Lecture Notes in Networks and Systems

Volume 232

The series "Lecture Notes in Networks and Systems" publishes the latest developments in Networks and Systems—quickly, informally and with high quality. Original research reported in proceedings and post-proceedings represents the core of LNNS.

Volumes published in LNNS embrace all aspects and subfields of, as well as new challenges in, Networks and Systems.

The series contains proceedings and edited volumes in systems and networks, spanning the areas of Cyber-Physical Systems, Autonomous Systems, Sensor Networks, Control Systems, Energy Systems, Automotive Systems, Biological Systems, Vehicular Networking and Connected Vehicles, Aerospace Systems, Automation, Manufacturing, Smart Grids, Nonlinear Systems, Power Systems, Robotics, Social Systems, Economic Systems and other. Of particular value to both the contributors and the readership are the short publication timeframe and the world-wide distribution and exposure which enable both a wide and rapid dissemination of research output.

The series covers the theory, applications, and perspectives on the state of the art and future developments relevant to systems and networks, decision making, control, complex processes and related areas, as embedded in the fields of interdisciplinary and applied sciences, engineering, computer science, physics, economics, social, and life sciences, as well as the paradigms and methodologies behind them.

Indexed by SCOPUS, INSPEC, WTI Frankfurt eG, zbMATH, SCImago.

All books published in the series are submitted for consideration in Web of Science.

More information about this series at http://www.springer.com/series/15179

Radek Silhavy · Petr Silhavy ·
Zdenka Prokopova
Editors

Software Engineering Application in Informatics

Proceedings of 5th Computational Methods in Systems and Software 2021, Vol. 1

Springer

Editors
Radek Silhavy
Faculty of Applied Informatics
Tomas Bata University in Zlín
Zlín, Czech Republic

Petr Silhavy
Faculty of Applied Informatics
Tomas Bata University in Zlín
Zlín, Czech Republic

Zdenka Prokopova
Faculty of Applied Informatics
Tomas Bata University in Zlín
Zlín, Czech Republic

ISSN 2367-3370 ISSN 2367-3389 (electronic)
Lecture Notes in Networks and Systems
ISBN 978-3-030-90317-6 ISBN 978-3-030-90318-3 (eBook)
https://doi.org/10.1007/978-3-030-90318-3

This Springer imprint is published by the registered company Springer Nature Switzerland AG
The registered company address is: Gewerbestrasse 11, 6330 Cham, Switzerland

Preface

This book constitutes the refereed proceedings of the Computational Methods in Systems and Software 2021 (CoMeSySo 2021), held in October 2021.

CoMeSySo 2021 conference intends to provide an international forum for the discussion of the latest high-quality research results in all areas related to systems, software, and networks in intelligent systems.

The addressed topics are the theoretical aspects and applications of software engineering and computational methods in engineering and science or artificial intelligence. The papers address topics as software engineering, cybernetics and automation control theory, econometrics, mathematical statistics or artificial.

CoMeSySo 2021 has received (all sections) 394 submissions and 165 of them were accepted for publication.

The volume Software Engineering Application in Informatics brings papers in which new approaches and methods applied to real-world problems are described. Furthermore, the basic research that describes novel approaches in the software engineering and informatics in the scope of the systems and software systems.

The editors believe that readers will find the following proceedings interesting and useful for their research work.

September 2021

Radek Silhavy
Petr Silhavy
Zdenka Prokopova

Organization

Program Committee

Program Committee Chairs

Petr Silhavy, Ph.D., Department of Computers and Communication Systems, Faculty of Applied Informatics, Tomas Bata University in Zlin, Czech Republic

Radek Silhavy, Ph.D., Department of Computers and Communication Systems, Faculty of Applied Informatics, Tomas Bata University in Zlin, Czech Republic

Zdenka Prokopova, Ph.D., Associate Professor at Department of Computers and Communication Systems, Tomas Bata University in Zlin, Czech Republic

Krzysztof Okarma, Ph.D., DSc, Faculty of Electrical Engineering, West Pomeranian University of Technology, Szczecin, Poland

Prof. Roman Prokop, Ph.D., Professor at Department of Mathematics, Tomas Bata University in Zlin, Czech Republic

Prof. Viacheslav Zelentsov, Doctor of Engineering Sciences, Chief Researcher of St.Petersburg Institute for Informatics and Automation of Russian Academy of Sciences (SPIIRAS), Russian Federation

Prof. Lipo Wang, School of Electrical and Electronic Engineering, Nanyang Technological University, Singapore

Silvie Belaskova, Head of Biostatistics, St. Anne's University Hospital Brno, International Clinical Research Center, Czech Republic

Roman Tsarev, Ph.D., Associate Professor, Department of Informatics, Siberian Federal University, Krasnoyarsk, Russia

International Program Committee Members

Organizing Committee Chair

Radek Silhavy, Ph.D., Tomas Bata University in Zlin, Faculty of Applied Informatics, email: comesyso@openpublish.eu

Conference Organizer (Production)

Silhavy s.r.o.
Web: http://comesyso.openpublish.eu
Email: comesyso@openpublish.eu

Conference Website, Call for Paper

http://comesyso.openpublish.eu

Contents

Relationship Between the Level of Economic Development and Education Level of the Employed Population

Yuriy A. Shamsutdinov(✉), K. V. Lebedev, and P. P. Bochkovskiy

RAO Institute of Educational Development Strategy, Moscow, Russia

Abstract. The sphere of education forms the structure of the country's population in terms of education level, which must meet the needs of real sectors of activity. Achieving a balance of supply and demand, considered at the level of demand for personnel of a certain level of education, presupposes finding an optimal match, at which the current demand is satisfied and at the same time there is a "positive" contradiction that stimulates further development. The problem of achieving a balance lies in the need to take into account this contradiction, the definition of which is not as obvious as if an assessment is carried out in order to satisfy only the current demand. Publication's review on this issue shows that the emphasis is on the benefits of a particular level of education both for individuals receiving it and for the whole society, while the assessment of compliance with the level of socio-economic development is carried out only in relation to a specific country. Conclusions that can be extended to other countries are recommendatory and they do not offer a universal method to assess the level of the achieved balance.

Method is proposed and tested in this article for determining the structure of the working-age population by levels of education, which is optimal from the point of view of both meeting the current demand for personnel of a certain level of education, and creating conditions for the development of real sectors of activity. By comparing the structure of the population by education level with the levels of development of economic spheres of the OECD countries, characterized by the indicator of the level of their technological development, the optimal structure of the population by education level was determined, which corresponds to the achieved high values of indicators of the level of development of the economic sphere and their high growth rate. The structure of the population by the level of education for the Russian Federation is analyzed for its proximity to the optimal structure. It is shown that in comparison with the optimal structure, the share of people with higher education is significantly exceeded, which does not turn out to be a "positive" contradiction that stimulates development.

The scientific novelty of the work lies in the proposal of a methodological approach to predicting the structure of the country's population by the level of education, optimal in terms of meeting the needs of real sectors of activity, including compliance with the conditions for their development, which can be applied to countries with different levels of socio-economic development.

The results of the work can be used in making decisions in the field of education.

R. Silhavy et al. (Eds.): CoMeSySo 2021, LNNS 232, pp. 1–14, 2021.
https://doi.org/10.1007/978-3-030-90318-3_1

Keywords: Technological structure · Level of technology development · Sphere of education · Level of education of the population · System of personnel training

1 Introduction

Any activity field requires the personnel employed certain qualifications in it. Abstracting from the achieving a matching problem between supply and demand for specific specialities and considering this connection in the broadest sense, the working-age population country's structure in education levels terms, determined by the existing education system, should correspond to the activity state's spheres aggregate demand, depending on their structure and development level. However, such a problem solution is faced with the "compliance" very concept defining the problem. If full compliance with the request is achieved, is this satisfactory for development? According to the dialectics' laws, development requires contradiction as a development source, and if we accept the contradiction presence as a necessary factor in the development, then what is the "positive" contradiction source - unsatisfied demand or excess supply? What imbalance level is optimal for development, and what are its maximum permissible limits? Limiting ourselves to internal criteria, it is impossible to solve such an optimization problem, since the socio-economic development level growth without comparison with the world countries, where on average it may turn out to be higher and the country's competitiveness, in this case, will decrease, may give an optimal compliance false idea. A correct assessment can be made by the considered parameters cross-country comparison method and their changes dynamics comparison. The criterion for optimal compliance, in this case, will be a socio-economic development indicators high level and their positive dynamics.

This work is devoted to the development and a method' testing for countries comparative assessment according to the compliance under consideration level.

2 Literature Review

There is an articles huge number on this topic, which indicates, on the one hand, about its relevance, on the other, about its consideration aspects wide range from the point of view of assessing and analysing the relationship between the population education level and the socio-economic development achieved level.

In [1], a growth dynamics comparative analysis in the adult population education level in the world developing and developed countries for the period 1960–1985 was carried out, which showed an "explosive" growth in developing countries, where, in particular, the people annual growth rate with higher education was 43%, while in the developed only 9.6%. Concerning the relationship between the population and economic growth educational attainment growth, it was concluded that the primary and secondary education development is preferable for the poorest countries, since the return on financial investments there is higher, and the educational growth in the higher education segment leads to an increase in the disproportion between private and social benefits in the former favour, while for these countries' development stage, social growth is a priority. For rich, economically developed countries, the increase in the people proportion with higher education has a greater effect, however, the costs for it are required significantly higher. That is, rich countries can "afford" to invest in higher education, while poor or developing countries, to catch up with the developed and focusing on the higher education development, do not do well. The connection under consideration issue, therefore, affects a socio-economic development aspects wider range. With all the education benefits obviousness (an increase in the human potential productivity, an improvement in the society quality, an increase in the labour force qualifications, etc.), investments by education levels have different profitability in economic growth terms, the improvement strategy depends on the development basic level, and in social terms it is, on the one hand, it leads to a reduction in poverty, on the other hand, potentially contributes to the social stratification.

Further research carried out in the world' various countries, in an attempt to find universal laws applicable to all countries, is often limited to general conclusions and, depending on the research methods used, results are obtained, which aggregate cannot form unambiguously interpreted conclusions. In works [2, 3], carried out in Zimbabwe and China (based on the data analysis for Guangzhou province), respectively, it is concluded that the population education level is a priority, which contributes to economic growth. In work [4], devoted to the European countries study, it is reported that only in Belgium, France, Ireland and Great Britain a causal relationship between the areas under consideration was revealed. In the countries rest, neither an increase in significant effect in the people proportion with higher education on economic growth nor the real GDP growth impact on growth in the people adult population share with higher education was revealed. A deeper analysis was carried out in [5]. Based on the data analysis for Pakistan, it was concluded that with the real physical capital presence, all education levels are the economic growth caused, while economic growth does not affect such education parameters as higher (university) education, including vocational education, and education spending. In work [6], on the data for Spain, the considered dependence nonlinearity was analysed taking into account the physical capital, expenditures parameters on education, the labour force structure. It is shown that if the linear model gives an unambiguous solution (an increase in the people share with secondary education by 10% leads to an increase in economic growth by 1.5%, while the same growth in higher education increases the economic growth rate by 0.9%, with other equal conditions), the non-linear model shows a greater economic effect due to the higher education growth in general, and growth due to secondary education depends on a previous period parameters number (GDP, physical capital, labour force, government spending). In the article [7], attention is paid to taking into account such a factor as the technologies' development, which is ignored in studies on this topic, judging by any quantitative estimates' absence or their characterizing parameters in various calculation models, but also in the work itself, its influence or dependence on education has not been investigated.

Among the Russian authors, one can single out a scientific report of the RPAN and PA [8] and a work similar to it [9], where the problem is considered quite versatile, from different aspects, but with rather general conclusions, among which the conclusion about the service sector growth in economies is developed countries attention-worthy, which employs up to 2/3 of the workers, and this forces a different approach to the "economic growth" concept content, which has traditionally been associated with industrial growth. The work [10] assessed the human capital contribution to the gross regional product growth in regions with development different levels. The conclusions obtained in the work: the human capital contribution is higher in more developed regions, which can be interpreted as a higher return on a highly educated population in regions with a development higher level.

The most productive approach was applied in the study [11]. (With an emphasis on analysis specifically for Russia, a similar study was carried out in [12]. In an earlier work [13] a similar approach was applied, but an attempt to find a direct connection between the population education level and the economic growth level on its basis is not entirely successful). The approach essence is to assess the investment feasibility in education from the return rate assessing standpoint on investment in education. According to the proposed approach, the profit rate from education (return on education) is estimated, which is determined by a person's earnings cost over a lifetime (we mean additional earnings generated from additional education, otherwise such a return rate calculation is incorrect - ed.) to education costs net present value. Since the costs' recoupment occurs in activity real areas, where the return rates usually express their development level and through a wages level differentiated offer or the income current level, depending on the in-demand personnel education level, the need for people with an education-specific level is determined. It is fairly assumed that, focusing on this expected income, people invest money and time in education, counting on its payback and profitability at the end, which forms the population structure in education terms that is adequate to the activity' real sectors needs. Figure 1 most integrally expressed the study conclusions, which boil down to the following.

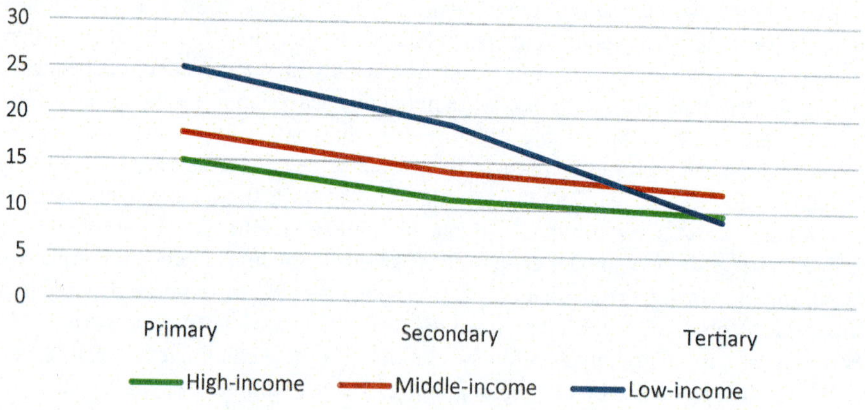

Fig. 1. The Structure of social returns.

Large social returns in countries with a development low economic level (compared to developed countries) from school and secondary vocational education. The return on higher education for all countries is always lower than the return on other education levels. In [12], when comparing indicators for Russia with other countries, it is shown that the return average rate on education in Russia decreased from 10% in 2006 to 5.6%

in 2018, while the average for all countries in 2018 it is estimated at 8–9%, and in the higher education field, the return in Russia is estimated at 8%, in European countries 10%, and in all countries, it is estimated at 15%.

An analysis of these and those close to them in research methods terms used and other works approaches shows that the more parameters are taken into account in the calculations, the more ambiguous and general the conclusions are, the different research methods applied to the same data leads to different results. The often-used economic growth estimate in the GDP growth rate terms per capita, as a criterion for assessing economic growth due to the increase in the population education level, is too rough, due to the latter isolating the contribution difficulty. Conclusions on the causal relationship are contradictory and do not provide unambiguous recommendations on how, based on the economy existing structure in physical capital volume indicators ratio terms, the human capital structure, the employment structure, to decide on priorities in development: to invest in machinery and equipment or in education.

The assessing idea attractiveness at the contribution level to economic growth is obvious, but this task seems unattainable if we expect its solution in the correspondence finding universal rules form that all countries must follow.

In this regard, a particular problem solution seems to be more productive, namely, the education level correspondence to a certain parameter that reflects or characterizes the economic development level quite integrally, but at the same time which connection with the education sphere is more direct. This parameter can be the country technological development level, which growth, as a rule, determines economic growth. The technology development level, in contrast to other macroeconomic indicators, is directly related to the population education level, since the human capital contribution is most significant here, and the latter structure largely determines their development level. This level can be expressed in the country's economy technological structure assessments [14, 15], the country's readiness degree for innovations [16], the potential for the technologies' introduction [17].

3 Methodology, Materials and Research Methods

The working-age population structure in education level terms was assessed under the country technological development level, as an activity economic sphere key parameter, which, in turn, is any state activity main sphere.

The country's economic development structure and level assessment are most integrally assessed by the technologies level used in it. One of the approaches is to determine the technologies used to share distribution, classified under the technological structures (In the world literature, the term techno-economic paradigm is used). The leading researcher in this area is recognized as Carlota Perez.), in which they are dominant [18–20]. Each technological structure puts forward its requirements for the personnel training level, while a higher level technological structure dominance presupposes an education people with a higher level greater proportion in the employment structure (Table 1) [21].

As you can see, the technologies' development largely determines the socio-economic progress dynamics and their development is directly related to population education. Unfortunately, technologies shares distribution quantitative estimates used,

Table 1. The relationship between technological waves and education as part of human capital.

Structure characteristics	Technological structures		
Dominance period (years)	I technological structure (1770–1830)	II technological structure (1830–1880)	III technological structure (1880–1930)
Core	Textile machinery manufacturing, textile manufacturing, iron and iron production, canal construction	Steam engines' production, steamships and steam locomotives, coal mining, iron and iron production, machine tools, railway construction	Electrical engineering manufacture, power lines and power plants construction, plastics and synthetic fabrics, automobiles manufacture
Key element	Textile machines	The steam engine	Electric motor
Main energy carrier	Water	Coal	Electrical Energy
Carrier industries	Textile engineering, textile industry, ferrous metallurgy, transport	Heavy industry, coal industry, ferrous metallurgy, transport	Electrical, heavy engineering, medium engineering, chemical industry, electric power
Changes in the employee's work content	Direct impact on the labour subject	Equipment technological maintenance	Managerial labour allocation into a separate function
Workers prevailing educational level	absent, training takes place in the workplace	Primary Education	Initial vocational training
Main educational institutions	absent, training takes place in the workplace	School	Educational institution
Dominance period (years)	IV technological structure (1930–1970)	V technological structure (1970–2010)	VI technological structure (2010–2050)
Core	Cars and tractors manufacture, household appliances, aluminium, polymeric materials, oil production and refining	Computing and fibre-optic equipment production, software development, composite materials' production, robots and CNC machines, gas production and processing, telecommunications development	Nanoelectronics, molecular and nanophotonics, nanomaterials and nanostructured coatings, nanobiotechnology, nanosystem technology

(*continued*)

Table 1. (*continued*)

Structure characteristics	Technological structures		
Dominance period (years)	I technological structure (1770–1830)	II technological structure (1830–1880)	III technological structure (1880–1930)
Key element	Explosion engine	Microelectronic components	Nanotechnology, cell technology
Main energy carrier	Oil	Gas	Nuclear fuel, hydrogen
Carrier industries	Medium engineering, non-ferrous metallurgy, chemical and petrochemical industry	The electronics industry, robotics, gas industry, information services	Electronic, nuclear and electrical industries, alternative energy, education, health care, information and communication sector, crop production
Changes in the employee's work content	Control over the production process	Collection, processing, classification, information transmission	Introducing a creative approach to solving production problems and mastering ways to enhance the inventive activity, increasing knowledge
Workers prevailing educational level	Secondary prof. degree	Higher vocational training	Lifelong education
Main educational institutions	Technical schools, colleges	University	University, enterprise

attributed to certain technological paradigms, are rather approximate and not systematized by country, which does not allow comparison. We can only note the data most frequent comparison for the USA and Russia (see, for example, work [17]), where the following estimates are given for 2015. In the USA: 20% of the technologies used belong to the fourth technological order, 60% - to the fifth, 5% - to the sixth; in Russia: 30% of the technologies used belong to the third technological order, 50% to the fourth, 10% to the fifth.

If the technologies share distribution used for the United States is taken as a typical distribution in economically developed countries, with the technologies' dominance for the high-tech goods' production, then in Russia the technologies for the raw materials extraction and processing are dominant, which suggests, according to the correspondences given in the table, the "dominant" demand for personnel with an average professional education.

Given the lack of statistical data characterizing the countries technological development level through the dominant technologies share distribution in them, the assessment is possible through the country's "technological readiness" index, determined for the OECD countries and presented in the Report by The Economist Intelligence Unit "Technological Readiness Ranking" [22]. This level (both the technologies' development and the prospects' availability for their development) was assessed by an indicators number taking into account the Internet development, the economic infrastructure digitalization (e-commerce, e-government, cybersecurity readiness) and openness to innovations (patenting, the research and development costs volume, etc., research infrastructure). The population education structure comparison with this index is presented in Table 2.

As can be seen from the data in the table, the average for OECD countries for the period 2014–2019 the people share with higher education has increased significantly, as well as the technological readiness index relative value, which is consistent with the relationship to qualitative assessment between technologies and the education level (see Table 1): a higher technological order technologies use requires a population education higher level. This conclusion is quite obvious, but in this table, it is not so much the absolute values and their growth rate that are informative, but the countries' distribution, which in the table are arranged in technological readiness index descending order. Here Norway determines the countries' division boundary relative to the technological readiness index average value and the countries located higher can be classified as those in which the population structure is consistent with the *current* query. If we rank countries by the change rate in the technological readiness relative index and divide them into two groups according to the above-average and below-average growth rate, then we can also conditionally classify countries with a rate above average as countries in which the population structure (in education terms) is consistent with the request *for development*. Then the countries in which both parameters turned out to be above average values can be attributed to those in which the population structure in education level terms is close to optimal in meeting the current demand terms and the development potential availability (Table 3).

4 Research Results

As can be seen from the data in Table 3, the optimal structure for countries that can be classified as those in which the dominant technologies belong to the 5th technological order (which, as noted earlier, is inherent in economically developed countries), is an approximately equal share of people aged 25–64 years old with secondary and higher education - (42–43)% each and the people remaining share without special education (15%).

In the Russian Federation, as can be seen from the data in Table 2, the people dominant share with higher education (56.7%) and the lowest among the countries under review, the people share without special education (4.8%). The technological readiness index value is in the bottom quarter of the countries, ranked in the index descending order; its relative growth rate is approximately in line with the average value for the OECD countries.

Table 2. Education level structure of the OECD countries and Russia in comparison with the relative index of technological readiness.

Countries	The population distribution by the educational level at the age of 25–64								Technology Readiness Index 2019	
	Shares of total headcount (2019)				Relative change for the period 2014–2019,%				Absolute value (2019)	Relative change for the period 2014–2019,%
	General	Average	Higher	Integral index[a]	General	Average	Higher	By integral index		
Australia	17.1	35.7	47.1	134.3	75	102	112	102.1	9,71875	106.1%
Sweden	16.4	39.7	44.0	133.9	89	93	114	101.3	9,71875	103.0%
Finland	9.7	44.3	45.9	136.3	72	99	110	101.5	9,43750	97.1%
France	19.6	42.5	37.9	131.7	78	100	118	102.2	9,43750	1098%
Germany	13.3	56.8	29.9	132.0	101	95	111	100.4	9,43750	103.1%
Netherlands	20.4	39.2	40.4	131.9	85	94	117	101.8	9,43750	103.1%
United States	9.2	42.5	48.3	136.9	88	94	109	100.9	9,43750	109.8%
Austria	14.4	51.8	33.8	132.4	89	96	113	101.0	9,15625	100.0%
Belgium	21.3	38.0	40.7	131.7	81	104	110	101.8	9,15625	110.2%
Korea	11.3	38.7	50.0	136.6	75	96	112	101.6	9,15625	103.2%
Canada	8.1	32.5	59.4	139.4	81	89	111	101.2	8,87500	103.3%
Denmark	18.4	41.2	40.4	132.5	90	94	113	101.2	8,87500	103.3%
Estonia	9.9	48.7	41.4	135.3	124	92	109	101.0	8,87500	106.8%
New Zealand	19.3	41.6	39.1	132.0	74	108	110	102.1	8,87500	103.3%
Switzerland	11.0	44.6	44.4	135.6	91	93	110	100.9	8,87500	106.8%
Israel	12.9	36.8	50.2	136.2	88	100	104	100.6	8,59375	100.0%
United Kingdom	32.3	20.5	47.2	129.7	84	106	112	102.2	8,59375	100.0%
Norway	**17.5**	**38.4**	**44.1**	**133.6**	**97**	**96**	**106**	**100.5**	**8,31250**	**100.0%**
Ireland	20.4	20.4	59.1	135.7	96	54	144	102.9	8,03125	111.7%
Spain	38.7	22.7	38.6	126.1	89	104	111	101.8	8,03125	107.5%
Czech Republic	6.2	69.5	24.2	133.0	92	97	113	100.6	7,46875	103.9%
Italy	37.8	42.5	19.6	122.6	93	100	116	101.2	7,46875	103.9%
Poland	7.4	60.6	32.0	134.2	78	95	119	101.2	7,46875	108.1%
Slovenia	11.2	55.5	33.3	133.3	78	97	116	101.4	7,18750	108.5%
Russian Federation	4.8	38.5	56.7	139.9	**90**	**95**	**104**	**100.5**	7,18750	**104.1%**
Chile	32.6	42.3	25.2	125.3	84	105	119	102.1	6,90625	104.2%
Portugal	47.8	25.9	26.2	120.8	84	120	121	103.0	6,90625	104.2%

(*continued*)

Table 2. (*continued*)

Countries	The population distribution by the educational level at the age of 25–64								Technology Readiness Index 2019	
	Shares of total headcount (2019)				Relative change for the period 2014–2019,%				Absolute value (2019)	Relative change for the period 2014–2019,%
	General	Average	Higher	Integral index[a]	General	Average	Higher	By integral index		
Slovak Republic	8.8	65.4	25.8	132.5	96	93	127	100.9	6,90625	125.6%
Hungary	15.0	59.0	26.0	130.7	89	99	111	100.8	6,62500	109.3%
Turkey	58.3	19.7	22.0	116.9	91	104	131	102.5	5,78125	110.8%
Greece	26.0	42.1	31.9	128.6	82	105	113	102.0	5,21875	100.0%
OECD average	**22.1**	**40.3**	**37.6**	**130.9**	**87.6**	**97.4**	**115.0**	**101.5**	**8,23085**	**105.4%**

[a] The integral index is defined as the shares weighted sum under consideration. The weights are determined under the average study time proportions by education levels: for general education - 1 (10 years), for secondary vocational - 1.3 (13 years), for higher education - 1.5 (15 years). As the population proportion with an education increases the higher level, the index increases and vice versa.

Table 3. Countries, where education level structure of population is close to optimal in terms of compliance with the level of technological development in terms of ensuring its high level and creating development potential

Countries	The population distribution by the educational level at the age of 25–64						Technology Readiness Index 2019	
	Shares of total headcount (2019)			Relative change for the period 2014–2019,%			Absolute value (2019)	Change for the period 2014–2019
	General	Average	Higher	General	Average	Higher		
Belgium	21.3	38.0	40.7	81	104	110	9.15	110.2%
France	19.6	42.5	37.9	78	100	118	9.43	109.8%
United States	9.2	42.5	48.3	88	94	109	9.43	109.8%
Estonia	9.9	48.7	41.4	124	92	109	8.87	106.8%
Switzerland	11.0	44.6	44.4	91	93	110	8.87	106.8%
Australia	17.1	35.7	47.1	75	102	112	9.72	106.1%
Average for selected countries	14.7	42.0	43.3	90	97	112	9.25	108.2%
OECD average	22.1	40.3	37.6	88	97	115	8.23	105.4%

5 Discussion

What conclusions follow from this? The people share with higher education in Russia is overestimated both relative to the average values for the OECD countries and the country's "technological readiness" index relatively low value. On the one hand, this is a positive factor, since the population educational level is high. On the other hand, this is not an optimal structure in the technological development achieved level terms. As discussed above, the optimal value for developed countries is (42–43) % and this applies to countries with a technological development high level. To analyze this moment in detail, let us turn to the internal statistics provided by the RSA on the country employed population *structure* at the age of 25–64 in terms of education level (Table 4) [23].

Table 4. Structure of the employed population of the Russian Federation by level of education (2019).

Indicators	Total	Including have education:		
		General	Secondary professional	Higher
Employed at the age of 25–64, a thousand people	66,549	13,149	29,987	23,413
Those employed shares at an education certain level,%		19.6	44.7	34.9

Based on the data published in the OECD and RSA statistical databases, it turns out that of 84,8 million people aged 25–64, 48,1 million people have higher education (56.7%), of which only 23,4 million are employed (34.9% of 66,5 million people). That is, there are 24,6 million unemployed people aged 25–64 with higher education in Russia, which is more than a third of the total employed, that is, a very impressive figure. It is obvious that they somehow earn a living, but it is also obvious that they take little part in the technologies' development (except for certain industries), since the technologies' development requires capital expenditures (in buildings, machines, equipment) and a relatively long time required for their creation and implementation. Both the first and the second implies a person registered as an employee.

The foregoing allows us to conclude that the highly educated people surplus in Russia is not a "positive contradiction" factor that stimulates economic growth, since neither the actual level achieved nor the growth prospects, according to the "readiness index", are estimated rather low. In general, the population structure in education level terms is suboptimal.

The solution to this problem is seen not in the people number reduction with higher education, but in the investments' growth in fixed assets and their optimal distribution by industry. But so far in Russia in the fifth technological order industry, there is only (11–12)% of the total investment, and the lion's share in the industry attributed to the third and fourth technological order [17].

6 Conclusion

The relationship analysis between the economic structure and the training personnel system in the education field, which forms the country's population structure in education terms, shows that the Russia population structure in the adult population education level terms can be characterized as "re-educated" and generally not optimal for the technology development current level in the country. In this current situation, the growth cause should be economic growth (technologies introduction), education - the consequence. As world countries' comparative analysis shows, with a population education lower level in economically developed countries, its structure is nevertheless optimal for achieving a technology development high level, while a people high proportion with higher education in Russia does not lead to an increase in technology, corresponding to such a population structure. As a result, investments in education with an orientation towards domestic demand turn out to be low profitable both on the part of the state financing higher education and in private investment terms, and the population's education growth social effect is underestimated.

This conclusion cannot be interpreted as a population poor educational structure in general, since a population higher educational level is better than a low level: this is a good potential for development, it has more qualified personnel, a population higher cultural level, etc. It should be interpreted as a country's adult population education level suboptimal correspondence to its economic development level, as a consequence, suboptimal conditions for economic growth, and this also characterizes the causal relationship in this ratio, which, as can be seen from the conclusions from the works, does not interpreted unambiguously since they depend on the reason why the optimal balance is not achieved.

A possible way to solve this problem in Russia is to correct the training personnel system in the education field towards an increase in the people number receiving secondary vocational education, and increased investment in fixed assets in high-tech sectors of the economy.

For other countries, the relationship between level of education and level of socio-economic development may be different, but which is also determined on the basis of principle of compliance with the optimal balance, in the sense that is attached to him in this article.

Acknowledgments. The work was performed within the framework of the state task of the Federal state budgetary institution "Institute of education development strategy of the Russian Academy of education" No. 073-00007-20-01 for 2020 and for the planning period of 2021 and 2022. The project «Conducting a statistical assessment of the state of the education sector, including an assessment of the level of technological development and the level of education in the constituent entities of the Russian Federation in accordance with the powers of the Ministry of Education of Russian Federation».

References

1. Jandhyala, B.G.: Tilak. Education and its relation to economic growth, poverty, and income distribution. In: World Bank Discussion Papers, vol. 46 (1989). http://documents1.worldbank.org/curated/en/357751468739481512/pdf/multi-page.pdf. Accessed 2 Feb 2021

2. Zivengwa, T.: Investigating the causal relationship between education and economic growth in Zimbabwe. Global J. Manag. Bus. Res. **12**(8), 107–117 (2012)
3. Liao, L., Du, M., Wang B., Yu, Y.: The impact of educational investment on sustainable economic growth in Guangdong, China: A Cointegration and Causality Analysis, Sustainability. https://www.mdpi.com/2071-1050/11/3/766/htm. Accessed 2 Feb 2021
4. Dudzevičiūtė, G., Šimelytė, A.: Education and economic development in the selected European Union Countries. Eur. J. Sustain. Dev. **7**(2), 14–28 (2018)
5. Afzal, M., Rehman, H.U., Farooq, M.S., Sarwar, K.: Education and economic growth in Pakistan: a cointegration and causality analysis. Int. J. Educ. Res. **50**, 321–335 (2011)
6. Marquez-Ramos, L., Mourelle, E.: Education and economic growth: an empirical analysis of nonlinearities. Appl. Econ. Anal. **27**(79), 21–45 (2019)
7. Canals, C.: Education and economic growth. https://www.caixabankresearch.com/en/economics-markets/labour-market-demographics/education-and-economic-growth. Accessed 2 Feb 2021
8. Bozhechkova, A.V., Klyachko, T.L., Knobel', A.Yu., Loshchenkova, A.N., Lyubimov, I.L., Sinel'nikov-Murylev, S.G.: Education and economy development. «Delo» RANEPA, Moscow, Russia (2019)
9. Klyachko, T.: Education development and economic growth in developed and developing countries, 17 April 2020. https://ssrn.com/abstract=3644290 or https://doi.org/10.2139/ssrn.3644290. Accessed 2 Feb 2021
10. Komarov, A.V., Papshok, O.V.: The estimation of the human capital's impact into the economic growth of the regions of Russian Federation (Using Mankiw – Romer – Wail's Model). Novosibirsk State Univ. Bull. Ser. Socio-economic Sci. **3**, 191–201 (2007)
11. Psacharopoulos, G., Patrinos, H.A.: Returns to investment in education. A Decennial Review of the Global Literature. Policy Research Working Paper 8402, Education Global Practice. http://documents1.worldbank.org/curated/en/442521523465644318/pdf/WPS8402.pdf. Accessed 2 Feb 2021
12. Melianova, E., Parandekar, S., Patrinos, H.A., Volgin, A.: Returns to education in the Russian Federation some new estimates. In: Policy Research Working Paper, vol. 9387, Education Global Practice. http://documents1.worldbank.org/curated/en/172031599588105838/pdf/Returns-to-Education-in-the-Russian-Federation-Some-New-Estimates.pdf. Accessed 2 Feb 2021
13. Belyakov, S.A., Klyachko, T.L.: Methodology for assessing the contribution of education to the socio-economic development of the Russian Federation and its subjects. «Delo» RANEPA, Moscow, Russia (2015)
14. Hongyu, Z., Li, X.: Routes for education reform at the turning point in human history. J. Natl. Acad. Educ. Admin. **7** (2014)
15. Ubushiev, E.V.: Economic security in various technological structures. Theor. Appl. Econ. **3**, 1–21 (2018)
16. Odior, E.S.O.: Government spending on education, economic growth and long waves in a CGE micro-simulation analysis: the case of Nigeria. Br. J. Econ. Finan. Manag. Sci. **1**(2), 74–87 (2011)
17. Prokhorova, I.S.: Technological potential for development of innovation economy in Russia: the limits and limitations of the sixth technological order. Bull. Univ. **2**, 68–75 (2020)
18. Glazyev, S.: The global economic crisis as a process of technological shifts. Probl. Econ. Transit. **52**(5), 3–19 (2014)
19. Cantwell, J., Santangelo, G.: Capitalism, profits and innovation in the new techno-economic paradigm. J. Evol. Econ. **10**, 131–157 (2000)
20. Yegorov, Y.: Long economic waves as innovation cycles. In: Proceedings of Simon Kuznets International Symposium. SKISD, Kiev, Ukraine (2011)

21. Potekhina, N.V., Shulinina, Yu.I.: The relationship of technological structures and education as part of human capital. Theory Pract. Soc. Dev. **3** (2016)
22. Technological Readiness Ranking. http://pages.eiu.com/rs/753-RIQ-438/images/Technological_readiness_report.pdf. Accessed 2 Feb 2021
23. Labour force, employment and unemployment in Russia – 2020, section "Number of employed by education level and status in 2019", https://rosstat.gov.ru/bgd/regl/B20_61/Main.htm. Accessed 2 Feb 2021

Replacing the Reinforcement Learning (RL) to the Auto Reinforcement Learning (AutoRL) Algorithms to Find the Optimal Structure of Business Processes in the Bank

Andrey A. Bugaenko(✉)

Data Science Director, Sberbank PJSC, Moscow, Russia
aabugaenko@sberbank.ru

Abstract. The paper provides a description of the algorithm allowing the shift from *RL* to *AutoRL* in the search for the optimal path in a banking business process grid graph. The article discusses which attributes of the environment are universal and which are not. It describes how it is necessary to change the *Reward* so that it becomes universal for any business process in the bank. #COMESYSO1120.

Keywords: Reinforcement learning · RL · Auto reinforcement learning · AutoRL · Environment creation · Process mining · Banking business process

1 Introduction

One of the big tasks that the analytical units are tackling in banks is seeking ways to optimize business processes. Classically, process mining models [1, 2] are favored to optimize business processes in those banks keeping employee action logs. Data is collected from primary systems, aggregated, and a business process grid graph is built on that basis. Unlike abstract representation in regulatory documents, this approach shows the realistic structure of a business process. Next, an analyst will look in detail at the grid graph of a business process using BI tools to address the questions: *"How does the business process actually work?"* and *"What are the maximum optimization opportunities?"* Another grand question is *"How should a business process be arranged optimally?"* However, BI-tools will not be able to manually analyze a grid graph of dozens and hundreds process stages and hundreds and thousands of stage-to-stage transitions, some of which are optional. *Reinforcement Learning (RL)* approaches are among the most effective ways to complete this task [3–7].

A specific feature of using *RL* in banks to optimize business processes is that Environment have to be easy to interpret. This feature pertains to economic and financial grounds of the optimal structure of a business process favored, i.e. reasoning for the optimal path in the grid graph, required by the bank's business units and the leadership. A company's TOP management may not need such reasoning, and may accept data-scientist deliverables without verifying them. It would of course be a green light for the use of non-interpretive environments to solve tasks [3–5]. Nonetheless, in reality, if

R. Silhavy et al. (Eds.): CoMeSySo 2021, LNNS 232, pp. 15–22, 2021.
https://doi.org/10.1007/978-3-030-90318-3_2

financial considerations cannot justify an *RL*-assisted solution, the bank's management can readily reject it for lack of validation. Authors of [6, 7] show a way to apply *RL* with the interpreted Environment. They suggest presenting the environment in the format:

- *Environment* - neting graph of business process;
- *State* - *node* (stage of process);
- *Action* - *edge* (transition between nodes);
- *Reward* – bank's profit from one transaction (one business process example) (or fictive profit);
- *Loss* – bank's operation expanses for every stage of process.

This approach has already proved to be highly successful in many large banks [6]. As such, this approach enables a analyst who pursues optimization of business processes to 'robotize' the most tedious jobs, reducing the task execution time pretty much.

Banks, however, have a huge number of versatile business processes requiring analysis and optimization. Although using *RL* is a common solution, a data scientist must have it fitted individually with each business process. Actually you have to make a lot of efforts to match the environment with each business process on a case-by-case basis.

2 RL – AutoRL Replacement Specifics

The solution is to use the *AutoRL* technology rather than *RL*, which actually enables the data scientist manual actions to fit *RL* with each particular business process to be automated.

Many machine learning tasks widely leverage auto machine learning (AutoML) approaches. They are favored by banks too. The essence is that the machine learning model is built 'by itself'. Besides the training and test set, some of the data is put aside as a 'validation' set to designate the best machine learning method, hyperparameters, and for 'feature selection' purposes. Accordingly, all possible 'method–hyperparameters–factors' combinations are searched by some selection method (usually greed search). The quality of the model is evaluated on a validation set by some criterion (for example: MAPE, MSE, R2). The best combination (method-hyperparameters-factors) is then selected for programming. The model is programmed and tested based on a test set.

AutoRL can function likewise. For a combination of three elements, automatic selection is needed here:

- *environment* (include: *State, Action, Loss, Reward*);
- *RL* - method;
- method - hyperparameters.

Obviously, each element of the method–hyperparameters–factors combination must be either generic and, therefore, not included in the grid search dictionary, or it can be selected by the grid search depending on each business process.

As *RL* methods with baseline hyperparameters, you can set the following in the grid search dictionary:

- q-learning:
  - ε - randomness coefficient,
  - learning rate,
  - number of cycles;
- cross entropy:
  - β,
  - ε,
  - number of cycles;
- genetic algorithm:
  - number of children,
  - share of mutations,
  - number of cycles.

The environment setting method in [6, 7] is generic for most elements.

The basic setting rule of a process grid graph, where *node* (stage of process) – *state*, *edge* (transition between nodes) - *action* is a generic rule that is fair without limitation for almost any banking business process in a bank branch and applies without a grid search.

The *Loss* setting method is also generic. Any stage of a business process can be represented as:

$$Loss_i = t_i \times Pr \times OPEX/HR\_OPEX \tag{1}$$

where t_i is the time to complete the ith process stage (s), I ε [1,h], where h is the maximum possible number of stages in a process, *Pr* is the staff (manager) labour second cost (rubles), *OPEX* is the operating expenses per business line, *HR_OREX* is the operating HR payroll expenses. Fully automatic processes are not covered here. OPEX determination rules for them require separate consideration. For any manual or automated processes, the *Loss* determination rule is generic.

A *Reward* setting rule is not that easy. There is no doubt that for business processes that have a fiscal effect, i.e. P&L per transaction (process instance) are positive, representing a *Reward* as equal to an P&L per transaction is an ideal solution, but there are business processes in bank branches that do not give a finance effect. Example: closing deposits, exchanging cash, banknote counterfeiting tests (examination of banknotes). This means that their *Reward* will be zero, so:

$$\sum\nolimits_{\overline{S}}(R + L) \rightarrow Max \Longleftrightarrow \sum\nolimits_{\overline{S}}|L| \rightarrow Min \Longleftrightarrow \sum\nolimits_{S_0}|L| \tag{2}$$

where $R \geq 0$ is the *Reward*, $L \leq 0$ is the *Loss*, $\overline{S}$ is the travel strategy on a grid graph opted for, S_0 is the strategy: log in and immediately log out, because any next step will bring:

$$\sum_{S_0} |L| \leq \sum_{S_i \in \overline{S}} |L| \quad (3)$$

Furthermore, banks sometimes have processes with a fiscal effect, but the P&L per transaction are less than the OPEX value per transaction, i.e.

$$\sum_{S_0} |L| \leq \sum_{S_i \in \overline{S}} (|L| - R) \quad (4)$$

Again, the optimal path in the grid graph of a business process will be: log in and immediately log out. This situation can be observed e.g. when a bank is bound to the state to process some incoming budget payments at an 0 bank fee and cannot cancel them.

In [7], this issue is proposed to be solved using a *'fictive' Reward*, but determining it is assumed to be a data scientist's duty because of using *RL* rather than *AutoRL* to optimize the business process, and rendering this criterion generic is not required.

The desired *'fictive' Reward* (R_a) must have two properties:

1)

$$\sum_{S_0} |L| \geq \sum_{S_i \in \overline{S}} (|L| - R_a) \quad (5)$$

2) R_a – must have a business interpretation.

The first possible solution is to break the task into 2 parts:

- Part 1: the search for the optimal path from start to *node* with *Reward*, and
- Part 2: the search for the optimal path from *node* with *Reward* to end.

Despite this, the overall problem statement here will not be the same as for a positive *Reward* value, and therefore this approach is not applicable if we have decided to use *AutoRL*.

Another approach may be setting the greatest possible *AutoRL* in the corresponding *node*, and then subtracting the *AutoRL* value, i.e.:

$$\sum_{\overline{S}} (R + L) \rightarrow Max \iff \left(\sum_{S_0} (R_a + L)\right) - R_a \quad (6)$$

This expression makes sense:

$$R_a \rightarrow \infty, for\ \forall L \geq 0 \Rightarrow \sum_{S_0} |L| \geq \sum_{S_i \in \overline{S}} (|L| - R_a) \quad (7)$$

However, setting ∞ as a *Reward* is obviously infeasible.

The third approach is to search for a R_a that obeys

$$max\left(\sum_{\overline{S_a}} |L|\right) \leq R_a \quad (8)$$

where $\overline{S_a}$ are the paths in the grid graph passing through *node* with *Reward*. However, a *Agent* is able to pass through the same edges of a grid graph several times, unrestricted by the number of returns from the same vertex back. In this case:

$$max(\sum\nolimits_{\overline{S_a}}|L|) \rightarrow \infty \Rightarrow max(\sum\nolimits_{\overline{S_a}}|L|) \leq R_a \iff \infty \leq R_a \tag{9}$$

which reduces the solution to expression (7), meaning that this approach is not applicable either.

Given (7) and (9), we can conclude that choosing a R_a that will be generic for any situation is infeasible. Nonetheless, these findings can apply only to a situation where the number of considered paths in the graph is $i \rightarrow \infty$. In reality, the number of these paths is limited by the method's hyperparameter 'number of cycles': i_{max}, i.e.

$$max(\sum\nolimits_{\overline{S_a}}|L|) \rightarrow L_{max} \tag{10}$$

Since we use *RL* methods to solve the issue, we must recognize a training model pattern, i.e.

$$\sum\nolimits_{S_j \subset \overline{S}}|L| \geq \sum\nolimits_{S_{j+1} \in \overline{S}}|L| \geq \sum\nolimits_{S_i \in \overline{S}}|L| \tag{11}$$

where $j < j + 1 < i$ are the ordinal numbers of the paths. Hence, we can conclude that:

$$\forall R_{a,i} = \sum\nolimits_{S_i \in \overline{S}}|L| \Rightarrow R_{a,i} \geq \sum\nolimits_{S_{i+1} \in \overline{S}}|L| \Rightarrow R_{a,1} \geq \sum\nolimits_{S_{i+1} \in \overline{S}}|L| \tag{12}$$

This suits us perfectly. In other words, assuming R_a as equal to the sum of *Loss* in the first programming cycle, we attain a value *Reward* obeying the problem situation. Naturally, after attaining $\sum_{\overline{S}}(R_a + L)$, the R_a must be subtracted. For clarity, any arbitrary large number cannot be taken as a R_a: it may be less than the sum of *Loss.*

3 AutoRL Usage Algorithm

Accordingly, the essence of the *AutoRL* usage method for finding the optimal way in the grid graph of a business process in bank branches will be the steps as below:

1) Set *Environment* - grid graph, which have got from process mining, *State* – node's, *Action* – edge's, *Loss* – bank's OPEX (1) for each process stage [7]
2) Set a *Reward* equal to the bank's gross profit from a single transaction (a process instance). If *Reward* $= 0$, go to step 4.
3) Search for the maximum value of $\sum_{\overline{S}}(R + L)$ for each combination of hyperparameter settings for methods: q-learning, cross entropy, genetic algorithm, and make sure the setting recommendations are obeyed. If $MAX(\sum_{\overline{S}}(R + L)) > 0$, go to step 6.
4) Define *MAX* $(\sum_{s_1 \overline{M}}|L|)$, where *M* is the vector of *RL* methods with various hyperparameters. The number of cycles is 1. In other words, repeat each method *1* time and take the maximum of the sum *Loss* as a *Reward.*

5) Search for the maximum value of $\sum_{\overline{S}}(R_a + L) - R_a$ for each combination of hyper-parameter settings for methods: q-learning, cross entropy, genetic algorithm, and make sure the setting recommendations are obeyed.
6) The optimal path in the grid graph will be S_0 for which

$$MAX(MAX(\sum\nolimits_{\overline{S}}(R + L)); MAX(\sum\nolimits_{\overline{S}}(R_a + L) - R_a)) \tag{13}$$

Then the algorithm was tested. The testing involved the business processes, namely: accepting payments from natural persons (NP); depositing/withdrawing cash to/from an account; issuing, exchanging and closing bank cards; selling coins and gold bars; NP's bank transfers; paying off regular loan payments; opening/closing/adding deposits; exchanging NP's currency; renting safety deposit boxes; security and certificate sale; cash change and damaged banknote exchange; banknote counterfeiting tests; indemnity and payouts to non-community bank depositors.

The resulting data showed that the algorithm for all business processes found a path in the grid graph in which the sum of *Loss* + *Reward* is greater than in the most likely path in the grid graph.

4 Conclusions

The paper provides a description of the algorithm allowing the shift from *RL* to *AutoRL* in the search for the optimal path in a banking business process grid graph. Notably, while *RL* itself helps to significantly reduce the efforts of analyst exploring process mining -derived grid graphs, data scientist's enabling the use of *RL* will also avail of shifting to *AutoRL* that requires fewer efforts.

The essence of the algorithm is confined to using the same principle as the use of *RL* [7]. Despite this, *RL* methods and hyperparameters are selected using '*greed search*'. In addition, a generic *'fictive' Reward*: $R_a = MAX(\sum_{s_1\overline{M}}|L|)$ equal to the sum of *Loss* of the first cycle per *RL* method must be set. For the sake of designating the best path in a grid graph, a target function changes from $MAX(\sum_{\overline{S}}(R + L))$ to $MAX(MAX(\sum_{\overline{S}}(R + L)); MAX(\sum_{\overline{S}}(R_a + L) - R_a))$.

This research has potential for furtherance. The method assumes that if a fiscal effect P&L of a single transaction is 0 in a single process instance, there is only one *Reward*, and this is true for the vast majority of process instances. Very rarely, however, it happens that there may be multiple events within a single instance that correspond to success, and therefore an agent would have to collect multiple *Reward*. For example: a client decided to change cash several times in a row in one visit to a the cash desk, or a client had several different deposits in the same bank and decided to close them all at once in one visit to a manager. Such events, though constituting a tiny fraction of all process instances, may well occur, and it would be interesting to analyze the likelihood of extending the algorithm to such exceptions. Alongside this, using 'greed search' to find the best *RL* method and, accordingly, hyperparameters requires very large computing capacity. The paper [7] analyses general principles of setting up hyperparameters of methods for a given task, but more appealing is searching for some settings that will be the best in any business process, which would allow them not to be included in ‘greed search’ and greatly reduce GPU(CPU) requirements in the end.

References

1. van der Aalst, W.M.P.: Process Mining: Discovery, Conformance and Enhancement of Business Processes. Springer, Heidelberg (2011). ISBN 978-3-642-19344-6, https://doi.org/10.1007/978-3-642-19345-3
2. van der Aalst, W.: Process Mining, Data Science in Action, 2nd edn. Springer, Heidelberg (2016). https://doi.org/10.1007/978-3-662-49851-4
3. Huang, Z., Van Der Aalst, W.M., Lu, X., Duan, H.: Reinforcement learning based resource allocation in business process management. Data Knowl. Eng. **70**(1), 127–145 (2011)
4. Silvander, J.: Business process optimization with reinforcement learning. In: Shishkov, B. (ed.) BMSD 2019. LNBIP, vol. 356, pp. 203–212. Springer, Cham (2019). https://doi.org/10.1007/978-3-030-24854-3_13
5. Arango, M., Foster, M., Muller, R., Vengerov, D.: Business Process Optimization via Reinforcement Learning (2017). https://doi.org/10.13140/RG.2.2.21505.04964
6. Panfilov, M., Goncharenko, I., Bugaenko, A.: Application of DS methods for solving applied task in finance. In: AI Journey Conference, Kaliningrad (2019)
7. Bugaenko, A.A.: Application of reinforcement learning to optimize business processes in the bank. Turkish J. Comput. Math. Educ. **12**(6), 1638–1644 (2021)
8. Sowjanya, K., Krishna Mohan, G.: Predicting heart disease using machine learning classification algorithms and along with tpot (Automl). Int. J. Sci. Technol. Res. **9**(4), 3202–3210 (2020)
9. Drozdal, J., Weisz, J., Wang, D., Ju, L., Su, H.: Trust in AutoML. In: International Conference on Intelligent User Interfaces, pp. 297–307. Proceedings IUI (2020)
10. Yu, C., Karlaš, B., Zhong, J., Zhang, C., Liu, J.: AutoML from service provider's perspective: multi-device, multi-tenant model selection with GP-EI. In: 22nd International Conference on Artificial Intelligence and Statistics, AISTATS 2019 (2020)
11. Tsiakmaki, M., Kostopoulos, G., Kotsiantis, S., Ragos, O.: Implementing autoML in educational data mining for prediction tasks. Appl. Sci. (Switzerland) **10**(1), 90 (2020)
12. Cai, H., Lin, J., Lin, Y., Zhu, L., Han, S.: AutoML for architecting efficient and specialized neural networks. IEEE Micro **40**(1), 8897011, 75–82 (2020)
13. Xu, P., Wu, C.-S., Madotto, A., Fung, P.: Clickbait? Sensational headline generation with auto-tuned reinforcement learning. In: 2019 Conference on Empirical Methods in Natural Language Processing and 9th International Joint Conference on Natural Language Processing, Proceedings of the Conference, pp. 3065–3075. EMNLP-IJCNLP 2019 (2020b)
14. John, I., Sreekantan, A., Bhatnagar, S.: Auto-scaling resources for Cloud Applications using reinforcement learning. Grace Hopper Celebration India GHCI **2019**, 9071835 (2019)
15. Bibal Benifa, J.V., Dejey, D.: RLPAS: reinforcement learning-based proactive auto-scaler for resource provisioning in cloud environment. Mobile Netw. Appl. **24**(4), 1348–1363 (2018). https://doi.org/10.1007/s11036-018-0996-0
16. Chiang, H.-T. L., Faust, A., Fiser, M., Francis, A.: Learning navigation behaviors end-to-end with AutoRL. IEEE Robot. Autom. Lett. **4**(2), 8643443, 2007–2014 (2019)
17. Allday, R., Hadfield, S., Bowden, R.: Auto-perceptive reinforcement learning (APRIL). CEUR Workshop Proc. **2487**, 103–112 (2019)

18. Wei, Y., Kudenko, D., Liu, S., Wu, L., Meng, X.: A Reinforcement learning based auto-scaling approach for saas providers in dynamic cloud environment. Math. Probl. Eng. **2019**, 5080647 (2019)
19. Chen, L., Lingys, J., Chen, K., Liu, F.: Auto: Scaling deep reinforcement learning for datacenter-scale automatic traffic optimization. In: Proceedings of the 2018 Conference of the ACM Special Interest Group on Data Communication, pp. 191–205. SIGCOMM (2018)

Features of Using Programs for Casting Processes Modeling

Viktor Kukartsev[1], Inga Kaposhko[1], Vladislav Kukartsev[1,2], Vadim Tynchenko[1,2,3(✉)], Anna Leonteva[1], and Igor Kartsan[3,4]

[1] Siberian Federal University, Krasnoyarsk, Russia
[2] Reshetnev Siberian State University of Science and Technology, Krasnoyarsk, Russia
[3] Marine Hydro-Physical Institute, Russian Academy of Sciences, Sevastopol, Russia
[4] Sevastopol State University, Sevastopol, Russia

Abstract. The article provides information on existing modeling program systems, their brief description and ability of practical application. A comparative analysis of the advantages and disadvantages of such simulation systems, which allowed in the framework of the tasks to reduce the number of systems under study to two were carried out. The calculation of the sprinkling system for casting from the alloy 20X5ML of the compressor wheel based on its model unit is per formed, while using the LVMflow modeling program, its most effective version of such a system was found. It is designed for the yield of suitable, the materials are selected to obtain the required chemical composition of the alloy, based on the condition that the smelting will be made on return and raw materials, the costs of their acquisition are determined. Further, this calculation option was used as the original for modeling in the Procast program, which made it possible to obtain more higher prime yield with suitable costs for raw melting stock.

Keywords: Casting modeling · Gating system · 20X5ML alloy · Prime yield · Melting stock · LVMflow · ProCAST

1 Introduction

The most important task of modern production of semi-finished products and products from foundry alloys is to reduce their labor intensity and an increase in the use of metal utilization. One of the ways to increase the coefficient of use of the metal is to apply modeling program systems.

Computer simulation becomes an integral part of the design processes of new parts and the design of the technological processes of their manufacture [1, 2]. It acquires the status of an important, and often decisive competitive advantage. Increasingly, customers on the foundry in the list of requirements for the manufacturer of this product put forward a requirement for the mandatory use of computer simulation [3, 4]. The advantages provided by the automated modeling systems of casting processes (AMS LP) are obvious. First of all, it is the possibility of working out the nuances of the foundry technology on the virtual prototype of the casting manufactured, which reduces or completely eliminates

R. Silhavy et al. (Eds.): CoMeSySo 2021, LNNS 232, pp. 23–30, 2021.
https://doi.org/10.1007/978-3-030-90318-3_3

the need for test castings, reduces the process of designing the technology and reduces the cost of castings. Visualization of the physical processes of foundry technology, such as filling by the mold, cooling and solidification of the metal, its warping under the action of thermal stresses makes it possible to better understand the features of these processes, and consequently, to more effectively manage them in order to reduce the marriage of castings and increasing the yield [5, 6].

To date, countries, the developers of the most common programs for modeling casting processes are Germany, France, USA, England, Finland and Russia. Brief available information about programs offered by these countries is presented in Table 1 [7–11].

Table 1. Foundry modeling programs.

Program name	Developer country	Developer	Functions	Solving method	Minimum price for 1 workplace, Euro
Magmasoft	Germany	MAGMA GmbH	HTSM	FDM	65000
WinCast	Germany	Simtec Inc	HTSM	FEM	65000
Procast	France	ESI Group	HTSM	FEM	60000
PAM-CAST	France	ESI Group	HTSM	FDM	30000
Flow3D	USA	Flow Science	HTS	MCV	28900
SolidCast	USA	Finite Solutions Inc	HTO	FDM	26500
CastCAE	Finland	CT-Castech Inc. Oy, Espoo	HTO	FDM	15000
LVMFlow	Russia	CJSC NPO MKM	HTS	MCV	29500
Poligon Soft	Russia	Sisoft Polygon Plus	HTSM	FEM	34600
Mavis-Flow	United Kingdom	Alphacast Software, Swansea, UK	HT	FDM	11000

Programs are characterized by the following possibilities:

- **H** - conducting the hydrodynamic calculation of filling the melted form.
- **T** - calculation of temperature fields during crystallization and detection of shrinkage defects.
- **S** - calculation of stresses in casting and residual deformation.
- **M** - structure modeling (grain structure, ferrite and perlite distribution, graphite inclusions size, etc.).
- **O** - parametric optimization of the grieving system in automatic mode.

The programs use the following methods for solving the simulation of casting processes:

- **FDM** - method of finite differences.
- **FEM** - method of finite elements.
- **MCV** - method of control volumes.

In Russia, today, the following modeling programs are widely spread:

- Magmasoft at 15 enterprises.
- WinCast at 3 enterprises.
- Procast at FSUE MMPP Salute (Moscow).
- SolidCast at OJSC Rostov Foundry Plant (Rostov-on-Don).
- LVMflow at 67 enterprises and 17 educational institutions.
- Poligon Soft at 8 enterprises and 10 educational institutions.

2 Materials and Methods

As a rule, a technology technologist uses the simulation program to verify the developed variant of the manufacture of casting. First of all, it determines the place of education of shrinkage defects and is looking for an option to eliminate them. Since the market dictates its own conditions for the execution of orders, this work technologist must fulfill this operation for 1–2 days (otherwise the favorable order will go to another manufacturer). This is especially true of the manufacture of castings from alloyed steels of complex configuration, the cost of which is high enough (at least 500 rubles/kg). For this reason, the enterprise has a serious task of choosing the necessary modeling program [12].

As practice shows, the best results are provided by the programs Poligon, ProCast, WinCast, which allow you to maximize the geometry of casting and identify even minor defects. However, the complexity of the application of method for solving the casting processes problem - FEM, leads to the preparation of the source data for computer modeling can take several days and even weeks for 1 calculation. In addition, it requires highly qualified personnel. For this reason, the mass distribution of foundry programs using the FEM, in Russia did not receive.

Today, the ProCast program has been successfully operating on FSUE Salyut, where the computer aided design (CAD) department was created, which completely optimizes the process of obtaining responsible castings, spending, at the same time, about a month per casting [13, 14].

In Russia, the only development that uses the FEM is the program Poligon, which is also very difficult to use and significantly more expensive programs that apply FDM. For many parameters, this program can be considered as a Russian analogue of the ProCast system. However, practical application at the foundry enterprises of Russia showed that the program Poligon cannot act as a daily program for the development of the technology of obtaining castings. The complexity of use, the duration of obtaining results, as well as the lack of qualified personnel, has reduced its use in ordinary enterprises to zero. This was also influenced that the duration of the training for using Poligon in production conditions ranges from six months to a year.

The method of control volumes combines the simplicity and factorization of FDM, as well as a good approximation of boundaries between different materials and various phases. This allows modeling as quickly as possible without losing the accuracy of calculations. LVMFlow is the only full-fledged foundry that uses this method and greatly wins among these programs in capabilities, has a Russian interface and a domestic database [15, 16]. In addition, the duration of learning to work with AMS LP LVMFlow does not exceed 16 h, after which the technologist can begin to independently simulate the casting processes.

Thus, the choice of a modeling program systems for an enterprise is a serious task affecting the preparation of the production of castings.

3 Experimental Results and Discussion

As part of the study, the adjustment of the grieving system (GS) was performed for casting from steel 20X5ML using LVMFlow and ProCast modeling programs. Then, based on the achieved prime yield, the material costs for raw charge materials were determined.

As an example, the casting of the compressor wheel with weight of 84 g was modelled (Fig. 1a).

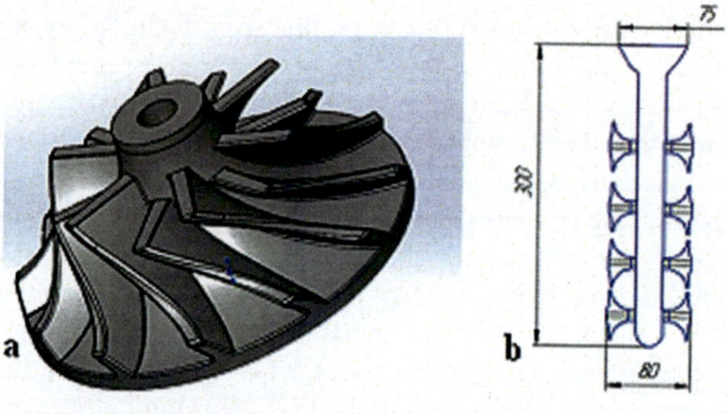

Fig. 1. Compressor wheel and model unit: a - compressor wheel; b - model block.

Considering the complexity of the configuration and the need to obtain sizes with minimal processing allowances, a method for manufacturing castings on smelt models was chosen. Using a methodology developed by M. L. Henkin, which provides the metal supply to the most massive casting nodes, the gating system of 1st type was calculated, and the model block was built with the following dimensions: diameter of the riser is 30 mm, feeder length is 8 mm, feeder diameter is 12 mm (Fig. 1 b.). As an initial, the simplest scheme of castings around the rising was accepted. The central risker serves as the basis for the creation of an integrated mechanized technological process of production of small castings. The use of a unified metallic frame as a supporting structure provides the convenience of link assembly of the model block and its high strength in the manufacture of the shell form. Already in the cast block, the central riser is securely fixed in the fixtures of the machines when cleaning and cutting the castings.

According to the sizes, a casting model with a spree-eating system was built in the SolidWorks program and the prime yield was calculated as 32.98%.

The formation of castings is a very complex process, so it is almost impossible to take into account all the factors affecting the crystallization process. Experimentally work out all the options for LPS for finding the best is a complex and long-term process. As a rule, the development of the most efficient version of the gating and supplying system to obtain an experimental casting takes several weeks, so to determine the correctness of the developed spruce system we applied the LWMFlow modeling program. The result of modeling has shown that the calculated maritime system does not provide a non-defective fill (Fig. 2a).

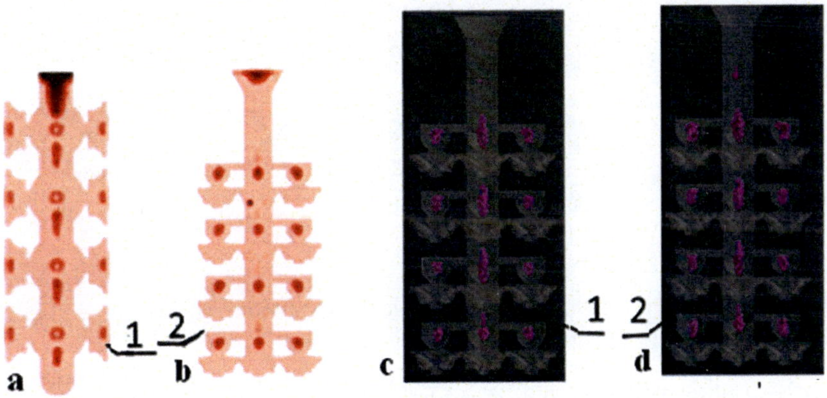

Fig. 2. Modeling results: a - defects detected during modeling in LWMFlow based on the initial grieving system; b - section of castings in a block showing the absence of defects (modeling in LWMFlow); c - modeling in ProCast with the sizes of the grieving system taken from the LWMFLow program; d - final result of modeling in the Procast program.

After a set of unsuccessful fillings, it was decided to change the design of the grieving system, including the location of the castings. As a result, it was possible to choose the most efficient variant of the spree-supplying system (Fig. 2b) with a riser diameter equal to 40 mm and a length of 320 mm. The prime yield was 30.76%.

The next stage of work was to carry out modeling using the ProCast program. The source data served as the results obtained using the LWMFlow program. Figures 2c and 2d show the results of this study.

After automatic optimization of the grid and 3D models of the casting, it was possible to reduce the diameter of the riser to 27 mm and its length up to 310 mm, which made it possible to increase the technological prime yield from 30.76% to 31.92%.

Further, on the basis of the obtained values of the outdoor casting, the metal balance required for the release of the "compressor wheel" cast was made of steel 20X5ML (Table 2).

Table 2. Estimated metal balance.

Balance item	Technology using LVMflow program		Technology using ProCAST program	
	%	kg	%	kg
Prime yield	30.86	1000	31.92	1000
Spoilage	4	129.618	4	125.313
Gates	58.64	1900.19	57.58	1803.88
Plums and splashes	2.5	81.011	2.5	78.3208
Irrecoverable losses	2	64.8088	2	62.6566
Waste	2	64.8088	2	62.6566
Metal filling	100	3240.4407	100	3132.83208

The balance of the metal is of great technical and economic importance. The consumption of metal, the cost of liquid metal and castings depend on it. The following indicators are of decisive importance: the type of melting unit and the technological level of production. When a new casting is launched into production, such indicators as: scrap of castings, waste (for induction furnaces), irrecoverable losses, waste and scrap are set based on those achieved in a similar production.

Having a balance of metal, we make the choice of charge materials, proceeding from the condition that the smelting will be carried out in an induction furnace with an acid lining, with full use of the return [17, 18]. The choice of materials is carried out in such a way that the chemical composition of the resulting alloy for each element has an average (permissible) value and meets the requirements of Russian State Standard (GOST) for steel 20X5ML. To do this, we calculate the charge using the selection method, which allows us to determine the required combination of starting materials. After that, we move on to determining the consumption of the selected charge materials and the cost of purchasing them. The results are shown in Table 3.

Table 3. Calculation of costs for raw batch materials (per one ton of prime yield).

Name of material	GOST or specification	Price, rubles / kg	Consumption and costs at prime yield of 30.86%			Consumption and costs at prime yield of 31.92%		
			%	kg	Rubles	%	kg	Rubles
Scrap 15X5	20072-74	20	32.5	1053.1	21 060	33.5	1049.5	20 990
Chrome metal X99	5905-2004	730	1.7	55.2	40 300	1.79	56.09	40 950
Manganese metal Mn 998	6008-90	140	0.186	6.04	850	0.185	5.84	820
Silicon metallic Cr 0	2169-69	140	0.188	6.07	850	0.194	6.1	854
Molybdenum metal MSh-2	TU 48-19-73-86	5500	0.217	7.04	38 740	0.186	5.84	32 140
Electrode break, grade A	TU 1911-109-73-2000	70	0.062	1.995	138	0.063	1.986	139
Total costs of raw materials			101 938			95 893		

4 Conclusion

The calculations show that when using the ProCAST casting process modeling program for developing a gating-feeding system, in comparison with the LVMflow program, it saves the cost of purchasing raw materials in the amount of 6045 rubles per 1 ton of suitable alloy steel 20X5ML. This enables the foundry to successfully compete in the market for castings from the specified alloy, at the expense of reducing their cost.

Thus, the article showed how the correct choice of a casting process modeling system can affect both the quality of such a process itself and the final cost of products and, accordingly, the competitiveness of an enterprise in the market.

Acknowledgements. This article was prepared as part of a research carried out with the financial support of the Russian science Foundation according to the research project No. 19–71-00028 within the framework of the Competition of 2019 “Conducting initiative research by young scientists”.

References

1. Gorski, F., Zawadzki, P., Hamrol, A.: Knowledge based engineering as a condition of effective mass production of configurable products by design automation. J. Mach. Eng. **16**(4), 5–30 (2016)
2. Gramegna, N., Corte, E.D., Poles, S.: Manufacturing process simulation for product design chain optimization. Mater. Manuf. Process. **26**(3), 527–533 (2011)
3. Grzegorzewski, P., Kochanski, A.: Data and modeling in industrial manufacturing. In: Grzegorzewski, P., Kochanski, A., Kacprzyk, J. (eds.) Soft Modeling in Industrial Manufacturing. SSDC, vol. 183, pp. 3–13. Springer, Cham (2019). https://doi.org/10.1007/978-3-030-03201-2_1

4. Kozłowski, J., Jakimiuk, M., Rogalewicz, M., Sika, R., Hajkowski, J.: Analysis and control of high-pressure die-casting process parameters with use of data mining tools. In: Hamrol, A., Kujawińska, A., Barraza, M.F.S. (eds.) MANUFACTURING 2019. LNME, pp. 253–267. Springer, Cham (2019). https://doi.org/10.1007/978-3-030-18789-7_22
5. Gramegna, N.: Smart casting process control and real time quality prediction. In: 9th VDI Conference with Specialist Exhibition on Casting Technology in Engine Construction: Potential for the Next generation of Vehicle Propulsion, vol. 2304, pp. 227–234 (2017)
6. Sika, R., Ignaszak, Z.: Cause-effect analysis using A&DM system for casting quality prediction. Arch. Foundry Eng. **20**(2), 5–12 (2020)
7. Patnaik, L., Saravanan, I., Kumar, S.: Die casting parameters and simulations for crankcase of automobile using MAGMAsoft. Mater. Today Proc. **22**, 563–571 (2020)
8. Kumar, R., Madhu, S., Aravindh, K., Jayakumar, V., Bharathiraja, G., Muniappan, A.: Casting design and simulation of gating system in rotary adaptor using procast software for defect minimization. Mater. Today Proc. **22**, 799–805 (2020)
9. Haldar, P., Sutradhar, G.: Simulation and validation of castings in shop floor. In: Casting Processes and Modelling of Metallic Materials. IntechOpen (2021)
10. Jones, C.A., et al.: A verification of thermophysical properties of a porous ceramic investment casting mould using commercial computational fluid dynamics software. In: IOP Conference Series: Materials Science and Engineering, vol. 861, no. 1, p. 012036 (2020)
11. Ogorodnikova, O.M., Yeltsin, S.V., Martynenko, S.V.: Reconstruction of thermo-physical properties to improve material database for casting simulation. In: IOP Conference Series: Materials Science and Engineering, vol. 971, no. 3, p. 032089 (2020)
12. Turishchev, V.V.: Modeling of foundry processes: what to choose? CADmaster **2**, 33–35 (2005)
13. Kozlowski, J., Sika, R., Gorski, F., Ciszak, O.: Modeling of foundry processes in the era of industry 4.0. In: Design. Simulation, Manufacturing: The Innovation Exchange, pp. 62–71. Springer, Cham (2018)
14. Abdullin, A.D.: New capabilities of software package ProCAST 2011 for modeling foundry operations. Metallurgist **56**, 323–328 (2012)
15. Dong, R., Feng, H., Wang, J.: Study on simulation and optimization of the casting method of mining up-tight wheel based on ProCast software. GAK gummi fasern kunststoffe **69**(13), 1086–1090 (2016)
16. Zaloznik, M., Sarler, B.: Modeling of macrosegregation in direct-chill casting of aluminum alloys: estimating the influence of casting parameters. Mater. Sci. Eng. A **413**, 85–91 (2005)
17. Kukartsev, V.A., Kukartsev, V.V., Tynchenko, V.S., Radionov, A.A.: Production technology of synthetic iron to ensure the efficiency of casting production. In: Materials Science Forum 989 MSF, pp. 600–605 (2020)
18. Kukartsev, V.A.: Cast iron and steel smelting in industrial frequency induction furnaces. Steel **5**, 26–28 (2016)

Development of an Information System to Help Identify Symptoms and Predict the Spread of COVID-19 Using Artificial Intelligence

E. V. Griva(✉), S. V. Konovalov, R. S. Kulshin, P. V. Senchenko, and A. A. Sidorov

Tomsk State University of Control Systems and Radioelectronics, 40, Lenina Avenue, Tomsk 634050, Russia

Abstract. The article discusses the development of an information system, the Main purpose of which is to help identify the symptoms of a new coronavirus infection COVID-19, predict the level of infection and mortality, using artificial intelligence technology, which is used for conducting a user survey and its subsequent analysis. Currently, predicting the rate of spread of COVID-19 is important for managing the situation and making decisions. Hospitals are overloaded and can't accept all potential carriers, and the category of people who are prone to panic due to the slightest changes in their health status only complicates the work of medical institutions. The use of the developed technology will reduce the number of false calls to hospitals, help to make a forecast of the spread of infection, and also model the social graph of transmission from carriers to healthy people.

Keywords: COVID-19 · Chat-bot · Risk of infection · Forecasting the spread of infection

1 Introduction

Currently, the issue of curbing the spread of COVID-19 is acute, and medical institutions are no longer able to cope with the flow of patients. People began to panic, and they did without knowing how to distinguish the symptoms of a common flu or other respiratory viral infection from COVID-19, began to call for medical help for any minor reason. The number of false calls to the ambulance and clinics has increased, which has created an additional burden.

At the same time, many people are constantly at real risk by contacting their friends and colleagues. Potential carriers of infection may not be aware of their infection until the last moment.

No less important is the problem of predicting the rate of infection both in General for countries and for individual regions. The lack of adequate forecasts leads to incorrect management decisions that can only exacerbate the previously described problems.

To solve previously identified problems, a user survey mechanism and a technology for predicting the probability of infection are being developed. It is proposed to predict

R. Silhavy et al. (Eds.): CoMeSySo 2021, LNNS 232, pp. 31–40, 2021.
https://doi.org/10.1007/978-3-030-90318-3_4

the rate of spread of COVID-19. Another task is to create a social graph that will notify users of a high risk of infection if someone from their environment confirms the fact of infection.

The proposed solution is implemented on the basis of chatbots for social networks and messengers, which allows you to ensure a low entry threshold, since the user does not need to go to third-party sites and download applications, you only need a social network or messenger account, all actions take place in the messenger where it is more convenient for the user to work. Data privacy is provided by the platforms on which the chatbot is based. A web application is also being created that allows you to view the COVID-19 spread forecast and visualize social graph for each region.

2 The Application of Neural Networks to Classification of Users Group at Risk

Currently determining the risk group for the disease is a classification task. There are many machine learning approaches and technologies available to solve this type of problem. The most popular ones are: decision trees, decision tables, Bayesian classifiers, and neural networks.

Decision trees are a tree-like hierarchical structure consisting of decision rules. To get the result, you need to pass from the root to the leaf and get the required value. Decision trees have a number of disadvantages that make it impossible to apply them to solve this problem. The main problems of decision trees are: the complexity of obtaining an optimal tree - making a decision at each local node and weak interconnection of branches, as well as the complexity of retraining the tree - generating a tree with a complex hierarchy that does not reflect the data sufficiently representatively [1].

Decision tables are a way to represent models with complex logic, establishing a relationship between conditions and subsequent actions, but at the same time presenting in a simple form a relationship between a large number of conditions and actions. The use of decision tables is not rational, since there are problems with changing the table when new data appears, as well as difficulties in establishing relationships between factors that are conditions [2].

Bayesian classifiers are based on the maximum a posteriori probability. For the classified object, the likelihood functions of each class are calculated, and then the posterior probabilities for each class are calculated and the class with the maximum probability is determined. The disadvantages of Bayesian classifiers are: a strong dependence on data representativeness – the use of distorted data at the training stage leads to irreversible errors in classes in the future [3].

Neural networks are mathematical models that implement networks of nerve cells in a living organism. A neural network is a system of interconnected neurons that transmit a signal to each other and make changes to it. Neural networks avoid the disadvantages of the methods described above, although they are not without others, but their use for classification of diseases is the most optimal solution [4, 5].

A neural network was developed to determine the risk group for the disease. The neural network is written in the Python programming language and the keras neural network library. Eleven criteria-factors are fed to the input of the neural network, according to which it must classify the user into five risk groups. The multi-layer perceptron was chosen as the network architecture. This architecture perfectly copes with the tasks of multi-class classification.

The input layer consists of eleven neurons that are criteria-factors. A sigmoid was used as the activation function. The sigmoid provides generation of non-linear functions at the output, and also provides a smooth gradient. When solving the classification problem, the value of the sigmoid tends to the sides of the curve, which allows you to define clearer classification boundaries.

The output layer consists of five neurons that correspond to five risk groups. The activation function on the output layer is softmax. Softmax perfectly copes with the task of non-binary classification, by applying it not to a single neuron as in the case of binary classifications, but to a layer in General. The softmax function at the output allows you to get a vector of probability values for all classes.

One of the varieties of stochastic gradient descent, the adaptive inertia method (Adam), was used as an optimization method. Adam copes well with a large number of criteria, is less susceptible to damping in local minima, and also shows good results very quickly. In the diagram, x's denotes the input layer consisting of the user's responses to the survey, and the output layer with y's denotes one of the five possible risk groups that the user will be assigned to base on the results of the neural network. The neural network diagram is shown in Fig. 1.

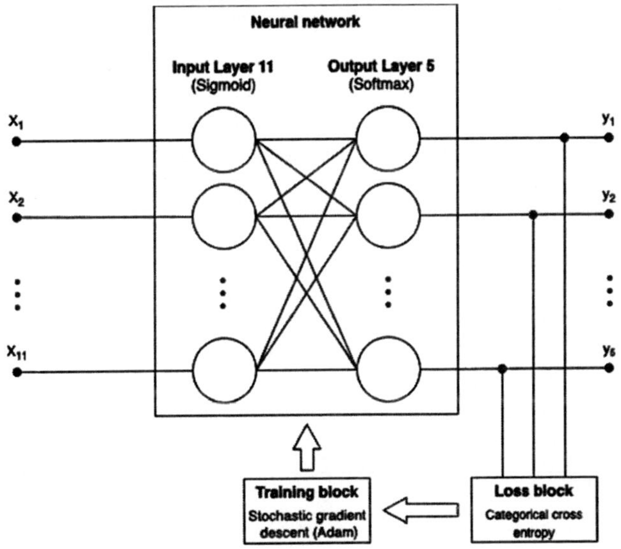

Fig. 1. Scheme of the neural network.

When designing a neural network, a categorical cross - entropy was chosen as the loss function. This function is intended for use in multi-class classification tasks. The developed neural network architecture provides stable and accurate work on classifying people into risk groups, and the probability of error is quite small. The network is constantly learning more from user data and the accuracy is getting higher.

After activating the chatbot, the user must complete a survey that will help determine whether the user has symptoms of COVID-19. The survey consists of 11 questions. The received responses are processed and sent to the input layer of the neural network, after which the assessment takes place and the user belongs to one of the five risk groups. Using a neural network allows you to adjust the accuracy of assigning users to different risk groups, based on the responses to the survey. To train the neural network, we used data from 1,000 people who were tested in the chatbot.

3 The Forecasting the Spread of Infection

Currently, there are many methods for predicting the spread of infection, such as Youyang Gu COVID model (YYG) [6], COVID-19 agent-based model (COVID-19 ABM) [7], and others [8]. But to use many of them, need a lot of computing power and data on a huge number of parameters. In most regions, the government provides a limited set of data, so the development uses the SIERD model (Susceptible, Infected, Exposed, Recovered, Dead) [9], a modification of the SIR model (Susceptible, Infected, Recovered) [10]. Unlike the SIR model the SIERD model takes into account parameters such as the incubation period of infection and the number of people who died, which are important parameters for modeling. In the SIERD model, the population is divided into several groups. The first group is people who are susceptible to infection, the second group is people who have the disease in the incubation period, the third group is sick people, the fourth is recovered people, and the fifth is dead. Each group is associated with a variable in the system differential equations, solving which you can predict the spread of COVID-19. This system of equations is presented below (1–5):

$$\frac{dS}{dt} = -\beta\frac{S}{N}I \quad (1)$$

$$\frac{dE}{dt} = +\beta\frac{S}{N}I - \sigma E \quad (2)$$

$$\frac{dI}{dt} = +\sigma E - \gamma I \quad (3)$$

$$\frac{dR}{dt} = (1 - f)\gamma I \quad (4)$$

$$\frac{dD}{dt} = f\gamma \quad (5)$$

Where S is people susceptible to infection, E — those who have the disease in the incubation period, I-sick, R-recovered, D-dead, N-the total population.

ββ is a constant that determines the likelihood of contact between a susceptible person and a sick person.

σ is a constant that determines the probability of infection.

f is the case fatality rate.

(1 − f)γ is the probability of recovery.

fγ is the probability of death.

The data for the system of equations were taken from reports from the World Health Organization [11].

This model was implemented in the Python programming language using the Pandas, SciPy, NumPy libraries. For clarity, a graph was built which displays data on the number of infected official statistics in Russia and displays data with the predicted model. This graph is shown in Fig. 2. The accuracy of the model was 99.2%. The following formula was used to obtain the accuracy of the model (6):

$$Accuracy = 100\% - \frac{\sum |A - B|}{\sum A} * 100\% \tag{6}$$

where A - true actual values, B - predicted values.

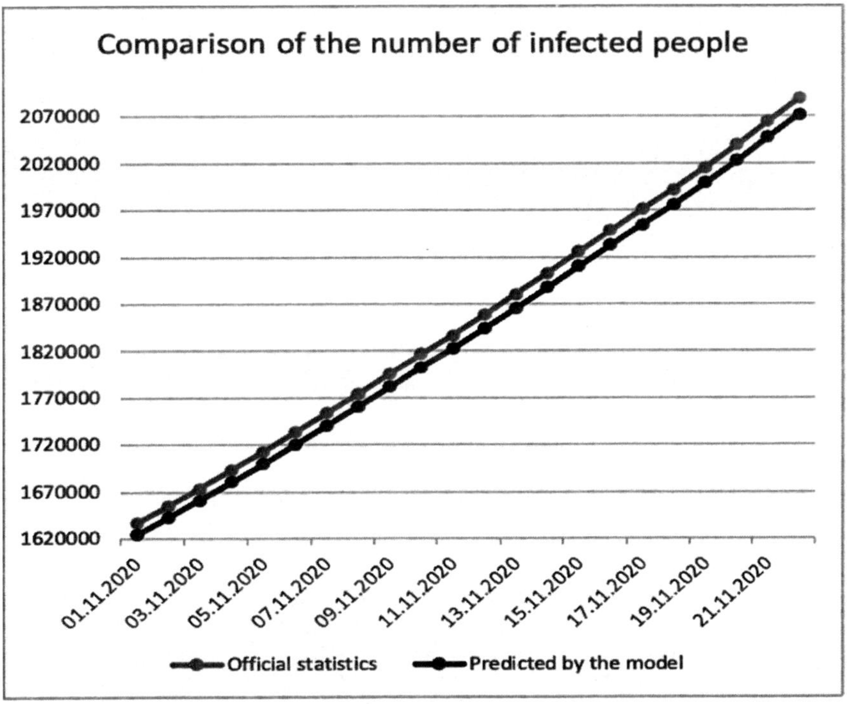

Fig. 2. The number of infected according to official statistics and predicted by the model.

4 The User Interaction with the Program

For comfortable interaction of users with the program, the interface of chatbots for social networks and messengers was chosen. During a small comparative analysis, the chatbot interfaces in the Russian social network Vkontakte and Facebook messenger were highlighted, as they provide a user-friendly interface and a functional application programming interface (API). Visualization of the results of forecasting and building social graphs is provided via the web interface.

To interact with the chatbot, the user only needs a social network account. To start a conversation with the chatbot, the user just needs to write a message. As soon as this happens, the chatbot sends the user information about the privacy policy and uses the social network interfaces to add a special keyboard with answer options and the necessary buttons to the dialog. The created keyboard is a set of special commands and keywords that the chatbot understands. When you click on one of the keys, a special command is sent to the chatbot and, depending on the selected action, the necessary information is provided. The chatbot operation scheme is shown in Fig. 3.

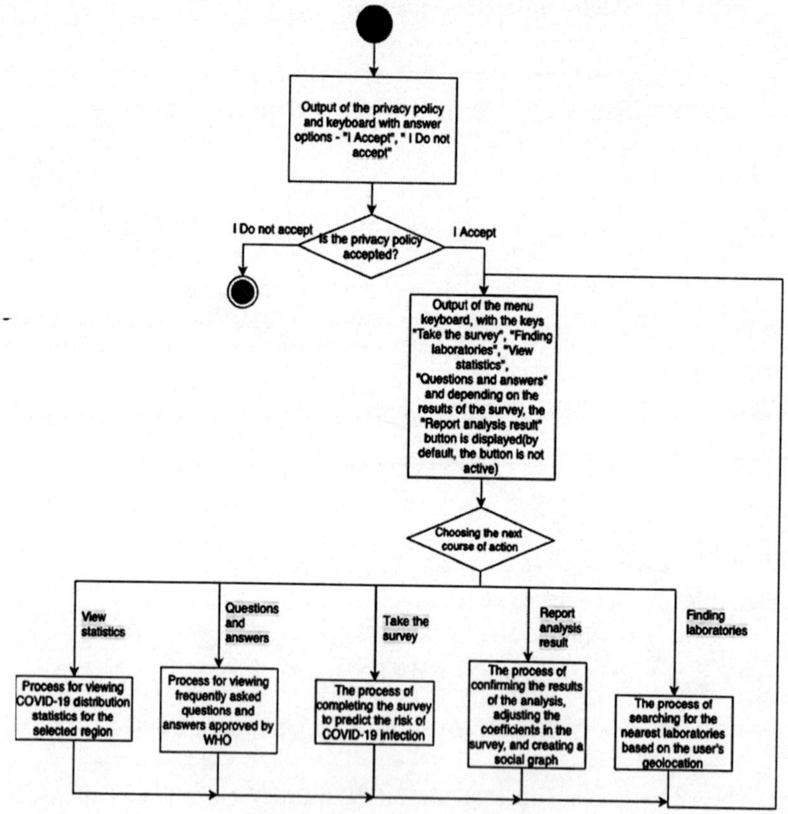

Fig. 3. Scheme chatbot works.

Chat-bot is fixated, it has no exit point after executing any action, the user still gets to the main menu. The chatbot always works and doesn't stop for a second. If the user no longer wants to use the bot, they can use the messenger tools to prevent sending messages to the chatbot. At any time, the user can return permission to send messages from the chatbot and the conversation will continue from where it left off.

The most important function is to pass a survey to calculate the risk of infection. The chatbot sends questions to the user in the format of a question-and-answer dialog and changes the keyboard, allowing them to choose different answers. The test contains 11 questions in total (Table 1). The survey is based on the recommendations of major medical organizations and officially approved methods.

Table 1. Questions included in the survey.

#	Question	Answer options
1	Have you recently visited other countries, regions, or cities?	Yes/No
2	Specify your current status?	Healthy/Quarantine/Home treatment/Hospitalization
3	Have you had any recent contact with people who are ill?	Yes/No/I don't know
4	Do you have a chronic medical condition (such as Asthma, diabetes, etc.)?	Yes/No
5	Do you visit places of a mass congestion of people (work, shops, etc.)?	Yes/No
6	Do you violate the self-isolation regime or do your roommates violate it?	Yes/No
7	Do you have difficulty breathing?	Yes/No
8	Do you have sneezing, coughing, and nasal congestion?	Yes/No
9	Do you experience pain in your muscles and chest?	Yes/No
10	Do you experience headaches and weakness?	Yes/No
11	What is your temperature?	Any number entered

From the questions included in the survey (Table 1), you can see different types of answers. To make it easier to work with them, each response is processed separately and assigned its own weight value. In questions with the answer type "Yes/No" answer weights can take the value 1 or 0. In the second question, the answers "Healthy/Quarantine/Home treatment/Hospitalization" correspond to weight coefficients in the form of fractional numbers in the range from 0 to 1. The same thing happens in the eleventh question, the more or less the temperature value deviates from normal values, the greater the weight coefficient in the range from 0 to 1 will be obtained (For example, at 37.6 °C, the weight coefficient value will be about 0.67).

After completing the survey, the user can view the probability of the risk of infection with COVID-19 and recommendations for further actions. When the user is familiar with the results of the survey, an additional button is added that will allow the user to specify the results of tests for COVID-19, if they passed them. This result will be taken into account and if the result is positive, the chatbot displays a keyboard with a list of friends. On this keyboard, the user can mark the people they have recently contacted. After confirming the selection, these users receive a notification from the chatbot, which says that one of their friends is sick and a list of recommendations indicating further actions. The notification does not disclose which of the friends became ill, in order to maintain anonymity.

After clicking the search button for the nearest laboratories, the chatbot offers to provide it with geodata, or enter the locality manually. The chatbot will analyze the list of all laboratories, and then display the nearest ones to the user. If you can't find a lab in a locality, you will automatically be offered another one in the nearest localities.

The button with recommendations for infection control and prevention is a list of frequently asked questions and answers provided by the world health organization.

The statistics button lets you select any region in alphabetical order. After selecting a region, the chatbot sends the user up-to-date statistics on the spread of infection, which includes: the number of cases, the number of deaths, and the number of people who recovered. Statistics are updated once an hour and are taken from official sources.

The Web interface provides two functions. The first is to view graphs compiled by the COVID-19 spread prediction model. The second function is visualization of social graphs of infection spread, with the condition that each element of the graph remains anonymous. The graph allows you to view a general picture of how the infection is transmitted from person to person. An example of a social graph with a small sample is shown in Fig. 4.

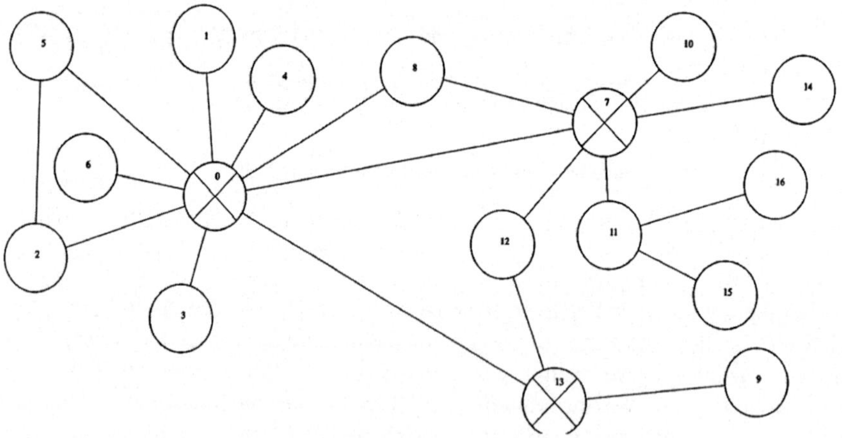

Fig. 4. Example of a social graph.

In the social graph, the data is anonymous, so all people are identified by a number. In Fig. 4, the dots numbered 0, 7, and 13 indicate people who have confirmed their positive diagnosis. At the same time, people with numbers 7 and 13 could get infected both from the person designated by the number 0, and from other people not included in the schedule. For example, the numbers 0 and 7 independently indicated that they were in contact with the person number 8. Patients with a negative or unverified diagnosis could also be in contact with each other, as in the case of numbers 5 and 2. The more people who fall into the social graph, the faster they can be notified of the potential risk of infection, which allows medical organizations to more accurately track COVID-19 outbreaks and foci.

5 Conclusion

Thus, the developed information system should help in the fight with COVID-19. The neural network will learn from the responses of new users and provide more accurate results for assessing the risk of infection. Using the program will help reduce the number of false calls to medical organizations, which will have a positive impact on their work and allow them to focus their efforts on really sick patients. Building a social graph and predicting the spread of infection will make it possible to make more informed management decisions.

After conducting mass testing on users, it will be possible to conduct an analysis that will make it clear whether the questions are correctly formulated, and whether the system correctly evaluates the significance of each of them. It is also planned to analyze the constructed social graphs and determine how correct the resulting graphs are and in which areas they can be applied.

In the future, it is also planned to improve the functionality of the information system, as well as add new functions. It is planned to use chatbot technology in other social networks and messengers in order to increase the reach of various user groups. It is planned to improve the algorithms for conducting a survey and building social graphs, to maximize the simplification of working with users.

Acknowledgments. This paper is designed as part of the state assignment of the Ministry of Science and Higher Education; project FEWM-2020–0036.

References

1. Patel, H., Prajapati, P.: Study and analysis of decision tree based classification algorithms. Int. J. Comput. Sci. Eng. **6**(10), 74–78 (2018)
2. Senchenko, P.V., Ekhlakov, Y.P.: Use of decision tables in monitoring of performance discipline. In: 13th IEEE International Conference on Application of Information and Communication Technologies (AICT), pp. 1–4 (2019)
3. Langley, P., Iba, W., Thompson, K.: An analysis of Bayesian classifiers. In: Proceeding AAAI, vol. 90, pp. 223–228 (1992)
4. Ardabili, S.F., et al.: Covid-19 outbreak prediction with machine learning. medRxiv 2020.04.17.20070094. https://ssrn.com/abstract=3580188 (2020). Accessed 01 May 2021
5. Zhang, P.: Neural networks for classification: a survey. IEEE Trans. Syst. Man Cybern. Part C (Appl. Rev.) **30**(4), 451–462 (2000)
6. Gu, Y.: YYG model. https://covid19-projections.com (2020). Accessed 01 May 2021

7. Wallentin, G., Kaziyeva, D., Reibersdorfer-Adelsberger, E.: COVID-19 intervention scenarios for a long-term disease management. Int. J. Health Policy Manage. **9**(12), 508–516 (2020)
8. Fong, S.J., Li, G., Dey, N., Crespo, R.G., Herrera-Viedma, E.: Finding an accurate early forecasting model from small dataset: a case of 2019-ncov novel coronavirus outbreak. Int. J. Interact. Multimedia Artif. Intell. **6**(1), 132–140 (2020)
9. Piccolomini, E.L., Zama, F.: Monitoring italian COVID-19 spread by an adaptive SEIRD model. medRxiv 2020.04.03.20049734. https://doi.org/10.1371/journal.pone.0237417. Accessed 04 May 2021
10. Gao, S., Teng, Z., Nieto, J.J., Torres, A.: Analysis of an SIR epidemic model with pulse vaccination and distributed time delay. J. Biomed. Biotechnol. https://doi.org/10.1155/2007/64870, Accessed 04 May 2021
11. World Health Organization: Coronavirus disease (COVID-19) pandemic. https://www.who.int/emergencies/diseases/novel-coronavirus-2019 (2021). Accessed 05 May 2021

Use of Analytic Hierarchy Process in Tasks of Knowledge Visualization

V. V. Mayorov, I. A. Shaytorova(✉), Violetta V. Grigorenko, N. B. Nazina, and A. A. Egorov

State-Funded Institution of Higher Education of Khanty-Mansiysk Autonomous District – Yugra, «Surgut State University», Okrug 628408, Russia

Abstract. The importance of knowledge visualization for better perception and acquisition of information in educational systems is obvious, because this facilitates the understanding of material studied. The quantity of visualization tools is high and selection of them is random. Article reviews the mechanism of selecting a knowledge visualization tool with the help of Saaty's analytic hierarchy process when developing remote courses for training specialists of oil and gas industry. Article presents hierarchical model, decision-making stages while selecting a knowledge visualization tool. Hierarchies of criteria and alternatives were formed, balance scale of criteria was established and according to this, the alternative assessment was performed in conformity with the method of linear convolution. Such an approach defines the relative significance of alternatives being studied for all criteria, which are in hierarchy. The example of selecting a tool of knowledge visualization, with the help of a hierarchical model, was given. Pairwise comparison matrices were built with regard to each criterion. In conclusion it is emphasized that Saaty's analytic hierarchy process can be used for objective selection of knowledge visualization tools in development of remote courses for training specialists of oil and gas industry.

Keywords: Analytic hierarchy process · Saaty's method · Decision-making · Knowledge visualization · Hierarchical model

1 Introduction

Competency management, qualification upgrade, requalification, assimilation of new technologies - these are the basic directions of remote training of employees in the oil and gas industry.

Remote education – is acquiring knowledge with the help of modern informational means, telecommunications, Internet.

One of the most important tasks of remote education of specialists, working in different subdivisions of the oil and gas mining branch, is the improvement of training quality. Informational saturation with knowledge generates a need for transferring the information in visually observable form. This can be achieved with the help of content-related compression of information for presenting it in convenient form, because 90%

R. Silhavy et al. (Eds.): CoMeSySo 2021, LNNS 232, pp. 41–47, 2021.
https://doi.org/10.1007/978-3-030-90318-3_5

of information is received by humans with the help of eyesight, 50% of brain neurons process the visual information, presence of images increases the wish to read the text in 80%. Human being perceives only 10% of information passing through his organs of hearing, 20% – of information read and 80% – of information seen. If the guidelines for using oil-and-gas production equipment don't contain illustrations, then the information is ingested only in 60%, if to add images, then the percent will grow up to 95%.

Intelligent visualization of knowledge will provide the following advantages: the better understanding of data; increase of decision-making speed; high involvement of the audience; accelerates perception of information; to explain difficult themes in accessible ways; to analyse large data sets; to create reports and analyse statistics. Knowledge visualization – is a graphic representation of data and knowledge, in the form of pictures, graphics, diagrams, animation etc.

Analytic hierarchy process - is mathematical tool, which helps to take decisions in difficult situations: when there are few options of task solution, when criteria, defining the measure of suitable solution, is specified, when the reasons and conditions, influencing the selection of end result, are known.

Analytic hierarchy method is highly universal and has the important advantage - is used for solving a wide variety of tasks (distribution of resources, formation of ratings, tool selection, analysis of possible scenarios of unfolding situations).

2 Stages of Applying Saaty's Analytic Hierarchy Method

Step 1: Selection of research goal.

Step 2: Defining the basic criteria and alternatives.

Step 3: Build-up of hierarchy: tree from the goal through criteria towards alternatives.

Step 4: Forming the matrix of pairwise comparison of criteria according to the goal and alternatives according to criteria.

Step 5: Using the methodology of analysing the matrices received.

Step 6: Defining the weights of alternatives by hierarchy system.

While preparing the material for training in traditional or remote form, knowledge visualization is important for better perception and acquisition of material. When using Saaty's analytic hierarchy method for selecting a knowledge visualization tool, it's important to highlight the basic criteria and alternatives.

Criteria – is defining to which extent this or another decision is suitable.

Alternatives – are an abundance of solution variants.

Because classification of knowledge visualization includes: information difficulty level, difficulty level of data entered, difficulty level of display, difficulty level considering the type of data [1–6], as a result of discussions it was decided to select five criteria according to the level of difficulty taking into account the type of data, which have to satisfy to decision on knowledge visualization tool selection.

The first step consists in choosing the goal of research. Goal of this research lies in choosing one of three knowledge visualization tools applicable to remote studies: animation, 3D – graphics, video.

Second step involves the selection of basic criteria and alternatives of knowledge visualization (Table 1).

Difficulty level of data representation, taking into account the type of data, was selected as criteria. This research considers such criteria as:

Qualitative data – is information, which can't be measured with the help of digits.

Quantitative data – is data collected by a group of people and processed with the help of different methods.

Difficult data – is data, which has a difficult structure. For example, in programming: array, record, file.

Simple data – is data, which do not contain the other types of data within themselves.

Ordered data (categorical data) – is data, within which the ordering relation is established, but the distances between categories are not defined, for example, health can be excellent, good, satisfactory.

Such knowledge visualization tools as - animation, 3D - graphics and video, serve as major alternatives.

Table 1. Selection of basic criteria and alternatives of knowledge visualization.

Major criteria of selecting knowledge visualization tool considering the type of data	
1	Qualitative data
2	Quantitative data
3	Difficult data
4	Simple data
5	Ordered data
Basic alternatives	
1	Animation (A)
2	3D-graphics (B)
3	Video (C)

Third step – build-up of a three level hierarchy tree based on the goal of research, then the second level – there are criteria, which define the alternatives (Fig. 1).

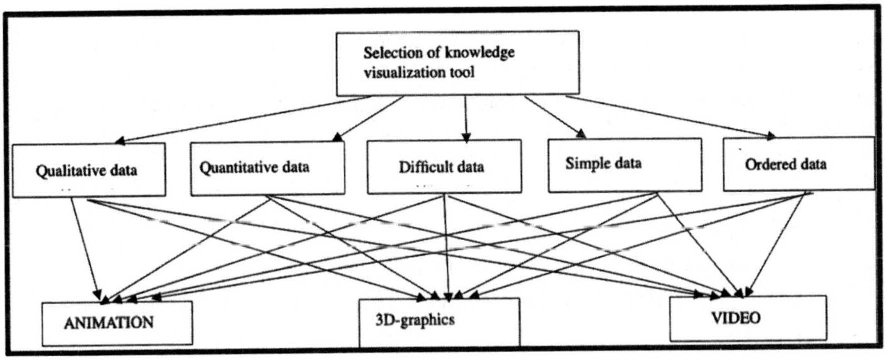

Fig. 1. Hierarchy composition: tree coming from the goal through criteria to alternatives.

At fourth step of research, the comparison matrix was built on the basis of Saaty's 1–9 scale of the relative importance [7] for comparing the criteria with regard to the selected knowledge visualization tool (Table 2) and for comparing the alternatives (knowledge visualization tools) according to criteria (Table 3).

Table 2. Comparison of criteria with regard to selection of knowledge visualization tools.

	Qualitative data	Quantitative data	Difficult data	Simple data	Ordered data
Qualitative data	1	3	1	½	5
Quantitative data	1/3	1	¼	1/7	2
Difficult data	1	4	1	1	6
Simple data	2	7	1	1	8
Ordered data	1/5	1/2	1/6	1/8	1

Table 3. Comparison of alternatives by criteria.

Qualitative data	A	B	C	Quantitative data	A	B	C	Difficult data	A	B	C
A	1	4	1/2	A	1	1/2	3	A	1	1	2
B	¼	1	1/5	B	2	1	4	B	1	1	3
C	2	5	1	C	1/3	1/4	1	C	1/2	1/3	1
Simple data	A	B	C	Ordered data	A	B	C				
A	1	1/3	4	A	1	2	1/5				
B	3	1	5	B	1/2	1	1/6				
C	¼	1/5	1	C	5	6	1				

At the fifth step we use methodology of analysing the matrices received. Calculation of criteria weight, from point of view of the goal set, is presented in Table 4, in Table 5 we made pair-wise comparison of alternatives according to the criteria. Calculation of weights by criteria is presented in Table 6. Matrix of alternatives weight by each criterion is formed in Table 7. In order to find the sum of elements of each column, we use the following formula (1):

$$S_j = a_{1j} + a_{2j} + \ldots + a_{nj}; \tag{1}$$

where S_j – sum of elements in each column of matrix; i – rows of matrix; j – matrix columns; a_{nj} – matrix element in n- row j-column.

Through the formula (2) we find scoring matrix, for this we divide all matrix elements by sum of elements in respective column:

$$A_{ij} = \frac{a_{ij}}{S_j}; \tag{2}$$

where A_{ij} – scoring matrix; S_j – sum of elements in each column of matrix; i – rows of matrix; j – columns of matrix; a_{ij} – *matrix* element, which defines the importance of criterion i with regards to criterion.

Table 4. Calculation of criteria weight from point of view of the goal established.

	Qualitative data	Quantitative data	Difficult data	Simple data	Ordered data	Weight in decimal fractions	Weight in %
Qualitative data	0.221	0.194	0.293	0.181	0.227	0.223	22.3
Quantitative data	0.074	0.065	0.073	0.052	0.091	0.071	7.1
Difficult data	0.221	0.258	0.293	0.361	0.273	0.281	28.1
Simple data	0.441	0.452	0.293	0.361	0.364	0.382	38.2
Ordered data	0.044	0.032	0.049	0.045	0.045	0.043	4.3

Table 5. Pairwise comparison according to criteria.

Qualitative data	A	B	C	Weight in decimal fractions	Weight in %	Quantitative data	A	B	C	Weight in decimal fractions	Weight in %
A	0.31	0.40	0.29	0.334	33.39%	A	0.30	0.09	0.71	0.366	36.56%
B	0.08	0.10	0.12	0.098	9.82%	B	0.60	0.18	0.06	0.280	28.02%
C	0.62	0.50	0.59	0.568	56.79%	C	0.10	0.73	0.24	0.354	35.42%
Difficult data	A	B	C	Weight in decimal fractions	Weight in %	Ordered data	A	B	C	Weight in decimal fractions	Weight in %
A	0.40	0.43	0.33	0.387	38.73%	A	0.15	0.22	0.15	0.174	17.41%
B	0.40	0.43	0.50	0.443	44.29%	B	0.08	0.11	0.12	0.103	10.33%
C	0.20	0.14	0.17	0.170	16.98%	C	0.77	0.67	0.73	0.723	72.25%
Simple data	A	B	C	Weight in decimal fractions	Weight in %						
A	0.24	0.22	0.40	0.284	28.42%						
B	0.71	0.65	0.50	0.619	61.94%						
C	0.06	0.13	0.10	0.096	9.64%						

Table 6. Criteria weights.

	Weight	Weight in %
Qualitative data	0.223	22.3%
Quantitative data	0.071	7.1%
Difficult data	0.281	28.1%
Simple data	0.382	38.2%
Ordered data	0.043	4.3%

Table 7. Matrix of alternatives weights by each criteria.

		Qualitative data	Quantitative data	Difficult data	Simple data	Ordered data
A	Animation	0.334	0.366	0.387	0.284	0.174
B	3D - graphics	0.098	0.280	0.443	0.619	0.103
C	Video	0.568	0.354	0.170	0.096	0.723

By multiplying the received matrix of alternatives weights, according to each criterion, by a column of criteria weights, we receive weights of alternatives from the point of view of achieving a goal:

$$\begin{pmatrix} 0.334\ 0.366\ 0.387\ 0.284\ 0.174 \\ 0.098\ 0.280\ 0.443\ 0.619\ 0.103 \\ 0.568\ 0.354\ 0.170\ 0.096\ 0.723 \end{pmatrix} \times \begin{pmatrix} 0.223 \\ 0.071 \\ 0.281 \\ 0.382 \\ 0.043 \end{pmatrix} = \begin{pmatrix} 0.325 \\ 0.407 \\ 0.267 \end{pmatrix} \quad (3)$$

As a result, in the table we get the weights of alternatives from the point of view of achieving the goal established – selection of knowledge visualization tool (Table 8).

Table 8. Weights of alternatives for selection of visualization tool.

		Weight	Weight in %
A	Animation	0.325	32.53%
B	3D - graphics	0.407	40.73%
C	Video	0.267	26.74%

3 Conclusion

In the course of research work the Saaty's analytic hierarchy process was applied for the purpose of selecting knowledge visualization tool while developing course of remote training of employees from oil and gas producing industry. Following on from the results of research, the percentage of selection of knowledge visualization tool, among the alternatives found, was calculated.

Specifically, calculation demonstrated that the 3D-graphics (41%) is more attractive for knowledge visualization in this subject area. If we will use various instruments of visualization, then further, by priority, animation (32%) and video (27%) will follow.

Suggested approach provides objective assessment of quality of selecting visualization tool while preparing course of remote training of staff in oil and gas mining branch.

Acknowledgements. Research was done in the frames of grant RFBR 18-47-860003 p_a «Mathematical modelling of the processes of oil refining and petro chemistry on the basis of dynamic models in terms of compositionally continuous mixtures».

References

1. Gavrilova, T.A., Alsufev, A.I., Grinberg, E.Ya.: Visualization of knowledge: critique of the St. Gallen school and analysis of modern trends. Bus. Inf. **3**(41), 7–19 (2017)
2. Gavrilova, T.A., Guliakina, N.A.: Visual methods of working with knowledge: attempt of review. Artif. Intell. Decis. Mak. **1**, 15–21 (2018)
3. Gavrilova, T.A., Leshcheva, I.A., Strahovich, E.V.: About usage of visual conceptual models in teaching. Vestnik of St. Petersburg State University. Manag. Ser. **4**, 125–151 (2011)
4. Gavrilova, T.A., Kudriavtsev, D.V., Leshcheva, I.A., Pavlov, Ya.Yu.: About one method of classification of visual models. Bus. Inf. **4**(26), 21–34 (2013)
5. Gavrilova, T.A., Leshcheva, I.A.: Cognitive style and formation of conceptual frameworks of knowledge. J. Psychologies **1**(13), 137–153 (2016)
6. Grigorenko, V.V, Gorbunov, S.V., Hvostov, D.Yu, Kasatkin, V.V.: Unstable systems: homogeneity problem of groups. Vestnik kibernetiki **1**(33), 67–75 (2019)
7. Nazin, A.S., Novichkov, N.A., Grigorenko, V.V., Nazina, N.B.: Algorithm of decision-making on selection of optimal abstract for scientific article in English language. Nat. Tech. Sci. **5**(143), 95–102 (2020)
8. Yau, N.: The Art of Visualization in Business. How to Present Complex Information in Simple Images. Ed. Yau Nathan, translation from English by Svetlana Kirova. Mann, Ivanov, Ferber, Moscow, p. 352 (2013)
9. Saaty, T.L.: Decision making. Analytic Hierarchy Process. Economic Theory of Money, Banking Operations and Financial Markets. Translation from English by R. G. Vachnadze. Radio i svyaz, Moscow, p. 278 (1993)

Blockchain-Oriented Software Development Issues: A Literature Review

Ammar AL-Ashmori[1(✉)], Shuib Basri[1], P. D. D. Dominic[1], Amgad Muneer[1], Qasem Al-Tashi[1], and Yasser Al-Ashmori[2]

[1] Department of Computer & Information Sciences, Universiti Teknologi PETRONAS Seri Iskandar, Perak, Malaysia
ammar_18003398@utp.edu.my
[2] Department of Civil & Environmental Engineering, Universiti Teknologi PETRONAS Seri Iskandar, Perak, Malaysia

Abstract. Blockchain-oriented software is a trend and a hot topic for discussion. The fitness of the common software engineering practices, tools, techniques, and methods for this new type of software has been discussed by many investigators. Furthermore, a new branch of software engineering known as blockchain-oriented software engineering has been announced as a new research field. However, the current issues of Blockchain-oriented software development have not been well defined yet. This study offers a systematic review procedure for software engineering, the stages pursued, the explanation of every one of them, along with its execution employed to the subject of the current issues in blockchain-oriented software development. The results of this study confirm that there 15 issues related to the exceptional properties of blockchain-oriented software affecting its development process.

Keywords: Blockchain-oriented software · Software process improvement · Development issues · Systematic review

1 Introduction

An anonymous figure known by Nakamoto published a paper in 2008 talking about the revolutionary idea of peer-to-peer electronic money exchanging system. This system allows financial participants to transfer payments to each other without centralized authority, such as banks avoiding the double-spending issue. Based on this, and in the same year, the first cryptocurrency appeared known as Bitcoin [1]. Presently, over one thousand Cryptocurrencies [2] are there. Bitcoin is the top cryptocurrency with over 100 billion USD market value. Still, many experts think blockchain technology should not be limited to cryptocurrency as it can be a public ledger for any transactions rather than digital currency [3]. Therefore, blockchain's evolution can be split into three generations, where the first generation is for digital currency. The second generation is for digital finance, and the third generation is for digital society [4].

R. Silhavy et al. (Eds.): CoMeSySo 2021, LNNS 232, pp. 48–57, 2021.
https://doi.org/10.1007/978-3-030-90318-3_6

Designing, developing, and deploying blockchain-oriented software (BOS) have technological innovation. Therefore, BOS developers should be careful around maliciousness, immutability, and distribution of BOS. BOS developers should be Carrying the inevitable scarcity of tools and resources for a new technology, which is considered a new development paradigm [5]. This new paradigm, Blockchain-oriented Software Engineering (BOSE) was proposed by [6] and [7] to address the considerable distinctions between traditional development for software and BOS development.

In this new development paradigm, errors fixing after the deployment is excessively hard or practically impossible without compromising the blockchain's integrity. This vulnerability was the motivation for most of the attacks on the blockchain platform. Some of them did succeed and caused an unexpected financial loss. MtGoX attack in 2014 was the largest with a loss of $460 million, and one more exceptional exploit was in 2016 that experienced by the DAO establishment [8].

Consequently, to avoid such a scenario, it was acknowledged the need to develop specific techniques and tools for BOS development. Best practices, quality metrics, specific modeling languages, development methods, design patterns, security strategies, requirement collection, task assignment, and testing tactics should be developed for BOS [6, 7, 9–14].

Although the new development paradigm suggests that BOS development is different from traditional software development, very few studies addressed the issues of BOS development partially, and none of them have focused on identifying these issues. Recognizing the BOS development issues will help the research community investigate and propose suitable solutions to support the process of designing, developing, and deploying BOS, and will help the people interested in joining BOS projects to prepare themselves with common issues for their projects. Therefore, this research seeks to identify the BOS development distinctive issues by gathering all the literature related to BOS and systematically reviewing them to comprehend the study topic, BOS development issues that serve the other researchers' investigations as a start point.

This study's background is discussed briefly in Sects. 2, and 3 describes the research design. The results and the discussion are presented in Sects. 4, and 5 concludes the paper and encourages future works.

2 Background

Lately, blockchain is of the hottest topics in the academic and industrial worlds. There is a high demand for BOS applications, and BOS projects are initiated almost every day spontaneously that never happened since the appearance of internet technology [12]. BOS is being used in a variety of areas containing but not restricted to the internet of things [15], data storing and management [16], licensing authentication [17], and educational record and reputation [18]. BOS term refers to any software or application built upon the effective realization of blockchain technology [6].

Blockchain is a peer-to-peer network protocol to store data in a distributed database which records all the transactions history that has ever happened on the network. A version of this database can be found at all the network nodes and all the blocks linked to each other via cryptographical hashing. A block refers to its predecessor block by keeping its cryptographical hashing, which builds a chain of blocks. Moreover, public-private key cryptography is being used by blockchain, so it is secured from cheating, stealing, and corruption. When a transaction is accepted and stored in the block, it cannot be modified or deleted, which leads to data integrity with mitigation of the double-spending issue [1].

A programming unit on blockchain called a smart contract can be considered as a trustworthy intermediary among non-trusting participants, which can be developed and deployed to be used by a network node easily via creating a transaction on a blockchain network [19]. Smart contract code is immutable, so it cannot be updated once it is contained in the blockchain [20]. As a result, the smart contract does not follow a standard software life cycle where the bugs can be fixed only by publishing a new edition of the smart contract [7].

BOS development projects are rapidly constructed and developed across several BOS applications. The development process of the BOS project has generally messy and rushed attributes. The competitive nature of the BOS project and the unofficial processes for the fast development of BOS applications do not guarantee the quality of software or the essential software engineering perceptions [21].

It has been suggested that useful testing, improving cooperation in teams, and accelerating the smart contracts development are significant factors to establish a development process for BOS in the future [6]. A development process based on Agile practices and the usage of the UML diagrams with additional notations of the specific characteristics of BOS was proposed by [22]; similarly, a suggestion to collaborate and improve UML class diagram and BPMN to design the development process of BOS was recommended by [23].

In short, many researchers agreed that the current software development process for BOS has many problems and should be enhanced to address the distinctive characteristics of BOS [6, 7, 9–14]. A few researchers mentioned some of the current software development process issues and introduced some suggestions for recent software development process improvement [22, 23]. Nevertheless, no systematic review has been conducted to collect the issues of the current software development process for BOS. Collecting these issues and recognize them will help the efforts to improve.

3 Research Design

This study research method is the systematic literature review, which is suggested by Kitchenham [24] to help to identify, assessing, and synthesizing the existing knowledge on a topic of specific concern being researched. This systematic review has the following phases:

Recognizing the Reason for This Review to Be Conducted. As many researchers agreed that the current development process for BOS should be improved, it is imperative to identify and understand the BOS development process's current issues. This will help to direct the software process improvement of BOS and evaluate them by solving the current issues of the BOS development process.

Formulating the Research Question. What are the issues of the current BOS development process?

Creating the Research String. The term used for this is Blockchain-oriented software.

Selecting the Data Resources. ACM, ProQuest, Springer, Web of Science, Emerald Insight, and IEEE Explore are the chosen database.

Preforming the Review. The research string was applied to the selected databases, and the results were summarized as shown in Table 1.

Table 1. Summary of research results

No	Database/Research engine	Results
1	IEEE Explore	39
2	Springer	13
3	ACM	15
4	ScienceDirect	5
5	ProQuest	21
6	Web of Science	7
7	Emerald insight	3
	Database/ Research Engine	**103**

Defining and applying criteria for the exclusion: there were three reasons to exclude papers, and a total excluded paper 30, as shown in Table 2.

Table 2. Excluded and included papers.

No	Database/Research engine	Results
1	Duplicated papers	11
2	Papers that are not in English	4
3	The author index, Contents, Front cover, Conference index, Conference Message, Table of contents, Keynote and Summary	15
	Excluded papers	**30**
	Included papers	**73**
	Total	**103**

Data analyzing and results reporting: full-text reading was performed on the selected 73 papers. The thematic analysis was used to determine and report the issues' BOS development process as the last step in this research. More details about this step in the coming section, Data analysis, and results. Figure 1 shows these research phases.

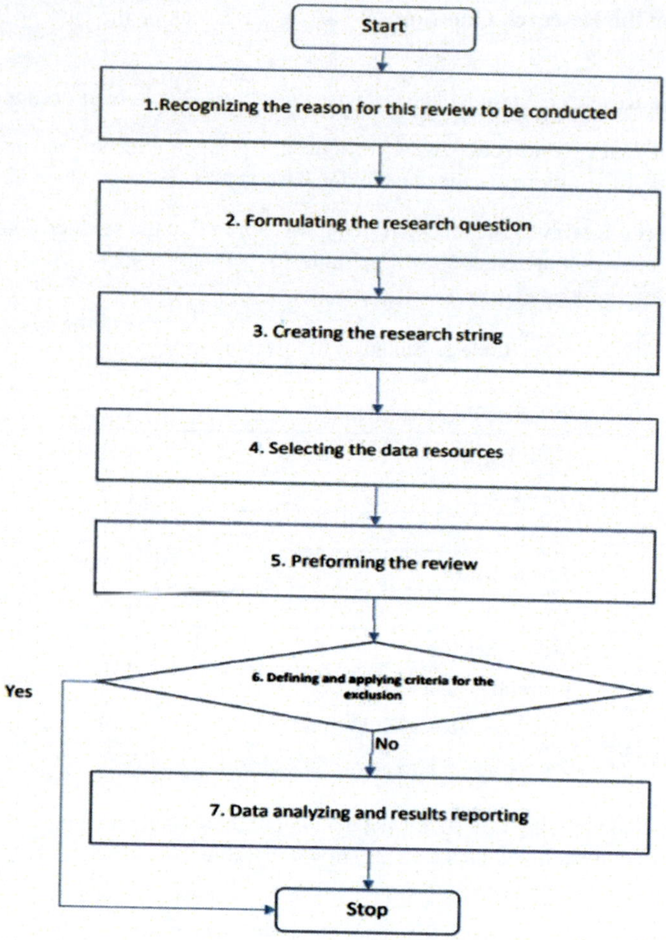

Fig. 1. Systematic review phases.

4 Data Analysis and Results

The thematic analysis was used to extract the data with an inductive approach, and this data established the themes as recommended by [25]. Firstly, the entire paper was read to understand the data, and secondly, contents related to the research question were coded and documented in an excel file to be analyzed later. In the third step, identifying patterns on the documented contents was performed to generate suitable themes. Each theme is given a meaningful name to represent the problem in the BOS development process. Fifteen issues were found as summarized and shown in Table 3, and by identifying these issues, this study question has been answered.

As shown in Fig. 2, the 15 issues were classified into the following categories. Technology, it refers to the issues related to blockchain technology. Data and smart contracts cannot be modified issue were mentioned in 27 papers as the highest mentioned

Table 3. Summary of research results

No	Database/Research engine	Total	%
1	Data and smart contract cannot be modified	5	6.85
2	Modeling diagrams do not have specific notations for BOS	4	5.48
3	Testing BOS is critical	4	5.48
4	Inadequate Programing language	3	4.11
5	BOS is slow	3	4.11
6	BOS development is a non-standard software life cycle	3	4.11
7	BOS development is rushed	2	2.74
8	Testing BOS is complex	2	2.74
9	BOS has a new development paradigm	1	1.37
10	BOS needs specialized professional figure	1	1.37
11	Inadequate testing tools	5	6.85
12	Inadequate development tools	4	5.48
13	Requirements identified mostly by a community discussion	4	5.48
14	Lack of standards and regulations	3	4.11
15	Lack of BOS characteristics understanding	3	4.11

issue with 36.99%. BOS slow issue was mentioned six times with 8.22%. The inadequate Programming language was also mentioned six times, with 8.22%.

People. It refers to the issues related to the people who are working on BOS projects. BOS needs specialized professional figure issue was mentioned in 3 papers with 4.11%. Lack of standards and regulations and lack of BOS characteristics understanding were mentioned one time for each one with a rate of 1.37%.

Process. It refers to the issues related to the process of BOS development. Modeling diagrams do not have specific notations for BOS issue was mentioned in 9 papers with 12.33%. Testing BOS is a critical issue that was mentioned seven times, with 9.59%. BOS development is a non-standard software life-cycle issue that was mentioned five times with 6.85%. BOS development is rushed issue was mentioned four times with 5.48%. Testing the BOS complex issue was mentioned four times, with 5.48%. BOS's new development paradigm issue was mentioned three times, with 4.11%. Inadequate testing tools issue was mentioned three times, with 4.11%. Inadequate development tools issue was mentioned twice with 2.74%. Requirements identified mostly by community discussion issue was mentioned twice with 2.74%.

In summary, most of the BOS development process issues are related to technology and process as both groups' issues were mentioned 39 times. The BOS development process issues related to people were noted in 5 papers only, as shown in Fig. 3.

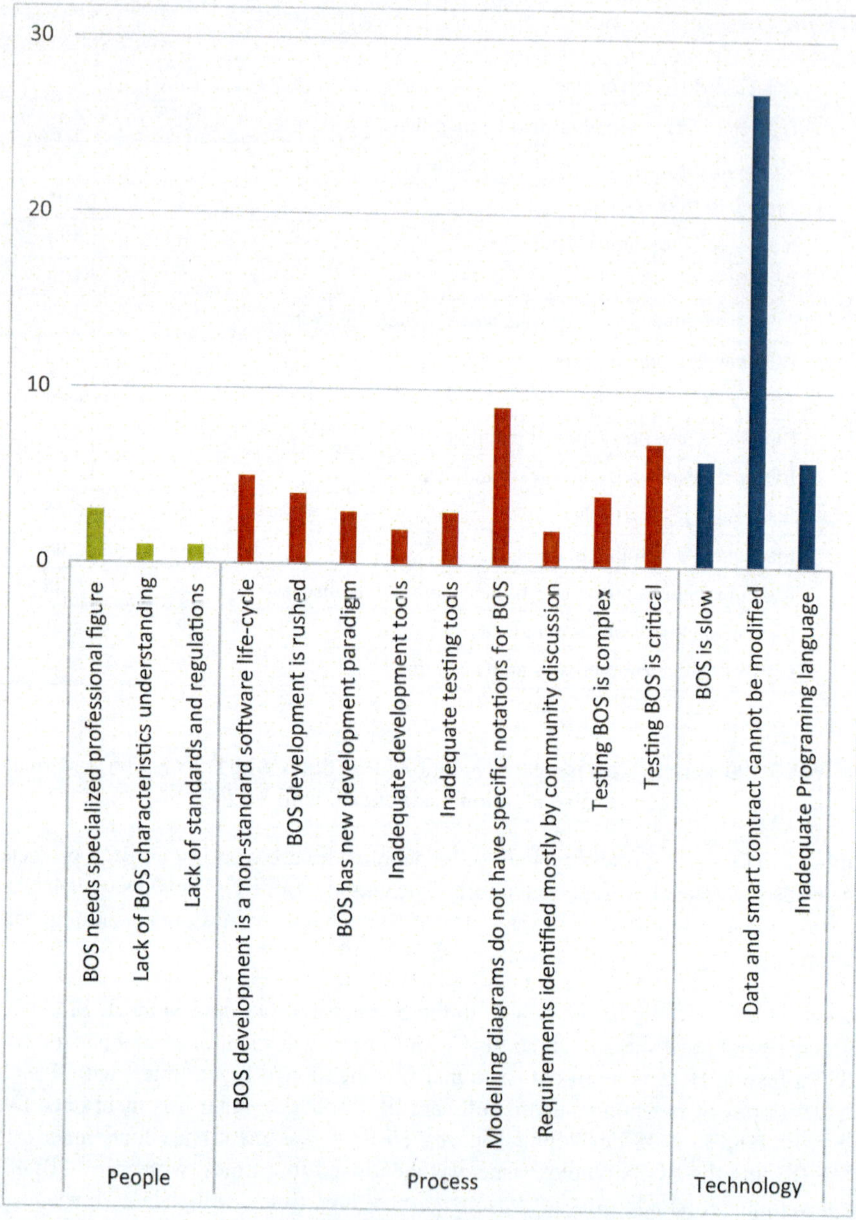

Fig. 2. Count of issues by group.

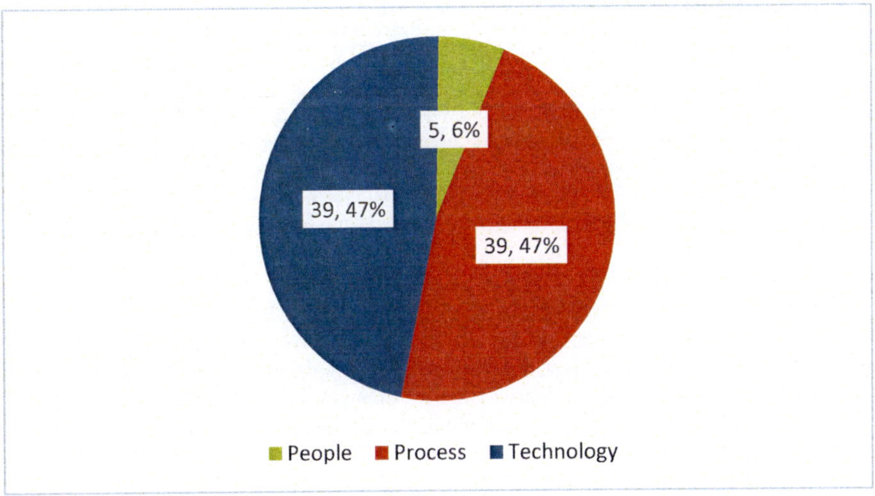

Fig. 3. Total issues by groups.

5 Conclusion and Future Works

This study presented a systematic review to identify the BOS development process issues from the current literature on BOS. There are 15 issues in the development process found in the literature of BOS, and these issues were divided into three groups people-related issues, technology-related issues, and process-related issues. It is observed that data and smart contracts cannot be modified issue was the most mentioned issue, and the issues related to people are the less mentioned issues. This study results were closed on BOS development process issues only from the available current literature on this study topic. Quite a few challenges arise as future work. Firstly, these issues found in this study should be confirmed by doing a confirmatory study as a survey or interview. Secondly, they should be further analyzed to determine the distinctive BOS characteristics causing these issues to happen. Finally, these issues should be solved or eased as part of BOS development process improvement, possibly a new SPI branch called Blockchain-oriented software process improvement BSPI.

Acknowledgment. This research was fully supported by Ministry of Higher Education Malaysia, under the Fundamental Research Grant Scheme (FRGS) with Ref. No. FRGS/1/2018/ICT04/UTP/02/4.

References

1. Nakamoto, S.: Bitcoin: a peer-to-peer electronic cash system. https://bitcoin.org/bitcoin.pdf. Accessed 27 Dec 2020
2. Silva, T.B., Morais, E.S., Almeida, L.F.F., Rosa Righi, R., Alberti, A.M.: Blockchain and industry 4.0: overview, convergence, and analysis. In: Rosa Righi, R., Alberti, A.M., Singh, M. (eds.) Blockchain Technology for Industry 4.0. BT, pp. 27–58. Springer, Singapore (2020). https://doi.org/10.1007/978-981-15-1137-0_2

3. Zhao, J.L., Fan, S., Yan, J.: Overview of business innovations and research opportunities in blockchain and introduction to the special issue. Financ. Innov. **2**(1), 1–7 (2016). https://doi.org/10.1186/s40854-016-0049-2
4. Swan, M.: Blockchain: Blueprint for a new economy. O'Reilly Media Inc., Newton (2015)
5. Koul, R.: Blockchain oriented software testing-challenges and approaches. In: 3rd International Conference for Convergence in Technology (I2CT) 2018, pp. 1–6., IEEE (2018)
6. Porru, S., Pinna, A., Marchesi M., Tonelli, R.: Blockchain-oriented software engineering: challenges and new directions. In: IEEE/ACM 39th International Conference on Software Engineering Companion (ICSE-C) 2017, pp. 169–171. IEEE (2017)
7. Destefanis, G., Marchesi, M., Ortu, M., Tonelli, R., Bracciali A., Hierons, R.: Smart contracts vulnerabilities: a call for blockchain software engineering? In: International Workshop on Blockchain Oriented Software Engineering (IWBOSE) 2018, pp. 19–25. IEEE (2018)
8. dos Santos, R.: Consensus algorithms: A matter of complexity? In: Blockchain Economics: Implications of Distributed Ledgers-Markets, Communications Networks, and Algorithmic Reality, vol. 1, p. 147. World Scientific Publishing Europe Ltd (2019)
9. Rocha, H., Ducasse, S., Denker M., Lecerf, J.: Solidity parsing using smacc: challenges and irregularities. In: Proceedings of the 12th edition of the International Workshop on Smalltalk Technologies, pp. 1–9. Universitätsverlag Potsdam (2017)
10. Ibba, S., Pinna, A., Lunesu, M.I., Marchesi M., Tonelli, R.: Initial coin offerings and agile practices. In: Future Internet, vol. 10, p. 103. Multidisciplinary Digital Publishing Institute (2018)
11. Pinna, A., Tonelli, R., Orru, M., Marchesi, M.: A petri nets model for blockchain analysis. In: The Computer Journal, vol. 61, pp. 1374–1388. Oxford University Press (2018)
12. Marchesi, M.: Why blockchain is important for software developers, and why software engineering is important for blockchain software (Keynote). In: International Workshop on Blockchain Oriented Software Engineering (IWBOSE) 2018. IEEE (2018)
13. Chakraborty, P., Shahriyar, R., Iqbal, A., Bosu, A.: Understanding the software development practices of blockchain projects: a survey. In: Proceedings of the 12th ACM/IEEE International Symposium on Empirical Software Engineering and Measurement, pp. 1–10. Association for Computing Machinery (2018)
14. Kazakzeh, S., Ayoubi, E., Muslmani, B., Qasaimeh, M., Al-Fayoumi, M.: Framework for blockchain deployment: the case of educational systems. In: 2nd International Conference on New Trends in Computing Sciences (ICTCS) 2019, pp. 1–9. IEEE (2019)
15. Zhang Y., Wen, J.: An IoT electric business model based on the protocol of bitcoin. In: 18th international conference on intelligence in next generation networks 2015, pp. 184–191. IEEE (2015)
16. Zyskind, G., Nathan, O.: Decentralizing privacy: using blockchain to protect personal data. In: IEEE Security and Privacy Workshops 2015, pp. 180–184. IEEE (2015)
17. Herbert, J., Litchfield, A.: A novel method for decentralised peer-to-peer software license validation using cryptocurrency blockchain technology. In: Proceedings of the 38th Australasian Computer Science Conference (ACSC 2015), p. 30. Australian Computer Society (2015)
18. Sharples, M., Domingue, J.: The blockchain and kudos: a distributed system for educational record, reputation and reward. In: Verbert, K., Sharples, M., Klobučar, T. (eds.) EC-TEL 2016. LNCS, vol. 9891, pp. 490–496. Springer, Cham (2016). https://doi.org/10.1007/978-3-319-45153-4_48
19. Fernandez-Vazquez, S., Rosillo, R., De La Fuente, D., Priore, P.: Blockchain in FinTech: a mapping study. In: Sustainability, vol. 11, pp. 63–66. Multidisciplinary Digital Publishing Institute (2019)
20. Alharby, M., Aldweesh, A., van Moorsel, A.: Blockchain-based smart contracts: a systematic mapping study of academic research. In: International Conference on Cloud Computing, Big Data and Blockchain (ICCBB) 2018, pp. 1–6. IEEE (2018)

21. Saghiri, A.M.: Blockchain architecture. In: Kim, S., Deka, G.C. (eds.) Advanced Applications of Blockchain Technology. SBD, vol. 60, pp. 161–176. Springer, Singapore (2020). https://doi.org/10.1007/978-981-13-8775-3_8
22. Marchesi, M., Marchesi, L., Tonelli, R.: An agile software engineering method to design blockchain applications. In: Proceedings of the 14th Central and Eastern European Software Engineering Conference Russia, pp. 1–8. ACM (2018)
23. Rocha, H., Ducasse, S.: Preliminary steps towards modeling blockchain oriented software. In: IEEE/ACM 1st International Workshop on Emerging Trends in Software Engineering for Blockchain (WETSEB) 2018, pp. 52–57. IEEE (2018)
24. Kitchenham, B., Brereton, O.P., Budgen, D., Turner, M., Bailey, J., Linkman, S.: Systematic literature reviews in software engineering: a systematic literature review. In: Information and Software Technology, vol. 51, pp. 7–15. Elsevier (2009)
25. Caulfield, J.: How to do thematic analysis (2019). https://www.scribbr.com/methodology/thematic-analysis. Accessed 11 Nov 2021

Identification of Magnetic Levitation System

Aleš Novotný(✉), Daniel Honc, and František Dušek

Faculty of Electrical Engineering and Informatics, University of Pardubice, náměstí Čs. legií 565, 530 02 Pardubice, Czech Republic
ales.novotny1@student.upce.cz

Abstract. Magnetic Levitation Systems (MLS) are usually highly non-linear systems with a great sensitivity to the size of the control input. Therefore, special emphasis should be placed on the correct identification of all unknown MLS parameters. This paper describes the principle and procedure for identifying laboratory plant CE 152 MLS with an emphasis on automatically processing identification data. Moving-Average Filter (MAF) and Fast Fourier Transform Filter (FFTF) methods are compared to filtering input data noise. Key parameters are then estimated using the Least Squares Method (LSM). The results are verified in simulation and real-world experiment using a simple PID controller.

Keywords: Magnetic levitation system · Non-linear system · Control input · Laboratory plant CE 152 · Fast Fourier Transform Filter · Moving-average filter · Least squares method · PID controller

1 Introduction

The aim of this work is to identify the CE152 laboratory model produced by Humusoft (see Fig. 1). This model is well described in the manual [1] and many other publications such as [2–4], etc. Each of the previously mentioned authors came up with a little different solution to this issue but always came to satisfactory results.

First in Sect. 2, the signal deprived by noise is processed. In this paper are presented results using Moving-Average Filter (MAF) [5] and using Fast Fourier Transform Filter (FFTF) [6]. The performances of both approaches are compared.

Next in Sect. 3, the system, equations and parameters are described. The parameters that are easy to determine are calculated. The remaining parameters are estimated using Least Squares Method (LSM) [7]. The parts of the system which have a negligible effect on the operation of the system are neglected, thereby the system is simplified.

Subsequently in Sect. 4, a simple control circuit with PID controller is designed [8,9]. The correctness of the calculations is verified by a simulation using MATLAB/Simulink environment. Finally, the accuracy of the procedure

R. Silhavy et al. (Eds.): CoMeSySo 2021, LNNS 232, pp. 58–75, 2021.
https://doi.org/10.1007/978-3-030-90318-3_7

Fig. 1. Laboratory plant CE152 (Source: https://www.humusoft.cz/data/img/models/ce152.jpg.).

is proved by comparing the simulation input-output trajectories with the real input-output trajectories of the laboratory plant.

2 Input Data Processing

2.1 Gathering Input Data

Firstly, it is necessary to measure directly measurable data describing the system as the distance between the coil and the sensor, the diameter and weight of the ball and the sensor output voltage when the ball is in both limit positions.

Next, it is necessary to determine how to measure indirectly measurable data describing the system. It is generally known that the best data is such that it contains the highest possible quantity of information. The same is also with data collection for this system. As this system consists of a coil, ball and distance sensor, it is good to get data where the ball is steady in as many distances as possible (see Fig. 2).

For keeping the ball in desired distances purpose, it is possible to use the proposed controller in the manual [1] or design your own feedback PID controller.

After selecting a suitable method for obtaining data, in most cases, data is affected significantly by "noise" (controller oscillations), as in this case (see Fig. 3). The noise needs to be suppressed as far as possible for the successful and

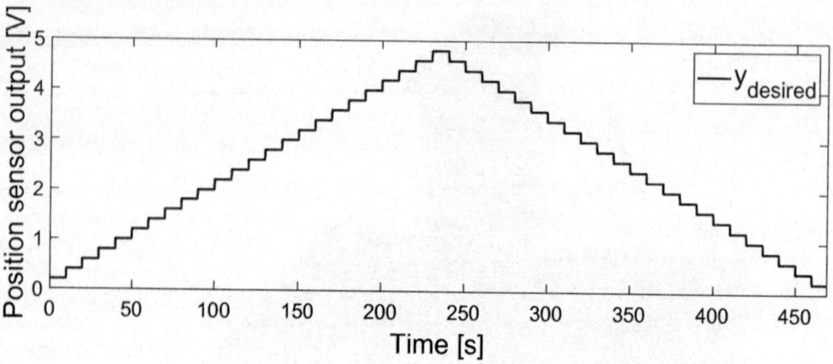

Fig. 2. Example of the method of acquisition of suitable data.

accurate identification of directly unmeasurable system variables. It is worth noting that the amplitude of the control input in the extreme ball positions significantly fluctuates and does not provide accurate information. These positions are good to neglect to achieve higher accuracy. As mentioned, two filtration techniques will be presented in the following two chapters and will subsequently be compared.

2.2 Input Data Processing Using Moving-Average Filter

One of the simplest filtering techniques is averaging using a fixed number of last values (see Eq. 1)

$$y(n) = \frac{\sum_{i=(n-w+1)}^{n} x(i)}{w}, \tag{1}$$

where $y(n)$ is the n-th result of averaging, w is window size, and x is input.

2.3 Input Data Processing Using Fast Fourier Transform Filter

More sophisticated access is to take into consideration the characteristics of noise and design a filter that fits this kind of noise. As shown in Fig. 3, the noise is random and mostly represented by pulses with a very short duration. Therefore, it is appropriate to make an assumption that noise frequency differs from the frequency of the desired signal. Thus, it is possible to use preferably a filter that transforms the input data to the frequency spectrum, select the desired number of amplitudes and convert it back to the time spectrum. The FFTF presented in [10] is used in this paper.

2.4 Comparison of the Input Data Processing Methods

A comparison of MAF and FFTF is shown in Fig. 4. It is evident from the figure that MAF (see Fig. 4a) allows you to generate a smoother output than FFTF

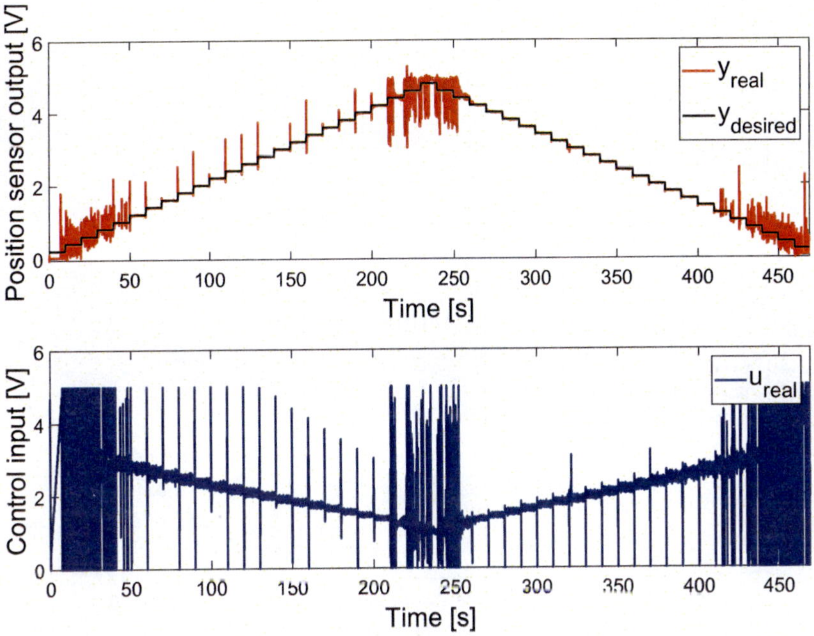

Fig. 3. Example of real data for system identification.

(see Fig. 4b) for bigger window sizes. On the other hand, the larger window values using MAF leads to greater distortion, which results from the nature of the averaging. In the FFTF case, the distortion is increasing when accepting a small number of amplitudes because a large number of amplitudes is needed to approximate the step transition between positions.

However, if we do not consider the visual impression and computational complexity, the results of both methods are approximately the same precise. Therefore, the techniques are not necessary to be combined or to look for additional data filtration methods. The outputs of both filtration methods will be further compared during system identification in the next Sect. 3.

3 First Principle Modeling

3.1 Introduction to CE 152 System

As shown in Fig. 5, the CE 152 system can be divided into D/A DAQ, U/I converter and coil system, Ball and coil system, Position sensor, A/D DAQ. The blocks are more described below.

In Sect. 3.2, the individual blocks and their equations are described in more detail, directly measurable and determinable parameters are found, and parts of the system with negligible influence are neglected.

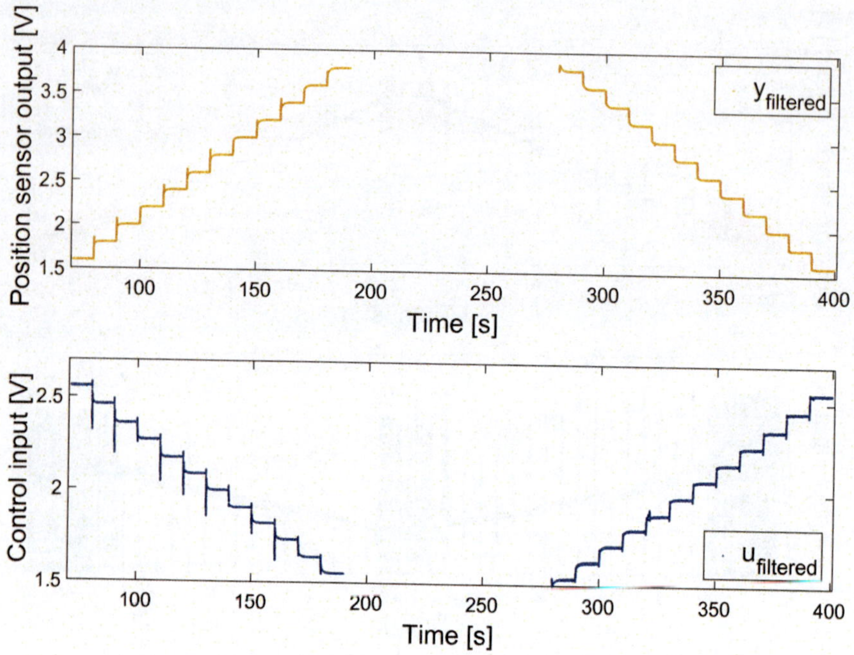

(a) Filtering by Moving-Average Filter (MAF) with $w = 200$.

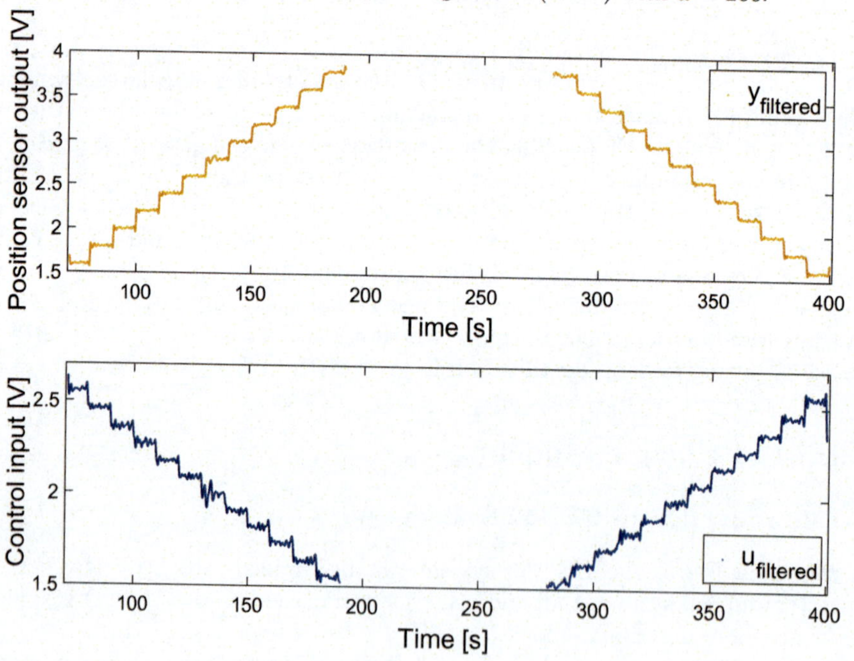

(b) Filtering by Fast Fourier Transform Filter (FFTF) when accepting 500 amplitudes.

Fig. 4. Comparison of filtering by MAF and FFTF.

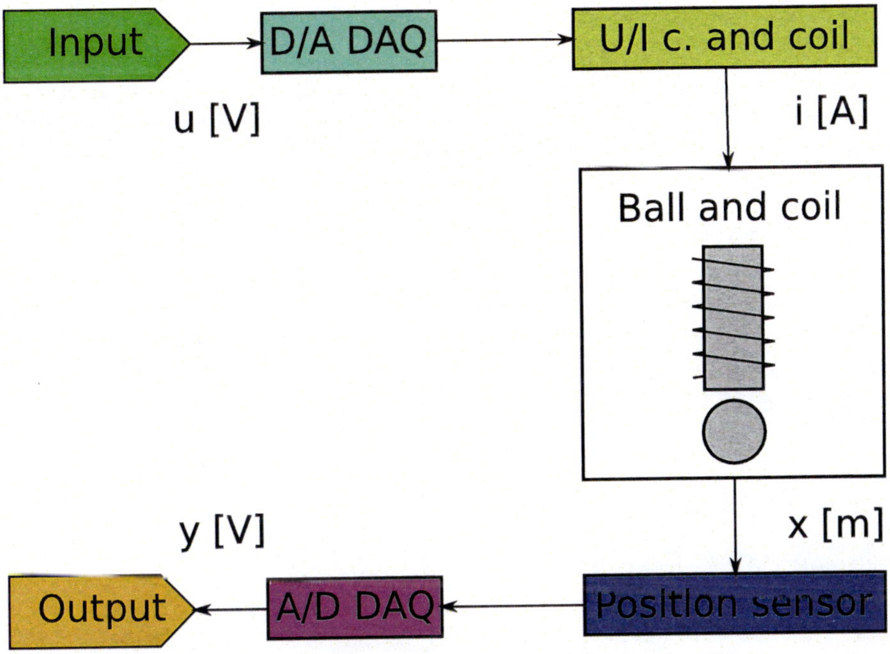

Fig. 5. Block diagram of CE 152 system.

3.2 Identification Directly Measurable Parameters

D/A DAQ. 14-bit D/A converters with range $\pm 10\,V$ are used in CE 152. The quantization step is then determined using Eq. 2.

$$\Delta_{DA} = \frac{u_{da\_max} - u_{da\_min}}{2^{da\_bits} - 1} = 1.2 \cdot 10^{-3}\ V \tag{2}$$

Since it is reasonable to generate a voltage at 0 to 5 V relative to the U/I converter input characteristic, it is possible to change the value with step of 0.024% maximum value. This step is precise enough, and unwanted D/A conversion effects is further neglected.

U/I Converter and Coil System. U/I converter together with the coil is the first order system that can be described after Laplace transform by the Eq. 3 [1].

$$I(s) = \frac{K_{um}}{R + K_{am}K_sR_s}\frac{U(s)}{\frac{L}{R+K_{am}K_sR_s}s + 1} = k_i\frac{U(s)}{T_as + 1} \tag{3}$$

Where k_i is the gain of the U/I converter and the T_a is system time constant and K_s, K_{am}, R, R_s, L are electric circuit element values. In order to calculate

k_i and T_a constants, the nominal values of the circuit element values given in the manual [1] are used.

Searched constants have approximate values

$$\begin{aligned} k_i &= 2.970 \cdot 10^{-1}\ A \cdot V^{-1}, \\ T_a &= 8.910 \cdot 10^{-5}\ s. \end{aligned} \tag{4}$$

As can be deduced from Eq. 4, the system output starts approaching the desired value of approximately 100 microseconds after the input change. The used computer can generate inputs and evaluate outputs with a minimum period of 1 ms. Thus, a transient of the system cannot be incorporated into calculations by the computer, so it is unnecessary to consider this transition phenomenon. Equation 5 is considered as a transfer equation for the U/I converter in the rest of the work.

$$i(t) = 2.970 \cdot 10^{-1} u(t) \tag{5}$$

Ball and Coil System. The main forces attracting the ball are mentioned in Eq. 6. Where F is total force, F_m is magnetic force generated by the coil, F_g is gravitation force, F_{fv} is air resistance. The output is $x(t)$ which is ball position in time, the input is control input voltage in time $u(t)$. It should be noted that the force directions are not involved in the initial considerations in order to keep the possibility to change simply the convention of ball position, velocity and acceleration by changing p_{dir}.

For example, a zero ball position can be assumed nearby the sensor or the coil. Furthermore, it is possible to measure the ball position either from the top side, the centre of gravity or the underside. With this preservation of generality, it is possible to use simply any of these conventions without changing the clarity of equations.

$$\begin{aligned} F &= \operatorname{sign}(p_{dir}) \cdot (F_m - F_g - F_{fv}) \\ F &= m_k \frac{d^2}{dt^2} x(t) \\ F_m &= k_c \left(\frac{k_i u}{(x + x_0)} \right)^2 \\ F_g &= m_k g \\ F_{fv} &= k_{fv} \frac{d}{dt} x(t) \end{aligned} \tag{6}$$

where constant p_{dir} is the convention. If the zero ball position is considered near the sensor, then $p_{dir} = 1$, else $p_{dir} = -1$. The value of k_i is known from (4). The gravitation constant g is generally known. The ball weight m_k, diameter D_k and space between coil and sensor l_0 are directly measured. The distance where the ball can move is calculated as the subtraction of l_0 and D_k.

The value of air resistance constant k_{fv} can be easily calculated from Stokes equation

$$k_{fv} = 6 \cdot \pi \cdot \eta \cdot \frac{D_k}{2} \tag{7}$$

where η is fluid viscosity. All above mentioned constant values can be seen in Table 1.

Table 1. Directly measurable system constants

Constant	Sym.	Value
U/I converter gain	k_i	$2.970 \cdot 10^{-1}\ A \cdot V^{-1}$
Gravitation constant	g	$9.807\ m \cdot s^{-2}$
Ball weight	m_k	$8.280 \cdot 10^{-3}\ kg$
Ball diameter	D_k	$1.270 \cdot 10^{-2}\ m$
Space between coil and sensor	l_0	$1.840 \cdot 10^{-2}\ m$
Ball manipulation space	l	$5.700 \cdot 10^{-3}\ m$
Dynamic viscosity (air 25 °C)	η	$1.710 \cdot 10^{-5}\ Pa \cdot s$
Air resistance const.	k_{fv}	$2.047 \cdot 10^{-6}\ kg \cdot s^{-1}$

It is worth noting that the k_{fv} value is a small number. The maximum speed of the ball is expected in units or tens of meters per second. Thus, the air resistance F_{fv} is negligible (see Fig. 6), and in the rest of this article is considered as zero.

Constants coil offset x_0 and coil constant k_c cannot be measured by direct measurement. The technique of finding these constants is described in Sect. 3.3.

Position Sensor. For sensor identification, it is necessary to measure the limit values of the sensor depending on the ball. The values are shown in Table 2.

Table 2. Sensor values in limit positions

Position of ball	Voltage on sensor
Ball nearby sensor	0.047 V
Ball nearby coil	4.920 V

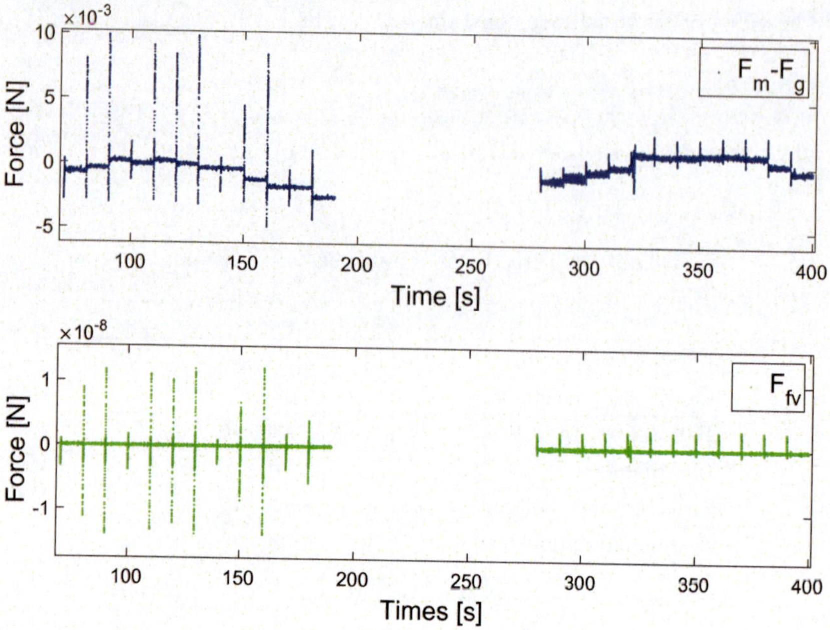

Fig. 6. Comparison of computed forces acting on the ball from rough data filtered using MAF with $w = 200$.

The sensor measuring the ball position is noted in [1] as approximately linear. Therefore, it is possible to approximate the relationship between voltage and position by a line equation

$$y = k_x x + y_0, \tag{8}$$

where y is sensor voltage, x is ball position. The constants k_x and y_0 depends on selected convention shown in Sect. 3.3.

A/D DAQ. 14-bit A/D converters with range $\pm 10\,V$ are used in CE 152. The quantization step is then determined using Eq. 9.

$$\Delta_{AD} = \frac{u_{ad\_max} - u_{ad\_min}}{2^{ad\_bits} - 1} = 1.2 \cdot 10^{-3}\ V \tag{9}$$

Since it is expected to receive a voltage at 0.047 to 4.920 V relative to the position sensor and ball characteristics, it is possible to sense 4060 positions. If the ball manipulation space l is took into consideration (see Table 1), the step is $1.404 \cdot\ 10^{-6}\ m$. This is precise enough, and unwanted A/D conversion effects is further neglected.

3.3 Identification Remaining Parameters

Selecting Position Convention. First of all, in order to further continue it is necessary to set one of the ball position conventions described in Sect. 3.2. In this paper, it is considered that the ball zero position is located nearby the sensor ($p_{dir} = 1$) and the position is counted from the top boundary of the ball. The values further identified constants depends on this decision.

Finding Sensor Constants. k_x and y_0. Due to information from Sects. 3.2 and 3.3 it is possible to express k_x and y_0 from two variables two linear equations problem. The values are shown in Table 3.

Table 3. Sensor constants.

Constant	Sym.	Value
Position sensor gain	k_x	$8.549 \cdot 10^2\ V \cdot m^{-1}$
Position sensor offset	y_0	$4.700 \cdot 10^{-2}\ V$

Finding Coil Constants x_0 and k_c. The most complex part of the identification is to identify the coil constants. These constants dependent not only on the choice of position convention (see Sect. 3.3) but also on the quality of input data filtering (see Sect. 2.4).

For identifying these constants from system equations (see Eq. 6), it is appropriate to consider velocity and acceleration as zero, which leads to eliminating total force, and it is possible to express sought constants better. Also, it is possible to convert the position variable into output sensor voltage using Eq. 8. This modification gives Eq. 10.

$$m_k g = k_c \left(\frac{k_i u}{(\frac{y-y_0}{k_x} + x_0)} \right)^2 \tag{10}$$

The output can be then expressed by

$$\frac{y - y_0}{k_x} = \sqrt{\frac{k_c}{m_k g}} k_i u - x_0 = au + b. \tag{11}$$

From Eq. 11 is deduced conclusion that if the result of the forces acting on the ball is zero, the relationship between the control voltage and the position (voltage on the sensor) is linear.

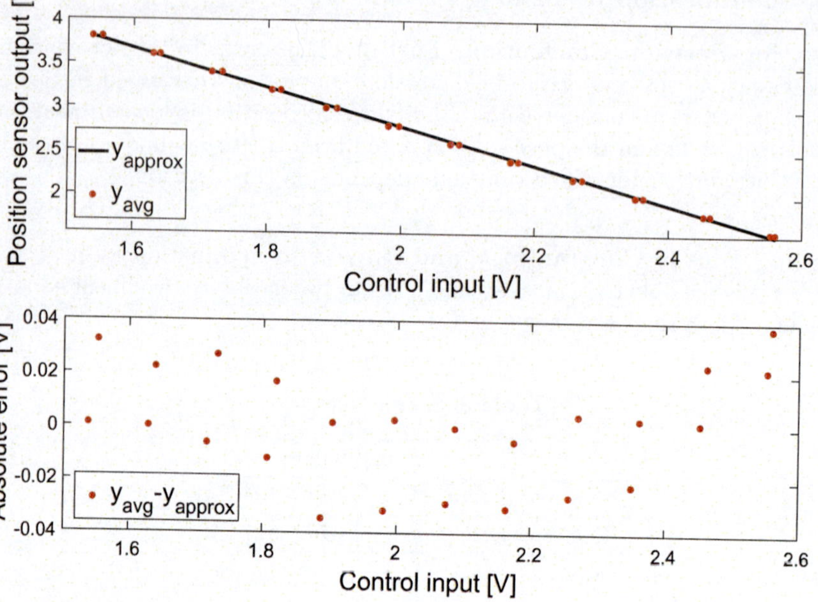

(a) Application of Least Squares Method (LSM) approximation on signal filtered by MAF with window size 200 ms.

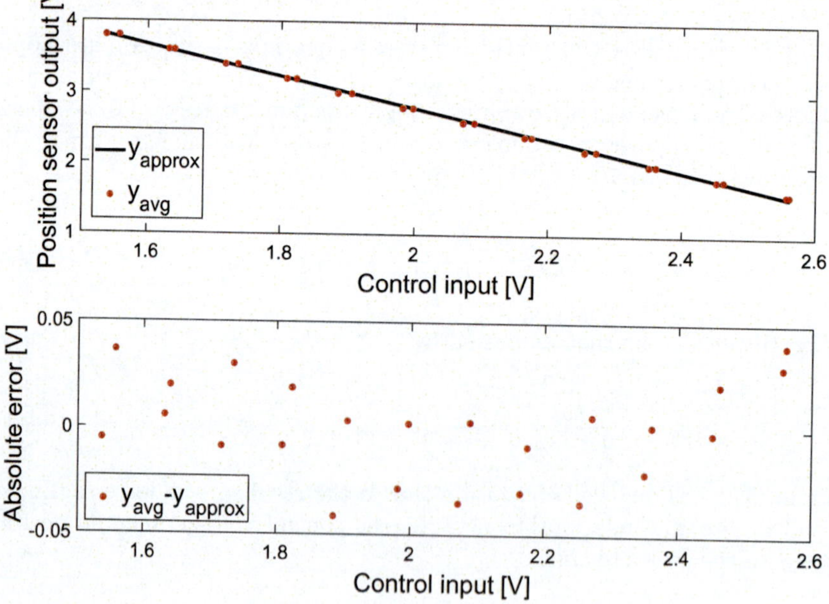

(b) Application of Least Squares Method (LSM) approximation on signal filtered by FFTF accepting 500 amplitudes.

Fig. 7. Comparison of filtering by MAF and FFTF.

For precise searching of constants, steady states are cut from filtered data (using FFTF and MAF). The data are averaged for each step (first dot is the way up and second is the way down) and then the LSM method is applied to points. Thus, a straight-line approximation is performed (see Fig. 7) and then the LSM method is applied to points. The x_0 is found and then k_c. The values are shown in Tables 4 and 5.

Below in Figs. 7a and 7b shows an average absolute error with which method works. It can be deduced from the figures that little inaccuracy arises for a straight-line approximation. For increasing accuracy, it is possible to select a more complex curve, such as hyperbole.

Table 4. Coil constants - FFTF accepting 500 amplitudes

Constant	Sym.	Value
Coil offset	x_0	$-8.335 \cdot 10^{-3}\ m$
Coil constant	k_c	$6.049 \cdot 10^{-6}\ N \cdot m^2 \cdot A^{-2}$

Table 5. Coil constants - MAF with window size 200 ms

Constant	Sym.	Value
Coil offset	x_0	$-8.333 \cdot 10^{-3}\ m$
Coil constant	k_c	$6.035 \cdot 10^{-6}\ N \cdot m^2 \cdot A^{-2}$

It is worth noting in Tables 4 and 5 is the distance coil offset x_0 negative number. It is the consequence of the positive notation in the denominator of F_m.

4 Verification of the Model

4.1 Design of Simply Control Circuit for the CE 152 System

For verifying the identification results, the control circuit is needed to be designed. The designed control circuit (see Fig. 8) consists of the PID controller, identified MLS and x/u converter. The PID constants are found manually, the values are in Table 6. The x/u converter is based on the linear relationship between y and u described in Sect. 3.3.

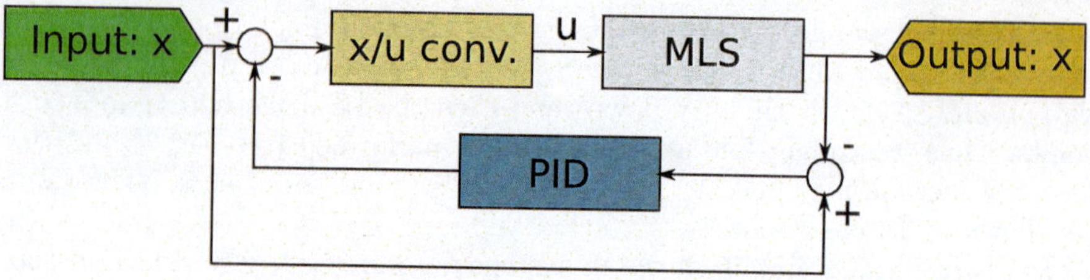

Fig. 8. Simple control circuit for CE 152 system.

Table 6. PID controller constants

Constant	Value
Proportional (P)	2
Integral (I)	3
Derivative (D)	0.03
Filter coefficient for derivative (N)	1000

4.2 Verification by Simulation and Real-World Experiment

A series of real experiments were performed to verify the correctness of the designed model. These experiments were then compared to the simulation results for the same control circuit (in the MATLAB/Simulink env.). To maintain consistency with previous images, regardless on the selected convention, position sensor voltage represents ball position.

First, the expected and real behaviours of the system were compared for steps from different initial positions (see Figs. 9, 10, 11 and 12). Subsequently, the behaviour of smooth transitions between states was tested using ascending and descending ramps (see Fig. 13). Finally, the system was subjected to a stress test when a sinusoidal signal was selected as its input (see Fig. 14).

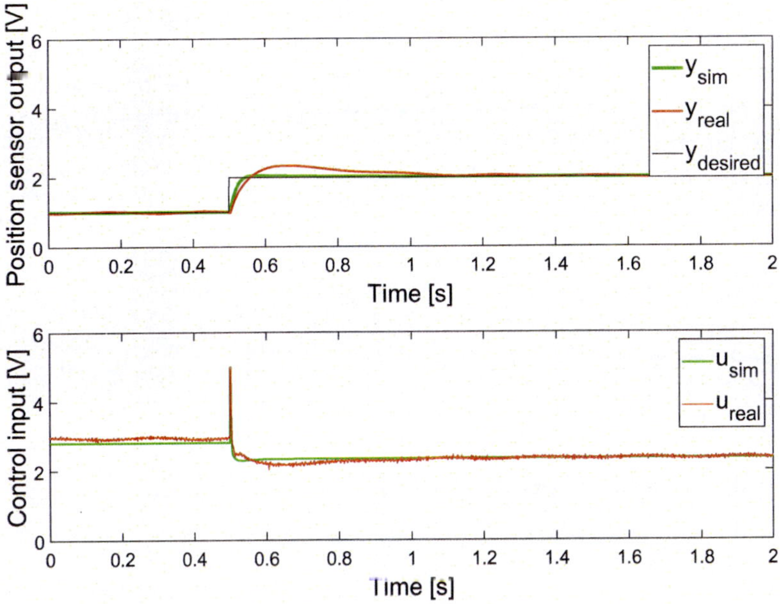

Fig. 9. Comparison of step 1–2 V response for plant model simulation and real-world experiment.

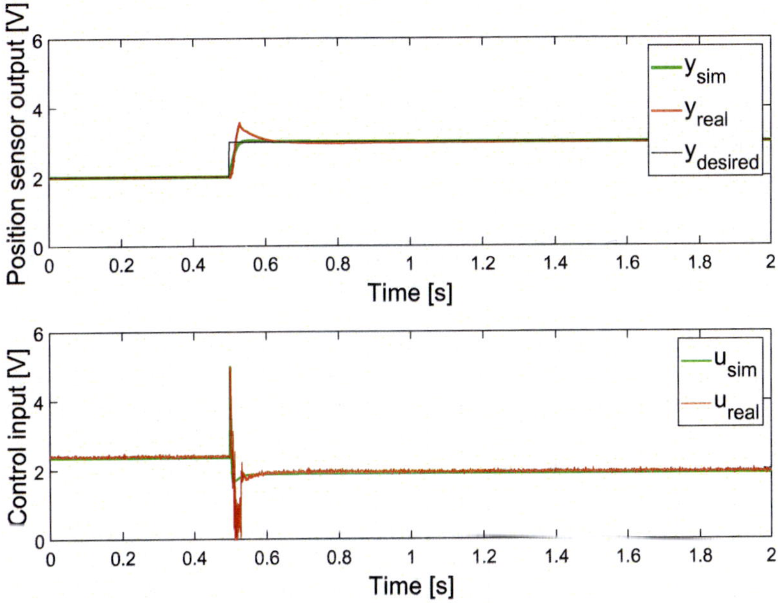

Fig. 10. Comparison of step 2–3 V response for plant model simulation and real-world experiment.

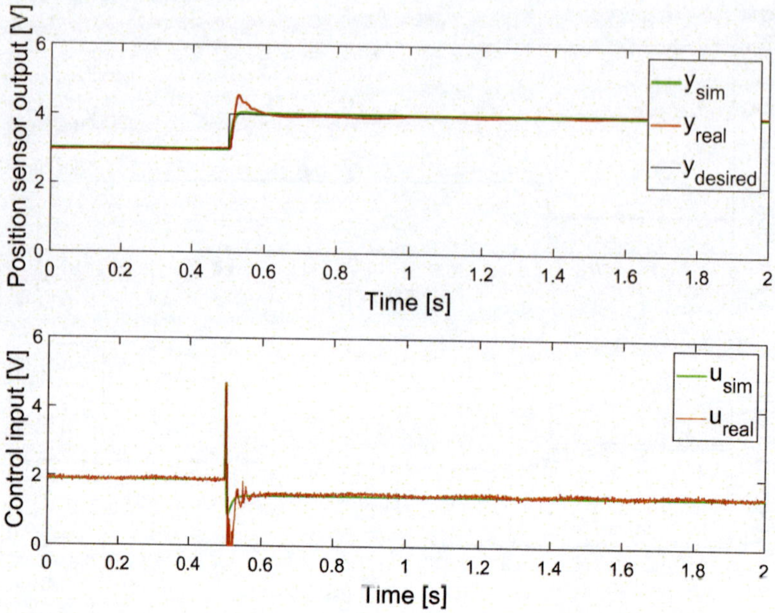

Fig. 11. Comparison of step 3–4 V response for plant model simulation and real-world experiment.

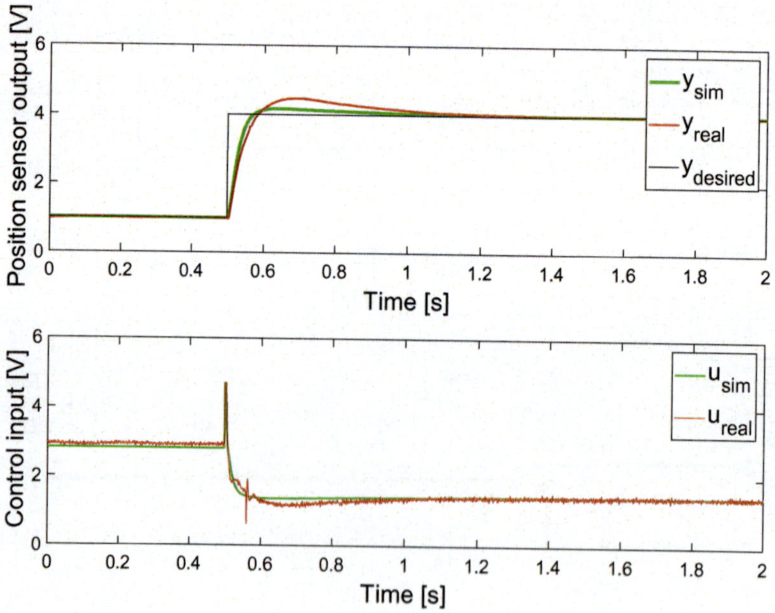

Fig. 12. Comparison of step 1–4 V response for plant model simulation and real-world experiment.

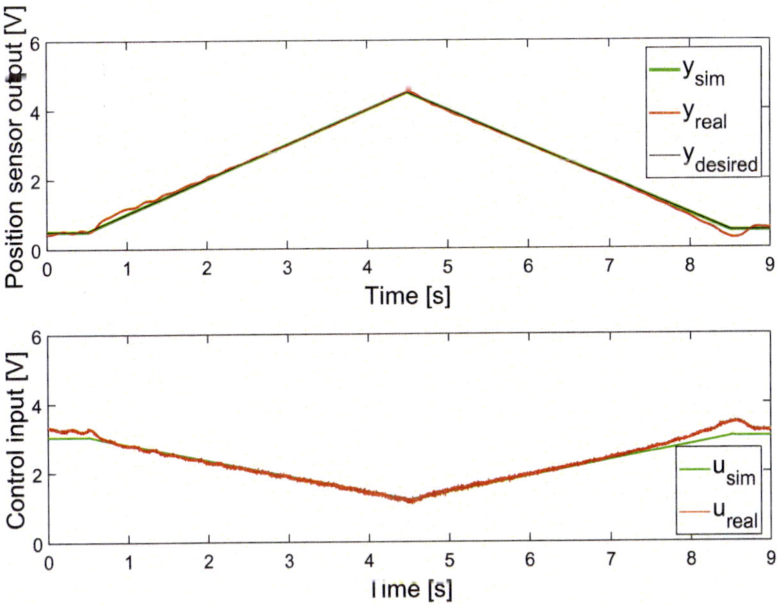

Fig. 13. Comparison of responses to ascending ramp 0.5–4.5 V and descending ramp 4.5–0.5 V for plant model simulation and real-world experiment.

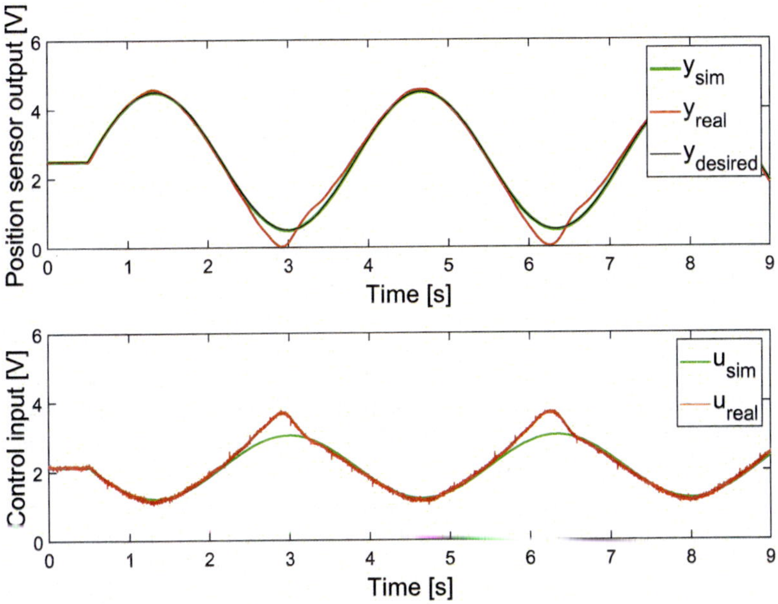

Fig. 14. Comparison of responses to sinus with offset 2.5 V, amplitude 2 V and frequency 0.3 Hz for plant model simulation and real-world experiment.

It is worth noting that some desired steady-state positions could not be achieved with this control circuit. The size of the minimum possible distance from the sensor and the coil in which the ball is possible to stabilize varied according to the PID controller constants.

In the case of steps, the expected system had smaller overshoot and a shorter settling time than real. In the case of a ramp, the trajectories y and u of the designed and real system were almost identical except for a small oscillation at the beginning and end of the signal. In the case of the sinusoidal input signal, the trajectories y and u were also very similar, except for positions when the ball was near the sensor. In this case, the controller was unable to stabilize the ball position with non-zero velocity and the ball dropped to the sensor for a moment.

From the verification results, it is possible to note that after the system settles, the desired y and u are reached, and for slowly changing continuous signal as the ramp is, the trajectories of the model were identical to the real system trajectories. Due to this fact, it is possible to conclude that the CE 152 model parameters were well determined because the static properties of the designed model corresponded to the static properties of the real plant.

On the other hand, when comparing the response to a faster change, such as step, it is possible to conclude that the dynamics of the proposed model could be better. The differences were probably caused by neglecting some parts of the system or not considering all phenomena affecting the system (heating the coil, electrical circuit components errors, etc.). These deficiencies can be easily removed by designing a more sophisticated or non-linear controller that will better control such a fast and highly non-linear system as CE 152.

5 Conclusion

In this work, the procedure for identifying the CE 152 system was outlined. Firstly, the technique for obtaining data from the system was described.

Next, the system was described and subsequently identified. The main characteristics of this system were introduced.

Finally, the simply PID control circuit was designed for testing the system model and the functionality was verified by comparison of simulations in the MATLAB-Simulink environment and real-word experiment.

The experiment showed that static properties (remaining in steady-state, following slowly changing continuous reference signal) of the proposed model corresponded to the real plant. On the other hand, the correspondence of dynamic properties (following step reference signal) could be better. In this case, the system is probably affected by a physical phenomenon that was not considered or neglected. It was found that some positions close to the sensor and the coil are unreachable. These zones sizes varied depending on the PID controller constants and is probably the result of linear control of such a fast and highly non-linear system as CE 152.

Another more sophisticated or non-linear control circuit is needed for better CE 152 control. Due to this fact, a new non-linear control circuit for this system will be designed in further work.

Acknowledgements. This research was supported by SGS 2021 grant at Faculty of Electrical Engineering and Informatics, University of Pardubice.

References

1. Humusoft: CE 152 Magnetic levitation model - educational manual (1996)
2. Honc, D.: Modelling and identification of magnetic levitation model CE 152/Revised. In: Silhavy, R. (ed.) CSOC2018 2018. AISC, vol. 765, pp. 35–43. Springer, Cham (2019). https://doi.org/10.1007/978-3-319-91192-2_4
3. Šuster , P., Jadlovská, A.: Modeling and control design of magnetic levitation system. In: 2012 IEEE 10th International Symposium on Applied Machine Intelligence and Informatics (SAMI), pp. 295–299 (2012)
4. Chalupa, P., Maly, M., Novak, J.: Nonlinear simulink model of magnetic levitation laboratory plant, pp. 293–299, June 2016
5. Safaei Pirooz, A.A., Flay, R.G., Minola, L., Azorin-Molina, C., Chen, D.: Effects of sensor response and moving average filter duration on maximum wind gust measurements. J. Wind Eng. Ind. Aerodyn. textbf206, 104354 (2020). https://www.sciencedirect.com/science/article/pii/S0167610520302646
6. González-Díez, A., Barreda-Argüeso, J., Rodríguez-Rodríguez, L., Fernández-Lozano, J.: The use of filters based on the fast fourier transform applied to dems for the objective mapping of karstic features. Geomorphology **385**, 107724 (2021). https://www.sciencedirect.com/science/article/pii/S0169555X2100132X
7. Strejc, V.: Least squares parameter estimation. Automatica **16**(5), 535–550 (1980). https://www.sciencedirect.com/science/article/pii/0005109880900771
8. Bojan-Dragos,, Preitl, S., Precup, R., Hergane, S., Hughiet, E.G., Szedlak-Stinean, A.: State feedback and proportional-integral-derivative control of a magnetic levitation system. In: 2016 IEEE 14th International Symposium on Intelligent Systems and Informatics (SISY), pp. 111–116 (2016)
9. Chacón, J., Vargas, H., Dormido, S., Sánchez, J.: Experimental study of nonlinear pid controllers in an air levitation system. IFAC-PapersOnLine **51**(4), 304–309 (2018). 3rd IFAC Conference on Advances in Proportional-Integral-Derivative Control PID 2018. https://www.sciencedirect.com/science/article/pii/S2405896318303781
10. Ben-Ezra, S.: FFT filter - clean your signals and display results! MATLAB central file exchange. https://www.mathworks.com/matlabcentral/fileexchange/25017-fft-filter-clean-your-signals-and-display-results (2021). Accessed 19 Mar 2021

The Optimal Model for Consumers' E-wallets Usage Intention

Dam Tri Cuong(✉)

Industrial University of Ho Chi Minh City, Ho Chi Minh City, Vietnam
damtricuong@iuh.edu.vn

Abstract. These days, the entire world is getting digitalized, and payments through cell phones have become the most straightforward payment method, supplanting cash-based payment methods. The technological progressions have made it essential for the present buyers to utilize their cell phones since it is quicker, simpler, and convenient to do the everyday exchanges. In addition, the reasons (bonus points and cashback) give to the client, deciding to pay through e-wallets, pulls the clients from all age gatherings to pay through e-wallets. This study aims to empirical research for searching the optimal model to clients' e-wallets usage intention. We implemented the Akaike information criterion (AIC) technique with RStudio software for selecting the optimal model. AIC is one of the most widely used model choice techniques in most statistical software packages. The finding reveals the best model includes determinants (effort expectancy, perceived ease to use, promotions, and perceived security and privacy) that significantly impact clients' e-wallets usage intention.

Keywords: Optimal model · E-wallet · AIC · Usage intention

1 Introduction

These days, the entire world is getting digitalized, and payments through cell phones have become the most straightforward payment method, replacing cash-based payment methods. The technological progressions have made it essential for the present buyers to utilize their cell phones since it is quicker, simpler, and convenient to do the everyday exchanges. In addition, the reasons (bonus points and cashback) give to the client, deciding to pay through e-wallets, pulls the clients from all age gatherings to pay through e-wallets. Furthermore, attributable to the convenience and safety of this payment method, it has become effectively a broadly utilized method in developed countries like Europe and the United States [1]. Besides, according to research by Boston Consulting Group, 49% of urban consumers in Southeast Asia who are commercial bank customers have used e-wallets, forecasting this rate will reach 84% by 2025. In reality, the e-wallet application accelerated rapidly after the Covid-19 pandemic, prompting a dramatic rise in the forms of digital payment and home delivery. The outbreak of the Covid-19 pandemic and its consequence will drive more users to accept digital payments. Although the bustle of the e-wallet playground is a good signal for the fintech

R. Silhavy et al. (Eds.): CoMeSySo 2021, LNNS 232, pp. 76–85, 2021.
https://doi.org/10.1007/978-3-030-90318-3_8

sector in Vietnam, this also signals the fierce competition of this market in the coming time. To see this clearly can make a comparison. At the same time, Vietnam has only more than 97 million people but 37 electronic wallets of payment intermediaries, not including e-wallets of banks. While China with more than 1.4 billion people but only a few e-wallets occupy the dominant market share. With the fierce competition for market share, many e-wallets have had to choose a new direction to survive and keep customers [2]. So, it is vital to identify determinant factors for clients' e-wallets usage intention. Base on that, managers will establish plans to make clients increase e-wallets usage intention.

On the other hand, many prior researchers involving consumers' e-wallets usage intention [3–6]. However, up to now, using the optimal model via Akaike information criterion (AIC) to clients' e-wallets usage intention has few studies in Vietnam. Therefore, this paper is trying to fill this gap.

2 Literature Review

2.1 E-wallets

A digital wallet or e-wallet is an electric payment tool that permits a person to make electronic exchanges. The e-wallet can cover buying things online with a computer or utilizing a cell phone to buy something at a shop. The bank account of a person can likewise be connected to the digital wallet [7]. The e-wallet fundamentally has worked as a physical wallet, which can be utilized to keep cash, client identification, credit card information, and other information when transacting on the web or e-commerce websites [4].

2.2 Unified Theory of Acceptance and Use of Technology (UTAUT)

Venkatesh et al. [8] created the unified theory of acceptance and use of technology or UTAUT to predict behavioral intention and use of technology. UTAUT theory was created by a comprehensive combination of former technology acceptance studies (Theory of reasoned action – TRA, technology acceptance model - TAM), motivational model - MM, theory of planned behavior - TPB, combined TAM and TPB - C-TAM-TPB, model of PC utilization - MPCU, innovation diffusion theory - IDT, and social cognitive theory – SCT) [8].

Davis first proposed the technology acceptance model (TAM) in 1986 to determine the clients' acceptance of using technology. The model uses perceived usefulness and perceived ease of use to anticipate clients' intention to use technology and clients' attitude to technology [9]. UTAUT model has four fundamental concepts (specifically, effort expectancy, performance expectancy, facilitating conditions, and social influence) that impact technology usage. UTAUT model differs from other theories as the model comprises external factors that enhance more accurate prediction than the TAM model [9, 10].

2.3 Usage Intention

Usage intention of technology, recognized as behavioral intention to utilize, is thought to be the motivational element that a person plans to make to the behavior [11]. Usage intention is the outcome of performance expectancy, effort expectancy, social influence, and facilitating conditions [8]. Moreover, prior studies stated that usage intention is also the promotion outcome [4] and perceived privacy/security [3]. As such, this paper analyzes the optimal model from determinants (performance expectancy, effort expectancy, social influence, facilitating conditions, promotion, perceived privacy/security) for consumers' e-wallets usage intention by using the AIC technique.

2.4 Performance Expectancy

Performance expectancy is portrayed as how clients utilizing the application will benefit clients in doing certain transactions [12]. To additionally explain, it is a condition where individuals will use new technology when they see it will help them execute their work [13]. Performance expectancy also is how much an individual believes that utilizing the application will help an individual improve the outcome of a business or exchange [14]. Some empirical said that performance expectancy had positively impacted clients' e-wallet usage intention [5, 13].

2.5 Effort Expectancy

Effort expectancy refers to comfort, which brings clients when utilizing an e-wallet as a payment approach. Effort expectancy is likewise described as the degree of effortlessness related to the utilization of payment [12]. Other scientists consider perceived effort expectancy as easy to use [6]. Some studied revealed effort expectancy as one of the most elements predicting clients' intention to utilize an e-wallet [13, 15].

2.6 Social Influence

Social influence is the degree to which buyers recognize that significant persons (For instance: loved ones, relatives, colleagues, etc.) consider utilizing a specific technology [12]. Social influences allude to the pressing of social networks to perform or not a particular behavior intention [16]. Previous empirical studies indicated social influence had related to predicting consumers' mobile payment intention to utilize [17, 18].

2.7 Facilitating Conditions

Facilitating conditions are illustrated as customers' thoughts of the resources and assistance ready to behave [12]. Regarding e-wallets, facilitating conditions are accessibility and affordability of cell phones and web associations, requiring knowledge about cell phones and security/privacy laws [19]. Prior studies demonstrated facilitating conditions had positively influenced clients' intention to utilize an e-wallet [20].

2.8 Promotion

Promotional programs involve advertisement, personal selling, public relations and publicity, sales promotion, and direct marketing. Promotion includes the process of informing, persuading, and motivating a purchasing decision [21]. Sales advertisements that enhance item attractiveness had positively affect customer behavior [4]. Sales promotions utilizing price decreases and coupons significantly affect consumer satisfaction when shopping at a store recognized with behavioral intentions [13]. Former researchers stated that promotions had impacted positively clients' intention to utilize an e-wallet [4].

2.9 Perceived Privacy and Security

Perceived security and privacy are characterized as the extent to which a client supposes that utilizing a specific e-payment process will be secure [22, 23]. Perceived security, which includes the utilization of technological progressions like cryptography, digital marks, and authentications, pointed toward guarding clients against the danger of cheating or hacking [24, 25]. Perceived privacy is the likelihood that online organizations gather information about clients and misuse them [26]. Other scholars disclosed that perceived security and privacy had related to positively clients' intention to utilize an e-wallet [22].

In this study, we have discovered empirical research of the optimal model for determinants of clients' intention to utilize an e-wallet in Vietnam by using the AIC technique as below:

$$INT = \beta 0 + \beta 1PER + \beta 2EFF + \beta 3SOC + \beta 4FAC + \beta 5PRO + \beta 6SEP + \varepsilon \tag{1}$$

Where:

- INT: Clients' e-wallet intention to utilize
- PER: Performance expectancy; EFF: Effort expectancy; SOC: Social influence; FAC: Facilitating conditions; PRO: Promotions; and SEP: Perceived security and privacy
- ε: Error

3 Research Methodology

3.1 Analytical Technique

Akaike [27] presented the Akaike information criterion (AIC), an information theory methodology for model (variable) choice, by Kullback-Leibler divergence. AIC is one of the most widely used model choice techniques in most statistical software packages [28]. Besides, the model choice criteria have become standard tools to use for particular regression models. One justification for this is that the scientist can utilize model choice rules precisely for picking between model other options, which doesn't expect them to be nested. They are simple to use, simple to explain, and accessible in statistical program packages. Likewise, model choice criteria are regularly used to search for an appropriate model for the data being scrutinized when building models in cases where the dependent or explained variables are thought to be functions of a few independent or explanatory factors [29]. Furthermore, the most common method for comparison and choice of models has penalized model choice criteria. One of the most generally utilized penalized model choice indexes is the Akaike information criterion (AIC) [27, 28].

In this study, the AIC technique by RStudio software was used to evaluate the optimal model. The best model is the model that has the lowest AIC value [30]. AIC is computed as follows [31]:

$$AIC = -2 \log L(\hat{\theta}) + 2k \quad (2)$$

Where:

- θ: the set (vector) of model parameters
- $L(\hat{\theta})$: the likelihood of the candidate model provided the data when estimated at the maximum likelihood estimate of θ
- k: the number of estimated parameters in the candidate model

3.2 Data and Sample

Data in this examination was gathered from customers who have used e-wallets in Ho Chi Minh City. A five-point Likert scale was utilized to measure the factors (from 1 = entirely oppose to 5 = entirely assent). A convenient methodology assembled the population sample. In the wake of disposing of the responses didn't finish information, and there were 368 responses acknowledged for the factual examination. Our population represented 129 (35.1%) males and 239 (64.9%) females. The age and income traits of clients as demonstrated in Tables 1 and 2.

Table 1. Age

Age	Frequency	Percent (%)
18 - 25	106	28.8
26 – 35	142	38.6
36 – 45	88	23.9
Above 45	32	8.7
Total	368	100.0

The sample comprises four age groups, in which the two-age groups are the most important in the age group from 26 to 35 with 142 customers (38.6%) and 18–25 with 106 customers (28.8%).

Table 2. Income

Income	Frequency	Percent (%)
Below 8 mil VND	15	4.1
8 mil – < 15 mil VND	185	50.3
15 mil – < 25 mil VND	135	36.7
25 mil – 35 mil VND	22	5.9
Above 35 mil VND	11	3.0
Total	368	100.0

There are five income groups, in which two groups of the income principally in the population are the group with income from 8 mils VND to below 15 mils VND, and 15 mils VND to below 25 mils VND, account for (87.0%).

4 Results and Discussion

4.1 Results

Akaike Information Criterion (AIC)
AIC criteria as presented in Table 3.

Table 3 exhibits the AIC results by each stage to look through the optimal model. The first stage includes six predictor factors with AIC index = −166.08. The subsequent step comprises five independent elements with AIC index = −168.07. The last step with four independent factors with AIC index = −169.28.

Table 3. AIC

Model	AIC
INT ~ PER + EFF + SOC + FAC + PRO + SEP	−166.08
INT ~ PER + EFF + SOC + PRO + SEP	−168.07
INT ~ PER + EFF + PRO + SEP	−169.28

The finding of the results with three steps to search the best model, RStudio stops at the model with four predictor variables. That implies, in this examination, the best model is the model with four independent variables since it has the smallest AIC index.

Coefficients of Predictors and Model Testing
As demonstrated in Table 4, the coefficients of predictors and model testing.

Table 4. Coefficients of predictors & model testing

	Estimate	Std. Error	t-value	p-value
(Intercept)	−0.37789	0.09960	−3.794	0.000173
PER	0.34899	0.02980	11.711	0.000000
EFF	0.33633	0.02907	11.571	0.000000
PRO	0.27318	0.02819	9.689	0.000000
SEP	0.11047	0.02901	3.808	0.000164

Multiple R-squared: 0.8417	Adjusted R-squared: 0.8399			
F-statistic: 482.4	p-value < 0.000000			

Table 4 presentations coefficients of predictor factors with a p-value < 0.05 and are statistically meaningful, so the assumption of predictors in the model is acknowledged. Besides, the R-squared in the optimal model is 0.8417, which means, 4 preditor variables (Effort expectancy, performance expectancy, promotion, and perceived security and privacy) explains 84.17% of the variance on the clients' e-wallets usage intention. In addition, as given in Table 4, the optimal model has statistically significant. Therefore, the best model is as follows:

$$Y = -0.37789 + 0.34899\ \text{PER} + 0.33633\ \text{EFF} + 0.27318\ \text{PRO} + 0.11047\ \text{SEP} + \varepsilon \quad (3)$$

Variance Inflation Factor (VIF) Testing

Variance inflation factor (VIF) testing in the model as manifested in Table 5:

Table 5. VIF index

	PER	EFF	PRO	SEP
VIF	1.716615	2.331537	1.961011	2.3177917

As given in Table 4, the VIF index of the predictor variable is lower than 5, so there is no multicollinearity between the predictors [32]:

Relative Importance for Linear Regression

The relative importance for linear regression was estimated through the LMP index (see Gromping [33]). Relative significance alludes to the measurement of a particular regressor to a multiple regression model. Each predictor's contribution is only the R^2 from univariate regression, and all univariate R^2-indexes sum up the entire model R^2 [33]. Relative importance metrics of predictor variable presentations in Table 6:

Table 6. Relative importance metrics

	lmp
PER	0.2115852
EFF	0.2553861
PRO	0.2110894
SEP	0.1636143

4.2 Discussion

The finding of this study indicates the optimal model as given in model (3). This model (3) shows that some determinants (Effort expectancy, performance expectancy, promotion, and perceived security and privacy) positively impact clients' e-wallets usage intention.

Moreover, as shown in Table 6, effort expectancy (EFF) the most significantly affects clients' e-wallets usage intention (lmp = 0.2553861). That means when effort expectancy (EFF) increases by 1 point, the clients' e-wallets usage intention will grow by 0.2553861. Next, performance expectancy (PER) is the second important role on clients' e-wallets usage intention (lmp = 0.2115852). That means when performance expectancy (PER) increases by 1 point, the clients' e-wallets usage intention will grow by 0.2115852. The third significant role that impacts clients' e-wallets usage intention is promotions (0.2110894). That implies when promotions (PRO) increase by 1 point, the clients' e-wallets usage intention will increase by 0.2110894. The fourth significant role that impacts clients' e-wallets usage intention is perceived security and privacy (0.1636143). That implies when perceived security and privacy (SEP) increase by 1 point, the clients' e-wallets usage intention will increase by 0.1636143.

5 Conclusions and Limitations

The finding of this examination assigns the best model as given in model (3). This model (3) recognizes that a few determinants (Effort expectancy, performance expectancy, promotion, and perceived security and privacy) positively influence clients' e-wallets usage intention. Therefore, practitioners should note these determinants and build action plans to enhance clients' e-wallet usage intention.

This study determined the optimal model for clients' e-wallets usage intention in Ho Chi Minh City. Future research should survey other cities in Vietnam to increase the extent of generalization of the model.

References

1. Nag, A.K., Gilitwala, B.: E-Wallet- factors affecting its intention to use. Int. J. Rec. Technol. Eng. 8(4) (2019). https://doi.org/10.35940/ijrte.d6756.118419
2. Hoang, T.: E-wallet market boom (Vietnamese). https://bankingplus.vn/bung-no-thi-truong-vi-dien-tu-99515.html

3. Singh, N., Sinha, N., Liébana-Cabanillas, F.J.: Determining factors in the adoption and recommendation of mobile wallet services in India: analysis of the effect of innovativeness, stress to use and social influence. Int. J. Inf. Manage. **50**, 191–205 (2020). https://doi.org/10.1016/j.ijinfomgt.2019.05.022
4. Ridaryanto, Firmansyah, R., Kartono, R., Sundjaja, A..: Factors affecting the use of E-Wallet in JABODETABEK area. Int. J. Adv. Trends Comput. Sci. Eng. **8**, 3645–3651 (2019)
5. Karim, M., Haque, A., Ulfy, M., Hossain, M., Anis, M.: Factors influencing the use of E-wallet as a payment method among malaysian young adults. J. Int. Bus. Manag. **3**, 1–11 (2020). https://doi.org/10.37227/jibm-2020-2-21/
6. Yang, M., Al Mamun, A., Mohiuddin, M., Nawi, N.C., Zainol, N.R.: Cashless transactions: a study on intention and adoption of e-wallets. Sustainability (Switzerland). **13**, 1–18 (2021). https://doi.org/10.3390/su13020831
7. Punwatkar, S., Verghese, M.: Adaptation of e-wallet payment: an empirical study on consumers' adoption behavior in central india. Int. J. Adv. Manag. Technol. Eng. Sci. **8**, 1147–1156 (2018)
8. Venkatesh, V., Morris, M., Davis, G., Davis, F.: User acceptance of information technology: toward a unified view. MIS Q. **27**, 425–478 (2003). https://doi.org/10.1016/j.inoche.2016.03.015
9. Davis, F.: Perceived usefulness, perceived ease of use, and user acceptance of information technology. MIS Q. **13**, 319–340 (1989). https://doi.org/10.5962/bhl.title.33621
10. Anwar, N.B.: Singaporean consumers' attitudes to technology usage. Int. J. Bus. Admin. Stud. **4**, 1–14 (2018). https://doi.org/10.20469/ijbas.4.10001-1
11. Ajzen, I.: The theory of planned behavior. Organ. Behav. Hum. Decis. Process. **50**, 179–211 (1991). https://doi.org/10.1016/0749-5978(91)90020-T
12. Venkatesh, V., Thong, J.Y., Xu, X.: Consumer acceptance and use of information technology: extending the unified theory of acceptance and use of technology. MIS Q. **36**, 157–178 (2012). https://doi.org/10.1109/MWSYM.2015.7167037
13. Angelina, C., Rahadi, R.A.: A conceptual study on the factors influencing usage intention of e-Wallets in Java. Int. J. Acc. Financ. Bus. **5**, 19–29 (2020)
14. Okumus, B., Ali, F., Bilgihan, A., Ozturk, A.B.: Psychological factors influencing customers' acceptance of smartphone diet apps when ordering food at restaurants. Int. J. Hosp. Manag. **72**, 67–77 (2018). https://doi.org/10.1016/j.ijhm.2018.01.001
15. Junadi, S.: A model of factors influencing consumer's intention to use e-payment system in Indonesia. Proc. Comput. Sci. **59**, 214–220 (2015). https://doi.org/10.1016/j.procs.2015.07.557
16. Lu, J., Yao, J.E., Yu, C.S.: Personal innovativeness, social influences and adoption of wireless Internet services via mobile technology. J. Strat. Inf. Syst. **14**, 245–268 (2005). https://doi.org/10.1016/j.jsis.2005.07.003
17. Yang, S., Lu, Y., Gupta, S., Cao, Y., Zhang, R.: Mobile payment services adoption across time: an empirical study of the effects of behavioral beliefs, social influences, and personal traits. Comput. Hum. Behav. **28**, 129–142 (2012). https://doi.org/10.1016/j.chb.2011.08.019
18. Khatimah, H., Susanto, P., Abdullah, N.L.: Hedonic motivation and social influence on behavioral intention of e-money: the role of payment habit as a mediator. Int. J. Entrepreneurship **23**, 1–9 (2019)
19. Chawla, D., Joshi, H.: Role of mediator in examining the influence of antecedents of mobile wallet adoption on attitude and intention. Glob. Bus. Rev. 1–17 (2020). https://doi.org/10.1177/0972150920924506
20. Peñarroja, V., Sánchez, J., Gamero, N., Orengo, V., Zornoza, A.M.: The influence of organisational facilitating conditions and technology acceptance factors on the effectiveness of virtual communities of practice. Behav. Inf. Technol. **38**, 845–857 (2019). https://doi.org/10.1080/0144929X.2018.1564070

21. Kotler, P., Keller, K.L.: Marketing Management. Pearson Education Inc, Upper Saddle River, New Jersey (2016)
22. Amoroso, D.L., Magnier-Watanabe, R.: Building a research model for mobile wallet consumer adoption: the case of mobile Suica in Japan. J. Theor. Appl. Electron. Commer. Res. **7**, 94–110 (2012). https://doi.org/10.4067/S0718-18762012000100008
23. Shin, D.H.: Towards an understanding of the consumer acceptance of mobile wallet. Comput. Hum. Behav. **25**, 1343–1354 (2009). https://doi.org/10.1016/j.chb.2009.06.001
24. Ranganathan, C., Ganapathy, S.: Key dimensions of business-to-consumer web sites. Inf. Manag. **39**, 457–465 (2002). https://doi.org/10.1016/S0378-7206(01)00112-4
25. Roca, J.C., García, J.J., de la Vega, J.J.: The importance of perceived trust, security and privacy in online trading systems. Inf. Manag. Comput. Secur. **17**, 96–113 (2009). https://doi.org/10.1108/09685220910963983
26. Jarvenpaa, S.L., Toad, P.A.: Consumer reactions to electronic shopping on the World Wide Web. Int. J. Electron. Commer. **1**, 59–88 (1996). https://doi.org/10.1080/10864415.1996.11518283
27. Akaike, H.: A new look at the statistical model identification. IEEE Trans. Autom. Control **19**, 716–723 (1974). https://doi.org/10.1109/TAC.1974.1100705
28. Chaurasia, A., Harel, O.: Model selection rates of information based criteria. Electron. J. Stat. **7**, 2762–2793 (2013). https://doi.org/10.1214/13-EJS861
29. Karlsson, P.S., Behrenz, L., Shukur, G.: Performances of model selection criteria when variables are ill conditioned. Comput. Econ. **54**(1), 77–98 (2017). https://doi.org/10.1007/s10614-017-9682-8
30. Bulteel, K., Wilderjans, T.F., Tuerlinckx, F., Ceulemans, E.: CHull as an alternative to AIC and BIC in the context of mixtures of factor analyzers. Behav. Res. Methods **45**(3), 782–791 (2013). https://doi.org/10.3758/s13428-012-0293-y
31. Fabozzi, F.J., Focardi, S.M., Rachev, S.T., Arshanapalli, B.G.: Appendix E: model selection criterion: AIC and BIC. In: The Basics of Financial Econometrics. pp. 399–403. John Wiley & Sons, Inc, New Jersey (2014). https://doi.org/10.1002/9781118856406.app5
32. Hair, J.F., Hult, G.T.M., Ringle, C.M., Sarstedt, M.: A Primer on Partial Least Squares Structural Equation Modeling (PLS-SEM). SAGE Publications Inc, Los Angeles, CA (2017)
33. Grömping, U.: Relative importance for linear regression in R: the package relaimpo. J. Stat. Softw. **17**, 1–27 (2006). https://doi.org/10.18637/jss.v017.i01

Issues and Possibilities of Personal Data Remote Processing in the COVID19 Pandemic Environment

Dmitry A. Mityushin(✉)

Institute of Information Science and Safety Technologies of Russian State University for the Humanities, Miusskaya sq. 6 GSP-3, 125993 Moscow, Russia

Abstract. In the spring of 2020, an unprecedented event happened – a massive transfer of employees to remote working, to work from home. As is well known, this is due to the outbreak of the COVID19 coronavirus pandemic. The Russian Government ordered the heads of companies, organizations and institutions to voluntarily and compulsorily send employees to remote work. We will not discuss the effectiveness of these measures and the impact that was inflicted as a result on both business and the economy as a whole. A similar thing happened in November 2020, which is associated with the second wave of coronavirus. In this regard, the management of companies and organizations faced a number of tasks that require a quick and efficient solution. One of the important tasks is to ensure the protection of sensitive information. Such information is personal data belonging to the category of restricted data. Therefore, it is necessary to consider the possibility of personal data processing in the remote working mode. This raises a number of questions. Is it technically possible to protect personal data from leaks? What do the regulatory documents of regulatory authority for the personal data say protection about this? How to organize and control the personal data operator employees' working and actions? The author will try to answer to these questions in this article. Due to the fact that the issues related to both the organization of remote working and the organization of the personal data processing in different countries differs, the author will rely on the regulatory legal acts of the Russian Federation.

Keywords: Personal data · COVID19 · Pandemic · Data security · Remote work · Personal data operator · Personal data processing

1 Statement of the Problem

It will be still about the recommendations of the country's leadership for transferring employees of companies, organizations and institutions (both state and non-state) to a working remotely mode (hereinafter WRM) in a pandemic.

Is it possible to organize work with personal data at home and what do the legislation and technical requirements dictate to us?

R. Silhavy et al. (Eds.): CoMeSySo 2021, LNNS 232, pp. 86–94, 2021.
https://doi.org/10.1007/978-3-030-90318-3_9

First, let's take a look at what personal data is and what their processing is.

The main regulatory legal document governing the working with personal data is the Federal Law of June 27, 2006 No. 152-FZ "On Personal Data" (hereinafter 152-FZ).

According to 152-FZ:

1) *personal data* (hereinafter PD) is any information relating directly or indirectly to a specific or identifiable individual (personal data subject);
2) *personal data operator* (hereinafter PDO, operator) is a state body, a municipal body, an juridical or individual person, independently or jointly with other persons organizing and (or) carrying out the personal data processing, as well as determining the purposes of personal data processing, the composition of PD to be processed, actions (operations) committed with PD;
3) *personal data processing* is any action (operation) or a set of actions (operations) committed using automation tools or without using such tools with PD, including collecting, recording, systematization, accumulation, storage, clarification (update, change), extraction, use, transfer (distribution, provision, access), depersonalization, blocking, deletion, destruction of personal data;

Thus, the Ministry of Labor and Social Protection of Russia recommended the transfer of employees to remote work. At the same time, the employer's responsibility to provide employees with equipment, software and hardware, and information protection means is not provided for by law [1]. Also, the law does not provide for the employer's responsibility to compensate the employee for communication and Internet costs. These issues can be settled by an agreement between the employee and the employer. At the same time, according to Rostrud[1], when transferring employees to the WRM, it is necessary to create conditions for such work – the ability to perform tasks and assignments of the employer, contact the company, as well as the ability to control the performance of the employee's functions.

Therefore, there is a problem: how to organize the work of the PDO at home and whether this is possible at all.

Thereat, it is necessary to consider technical, organizational and legal aspects of solving this problem.

First, let's consider the requirements for the personal data protection.

2 Requirements of Regulators for the Personal Data Protection

General Data Protection Regulation of the European Union (hereinafter GDPR) requires encryption of the transmitted data and data stored on the user's devices (Recital 83 and Article 32), which can be employees' Android, iOS, macOS, and Windows devices. It's recommended using also third party hard drive encryption software, like VeraCrypt, for data protection on users' devices [2].

GDPR recommends using a corporate VPN to encrypt PD transfers.

[1] Federal Service for Employment and Labor Relations.

In Russia the regulator in the field of personal data protection is the Federal Service for Technical and Export Control of the Russian Federation (hereinafter FSTEC of Russia). In Order No. 21 of February 18, 2013, the FSTEC of Russia specifies requirements for personal data information systems (hereinafter PDIS). This order says that the composition of measures to ensure the PD security, implemented within the framework of the personal data protection system, taking into account current threats to the PD security and applied information technologies, includes fifteen types of measures:

1) Identification and Authentication of Access Subjects and Access Objects;
2) Access Control of Access Subjects to Access Objects;
3) Software Environment Limitation;
4) protection of data digital media on which personal data are stored and (or) processed (hereinafter personal data digital media);
5) security events logging;
6) anti-virus protection;
7) Intrusions Detection (Prevention);
8) Personal Data Security Control (Analysis);
9) Ensuring the Integrity of the Information System and Personal Data;
10) Ensuring the Personal Data Availability;
11) Virtualization Environment Protection;
12) Technical Equipment Protection;
13) protection of the information system, its means, communication and data transmission systems;
14) identification of incidents (one event or a group of events) that can lead to failures or disruption of the information system functioning and (or) the emergence of threats to the personal data security (hereinafter incidents), and responding to them;
15) configuration management of the information system and the personal data protection system.

Let's consider the possibility of implementing the above requirements.

3 Technical Aspects of the Remote Personal Data Processing

In this section, we will consider the technical issues of solving the problem of remote personal data processing. It goes without saying that all three groups of issues (technical, organizational and legal) should be solved together.

Obviously, when a person receives any central and local government services, using the corresponding online services, there can hardly be a question of the personal data processing, as in this case, the person is transferring (or previously had transferred) to the service his personal data.

Data is transmitted over the Internet using the secure https protocol.

It is quite another matter when this personal data is processed by the PDO employees.

In this case, we have a computer network, which includes a server where personal data is stored, automated workstations (hereinafter AWP) of the PDO employees, high-speed uninterrupted connection and access to the operator's database.

The PD server can be located both in the same building where the operator's subdivisions are located, and in another one, in the case of a geographically distributed infrastructure of the PDO. In addition, the PDO can exchange personal data with other operators, with federal executive and municipal authorities. Of course, PD should be transmitted via a secure communication protocol (not only https) or via a dedicated line.

Of course, the operator must also use DLP or/and SIEM systems.

Now let's imagine the PDO, following the instructions of the government, is forced to transfer his employees, everybody or part of them, to work remotely from home.

In this case, the employer must provide the employee with a workstation. This is because not all the requirements of Order No. 21 dated February 18, 2013 of the FSTC of Russia (hereinafter Order 21) can be fulfilled using a home computer for work.

The use of DLP systems on a home computer can threaten the system will either block or analyze personal information of an employee. Which, again, is unacceptable and what was said more than once at the conference on December 16 [3].

In addition, the processed personal data may be stored in a file or swap area of a home computer.

In this regard, few people want to make using their home PC much more complicated. S. Konyavskaya speaks about this in her article [4].

So, the first solution is the workstation (PC or laptop, or other device) must be provided by the employer. This entails additional financial costs for the employer.

The second is the architecture of this workstation.

Most of the Order 21 requirements can be satisfied if a thin client, slightly thickened, is used as a workstation. There should be no hard drive. The connection between an employee's home workstation with the PD server should be carried out through the corporate VPN.

Another solution that will allow meeting most of the regulator's requirements is the use of Remote Virtual Desktop Infrastructure (RVDI) technology.

In this case, the employer (PD operator) has another problem – the creation of a virtualization environment, including a virtualization server for creating employees' remote desktops, as well as the need to fulfill the requirements of paragraph 11 (protection of the virtualization environment) of the above measures. This will incur additional costs for the organization if such an environment has not been established previously.

Thus, the optimal solution would be a workstation (thin client) issued to an employee with the necessary security equipment, with a case under seal, operating over a VPN channel using RVDI technology with only the necessary software installed.

4 Organizational and Legal Aspects of the Remote Personal Data Processing

Now we will consider the organizational and legal issues of fulfilling the regulator requirements during remote personal data processing.

The GDPR developers attach everything to the following four main steps [2]:

- Update the cybersecurity policy to reflect the new "working from home" reality.
- Train the employees and make sure your cybersecurity team is ready to support them.
- Keep data encrypted in transit and at rest.
- Limit access to sensitive data and keep the connections secure with a corporate VPN.

Most of the requirements of Order 21 are disclosed in detail in a document called "Information Protection Measures in State Information Systems" and approved by the FSTEC of Russia on February 11, 2014.

Consider some of the fifteen types of measures to ensure the personal data security listed above.

4.1 Identification and Authentication of Access Subjects and Access Objects (IAF)

This is the 1st group, which includes six measures to ensure the PD security. Five of the six measures should be applied in the personal data information systems (hereinafter PDIS) of all four PD protection level. And only the second measure *IAF.2 Identification and authentication of devices, including stationary, mobile and portable* is applied on PD protection levels 1 and 2.

Requirements *IAF.3 Management of identifiers, including the creation, assignment, destruction of identifiers* can also be performed, but in any case, the administrator responsible for this measure should not work remotely.

In general, the requirements of measure IAF.4 *Management of authentication means, including storage, issuance, initialization, blocking of authentication means and taking action in case of loss and (or) compromise of authentication means* can be fulfilled. However, there is a significant increase in the probability of authentication means compromise by worker's family.

If the cybersecurity policy of the PDO determines personal hardware identifiers (USB, eToken, etc.) at the end of the working day must be handed over by employees to the head of the department, then in case of remote work, the probability of their loss or theft increases significantly.

In general, the measures of group 1 can be implemented. But the personal data compromising risk is increasing.

4.2 Access Control of Access Subjects to Access Objects (UPD)

The 2nd group includes seventeen measures to ensure personal data security. Thirteen of them must be applied in the PDIS of any PD protection levels. The measures *UPD.10 Blocking an access session to the information system after a specified time of inactivity of the user or at his request* and *UPD.11 Permission (prohibition) of user actions allowed before identification and authentication* are applied on the PDIS of protection levels 1, 2 and 3. Measure *UPD.17 Provision of trusted loading of computer equipment* is applied only on PDIS of protection levels 1 and 2. Measure *UPD.13 Implementation of secure remote access of access subjects to access objects through external information and telecommunication networks* can be implemented by with the VPN technology. The requirements of the measure *UPD.14 Regulation and control of the use of wireless access technologies in the information system* are rather difficult to fulfill. Therefore, we recommend excluding the use of a wireless network at the workstation. Similarly, for the *UPD.15 Regulation and control of the mobile technical means use in the information system* there should be a complete prohibition (or physical exclusion of such a possibility) of the use of removable computer storage media, portable computing devices and communication devices with the ability to process information (laptops, netbooks, tablets, cell phones, digital cameras, sound recording and other devices).

In general, the measures of group 2 can be performed with consideration to the listed restrictions.

4.3 Software Environment Limitation (OPS)

The 3rd group of measures to ensure the personal data security includes four ones that are applied only on PDIS of protection levels 1 and 2.

The measures of this group can be implemented when working remotely.

4.4 Protection of Data Digital Media (ZNI)

The 4th group of measures to ensure the personal data security includes eight ones. To fulfill the measure *ZNI.1 Accounting for machine data carriers* the employee must receive a workstation under signed receipt and be responsible for its safety. Only an authorized officer should deliver the workstation to the employee's home (to fulfill the *ZNI.3 Control of the data digital media movement outside the controlled area*).

Measures of this group can be implemented, including in the absence of the physical ability to connect removable devices to the workstation.

4.5 Security Event Logging (RSB)

The 5th group includes eight measures to ensure the personal data security. All of them should be used on any PDIS, and only the *RSB.5 Monitoring (viewing, analysis) of the results of security events registration and response to them* is applied on PDIS of protection levels 1 and 2.

In general, the measures of group 5 can be implemented when working remotely.

4.6 Antivirus Protection (AVZ)

The 6th group includes two measures. All of them must be applied on PDIS of any protection level.

The requirement of the measure *AVZ.2 Updating the signs of malicious computer programs (viruses) database*, as well as updating programs and software components, must be carried out using the technology of remote virtual desktops.

In general, the measures of group 6 can be implemented with remote work.

4.7 Intrusions Detection (Prevention) (SOV)

The 7th group also includes two measures, and all of them must be applied on the PDIS of protection levels 1 and 2.

In general, the measures of group 7 can be implemented when working remotely.

4.8 Personal Data Security Control (Analysis) (ANZ)

The 8th group includes five measures, one of which – *ANZ.2 Control of software updates installation, including software update of data protection means* – should be applied on PDIS of any protection level. The measure *ANZ.5 Control of the rules for generating and changing user passwords, creating and deleting user accounts, implementing the rules for differentiating access, user rights in the information system* should be applied on the PDIS of protection levels 1 and 2, the rest – on PDIS of protection levels 1, 2 and 3.

Implementation of the measure *ANZ.4 Control of the composition of technical means, software and data protection means* may require the deployment of a video surveillance system at the remote employee workplace, including for monitoring his presence in the workplace. In this case, the image from the workstation monitor should not fall into the video camera field of view. To transmit information from a video camera, it is advisable to use a separate information stream, isolated from the PDIS ones.

At the same time, the location of the video camera should not lead to a violation of the constitutional rights to privacy of the employee's folks.

In general, the implementation of the measures of this group can be carried out; one measure can be implemented with significant difficulties.

4.9 Ensuring the Integrity of the Information System and Personal Data (OCL)

The 9th group includes eight measures.

In general, the measures of this group can be implemented when working remotely.

4.10 Ensuring the Personal Data Availability (ODT)

The 10th group includes five measures. All measures are intended to be implemented on the PDIS of protection levels 1 and 2. However, carrying-out of requirements of *ODT.2 Reservation of hardware, software, information transmission channels, means of ensuring the functioning of the information system* can cause serious difficulties and turn out to be economically unprofitable.

Thus, not all measures to ensure the security of personal data of this group can be implemented when working remotely.

4.11 Virtualization Environment Protection (ZSV)

The 11th group includes 10 measures. Since RVDI technology is used, the implementation of this group of measures is mandatory.

In general, the measures of this group can be implemented when working remotely.

4.12 Technical Equipment Protection (ZTS)

The 12th group includes five measures. Most of them should be implemented in PDIS of any protection level. The main problems with compliance with the regulator requirements arise here.

ZTS.1 Protection of processed by technical means information from its leakage through technical channels. Technically, it is possible to implement this measure at home, although this can cause a number of difficulties, but this will have to be done by almost every employee at home.

ZTS.2 Organization of a controlled area, within which stationary technical means processing information and means of protecting information, as well as means of ensuring functioning, are permanently located. This measure cannot be implemented without violating the legal rights of the employee's family members. The same applies to the *ZTS.3* measure, indicating the need to control and manage physical access to technical means, data protection means, means of ensuring functioning, as well as to the premises and structures in which they are installed, excluding unauthorized physical access to information processing means, data protection and means of ensuring the functioning of the information system and the premises and structures in which they are installed. In enhanced mode, this measure involves the establishment of a physical access control system and video surveillance.

ZTS.4 Placement of information output (display) devices, excluding its unauthorized viewing. And in this case, the risk of personal data leakage from the employee's family members is seriously increasing.

In principle, we can stop at this.

5 Conclusion

The process of transferring the processing personal data operator's employees to remote work is associated with a number of serious problems, both technical and organizational and legal. First, the number of internal information security violators increases significantly at the expense of the employee's family members. Secondly, not all the requirements of the personal data protection regulator are carried out. Third, the constitutional rights of the employee's family members may be significantly violated. In addition, we did not say anything about possible failures in the system, as well as data transmission channels.

Thus, the transfer of personal data processing employees to remote work is impractical, since the requirements of the regulator may be violated. This can only be allowed in extreme cases. But how will the regulator look at it?

References

1. Letter of the Ministry of Labor and Social Protection of the Russian Federation of April 23, 2020 No. 14–2/10/P-3710 "On sending Recommendations on the use of flexible forms of employment in the preventing the spread of a new coronavirus infection environment in the territory of the Russian Federation", http://www.consultant.ru/document/cons_doc_LAW_351512/. Accessed 22 May 2021

2. Data protection and working remotely. https://gdpr.eu/working-remotely-data-security/. Accessed 22 May 2021
3. DLP in conditions of remote work. Online conference. December 16, 2020. https://www.youtube.com/watch?v=5g_EnNjqYzg. Accessed 22 May 2021
4. Konyavskaya, S.: Self-isolation will not protect the information system from infection. Information Security/Informatsionnaya bezopasnost, **2**, 6 (2020). https://www.itsec.ru/articles/samoizolyaciya-ne-zashchitit-ot-zarazheniya-informacionnuyu-sistemu. Accessed 22 May 2021

Oil Consumption Analysis in the USA in Current Time

A. A. Vysotskaya, Roman I. Dzerzhinsky(✉), and E. N. Pronina

Applied Mathematics Department 78, MIREA – Russian Technological University, Vernadsky pros, Moscow 119454, Russia

Abstract. This article examines the oil consumption dynamics from 1965 to 2020 in the United States of America. The corresponding time series fluctuations and trend components analysis is carried out. The oil consumption dynamics a short-term forecast is also built. The trend is preliminarily eliminated based on the proportions' theory for time series fluctuations further analysis. Near-period values are determined using shift and autocorrelation functions. Based on the near-period obtained value, a development cell is built. Presumably, in the period from 2021 to 2042, the oil consumption dynamics in the United States will be distinguished by steady, slow growth with an average annual rate of about 1%, which will be interrupted by short-term recessions. After 2042, an unstable development phase will begin until 2061. During this period, a qualitative restructuring and a change in the oil consumption dynamics structure in the United States will take place. #COMESYSO1120.

Keywords: US oil consumption · Proportions theory · Alter-Johnson function · Autocorrelation function · Development cell

1 Introduction

A specific time series one example is the mining volume, incl. oil. Affected by political and economic factors, which at first glance is random and always determined by final demand, the oil consumption value is the economic stability indicator and the new methods' application volume in the energy sector.

The factors same set leads to the need to study the processes' interaction representation using fundamentally different time scales. This is achieved by dividing the initial time series into components that characterize the so-called fast and the process slow components. As a result, the fast movements' analysis makes it possible to determine the time intervals' hierarchy that are closest to periods (near-periods).

Knowing the near-periods allows us to use them as the initial series smoothing intervals to determine the trends' hierarchy corresponding to the near periods obtained values and to identify the so-called development cell that determines the prognostic observation period.

R. Silhavy et al. (Eds.): CoMeSySo 2021, LNNS 232, pp. 95–106, 2021.
https://doi.org/10.1007/978-3-030-90318-3_10

In addition, there is a near-proportions problem, which can be identified both with the algorithms help for finding near-proportions and can be reproduced through the near-periods proportions [1–3].

The application of such an analytical ideology makes it possible to analyze markets and systems of different nature [4, 5]. The choice of information processing methods is especially relevant when modeling complex technical and socio-economic systems [6, 7].

This paper examines the oil consumption dynamics in the period from 1965 to 2020 in the United States of America [8, 9].

The study aim is to determine the process main characteristics and determine its development phase at a given point in time.

2 The Exponential Trend Highlighting Areas

Let us consider the time series characterizing the oil consumption dynamics in the USA [2] in quadrillion British thermal units, quadBtu (Btu - British thermal unit, 1 kWh = 3412 Btu, 1 quadrillion = 1015), on the time interval 1965–2020. The data is presented on a semi-logarithmic scale (Fig. 1).

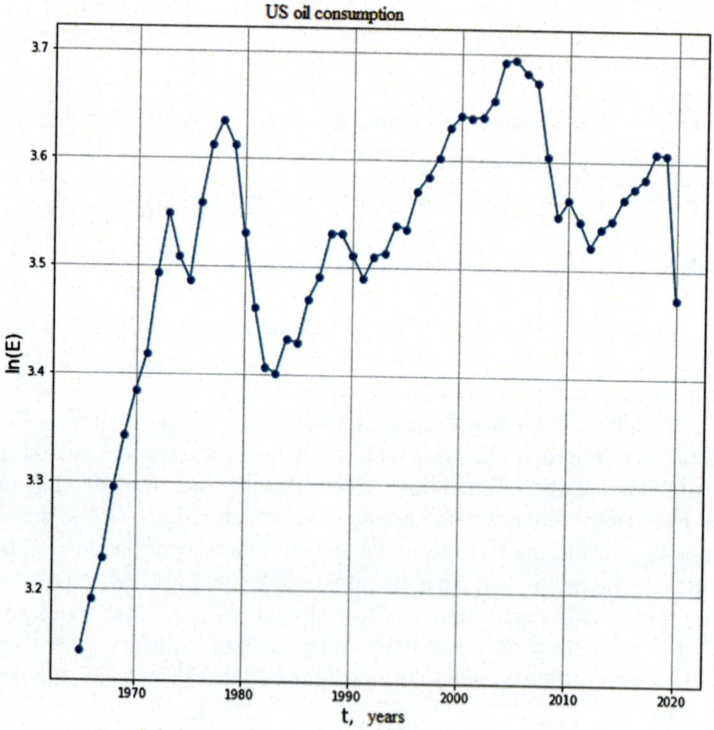

Fig. 1. Oil consumption dynamics in the USA, 1965–2020.

A semi-logarithmic coordinate system is a natural way to identify exponential trend areas, which is characterized by constant relative increments. Growth rates are calculated using the following formula:

$$r(t) = \frac{\ln \frac{E_2}{E_1}}{t_2 - t_1}$$

Consider two sections that are straightened on a semi-log scale, the first for the time interval from 1965 to 1973, the second from 1983 to 2005 (Fig. 2).

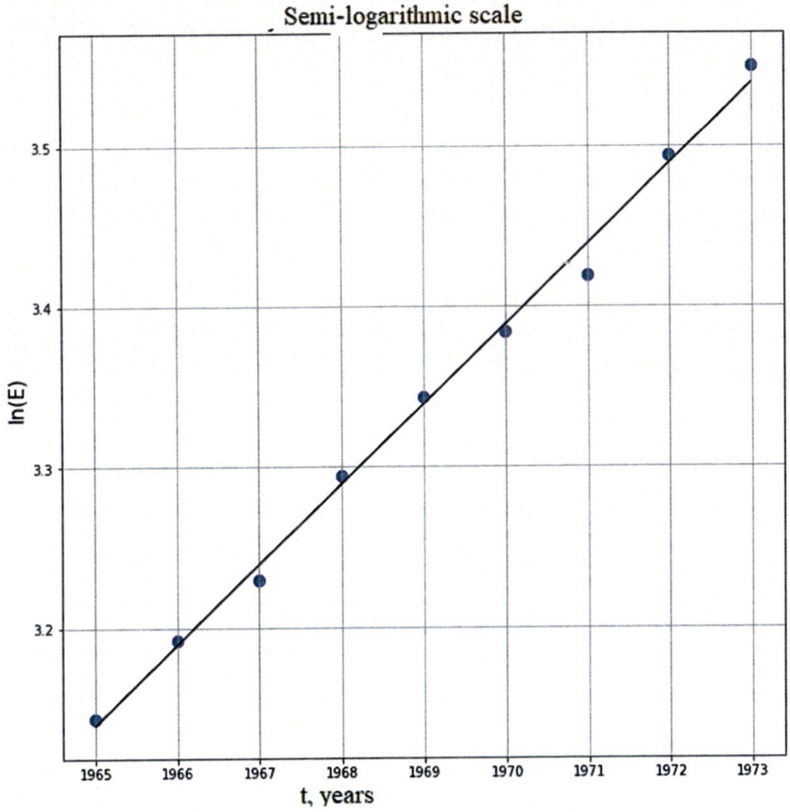

Fig. 2. The first straightening section, 1965–1973.

Linear trend equation $y(t) = 0.05 \times -95.11$.

This oil consumption dynamics interval in the United States is characterized by the relative growth rates $r = 0.05$. The process characteristic time in this interval is determined by the relative increments reciprocal and is equal to 20 years. In the initial coordinates, this section corresponds to an exponential increasing dependence (Fig. 3).

$$E(t) = 23.18 \cdot e^{0.05(t-1965)}.$$

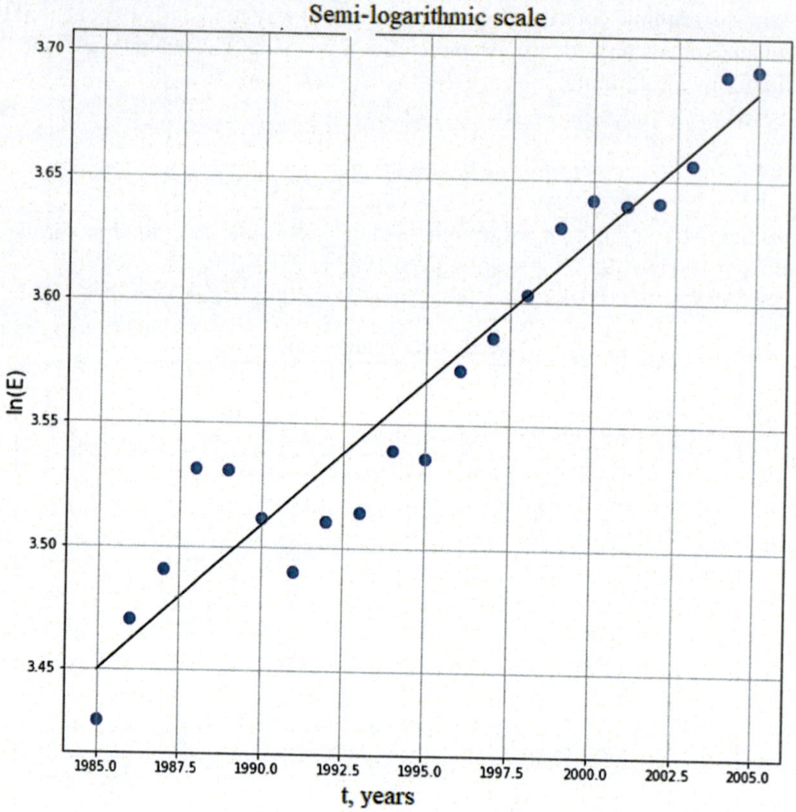

Fig. 3. Second straightening section, 1983–2005.

Linear trend equation y(t) = 0.012 × −20.5. In the initial coordinates, this section corresponds to an exponential increasing dependence.

E(t)=29.997·e^0.012(t-1983)

The exponential growth next segment is located in the interval 2012–2020. The oil resources consumption dynamics curve during this period again returns to the previous rate, which value is approximately 1.2% per year,

E(t)=33.83·e^(0.0122(t-2012))

3 Fluctuations Analysis Manifested in the Oil Consumption Dynamics

It is quite problematic to analyse fluctuations since the initial data contain a trend component. It distorts the result. Therefore, we first exclude the trend from the initial data. To exclude a trend, the empirical data transformation in coordinates is used

$$\ln\left(\frac{y_{t-\Delta t} + y_{t+\Delta t}}{2 \cdot y_t}\right) \sim t, \tag{1}$$

$$\ln\left(\frac{y_{t-\Delta t} \cdot y_{t+\Delta t}}{{y_t}^2}\right) \sim t. \tag{2}$$

This method is based on arithmetic and geometric proportions.

Applying transformation (1) to the oil consumption time series with different shifts Δt, we obtain the results shown in Fig. 4, 5.

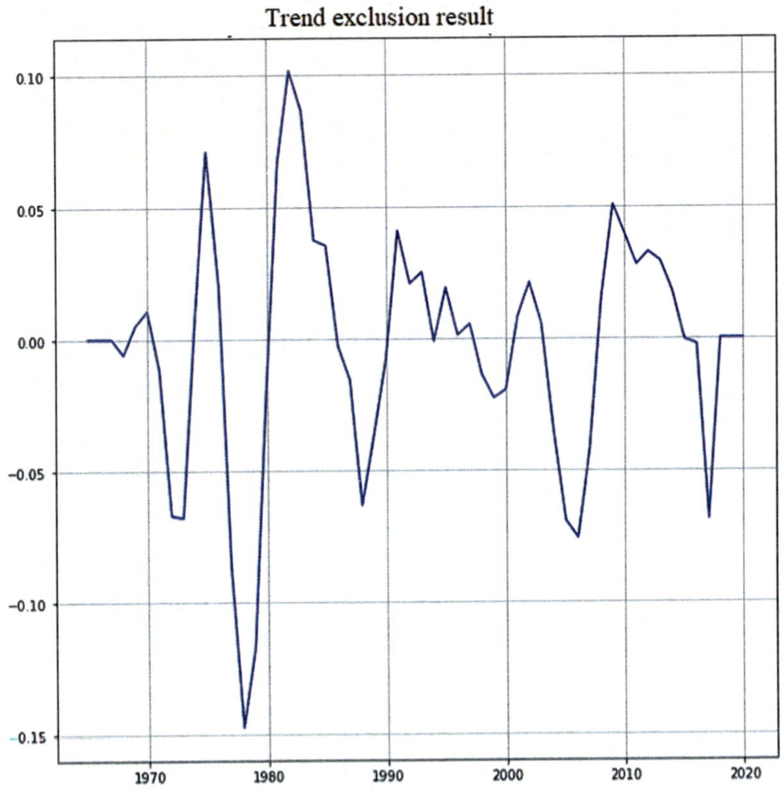

Fig. 4. Excluded trend, $\Delta t = 3$.

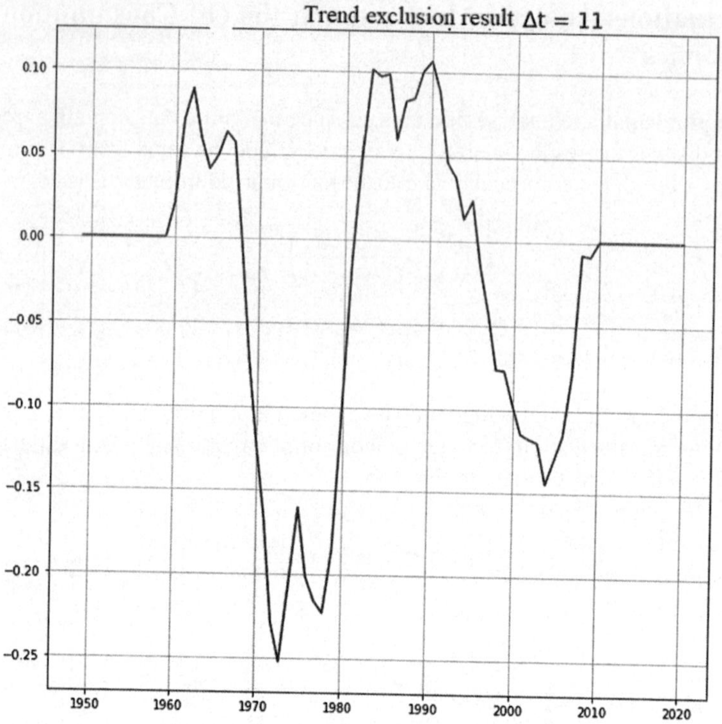

Fig. 5. An Excluded trend, $\Delta t = 11$.

4 Search Shift Function

The shift function or Alter-Johnson function for determining near-periods relies on a function period fundamental characteristic property, which is the function values repetition at equal time intervals.

Since real data contains non-linear fluctuations, the periods' true values are quite rare. With the shift function help, we can determine only the values that are as close as possible to the periods. Such values are called near-periods. Knowing the near-periods allows us to use them as smoothing intervals to clarify the time series trend components.

To determine the near-periods, we introduce the following metric:

$$a(\tau) = \frac{1}{n - \tau}\sum_{t=1}^{n-\tau}|f(t + \tau) - f(t)|.$$

By the local minima location, which corresponds to the closest to the period values τ, the system near-period is determined.

Let us consider the shift function application to the analysed time series with the previously excluded trend component (Fig. 6, 7).

The shift functions local minima position indicates near-periods: $\tau 1 = 9$ years and $\tau 2 = 27$ years. Since τ is 2 multiples of $\tau 1$, we can assume that there is only one near-period in the original series, equal to 9 years.

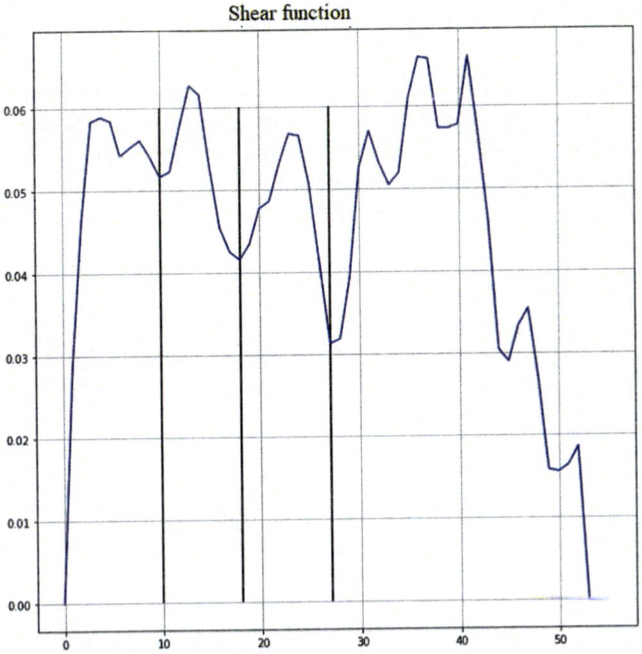

Fig. 6. Shear function for the data in Fig. 4.

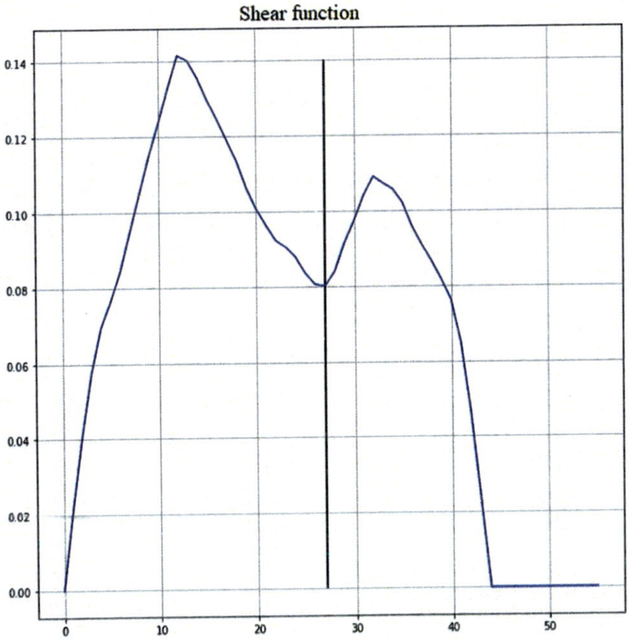

Fig. 7. Shear function for the data in Fig. 5.

5 Autocorrelation Function Definition

To check the results obtained reliability, we will use one more function that can be used to determine the near-periods values. It is an autocorrelation function (ACF). The autocorrelation function allows finding time series repeating portions or to determine the carrier frequency, which may be hidden due to the noise or fluctuations superposition at other frequencies. Defines repeating areas by superimposing a row on itself, but with different offsets τ in time.

The autocorrelation function is calculated by the formula:

$$a(\tau) = \frac{1}{n-\tau}\sum_{t=1}^{n-\tau} f(t+\tau)\cdot f(t).$$

Autocorrelation functions built after excluding the trend are shown in Fig. 8, 9.

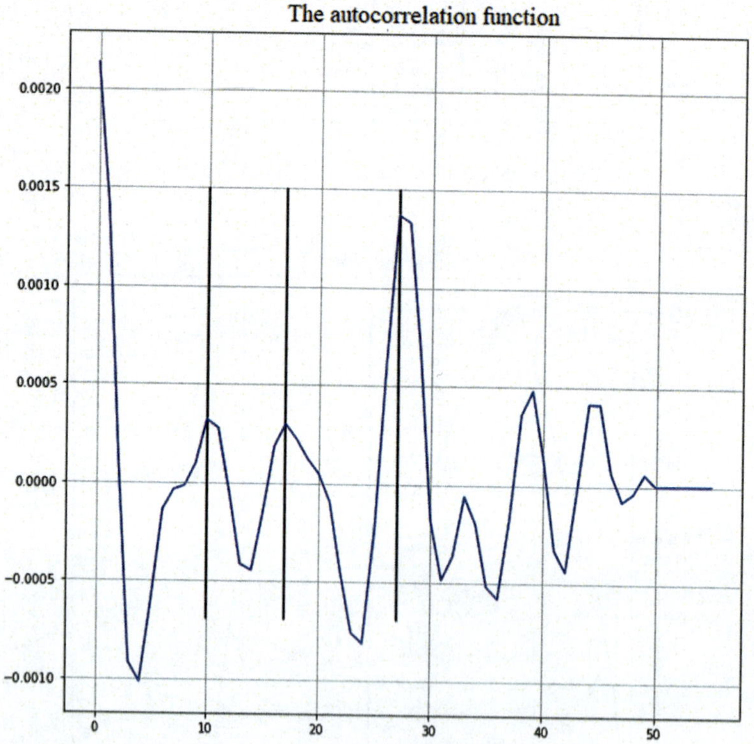

Fig. 8. Autocorrelation function for the data in Fig. 4.

The autocorrelation function local maxima arguments coincide with the near-periods obtained from the shift function. Thus, it can be expected that the oil consumption dynamics in the United States will show a temporal rhythm with a duration of 9 years.

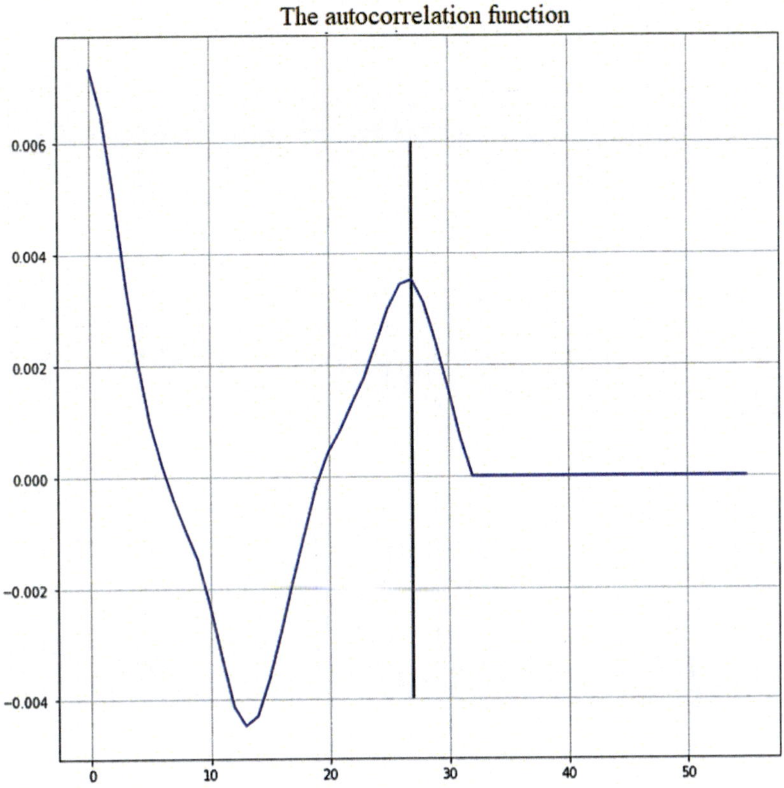

Fig. 9. Autocorrelation function for the data in Fig. 5.

6 Examining Critical Boundaries

The conditions for the rhythms' coordination, built according to the arithmetic progression law, and rhythms following a geometric progression, are formulated in the critical phenomena theory. The corresponding structure was named the development cell [1].

Let us construct a development cell with a near-period of 9 years for the time interval from 1983 to 2012. During this period, US oil production entered a mature phase. After passing the production peak in 1970, the United States is turning into the world's largest oil importer, especially since forty years ago it was more profitable to import cheap Arab oil than to increase investments in its own fields' development. During these years, oil consumption in the United States exceeded its production by 3.6 times. Fuel and energy resources are becoming a factor limiting the further increase in energy production and consumption.

Let's take the local minimum point (1983) on the oil consumption dynamics curve in the United States as the starting point. The development cell remaining boundaries are shown in Fig. 10.

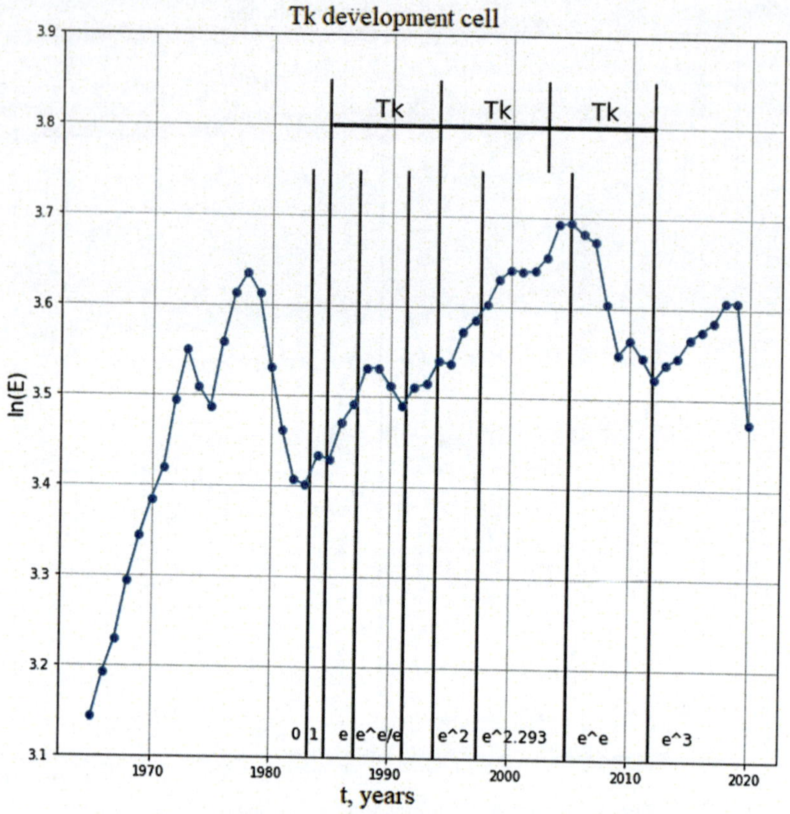

Fig. 10. The rhythmic structure manifested in the oil consumption dynamics in the period from 1983 to 2012.

Attention is drawn to the boundary coincidence e^e with the maximum consumption during the entire analysed period, followed by the development' an unstable phase (restructuring phase), on the graph, it is shown as another decline. During this period, the US government policy changed, a decision was made to reduce oil imports, which prices had become significant by that time.

Thus, the critical boundaries obtained by the calculation according to the development cell indicate significant changes in the oil consumption dynamics: temporary stops in growth, local minimums, maximums and trend changes.

The system such a parameter knowledge as the cycle length allows predicting the oil resources consumption dynamics further development. To do this, let's build a next-level development cell $T_k + 1$ (Fig. 11). As the basis for calculating the cell structure boundaries, let us take the duration equal to $e \cdot T_k \approx 24$ years (Fig. 11).

Significant changes in the real dynamics curve, which are manifested as local minimums and maximums in 1991, 2005 and 2012, corresponding to the critical boundaries $\frac{e^e}{e}, e^e, e^3$.

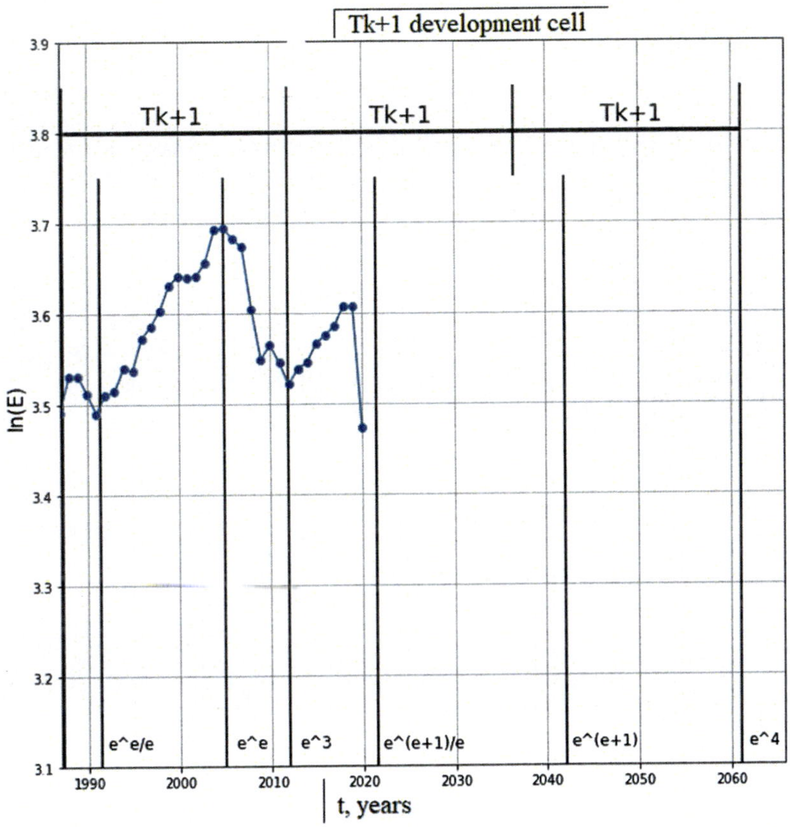

Fig. 11. Development Cell $T_k + 1$.

7 Conclusion

Based on the results obtained above, it can be concluded that the oil consumption annual dynamics in the United States has reached its maximum level, the historical maximum has been passed in 2004–2005.

Presumably, in the period from 2021 to 2042, the oil consumption dynamics in the United States will be distinguished by steady, slow growth with an average annual rate of about 1%, which will be interrupted by short-term recessions. After 2042, an unstable development phase will begin until 2061. During this period, a qualitative restructuring and a change in the oil consumption dynamics structure in the United States will take place.

References

1. Veniaminov, S.S.: Revealing Latent Structure Laws in Processes and Signals. M., URSS p. 216 (2014)

2. Kuzmin, V.I., Gadzaov, A.F.: Method of construction of models on empirical data. The manual. M., MGTU MIREA 94 (2012)
3. Serebrennikov, M.G., Pervozvansky, A.A.: Revealing of the Latent Periodicity. M., Science p. 244 (1965)
4. Dzerjinsky, R.I., Pronina, E.N., Dzerzhinskaya, M.R.: The Structural analysis of the world gold prices dynamics. In: Silhavy, R. (ed.) CSOC 2020. AISC, vol. 1225, pp. 352–365. Springer, Cham (2020). https://doi.org/10.1007/978-3-030-51971-1_29
5. Paramonov A.A, Kuzmin V.I., Dzerjinsky R.I.: Analysis of almost-periodic and almost-proportional characteristics of a representative sample local minima time series. In: IOP Conference Series: Materials Science and Engineering, vol. 1047, p. 012045 (2020).https://doi.org/10.1088/1757-899X/1047/1/012045.
6. Semenenko, M.G., et al.: How to use neural network and web technologies in modeling complex technical systems. In: IOP Conference Series: Materials Science and Engineering, vol. 537, 032095 (2019)
7. Gruzenkin, D.V., Sukhanova, A.V., Novikov, O.S., Grishina, G.V., Rutskiy, V.N., Tsarev, R.Y., Zhigalov, K.Y.: Neural networks to solve modern artificial intelligence tasks. J. Phys. Conf. Series **1399**, 033058 (2019)
8. Our world in data. https://ourworldindata.org/search?q=consumption+oil. Accessed 8 Aug 2020
9. Machinelearning. http://www.machinelearning.ru/. Accessed 8 Aug 2020

The Nestle Share Price Dynamics Analysis and Forecasting

Roman I. Dzerjinsky(✉)

Applied Mathematics Department, MIREA – Russian Technological University, 78, Vernadsky pros, Moscow 119454, Russia

Abstract. The article presents the latent periodicities' study results in the Nestle shares dynamics using anamorphosis for the Gompertz models. The self-similar growth ranges boundaries are interconnected with the functions local minimums and together with them determine the world prices spectral composition for Nestle shares. The series structural analysis showed the growth possibility in the shares' value until 2025, and after reaching its global maximum at the 2025 end, its fall and reaching its minimum in the 2026 third quarter. The development unstable phase (the restructuring phase) falls on the period from 2028 to 2037. #COMESYSO1120.

Keywords: Intensive growth models · Anamorphosis · Model Gompertz · Near-periods · Dynamics predictive assessment · Self-similarity

1 Introduction

All the stock exchange price is a specific time series typical example. Their feature is the ability to represent the processes' interaction using fundamentally different time scales. This is achieved by dividing the initial time series into components that characterize the so-called fast and the process slow components. As a result, the fast movements' analysis makes it possible to determine the time intervals' hierarchy that are closest to periods (near-periods).

Knowing the near-periods allows us to use them as smoothing intervals for the initial series to determine the trends' hierarchy corresponding to the near-periods obtained values. Slow movements, as the main trends, are also analysed directly, without the fast movements' exception, on the anamorphosis basis, which bijectively maps the initial data into piecewise linear dependencies. The resulting synthetic construction is determined by comparing the near-periods characteristics and trends.

In addition, there is a near-proportions problem, which can be identified both with the algorithms help for finding near-proportions and can be reproduced through the near-periods proportions [1–3].

The application of such an analytical ideology makes it possible to analyze markets and systems of different nature [4, 5]. The choice of information processing methods is especially relevant when modeling complex technical and socio-economic systems [6, 7].

R. Silhavy et al. (Eds.): CoMeSySo 2021, LNNS 232, pp. 107–127, 2021.
https://doi.org/10.1007/978-3-030-90318-3_11

This paper examines the company Nestle share price dynamics-a transnational company headquartered in Switzerland, one of the world market leaders. As data, the share price values in trading days were taken for the period from 11.13.1996 to 11.13.2020 [8, 9].

The study purpose is to determine the process main characteristics, as well as to predict this company shares value and determine its development phase at the time moment.

2 The Nestle Shares Exchange Rate Analysis

To analyse the Nestlé stock price, you will need the initial data - the MICEX Index throughout the entire analysed interval. To make it convenient to work with the charts, the days starting from the first one will be plotted on the abscissa axis (Fig. 1). But to further correlate the required trading day with the real date, a figure is also presented, where dates are plotted on the abscissa axis (Fig. 2).

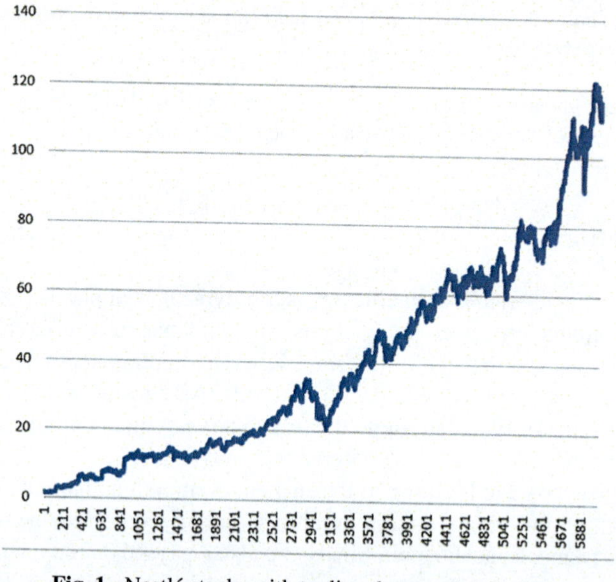

Fig. 1. Nestlé stocks with trading days on the abscissa.

Figure 2 clearly shows the general upward trend in Nestlé shares. For the initial data a detailed analysis, as well as identifying the critical levels corresponding to the main boundaries' formation, it is necessary to construct the initial data in semi-logarithmic coordinates, that is, to log the values along the ordinate axis (Fig. 3).

This figure clearly shows that at certain moments the trend changed the inclination angle, and you can also see the channels in which the process is taking place. Note that the graph begins with a 0.4 value along the y axis (Fig. 3), then values greater than one follow, then rises to the 0.9 level. After that, the graph gradually rises to the value

Fig. 2. Nestlé stocks with dates on the abscissa.

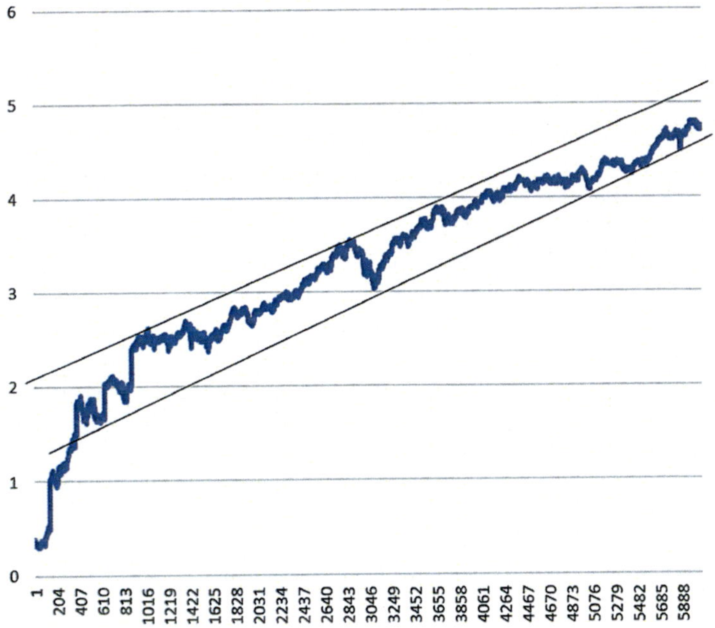

Fig. 3. Initial data on a semi-logarithmic scale.

3. Fluctuating about the value 3, the graph begins to tend to the value 4 on the axis y. Further, the chart from 4 tends to 5, fluctuating around this level.

A change in the unit price logarithm value will correspond to a change in the original data by a times factor. Parallel axis lines of x corresponding to share price 1, 2, 3 and 4 natural logarithm values will be support and resistance lines for the share price respective growth sites (fall). Resistance line - a level that does not allow rising; support line - a level that does not allow stocks to fall. Support and resistance lines pass through at least 3 points on the chart and form a corridor for an upward or downward trend.

3 Intensive Growth Model

3.1 First Growth Rate Model

The model general structure is as follows:

$$\frac{dy}{dt} = f(t, y)y,$$

moreover, k can take any function kind. In the case when it is necessary to determine the intensive growth on the graph, k has the form k = a * y, and the general equation will have the form $y' = a * y^2$. The solution to this equation will look like this: $\frac{dy}{y^2} = a * dt$ and integrate, we get the following form equation:

$$\frac{1}{y} = -a * t + C.$$

To get anamorphosis, straightening the graph in those places where there is intensive growth, it is necessary to build a graph $\frac{1}{y} \sim t$ and mark a trend line on it (see Fig. 4).

Let's Any intensive growth has its limit to which it will grow. On a graph with the original data, this would look like a quadratic function tending to the vertical asymptote. To find this asymptote, it is necessary to know the intensive growth period end. To do this, you need to draw lines in all areas where you can see the graph a clear straightening.

Since the graph in Fig. 4 falls rather sharply, varying in a wide range, it will be difficult to evaluate it based on intensive growth. For intensive growth a correct assessment, it is necessary to consider in more detail each section on the graph where straightening occurs, that is, take it in certain intervals (Fig. 5, 6 and 7).

The sought values are respectively equal to 540, 1580, 1980, 3600 and 4000. Note the obtained values on the original semilogarithmic graph and draw straight lines from them parallel to the y axis, these will be the sought asymptotes (Fig. 8).

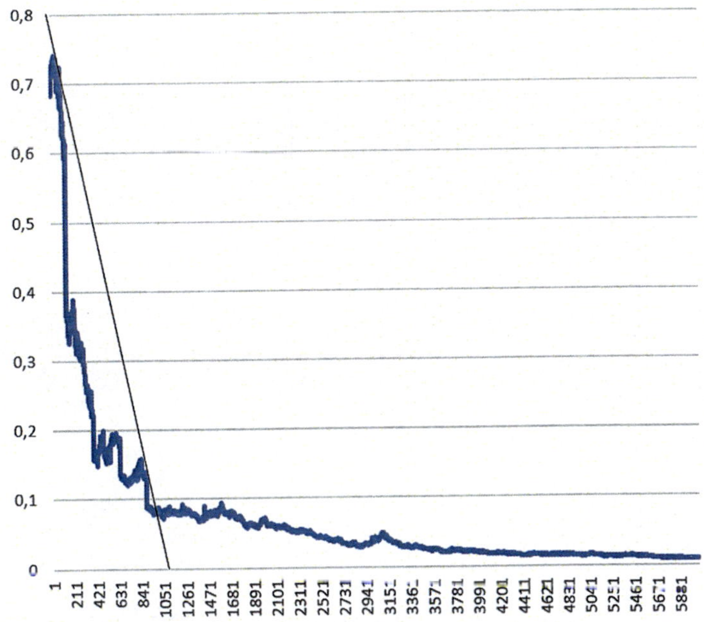

Fig. 4. Anamorphosis to intensive growth in coordinates.

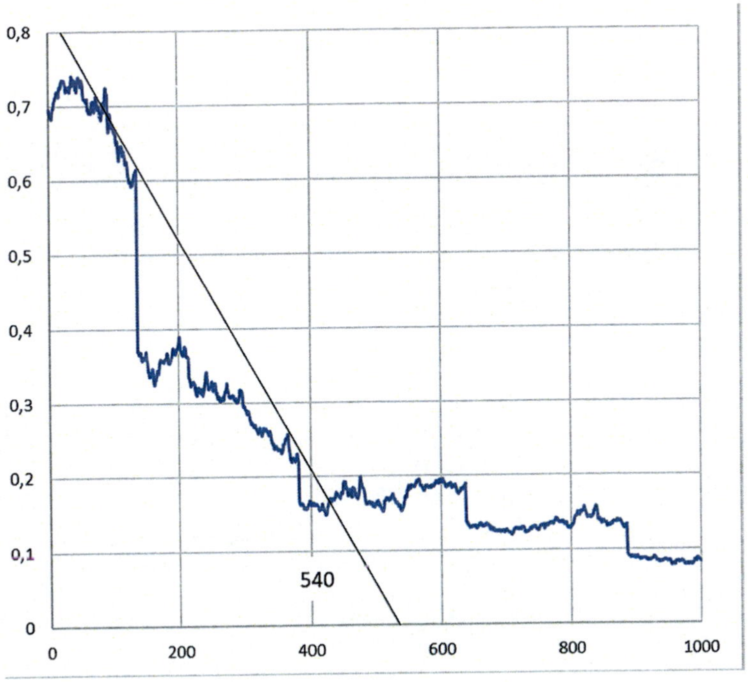

Fig. 5. Anamorphosis to intensive growth in the range from 0 to 1000.

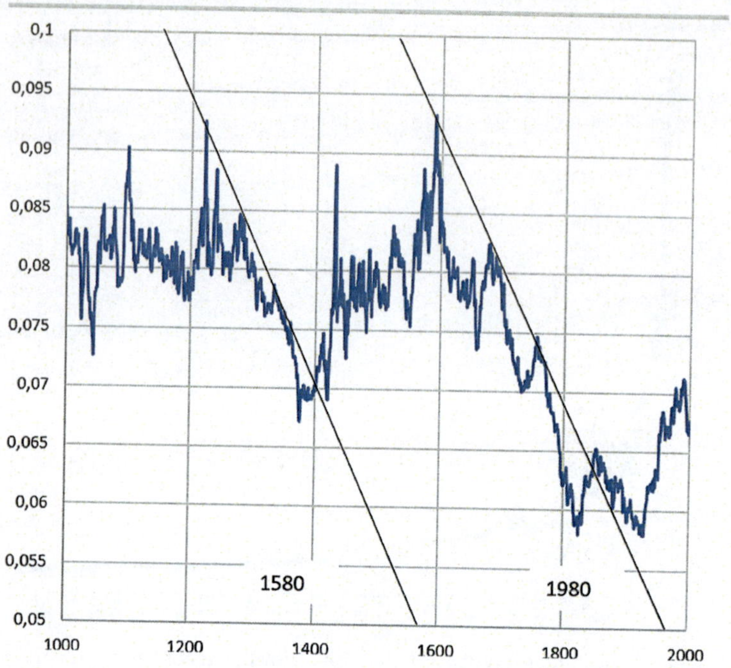

Fig. 6. Anamorphosis to intensive growth in the range from 1000 to 2000.

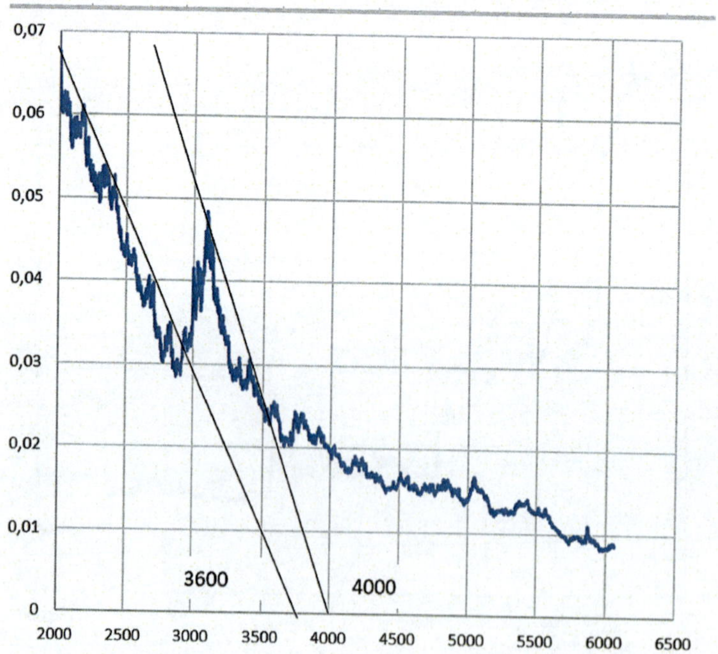

Fig. 7. Anamorphosis to intensive growth in the range from 2000 to 6500.

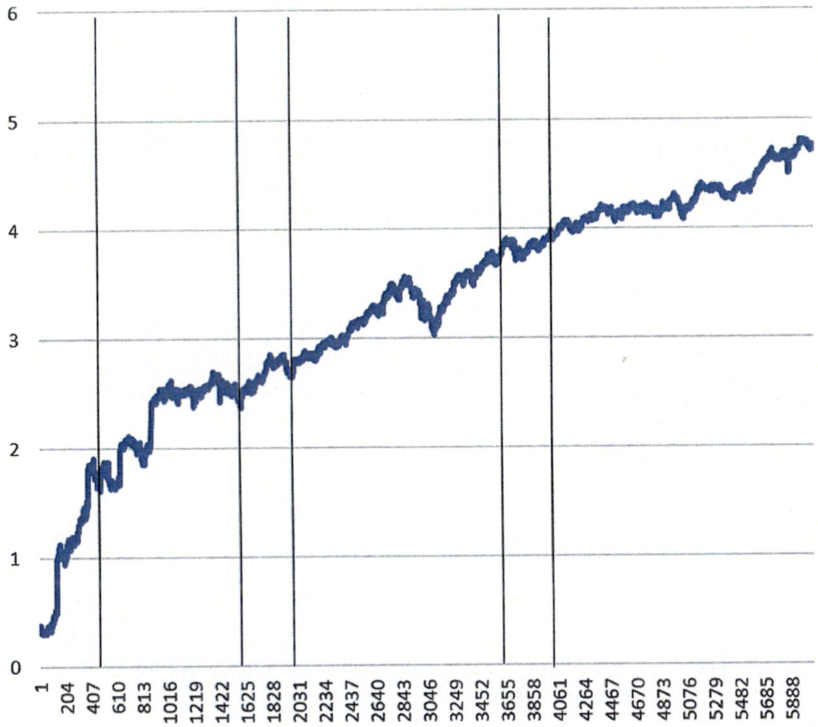

Fig. 8. Initial graph (y) $\sim t$, with marked initial asymptotes.

3.2 Second Growth Rate Model

After determining the intensive growth period, it is necessary to determine the cutoff levels, that is, those values *ln (y)*, at which these processes will end. To do this, you need to represent $f(t, y) = a * t$, and solve this equation.

In the equation $y' = at * y^2$, divide the variables $\frac{dy}{y} = at * dt$ and integrating both sides we get:

$$\ln(y) = \frac{a * t^2}{2} + \ln(C)(t - t_0),$$

where trading days and t_0 is the trading day at the countdown start.

To find the intensive growth limit, it is necessary to substitute the found t^* on the abscissa axis, find the intersection with the plotted graph and reflect it on the ordinate axis. Then we will find the desired value, but only under the logarithm. So for a final answer, we need to put it as an exponent degree. To do this, you need to build an anamorphosis in the coordinates *ln (y)~*t^2 and substitute the found $t*$ (Fig. 9, 10).

Now let's find the height limits for each age point $t*$ using Fig. 2.2.2. For $t* = 540$ it equals $e^{0,31} = 1.363$, for $t* = 1580$ it equals $e^{0.297} = 1.345$, for $t* = 1980$ it is equal to $e^{0,318} = 1.374$, for $t* = 3600$ it is $e^{0.336} = 1.399$, for $t^* = 4000$ it is equal to $e^{0.37} =$ 1.447.

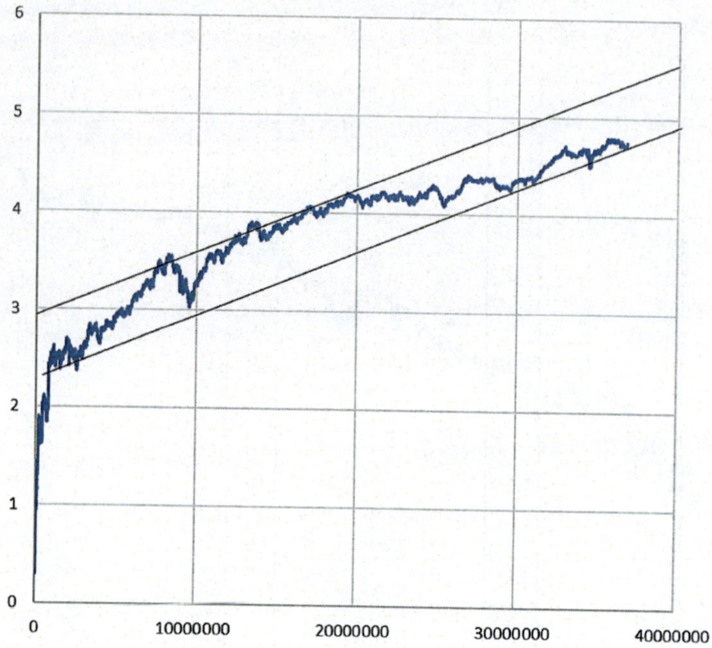

Fig. 9. Anamorphosis to intensive growth in coordinates ln(y)~t^2.

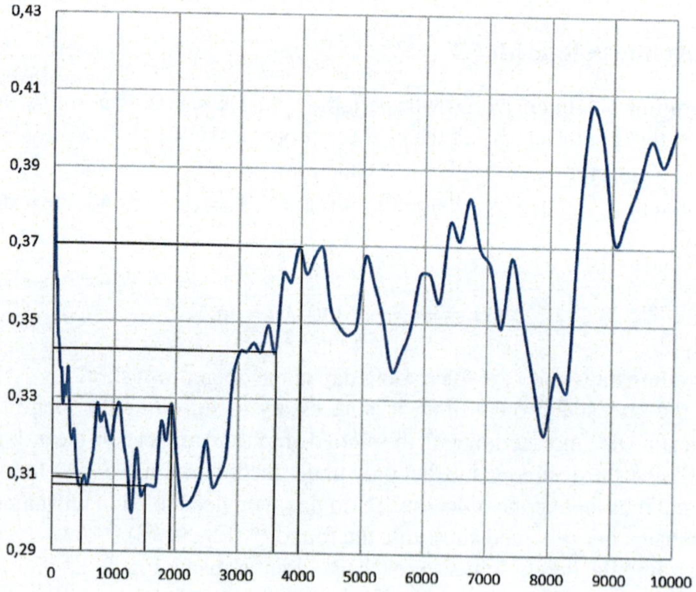

Fig. 10. Anamorphosis to intensive growth in coordinates ln(y)~t^2 in the range from 0 to 4000.

3.3 Third Growth Rate Model

The growth rate increases exponentially with the age of the system $f(t, y) = A * e^{kt}$, where t is the system age.

Then, $y' = A * e^{kt} * y$; $\frac{dy}{y} = A * e^{kt} dt$.

Therefore, this anamorphosis to intensive growth has the form:

$$\ln\left(\ln\left(\frac{y}{C}\right)\right) = \ln\left(\frac{A}{k}\right) + kt.$$

Straightening is achieved in coordinates $\ln\left(\ln\left(\frac{y}{C}\right)\right) \sim t$. By varying the constant *with*, the original data best straightening is achieved. In this case, $C = 0.1$ (Fig. 11).

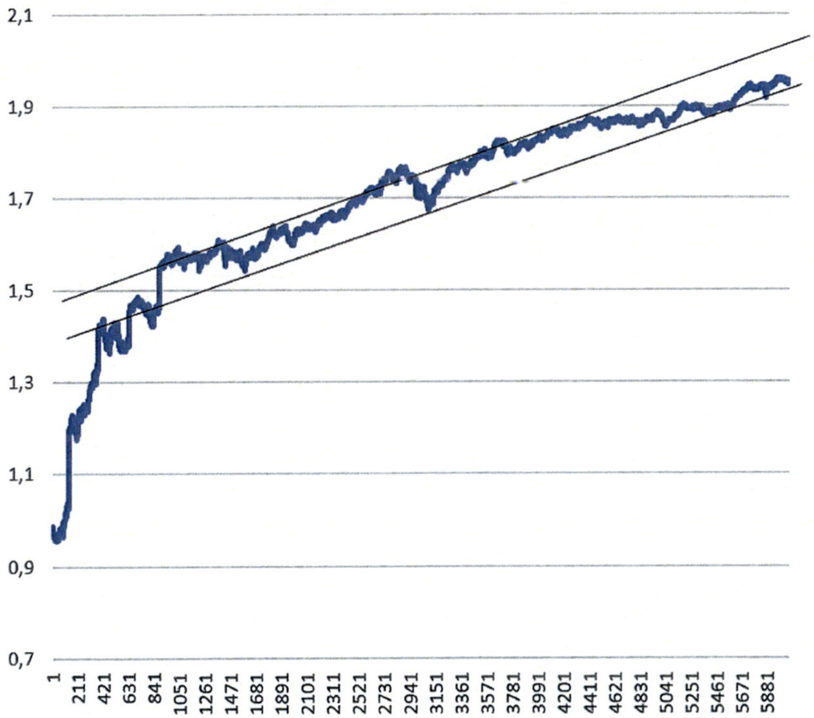

Fig. 11. Anamorphosis to intensive growth in coordinates.

But since the growth rates do not exponentially depend on the system age, by varying the constant C it is impossible to achieve the rectification best angle.

4 Limited Height Model

4.1 Gompertz Model

4.1.1 Gompertz First Model

For the Gompertz model.

$$\frac{dy}{y} = A * e^{-kt} y.$$

Divide both parts by yand integrate, we get the equation:

$$\ln\left(\frac{1}{y}\frac{dy}{dt}\right) = \ln(A) - kt.$$

Let's construct this anamorphosis in coordinates $\ln\left(\frac{y'}{y}\right) \sim t$ (see Fig. 12).

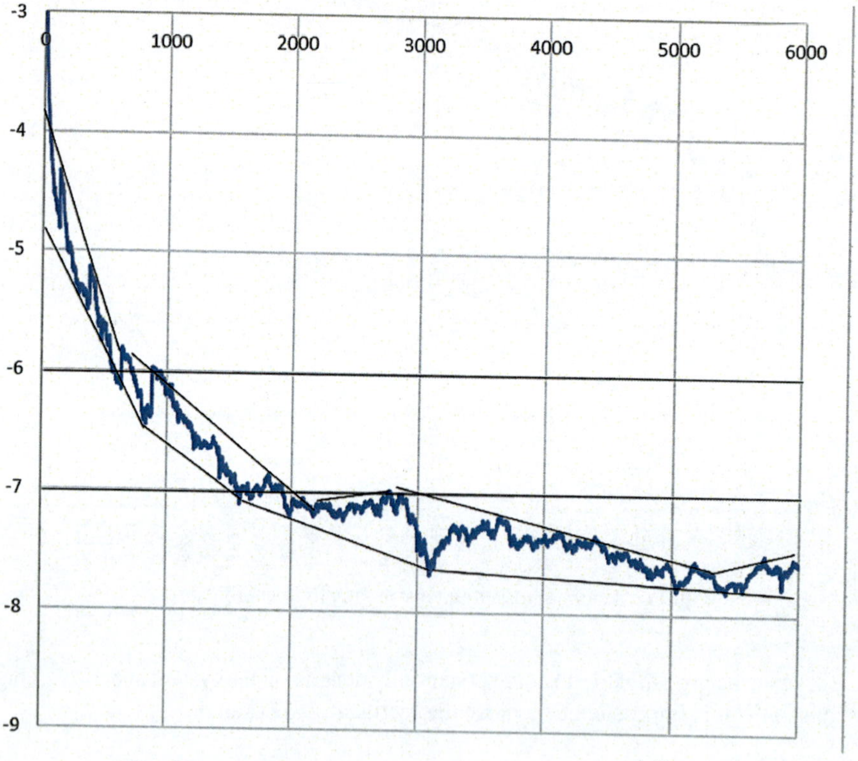

Fig. 12. Anamorphosis to Gompertz in coordinates $\ln\left(\frac{y'}{y}\right) \sim t$.

Let's define 2 areas for analysis and consider them in more detail (Fig. 13, 14).

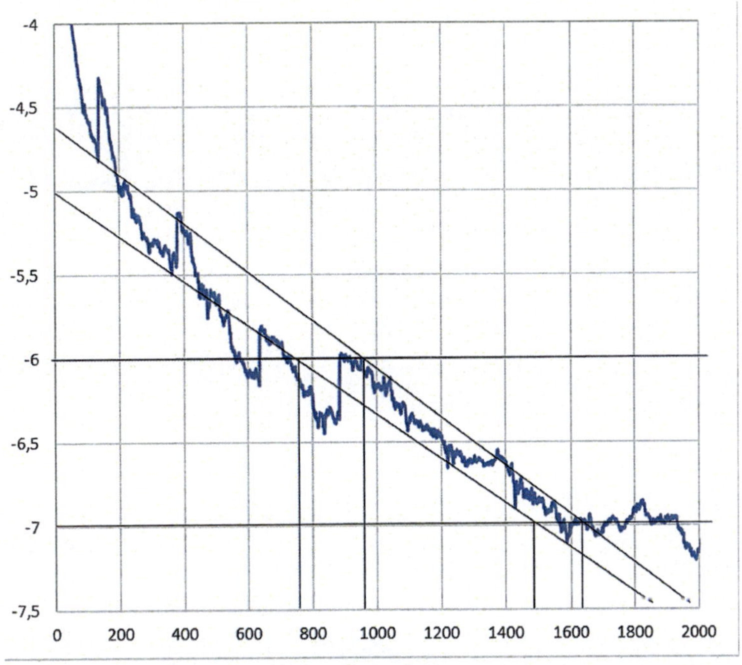

Fig. 13. Anamorphosis to Gompertz in coordinates in the range from 0 to 1000.

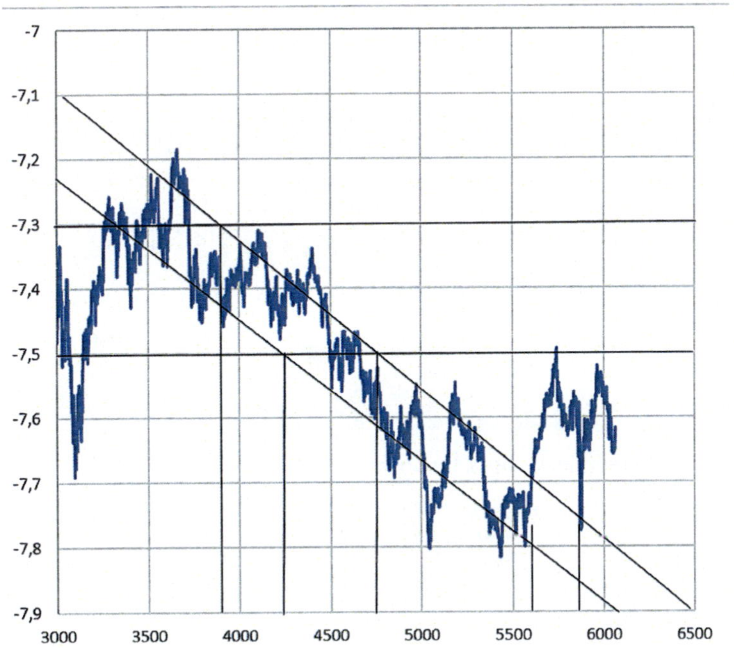

Fig. 14. Anamorphosis to Gompertz in coordinates in the range from 3000 to 6500.

From these graphs, you can find the parameter k. For $\ln(k) = -6$,

$$k = e^{(-6)} = 0,00248; \text{ for } \ln(k) = -7, k = e^{(-7)} = 0,000913; \text{ for } \ln(k) = -7,3, k = e^{(-7,3)} = 0,000676, \ \ln(k) = -7,5, k = e^{(-7,5)} = 0,000566.$$

Using the drawn lines that are parallel to the *x*axis from the points and correspond to the found $\ln(k)$, we find their intersection points with the trend lines. These points coordinate along the *x-axis* are 760, 1490, 3850 and 4750, respectively. Let's mark the resulting values on the original semi-logarithmic graph, draw straight lines from them, parallel to the *y* axis, they fall into local maximums - these will be the limited growth limits (Fig. 15).

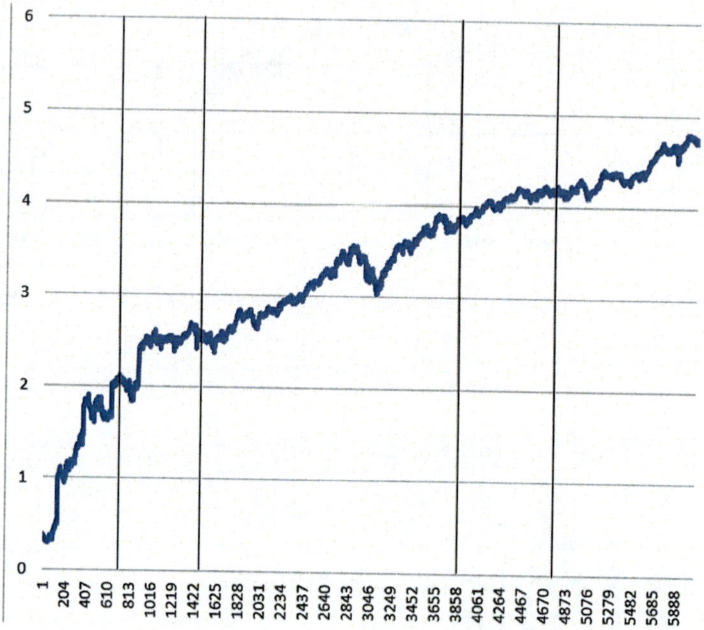

Fig. 15. The original graph in coordinates ln (y)~t, with highlighted local maxima.

4.1.2 Second Gompertz Model

The following anamorphosis looks like this:

$$\frac{1}{y}\frac{dy}{dt} = k(\ln(y_\infty) - \ln(y)).$$

That is, to determine the areas belonging to the Gompertz distribution, it is necessary to construct the initial data in the coordinates $\frac{y'}{y} \sim \ln(y)$ (Fig. 16).

To determine y_∞, it is necessary to extend the straight lines that are plotted on the areas that are straightened in these coordinates to the abscissa axis. In this case, we will define the value k * ln (y_∞), that is, to determine the parameter y_∞itself, we need to put this value as the exponent power. But in Fig. 3.1.2.1, no straight sections are visible, since the first points are very much apart from the others, which are in the range from 0 to 0.001. Therefore, it is necessary to detail a certain area (Fig. 17).

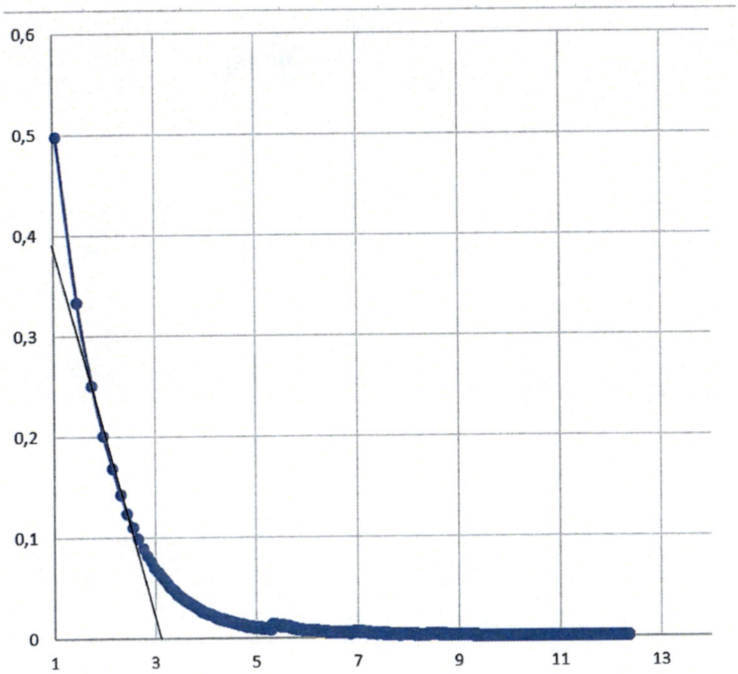

Fig. 16. Anamorphosis to Gompertz in coordinates.

With Further, according to the graph shown in Fig. 17, we can find y_∞. $y_\infty = e^{9.67} = 14835$.

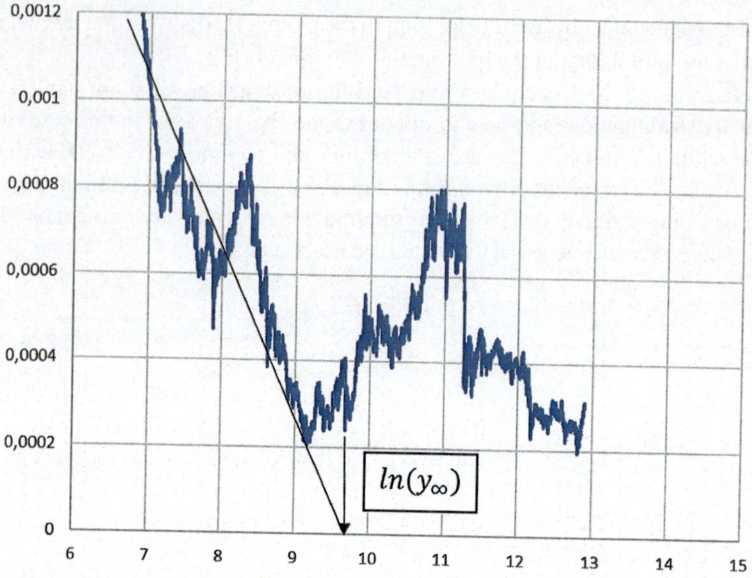

Fig. 17. Anamorphosis for logistic distribution in coordinates 〖(y`)/y~ln(〗 ▯ 〖y)〗 in the interval from 6 to 13 along the abscissa axis.

4.1.3 Gompertz's Third Model

Let's define an additional identification method $\ln(y)$ Integrate both sides of the equation $\frac{dy}{dt} = -\text{ky} * \ln\left(\frac{y}{y_\infty}\right)$ and get the equation $\ln(y) = -\frac{A}{k}e^{-kt} + \ln(C)$. Considering that $t \to \infty$, and $= y_\infty$, write $\ln\left(\frac{y}{y_\infty}\right) = -\frac{A}{k} * e^{-kt}$. Taking the equation both sides logarithm, we get:

$$\ln\left(\ln\left(\frac{y\infty}{y}\right)\right) = \ln\left(\frac{A}{k}\right) - kt.$$

Let's construct this anamorphosis in coordinates $\ln\left(\ln\left(\frac{y\infty}{y}\right)\right) \sim t$ (see Fig. 18).

From this graph, it follows that the slope coefficient is k = 0.000668. This value practically coincides with the previously found parameter k, which means that the calculations were performed correctly.

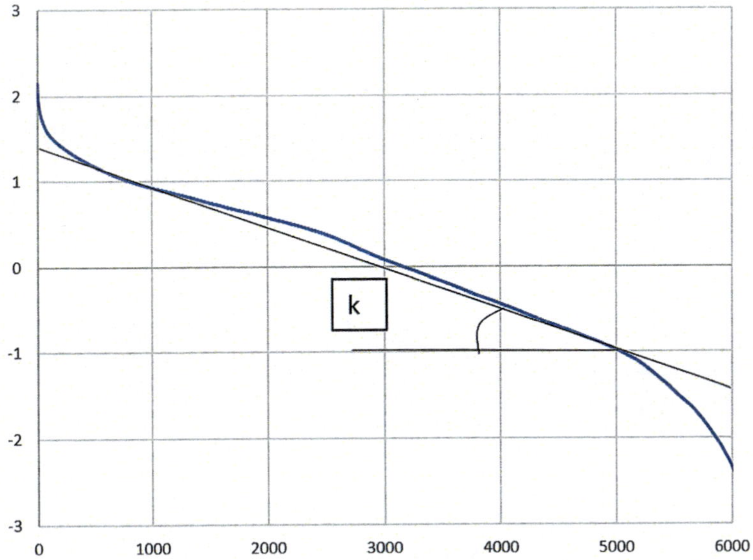

Fig. 18. Anamorphosis to Gompertz in coordinates.

4.2 Logistic Model

The simplest widespread limited growth model is the form logistic equation:

$$\frac{dy}{dt} = ky * \left(1 - \frac{y}{y_\infty}\right).$$

Let's build this logistic anamorphosis in coordinates $\frac{y'}{y} \sim y$ Fig. 19.

Since we do not observe any straight sections, the first points are very different from the others, which are in the range from 0 to 0.001. Therefore, it is necessary to detail a certain section (Fig. 20).

It follows from the figure that the y value $_\infty$ is approximately equal to 14835. Therefore, the value y $_\infty$ was found correctly.

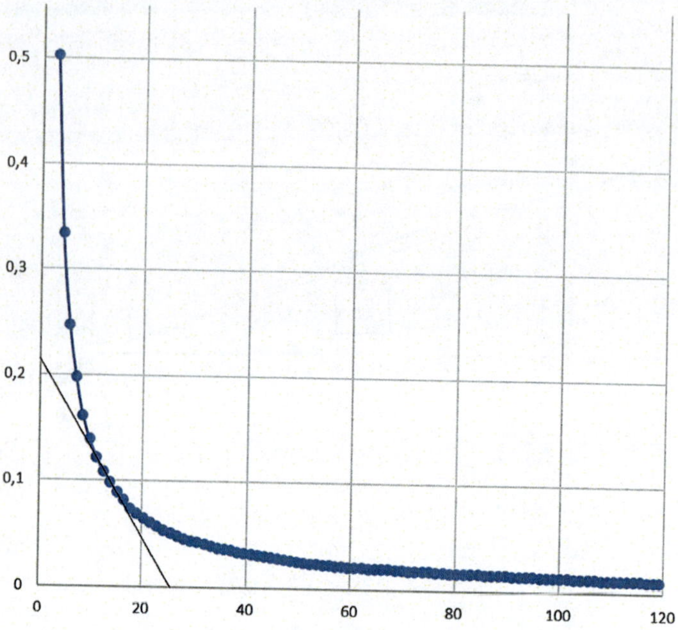

Fig. 19. Anamorphosis for logistic distribution in coordinates.

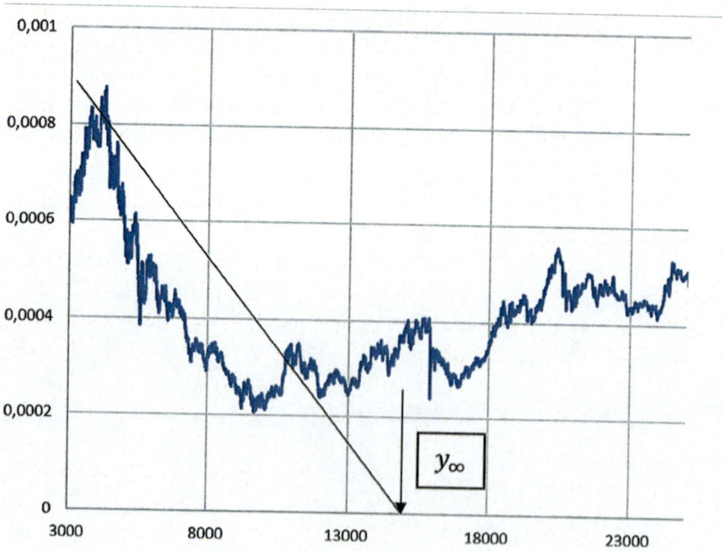

Fig. 20. Anamorphosis to logistics in the interval from 0 to 14000 along the abscissa axis.

5 Near-Periods Definition

5.1 Finding Near-Periods

In In any process, an important characteristic is its frequency, which is when a certain section will be repeated over and over again. But in real processes there is practically no period in its pure form, therefore there is such a thing as a near-period. To determine it, you first need to remove the trend component from the initial data, since the near-period refers to the system internal fluctuations.

You can remove a trend based on different metrics, that is, you can use arithmetic, geometric or any other proportion. In our case, the geometric proportion will be used, after which we will logarithm it:

$$R=\ln((y_{(t-\Delta t)}*y_{(t+\Delta t)})/(y_t^2)).$$

That is, R is the new value on the ordinate axis. We select the value Δt, which can affect both the trend harvesting quality and the near-period value truth. Therefore, we take the value $\Delta t = 500$. As a check that the trend has been removed completely (and there are times when it is not possible to remove it the first time), it is necessary to take the average of the resulting graph without a trend, that is, add all the points and divide by their total number. If the value is close to zero, then the result is satisfactory (Fig. 21).

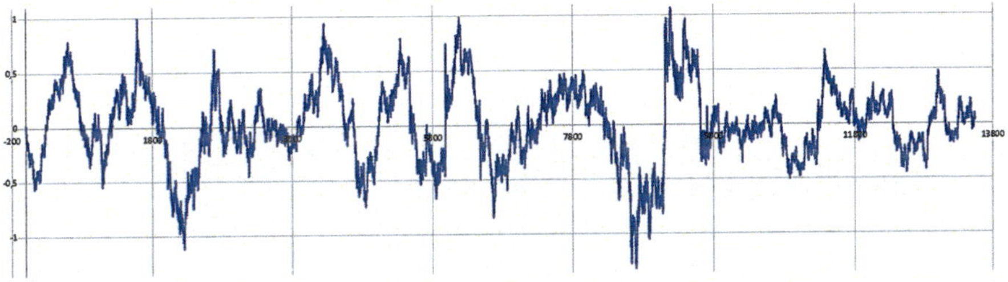

Fig. 21. Data without trend at $\Delta t = 500$.

To check, we use the following equation (Fig. 22):

$$R = \ln\left(\frac{yt - \Delta t + yt + \Delta t}{yt * 2}\right).$$

Now you need to calculate the almost-periods based on each chart and compare them. The shear function is calculated by the formula:

$$a(\tau) = \frac{1}{n-\tau} * \sum\nolimits_{i=1}^{n-\tau} |f(t+\tau) - f(t)|,$$

where τ should be changed from 1 to approximately 2/3 of the points total number from the original data since the next is an incorrect estimate due to the remaining points very

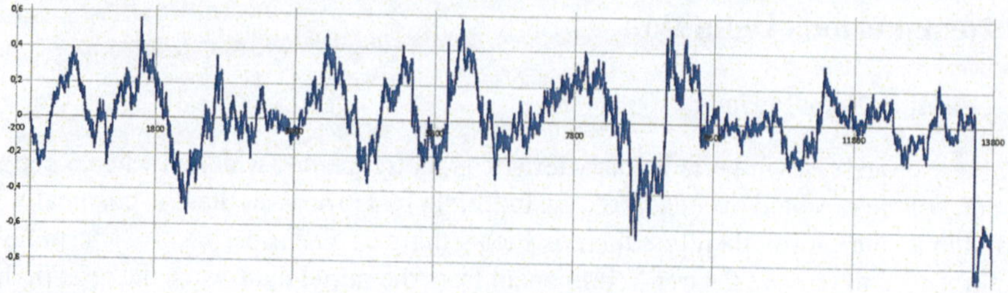

Fig. 22. Data without trend at $\Delta t = 500$.

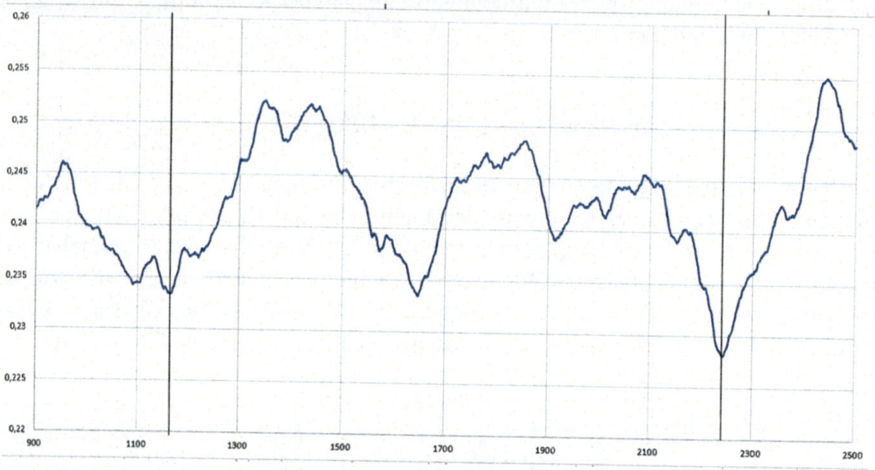

Fig. 23. Shear function.

small number. In the resulting chart, select the interval where the near-period is visible, and select the minimum points on this interval (Fig. 23).

Figure 23 shows that the second minimum point is the first point multiple twice: $\frac{2244}{1122} = 2$ Therefore, it is necessary to build only one boundary synchronization schedule.

5.2 Finding the Restructuring Phase

The relationship between the level 1 uniform intervals and the level 2 geometric progression members in the critical constants' hierarchy (see Fig. 24) is determined by their boundaries' synchronization under the relation:

$$X_k = \left(\frac{2e}{e-1}\right)T_k,$$

where is the arithmetic progression uniform unit clock cycles (level 1) - the geometric progression boundary with the module e (level 2).

The most significant border is also the line. In the interval between borders e^e and border, e^3 there is a zone in which systems change their structure (restructuring phase), (Fig. 25).

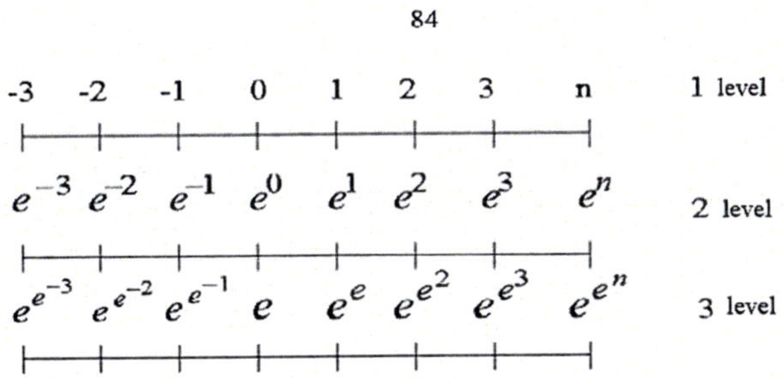

Fig. 24. Critical constants hierarchy determined by Neper's number.

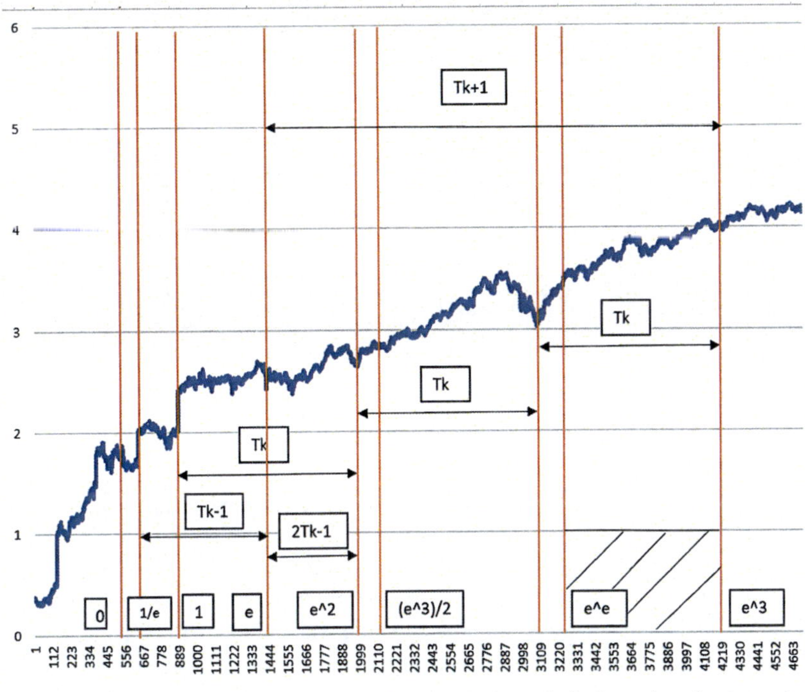

Fig. 25. The restructuring phase determination by the boundaries' synchronization values.

The vertical lines mark the development' cell boundaries. T_k is equal to the near-period, therefore equal to 1122, which is approximately 5 trading years.

6 Forecast

The analyzed process is characterized by cycles T_k = 1122, T_(k + 1) = 2685, T_(k − 1) = 794. Knowing these indicators, you can predict the rise or stocks fall. In the case of Nestlé, between 870 and 2800 trading days, T_k corresponds to global highs. Also, in the sections from 2800 to 3100, the structure changes and at point 4215, T_k and T_(k + 1) are synchronized.

Using all the obtained indicators, let's assume how the boundary synchronization graph will look like and predict a new restructuring phase (Fig. 26).

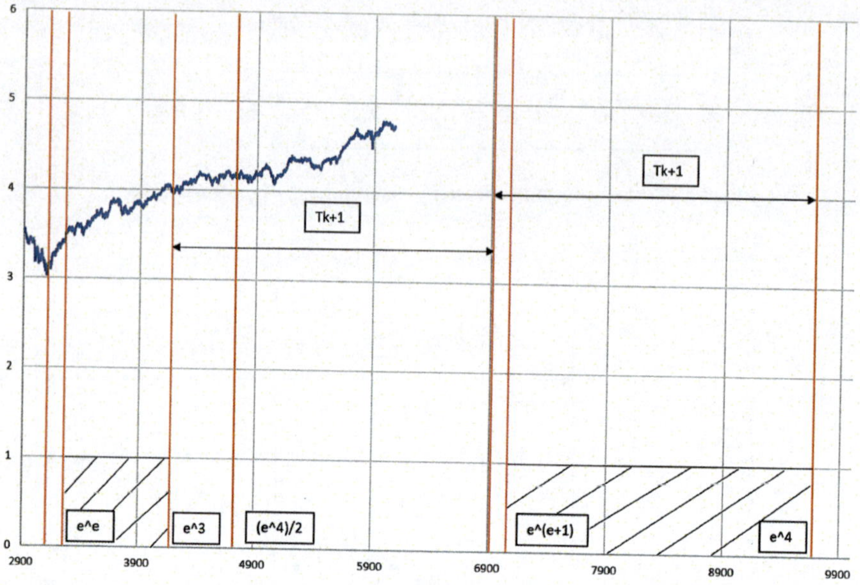

Fig. 26. A new restructuring phase prediction based on the boundaries' synchronization values.

Figure 5.1 shows that at the critical constants' hierarchy next level $e^4 = T_k * 2 = 4800$, $\frac{e^4}{2} = 9600$, $e^{e+1} = 7043$. This means that the new restructuring phase will last from 7043 to 9600 trading days, that is, 2557 trading days, which roughly corresponds to 9 real-time years.

Knowing that the trading day value e^3 corresponds to November 2011, it can be predicted that the rise in the shares' value will rise during 2025, and after reaching its global maximum at the 2025 end, the shares' value will fall and will reach its minimum in the 2026 third quarter. This will be followed by a new restructuring.

Consequently, the new restructuring phase will be roughly between 2028 and 2037.

Summing up, it can be noted that today Nestle is one of the world's largest food manufacturers. In the medium and long term, Nestlé shares are very stable, the company's business is diversified, and therefore dependence on crises' various kinds is not very high.

References

1. Veniaminov, S.S.: Revealing Latent structure laws in processes and signals. In: URSS, p. 216 (2014)
2. Kuzmin, V.I., Gadzaov, A.F.: Method of construction of models on empirical data. The manual. In: MGTU MIREA, p. 94 (2012)
3. Serebrennikov, M.G., Pervozvansky, A.A.: Revealing of the latent periodicity. In: Science, p. 244 (1965)

4. Dzerjinsky, R.I., Pronina, E.N., Dzerzhinskaya, M.R.: The structural analysis of the world gold prices dynamics. In: Silhavy, R. (ed.) CSOC 2020. AISC, vol. 1225, pp. 352–365. Springer, Cham (2020). https://doi.org/10.1007/978-3-030-51971-1_29
5. Paramonov, A.A, Kuzmin, V.I., Dzerjinsky, R.I.: Analysis of almost-periodic and almost-proportional characteristics of a representative sample local minima time series. In: IOP Conference Series: Materials Science and Engineering, vol. 1047, p. 012045 (2020). https://doi.org/10.1088/1757-899X/1047/1/012045
6. Semenenko, M.G.,et al.: How to use neural network and web technologies in modeling complex technical systems. In: IOP Conference Series: Materials Science and Engineering, vol. 537, p. 032095 (2019)
7. Gruzenkin, D.V., et al.: Neural networks to solve modern artificial intelligence tasks. J Phys. Conf. Series **1399**, 033058 (2019)
8. Training dataset. www.kitco.com/charts, Accessed 08 Aug 2021
9. Finance. Yahoo – service for obtaining data on the sale of Nestle shares. https://finance.yahoo.com/, Accessed 08 Aug 2021

The Procter and Gamble Stock Price Dynamics Analysis and Forecasting

K. V. Gusev, Roman I. Dzerjinsky(✉), and P. A. Palamarchuk

Applied Mathematics Department, MIREA – Russian Technological University, 78, Vernadsky pros, Moscow 119454, Russia

Abstract. The Procter & Gamble stock price analysis is carried out. The anamorphosis almost periodic analysis application for the latent periodicities study in the company's shares dynamics is proposed under the Gompertz models applicability assumption. The series structural analysis showed the stable growth possibility in the company's share price in the period from 2021 to 2035. Between 2038 and 2051, an unstable developmental phase can be expected, known as the restructuring phase.

Keywords: Intensive growth models · Anamorphosis · Model Gompertz · Near-periods growth rate model · Dynamics predictive assessment

1 Introduction

The stock exchange price is traditional for the time-series study. Such objects mathematical study is reduced to their representation on fundamentally different time scales. This is achieved by dividing the initial time series into components that characterize the so-called fast and the process slow components. As a result, the fast movements' analysis makes it possible to determine the time intervals' hierarchy that are closest to so-called near-periods.

To determine the trends' hierarchy, understanding near-periods allows you to use them as smoothing intervals for the original series. The initial data presentation in the piecewise linear dependences forms for slow movements, as the main tendencies, makes it possible to analyse directly, without excluding fast movements, based on anamorphosis.

In the study course, almost-proportions are determined, which can be identified both by specialized algorithms means and by analysing the near-periods ratio [1–3].

The consideration' such an ideology application makes it possible to analyse different nature markets [4, 5].

This paper examines the Procter & Gamble share price dynamics in the period from February 1962 to October 2020 [6, 7].

R. Silhavy et al. (Eds.): CoMeSySo 2021, LNNS 232, pp. 128–136, 2021.
https://doi.org/10.1007/978-3-030-90318-3_12

The study purpose is to determine the process main characteristics, as well as to predict this company shares value and determine its development phase at the time moment.

2 The Procter and Gamble Shares Price Characteristics Analysis

To analyse the Procter & Gamble stock price, you will need the initial data - the MICEX Index throughout the entire analysed interval [6, 7]. The initial data on a semi-logarithmic scale is shown in Fig. 1.

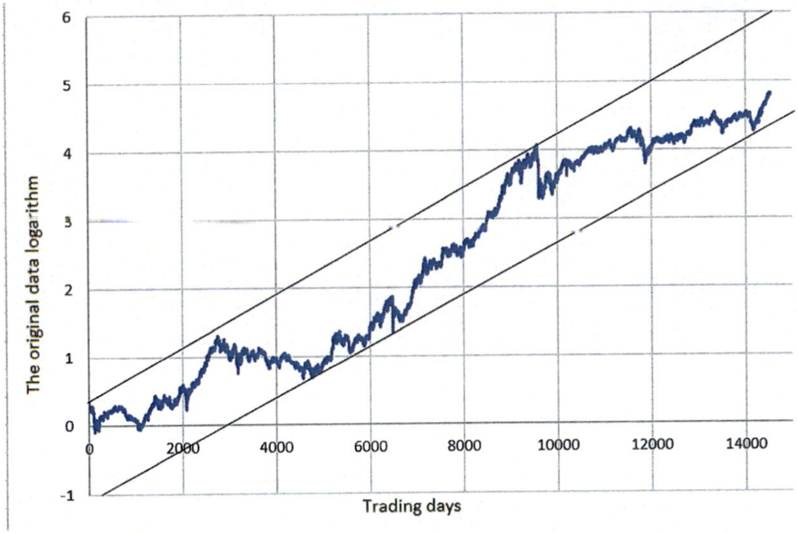

Fig. 1. Initial data on a semi-logarithmic scale.

3 Intensive Growth Model

To analyse the initial data, the intensive growth modelling method was used. The first growth rate model is calculated using the following formula [1]:

$$\frac{dy}{dt} = f(t, y)y, \tag{1}$$

Any intense growth has a limit to which it will grow. In the original graph, this limit is a quadratic function that tends to vertical asymptote. To mark these asymptotes were found values on the greatest straightening sites, which respectively equal to 1900, 2900, 7700 and 9600 (Fig. 2, 3, 4).

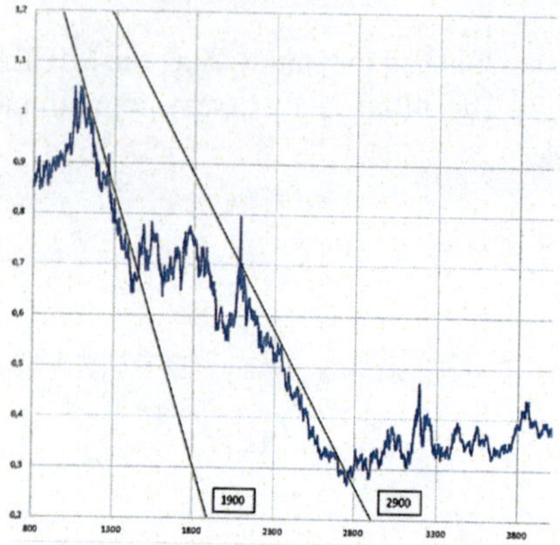

Fig. 2. Anamorphosis to intensive growth in the range from 800 to 4000

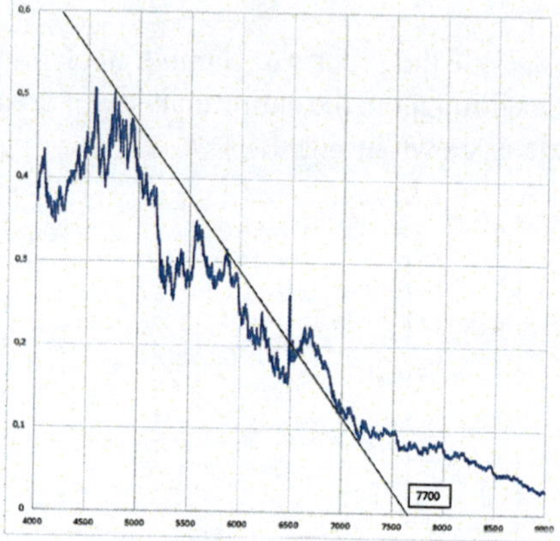

Fig. 3. Anamorphosis to intensive growth in the range from 4000 to 9000

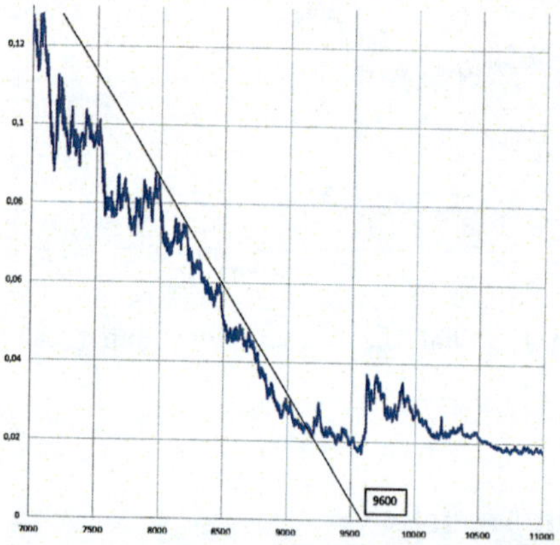

Fig. 4. Anamorphosis to intensive growth in the range from 9600 to 11000.

The sought values are respectively equal to 540, 1580, 1980, 3600 and 4000. Note the obtained values on the original semilogarithmic graph and draw straight lines from them parallel to the y axis, these will be the sought asymptotes (Fig. 5).

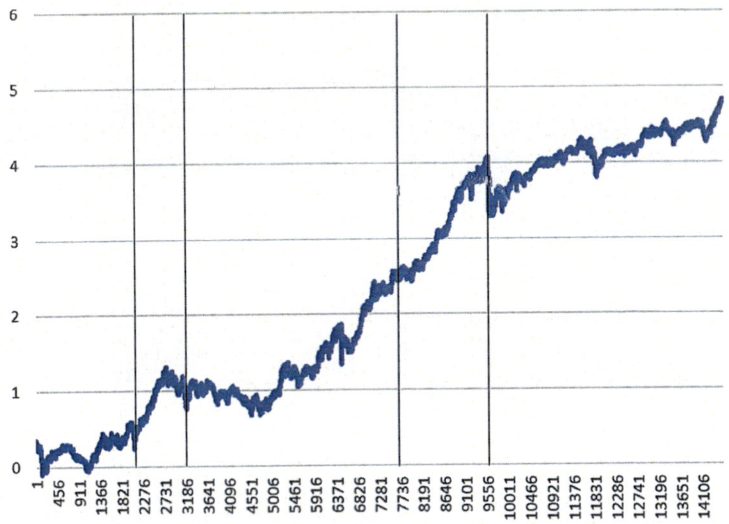

Fig. 5. Original graph with data with marked asymptotes.

4 Second Growth Rate Model

After determining the intensive growth period, it is necessary to determine the cutoff levels, that is, those values *ln (y)*, at which these processes will end. To do this, initial growth a second model is built according to the following formula [1]:

$$\ln(y) = \frac{a * t^2}{2} + \ln(C)(t - t_0) \tag{2}$$

On this basis, anamorphosis is built to an intense growth in the $ln(y) \sim t^2$ coordinates (Fig. 6).

The graph shows the early values t^* found on the abscise axis, lines from these points to the graph are drawn, and lines reflected from the intersection point with the graph on the ordinate axis.

Now we find the growth limits for each age point t^* using Fig. 6. For $t^* = 1900$ it is equal to $e^{0.219} = 1{,}244$, for $t^* = 2900$ it is equal to $e^{0.255} = 1{,}289$, for $t^* = 7700$ it is equal to $e^{0.250} = 1{,}283$, for $t^* = 9600$ it is equal to $e^{0.202} = 1{,}223$.

To study the model limited growth, the method is used as a limited growth model a method by Gomperc.

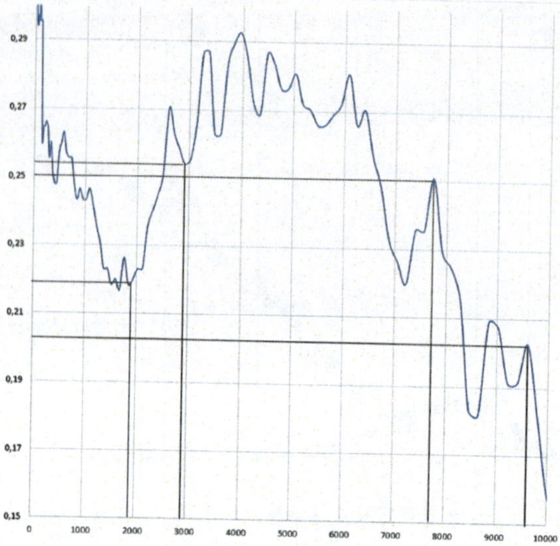

Fig. 6. Anamorphosis to intensive growth in coordinates *ln(y)* ~ t^2 in the range from 0 to 10000.

5 Limited-Growth Model (Gomperc)

The Gomperc limited growth model is built according to the following formula [1]:

$$\frac{dy}{y} = A * e^{-kt} y \tag{3}$$

Anamorphosis to Gomperts in coordinates is represented in Fig. 7, 8.

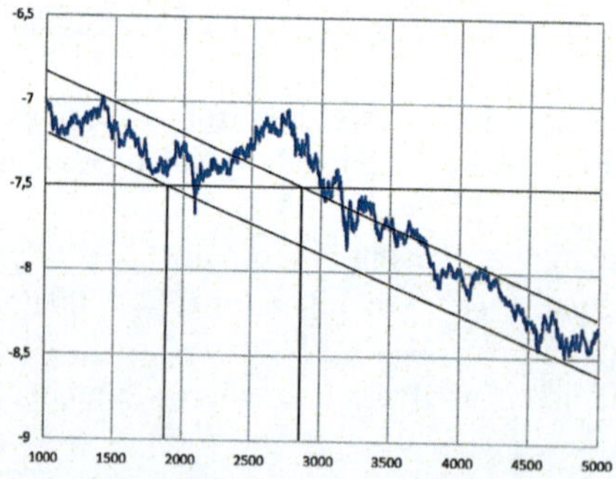

Fig. 7. Anamorphosis to Gompertz in the range coordinates in the range 1000 to 5000.

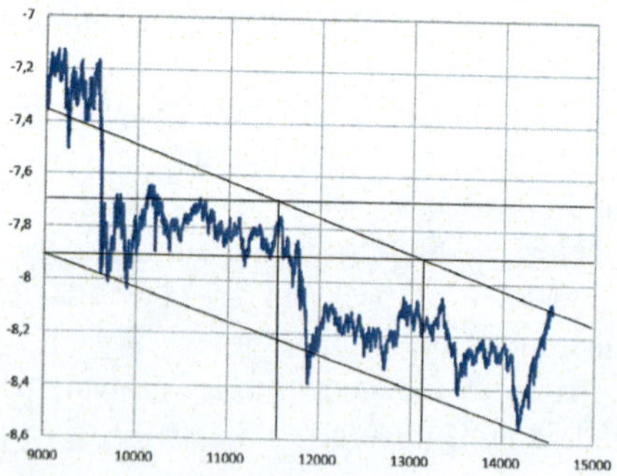

Fig. 8. Anamorphosis to Gompertz in coordinates in 9000 to 15000

From these graphs, you can find the parameter k. With the parameter k, you can estimate the tact length in this system. For $\ln(k) = -7.5$,

$$k = e^{(-7.5)} = 0.000566; \text{ for } \ln(k) = -7.7, k = e^{(-7.7)} = 0,000464;$$
$$\text{for } \ln(k) = -7,9, k = e^{(-709)} = 0,00038.$$

Using the drawn lines that are parallel to the x axis from the points and correspond to the found $\ln(k)$, we find their intersection points with the trend lines. These points coordinate along the *x-axis* are 2800, 11600, and 13100, respectively. Let's mark the resulting values on the original semi-logarithmic graph, draw straight lines from them, parallel to the y axis, they fall into local maximums - these will be the limited growth limits (Fig. 9).

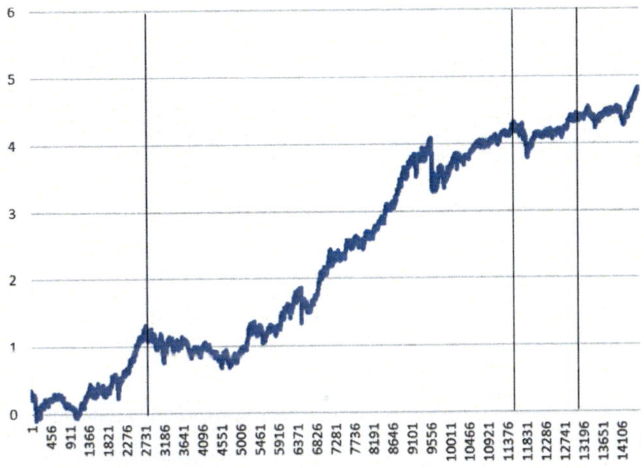

Fig. 9. The original graph in coordinates y ~ t, with highlighted local maxima.

6 Near-Periods Definition

In any process an important characteristic is its frequency, that is, when a certain section will be repeated over and over again. But in real processes there is practically no period in its pure form, therefore there is such a thing as a near-period. To determine it, you first need to remove the trend component from the initial data, since the near-period refers to the system internal fluctuations.

You can remove a trend based on different metrics, that is, you can use arithmetic, geometric or any other proportion. In our case, the geometric proportion will be used, after which we log it using the following formula [3]:

$$R = \ln\left(\frac{yt - \Delta t * yt + \Delta t}{y_t^2}\right). \tag{4}$$

For the near-periods calculation, a shift function according to the formula (5) is considered [3]:

$$a(\tau) = \frac{1}{n - \tau} * \sum\nolimits_{i=1}^{n-\tau} |f(t + \tau) - f(t)|, \tag{5}$$

According to the calculations received, a shift graph is built, and an interval is allocated, where the near-period (Fig. 10) is visible.

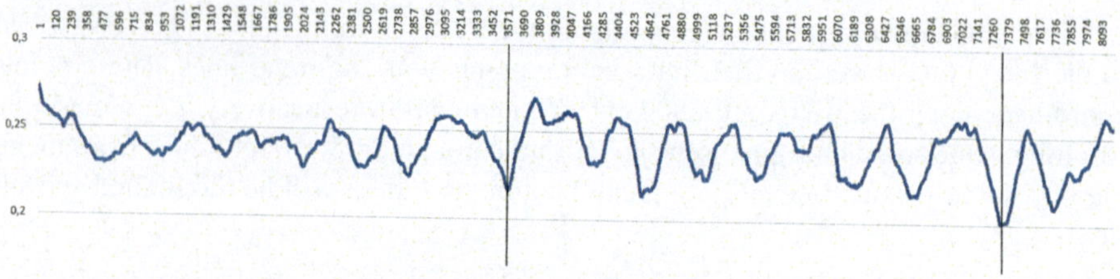

Fig. 10. Shift function graph.

Figure 10 shows that the minimum second point is the first point twice multiple: $\frac{7140}{3570} = 2$. Therefore, it is necessary to build only one timetable for synchronizing the boundaries to predict the shares' growth.

7 Finding a Development Unsustainable Phase

Synchronized boundary under the ratio [1]:

$$X_k = \left(\frac{2e}{e-1}\right)T_k, \tag{6}$$

where T_k - the arithmetic progression uniform unit clock cycles (level 1), X_k - the geometric progression boundary with the modulus e (level 2), the relationship between the level 1 uniform intervals and the level 2 geometric progression members in the critical constants' hierarchy is determined.

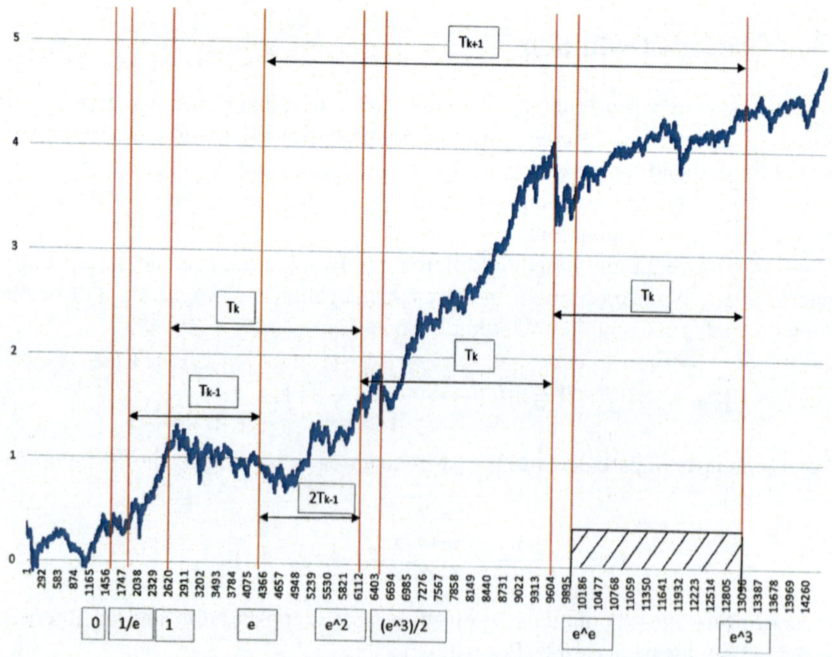

Fig. 11. The restructuring phase determination by the boundaries' synchronization values.

The most significant border is also the line e. In the interval between the boundary e^e and the boundary e^3, there is a zone in which the systems change their structure, that is, there is a restructuring phase (Fig. 11).

The vertical lines mark the development' cell boundaries. T_k is equal to the near-period, therefore equal to 3570, which is approximately 13 trading years.

The analyzed process is characterized by cycles $T_k = 3570$, $T_{k+1} = 8860$, $T_{k-1} = 2762$. Knowing these indicators, you can predict the rise or stocks fall. In the Procter & Gamble case in the areas from 2620 to 9600 trading days, the period T_k corresponds to global highs. Also, in the sections from 9600 to 9862, the structure changes and at point 9630, and are synchronized T_k and T_{k+1}. This point corresponds to the global minimum.

Using all the obtained indicators, let's assume how the boundary synchronization graph will look like and predict a new restructuring phase (Fig. 12, 13).

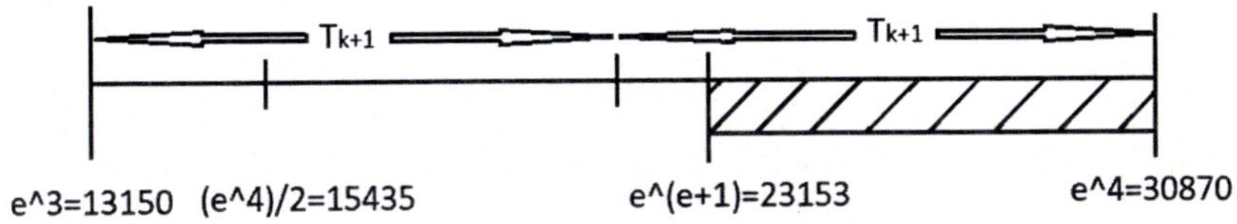

Fig. 12. A new restructuring phase schematic forecast.

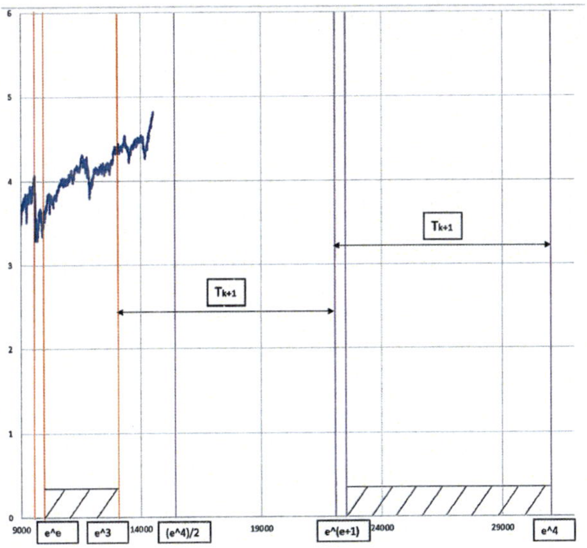

Fig. 13. A new restructuring phase prediction based on the bound ries' synchronization values.

Figure 13 shows that at the critical constants' hierarchy next level $e^4 = T_k * 2 = 30870$, $\frac{e^4}{2} = 15435$, $e^{e+1} = 30870 * \frac{3}{4} = 23153$. This means that the new restructuring phase will last from 23153 to 30870 trading days, that is, 7717 trading days, which roughly corresponds to 30 real-time years.

8 Conclusion

Knowing that the trading day e^3 value corresponds to November 1999, we can predict that the growth in the shares' value will increase during 2035, and after reaching its global maximum at the 2035 end, the shares' value will fall and reach its minimum in the 2036 third quarter. This will be followed by a new restructuring.

Consequently, the new restructuring phase will be roughly between 2038 and 2051.

Summing up, it can be noted that today P&G is one of the cartels' world's largest producers. In the medium and long term, P&G shares are very stable, the company's business is diversified, and therefore the dependence on crises' various types is not very high.

References

1. Veniaminov, S.S.: Revealing Latent structure laws in processes and signals. In: URSS, p. 216 (2014)
2. Kuzmin, V.I., Gadzaov, A.F.: Method of construction of models on empirical data. The manual. In: MGTU MIREA, p. 94 (2012)
3. Serebrennikov, M.G., Pervozvansky, A.A.: Revealing of the latent periodicity. In: Science, p. 244 (1965)
4. Dzerjinsky, R.I., Pronina, E.N., Dzerzhinskaya, M.R.: The structural analysis of the world gold prices dynamics. In: Silhavy, R. (ed.) CSOC 2020. AISC, vol. 1225, pp. 352–365. Springer, Cham (2020). https://doi.org/10.1007/978-3-030-51971-1_29
5. Paramonov, A.A., Kuzmin, V.I., Dzerjinsky, R.I.: Analysis of almost-periodic and almost-proportional characteristics of a representative sample local minima time series. In: IOP Conference Series: Materials Science and Engineering, vol. 1047, p. 012045 (2020). https://doi.org/10.1088/1757-899X/1047/1/012045.
6. Training dataset. www.kitco.com/charts
7. Finance. Yahoo – service for obtaining data on the sale of Procter & Gamble shares. https://finance.yahoo.com/

Measuring the Adoption of Biometric Verification Technique Implementations on Mobile Devices

William Ratjeana Malatji[1(✉)], Rene VanEck[2], and Tranos Zuva[2]

[1] Vaal University of Technology, 1911 Andries Potgieter Blvd, Vanderbijlpark, South Africa
[2] Department of ICT, Vaal University of Technology, 1911 Andries Potgieter Blvd, Vanderbijlpark, South Africa
{Rene,Tranosz}@vut.ac.za

Abstract. Mobile device growth is rapid, and it has a significant impact on our private and professional lives. All mobile users want to be guaranteed that their data is safe, which is why biometrics exists for mobile devices. Countless surveys have been performed on the adoption of biometric verification techniques, however, only a limited number of these surveys have concentrated on mobile device-based biometrics, including this current research. This quantitative study's overall model was put to test on 302 mobile users via survey questionnaire distribution and analyzed utilizing the Statistical Package for Social Science (SPSS). Validity and reliability tests were performed and the model proved to be fit. The findings showed that the SSN variable of the proposed model was not supported, indicating that more research is required. In addition, the research model's functional elements have a greater impact on the participants' desire to use the mobile biometric system than the social elements. The study adds to academic knowledge by proposing new constructs that combine MBTAM to assess the likelihood of mobile device users adopting biometric verification techniques.

Keywords: Mobile biometric technology adoption model (MBTAM) · Perceived social presence (PSP) · Perceived humanness (PH) · Perceived interactivity (PI) · Statistical package for social sciences (SPSS)

1 Introduction

Something that has an impact on a large population inevitably has an impact on society, and technology is not an exception. A lot of research has been done about how and why people adopt or reject inventions [1]. The adoption of technology is more critical than the inventor, the first company to market, or even the technology itself [2]. The willingness of a country to adopt new technology is critical, as it affects everything from the cost of capital goods to per capita income [2].

Many people seem to quickly adopt fashion as mobile devices have come to be the preferred personal information processing interface [3]. However, there is no research into how early adopters of mobile devices overcame technological uncertainties [3].

R. Silhavy et al. (Eds.): CoMeSySo 2021, LNNS 232, pp. 137–148, 2021.
https://doi.org/10.1007/978-3-030-90318-3_13

These devices are now being used to perform daily business activities. This makes users start questioning the safety or security of their data [4]. For several years, the password was the only reliable method of authentication, however, a huge increase in the use of biometrics to replace passwords have been seen recently since the technology has proved to be much simpler and time-saving [4]. As a result, in the consumer market, there is now winning mobile devices and biometrics combination allowing the technology to become even more widely adopted [1].

According to [5], the utilization of biometric verification techniques on mobile devices such as smartphones and tablets is referred to as mobile biometrics. A biometric device can determine or verify an individual's identity by measuring physical or behavioral characteristics or DNA information [6]. The biometric system is growing due to rising adoption in both the private and public sectors. Moreover, transportation and immigration represent the largest end-user segments for biometric technology [7]. Biometrics adoption, whether in the form of fingerprint recognition, voice, or multi-factor biometrics, has changed and people are now adjusting to the new technology [8]. By 2021, the biometrics industry will worth $30 billion and much of this growth is being fuelled by improved smartphone capabilities (ABI, 2016).

There were a very limited number of researches that focused on the adoption of biometric authentication invention on mobile devices. Thus, this research aims to measure the adoption of biometric authentication inventions on mobile devices. The plan of this study is organized as follows: Related work, methodology, findings, discussion, study limitations and suggestions for future research and conclusion.

1.1 Statement of the Problem, Research Questions and Objective of the Study

In recent years, it has become common to access information via mobile devices in corporate and individual environments. Technically, the world is more advanced, and all users of mobiles need to know that their data is safe [4]. Biometrics are best used to improve the security of mobile devices and the efficiency of wireless networks [9].

Many studies have been carried out on biometric devices and their application adoptions, consumer attitudes in the direction of these devices and measurement of their performance impact. In other words, only limited studies have put their focus on the factors affecting the adoption and acceptance of biometric devices [10]. According to the research of [11], there are only a few studies involving the problems faced by users in the adoption of biometrics, and some studies have been conducted to evaluate the adoption of biometrics.

Only a few studies put their focus on the adoption of biometric verification technique implementations on mobile devices. Now that mobile devices are utilized for both commercial and own purposes, this study attempted to determine the adoption of biometric verification technique implementations on mobile devices by providing answers to the following research questions:

- What has been done to measure the user adoption of technology, according to the literature?
- What model can be suggested to evaluate the adoption of biometric verification techniques on mobile devices?

- How does one measure the effect of every variable of the model in the adoption of biometric verification technique implementations on mobile devices?

The above research questions will be answered through the following research objectives:

- To study and determine what was done, according to the literature, to measure the user adoption of technology.
- To suggest a model that could be used to evaluate the adoption of biometric verification techniques on mobile devices.
- To measure the effect of every variable of the model in the adoption of bio- metric verification technique implementations on mobile devices.

2 Related Work

Many research studies on the adoption of biometric technology have been published in the literature. Just a few of these studies, however, concentrated on the adoption of biometrics on mobile devices. A report that was provided by [12] of the study on the growing adoption of biometric solutions due to the COVID-19 pandemic. The findings of the study revealed that biometrics are used by 73% of survey respondents to maintain a positive customer and user experience, 55% use biometric verification technology, and three-quarters (75%) use biometrics for identity verification. According to the report, there has been a 45% rise in biometrics adoption due to COVID-19.

A study carried out by [13] on the evaluation and performance comparison of the model for the adoption of biometrics in online banking. The findings revealed that self-efficacy, perceived behavioral control, reliability, appearance, support, content, safety, and privacy are critical criteria for bank management to consider when considering the use of biometrics in online banking in Kuwait.

According to [14], only a few examinations on mobile biometric devices have been performed, and the benefits and drawbacks of such devices have been addressed. There has been no clarity on the factors influencing the use of biometric techniques on mobiles, so research was conducted on both the advantages and disadvantages of the technique. These factors have an effect on the workplace, schooling, government, and other industries. According to this article, there is a technological phenomenon of user adoption.

A survey carried out by [15] on biometric verification in financial institutions and the willingness of banks to adopt powered ATM biometrics. The results revealed that perceived benefit and external pressure affect the adoption of powered ATM biometrics positively, although the perceived risk has a negative effect, according to the empirical findings. Organizational readiness, on the other hand, was found to be negligible in this analysis.

A survey on the adoption of biometric authentication systems was conducted by [16]. The study looked at the relationship between three variables: readiness, perceived benefit, and external pressure, and the intention to use biometric verification in credit unions. Factors that are competitive, the perceived benefits, and the financial- resource readiness of the organization related to the invention (including regulatory meeting and requirements of the security) were found to be linked to adoption intent, but management support was not an important determinant of readiness and thus of adoption intent.

The study conducted by [17] on the adoption of biometric techniques in online applications revealed that perceived credibility has the greatest predictive power for actual use, followed by perceived ease of use, personal innovativeness, perceived risk of a new invention and computer self-efficacy.

2.1 Open Issues on the Adoption of Biometric Authentication Techniques

Aside from the numerous benefits that can be obtained through the implementation of biometric technologies, several factors that have been listed in several studies deter the adoption and use of the technology. The study conducted by [18] denoted that challenges to the implementation of the biometric fingerprint system include intermittent power supply, computer breakdowns, non-recognition of certain fingerprints, and long lines in front of the machine at work closing time.

The study of [18] on the opportunities and challenges for the biometric system, identified several barriers to biometric adoption, including anonymity, fear of harm from using the device, and general user anxiety. Several obstacles to biometric technology adoption were identified by [19] in their study of challenges to biometric technology adoption which includes the absence of dependable biometric security innovation providers, finances to support the rollout of biometric security, acceptance of stakeholders or users, energy to power the gear, operational difficulties resulting from weak cooperation between suppliers and government, challenges of logistics in spreading out the concept of biometric security delivery, and so on.

The company must understand the related difficulty involved in enrolling new users and administering usage training in addition to hardware and software costs [1]. There are also concerns about the accuracy of biometric technology use, as incorrect matches in authentication and identification systems may give unauthorized users access to facilities or services [1].

3 Methodology

This section provides a brief explanation of the methodological procedure followed in this study. The model in Fig. 1 derived from [20] for this quantitative study contains nine constructs which include social elements (perceived humanness, perceived interactivity and perceived social presence), functional elements (perceived ease of use, perceived usefulness and subjective social norm), reliability, trust as well as actual use. The hypotheses are shown in Fig. 1 as H_1, H_2, H_3, H_4, H_5, H_6, H_7, H_8, H_9, H_{10} and H_{11}.

The items in the survey questionnaire for this study were derived from other relevant research studies as shown in Table 1. The questionnaire was categorized into two sections: section one and section two. The first section covers the question's specifics, such as internet use, technological skills, demographics, and knowledge of internet scams. Questions about estimating the importance of mobile biometrics (suitable use of various biometrics) make up the final section. This research used a five-point Likert-scale style calculation, with one being "strongly agree" and five being "strongly disagree."

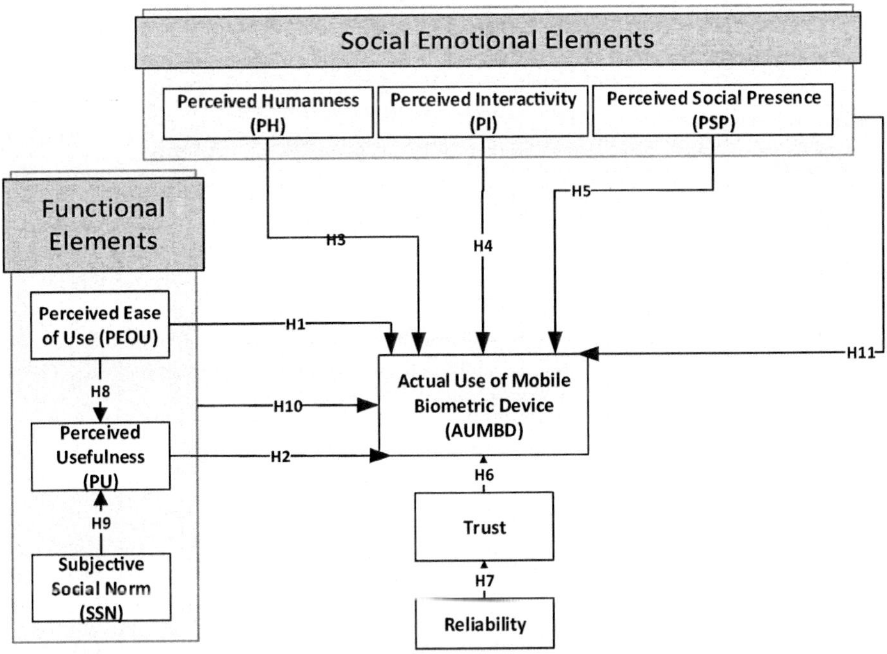

Fig. 1. Proposed mobile biometric technology acceptance model (Source: [21])

Table 1. Construct, number of items, and source.

Constructs	Number of items	Source-citations
Perceived Ease of Use (PEOU)	4	[21]
Perceived Usefulness (PU)	4	[21]
Subjective Social Norm (SSN)	4	[22]
Perceived Humanness (PH)	3	[23]
Perceived Interactivity (PI)	3	[24]
Perceived Social Presence (PSP)	3	[23]
Actual Use of Mobile biometric device (AUMBD)	3	[25]
Trust	4	[26]
Reliability	1	[27]

A pilot study was conducted using 30 participants (which is 10% of the sample size) who were randomly selected to ensure that the questions were clear, the instruments were long enough, and the content was complete. Finally, questionnaires were distributed online and only 302 were returned and usable.

Statistical Package for Social Science version 26.0 was used for this study. Reliability and validity were measured. Regression analysis was used to test the hypotheses at 0.05 level significant.

4 Findings

Only 302 responses were returned and usable. This shows that 99% of the questionnaires were returned which is good for this type of study [28]. All usable data was cleaned to remove the outliers before use. The reliability test was conducted in this research, and it was discovered that the analysis of reliability for all constructs was reasonably strong, indicating that the internal consistency between constructs was good and robust. These

Table 2. Demographic information of the respondents.

Variable	Frequency	Percet
Gender		
Male	186	61.6%
Female	116	38.4%
Age		
19 and Below	14	4.6%
20-30	211	69.9%
31-40	53	17.5%
41-50	17	5.6%
51-60	7	2.3%
61 and Above	0	0%
Race		
Black	198	65.6%
White	87	28.8%
Other	17	5.6%
Employment status		
Self-employed	26	8.6%
Employed	72	25.5%
Retired	3	1.0%
A student	188	62.3%
Other	8	2.6%
Level of study		
Primary	1	.3%
Secondary	11	3.6%
Undergraduate	117	38.7%
Postgraduate	97	32.1%
Other	76	25.2%
Do you have a mobile device of your own?		
Yes	294	97.4%
No	3	1.0%
Owned it before	5	1.7%
Have you utilized the biometric verification technique before?		
Yes	215	71.2%
No	87	28.8%
Would you prefer to utilize a mobile biometric device?		
Yes	256	84.8%
No	15	5.0%
Not sure	31	10.3%
I have gained access to the internet using a mobile biometric device		
Yes	141	46.7%
No	161	53.3%
Total	**302**	**100%**

tests produced values above 0.70 which is acceptable for this study [29]. Normality tests were conducted and the results proved that the degree of closeness to normality was excellent. The obtained results revealed that the dependent variables do not deviate from the linearity assumption. Furthermore, there is no evidence of homoscedasticity or multicollinearity, according to the findings. This suggests that the statistical inferences made about the data may be accurate.

The demographic results of this study can be seen in Table 2 as well as answers to the questions that aimed to gather information related to internet scams from the study participants.

According to the study, there were 186 male respondents and 116 female respondents, indicating that the number of male respondents is higher than the number of female respondents. The largest group of respondents (69.9%) is between the ages of 20 and 30, while the smallest group (2.3%) is between the ages of 51 and 60. Given the existence of mobile devices, this disparity is understandable, since the majority of mobile users are typically teenagers [30]. The results revealed that 198 respondents were black, 87 were white, and 17 were of other races among the various races who took part in the survey. The findings indicate that 8.6% of the participants were self- employed, 25.5% were employed, and 1.0% were retired, while 62.3% were students, and 2.6% were other. The findings also revealed that, at the study's level, the majority of the respondents were undergraduate students (38.7%), with primary students accounting for just 3%.

4.1 Regression Analysis

This section of the study will outline the results of the regression analysis in the order that they were obtained.

Table 3. Regression findings for PU, PEOU, SSN, trust, PH, PI, PSP and actual use.

Model	Unstandardized coefficients		Standardized coefficients	t	Sig.
	B	Std. error	Beta		
1 (Constant)	.026	.166		.154	.877
PU	.350	.066	.324	5.278	.000
PEOU	.195	.063	.162	3.054	.011
SSN	−.054	.038	−.070	−1.418	.157
Trust	.350	.066	.311	5.103	.000
PH	.132	.060	.126	2.196	.029
PI	.196	.064	.166	3.070	.002
PSP	.211	.052	.229	4.030	.000
a. Dependent Variable: Actual Use					

The findings displayed in Table 3 above indicate that perceived ease of use ($\beta = 0.162$; $p < 0.05$), perceived usefulness ($\beta = 0.324$; $p < 0.01$), perceived humanness ($\beta = 0.126$, $p < 0.05$), perceived interactivity ($\beta = 0.166$; $p < 0.05$), perceived social presence ($\beta = 0.229$; $p < 0.01$) and trust ($\beta = 0.311$; $p < 0.01$) were supported. It was further revealed by the above findings that subjective social norm was not supported with ($\beta = -0.070$; $p > 0.5$). The formal equation for the supported variables is shown in Eq. (1):

$$AU = 0.324PU + 0.162PEOU + 0.311Trust + 0.126PH + 0.166PI + 0.229PSP \quad (1)$$

Table 4. Regression findings for PEOU, SSN and PU

Model	Unstandardized coefficients		Standardized coefficients	t	Sig.
	B	Std. error	Beta		
1 (Constant)	.427	.132		3.228	.001
PEOU	.589	.048	.576	12.340	.000
SSN	.128	.032	.180	3.961	.000
a. Dependent Variable: PU					

The above results in Table 4 show that perceived ease of use ($\beta = 0.576$; $p < 0.01$) and subjective social norm ($\beta = 0.180$; $p < 0.01$) are good determinants of perceived usefulness. These two variables were found to be supported and the formal equation for this results is shown in Eq. (2):

$$PU = 0.576PEOU + 0.180SSN \quad (2)$$

Table 5. Regression findings for reliability and trust

Model	Unstandardized coefficients		Standardized coefficients	t	Sig.
	B	Std. error	Beta		
(Constant)	1.207	.182		6.634	.000
Reliability	.466	.052	.468	8.971	.000
a. Dependent Variable: Trust					

Table 5 indicates that reliability is a good determinant of trust. This variable was supported with ($\beta = 0.468$; $p < 0.01$) and the formal equation for this findings is shown in Eq. (3):

$$Trust = 0.468R \tag{3}$$

The final model with supported variables is shown in Fig. 2.

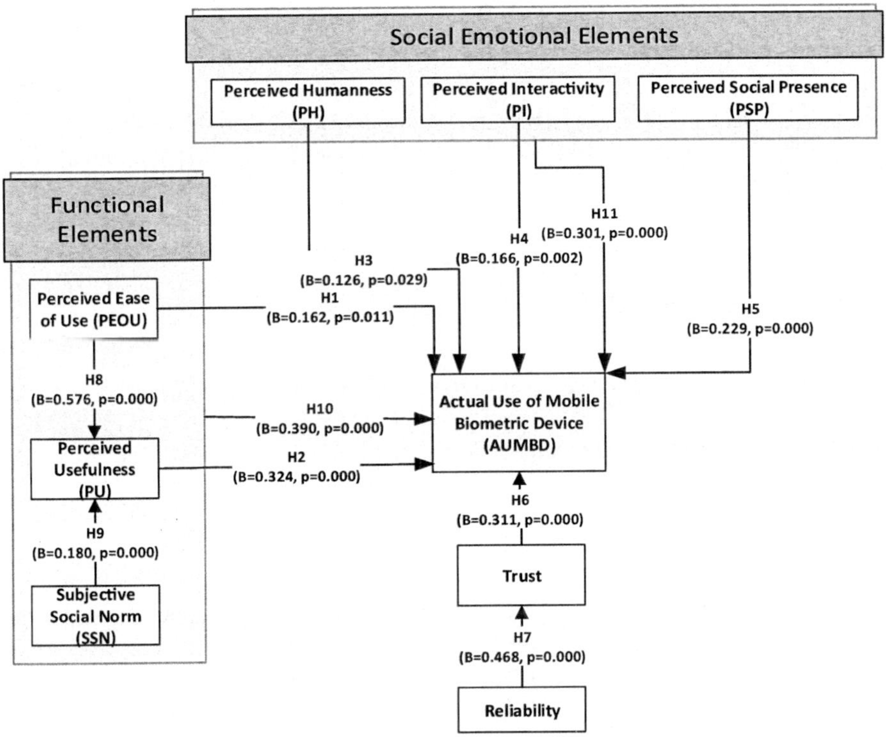

Fig. 2. The final model for this research

The overall formal equation for the supported variables in the final model of this study is shown in Eq. (4):

$$AU = 0.349PEOU + 0.058SSN + 0.1468R + 0.126PH + 0.166PI + 0.229PSP \tag{4}$$

5 Discussion

This study's proposed overall mobile biometric adoption model has been validated. Starting with the model's functional elements, the results showed that PEOU has a positive impact on customers' actual use of mobile biometric devices, with [17], confirming these findings. This indicates that as PEOU increases, so will the actual use. The findings indicated that the PU obtained has an effect on the optimistic plan to consider mobile use. However, SSN was not found to be effective on its own in this review. The same findings were found in [31]'s research on "factors determining the behavioral intention to use mobile learning: an application and extension of the ATAUT model." PEOU and PU can be retained and utilized in future research to test the adoption of biometric verification technologies on mobile devices, based on the findings. The findings revealed that SSN is not supported, however, this variable on its own has an impact on PU. The previous study by [20], found that SSN variable was supported. Finally, the addition of all functional elements shows a significant impact on the actual use.

The proposed model's social elements are all supported. The results show that PI, PH, and PSP have a relationship with actual use, and [32], support the findings of these three variables. As a result, these variables can be retained and utilized in future studies to predict consumer adoption of biometric verification implementations on mobile devices. Reliability, in and of itself, has a significant impact on trust. These findings are consistent with [11]'s study. In view of these discoveries, it can be decided that trust, reliability and actual use should be retained and used in future research to assess biometric authentication techniques adoption on mobile devices.

6 Study Limitations and Suggestions for Further Research

The following are the two shortcomings that were the subject of this research. To begin with, the research was limited to the adoption of biometric verification technology on mobile devices. Except for mobile devices, further research is required into the adoption of biometric verification techniques on other recent technologies. The gender and age of the respondents were the study's second major limitation. The majority of the respondents who took part in this study were men, and the average age was between 20 and 30. As a consequence, there is a problem of unbalanced results. In general, mobile devices are used by both males and females of various age groups today. The adoption of biometric authentication techniques on other recent innovations should be evaluated using the conceptual framework used in this report. This study contributed by producing a mathematical model that can be used to measure the adoption of mobile biometrics. This can be achieved also by the following formulae derived from the supported variables of the model for this study by the researchers (Eq. (4)): $AU = 0.349PEOU + 0.058SSN + 0.1468R + 0.126PH + 0.166PI + 0.229PSP$. It is hoped that the produced formulae will be useful in future studies.

7 Conclusion

This study aimed to see how well biometric authentication technique implementations on mobile devices were received. It has also addressed the user adoption issues identified in the research question to increase the user adoption of mobile biometric innovation dependent on security. Biometric authentication protection technology is becoming more prevalent in new devices; however, few studies have examined consumer adoption of such innovation on mobile devices. The framework utilized in this study was found to be accurate, appropriate, and well-supported. The majority of the participants agreed or are prepared to consider biometric verification techniques for use as protection on mobile devices, according to the findings of this study. This study has significantly contributed by adding to the consumers' understanding of the biometric verification reformation on mobile devices. The results of this survey aid the decision-makers in being mindful of the concerns that influence users' decisions to adopt and use a particular system, allowing them to consider them during the implementation stage. Future researchers may benefit from this study because it provides useful knowledge about biometric validation procedures on mobile devices, and a portion of their examination questions might be addressed. In addition, further study according to the researcher should be done, especially with the constructs that were found to be supported in the framework of the current study.

Acknowledgement. This work was authored by Malatji W.R. The author would like to thank Doctor Rene Van Eck and Professor Tranos Zuva for their contributions, helpfulness and support during this study.

References

1. Malatji, W.R., Zuva, T., VanEck, R.: Acceptance of biometric authentication security technology on mobile devices. In: Book Acceptance of biometric authentication security technology on mobile devices' (Springer Nature, (2020), edn.), pp. 145–156
2. https://www.qualitymag.com/articles/96205/technology-adoption. Accessed 1 April 2021
3. Tscherning, H., Mathiassen, L.: Early adoption of mobile devices-a social network perspective. J. Inf. Technol. Theory Appl. **11**(1), 23–42 (2010)
4. Kadena, E., Ruiz, L.: Adoption of biometrics in mobile devices. In: Book Adoption of Biometrics in Mobile Devices, edn., pp. 140–148 (2017)
5. Ashbourn, J.: Biometrics in the world: The cloud, mobile technology and pervasive identity. Springer International Publishing (2018)
6. Jain, A.K., Ross, A.A., Nandakumar, K.: Introduction to Biometrics. Springer (2011)
7. Dhraief, M.Z., Bedhiaf-Romdhania, S., Dhehibib, B., Oueslati-Zlaouia, M., Jebali, O., Ben Youssef, S.: Factors affecting the adoption of innovative technologies by livestock farmers in Arid area of Tunisia. FARA Res. Rep. **3**(5), 22 (2018)
8. https://www.enterpriseedges.com/biometric-banking-authentication-industry. Accessed 1 April 2021
9. Clarke, N.L., Furnell, M.: Authentication of users on mobile telephones—a survey of attitudes and practices. Comput. Security **24**, 519–527 (2005)
10. James, T., Pririm, T., Katherine, B., Reithel, B., Barkhi, R.: Determining the intention to use biometric devices: an application and extension of the technology acceptance model. J. Organ. End User Comput. (JOEUC) **18**(3), 24 (2006)
11. Chau, A., Stephens, G., Jamieson, R.: Biometrics acceptance - perceptions of use of biometrics. In: Book Biometrics Acceptance - Perceptions of Use of Biometrics (edn.) (2004)
12. https://www.biometricupdate.com/202104/Biometrics-adoption-boom-from-pandemic-expected-to-continue-goode-and-ID-R&G-survey. Accessed 01 May 2021
13. Abd-El-Barr, M., Qureshi, K., Aljanahi, N.: Evaluation and performance comparison of a model for adoption of biometrics in online banking. Kuwati J. Sci. **48**(2) (2021)

14. Vrana, R.: Acceptance of mobile technologies and m-learning in higher education learning: an explorative study at the Faculty of Humanities and Social Science at the University of Zagreb. Croatian Soc. Inf. Commun. Technol. Electron. Microelectron. **41**, 814–819 (2018)
15. Mandari, H., Koloseni, D.: Biometric authentication in financial institutions: the intention of banks to adopt biometric powered ATM. Adv. Comput. Sci. Int. J. **5**(4), 9–17 (2016)
16. Laux, D., Luse, A., Mennecke, B., Townsend, A.M.: Adoption of biometric authentication systems: implications for research and practice in the deployment of end- user security systems. J. Organ. Comput. Electron. Commer. **22**(4) (2012)
17. Soh, K.L., Wongand, W.P., Chan, K.L.: Adoption of biometric technology in online applications. Int. J. Bus. Sci. Appl. Manage. **3**(2), 121–146 (2010)
18. Boateng, M.S., Asiamah, K.O., Lamptey, R.B.: Impact assessment of biometric fingerprint application for time keeping at KNUST library. J. Appl. Thought **4**, 256–273 (2015)
19. Cristian, M.: Opportunities and challenges for biometric systems in travel: a review. Travel Tourism Res. Assoc. Adv. Tourism Res. Globally **61**, 1–120 (2016)
20. Erastus, L.R., Jere, N., Shava, F.B.: Exploring challenges of biometric technology adoption: a Namibian review. In: Proceedings of the International Conference on Emerging Trends in Network and Computer Communication, Windhoek (2015)
21. Ho, G., Stephens, G., Jamieson, R.: Biometric authentication adoption ıssues. In: Proceedings of the 14th Australasian Conference on Information Systems, Perth, Western Australia (2003)
22. Johnson, E.M., Howard, C.: A library mobile device deployment to enhance the medical student experience in a rural longitudinal integrated clerkship. J Med Libr Assoc **107**(1), 30–42 (2019)
23. Apostolou, B., Bélanger, F., and Schaupp, L.C.: Online communities: satisfaction and continued use intention. Inf. Res. **22** (2017)
24. Lankton, N., McKnight, D., Tripp, J.: Technology, humanness, and trust: rethinking trust in technology. J. Assoc. Inf. Syst. **16**, 880–918 (2015)
25. Gao, Q., Rau, P.-L., Salvendy, G.: Perception of interactivity: affects of four key variables in mobile advertising. Int. J. Hum.-Comput. Interact. **25**, 479 (2009)
26. Asiimwe, E.N., Grönlund, A.: MLCMS actual use, perceived use, and experiences of use. ijEDict – Int. J. Educ. Dev. Inf. Commun. Technol. **11**(1), 101–121 (2015)
27. Cheng, X., Fu, S., Sun, J., Bilgihan, A., Okumus, F.: An investigation on online reviews in sharing economy driven hospitality platforms: a viewpoint of trust. Tour. Manage. **71**, 366–377 (2019)
28. Tuunainen, V., Pitkänen, O., Hovi, M.: Users Awareness of Privacy on Online Social Networking sites – Case Facebook (2009)
29. Taherdoost, H.: Sampling methods in research methodology; How to choose a sampling technique for research. Int. J. Acad. Res. Manage. **5**, 18–27 (2016)
30. Taber, K.S.: The use of Chronbach's Alpha when developing and reporting research instruments in Science Education. Springer **48**, 1273–1296 (2017). https://doi.org/10.1007/s11165-016-9602-2
31. Hosokawa, R., Katsura, T.: Correction: association between mobile technology use and child adjustment in early elementary school age. PLoS ONE **13**, 12 (2018)
32. Chao, C.M.: Factors determining the behavioral intention to use mobile learning: an application and extension of the UTAUT model. Front. Psychol **10**, 1–14 (2019)
33. Lankton, N.K., Mcknight, D.H., Tripp, J.F.: Technology, humanness, and trust: Rethinking trust in technology. J. Assoc. Inf. Syst. **16**(10), 880–918 (2015)

Thermal Processes with Thermal Feedback in Electronic Systems and Their Modeling by Nonlinear Matrix Equations

Alexander G. Madera(✉) and Ekaterina K. Grebennikova

Scientific Research Institute of System Analysis of Russia Academy of Sciences, 36, k.1, Nachimovskiy Prosp, 117218 Moscow, Russia

Abstract. Thermal processes in electronic systems are subjected to thermal feedback is described by matrix nonlinear equations with polynomials on the right side. The solving such nonlinear matrix equations is very difficult task. This article describes a method for solving such equations. According to the developed method, the system of nonlinear interconnected equations is reduced to a system of independent algebraic equations, each of one is easy to solve. The method has a significant scientific novelty and has no analogues in the existing literature currently. This article describes a mathematical model of an electronic system with electronic components which have temperature-dependent power consumption and thermal feedback. Comparison of the results obtained by the developed and numerical methods show their complete match.

Keywords: Mathematical modeling · Nonlinear matrix equation · Polynomial functions · Spectral decomposition · Similarity transformation · Similarity matrix

1 Introduction

The effect of thermal feedback occurs in electronic systems (integrated circuits, resistors, electro-radio cells) during its operation. Thermal feedback arises in all electronic systems in which the operating parameters and power consumption depend on the temperature of self-heating, and the temperature, in turn, depend upon the consumed power [1–5]. The resulting interaction of the parameters of the functional and thermal modes of the electronic system with each other, which returns new values of the parameters along the feedback loop, leads to thermal feedback. The effect of the thermal feedback is manifested in any electronic systems and can significantly affect their performance and reliability and is also constitute a serious obstacle to the creation of competitive electronic systems.

Thermal processes in electronic systems affected by thermal feedback described by the mathematical model in the form of a nonlinear matrix equation for unknown temperatures of electronic system elements in the form

$$HT = P(T), \tag{1}$$

R. Silhavy et al. (Eds.): CoMeSySo 2021, LNNS 232, pp. 149–156, 2021.
https://doi.org/10.1007/978-3-030-90318-3_14

where H is square nondegenerate $N \times N$-matrix of the electronic system thermal model graph [1, 6] with constant elements h_{lm}, $l, m = 1, 2, \ldots, N$; $T = (T_1, T_2, \ldots, T_N)^T$ is N-vector of temperatures T_i of electronic system elements $i = 1, 2, \ldots, N$ in N graph nodes; $P(T) = (P_1(T_1), P_2(T_2), \ldots, P_N(T_N))^T$ is nonlinear vector function of the power consumption $P_i(T_i)$ of electronic elements i that depend nonlinearly on the temperature of the elements T_i.

Nonlinearly temperature-dependent power consumption $P_i(T_i)$ can always be approximated either by k-degree polynomials, or decomposed into Taylor power series. In other words, power consumption temperature dependence $P_i(T_i)$ of electronic element i can be presented as a k-degree polynomial $P_i(T_i) = p_{0i} + p_{1i}T_i + p_{2i}T_i^2 + \cdots + p_{ki}T_i^k$, where p_{li}, $l = 1, 2, \ldots, k$, $i = 1, 2, \ldots, N$ are constant coefficients. Then Eq. (1) with the nonlinear polynomial right-hand side $P(T)$ takes the form

$$HT = \begin{pmatrix} p_{01} + p_{11}T_1 + p_{21}T_1^2 + \cdots + p_{k1}T_1^k \\ p_{02} + p_{12}T_2 + p_{22}T_2^2 + \cdots + p_{k2}T_2^k \\ \vdots \\ p_{0n} + p_{1n}T_n + p_{2n}T_n^2 + \cdots + p_{kn}T_n^k \end{pmatrix} \tag{2}$$

Currently, therc are no analytical matrix methods for obtaining solutions for nonlinear matrix Eqs. (2). The application of numerical methods to the solution of Eqs. (2) encounters problems of convergence, step selection, convergence rate, accuracy and calculation errors, and these problem solutions are a priori unknown.

For the solving Eq. (2), as a rule, make various and unfounded assumptions, that reduce a nonlinear matrix equation to a linear one. For example, to the assumption that nonlinear functions $P_i(T_i)$, $i = 1, 2, \ldots, N$ can be approximated by linear ones, or that the nonlinear function decomposition into a Taylor power series with leaving only the first-order terms of smallness is enough. All these assumptions lead to inadequate modeling results and errors in design of electronic systems.

The article considers an analytical method for solving nonlinear matrix equations in the form (1) with a polynomial right-hand side (2). The proposed method is based on the following fundamental aspects: (a) the transformation of the original matrix-vector equation with unknown vector to a matrix equation with unknown matrix, (b) spectral decomposition of the matrix H in Eq. (2) and its reduction to a diagonal identity matrix, and (c) reduction of interrelated dependent nonlinear equations system to a system of unrelated independent polynomial equations, the roots of which determine the desired solution of the original nonlinear matrix-vector Eq. (2). The general method is described in [7].

The method has been tested for mathematical modeling of heat processes with thermal feedback in real electronic systems. A comparative analysis of the solutions obtained by the developed method and by the known numerical methods have showed a complete coincidence of these solutions.

2 Proposed Method for Solving Nonlinear Matrix Equations with a Polynomial Right-Hand Side

Let us transform the original matrix-vector Eq. (1) with an unknown vector T to a matrix equation with an unknown matrix T [7].

For this purpose the vector $T = (T_1, T_2, \ldots, T_N)^T$ represents as the product $T = \mathcal{T}j$ of a diagonal $N \times N$ -matrix $\mathcal{T} = diag\{T_1, T_2, \ldots, T_N\}$ and an identity N-vector $j = (11\ldots1)^T$. It is easy to verify the validity of the following relation for the degrees of the matrix $\mathcal{T}^m = diag\{T_1^m, T_2^m, \ldots, T_N^m\}, m = 1, 2, \ldots k$, namely,

$$(T_1^m, T_2^m, \ldots, T_N^m)^T = diag\{T_1^m, T_2^m, \ldots, T_N^m\}j = \mathcal{T}^m j.$$

Entering the polynomial vectors $\mathcal{P}_0 = (p_{01}, p_{02}, \ldots, p_{0N})^T$, $\mathcal{P}_1 = (p_{11}, p_{12}, \ldots, p_{1N})^T$, ..., $\mathcal{P}_k = (p_{k1}, p_{k2}, \ldots, p_{kN})^T$, and substituting them into the matrix-vector Eq. (2), we obtain the matrix equation with the unknown diagonal $N \times N$-matrix $\mathcal{T}$

$$H\mathcal{T}j = \mathcal{P}_0 + \mathcal{T}\mathcal{P}_1 + \mathcal{T}^2\mathcal{P}_2 + \cdots + \mathcal{T}^k\mathcal{P}_k. \tag{3}$$

Let us introduce a nonsingular transforming $N \times N$-matrix X that make the similarity transformation $\mathcal{T} = XDX^{-1}$ between $N \times N$-matrices $\mathcal{T}$ and D. Since for the degrees of the matrix $\mathcal{T}$ the relations

$$\mathcal{T} = XDX^{-1}, \mathcal{T}^2 = XD^2X^{-1}, \ldots, \mathcal{T}^k = XD^kX^{-1},$$

are valid, Eq. (3) will take the following form:

$$HXDX^{-1}j = P_0 + XDX^{-1}P_1 + XD^2X^{-1}P_2 + \cdots + XD^kX^{-1}P_k. \tag{4}$$

Reduce the matrix H of the system of Eqs. (4) to a diagonal identity form.

Since the $N \times N$-matrix H in Eq. (1) is nonsingular and positive definite, there is a spectral decomposition $U^{-1}HU = \Lambda_H$ for it with a transforming similarity $N \times N$-matrix U and a diagonal $N \times N$-matrix of its eigenvalues Λ_H [7–11]. Let us present the transforming similarity matrix X in the form $X = U\Lambda_H^{\frac{-1}{2}}V^{-1}$ with still unknown transforming similarity $N \times N$-matrix V and multiply Eq. (4) on the left by the matrix $V\Lambda_H^{\frac{-1}{2}}U^{-1}$.

Then, considering the relation $V\Lambda_H^{\frac{-1}{2}}U^{-1}X = V\Lambda_H^{\frac{-1}{2}}U^{-1}U\Lambda_H^{\frac{-1}{2}}V^{-1} = V\Lambda_H^{-1}V^{-1}$, we get the following equation instead of (4):

$$\begin{aligned} &V\Lambda_H^{-\frac{1}{2}}U^{-1} \cdot H \cdot U\Lambda_H^{-\frac{1}{2}}V^{-1}DX^{-1}j \\ &= V\Lambda_H^{-\frac{1}{2}}U^{-1}\mathcal{P}_0 + V\Lambda_H^{-1}V^{-1}DX^{-1}\mathcal{P}_1 + V\Lambda_H^{-1}V^{-1}D^2X^{-1}\mathcal{P}_2 + \cdots + V\Lambda_H^{-1}V^{-1}D^kX^{-1}\mathcal{P}_k. \end{aligned} \tag{5}$$

As long as $V\Lambda_H^{\frac{-1}{2}}\underbrace{U^{-1} \cdot H \cdot U}_{\Lambda_H}\Lambda_H^{\frac{-1}{2}}V^{-1} = VV^{-1} = I$, the matrix H of the equation system (2) is reduced to a diagonal form with the identity matrix $I = diag\underbrace{\{1, 1, \ldots, 1\}}_{N}$.

Multiplying Eq. (5) on the left by the matrix $V\Lambda_H V^{-1}$, we get

$$V\Lambda_H V^{-1}DX^{-1}j = V\Lambda_H^{\frac{1}{2}}U^{-1}\mathcal{P}_0 + DX^{-1}\mathcal{P}_1 + D^2X^{-1}\mathcal{P}_2 + \cdots + D^kX^{-1}\mathcal{P}_k. \quad (6)$$

Let us introduce the vectors $U^{-1}j = \alpha = (\alpha_1, \alpha_2, \ldots, \alpha_N)^T$ and $U^{-1}\mathcal{P}_l = \beta_l = (\beta_{l1}, \beta_{l2}, \ldots, \beta_{lN})^T$, $l = 0, 1, 2, \ldots, k$, and consider that $X^{-1} = V\Lambda_H^{\frac{1}{2}}U^{-1}$. Then we can write the following relations: $X^{-1}j = V\Lambda_H^{\frac{1}{2}}U^{-1}j = V\Lambda_H^{\frac{1}{2}}\alpha$, $X^{-1}\mathcal{P}_l = V\Lambda_H^{\frac{1}{2}}U^{-1}\mathcal{P}_l = V\Lambda_H^{\frac{1}{2}}\beta_l$, $l = 0, 1, 2, \ldots, k$.

Substituting the last expressions in the Eq. (6) we obtain

$$V\Lambda_H V^{-1}DV\Lambda_H^{\frac{1}{2}}\alpha = V\Lambda_H^{\frac{1}{2}}\beta_0 + DV\Lambda_H^{\frac{1}{2}}\beta_1 + D^2V\Lambda_H^{\frac{1}{2}}\beta_2 + \cdots + D^kV\Lambda_H^{\frac{1}{2}}\beta_k. \quad (7)$$

Let us demand that the matrix V is to be a transforming similarity matrix for the matrix D, so that its spectral decomposition has the form $V^{-1}DV = \Lambda_D$, where $\Lambda_D = diag\{\lambda_{D1}, \lambda_{D2}, \ldots, \lambda_{DN}\}$ is the diagonal matrix of the matrix D eigenvalues.

Multiplying, finally, Eq. (7) on the left by the matrix V^{-1}, considering the spectral decomposition of the matrix D, as well as the diagonal structure of the matrices Λ_D and Λ_H, we obtain finally the matrix-polynomial equation that the eigenvalue matrix Λ_D must satisfy

$$\beta_0 + \Lambda_D(\beta_1 - \Lambda_H\alpha) + \Lambda_D^2\beta_2 + \cdots + \Lambda_D^k\beta_k = 0. \quad (8)$$

The matrix Eq. (8), due to the diagonal nature of the matrices Λ_H and Λ_D, decomposes into N independent unrelated polynomial equations of k-th order with unknown eigenvalues λ_{Di}, $i = 1, 2, \ldots, N$, of the matrix D, namely,

$$\beta_{0i} + (\beta_{1i} - \lambda_{Hi}\alpha_i)\lambda_{Di} + \beta_{2i}\lambda_{Di}^2 + \ldots + \beta_{ki}\lambda_{Di}^k = 0 \quad (9)$$

The resulting equations are the conditions that must be obeyed by the eigenvalues λ_{Di}, $i = 1, 2, \ldots, N$, of the matrix D.

The desired temperature vector T, that is the solution of the matrix Eq. (1) with the polynomial right-hand side (2), is determined by the relations:

$$T = \mathcal{T}j = XDX^{-1}j = U\Lambda_H^{-\frac{1}{2}}\underbrace{V^{-1}DV}_{\Lambda_D}\Lambda_H^{\frac{1}{2}}U^{-1} = U\Lambda_D U^{-1}j \quad (10)$$

In expression (10), the transforming similarity matrix U is known from the spectral decomposition of the matrix H of the original equation system (1), and the elements λ_{Di}, $i = 1, 2, \ldots, n$, of the diagonal matrix Λ_D are determined from the solution of algebraic polynomial equations of the k-th order (9).

The developed method allows to obtain a vector solution T of the matrix nonlinear Eq. (1) with a polynomial right-hand side (2).

The method application contains the following steps:

- the spectral decomposition $U^{-1}HU = \Lambda_H$ of the matrix H of the original equation system (1), the transforming similarity $N \times N$-matrix U and the diagonal $N \times N$-matrix

$\Lambda_H = diag\{\lambda_{H1}, \lambda_{H2}, \ldots, \lambda_{HN}\}$ of the matrix H eigenvalues, are obtained [8, 9]. Since the matrix H is nondegenerate and positive definite, its spectral decomposition exists;

- the k-th order N independent and unrelated algebraic polynomial Eqs. (9) are forming. From this solution the matrix D eigenvalues λ_{Di}, $i = 1, 2, \ldots, N$, are determined;
- the k roots of k-th degree algebraic Eqs. (9) generally contain both complex and real roots. The complex roots are excluded from consideration due to the physical and mathematical nature of temperatures. As regards the real roots, it should be analyzed and those of them that have a physical and informative meaning for the problem being solved should be selected;
- the desired vector T of the original nonlinear Eq. (1) solution with the polynomial right-hand side (2) is determined according to the expression (10), namely, $T = U\Lambda_D U^{-1}j$.

3 Example of Using the Method

Consider the steady-state thermal process in an electronic system containing two integrated circuits IC1 and IC2, placed on a common printed circuit board (PCB, Fig. 1, *a*) and being in a state of heat exchange between themselves and the environment.

The power consumptions $P_1(T_1)$, $P_2(T_2)$ of integrated circuits IC1 and IC2 depend nonlinearly on the temperature values T_1 and T_2 of IC1 and IC2 chips, as second-degree polynomials (Fig. 1, *b*).

$$P_1(T_1) = p_{01} + p_{11}T_1 + p_{21}T_1^2 = 5 - 0,001T_1 + 0,0002T_1^2,$$

$$P_2(T_2) = p_{02} + p_{12}T_2 + p_{22}T_2^2 = 8 - 0,003T_2 - 0,0003T_2^2.$$

Due to the temperature dependence of the power consumption $P_1(T_1)$, $P_2(T_2)$, and the thermal interaction by the conductive heat transfer on the printed circuit board integrated circuits IC1 and IC2 are affected by thermal feedback [1]. Note that the temperatures T_1 and T_2 simulate the excess of the temperature of integrated circuits IC1 and IC2 over the environment temperature.

The mathematical model describing the thermal interaction between integrated circuits (Fig. 1), is a matrix Eq. (1), and defines an unknown desired temperature vector $T = (T_1, T_2)^T$, and has the form

$$HT = \begin{pmatrix} h_{11} & h_{12} \\ h_{21} & h_{22} \end{pmatrix} \begin{pmatrix} T_1 \\ T_2 \end{pmatrix} = \begin{pmatrix} p_{01} + p_{11}T_1 + p_{21}T_1^2 \\ p_{02} + p_{12}T_2 + p_{22}T_2^2 \end{pmatrix},$$

where $h_{11} - 0,31422$, $h_{12} = h_{21} = -0,0065$, $h_{22} = 0,66675$.

Note that the matrix H is symmetric, nondegenerate and positive definite, and its determinant is equals to $det\{H\} = 0,209$.

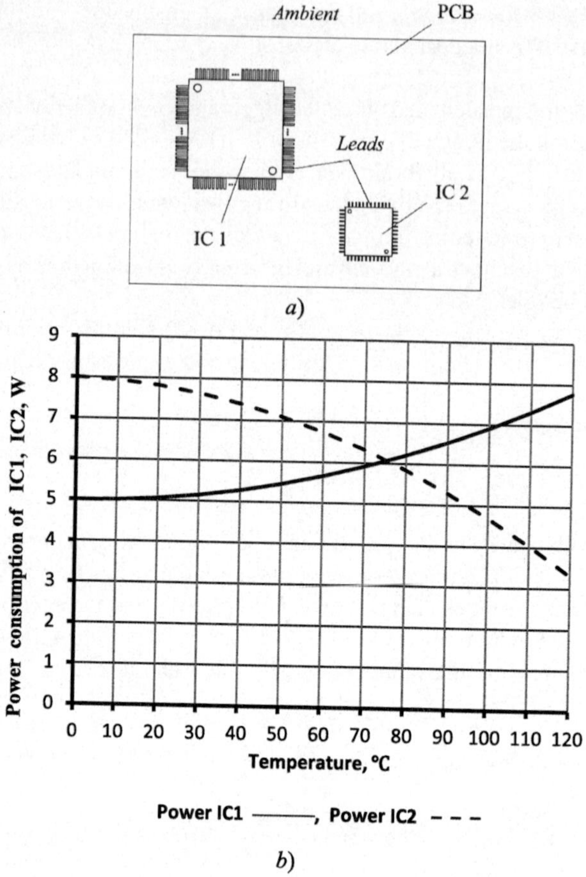

Fig. 1. Dependences of power consumption (W) of integrated circuits IC1 and IC2 on temperature (°C)

The matrix equation of the consideration mathematical model by applying the transformation $T = \mathcal{T}j$ with a diagonal matrix $\mathcal{T} = diag\{T_1, T_2\}$ and the identity vector $j = (11)^T$ (see point 2), and introducing the vectors $\mathcal{P}_0 = (p_{01}, p_{02})^T, \mathcal{P}_1 = (p_{11}, p_{12})^T$, $\mathcal{P}_2 = (p_{21}, p_{22})^T$, is written as the following matrix Eq. (3)

$$HTj = P_0 + TP_1 + T^2P_2.$$

According to the applying procedure for the developed method (see point 2), the spectral decomposition of the matrix $H = U\Lambda_H U^{-1}$ is determined at first

$$\begin{pmatrix} 0,31422 & -0,00605 \\ -0,00605 & 0,66675 \end{pmatrix}$$

$$= \begin{pmatrix} 58,3 & -0,0172 \\ 1 & 1 \end{pmatrix} \begin{pmatrix} 0,314 & 0 \\ 0 & 0,667 \end{pmatrix} \begin{pmatrix} 0,0172 & 0,000294 \\ -0,0172 & 1,0 \end{pmatrix},$$

and from it the transforming similarity matrix U, its inverse matrix U^{-1}, and the diagonal matrix of the eigenvalues Λ_H of the matrix H, are determined as following:

$$U = \begin{pmatrix} 58,3 & -0,0172 \\ 1 & 1 \end{pmatrix}, \Lambda_H = \begin{pmatrix} 0,314 & 0 \\ 0 & 0,667 \end{pmatrix}, U^{-1} = \begin{pmatrix} 0,0172 & 0,000294 \\ -0,0172 & 1,0 \end{pmatrix}$$

After this, according to (9), the second-degree algebraic polynomial equations determining the elements λ_{D1}, λ_{D2} of the diagonal matrix Λ_D, are formed:

$$3,33 \cdot 10^{-6}\lambda_{D1}^2 - 0,00548\lambda_{D1} + 0,08785 = 0,$$

$$-0,0003\lambda_{D2}^2 - 0,6586\lambda_{D2} + 7,91,45 = 0.$$

The roots of these equations are equal: the 1-st pair of roots $(\lambda_{D1}; \lambda_{D2}) = (1644,176; -2207,21)$ and the 2-nd pair of roots $(\lambda_{D1}; \lambda_{D2}) = (16,34; 11,95)$. In these pairs of roots, only the 2-nd pair corresponds to the real physical and informative meaning of the mathematical model under consideration, namely $(\lambda_{D1}; \lambda_{D2}) = (16,34; 11,95)$.

The desired solution vector $T = (T_1, T_2)^T$ of the considered nonlinear mathematical model is determined by the expression (10):

$$T = U\Lambda_D U^{-1}j$$
$$= \begin{pmatrix} 58,3 & -0,0172 \\ 1 & 1 \end{pmatrix}\begin{pmatrix} 16,34 & 0 \\ 0 & 11,95 \end{pmatrix}\begin{pmatrix} 0,0172 & 0,000294 \\ -0,0172 & 1,0 \end{pmatrix}\begin{pmatrix} 1 \\ 1 \end{pmatrix} = \begin{pmatrix} 16,46 \\ 12,03 \end{pmatrix}.$$

It follows that the excess temperature of integrated circuits IC1 and IC2 over the environment temperature is 16.46 °C and 12.03 °C, respectively.

The solution obtained by applying the Newton numerical method is

$$T = \begin{pmatrix} 16,27 \\ 12,03 \end{pmatrix}.$$

Thus, a comparison of the solutions for the temperature vector T obtained by the developed method and calculated by the Newton method shows their almost an exact match, their relative error is under than 1.16%.

4 Conclusion

A mathematical model of a steady-state thermal process in an electronic system with thermal feedback, in which the element power consumption depends nonlinearly on the temperature of the elements themselves, is a nonlinear matrix equation with polynomial right-hand side. The method presented in the article is a general analytical method, which does not contain limitations on the type and degree of the polynomial and allows to obtain a closed analytical solution of the matrix nonlinear equation exclusively in the matrix form without usage of numerical methods. The developed method has no limitations and disadvantages inherent in numerical methods, such as convergence, step selection, convergence rate, calculation accuracy, rounding errors, etc.

The application of this method is considered on the example of thermal steady-state process in an electronic system containing two integrated circuits placed on a printed circuit board under the thermal feedback. The power consumption of microchips depends on the values of their temperature and describes by quadratic functions (Fig. 1, *b*). A comparison of the results obtained by the developed method and ones calculated by the numerical Newton method have showed their almost complete coincidence with their relative error not exceeding 1.16%.

Publication is made as part of national assignment for SRISA RAS (fundamental scientific research 47 GP) on the topic No.0580-2021-0001 (121031300047-6).

References

1. Madera, A.G.: The effect of thermal feedback and statistical technological dispersion of microcircuits parameters on their thermal mode. In: IOP Conference Series: Materials Science and Engineering, vol. 862, no. 32, 27 May 2020; 032106, 8 p. (2020). https://doi.org/10.1088/1757-899X/862/3/032106
2. Madera, A.G., Kandalov, P.I.: Thermal processes in electronic equipment at uncertainty. J. Eng. Thermophys. **29**(1), 170–180 (2020). https://doi.org/10.1134/S1810232820010129
3. Madera, A.G.: A method for explicitly solving matrix differential equations in second-order ordinary derivatives based on diagonalizing the matrices of the equation using spectral decomposition and Kronecker matrix algebra. Differencialnie Uravnenia i Protsesy Upravlenia (Differential Equations and Control Processes), no. 3, pp. 10–24 (2021)
4. Zhang, L., Song, D., Xiao, Y., Lin, X., Chan, M.: On the formulation of self-heating models for circuit simulation. IEEE J. Electron Devices Soc. **6**, 291–297 (2018). https://doi.org/10.1109/Jeds.2018.2801301
5. Park, J., Yun, D., Kim, S., Choi, Y.: Suppression of self-heating effects in 3-D V-NAND flash memory using a plugged pillar-shaped heat sink. IEEE Electron Device Lett. **40**(2), 212–215 (2019). https://doi.org/10.1109/LED.2018.2889037
6. Salamin, S., Van Santen, V.M., Rapp, M., Henkel, J., Amrouch, H.: Minimizing excess timing guard banding under transistor self-heating through biasing at zero-temperature coefficient. IEEE Access **9**, 30687–30697 (2021)
7. Camarchia, V., Cappelluti, F., Pirola, M., Guerrieri, S.D., Ghione, G.: Self-consistent electrothermal modeling of class A, AB, and B power GaN HEMTs under modulated RF excitation. IEEE Trans. Microw. Theory Tech. **55**(9), 1824–1831 (2007). https://doi.org/10.1109/TMTT.2007.903839
8. Lankaster, P.: Theory of Matrix. Academic Press, New York – London (1969)
9. Horn, R.A., Johnson, C.R.: Matrix Analysis. Cambridge University Press, Cambridge (1986)
10. Bellman, R.: Introduction to Matrix Analysis. McGraw-Hill, New York (1960)
11. Gantmacher, F.R.: The Theory of Matrix. Chelsea Publishing Company, New York (1960)

Software for Determining Functional Properties of Tribological Conjugations

Andrey Rachishkin(✉)

Tver State Technical University, Tver, Russian Federation

Abstract. Article describes computer aided simulation of patterns of surfaces with roughnesses. The model is based on experimental data of interacting surfaces' morphology, including those with coatings. The assumptions made do not have a considerable influence on the simulated surface credibility. It is designed to model contact interaction and evaluate the frictional parameters of the friction units, anticipate thermal resistance. A flexible algorithm and an independent segmented mathematical logic are developed for the optimal calculation of simulated processes using numerical methods. This software combines modern information technologies and analytical developments of generation of rough surfaces.

Keywords: Computer simulation · Roughness · Topography · Contact interaction · Frictional interaction · Thermal resistance

1 Introduction

Improving the reliability and durability of mechanical systems is one of the main issues in modern engineering. The main reason for the failure of machines is not their breakdown, but the wear of their tribological conjugations. It is necessary to optimize the functional properties of mechanical systems according to the criteria for reducing wear and friction losses in tribological couplings at the design stage considering specific conditions of their operation.

Microgeometry modeling of technical surfaces and contact interaction is one of the effective approaches in tribology [1–4]. Combining theoretical methods based on the CAD approach and building their interaction allows us to design a system of discrete event-driven computer simulation for virtual R&D to calculate various tribological conjunctions.

The development of the system for modeling a morphology of technical surfaces, taking into account the presence of coatings and their anisotropy, the calculation of the characteristics of contact and friction interaction, the thermal resistance of joints, allows us to solve a wide range of engineering practice problems.

Development of the software for computer simulation of interactions of roughness of technical surfaces should incorporate existing approaches to forecasting the operational characteristics of contact of rough surfaces and modern computer technologies [5]. Initial implementation includes two fundamental levels that should be distinguished in the simulated system: modeling the morphology of technical surfaces and modeling

R. Silhavy et al. (Eds.): CoMeSySo 2021, LNNS 232, pp. 157–162, 2021.
https://doi.org/10.1007/978-3-030-90318-3_15

physical processes occurring on the places of real contact. Then, it is necessary to determine relatively autonomous parts (modules) of the given processes and design software modules for them. This approach allows adding, modifying and debugging of the modules without a restructuring of the software. The use of unified logic makes it possible to expand the computer model and modernize existing components.

System includes a user input verification tool, detecting human errors such as creation of logically impossible situations and improving UX in general. Initial data includes microgeometric and physico-mechanical parameters received during experiments. Altering one input parameter does not necessarily require recalculation of the entire system. It is sufficient to apply corrective modelling of parameter dependent modules. Such an approach improves the performance of the system.

Next step of simulation is where the simulation of the contact interaction of roughnesses and processes at real contacts take place. All generated roughnesses and local physical and mechanical properties of the simulated surfaces are properly checked. At the verification stage roughness is excluded from the modelling if there is no contact. During contact interaction the presence and thickness of functional coatings and the nature of the resulting deformations are analyzed. The mathematical model is built considering possible algorithmic situations. When all contacting roughnesses are processed the output is reduced to the resulting parameter over the entire surface at this iteration of the introduction increment. Then the system calculates next iteration bringing it to the value corresponding to a given load. When the final value is reached, the general results of the modeling process are displayed in a given format.

2 Software Design

The morphology generation module of the technical surface. Morphology of technical surfaces defined by their topography, physical and mechanical properties, the presence of functional coatings and surface films.

Surface 3D model describes the topography of the surfaces. During model design we should acknowledge that for most contact interaction cases we take into account only the part of the layer with protrusions [2]. Mathematical apparatus created uses identically oriented segments of rotation ellipsoids to form a model of the roughness of a surface. Experimentally obtained standard microgeometry parameters allowed to develop an algorithm of distribution of rougnessess on the nominal plane. In a given range the heights and radiuses of roughnesses are distributed. The model of roughness is of the shape of an ellipsoid of revolution segment. Given model recognizes the anisotropy of the contacting surfaces. Model can be set of the shape of a spherical segment, which is widely used in studies of contact interaction [1–4].

Local physical and mechanical properties of the surface and presence and thickness of functional coatings can be changed according to one of the laws of distribution of random values or set as constant.

Module of Contact Interaction. A 3D model of a rough surface is a foundation for modelling contact interaction. Mathematical model handles the variant analysis of contact interaction with an elastic half-space having an elastoplastic coating and with an elastoplastic half-space without a coating. The resulting deformations are verified and calculated as either elastic or elastoplastic. Thus, the calculation of the total load for all irregularities in the contact can be expressed as follows:

$$N = \sum_{i=1}^{j} N_{1ie} + \sum_{i=1}^{k} N_{1iep} + \sum_{i=1}^{l} N_{2ie} + \sum_{i=1}^{q} N_{2iep} \quad (1)$$

where: N – normal load applied to the rough surface; N_{1ie} – loads applied to segments that contact coating in the case of its elastic deformation; N_{1iep} – loads applied to segments that contact coating in the case of its elastoplastic deformation; N_{2ie} – loads applied to segments that contact base in the case of its elastic deformation; N_{2iep} – loads applied to segments that contact base in the case of its elastoplastic deformation; j – number of segments that contact coating in the case of its elastic deformation; k – number of segments that contact coating in the case of its elastoplastic deformation; l – number of segments that contact base in the case of its elastic deformation; q – number of segments that contact base in the case of its elastoplastic deformation.

This task can be resolved by using numerical method [6] accommodating step increment of implementation. Module test results are well aligned with experiments. In Fig. 1 shows the results of computer simulation of the contact interaction of rough surfaces having coatings and the corresponding experimental data [7, 8]. Points correspond to experimental data, lines correspond to computer simulation results.

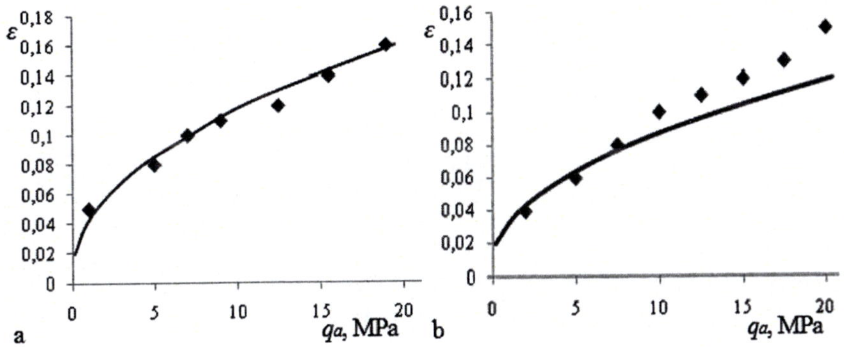

Fig. 1. Graphs of relative penetration for rough steel specimens in contact with steel specimens covered with functional or solid lubricating coating: a) gold, $\delta = 16\ \mu m$; b) silver, $\delta = 30\ \mu m$;

Module of Friction. The module can be plugged-in and will inherit modelling steps [9]. The total force is calculated as follows:

$$F = \sum_{i=1}^{j} F_{1ie} + \sum_{i=1}^{t} F_{1iep} + \sum_{i=1}^{l} F_{2ie} + \sum_{i=1}^{q} F_{2iep} \quad (2)$$

where: F – force of friction; F_{1ie} – forces of friction on segments that contact coating in case of its elastic deformation; F_{1iep} – forces of friction on segments that contact coating in case of its elastoplastic deformation; F_{2ie} – forces of friction on segments that contact base in case of its elastic deformation; F_{2iep} – forces of friction on segments that contact base in case of its elastoplastic deformation; j – number of segments that contact coating in case of its elastic deformation; t – number of segments that contact coating in case of its elastoplastic deformation; l – number of segments that contact half-space in case of its elastic deformation; q – number of segments that contact half-space in case of its elastoplastic deformation.

User input acts as main parameters and at this stage of modelling the parameters of microgeometry, materials and coatings physico-mechanical properties and the laws of their statistical distribution are already present.

Module testing was carried out using experimental data [8] (Fig. 2). In these experiments, samples with the following SLCs were used: VNIINP-230 (Vserossijskij Nauchno-Issledovatelskij Institut po Pererabotke Nefti (Russian Oil Refinment Research Institute)) 0,03 mm thick (rhombic symbols), VNIINP-212 0,03 mm thick (square symbols), and VNIINP-212 0,005 mm thick (triangle symbols). The experimental data show satisfactory agreement with the results of computer simulation shown by blank symbols.

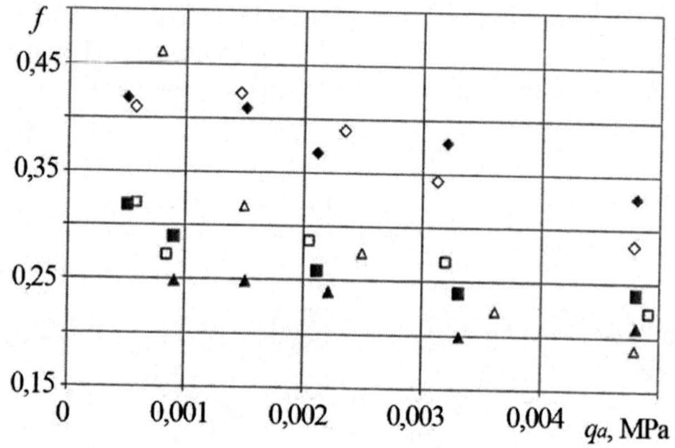

Fig. 2. Comparison of experimental values of the constant of friction for solid-lubricating coatings with computer simulation results.

Module of Thermal Resistance. The module can be plugged-in and will inherit modelling steps. Input data is thermophysical properties of the contacting materials and coatings.

The thermal resistance of a discrete contact can be represented as aggregation of parallel thermal resistances of single contact spots formed under the action of an external load. Then, thermal resistance R_{rh} for a thermally stressed rough contact can be defined as:

$$\frac{1}{R_{rh}} = \sum_{i=1}^{n_r} \frac{1}{R_{rhi}} \tag{3}$$

Where: n_r – number of the roughnesses that have come in contact with the coating; R_{rhi} – thermal resistance of a single contact spot with a coating.

For the mathematical model of calculating the thermal resistance of a single contact, the results obtained in the work [10].

Modelling results were tested by experimental data [11] (Fig. 3). Contour markers show the results of computer modeling, black - experimental data.

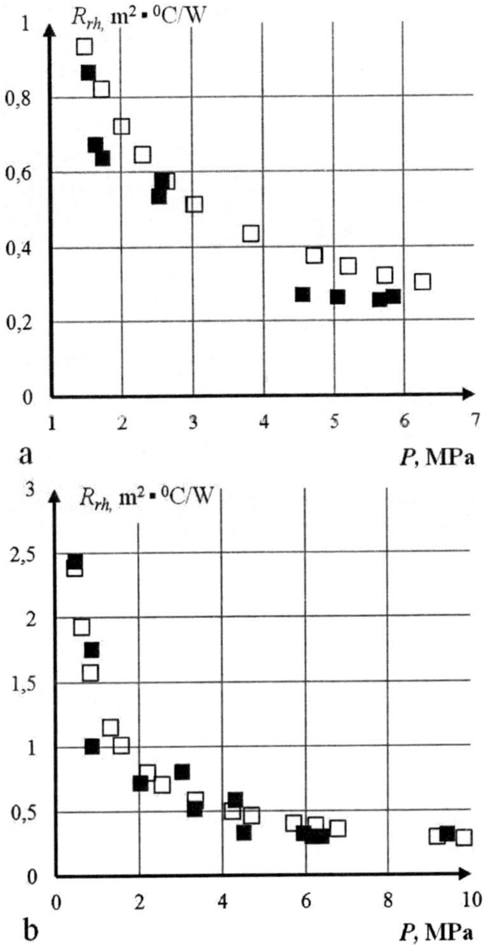

Fig. 3. Comparison of the experimental data with the modeled AISI grade 304 stainless steel information: a) finished grinding and b) regrinding.

The results show good convergence of the experimental and simulated data.

3 Conclusion

Computer program with modular design has been developed and implemented, allowing to computer-based experiments for tribological conjugations for various purposes.

Optimization of data processing and orientation to the object structure allows adequately describe the morphology of technical surfaces and to simulate various physical and mechanical processes occurring on spots of contact. Changing the input simulation parameters in wide ranges allows us to optimize the functional properties of tribojunctions at the design stage.

Aknowledgements. This work is supported by the research grant of the President of Russian Federation for state support of young Russian scientists MK-326.2021.4.

References

1. Greenwood, J.A., Williamson, J.B.P.: Contact of nominally flat surfaces. Proc. Roy. Soc., Ser. A **295** (1442), 300–319 (1966)
2. Demkin, N.B., Ryzhov E.V.: Kachestvo poverkhnosti i kontakt detalei mashin (Surface Quality and Machine Detail Contact), Moscow: Mashinostroenie (1981). (in Russian)
3. Goryacheva, I.G.: Mechanics of discrete contact. Tribol. Int. **39**, 381–386 (2006)
4. Bhushan, B.: Contact mechanics of rough surfaces in tribology: multiple asperity contact. Tribol. Lett. **4**, 1–35 (1998)
5. Nagel, C.: Professional C# 7 and .NET Core 2.0. Wiley (2018). https://doi.org/10.1002/9781119549147
6. Sutyagin, O.V., Bolotov, A.N., Rachishkin, A.A.: Computer simulation of contact interaction of rough surfaces. J. Frict. Wear **37**(3), 198–203 (2016)
7. Alekseev, N.M.: Metallicheskie pokrytiya opor skol'zheniya (Metallic Coatings of Sliding Supports). Nauka, Moscow (1973).(in Russian)
8. Sutyagin, O.V., Bolotov, A.N., Vasil'ev, M.V.: Kontakt sherokhovatykh tel s tverdosmazochnymi pokrytiyami (Rough Body Contact with Solid-Lubricating Coatings), Tver: Tver. Gos. Tekhn. Univ. (2014) (in Russian)
9. Sutyagin, O.V., Bolotov, A.N., Rachishkin, A.A., Burdo, G.B.: Computer simulation of the friction of rough surfaces. J. Frict. Wear **37**(4), 330–336 (2016). https://doi.org/10.3103/S1068366616040188
10. Korotkov, M.A., Sutyagin, O.V., Demenkov, E.A.: Effect of a coating on the electrical and thermal resistance of a plane single contact for a two-layer half-space. J. Eng. Phys. Thermophys. **67**(1–2), 823–830 (1994)
11. Thomas, T.R., Probert, S.D.: Correlations for thermal contact conductance in Vacuo. J. Heat Transf. **94**(3), 276–280 (1972). https://doi.org/10.1115/1.3449928

Algorithmic Procedure for the Reduction of the Classifier Based on the ε-, δ-criterion in the Method of Logical Analysis of Data

Roman I. Kuzmich(✉)

Siberian Federal University, Krasnoyarsk, Russia
rkuzmich@sfu-kras.ru

Abstract. The selection of logical patterns for a new classifier by reducing the original classifier based on the ε-, δ-criterion is proposed. The implementation of the algorithmic procedure for the reduction of the classifier consists of a number of steps: the formation of the initial classifier from a set of partial patterns with the application of the increasing procedure to them, the assessment of the suitability of each pattern in the set using the ε-, δ-criterion, the reduction of the classifier based on the assessment of suitability. Numerical experiments on real problems which make it possible to assess the feasibility of the proposed algorithmic procedure are presented.

Keywords: Pattern · Classifier · Reduction · ε- · δ-Criterion

1 Introduction

This paper considers one of the methods of inductive learning – the method of logical analysis of data. The main principle of the method is the formation of patterns (logical rules) based on the initial data. Note that each pattern is a conjunction of attributes or their negations. As the properties of a pattern, we can note: coverage and degree. Coverage is the number of satisfied its conditions observations of its class. Degree is the number of variables that it contains.

From the obtained patterns, it is necessary to build a classifier, on the basis of which a decision about the belonging of a new observation (observation of a test sample) to a particular class is made.

For clarity, we will consider the problem of binary classification, the initial data sample in which consists of disjoint sets $\Psi+$ (a set of observations of a positive class) and $\Psi-$ (a set of observations of a negative class) of n-dimensional vectors. For real problems, vector elements can take different types of values (binary, nominal, quantitative).

At the stage of constructing a classifier there is a problem related to the selection of informative patterns in the classifier. Informativity is the quality of the formed logical rule. Note that the larger the coverage of observations of the class on the basis of which the pattern is built, and the smaller the coverage of observations of another class, the higher its informativity.

R. Silhavy et al. (Eds.): CoMeSySo 2021, LNNS 232, pp. 163–170, 2021.
https://doi.org/10.1007/978-3-030-90318-3_16

Thus, at the stage of selecting patterns in the classifier, it is necessary to remove patterns that are unsuitable for classification. Literature analysis has shown that there are two approaches to the selection of patterns in the classifier [1–5]. The first approach is associated with hard constrains at the stage of building patterns, which initially reduces the number of patterns in the set. Its disadvantage is a decrease in the generalizing abilities of the classifier (retraining). The second approach is based on the formation of a set of various patterns for the classifier and its subsequent reduction based on the proposed concepts. For example, in [1], the concept of obtaining a minimum set of rules covering all observations of the training sample is proposed. In practical use, the disadvantage of the concept is the increase in the number of uncovered observations of the test sample, which is associated with an insufficient number of rules in the set.

In this paper, we propose a concept of classifier reduction based on the application of the ε-, δ-criterion for assessing and selecting patterns in a new classifier. Using the ε-, δ-criterion for the reduction (reduction of the number of patterns) of the classifier, we can increase its interpretability (reduce the number of rules in it) and generalization (reduce the number of errors on the test sample).

2 Materials and Methods

Let introduce φ: $Z \rightarrow \{0, 1\}$. It is a predicate defined on the set of observations Z. If the condition $\varphi(z) = 1$ is satisfied, then the predicate φ covers the observation z. If a predicate covers a sufficient number of observations of one class and practically does not cover observations of other classes, then it can be called a pattern [6].

Each constructed pattern classifies only a certain subset of observations from the set Z. Therefore, it is necessary to combine a certain number of patterns into a composition in order to obtain a classifier capable of classifying any observations from the set Z.

A class is called its own class if a pattern is built on the basis of its observations. Observations of one class are also called positive and another – negative.

To determine the ε-, δ-patterns, the following notation is introduced [6]:

P_k is the number of observations of a positive class k in the sample Z^ℓ, where $P_k > 1$, $P_k + N_k = \ell$;

$p_k(\varphi)$ is the number of observations for which the condition $\varphi(z) = 1$ is satisfied;

N_k is the number of observations of all other classes in the sample Z^ℓ, where $N_k > 1$;

$n_k(\varphi)$ is the number of observations for which the condition $\varphi(z) = 1$ is satisfied.

According to the indicated notations, the task of constructing an informative pattern consists in optimization according to two criteria: $p_k(\varphi) \rightarrow \max$ and $n_k(\varphi) \rightarrow \min$. Two types of patterns are unsuitable for classification. The patterns of the first type cover too few observations, the patterns of the second type cover positive and negative observations approximately in the proportion in which they are presented in the training sample.

Let introduce the notation E_k for the fraction of negative observations among all covered observations and D_k for the fraction of covered positive observations:

$$E_k\left(\varphi, Z^{\updownarrow}\right) = \frac{n_k(\varphi)}{p_k(\varphi) + n_k(\varphi)};$$

$$D_k\left(\varphi, Z^{\updownarrow}\right) = \frac{p_k(\varphi)}{p_k(\varphi) + n_k(\varphi)}.$$

Based on the introduced notation, we define the concept of logical ε-, δ-patterns for the class k. If $E_k(\varphi, Z^\ell) \leq \varepsilon$ and $D_k(\varphi, Z^\ell) \geq \delta$ for given sufficiently small ε and sufficiently large δ from the interval [0, 1], then the pattern φ is called a logical ε-, δ-pattern.

Before starting to work with attributes, it is necessary to binarize them since they can take on different types of values in the initial sample [7]. The resulting set of binary attributes can be large, therefore, to reduce the complexity of the problem being solved, it is necessary to choose a subset of attributes S capable to separate the sets of observations of positive and negative classes with high accuracy. The projections Ψ_S^+ and Ψ_S^- of the sets Ψ^+ and Ψ^- on S are used for the further work of the method [1].

At the next stage of the method, it is necessary to bring optimization models that make it possible to form patterns. Let us present optimization models using the example of positive patterns. Similar reasoning can be applied to negative patterns.

A positive α-pattern for $\alpha \in \{0,1\}t$ is a pattern built on the basis of the α observation. The task is to form for each observation $\alpha \in \Psi_S^+$ the maximum α-pattern, that is, capturing the maximum number of observations of a positive class. For this purpose, we define an α-pattern using binary variables $X = (x_1, x_2, \ldots, x_t)$:

$$x_k = \begin{cases} 1, \text{if the } k - \text{th attribute is in the rule;} \\ \quad 0, \text{ otherwise.} \end{cases}$$

For pure patterns, $n_k(\varphi) = 0$. Therefore, for each observation of negative class $\beta \in \Psi_S^-$, the variable xk must be equal to 1, at least for one index k for which $\beta_k \neq \alpha_k$:

$$\sum_{\substack{k=1 \\ \beta_k \neq \alpha_k}}^{t} x_k \geq 1 \text{for any } \beta \in \Psi_S^-.$$

In order to increase fault tolerance in the constraint function, you can change the number 1 to another positive integer d.

On the other hand, the α-pattern covers the observation of a positive class $\sigma \in \Psi_S^+$ when the value of the variable xk is 0 only for indices k that are not fixed in the pattern. Hence, the objective function of the optimization model is the total number of observations of a positive class captured by the α-pattern, and can be calculated as:

$$\sum_{\sigma \in \Psi_S^+} \prod_{\substack{j=1 \\ \sigma_k \neq \alpha_k}}^{t} (1 - x_k).$$

By combining the objective function and the constraint function, we obtain an optimization model for finding patterns [8]:

$$\sum_{\sigma \in \Psi_S^+} \prod_{\substack{j=1 \\ \sigma_k \neq \alpha_k}}^{t} (1 - x_k) \to \max \tag{1}$$

$$\sum_{\substack{k=1\\ \beta_k \neq \alpha_k}}^{t} x_k \geq d \text{ for any } \beta \in \Psi_S^-, x \in \{0, 1\}^t. \tag{2}$$

Based on this optimization model (1–2), we will obtain pure patterns ($n_k(\varphi) = 0$).

In real classification problems, the original data samples are often incomplete and inaccurate (e.g., medical or economic problems). Consequently, for such problems, partial patterns ($n_k(\varphi) > 0$) are better suited, the degree of which, as a rule, is less, and the coverage is greater than that of pure patterns [6].

Limiting function of the optimization model for the formation of partial patterns can be represented as follows:

$$\sum_{\sigma \in \Psi_S^-} \eta_\beta \leq D, \tag{3}$$

$$\text{where } \eta_\beta = \begin{cases} 0, \text{ if } \sum\limits_{\substack{j=1\\ \beta_k \neq \alpha_k}}^{t} x_k \geq d, \\ 1, \text{ otherwise;} \end{cases}$$

where D is the number of observations of the opposite class that can be covered by the pattern (positive integer).

Using model (1, 3), we obtain patterns with a small degree and maximum coverage. Logical rules with a small degree correspond to large areas in the attribute space, therefore, they capture more observations of the opposite class of the training sample ($n_k(\varphi)$ increases). An increase of $n_k(\varphi)$ negatively affects the value of the informativity of the rule being formed. On this basis, an algorithmic procedure for building patterns is proposed, which makes it possible to increase the value of informativity. Its essence is the maximum increase in the degree of already constructed patterns while maintaining the number of covered observations of its class [9]:

$$\sum_{k=1}^{t} x_k \to \max,$$

$$fc(X) = fc'(X)$$

where fc(X) is the coverage of observations of its class for a logical rule before the increasing procedure, fc'(X) is the coverage of observations of its class for a logical rule after the increasing procedure.

The result of the pattern increasing procedure is a set of logical rules with maximum coverage and a degree higher than the initial one. The generalizing ability of a classifier composed of such a set of informative logical rules should be improved. We emphasize that an increase in the informativity of the rules being formed is associated with a decrease in the value of the indicator $n_k(\varphi)$.

After building the initial classifier, which consists of all the patterns built on the basis of each observation of the training sample, we proceed to the algorithmic procedure for its reduction. The procedure for reducing of the classifier is needed in the case of a large volume of the training sample in order to build an interpreted classifier.

Algorithmic procedure for reducing the classifier is proposed to be carried out on the basis of the ε-, δ-criterion. First, it is proposed to fix the parameter ε at a certain value (in this work, 0.2). The constructed patterns, for which $E_k > 1/5$, will be removed from the classifier. Second, it is proposed to form a classifier by removing patterns, the value of the parameter D_k for which is lower than that of the remaining logical rules in the set. For this, the following algorithmic procedure for reducing the classifier is proposed:

1. Determine the values of the parameter D_k for each pattern in the original classifier;
2. Rank the patterns by the value of the parameter D_k in decreasing order for each class.
3. Remove patterns with a minimum value of the parameter D_k from the set of each class, leaving 87.5%, 75%, 67.5% and 50% of patterns from the initial number in the set and forming new classifiers from them.
4. Check the obtained classifiers on the test set of observations.

The criterion for identifying the optimal set of patterns from the presented ones can be: an increase in the number of uncovered observations of the test set (it indicates the lack of logical rules in the classifier) or a sharp decrease in the classification accuracy on the test set.

The main idea of this paper is to determine the appropriateness of using the ε-, δ-criterion for the reduction of the original classifier.

3 Results

An experimental study of the proposed algorithmic procedure for reducing the classifier is carried out on the problem of the results of radar scans of the ionosphere and the problem of SPAM detection by email [10].

To find the rules, the optimization model (1, 3) was used. Subsequently, for each pattern, an algorithmic increasing procedure was applied. Optimization model (1, 3) is a conditional pseudo-Boolean optimization problem, the solution of which is necessary in the formation of each pattern. Taking into account that the functions of the optimization model (1, 3) are monotonic, specially developed optimization algorithms are proposed [11].

In all the presented practical problems, 20% of the sample was used to test the classifier, and 80% of the sample was used as a training sample.

When implementing the algorithmic procedure for reducing the classifier, studies were carried out on five sets of patterns. The first set of patterns is the original (without reduction), the second set is 87.5% of patterns from the initial set (reduction 12.5%), the third set is 75% of patterns from the original set (reduction 25%), the fourth set is 62.5% of patterns from the initial set (reduction 37.5%), the fifth set is 50% of patterns from the initial set (reduction 50%).

The classification results for the problems under consideration are presented below.

3.1 Classification of the Results of Radar Scans of the Ionosphere

Classification of the results of radar scans of the ionosphere. The initial data sample consists of 225 observations of a positive class (detection of a structure in the ionosphere) and 126 observations of a negative class (lack of structure). The results for all sets of patterns are presented in Table 1.

Table 1. Accuracy to each set of patterns in solving the problem of classification of the results of radar scans of the ionosphere the results of radar scans of the ionosphere.

Set number	Set of rules	Number of rules	Coverage of negative observations	Coverage of positive observations	Average degree of rule	Average share of coverage obs. D_k	Classification accuracy, %	Total accuracy,%
1	Neg	95	33	7	3	0.82	84	91
	Pos	186	4	100	6	0.96	95	
2	Neg	83	35	7	3	0.84	84	91
	Pos	163	4	104	6	0.97	95	
3	Neg	71	36	7	3	0.85	84	91
	Pos	140	4	104	6	0.97	95	
4	Neg	59	37	6	3	0.86	84	91
	Pos	116	3	104	6	0.97	95	
5	Neg	48	35	5	3	0.87	84	91
	Pos	93	3	102	6	0.97	95	

According to the obtained results (Table 1), it can be noted that when patterns are removed from the initial set, the average coverage decreases by observation rules of the opposite class, which indicates an increase in the informativity of the set of patterns for the classifier. This allows maintaining high classification accuracy while removing rules from the original classifier. The absence of unclassified observations for all sets allows us to consider the set of patterns No 5 as optimal (the minimum number of patterns).

3.2 SPAM Detection by Email

The original data sample consists of 181 observations of the positive class (SPAM) and 279 observations of the negative class (non-SPAM). The results for all sets of patterns are presented in Table 2.

According to the results (Table 2), it can be noted that when the patterns are removed from the set, the average value of the index $p_k(\varphi)$ for the set of patterns in each class increases, and the average degree of logical rules also slightly decreases. As a result, this leads to an increase in the informativity of a set of patterns for the classifier, as well as to an improvement in its generalizing ability. The absence of unclassified observations for all sets allows us to consider the set of patterns No 5 as optimal (the minimum number of patterns).

Table 2. Accuracy to each set of patterns when solving the problem of SPAM detection by email.

Set number	Set of rules	Number of rules	Coverage of negative observations	Coverage of positive observations	Average degree of rule	Average share of coverage obs. D_k	Classification accuracy, %	Total accuracy,%
1	Neg	234	99	5	5	0.95	98	85
	Pos	134	4	49	4	0.92	72	
2	Neg	205	104	5	5	0.96	98	85
	Pos	117	4	54	4	0.93	72	
3	Neg	176	107	5	5	0.96	98	85
	Pos	101	4	56	4	0.94	72	
4	Neg	146	110	5	5	0.96	98	85
	Pos	67	4	59	3	0.95	72	
5	Neg	117	112	5	5	0.96	98	85
	Pos	67	3	60	3	0.95	72	

4 Conclusions

According to the obtained results, it can be noted that the idea of using the ε-, δ-criterion for the reduction of the original classifier is reasonable. As a result of the implementation of the algorithmic procedure for the reduction of the classifier based on the ε-, δ-criterion, we obtain an interpreted classifier with a high recognition accuracy.

According to the Tables 1–2, the high classification accuracy during its reduction is primarily due to a decrease in the average value of the indicator $n_k(\varphi)$, as well as an increase in the average value of the indicator $p_k(\varphi)$ for a set of patterns of each class.

An increase in the interpretability of the classifier occurs due to its reduction, the stopping criterion of which is an increase in the number of uncovered observations of the test set.

Acknowledgements. The research was supported financially by a grant from the President of Russian Federation (Project No. MK-1376.2020.9).

References

1. Hammer, P.L., Bonates, T.: Logical analysis of data: from combinatorial optimization to medical applications. RUTCOR Res. Rep. **10**, 1–27 (2005)
2. Weinzweig, M.N.: The learning algorithm for pattern recognition “kora”. Learn. Pattern Recog. Algor. 110–116 (1973)
3. Bonates, T., Hammer, P.L., Kogan, A.: Maximum patterns in datasets. RUTCOR Res. Rep. **9**, 1–18 (2006)
4. Boros, E., Hammer, P.L., Kogan, A., Crama, Y., Ibaraki, T., Makino, K.: Logical analysis of data: classification with justification. RUTCOR Tech. Rep. **5**, 1–34 (2009)
5. Antamoshkin, A.N., Masich, I.S.: The choice of logical patterns for constructing the decisive recognition rule. Vestnik SibSAU **5**(57), 20–25 (2014)

6. Vorontsov, K.V.: Lectures on logical classification algorithms (2007). http://www.ccas.ru/voron/download/LogicAlgs.pdf
7. Rastrigin, L.A., Freimanis, E.E.: Solving multi-scale optimization problems by random search methods. Prob. Rand. Search **11**, 9–25 (1988)
8. Masich, I.S., Kraeva, E.M., Kuzmich, R.I., Gulakova, T.K.: Comparative analysis of data classification methods on practical problems of forecasting and diagnostics. Control Syst. Inf. Technol. **1**(43), 20–25 (2011)
9. Kuzmich, R.I., Masich, I.S., Stupina, A.A.: Patterns of formation of patterns in the method of logical analysis of data. Control Syst. Inf. Technol. **1**(67), 33–37 (2017)
10. UCI Machine Learning Repository. http://archive.ics.uci.edu/ml/index.html
11. Antamoshkin, A. N., Masich, I.S.: Pseudo-Boolean optimization in case of unconnected feasible sets. In: Models and Algorithms for Global Optimization. Series: Springer Optimization and Its Applications vol. 4, pp. 111–122 (2007). https://doi.org/10.1007/978-0-387-36721-7_7

Information Security Audit Method Based on the Use of a Neuro-Fuzzy System

V. Lakhno[1], B. Akhmetov[2], V. Chubaievskyi[3], Alona Desiatko[3](✉), K. Palaguta[3], A. Blozva[1], and Y. Chasnovskyi[1]

[1] Department of Computer Systems and Networks, National University of Life and Environmental Sciences of Ukraine, Kyiv, Ukraine
andriy.blozva@nubip.edu.ua

[2] Caspian University of Technology and Engineering named after Sh.Yessenov, 32 microregion, Aktau, Republic of Kazakhstan

[3] Department of Software Engineering and Cybersecurity, Kyiv National University of Trade and Economics, Kyiv, Ukraine
{chubaievskyi_vi,palagutae}@knute.edu.ua

Abstract. The article proposes measures to improve information security audit procedures (ISA) for various objects of informatization (OBI). It is shown that it is advisable to assess the level of information security (IS) for OBI based on evaluating the effectiveness of a set of criteria for the method of analysis of hierarchies (MAH). At the same time, such an assessment of the degree of IS and all associated ISA audit procedures are most effective for a multilateral assessment of IS OBI. Both standard numerical information security metrics and metrics proposed by information security experts and agreed with the OBI management can be used as assessment metrics.

A modified method for the analysis of hierarchies is proposed, based on the application of the apparatus of the theory of fuzzy sets and neural networks. This method enables management to make informed management decisions in the field of information security OBI. The solutions obtained are aimed at improving not only the OBI IS itself, but also ultimately optimize the OBI control system, reduce costs and increase the efficiency of OBI business processes as a whole. It is shown that the use of the mathematical apparatus of the Moscow Aviation Institute and the corresponding software, in particular, the developed intellectual system, makes it possible to increase the degree of reliability of the results of a comprehensive audit of IS OBI. Moreover, this statement is true for the procedures of the internal ISA OBI and the external ISA.

Keywords: Audit of information security · The object of informatization · Method of analysis of hierarchies · Intelligent information system · Neural network · Fuzzy logic

1 Introduction

The dynamics of the increase in the number and complexity of cyberattacks on various objects of informatization (OBI) only in the last few years [1, 2] shows that, despite all

R. Silhavy et al. (Eds.): CoMeSySo 2021, LNNS 232, pp. 171–184, 2021.
https://doi.org/10.1007/978-3-030-90318-3_17

the efforts of the defense side, to oppose the attackers with more and more technically advanced hardware and software information security (IS) and to this day, the problem of obtaining current and forecast estimates of the level of information security OBI does not lose its relevance. This task has become especially relevant for the objects of critical infrastructure (CI) of the state [3]. After all, unauthorized interference in computer systems (CS) of which can cause disruptions in business processes and affect the safety of people. For example, even a short-term failure of a compressor station, in particular, of utility companies, can cause interruptions in the supply of electricity, water, interruptions in supply to retail networks, etc.

The more complex the OBI structure and the more complex the information technologies (IT) used on OBI, the more difficult it is to build for it an information security management system (ISMS) that meets modern requirements. Any important object of informatization a priori suggests the need to have a modern management system, in particular, in matters related to information security. Systems of this kind have become an integral part of management systems (MS) today. And the integration of such MSs into the tasks of ensuring the OBI information security implies the need to build a system for conducting a periodic audit of the OBI security level. This, in particular, is possible using expert or computer obtaining estimates (metrics) of information security in the ISMS.

To build an effective ISMS, it is necessary to adhere to the following algorithm of actions. First, the design of the OBI ISMS should be based on risk-based standards. Secondly, the formation of requirements for OBI business processes from the point of view of ensuring information security should be carried out based on the use of clear measurable information security metrics. Thirdly, the task of an OBI IS audit should be considered comprehensively, and in such a formulation of this task, it is difficult to do without computer support for decision-making.

All of the above and determines the need for new research in the tasks of increasing the efficiency of information security audit procedures (AIS) based on the integrated application of intelligent systems.

2 Review and Analysis of the Literature

Quite a lot of works have been devoted to the problem of conducting an IS audit for various OBIs. In this series, we can mention the work that has become classic for solving the problems of ISA [4, 5].

The work [6] shows the importance of analyzing information flows for the correct conduct of audit procedures in OBI information systems. But the author does not consider the potential of using intelligent systems to improve the quality of information security audit procedures.

The work [7] analyzes the relationship between the procedures for internal audit of information security and external audit. However, the authors did not take into account the constant development of IS protection systems. Note that the introduction of new, more advanced information security systems into the IS contours can change the list of basic IS metrics adopted in the organization.

In [8], the features of conducting an IS audit for "zero-day" threats are analyzed. In particular, the authors noted that providers of information security tools, as a rule, can only offer an option for continuous development and improvement of technical means of information security (MIS).

In [9, 10], the authors also noted that, although continuous improvement of the MIS is necessary, it is "beneficial" mainly for ESS manufacturers. And only one improvement of MIS is not able to independently solve the problem of constant confrontation between MIS and IS threats. Moreover, as shown in [11, 12], if the defense side encounters a targeted (targeted) attack, then it would be a mistake to rely only on the MIS.

In this regard, many experts [8, 13, 14] note the need to apply not only technical approaches (use of MIS) to counter cyber threats but the introduction of combined methods. The authors did not disclose in detail the concept of combined methods, but they mention the need to build them based on the ISO 27001 and 19011 families of standards [15–19].

The scientific problem that this study touches can be formulated as follows. It is necessary to overcome the contradiction between the state of the theory of information security, in the part that regulates the requirements for conducting audits of information security OBI and the established practices of ensuring information security OBI that depend on the landscape of cyber threats.

The solution to this problem, in particular, presupposes the need to revise the existing static models of information security management. The need to improve the ISA system should be mentioned as one of the subtasks. Consequently, the decision-making procedure by a decision-maker with a fixed set of alternatives implies the need to create new approaches to IS audit procedures.

Note that the methodology for conducting an IS audit is well known and worked out by specialists, but the issues related to the introduction of intelligent decision support systems (IDSS) into audit procedures are not yet fully developed. At the same time, the increasing requirements for the quality of IS audit procedures dictate the need to use the potential of IDSS in the course of a prompt response to identified threats in information systems (IS). And this makes the task of increasing the degree of security, as well as obtaining current and forecast estimates of IS OBI relevant.

3 Statement of the Problem

The solution to the problem of increasing the degree of security, as well as obtaining the current and predictive assessment of IS OBI is most expedient by using accurate numerical estimates - IS metrics [4, 5].

This is consistent with the main provisions of the "basic" standard of the information security management system 27004: 2009 [5]. The following information can be used as data sources during the implementation of IS audit procedures (both internal and external) [6]:

- results of analysis and assessment of risks for IS OBI;
- reports of previous ISA procedures;
- logs of registration of information security incidents
- reports of intrusion detection systems and/or such software category as Security information and event management (SIEM);

- personnel reports on information security incidents;
- the results obtained in the course of testing the functional subsystems of the OBI CS;
- results obtained in the course of IS training for OBI personnel;
- and etc.
- Thus, we will formulate the following formulation of the research tasks:
- development of the ISMS method for auditing IS OBI and obtaining numerical current and/or predictive estimates of the degree of its security in a dynamic confrontation with the attacker);
- development and testing of an intelligent decision support system aimed at increasing the degree of IS with the ability to synthesize a numerical assessment of the effectiveness of an OBI IS audit.

4 Methods and Models

When developing an IS audit program (hereinafter PISA), not all links between ISA certificates can be taken into account in a particular situation. This is primarily due to the lack of necessary information.

When conducting an IS audit of large companies or enterprises, it is quite difficult to fully consider the audit object in its entirety. It is more advisable for auditors to choose the most important informative audit evidence (hereinafter EA) or information security metrics. These selected metrics and EAs will be of great importance and, at the same time, the cost of obtaining them will be low.

Usually, to build a model of the ISA object (hereinafter referred to as OISA), it is advisable to use the weighting factors of the significance of the audit evidence.

As the real practice of conducting IS audits has shown, taking into account the significance of the EA is a difficult task. At the same time, an important factor is the experience of the auditor, and, first of all, the person responsible for drawing up the ISA program and the systematic analysis of the results obtained in the audit process. Incorrect formulation of the initial tasks of the ISA can nullify the main goal of the OBI carried out by the ISA or give unreliable results. All of the above determines the effectiveness of the combination of expert and mathematical methods of processing the received expert assessments. As the analysis of literature sources [10, 20] has shown, the following methods can be used to solve the above problem: pairwise comparisons; ball marks; vectors of preferences; analysis of hierarchies (MAH), etc. A sufficiently detailed analysis of the effectiveness of the application of these methods is presented in [20].

Considering that the selection of IS metrics for each OBI has its characteristics, dictated both by the OBI industry and by the degree of its criticality, we will further formalize the typical problem of selecting ISA metrics. In this case, it is proposed to be guided by the following algorithm, see Fig. 1.

In the modified algorithm for conducting an information security audit, in comparison with the basic procedures, it is necessary to take into account the following new circumstances:

1. You should focus, first of all, on the priority information security metrics selected by the auditors. These metrics selected by the auditors might not have been met before during the previous cycle of the OBI IS audit procedure;
2. The emerging deviations in the quality characteristics of the selected IS metric should be monitored.
3. Rules for the selection and replacement of previously selected IS metrics should be formed, for example, guided by the results of previous IS audits, or in connection with a change in the threat landscape for OBI information systems.

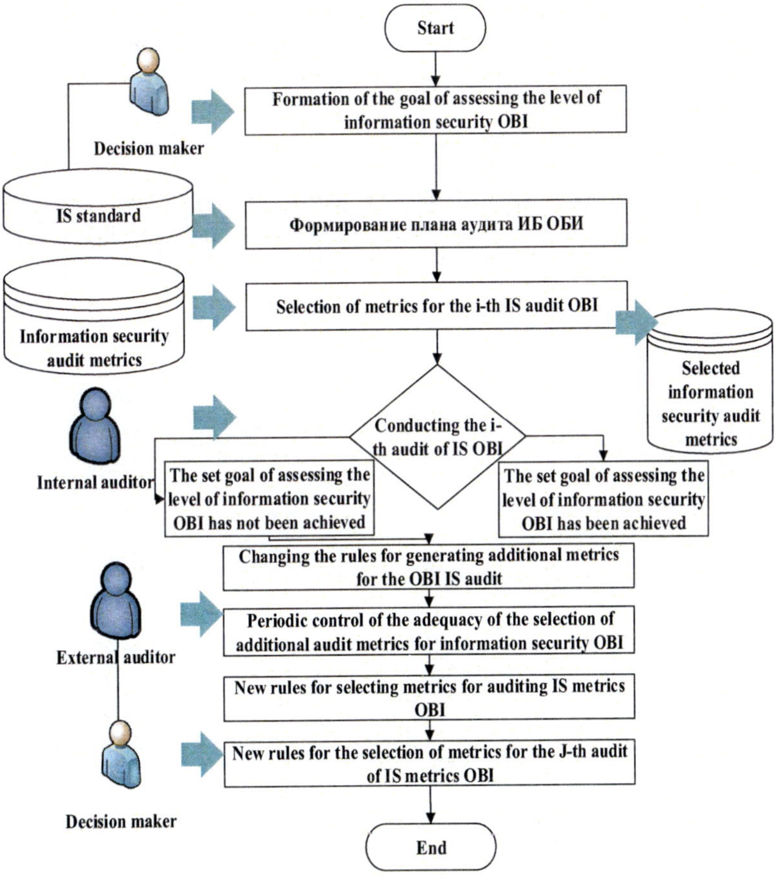

Fig. 1. Block diagram of the algorithm for selecting individual criteria

(information security metrics)

Building a hierarchy of information security metrics during the audit is as follows:
Stage 1. Determine the sub-goals of the ISA. These can be private tasks, for example:

- assess the availability, confidentiality, integrity of information in the IS;
- justify the set of EAs that are included in the ISA program (PISA);
- withdraw appropriate audit evidence that is required by the PISA. ISA certificates are usually calculated by information security analysts.
- Stage 2. We select the factors that are important for the 2nd level of the PISA hierarchy. At this stage:
- collect objective evidence from the ISA. These include, for example, facts that are important from the point of view of the audit criteria;
- focus the forces and means of the ISA. For example, this includes audit teams, individual information security auditors, etc.

Note that for each OBI, the IS audit task has its specifics, which are determined by the degree of criticality of information processes in the organization's business processes.

Taking into account the publications [23], modify the MAH through the additional application of the following steps:

Step 1. We estimate the stability of local rankings based on the vectors of changes in the elements of matrices of pairwise comparisons;

Step 2. Find expert assessments of pairwise comparisons, which to the greatest extent affect the change in the local rankings of decision alternatives and the change in the levels of consistency of the set of assessments;

Step 3. Evaluate the sensitivity of the global ranking of decision alternatives to changes in the weights of the hierarchy elements;

Step 4. Find the most sensitive and stable elements for each level of the hierarchy.

In fact, MAH is necessary at the stage of calculating intermediate indicators and determining the final rank of objects. This is similar to the procedures for assigning membership functions of fuzzy sets used to describe objects of observation and requirements for them.

In MAH, to find the rank of an object, they use the formula:

$$p_i = \sum\nolimits_{j=1}^{n} g_j \cdot v_{ij}, \tag{1}$$

where: $n-$ the number of criteria (IS metrics), g_j- an indicator of the criterion importance (IS metrics), $v_{ij}-$ an indicator of the preference of the $i-$ th object according to the $j-$ th criterion.

It is not difficult to notice a certain similarity with the simplest model of a unidirectional neural network. In the theory of neural networks, the following formula is used for this [24]:

$$y = f\left(\sum\nolimits_{i=0}^{N} w_i \cdot u_i\right), \tag{2}$$

where $N-$ is the number of inputs of the neuron, w_0 = the threshold value, $u_0 = -1$, u_i- the input signal of the $i-$ neuron, w_i- the synaptic weights of the inputs, $y-$ the output signal of the neuron.

The similarity of these formulas is obvious. In this case, the indicator of the object's preference according to various IS metrics v_{ij} is put following the synaptic weights of the neuron inputs w_i. Then, the process of calculating the parameters of the object's preferences for various requirements performs the functions of training neurons.

Taking into account the specifics of the task of conducting an IS audit, such changes are proposed for modifying the MAH. This allows not only to take into account the specifics of the subject area of the ISA OBI, but also the subsequent adaptation of the proposed changes for the practical implementation of intelligent DSS using object-oriented design methods.

First, we introduce restrictions on the original data. This is due to the peculiarities of the organization of the ISA OBI procedure. We will divide the set of criteria for assessing the degree of cybersecurity of OBI into two parts. These are, respectively, general and individual criteria (information security metrics).

General criteria are the criteria that apply to any OBI when conducting ISA. This set does not depend on the purpose and functionality that are implemented by the OBI information systems. We believe that many of the general criteria, for example, reliability, cost, are limited and constant.

At the same time, each ISA OBI procedure should take into account its specifics. We take into account these individual criteria (information security metrics) in a separate set. Individual criteria are extremely important for a particular OBI. If at least one of the criteria is not met, then the OBI protection state does not meet the required IS level.

Second, the following assumption was made in the process of modifying the MAH. A comparative assessment of the importance of common criteria can be performed using classical peer review. In this situation, there is no need to resort to using paired comparisons of the MAH. This becomes possible because the number of general criteria for ensuring IS OBI is relatively small. As practice has shown, for most IS practices, the selected four to five criteria (IS metrics) become decisive.

Third, we will make adjustments to the algorithm for calculating the synaptic weights of the neuron inputs. The values of the synaptic weights of the inputs of neurons, which correspond to each compared object, are calculated using a system of fuzzy rules such as IF-THEN. The rules are based on the application of the Takagi-Sugeno method. As the initial data of such a system, we use the OBI IB criteria, which correspond to a given neuron. And besides, we take into account the individual criteria of IS OBI.

The following initial data were accepted for the modified MAH, which can be used in the OBI IS audit procedures:

1. a set $\{Y_i\}$, $i \in [1, n]$, that contains expert assessments of the importance of each of the IS metrics, the number of criteria (IS metrics);
2. a set $\{Z_j\}$, $j \in [1, m]$, that contains individual criteria (IS metrics) OBI, the number of individual criteria.

Performing a comparative assessment of the importance of common criteria involves the following steps:

Stage 1. Guided by the required level of IS OBI expert, it is important for all general IS criteria in the form of a set $\{Y_i\}$, $i \in [1, n]$,, for example, for$Y \in [1, 0]$, here "0" corresponds to a situation where there are no IS requirements for the audited object, and "10" is the maximum importance of the criterion (by analogy with MAH T. Saati);

Stage 2. We transform a set $\{Y_i\}$ into a set $\{u_i\}$. The transformation is implemented by normalizing the elements $\{Y_i\}$ to an interval [0, 1] :

$$u_i = \frac{Y_i}{\sum_{j=1}^{n} Y_j}. \tag{3}$$

The resulting set $\{u_i\}$ will contain comparative indicators of the importance of general IS criteria, which are analyzed during the ISA OBI.

The neural network (NN) used in calculating the ranks of audit objects will contain the number of neurons h, which is equal to the number of objects l potentially acceptable in the information security and cybersecurity circuits of OBI. Each of the neurons has several inputs equal to the number of common requirements n. At the output of neurons, a value will be formed, which will determine the rank of the corresponding audit object.

The base of fuzzy rules (BFR) for the selection of individual metrics of the IS audit, and allowing to calculate the synaptic weights of the inputs of each of the neurons, will include rules of the form:

$$\left\{R^k\right\} : \textit{IF } (x_k \textit{ this } A_k) \textit{ THEN } w_k = c_k, \tag{4}$$

$$w_i = \frac{\sum_{k=1}^{K} \mu_{A_k} \cdot (x_k) \cdot w_k}{\sum_{k=1}^{K} \mu_{A_k} \cdot (x_k)} \cdot \prod_{j=1}^{m} \mu_{z_j} \cdot (z_j), \tag{5}$$

where: $\left\{R^k\right\}$, $k \in [1, K]$– BFR, which contains fuzzy rules;

c_k, $k \in [1, K]$– a constant that depends on the specific rule $c_k \in (0, 10]$;,

$A_k = \{x_k, \mu_{Ak}(x_k)\}$, $k \in [1, K]$ - fuzzy sets, which are specified by membership functions $\mu_{Ak}(x_k)$ on the set of possible values of the characteristics of an IS audit object, and corresponding to general IS metrics;

x_k, $k \in [1, K]$– the values of the variables that characterize the properties of the IS audit object and meet the implementation of the general criteria for IS OBI;

$z = \left\{z_j, \mu_{zj}\left(z_j\right)\right\}$, $j \in [1, m]$ - the classical set, set by membership functions $\mu_{Zj}\left(z_j\right)$ equal to 0 or 1. This set describes the values of the properties of the IS audit object, which are responsible for the implementation of individual criteria (IS metrics);

z_j, $j \in [1, m]$– variables that characterize the properties of the IS audit object, following individual criteria;

i, $i \in [1, n]$– input number of the corresponding neuron.

The functions located in the part with the rule operator *THEN* are defined as constants. Then, these functions will take maximum values in cases where the properties of the IS audit objects correspond to fuzzy sets. Such fuzzy sets include, for example, the following: information on the effectiveness of "instant audits" of information security; information on the effectiveness of all types of audits; information about information security incidents; information about new preferences in the information security policy of the decision-maker, etc. In the course of the research, it was found that the use of only five, six terms for the linguistic assessment of the properties of the IS audit object will allow the object to be evaluated with a sufficiently large degree. At the same time, we keep the simplicity and clarity of the model, which improves the classic MAH.

We set the set of individual ISA criteria Z on the set of group and individual properties of all objects that are included in the OBI IB contours. The membership function $\mu_{zj}(z_j)$ of the set Z will be equal to one for those properties of ISA objects that ensure the implementation of individual criteria. Accordingly, the zero value will be in the case of all other properties.

The sets z of individual criteria that are assigned to individual ISA objects are subsets of the set Z. In the set that corresponds to a specific object of the IS audit, we include only those individual ISA criteria that can be presented to this object of the OBI IS contours.

In our opinion, in comparison with the classical MAH by T. Saaty, the use of such a fuzzy neural system in IS audit procedures has several advantages. First, it will speed up and simplify the computation of synaptic weights. Secondly, it will be possible to fairly accurately take into account the individual criteria (IS metrics) characteristic of various objects of IS audit. Thirdly, a fuzzy system will allow to take into account and evaluate quantities of a quantitative and qualitative nature for various metrics. This primarily concerns metrics that are described as numerical information security metrics.

As a result, the modified MAH can be conceptually implemented in the form of the following neuro-fuzzy system, see Fig. 2. Such a scheme implies the unification of a neural network, in which comparisons of objects of IS contours are carried out, as well as a fuzzy system, which is based on the application of the base of fuzzy rules described above.

The fuzzy system, following the developed scheme, will calculate the synaptic weights of the w_i neuron inputs. At the same time, individual criteria (see Table 1) selected for the IS audit procedures of a particular OBI are also taken into account. At the input of the neuro-fuzzy system, expert assessments of the importance of the criteria for a specific OBI will be submitted, and at the output, the ranks $p_1, \ldots, p_n$ of the objects of the IS contours will be read.

5 Software Implementation of Intelligent DSS for Information Security Audit Based on the Use of a Neuro-Fuzzy System

The described modified MAH is implemented in the Visual Studio 2019 system (C# programming language) in the form of an intelligent DSS to improve the efficiency of IS audit procedures, see Fig. 3. Intelligent DSS ensures the input of individual criteria and expert assessments of the importance of general criteria, which are checked during the audit of IS OBI. The results of object comparison are presented in the form of a graph (see Fig. 4).

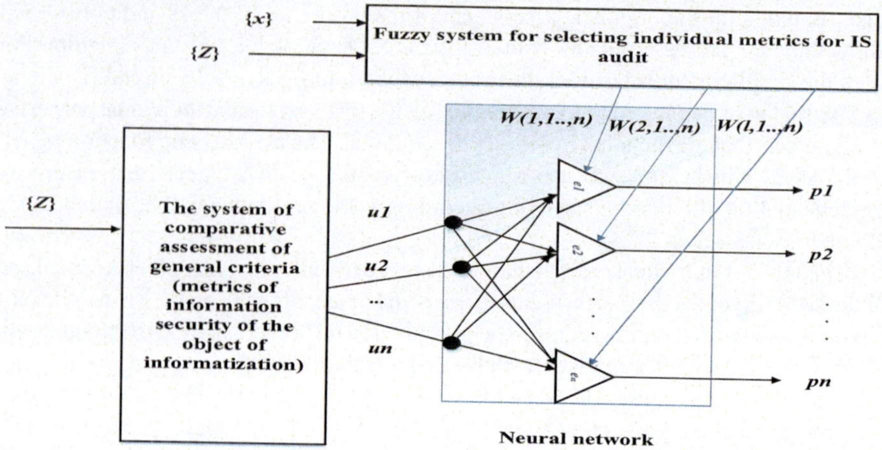

Fig. 2. The conceptual structure of a neuro-fuzzy system for a modified hierarchy analysis method

Table 1. An example of the formation of general and individual metrics when conducting an IS audit

General information security metrics	
1	Metrics characterizing hosts and their connectivity
2	Percentage of critical applications
3	Average time to fix a vulnerability
	…
N	Total Gain and Expected Annual Loss
Individual IS metrics (selected IS audits for a specific OBI)	
1	A probabilistic measure of vulnerability, indicating how likely a zero-day vulnerability is to occur over a specified period of time
2	Ensuring the maximum completeness of the list of information assets, in terms of additional information about IS threats
3	Determination of the degree of implementation of measures (means) to ensure information security
	…
M	Specified business risk

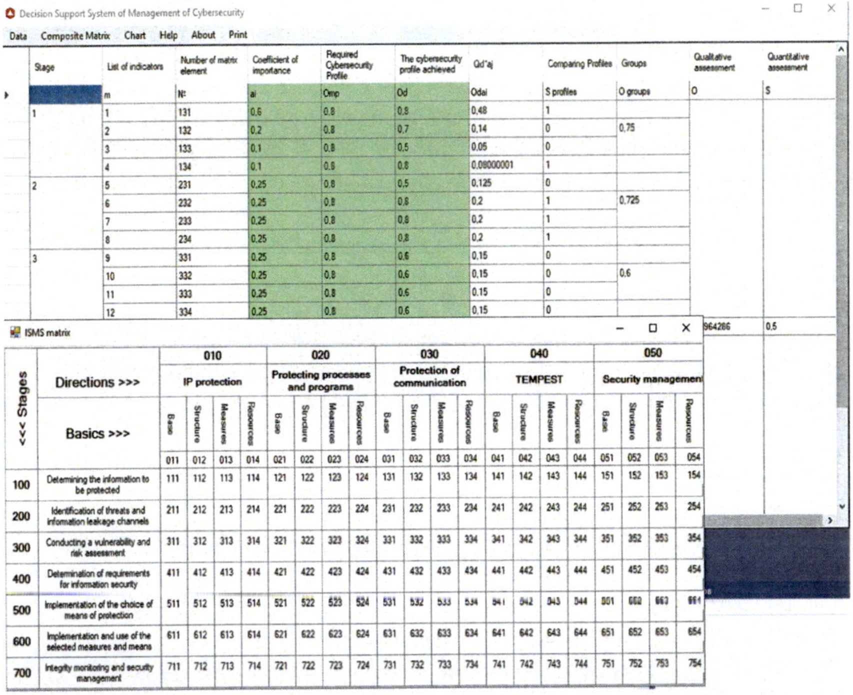

Fig. 3. General view of an intelligent DSS for information security audit based on the use of a neuro-fuzzy system

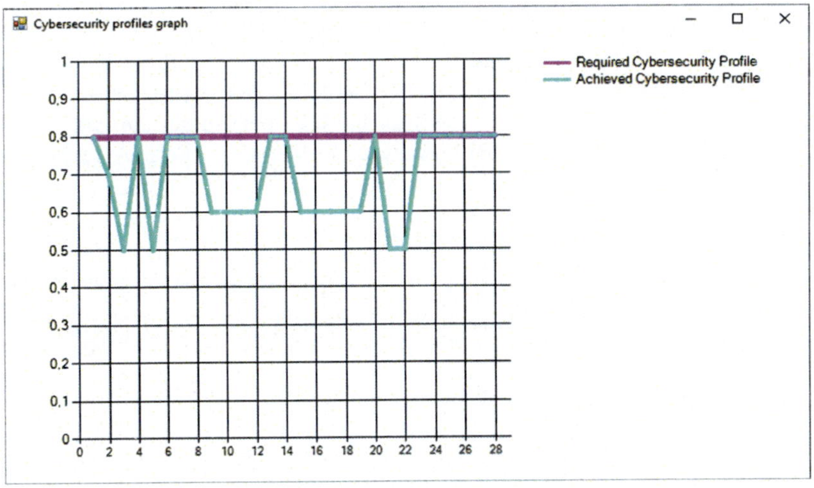

Fig. 4. Results of IS audit OBI and comparison of reference requirements for IS metrics and the achieved level

6 Discussion of the Results

The research carried out within the framework of this article was based on ideas that made it possible to harmoniously combine the theory of neural networks and fuzzy sets, decision-making methods and a method for analyzing hierarchies for conducting and improving procedures for auditing information security of various objects of informatization.

As can be seen in the diagram shown in Fig. 4 graph, the use of the modified MAH made it possible to obtain a graph of the current state of information security of the surveyed object of informatization (blue line). Moreover, the selected IS criteria are approximately 25–30% lower than the reference values. Although the application of the classical MAH did not give such discrepancies. The method has been tested in the course of performing IS audits of several enterprises in Ukraine and Kazakhstan.

7 Conclusion

The research carried out allows us to draw the following conclusions:

1. It is shown that the assessment of the level of the degree of information security (IS) for objects of informatization (OBI) should be carried out based on assessing the effectiveness of the set of criteria of the method of analysis of hierarchies (MAH). At the same time, such an assessment of the degree of IS and all related IS audit procedures (ISA) are most effective for a multilateral assessment of IS OBI. Both standard numerical information security metrics and metrics proposed by information security experts and agreed with the OBI management can be used as assessment metrics.
2. A modified method for the analysis of hierarchies is proposed, based on the application of the apparatus of the theory of fuzzy sets and neural networks. This method enables management to make informed management decisions in the field of information security OBI. The solutions obtained are aimed at improving not only the OBI IS itself, but also ultimately optimize the OBI control system, reduce costs and increase the efficiency of OBI business processes as a whole. An example of calculating indicators of the level of information security of a conditional object of informatization is presented. This example illustrates the adequacy of the modified MAI.
3. It is shown that the use of the mathematical apparatus of the Moscow Aviation Institute and the corresponding software, in particular, the developed intellectual system, makes it possible to increase the degree of reliability of the results of a comprehensive audit of IS OBI. Moreover, this statement is true for the procedures of the internal ISA OBI and the external ISA.

Acknowledgments. The work was carried out as part of the grant study AP08855887-OT-20 "Development of an intelligent decision support system in the process of investing in cybersecurity systems."

References

1. Lallie, H.S., et al.: Cyber security in the age of COVID-19: a timeline and analysis of cyber-crime and cyber-attacks during the pandemic. Compute. Secur. **105**, 102248 (2021)
2. Miao, Y., Chen, C., Pan, L., Han, Q. L., Zhang, J., Xiang, Y.: Machine learning based cyber attacks targeting on controlled information: a survey. arXiv preprint arXiv:2102.07969 (2021)
3. Yamin, M.M., Ullah, M., Ullah, H., Katt, B.: Weaponized AI for cyber attacks. J. Inf. Secur. Appl. **57**, 102722 (2021)
4. Golyash, I., Sachenko, S., Rippa, S.: Improving the information security audit of enterprise using XML technologies. In: Proceedings of the 6th IEEE International Conference on Intelligent Data Acquisition and Advanced Computing Systems, vol. 2, pp. 795–798. IEEE, Sept 2011
5. Steinbart, P.J., Raschke, R.L., Gal, G., Dilla, W.N.: The influence of a good relationship between the internal audit and information security functions on information security outcomes. Acc. Organ. Soc. **71**, 15–29 (2018)
6. Griffiths, P.: Where next for information audit? Bus. Inf. Rev. **27**(4), 216–224 (2010)
7. Steinbart, P.J., Raschke, R.L., Gal, G., Dilla, W.N.: The relationship between internal audit and information security: an exploratory investigation. Int. J. Account. Inf. Syst. **13**(3), 228–243 (2012)
8. Kaur, R., Singh, M.: A survey on zero-day polymorphic worm detection techniques. IEEE Commun. Surv. Tutorials **16**(3), 1520–1549 (2014)
9. Steinbart, P.J., Raschke, R.L., Gal, G., Dilla, W.N.: Information security professionals' perceptions about the relationship between the information security and internal audit functions. J. Inf. Syst. **27**(2), 65–86 (2013)
10. Kayworth, T., Whitten, D.: Effective information security requires a balance of social and technology factors. MIS Q. Exec. **9**(3), 2012–2052 (2010)
11. Jarison, J., Morris, L., Wilkinson, C.: The future of cyber security in internal audit. Disponibil online la. Ashx (2018). www.crowe.com/-/media/Crowe/LLP/foliopdf/The-Future-of-Cybersecurity-in-IA-Risk-18000-002A-update
12. Suduc, A.M., Bîzoi, M., Filip, F.G.: Audit for information systems security. Informatica Economica **14**(1), 43 (2010)
13. Herath, H.S., Herath, T.C.: IT security auditing: a performance evaluation decision model. Decis. Support Syst. **57**, 54–63 (2014)
14. Atymtayeva, L.B., Bortsova, G.K., Inoue, A., Kozhakhmet, K.T.: Methodology and ontology of expert system for information security audit. In: The 6th International Conference on Soft Computing and Intelligent Systems, and The 13th International Symposium on Advanced Intelligence Systems, pp. 238–243. IEEE, Nov 2012
15. ISO/IEC 27001:2013: Information technology. Security techniques. Information security management systems. Requirements, International Organization for Standardization, p. 23 (2013)
16. ISO/IEC 27000:2014: Information technology. Security techniques. Information security management systems. Overview and vocabulary, International Organization for Standardization, p. 31(2014)
17. ISO/IEC 27004:2009: Information technology. Security techniques. Information security management systems. Measurement, International Organization for Standardization, p. 55 (2009)
18. ISO/IEC 27005-2011: Information technology. Security techniques. Information security management systems. International Organization for Standardization, p. 68 (2011)
19. ISO 19011:2011: Guidelines for auditing management systems. International Organization for Standardization, p. 44 (2011)

20. Voevodin, V.A.: Etalonnaya model' ob"ekta audita informacionnoj bezopasnosti [Reference Model of an Information Security Audit Object]. Modern Science: actual problems of theory and practice. Ser. Nat. Tech. Sci. (9), 56–60 (2019) (in Russian)
21. Voevodin V.A.: Method of the study of privacy protection in information. Am. Sci. J. **2**(32), 47–51 (2019) (in Russian)
22. Voevodin, V.A.: Conceptual model of information security auditobject. Comput. Nanotechnol. (3), 92–95 (2019). https://doi.org/10.33693/2313-223X-2019-6-3-92-95 (in Russian)
23. Aguarón, J., Escobar, M.T., Moreno-Jiménez, J.M.: Consistency stability intervals for a judgement in AHP decision support systems. Eur. J. Oper. Res. **145**(2), P.382–393 (2003)
24. De Wilde, P.: Neural Network Models: Theory and Projects. Springer, Heidelberg (2013). https://doi.org/10.1007/978-1-84628-614-8

Detection of Steel Surface Defects Using U-Net with Pre-trained Encoder

Ammar Alhaj Ali, Bronislav Chramcov, Roman Jasek, Rasin Katta(✉), Said Krayem, and Mohammad Kadi

Faculty of Applied Informatics, Tomas Bata University in Zlin, Zlin, Czech Republic
{alhaj_ali,chramcov,jasek,Katta,krayem}@utb.cz

Abstract. Steel is one of the most widely building materials of modern times. Automatic detection of manufacturing defects in steel surfaces is a very important step for product quality control in the steel manufacturing industry. Traditional inspection systems require high maintenance costs and allow little flexibility. In this paper, we propose using the deep learning network U-Net for detecting defects. U-Net is a type of Convolutional Neural Networks that perform semantic segmentation of images, it was originally developed to achieve biomedical image segmentation, as U-Net requires smaller training sets compared with typical CNNs. We will present that segmentation can be done with promising results when using the U-Net model with pre-trained encoder; as transfer learning increases the accuracy and stability of the object detection models.

Keywords: Steel surface defect detection · U-Net · Transfer learning · CNN · Deep learning

1 Introduction

The steel manufacturing industry is obliged to meet high-quality standards. Sometimes, a single failure can lead to enormous financial damage. Steel surface defects such as scratches, patches, and inclusions have a harmful influence on material properties, i.e., fatigue strength and corrosion resistance, as well as the appearance [1]. To guarantee high-quality standards, manufacturers employ surface inspection systems, which are either done through human experts or with machine vision systems [2]. The first option leads to human errors, especially after long hours of work on the arduous inspection task. Currently, steel manufacturers mostly adopt manual detection methods, but they are looking to apply machine learning to increase efficiency, improve automation, and maintain high quality in their production [3]. The automated surface inspection system ASIS of the steel basically consists of the following components: image acquisition, rapid detection, and defect classification [4, 5], see Fig. 1.

Last decade, computer vision applications made impressive progress in objects detection and classification in various domains such as image search, self-driving cars, handwriting recognition, and face detection [6]. And convolutional neural networks CNNs

R. Silhavy et al. (Eds.): CoMeSySo 2021, LNNS 232, pp. 185–196, 2021.
https://doi.org/10.1007/978-3-030-90318-3_18

have been using to get an advanced performance in image segmentation and classification, these kinds of networks consist of multilayers that could learn the understructure of the information from data, CNNs also save time and effort to extract features by themselves without any need to be designed by a human [7], in addition, semantic segmentation tasks have also been revolutionized by convolutional networks.

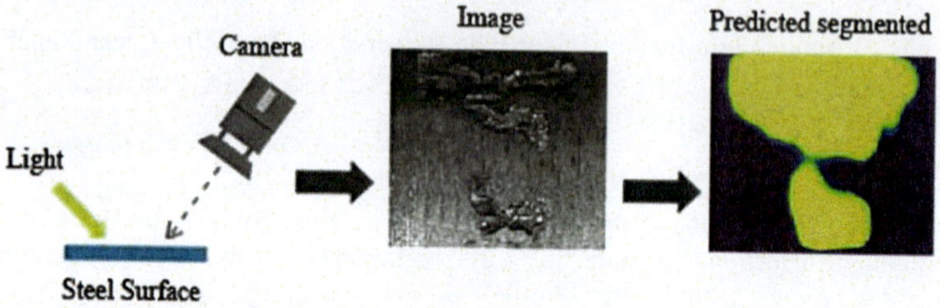

Fig. 1. Automated vision system for quality inspection

Despite the great breakthrough in computer vision tasks, a major drawback of CNNs is that they require huge volumes of training data. Unfortunately, not only the acquisition of images is complicated and expensive, but also the accuracy of training data annotation adds more complexity. Nevertheless, in recent years CNNs have significantly achieved promising results in image segmentation, and most of the credit goes to U-Net [8].

Usually, convolutional neural networks are used to classify objectives of images, where the output to an image is a class label, in addition, some visual tasks required output should include localization [9]. U-Net has been offering impressive potential in segmenting images, to an extent that it has become the standard in image segmentation like the medical images, even with a small amount of labeled training data. U-Net - called in this way due to its U shape - is a type of convolutional neural network that performs the task of semantic segmentation; it has symmetric architecture, having encoder that extracts spatial features from the images, and a decoder that forms the segmentation map from the encoded features [8], see Fig. 2.

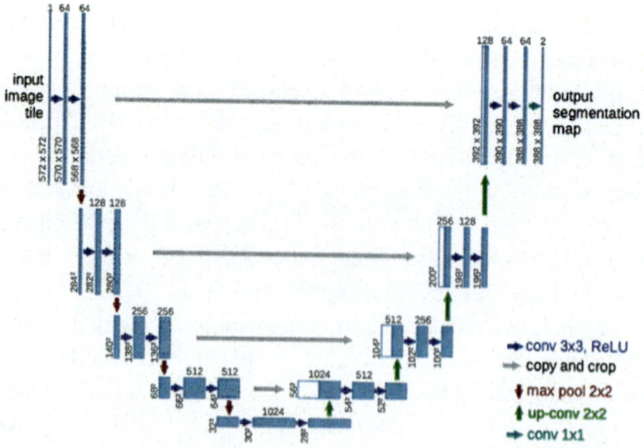

Fig. 2. Architecture of U-Net. From [9]

On another side, The CNN models trained for image classification contain meaningful information and have been used to obtain an advanced performance in image segmentation. The performance of U-Net can be improved using the convolution layers of the pre-trained models in the encoder layers of the segmentation model [10]. U-Net models can use various backbones such as ResNet, VGG, and EfficientNet that are pre-trained on a large dataset like ImageNet, this shows better convergence than those trained from scratch on a small dataset.

The contributions of this study can be summarized as follows:

1. Presenting a deep learning model to detect the defects of steel surfaces properly and improve the capability of quality control in the steel industry.
2. Demonstrating how the U-Net architecture can be improved by using the pre-trained encoder.
3. Comparing the performance metrics of the U-Net model using various pre-trained backbones to detect steel surface defects.

The rest of the paper is organized as follows: Sect. 2 introduces image segmentation using U-Net. Section 3 presents the proposed method like network architecture and evaluation metrics. Section 4 analyzes the dataset used. Section 5 provides the training specifications and the experimental results. Finally, Sect. 6 summarizes the study and draws a conclusion.

2 Image Segmentation Using U-Net

Image segmentation is a way of separating an image into multiple regions according to the different properties of pixels to extract meaningful information. In other words, image segmentation describes the process of linking each pixel of the input image to a class label. In this case, the image pixels are labeled in a way that each pixel in an image shares certain characteristics such as color, texture, intensity, etc. [11]. Broadly image segmentation is divided into two types: 1-semantic segmentation, 2-instance segmentation. Semantic segmentation handles multiple objects of the same class as a single entity. On the other hand, instance segmentation handles multiple objects of the same class as distinct individual object (or instance).

CNN's are powerful visual models and can produce a segmentation map for an entire input image. One of the most successful deep learning methods is based on the Fully Convolutional Networks (FCN). The typical convolution neural network (CNN) often contains fully connected layers that do not perform the convolution operation this makes it not fully convolutional. The main idea of FCN is to use CNN as a powerful feature extractor by substituting the fully connected layers with convolution layers to produce spatial feature maps instead of classification scores. Simplify, a fully convolution network (FCN) is a neural network that only performs convolution operations. Equivalently, an FCN is a CNN without fully connected layers (dense layers) and stack a bunch of convolutional layers in an encoder-decoder fashion.

In the first half of FCN, the encoder downsamples the spatial resolution of the image by expanding feature maps carefully. At this stage, we get highly efficient discrimination between different classes. Then to recover the location information, downsampling is followed by an upsampling, where the decoder upsamples the image by taking multiple lower resolution images (feature maps) as input to produce dense pixel-wise output (segmentation map) [12, 13].

In U-Net, the encoder extracts the features where the decoder uses features extracted from the input images to produce the desired segmentation map. The encoder can be replaced by a pre-trained network such as VGG, ResNet, MobileNet, or Efficientnet, etc. These pre-trained models are already trained on the ImageNet dataset and have the necessary feature extraction capabilities [14–17] Fig. 3 shows U-Net uses a pre-trained network (VGG16) as an encoder, each box represents a multi-channel features map passing through a series of transformations, and the blue arrows denote skip-connections where information is transmitted from the encoder to the decoder [12, 15].

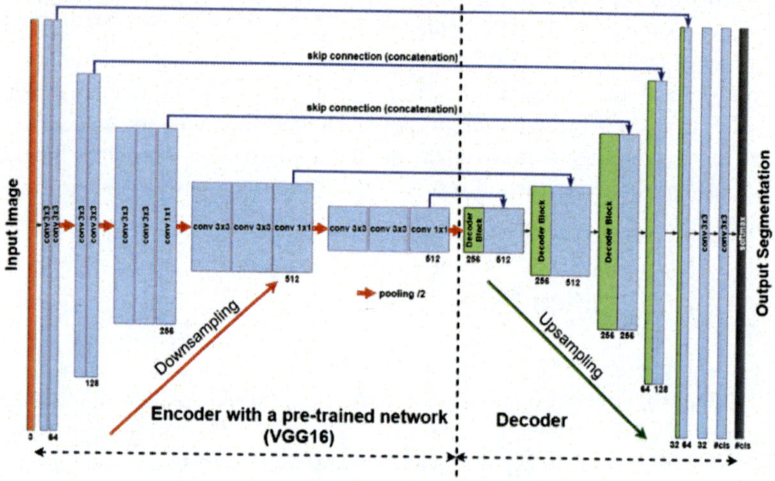

Fig. 3. U-Net with a pre-trained network as an encoder. From [15]

3 Proposed Method

In this section, we briefly introduce an overview of the research, followed by details of deep learning-based proposed architecture using a U-Net and pre-trained networks as an encoder, as well as evaluation metrics used in the experiment.

3.1 Overview and Motivation

The importance of image segmentation in some domains became inevitable and hence its accuracy and performance are a priority. Deep learning has efficiently paved the way for objects detection and diagnose in various fields of applications, yet the amount of data used is an essential factor. The U-Net model for segmentation can be trained in two ways; either the model can be trained from scratch or by using transfer learning. Training a model from scratch is weary and the amount of data required is so huge, while in transfer learning, the encoder part is replaced by a specified pre-trained weight of an already trained model also known as a backbone; this technique is called transfer learning. This study investigates the performances of the U-Net model using various pre-trained backbones to detect steel surface defects and evaluates the performance of models by several evaluation metrics. Figure 4 summarizes the complete evaluation process.

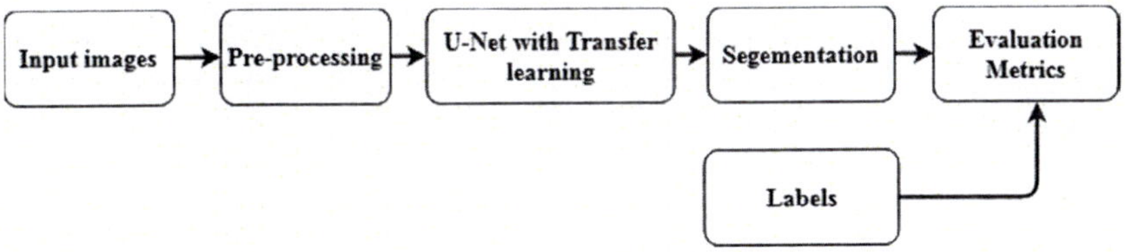

Fig. 4. Segmentation evaluation procedure

3.2 Network Architecture

In semantic segmentation, there are several things that should be taken into account when choosing the model architecture such as the number of training images, the size of the images, and the domain of the images. The U-Net architecture adopts an encoder-decoder framework with skip connections, the encoder and decoder layers are symmetrical to each other.

Usually, for many applications, selecting a model pre-trained on ImageNet is the best choice, VGG, ResNet, or MobileNet is chosen as standard model for the base network (backbone). In the U-Net segmentation model, the encoder will be substituted by specified weights of a pre-trained model. The backbone model (encoder) is used as a feature extractor without the dense layers, and the decoder is built depending on the specified features of the selected encoder.

3.3 Evaluation Metrics

The evaluation metrics are useful to set the quality of a model. In semantic segmentation, a successful prediction is the one that maximizes the overlap between the predicted segmented and the true object (ground-truth). The Dice and Jaccard coefficients are two metrics used for evaluating the overlap. The Jaccard index is also known as Intersection over Union (IoU). Both metrics are described in the formulas below:

$$Dice\ Score\ (A, B) = \frac{2 \times |A \cap B|}{|A| + |B|}$$

$$Jaccard\ (A, B) = IoU\ (A, B) = \frac{|A \cap B|}{|A \cup B|}$$

Where A and B are the predicted segmented and the ground-truth. Where ∩, ∪ denote the intersection and union operator, respectively.

Also, to evaluate the effect of the model, this study selects Recall, Precision, F1-score, and Accuracy to compare the models;

Figure 5 shows segmentation errors representation (confusion matrix) presenting predicted and actual values.

- True positive (TP) is the score of identifying ground truth pixels exactly.
- True negative (TN) is the score of identifying background pixels exactly.
- False positive (FP) is score of failing in identifying ground truth pixels.
- False negative (FN) is score of failing in identifying background pixels into ground truth pixels.

Actual	Predicted	
	Positive	*Negative*
Positive	TP	FN
Negative	FP	TN

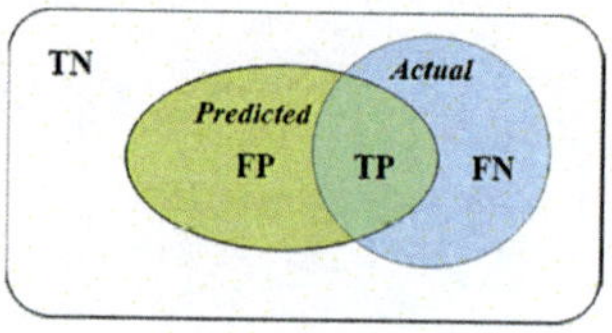

Fig. 5. Segmentation errors representation

Accuracy, Recall, Precision, and F1score are described in following formulas:

$$Accuracy = \frac{TP + TN}{TP + TN + FP + FN}$$

$$Recall = \frac{TP}{TP + FN}$$

$$Preicision = \frac{TP}{TP + FP}$$

$$F1\ Score = 2 \times \frac{Recall \times Preicision}{Recall + Preicision}$$

Recall indicates the proportion of positive samples that are correctly identified, and Precision indicates the proportion of real positive samples among the identified positive samples.

4 Steel Defect Dataset

The dataset used in this study comes from the Kaggle competition, Severstal: Steel Defect Detection [18], with a total of 12,568 steel sheet images having the size of 1600 × 256. There are four classes of defects, see Fig. 6. Images may have no defects, a defect of a single class, or defects of multiple classes, see Fig. 7. While Fig. 8 shows the defects distribution graph.

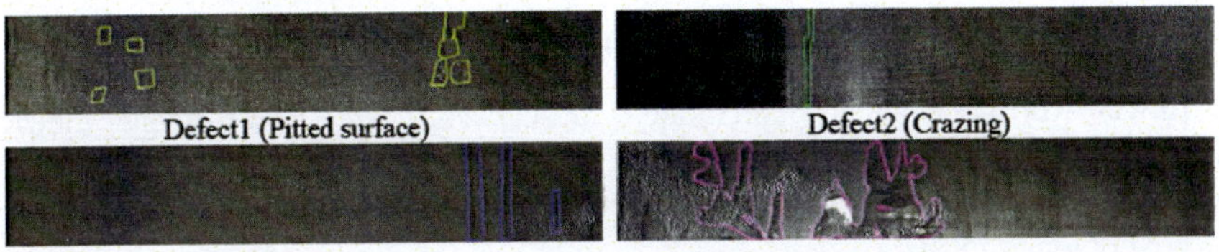

Fig. 6. Classes of the surface defects

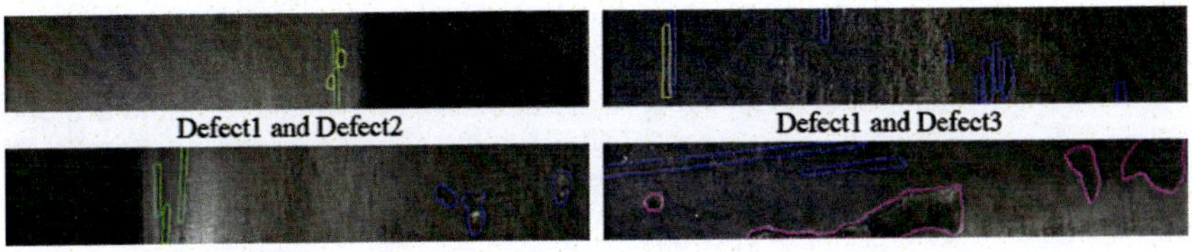

Fig. 7. Multiple classes of defects in the same image

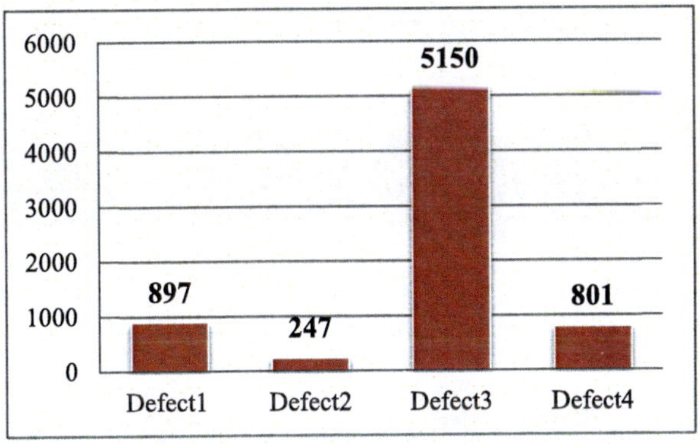

Fig. 8. The number of defects in dataset

5 Training and Experimental Results

5.1 Training and Implementation

Transfer learning is used to increase the accuracy and stability of the object detection models [19]. In this study, the experiments are implemented using the Keras framework, and segmentation was done on U-Net models with ResNet34, VGG16, MobileNet, SeResNet34, Densenet121,and Efficientnetb0 backbones, all various backbones were first pre-trained on the ImageNet dataset, and then the U-Net model training was done in 100 epochs with a batch size of 16, all of the models were optimized via Adam optimizer [20] with a learning rate of 0.001, and Binary Cross-Entropy was used as a loss function [21]. The dataset was split into two sets: a training set and a validation set with a ratio of 80:20, and images were resized by a factor of 0.5 from 256 × 1600 to 128 × 800.

5.2 Experimental Results and Discussion

Experiments were done on the U-Net model with six backbones; the performance metrics were calculated and tabulated in Table 1. Figures 9 and 10 show validation and training dice coefficient respectively, while Figs. 11 and 12 show validation and training accuracy respectively for proposed models.

Figure 13 shows the input image and corresponding ground truth mask, and the segmented output for some images from the validation dataset.

The performance metrics results in Table 1 shows that all the U-Net models have performed good well, and the U-Net model with Efficientnetb0 as the backbone has outperformed the other models with a dice score of 0.7305, IoU of 0.6421, and accuracy of 0.9963 followed by the SeResNet34 model.

Furthermore, increasing the accuracy of the models could be done by analyzing the findings using a larger number of epochs or different activation functions, as well; more types of backbones for the U-Net could also be used such as EfficientnetB7.

Table 1. Performance metrics of segmentation models experimented

U-Net Backbone	metrics					
	Dice	IoU	Acc	Recall	Precision	F1 Score
VGG16	0.6681	0.5516	0.9955	0.7738	0.6109	0.6827
ResNet34	0.6932	0.6158	0.9958	0.7723	0.6610	0.7123
MobileNet	0.7007	0.6269	0.9959	0.7760	0.6704	0.7194
Densenet121	0.7065	0.5785	0.9960	0.7692	0.6915	0.7283
SeResNet34	0.7113	0.6210	0.9960	0.7471	0.7251	0.7359
Efficientnetb0	0.7305	0.6421	0.9963	0.7800	0.7267	0.7524

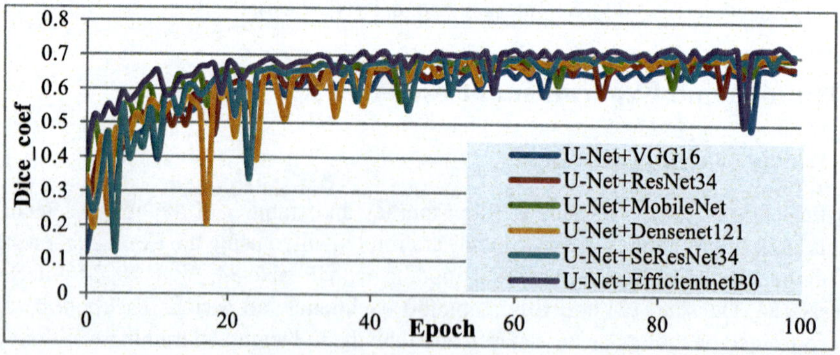

Fig. 9. Validation dice coefficient of the proposed models

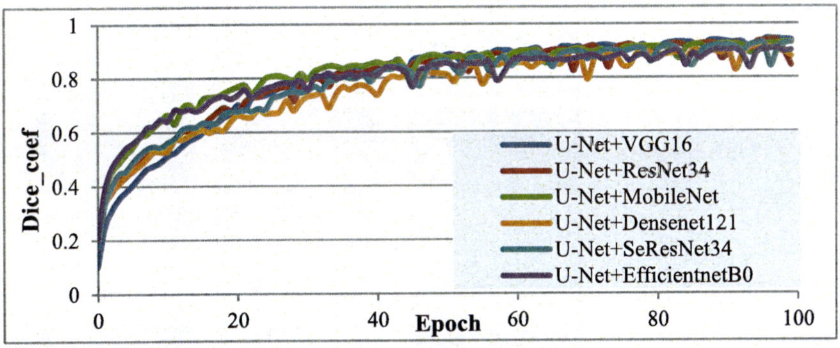

Fig. 10. Training dice coefficient of the proposed models

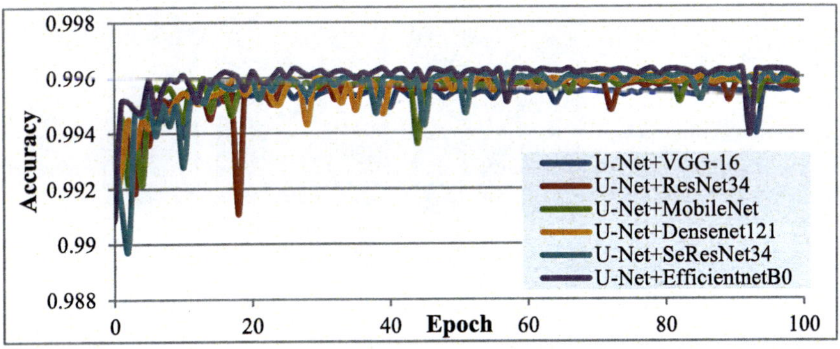

Fig. 11. Validation accuracy of the proposed models

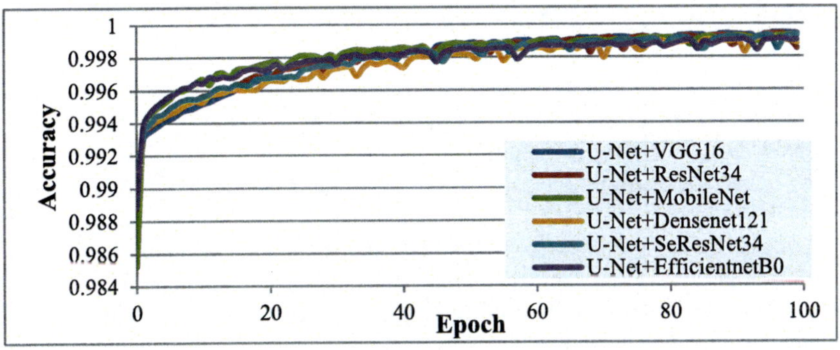

Fig. 12. Training accuracy of the proposed models

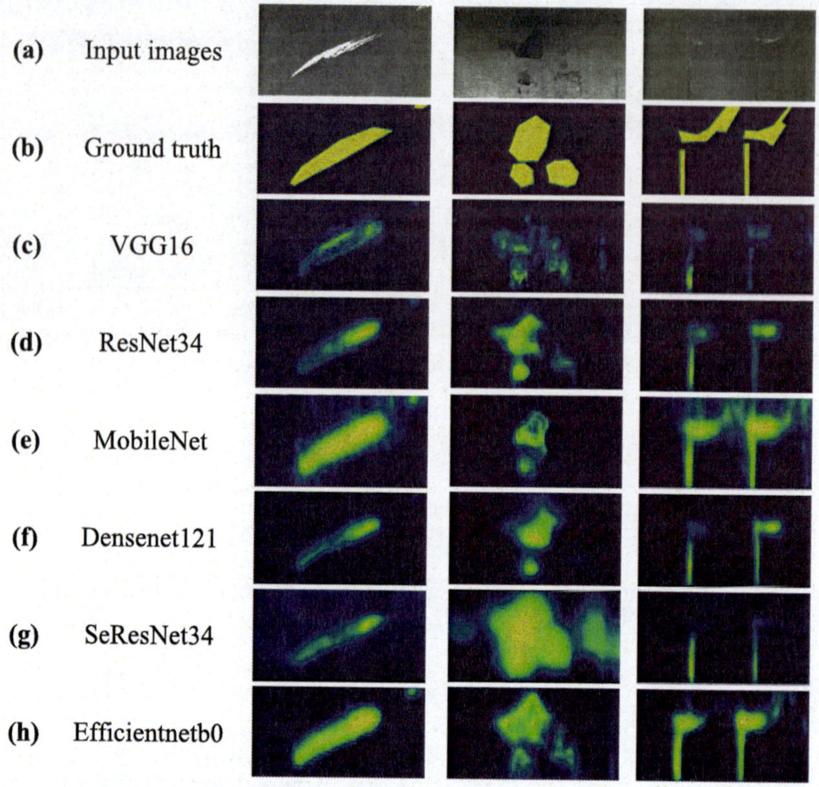

Fig. 13. The experimental results of the proposed models, where (a) input images, (b) ground truth, (c) the outputs of U-Net + VGG16 model, (d) the outputs of U-Net + ResNet34 model, (e) the outputs of U-Net + MobileNet model, (f) the outputs of U-Net + Densenet121 model, (g) the outputs of U-Net + SeResNet34 model, and (h) the outputs of U-Net + Efficientnetb0 model

6 Conclusions

In this paper, we proposed a steel surface defect detection model based on a U-Net deep learning model with various pre-trained backbones, the model was trained on a given dataset during 100 epochs since it was enough to test Dice and Jaccard score. As well, many evaluation metrics were used to set the quality of the model.

Usually, Deep learning techniques fail to perform effectively when using small datasets, transfer learning is one of the most important techniques to overcome this problem. U-Net can be improved by replacing the encoder part with networks pre-trained on the ImageNet dataset. Furthermore, pre-trained networks basically reduce training time and help to prevent overfitting. Transfer learning is the perfect choice for faster convergence that would be useful when high-performance computing resources are limited.

The experiments results showed that the U-Net performs segmentation with remarkable accuracy, and the proposed architecture has performed well and increased the accuracy and stability of the object detection model. These results can be further improved by extra hyperparameter tuning combined by training time increase.

References

1. Lee, S.Y., Tama, B.A., Moon, S.J., Lee, S.: Steel surface defect diagnostics using deep convolutional neural network and class activation map. Appl. Sci. **9**(24), 5449 (2019). https://doi.org/10.3390/app9245449
2. Prappacher, N., Bullmann, M., Bohn, G., Deinzer, F., Linke, A.: Defect detection on rolling element surface scans using neural image segmentation. Appl. Sci. vol. **10**(9) (2020). https://doi.org/10.3390/app10093290
3. Wang, S., Xia, X., Ye, L., Yang, B.: Automatic detection and classification of steel surface defect using deep convolutional neural networks. Metals (Basel) **11**(3), 1–23 (2021). https://doi.org/10.3390/met11030388
4. Fu, J., Zhu, X., Li, Y.: Recognition of surface defects on steel sheet using transfer learning. arXiv preprint arXiv:1909.03258. 7 Sept 2019
5. Neogi, N., Mohanta, D.K., Dutta, P.K.: Review of vision-based steel surface inspection systems. J. Comput. Higher Educ. **2014**(1), 50. Springer, New York (2014). https://doi.org/10.1186/1687-5281-2014-50
6. Soukup, D., Huber-Mörk, R.: Convolutional neural networks for steel surface defect detection from photometric stereo images. In: Bebis, G., et al. (eds.) ISVC 2014. LNCS, vol. 8887, pp. 668–677. Springer, Cham (2014). https://doi.org/10.1007/978-3-319-14249-4_64
7. Khanh, T.L.B., et al.: Enhancing U-Net with spatial-channel attention gate for abnormal tissue segmentation in medical imaging. Appl. Sci. **10**(17), 5729 (2020). https://doi.org/10.3390/app10175729
8. Ibtehaz, N., Rahman, M.S.: MultiResUNet: Rethinking the U-Net architecture for multimodal biomedical image segmentation. Neural Netw. **121**, 74–87 (2020)
9. Ronneberger, O., Fischer, P., Brox, T.: U-Net: convolutional networks for biomedical image segmentation. In: Navab, N., Hornegger, J., Wells, W.M., Frangi, A.F. (eds.) MICCAI 2015. LNCS, vol. 9351, pp. 234–241. Springer, Cham (2015). https://doi.org/10.1007/978-3-319-24574-4_28
10. Baheti, B., Innani, S., Gajre, S., Talbar, S.: Eff-UNet: a novel architecture for semantic segmentation in unstructured environment (2020). https://cvit.iiit.ac.in/ncvpripg19/. Accessed 13 Mar 2021
11. Sultana, F., Sufian, A., Dutta, P.: Evolution of image segmentation using deep convolutional neural network: a survey. Knowl.-Based Syst. **201**, 106062 (2020)
12. Iglovikov, V., Shvets, A.: TernausNet: U-Net with VGG11 encoder pre-trained on imagenet for image segmentation. https://github.com/ternaus/TernausNet. Accessed 13 Mar 2021
13. Shelhamer, E., Long, J., Darrell, T.: Fully convolutional networks for semantic segmentation. IEEE Trans. Pattern Anal. Mach. Intell. **39**(4), 640–651 (2017). https://doi.org/10.1109/TPAMI.2016.2572683
14. Alam, S., Tomar, N.K., Thakur, A., Jha, D., Rauniyar, A.: Automatic polyp segmentation using U-Net-ResNet50 (2020). http://arxiv.org/abs/2012.15247. Accessed 13 Mar 2021
15. Shvets, A.A., Iglovikov, V.I., Rakhlin, A., Kalinin, A.A.: Angiodysplasia detection and localization using deep convolutional neural networks. In: Proceedings - 17th IEEE International Conference on Machine Learning and Applications, ICMLA 2018, pp. 612–617, Jan 2019. https://doi.org/10.1109/ICMLA.2018.00098

16. Abdur Rehman, N., Saif, U., Chunara, R.: Deep landscape features for improving vector-borne disease prediction. In: Proceedings of the IEEE/CVF Conference on Computer Vision and Pattern Recognition Workshops. pp. 44–51 (2019)
17. Alex, D.M., Abraham Chandy, D.: Investigations on performances of pre-trained U-Net models for 2D ultrasound kidney image segmentation. In: Miraz, M.H., Excell, P.S., Ware, A., Soomro, S., Ali, M. (eds.) iCETiC 2020. LNICSSITE, vol. 332, pp. 185–195. Springer, Cham (2020). https://doi.org/10.1007/978-3-030-60036-5_13
18. Severstal: Steel Defect Detection ,Kaggle. https://www.kaggle.com/c/severstal-steel-defect-detection. Accessed 13 Mar 2021
19. Ali, A., Chramcov, B., Jasek, R., Katta, R., Krayem, S.: Classification of Plant Diseases Using Convolutional Neural Networks. In: Silhavy, Radek (ed.) CSOC 2021. LNNS, vol. 229, pp. 268–275. Springer, Cham (2021). https://doi.org/10.1007/978-3-030-77445-5_24
20. Kingma, D.P., Ba, J.L.: Adam: a method for stochastic optimization, Dec. 2015. https://arxiv.org/abs/1412.6980v9. Accessed 13 Mar 2021
21. Ho, Y., Wookey, S.: The real-world-weight cross-entropy loss function: modeling the costs of mislabeling. IEEE Access **8**, 4806–4813 (2020). https://doi.org/10.1109/ACCESS.2019.2962617

Free Software at School and the Educational Process: Current State, Systematization of Typical Problems

Alexander Fedosov[1](✉), Mikhail Markushevich[2], and Tatiana Gubina[3]

[1] Russian State Social University, Moscow, Russia
[2] School 1352, Moscow, Russia
[3] Basalt SPO" Company, Moscow, Russia

Abstract. The article presents the genesis of the problem of using free software in a Russian general education school, reveals a range of problems that educational institutions face when using (choosing) free software. The authors describe the general set of problems characteristic of the migration process to free software, and also consider in detail the relevance of each of them in relation to the field of education in the Russian Federation over the past 12 years, starting in 2008, when the first attempt was made to translate the educational process general education Russian school for free software. The article also provides possible solutions to some of the identified problems and examples of their implementation .

Keywords: Free software · Communitarianism · Pupils learning · Educational pro cess · Open source free software

1 Introduction

The history of the massive implementation of free software in schools in the Russian Federation began in 2007, when, as part of realization of the project "First Aid", a free software package (FSP) was implemented in three pilot regions of the Russian Federation: the Perm Territory, the Republic of Tatarstan, Tomsk Region. The approbation was recognized as successful, and in 2007–2009 in accordance with the decision of the Government of the Russian Federation No. 1447-r of October 18, 2007, the Federal Education Agency has delivered the FSP to all educational institutions of Russia. This package contained the entire range of programs and teaching materials necessary to ensure the educational process and functionally fully corresponded to the project of the Federal Agency for Education of the Russian Federation for providing schools with licensed software products implemented in 2007–2009 - "First Aid", and it also contained dozens of additional programs for educational institutions.

It should be noted here that simultaneously with the FSP, "First Aid" included a set of proprietary programs, which was also actively used in schools.

Thanks to the Federal program, teachers from many educational institutions have appreciated the benefits, advantages and disadvantages of free software over proprietary

R. Silhavy et al. (Eds.): CoMeSySo 2021, LNNS 232, pp. 197–207, 2021.
https://doi.org/10.1007/978-3-030-90318-3_19

software, and now they actively continue to use, if not the entire set of FSP, then at least part of the programs from its set.

The debates on the merits and demerits of proprietary and free software continues to this day. Each teacher, instructor tries to select those tools (software products) that will create a modern subject-oriented educational environment, aimed at the implementation of basic educational programs in accordance with the position held.

It is necessary to make an important note: software can be used by a teacher, an instructor as a teaching tool, or it can act as a subject for students to study. And it is the second option that arouses the greatest interest and subject for discussion. In this connection, the further presentation of the article will mainly be conducted in the context of the applicability and possibility of using open source software within the frame- work of the subjects of the information technology cycle: computer science and ICT, technology, programming, computer graphics, robotics, information technology.

What software products should be used to build laboratory and practical work in the school course of computer science and ICT?

Learning kits and coursebooks for the school computer science course from different authors' teams give the teacher the opportunity to choose software. For example, the learning kit of Bosova L.L. for the grades 5–9 contains presentations for the lessons, files-templates for coursebooks on computer science, templates for a computer practice, which can be used both when working in the office suite Microsoft Office, and in LibreOffice [1–3]. In addition, it is proposed to build the study of the fundamentals of algorithmization and programming on free programs Scratch [4], Kumir [5].

The fact that free software has less functionality and displeases users with graphic design in comparison with similar proprietary software also has a right to life.

So, for example, there are regular discussions on the choice of a graphics editor: GIMP or Photoshop? When studying the section "Technologies for the creation and processing of raster graphic images" in a general education school, the functionality and capabilities of the free raster graphic editor GIMP are quite enough, but, according to experts, to train professional designers in secondary specialized and higher institutions, it is necessary to master the functionality of a proprietary professional graphic editor Adobe Photoshop. This opinion is supported by the fact that knowledge of Adobe Photoshop is a standard requirement in the labor market on the part of an employer when applying for a job as a designer.

What is to be done? Should we start teaching schoolchildren to work in Photoshop right away, or should we still opt for GIMP or another free editor?

The authors of the article decided to take a short excursion into the history of the issue and analyze all the pros and cons of free software.

We should note that this paper does not pretend to be a detailed analysis of the problems that educational institutions face when using/choosing free software.

The authors focus on the task of only describing a certain general stack of problems characteristic of the migration to open source software, as well as a detailed consideration of the relevance of each of the difficulties under consideration in relation to education in the Russian Federation, over the past 12 years, starting in 2008, when the first attempt of the transfer of the educational process of a general education Russian school to free software was made.

It is obvious that the process of transition to the use of open source software in each of the branches of the modern economy (education, public sector, oil and gas industry, etc.) will be specific, but, nevertheless, from the authors' point of view, a stack of problems that are actualized in the process of the considered possible transition will be approximately the same, which is shown by the corresponding analysis of the literature on this issue and carried out in this work.

2 Review of the Literature Devoted to the Problems of Transition to Free Software (Back to the Background)

The main advantages of using free software in the educational process of a general education school include the following:

1. Free software licensing policy, which gives 4 degrees of freedom, one of which is free licenses for use by anyone and anywhere [6], which cannot be said about non-free software;
2. Cross-platform nature of free software, and what is especially important is the availability of versions of application software for the GNU / Linux OS, which is often not typical for proprietary software;
3. The opportunity for all participants in the educational process to install and use open source software on home computers without the need to purchase licenses for the right to own the software product and, as a result, encourage students to observe copyright [7];
4. Relatively low system requirements, typical for the majority of programs classified as free software;
5. The communitarian ideology of free software [8];
6. Open source free software;
7. Open Bug Tracking System, which allows any registered user to write about bugs that he found while using the software.

However, despite such significant advantages of free software in comparison with proprietary software, one can state the presence of both objective and subjective difficulties on the path of migration of the educational process of educational organizations that implement general education programs to free software.

By 2008 it was obvious that the use of proprietary software in the educational process of a Russian school is not the only and, perhaps, not the most optimal way, and that its alternative is the use of free software. To implement this solution, a free software package (FSP) based on the domestic Alt Linux operating system was sent to schools in 2008 for testing and use. The package contained a number of distributions for various purposes: desktop operating systems for computers with different performance "Alt Linux School Junior" and "Alt Linux School Master", a version of the

server system "Alt Linux School Server", distributions of free software for various purposes.

At the same time, the first experience of using open source software in educational institutions appeared, and, what is noteworthy, not only general education, but also higher and secondary specialized ones. Teachers actively began to share their experience in building the educational process on the basis of SPE at conferences, in scientific journals and collections of works.

So, V.G. Slugin, who in 2010 was engaged in a partial translation of the educational process at the Nizhny Novgorod Radio Engineering College to free software, notes the following difficulties that he encountered during this process.

Incompatibility of some peripheral equipment (printers, scanners) with free operating systems;

The need for preliminary training of staff and teachers in the use of new software;

Many on-line data collection systems about educational institutions are focused on working only with the Internet Explorer web browser and do not perform their functions correctly in other browsers;

Higher and supervisory authorities correspond exclusively in the formats of proprietary programs (doc, docx, xls, xlsx) [9].

In the same period of time E.R. Alekseev and I. Shevchenko identified the following problems hindering the migration of the educational process to free software [10]:

1. Insufficient amount of educational and methodological literature;
2. Lack of departments specializing in free software in CIS universities. Even at that time, the author suggested to think seriously about organizing a free software department in one of the CIS universities;
3. Passivity of thinking of teachers of IT disciplines.

E.V. Andropova and T.N. Gubina, who actively implemented open source software at the Yelets State University named after I.A. Bunin, point to the following difficulties that hinder the successful migration of the educational process to free software:

The passive behavior of many teachers, instructors and engineers who are directly related to this issue, waiting for the "command "from upstairs". If such a command is not received, then they try to delay the onset of events as far as it is possible" [11, p. 32];

The factor of the teachers' habit of using certain, previously used software, most often proprietary, and the prospect of switching to "unfamiliar" software causes a negative reaction.

P.B. Berestov who was involved in the implementation of open source software in educational institutions of the Sverdlovsk region described similar difficulties in his article [12]:

1. Low level of knowledge of information and communication technologies by users, even when working with familiar programs;
2. Opposition to the implementation by heads and key employees. The author notes, "The administrative resource is the main component of success in the transition to open source software" [12, p. 64].

Formulated differently, the initiative to migrate the educational process to free software, according to practicing teachers, should be supported at the level of the administration of the educational institution, by educational and methodological materials, training programs, and an attempt to organize it by the efforts of enthusiastic teachers is doomed to failure: all the entire system of migration is crumbling with a change of place the teacher's work.

Despite the described difficulties, many teachers, having felt all the degrees of freedom and advantages of open source software, continued to implement it into the educational process, but not on such a scale as originally planned. And there were enough reasons for this. So, N.V. Klyuchko, a teacher from the Moscow Metropolitan Governance Yury Luzhkov University, pointed at "professional and aggressive competitive marketing activity" [13, p. 62] of foreign corporations developing proprietary software in the list of reasons that create difficulties with the implementation of free software in the educational process. And this is true, and it is still relevant today, since large corporations can afford to finance marketing activities, injecting money into the rapid development of software, into improving graphic design and user interface, which cannot be said about the free software community.

The marketing policy of large corporations contributed to the fact that the Internet was flooded with articles, video tutorials, in which teachers were taught to work with proprietary software.

S.A. Dochkin from "Kuzbass State Technical University" identified as the main problems:

1. Low awareness of teachers and administration in the field of application of free software in the educational process and, as a result, its narrow distribution in educational organizations. In particular, he notes, "The analysis showed that only about 10% use operating systems based on GNU / Linux. Free applications Firefox, OpenOffice.org, GIMP and others have already been mastered and installed on at least 7–8% of computers..." [14, p. 134];
2. Lack of qualified teaching staff capable of using free software products for educational purposes competently;
3. Professional development programmes and retraining of teachers of vocational education institutions of the Kemerovo region are focused on the use of products and technologies based on the Windows operating system;
4. The audience of students at professional development programmes on free software, held in the Kemerovo region, is not actively following the path of mastering open source software;
5. Lack of a legal and organizational mechanism capable of ensuring the implementation of free software in vocational education institutions of the Kemerovo region, and a methodology for assessing the competencies of teachers, administrative and technical workers in relation to the implementation and use of open source software in the educational process;
6. Availability of free software products for a price and the need to pay for services to support the implementation of open source software in the educational process [14].

Thus, the author draws attention to one of the main problems of the transition period which is the lack of professional development programmes, organizational and methodological support for teachers.

According to the author of the article V.E. Veichko [15] the creation of an information and educational environment at the Luhansk Taras Shevchenko National University on the basis of free software did not carry neither systematic nor consistent character for the following reasons:

A psychological problem based on the myth that free software is intended for specialists in the field of information technology and its use will be beyond the strength of an ordinary user;

Methodological problem associated with the fact that until now (2014) there is no purposeful orientation regarding the maintenance of educational materials that would rely on free software;

A financial problem due to the fact that nevertheless, funds are needed to implement free software, which is often free in the educational and administrative process. These funds are required for organizing professional development programmes, for the development, adaptation and implementation of new curricula, as well as for the popularization of open source software among the administration, teachers and students.

The author notes, "The analysis of the demands of the information and educational environment, carried out according to the structure of its components, for software shows that modern free software fully satisfies both quantitatively and qualitatively the ever increasing demands of university education" [15, p. 127].

V. Dagiene [16] notes that one of the main challenges in using open source software is the need to engage educators and policymakers to effectively disseminate information about open source, justify the economic and pedagogical feasibility of using open source software in schools, and bringing together students and teachers to improve open source.

Returning to the current realities, more than 10 years after the beginning of the massive transition to free software, we see that even large corporations are increasingly looking towards open source software. A.E. Novodvorskiy also speaks about the validity of this approach, "the creation and development of complex software products "requires colossal resources either human ones, time or financial. The international community of open source software developers distributes this load, and thus we all significantly save resources and release new versions of products faster. The components of free software are used today by almost all developers from small firms to such world giants as Google and Microsoft" [17].

Many teachers, professors, methodologists of our days understand the importance of acquaintance of the younger generation with free software and its demand.

SBI SVE "St. Petersburg Center for Evaluation of the Quality of Education and Information Technologies" annually holds a scientific and practical conference "Problems and prospects for the implementation of free software (FS) in educational institutions of St. Petersburg" [18].

In 2020, the conference participants again discussed the problems of implementing free software into educational organizations in St. Petersburg. I.A. Tumanov in his report highlighted the important, most urgent problems of present time that complicate the refusal of educational organizations from proprietary software in favor of

open source software [19]. On the one hand, these are actually the same problems, such as a different user interface of both system and application open source software; lack of real and effective support from the state. However, the emphasis is also placed on more pressing problems in modern political and economic conditions. And here it is important to note that it is not just about proprietary software, but about domestic and imported proprietary software; about the difficulties that system administrators in schools have to face when trying to integrate free software into the existing infrastructure of an educational institution (including the inability to use special software, such as GIS, under free operating systems); about the inactivity of government agencies waiting for real problems with proprietary imported software (for example, its blocking as a result of the implementation of a new package of sanctions from foreign countries, on the territory of which the headquarters of firms developing proprietary software are located), and only then ready to act; about insufficient active position of domestic developers and non-profit associations of developers (communities) in advocating the use of open source software.

The introduction of free software into the educational process is increasingly becoming an element of government policy in the field of education. Brazil and Mexico is a significant examples [20, 21].

3 Materials and methods

The authors of this article carried out a frequent analysis of the mention of various problems of migration to free software, indications of which can be found in the review of articles described above on the topic we are considering. Thus, we can highlight the following main problems that arise during migration to free software and the corresponding number of references to these problems in scientific literature:

1. Lack of a plan for systematic work on the implementation of free software in edu cational organizations (S.A. Dochkin);
2. Passive behavior of teachers and engineers who are directly related to this issue, waiting for the "command from upstairs" (E.V. Andropova, T.N. Gubina, P.B. Berestov);
3. Lack of large-scale permanent state support (P.B. Berestov, I.A. Tumanov);
4. Lack of a high-quality, effective, targeted professional development programme for teaching and technical personnel of educational institutions and, as a result, lack of a sufficient number of qualified personnel (V.E. Velichko, I.A. Khakhaev, V.G. Slugin, dos Santos E.E.F., da Silva Figueira-Sampaio A.);
5. The presence of inactivity in terms of the use of familiar proprietary software by employees of educational organizations and unwillingness to make efforts to master new software tools, which, in particular, is due to the difference in the interfaces of free and proprietary software (A. Gonzales, V.E. Velichko, G. Zlobin, I.A. Khakhaev, E.V. Andropova, T.N. Gubina, E.R. Alekseev, I. Shevchenko,I.A. Tumanov);
6. Lack of a sufficient amount of high-quality methodological literature to support the process of migration to free software (V.E. Velichko, E.R. Alekseev, I. Shevchenko);
7. Problems of compatibility of text documents, spreadsheets and presentations when processing documents received from other users or external contractors (I.A. Khakhaev, V.G. Slugin);
8. Integration with internal corporate information systems, including incompatibility between special software and free operating systems based on the Linux kernel (I.A. Khakhaev, V.G. Slugin, I.A. Tumanov);

9. Aggressive competitive marketing activity of foreign corporations developing proprietary software and insufficient activity on the part of domestic free software vendors (N.V. Klyuchko, I.A. Tumanov);
10. Incompatibility of some peripheral equipment (printers, scanners) with free operating systems (V.G. Slugin).

4 Results

Let us present the frequency of mentioning various problems arising in the process of transferring the educational process of educational organizations to free software in Fig. 1.

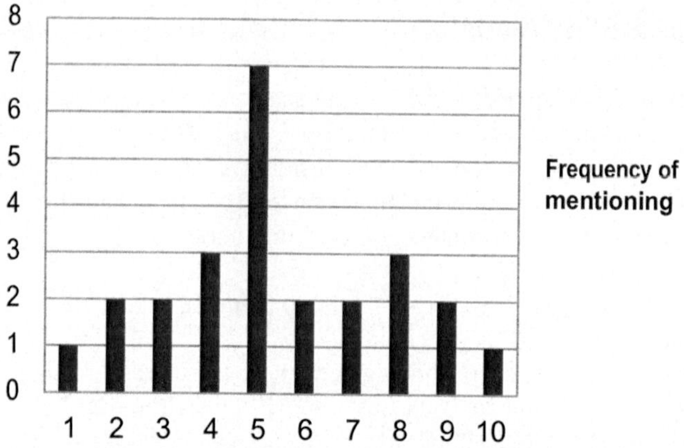

Fig. 1. Frequency of mentioning various problems arising in the process of migration to open source software

5 Discussion and conclusion

As a result of the analysis of the histogram shown in Fig. 1, it should be noted that more often than others, researchers note problems No. 4, 5 and 8, namely, the lack of a high-quality, effective, targeted professional development programme for teaching and technical personnel of educational institutions, the presence of inactivity in terms of the use of familiar proprietary software by employees of educational organizations, as well as Integration with internal corporate information systems.

In terms of finding and implementing ways to solve problem No. 4, we would like to note the serious contribution of LLC "Basealt FS", a domestic developer of operating systems for the ALT family, including for educational organizations. The distribution kit "Alt Education" includes a set of programs, including educational ones, from the category of "free software". These are applied programs that have been success- fully used in educational institutions for many years. In 2017, to help teachers, the company created an Internet portal located at kurs.basealt.ru for remote methodological support of teachers who use open source software in their professional work.

The remote support portal contains:

Training courses that allow teachers to master the technology of using operating systems of the Alt family, as well as free applied software, to solve professional pedagogical problems (general user component of teachers' ICT competence);

Methodological courses that consider the issues of methods of constructing the educational process in a general education school based on Alt software solutions (general pedagogical component of teachers' ICT competence);

Specialized training courses aimed at considering the methods of using applied and special software in the educational process, including programming systems [22].

As for the solution of the problem No. 5 in terms of overcoming the inactivity of the use of the usual proprietary software by employees of educational institutions, here, from the authors' point of view, serious efforts of the entire Russian pedagogical community are required, aimed, first of all, at popularizing the use of free software in teaching and educational process of educational organizations.

As an example of movement in this direction, we can cite the organization, within the framework of the Foresight session of informatics teachers "Looking into the Future", organized in September 2020 on the basis of the "TemoCenter" State Autonomous Educational Institution of Supplementary Vocational Education, the "Free Software" section [23], where the workshops on the use of open source software in teaching informatics in a secondary school were held. This event aroused genuine interest among teachers of computer science in Moscow.

In addition, it is important to note the fact that the difficulties faced by an organization providing services in the use of regulatory documents (the Consortium "Kodeks") [24], when switching to the use of free software, refer to the list of difficulties we have considered above No. 4, 5, 7, 8, typical for the migration of the educational process to open source software in educational organizations, which allows us to make a preliminary assumption that the stack of problems that are actualized in the process of the considered possible transition will be approximately the same for different sectors of the economy. Naturally, for different sectors of the economy, the ranks of the relevance of various difficulties will differ, which requires further research in this direction.

In the case of general education at all its levels, the goal is to form a diversified, creative personality capable of realizing creative potential in dynamic socio-economic conditions both in their own vital interests and in the interests of society (continuation of traditions, development of science, culture, technology, strengthening historical continuity of generations). Thus, we can talk about familiarizing pupils with various modern information technologies, but not about mastering them at the level of professional use, for which the use of free software can be considered as the most optimal direction.

References

1. Electronic supplement to the textbook "Informatics" for grade 7. https://lbz.ru/metodist/authors/informatika/3/eor7.php
2. Author's workshop L.L.Bosova. https://lbz.ru/metodist/authors/informatika/3/

3. Blanks for a computer workshop grade 7-9, https://lbz.ru/files/18943/
4. Informatics. Grade 5-6: Workshop on programming in the Scratch environment. https://lbz.ru/books/1072/10579/
5. Informatics. 5-6 grades: we study algorithms. My Kumir. https://lbz.ru/books/1072/10366/
6. Free software in government agencies. https://digital.gov.ru/ru/activity/directions/106/
7. Fedosov, A., Markushevich, M.V.: Methodology of choice between system and general applied proprietary and free software for educational process implementation in distance form. Inf. Soc. Educ. Sci. Cult. Technol. Future **1**, 114–130 (2017). (in Russian)
8. Fedosov, A., Markushevich, M.V.: Community ideology of free software as the basis for the formation of components of moral and ethical culture, life and personal self- determination of the student. Int. J. Open Inf. Technol. **8**(11), 42–48 (2020). (in Russian)
9. Slugin, V.G.: Implementation, use and maintenance of open source software in an educational institution on the example of FGOU open source software NRTK In: Fifth Conference Free Software in Higher Education, pp. 27–30. Alt Linux, Pereslavl (2010). (in Russian)
10. Alekseev, E.R.: Choice of free software for the educational process (on the example of the Department of Military and Industrial Engineering of DonNTU). In: Fifth Conference Free Software in Higher Education, pp. 25–27. Alt Linux, Pereslavl (2010). (in Russian)
11. Gubina, T.N.: Free software as a factor in the process of forming information and technological competencies: achieved results. In: Sixth conference Free software in higher education, pp. 32–34. Alt Linux, Pereslavl (2011). (in Russian)
12. Berestov, P.B. Experience of implementation of ALTLinux PSPO in educational institu- tions. In: Sixth conference Free software in higher education, pp. 63–65. Alt Linux, Pereslavl (2011). (in Russian)
13. Klyuchko, N.V.: Problems of using free software in practical training in a higher educational institution. In: Eighth Conference Free Software in Higher Education, pp. 60–64. Alt Linux, Pereslavl (2013). (in Russian)
14. Dochkin, S.A.: Features of the implementation of free products in vocational education: a regional aspect. In: Ninth Conference Free Software in Higher Education, pp. 133–137. Alt Linux, Pereslavl (2014). (in Russian)
15. Velichko, V.E.: Building an information and educational environment of the university based on free software. In: Ninth Conference "Free Software in Higher Education", pp. 127-125. Alt Linux, Pereslavl (2014). (in Russian)
16. Dagiene, V.: Research on Open Source Software Intended to Promote its Usage in Educa- tion. In: Kumar D., Turner J. (eds) Education for the 21st Century — Impact of ICT and Digital Resources. IFIP WCC TC3 2006. IFIP International Federation for Information Processing, vol. 210. Springer, Boston, MA (2006). https://doi.org/10.1007/978-0-387-34731- 8_33
17. Interview of TAdviser: Aleksey Novodvorskiy, "Basalt SPO" - about Linux in the "post- Intel" era. https://www.tadviser.ru/index.php/%Article:Novodvorskiy_Aleksey%2S_Deputy_General_Director_Basalt_SPO
18. Results of the XIII Conference "Problems and Prospects for the Implementation of Open Source Software in OO SPb 2020". https://umr.rcokoit.ru/news/194-itogi-xiii-konferencii-problemy-i-perspektivy-vnedrenija-spo-v-oo-spb-2020.html
19. Tumanov, I.A.: Problems of introducing domestic and free software in educational organizations. https://drive.google.com/file/d/1KXSqyTtOAQ_Sv6dWOOb9ku0sVCxsyaL/view?usp=sharing
20. dos Santos E.E.F., da Silva Figueira-Sampaio A.: Teacher education in Brazil: the use of free software in math education. In: Peters, M. (eds) Encyclopedia of Teacher Education. Springer, Singapore (2020). https://doi.org/10.1007/978-981-13-1179-6_401-1
21. Angel, G.: Mexican schools embrace Windows. https://www.wired.com/2001/08/mexican-schools-embrace-windows/?%20%20currentPage=all&%2520%2520currentPage=all

22. Gubina, T.N.: Methodological support of teachers for the transition to Russian operating systems of the Alt family: directions of work, results achieved, plans for the future. In: Four-tenth Conference "Free Software in Higher Education", pp. 25-29. MAKS Press, Moscow (2019). (in Russian)
23. Foresight session of teachers of informatics, Look into the future. http://foresight.itteachers.ru/
24. Khakhaev, I.A.: LibreOffice Teaching methodology based on the experience of implementing an office suite in the codex consortium. In: Twelfth Conference "Free Software in Higher Education", pp. 34–37. Basealt, Pereslavl (2017). (in Russian)

The Relationship Between the Gender; Age and Home Educational Resources and Mathematics Achievement in TIMSS 2015: Multilevel Analysis

Khalid Boulifa[1(✉)] and Kaaouachi Abdelali[2]

[1] Laboratory of Mathematics and Applications, Tangier, Morocco
[2] Research on Education and Training Lab, High School of Technology, University Mohammed First, Oujda, Morocco

Abstract. This paper studies the effect of gender, age and Home Educational Resources index (HER) on the performance of the Moroccan's eighth grade students in mathematics as part of the TIMSS 2015 survey (Trends international Mathematics and Science Study). Multilevel analysis results show that the variance explained at the school level is 26.43%. They also show that the HER index and students' age are positively correlated with students' achievement in mathematics and boys are favored over girls.

Keywords: Multilevel model · Home Educational Resources (HER) index · Age · PCM model · TIMSS 2015 · R software

1 Introduction

it is well known that the student's achievement has been widely investigated from the beginning of 1960, especially in the United States after the publications of the Coleman report [1]. A study made by [2], shows that the intimidation has a negative effect on student achievement in science. According to [3–6] the variations in school characteristics are associated with variations in student achievement. Another study by [6,7] shows that a good school climate has a positive effect on student achievement in science. Studies by [6,8] show that students' academic performance is affected if they perceive their school as "dangerous".

This study evaluates the impact of gender, age and Home Educational Resources index on mathematics achievement by using TIMSS 2015 eight grade data base. For this purpose, we assume that all factors are controlled, and we will try to find answers to the following research questions:

(1) Do schools make a difference in Mathematics achievement?
(2) What is the difference between girls and boys in Mathematics achievement?
(3) What is the effect of the Home Educational Resources (HER) index on student achievement in mathematics?
(4) What is the effect of students' age in mathematics achievement?

R. Silhavy et al. (Eds.): CoMeSySo 2021, LNNS 232, pp. 208–215, 2021.
https://doi.org/10.1007/978-3-030-90318-3_20

2 Method

Our study relates to 345 schools with 12992 students (6212 girls and 6780 boys), for which the average age is 14.49 years. We use two statistical software packages in our application: R Core Team and SPSS 20. To answer the first question, we apply the "One-way random effect ANOVA" model and for other questions, we use the "Intercepts and Slopes as Outcomes Model" [9]. The parameters of these models are estimated by the *lme4 package* [10], the significance of fixed and random parameters is tested using the *lmerTest package* [11]. The graphs are created by the *ggplot2 package* [12]. The partial credit model (PCM) [13] is used to scale the HER index. The missing data are imputed by the Multivariate Imputation by Chained Equation Method with the R *mice package* [14].

2.1 TIMSS Survey

The TIMSS survey has been conducted every four years since 1995. It concerns pupils continuing their studies in 4th and 8th grades of the compulsory education. This survey uses two-stage stratified cluster sampling: the first step is to sample schools using the "systematic probability-proportional to size (PPS)" approach [15].

Models Used in the TIMSS Survey. Item response models have been developed to allow a reliable estimate of the parameters of the items and the skill parameters of the students without these estimates being disturbed by the missing data. The model used to estimate item parameters is a development of the Rasch model, the partial credit model [13].

TIMSS 2015 Mathematics Framework: Eighth Grade. The TIMSS evaluation framework is divided into two dimensions: Content domains (Number, Geometry, Data and chance) and Cognitive domains (Applying, Knowing, Reasoning) [16] (see Table 1).

Table 1. Proportion of the TIMSS 2015 mathematics assessment: content and cognitive domains in the eighth grade

Content domains	Percentages	Cognitive domains	Percentages
Number	30	Knowing	35
Algebra	30	Applying	40
Geometry	20	Reasoning	25
Data and chance	20		

Sources data: [16]

2.2 Measures and Covariates

Dependent Variables. The dependent variable is the average scores of student's achievement in mathematics (from the five plausible values) obtained by the methods of item response theory [17].

Explanatory Variables. There are three variables: gender, age and the Home Educational Resources (HER) index. This last variable is based on responses to the following questions in the students' Questionnaire [18].

A- Number of books at home: 1 = "0–10"; 2 = "11–25"; 3 = "26–100"; 4 = "101–200"; 5 = "More than 200".

B- Number of home study supports: 1 = None ; 2 = Internet connection or own room; 3 = Both.

C- Highest level of education of either parent: 1 = Finished lower primary or lower secondary or did not go to school;

2 = Finished lower secondary; 3 = Finished upper secondary; 4 = Finished post-secondary education; 5 = Finished university or higher.

Scaling the HER Index. We use the unidimensional PCM model [13] and the R *TAM package* [19]. The HER index is estimated by the Weighted Likelihood Estimation method. Table 2 represents the *Infit* and *Outfit* indexes, we note that the values of these indexes are between 1 and 0.99. These results show that the model adequately fits the items [20].

Table 2. Outfit and infit for the home educational resources items.

Items	Outfit	Outfit	Infit	Infit
BSBG04	1	0.03199	1	0.0319
BSDG06S	1	0.00256	1	0.0026
BSDGEDUP	0.99	−0.0360	0.9999	−0.0360

3 Multilevel Analysis

3.1 Schools Difference in Mathematics Achievement

The first question concerns the study of the variability of student scores between schools. This question can be addressed using a model known by the unconditional multilevel model [9]:

Level 1 model

$$MathAch_{ij} = \beta_{0j} + \varepsilon_{ij};\ \varepsilon_{ij} \sim\ N(0, \sigma^2), \tag{1}$$

Level 2 model

$$\beta_{0j} = \gamma_{00} + \mu_{0j};\ \ \mu_{0j} \sim\ N(0, \tau_{00}^2), \tag{2}$$

Full model M_0

$$MathAch_{ij} = \gamma_{00} + \mu_{0j} + \varepsilon_{ij}, \tag{3}$$

Intraclass Correlation Coefficient

$$ICC = \tau_{00}^2/(\sigma^2 + \tau_{00}^2). \tag{4}$$

$MathAch_{ij}$: Mathematics achievement score of student i nested in school j; γ_{00} the grand mean Mathematics achievement; β_{0j} : the mathematics achievement mean for school j; τ_{00}^2: between-school variance and σ^2: Within school variance.

3.2 Variation of Student's Achievement as a Function of HER "(a) at the Student's Level and (b) Across Schools

Figure 1b represents 345 regression lines of mathematics on HER (weak lines) and the average line corresponding (solid line). From this Figure, the slopes and intercepts are found to be different. So, to answer the question of variation of student's achievement as a function of HER, we use the "Intercepts and Slopes as Outcomes" Model, The variable HER will be centered with the group-mean $HER_{.J} = \frac{1}{n_j}\sum_{i=1}^{n_j} HER_{ij}$ (group-mean centering) [9].
Level 1 model

$$MulhAch_{ij} = \beta_{0j} + \beta_{1j}(HER_{ij} - HER_{.j}) + \varepsilon_{ij};\ \varepsilon_{ij} \sim\ N(0, \sigma^2), \tag{5}$$

Level 2 model

$$\beta_{0j} = \gamma_{00} + \mu_{0j};\ \mu_{0j} \sim\ N(0, \tau_{00}^2) \tag{6}$$

$$\beta_{1j} = \gamma_{10} + \mu_{1j};\ \ \mu_{1j} \sim\ N(0, \tau_{11}^2) \tag{7}$$

Full model M_1

$$MathAch_{ij} = \gamma_{00}+\gamma_{10}(HER_{ij}-HER_{.j})+\mu_{1j}(HER_{ij}-HER_{.j})+\mu_{0j}+\varepsilon_{ij}. \tag{8}$$

β_{0j}: is the unadjusted mean for school j [9]; μ_{0j} and μ_{1j} are the random effects; γ_{01}; γ_{10} and γ_{00} are the fixed effects.

3.3 Difference Between Girls and Boys in Mathematics Achievement

To study the difference between girls and boys in Mathematics achievement, We add the variable ITS EX in the M_1 model (Eq: (8)).
Level 1 model

$$MathAch_{ij} = \beta_{0j} + \gamma_{01}ITSEX + \beta_{1j}(HER_{ij} - HER_{.j}) + \varepsilon_{ij};\ \varepsilon_{ij} \sim\ N(0, \sigma^2), \tag{9}$$

Level 2 model

$$\beta_{0j} = \gamma_{00} + \mu_{0j};\ \mu_{0j} \sim\ N(0, \tau_{00}^2), \tag{10}$$

$$\beta_{1j} = \gamma_{10} + \mu_{1j};\ \ \mu_{1j} \sim\ N(0, \tau_{11}^2), \tag{11}$$

Full model M_2

$$\begin{aligned} MathAch_{ij} &= \gamma_{00} + \gamma_{01}ITSEX + \gamma_{10}(HER_{ij} - HER_{.j}) \\ &\quad + \mu_{1j}(HER_{ij} - HER_{.j}) + \mu_{0j} + \varepsilon_{ij}. \end{aligned} \tag{12}$$

3.4 Variation of Students Achievement as Function of Age: (a) at the Student's Level and (b) Across Schools

Figure 1a represents 345 regression lines of mathematics on Age (Weak lines) and the average line corresponding (solid line). From this Figure, the slopes and intercepts are found to be different. So to answer the question of variation of students achievement as a function of Age, we use the "Intercepts and Slopes as Outcomes" Model. The variable Age will be centered with the grand-mean (grand-mean centering) [9].

In the M_2 model (Eq. (12)), the test the test $H_0 : \tau_{11}^2 = 0 \; vs \; H_a : \tau_{11}^2 > 0$. is not significant at the 0.05 level. Thus we eliminate the random part $\mu_{1j}(HER_{ij} - HER_{.j})$ of the M_2 model (Eq. (12)). *Level 1 model*

$$Mathach_{ij} = \beta_{0j} + \gamma_{01} ITSEX + \beta_{1j} Age + \gamma_{10}(HER_{ij} - HER_{.j}) + \varepsilon_{ij}; \varepsilon_{ij} \sim N(0, \sigma^2), \quad (13)$$

Level 2 model

$$\beta_{0j} = \gamma_{00} + \mu_{0j}; \;\; \mu_{0j} \sim N(0, \tau_{00}^2), \quad (14)$$

$$\beta_{1j} = \gamma_{02} + \mu_{1j}; \;\; \mu_{1j} \sim N(0, \tau_{11}^2), \quad (15)$$

Full model M_3

$$Mathach_{ij} = \gamma_{00} + \gamma_{01} ITSEX + \gamma_{02} Age + \gamma_{10}(HER_{ij} - HER_{.j}) + \mu_{1j} Age + \mu_{0j} + \varepsilon_{ij}. \quad (16)$$

β_{0j}: is the unadjusted mean for school j [9]; μ_{0j}: a random effect that measures the difference between γ_{00} and the β_{0j} and its variance τ_{00}^2 measure the heterogeneity of schools in math achievement; μ_{1j}: a random effect which measures the difference between the average slope γ_{01} and slope β_{1j}, its variance τ_{11}^2 measure the heterogeneity of the students' age effect across schools.

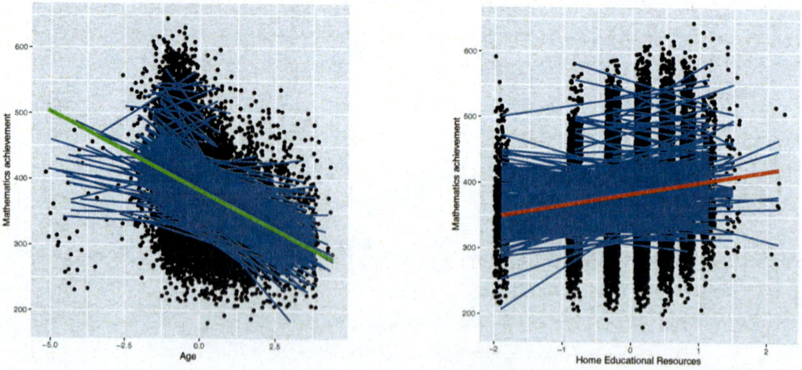

(a) The test $H_0 : \tau_{11}^2 = 0 \; vs \; H_a : \tau_{11}^2 > 0$. is significant at the 0.05 level.

(b) The test $H_0 : \tau_{11}^2 = 0 \; vs \; H_a : \tau_{11}^2 > 0$. is not significant at the 0.05 level.

Fig. 1. 345 regression lines of mathematics on Age (Fig. 1a) and HER index (Fig. 1b) (weak lines) and the average lines corresponding (solid lines)

4 Results

The results of models M_0; M_1; M_2 and M_3 are shown in Table 3.

Table 3. Parameter estimates for alternative multilevel achievement models.

Model	M_0	M_1	M_2	M_3
Fixed effect				
Intercept (γ_{00})	385.64 (2.179)^{s}	385.64 (2.22)^{s}	385.3 (2.22)^{s}	380.02 (1.95)s
HER (γ_{10})		8.57 (.93)^{s}	8.71 (0.99)^{s}	7.24 (0.98)^{s}
ITSEX (Ref = girls) (γ_{01})			0.56 (1.13)^{ns}	9.30 (1.08)^{s}
Age				−21.80 (0.56)^{s}
Variance components				
Level 1-student (σ_{00})	63.76	63.56	63.46	59.51
Level 2-school (τ_{00})	38.22^{s}	38.25^{s}	38.26^{s}	33.23^{s}
Level 2-school (τ_{11})		5.70^{ns}	5.70^{ns}	4.72^{s}
$Cor(\tau_{00}, \tau_{11})$		0.03	0.03	−0.58
ICC	26.43%			
PRV^{PRV}		−0.16%	−0.21%	24.41%

s: Test significant at the 0.01 level;
ns: Test not significant at the 0.01 level;
PRV: Proportional Reduction in Variance:
$PRV = \frac{\tau_{00}^2(null\ \ model) - \tau_{00}^2(full\ \ model)}{\tau_{00}^2(null\ \ model)}$ [9,21].

A-Schools Difference. The results of the M0 model show that the share of the variance explained at school level is 26.43%. The average score in mathematics is $3385.64, SD = 2.18$. The test $H_0 : \tau_{00}^2 = 0\ vs\ H_a : \tau_{00}^2 > 0$. is significant at the 0.05 level. Therefore, the students' achievement varies significantly across schools.

B-Effect of the HER Index. *(a) At the student's level:* The HER index is positively correlated with the students' achievement and is significant at the 0.01 level. When HER index increases by one unit the $Mathach$ increases by 8.57 units ($\gamma_{10} = 8.57, SD = 0.99, p < .001$).
(b) Across schools: The test of random slope:$H_0 : \tau_{11}^2 = 0\ \ vs\ \ H_a : \tau_{11}^2 > 0$ is not significant at the 0.01 level, therefore, there is a homogeneity of the HER effect through schools.

C-Gender Differences. The results of the M_2 model show that the boys are favored over girls by 9.30 points. this difference is significant ($\gamma_{10} = 9.30,\ SD = 0.98,\ p < 0.01$).

D-Variation of Students Achievement as a Function of Age *(a) At the student's level:* The students' achievement is negatively correlated with students'

Age and is significant at the 0.01 level. When Age increases by one unit the $Mathach$ decreases by 21 units ($\gamma_{02} = -21.80, SD = 1.08, p < .001$).

(b) Across schools: The test of random slope: $H_0 : \tau_{11}^2 = 0 \;\; vs \;\; H_a : \tau_{11}^2 > 0$ is significant at the 0.01 level, therefore, the Age effect varies through schools.

5 Conclusion

The results of the M_2 model (Eq: (16), Table 3, column 5) show that the Home Educational Resources has a positive and significant effect on the performance of students in Mathematics and the HER effect does not vary significantly across schools. After adding the Age variable to the M_2 model, the difference between the two sexes increases from 0.56 point to 9.30 points ($\gamma_{01} = 0.56, \;\; SD = 1.13 \;\; versus \;\; \gamma_{01} = 9.30, \;\; SD = 1.08$). Additionally, the effect of HER changes slightly as: $\gamma_{10}(M_3) - \gamma_{10}(M_2) = 1.47$. Additionally. the proportion of total variance at school level decreases by 24.41% (Proportional Reduction in Variance). These results lead us to conclude that the students Age provides an improvement of students' performance and affects negatively their performance in Mathematics. This study examines the effect of Age and Home Educational Resources on students' achievement in mathematics. The research findings show that: (a) The boys are favored over the girls (b) The HER index is positively correlated with the students' performances in mathematics (C) The Age is negatively correlated with the students' performances in mathematics.

References

1. Coleman, J.S., et al.: Equality of educational opportunity. Government Printing Office, Technical report, Washington, DC (1966)
2. Boulifa, K., Kaaouachi, A.: The link between the home resources for learning index and mathematics achievement in Moroccan context: a multilevel analysis. Appl. Math. Sci. **9**(13), 637–652 (2015)
3. Card, D., Krueger, A.B.: School resources and student outcomes: an overview of the literature and new evidence from north and south Carolina. J. Econ. Perspect. **10**(4), 31–50 (1996)
4. Greenwald, R., Hedges, L.V., Laine, R.D.: The effect of school resources on student achievement. Rev. Educ. Res. **66**(3), 361–396 (1996)
5. Lockheed, M.E., Verspoor, A.: Improving Primary Education in Developing Countries. Oxford University Press, Washington, D.C. (1991).Published for the World Bank
6. Boulifa, K., Kaaouachi, A.: The relationship between students' perception of being safe in school, principals' perception of school climate and science achievement in TIMSS 2007: a comparison between urban and rural public school. Int. Educ. Stud. **8**(1), 100–112 (2014). Published by Canadian Center of Science and Education
7. Parcel, T.L., Dufur, M.J.: Capital at home and at school: effects on student achievement. Soc. Forces **79**(3), 881–911 (2001)
8. Milam, A., Furr-Holden, C., Leaf, P.: Perceived school and neighborhood safety, neighborhood violence and academic achievement in urban school children. Urban Rev. **42**(5), 458–467 (2010)

9. Raudenbush, S., Bryk, A.: Hierarchical Linear Models: Applications and Data Analysis Methods. SAGE Publications (2002)
10. Bates, D., Mächler, M., Bolker, B., Walker, S.: Fitting linear mixed-effects models using lme4. J. Stat. Softw. **67**(1), 1–48 (2015)
11. Kuznetsova, A., Bruun Brockhoff, P., Haubo Bojesen Christensen, R.: lmerTest: tests for random and fixed effects for linear mixed effect models (lmer objects of lme4 package) (2013). R package version 2.0-0
12. Wickham, H.: ggplot2: Elegant Graphics for Data Analysis. Springer, New York (2009). https://doi.org/10.1007/978-0-387-98141-3
13. Masters, G.: A rasch model for partial credit scoring. Psychometrika **47**(2), 149–174 (1982)
14. Buuren, S., Groothuis-Oudshoorn, K.: MICE: multivariate imputation by chained equations in R. J. Stat. Softw. **45** (2011)
15. Foy, P., Olson, F.: TIMSS 2007 User Guide for the International Database, Supplement3. Teacher, and School Questionnaire Data. TIMSS and PIRLS International Study Center, Variables Derived from the Student (2009)
16. Michael, O., Martin, I.V.M., Hooper, M.: TIMSS 2015 mathematics framework. In: Liv Sissel, G., Mary, L., Alka, A., Mullis, I.V. (eds.) TIMSS 2015 Assessment Frameworks. Number 2016919196. TIMSS & PIRLS International Study Center, Lynch School of Education, Boston College and International Association for the Evaluation of Educational Achievement (IEA) (2016)
17. Olson, F., Martin, O., Mullis, V.: TIMSS 2007 technical report. Technical report, TIMSS and PIRLS International Study Center, Lynch School of Education, Boston College (2008)
18. Ina V.S.M., Michael, O.M., Shirley, G., Kerry, C.: TIMSS 2015 encyclopedia: education policy and curriculum in mathematics and science. Boston College, TIMSS & PIRLS International Study Center (2016)
19. Kiefer, T., Robitzsch, A., Wu, M.: TAM: test analysis modules. R package version 1.0-2 (2014)
20. Smith, R., Schumacker, R., Bush, M.: Using item mean squares to evaluate fit to the rasch model. J. Outcome Meas. **2**(1), 66–78 (1998)
21. Singer, J., D.Willett, J.B.: Applied Longitudinal Data Analysis: Modeling Change and Event Occurrence. Oxford University Press, New York (2003)

Development of a Business Architecture Model and Its Role in Constructing an Organizational Architecture

Kamelia Shoilekova(✉)

Angel Kanchev University of Ruse, 8 Studentska Street, 7000 Ruse, Bulgaria
kshoylekova@uni-ruse.bg

Abstract. This paper describes the interrelation between business and enterprise architectures. The main purpose is to highlight in what way a carefully planned business architecture in supplement to the enterprise architecture units is able to reduce the risks faced by an organization and to remove the occurrence of unusable data islands.

The paper presents several aspects of the bridge between Business and Software Architecture and the circumstances leading to it. The research described in this paper confirms the existence of different aids and methods helping the convergence of the business architecture and software architecture, which strengthens the company's positions and benefits. Their advantages and disadvantages are pointed out.

Keywords: Business architecture · Enterprise architecture · Organizational architecture · Software architecture

1 Introduction

The dynamics of contemporary business environment sets modern enterprises to challenges demanding a prompt response to all changes and fluctuations, and consequently, raises the necessity of appropriate tools for implementing flexible management.

The concept of enterprise architecture is developed as such an instrument. It is built on the perception that the management should have a "view from the top" to highlight the entire organization with all its units in order to observe how they work, interact and synchronize. Besides, the significant role of the information technologies (IT), which enhance the enterprise adaptability and its successful performance, is a generally acknowledged fact. However, the management often confronts situations when the existing ITs are not synchronized with the organization goals in result of which "islands of information" are being created. Increasingly, the achievement of accordance between business processes and IT is laid in the focus because it directly effects the organization flexibility and its adaptability to the dynamic environment [4, 6].

R. Silhavy et al. (Eds.): CoMeSySo 2021, LNNS 232, pp. 216–223, 2021.
https://doi.org/10.1007/978-3-030-90318-3_21

2 Business Architecture as an Element of the Organization Architecture

To understand the essence and practical applicability of the Enterprise Architecture (EA) it is needed to specifically clarify its structural units. EA can be presented as a composition of four interconnected elements: Technology Architecture, Information Architecture, Application Architecture and Business Architecture (Fig. 1). Introducing each of these architectural structures contributes to improving the activities within the separate enterprise units, and to the successful performance of the enterprise as a whole.

Business architecture is the bridge between a business model and the enterprise strategy on one side, and the economic functionality of the enterprise on the other side. It interconnects the functional structure of an enterprise in terms of business services and business information. The business architecture defines the business strategies, goals, processes, organizational structure, roles and responsibilities as well as their interconnections. It is the basis for specifying the requirements to the information systems and technologies necessary for implementing the business activities [7, 9, 11].

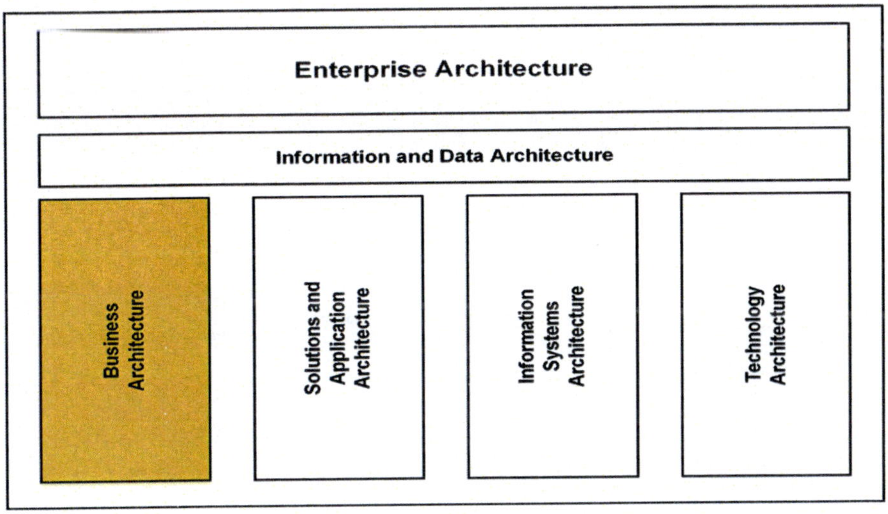

Fig. 1. Business architecture related to enterprise architecture

3 Relationship Between the Development Stages of Business and Software Architectures

Figure 2 shows the development stages of both the business architecture and software architecture. The diagram illustrates the relationships between the two architectures and the stages, which are developed in parallel.

The stage of defining the requirements to the business architecture can be overlaid on the stage that specifies the requirements to the software architecture. The aim of applying one architecture over another is to create a document describing the project implementation (respective modification). Once the document is worked out, it is analyzed, and then the specification of the technical project details is made including identification and description of the basic business rules, business processes, and participators. Most often, the description is implemented with the help of UML diagrams. The diagrams are precisely compiled and analyzed by the teams responsible for the business and software architectures. At this stage, it is essential that both teams use the same notations in the process of architecture development [5, 9].

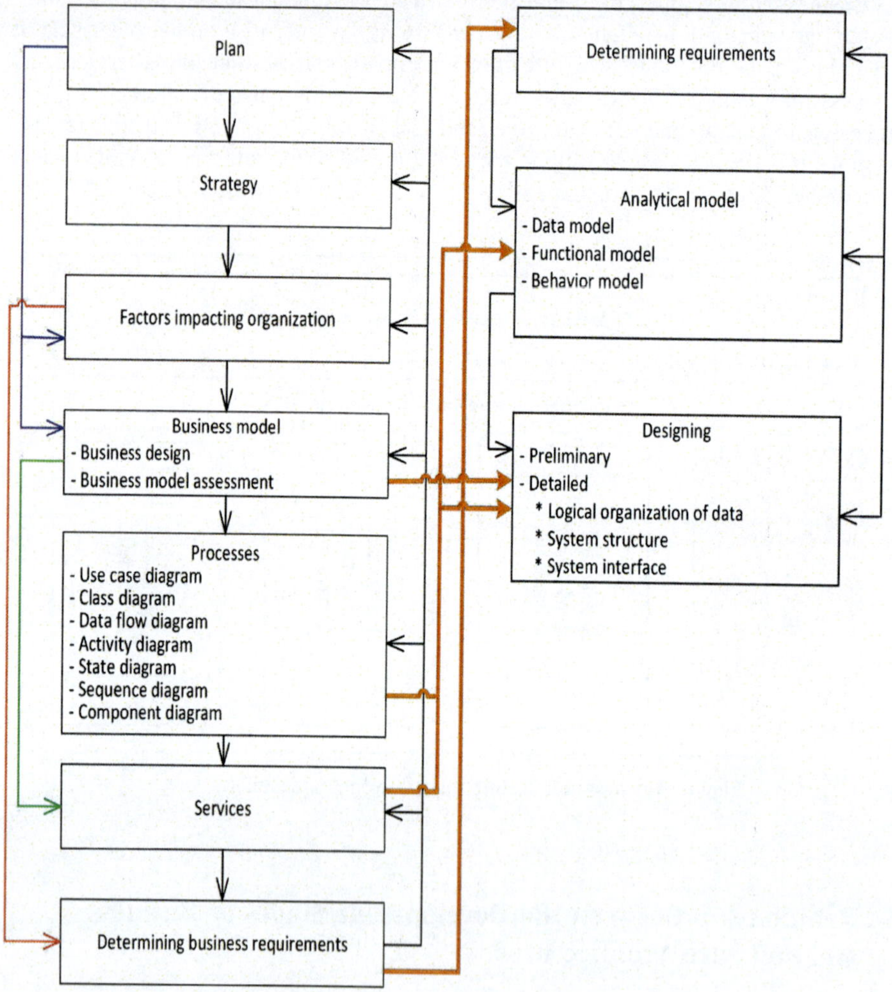

Fig. 2. Development stages of business and software architectures

At the phase of architectures' overlaying (Fig. 2) it is observed that the business requirements are applied on the stage of defining the requirements to the software architecture. A low point in developing the software architecture may appear when the requirements to it have been not clearly formulated. This drawback can be avoided if a model of the business architecture is created in advance. Therein the requirements to the software architecture are initially defined. Once they have been described and analyzed, the probability of inaccuracy or mistakes in the phase of re-use is much lessened. Another problem connected with the description of the requirements to a system is that they are dependent on the viewpoints of different participators in the process of software development. The core reason is the mismatches between the descriptions made by the specialists in a given subject domain (users) and the software developers who are responsible for the initial setting up of the system framework, strategy and goals (system analysts or designers). Most commonly, users define their demands and expected results unclearly and imprecisely. Being not aware of the computer system capabilities, they miss important details in their description. Analysts, on the other hand, are not familiar with the peculiarities of the activities, which have to be automated, and cannot estimate the importance of the raised problems. That is why the requirements should be determined in an interactive process with the participation of properly selected representatives of all interested parties [8].

The key component for settling the requirements to the business architecture is the set of factors, which put pressure on the organization (Fig. 2). After these factors are specified, the business model must be re-worked. The transformation of the business model is made in compliance with these factors, as well as the competitive enterprises. The transformation reflects on the business processes and services on one side, and on the other side, each transformation of the business process may lead to changing the number of the pressure factors, and to modification of the business model, processes and services. After the requirements to the business architecture are being changed, it is recommended to provide a concise analysis of the organization in order to estimate whether relevant amendments of the business plan and strategy should be implemented. If it is not necessary to make amendments, then the stage of specifying the requirements to the business architecture is applied on the stage defining the requirements to the software architecture. Otherwise, the requested changes are made on the stages concerning preliminary business plan and strategy, and after that due corrections are inserted on the remaining stages. The modifications are discussed with the enterprise management and after being confirmed, they are applied on the stage that specifies the requirements to the software architecture.

To develop an integrated view of the research, it is necessary to examine the organization structure and activities from different views. The first view concentrates on the idea that has provoked an owner to establish their enterprise (Fig. 3) [3]. For each enterprise, the idea is implemented with the help of five key elements:

- Management – for planning the personnel and their qualification in order to implement the idea. Is there a need for collaboration?
- Information - the task is to examine whether the idea deserves to be realized;
- Calculations - preliminary evaluation of the investment that is required for implementing the idea and in what time it shall be 100% refunded;

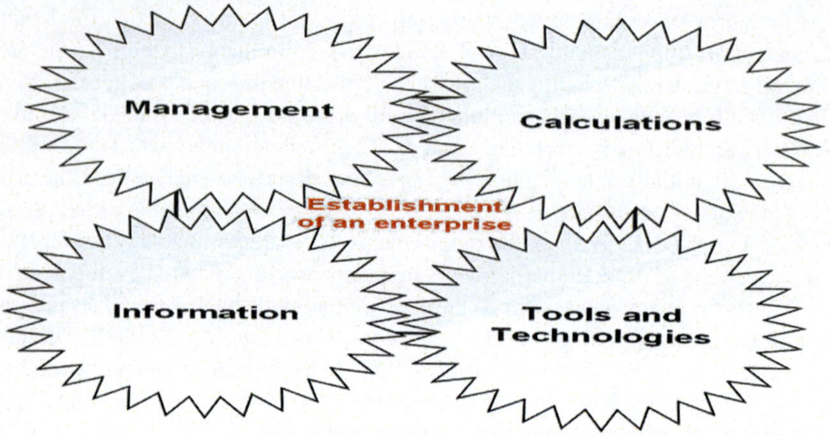

Fig. 3. Idea of enterprise foundation

- Facilities and instruments - what facilities should be provided for implementing the idea;
- Technologies - what technologies should be introduced for implementing the idea.

Understanding the enterprise architecture (EA) in substance and its practical applicability demands a more concrete elucidation of its structural components as well as an exact and clear determination of the aims and strategies for enterprise future development. EA can be presented as a structure of four interrelated components, namely: technology architecture, information architecture, application architecture and business architecture (Fig. 4). Introducing each of these architectural structures contributes to improving the activities of the enterprise units [12, 13].

Each enterprise can be presented by a structured approach to changes establishment in an organization. The approach aims to determine the critical points of an enterprise, which are examined so as to protect the company from plenty of turbulent events. The structured approach to changes establishment uses a change management tree for the enterprise modifications. Changes have an impact on the enterprise business processes, which on their side effect the promptness in establishing the pluses and minuses within an organization. The goal is to create a mechanism for an effective reassessment of the changes in the business organization. In this context, the most appropriate approach is to use a rank scale that allows to arrange the assessment elements' values by comparing them with specified scores as E (excellent), G (good), B (bad).

The priority of the participants in the structured approach to changes establishment is analyzed from two aspects:

- Significance of each scenario for the system success;
- Degree of difficulty provoked by the scenario implementation.

In the process of the structured approach to changes establishment performance due analysis is conducted to find out the critical points in the enterprises and solutions for

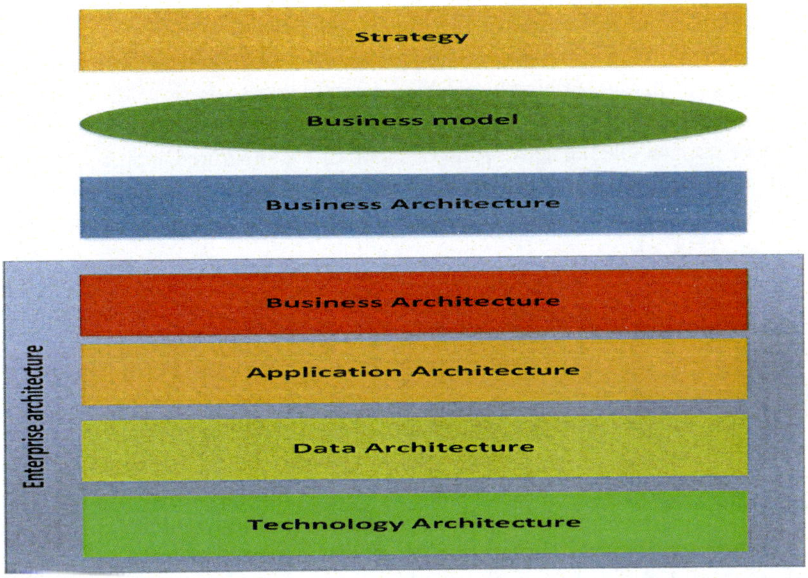

Fig. 4. Business Architecture in an EA context

their elimination are proposed. Figure 5 shows the tree revealing the main critical points of each enterprise. A company is characterized with two basic critical points:

- Business architecture – it is the handbook of the IT experts. The handbook must contain a very detailed route that has to be followed by the software architecture development team.
- Software architecture – it is the key to the company's successful performance [1, 2, 10].

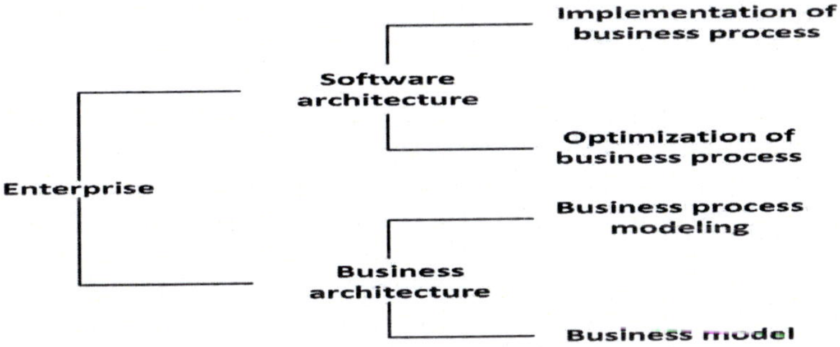

Fig. 5. Management of priorities in the scenarios of each company

After determining the critical points encountered in enterprises of different types, the findings ascertained are related to:

1. The essential role of the architecture in building a completed company's model;
2. The business architecture design's impact for eliminating the critical points yet at the phase of software architecture design.

It can be noticed that the above stated method of the structured approach to changes establishment incorporates a relation between the business architecture and software architecture. The relation is most clearly distinguished at the stage of designing both architectures (Fig. 2).

4 Conclusion

Today's economic realities, the global interactions and collaborations of companies place the business in extremely intensive conditions. The amount of activities that are performed per unit of time is continually growing and companies have to change their operating manner in order to adapt to the dynamics of work. The business architecture design aims to incorporate the solutions of a great number of problems that have to be later implemented by the software architecture. The research described in this paper confirms the existence of different aids and methods helping the convergence of the business architecture and software architecture, which strengthens the company's positions and benefits.

The enterprise architecture has one basic purpose: to ensure the successful performance of all the processes in the enterprise that shall raise its effectiveness. This is why a key principle in EA development is the synchronization of IT with the mission, vision, and goals of the organization. The synchronization can be achieved by developing a detailed and precise business architecture which projection is applied on the enterprise architecture.

The development of organization architecture proceeds in two main stages: design and implementation. Organization architecture design is an object of investigation firstly to business architecture and software architecture.

References

1. Enterprise Architecture Solutions, Business Modeling (2011). http://www.enterprise-architecture.org/business-architecture-tutorials/55-business-architecture-overview
2. Erl, T.: Service-Oriented Architecture: Concepts, Technology, and Design. Prentice Hall Publishers (2005)
3. Krastev, G., Voinohovska, V.: Smart mobile application for public transport schedules - logical model. TEM J. (2020). ISSN 2217-8309. 10.18421/TEM92-15
4. Rivkin W.: Closing the Business-IT Gap Once and for all. http://bpminstitute.org/articles/article/article/closing-the-business-it-gap-once-and-forall.html
5. Cullen, A.: Closing the Business Technology Alignment Gap. http://forrester.com/Research/Document/Excerpt/0,7211,46388,00.html
6. Jackowski, Z.: Bridging the IT-Business Gap with BPM and SOA. http://www.ebizq.net/topics/soa_security/features/7026.html?&pp=1
7. Basu A.: More on the IT/Business Alignment Gap. http://businessfoundation.typepad.com/bf_blog/2008/03/more-on-the-itb.html

8. Malveau, R.: Bridging the Gap: Business and Software Architecture. http://www.cutter.com/research/2004/edge040113.html
9. Penker, M.: Business Modeling with UML, 1st edn. (January 2000)
10. IBM, Getting Started with SOA. Student Notebook (2007)
11. Jarnagin, Ch.: How to Tap IT's Hidden Potential
12. Scott, J.:, Approaches to Business Architecture, Forrester Research (2008)
13. Atanasova, G.: Business Process Simulation Essence and Component Based Conceptual Architecture, International Journal of Computer Science and Information Security (IJCSIS) (2018). ISSN 1947-5500

Stacking-Based Ensemble Learning Method for House Price Prediction

Yuanning Liu[1], Yifan Wu[2], Linlin Su[3], Wenxuan Li[4], and Jianjun Lei[1](✉)

[1] School of Computer Science and Technology,
Chongqing University of Posts and Telecommunications, Chongqing, China
leijj@cqupt.edu.cn
[2] School of Economics, Shenzhen University, Shenzhen, China
[3] School of Mathematics and Physicial Science, University College London, London, UK
[4] School of Finance and Economics, Qinghai University, Xining, China

Abstract. It is difficult for the empirical prediction to provide accurate prediction results for house price due to its frequent fluctuation. Motivated by recent developments and advantages of emerging machine learning algorithms, this paper proposes a stacking model to improve the prediction accuracy of house price, which merges several outstanding base models, Bagging regression, Extra-Trees regression, XGBoost and LightGBM. Meanwhile, we analyze all the factors affecting house price and present a more practical and complex data preprocessing method to select the most contributing features by combining a creative feature engineering method. Moreover, we also use linear model as the meta model. Experimental results reveal that the proposed stacking model has high prediction accuracy and obtains the best performance over all base models and the meta model especially in predicting extreme values.

Keywords: House price prediction · Machine learning · Stacking ensemble model · Regression model

1 Introduction

In recent years, due to the geographical location, the number of bedrooms, construction time and other characteristics, the house price has undergone great changes [1]. For most people, house price affects their lifestyle, and the "house buying boom" follows. Therefore, it is valuable to predict the house price in the future through a specific algorithm. However, we cannot intuitively predict house prices according to the characteristics provided. Previous studies adopted fully Convolutional Neural Networks (CNN), Ordinary Least Square (OLS) regression and non-parametric regression methods to predict house price [2], but most of them use existing single model, which cannot avoid some defects existing in the single model.

In general, the house price index represents a snapshot of changes in house prices. However, a single model cannot make full use of all the factors affecting

R. Silhavy et al. (Eds.): CoMeSySo 2021, LNNS 232, pp. 224–237, 2021.
https://doi.org/10.1007/978-3-030-90318-3_22

house price. In this paper, a more practical and complex data preprocessing method, a creative feature engineering method, are studied for limited datasets and data features [3]. It is a stacking Model formed by integrating the Linear meta model, Bagging regression model [4], Extra-Trees regression model [5], XGBoost model [6] and LightGBM model [7]. Our staking model has many advantages. First, staking can further improve the accuracy and reduce the time cost when the accuracy of a single model is already high. Then, this new model does not need to adjust the parameters and other problems. Finally, staking model is good and stable in predicting extreme values.

The following sections of this paper are constructed as follows: related work reviews past work on house price prediction using different machine learning techniques. Section 3 explains the dataset and how to convert it into cleaned data. Section 4 introduces the principle of our stacking model and the characteristics of several sub-models. The model validation and experimental results will be described in Sect. 5 and conclusions will be derived in Sect. 6.

2 Related Work

The factors that affected the real estate market in the past were more straightforward than the present, and relatively accurate predictions of house price can be made through experience and simple regression models such as Linear regression. But as the factors affecting house prices gradually increase, for example, the number of bedrooms and bathrooms and the house's location are all highly correlated with the cost of the house, house price predicting starts to confuse researchers in investigation.

Thus, researchers choose to use more complex methods such as Bagging and Boosting algorithms to solve multivariate regression problems. C. R. Madhuri [8] attempts to model predictions by training and splitting data and used various machine learning techniques for prediction, such as Multiple Linear regression, Elastic Network regression, Gradient Boosting and Adaboost regression, with an emphasis on comparing different machine learning algorithms. Compared with other algorithms, the Gradient Boosting algorithm has higher accuracy in predicting house price. Pow [9] uses four regression techniques, namely Linear regression, Support Vector Machine (SVM), K-Nearest Neighbor (KNN) and Random Forest regression, as well as an integrated method combining KNN and Random Forest techniques to predict the price value of the real estate. The integration method has the least price prediction error, which is 0.0985. The application of Principal Component Analysis (PCA) does not impair the forecast results. Cheng Chen [10], in order to perform effective feature fusion and eliminate redundant and noise information in the initial feature vector, Elastic Network is used to determine the vital subset of features. In terms of PPIs prediction, the researchers use LightGBM to construct a PPIs prediction model. Five-fold cross-validation shows that the prediction performance of LightGBM is better than SVM and Random Forest.

Moreover, not satisfied with the simple use of a single algorithm, investigators combine some dimension reduction and feature screening technologies to make the model more efficient. Therefore, more and more investigators are trying to use the fusion model. For example, G. Srirutchataboon [11] evaluates the effectiveness of stacking different types of models, namely CNN-Random Forest, CNN-XGBoost and CNN-Adaboost, looking for the most effective stacking model for house price prediction. H. Fan [12], using XGBoost as a super model, integrates the Random Forest, Adaboost and LSTM models to obtain the final stacking model, which provides an essential reference for the formulation of natural rubber procurement strategy.

This article will also discuss the Linear regression, Bagging regression, Extra-Trees regression, XGBoost, LightGBM and stacking model based on these basic models to analyze house price.

3 Data Analysis and Pre-processing

3.1 Original Data

In this research, the dataset provided by the Kaggle website [13] is used. This dataset contains the sales of houses in King County, including Seattle, the United States, from May 2014 to May 2015. The original data includes 21613 observations and 21 variables. The meaning and type of each feature are shown in Table 1.

Table 1. Variables Description

Name	Descriptions	Type
Id	Identification number	int64
Date	When the house was sold	object
Price	House price(prediction target)	float64
Bedrooms	Number of Bedrooms	int64
Bathrooms	Number of bathrooms	float64
sqft_living	Square footage of the living room area	int64
sqft_lot	Square footage of the lot	int64
Floors	Total floors in house	int64
Waterfront	Whether the house has a view to a waterfront	int64
View	Has been viewed or not	int64
Condition	How good the condition is	int64
Grade	Grade given to the residential unit, based on King County Grading system	int64
sqft_above	Square footage of house apart from basement	int64
sqft_basement	Square footage of the basement	int64
yr_built	Built year	int64
yr_renovated	When the house was renovated	int64
zipcode	Zipcode	int64
lat	Latitude	float64
long	Longitude	float64
sqft_living15	Square footage of the living room area in 2015	int64
sqft_lot15	Square footage of the lot area in 2015	int64

These data cover most of the information for building a housing price prediction model. The housing price that needs to be predicted is a constant value, and some other numerical data are discretely distributed. The characteristic value of 'date' is a non-numeric type.

3.2 Data Correlation Analysis

Above all, an investigation of if there are any missing values in the dataset is necessary. According to statistics through scikit-learn, all observations are non-null.

It could be found that the feature contains the construction time (yr_built) and renovation time (yr_renovate). These two characteristics can show the house's present condition, so we use Eq. (1) to calculate the age of the house and add age as a new feature to the dataset. Similarly, 'zipcode' represents the geographic location of the house, which the latitude (lat) and longitude (long) can replace in the feature, so the 'zipcode' column is deleted.

$$age = \begin{cases} 2015 - yr\_built, & yr\_renovated = 0 \\ 2015 - yr\_renovated, & yr\_renovated \neq 0 \end{cases} \tag{1}$$

We also delete 'id' and 'date' that are not much related to establishing the model. After replacing and deleting steps, there are 19 features, including housing price.

To better understand the relationship between various characteristics and 'price', a heat map is drew to develop a correlation coefficient matrix. Through Fig. 1, factors that are most relevant to the housing price can be found.

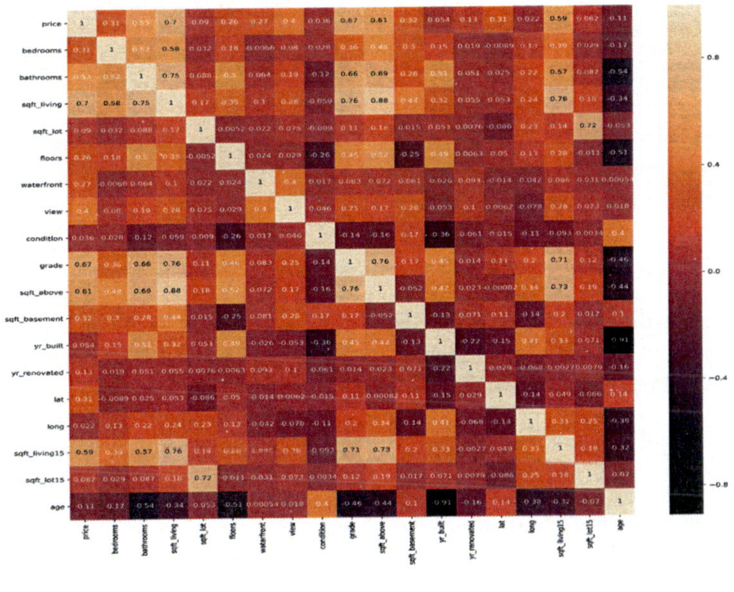

Fig. 1. Correlation matrix drawn by heat map

Table 2 lists the five features that have the strongest correlation with 'price'. Other components that are not mentioned have correlation coefficients with 'price' below 0.5. Figure 2 shows the linear relationship between 'price' and the five features with the largest correlation coefficient.

Table 2. Top 5 features with the highest correlation coefficients with 'price'.

Name	Correlation coefficient
sqft_living	0.702035
grade	0.667434
sqft_above	0.605567
sqft_living15	0.585379
bathrooms	0.525138

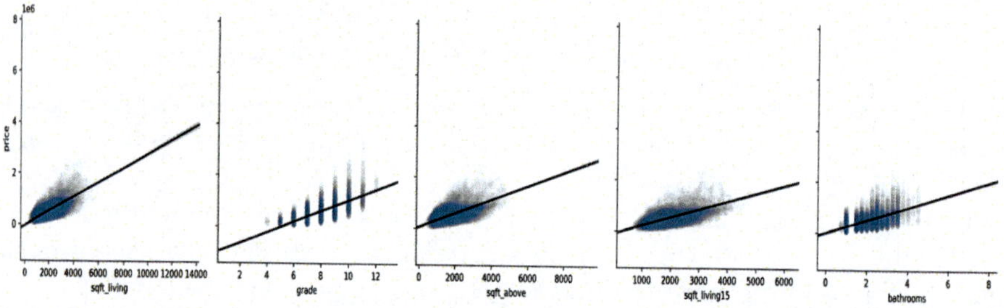

Fig. 2. The linear relationship between 'price' and the five features with the largest correlation coefficient.

3.3 Data Pre-processing

We observe the statistical description of the data, such as their mean value, standard deviation, maximum value, minimum value, and quartile points, found that most of the features have outliers. For example, the ultimate value of feature 'bedrooms' is 33, much larger than the average of 3.37. Outliers will affect the training of models, so Box Plot Diagram is applied to identify these outliers.

- A box plot displays five-number summary of a set of data: minimum, first quartile (Q1/25th Percentile), median, third quartile (Q3/75th Percentile), and maximum.
- Calculate Interquartile range (IQR) based on Q1 and Q3.

$$IQR = Q3 - Q1. \tag{2}$$

- Calculate lower limit and upper limit.

$$\begin{cases} Lower\ Limit = Q1 - 1.5IQR \\ Upper\ Limit = Q3 + 1.5IQR \end{cases} \tag{3}$$

- The data that lies within lower and upper limit are statistically considered normal and thus can be used for further observation.

As a result, the cleaned dataset has 13868 observations. Figure 3 compares the outcomes of 'price' and its five most relevant features before and after the outliers applying the box plot diagram.

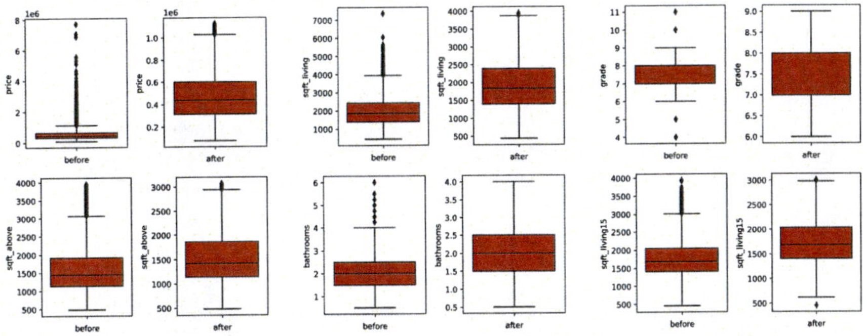

Fig. 3. Comparison before and after the box plot processing.

According to the latitude and longitude information provided by the dataset, the distribution of housing prices based on geographic location is visualized. After the purification, data points were more evenly distributed, convenient for exploring the mutative trend of house prices with geographic positions. The result is displayed in Fig. 4.

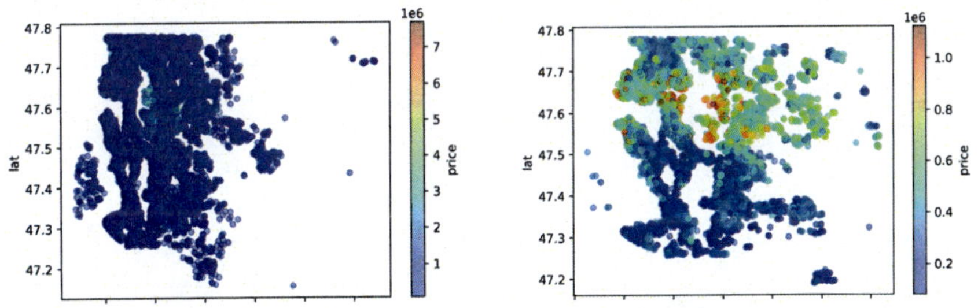

(a) Before purification

(b) After purification

Fig. 4. Distribution of house prices according to latitude and longitude.

At this point, the preprocessing of the data is basically accomplished. After that, the dataset is split into two parts: a training set and a test set. The training set accounts for 80% of the total data, which has 11094 observations, and the remaining 20% is used as the test set.

It could be seen in Fig. 5(a) that price's distribution in the training set is skewed. For its skewness, which is helpful to measure the asymmetry of a real-valued random variable's probability distribution about its mean [14] and its kurtosis, which describes the "tailedness" of the distribution [15], there is still a particular gap between it and the normal distribution. After applying log transformation to the price variable, the kurtosis and skewness are reduced from 0.877658 and 0.624419 to −0.098020 and −0.291361 respectively, makes it approximately distributed in a normal distribution shown in Fig. 5(b). In the results comparison stage, we need to restore the smoothed 'price' value.

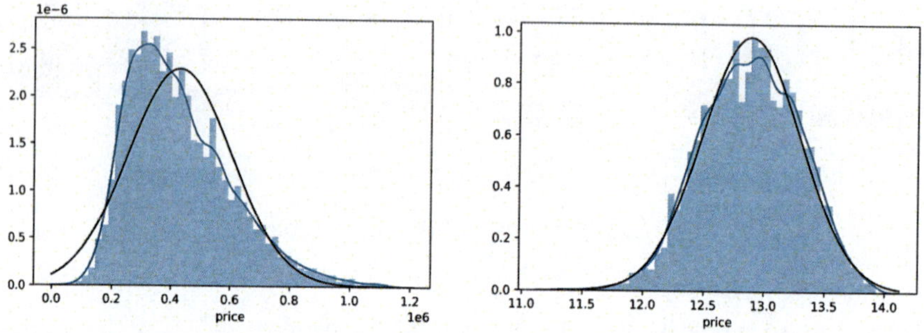

(a) Before log transformation

(b) After log transformation

Fig. 5. Distribution of price in training data set.

4 Ensemble Learning Based Prediction Model

In the early stage of the experiment, to understand what algorithms work better for our dataset, many different algorithms are applied, from which we screen four well-performing models: Bagging Regression (BR), Extra-Trees Regression (ER), XGBoost (XG) and LightGBM (LG). BR and ER belong to the Bagging algorithm in machine learning, while LG and XG belong to the Boosting algorithm. These four models all have excellent prediction results, and RMSE is basically around 0.15. Because the model is already well fitted, further improvement in accuracy with a single model is more complicated. Therefore, we propose a fusion model, stacking combining BR, ER, LG, and XG, to improve prediction precision continually. Figure 6 presents the flow framework of the stacking algorithm.

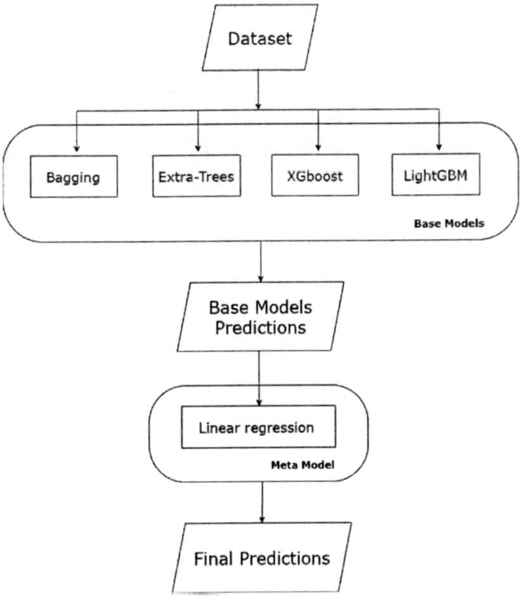

Fig. 6. The stacking ensemble scheme.

As shown in Fig. 6, the stacking model has two phases. The first stage is to train the four previously trained models as the base model of stacking, whose output will be passed as input into the second stage meta model that learns the performance of the first stage sub-models and gives them different weights according to the error, ultimately predicting the results and calculating the difference between the actual values. Here, we have tried algorithms with varying degrees of complexity to assume meta model, find that on our dataset, the more complex the algorithm does not mean higher prediction accuracy. On the contrary, some simple regression methods, such as Linear regression, can better fit the results.

4.1 Base Model Analysis

BR, ER, LG, and XG are selected as the stacking base models not only because they have superior prediction accuracy over other models, but also because they belong to two of the popular classes of machine learning algorithms, Bagging (BR, ER) and Boosting (LG, XG), used together to compensate for their respective shortcomings.

The principles of BR, ER, LG, and XG will not be covered in this article, we will mainly focus on the characteristics of each algorithm.

Bagging Regression. Compared with single regression tree, Bagging shows better prediction accuracy [16]. The most crucial point about the functionality of Bagging is that it can balance out the impact of these particular observations in the presence of "bad" outliers, but when there are some "good" outliers being underweighted, this can harm the forecast [17]. Bagging can effectively alleviate the model overfit case, and it can perform well when the data dimensions are large. However, its limitations are insufficient in improving prediction accuracy when the data is biased or underfitted. Bagging also consumes a lot of computing power causing it to be inapplicable in some specific cases.

Extra-Trees Regression. Extra-Trees is also a Bagging algorithm that can productively reduce the variance to prevent the emergence of overfitting. Unlike BR, it can also cope with the problem of rising bias and significantly save operational time. Nevertheless, when randomization increases above the optimal level, the variance decreases slightly, while the bias usually increases considerably [18].

XGBoost. Like other Boosting algorithms, XGBoost prevents the production of overfitting. XGBoost is highly scalable, and its strong scalability stems from some systematic and algorithmic optimizations that make XGBoost run faster and consume fewer resources than other Boosting models when the amount of data to be processed is enormous [6]. The weaknesses of XGBoost are reflected in that it is susceptible to outliers and cannot handle issues involving categorical variables.

LightGBM. LightGBM has faster running speed and less memory consumption than XGBoost without loss of precision [7]. LightGBM can accelerate the training process and improve the operating efficiency when determining the split points [10]. Moreover, it can also handle datasets with categorical features, which XGBoost cannot do [19]. The disadvantage of LightGBM is that it may generate a relatively deep decision tree leading to overfitting and being more sensitive to noisy points.

4.2 Stacking Ensemble Model

Different from the familiar Bagging and Boosting fusion algorithms, which fuse "weak" learners together, stacking integrates "strong" yet distinct models, further improving the accuracy of prediction on the original basis and without adding much computing cost. Figure 7 explains the operation process of our stacking model in detail.

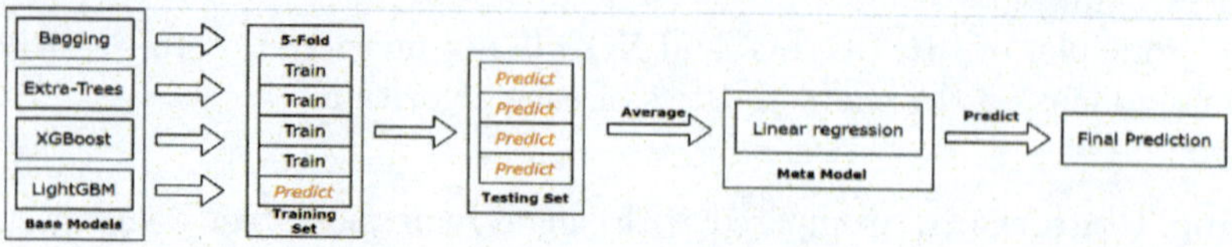

Fig. 7. The stacking ensemble scheme.

Firstly, the training set is divided into five parts and input into the base model. One is a test set, and the remaining four are training sets, and the training is carried out using 5-fold cross-validation. The five prediction results of the primary model output will be combined to obtain the average value and then be trained in the meta model, whose output is the final result of the stacking model. The detail of the stacking model is described by Algorithm 1.

Algorithm 1: Algorithm of Stacking Ensemble

Input: Dataset $D = \{(x_1, y_1), (x_2, y_2), \ldots, (x_m, y_m)\}$;
Base learning algorithms: $L_1, L_2, L_3, L_4 =$
$\{$*Bagging, Extra-Trees, XGBoost, LightGBM*$\}$;
Meta learning algorithm $L = \{$*Linear regression*$\}$;
Output: A stacking ensemble regressor H
Step 1 Learn base-level regressors
for $t = 1, ..., T$ **do**
  $h_t = L_t(D)$
end
Step 2 Construct a new dataset D' from D
$D' = \phi$
for $i = 1, ..., m$ **do**
  for $t = 1, ..., T$ **do**
    $x_{it}^{new} = h_i(x_i)$ % Use h_t to predict training example x_i and get new training set x_{it}^{new}
  end
  $D' = D' \cup \{((x_{1t}^{new}, x_{2t}^{new}, \ldots, x_{mT}^{new}), y_i)\}$
end
Step 3 Learn a meta regressor
$h^{new} = L(D\prime)$
Return $H(x) = h^{\text{new}}(h_1(x), h_2(x), \cdots, h_T(x))$

---

5 Results

In the validation phase, Linear regression will be used as the baseline for model evaluation, and the prediction results of other models will compare with it. Root Mean Square Error (RMSE) is used to tell how concentrated the data is around the line of best fit. It is calculated as:

$$RMSE = \sqrt{\frac{1}{N}\sum_{i=1}^{N}(\log y_i - \log \hat{y}_i)^2} \tag{4}$$

In this equation, N represents the number of observations, y_i indicates the actual value, and $\hat{y}_i$ expresses the predicted value. Because log transformation have been applied to the price variable in the training set, to maintain the order of magnitude consistency, the price variable in the test set also needs to use the same processing method separately.

In the model's training, 5-fold cross-validation is applied firstly on the training set to adjust parameters and select the model with the best prediction effect. Table 3 shows the RMSE value of each model on the training set and the test set. Eval. Ratio represents the evaluation RMSE of each model divides to the evaluation RMSE of Linear regression. For these three evaluation indicators, the smaller the value, the higher the model's prediction accuracy.

Table 3. Prediction result

Model	Train RMSE	Eval. RMSE	Eval. Ratio
Linear regression	0.229759	0.222157	1.000000
BaggingRegression	0.166787	0.162618	0.731996
ExtraTreesRegression	0.158226	0.151100	0.680149
LightGBM	0.149716	0.142748	0.642555
Stacking	0.149151	0.142491	0.641398

From Table 3, it can be seen that the result of the other four models are obviously better than that of the Linear regression model. Meanwhile, the best model evaluated by RMSE is undoubtedly the stacking model, though LightGBM also generates a close outcome. All models perform better on the test set than the training set, meaning there is no over-fitting issue.

Measuring the quality of a model is more than just checking its accuracy. It is also necessary to pay attention to the model fitting time, which may go up as the complexity of the model increases.

The fitting time of Linear regression is the shortest, and that of the stacking model is the longest of course. One thing worth to be mentioned is the LightGBM, as we can see from Table 4, its fitting time is the second shortest ranking just after that of the Linear regression model. Besides, from the perspective of accuracy, it beats almost all of the models except for stacking model, with only a tiny difference to win. Nevertheless, the fitting time of the stacking model is much longer than that of LightGBM, which is indicated in Table 4. After all, sacrificing the time, stacking can offer us better result than any other models in our experiment even for LightGBM.

In Fig. 8(a), the model that has the best prediction effect—stacking is compared with baseline model—Linear regression. Whether it is a house price near the average value or a house price that deviates far from the average value, the prediction result of stacking is much better than that of the baseline. Then we contrast LightGBM and stacking model, it is observed in Fig. 8(b) that although LightGBM demonstrates slightly better performance in predicting extreme values, it is also apparent that it sometimes exaggerate the extremes, which is not appealing compared with the stacking model, and this will also overstate the

Table 4. Fitting model runtime

Model	Fitting time (sec.)
Linear regression	0.0076
BaggingRegression	0.5942
ExtraTreesRegression	3.5942
LightGBM	0.3421
Stacking	22.9733

volatility. We consider that this might be one of the reasons why the accuracy of LightGBM is a little bit lower.

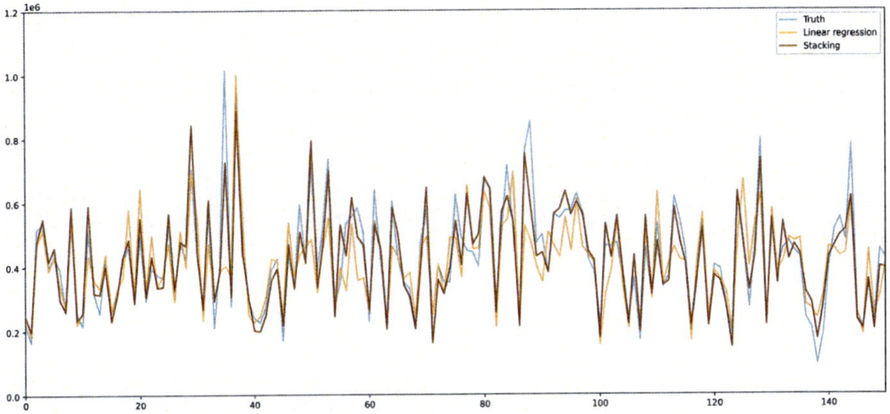

(a) Stacking & Linear regression

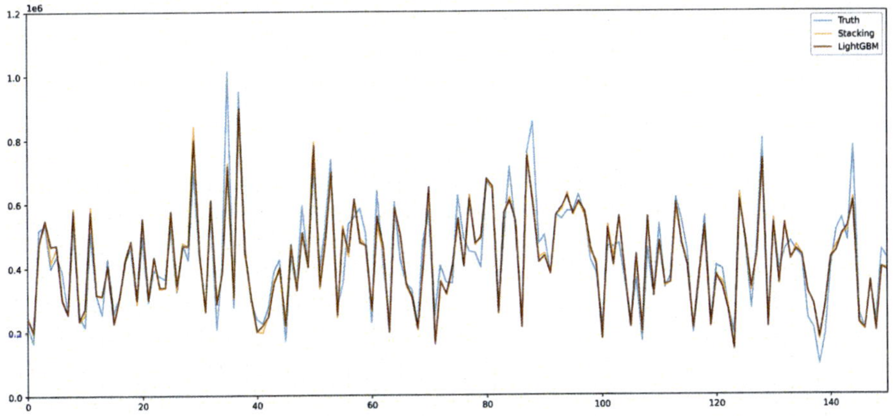

(b) Stacking & LightGBM

Fig. 8. Fitting result comparison.

In further experiments in the future, stacking is more suitable to make more precise predictions especially for extreme values. First, stacking eliminates many parameter tuning steps that can be cumbersome and time-consuming when training the model. Second, stacking can effectively neutralize models with poor predictions, such as Bagging and Linear in this article, which do not perform well when used alone. Stacking and LightGBM are much alike in prediction accuracy, but the exaggeration of LightGBM on extreme values is a key flaw that shouldn't be neglected.

6 Conclusion

In this study, a stacking model is proposed to make house price prediction for the King County's house dataset provided by Kaggle. 'sqft _living', 'grade' and 'sqft_above' are found to be several factors more correlated to house price, which is the same as the reality. Several outstanding single models, Bagging regression, Extra-Trees regression, XGBoost and LightGBM, are used as the base model of the stacking model, and Linear is used as the meta model. Experimental results reveal that the proposed stacking model has the best performance over all base models and the meta model. It is evident that the proposed stacking model has high prediction accuracy and is expert in predicting extreme values. Moreover, stacking's convenience in tuning parameters makes it a more competitive prediction model. However, there are still some limitations that need to be addressed in future studies. First, the prediction accuracy of stacking does not show significant advantages, and a more suitable model combination can be found by trying to fuse different regression prediction models and compare their results. Second, some data dimension reduction methods such as PCA, Linear Discriminant Analysis (LDA) can be used in the data preprocessing stage to select the most contributing elements to increase the running speed of stacking.

References

1. House Prices - Advanced Regression Techniques. https://www.kaggle.com/c/house-prices-advanced-regression-techniques
2. Beracha, E., Gilbert, B.T., Kjorstad, T., Womack, K.: On the relation between local amenities and house price dynamics. Real Estate Econ. **46**, 612–654 (2018)
3. Bogin, A.N., Doerner, W.M.: Property renovations and their impact on house price index construction. J. Real Estate Res. **41**(2), 249–284 (2019)
4. McMillen, D.P., Redfearn, C.L.: Estimation and hypothesis testing for nonparametric hedonic house price functions. J. Reg. Sci. **50**, 712–733 (2010)
5. Lu, B., Charlton, M., Harris, P., Stewart Fotheringham, A.: Geographically weighted regression with a non-Euclidean distance metric: a case study using hedonic house price data. Int. J. Geograph. Inf. Sci. **28**(4), 660–681 (2014)
6. Chen, T., Guestrin, C.: XGBoost: a scalable tree boosting system. In: Proceedings of the 22nd ACM SIGKDD International Conference on Knowledge Discovery and Data Mining (KDD 2016), pp. 785–794. Association for Computing Machinery, New York (2016)

7. Ke, G., et al.: LightGBM: a highly efficient gradient boosting decision tree. In: Proceedings of the 31st International Conference on Neural Information Processing Systems (NIPS 2017), pp. 3149–3157. Curran Associates Inc., Red Hook (2017)
8. Madhuri, C.R., Anuradha, G., Pujitha, M.V.: House price prediction using regression techniques: a comparative study. In: 2019 International Conference on Smart Structures and Systems (ICSSS), pp. 1–5 (2019). https://doi.org/10.1109/ICSSS.2019.8882834(2019)
9. Pow, N.: Applied Machine Learning Project 4 Prediction of real estate property prices in Montréal (2014)
10. Chen, C., Zhang, Q., Ma, Q., Yu, B.: LightGBM-PPI: predicting protein-protein interactions through LightGBM with multi-information fusion, Chemometrics and Intelligent Laboratory Systems, vol. 191, pp. 54–64 (2019). ISSN 0169-7439
11. Srirutchataboon, G., Prasertthum, S., Chuangsuwanich, E., Pratanwanich, P.N., Ratanamahatana, C.: Stacking ensemble learning for housing price prediction: a case study in Thailand. In: 2021 13th International Conference on Knowledge and Smart Technology (KST), pp. 73–77 (2021). https://doi.org/10.1109/KST51265.2021.9415771
12. Fan, H., Luo, X., Sun, Z.-H., Yuan, X., Qiu, S.: Multi-model fusion based on stacking: a predictive model for the price trend of natural rubber. In: 2020 IEEE 16th International Conference on Automation Science and Engineering (CASE), pp. 1237–1242 (2020). https://doi.org/10.1109/CASE48305.2020.9216888
13. House Sales in King County, USA data. https://www.kaggle.com/harlfoxem/housesalesprediction
14. Skewness. https://en.wikipedia.org/wiki/Skewness
15. Kurtosis. https://en.wikipedia.org/wiki/Kurtosis
16. Bühlmann, P., Bin, Yu.: Analyzing bagging. Ann. Stat. **30**(4), 927–961 (2002)
17. Grandvalet, Y.: Bagging equalizes influence. Mach. Learn. **55**, 251–270 (2004)
18. Geurts, P., Ernst, D., Wehenkel, L.: Extremely randomized trees. Mach. Learn. **63**, 3–42 (2006)
19. Jin, D., Lu, Y., Qin, J., Cheng, Z., Mao, Z.: SwiftIDS: real-time intrusion detection system based on LightGBM and parallel intrusion detection mechanism. Comput. Secur. **97**, 101984 (2020). ISSN 0167-4048

Methodology for the Development of Adaptive Training Systems Based on Neural Network Methods

Artem Obukhov(✉), Victoria Vostrikova, Alexandra Nazarova, Daniil Teselkin, and Ekaterina Surkova

Tambov State Technical University, 392000 Tambov, Russian Federation

Abstract. The development of modern adaptive training systems (ATS) based on virtual reality technologies is an expensive process. To increase the efficiency of the development of such adaptive systems, a methodology is proposed that includes a set of neural network methods. This made it possible to automate the processes of analysis, information processing, decision-making, which reduced the software complexity of the ATS implementation and the cost. The article describes in detail the main stages of the methodology, the tools used in the course of its implementation. The methodology differs from existing approaches by the transition to neural network architecture and the addition of a stage for the implementation of neural network components. The proposed methodology has been successfully tested in the subject area of adaptive training systems. The application of the methodology and neural network methods in the implementation of the ATS for the training of miners made it possible to reduce economic costs by 18.9% and the complexity of software by 19.6%, to increase adaptability (by 20%), quality (by 12%) and productivity (by 5.5%) relative to the system developed according to RAD methodology and MVC architecture.

Keywords: Methodology for the development of adaptive systems · Adaptive training systems · Neural network methods · Automation of the development of information systems · Machine learning

1 Introduction

Existing methodologies for the development of adaptive information systems (cascade model, CASE-technologies (computer-aided software engineering), agile methodologies, RUP (rational unified process), RAD (rapid application development), as well as various architectural patterns) allow to implement software of varying complexity and structure [1–5]. However, despite the wide distribution and effectiveness of these methodologies, the task of automating the process of developing information systems is not solved in them. Their use makes it possible to reduce the cost of material or time resources, to optimize the processes of implementing systems, but they are not universal and uniquely applicable approaches, since each task and subject area imposes its own limitations [6].

R. Silhavy et al. (Eds.): CoMeSySo 2021, LNNS 232, pp. 238–253, 2021.
https://doi.org/10.1007/978-3-030-90318-3_23

In the subject area of adaptive training systems (ATS), improving the development efficiency is an urgent scientific problem of great economic importance. This is due to the high cost of the used hardware components (virtual reality systems, simulation equipment). In addition, the process of developing such training systems is lengthy and resource-intensive. During the implementation of simulators, their adaptability is of great importance, namely the ability to adapt to the characteristics of the user in order to increase the comfort of being in virtual reality. Therefore, an urgent task is to create a methodology for the development of such adaptive systems using modern technologies - neural networks, which will reduce economic costs, the complexity of software implementation, increase the adaptability, quality and performance of the system.

However, the existing methodologies and software architectures do not sufficiently solve the problems listed above and increase the efficiency of the development of adaptive information systems (including adaptive training systems). Therefore, in the process of work, a new methodology is considered based on the use of neural network technologies and methods for automating control processes, analysis and information processing.

2 Methods

2.1 Analysis of Existing Development Methodologies and Architectures of Adaptive Systems

At the first stage of the research, an analysis of existing methodologies for the development of adaptive systems was carried out, which made it possible to draw the following conclusion. Despite the widespread use and effectiveness of existing methodologies (cascade model, CASE-technologies, agile methodologies, RUP, RAD [1–5]), the task of automating the process of developing information systems and integrating adaptability functions in them is not solved. Their use makes it possible to reduce the cost of material or time resources, to optimize the processes of implementing systems, but they are not universal and unambiguously applicable approaches, since each task and subject area imposes its own limitations.

The RAD methodology, however, in its principles corresponds to the direction of the problem being solved to increase the efficiency of the development of adaptive systems, but it will require modernization and changes. Significant disadvantages of the RAD methodology are the need for a large development team; the system must be decomposed into separate modules; attraction of the user (customer) at all stages of development; high requirements for the qualifications of developers; tight time constraints; high risks of project failure [4]. However, some modifications of this methodology will make it possible to neutralize the effect of these disadvantages.

Further, an analysis of information systems architectures was carried out, which made it possible to identify some key features that can be useful in automating their development.

First, it is necessary to note the service-oriented architecture, which contains effective solutions for data processing problems and the implementation of system modules. Some of the services can be further replaced by automated components with the development of methods and technologies for data processing, which will allow using already working, functioning systems as a basis [7].

Secondly, the existing architectural patterns (MVC, Mediator, Abstract Factory, and others) can be actively used to solve individual problems in the implementation of some modules and be integrated into the overall architecture of the system. This will significantly increase its flexibility and efficiency through the use of optimal approaches. [8].

Finally, the analysis showed that there is no ideal architecture that can be applied to the implementation of most information systems. Many architectures have serious problems with scalability and flexibility, which is a significant disadvantage for an adaptive system.

Therefore, a promising solution is the synthesis of a new architecture that takes into account the strengths of the considered approaches (for example, MVC) and is focused on automating the processes of analysis, processing and data transmission.

In this way, in the course of the analysis, it was found that the most promising for solving the problem of developing adaptive training systems is the rapid application development (RAD) methodology, which has a number of shortcomings associated with the integration of machine learning technologies, adaptation of the system to user characteristics and the influence of external Wednesday. As a basis for the system architecture, it is proposed to use the Model-View-Controller (MVC) pattern, which has proven its effectiveness in the development of many information systems.

2.2 Statement of the Research Problem

The aim of the work is to increase the efficiency of the development of adaptive training systems, expressed in reducing the complexity, cost of the software implementation process, and increasing adaptability.

In formalized form this problem belongs to the class of problems of structural-parametric synthesis and has the following form:

It is necessary to determine such a combination of structural elements, connections between elements, control methods, analysis, processing, generation and transmission of information, as well as a set of parameters for each element of the structure of the adaptive training system, at which the target function of the ATS efficiency reaches a maximum:

$$\{NNA^*, PRM^*\} = \underset{NNA,PRM}{\arg\max}\,(R), \tag{1}$$

$$NNA^* = (E^*, C^*, M^*, V^*, NN^*, D^*, MTD^*). \tag{2}$$

while doing:

- set of restrictions SO on the link between structure elements NNA:

$$SO = \{so_i\}, so_i \to \gamma_i(D) \subseteq D^*, \tag{3}$$

- sets of constraints PO to the range of parameter values:

$$PO = \{po_i\}, po_i \to \begin{cases} \varphi_i(PRM^*) = 0, \\ \varphi_i(PRM^*) < 0, \\ \varphi_i(PRM^*) \in PRM_i, \end{cases} \quad (4)$$

– sets of constraints RO on the maximum and minimum boundaries of the values of the quality estimates of the adaptive system:

$$\begin{aligned} &RO = \{ro_i\}, \\ &ro_i \to \alpha_i^{\min} \le \phi_i(R) \le \alpha_i^{\max}, \end{aligned} \quad (5)$$

where E^*, C^*, M^*, V^*NN^*, D^*, MTD^*- the selected elements of the Entities Environment, Control, Model, View, Neural Networks, as well as links and information processing methods, respectively.

$\gamma_i(D)$ - a function that forms a subset of links in accordance with the condition so_i, that must be included in the generated set $D*$;

$\varphi_i(PRM^*)$ - a function that transforms the parameters of the system into a form in which they can be used in equalities, inequalities, or to check whether they are in some subsets PRM_i;

$\phi_i(R)$ - a function that transforms the objective function of efficiency or its individual components to check compliance with the minimum $\alpha_i^{\min}$ and maximum $\alpha_i^{\max}$ admissible limits.

By the ATS efficiency function R in general, we mean a complex assessment of the system of the following form:

$$R = FR(R_V, R_D, R_A, R_Q, R_P) \quad (6)$$

where FR - a method for finding the optimal solution in a multicriteria optimization problem;

R_V - estimation of economic costs for the development of the ATS [9];
R_D - software complexity assessment of the ATS [10];
R_A - adaptability assessment of the ATS [11];
R_Q - quality control of the ATS [12];
R_P - performance evaluation of the ATS [13].

To solve this problem, we have developed a new methodology based on neural network methods.

2.3 Methodology for the Development of Adaptive Training Systems

Analysis of existing methodologies for the development of information systems made it possible to formulate a list of requirements and restrictions for the developed methodology [1–5]. Their implementation makes it possible to solve the set tasks related to the automation of the process of structural-parametric synthesis of information systems. In this way, the following principles of the developed methodology are formulated:

The principle of the neural network organization of the structure of the ATS determines that the structure of the ATS is represented as a set of isolated and independent entities, grouped into 5 categories: Environment, Neural Networks, Model, View and Control. Each element of the ATS uses neural network data channels to interact with other components and transfer information [14].

The principle of preparation and correspondence of information is that for the implementation of neural network methods of analysis, processing, generation and transmission, functioning on the basis of neural networks, in the ATS it is necessary to provide preliminary processing, correctness, sufficiency and correspondence of the initial data for the subsequent training of neural networks and artificial algorithms intelligence.

The principle of applicability of the neural network architecture is in the possibility of using methods and models based on the approaches of the theory of artificial intelligence, and forms a list of requirements for the sufficiency of the initial data, formalization of information processing processes, the possibility of using approximating functions and the need to implement adaptively functions in the chosen subject area.

Following the above principles allows formulating a general methodology for structural and parametric synthesis of the ATS based on neural network technologies. This methodology covers a wide class of systems of varying complexity. This is achieved through the development of separate universal automated methods and algorithms for analysis, processing, generation and transmission of information.

In the process of developing the methodology, the most relevant and universal approaches to software design, terms and concepts adopted in the field of programming were taken into account. However, the use of neural networks as a central element of the process of structural-parametric synthesis of the ATS, an integrated approach to automating the processes of analysis, processing, generation and transmission of information using neural networks is proposed for the first time.

The proposed methodology is based on the concept of rapid application development (RAD) [4], since its basic principles fully meet the stated goal: reducing the cost and time of developing information systems, focusing on small teams. The methodology includes all the main stages of the ATS, from planning the structure of the system and formalizing the subject area, ending with the development, optimization and modernization of the ATS.

The methodology implements the following new methods, discussed in detail and tested in previous studies:

- method of information flow formalization based on multi-level graph models to represent information processing processes, study the patterns of movement and life cycle of information objects, analysis and synthesis of the structure of information flows of the system [15];
- neural network method of information processing and transmission for automated organization of communication between the components of the neural network architecture based on the implementation of neural network data channels that perform analysis, processing and transmission of data using artificial intelligence methods [16];

- neural network method of automatic data generation based on the use of neural networks to form information objects with a given structure and properties with automatic search for optimal parameters of neural networks, used to obtain a compressed state of the object, recovery of lost data, prediction [17];
- A neural network method for automatic forwarding of information between ATS users, featuring the use of neural networks to determine the optimal performer of an operation based on the prediction of time and probability of its successful execution, enabling the routing of information;
- neural network method of data classification and distribution to organize their automated storage, retrieval, verification and comparison;
- A neural network method of adapting the information system to user characteristics, based on artificial intelligence, big data analysis and machine learning technologies [18];
- neural network-based control method based on the application of neural networks to automate decision support and control processes in ATS [19].

Let us consider the main steps of the proposed methodology.

A1. Planning. At this stage a systematic analysis of the subject area, aimed at obtaining the requirements for the information system.

The first step is a systematic analysis of the main objects, subjects and processes belonging to the subject area. ATS development team, the customer and future users determine the main objectives of the project by decomposing the subject area and analyzing the objects, subjects and processes occurring in it, after which a preliminary assessment of the complexity of the project is carried out. This allows you to identify the complexity and scope of the project, to assess the cost and time of its implementation, the possible technical requirements. Next, the necessary functions of the adaptability of the system are identified: the variability of interface and navigation, the ability to customize the design of the system interface, cross-platform, etc.

At the end of the stage on the basis of the information collected, the terms of reference for the design of ATS is formed.

A2. Formalization. The second stage is an unambiguous and detailed description of the information system architecture using various mathematical apparatuses and methods of system analysis.

To formalize the objects and subjects of this subject area, the apparatus of theories of sets and graphs, as well as the method of formalizing information flows based on the model of multi-level graphs are used. The regularities of information processing processes in a subject area can be formalized in the form of proofs of lemmas, theorems, and corollaries.

The resulting graph-theoretical model of objects and process diagrams of the subject area are integrated into the ATS components in accordance with the neural network architecture. Part of the processes are replaced by neural network data channels corresponding to them in functionality and type. Objects and subjects are allocated to neural network architecture entities (Environment, Control, Model and View). The obtained formalized scheme of the system elements, mathematical and functional models describing objects, subjects and processes of the subject area are combined into a single theoretical and graphical model of the ATS structure.

On the basis of this model a formalized formulation of the problem of structural-parametric synthesis of ATS is carried out. The solution to this objective will be carried out at the following stages of the methodology.

A3. Implementation of neural network components is the stage of analysis and processing of data of the subject area, formation of the structure and training of neural networks used in the course of automated solution of information processing tasks.

Based on the formalized model of ATS structure, properties of objects, subjects and processes of the subject area, a list of components based on neural network technologies is formed. Their totality will constitute the essence of Neural Networks in the proposed ATS architecture.

After determining the set of necessary components, the analysis and preprocessing of raw data is carried out. Data sampling is carried out in accordance with the requirements of the terms of reference.

The next step is to select the optimal architecture and parameters of neural networks for each neural network component of the ATS. This objective can be solved analytically by developers or machine learning experts, as well as using genetic algorithms and various software tools (For example, AutoKeras, AutoGAN, Google AutoML and so on [20–23]).

The trained neural networks are evaluated. If the accuracy, quality or other characteristics of the neural networks defined at this step are found to be unsatisfactory, then stage A3 begins again. If the obtained neural networks satisfy the thresholds for the selected metrics, then the transition is made to the stage of implementation of neural network components.

Next, the process of implementing neural network components based on the neural network methods developed within the methodology [14–19]. These methods allow to automate the processes of analysis, processing, generation, transfer of information, as well as to implement control algorithms using neural network technologies. The possibility of solving such problems using neural networks is confirmed by numerous studies [24, 25].

Step A3 is iterative in nature and is carried out to find the optimal set of input data and neural network structures. If new tasks and conditions arise, the corresponding neural network components are implemented and upgraded.

A4. Design and optimization are the main stage of ATS development. At this stage, the developers integrate the developed neural network components into the ATS, as well as make the necessary corrections to the system based on the comments generated by users in the next stage. Thus, this stage is also repeated many times, at the first iteration the primary prototype is formed, at subsequent iterations it is finalized to obtain the final version of the ATS.

In this stage the problem of structural-parametric synthesis of ATS is solved. Since in solving the optimization problem applies several metrics that are part of a complex criterion for ATS optimization, it is necessary to use one of the methods of solving multi-criteria problems [26]. After the successful solution of the problem the structure and parameters of the ATS are formed. They are analyze and adapt to the neural network architecture of the ATS. Then begins the development of the ATS itself. First, the readiness of the ATS prototype for implementation is assessed, after which the developers implement the ATS modules, design, and integration of neural network components obtained at the previous stage into the ATS prototype. Development is carried out in accordance with existing programming methods and paradigms within the framework of the previously developed module structure. The ATS prototype obtained at each iteration is sent to the users for analysis (in the next step).

A5. User rating. The stage implements the interaction of developers and users in order to get the best in terms of quality and convenience ATS.

The first step is a check for compliance of the current version of the ATS prototype with the requirements of the terms of reference, after which it is evaluated by quantitative and qualitative metrics of the complex criterion of ATS optimization. This assessment allows you to monitor the progress of the project and in case of exceeding the threshold for going beyond the permissible area to make the necessary adjustments. Next, the user evaluation of the ATS prototype is carried out, as well as its testing [27]. This allows you to identify inconsistencies between the views of developers and users on the project, as well as software bugs, inconvenient interface or operation complexity in program. Thus, the ATS prototype is subjectively evaluated for quality, performance and degree of adaptability, which allows to collect and consider the features and wishes of users.

The final stage of the user evaluation is consolidation of the collected information on non-compliance with the technical specification, exceeding the limits of quantitative and qualitative metrics of ATS evaluation, users' comments. The collected list of corrections in ATS is transferred to the previous stages (A3 and A4) For making changes to the structural-parametric synthesis problem, implementation of neural network components, software modules, interface, and so on.

The users are thus involved in the ATS development process and carry out its correction, which allows you to immediately eliminate inconsistencies in the ATS. This step is quite long and is repeated many times when the next iteration of the ATS prototype appears.

Stages A3–A5 are closely interconnected and cyclically follow each other (or run in parallel).

After receiving an acceptable assessment from users, completing the implementation of all neural network components, matching the ATS metrics of the complex optimization criterion and proving its readiness, the transition to the final stages of the ATS structural-parametric synthesis is made. After successful completion of the user evaluation phase, it is possible to determine the final values of all metrics and evaluate the optimality of the resulting solution for criteria such as adaptability and quality.

A6. Implementation is the final stage of the project, which includes final testing, implementation of the information system and user training.

Detected technical errors and comments from ordinary users (at this stage they include all users interacting with the ATS, not a separate group, as at stage A5) are corrected.

A package of accompanying documentation is prepared for the tested and operable ATS under the guidance of a group of methodologists. Methodologists also organize the process of training users to work with the ATS. At this stage it may be necessary to make small adjustments to facilitate the interaction of users with the system or prepare additional reference material.

After user training is completed, the ATS is implemented in the subject area in normal operation mode.

A7. Operation and modernization - the stage of operation in normal mode and support of the ATS.

Since ATS can change under the influence of structural changes, external influences, the emergence of new technologies, hardware and software, the analysis of changes in the structure of information objects, subjects, processes in the subject area is periodically carried out. The monitoring of external factors influencing the functioning of ATS is also carried out: failures, technical changes, load on the system, and so on. Thus, a list of necessary changes and adjustments to the ATS is formed, transferred to stages A3 and A4, on the basis of which the ATS is refined according to an abbreviated scenario, taking into account all the previously obtained results in the form of models, methods, software and hardware solutions.

This completes the decomposition of the stages of the methodology of structural-parametric synthesis of the ATS based on the neural network architecture. The methodology is based on existing approaches in the field of information systems development (RAD) [4], but its scientific novelty lies in separating neural network components into a separate stage of implementation and focusing on the use of neural network technologies, methods and components for solving problems of automating processes of analysis, processing, generation and transmission of data.

Next, we will consider the approbation of the developed methodology on the example of a specific adaptive training system for training miners.

3 Results

In accordance with the developed methodology, at the first stage, it is necessary to carry out planning, including a systematic analysis of the main objects, subjects and processes belonging to the subject area.

The main object in the ATS is a component of the training complex, which can be represented as software (virtual object or software module) or hardware (sensor, controller, simulation system, etc.) software or system. The subjects in the ATS are the users of the system who carry out many interactions with the virtual reality or the hardware components of the simulator. Information processing processes include: processing and recovery of lost data from sensors, making decisions in the system for imitating physical loads to adapt its speed to the nature of a person's movement; processing data on physical or chemical phenomena in virtual reality.

The functions of adaptability include: adaptation of the components of the simulator; adaptation of the speed of the system of imitation of physical activity to the actions of the user.

After analyzing the subject area, we will move on to solving the problem of structural-parametric synthesis of an adaptive simulator for training miners, which has the following form: it is necessary to determine the structure NNA^* and parameters PRM^* of ATS, at which the objective function of the complex optimality criterion reaches an extremum:

$$\{NNA^*, PRM^*\} = \underset{NNA,PRM}{\arg\min}\,(R_V) \quad (7)$$

subject to the following restrictions:

$$R_V \leq R_V^{MAX}, R_D \leq R_D^{MAX}, R_A \leq R_A^{MIN}, R_P \leq R_P^{MIN}, R_Q \leq R_Q^{MIN}, \quad (8)$$

where $R_V^{MAX} = 2000000$ is the maximum costs for the implementation of the ATS;
$R_D^{MAX} = 0.75$ is the maximum estimation of software complexity;
$R_A^{MIN} = 0.5$ is the minimum score for adaptability;
$R_P^{MIN} = 0.75$ is the minimum possible performance of the ATS;
$R_Q^{MIN} = 0.75$ is the minimum acceptable software quality assessment.

The dimension of the problem of structural and parametric synthesis of automatic telephone exchanges depends on the number of input variables, which include:

- 7 hardware parameters of the equipment on which the ATS operates (processor, video accelerator, storage device and its characteristics, RAM, etc.);
- 3 hardware parameters for interacting with virtual reality (means of displaying virtual space, type of controller, means of moving in virtual space);
- 1 parameter of imitation equipment (the presence of imitation of physical activity and breathing apparatus);
- 3 software parameters: the chosen development environment, DBMS and software implementation (technology stack for the development of software modules);
- 1 a parameter that determines the selected method of analysis and processing of information when conducting research and development for the implementation of software modules;
- 10 variables for each neural network component that determine the characteristics and structure of neural networks for solving specific information processing problems.

Thus, we have the dimension of this problem, equal to 55 (taking into account 4 neural network components for the implementation of the necessary functions of adaptability and automation of information processing processes).

Next, we will consider the area of variation of the input parameters, their possible values.

Hardware parameters:

1. Processor (20 values): ranging from 4 cores with a clock frequency of 3000 MHz to 16 cores with a frequency of 3500 MHz.
2. Video accelerator (20 values): variable from 3 GB to 16 GB of video memory with a GPU clock speed from 1000 to 1800 MHz.

3. Storage (20 values): ranging from 500–10000 GB hard disk drive (HDD) to 120–2000 GB solid-state disk (SSD).
4. RAM (8 values): ranging from 4 GB to 128 GB.
5. Motherboard: 10 values corresponding to different processor sockets.
6. Network card (4 options): with speeds from 10 Mbps to 10 Gbps.
7. Power supply and case (10 options): power supply capacity from 300 W to 1500 W.
8. Means of displaying the virtual space (10 values): ranging from the monitor to various models of virtual reality helmets.
9. Controller type (3 values): keyboard + mouse, virtual reality controllers, virtual reality gloves.
10. A means of movement in virtual space (3 values): keyboard, virtual reality controllers, treadmill.
11. Simulation equipment: 4 values reflecting all combinations of the presence or absence of simulation equipment to simulate physical exertion and respiratory activity.

Software parameters:

1. Development environment (5 options): Visual Studio, MonoDevelop, Android Studio, Pycharm, UnrealEd.
2. DBMS (5 options): (Oracle, MySQL, SQL, PostgreSQL, MongoDB).
3. Software implementation: 5 options, depending on the programming language and the selected engine for the implementation of virtual reality, Unity, Unreal, CryENGINE, PyQt, Android application.

The method of analyzing and processing data when performing scientific research works: 2 values, a classical analytical method for preparing data for their further use in the algorithmic support of software modules and a method for preprocessing data for training neural networks.

Neural network parameters (for each neural network component):

1. Neural network architecture (2 options): sequential or functional.
2. Type of neuron layers used (10 options): various workable combinations of dense layers for numerical values, convolutional for images, recurrent for sequences, normalization layers to reduce the effect of overfitting.
3. The number of dense layers of neurons (20 values): from 1 to 20.
4. Number of dropout layers (10 values): from 0 to 9.
5. The number of specialized layers (convolutional, recurrent, etc.) (10 values): from 1 to 10.
6. Number of neurons in a layer (50 values): from 5 to 2000.
7. Type of activation functions (5 values): linear, sigmoidal, hyperbolic tangent, linear rectifier (ReLU), scaled exponential linear function (SELU).
8. Optimizer (5 values): SGD, Adam, Adamax, Nadam, RMSprop.
9. Loss function (5 values): squared (MSE), mean absolute error (MAE), mean absolute percentage error (MAPE), cross-entropy, binary entropy.
10. Number of training epochs (10 values): from 3 to 300.

The maximum number of possible combinations of parameters and structure for each neural network component is also 500 million (for specific modules from 50 to 250 million).

The number of combinations of software and hardware parameters was 125 and 9.21 billion, respectively, but the real number is lower due to restrictions on the combination of components and software compatibility.

When solving the problem of structural-parametric synthesis, it is not required to carry out a full enumeration of all possible combinations, since the use of the developed neural network methods aimed at automating the process of selecting the structure and parameters of neural networks for solving information processing problems allows us to unambiguously determine the optimal parameters for each neural network component. Further, on the basis of checking the constraints, from the total number of admissible combinations of software and hardware parameters of ATS, an optimal combination is selected that provides the best value of economic costs.

Thus, among the set of all combinations of structures and parameters of software, hardware, neural network components, two options were chosen:

1. The best solution based on the RAD methodology and MVC architecture without the use of neural networks, designated as the classic ATS implementation.
2. Neural network implementation of ATS.

For both options, the same software and hardware is used, which will allow you to correctly compare their performance and cost. The results of this comparison are shown in Table 1.

Table 1. Comparison of classical and neural network ATS implementation.

Metrics	Classic implementation	Neural network implementation
Economic development costs R_V, rub	1685842	1366651
Complexity of the software R_D	0.61	0.49
Adaptability assessment R_A	0.68	0.88
Quality control R_Q	0.75	0.87
Performance R_P	0.885	0.94

After applying the methodology, the overall economic costs and software complexity are reduced, the quality, adaptability and performance of the system are increased. The conditions defined in the formulation of the structural-parametric synthesis problem have been met (Fig. 1).

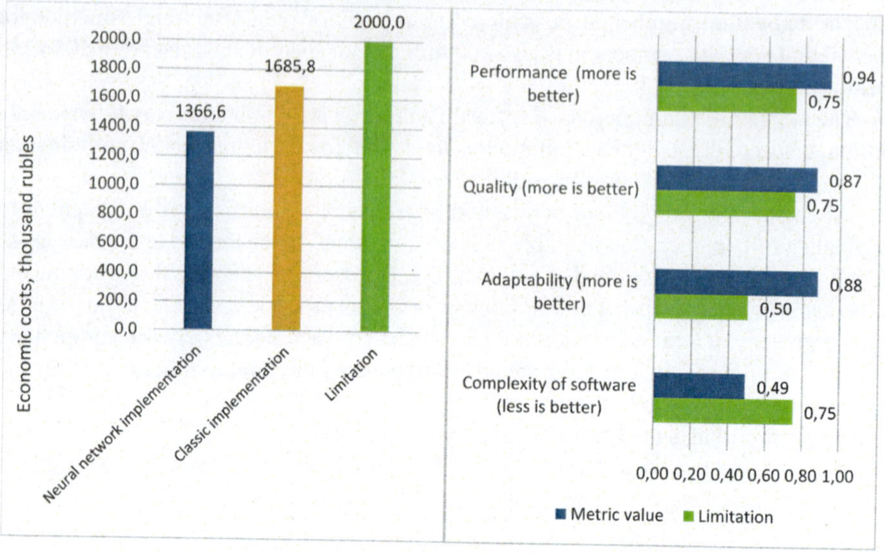

Fig. 1. Results of solving the problem of developing ATS.

As a result of the application of the methodology, a decrease in economic costs (by 18.9%), the complexity of software implementation (by 19.6%), an increase in adaptability (by 20%) and system quality (by 12%), as well as an increase in productivity (by 5.5%) were achieved.

The use of the methodology and neural network methods made it possible to implement the necessary adaptability functions: adaptation of the simulator components when replacing them and determining the speed of the running platform depending on the actions of a particular user. The developed ATS includes an adaptive running platform

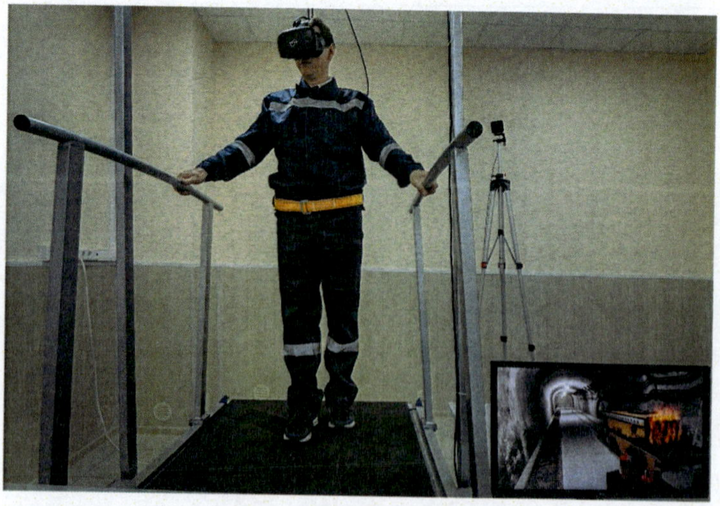

Fig. 2. Adaptive running platform for movement in virtual reality.

(Fig. 2) and a module for modeling respiratory processes to simulate smoke and air pollution (Fig. 3).

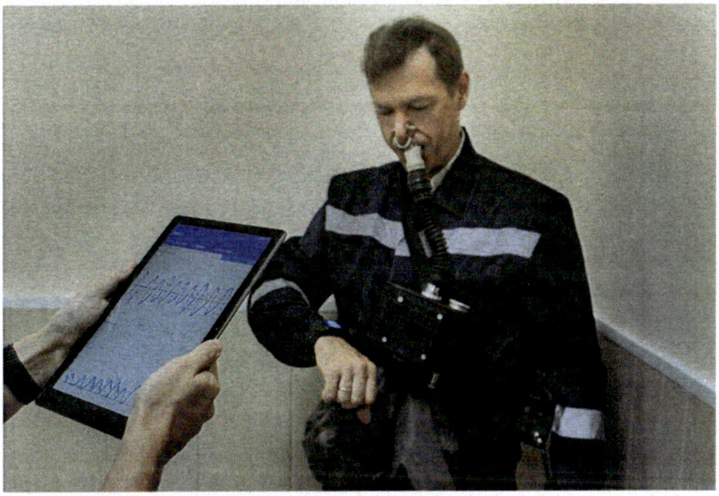

Fig. 3. A module for modeling respiratory processes and tracking physical indicators.

Thus, the proposed methodology made it possible to achieve the set goal - to increase the efficiency of the ATS development.

4 Conclusion

The paper considers an urgent scientific problem of great economic importance - increasing the efficiency of the process of developing adaptive information systems through the development and development of theoretical foundations and software tools based on machine learning technologies and neural networks. This problem is formalized in the form of the problem of structural-parametric synthesis of the ATS.

To solve it, a new methodology of structural-parametric synthesis based on neural network technologies is proposed. The basic principles of the methodology, the general structure, and the methods used are formulated. The stages of the methodology are formalized with a detailed description of each step in solving the problem of structural-parametric synthesis of the ATS. Within the framework of the methodology, a set of neural network methods has been implemented, aimed at reducing the complexity of the software implementation of the processes of developing ATS modules and information processing.

Approbation of the methodology in two subject areas has shown its high efficiency. Further research will be associated with the application of the developed methodology and neural network methods in new subject areas for the automation of processing, information analysis, and control.

This work was supported by the Ministry of Science and Higher Education of the Russian Federation within the grant of the President of the Russian Federation, MK-74.2020.9. The work was carried out on the basis of the Laboratory of medical VR simulation systems for training, diagnosis and rehabilitation.

References

1. Volkova, V.N., Vasiliev, A.Y., Efremov, A.A., Loginova, A.V.: Information technologies to support decision-making in the engineering and control. In: 2017 XX IEEE International Conference on Soft Computing and Measurements (SCM), pp. 727–730. IEEE, Saint-Petersburg (2017). https://doi.org/10.1109/scm.2017.7970704
2. White, B.E.: A complex adaptive systems engineering (CASE) methodology — the ten-year update. In: 2016 Annual IEEE Systems Conference (SysCon), pp. 1–8. IEEE, Orlando (2016). https://doi.org/10.1109/syscon.2016.7490547
3. Al-Zewairi, M., Biltawi, M., Etaiwi, W., Shaout, A.: Agile software development methodologies: survey of surveys. J. Comput. Commun. **5**(05), 74–97 (2017). https://doi.org/10.4236/jcc.2017.55007
4. Naz, R., Khan, M.N.A.: Rapid applications development techniques: a critical review. Int. J. Softw. Eng. Appl. **9**(11), 163–176 (2015). https://doi.org/10.14257/ijseia.2015.9.11.15
5. Asgar, T.S., King, T.M.: Formalizing Requirements in ERP Software Implementations. Lect. Notes Softw. Eng. **4**(1), 34–40 (2016). https://doi.org/10.7763/lnse.2016.v4.220
6. Jipp, M., Ackerman, P.L.: The impact of higher levels of automation on performance and situation awareness. J. Cogn. Eng. Decis. Mak. **10**(2), 138–166 (2016). https://doi.org/10.1177/1555343416637517
7. Rojas, H., Arias, K.A., Renteria, R.: Service-oriented architecture design for small and medium enterprises with infrastructure and cost optimization. Procedia Comput. Sci. **179**, 488–497 (2021). https://doi.org/10.1016/j.procs.2021.01.032
8. Chelliah, P.R., Subramanian, H., Murali, A.: Architectural Patterns: Uncover Essential Patterns in the Most Indispensable Realm of Enterprise Architecture. Packt Publishing Ltd., Birmingham (2017)
9. Sanchez, O.P., Terlizzi, M.A.: Cost and time project management success factors for information systems development projects. Int. J. Project Manag. **35**(8), 1608–1626 (2017). https://doi.org/10.1016/j.ijproman.2017.09.007
10. Kuznetsov, M.A., Surkov, V.O.: Analysis of complexity metrics of a software code for obfuscating transformations of an executable code. IOP Conf. Ser. Mater. Sci. Eng. **155**, 012008 (2016). https://doi.org/10.1088/1757-899x/155/1/012008
11. Bastien, J.M.C., Scapin, D.L.: Evaluating a user interface with ergonomic criteria. Int. J. Hum.-Comput. Interact. **7**(2), 105–121 (1995). https://doi.org/10.1080/10447319509526114
12. Hadji, B., Degoulet, P.: Information system end-user satisfaction and continuance intention: a unified modeling approach. J. Biomed. Inform. **61**, 185–193 (2016). https://doi.org/10.1016/j.jbi.2016.03.021
13. Al Rawajbeh, M.: Performance evaluation of a computer network in a cloud computing environment. ICIC Exp. Lett. **13**(8), 719–727 (2019)
14. Obukhov, A.D., Krasnyansky, M.N.: Neural network architecture of information systems. Vestnik Udmurtskogo Universiteta. Matematika. Mekhanika. Komp'yuternye Nauki **29**(3), 438–455 (2019). https://doi.org/10.20537/vm190312
15. Obukhov, A.D., Krasnyansky, M.N., Dedov, D.L., Karpushkin, S.V.: Mathematical model of information processing in electronic document management system. Int. Rev. Autom. Control (IREACO) **11**(6), 336–345 (2018). https://doi.org/10.15866/ireaco.v11i6.15305

16. Obukhov, A.D., Krasnyansky, M.N.: Neural network method of data processing and transmission in adaptive information systems. Vestnik Udmurtskogo Universiteta. Matematika. Mekhanika. Komp'yuternye Nauki **31**(1), 149–164 (2021). https://doi.org/10.35634/vm210111
17. Obukhov, A.D., Krasnyanskiy, M.N.: Neural network method for automatic data generation in adaptive information systems. Neural Comput. Appl. **33**, 1–23 (2021). https://doi.org/10.1007/s00521-021-06169-2
18. Obukhov, A., Krasnyanskiy, M., Nikolyukin, M.: Algorithm of adaptation of electronic document management system based on machine learning technology. Progr. Artif. Intell. **9**(4), 287–303 (2020). https://doi.org/10.1007/s13748-020-00214-2
19. Obukhov, A., Siukhin, A., Dedov, D.: The model of the automatic control system for a treadmill based on neural networks. In: 2020 International Multi-Conference on Industrial Engineering and Modern Technologies (FarEastCon), pp. 1–5. IEEE, Vladivostok (2020). https://doi.org/10.1109/fareastcon50210.2020.9271589
20. Jin, H., Song, Q., Hu, X.: Auto-Keras: an efficient neural architecture search system. In: Proceedings of the 25th ACM SIGKDD International Conference on Knowledge Discovery & Data Mining, pp. 1946–1956. Association for Computing Machinery, New York (2019). https://doi.org/10.1145/3292500.3330648
21. Budjač, R., Nikmon, M., Schreiber, P., Zahradníková, B., Janáčová, D.: Automated machine learning overview. Res. Pap. Fac. Mater. Sci. Technol. Slovak Univ. Technol. **27**(45), 107–112 (2019). https://doi.org/10.2478/rput-2019-0033
22. Elsken, T., Metzen, J.H., Hutter, F.: Neural architecture search. In: Hutter, F., Kotthoff, L., Vanschoren, J. (eds.) Automated Machine Learning. TSSCML, pp. 63–77. Springer, Cham (2019). https://doi.org/10.1007/978-3-030-05318-5_3
23. Gong, X., Chang, S., Jiang, Y., Wang, Z.: Autogan: neural architecture search for generative adversarial network. In: Proceedings of the IEEE International Conference on Computer Vision, pp. 3224–3234. IEEE (2019). https://doi.org/10.1109/iccv.2019.00332
24. Cioffi, R., Travaglioni, M., Piscitelli, G., Petrillo, A., De Felice, F.: Artificial intelligence and machine learning applications in smart production: progress, trends, and directions. Sustainability **12**(2), 492 (2020). https://doi.org/10.3390/su12020492
25. Dimiduk, D.M., Holm, E.A., Niezgoda, S.R.: Perspectives on the impact of machine learning, deep learning, and artificial intelligence on materials, processes, and structures engineering. Integr. Mater. Manuf. Innov. **7**(3), 157–172 (2018). https://doi.org/10.1007/s40192-018-0117-8
26. Aruldoss, M., Lakshmi, T.M., Venkatesan, V.P.: A survey on multi criteria decision making methods and its applications. Am. J. Inf. Syst. **1**(1), 31–43 (2013). https://doi.org/10.12691/ajis-1-1-5
27. Harte, R., et al.: A human-centered design methodology to enhance the usability, human factors, and user experience of connected health systems: a three-phase methodology. JMIR Hum. Fact. **4**(1), e8 (2017). https://doi.org/10.2196/humanfactors.5443

Check for updates

Multi-agent Simulation of Agile Team Dynamics: Experiments on Team Strategies Comparisons

Zhe Wang(✉)

Lincoln University, Lincoln, New Zealand
Zhe.wang@lincolnuni.ac.nz

Abstract. Scrum is an agile process that incrementally, iteratively and continuously delivers software based on sprints. It is comprised of user stories stored in product backlogs and delivered through sprints by a Scrum team consisting of developers, a Scrum Master and a Product Owner. The performance of a Scrum team is largely dependent on the capabilities of team members and the technical practices they adopt. One such practice, pair programming has been studied in a variety of contexts but not extensively in a Scrum context. Pair programming is a technique where two programmers work side by side to design, code, and test their program.

A multi agent system is used to simulate a scrum environment where a team (with varying team members' capability) work on delivering user stories (which consists of multiple tasks with varying complexities) in multiple sprints. Using this simulated environment, various strategies of compulsory pairing and voluntary pairing are investigated. Impact is measured based on the team's work efficiency, completion time, effort time and idle time.

Experiments were performed to test these strategies in varying environments and results showed that a hybrid pairing strategy performed the best in fixed environments as it avoided negative pairing situations. An adaptive strategy (which changes strategy depending of the composition of the team and the tasks to be completed) performed best in the random setting as it was able to use the best strategy based on the current environment.

Keywords: Scrum · Team dynamics · Agent-based modelling · Multi-agent system · Team strategies · Solo programming · Pair programming

1 Introduction

Software is created by people using technologies and tools which are integrated in a process known as the software development life cycle (SDLC). They are two major approaches in software development; the waterfall model and the Agile approach. The waterfall model is a non-iterative sequential process which progresses from requirements analysis through system design, and coding, to testing and maintenance. Such an approach provides limited interactions between the customer and the development team due to its sequential nature where interactions only happen during requirements analysis and in the testing and maintenance phases.

R. Silhavy et al. (Eds.): CoMeSySo 2021, LNNS 232, pp. 254–286, 2021.
https://doi.org/10.1007/978-3-030-90318-3_24

Agile software development is an approach to creating software through an iterative and incremental process. (Beck and Fowler 2001) describe Agile as: "Individuals and interactions over processes and tools; Working software over comprehensive documentation; Customer collaboration over contract negotiation; Responding to change over following a plan." Scrum is one of the agile processes that realises the Agile manifesto (Kent et al. 2020) It splits users' stories into several parts and aims to achieve each part within a time period, which can last from one day to thirty days, and is known as a sprint. Within each sprint, the process is similar to the waterfall model. However, there are more opportunities for the Scrum team and the customer (also known as the product owner) to discuss the software requirements during the process because there are multiple sprints in an Scrum project and each sprint contains requirement analysis and this allows the requirements to be refined during each sprint. Each sprint is aims to deliver working software and therefore a higher project success rate compared to the waterfall approach because the size of the software is smaller, its design goal is clearer, and the Scrum team is fully focused on the project (Schwaber and Sutherland 2017). The Scrum process is composed of user stories, tasks, sprints, sprint meetings, deliverable software and conducted by the Scrum team. User stories are designed based on the users' requirements. Each user story contains one or more tasks that need to be completed by the Scrum team during one sprint or across several sprints. A Scrum team comprises several members who have different skills and capabilities, such as designers, developers, database administrators and testers. Scrum team members need to interact and collaborate with each other to achieve the goals of the team, through the sprints. A Scrum team will have daily sprint meeting (a scrum) to discuss what has been done and what needs to be done.

Although the Scrum approach is proposed as an improvement over the waterfall model, it can suffer from several problems. One such problem is team dynamics, which largely affect the quality of the produced software, risk of project failure and project completion time (and therefore cost). Team dynamics refers to the team's composition, task allocation, interactions between team members and how they work together. Song et al. (2015) define effective team dynamics according to the following criteria, as also indicated by (Nadler et al. 1979):

Team performance (i.e., the product of teamwork meets the expectations of those who use it);

Member satisfaction (i.e., each team member's experience contributes to his or her personal well-being and development); and.

Team adaptation (i.e., the team experience enhances each member's capability to work and learn together in the future).

A team comprising of experienced and highly skilled members will normally perform better than a junior team that is less experienced and skilled. In a Scrum environment, the composition of the team will affect the performance of the team, because a Scrum team needs high level of cooperation among its team members to achieve the sprint goal, and the skills, experience and capabilities of the team members affect the performance of the team. Varying methods of tasks allocation may result in different outcomes and may affect the delivery of the software.

1.1 Agent-Based Modelling

The main idea of agent based modelling is that many real-world phenomena can be effectively modelled as an agent, an environment and as a description of agent-agent interactions and agent-environment interactions (William and Uri 2015). Modelling the real world involves developing a computational model that comprises of agents and the environment in which the main actors are the agents (Gilbert and Troitzsch 2005; Mehmood et al. 2019).

Agent-based modelling can be used to model social behaviour and interactions. (Boulahbel-Bachari and El Saadi 2019) used agent-based modelling to model the behaviour of migration and self-selection of people. (Joslin and Poole 2005; Macal and North 2006) provide an overview of how to use agents to model software project planning by modelling the developer team. In this work, agents are used to model and simulate human developers, such that an agent can have capability, skills and experience. Agents can pick up a task and work on it and deliver the product. Different agents can have different skill sets and motivation. Task modelling is important when constructing the scale of the overall project. Agents' interactions with an agent or a task can be captured using agent-based modelling, and such modelling can model the process of how each agent completes the task.

(Agarwal 2007; Macal and North 2008; Phillips 2006; Yilmaz and Phillips 2007) (Agarwal and Umphress 2010) indicate that agent-based modelling is suitable for undertaking team-based research to capture team behaviour and interaction when carrying out software project estimations and optimization. Team-based modelling uses multi-agents, which are a group of agents that interact with each other in a team. The shared goal of these agents is to complete all the tasks. In the real world, a software developer team usually contains more than one developer, and an agent can be used to model and simulate each developer and to group them as a team. The reasons for this are based on an agent's unique characteristics, especially its features of human behaviour and decision-making. Agent-based modelling can be used to model a group of agents and verify their impacts on team goal realization and achievement. Such features can be used to undertake team-based analysis in particular organizational-based contexts.

Abar et al. (Abar et al. 2017) specifically reviewed all the tools that have been developed for agent-based modelling and simulation, classifying them by their functionality and scales of computing that can be applied. These tools can be distributed for various modelling purposes, such as human behaviour modelling, animal behaviour modelling and cell modelling in biology. One of the more popular tools is NetLogo (William and Uri 2015), which is used to simulate natural and social phenomena and is well suited for modelling complex systems over time. NetLogo allows developers to explore the various agents' behaviour under a variety of settings.

For the purpose of this study, a multi-agent system was used to simulate the Scrum environment to enable agents to make autonomous decisions and interact with one another using a standard agent communication language. The next sections describe the agent and the multi-agent systems in some detail.

1.2 What is an Agent and Its Characteristics?

Agent is a special software component that can make decisions autonomously based on its inner rules and logic reasoning (Wooldridge and Jennings 1995). An agent can be defined as a computer system that is situated in some environment that is capable of autonomous actions in order to meet its design objectives. Agents can be used to model human behaviour or to construct an intelligent system (Wooldridge and Jennings 1995).

Wooldridge and Jennings (1995) noted that computer agents typically have the following properties:

- Autonomy – agents operate without having direct control over their actions and internal state;
- Social ability – agents interact with other agents through some kind of 'language' (a computer language, rather than a natural language);
- Reactivity – agents are able to perceive their environment (which may be the physical world, a virtual world of electronic networks or a simulated world that includes other agents) and respond to it; and
- Proactivity –agents are also able to take the initiative and engage in goal-directed behaviour.

Agents can communicate with each other through a common language structure, such as Foundation for Intelligent Physical Agent – Agent Communication Language (FIPA ACL), which is an agent language protocol standard, to exchange information.

1.3 What is Multi-agent System?

Agents are special software than can act autonomously on behalf of the users to achieve specific goals. Agents often need to work together to solve more complex problems. A multi-agent system (MAS) is a loosely coupled network of agents that interact to solve problems that a single agent will not be able to solve alone (Ali et al. 2019; Tyrychtr et al. 2019). Agents may need to cooperate, coordinate, collaborate and negotiate with each other to achieve their goals in a MAS environment. The ability to interact is made possible by the presence of a standard agent communication language (ACL), such as FIPA-ACL (Bellifemine 2007). Various MAS applications have been developed to solve a variety of problems in multiple domains, such as e-commerce (Tyrychtr et al. 2019), the military (Han et al. 2019), networking (Boulahbel-Bachari and El Saadi 2019), logistics and supply chains (Ivanov et al. 2019) and disaster rescue (Hashimoto et al. 2019).

MAS is ideal for this research as each individual agent can be programmed to possess characteristics that embody the different roles in Scrum, such as the Scrum Master and developers. Each individual agent is able to make its own decisions about whether to take on a specific task, depending on its capability, as well as being able to make decisions about which agent to pair to gain an additional advantage. These developer agents can communicate with each other and the Scrum Master agent. They can also share information with one another either by means of communication or a shared blackboard. The Scrum MAS environment is described in the next section.

1.4 Scrum Background

There are three components in Scrum, as shown in Fig. 1, the roles, the events and the artefacts.

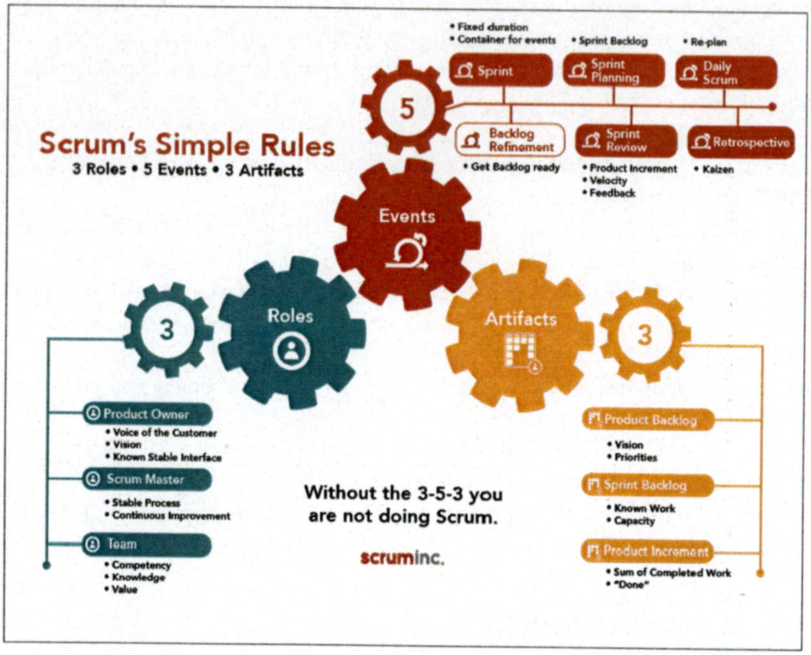

Fig. 1. Scrum model in brief (Justice 2018)

Roles

There are three roles in a Scrum team; Product Owner, Scrum Master, and Development team (including developers, designers, testers, database administrators, user experience experts). The product owner represents the client stakeholders (Stober and Hansmann 2010). The product owner must ensure that he or she is representing the interests of all stakeholders. The product owner also provides the requirements, authorises expenditure for the project, and signs off on any deliverable. The development team is responsible for developing and testing the project deliverables. Developers usually hold several key skills to solve user story problems, such as programming capability, software system design, software testing and database programming. These skills are matched with a problem-solving background. For example, an Android development project would require the developer to have Java programming skills much more than C#. Finally, the Scrum Master is responsible for the Scrum process, adjusting the Scrum to best fit the project and the organisation, as well as ensuring that any issue or problem gets resolved so the Scrum team can be as effective as possible.

Scrum Events
Scrum events are referred to as Scrum ceremonies and include sprint planning, daily Scrum (stand-up), sprint review and sprint retrospective. At the beginning of the Scrum process, the user stories are defined, prioritised and estimated, then stored in the product backlog. At the beginning of each sprint, there is a sprint planning session where the team decides what user stories are worked on during that sprint. Each sprint is, typically, time boxed at two to four weeks. Each day, the team has a daily stand-up meeting (daily Scrum) where they discuss their progress, plan forward and identify problems that may arise. At the end of the sprint, a sprint review is conducted with the product owner to report on progress, to inspect the increments and to discuss anything (if necessary) that may require changes to the product backlog. A sprint retrospective will be conducted at the end of the sprint by the Scrum team with the Scrum Master in order to determine how the previous sprint could have been done better, the defects in the previous sprints and how to improve them.

Artefacts
The artefacts of Scrum include the product backlog (Alsalemi and Yeoh 2015), sprint backlog, and increments (Schwaber and Sutherland 2017). Requirements (user stories) are gathered and prioritised in a list called the product backlog (Stober and Hansmann 2010). The product backlog is used to hold user stories defined by the product owner and are ordered by their priority. This gives the product owner an advantage in defining the user stories (Mahnič and Hovelja 2012) iteratively and incrementally. The highest priority user story always goes to the top of the product backlog. The Scrum team will then choose the user stories with the highest priorities to work on in each sprint. A sprint can deliver one or more user stories, depending on its size, complexity of the user stories and capability of the team. Several sprints are typically necessary to complete enough user stories for the initial delivery of the software. An incremental delivery model is a key point in encouraging frequent discussions between the team and the product owner. Regular reviews help to guarantee the quality of the product and the increment delivered in each sprint. The process runs iteratively until all the user stories in the product backlog are completed by the Scrum team or the product owner decides to stop the development of the software.

1.5 Pair Programming

In pair programming, two programmers work side-by-side at one computer to collaborate on the same design, algorithm, code or test. This kind of programming has been practiced in industry with great success for years (Williams et al. 2000). Pair programming can enhance the performance of developers to deliver quality code in less time compared with solo. The benefit of pair programming is effective communication, knowledge sharing/transfer, enjoying work, increase satisfaction and confidence, reduce code defect and increase the product quality, enhance team performance.

1.6 Research Motivation

Teamwork is considered as the most critical factor (da Silva et al. 2013) that could affect the success rate of software projects. The Scrum guide indicates that the Scrum team plays an important role in the software development project. However, the team composition and the how the team should work together (in pair or solo) have not been fully investigated.

Pair programming is widely used in industry, because it can support two programmers working together, to share knowledge and enhance the software delivery capability. However its positive and negative impacts on team performance that could affect its software delivery under scrum context has not been fully investigated. This research investigates the use of pair programming in a Scrum team and how it can affect the performance of the team through multi-agent simulation. A multi-agent approach is adopted in this research to model each developer in the Scrum team and to facilitate communication between agents. Moreover, there is very limited work using multi-agents system to model scrum team behaviour, especially on how to use such an approach to address Scrum team strategy design and implementation (Ramanujam and Lee 2011) (Zhen et al. 2018).

2 Literature Review

The study by (Silva et al. 2015) proposed a reference model outlining what should be considered in the model from the agile process point of view. In this study, it was suggested that the developer's skills should be considered in the agent model, such as programming skills in Java. The simulation of an agile process can be based on the Scrum model which includes the product backlog, sprint backlog, sprint, the role of Scrum Master and the activities of Scrum.

(Zualkernan et al. 2008) worked on an agent-based simulation project to model Scrum. They simulated the process of Scrum in order to train students to understand the correct steps in the Scrum, such as the role of members and the progress of the Scrum. This training was very useful as it helped these students avoid making fundamental mistakes when applying Scrum in their projects. This is also a type of modelling and simulation that is used for training purpose. This research found that the training system can help people understand what Scrum is and how to practice it, as most of the problems for Scrum occurs during the adoption of Scrum. The research practised the concept of Scrum through the analysis of roles, such as the correct person performing the correct role. For example, who should prepare the product backlog, the Scrum Master or the product owner? For Scrum activity, should the team perform a daily meeting or not? For the sequence of Scrum activity, such as the estimation of the user story points, should it happen before or after the product backlog preparation? For the location of the Scrum activity, should some of the distributed team hold their meetings online? These questions help Scrum practitioners to fully understand the Scrum concepts.

The study by (Jun 2013) addressed task allocation in Scrum projects where he designed an algorithm to allocate tasks to team members and then verified the time delay of the project based on the task allocation methods designed. He found that the task allocation should be both average and focused, which means that all team members should be working instead of idling. However, the strong members should not work too much, so as not to cause any further pressure on the others, as this can enhance the team's work efficiency. When the Scrum process is completed, the delay rate of the tasks is calculated and compared to the accept-when-requested task allocations. He found that the algorithm was more stable and performed better for task delay rates when most tasks were completed within the estimated time limitations. The main purpose of this algorithm is to ensure a balance between the considerations for quality and timeliness when working on the tasks, allocating tasks to highly capable agents, and the on-time completion of the tasks, so that there should be low number of agents that are idle. The system was constructed using a multi-agent-based simulation.

The study by (Noori and Kazemifard 2015) developed an agent-based model on pair programming for an agile project. They argued that the different personalities and characteristics will affect pair programming. They modelled the developer through agents and found that some good combinations of specific personalities were well suited for pair programming. The results showed that personality played an important role in the formation and utility of a pair. For example, when the expertise of both individuals is high, the best pairing is introvert-extrovert. When both individuals are extrovert, the best pairing is low-high or medium-high expertise. This work focused on the team selection in Agile Scrum.

A study by (Joslin and Poole 2005) proposed several key ideas on how to apply agent-based simulations for software projects. The goal of agent-based software project simulation is for decision support in software project management. The agent can be used to simulate how developers work and how resources are allocated. The agents' strategy can be separated from its algorithmic design. This strategy mainly focused on high level planning while the algorithm focused more on specifics in its realization, such as the agent's behaviour and activity. The search strategy can be optimised using genetic algorithms. There should be initial simulation parameters, which will be updated at each sprint. The duration of the sprint can be defined based on the Scrum Master and the development team. At the end of each sprint, the initially estimated parameters will receive their final value. They also indicated that the task allocation for multi-agents was a critical part of the software development process.

The research by (Tamburri et al. 2012) explores various time zone-based Scrum teams and how time can affect task completion as, obviously, the time zone will affect the tasks assigned to the team who can be in different locations (global software engineering). Delays or errors in the tasks caused by communication are affected by the time zone, which needs to be considered and optimised through Scrum and adapted to be distributed for worldwide team performance. The stakeholder is also located in a different location and they have no information on the whereabouts of the team members. These tasks are dependent on each other, which makes the delivery process even more complex. The time zone gap is one of the major difficulties that prevents team members from communicating effectively and in a timely fashion. An error occurring in one task may be informed about, or detected by, a team in another place, so two teams can be affected by this error that was shared between tasks. The research further adopts the scrum and agile service network simulation to understand the performance of the model, and found that the agile service network is more feasible in supporting global software engineering.

The study by (Griffith et al. 2014) provides a discrete-event based simulation of the Scrum process with a particular focus on defects and technique debt creation. They updated the Scrum process by considering technical debt creation. The purpose of this simulation is to investigate how technical debt could affect the Scrum process, including the impact on the software products delivered, as the team effort is a limited resource. However, the technical debt can be almost unlimited depending on the type of debt. The estimation of the time needed to complete a task is based on the developer's capability, where juniors need 2.0 time, middle level developers need a1.0 time and seniors need 0.5 time to complete a given task. This shows that the more technical debt that exists the lower the team performance. If the error size is the same as the system size then the team productivity is zero. The smaller the error size the higher the team productivity. This study focusses on the effect of technical debt in Scrum. It further simulates the debt process into the Scrum process, so technical debt needs to be solved before the project can move on. However, the Agile team cannot pay too much attention to every debt, otherwise the budget of the project will be increased.

(Lunesu et al. 2018) used an event-based simulation to model and simulate software project delivery through Scrum. They compared the size of the project and the number of developers to predict the effort and time to complete a project through event-based simulation as well as the differences between the Scrum and Kanban models in delivering software through the cloud platform. They found that project teams faced problems with the communication and organization of distributed projects that affect teams' productivity and increases the time required to achieve the project's goals.

A study by (Orłowski et al. 2014) proposed a framework to simulate the Scrum process. The system simulates the Scrum methodology, including its management processes and project roles. For the implementation of Scrum processes, a Scrum ontology is proposed for the competencies of the roles of project team members and a fuzzy-logic representation is provided. As a result, they presented the hybrid fuzzy-ontological system. The framework of the design processes proposed in the article was verified based on the real course of project management processes in a large IT company.

(Košinár and Štrba 2013) used an event-based simulation to provide an estimation for the time and effort needed to complete healthcare information projects through Scrum modelling and simulation. They simulated the number of tasks and the size of the development team in order to estimate the completion time of the project using a burndown chart. Even though the simulation was very close to the real world, not all simulations are precise enough due to external factors such as unpredicted obstacles or changes to the teams.

2.1 Task Allocation Research

Task allocation was studied by (Masood et al. 2017) who analysed real cases in software companies and found that the majority of developers in the software team considered their skills and matched with the task allocated. The goals of the whole team were often neglected by the team, and this prevented other team members from showing their options and influencing each other in a positive way. (Jun 2013, 2015; Jun et al. 2014; Lin et al. 2014) proposed a task allocation algorithm for the Scrum team based on context aware methods. This method balances the workload of the task and agent status, such as the requirement for task quality, completion efficiency and an agent's psychology and pressure. (Jun et al. 2014; Lin et al. 2014) conducted survey-based research about the developer's self-assessment of confidence in completing the task to find whether a correlation between the confidence and completion time of the task is correct or not for task allocation decision-making adjustments. The confidence scale was from 0 to 11 points to indicate from not confidence to very confident. (Jun et al. 2014) analysed task allocation in the Scrum context by considering the developer's morale. This research had 125 undergraduate software engineering students complete a 12-week course work using the agile software development method. The key findings about the Scrum-based Adaptive Software Development (ASD) process practised by novice teams from this study were: 1) task allocation in agile teams positively correlated with the students' technical productivity; 2) collaboration is negatively correlated with the teams technical productivity, team morale, and team score; and 3) team morale was positively correlated to their technical productivity.

3 Method

In this section, we describe three different strategies for allocating developers to tasks; a solo strategy that can be used in a situation where a single task must be worked on by a single agent. We also describe two pairing strategies where a pair of agents must work as pair to complete a task (must pair) and where a task may be worked on by a pair of agents or a single agent (voluntary).

3.1 Strategy Design

To evaluate the impact of pairing in a Scrum software development project, we developed one solo strategy and two pairing strategies (must pair and voluntary pairing) that can be used by the scrum team to complete all the user stories in the sprint backlog. The must pair strategy dictates that the agents must work in pair to complete a task whereas with the voluntary pairing strategy a task may be worked by a single agent or a pair of agents. The purpose of introducing these two types of pairing is to investigate the impact of compulsory pairing as opposed to voluntary pairing. Voluntary pairing is designed to maximise the benefits of pairing, so we only pair agents when it is beneficial to pair and allows them to work solo when no acceptable pair is found. We further sub-divided the must pair strategy and voluntary pair strategy into three sub-categories based on the capability of the pair.

These strategies are evaluated in a single sprint where a sprint contains multiple user stories each containing multiple tasks. Each task has a size which ranges from 1 to 10, where 1 is the easiest task and 10 is the most difficult task. Tasks between 1 and 4 are categorised as easy tasks, 5–7 are intermediate tasks and 8–10 are complex tasks. These tasks will be worked on by the developer agents (the Scrum team) with varying capabilities. The developer's capability ranges from 1 to 10 where 1–4 is considered novice, 5–7 intermediate and 8–10 expert. Ideally, a developer agent should only work on task with complexity equal to its capability. However, in situations where this is not possible, other options can be explored such as pairing with another agent or working alone on the task with a penalty. It is also assumed that the team composition and the tasks are dynamic (these configurations vary in each sprint) and that certain strategy may work well for certain environment. These strategies are described in detail in the next sections.

3.2 Solo Strategy

The solo strategy allocates a task to a single agent based on the capability of the available agents. The task is assigned to the agent with the closest capability to the complexity of the task. A preference value (agent capability – task complexity) is used to determine which agents can undertake a task within the simulation (Wang 2019). Here, an agent is only permitted to work on tasks where the preference value is greater than or equal to −3. The time required to complete a given task is dependent on the agent's capability and the task's complexity. If the agent's capability is greater than or equal to the task complexity, then the time required to complete the task is equal to the task size (Eq. 1).

$$TimeRequired = TaskSize \tag{1}$$

otherwise, if the agent capability is lower than the task's complexity, the time required to complete a task is calculated as:

$$TimeRequired = |AgentCapability - TaskComplexity| * 0.33 * TaskSize + TaskSize \tag{2}$$

Here, the time required to complete the task is longer since an agent with a lower capability will need more time to work on the given task. Since a gap of −3 is allowed, 0.33% is added to the time required to complete the task for each gap value (−1, −2 and −3) based on (Griffith et al. 2014) where, for a given task that can be worked in a 1.0 unit of time by a middle level worker, it will take a 0.5 time unit for the senior to complete and 2.0 units of time for the junior to complete the task.

As discussed previously, a developer agent's capability can be categorised as novice, intermediate or expert agents such that same level of pairings (expert-expert, intermediate-intermediate, and novice-novice) and cross level pairings (expert-intermediate, expert-novice, and intermediate-novice) are possible. However, based on the literature (Arisholm et al. 2007; Dybå et al. 2007; Lui and Chan 2006; Nilsson 2003) the effectiveness of the pairing is dependent on the level of expertise of the individuals in the pair and the task to be worked on. The following section provides an analysis of the different pairings.

3.3 Same Level Pairing

Same level pairing refers to pairing of developers with the same capability to work together on a specific task. This is the original idea for pair programming, which aimed to enhance the quality of code cooperation between developers who enjoyed the work of coding to promote product delivery.

Expert-Expert Pair
This type of pair is considered as negative pairing as both experts may have disagreement on software design, coding and analysis. (Dybå et al. 2007) supports two experts working on complex tasks as a good choice but suggests that experts should not work in pair when working on an easy task. Similarly, (Arisholm et al. 2007) also supports two experts pairing to work on complex tasks but not on easy tasks. In addition, (Lui and Chan

2006) shows that once a novice pair moved up to expert rank, it is no longer useful to pair them.

Intermediate-Intermediate Pair
This type of pairing is only effective when the pair is working on a complex task. (Dybå et al. 2007) supports two intermediate developers working on complex tasks as a good choice, but a bad one when working on an easy task. (Arisholm et al. 2007) also supported two intermediate developers working on complex tasks but not on easy tasks. A single intermediate developer is not able to work on a complex task. However, pairing two intermediate developers can enhance the performance of each developer. This pairing is not recommended when working on intermediate tasks as an intermediate task can be worked on by a single intermediate expert (Arisholm et al. 2007; Dybå et al. 2007; Hannay et al. 2009; Lui and Chan 2006).

Novice-Novice Pair
This type of pairing is not recommended when working on complex tasks as it is not possible for two novices to solve complex problems due to gaps in knowledge. A novice-novice pairing is effective when working on easy and intermediate tasks. Based on (Dybå et al. 2007; Arisholm et al. 2007) it is possible for two novices to work on all tasks.

3.4 Cross Level Pairing

The purpose of cross level pairing is to facilitate learning and knowledge transfer between the pairs (Zieris and Prechelt 2014). Good pairs can transfer knowledge, as supported by (Plonka et al. 2015). Many researchers indicate that knowledge transfer between developers exists in an organization (Argote and Fahrenkopf 2016) (Lin and Zhang 2019) (Wang et al. 2019). By doing this, the lower level developer can learn from the high-level developer. Knowledge transfer occurs when pairings, such as when a student works together with a supervisor.

Expert-Intermediate Pair
This type of pair is effective as the intermediate can learn from the expert. This pairing is highly recommended when working on a complex task as this allows the intermediate to learn from the expert (Plonka et al. 2015). It is not recommended for this kind of pairing to work on easy and intermediate tasks as no learning can take place and the combined capabilities of the intermediate and the expert are not maximised.

Expert-Novice Pair
The expert-novice pair has the largest gap between the two developers, and this is considered an effective pairing as the novice can learn from the expert. A novice developer will not be able to work on an intermediate or complex task alone. However, by pairing with an expert, the novice has the opportunity to work on tasks at all levels.

Intermediate-Novice Pair
The intermediate-novice pair provides the opportunity for the novice to work on easy and intermediate tasks. The novice will also be able to learn from the intermediate. This pairing is not recommended when working on a complex task as the intermediate might not have the knowledge to tackle the task.

Table 1 summarises the different pairings for the varying task complexities used in this simulation. Positive pairing is the preferred pairing as it results in better code quality and allows junior developers to learn from senior developers.

Table 1. Pair programming modelling assumptions

Pair	Easy task	Intermediate task	Complex task
Novice-novice	Positive	Positive	Negative (not used)
Intermediate-intermediate	Negative	Negative	Positive
Expert-expert	Negative	Negative	Negative
Novice-expert	Positive	Positive	Positive
Intermediate-expert	Negative	Negative	Positive
Intermediate-novice	Positive	Positive	Negative (not used)

There are several situations where pairing is not recommended (negative pairing):

- Novice-novice working on a complex task
- Expert-expert (all cases)
- Intermediate-intermediate working on intermediate and easy tasks
- Expert-intermediate working on intermediate and easy tasks
- Intermediate-novice working on a complex task

In our simulation, the time required to complete a task worked on by a negative pair will be inflated by 15% to accommodate the overheads (misunderstandings and disagreements) from this pairing. (Arisholm et al. 2007).

When pairing, the pair capability is based on the capability of the stronger agent (referred to as the lead agent). The time required to complete a task for a pair is half the time taken by a solo developer as there are two developers working on the same task. For negative pairing, a penalty is imposed on the pair that increases the time required for the pair by 15%. Negative pairing may also affect the quality of the completed task. In our modelling we only considered the time penalties required, as the negative pair can result in low team performance that reduces the team's efficiency. The time required to complete a given task is as follows:

$$TimeRequired = \begin{cases} TS, LAC \geq TC \\ (|LAC - TC| * 0.33 * TS) + TS, LAC < TC \end{cases} \tag{3}$$

where LAC is the lead agent capability, TS is the task size and TC is the task complexity. As before, 0.33 is the additional time imposed by a developer working on a task that has a complexity higher than the developer's capability (0.33 for a gap of 1, 0.66 for a gap of 2 and 1.0 for a gap of 3). In this scenario, the time required to work is longer since the agent with the lower capability needs more time to work on the given task.

The time required is updated based on the nature of the pairing where negative pairing has a penalty imposed as follows:

$$TimeRequired = \begin{cases} TimeRequired, if \ \ no\ penalty \\ \\ 1.15 * TimeRequired, with\ penalty \end{cases} \tag{4}$$

The time required for each individual agent in the pair is:

$$TimeRequired_A = \frac{TimeRequired}{2} \tag{5}$$

as each agent can split the work equally. In (Lui and Chan 2006), they indicated that pairs will use half the time to complete the same task compared with a solo.

3.5 Strategy Types

In our simulation, we modelled our pairing based on the following:

- Type 1 – the lead agent must have a capability equal to or higher than the task complexity. In this setup, the lead agent can choose who to pair with depending on the task complexity.
- Type 2 – the lead agent must be in the same category or higher than the task complexity. This means that a novice agent can work on any easy task, an intermediate agent can work on medium and easy tasks, and an expert agent can work on complex, medium and easy tasks.
- Type 3 – similar to type 2, but the lead agent with a gap (agent capability – task complexity) of −3 or greater can work on the task. By allowing a gap of 3, it is possible for two intermediate agents to work on a complex task (Dybå et al. 2007). Also, in this setup, a pair of novice agents can work on a medium task.

These strategies were set up to observe how they affect the team's efficiency, completion time and idle time. In Type 1, the pair can only work on a task equal to or lower than the pair's capability. The Type 2 strategy allows the pair to work on a task in the same category as the capability of the pair, while Type 3, allows tasks in a higher category to be worked on by the pair (e.g. an intermediate task can be worked by a novice pair and a complex task can be worked by an intermediate pair).

3.6 Must Pair Strategy

In the must pair strategy, each task must be undertaken by a pair of agents and no agent can work on a task alone. In this strategy, each task needs to be verified by a pair of agents who have not worked on the task previously. To accommodate the must pair, a single agent must wait for another agent to pair with before working on another task. This also means that some pairings may be ineffective. The must pair strategy has three sub-types as shown in Tables 2 and 3.

Must Pair Type 1 (MP1)

Table 2. Must pair type 1

Task	First choice	Second choice	Third choice	Fourth choice
Easy	Novice	Novice	Intermediate	Expert
	Novice	Any agent	Any agent	Any agent
Intermediate	Intermediate	Intermediate	Expert	Expert
	Novice	Any agent	Novice	Any agent
Complex	Expert	Expert	Expert	
	Novice	Intermediate	Expert	

For easy tasks in this strategy, a novice who has a capability equal to or higher than the task complexity is the lead agent. In this situation the novice agent should pair with another novice agent. If no novice agent is found, then it may pair with any agent. If no novice agent is available to work on the easy task, an intermediate agent can be nominated as the lead agent and may pair with any other agent. Finally, when there are no novice or intermediate agents available, an expert, takes the role of a lead agent and can pair with any other agent.

For an intermediate task, an intermediate agent who fulfils the conditions will be the lead agent and must first pair with a novice (to provide an opportunity for the novice to learn), followed by any agent. If an intermediate agent is not available, the expert agent will be the lead agent who will then pair with a novice first (if available), followed by any agent.

For a complex task, an expert agent with the right capability is the lead agent and may first pair with a novice, followed by an intermediate and an expert.

Must Pair Type 2 (MP2)
The MP2 strategy is similar to MP1, so the pairing decisions are as in Table 2. However, the main difference in this strategy is that we allowed some flexibility on what task the agents can work on. A pair of agent can work on a task as long as the task complexity is in the same category as the lead agent's capability (i.e. an easy task can be worked on by a novice agent, a medium task can be worked on by any intermediate agent, and a complex task can be worked on by any expert agent).

Must Pair Type 3 (MP3)
MP3 is another variation of MP1 where, for this strategy, a novice-novice pair is allowed to work on an intermediate task and an intermediate-intermediate pair is allowed to work on a complex task.

This strategy allowed the intermediate and novice to work on higher level tasks and this is very useful when the team's maximum capability is lower than the task required.

Table 3. Must pair type 3

Task	First choice	Second choice	Third choice	Forth choice	Fifth choice
Easy	Novice	Novice	Intermediate	Expert	
	Novice	Any agent	Any agent	Any agent	
Intermediate	Intermediate	Intermediate	Expert	Expert	Novice P-value $\geq$ 4
	Novice	Any agent	Novice	Any agent	Novice
Complex	Expert	Expert	Expert	Intermediate P-value $\geq$ 4	
	Novice	Intermediate	Expert	Intermediate	

3.7 Intelligent Pair Strategy

The intelligent pair strategy (voluntary pair) ensures that agents only pair if the pairing is positive and, in situations where pairing is not possible, the task will be assigned to a solo agent with sufficient capability. Each task needs to be verified by an agent who has not worked on the task. The intelligent pair strategy has three sub-types that are shown in Tables 4 and 5.

Intelligent Pair Type 1 (IP1)

Table 4. Intelligent pair Type 1

Task	First choice	Second choice	Third choice	Fourth choice
Easy	Novice	Novice solo	Intermediate solo	Expert solo
	Novice			
Intermediate	Intermediate	Intermediate solo	Expert	Expert solo
	Novice		Novice	
Complex	Expert	Expert	Expert solo	
	Novice	Intermediate		

In this strategy (see Table 4), effective pairing has a higher priority. For an easy task, if there is a novice lead agent available, it will try to pair with another novice, followed by an intermediate and then an expert. If there are no agents available to pair with at that time tick, the novice agent will work on the task independently. If there is no novice agent, then a solo intermediate agent is preferred over an expert agent.

For an intermediate task, the intermediate agent is the lead agent and will pair first with a novice agent. If this pairing is not possible, a solo intermediate is the next choice, followed by an expert-novice pair and then an expert solo.

For a complex task, the expert agent is always the lead agent with the novice agent as the pair, followed by the intermediate. If novice and intermediate agents are not available, an expert agent will work on the task individually.

Intelligent Pair Type 2 (IP2)

The IP2 strategy is similar to IP1, so the pairing decision is the same as in Table 4. As before, the main difference is that in this strategy, we allow some flexibility on what task the agents can work on. An agent can work on a task as long as the task complexity is in the same category as the agent's capability (i.e. easy tasks can be worked on by a novice agent, a medium task can be worked on by any intermediate agent, and a complex task can be worked on by any expert agent).

Intelligent Pair Type 3 (IP3)

IP3 as shown in Table 5 is another variation of IP1 where in this strategy, a novice-novice pair and a novice solo are allowed to work on an intermediate task and an intermediate-intermediate pair and an intermediate solo pair are allowed to work on a complex task. This strategy is similar to MP3 in that it allows intermediates and novices to work on higher level tasks.

Table 5. Intelligent pair Type 3

Task	First choice	Second choice	Third choice	Fourth choice	Fifth choice	Sixth choice
Easy	Novice Novice	Novice solo	Intermediate solo	Expert solo		
Intermediate	Intermediate Novice	Intermediate solo	Expert Novice	Expert solo	Novice P value ≥ 4 Novice	Novice solo P value ≥ 4
Complex	Expert Novice	Expert intermediate	Expert solo	Intermediate P value ≥ 4 Intermediate	Intermediate solo P value ≥ 4	

3.8 Evaluation Metrics

To evaluate the performance of the various strategies, five different metrics are defined:

Completion Time

This is the elapsed time used by the team to complete all the tasks (for the pre-defined user stories) from the start of the sprint until the end of the sprint when all tasks in the sprint backlog were completed, verified and marked as completed.

Person Hours

$$PH = CT \times NA \tag{6}$$

The person hours can be used to calculate the cost for the team to complete the product backlog. It has an economic purpose. This is calculated by multiplying the completion time with the number of agents in the team as shown in Eq. 6 where PH is person hours, CT is completion time and NA is the number of agents in the team.

Idle Time
The idle time indicates the amount of time that an agent was available but not actively working on a task. The total idle time is collected for a Scrum team in each run.

Effort Time
This indicates the amount of time an agent was working and busy. The total effort time is collected for a Scrum team in each run.

Working efficiency
The working efficiency is the teamwork efficiency. This is obtained by dividing the completion time with the average workload of the agents (the lower the better). Here, working efficiency (WE) is defined as:

$$\mathbf{WE} = \frac{\mathbf{CT}}{\left(\frac{\mathbf{WL}}{\mathbf{NA}}\right)} \tag{7}$$

where CT is the completion time, WL is the total workload (summation of the all the tasks in the sprint backlog) and NA is the number of agents in the team.

3.9 Scrum Environment, Agents and User Stories

Figure 2 shows an overview of the Scrum simulation using a multi-agent system. In this simulation, there are two types of agents, the Scrum Master agent and the developer agents. Following the Scrum guidelines, there are between four and eight developer agents and one Scrum Master agent (Schwaber and Sutherland 2017). This team of agents work through sprints to complete all the user stories in the product backlog. In each sprint, the agents' objectives are to work collaboratively to complete the user stories (which are comprised of tasks) in the sprint backlog. At the beginning of each sprint, user stories from the product backlog are moved into the sprint backlog. This will be repeated until all user stories in the product backlog are completed, at which time, the software development project will be deemed to be completed. Agents share and exchange information through the use of the Scrum board where agents can view information about the status of another agent, the task and the user stories.

Scrum Master Agent
The Scrum Master agent is designed to communicate with the developer agents during task allocation, to allocate tasks and to manage the Scrum board. The Scrum Master agent communicates with all developer agents to check their status, get their preferences for

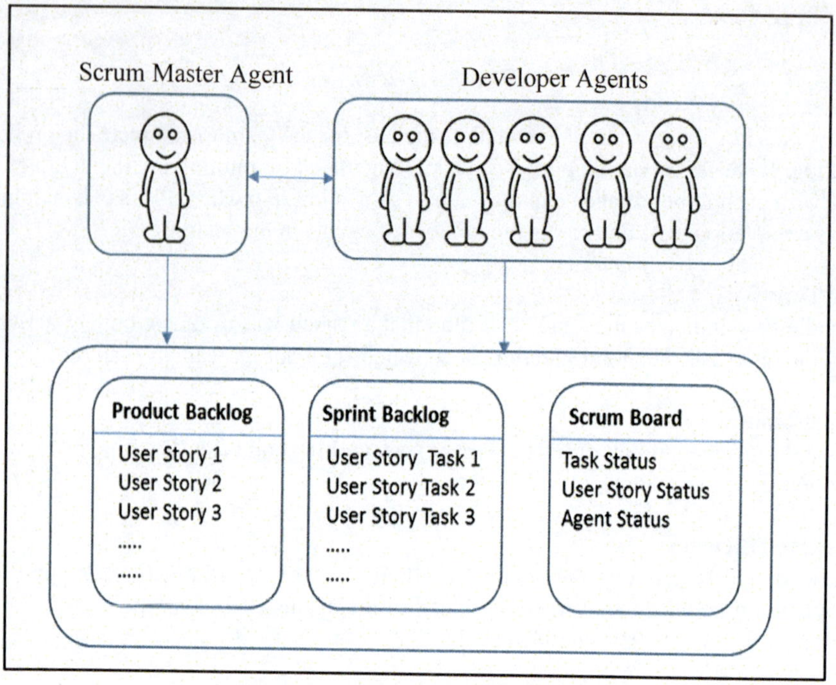

Fig. 2. Overview of scrum simulation

particular tasks, allocate tasks to each agent, form a pair based on the task requirements and the capability of the available agents, and allocate tasks to the pairs. The number of agents and the number of user stories can be defined in the simulation. At the start of the simulation, agents (with different capabilities) and tasks (with different levels of complexities and sizes) are generated.

At the beginning of the sprint, the Scrum Master will coordinate the task allocations for the developer agents by sending a list of tasks that are available. Each agent sends its preferences to the Scrum Master and the Scrum Master decides the task allocations based on the agents' preferences[1]. At each time tick thereafter, each developer agents sends its status to the Scrum Master and in turn the Scrum Master updates the status of each agent and task. When an agent completes a task and becomes available, the Scrum Master allocates a new task to the agent, which the agent may accept or not depending on the task complexity and the agent's capability. This is repeated until all the tasks in the sprint backlog are cleared and marked as completed.

A task can be allocated to a single agent (referred to as a solo programming) or it can be allocated to a pair of agents (referred to as pair programming). For solo programming, the Scrum Master allocates the tasks to agents based on task complexity and agent

[1] This is a slight variation from the Scrum setting where developers (including the Scrum Master) negotiate on which tasks they want to work on. In this implementation, we have made Scrum Master a coordinator agent to reduce communication threads between all the agents. However, all tasks are allocated based on the agents' preferences.

preference, whereas, to allocate the task to a pair it chooses the two most appropriate agents to work on the task based on a pairing strategy.

The Scrum Master also initiates the verification of the task once it is completed by adding a verification task in the sprint backlog. When a user story is completed (all the tasks for that user stories are completed and verified), the Scrum Master initiates a user story verification task, which also gets added to the sprint backlog. The Scrum Master also manages the Scrum board where it updates all the information relating to the sprint, such as the agent's status, the agent's idle time, the agent's working time, task status, task start time, task end time, task verification time, user story status, the current time tick and the sprint completion time.

Developer Agents

The main role of the developer agents are to work on completing the tasks assigned to them. They communicate with the Scrum Master to provide information about their status, the status of the task they are working on and their task preference. Each developer agent has a name, capability and status. The capability represents technical capability ranging from 1 to 10 where 1–4 is considered novice, 5–7 intermediate and 8–10 expert2. This capability value is then used to determine how much time is required by the individual agent to complete a given task. Each agent has a status to indicate their current activity (working, verifying, idle). Working indicates that the agent is currently working on a task, verification indicates that the agent is verifying a task that has been completed by another agent and idle indicates that the agent is free and available to take on work.

User Stories and Tasks

Each user story is composed of multiple tasks. In this simulation, each user story is assumed to have five tasks. This is based on data collected from a number of software companies in Christchurch, New Zealand. Each task has a size, which indicates how much effort is needed to complete the task (the more complex the task, the larger its size). The task sizes range from 1 to 10, where 1 is the easiest task and 10 is the most difficult task. Tasks with size 1–4 are categorised as easy tasks, 5–7 are intermediate tasks and 8–10 are complex tasks. In this model, it is assumed that each task can be worked on by a single developer agent or a pair of developer agents. A task has a state with four possible values: to do (task has not been worked yet); in progress (task currently being worked); to be verified (task completed but has not been verified); and done (task is completed and verified). A task that has been completed must not be verified by the same agent(s) who worked on it.

The Scrum Board

The Scrum board shows the progress of the sprint at each time tick by showing the status of the agents, the tasks and the user stories. The Scrum board provides a platform for all agents to share and exchange information about the current sprint. The Scrum board helps the team keep all the progress transparent and is maintained by the Scrum Master.

Task Allocation Process

At the beginning of the sprint, the Scrum Master selects tasks from the sprint backlog and sends these tasks to all the developer agents. It is assumed that the tasks in the sprint backlog are already prioritised such that the most important task is at the top of the list. Each developer agent inspects the tasks and gives its preferences for these tasks. A preference value is then used to determine which agents are able to undertake a task. This preference value is the gap between the task complexity and the agent's capability (agent capability – task complexity). Once the Scrum Master has received the developer agents' preferences, task allocation is then decided. A developer agent may work on the task independently (solo) or a pair of agents (pair) may work on the task. If the work scheme is solo, the developer agent who has indicated a preference for the task will be selected. If more than one agent is keen to work on the task, then the task is allocated to the agent with the lowest positive preference value. If agents are working in pairs, then the agent with the lowest positive preference value can work with any other agent (as discussed in strategy design.)

If the work scheme is a pair, then a lead developer will be identified to work on the task (based on preference value) and an appropriate pair will be selected (as described in detail in Strategy Design). In both cases, the task is identified before deciding which agent(s) take(s) the task. In this work, there are three main types of working schemes: agent working on a task independently (solo); two agents must work on a single task (must pair); and agents that can work either solo or in pairs (intelligent/voluntary pair).

During the sprint, when a task is completed and agents become idle, the Scrum Master advertises the next task to the idle agents and the task allocation process is repeated. The Scrum Master updates the Scrum board at each time tick. The task verification must not be verified by the same developer agent and the user story must be verified by a product owner.

Simulating a Single Sprint

1. Before the start of a sprint, the agents and the user stories are generated. The user stories are then broken into smaller tasks, prioritised and stored in the sprint backlog. Assuming that a single agent always works on a single task, the single sprint works as follows: When a sprint starts, the Scrum Master agent selects tasks from the sprint backlog and allocates the tasks to the developer agents based on their preferences. Each agent updates its status at every time tick. When an agent has completed a task, a task is marked as 'to be verified' and the agent's status is set to idle. At this stage, the agent can request another task from the Scrum Master. The Scrum board is updated at every time tick to show the status of the agents and tasks. This process is repeated until the end of the sprint. During the sprint, when all tasks belonging to a user story are completed, the user story needs to be verified by the product owner who is external to the developer agent team. If the product owner is satisfied with the user story, it will be marked as completed. Otherwise, if the product owner is not satisfied with the user story, the tasks that belong to that user story will be put back into the sprint backlog. Each will then be given a remedial task, named as a review task (number) where the size of the remedial task will be smaller than its original task size. At the end of the sprint, any remaining tasks in the sprint backlog will be moved to the product backlog to be worked on in the next sprint. If agents are working in pair, an available agent has to wait for another agent to become available

before they can work on the task. In this simulation, it is possible for tasks to be worked by individual agents and paired agents.
2. To simulate the software development project, multiple sprints will be run until all the user stories in the product backlog are verified and marked as completed.

Implementation

The simulation system is developed based on the Java Agent Development framework (JADE). This framework defines agents based on their behaviour, where each agent can be designed to have various behaviours and attributes (Bellifemine 2007). JADE is a Java-based platform that can be used to develop multi-agent system. It is widely used in telecommunication industry and is not limited to modelling and simulation only.

The communication between the developer agents and the Scrum Master agent is facilitated through the message passing using the standard FIPA-ACL. The Scrum Master agent can communicate with all working agents to identify whether their working status is busy or idle, know the preference value for the task, allocate tasks to each agent, form pairs based on task requirements and allocate a task to the pair. This simulation is flexible as it allows the designer to specify the number of agents, agent's capabilities, user stories, tasks and its complexity, duration of the sprint and whether task should be worked individually or in pair. Using this simulation, various information can be collected at each time tick such as the time taken to complete a task, who worked on the task, when the task was started, when the task was completed. Information about agent working time and idle time can also be obtained.

4 Experiments

The purpose of this experiment is to investigate the performance of strategies under various settings namely fixed environment and random environment. In the fixed environment, the number of team is set to five or six agents with varying capabilities such as a novice dominated team, an intermediate dominated team, an expert dominated team and an evenly distributed team. Five or six agents are chosen as they are taken as the midpoint (the Scrum guide recommends a team to have between four and nine members). Each sprint comprised of 50 tasks (10 user stories and five tasks for each story), with varying complexities. The tasks to be worked by the team in each sprint can be classified as many easy tasks, many intermediate tasks, many expert tasks and evenly distributed tasks. In the random environment, the number of agents, the agents' capabilities and the task complexities are generated randomly to mimic the real-world setting. Here, only one of the strategies is tested.

4.1 Fixed Testing

In the fixed environment, we fixed the number of agents in a team between five and six agents. There are eight types of team setup: five agents with many novices; five agents with many intermediates; five agents with many experts; five agents with an evenly distributed capability; six agents with many novices; six agents with many intermediates;

six agents with many experts; and six agents with an evenly distributed capability. The setup for these eight types of team is enough for us to observe the performance of these seven strategies and to understand the effect the team composition may have on these strategies. Table 6 shows the team compositions for the eight teams and the agents' capabilities in each team.

Table 6. Team for fixed testing

Agent Capability	Five developers				Six developers			
	Many Novices	Many intermediate	Many experts	Evenly distributed	Many Novices	Many intermediates	Many experts	Even
1	One			One	One			One
2	One			One	One			One
3	One				One			
4		One	One			One	One	
5	One	One			One	One		
6		One		One	One	One	One	One
7		One	One	One		Two	One	One
8	One	One	One	One	One	One	One	Two
9			One				One	
10			One				One	

The tasks are distributed based six scenario shown in Table 7. The Type 1 scenario set has 10 easy tasks, 20 intermediate tasks and 20 complex tasks. The Type 2 scenario has 20 easy tasks, 10 intermediate tasks and 20 complex tasks. The Type 3 scenario has 20 easy tasks, 20 intermediate tasks and 10 complex tasks, the Type 4 scenario has 30 easy tasks, 10 intermediate tasks and 10 complex tasks, the Type 5 scenario has 10 easy tasks, 30 intermediate tasks and 10 complex tasks. The Type 6 scenario has 10 easy tasks, 10 intermediate tasks and 30 complex tasks. Scenario 1, 2, and 3 are variations of the evenly distributed tasks, while Scenario 4 has more easy tasks, Scenario 5 has more intermediate tasks and Scenario 6 has more complex tasks. We set up these various task sets in order to see how the teams reacted to those different task sets based on the use of different strategies.

Each testing is made up of 10 user stories where each user story consists of five tasks with varying levels of complexity. The distribution of the task complexities for each user story are shown in Table 8.

4.2 All Possible Combinations

There are 48 environments in total (8 variations of agents' capabilities shown in Table 6 and 6 variations of task complexities shown in Tables 7 and 8). In each of these environments, the seven strategies are tested to see how they performed and, based on that, we can analyse how each strategy performed under each environment.

Table 7. Task set categories

Scenario	Description
One	10 easy tasks, 20 intermediate tasks, 20 difficult tasks. Each user story with 30 as the workload. Total of 50 tasks
Two	20 easy tasks, 10 intermediate tasks, 20 difficult tasks. Each user story with 28 as the workload. Total of 50 tasks
Three	20 easy tasks, 20 intermediate tasks, 10 difficult tasks. Each user story with 26 as the workload. Total of 50 tasks
Four	30 easy tasks, 10 intermediate tasks, 10 difficult tasks. Each user story with 25 as the workload. Total of 50 tasks
Five	10 easy tasks, 30 intermediate tasks, 10 difficult tasks. Each user story with 30 as the workload. Total of 50 tasks
Six	10 easy tasks, 10 intermediate tasks, 30 difficult tasks. Each user story with 34 as the workload. Total of 50 tasks

Table 8. User stories for each test case

Task Complexity in each user story	Scenario 1	Scenario 2	Scenario 3	Scenario 4	Scenario 5	Scenario 6
1						
2	One	One	One			
3		One	One	One		
4				Two	One	One
5	One		One		One	One
6	One	One		One	One	
7			One		One	
8	One	One		One	One	Two
9	One	One	One			One
10						

4.3 Random Testing Setup

The purpose of random testing is to simulate a real-world situation where we could not predict the team composition or what type of task set the team is going to work on. We also wanted to observe which strategy worked best in this random situation. These random tests were repeatedly run, up to 100 times, to explore the various combinations of team and task. In each run the number of agents generated is between five and six. We then observed how these strategies performed under unknown situations. The task complexity is randomly generated between 1 and 10. Similarly, the agent capability is also randomly generated between 1 and 10. This is repeated 100 times for all strategies.

Figures 3 and 4 show the distribution for the tasks and agents for 100 runs. We also introduced an additional strategy (referred to as adaptive strategy) where the strategy used is dependent on the environment it is in. For example, if the current environment has 6 agents in a team that needs to work on many complex tasks, then the strategy to be used is the strategy that performed well in that fixed environment. In every run, the strategy to be used is changed accordingly.

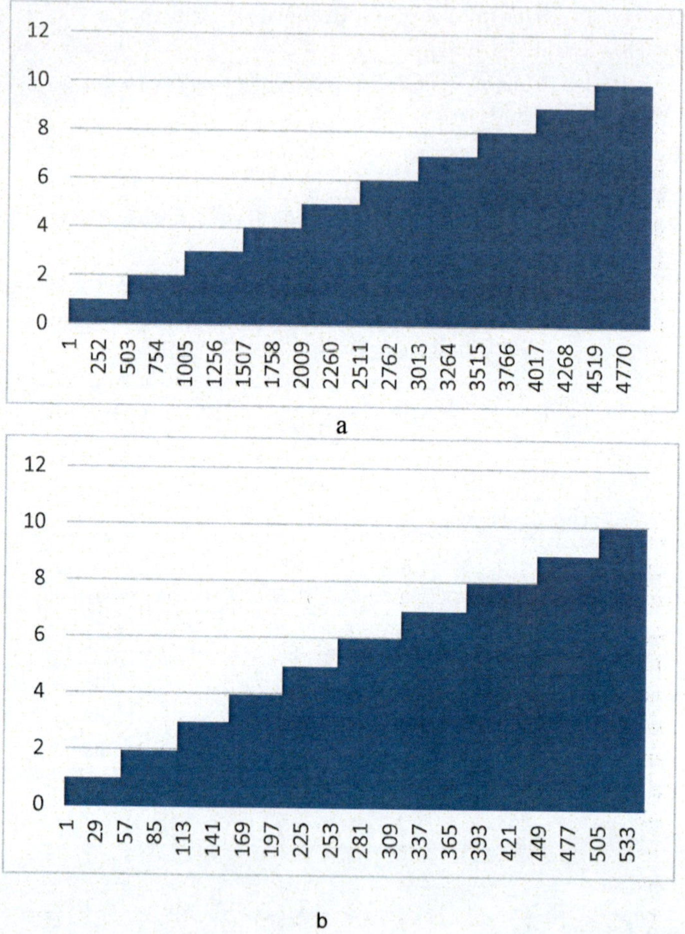

Fig. 3. Task complexity distribution(a) and agent capability distribution (b)

4.4 Experimental Result

The summary result for the fixed environment are shown in Table 9. Here, the top three strategies are shown based on its average work efficiency. First of all, we can observe that the intelligent stategies (IP1, IP2 and IP3) performed well in almost all environments. In this fixed environment, MP3 performed slighty better than IP3 and IP2. The intelligent strategies worked well because it avoided negative pairing and opted to work solo when positive pairing is not possible. Moreover, by not waiting for an agent to pair, the idle time for any agent is reduced thereby maximising the agent's productivity. On the other hand, MP3 did very well in the environment where majority of the agents are novice but working on complex tasks. The must pair strategy performed particularly well when there are even number of agents. In this fixed environment, the must pairs that took place are positive pairing where an expert or and intermediate agent is paired with a novice agent (for 5 agents, there are 3 novices, 1 intermediate and 1 expert and for 6 agents, there are 3 novices, two intermediate and one experts as shown in Table 6). This cross pairing enables the novice agents to learn from the intermediate and expert agents. More importantly, the cross pairing allows a novice agent to work on more complex tasks.

It can also be observed that solo strategy performed better than must pair strategies but never better than the intelligent strategies as shown in Table 9. These results also suggest solo strategy can also be used in an environment where the developers are working on tasks with evenly distributed complexities. This result indicates that even though pairing brings a lot of benefit, it is not recommended to pair all the time and that pairing should only take place when the pairing is positive, which in turn maximises the benefit of pairing in that it delivers higher quality software (fewer errors) in a shorter time.

Table 9. Results summary for five and six agents

Agent/task	Evenly distributed tasks	Easy tasks	Intermediate tasks	Complex tasks
Five or six where majority are novice agents	IP2 (2.04) IP3(2.05) MP3(2.20)	IP1(1.62) IP2(1.65) IP3(1.77)	IP2(1.72) IP3(1.83) IP1(1.86)	MP3(2.16) IP3(2.21) IP2(2.36)
Five or six where majority are intermediate agents	IP3(1.67) SOLO(1.74) IP2(1.93)	IP1(1.41) IP2(1.43) IP3(1.44)	IP1(1.38) IP2(1.39) IP3(1.44)	IP3(1.70) SOLO(1.73) MP3(1.92)
Five or six where majority are expert agents	IP1(1.31) IP2(1.34) SOLO(1.42)	IP1(1.28) IP2(1.30) Ip3(1.34)	IP1(1.27) IP2(1.27) IP3(1.32)	IP1(1.29) IP2(1.30) IP3(1.37)
Five or six where agents capabilities are evenly distributed	IP2(1.70) IP3(1.70) SOLO(1.86)	IP1(1.43) IP2(1.50) IP3(1.58)	IP1(1.37) IP2(1.45) IP3(1.55)	IP3(1.78) IP2(1.84) MP3(1.89)

4.5 Random Testing

Table 10 shows the result when running the seven strategies and the adaptive strategy defined in Sect. 3. It can be seen that in this environment, only Solo, MP3, IP3 and the adaptive strategies were able to complete all the tasks (labelled as 100 runs). IP1 and MP1 only managed to complete 48 runs out of 100, IP2 and MP2 were able to complete 93 of 100 runs. This means that Type 1 and Type 2 strategies can only work in some situations, whereas Type 3 strategies can be used for all situations. Similarly, the adaptive strategies work for all situations as it was tuned to select the best strategy to used based on the current environment. The average person hours (APH) is added to the table to show the total hours spent by the team.

Table 10. Random testing results

Strategy	ACT	AET	AIT	APH	Number of successful run	Average WE
Solo	**99.32**	**431.30**	**63.39**	**542.20**	**100**	**1.97**
MP1	94.98	410.93	84.39	517.59	48	1.89
IP1	76.58	333.81	64.45	421.66	48	1.53
MP2	105.99	442.68	109.64	573.55	93	2.08
IP2	86.63	377.19	70.62	472.60	93	1.72
MP3	**107.60**	**466.58**	**94.97**	**582.65**	**100**	**2.12**
IP3	**89.86**	**425.40**	**32.46**	**490.69**	**100**	**1.78**
Adaptive	**89.77**	**372.48**	**90.48**	**489.95**	**100**	**1.77**

As IP1, IP2, MP1 and MP2 were not able to complete all the runs successfully, we will only compare the four strategies that have completed all runs successfully (Solo, IP3, MP3 and Adaptive). As expected, the adaptive strategy recorded the best average work efficiency followed by IP3, solo and MP3. MP3 was worse off than solo due to the negative pairings that might have taken place. The performance of MP3 is also dependent on whether the number of agents in team is even or odd. The adaptive strategy also recorded the shortest average completion time, the lowest average effort time, and the lowest average person hours. If this is translated to the cost of paying an individual person for their hours worked on the sprint, both the adaptive strategy and IP3 will have the least cost. IP3 performed very closely to the adaptive strategy and recorded the lowest average idle time. This means, in the absence of an adaptive strategy, IP3 can be used as its performance is acceptable. Figure 4 shows the comparisons of the four strategies based on average completion time, average idle time, average effort time and person hour.

Based on this results, both the adaptive strategy and the IP3 performed well in this environment. The adaptive strategy performed best as it changes its strategy between IP1, IP2 and IP3 depending on the type of environment it is in. This strategy is useful and can be used only if the environment can be accurately identified at the beginning of a software development project. However, in a real world setting this is not always true as the user requirements are dynamic and the categorisation of the task complexities and the agent capabilities may not be accurate. An alternative to this adaptive strategy is IP3 as it provides some level of flexibilities for pair with lower capabilities to work on task

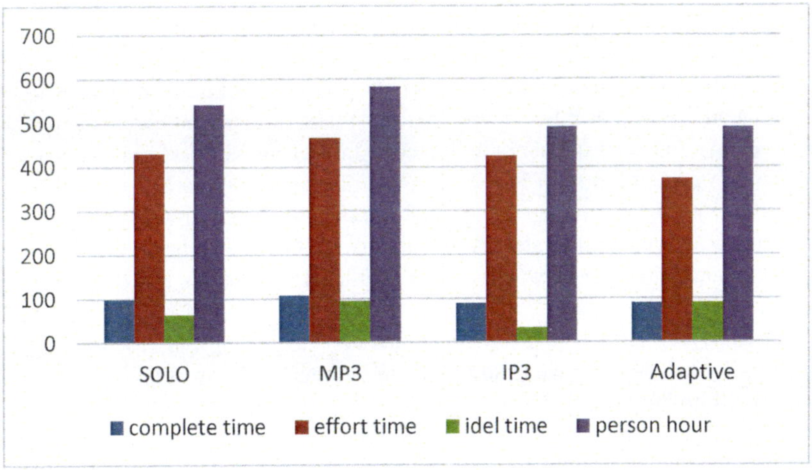

Fig. 4. Comparison of strategies for completion, effort, person hours and idle time

with higher level of complexities (with some penalties). The result obtained suggests that pairing should only take place when it results in a positive pairing and when this is not possible, it is not necessary to pair.

5 Summary

In this experiment, pair programming is used in both intelligent pair and must pair, however in the intelligent pair, only positive pairing is considered. In order to evaluate the impact of pair programming (either MP or IP), we compared their performance solo programming with pair programming, and we found that must pair is not always better than solo, but intelligent pair is always better than solo. Based on the result obtained, the must pair only performed well in one of the 48 environments that were tested. In contrast, the intelligent pair performed well in 47 environments. Moreover, the average effort time recorded by the intelligent pair are lower than that of the solo strategy. The use of intelligent strategy also suggests that pairing is recommended when it results in a positive pairing and when no positive pairing is possible, solo programming is recommended to reduce the team's average idle time.

The strategy chosen will impact the efficiency of team working performance. We show that pairing was better than solo (except when there was negative pairing). Then we investigated the pairing schemes (MP and IP) in more detail and what kind of tasks were taken by pairs (Type 1, Type 2 and Type 3). Seven strategies were designed and implemented, intelligent pair (IP1, IP2 and IP3), must pair (MP1, MP2 and MP3)) and solo. Type 1 strategy dictates that a pair of agents can only work on a task that has the same complexity as the pair capability. Type 2 allows a pair of agents to work on task in the same category of complexity as the pair capability (i.e. novice working on easy task, intermediate working on intermediate task and expert working on complex task). Type 3 is the most flexible strategy as it allows a pair of agents to work on task in a category higher than the pair capability (novice working on intermediate task, intermediate working on complex task) with a penalty.

In the must pair strategy, negative pairing is allowed as all tasks must be worked by a pair of agents. As such, it was found that must pair strategy only worked well when there are even number of members in the team. Intelligent pair performed well in most of the environments because it avoids negative pairing. This strategy worked for both odd numbered and even numbered team. Solo strategy is used to compare with must pair strategy and intelligent pair strategy. Solo strategy will be used when positive pairing is not possible and when the task to be completed has the same complexity as the agent capability.

After analysing all possible environments and identifying the best strategy to be used for each environment, we tested all the 7 strategies and an adaptive strategy in a random environment. The adaptive strategy is introduced to investigate whether what happen if we tuned the strategy to the prevailing environment. Using adaptive strategy, we assumed that the environment can be accurately identified and hence we can select the best strategy that can be used for this environment. Based on the result, we can conclude that the adaptive strategy is the best strategy and should be used if we can accurately identify the environment it is in. However, when this is not possible (as requirement changes dynamically and we may not be able to accurately categorise the pair capability and the task complexity), IP3 can be used as an alternative strategy.

6 Conclusion and Future Work

A comprehensive investigation was undertaken into pair programming in a Scrum team by providing an advanced simulation tool that was developed using a multi-agent system. Seven strategies were designed and provided to compare the performance of various kinds of pair and solo programming. The concept of negative pairing, positive pairing, cross level pairing and same level pairing were all considered in the design of these strategies. Experiments were undertaken to test those strategies in fixed and random environments. The intelligent pair strategies were observed to give the best performance in all contexts and to provide stability in real world simulations. The adaptive strategy also performed well in the random environment where it was able to change its strategy based on the prevailing environment and would work well if the environment can be accurately identified. This research provides a comprehensive analysis of different kind of pairings; must pair and voluntary pair and suggests how pairing should be undertaken based on varying situations (varying agent capabilities and varying task complexities).

The current simulation models developer agents with a single capability value, a future extension to this would be providing agents with per skill capabilities and describing task complexity in terms of skills required to complete. This would allow the exploration of specialist agents versus representations of cross-functionally skilled team members. A further extension to the developer agents, would see the addition of personality characteristics to allow the modelling of the interaction of differing personality types when pairs are formed. Team members in a development team, over time, increase their skills and capabilities through external learning and also by working on tasks within projects. Currently this project learning is not modelled in the simulation and would be an addition that would help provide insight into how teams evolve over time, and the impact that pairing has on learning rates as junior developers are able to tackle more complex tasks alongside more senior developers increasing their exposure to complex problems.

Acknowledgments. Sincerely, Thanks for Dr Patricia Anthony and Dr. Stuart Charters at Lincoln University, New Zealand to support this PhD research, also thanks for Pro. Guojian Cheng at Xi'an Petroleum University, China. Dr Gang Li at Deakin University, Australia and Professor Longbing Cao at University of Technology Sydney provides funding in the related data analysis and machine learning research which I was doing my invited research at UTS, Australia. I also thanks to Edinburgh Napier University, United Kingdom where I get my MSc in Advanced Software Engineering. They are all my best Supervisors support me to growth and become more and more professional.

References

Abar, S., Theodoropoulos, G.K., Lemarinier, P., O'Hare, G.M.P.: Agent based modelling and simulation tools: a review of the state-of-art software. Comput. Sci. Rev. **24**, 13–33 (2017). https://doi.org/10.1016/j.cosrev.2017.03.001

Agarwal, R.: A flexible model for multi-agent based simulation of software development process. Auburn University (2007)

Agarwal, R., Umphress, D.: A flexible model for simulation of software development process. In: Paper presented at the Proceedings of the 48th Annual Southeast Regional Conference, Oxford, Mississippi (2010)

Ali, S., et al.: Multi-agent system using scrum methodology for software process management, Singapore (2019)

Alsalemi, A.M., Yeoh, E.: A survey on product backlog change management and requirement traceability in agile (Scrum). In: Paper presented at the 2015 9th Malaysian Software Engineering Conference (MySEC) (2015)

Argote, L., Fahrenkopf, E.: Knowledge transfer in organizations: The roles of members, tasks, tools, and networks. Organ. Behav. Hum. Decis. Process. **136**, 146–159 (2016). https://doi.org/10.1016/j.obhdp.2016.08.003

Arisholm, E., Gallis, H., Dyba, T., Sjoberg, D.I.K.: Evaluating pair programming with respect to system complexity and programmer expertise. IEEE Trans. Softw. Eng. **33**(2), 65–86 (2007). https://doi.org/10.1109/TSE.2007.17

Beck, K., Fowler, M.: Planning Extreme Programming (2001)

Bellifemine, F.: Develpoing Multi-Agent Systems with JADE (2007)

Boulahbel-Bachari, S., El Saadi, N.: Migration and self-selection: an agent-based model. In: Silhavy, Radek, Silhavy, Petr, Prokopova, Zdenka (eds.) Computational Statistics and Mathematical Modeling Methods in Intelligent Systems: Proceedings of 3rd Computational Methods in Systems and Software 2019, Vol. 2, pp. 288–303. Springer International Publishing, Cham (2019). https://doi.org/10.1007/978-3-030-31362-3_28

da Silva, F.Q.B., et al.: Team building criteria in software projects: a mix-method replicated study. Inf. Softw. Technol. **55**(7), 1316–1340 (2013). https://doi.org/10.1016/j.infsof.2012.11.006

Dybå, T., Arisholm, E., Sjøberg, D.I.K., Hannay, J.E., Shull, F.: Are two heads better than one? on the effectiveness of pair programming. IEEE Softw. **24**(6), 12–15 (2007). https://doi.org/10.1109/MS.2007.158

Gilbert, N., Troitzsch, K.G.: Simulation for the Social Scientist. Open University Press (2005)

Griffith, I., Taffahi, H., Izurieta, C., Claudio, D.: A simulation study of practical methods for technical debt management in agile software development. In: Paper presented at the Proceedings of the Winter Simulation Conference (2014)

Han, J., Liu, J., Lei, H.: Agent-based simulation of emergency response of urban oil and gas pipeline leakage. In: Paper presented at the Proceedings of the 11th International Conference on Computer Modeling and Simulation, North Rockhampton, QLD, Australia (2019)

Hannay, J., Dybå, T., Arisholm, E., Sjøberg, D.: The effectiveness of pair programming: a meta-analysis. Inf. Softw. Technol. **51**(7), 1110–1122 (2009). https://doi.org/10.1016/j.infsof.2009.02.001

Hashimoto, A., Takata, K., Ito, N., Matoba, R., Tani, K., Maeda, Y.: Study of the influence of an obstacle on the evacuation behavior using multi-agent simulation where the intimate space around each agent is considered. In: Paper presented at the Proceedings of the 11th International Conference on Computer Modeling and Simulation, North Rockhampton, QLD, Australia (2019)

Ivanov, D., Kapustyan, S., Kalyaev, A., Korovin, I.: Decision support systems for the oil fields with cloud multiagent service. In: Silhavy, Radek, Silhavy, Petr, Prokopova, Zdenka (eds.) CoMeSySo 2019 2019. AISC, vol. 1047, pp. 16–23. Springer, Cham (2019). https://doi.org/10.1007/978-3-030-31362-3_3

Joslin, D., Poole, W.: Agent-based simulation for software project planning. Paper presented at the Proceedings of the Winter Simulation Conference (2005)

Jun, L.: Context-aware task allocation for distributed agile team. In: Paper presented at the Proceedings of the 28th IEEE/ACM International Conference on Automated Software Engineering, Silicon Valley, CA, USA (2013)

Jun, L.: Human Factors in Agile Software Development (2015). ArXiv, abs/1502.04170

Jun, L., Han, Y., Zhiqi, S.: An Empirical Analysis of Task Allocation in Scrum-based Agile Programming (2014). ArXiv, abs/1411.6201

Justice, J.: The 3–5–3 of Scrum (2018). https://www.scruminc.com/the-3-5-3-of-Scrum/

Kent, B., James, G., Robert, C.M.: Manifesto for Agile Software Development (2020). https://agilemanifesto.org/

Košinár, M., Štrba, R.: Simulations of agile software processes for healthcare information systems development based on machine learning methods. IFAC Proc. Vol. **46**(28), 175–180 (2013). https://doi.org/10.3182/20130925-3-CZ-3023.00028

Lin, J., Han, Y., Zhiqi, S., Chunyan, M.: Studying task allocation decisions of novice agile teams with data from agile project management tools. In: Paper presented at the Proceedings of the 29th ACM/IEEE international conference on Automated software engineering, Vasteras, Sweden (2014)

Lin, M., Zhang, Q.: Time scales of knowledge transfer with learning and forgetting. Physica A **525**, 704–713 (2019). https://doi.org/10.1016/j.physa.2019.03.084

Lui, K., Chan, K.: Pair programming productivity: Novice–novice vs. expert–expert. Int. J. Hum.-Comput. Stud. **64**(9), 915–925 (2006). https://doi.org/10.1016/j.ijhcs.2006.04.010

Lunesu, M.I., Münch, J., Marchesi, M., Kuhrmann, M.: Using simulation for understanding and reproducing distributed software development processes in the cloud. Inf. Softw. Technol. **103**, 226–238 (2018). https://doi.org/10.1016/j.infsof.2018.07.004

Macal, C.M., North, M.J.: Tutorial on agent-based modeling and simulation PART 2: how to model with agents. In: Paper presented at the Proceedings of the 2006 Winter Simulation Conference (2006)

Macal, C.M., North, M.J.: Agent-based modeling and simulation: ABMS examples. In: Paper presented at the Proceedings of the 40th Conference on Winter Simulation, Miami, Florida (2008)

Mahnič, V., Hovelja, T.: On using planning poker for estimating user stories. J. Syst. Softw. **85**(9), 2086–2095 (2012). https://doi.org/10.1016/j.jss.2012.04.005

Masood, Z., Hoda, R., Blincoe, K.: Motivation for self-assignment: factors agile software developers consider. In: Paper presented at the 2017 IEEE/ACM 10th International Workshop on Cooperative and Human Aspects of Software Engineering (CHASE) (2017)

Mehmood, S., Ahmed, S., Kristensen, A.S.: Application of integrated model of evacuation psychology in an agent-based simulation. In: Paper presented at the Proceedings of the 11th International Conference on Computer Modeling and Simulation, North Rockhampton, QLD, Australia (2019)

Nadler, D., Hackman, J.R., Lawler, E.E.: Managing organizational behavior (1979)

Nilsson, K.: Increasing quality with pair programming - an investigation of using pair programming as a quality tool (2003)

Noori, F., Kazemifard, M.: Simulation of pair programming using multi-agent and MBTI personality model. In: Paper presented at the 2015 Sixth International Conference of Cognitive Science (ICCS) (2015)

Orłowski, C., Bach-Dąbrowska, I., Kapłański, P., Wysocki, W.: Hybrid fuzzy-ontological project framework of a team work simulation system. Procedia Comput. Sci. **35**, 1175–1184 (2014). https://doi.org/10.1016/j.procs.2014.08.214

Phillips, J.: Team-RUP: an agent-based simulation study of team behavior in software development organizations (2006)

Plonka, L., Sharp, H., van der Linden, J., Dittrich, Y.: Knowledge transfer in pair programming: an in-depth analysis. Int. J. Hum. Comput. Stud. **73**, 66–78 (2015). https://doi.org/10.1016/j.ijhcs.2014.09.001

Ramanujam, R., Lee, I.: Collaborative and competitive strategies for agile scrum development. Paper presented at the The 7th International Conference on Networked Computing and Advanced Information Management (2011)

Schwaber, K., Sutherland, J.: The Scrum Guide (2017)

Silva, I.J.D., Rayadurgam, S., Heimdahl, M.P.E.: A reference model for simulating agile processes. In: Paper presented at the Proceedings of the 2015 International Conference on Software and System Process, Tallinn, Estonia (2015)

Song, H., et al.: Development and validation of the primary care team dynamics survey. Health Serv. Res. **50**(3), 897–921 (2015). https://doi.org/10.1111/1475-6773.12257

Stober, T., Hansmann, U.: Considerations on teaming and leadership. in agile software development: best practices for large software development projects, pp. 75–92. Springer, Heidelberg

Tamburri, D.A., Razo-Zapata, I.S., Fernández, H., Tedeschi, C.: Simulating awareness in global software engineering: a comparative analysis of Scrum and Agile Service Networks. In: Paper presented at the 2012 4th International Workshop on Principles of Engineering Service-Oriented Systems (PESOS) (2012)

Tyrychtr, J., Pelikán, M., Kvasnička, R., Ander, V., Benda, T., Vrana, I.: Multi-agent system in smart econometric environment. In: Silhavy, Radek, Silhavy, Petr, Prokopova, Zdenka (eds.) CoMeSySo 2019 2019. AISC, vol. 1046, pp. 434–442. Springer, Cham (2019). https://doi.org/10.1007/978-3-030-30329-7_38

Wang, X., Wang, J., Zhang, R.: The optimal feasible knowledge transfer path in a knowledge creation driven team. Data Knowl. Eng. **119**, 105–122 (2019). https://doi.org/10.1016/j.datak.2019.01.002

William, R., Uri, W.: An introduction to agent-based modeling: modeling natural, social, and engineered complex systems with NetLogo (2015)

Williams, L., Kessler, R.R., Cunningham, W., Jeffries, R.: Strengthening the case for pair programming. IEEE Softw. **17**(4), 19–25 (2000). https://doi.org/10.1109/52.854064

Wooldridge, M., Jennings, N.: Agent theories, architectures, and languages: a survey. In: Wooldridge, Michael J., Jennings, Nicholas R. (eds.) ATAL 1994. LNCS, vol. 890, pp. 1–39. Springer, Heidelberg (1994). https://doi.org/10.1007/3-540-58855-8_1

Yilmaz, L., Phillips, J.: The impact of turbulence on the effectiveness and efficiency of software development teams in small organizations. Softw. Process Improve. Pract. **12**(3), 247–265 (2007). https://doi.org/10.1002/spip.318

Zhen, Y., Wanpeng, Z., Hongfu, L.: Artificial intelligence techniques on real-time strategy games. In: Paper presented at the Proceedings of the 2018 2nd International Conference on Computer Science and Artificial Intelligence, Shenzhen, China (2018)

Zieris, F., Prechelt, L.: On knowledge transfer skill in pair programming. In: Paper presented at the Proceedings of the 8th ACM/IEEE International Symposium on Empirical Software Engineering and Measurement, Torino, Italy (2014)

Zualkernan, I.A., Darmaki, H.A., Shouman, M.: A methodology for building simulation-based e-learning environments for Scrum. In: Paper presented at the 2008 International Conference on Innovations in Information Technology (2008)

Comparison of Software Simulation and Network Testbed of DDoS Attacks for IPv4 and IPv6 Networks

Ladislav Huraj(✉), Marek Simon, and Matus Svrcic

Department of Applied Informatics, University of Ss. Cyril and Methodius in Trnava, 91701 Trnava, Slovak Republic
{ladislav.huraj,marek.simon,svrcic2}@ucm.sk

Abstract. DDoS attacks have now become a common threat that every organization connected to the Internet must face. Scientists are trying to come up with new proposals for protection against DDoS attacks, for which they need to test individual defense mechanisms. One of the testing options is a software simulator such as the OMNeT++ simulator. An INET framework containing a standard library of communication protocols and easily enabling the simulation of various communication network scenarios, such as UDP protocols, TCP, IPv4, IPv6, MANET routing protocols, etc., can extend the network OMNeT++ simulator.

The aim of the presented research was to find out whether the software tool OMNeT++ together with the INET framework is suitable for simulating DDoS attacks from the point of view of the impact on the computer network. The article presents a comparison of a real DDoS attack based on a testbed environment over UDP and ICMP protocol with simulated DDoS attacks in the OMNeT++ software simulator. In addition to comparing different protocols such as UDP and ICMP, experiments were performed and different IP protocols were compared, and attacks were performed on both IPv4 and IPv6 networks.

Keywords: DDoS · OMNeT++ · Simulation · Testbed

1 Introduction

With the advent of the Internet of Things, a wide range of devices can be observed that can be used with an Internet connection. The user can very easily control them via smartphones, tablets or computers. Modern Internet technologies make everyday life easier, but on the other hand, they also have their downsides in the form of various types of attacks on Internet services. An attacker also has easier access to a huge number of vulnerable devices, which increases the risk of potential threats. One of the most common threats exploiting the availability of an enormous number of vulnerable devices is the Distributed Denial of Service (DDoS) attack.

DDoS attacks use a number of different distributed nodes (named bots or zombies) involved in the so-called botnet in an effort to exhaust the victim's resources and reduce the quality of service. These are often devices where the device owner does not know

R. Silhavy et al. (Eds.): CoMeSySo 2021, LNNS 232, pp. 287–295, 2021.
https://doi.org/10.1007/978-3-030-90318-3_25

that his computer is being exploited for a DDoS attack. Such a misused computer or other device is involved in a DDoS attack and executes malicious commands sent to it by an attacker controlling the botnet. The usual impact of a DDoS attack on an organization is a reduction in quality or even a complete loss of the availability of the service that the organization operates for its own needs or provides to its customers. This will cause a loss of reputation and may lead to direct financial losses. Therefore, it is essential that effective mechanisms be devised to help effectively detect and defend against DDoS attacks. Increased use and reliance of society on Internet technologies has contributed to a significant increase in the company's vulnerability. Areas most affected by DDoS attacks are IT and cloud services, public and financial services, and the entertainment and media industries [1–3].

An important step in implementing a defense mechanism is to test the proposed method on specific attack scenarios. Testing can take place directly in the real network, or a software simulator can be used for testing, which facilitates and simplifies the performed experiment and reduces the costs of the testing process. Another way of testing is to use existing datasets from already implemented and recorded attacks and use this data to form new defense systems [4]. However, the datasets used also had to be created on the basis of real attacks or on the basis of data generated within the simulation. Therefore, when using a software simulator, it is important to know its accuracy, as well as the functional limits of the used software simulator compared to the implementation of real DDoS attacks.

2 Software Simulation vs. Network Testbed

When designing a defense mechanism against a DDoS attack, but also when testing the behavior of a DDoS attack, the experiment must be carried out under the control of researchers with the possibility of examining properties and changing parameters during the attack. In general, it is not possible to use a real computer-operating network for testing and conduct an attack directly on this network. For this reason, network simulators/emulators and testbeds are used to test DDoS attacks, which allow to test different scenarios in a controlled environment and obtain reproducible results [4].

The simulator does not represent the actual implementation of the simulated system, but is only software that predicts the behavior of the system under given conditions. Analytical methods are used in the simulator to model the behavior of the system. The behavior of network devices, links, protocols and applications running on the network and network transmission are modeled and simulated in a network simulator. Only the result of the final number of events that occur in the network in a certain period of time is monitored [4].

The experiments presented in this article focused on the OMNeT++ network simulator. Various authors use the OMNeT++ software simulation tool to test DDoS attacks or to create DDoS datasets where OMNeT++ is used as a traffic generator for DDoS attacks [5–7].

However, as in all simulation systems, the validation of component simulation is questionable, moreover, if there is no reference system for comparison.

In addition, due to the fact that the overall simulation system consists of many subsystems, which may have been individually tested and may show unprecedented behavior, the overall simulation can provide only limited reliability [4].

OMNeT++ is defined as a simulation library and framework that is designed to create network simulations. Simulators can be used to create wired and wireless networks and domain-specific features, such as support for sensor networks, wireless ad hoc networks, Internet protocols, network performance modeling, and more. The simulator provides model frameworks created as projects by independent developers and can use the INET framework.

INET framework is an open-source model library for the OMNeT++ simulation environment. Provides protocols, agents called intermediaries and other models for researchers working with communication networks. INET supports a wide class of communication networks including fixed, wireless, mobile, ad hoc and sensor networks and a wide range of network protocols such as TCP, UDP, IPv4, IPv6, OSPF, Ethernet, PPP, IEEE 802.11, etc. [8].

Unlike a simulator, cyber-physical testbeds using real networks create a more realistic environment for testing attacks. Testbed is a platform used for consistent and repeatable experimentation, in this case with DDoS attacks. The test environment must provide fidelity, repeatability, measurement accuracy and safe execution for the scientific experiment [9].

A network testbed using Ansible orchestration was used in a real network to compare DDoS attacks with the OMNeT++ software simulator. Ansible is an open-source IT automation tool that offers both configuration management and state-of-the-art orchestration. Ansible can be deployed on multiple nodes and run tasks remotely, with a simple, model-driven configuration management system [10]. Ansible is a powerful tool that simplifies the implementation of a test environment. In addition, no special hardware is required to perform attacks using the Ansible tool; the testbed created uses the existing infrastructure in the organization. Our previous studies with the implementation of a testbed using an Ansible environment show the ability of Ansible to create a network testbed comparable to a botnet with tens of thousands of bots. The testbed created in this way meets the requirements for a test environment for a scientific experiment. In addition, the testbed created over this tool can be used to implement DDoS attacks on specific target servers in both IPv4 and IPv6 networks [11].

3 Experimental Scenarios and Results

DDoS is a large-scale cyber attack in which an attacker uses more than one attack vector to generate a large amount of malicious traffic from different sources, flooding network services or resources with a large amount of network traffic. At the network layer, an attacker uses the Internet Protocol (IP) and the Internet Control Message Protocol (ICMP) to generate attacks. At the transport level, attackers use Transmission Control Protocol (TCP) and User Datagram Protocol (UDP) to perform flood attacks, and at the application level, attackers typically use Hypertext Transfer Protocol (HTTP) to disrupt Web server services [12]. The most common DDoS attacks take place via the UDP protocol [13].

Based on the above, the UDP and ICMP protocols were selected for pilot testing and several identical scenarios were tested for the OMNeT++ simulation environment as well as for the network testbed.

The network topology for simulations of DDoS attacks in the OMNeT++ simulator on networks with IPv4 protocol was compiled by connecting thirty INET *StandardHosts* attacking nodes. The attacking nodes were divided into botnets of ten nodes and connected by a *Router* node, which was then connected to the target server being attacked, Fig. 1. A similar infrastructure for simulation of DDoS attacks in the OMNeT++ simulator can be found e.g. in [14]. Nodes in the IPv6 network were represented in the simulation environment by the *StandardHost6* module from the INET framework, which has direct support for the Internet protocol version 6. Routers in the network were represented by the *Router6* module with support for the IPv6 protocol. Devices throughout the network were interconnected via the *DatarateChannel* channel. The connection speed was set to 10Mbps with a delay of 0.1 microsecond. The simulation time of the experiment was 20 s.

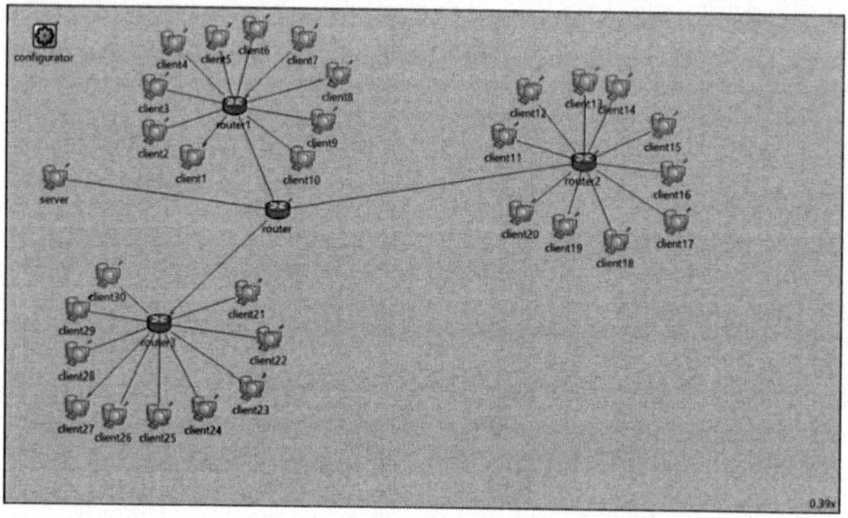

Fig. 1. Network topology for 30 attacking nodes in the OMNeT++ simulation environment.

During the experiments, individual nodes sent 32-byte packets to the target node. The intensity of sending (*sendInterval*) packets varied depending on the simulation. Packet transmission intensity values were designed to be 100 ms, 50 ms, 10 ms, 5 ms, 4 ms, 3 ms, 2 ms, 1 ms, and 0.5 ms. User interaction was not taken into account in the study. All designed and processed simulations in the work were combined in the development environment OMNeT++ version 6.0 with the addition of modules of the INET framework version 4.0.

The same infrastructure, number of nodes, experiment length and transmission intervals were also used in the implementation of experiments in the network testbed. The experiments were performed for 30 nodes sending packets to the destination server for UDP and ICMP protocols for both IPv4 and IPv6 networks. On attack machines,

OS Centos7 ran with default settings of network parameters. On the target server, the Debian10 OS ran on which the size of the send buffer (SO_SNDBUF socket option) and receive buffer (SO_RCVBUF socket option) was increased to 256 kilobytes. UDP uses such a receive buffer to store the received data so that the application can read it later. Because UDP does not have flow control, if a datagram arrives that no longer fits in the receiving socket buffer, that datagram is discarded. As a result, the fast sender can overhelm the slow receiver, causing datagrams to be dropped using UDP. In addition, the fast sender can also overpower its own network interface, causing the sender to drop the datagrams itself.

In a DDoS flood attack, the attacking nodes send non-existent port requests to the victim's server; the attacks carried out in the performed experiments were slightly different. Scenarios for the performed case study were designed in such a way that it was possible to perceive the implemented attack from the client's point of view, i.e. so that attacking computers can be viewed not only as attacking zombies but also as legitimate nodes requesting delivery of a response to a request sent to the server. For this reason, the *UdpEchoApp* application was assigned to the server during simulations in OMNeT++. UdpEchoApp "echoes" received packets back to the sender [15]. Also in the real simulation was used Echo Protocol where the server listens for UDP datagrams and when a datagram is received, the data from it is sent back in an answering datagram.

All executed scenarios for both the OMNeT++ simulator and the network testbed ran for 20 s. This time was sufficient to demonstrate the range of network overloading by sent packets. Table 1 and Table 2 show the average values and median values for the individual experiments performed. The tables show the value as a result of successfully returned packets during a DDoS attack in the OMNeT++ simulation and in the network testbed for different packet sending intervals for both IPv4 and IPv6 protocols. The course of the average values is also visualized in Fig. 2.

Table 1. Mean/median values for simulation and real results of individual scenarios for the UDP protocol performed for different intervals of sending packets from attacking nodes

		0.5 ms	1 ms	2 ms	3 ms	4 ms	5 ms	10 ms	50 ms	100 ms
IPv4 OMNeT++	Average	17.02%	34.03%	68.06%	80.06%	80.08%	80.10%	80.20%	81.00%	82.00%
	Median	6.91%	2.94%	76.12%	80.06%	80.08%	80.10%	80.20%	81.00%	82.00%
IPv4 Testbed	Average	0.15%	0.29%	0.59%	0.91%	1.21%	1.52%	3.03%	11.63%	23.50%
	Median	0.15%	0.30%	0.58%	0.89%	1.23%	1.56%	3.18%	11.63%	23.50%
IPv6 OMNeT++	Average	5.87%	18.08%	36.15%	60.31%	80.05%	80.08%	80.21%	81.20%	82.45%
	Median	3.17%	2.70%	46.44%	58.40%	80.10%	80.13%	80.25%	81.25%	82.50%
IPv6 Testbed	Average	0.30%	0.61%	1.20%	1.82%	2.40%	3.04%	6.07%	26.60%	51.68%
	Median	0.31%	0.61%	1.20%	1.85%	2.44%	3.10%	6.35%	26.50%	51.50%

As can be seen from Fig. 2, in all implemented attacks there was a partial or complete flooding of the network during the DDoS attack. It is natural that due to the higher frequency of sent packets, the network was flooded to a greater extent and the success of received packets decreased. The given effect can be seen in both implementations,

Table 2. Mean/median values for simulation and real results of individual scenarios for the ICMP protocol performed for different intervals of sending packets from attacking nodes

		0.5 ms	1 ms	2 ms	3 ms	4 ms	5 ms	10 ms	50 ms	100 ms
IPv4 OMNeT++	Average	17.02%	34.03%	68.06%	80.06%	80.08%	80.10%	80.20%	81.00%	82.00%
	Median	6.91%	2.94%	76.12%	80.06%	80.08%	80.10%	80.20%	81.00%	82.00%
IPv4 Testbed	Average	12.27%	21.28%	35.09%	45.29%	53.13%	59.25%	74.96%	94.39%	97.47%
	Median	13.11%	20.23%	33.32%	43.88%	52.14%	58.55%	74.75%	94.50%	97.50%
IPv6 OMNeT++	Average	5.87%	18.08%	36.15%	60.31%	80.05%	80.08%	80.21%	81.20%	82.45%
	Median	3.17%	2.70%	46.44%	58.40%	80.10%	80.13%	80.25%	81.25%	82.50%
IPv6 Testbed	Average	11.67%	20.41%	33.88%	43.82%	51.62%	57.77%	73.79%	94.14%	97.17%
	Median	12.76%	19.22%	32.53%	42.26%	50.50%	57.20%	73.35%	94.00%	97.00%

in the software simulator as well as in the real testbed. From the point of view of IPv4 and IPv6 protocols, the course of the attack was comparable for protocols, both for the OMNeT++ software simulation and for the network testbed. In both implementations, the IPv6 protocol shows a slightly better success rate of returned packets in a UDP attack compared to the IPv4 protocol, and thus also a slightly better resistance to DDoS attacks.

The differences can be seen when comparing the results from the point of view of software simulation and real testbed. While in ICMP attacks the simulation and testbed results are comparable and the overall success rate has increased steadily with decreasing attack frequency in both implementations, UDP attacks show a significant difference in packet return success. In the software simulation, there was extensive flooding of the network and insufficient return of responses to requests only at a higher frequency of sending packets by attacking nodes from the interval of 2ms; in the network testbed, packet return was low at each attack frequency. As can be seen from Fig. 2, the overall course of the attack using the UDP flood was different for IPv4 and IPv6 networks in the OMNeT++ software simulator and in the real network testbed.

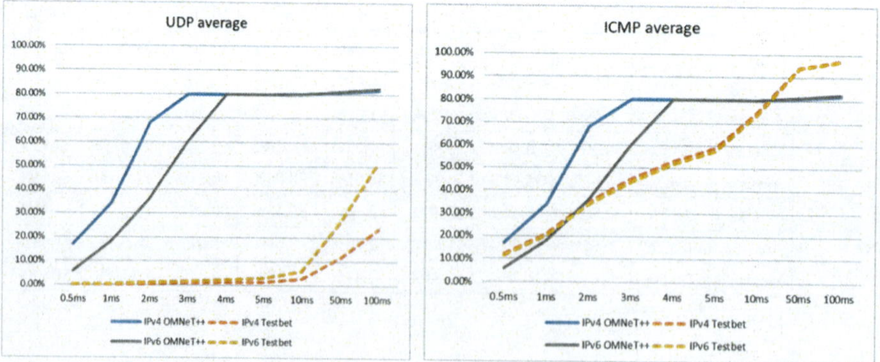

Fig. 2. Average results of successfully returned packets during DDoS attack in OMNeT++ simulation vs. network testbed for different interval of sending packets.

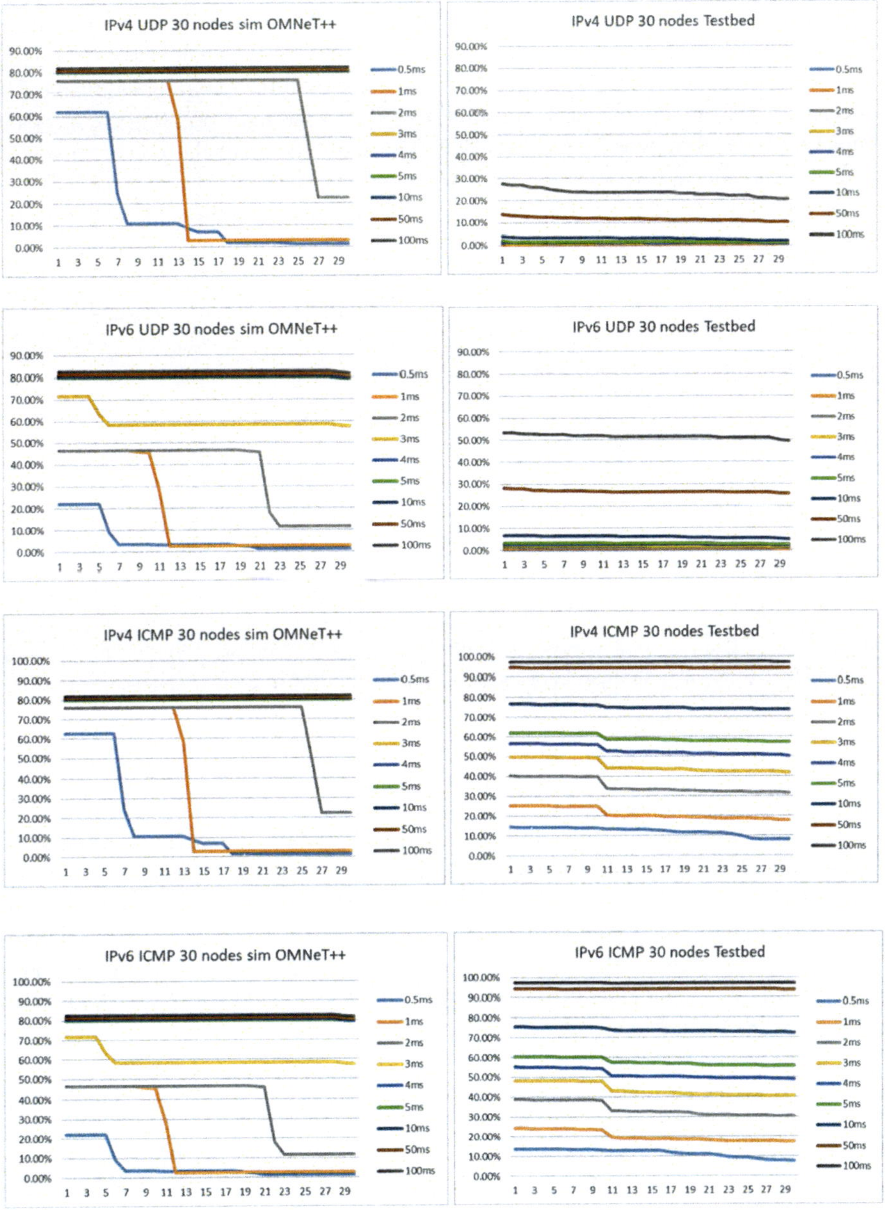

Fig. 3. Results of successfully returned packets during DDoS attack for individual nodes (descending order) in OMNeT++ vs. network testbed for different interval of sending packets.

In addition, during the software simulation in the OMNeT++ environment, there were some anomalies from the point of view of the attacking nodes. As can be seen in Fig. 3, the communication of some nodes had a higher priority in the environment than other nodes. From the descending graph it can be seen that while some nodes in the software simulation in IPv4 networks communicated with more than half the success, most of the others had almost zero success. For IPv6 networks, the course was similar, with the number and success of priority nodes being slightly lower. Compared to experiments in a real network testbed, the distribution of communication with individual nodes was even. Therefore, it is necessary to take the value of the total diameter or median of all nodes involved in the experiment as a telling result in the software simulation. For a software simulation environment, it is not possible to take only values from one node as a telling value, but it is necessary to take measurements on several in order to avoid incorrect results from possible anomalies. In the real network testbed, such an anomaly did not occur.

4 Conclusion

The aim of the article was to compare the results for a DDoS attack in the OMNeT++ software simulation environment and in a real network testbed. The case study shows that although the experiments were performed on the same structure and the same packet sizes and sent intervals, the results of the software simulation and the real network testbed differ.

From the point of view of performing the simulation with the OMNeT++ simulator, it can be stated that the simulator statistically gives more favorable results on the resistance of the network against DDoS attack compared to the network testbed. However, the implementation in the software simulation had shortcomings, which could be seen mainly in the distribution of sending packets by individual attacking nodes during the simulation, when the software significantly preferred the sending of packets from some nodes.

When further testing and simulating network attacks, it is possible to focus on comparisons with other simulators or take into account other types of DDoS attacks as well as the range of tested networks.

Acknowledgements. The work was funded by the grant KEGA 012UCM-4/2021 *Modern technologies and innovations in network security education*, and by the grant APVV-17–0116 *Algorithm of collective intelligence: Interdisciplinary study of swarming behavior in bats.*

References

1. Özçelik, İ, Brooks, R.R.: Deceiving entropy based DoS detection. Comput. Secur. **48**, 234–245 (2015)
2. Machaka, P., Bagula, A.: Statistical properties and modelling of DDoS attacks. In: Vinh, P.C., Rakib, A. (eds.) ICCASA/ICTCC -2020. LNICSSITE, vol. 343, pp. 44–54. Springer, Cham (2021). https://doi.org/10.1007/978-3-030-67101-3_4

3. Voštinár, P., Horváthová, D., Mitter, M., Bako, M.: The look at the various uses of VR. Open Comput. Sci. **11**(1), 241–250 (2021)
4. Brooks, R.R., Özçelik, I.: DDoS research: testing. In: Distributed Denial of Service Attacks, pp. 93–106. Chapman and Hall/CRC, Boca Raton (2020)
5. Alzahrani, S., Hong, L.: Generation of ddos attack dataset for effective ids development and evaluation. J. Inf. Secur. **9**(04), 225 (2018)
6. Khan, M.T., et al.: Simulation and analysis of DDoS attack on connected autonomous vehicular network using OMNET++. In: 2018 9th IEEE Annual Ubiquitous Computing, Electronics & Mobile Communication Conference (UEMCON). IEEE (2018)
7. Alhaidari, F., Alrehan, A.: A simulation work for generating a novel dataset to detect distributed denial of service attacks on Vehicular Ad hoc NETwork systems. Int. J. Distrib. Sensor Netw. **17**(3), 155014772110002 (2021)
8. OMNET++ Discrete Event Simulation System. https://omnetpp.org/, Accessed 06 Aug 2021
9. Siaterlis, C., Genge, B., Hohenadel, M.: EPIC: A testbed for scientifically rigorous cyber-physical security experimentation. IEEE Trans. Emerg. Top. Comput. **1**(2), 319–330 (2013)
10. Masek, P., et al.: Unleashing full potential of ansible framework: university labs administration. In: The 22nd Conference of Open Innovations Association (FRUCT), IEEE, Finland, pp. 144–150 (2018)
11. Čerňanský, M., Huraj, L., Šimon, M.: Controlled DDoS attack on IPv4/IPv6 network using distributed computing infrastructure. J. Inf. Organ. Sci. **44**(2), 297–316 (2020)
12. Kshirsagar, D., Kumar, S.: A feature reduction based reflected and exploited DDoS attacks detection system. J. Ambient Intell. Human. Comput., 1–13 (2021)
13. Furfaro, A., Pace, P., Parise, A.: Facing DDoS bandwidth flooding attacks. Simul. Model. Pract. Theory **98**, 101984 (2020)
14. Kaur, R., Sangal, A.L., Kumar, K.: Modeling and simulation of DDoS attack using Omnet++. In: 2014 International Conference on Signal Processing and Integrated Networks (SPIN), pp. 220–225. IEEE (2014)
15. Mészáros, L., Varga, A., Kirsche, M.: Inet framework. In: Virdis, Antonio, Kirsche, Michael (eds.) Recent Advances in Network Simulation: The OMNeT++ Environment and its Ecosystem, pp. 55–106. Springer International Publishing, Cham (2019). https://doi.org/10.1007/978-3-030-12842-5_2

Benchmarking NewSQL Cloud-Native or Cloud-Hosted Databases

Ionela Chereja(✉), Sarah Myriam Lydia Hahn, and Oliviu Matei

Computers and Information Technology, Technical University Cluj Napoca, Cluj-Napoca, Romania
ionela.chereja@ieee.org, sarah-hahn@gmx.net, oliviu.matei@holisun.com

Abstract. Evaluating the performance of high-impact components when building a solution is the process that provides an informed and fitting choice. Choosing a database system and the proper infrastructural configuration underneath it is however a process that can prove to be extremely complex. In the context of the highly volatile NewSQL database systems and Cloud hosting, choosing the fitting combination is not only difficult but can be daunting. This paper provides an overview of the implications of creating reliable benchmarking for NewSQL Cloud hosted databases. It includes the steps involved in accurate performance evaluations, existing tools, proposes a systematic approach for narrowing the number of combinations to test for and lists the standard benchmarks as well as alternatives. With this work we aim to introduce clarity and structure for the database benchmarking process, based on existing research and the current gaps therein.

Keywords: NewSQL · Performance analysis · Database systems · Benchmarking · Cloud · Big data · TPC

1 Introduction

Increasing attention towards the data in its different formats and various uses drives the requirement for performant data management. Depending on applicability of the data, different database management systems (DBMS) are involved in the process. SQL, NoSQL or NewSQL offer the choice of using the fitting DBMS category depending on how the data is structured, what constraints are in place from a functional perspective, compliance to certain business needs and trade-offs regarding availability and serving data performance. Several comparisons between the three are available [5,29,40,42,54,57], providing an overview on both the advantages and disadvantages for each category and providing guidance on how a choice can be made for a use case. With the development of Internet of Things (IoT) and Big Data, the shift was made more towards the two latter categories, NoSQL and more recently NewSQL systems. NewSQL systems were introduced as a separate category with the coining of the term [3] after a new type of systems became available to reconcile some of the gaps between SQL and NoSQL systems. These DBMS aim to combine the relational

R. Silhavy et al. (Eds.): CoMeSySo 2021, LNNS 232, pp. 296–310, 2021.
https://doi.org/10.1007/978-3-030-90318-3_26

data model and ACID (Atomicity, Consistency, Isolation, Durability) support of the traditional Relational Database Management Systems (RDBMS) with the high performance and high scalability of NoSQL data stores [8,28].

In the context of data analysis and the increasing demand for real-time analytics under ACID constraints, NewSQL systems offer the option of serving hybrid workloads and the use of embedded analytics support [10]. However, the choice of the actual DBMS within that category is still broad, spanning over multiple considerations [9,27].

Cloud computing developed along the development of Big Data. The increasing need for performant, scalable and elastic solutions for storing and processing data offered the proper opportunity to store these DBMS in the Cloud. While choosing the category of DBMS can be achieved by using the surveys and comparisons available, choosing the suitable DBMS [1] within the category prompts several challenges. Furthermore, as Cloud computing provides on-demand resources with scalable configurations, the choice of a platform becomes increasingly complex. The components involved in the choice of a Cloud-hosted DBMS then are: the category of the database, the database engine, the Cloud provider and the resources needed in terms of sizing and nodes. Each of these components have an influence on each other [51]. Disregarding this connection in making the choices might eventually lead to less than optimal performance.

An established approach to selecting fitting hardware and software platforms is to use benchmarking. Database benchmarking implies a systematic approach [6], yet it is limited to providing reproducible results in the context of using the same underlying hardware architecture and configuration. In the context of benchmarking database systems in the Cloud, several other components have to be involved so that the volatility of the underlying setup becomes part of the bechmarking results.

This paper has the objective to create an overview of the facets involved in benchmarking NewSQL database management systems in a Cloud-native or a Cloud-hosted infrastructure, as well as the available tools and existing limitations. The approach used is to define the concepts, the taxonomy of the involved components and the steps necessary in the process.

2 Related Work and Definitions

2.1 Cloud Computing

The term "Cloud computing" has been defined in numerous manners [23]. It is preponderantly used in the context of expanding capacity and resources on demand, without needing to invest in infrastructure, licensing new software or training new staff [34]. It is referred to in regards to pay-per-use services or subscriptions that have the ability to extend infrastructural or IT service capabilities in real time, over a network connection.

The National Institute of Standards and Technology (NIST) defines Cloud computing as follows [39]: "Cloud computing is a model for enabling ubiquitous, convenient, on-demand network access to a share pool of configurable computing resources (e.g., networks, servers, storage, applications and services) that can be rapidly provisioned and released with minimal management effort or service provider interaction". The definition also includes five characteristics of Cloud computing, three service models and four deployment models, making it a more encompassing understanding of the term. The types of service models are infrastructure, platform and software. The models of deployment list private, public, community and hybrid.

These characteristics provide advantages to the end-user, including higher scalability and elasticity, lower operational costs and easy access over the internet [61]. The broad definition with the particular subcategories also create a wide number of possible combinations, and therefore a varying extent to which specific requirements (such as performance, scalability, elasticity and others) will be delivered through a chosen configuration [59].

2.2 Big Data Management

The term big data is often used alongside terms such as Artificial Intelligence, Analytics and Cloud Computing [7]. It refers to massive amounts of information such as, but not limited to, social media feeds, streaming data and historical data. If previously the standard data management systems were relational databases [13], the changes in volume and data formats posed performance and storage challenges [44].

A re-architecture of database management systems has been implemented through the NoSQL data stores. These have developed next to Cloud computing and the internet, driving a high-performance design that comprised big data storage that was more efficient, lower costs, low latency and high scalability and concurrency [29]. Built with a schema-free architecture, NoSQL systems could primarily host unstructured and semi-structured data, but missing the ACID properties of relational databases. As enterprises still used systems that requested the transactionality and reliability of traditional database management systems, NoSQL was not a good fit for those scenarios [37].

A generation of database management systems started to be developed which integrated aspects from both SQL and NoSQL systems. By this approach, advantages from both categories were included and disadvantages minimized [5]. These systems are called NewSQL databases. They provide the ACID properties of relational SQL systems and also implement the versatility and high performance of NoSQL data stores [8].

2.3 Benchmarking

Performance evaluation is a need that is not bound to one specific industry or area, yet it has become particularly relevant in the IT industry. With the increase in variety for IT systems, the need for a more standardized way to compare and assess performance became prominent [36]. Quality management systems is a subject that became of wide concern through the introduction of ISO 9000 standard in 1987. The International Organization for Standardization introduced with the ISO 9000 a range of quality management systems standards [60].

Benchmarking is the process of measuring and then comparing distinct performance elements with predefined standards or against other systems. It can be used for both technical products and business processes. The three main steps involved in benchmarking are [19]:

1. Define and design model
2. Measure options
3. Analyze results

2.4 Database Benchmarking

Database benchmarking was addressed in 1993 by Jim Gray in a paper on evaluating the performance of database systems [26], which later on became a reference point for setting the specific database benchmark criteria. Gray uses four identified criteria for successfully benchmarking a database: a) the relevance to an application domain; b) portability of the solution to allow the benchmarking of different systems; c) ability to scale in order to support large systems benchmarking; d) simplicity in rendering results so that they are understandable.

Transaction Processing Performance Council (TPC) was founded in 1988 and its focus was to define benchmarks for database domains and for transaction processing [41]. Traditional database benchmarking focuses on measuring the throughput of the database on each infrastructural configuration for a specific workload. Based on these, the analysis is comprised of comparing results and choosing the best combination.

TPC defines benchmarks used in the academia and in the industry to measure the performance of database management systems [21]. A relevant part of defining a benchmark is represented by the workload. The workload is a series of ordered operations executed on a database. The way that these are processed vary, either by their sequence or by time. To create a workload, a distinction exists between a synthetic load and traced data. Traced data (or production database workloads) is captured from existing systems, while the synthetic load is generated based on patterns and limited by configurable constraints [30].

A set of standard database performance benchmarks are offered by TPC [55]:

- Transaction Processing: TPC-C, TPC-E
- Decision Support: TPC-H(V2/V3), TPC-DS (V2/V3), TPC-DI
- Virtualization: TPCx-V, TPCx-HCI
- Big Data: TPCx-HS, TPCx-BB
- Internet of Things:TPCx-IoT (V1/V2)
- Artificial Intelligence: TPCx-AI

In the context of NewSQL database management systems and particularly in the case of real-time analytics, NewSQL implementations offer Online Transaction Processing (OLTP) capabilities, Online Analytical Processing (OLAP) capabilities or Hybrid Online-Analytical Processing (HTAP) [10]. This creates the need for a reliable benchmarking option for each of the traditional two approaches, but also for the hybrid option [21].

NewSQL performance evaluations have been proposed in previous works. These evaluations are however mainly limited to only a few non-Cloud installations [32,33]. Another approach is to evaluate NewSQL DBMS over a specific characteristic, such as fault-tolerance [4]. In designing the benchmark, the first step requires the definition of the context for which the analysis is done. Currently, analyzing multiple systems across multiple dimensions is a too complex of an operation, as well as a too dynamic of a configuration. This is due to the ever evolving nature of the systems. Making a choice of the DBMS systems that are going to be involved in the benchmarking tests has to be a process that is done before the actual design of the benchmarks. This already involves a set of knowledge-based choices. This step is necessary particularly because of the time constraint involved in setting up the configuration and then the execution of tests on multiple different systems.

2.5 Cloud Benchmarking

A different component that is researched to influence the accuracy and reliability of the benchmarking results of analyzed database systems is the supporting infrastructure. In a non-Cloud installation, the tests can be performed over a preset infrastructural configuration that will not change between tests or systems. In a Cloud installation, database benchmarking results can differ significantly on different Cloud providers even on comparable configuration [15,16,35,53].

Cloud benchmarking is a multi-faceted procedure aimed to assess the performance of software stacks and Cloud infrastructures. Lacking a standardized approach to benchmarking Cloud configurations and providers, initial benchmarking designs were based on existing standards from transaction processing [38]. Yet considering the volatility of Cloud offerings, approaches to Cloud Benchmarks can quickly become obsolete [20,24]. Moreover, the readily-available Cloud benchmarks do not always provide the background information and justification of the decisions taken in designing the tests. Relevant server workloads and performance measurement methodologies were addressed in previous research [18]. What became apparent was the lack of transparency into the decisions, so new implementations were created to address that [47]. As with the development velocity of the Cloud infrastructure and offered services, such implementations become obsolete if not maintained and further documented.

The following section will provide the considerations that are involved in benchmarking NewSQL DBMS that are Cloud-hosted or Cloud-native, steps involved in creating them and a view over the existing tools to execute them in a reliable and reproducible way.

3 Methods

3.1 Considerations

Benchmarking NewSQL database systems hosted in the Cloud pose not only the question of choosing the list of databases and defining the workload, but also the challenge of including the Cloud provider and Cloud configuration. Two different benchmarking domains need to be combined for producing relevant results. The research on this topic shows a few works that have begun to arise only in the last years [46,53]. In the industry, Cloud database benchmarking is limited to only a few choices in terms of both resources and DBMS. A few companies have developed support for their analysis, yet no one commercial solution is available or extensive.

Finding the best fitted combination of Cloud resources and database system also poses an ongoing evaluation need. Because of the continuous changes in the Cloud resources and the continuous releases of upgraded versions of the DBMS, recurrent checks of the performance may become a requirement. The evolution of the workload and understanding it better also will result in more accurate results for the benchmarks. Contingently, adjustments are also usually made on the configuration of both infrastructure and database to reduce identified problems. Over time, these four components: Cloud infrastructural and services changes, database systems upgrades, configuration adjustments and workload changes can have an impact on the overall performance of the environment. Having a reproducible set of benchmarks and the ability to update them to latest requirements can become a routinely check.

3.2 Phases

The implications of creating reliable Cloud-hosted database benchmarking results include several phases [52]. First the design phase, secondly the execution of the pre-designed benchmarks and finally the process of analyzing the results to provide a clear and understandable overview.

3.2.1. Designing the benchmarking process is aimed to set the objectives of the performance tests, includes defining all the necessary steps and the targeted measurements as well as takes into account all restrictions known.

During the design process the definition of the intended results overview is created and a set of questions have to be answered. These include the following components, yet are not limited only to:

- Workload definition - what is the use case for which the performance has to be evaluated? What are the known details of the workload such as the data size, the distribution across the read and write operations and the number of operations.

- Database systems - which data model does the database have to support, what is the format in which it needs to be supported, what are the minimum known requirements for the database engine? Which DBMS covers the requirements and can be considered a candidate for the tests? This sub-step implies narrowing the testing sets by pre-filtering relevant database engines. An overview of the involved dimensions for NewSQL systems, based on previous work [9] can be seen in Fig. 1.
- Benchmark - which can be used for the defined workload and use case?
- Cloud systems - which Cloud providers and configurations are a candidate for the chosen DBMS?
- What are the metrics targeted, over how many dimensions and what is the intention for each?
- How many iterations of executions have to be done in order to provide reliable results?
- What kind of expertise is required for each of the involved steps (pre-design, design, execution and analysis)?
- What is the order in which the tests will be executed, what components can be configured for automation, what manual steps are involved and how are the manual steps documented and prove their accuracy?

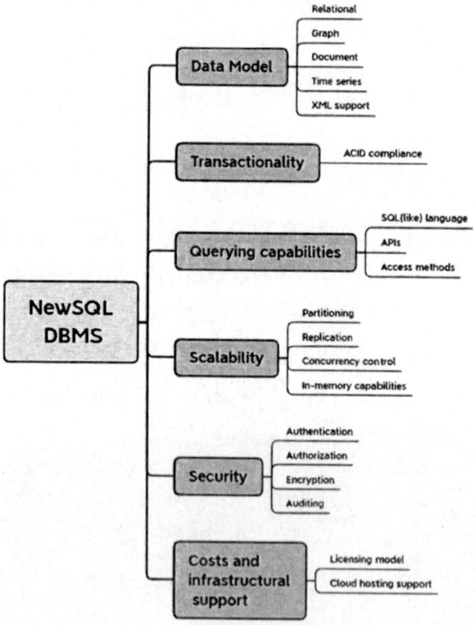

Fig. 1. NewSQL DBMS dimensions to be considered for pre-filter, based on the Multidimensional analysis paper [9]

The design phase is arguably the most important, having a high decisional impact for the rest of the process. Due to the extremely large choice in both domains, this phase requires domain specific knowledge for both the database management systems and the Cloud providers and potential configurations. This selection space needs to be narrowed already at this step so that the effort and in-depth knowledge gap is reduced to the only relevant choices from each perspective. The number of combinations that will be tested is directly impacted by the defined time component. This is part of the constraints that have to be defined.

Furthermore, approaching the process as an iterative procedure can yield the benefit of adjusting elements that were not known at the initial design phase. During the execution and analysis phase, domain specific information will become available, making further iterations more accurate. Properly documenting all the decisions and configurations will provide the reliability of the benchmark for future executions. Automation of parts of the process will provide an even less fault-prone approach.

3.2.2. Execution is the phase that relies strongly on the decisions already taken in phase one. This phase includes the actual implementation of needed configuration setups and then running the designed tests.

The execution phase requires technical expertise on the Cloud and database domains. The resources setup and installation and/or deployment is done at this point. Automation of the testing process or parts of the process is also part of this step. Depending on the extent of the tests, gaining knowledge on the chosen systems and the tools involved should be preceding it.

A major component of the test execution is isolating the designed tests from the artifacts of the tests themselves. No data that results from the test execution should be stored or combined in any way with the benchmarking data. Resources involved in the test execution have to remain separated from the actual benchmarked resources. No tools or benchmarking software should share any type of resource with the tested systems and configurations.

3.2.3. Analysis or Evaluation of the tests results also require data visualization technical expertise and even potentially data science expertise. The results of the benchmarking tests still provide at this point raw data that has to be put into perspective by doing the correct comparisons between the systems and the configurations.

3.3 Taxonomy

3.3.1 Cloud and Database Providers

Available choices in both Cloud providers and NewSQL database providers are broad. Including all providers for both or either of the domains would not provide an accurate view, as they are both extremely dynamic fields and the overview would become obsolete too quickly. In the present work we take a systematic approach for both domains.

An overview of the current offering for Cloud is presented in Fig. 2, specifying the Cloud types and examples of existing providers. The included public providers are based on the latest Gartner Magic Quadrant for Cloud Infrastructure and Platform Services [22] for the year 2020. The private Cloud section also mentions OpenStack [45] to include the on-premise hosted private Cloud. Hybrid Cloud represents a separate type, however it is a combination of the Private and Public Clouds (either on-premises or hosted).

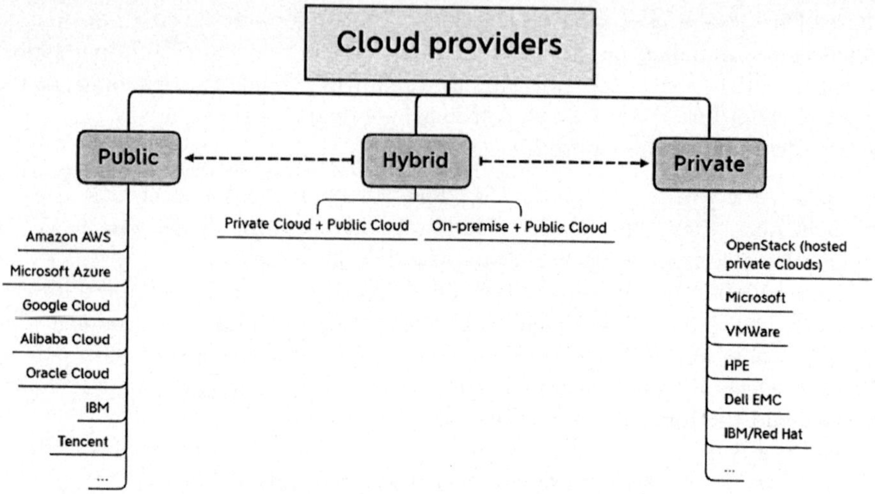

Fig. 2. Overview of Cloud types and providers

NewSQL database providers are also a volatile category. These are systems that continue to emerge and be adjusted, so providing a holistic list is not feasible. Similarly to the selection of the Cloud providers, we use a two-dimensional approach for choosing the database systems. The components for the system which supports the filtering of the engines are:

1. DB-Engines Ranking [25] score
2. Mentions of the DBMS in scientific papers in Google Scholar
3. Mentions of the DBMS in scientific papers in WebOfKnowledge

These dimensions are used to assert the stability of the included system and its maturity in the industry,as well as presence in research in the scientific community. Resulting from this initial analysis we limit the incorporated NewSQL database systems to the following: **SAP HANA** [50], **VoltDB** [58], **CockroachDB** [12], **Google Spanner** [31], Citus [11], **NuoDB** [43], **TiDB** [48], **FaunaDB** [17] , **Altibase** [2].

Furthermore, an overview of the licensing and infrastructural model can be seen in Table 1.

Table 1. Licensing and Infrastructural model

Database name	Open source license	Enterprise license	Cloud native
SAP HANA		Commercial	No
VoltDB	AGPL3	Commercial	No
CockroachDB	BSL and CCL	Commercial	No
Google Spanner		Commercial	Yes
Citus	GNU AGPLv4	Commercial	No
NuoDB		Commercial	No
TiDB	Apache 2.0		No
FaunaDB		Commercial	Yes
Altibase	GNU AGPLv3		No

3.3.2 Benchmark Types

For a reliable set of results the choice of the benchmarks is made from the standard available ones. The options available span over three categories, available in Fig. 3.

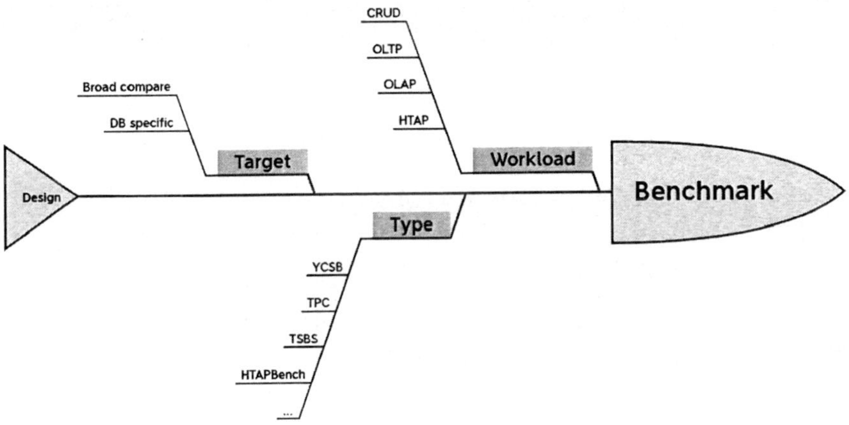

Fig. 3. Overview of database benchmark types and categories

3.3.3 Settings

Adding to the complexity of the components already mentioned, further impact will be leveraged by the configuration of settings on each level in the proposed solution. Any setting change at the level of database, Cloud configuration or benchmark configuration can impact the result of the execution of the tests. In Fig. 4, a view of available settings is shown.

3.3.4 Tools and Solutions

Currently there is a limited amount of solutions and/or tools that can be used in the process of evaluating the performance of such a complex configuration. The ones that are available cannot be all used in the context of NewSQL Cloud-hosted database systems due to either limitations in capabilities or configuration and cost constraints.

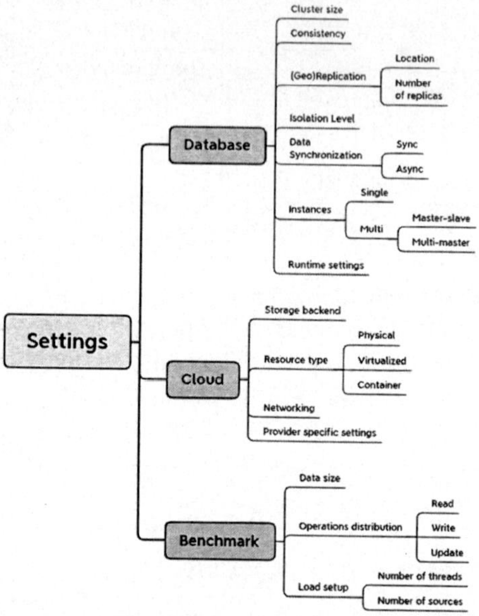

Fig. 4. Overview of available NewSQL Cloud benchmarking settings

There are currently three publicly available options:

- TPC (Transaction Processing Performance Council)
- YCSB (Yahoo! Cloud Serving Benchmark)
- SPEC-RC: Mowgli (The Standard Performance Evaluation Group - research Group)

For this work, we explicitly do not include Github database benchmarks or database provider benchmarks. For the first one, their relevance for academic research would have to be checked. For the second option we exclude provider benchmarks it because most are not freely available and the transparency into the solution might not be sufficient to deliver a reliable result.

TPC benchmarks, as mentioned previously, are standard ones that are generally used for testing transactional systems [21]. YCSB are benchmarks used particularly with NoSQL database systems and have been used in evaluating performance in the Cloud [14].

Mowgli is another available tool that has been developed through academic research [52] and has been made generally available through SPEC-RC [56].

Mowgli supports TPC-C and YCSB benchmarks and its use was primarily intended for NoSQL Cloud database benchmarking [52]. A more flexible and encompassing benchmarking tool that is based on the Mowgli framework is BaaS-Project: Cloud Database Decision Platform. This is not yet publicly available, but promises a fully automated and configurable solution for Cloud database benchmarking under several different licensing tiers [49].

4 Discussions

4.1 Conclusions

This paper is aimed at providing a systematic overview of the considerations, processes, steps and knowledge areas involved in reliably benchmarking NewSQL databases in the Cloud. It also provides a list of standard available benchmarks and tools. The research is intended to point out the existing gap in doing standard performance evaluations for database systems, particularly focusing on two volatile domains: NewSQL database management systems and Cloud hosting. While each of these, independently, pose a difficulty in creating reproducible and accurate performance tests results, the usage of both domains create a much more complex scenario. The research available shows how the performance of the end system, the database, is impacted by comparable yet different infrastructural configurations in different Cloud providers. Understanding what the intention is from performing the tests, what's the overview it should provide, what are the necessary prerequisites and the available tools, help in reducing the designing complexity of the evaluation.

4.2 Opportunities and Expansion Areas

Cloud database benchmarking is not a clearly established process that has cross-Cloud-provider and cross-database-provider tooling and portability. Multiple challenges come with implementing a performance evaluation for these domains. Some of the challenges include the rapid pace of changes in both NewSQL and Cloud technology. Version updates on either of the two can have an effect on the end-resulting performance and is therefore a requirement to keep adusting the tooling and configuration of the tests in order to have an accurate view of the available options when choosing a new system configuration.

The interpretation of the results from recurrent testing iterations can be a data analysis and statistics expertise area. Due to the high amount of combinations in settings and platforms, to give a proper overview of all the tested configurations, a clear and mathematically accurate picture comes into play. Furthermore, for allowing reproducible iterations of identical configurations in different environments, a highly automated solution is necessary. That involves a big up-front setup effort for the systems.

In every of the aforementioned areas there are gaps and opportunities for providing solutions. This paper identified, to name a few, the need for highly-configurable benchmarking tools, reliable and approved benchmarks for specific data usage patterns and visualization and analysis options for the accurately interpretation of the tests' results.

Acknowledgments. This research was made possible by funding from the ICT-AGRI-FOOD 2020 Joint Call. This work was supported by a grant of the Romanian National Authority for Scientific Research and Innovation, CCCDI - UEFISCDI, project number COFUND-ICT-AGRI-FOOD-MUSHNOMICS 205/2020, within PNCDI III.

References

1. Abadi, D., et al.: The Beckman report on database research. Commun. ACM **59**(2), 92–99 (2016)
2. Altibase: Altibase (2020). https://altibase.com/
3. Aslett, M.: What we talk about when we talk about newSQL, November 2020. https://blogs.451research.com/information_management/2011/04/06/what-we-talk-about-when-we-talk-about-newsql/
4. Azqueta-Alzúaz, A., Martinez, M.P., Vianello, V., Péris, R.J.: Fault-tolerance evaluation of a new SQL database. In: 2018 14th European Dependable Computing Conference (EDCC), pp. 81–86. IEEE (2018)
5. Binani, S., Gutti, A., Upadhyay, S.: SQL vs. NoSQL vs. NewSQL-a comparative study. Database **6**(1), 1–4 (2016)
6. Bitton, D., DeWitt, D.J., Turbyfill, C.: Benchmarking database systems-a systematic approach. Technical report, University of Wisconsin-Madison Department of Computer Sciences (1983)
7. Buhl, H.U., Röglinger, M., Moser, F., Heidemann, J.: Big data (2013)
8. Chandra, U.: A comparative study on: NoSQL, NewSQL and Polygot persistence. Int. J. Soft Comput. Eng. (IJSE) **7** (2017)
9. Chereja, I., Hahn, S.M.L., Matei, O., Avram, A.: Multidimensional analysis of NewSQL database systems. In: Silhavy, R. (ed.) CSOC 2021. LNNS, vol. 230, pp. 221–236. Springer, Cham (2021). https://doi.org/10.1007/978-3-030-77442-4_19
10. Chereja, I., Hahn, S.M.L., Matei, O., Avram, A.: Operationalizing analytics with NewSQL. In: Silhavy, R. (ed.) CSOC 2021. LNNS, vol. 230, pp. 249–263. Springer, Cham (2021). https://doi.org/10.1007/978-3-030-77442-4_21
11. CitusData: Citus (2020). https://www.citusdata.com/
12. Cockroachlabs: Cockroachdb (2020). https://www.cockroachlabs.com/
13. Codd, E.: An evaluation scheme for database management systems that are claimed to be relational. In: 1986 IEEE Second International Conference on Data Engineering, pp. 720–729. IEEE (1986)
14. Cooper, B.F., Silberstein, A., Tam, E., Ramakrishnan, R., Sears, R.: Benchmarking cloud serving systems with YCSB. In: Proceedings of the 1st ACM Symposium on Cloud Computing, pp. 143–154 (2010)
15. Copie, A., Fortiş, T.F., Munteanu, V.I.: Benchmarking cloud databases for the requirements of the internet of things. In: Proceedings of the ITI 2013 35th International Conference on Information Technology Interfaces, pp. 77–82. IEEE (2013)
16. Curino, C.A., Difallah, D.E., Pavlo, A., Cudre-Mauroux, P.: Benchmarking OLTP/Web databases in the cloud: the OLTP-bench framework. In: Proceedings of the Fourth International Workshop on Cloud Data Management, pp. 17–20 (2012)
17. Fauna: Fauna (2020). https://fauna.com

18. Ferdman, M., et al.: Clearing the clouds: a study of emerging scale-out workloads on modern hardware. ACM SIGPLAN Not. **47**(4), 37–48 (2012)
19. Ficco, M., Rak, M., Venticinque, S., Tasquier, L., Aversano, G.: Cloud evaluation: Benchmarking and monitoring. In: Quantitative Assessments of Distributed Systems, pp. 175–199 (2015)
20. Folkerts, E., Alexandrov, A., Sachs, K., Iosup, A., Markl, V., Tosun, C.: Benchmarking in the cloud: what it should, can, and cannot be. In: Nambiar, R., Poess, M. (eds.) TPCTC 2012. LNCS, vol. 7755, pp. 173–188. Springer, Heidelberg (2013). https://doi.org/10.1007/978-3-642-36727-4_12
21. Funke, F., Kemper, A., Neumann, T.: Benchmarking hybrid oltp&olap database systems. Datenbanksysteme für Business, Technologie und Web (BTW) (2011)
22. Gartner: Gartner, November 2020. https://www.gartner.com
23. Geelan, J., et al.: Twenty-one experts define cloud computing. Cloud Comput. J. **4**, 1–5 (2009)
24. Gillam, L., Li, B., O'Loughlin, J., Tomar, A.P.S.: Fair benchmarking for cloud computing systems. J. Cloud Comput. Adv. Syst. Appl. **2**(1), 1–45 (2013)
25. solidIT consulting & software development GmbH: Db-engines (2020). https://db-engines.com/
26. Gray, J.: Database and Transaction Processing Performance Handbook (1993)
27. Hahn, S.M.L., Chereja, I., Matei, O.: Analysis of transformation tools applicable on NewSQL databases. In: Silhavy, R. (ed.) CSOC 2021. LNNS, vol. 230, pp. 180–195. Springer, Cham (2021). https://doi.org/10.1007/978-3-030-77442-4_16
28. Hajoui, O., Dehbi, R., Talea, M., Batouta, Z.I.: An advanced comparative study of the most promising NoSQL and NewSQL databases with a multi-criteria analysis method. J. Theor. Appl. Inf. Technol. **81**(3), 579 (2015)
29. Han, J., Haihong, E., Le, G., Du, J.: Survey on NoSQL database. In: 2011 6th International Conference on Pervasive Computing and Applications, pp. 363–366. IEEE (2011)
30. Hsu, W.W., Smith, A.J., Young, H.C.: Characteristics of production database workloads and the TPC benchmarks. IBM Syst. J. **40**(3), 781–802 (2001)
31. Google Inc: Spanner (2020). https://cloud.google.com/spanner
32. Kaur, K., Sachdeva, M.: Performance evaluation of NewSQL databases. In: 2017 International Conference on Inventive Systems and Control (ICISC), pp. 1–5. IEEE (2017)
33. Knob, R.R., et al.: Análise e benchmarking das soluções newsql CockroachDB, MemSQL, NuoDB e VoltDB (2018)
34. Knorr, E., Gruman, G.: What cloud computing really means. InfoWorld **7**(20–20), 1–17 (2008)
35. Kuhlenkamp, J., Klems, M., Röss, O.: Benchmarking scalability and elasticity of distributed database systems. Proc. VLDB Endowment **7**(12), 1219–1230 (2014)
36. Lankford, W.M.: Benchmarking: understanding the basics. Coastal Bus. J. **1**(1), 57–62 (2000)
37. Leavitt, N.: Will NoSQL databases live up to their promise? Computer **43**(2), 12–14 (2010)
38. Lehrig, S., Sanders, R., Brataas, G., Cecowski, M., Ivanšek, S., Polutnik, J.: CloudStore-towards scalability, elasticity, and efficiency benchmarking and analysis in cloud computing. Futur. Gener. Comput. Syst. **78**, 115–126 (2018)
39. Mell, P., Grance, T., et al.: The NIST definition of cloud computing. NIST Special Publication, pp. 800–145 (2011)
40. Mohamed, M.A., Altrafi, O.G., Ismail, M.O.: Relational vs. NoSQL databases: a survey. Int. J. Comput. Inf. Technol. **3**(03), 598–601 (2014)

41. Nambiar, R., Wakou, N., Carman, F., Majdalany, M.: Transaction processing performance council (TPC): state of the council 2010. In: Nambiar, R., Poess, M. (eds.) TPCTC 2010. LNCS, vol. 6417, pp. 1–9. Springer, Heidelberg (2011). https://doi.org/10.1007/978-3-642-18206-8_1
42. Nayak, A., Poriya, A., Poojary, D.: Type of NoSQL databases and its comparison with relational databases. Int. J. Appl. Inf. Syst. **5**(4), 16–19 (2013)
43. NuoDB: Nuodb (2020). https://nuodb.com/
44. Ohlhorst, F.J.: Big Data Analytics: Turning Big Data into Big Money, vol. 65. Wiley (2012)
45. OpenStack: Openstack (2021). https://www.openstack.org/
46. Osborne, J.B.: A quantitative evaluation of the performance impact of Type-I virtualization on a NewSQL relational database management system. Ph.D. thesis, Colorado Technical University (2020)
47. Palit, T., Shen, Y., Ferdman, M.: Demystifying cloud benchmarking. In: 2016 IEEE International Symposium on Performance Analysis of Systems and Software (ISPASS), pp. 122–132. IEEE (2016)
48. PingCap: Tidb (2020). https://pingcap.com/
49. Project, B.: BaaS: benchmarking-as-a-service (2021). https://baas-project.de/
50. SAP: SAP HANA (2020). https://www.sap.com/products/hana.html
51. Seybold, D., Domaschka, J., et al.: Is distributed database evaluation cloud-ready? In: Kirikova, M. (ed.) ADBIS 2017. CCIS, vol. 767, pp. 100–108. Springer, Cham (2017). https://doi.org/10.1007/978-3-319-67162-8_12
52. Seybold, D., Keppler, M., Gründler, D., Domaschka, J.: Mowgli: finding your way in the DBMS jungle. In: Proceedings of the 2019 ACM/SPEC International Conference on Performance Engineering, pp. 321–332 (2019)
53. Seybold, D., Wesner, S., Domaschka, J.: King Louie: reproducible availability benchmarking of cloud-hosted DBMS. In: Proceedings of the 35th Annual ACM Symposium on Applied Computing, pp. 144–153 (2020)
54. Stonebraker, M.: SQL databases v. NoSQL databases. Commun. ACM **53**(4), 10–11 (2010)
55. TPC: Transaction processing performance council, June 2021. http://tpc.org/
56. Van Eyk, E., Scheuner, J., Eismann, S., Abad, C.L., Iosup, A.: Beyond microbenchmarks: the SPEC-RG vision for a comprehensive serverless benchmark. In: Companion of the ACM/SPEC International Conference on Performance Engineering, pp. 26–31 (2020)
57. Vatika, S., Meenu, D.: SQL and NoSQL databases. Int. J. Adv. Res. Comput. Sci. Softw. Eng. **2**(8), 20–27 (2012)
58. VoltDB: Voltdb (2020). https://www.voltdb.com/
59. Wang, L., et al.: Cloud computing: a perspective study. N. Gener. Comput. **28**(2), 137–146 (2010)
60. Zairi, M.: Effective Benchmarking. Springer, Dordrecht (1996)
61. Zhang, Q., Cheng, L., Boutaba, R.: Cloud computing: state-of-the-art and research challenges. J. Internet Serv. Appl. **1**(1), 7–18 (2010)

Multifactorial Analysis in Problems of Group Pursuit

Alexander A. Dubanov(✉)

Buryat State University, Ulan-Ude, Russian Federation

Abstract. This article is devoted to how in the task of group pursuit to achieve the simultaneous achievement of the pursuers of their targets. The proposed pursuit model is based on the fact that the pursuer tries to follow the predicted trajectory of movement. The predicted trajectory of movement is built at each moment of time. This path is a compound curve that respects curvature constraints. The time to reach the target of each pursuer is a dependence on the speed of movement and the minimum radius of curvature of the trajectory. The analysis of multidimensional moduli of velocities and minimum radii of curvature of the trajectories of each of the pursuers for the simultaneous achievement of their goals is the goal of the research discussed in this article.

Keywords: Multifactorial analysis · Radishchev diagrams · Target · Pursuer · Trajectory · Radius of curvature

1 Introduction

Let us consider a model for calculating the trajectory of the pursuer on a plane, where at each moment of time from the pursuer to the target a predicted trajectory is constructed, which the pursuer will try to adhere to (Fig. 1).

Curves $l_1(s)$, $l_2(s)$, $l_3(s)$ (Fig. 1) consist of a circular arc segment and a straight line segment. The radius of the circles is the limitation on the curvature of the predicted trajectories of the pursuers in the model. If the pursuer was at the moment of the beginning of the pursuit at the point P_i with the velocity vector V_i, then the center of the circle C_i of the radius r_i will be at the following point:

$$C_i = P_i \pm r_i \cdot \frac{\begin{bmatrix} -V_{iy} \\ V_{ix} \end{bmatrix}}{|V_i|}.$$

Then from the point of the target position T_i the tangent to the circle is drawn (C_i, r_i). The combination of the tangent and the circle will be the baseline of the predicted trajectory of the pursuer $l_i(s)$. Note that in the baseline equation from a one-parameter set of predicted trajectories, parameterization is performed on its arc length.

R. Silhavy et al. (Eds.): CoMeSySo 2021, LNNS 232, pp. 311–319, 2021.
https://doi.org/10.1007/978-3-030-90318-3_27

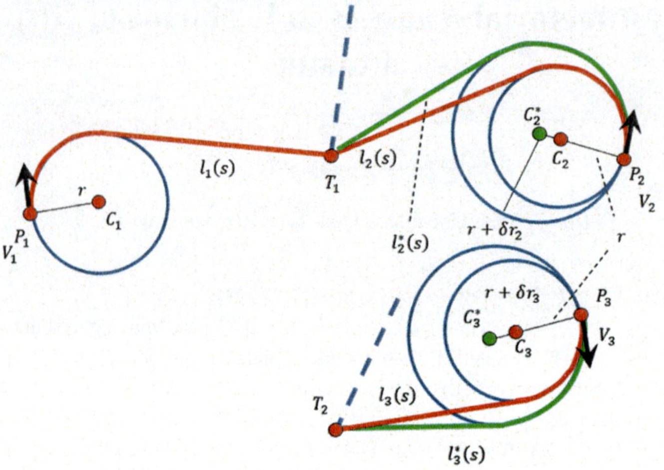

Fig. 1. Group pursuit of the set of targets.

With the new target position T_i the line $l_i(s)$ shifts, remaining parallel to itself (Fig. 2).

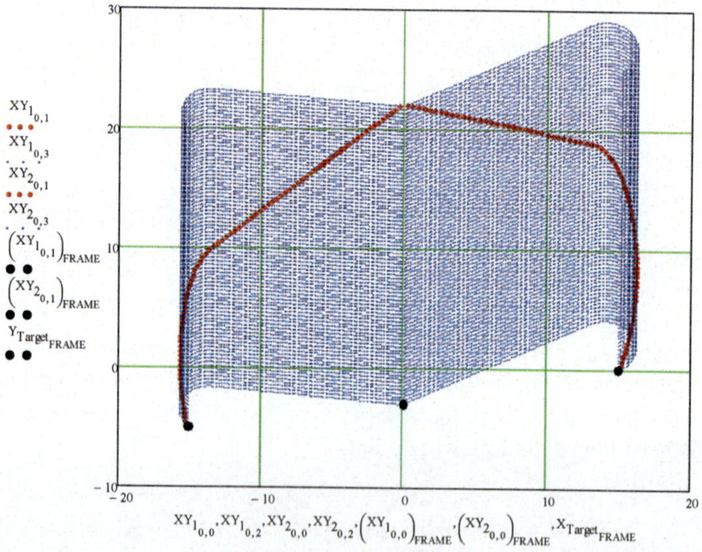

Fig. 2. One-parameter networks of predicted trajectories of pursuers.

If the i-th the pursuer is at the point P_{ij} (Fig. 3) at time t_j, while having a predicted trajectory of movement $l_{ij}(s)$, connecting to the current target position T_{ij}, then the next point of the pursuer trajectory will be the point P_{ij+1}.

The point P_{ij+1} is a point of intersection of the line $l_{ij+1}(s)$, appropriate to the target T_{ij+1} at the next moment in time t_{j+1}, and a circle centered at the point P_{ij} and radius $|V_{ij}| \cdot \Delta t$, $\Delta t = t_{j+1} - t_j$. This is the model for constructing the pursuer's trajectories.

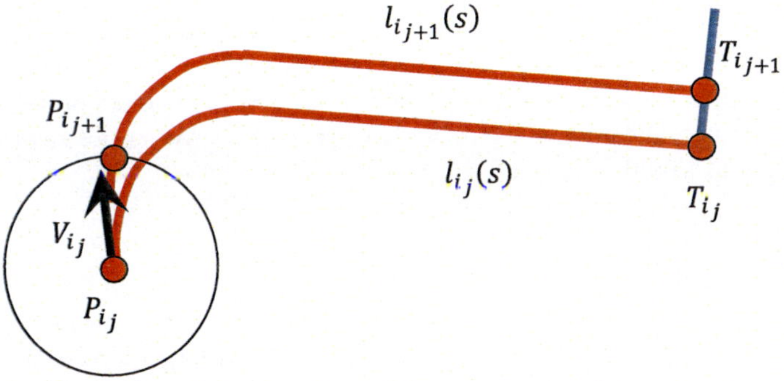

Fig. 3. Iterative process of calculating of the pursuer's trajectory.

2 Formulation of the Problem

Consider the group pursuit problem when a group of pursuers catches up with a group of targets. We will assume that each pursuer P_i strives to achieve its target T_i, although some pursuers may have the same targets, as shown in Figs. 1 and 2.

Moreover, the pursuer P_i reaches the target T_i for a certain time t_i, moving at the certain speed V_i. For the simultaneous achievement of targets, the equality of all values is necessary t_i to a certain value.

Figure 1 shows that to change the length of the baseline, you can change the radius of the tangent circle. The tangent circle was introduced so that the pursuer could smoothly transition to a straight line. If this were the case, then the problem would be reduced to pursuit by the parallel approach method.

Since the initial speed of the pursuer is directed arbitrarily, then with the help of a composite baseline, which moves as the target moves, remaining parallel to itself, a smooth transition to the parallel approach method takes place, observing curvature constraints (Fig. 2).

Figure 2 is supplemented with a link to an animated image, where you can see a smooth transition to a parallel approach.

The purpose of this article is to describe the method in which the pursuer reaches the target at the appointed time from the allowed values. As a consequence, one can consider the simultaneous achievement of their targets by a group of pursuers.

3 Theory

Based on the research results, a test program for the simultaneous achievement of targets by the pursuers has been developed, which can be viewed on the resource [6]. This program implements an algorithm that implements the iterative scheme for calculating the trajectory of the pursuer shown in Fig. 3.

In the model, it is assumed that there is a dependence for the pursuer P, who achieves its target T, during t: $t = F(P_s, T_s, n_P, n_T, V_P, V_T, R)$, where P_s, T_s are the coordinates of the points of position of the pursuer and the target at the moment of the beginning of the pursuit process, n_P, n_T are the unit vectors of the direction of the pursuer and the target at the start of the pursuit process, V_P, V_T are speed modules of the pursuer and targets in the pursuit process, R is the radius of the circle, the meaning of which is shown in the Figs. 1 and 3. In fact, the model calculates the number of steps in which the pursuer reaches the target. The number of steps, with a known discrete time interval, can be compared with the real time.

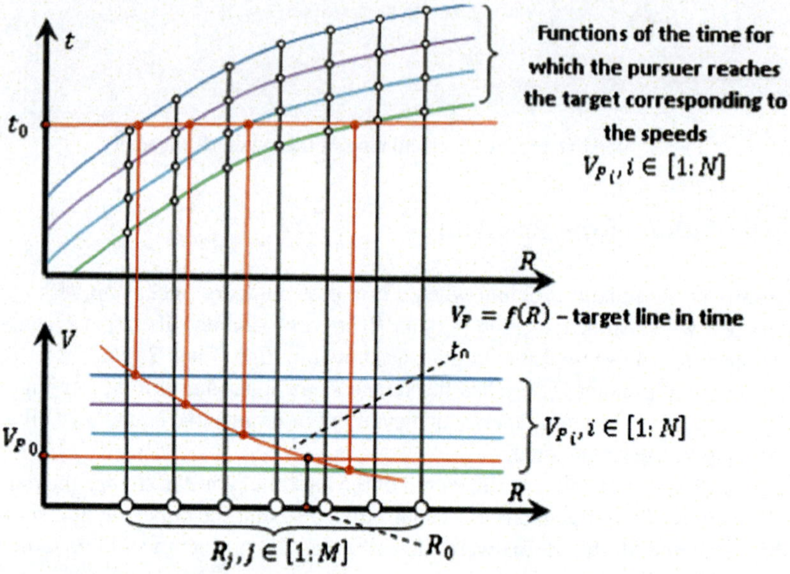

Fig. 4. Determination of the circle radius on the Radishchev diagrams.

If the target moves in a straight line and uniformly, then the dependence of the time to reach the target in the already begun iterative process can be considered a function of two variables: $t = F(V_P, R)$, from the pursuer speed module and from the curvature radius of the circle.

Although, in the model, it is believed that the pursuer moves at a constant speed V_P, nothing prevents us from changing the values of the modulus of speed, as well as the curvature radius. Let us assume that the speed modulus takes discrete values from the series V_{Pi}, $i \in [1:N]$, and the radius of the circle from the Figs. 1 and 3 takes values $R_j, j \in [1:M]$. For further research, Radishchev diagrams are used [1], where the coordinate planes (R, V) and (R, t) are used (Fig. 4).

Figure 4 shows experimental plotting of time dependences $t_{i,j} = F\left(V_{Pi}, R_j\right)$. The graphs of the time dependences of the pursuer reaching the target on the circle radius R at the fixed speed V_P are schematically shown on surface (R, t).

The equality $t = t_0$ is chosen as one of the optimizing factors [1] on the surface (R, t), where t_0 is required time to reach the target. Further, to solve our problem, on the plane (R, V) the equality $V_P = V_{P0}$ is chosen as the second optimizing factor, where V_{P0} it is the constant speed of the pursuer.

Although in the formulation of the problem it is said that the pursuer speed module is unchanged, the constructed series of velocities is necessary to calculate the circle radius R_0 on the projection plane (R, V).

Along the communication lines on the plane of projections (R, V) the corresponding intersection points with the lines of the velocity level V_{Pi} are found (Fig. 4). Polynomial regression is performed on the obtained points in the test program and, as a result, we obtain the function of the dependence of the pursuer's speed on the circle radius, at which the target is achieved in time t_0.

Then the intersection point of the function $V_P = f(R)$ (Fig. 4) with the level line $V_P = V_{P0}$ are searched for. Intersection point abscissa R_0 is the required radius of the circle to reach the target T by the pursuer P for the time t_0 with the speed V_{P0}.

The calculation is carried out on the condition that the target moves uniformly and in a straight line. If the target changes direction or speed, then a new circle radius of the compound baseline is calculated (analogous to the line of sight of the parallel approach method) to reach it and a new time to reach is set at the same pursuer's speed.

The lowest time limit for reaching the target with its uniform and rectilinear motion will be reached when the pursuer's speed is directed to the point K on the circle of Apollonius [2–5] (Fig. 5).

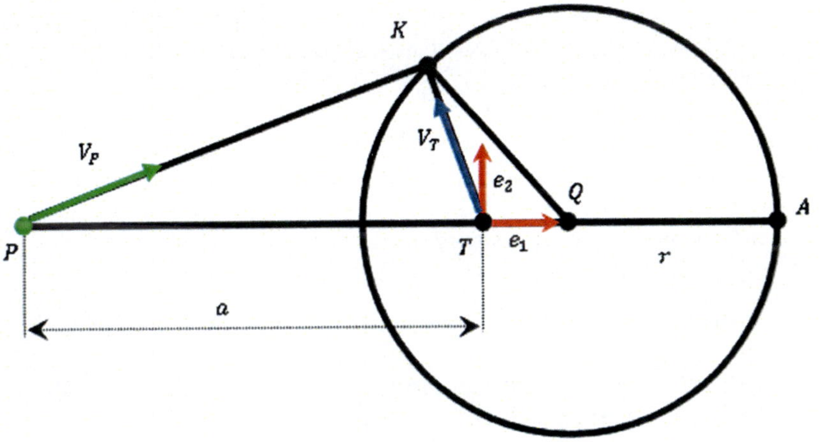

Fig. 5. Circle of Apollonius.

The circle of Apollonius is the locus of points, the distances ratio from which there is a constant value to two given points, not equal to one. Figure 5 illustrates this as: $|PK|/|TK| = |V_P|/|V_T|$.

When considering multiple pursuit of the targets group, the test program performs a preliminary calculation of the trajectories of the pursuers for the given initial parameters. Of the times when the targets were achieved, the longest time is selected to calculate the simultaneous achievement. And this time is chosen as a criterion for calculating the trajectories of the remaining pursuers. This moment is illustrated in Fig. 2, which is supplemented with a link to an animated image, where you can look at the simultaneous achievement of the target by two pursuers.

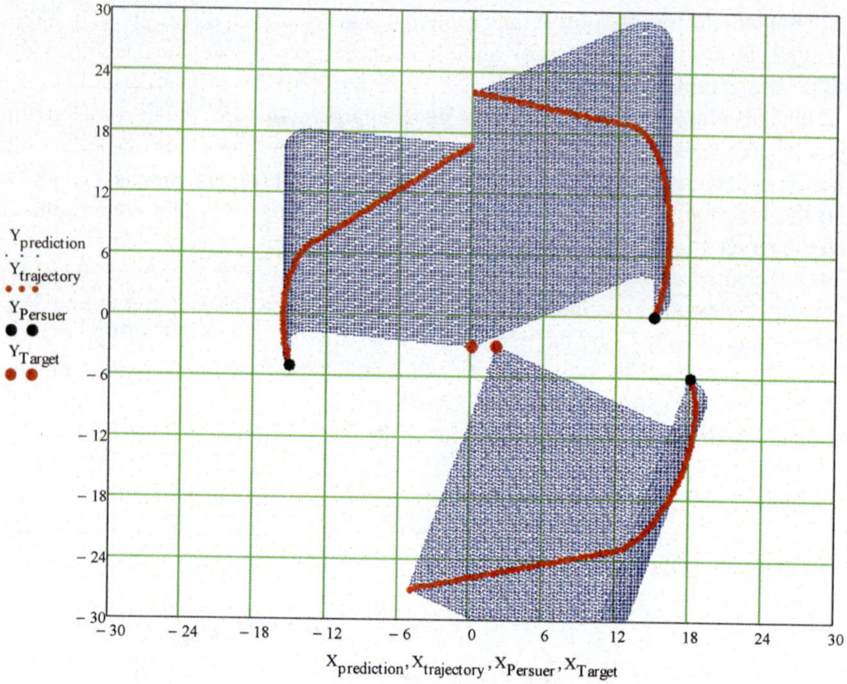

Fig. 6. Achievement of targets at different designated times.

Figure 6 shows how a shorter target time was set for one of the pursuers. Figure 6 is also supplemented with a link to an animated image, where you can watch the achievement of targets at various assigned times.

4 Experimental Results

Figure 7 shows some of the results of multivariate analysis in the problem of simultaneous target achievement by two pursuers. The target moves in a straight line and evenly. A set of allowable speeds was built for each pursuer. The range of acceptable values for the circle radius is varied using discrete variables dR. In the graphs in Fig. 7, it is the scale dR.

On the projection plane (dR, t) we build a one-parameter network of lines. Each line corresponds to a certain speed value and expresses the dependence of the time to reach the target on the increment in the circle radius. The graph in Fig. 7 shows a one-parameter network of velocity lines of one of the pursuers. A similar network was built for the second multivariate analysis in the test program.

The first optimizing factor is in charge of simultaneous achievement and is chosen for each pursuer [1], $t = t_0$. Where t_0 is the longest of the times of reaching the target, if the pursuers independently overtook the target with the same initial conditions. On the surface (dR, t) points of intersection with the level line $t = t_0$ are searched with velocity lines of a one-parameter network. Intersection points are found using built-in equation solving routines. In the system of computer mathematics MathCAD it could be a procedure *root*. The values dR and V correspond to the found intersection points on the projection plane (dR, V).

The built-in polynomial regression procedure is applied to the obtained points on the projection plane and the characteristic curve of the dependence of the speed on the circle radius of the composite baseline is found, which are shown in Fig. 1.

Figure 7, on the projection plane (dR, V) shows the same characteristic line of velocity dependence for another pursuer. Next, the second optimizing factor $V_1 = V_2 = V_0$ is applied. In the test program, objects move at the same speed. Built-in computer mathematics searches for points of intersection with the level line $V = V_0$. These points correspond to the values dR_1 and dR_2.

When starting an iterative process with a given value of the time to reach the target t_0, with the given modules of movement speeds V_0,with the found values of the increments dR_1 and dR_2 to the initial circle radius, the simultaneous achievement of the target by two pursuers has been achieved. This fact is demonstrated by the addition to Fig. 2.

5 Discussion of Results

This article proposes a method for achieving a set of targets by a group of pursuers, in which the time to reach can be assigned. The simultaneous achievement of targets is a particular result of this method. This method is a development of the parallel approach method. If this method is implemented in space, then it is necessary to ensure that the vectors of the pursuer and the target are in the same plane.

If the pursuit problem occurs in three-dimensional space, and we want to reduce it to the method of parallel approach but the pursuer's speed is directed arbitrarily, then the baseline of the predicted trajectories of the pursuer's movement should be built in the plane formed by the line of sight and the pursuer's speed.

The next step of the pursuer will be the intersection point of the sphere with the radius equal to the step of the pursuer and the base line, parallel to the transferred so that one of its ends coincides with the target position point.

As for the question of finding the circle of Apollonius and the point K in three-dimensional space: the circle itself will be in the plane formed by the line of sight and the target speed. The parameters of Circle of Apollonius such as circle center (p. Q, Fig. 5), circle radius r, Apollonius point (p. A, Fig. 5) and point K are determined by the target's velocity vector, the pursuer's velocity module and the pursuer's and target's positions. There is an analytical solution to this in the plane coordinate system shown in Fig. 5. The coordinate center is at the point of the target position, the abscissa vector will be a unit vector along the line of sight connecting the pursuer and target positions, and the ordinate vector will be perpendicular to the abscissa vector, but in the plane formed line of sight and target velocity vector.

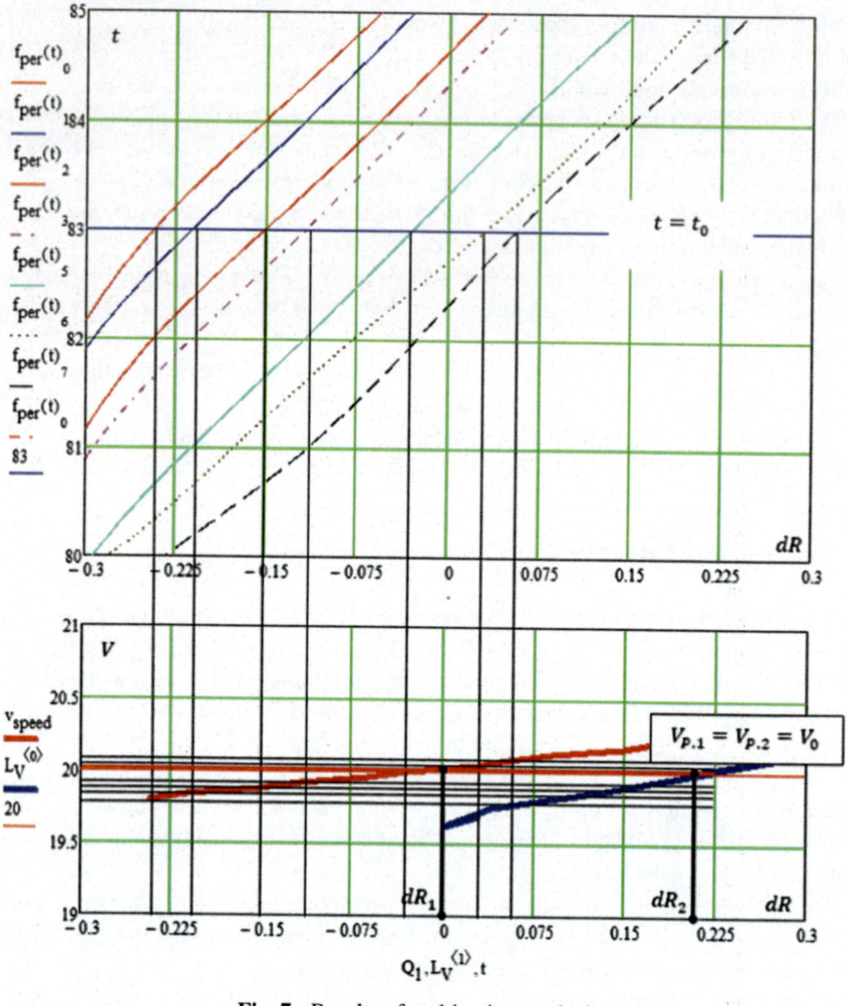

Fig. 7. Results of multivariate analysis.

6 Conclusion

The research results presented in this article may be in demand by the developers of loitering unmanned aerial vehicles that perform group coordinated tasks. The role of the guidance operator can be reduced to specifying targets and monitoring the execution of tasks.

Acknowledgments. This work was supported by an innovative grant from the Buryat State University in 2021 "Control of a four-link manipulator based on signals received from a neurointerface", Scientific adviser Dubanov A.A.

References

1. Volkov, V.Ya., Chizhik, M.A.: Graphic optimization models of multifactor processes. Monograph, Omsk **9**, 101 (2009). Publishing and Printing Center OGIS, 644099, Omsk, st. Krasnogvardeyskaya
2. Isaacs, R.: Differential Games. Mir, Moscow (1967)
3. Pontryagin, L.S.: Linear differential game of evasion, Tr. Steklov Math. Inst. USSR. **112**, 30–63 (1971)
4. Krasovsky, N.N., Subbotin, A. I.: Positional differential games. Science (1974)
5. Petrosyan, L.A.: Differential Pursuit Games, p. 222. Publishing house of Leningrad State University (1977)
6. http://dubanov.exponenta.ru Section "Simultaneous achievement of the goal on the plane" (Date of treatment 22 May 2021)
7. Video, the results of the simulation program for the simultaneous achievement of the goal. https://www.youtube.com/watch?v=7VNHNwCbWrg. (Date of treatment 22 May 2021)
8. Video, simulation results of the simultaneous achievement of two goals by three pursuers with visualization of a network of lines of predicted trajectories. https://www.youtube.com/watch?v=NNJDJOJT34I (Date of treatment 22 May 2021)
9. Video, simulation results of the simultaneous achievement of two goals by three pursuers without visualizing the network of lines of predicted trajectories. https://www.youtube.com/watch?v=tdbgoNoby3A. Accessed 22 May 2021
10. Video, simulation results of the simultaneous achievement of two goals by three pursuers at designated times. https://www.youtube.com/watch?app=desktop&v=F6MTsWZL2BY&feature=youtu.be (Date of treatment 22 May 2021)
11. Vagin, D.A., Petrov, N.N.: The problem of pursuing coordinated fugitives. Izvestiya RAN. Theory Control Syst. **5**, 75–79 (2001)
12. Bannikov, A.S.: Some non-stationary problems of group pursuit. In: Proceedings of the Institute of Mathematics and Informatics of UdSU, vol. 1(41), pp. 3–46 (2013)
13. Bannikov, A.S.: The nonstationary problem of group pursuit. In: Proceedings of the Lobachevsky Mathematical Center, vol. 34, pp. 26–28. Publishing house of the Kazan Mathematical Society, Kazan (2006)
14. Izmestiev, I.V., Ukhobotov, V.I.: The problem of pursuing low-maneuverable objects with a terminal set in the form of a ring. In: Proceedings of the International Conference "Geometric Methods in Control Theory and Mathematical Physics: Differential Equations, Integrability, Qualitative Theory" Ryazan, September 15–18, 2016, Results of Science and Technology. Topic. Obz., vol. 148, pp. 25–31. VINITI RAN, Moscow (2018)
15. Video, simulation results of simultaneous achievement of two goals by three pursuers with visualization of a network of lines of predicted trajectories. https://www.youtube.com/watch?v=NNJDJOJT34I
16. Video, the results of modeling the simultaneous achievement of two goals by three pursuers without visualizing the network of lines of predicted trajectories. https://www.youtube.com/watch?v=tdbgoNoby3A
17. Certificate of state registration of a computer program No. 2020665641. Kinematic model of the parallel approach method

Personalized Diet Web-Service: Business Modeling and Development

Vyacheslav D. Rudenko, Alexandr O. Ruban, Mikhail V. Kolesnikov(✉), Alexey N. Arseniev, and Tatyana G. Maximova

ITMO University, St. Petersburg, Russia
{rudenkov,aoruban,kmv,anarseniev,tgmaximova}@itmo.ru

Abstract. It is stated that the FoodTech web-services market is actively developing, providing a wide variety of formats for fulfilling the client's needs in individual diet planning. The aim of the work is an overview of the process of business modeling and designing a web service that provides personalized recommendations on an individual diet planning based on the results of the DNA test interpretation. Within the analysis of the adjacent markets, potential competitors were carried out, as well as the target audience of the web service was identified. A general model of the web service and the format for providing services have been obtained, key advantages in comparison with competitive services have been identified. The description of the minimum viable product (MVP) is presented in the form of a business model, a work plan, a list of necessary resources, and required technologies.

Keywords: Business model · Bioinformatics · Design · Web service · Minimum viable product · Market analysis · Personalized nutrition · Digital transformation · Foodtech · Business process · Genetic data

1 Introduction

According to the J'son and Partners report [1], in 2018 the value of the global market of FoodTech was estimated at 201.7 billion US dollars. The segment of personalized nutrition, including dietary services based on DNA analysis, diagnostic services, functional food, dietary supplements was also valued up to 7.2 billion USD. The volume of the Russian market for fitness services, which is quite nearly to the market of personalized food by the target audience and has a few similar products, in 2018 was estimated to 117.7 billion rubles or 1.69% of the global market. Comparing the market of personalized nutrition with the market of fitness services, we can see the prospects of the rapid growth of the Russian FoodTech market, including the personalized nutrition market, in the next 5–10 years.

Currently, the market of services providing food ration based on the results of DNA analysis is actively developing in the world, which significantly increases the degree of personalization of this type of service. Such format is just at the

R. Silhavy et al. (Eds.): CoMeSySo 2021, LNNS 232, pp. 320–332, 2021.
https://doi.org/10.1007/978-3-030-90318-3_28

beginning of its development in the domestic market, so the actual task is to study the needs of customers from Russia and CIS countries in IT solutions for the creation of personalized food ration, as well as the formation of a business model for launching such a service on the Russian market.

The goal of the project is to identify trends in the Russian personalized diet market, form a business model, and perform design and modeling of the web service that can provide personalized dietary recommendations based on the results of DNA tests. To achieve this goal, the objectives of analysis of the market, competitors, and potential target audience, preparation of a business model of service delivery, and preparation for the design of the web service are set. Within the work materials were used from J'son and partners' personalized nutrition market research, results of self-analysis of competitors in foreign and domestic markets, and the results of a survey of potential target audiences. The formed vision of the market was compared to competitors from the perspective of the marketing and positioning criteria. The idea of the target audience and potential goals of the client was formed. An idea of a unique trading offer of the service was prepared, the format of service delivery was described and a description of the business process of working with the client was prepared. Besides, key potential partners were identified, and the format of cooperation was described.

The results are presented in three sections of this paper. The "Market Research" section describes the results of the personalized nutrition market assessment and provides a comparative table of the types of competitors and specific competitive services based on the chosen criteria. The "Business Modeling and Software Engineering" section includes a detailed description of the resulting business model of a personalized diet web service. This section gradually describes the steps and results of segmenting the target audience, the description of the trade offer, the description of potential partners and formats of cooperation, as well as the description of the key business processes of the service in the form of the overview of the minimum viable product. The "Discussion" section summarizes the results, provides a vision on further plans, and describes options for the potential socioeconomic effect of the widespread introduction of this service.

2 Market Research

The FoodTech market has been developing gradually over time, but the idea of personalized nutrition has begun to enter the digital services market relatively recently. Due to the novelty of this area, designing a product for the Russian market requires the study of conjugated markets. The following categories of competitive markets are highlighted for this purpose:

- Personalized Nutrition Services
- Nutritionists and fitness trainers who provide nutrition planning services;
- Centers of Nutrition;
- Dietary services, including calorie, protein, fat, carbohydrates and/or calories spent;

- DNA analysis laboratories;
- Food delivery services.

Nutritionists and fitness trainers provide services to help with overweight, weight gain, or health support goals. Their target audience is represented by people aged 20–30 years. Among the interests of this audience can be distinguished: maintaining a healthy lifestyle, losing weight to obtain a good body shape [4], gaining muscle mass for sporting purposes or overall beauty. The representation of such services is usually limited to social networks, such as Instagram. For this segment, the product is a consultation expert in fitness and/or nutrition from specialists. The consultation is aimed at mastering the client's principles of diet and supporting it, as well as maintaining the client's interest in regular training.

Another close category of competitors is dietary centers that help customers with the problem of overweight. The target audience in this segment includes people over 30 years old. Clients of dietary centers solve the problem of excess weight to improve health, solve problems with diseases, and cure skin problems. This difference in the target audience leads to differences in the methods of providing the services. Before starting the weight loss procedure, the client often undergoes a diagnostic examination, passes many tests, and visits doctors such as an endocrinologist, therapist, psychologist. As a result, the nutritionist approach of working with the client is more personalized. When planning meals and training, the health status and other features of the client are also considered. Social networks and corporate websites are widely used as promotions in this market. In addition, such services are often provided by nutrition departments at paid health facilities, which indicates the presence of offline traffic. Among the popular centers of dietetics in Russia a center of weight loss "Dr. Bormental" and a clinic of weight correction "Dr. Kovalkov" can be mentioned.

To compare selected competitors on the DNA laboratories market, the criteria of business-model were chosen: segments of customers, sales channels, value propositions, key activities, and revenue streams. The results of the comparison of DNA laboratories, which have revenue from sales of DNA tests, are presented in Table 1. "23andMe" unlike the other competitors identifies the narrower segment of competitors and has more family clients. "FitnessGenes" gives to customers additional value propositions such as personalizing fitness or health activities and tracking service.

The results of the comparison of services, which have revenue from providing the food rations using the DNA analysis results, are presented in Table 2. It is worth mentioning that "Genopalate" addresses people with health problems as to the main customers, still providing the same value propositions as the other competitors. Blood tests are relevant except for DNA tests. Unlike the competitors, "Otri" focuses on helping to obtain healthy habits.

The results of the comparison of apps, which have revenue from mobile application features related to the diet process, are presented in Table 3. This category differs from the previous in the level of personalization of the diet and the list of available features in the application. Direct competitors form a personalized diet based on blood tests and/or DNA analysis, whilst mobile applications form

Table 1. DNA laboratories market research

Company	General	Customer	Organization
Genotek	Russia, cap: $25M	**Segments:** healthy lifestyle, athletes, extra weight, genetic diseases **Channels:** lectures and longreads in social networks, context marketing	**Value proposition:** predispositions to sport and diseases **Key activities:** DNA laboratory
Atlas	Russia	**Segments:** healthy lifestyle, athletes, extra weight, genetic diseases **Channels:** marketing in social networks, context marketing	**Value proposition:** predispositions to sport and diseases **Key activities:** DNA laboratory
Muhdo	UK	**Segments:** healthy lifestyle, athletes, extra weight, genetic diseases **Channels:** marketing in social networks, context marketing	**Value proposition:** predispositions to sport and diseases **Key activities:** DNA laboratory
FitnessGenes	UK	**Segments:** healthy lifestyle, athletes, extra weight **Channels:** marketing in social networks, context marketing	**Value proposition:** personalizing of fitness and health **Key activities:** DNA laboratory and fitness tracking service
23 and Me	USA, cap: $3.5bn	**Segments:** family healthy lifestyle and family genetics interests **Channels:** marketing in social networks, context marketing	**Value proposition:** predispositions to diseases **Key activities:** DNA laboratory

a diet based on height, weight, age, goal, etc. Thus, mobile applications provide less personalized services. The premium subscription in the application, which involves the preparation of a diet, has more advanced functionality, unlike the previous category of competitors: it has a counter for calories, proteins, fats, and carbohydrates, as well as meal planning tools, access to a recipe database, and tracking food and water consumption.

The results of the comparison of food delivery services are presented in Table 4. Not all competitors help to lose weight. For example, My-food positioning is focused on a simple delivery of ready-made products without pursuing the idea of healthiness. There are customers, who can't cook or have neither willing nor enough time to cook.

Unlike the previous categories of competitors, delivery services do not provide personalization of the diet, but rather imply that the client already knows his diet norms. From the analyzed list of competitors, only "Taste-Balance" provides

Table 2. Individual food ration delivery companies market research

Company	General	Customer	Organization
InsideTracker	UK	**Segments:** healthy lifestyle, athletes, extra weight **Channels:** social networks, context marketing	**Value proposition:** Personalized fitness and nutrition based on DNA and blood tests **Key activities:** Laboratory and Mobile app
DNAfit	UK	**Segments:** healthy lifestyle, athletes, extra weight **Channels:** social networks, context marketing	**Value proposition:** personalizing and control of diet and nutrition **Key activities:** Mobile app
Genopalate	USA	**Segments:** problems of health **Channels:** social networks, context marketing	**Value proposition:** Personalized nutrition based on DNA test **Key activities:** DNA laboratory and Mobile app
Otri	UK	**Segments:** healthy lifestyle, extra weight **Channels:** social networks, context marketing	**Value proposition:** formating of nutrition and healthy habits based of blood test **Key activities:** Laboratory and Mobile app
Athletigen	USA	**Segments:** healthy lifestyle, extra weight **Channels:** social networks, context marketing	**Value proposition:** Personalized fitness based on DNA test **Key activities:** DNA Laboratory

additional functionality for calculating calories, proteins, fats, and carbohydrates and offers ready-made rations based on such calculations.

There are only two clear leaders in the field of DNA testing services in Russia: Genotek and Atlas [2,3]. This market in Russia is still relatively young, so the leaders determine the state of the art on this market. Among the many types of analyses, the analysis, aimed at features in the diet, remains popular. It can be used to identify intolerance to foods, vitamin deficiency, and other features of the body. It is worth noting that the use of genome research methods in the personalized nutrition market is still in progress due to low level of trust on the market, which leads to an active position in promoting, offering potential customers lectures, articles, and visits to open houses. Dietary services are direct competitors in terms of providing individual diet programs, There are two types of dietary services: a web service or a mobile app. Using these services, it is possible to calculate the calories consumed, proteins, fats, and carbohydrates,

Table 3. App for weight loss market research

Company	General	Customer	Organization
FatSecret	Australia	**Segments:** healthy lifestyle, athletes, extra weight **Channels:** social networks, context marketing	**Value proposition:** Weight control service **Key activities:** Mobile app
Yazio	Germany	**Segments:** healthy lifestyle, athletes, extra weight **Channels:** social networks, context marketing	**Value proposition:** Calorie counting and food journal **Key activities:** Mobile app
Lifesum	UK	**Segments:** healthy lifestyle, athletes, extra weight **Channels:** social networks, context marketing	**Value proposition:** Calorie counting and food journal **Key activities:** Mobile app
MyFitnessPal	USA	**Segments:** healthy lifestyle, athletes, extra weight **Channels:** social networks, context marketing	**Value proposition:** Calorie counting and food journal **Key activities:** Mobile app

track the results of weight loss or weight gain, independently form a nutrition diary, keep a diary of exercises in a convenient way.

The following organizations were identified as a result of the analysis of direct competitors: DNAFit (UK), GenoPalate, Athletigen, Vitagene, and FitnessGenes. All data was collected from open sources such as official websites, blogs, social media materials, and the "Crunchbase" open database. Thus, the main direct competitors were analysed, their products and business models have been studied. Since direct competitors are not represented in the Russian market, a few related categories of competitors are analyzed, which may be close to the market under study by the service type or business model. Based on the analyzed data, it can be concluded that when working with a young segment of the audience (age 20–30 years) it is important to focus on fitness, while when working with a more adult audience it is important to pay attention to nutrition and configuration of a diet aimed at a healthy lifestyle. Long-term monitoring and a personal approach are important for all segments of the audience. Competitors mainly use Instagram as a source of incoming traffic.

Table 4. Food delivery services market research

Company	General	Customer	Organization
Level Kitchen	Russia	**Segments:** healthy lifestyle, athletes, extra weight, no time to cook **Channels:** social networks, context marketing	**Value proposition:** proper nutrition delivery **Key activities:** Kitchen and logistics
Taste-Balance	Russia	**Segments:** healthy lifestyle, athletes, extra weight, no time to cook **Channels:** social networks, context marketing	**Value proposition:** calculation of diet **Key activities:** kitchen and logistics
Grow Food	Russia	**Segments:** healthy lifestyle, athletes, extra weight, no time to cook **Channels:** social networks, context marketing	**Value proposition:** proper nutrition delivery **Key activities:** kitchen and logistics
M-Food	Russia	**Segments:** no time to cook **Channels:** social networks, context marketing	**Value proposition:** food delivery **Key activities:** kitchen and logistics
P-Food	Russia	**Segments:** healthy lifestyle, athletes, extra weight, no time to cook, premium segment **Channels:** social networks, context marketing	**Value proposition:** food delivery **Key activities:** Kitchen and logistics

3 Business Modeling and Software Engineering

The main tool for strategic planning and business model description is the Canvas model by Osterwalder and Pigneur. Creating such a model allows seeing a big picture of the future company model, forming a clear view of the target audience, key values, distribution channels, revenue streams, and cost structure [5]. The model was filled based on the analyzed information on the characteristics of competitors noted in the previous section, as well as the results of the survey prepared for the target audience. The resulting business model is depicted in Fig. 1.

In terms of the survey, the key goal was to identify the needs of the target audience in nutrition planning. For this, questions were asked about the services and methods used in the preparation of the diet, the difficulties in meal planning and cooking, as well as the difficulties encountered by the interviewees in following the chosen diet. A few questions were asked aimed at identifying trends in nutrition planning and general information about the correspondents. The target audience is represented by residents of Russia, mostly middle-aged, 25–45 years old. People under the age of 25 mostly won't be able to afford DNA tests, whilst for people over 45 years the idea of using digital services for weight loss is not that acute. Key needs are highlighted for the target audience, encouraging the use of services: maintaining health to improve well-being, increase immunity, treat, and prevent diseases, muscle mass gain, weight loss, development in professional and amateur sports.

In the process of building a business model, a unique trading proposal has been formulated, which includes the configuration of a diet that considers the personal characteristics of the client's body. Customer Interaction Platform: a web service and mobile app that will implement free functionality to help keep users active in the app, dieting and editing sections, personal profile, and paid subscription. The uniqueness of the trade offer lies in the personalization of the diet based on the results of DNA analysis. The diet becomes more effective by incorporating more suitable foods and eliminating foods that can cause various diseases.

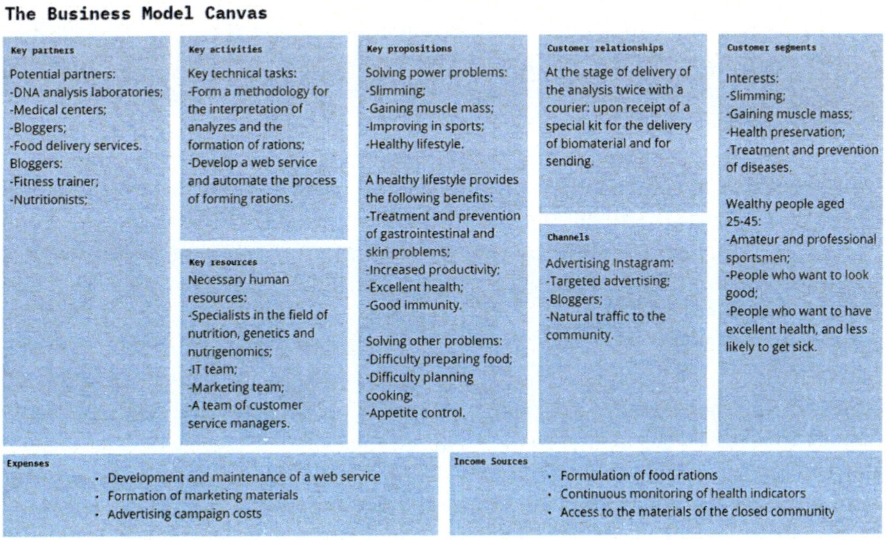

Fig. 1. Business model of the MVP in the form of canvas diagram

Among the key specialists required to develop the web service are nutritionists, genetics, IT specialists, marketing and customer service managers. Nutritional and genetic specialists are designated to prepare basic diets, develop and optimize the methodology for diet formation, research, and introduction of the new technologies into the ration adjustment process. The IT team is responsible for developing and supporting the web and mobile application, as well as ensuring security when dealing with personal data. The main tasks of marketing specialists are in broadening brand awareness, attract incoming traffic, work with potential B2B clients, and establish cooperation with potential partners. The sales team handles the traffic generated, is responsible for sales and customer management. The channels that the marketing and sales managers will work with include targeted advertising, working with bloggers, and running corporate pages on social networks, as well as in the networks of partners whose target audience coincides with the described one.

Further potential partners can be divided into three categories that include DNA analysis laboratories, experts in the field of nutrition, and food delivery services. Among the laboratories, there are three promising partners: "Atlas", "MyGenetics", and "Moi gen". Four possible partners are identified among food delivery services: "Level Kitchen", "Grow Food", "My Food" and "Yabisel". Possible formats of partnership work with DNA testing laboratories include the implementation of DNA testing services based on the partner's lab, as well as the offer of nutrition services to laboratory clients. Possible formats of partnership work with expert bloggers in the fields of nutrition science include joint events, exchange of auditory, a publication of ready-made diets, as well as the provision of promo codes for discounts. Possible forward-looking partnership format with delivery services includes delivery of prepared rations from the menu provided by the delivery service to customers. Thus, the clients will be able to get an optimal individual diet weekly at home without the need to purchase products on their own. Preparation of dishes and delivery is planned by the partner delivery service.

As a result, the main source of income will be a service to form a personalized diet. It is also worth highlighting as additional sources of revenue the service of constant monitoring of user indicators in the format of premium subscriptions and access to the closed community of users of the service. The main items of expenses will be the development and maintenance of the service, as well as advertising.

To create a service that could provide the customer with a personalized diet based on the data entered the system [10], it is necessary to form functional requirements for the system, such as compiling a recipe base, creating an algorithm for calculations of the calories, proteins, fat or carbohydrates, and macronutrients for each client. The architecture of the designed web services was described in the form of a structural model (Fig. 2). Multiple copies of the production server are required for product resiliency and hardware maintenance. Load balancer is distributing loads between production servers. Also for appending new functional, developers necessary personal equipment for testing programs. As inputs, the parameters that determine the work of the model are highlighted:

- Weight, height, purpose of a person, food preferences and difficulties of cooking;

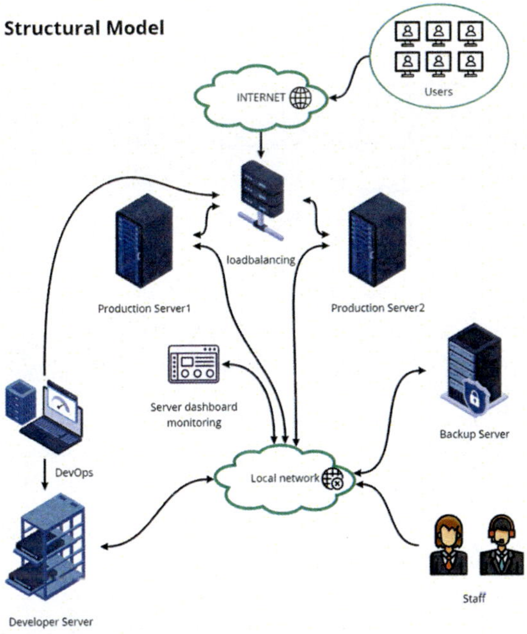

Fig. 2. Structural model of the MVP web-service

- .vcf genetic test results;
- Food preferences and the complexity of the user's cooking.

After receiving the necessary data from the user, the body mass index is calculated [8], and the amount of the main exchange based on which the rate of calorie intake is calculated using the physical activity factor [6,9]. Depending on the user's goals, the number of calories in the daily diet is proportionally increased or decreased. The output of the process is the limits on the rate of calories consumed and macronutrients for the user, considering the norms of physiological requirements in energy and food substances for different groups of the population. After obtaining genetic data in ".vcf" format [7], when a single-nucleotide replacement is detected, indicating a mutation in the allele of the diet-related gene, restrictions are formed in the set of products. The selection of dishes and products is the process of compiling the optimal diet from the base of dishes and recipes of the web service based on the input process of restrictions, user preferences in products for consumption, and complexity of cooking a meal. The process is shown in more detail in Fig. 3. On the Use Case diagram (Fig. 4) the key features that can be used by users of the MVP are shown. This diagram allows seeing the internal and external services as well as the context of the system services functionality.

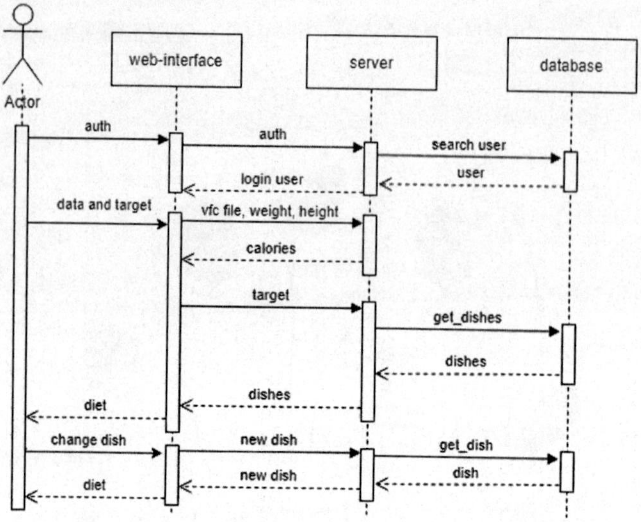

Fig. 3. Sequence model of the MVP web-service

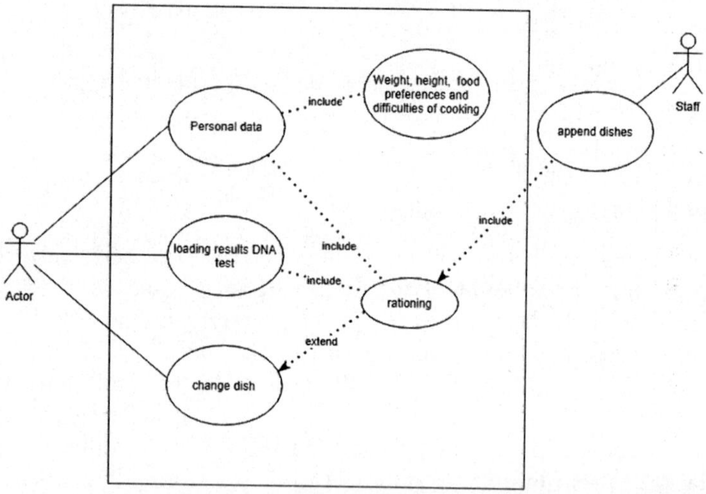

Fig. 4. Use case diagram of the MVP web-service

4 Discussion

As a result of the market analysis, it was concluded that the personalized food market is actively developing in foreign countries. The potential target audience was studied, the categories of indirect competitors were analyzed, and the main players were identified. The list of key competitors and potential partners includes "DNAfit", "GenoPalate", "Athletigen", "Vitagene", "FitnessGenes". At the moment, not all categories of competitors are represented in the Russian market. So, due to the lack of web services for the formation of a personal diet on the domestic market, the target audience was identified and the analysis of the request of potential customers was carried out. As a result of the work done, a detailed description of the business model of the service that can be implemented in the Russian market was obtained.

The business model of the minimum viable product includes a limited set of functions: the selection of a diet based on data on gender, height, weight, and the results of genetic analysis. These data are used to calculate the limits using body mass index indicators, the value of the metabolic rate, the coefficient of physical activity on which an individual diet is compiled, which helps to approach the "optimal" body mass index. The tasks of ensuring the reliability of personal data protection, unauthorized access, and fault tolerance of the service systems remain relevant and will be solved at the service development stage.

The service of forming a diet for selected segments of consumers will be the main source of income for the service. Additional sources are constant monitoring of user indicators in the premium subscription format and access to a closed community. The main items of expenditure will be the development and maintenance of the service and advertising.

The service can have a features for continuous monitoring of key health indicators using genetic analysis, blood tests, etc. The initial diet based on personal data (weight, height, activity, etc.) with the calculated amount of calories, protein, fat, and carbohydrates consumed will be adjusted based on the results of regular tests. Thus, the ratio between nutrients will be adjusted depending on the identified features in the metabolism of the human body.

Further research on the development of a business model for providing personalized nutrition based on genetic data will focus on calculating the possibility and cost-effectiveness of integration with ready-to-eat delivery services, which will solve the problem of planning and preparing meals. It is also planned to analyze the possibility of implementing the functionality of composing personal physical training aimed at helping you lose weight or gain weight. This will allow us to offer customers additional services to solve the problem of losing weight, gaining weight, or maintaining a healthy lifestyle.

References

1. Research of the Russian and world FoodTech market: key trends, limitations and prospects (in Russian). https://json.tv/ict_telecom_analytics_view/issledovanie-rossiyskogo-i-mirovogo-rynka-foodtech-klyuchevye-trendy-ogranicheniya-i-perspektivy-20200527010059
2. Genotek Medical Center - Laboratory of DNA analysis. https://www.genotek.ru/
3. Atlas Laboratory webpage - Laboratory of DNA analysis. https://www.atlas.ru/
4. Whelan, K.F.: Advanced Nutrition and Dietetics in Obesity. Wiley, Hoboken (2018)
5. Osterwalder A., Pigneur Y.: Business model generation. https://tudelft.openresearch.net/image/2015/10/28/business_model_generation.pdf
6. Federal Center for Hygiene and Epidemiology of Rospotrebnadzor: Norms of physiological needs for energy and nutrients for various groups of the population of the Russian Federation (in Russian)

7. Shin, D.M.: GEN2VCF: a converter for human genome imputation output format to VCF format. Genes Genomics **42**(10), 1163–1168 (2020)
8. Mokdara, T., Priyakorn, P., Jaturon, H.: Personalized food recommendation using deep neural network. In: 2018 Seventh ICT International Student Project Conference (ICT-ISPC), IEEE (2018)
9. Freyne, J., Berkovsky, S.: Intelligent food planning: personalized recipe recommendation. In: The 15th Annual Meeting of the Intelligent User Interfaces Community, pp. 321-324 (2010). 7–10 February 2010
10. Trevisiol, M., Chiarandini, L., Baeza-Yates, R.: Buon appetito - recommending personalized menus. In: 25th ACM Conference on Hypertext and Social Media, pp. 327-239 (2014). 1–4 September 2014

Designing a Wearable Wrist Body Thermometer Using Resistance Temperature Detector with Wireless Data for Non-contact Measurement

Febus Reidj G. Cruz, Rod Joervie P. Adante, and Rex Jose O. Bagain(✉)

Mapua University, Intramuros, Manila, Philippines
frgcruz@mapua.edu.ph, {rjpadante,rjobagain}@mymail.mapua.edu.ph

Abstract. The Coronavirus disease is an infectious disease caused by a recently discovered coronavirus originating from Wuhan, China. One of the main symptoms for COVID-19 is fever, specifically having a temperature equal to or greater than 38 °C. This study aims to design and develop a wearable wrist body thermometer using a resistance temperature detector (RTD) with wireless data transmission for non-contact measurement. Using different components, the wearable wrist thermometer was made and can upload data to the cloud. The device functions properly in terms of data acquisition and transmission. The device correctly outputs the correct LED light indicator based on a set temperature range. The device was also able to correctly transmit the data in the MCU to the cloud database with a lapsed time of about 12.3 s. Comparing the mean value of the wrist temperature taken for ten trials using the wireless thermometer and the TN-433L infrared thermometer indicates no significant difference between the two. From the ANOVA test for the precision testing of the device when taking the wrist temperature, the test shows that there is no significant difference between the readings of the temperature in these certain conditions. The device has better power efficiency due to the different power mode. It is also calibrated to be accurate, precise, and more efficient. It is characterized to have a battery life of 5 days while continuously recording wrist temperature every 30 min automatically without lapse but is only accurate for the first 24 h.

Keywords: Resistance temperature detector · Wrist temperature · COVID-19 · AD623 operational amplifier · ESP32 microcontroller

1 Introduction

COVID-19, or the Coronavirus disease, is an infectious disease caused by a recently discovered coronavirus originating from Wuhan, China and is now a global crisis or pandemic as of the year 2020. In response to lessening the spread of the disease, malls are closed, public gatherings are prohibited, lockdowns and other restrictions were implemented to reduce physical contact with other people. However, more people in the Philippines are getting infected and dying because of the disease. One of the main symptoms

R. Silhavy et al. (Eds.): CoMeSySo 2021, LNNS 232, pp. 333–347, 2021.
https://doi.org/10.1007/978-3-030-90318-3_29

for COVID-19 is fever, specifically having a temperature of equal or greater than 38 °C [1]. Health care workers always take care of watching over the patients, and they have a higher chance of getting infected with a disease. As COVID-19 spreads, it showed that healthcare workers have three times the risk of being infected compared to that of an average person [2]. To ensure safety, people are required to remain 6 feet apart from each other. Even with this rule, people are still getting infected due to the virus being capable of remaining on a surface for several hours [3]. It is also found that a wearable thermometer is viable and can be used in detecting Basal Body Temperature with the help of a Schmitt-trigger comparator [4]. The body has several points to measure the temperature accurately. Since the hypothalamus controls our temperatures, there are different readings of temperatures all around the body [5]. Differences in body temperatures range from 0.1 °C up to 6 °C depending on the location of measurement. In comparing the temperature of the wrist to the temperature of the upper arm, it has a temperature difference of about 5 °C [6]. Temperature from the wrist and forehead can be a good substitute for fever screening. Wrist measurement shows an average of −0.96 °C to −0.61 °C from tympanic measurement [7].

Due to face-to-face check-ups of the health workers and the persons under investigation (PUI) or persons under monitoring (PUM), which are possible COVID-19 positive patients, increases the risk of getting infected with the health workers COVID-19 increases. The contact between the health workers and the PUIs or PUMs cannot be avoided when taking their body temperatures using the common methods of axillary measurement.

This study aims to design and develop a wearable wrist body thermometer using a resistance temperature detector (RTD) with wireless data transmission for non-contact measurement. Specifically, this study aims to: (1) design the analog front-end circuit of RTD for a temperature reading resolution of 0.1 °C; (2) implement the thermometer system comprising of analog front-end circuit, microcontroller, and light color output as a wearable device incorporating different power modes such as active, ready and sleep; (3) develop the embedded instruction codes of the wearable thermometer with wireless data transmission to cloud-based on IEEE Std 11073-10408; and (4) characterize the performance of wearable wrist thermometer over a continuous use of the device.

With the wireless technology of wearable wrist thermometer devices, community health workers can avoid close contact with COVID-19 patients, thereby reducing the risk of getting infected. The design of this device gives many possibilities for medical applications. This research is also beneficial for the device users since it will provide a record of their body temperatures in a cloud which provides its users with an accessible and hassle-free way to monitor their past and present body temperatures.

The wearable wrist thermometer will be using the resistance temperature detector (RTD), such as the PT100 sensor. The resolution of the temperature reading is 0.1 °C. The system will be using the ESP32 microcontroller with a built-in analog-to-digital converter (ADC), Wi-Fi feature, real-time clock (RTC), and power management. Different power modes will be implemented – active, ready, and sleep, towards low power consumption of the proposed device. In the active mode, the Wi-Fi is working. In the ready mode, the Wi-Fi is disabled while the CPU computes for the wrist temperature from the ADC. In the sleep mode, only the RTC is operational while all the other parts are turned off.

The temperature measurement will be indicated using a light indicator. The light will turn to one color when the temperature is within the normal temperature. It will turn into another color when the temperature indicates a fever condition and blinks when the temperature goes into alarming fever condition. The accuracy of the proposed wearable wrist thermometer device will be determined by the compared values of the device to the axillary measurement of body temperature. The precision will be determined by the recorded data in different scenarios. The scenarios to be tested in recording the temperature are at air-conditioned environments, at room temperature, while moving, and stationary. The device will utilize Wi-Fi technology to transmit the data to the cloud database. This means an internet connection is required for the cloud service to function. There will be a website that will serve as an access point for the users. The website will contain the data sent by the device and accessed by the user and health workers. Since it uses software to record the data, all the records of body temperatures can only be accessed by the health worker and the user itself; this can be done through user accounts. To avoid privacy issues for the device users, an agreement between the user and the community health center should be made to give the user approval for the health workers to keep track of their temperature. The user first signed an agreement form as proof that they know that their body temperature is being monitored regularly. To ensure the interoperability of the device, IEEE Std 11073-10408 will be followed.

2 Methodology

2.1 Conceptual Framework

The device will have two main parts, the temperature sensor, which contains the PT100 RTD for one part, and the ESP32 microcontroller for the other. The PT100 is a sensor which varies its resistance which is directly proportional to the change in temperate [8]. This will be used in creating a readout circuit for the device. The conceptual framework for this research is shown in Fig. 1. For the input, the wrist temperature of the respondents will be used as the data while following the protocol for measuring body temperature and the IEEE standards for the thermometer. The process starts by designing the circuit of RTD for 0.1 °C resolution for accuracy. Then, each of the components is integrated into the wearable wrist thermometer. The device will be designed for the low power consumption of wireless sensors. The IEEE Std. 11073-10408 of the wireless data transmissions will also be applied in doing the cloud database. The device will be tested for its functionality. The final output of the study is a functional wearable wrist thermometer with wireless data to the cloud, which is accurate, precise, and power-efficient.

2.2 Design of Analog Front-End for Temperature Sensor

Operational Amplifiers or op-amps are among the most critical components for analog and mixed-signal (AMS) circuits [9]. A Resistance Temperature Detector or RTD, specifically the PT100, will be used as the device's temperature sensor for this research. It is a type of temperature sensor that comprises a resistor that changes its value directly proportional to the change in the value of the temperature. Regarding sensitivity, the

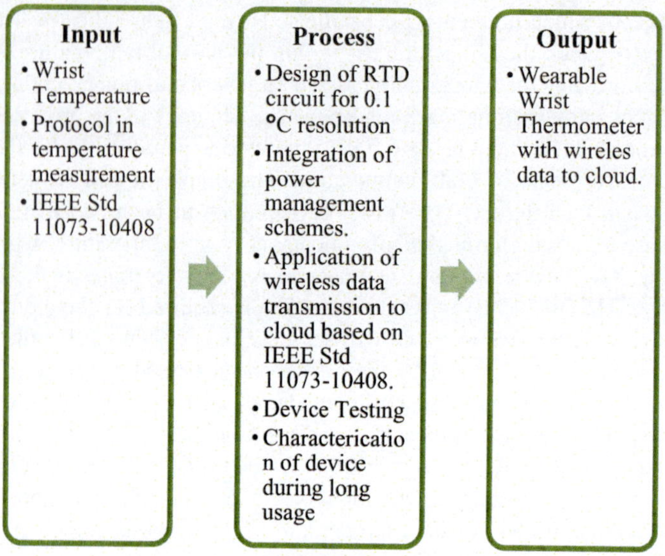

Fig. 1. Conceptual framework of the study

PT100 sensor's resistance increases by 0.385 Ω when temperature changes from 0 °C to 1 °C. The PT100 will be and should contact the person's skin for a more accurate reading of the temperature. The body and the PT100 will eventually have the same temperature, and the PT100 collects the data and varies its resistance directly proportional to the change in temperature from the cable. The output of this amplifier will be connected to the microcontroller for analog reading [10] (Fig. 2).

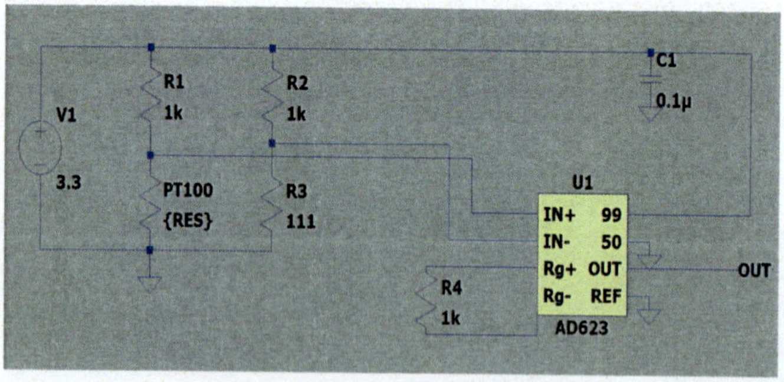

Fig. 2. Device schematic diagram

2.3 Implementation of Wearable Wrist Thermometer

Figure 3 shows the proposed setup of the whole system. The wrist temperature is taken using the PT100 of the wearable wrist thermometer worn on the wrist of a person (which

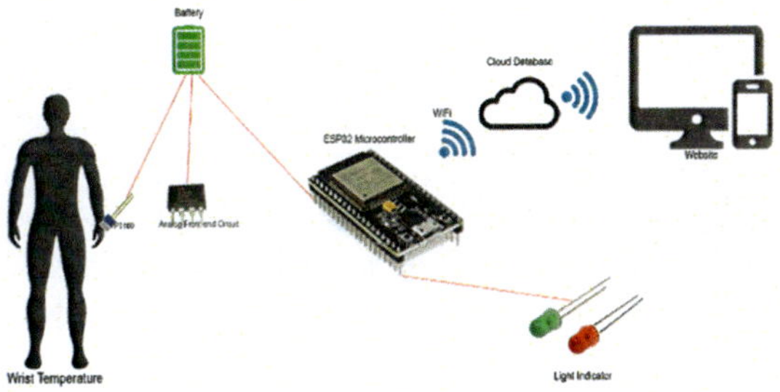

Fig. 3. Proposed setup of the whole system

can either be on the left or right hand). The wearable wrist thermometer is powered by a lithium-ion battery, where it powers all of the PT100, analog front-end circuit, and the ESP32 microcontroller. The Analog front-end circuit simply converts the change in resistance of the PT100 as a function of wrist temperature onto a voltage range that would allow the built-in 12-bit ADC in the microcontroller to provide a reading with a resolution of 0.1 °C. A light indicator is connected to the ESP32 microcontroller, where it is used as a primary temperature indicator based on the read temperature. The ESP32 microcontroller uploads the reading (using Wi-Fi) onto a cloud database of the website (http://esp-wearable-thermometer.com/login.php) created to store data such as the reading number wrist temperature, and time when the temperature is taken (Fig. 4).

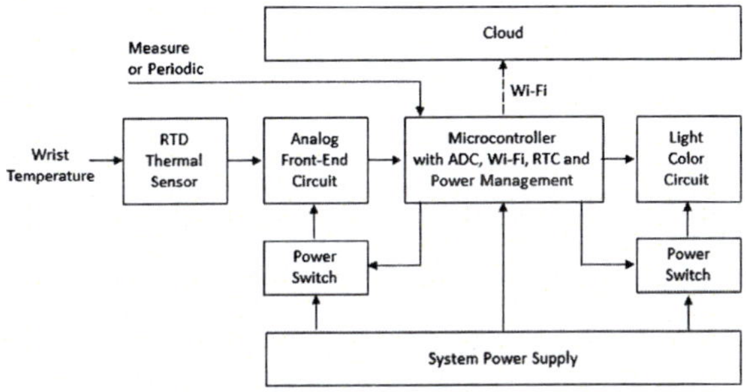

Fig. 4. Block diagram of the wearable wrist thermometer with wireless data to cloud

The block diagram shows how the wearable wrist thermometer with wireless data to the cloud will work. The wrist temperature is the input taken from the respondents. The wrist temperature is measured by the PT100 RTD thermal sensor in terms of change in resistance. The Analog front-end circuit then converts the change in resistance of the PT100 as a function of wrist temperature onto a voltage range that would allow the built-in 12-bit ADC in the microcontroller to provide a reading with a resolution of 0.1 °C. Noting that the wrist temperature is taken or recorded with 30-min intervals from each reading. The microcontroller is programmed with ADC, Wi-Fi, RTC, and power management which would then sends the recorded data into wrist temperature reading to the cloud. The analog front-end circuit, ESP32 microcontroller, and the light color circuit are turned on/off by a switch powered by the system power supply. The

LED of the light color circuit turns green when the measured wrist temperature does not indicate any fever (<37.5 °C) and turns red when the measured wrist temperature indicates mild fever condition (37.6 °C to 37.9 °C) and turns into blinking red when the measured wrist temperature goes into alarming fever level (>37.9 °C). The data stored in the cloud contained the wrist temperature measurement with its time and date when it was recorded. It can only be accessed through user login. Power modes will be implemented on the device for better power efficiency and longer battery life (Fig. 5).

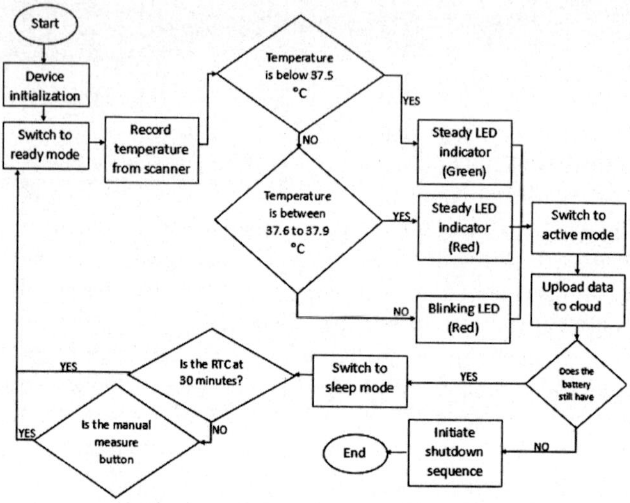

Fig. 5. System flowchart of the device

The figure above shows the system flowchart of the device on how the wearable wrist thermometer will work. After the device is turned on, it starts initializing then goes to ready mode (Wi-Fi is not yet turned on). It takes the person's wrist temperature and determines whether the person has a normal temperature, mild fever, or fever which will be indicated by a green light on the LED, red light, or blinking red light. The device will then switch to active mode to send or upload the recorded data to the cloud using Wi-Fi. After uploading the data, the device will go to sleep mode, leaving only the real-time clock operational. If the device ran out of battery, it would initiate the shutdown sequence or turn OFF; if there is still battery, it will then be in sleep mode. If the RTC detects 30 min after the last reading, it will switch to ready mode again to start taking the wrist temperature and repeat the process until going back to sleep mode again. Suppose the power button is pressed anytime between the 30-min interval. In that case, it will immediately switch to ready mode and repeat the data gathering and uploading cycle until it goes back to sleep mode again.

2.4 Development of Wireless Data Transmission to Cloud-Based on IEEE Std 11073-10408

The device will follow the data transmission standards for thermometers according to IEEE Std 11073-10408. For data transmission, there will be an interaction between the agent and the manager. The agent would represent the wearable thermometer, while the manager would represent the cloud server or database. This interaction would start with the agent sending its configuration, which is for the first time use, to the manager. Once the manager receives this configuration, the state would change into an operating stage, meaning that the agent device would start obtaining data (measuring body temperature). The measured data would then be sent to the manager. Once the manager receives the data, it would reply with a confirmation that the data was successfully received by the manager and is ready to end the measurement session. On the next measurement sessions, the configuration of the agent device would now be known by the manager and can now directly go to the operating state and start to send data. This process will be repeated over and over for each new device that will be created.

2.5 Device Calibration

Since the ESP32 can be programmed using the Arduino IDE environment. The calibration for the temperature data that will be obtained will be programmed in this environment as well. For calibration, it would first start by obtaining the analog value for the device's temperature using the analogRead() command; this is different from the voltage output of the readout circuit. The analog value will range from 0 to 4095 since the ESP32 has a 12-bit analog to digital converter. Once the analog value is obtained, the current temperature of the PT100 will be measured. The analog value and temperature will be recorded and will serve as one point for the slope. A temperature change will be applied to the PT100, which would change the reading from the readout circuit. The analog value and temperature will again be obtained. Finally, the map() command will create a linear slope obtained from the two points. The map command is a straightforward command for a 2-point from for a linear function. The experimentation also considers having a fixed offset to obtain the desired calibrated temperature readings of the device.

2.6 Device Testing

The device testing consists of: device functionality testing, where the data acquisition and transmission are tested; accuracy testing where the device is calibrated as compared to the TN-433L infrared thermometer; precision testing where the device is tested under airconditioned and room temperature environment while sitting down or walking; efficiency testing to lessen the power consumption; and characterization of the device for long term use where its data acquisition, data transmission, and battery life is determined.

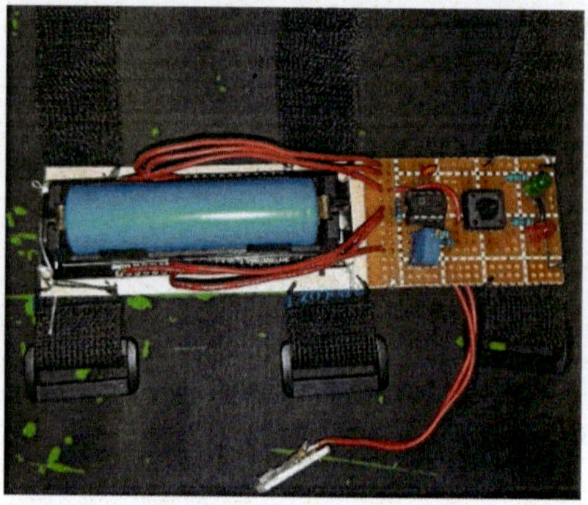

Fig. 6. Device prototype

3 Results and Discussion

3.1 The Wrist Body Thermometer Using RTD with Wireless Data for Non-contact Measurement

Overall Device

The figure above shows the overall design of the wrist body thermometer using RTD with wireless data being sent to the cloud. The wrist temperature that is being read by the device is stored on the website. The device is programmed to automatically take the user's wrist temperature every 30 min. It automatically turns off when it is not time to take the temperature (only the RTC is active) and turns on only when it takes the wrist temperature. The device also has a button where it can be pressed anytime to manually take the wrist temperature of the user. The device also has a primary temperature indicator where the LED lights up a specific color that indicates whether the user has no fever (green light), mild fever (red light), and alarming fever (blinking red light) temperatures before with the uploading of the data to the cloud (Fig. 6).

ESP32 Website/Cloud Database

The website prompts the user to enter a unique username and password provided to them for the login; the username and password for this study used aiwjurmy_user1 and P@ssword123, respectively. Access the website using: http://esp-wearable-thermometer.com/login.php The website contains the reading ID number with its corresponding wrist temperature value and timestamp (date and time). The user can also choose the number of readings to be displayed by the website. It will also show the minimum, maximum, and average reading of the displayed number of readings, wherein the figure above was for the last 20 readings (which is the default value) (Fig. 7).

3.2 Test for Voltage Outputs of the Front-End Readout Circuit

The test for voltage outputs of the front-end readout circuit will show if the voltages that will come from the AD623 instrumentation amplifier would change depending on

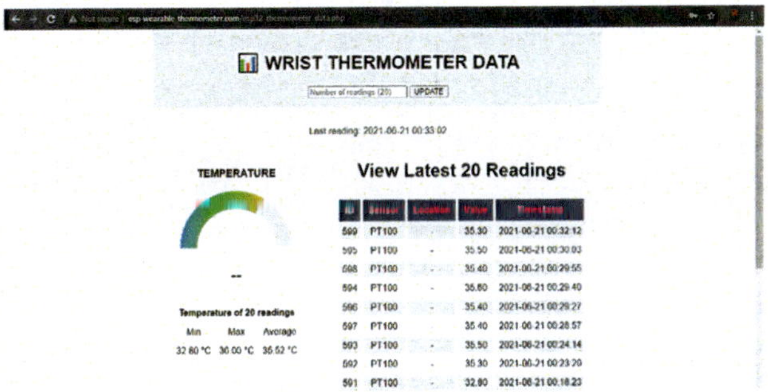

Fig. 7. Cloud database of the device

the resistance value of the temperature sensor. Figure 8 shows the obtained slope of the output voltage of the readout circuit with changing input temperature.

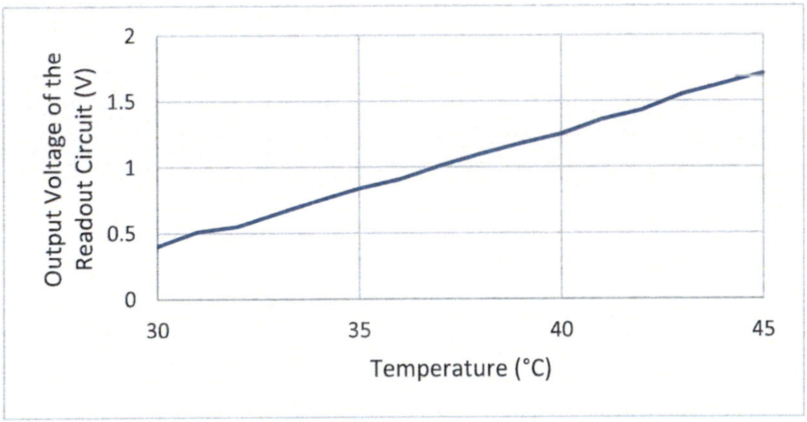

Fig. 8. Change of the output voltage of the readout circuit to temperature

3.3 Device Functionality Testing

For device functionality test, it focuses on the overall working state of the device; it would check if the voltage readings on the circuit should show that respective ADC value and output a temperature reading and a LED light color accordingly. The tests that were done for device functionality are shown in Table 1.

The gathered data for the functionality testing for data acquisition covers the wrist temperature taken by the RTD sensor, conversion to the ADC, then to MCU, and lastly to the LED light indicator, where it is used to determine whether the device is functional regardless of the accuracy, precision, and efficiency. The table shows the corresponding output voltage equivalent by the RTD circuit and its ADC equivalent value for a measured input temperature, and it also shows the output temperature of the MCU together with

Table 1. Functionality testing for data acquisition

Trials	Input temperature (°C)	Output voltage of RTD circ. (V)	ADC value (internal)	MCU output temperature (°C)	LED light color
1	33.2	0.710	674	33.5	Green
2	33.2	0.697	656	33.4	Green
3	37.8	1.07	1230	37.6	Red
4	39	1.12	1289	38.2	Blinking Red

its corresponding LED light indicator. The temperature values are converted correctly. The LED light indicator lights appropriately to the set temperature intervals, indicating no fever, mild fever, and high fever with green, red, and blinking red LED lights.

For data transmission, it covered the uploading of the recorded temperature (with time and date) of the MCU, using Wi-Fi, to the cloud, where it was used to determine if the device can send the data from the MCU to the cloud; the time lapsed for uploading the data to the cloud was used to determine how long the device sends the data to the website (which also supports the that the device can send gathered data to the cloud). The recorded data in the MCU were successfully uploaded to the cloud database with a lapsed time of about 12.3 s. The functionality test for data acquisition is summarized on Table 2.

Table 2. Functionality testing for data acquisition

Trials	Data in the MCU (Temp., Time, and Date)	Time Lapsed for uploading the data to the cloud
1	33.4 °C, 18:01:50, April 23 2021	12.360 s
2	33.2 °C, 18:02:38, April 23 2021	12.247 s
3	33.3 °C, 18:03:23, April 23 2021	12.311 s

3.4 Device Accuracy Testing

The data for testing the device's accuracy is used to determine whether the device has accurate wrist temperature measurements compared to the modern thermometer commonly used in public. The modern thermometer used for comparison was the TN-433L which is a non-contact infrared thermometer (NCIT). The average temperature for 10 trials will be used to compute the T-value of the device, which will determine whether there is (or no) significant difference between the temperature readings. No significant difference between the device and the modern thermometer would mean that the device accuracy is on point. The statistical treatment of the two-tail t-Test showed a p-value of 0.1446, which would mean that the null hypothesis indicates no significant difference between the readings of the wearable thermometer to the readings of the modern thermometer, which is not rejected, showing that the wearable thermometer is accurate (Fig. 9).

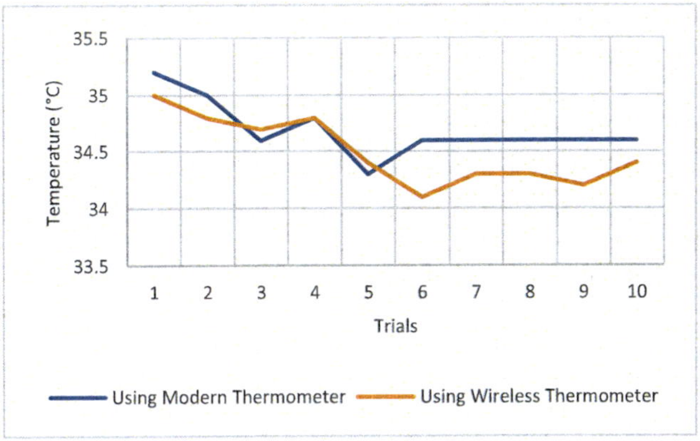

Fig. 9. Device accuracy testing results

The device was compared with the digital thermometer and mercury thermometer to see the difference in the accuracy of the wearable thermometer and other thermometers. The tests were shown in 10. From this test, both thermometers were taking the axillary temperature while the wearable thermometer. At the same time, the wearable thermometer is taking temperature from the wrist. The temperature difference from the results is because of the difference between axillary and wrist temperature and the overall accuracy of the device (Fig. 10).

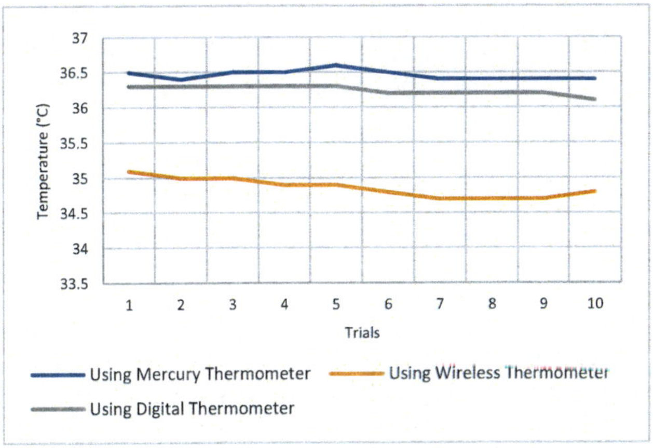

Fig. 10. Comparison of the device to other thermometers

3.5 Device Precision Testing

The data for testing the device's precision is used to determine whether the device has a precise reading given different environments. The obtained data showed the temperature readings of the device from the different environments and situations that it was put into. These parameters are for each of the given environments that the wearable thermometer was used. There were multiple ANOVA tests done to see the different p-values for the given environments and situations. For this test, a p-value greater than 0.05 would indicate that the wearable thermometer is precise given the various environmental changes. The calculated p-values obtained from the ANOVA test for the precision testing of the device when taking the wrist temperature under a room temperature and air-conditioned environment, regardless of whether the user is stationary or while walking, is 0.1198 and 0.9999; while the p-values for the precision testing of the device while the person is stationary or while walking, regardless of whether the user is in an air-conditioned environment or at room temperature, is 0.0962 and 0.7222 which means that there is no significant difference between the readings of the temperature in these certain conditions since all the p-values are greater than 0.5 (Fig. 11).

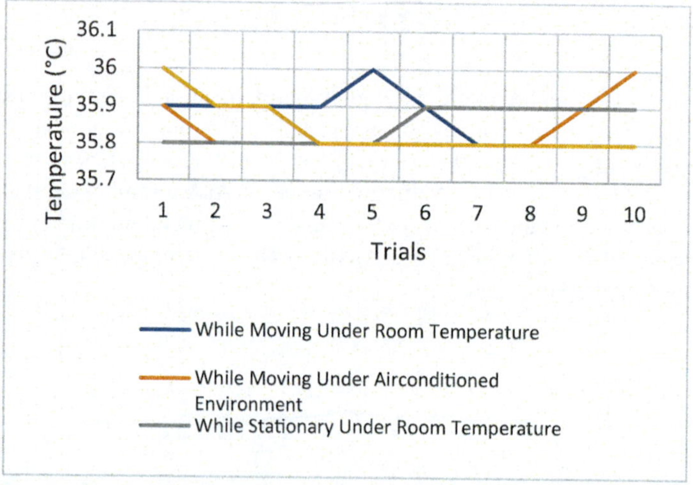

Fig. 11. Device precision testing results

3.6 Device Efficiency Testing

For device efficiency, the current draw for the different power modes was tested. This was done to see how long the battery, or the device would last before recharging. Shows the results for the efficient testing of the device. For the current draw, it would act as it is always in active mode without implementation of the power mode. This explains why it has the same current draw when the device is in active mode. From the obtained data, the average current draw for each power mode of the device is shown in Table 3; it shows that the current draw of the device without the implementation of the power mode would show a current draw similar to active mode. Using the device for like an hour straight would consume more power than having different power modes; the device would lessen its power consumption. It would not have a continuous average current draw around 169.26 mA, but instead, it would have a lower power consumption.

Table 3. Test for efficiency and power consumption

	Average current draw (mA)
Without implementation of power modes	169.26
Sleep mode	0.055
Ready mode	119.39
Active mode	170.93

3.7 Device Characterization

From the results, the device lasted for 5 days or 120 h without charging with no lapse time in the 30-min interval or taking the temperature. The results also show that the device is only accurate for the first 24-h of usage; based on the two-tailed T-test with a confidence level of 95%, the first 24 h has the only p-value amongst the 5 days, which means that there is no significant difference between the device's wrist temperature reading compared to the TN-433L wrist temperature readings (Fig. 12).

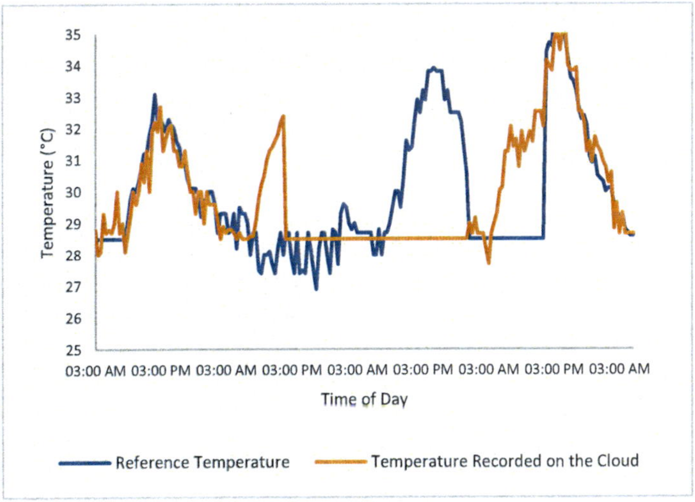

Fig. 12. Device characterization

4 Conclusion and Recommendations

4.1 Conclusion

The wearable wrist body thermometer using a resistance temperature detector was able to record the wrist temperature of a person without any contact with another person by simply accessing the uploaded data of the device in a cloud database which was wirelessly transmitted. The device uses a PT100 as the RTD sensor. It was specifically designed with a front-end circuit using AD623 as the operational amplifier to provide a reading resolution of 0.1 °C. The ESP32 microcontroller was programmed to read the wrist temperature readings, display different LED light indicators, upload data to the cloud, and incorporate its power modes. The code in the ESP32 for wirelessly transmitting data to the cloud followed the IEEE Std 11073-10408. The cloud database has user profiles to provide privacy to the users of the device.

The device is fully functional as its functionality testing for data acquisition and transmission shows that the device can gather wrist temperature data and use the LED to indicate the determined wrist temperature data whether it is within the set temperature range. The device correctly outputs a green LED light when the read temperature was 37.5 °C and below, a red LED light when the temperature was within 37.6 °C and 37.9 °C, and a blinking red LED light when the temperature was 38 °C and above. The device was also able to correctly transmit the data in the MCU to the cloud database with a lapsed time of about 12.3 s.

The device is calibrated that it can accurately measure the wrist temperature of a person as compared to the TN-433L, which is a commercial non-contact infrared thermometer. It can also precisely measure the wrist temperature of a person while the person is moving or not in an airconditioned environment and at room temperature. The device has better power efficiency due to the different power modes (such as active, ready, and sleep modes), significantly lessening its average current draw. Thus, it decreases its power consumption. The device can automatically record wrist temperature data every 30 min and upload the data in the cloud database for 5 days without charging it. However, the device is only accurate for the first 24 h of usage.

4.2 Recommendations

The researchers would recommend the future researchers test for the device's accuracy by comparing it to other wrist or body temperature devices commonly used for detecting fever. Also, find a way to maintain the device's accuracy for more than 1 day. Apply improvements to the device's aesthetics or size while maintaining its accuracy and precision. Lastly, have an easier way of getting the device to change Wi-Fi connections without reprogramming it.

References

1. McCallum, L., Higgins, D.: Measuring body temperature. Nurs. Times **108**(45), 20–22 (2012)
2. Nguyen, L., et al.: Risk of COVID-19 among frontline healthcare workers and the general community: a prospective cohort study. Lancet Publ. Heal **5**(9), e475–e483 (2020)

3. CDC Website. https://www.cdc.gov/coronavirus/2019-ncov/preventgetting-sick/social-distancing.html. Accessed 21 Nov 2021
4. Cruz, F.R., Balbin, J.J., Esporlas, M., Gacutan, L., Llapitan, I., Navale, D.: Schmitt trigger for ovulation detection using raindrop splash algorithm in basal body temperature. In: IEEE 10th International Conference on Humanoid, Nanotechnology, Information Technology, Communication and Control, Environment and Management HNICEM 2018, pp. 5–8 (2019)
5. Sund-Levander, M., Grodzinsky, E.: Assessment of body temperature measurement options. Br. J. Nurs. **22**(16), 942–950 (2013)
6. Choi, J.H., Yeom, D.: Study of data-driven thermal sensation prediction model as a function of local body skin temperatures in a built environment. Build. Environ. **121**, 130–147 (2017)
7. Chen, G.: Validity of Wrist and Forehead Temperature in Temperature Screening in the General Population During the Outbreak of 2019 Novel Coronavirus: A Prospective Real-World Study (2020)
8. Saifizi, M., Zakaria, M.S., Yaacob, S., Wan, K.: Development and analysis of hybrid thermoelectric refrigerator systems. IOP Conf. Ser. Mater. Sci. Eng. **318**, 012036 (2018)
9. Fakhfakh, M., Tlelo-Cuautle, E., Siarry, P. (eds.): Computational Intelligence in Analog and Mixed-Signal (AMS) and Radio-Frequency (RF) Circuit Design. Springer, Cham (2015). https://doi.org/10.1007/978-3-319-19872-9
10. Indestructables Website. http://www.instructables.com/id/Reading-Temperature-From-PT100-Using-Arduino/. Accessed 02 Aug 2020

Internet of Things and Global System for Mobile Based Temperature and Humidity Monitoring System

Kim John A. Ilagan, Diana Erika V. Jasareno(✉), and Flordeliza L. Valiente

Mapua University, Intramuros, Manila, Philippines
{kjailagan,devjasareno}@mymail.mapua.edu.ph,
flvaliente@mapua.edu.ph

Abstract. Server rooms are essential in having or building businesses. On other occasions, it can be seen in the home or mini workstations. In some instances, when the temperature in a server room goes too high or cannot maintain the ideal room temperature, the servers will shut down automatically due to overheating. Internet of Things and Global System for Mobile Based Temperature and Humidity Monitoring System provides real-time checking of the temperature and humidity inside the server room. The device offers various commands that the user can send to achieve the accepted values. Moreover, the gathered data in real-time is sent to the webserver ThingSpeak. In addition to that, it has a quarterly update in both SMS and the Telegram app. It utilizes the strength of both IoT and GSM in order to cater to the Filipino's needs. The heart of this project is the AHT20 sensor, one of the recently released sensors that surpass the specs of the DHT sensors. Meanwhile, its brain is the RasPi and Arduino Mega module. RasPi handles the IoT and calibration of the project while Arduino Mega is tasked with the IoT, GSM and acts as an actuator for the Air Conditioner. The project's limit, however, is that it is AirCon specific. The researchers recommend that future researchers find a board with higher storage capacity to control more brands and types of Air Conditioners.

Keywords: Internet of Things · Global system for mobile · Temperature and humidity monitoring system · Real time

1 Introduction

Server rooms are essential in having or building businesses. On other occasions, it can be seen in the home or mini workstations. Server rooms provide a single location where networking and other activities are made easier and more effective. Furthermore,

R. Silhavy et al. (Eds.): CoMeSySo 2021, LNNS 232, pp. 348–363, 2021.
https://doi.org/10.1007/978-3-030-90318-3_30

they provide an environment where operations of computer equipment can be done safely. The size of a server room is not uniformed. It can be in any shape or size. However, its shape can also be a contributing factor for the planning and environmental management, including cooling and airflow, the two important factors that can greatly affect the server room. Other aspects to consider in a server room design are its precise environmental control, airflow planning, fire suppression system, cable management solutions, redundant power sources, and physical security [1]. The usual equipment seen in a common server room is server racks, computer servers, routers, switches, network cabling, and cable management equipment. This equipment generates lots of heat, so cooling systems are installed to maintain the low temperature to guarantee the safety of the machinery. In some instances, when the temperature in a server room goes too high or cannot maintain the ideal room temperature, the servers will shut down automatically due to overheating [2]. This case is what Internet Technicians do not want to happen since the workstations heavily rely on server rooms. In this pandemic, work from home and schools rely on data centers or server rooms thru remote or VPN access. Hence, the importance of monitoring the temperature of server rooms is high and overheating must be prevented at all costs.

According to [7], Raspberry Pi and Arduino Module are effective means in taking the humidity and temperature in a server room. In addition, the use of the sensor DHT22 is effective since it was able to collect experimental data not far from the actual data. The experimental data refers to the gathered temperature and humidity data using the prototype in their research, while the actual data is the gathered temperature and humidity through measuring devices. Moreover, their receiver is Telegram, while the transmitter is the prototype. In their conclusion, it was specified that the Internet of Things is very helpful in maintaining the accurate temperature needed in a server room with the help of ThingSpeak, a web server that stores the data gathered by the sensor, in which a user can monitor the real-time temperature and humidity in a server room.

Data centers are considered to be the data holders of digital information of the world. Companies now in the modern world rely on data centers cloud and services to support and optimize e-commerce and online application [3]. Computer Business Review compiles multiple incidents. On January 6, 2015, the iiNet, the second-largest internet provider in Australia, was left offline due to overheating; the last recording was 45 °C. An iiNet representative, Christopher Taylor, posted in a forum post that the iiNet company had multiple air conditioners failing on-site, causing a rapid temperature increase. In March 2013, another incident happened due to overheating. Microsoft's email services were shut down for 16 h which affects Skydrive. Microsoft stated that due to the update failure of the software, the temperatures rose rapidly, reaching the breaking point of the data centre [4]. As per the study in [7], it is mostly applicable to Indonesia since the researchers are Indonesians. This paper aims to create a device that caters to the Filipinos using the concept made in [7]. The previous research can monitor the temperature and humidity thru ThingSpeak, and they get updates via the application Telegram. Their notification updates lack information and appeal to the eyes. It simply gives updates based on the information gathered in ThingSpeak. On the other hand, this research aims to create a better version of the previous research by incorporating a GSM Module and a more informative notification message. Furthermore, since the Philippines' climate is hot in general, an emergency text message is sent whenever the temperature and humidity go out of the accepted values. Next, the actuator in the previous research is automatic; hence there is no way of knowing whether the temperature of the aircon was truly adjusted since the communication is one-way. This research wants to give the users the capability to know whether the aircon is truly adjusted by integrating buttons in Telegram and keywords in SMS. The buttons represent different functions such as TURN UP and TURN DOWN, imitating the command in an actual remote control. The SMS feature has the same concept but with the use of keywords. Lastly, the previous research made use of only one sensor, so this current research wants to improve that by having two sensors connected in the circuit, and those two sensors will be calibrated to maintain the quality of the device since sensors are sensitive and are prone to errors once it touches moist or experienced overheating. The calibration technique will enable the researchers to protect the device and to detect the deterioration of sensors as soon as possible.

The purpose of this study is to develop an IoT and GSM Based Temperature and relative humidity Monitoring System, a device that can measure the temperature of server rooms or data centers with an alert response. Specifically, this study aims to: (1) design the circuit and program of the device; (2) develop a device with the utilization of Arduino and raspberry pi; (3) acquire the relative humidity and temperature of server rooms in real-time; (4) send feedbacks and real-time updates via the messaging app Telegram and SMS; (5) make the device act as an actuator to control the up and down button of the air conditioner.

The development of this study, IoT and GSM Based Temperature and relative humidity Monitoring System, is significant for data centers or server rooms since it can measure the room's temperature. It aims to make the job of IT staff easier by providing updates without the physical presence in the server room. This is found useful, especially today in times of pandemics. With the Wi-FI module, the individual can monitor the temperature in real-time through Telegram. The SMS feature is used for critical events, which are overheating detection or shutting down of servers. Furthermore, internet connection in the Philippines still needs improvement; hence GSM-based alerts were implemented. The IoT and SMS feature sends information or status to the individual that handles the data center or server rooms. Through this SMS, the individual can be aware of the temperature of the rooms for them to be able to maintain the room's ideal temperature. In addition, the device acting as an actuator for the air conditioner in critical situations can maintain the ideal temperature, hence avoiding equipment damage and preventing circumstances that affect the workforce that relies on data centers or server rooms.

The study focused on developing IoT and GSM Based Temperature and Monitoring systems for data centers or server rooms. The study used an Arduino and Raspberry Pi open-source electronics platform. The device can measure the temperature and relative humidity of a server room. The measured temperatures are categorized into three categories: High, Ideal, and Low. Temperatures above 27 °C are considered High, while below 18 °C are considered Low. The ideal temperature is between 18 °C to 27 °C (64.4 °F to 80.6 °F), with relative humidity anywhere between 40% and 60%. Furthermore, this project is AirCon specific and only works in Panasonic Air Conditioners released in December 2020 and up. The room where testing will be conducted has a dimension of 150″ × 88.5″ × 102″. There was an initial plan to conduct the testing in a company's server room, but since there are restrictions in this pandemic, it will be conducted in a personal workspace server room classified as Type A2 by ASHRAE.

2 Methodology

2.1 Conceptual Framework

The input for the system will be the acquired room temperature, and relative humidity gathered from the sensors. The process has two stages: (1) to transmit the data from RasPi to the database server and (2) to check the processed data whether is within the range of accepted values. There are two ways to receive the output: (1) via the Telegram app and (2) via SMS. Both notifications will be sent to the Telegram app and SMS. Meanwhile, by responding to the notification with a command, the same goes for the Telegram, the user can activate the actuator to control the air conditioner (Fig. 1).

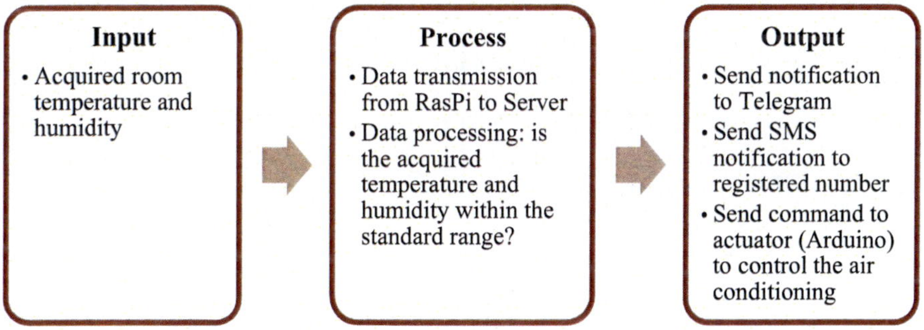

Fig. 1. Conceptual framework of the study

2.2 System Process Flowchart

The design process of the Internet of Things and Global System for Mobile Based Temperature and relative humidity Monitoring System starts with the sensors' acquisition of room temperature and relative humidity. Once the temperature and relative humidity is detected, the data will be transferred and stored to the database server from the Raspberry Pi. The device will then check if the acquired data is within the accepted values: 18 °C to 27 °C and 40% to 60% rH [5]. If the data is within the accepted values, real-time updates will be sent via the Telegram app and the operation will be terminated then restarted again. Else, the device will alert the user of the temperature and relative humidity straying from the accepted values via the Telegram app and SMS. The user also has the choice to send a command in either of the two modules. The command sent can activate the actuator and control the air conditioner (Fig. 2).

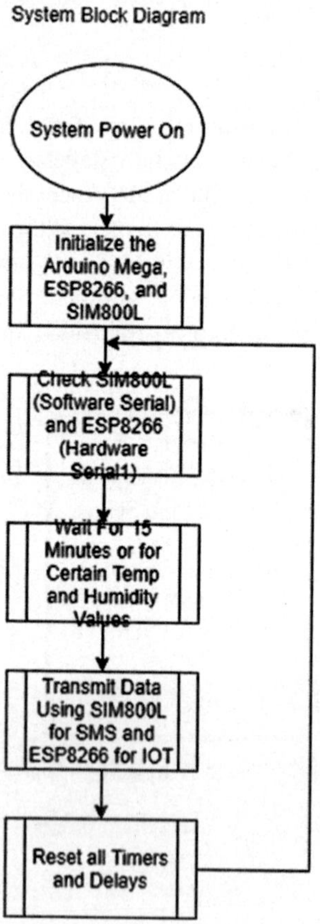

Fig. 2. System block diagram

2.3 IoT and SMS Notification

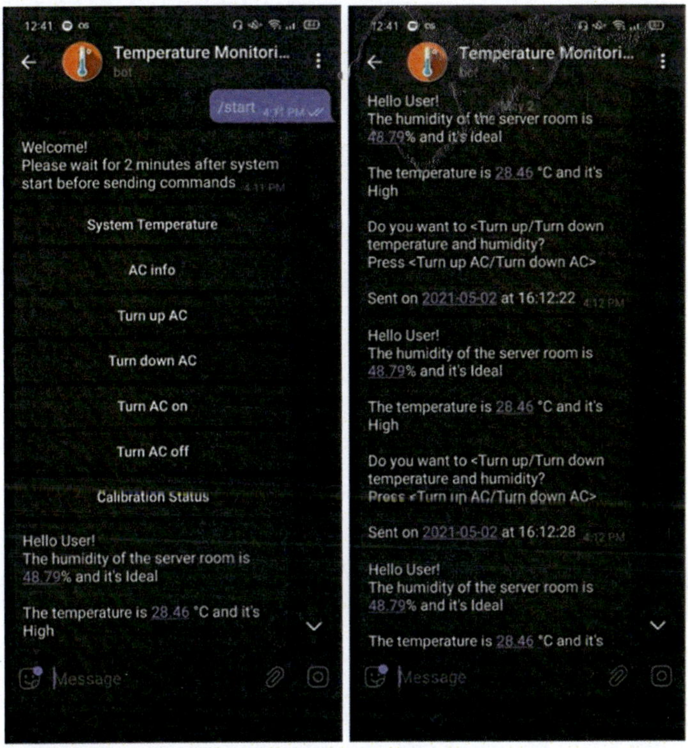

Fig. 3. Notifications sent in Telegram

In the figure above, we can see a sample notification received in the Telegram app. The telegram bot has six commands which are the following: system temperature, AC info, turn up AC, turn down AC, turn AC on, turn AC off, and calibration status. The temperature and relative humidity acquired from the telegram bot are from the AHT20 attached in the Raspberry Pi 4. The client can choose the calibration status command to determine if the two AHT20 are less than 2 degrees Celsius. On the right photo, it can be observed that there is also a quarterly update, just like in the SMS feature (Figs. 3 and 4).

In the figure above, we can see a sample notification sent in SMS. The notification shows the current temperature and relative humidity. These values are also processed whether it is within the ideal range of temperature of 18 °C to 27 °C and of relative humidity of 40% to 60%. These messages are the quarterly update programmed by the researchers. At the start of the Arduino Mega, it sends SMS, on the left photo, to the client that it is turning on. Commands can be sent via SMS and make the actuator control the air conditioner to achieve the ideal room temperature and relative humidity.

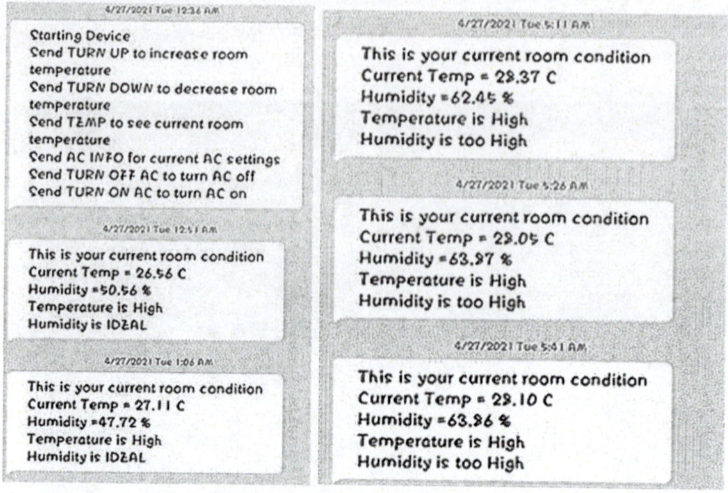

Fig. 4. Notifications sent in SMS

2.4 Setup

Fig. 5. Actuator placed beside the AirCon, Actuator zoomed in, RasPi placed above a table (from left to right)

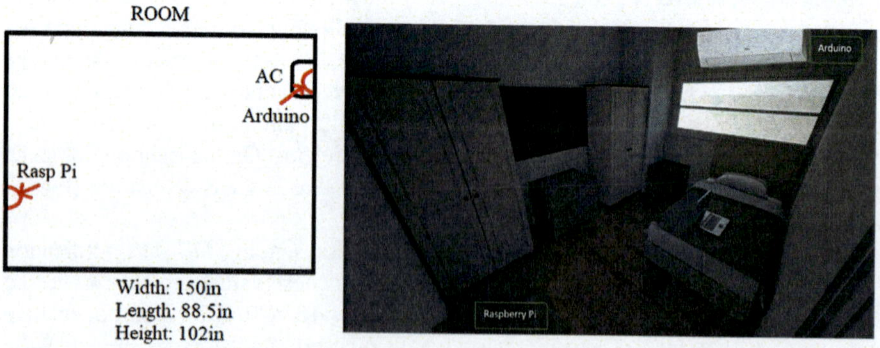

Fig. 6. Room dimension and 3D setup of the design

The figure above is the 3D setup of the design. It can be seen that there is a Smart TV, gaming laptop, and other electronic gadgets that emit heat. The Arduino is seen to be placed beside the AC while the Raspberry Pi is placed on the other side of the room (Figs. 5 and 6).

The testing room used in this research is a closed room with a dimension of 150″ × 88.5″ × 102″. Based on the standards set in [6] by ASHRAE, the size and temperature of this room are that of personal workspace. The data format of temperature reading of buildings, as stated in [6] by ASHRAE, has four different types varying from A1 to A4. A1 is a server room with mission-critical activities and tightly controlled environmental parameters (dew point, temperature, and relative humidity); enterprise servers and storage systems are typical commodities created for this environment. A2 is an information technology space, office, or lab environment with some environmental control (dew point, temperature, and relative humidity); volume servers, storage goods, personal computers, and workstations are typical products built for this environment. A3/A4 is an information technology space, office, or lab environment with some environmental control (dew point, temperature, and relative humidity); volume servers, storage goods, personal computers, and workstations are typical products built for this environment. The researchers used their guide in deciding the placement of the prototypes. The Arduino is the actuator; hence it is placed near the AirCon to detect failures immediately and to control the AirCon unit immediately. On the other hand, the RasPi is placed opposite to the Arduino and near to the server to record the surrounding temperature. There is a calibration of ±2 °C and ±5% for the rH to check whether the sensors are accurate and working. With that said, this prototype is efficient if placed in a mini-company server room with an area of 8.56 m^2.

As for the environmental setting, the room contains basic electronic gadgets that induce heat, such as chargers, cables, a mobile phone, a laptop, and the prototype circuits. When another server or computer is added, the rate of change of the environment vs the temperature of AirCon increases by approximately one °C per 10 s, unlike if the server is the only one with a rate of change of 0.13 °C. From this data, we can conclude that the more servers in a server room, the higher the heat. Therefore, the AirCon must be set to a lower temperature in order to avoid overheating.

2.5 Hardware

Figure 7 shows the schematic for the Arduino Mega. Coming from a 9 V input, a buck converter was used to step down the voltage input to 5 V to power up the GSM Module, Sim800L. Underneath the Sim800L is where the sim card is attached to the sim cardholder. There are also IR Transmitter and Receiver, IR LED 950 nm and TSOP38238, to turn up or down the switch of the air conditioner. The ESP8266 is attached to the Arduino Mega to gain connectivity to the Router.

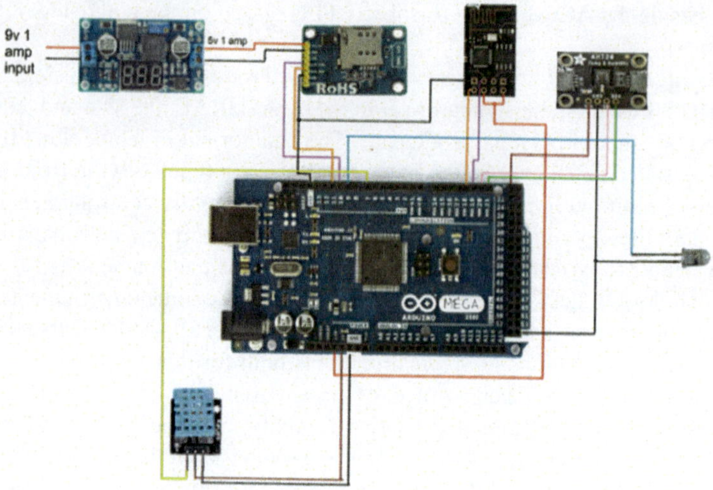

Fig. 7. Arduino mega schematic

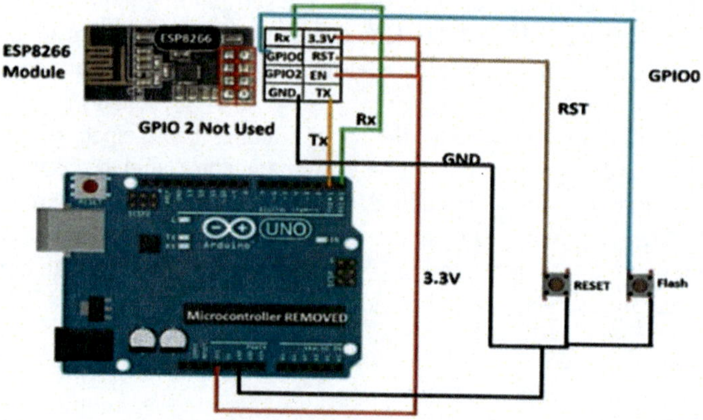

Fig. 8. ESP8266 module

Figure 8 shows the circuit for programming the ESP8266 Module. The Arduino IDE was used to program the Module. Following the circuit, the microcontroller was removed in the Arduino Uno to code the ESP8266. There are two pushbuttons, labelled Reset and Flash. To input the code into the ESP8266, the following steps are done while uploading the code: 1) press and hold the Flash button, 2) press the Reset button once, 3) leave the flash button. Once it is done, the uploading of code shall continue.

Figure 9 shows the schematic for the Raspberry Pi. The pins of the AHT20 are attached directly to the pins of the Raspberry Pi 4. The input voltage of the AHT20 was connected to the 5V power pin of the Raspberry Pi 4. The Ground, SDA, and SCL are connected to the GND, SDA, and SCL of the Raspberry Pi 4 as well, respectively.

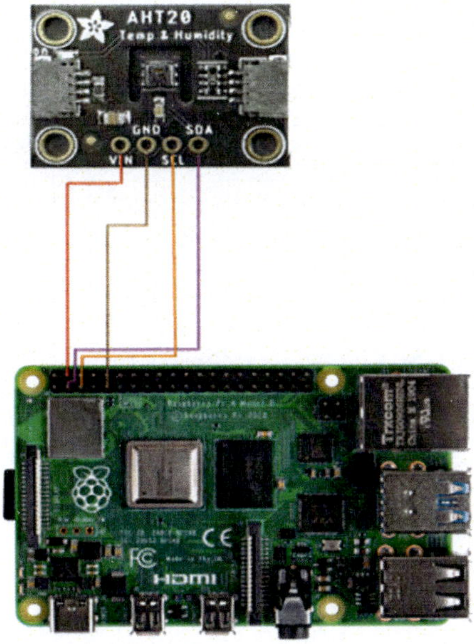

Fig. 9. Raspberry Pi schematic

2.6 Architectural Design of the Study

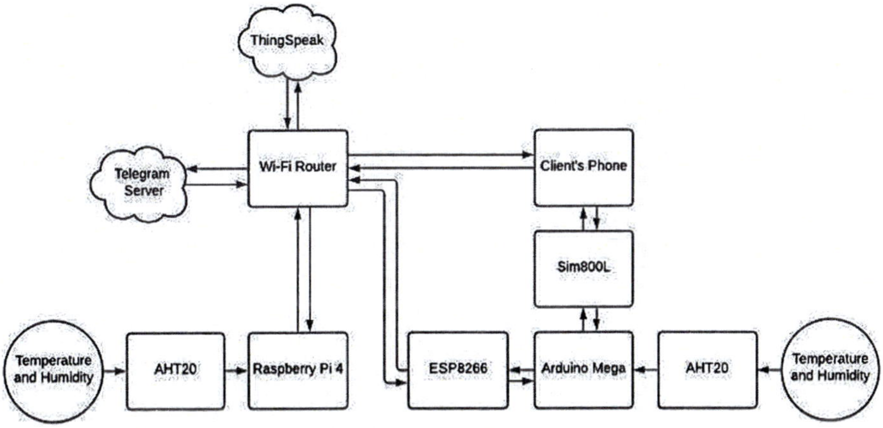

Fig. 10. Architectural design of the study

There are two microcontrollers used in the project, Arduino Mega and Raspberry Pi 4. Arduino Mega handles the SMS feature during the critical event of the system and the 15-min update feature. It has AHT20 for acquiring temperature and relative humidity. The ESP8266 attached to it was connected to the Wi-Fi to send the data to Thingspeak. It also handles turning up/down and on/off the feedback of the air conditioner in the server room. Raspberry Pi 4 handles a Wi-Fi feature that sends messages to Telegram via the bot feature. These messages are based on the data acquired via AHT20 sensors connected in the Raspberry Pi 4, temperature and relative humidity in the server room (Fig. 10).

3 Results and Discussion

3.1 Temperature and Relative Humidity Processing

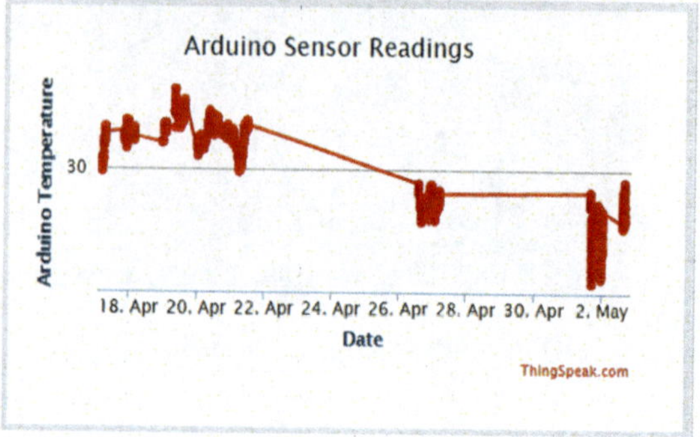

Fig. 11. Real time updates in ThingSpeak (temperature)

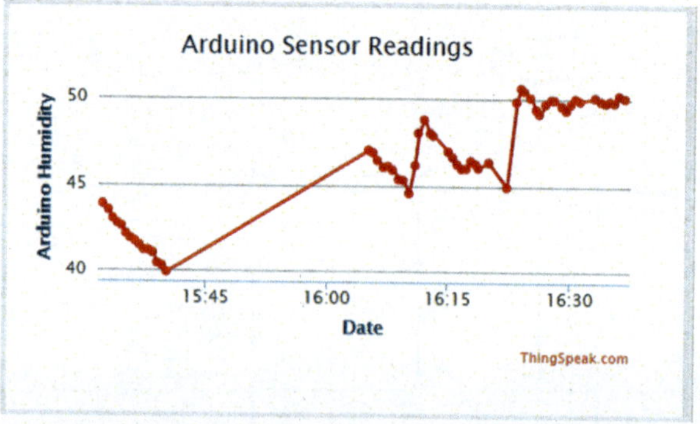

Fig. 12. Real time updates in ThingSpeak (humidity)

Each graph shows points and lines. Each point shows the values, which are the temperature and relative humidity, acquired, as well as the time and date it was acquired. The resulting straight lines are the times when the device is turned off. Since the device updates the database server realtime, the user can timely monitor the room (Figs. 11 and 12).

Figure 13 shows the sample data that was exported from the ThingSpeak. It can be observed that each entry has an interval of less than a minute. In the column 'C'

	A	B	C	D	E	F
3894	2021-05-02 07:18:11 UTC	3893	26.2	51.91	25	1
3895	2021-05-02 07:18:41 UTC	3894	26.33	51.77	25	1
3896	2021-05-02 07:19:17 UTC	3895	26.65	51.02	25	1
3897	2021-05-02 07:19:48 UTC	3896	26.93	50.49	25	1
3898	2021-05-02 07:20:19 UTC	3897	27.18	50.09	25	1
3899	2021-05-02 07:20:54 UTC	3898	27.53	49.87	25	1
3900	2021-05-02 07:21:28 UTC	3899	27.76	49.39	25	1
3901	2021-05-02 07:22:03 UTC	3900	28.08	49.14	25	1
3902	2021-05-02 07:22:41 UTC	3901	28.39	49.1	25	1
3903	2021-05-02 07:23:11 UTC	3902	28.44	47.89	25	1
3904	2021-05-02 07:23:46 UTC	3903	28.07	51.19	25	1
3905	2021-05-02 07:24:20 UTC	3904	27.88	52.77	25	1
3906	2021-05-02 07:24:52 UTC	3905	27.57	53.58	25	1
3907	2021-05-02 07:25:32 UTC	3906	27.24	53.87	25	1
3908	2021-05-02 07:26:10 UTC	3907	26.97	53.51	25	1
3909	2021-05-02 07:26:45 UTC	3908	26.71	52.59	25	1
3910	2021-05-02 07:27:20 UTC	3909	26.67	51.73	25	1
3911	2021-05-02 07:27:51 UTC	3910	26.87	50.15	25	1
3912	2021-05-02 07:28:21 UTC	3911	26.78	49.01	25	1
3913	2021-05-02 07:29:01 UTC	3912	26.81	47.84	25	1
3914	2021-05-02 07:29:32 UTC	3913	26.71	46.82	25	1
3915	2021-05-02 07:30:07 UTC	3914	26.65	46.01	25	1
3916	2021-05-02 07:30:37 UTC	3915	26.54	45.71	25	1
3917	2021-05-02 07:32:16 UTC	3916	26.2	43.87	25	1
3918	2021-05-02 07:32:57 UTC	3917	26.01	43.52	25	1
3919	2021-05-02 07:33:32 UTC	3918	26.09	43.04	25	1
3920	2021-05-02 07:34:07 UTC	3919	26.11	42.77	25	1
3921	2021-05-02 07:34:37 UTC	3920	26.1	42.61	25	1
3922	2021-05-02 07:35:12 UTC	3921	26.03	42.17	25	1
3923	2021-05-02 07:35:47 UTC	3922	26.01	41.91	25	1
3924						

Fig. 13. Sample data acquired in ThingSpeak

are the temperature acquired by the AHT20 sensor connected to the Arduino Mega; relative humidity is also acquired. In the column 'E' is the current temperature of the air conditioner while in column 'F' indicates if the air conditioner is turned-off, which is set to 0, or turned-on, which is set to 1.

3.2 Feedbacks and Real Time Updates

The Associating to the temperature and relative humidity processing, these data are sent to the client both via SMS and Wi-Fi. In Wi-Fi, the acquired values of temperature and relative humidity are sent, with different intervals of 14 min, 15 min, 16 min, and 17 min, to the client. The client receives the data via Telegram utilizing the bot-feature of the social media platform. In critical events, whereas no internet or the data values exceeded the ideal temperature, the device sends a message via the Sim800L, a GSM Module. Since it utilizes a sim card, the sim card requires load to be able to send messages to the client.

The design project using the Arduino Mega is capable of sending feedback in both the SMS-feature and the IoT-feature. In contrast of the AC to the design project to its temperature, the average percentage error is less than 1% which is 0.86291. Furthermore, it is also capable in triggering the emergency feature of the Arduino Mega in which it sends sms to the client in which alternate back and forth with the interval of 30 s and 29 s.

3.3 Switch Feature

Utilizing the Telegram and the SMS feature in Arduino Mega, the client is able to control the temperature button of the air conditioner of the server room. For the telegram, this is possible by setting a command to the telegram bot to send data to control the aircon. This command is triggered by the client whenever the client sends the command to the telegram bot. The telegram app also acts as a remote not just a notification platform.

3.4 Quarterly Update Monitoring

Based on the gathered data for the quarterly update for both the Telegram bot and the SMS feature, the quarterly update was successful as there is no failed notification, having a total of 10 received messages for each of the devices.

3.5 Rate of Change of Temperature

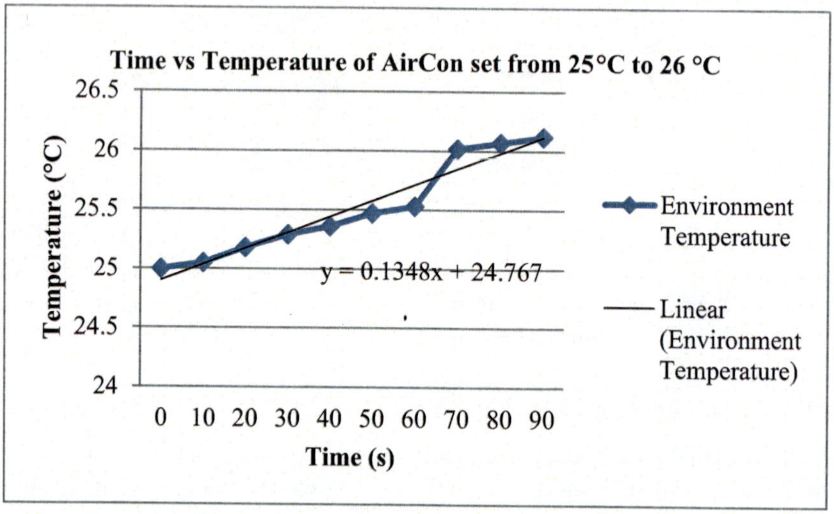

Fig. 14. Time vs Temperature of AirCon set from 25 °C to 26 °C

In this figure, we can see its relationship. The AirCon was adjusted from 25 °C to 26 °C. In a span of 90 s, the environmental temperature went up from 25 °C to 26.12 °C. It has a rate of change of 0.1348 °C per 10 s (Fig. 14).

3.6 Output Accuracy Test

The output accuracy test includes two parameters which are the temperature and the relative humidity that will be acquired in a server room, involved in this study. The temperature, and relative humidity acquired from the device will be compared to the data from actual measuring devices. This will determine its accuracy to measure the accuracy of the outputs. Thermo Hygrometer will be used for the expected values. Percent error and the accuracy will be computed based on the parameters' means.

$$\delta = |(\overline{x}A - \overline{x}E)/(\overline{x}E)| \cdot 100\% \tag{1}$$

$$\text{Accuracy} = 100\% - \delta \tag{2}$$

Table 1. Temperature and relative humidity accuracy data test

Time (hour)	Device output		ThermoHygrometer		% Error of temp.	% Error of rH
	Temp.	rH	Temp.	rH		
1	25.56	51.84	25.5	52	0.23529	0.307692
2	25.64	52.7	25.6	52	0.15625	1.346153
3	25.61	53.48	25.7	53	0.35019	0.905660
4	25.98	53.86	26	55	0.07692	0.254545
5	26.39	59.16	26.4	59	0.03787	0.271186
6	25.24	57.74	25.7	58	1.78988	0.448275
7	25.87	55.68	26.5	55	1.39622	1.236363
8	26.75	57.9	27.1	58	1.29151	0.172413
9	26.57	56.5	27.2	56	1.36029	0.892857
10	26.8	55.21	26.3	57	0.76045	1.385964
Average					0.74549	0.722111
Accuracy	99.25450865		99.27788862			

Table 1 shows the actual value, which is the prototype value, and the expected values, which are the thermohygrometer values. The percentage error was acquired for each trial. The average value of the percentage error was then acquired for the temperature, and relative humidity. With these parameters, the average percent error for both the temperature and relative humidity was less than 1%, hence, the accuracy was more than 99%.

3.7 Statistical Treatment

The Two systems are proposed to be used for output accuracy tests: the device output and the actual measuring device output as the reference system. The actual measuring device is a thermo hygrometer that measures temperature and relative humidity. Both systems will be used simultaneously for every sample.

The Bland-Altman analysis was performed to assess the agreement between the temperature acquired from the prototype and the thermohygrometer. Bland-Altman Analysis will be done using statistical software, NCSS. NCSS software is used by researchers around the globe in many industries. It is renowned for its accuracy and ease-of-use, graphical excellence and affordability [8].

It is found that the reports of the Bland-Altman Analysis are that the value acquired in both the prototype and the thermohygrometer are in agreement.

3.8 Application to Actual

The design is done in a mini server room. In application of the study, it gives a new concept of handling and monitoring the AC which are placed in the server rooms. The study provides the application of IoT which makes efficient operation management. Including the SMS feature, it allows the main user to be alerted realtime with its emergency feature. In controlling the temperature of the room itself, an AC inverter was used to maintain the temperature set by the user avoiding overcooling and overheating. Since the device is able to control the temperature of the AC, if an actual server room wasn't able to achieve the temperature it needs, the AC itself was the problem. Since the AC is an inverter type, the prototype didn't made to be automatic as the AC itself adjusts the load depending on the temperature in the environment to achieve the set temperature by the user.

4 Conclusion and Recommendations

4.1 Conclusion

This paper concludes that data acquisition such as temperature and relative humidity data is possible thru the incorporation of both Internet of Things and Global System for Mobile. Real-time monitoring of server room is possible through this research paper since it highlights the contactless controlling of the Air Conditioner and monitors the temperature and humidity in a server room. The key factor that enabled this is the webserver ThingSpeak which contains all the data acquired thru the AHT20 sensor. The objectives of this paper were met, such as designing the circuit and program of the device in both Raspberry Pi and Arduino Mega, acquiring the relative humidity and temperature of server rooms in real-time and without contact, receiving feedbacks and real-time updates via the messaging app Telegram and SMS, and making the device act as an actuator to control the up/down and on/off button of the air conditioner.

4.2 Recommendations

The proponents recommend considering the mobile service provider in the area to avoid delays in sending SMS. Since the design project was limited to a specific air conditioner, the researchers recommend that future researchers find a board with a higher storage capacity to control more brands and types of the air conditioner as the new generation of air conditioners have a more complex code. Also, with the increase of storage capacity, many new things can be added for future research. Moreover, the testing room used here is that of a mini-server room only, so there was no need for a transistor. It is suggested to implement a transistor for applications that require the prototype to be placed a distance from the air conditioner. Then, it is best to add an IR sensor to the Raspberry Pi to control the AirCon instead of relying only on the Arduino Mega's IR. Lastly, it is highly recommended to use an Ethernet cable for a more stable internet connection.

References

1. Gatlin, H.: Everything you should know about server rooms- RackSolutions. RackSolutions (2020). [Online]. https://www.racksolutions.com/news/blog/server-room-explained/?akamai-feo=off. Accessed 12 Sep 2020
2. Why Is Monitoring Server Room Temperature Important? Vertiv.com (2020). [Online]. https://www.vertiv.com/en-asia/about/news-and-insights/articles/educational-articles/why-is-monitoring-server-room-temperature-important/. Accessed 12 Sep 2020
3. Matthews, K.: Why Data Centers Are Important for the Retail Industry. Data Center Frontier (2018). [Online]. https://datacenterfrontier.com/why-data-centers-are-important-for-the-retail-industry/. Accessed 12 Sep 2020
4. Lima, J.: 10 incidents which prove the power of the data centre - Computer Business Review. Computer Business Review (2015). [Online]. https://www.cbronline.com/data-centre/10-incidents-which-prove-the-power-of-the-data-centre-4581872/. Accessed 12 Sep 2020
5. ASHRAE 2008 Applying 2011 ASHRAE Data Center Guidelines to HP Proliant-Based Facilities
6. Recommended standards for monitoring server rooms and data centers. ServersCheck (2014). [Online]. https://serverscheck.com/sensors/temperature_best_practices.asp. Accessed 01 Oct 2020
7. Alvan Prastoyo Utomo, M., Aziz, A., Winarno, Harjito, B.: Server Room Temperature & Relative Humidity Monitoring Based on Internet of Thing (IoT). J.Phys. Conf. Ser. **1306**, 012030 (2019). https://doi.org/10.1088/1742-6596/1306/1/012030
8. Cdn-learn.adafruit.com. AHT20 Product Manuals (2019). [online]. https://cdn-learn.adafruit.com/assets/assets/000/091/676/original/AHT20-datasheet-2020-4-16.pdf?1591047915

Reliability Models for Smartphone Applications

Sonia Meskini[1], Ali Bou Nassif[2], and Luiz Fernando Capretz[1](✉)

[1] Department of Electrical and Computer Engineering, Western University, London, ON N6A 5B9, Canada
lcapretz@uwo.ca

[2] Department of Electrical and Computer Engineering, University of Sharjah, P.O. Box 27272, Sharjah, UAE

Abstract. Smartphones have become the most used electronic devices. They carry out most of the functionalities of desktops, offering various useful applications that suit the user's needs. Therefore, instead of the operator, the user has been the main controller of the device and its applications, therefore its reliability has become an emergent requirement. As a first step, based on collected smartphone applications failure data, we investigated and evaluated the efficacy of Software Reliability Growth Models (SRGMs) when applied to these smartphone data in order to check whether they achieve the same accuracy as in the desktop/laptop area. None of the selected models were able to account for the smartphone data satisfactorily. Their failure is traced back to: (i) the hardware and software differences between desktops and smartphones, (ii) the specific features of mobile applications compared to desktop applications, and (iii) the different operational conditions and usage profiles. Thus, a reliability model suited to smartphone applications is still needed. In the second step, we applied the Weibull and Gamma distributions, and their two particular cases, Rayleigh and S-Shaped, to model the smartphone failure data sorted by application version number and grouped into different time periods. An estimation of the expected number of defects in each application version was obtained. The performances of the distributions were then compared amongst each other. We found that both Weibull and Gamma distributions can fit the failure data of mobile applications, although the Gamma distribution is frequently more suited.

Keywords: Smartphone applications · Software reliability · NHPP model · Software Reliability Growth Models · SRGM

1 Introduction

Smartphones are now so useful that many people prefer them over desktop or laptop computers. These mobile devices are overselling desktop and laptop computers. They are becoming a necessary commodity to many users and their prices have been decreasing. Hundreds of applications, usually suited to desktop or laptop computers, have been adapted to and carried out by these smartphones. Owing to their small size, other specific applications have been built in, ranging from simple applications to very critical ones. The high usage and trust placed in these devices and their applications make their reliability

R. Silhavy et al. (Eds.): CoMeSySo 2021, LNNS 232, pp. 364–388, 2021.
https://doi.org/10.1007/978-3-030-90318-3_31

a critically important goal to achieve [1]. Thus, manufacturers are competing to release the most reliable devices, and they have successfully achieved high reliability, in terms of hardware, by applying traditional and enhanced Hardware Reliability Growth Models (HRGM) [2]. These HRGMs have been useful for classic mobile phones. On the other hand, owing to their highly integrated software, smartphones are far more advanced devices and their functionalities far exceed those of the classic mobile phones. Therefore, increased attention is now being paid to the reliability and security of these devices.

Many companies in the mobile business, as they expand rapidly under pressure from the market and competition, do not use appropriate software engineering methods to develop their products and services [3, 4]. As a result, their software is less reliable and even more expensive than it should be. Therefore, the reliability issue is becoming as acute in the mobile area [5, 6] as for desktop and laptop computers. Furthermore, owing to the peculiarities of the Development Life Cycle (DLC) of mobile application software, the reliability issues [7] in the mobile area are likely to differ from those in the desktop or laptop area.

Software Reliability Growth Models (SRGMs) are among the tools that deal with the reliability of software applications; they have been constructed and successfully applied to desktop (classic/standard) applications. In recent work [8], we thoroughly investigated the applicability of these SRGMs to the mobile area. We applied three of the most used SRGMs to the collected failure data of three smartphone applications; our main conclusion was that none of the selected models was able to account for the observed failure data satisfactorily. Their failure was traced back to the specific features of mobile applications compared to desktop applications. Pursuing our investigation, and after collecting further failure data from many versions of the same applications, we addressed the following research questions:

(1) How do the existing successful reliability models, used to assess the desktop/laptop applications, perform when applied to the mobile area?
(2) What are the best non-linear distributions that fit smartphone application failure data?
(3) What useful information can be gained from this approach?

The rest of the paper is organized as follows: in Sect. 2 we present a brief comparative study of smartphones versus desktop or laptop computers in order to stress the main differences and their incidence for the reliability of mobile applications. In Sect. 3 we provide a short roadmap of the existing models that we will use, we describe our dataset collection, and we test the applicability of existing software reliability models. Finally, in Sect. 4, we carry out an analysis of the failure data with model distributions followed by a discussion in Sect. 5 and the threats to validity in Sect. 6. We present our conclusions in Sect. 7 and outline future work possibilities.

2 Reliability for Smartphone Applications

Smartphones have become personal devices that are used almost anywhere, at any time, and for practically everything. High expectations and trust are placed on this mobile device, and it is used as more than just a phone and the usage exceeds the functionalities of only sending SMS or making voice calls.

A few years ago, smartphone usage was limited to business. Lately, thanks to network and mobile technology improvements and progress, those smart devices have begun to gain traction and they have received remarkable acceptance in the user market. Since then there has been a significant increase in smartphone technology. Countless smartphone applications have been designed and developed. Along with this, the market has been growing rapidly, and market analysis has confirmed that it will continue growing to the point that it will exceed growth in desktop or laptop computers and oversell many other electronic devices, in particular laptop computers.

As a result, various types of customer needs must be satisfied to ensure continuing growth in this competitive market because previous studies have shown that those needs differ from one region to another and from one user to another [1]. In this situation the reliability of smartphone applications will play an important role to keep user trust in their devices.

Nevertheless, the quality and achievement levels of smartphone applications are not keeping pace with the increase in number of applications being developed. This is a consequence of the "time to market" strategy that the companies developing smartphone applications have adopted. For developers to meet project deadlines in designing these new applications, they have overlooked some development phases (especially the design phase, which is considered as the most important stage) of the DLC of the application [9]. During the design phase, multiple defects and bugs might be avoided. Hence, following a strategy of overlooking phases could result in many problems later on that might take more time and effort to solve than was used to design the entire application.

In step with this, for companies to be competitive it is important to study the market using surveys and other analysis and to understand that location is one of the important factors in the variety of user requirements. Therefore, a competitive smartphone application must meet these requirements before, during and after the DLC.

However, anticipating and meeting some of those requirements might be a challenge for developers in the design stage; the resulting failures might be difficult to solve in the execution stage since identifying and fixing the cause of the failure can be more difficult at that stage. The reason for this is that there are many factors that could result in the unreliability of the application or its failure – these include the nature of the technology used, the platform, the version of the Operating System (OS), and many other internal or external factors [9–11].

In the following discussion, we will provide a short road map of the most famous SRGMs that have been successfully applied in desktop and laptop applications, and we will check whether they will achieve the same success in the mobile area.

3 SRGMs Applied to Smartphone Applications

This section is devoted to a presentation of the application of three SRGMs that are known to be successful in both the desktop and laptop area to three concrete cases of smartphone applications.

The SRGMs used later in our experiments are: the NHPP – Crow – AMSAA model (also termed the NHPP-Power Law model), the Musa-Basic execution time model (or the exponential model), and the Musa-Okumoto model (or the Logarithmic Poisson model). The applications that have been chosen are Skype, Vtok, and a private Windows phone application.

- For the NHPP model the mean value function is:

$$\mu(t) = N((1 - \mathrm{e}^{-\mathrm{bt}}); \; where\{b, N\} > 0. \tag{1}$$

N: total number of faults and b: fault removal rate.

- The Musa-Basic model, also termed the exponential model, is given by the following mean value:

$$\mu(t) = \beta_0^E\left(1 - e^{-\beta_1^E t}\right). \tag{2}$$

where β_0^E is the expected number of failures, and β_1^E is the hazard rate.

- The Musa-Okumoto model, also termed the logarithmic model, is given by the following mean value:

$$\mu(t) = \beta_0^L Ln\left(1 + \beta_1^L t\right). \tag{3}$$

where β_0^L is the expected number of failures, and β_1^L is the hazard rate.

The reasons for our choice of these SRGMs are: (1) these are based on a few simple and reasonable assumptions, (2) they are simple to understand on physical grounds, and (3) they are implemented in reliability tools like RGA7 and SMERFS.

We present the procedure devised to collect the failure data for each application followed by the results of the application of the chosen SRGM to failure data for each application, and, finally, an analysis of the observed results.

3.1 Existing Reliability Models

As a matter of fact, hundreds of models exist but those models were chosen because, according to a Software Reliability modeling survey [5, 12], they are the most useful and successful models in the computer applications domain. Hence, one of our research goals is to check if those models will achieve the same success when applied to smartphone applications failure data as was the case for the desktop and Hardware Growth Reliability Models [2].

The purpose behind developing models is to measure and estimate software reliability. This has become an important and major target for companies because reliability has a significant effect in each stage of the DLC of an application from the design to the maintenance stages [13, 14].

The first SRGM was developed in 1972 [5, 13]. An SRGM usually results in a set of mathematical equations that accurately fit the collected failure data [12]. Any model relies on simplifying assumptions; however, some of these assumptions may not be useful in real situations [12].

Due to the limited space in this paper we will not present all of the assumptions of the different models used later but only the basic ones that are shared by all of the models. The detailed list of the different assumptions of each model can be found in [5].

The common assumptions are: "*1) The software is operated in a similar manner as that in which reliability predictions are to be made, 2) Every fault has the same chance of being encountered within a severity class as any other fault in that class,* and *3) The failures, when the faults are detected, are independent*" [5].

3.2 Datasets

We used Apple devices (iPhone, iPad and iPod Touch) crash files as well as a Windows Phone crash file as our "experimental" data. These crash files are not public, but are confidential. Hence, we will focus more on the Apple devices crash files since it was easier to collect them through a survey that was sent to different people from different parts of the world. Many of those to whom we sent surveys agreed to send us their failure data, whereas other did not.

For the Windows Phone case, we could only get the crash file report of one application due to confidentiality policies. Collecting the data was, and still is, a challenge. Figure 1 presents an example of the Apple devices crash log.

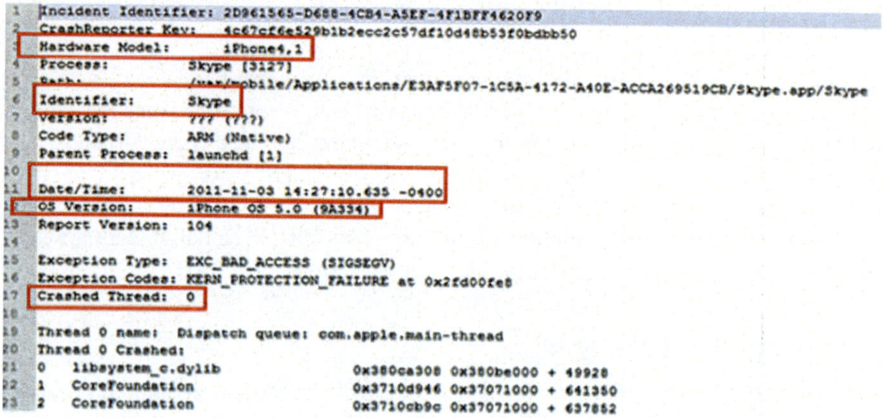

Fig. 1. Apple crash file.

The crash log is a long text file full of symbols and information that we ordinarily do not need, yet it contains information that we used to create our failure dataset. To achieve that, we developed a program in JAVA that we run each time we synchronize the devices or receive log folders from other users to update our dataset.

Figure 2 shows an example of the output file of the JAVA program developed for the purpose of extraction; in this example *Identifier* is the name of the application, *Date/Time* is the date and time of the crash, and *Crashed Thread* is the number of the thread that caused the crash.

3.3 Experiments

The reliability demonstration of smartphone applications was carried out through traditional testing, failure data collection, and the application of the most used SRGMs for

```
Identifier:         Skype
Date/Time:          2012-02-25 01:58:19.603 +0300
Crashed Thread:     0
Identifier:         Skype
Date/Time:          2012-02-26 00:15:58.353 +0300
Crashed Thread:     0
Identifier:         Skype
Date/Time:          2012-02-26 00:16:50.428 +0300
```

Fig. 2. Output of the JAVA program.

standard applications to observe and check the adequacy of these models in the mobile area.

For this purpose we used two applications for iOS and one for Windows mobile phone to test the models with different platforms. We could not collect enough data from Android phones but we are still collecting to eventually have enough data to test the models on Android applications.

The first iPhone application studied was Skype, which had been tested and used for one year (from November 1, 2011 to November 11, 2012). Hence, the data has been collected during this year with some missing values due to the occasional non-use of the application. We were, however, able to collect 39 data points for the Skype application.

The second application studied was Vtok (an application for Google talk). This application was used continuously every day for two months (from September 19, 2012 to November 25, 2012). Hence, we were able to collect failures every day (81 data points).

Each of the above mentioned SRGM models was applied to Skype and Vtok failure data, which represent two different situations: the Skype application used during one year but with some missing values, and the Vtok application used every day for two months, with the possibility of collecting more than one failure per day. This is an instance of testing the efficiency and accuracy of the models in different situations with different types of data.

During these periods, both the Skype and the Vtok applications were upgraded when new versions were released.

On the other hand, the Windows phone application was used and tested continuously for six months (from March 2012 to August 2012) by different users located in different parts of the world (more than 100 users). The crash count of the application is illustrated in Fig. 3.

We noted that June, July, and August are the months with the highest crash rate. Since this application is developed for the purpose of locating bicycle stations, it receives more use during the summer than in winter, which explains the high crash rate during the hot season. This highlights the fact that the type of application and its usage play an important role in its reliability. From this graph we extracted failure data during six months (177 data points).

We used two Software Reliability tools for this application to double check the results. The first tool is RGA 7 from ReliaSoft and the second one is **S**tatistical **M**odeling and **E**stimation of **R**eliability **F**unctions for **S**oftware (SMERFS). We configured our tools as follows: we choose 1 for the severity level of all failures and one hour for the unit.

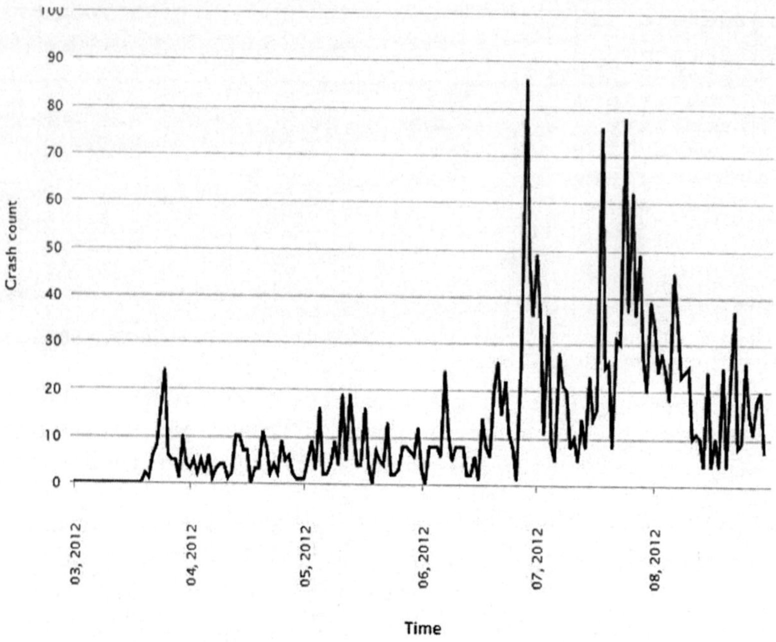

Fig. 3. Windows Phone crash count.

As the time scales of the three applications are very different, we choose to normalize our data between 0 and 1.

As the RGA tool does not accept zero value as a time to event, we entered 0.001 instead of 0 as the first value to be able to have results. For the severity, 1 was selected because the applications used are not going to cause harmful consequences if they fail. But this is not the case with other applications. When working with applications such as online banking, health, and stock exchange, etc. the severity of the failure has to be taken into consideration.

3.4 Evaluation

Smartphone application reliability is a challenge. Thus, it is a necessity for reliability methods being used elsewhere to be evaluated and to assess their validity in the mobile area. One of the main goals of this work is to check if the most accurate and most used SRGMs for desktop applications have the same success and accuracy when applied to smartphone applications.

Figure 4 presents the cumulative number of failures per time for the Skype application when applying the NHPP model. The RGA tool indicates an evident failure.

Figure 5, represents the cumulative number of failures per time for the Vtok application. Again the NHPP model failed to fit the data.

As mentioned in the previous section, we used Skype for a year and collected failure data that contains some missing values. Failure data for the Vtok application that was used for two months showed more than one failure per day. However, the NHPP still

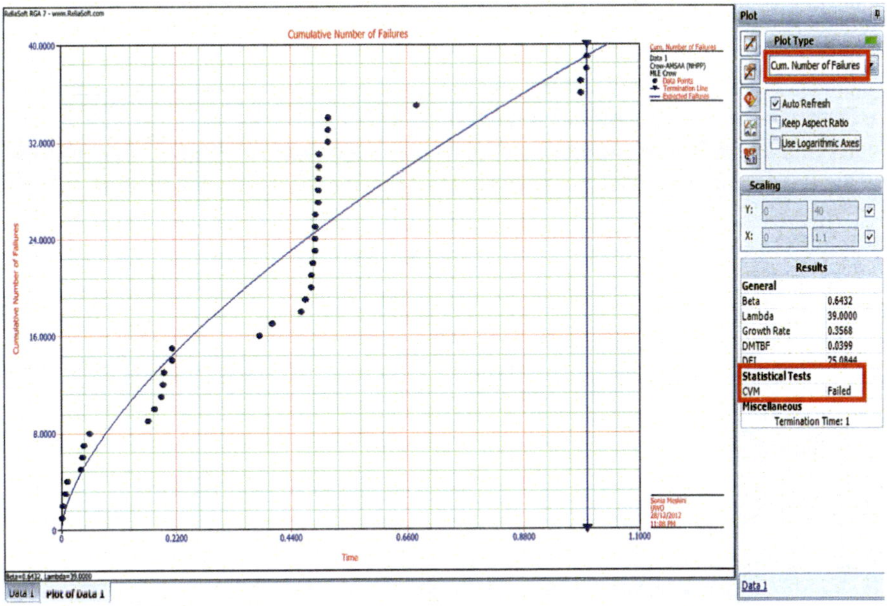

Fig. 4. Cumulative number of failures per Time (Skype).

fails to fit these two different types of data. One reason is that failure data is a dynamic

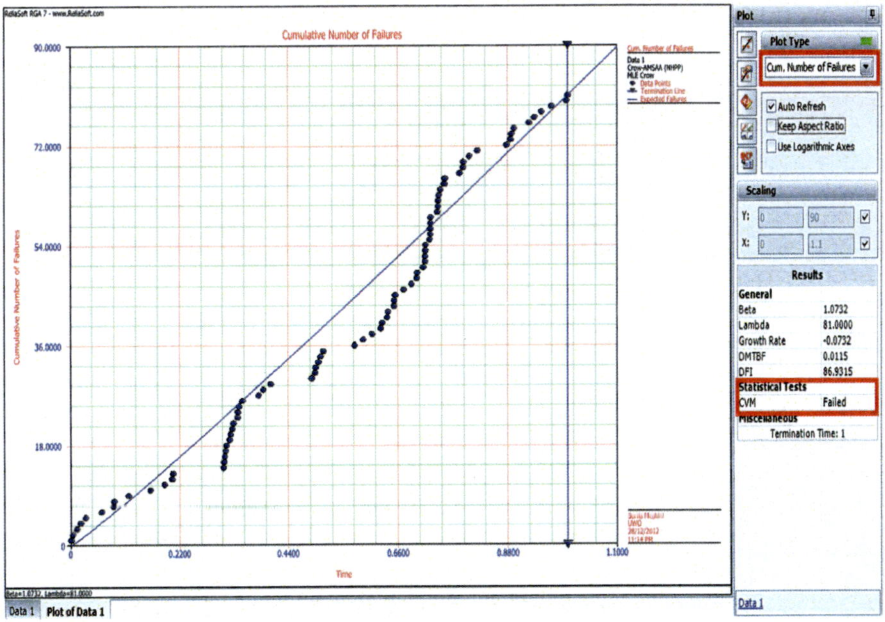

Fig. 5. Cumulative number of failures per Time (Vtok).

process for mobile applications, which means that the occurring number of failures is unpredictable, sometimes decreasing and sometimes increasing. For example, in Fig. 4 from t = 0.2076 until t = 0.3097 the application did not experience a failure and from t = 0.3097 until t = 0.3484 an important number of failures occurred.

In order to confirm our results we used a second tool, SMERFS, and we applied the NHPP model to the same data points. The result was the same – the failure of the model each time.

Figures 6 and 7, for Skype failure data, and Fig. 8, for Vtok failure data, show the results when applying the Musa-Basic and Musa-Okumoto models. Each time the models failed to fit the data. In fact, the models failed completely to fit the Vtok failure data (Fig. 8).

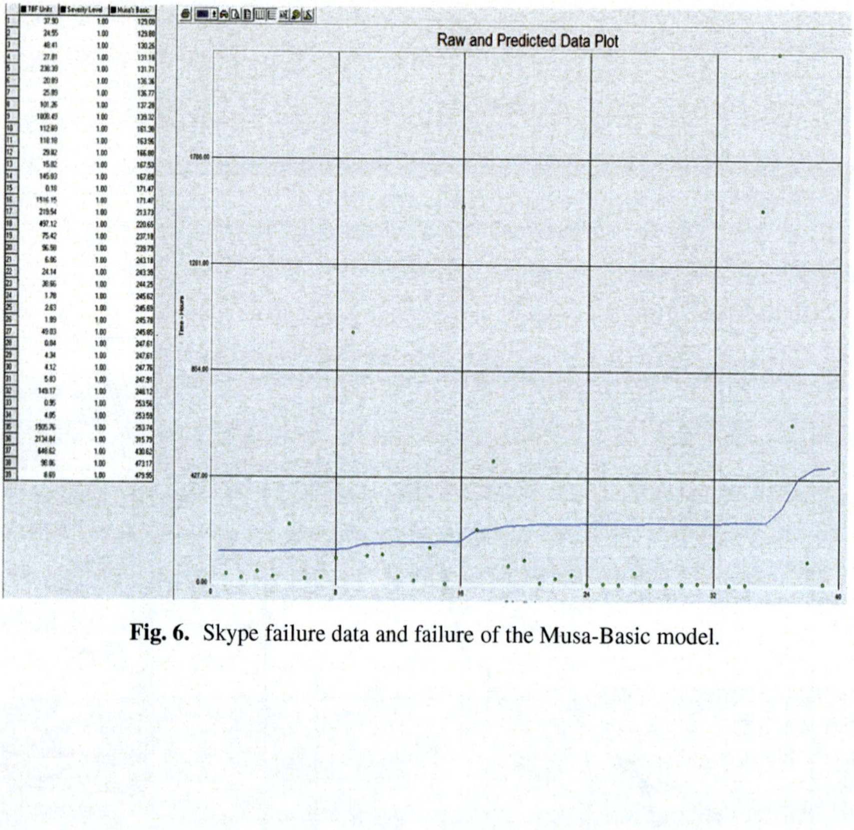

Fig. 6. Skype failure data and failure of the Musa-Basic model.

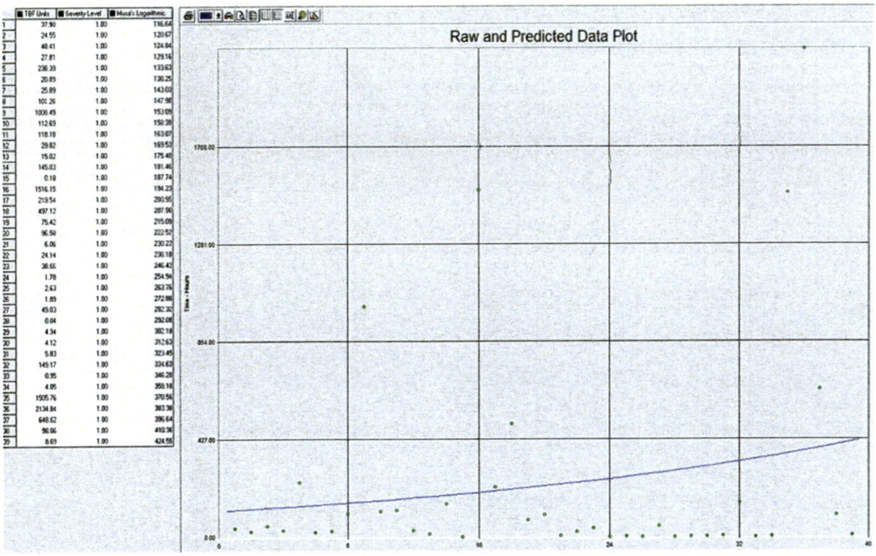

Fig. 7. Skype failure data and failure of the Musa-Okumoto model.

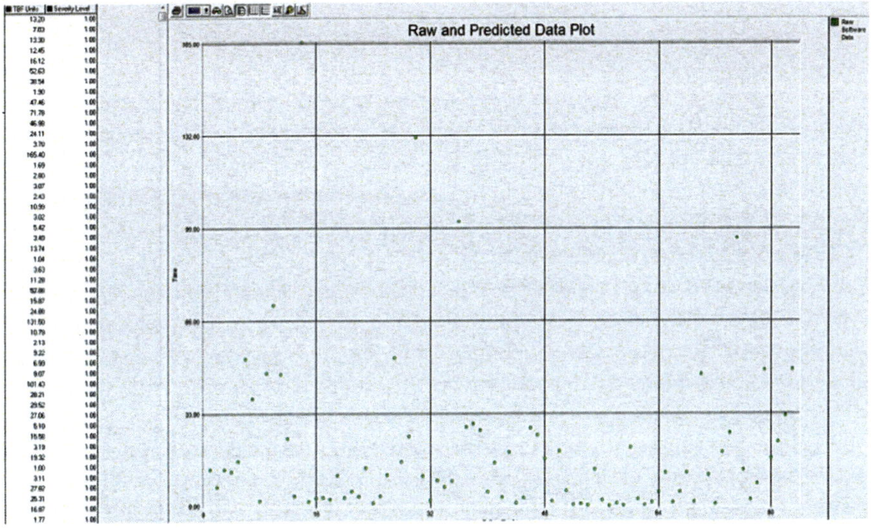

Fig. 8. Vtok failure data and failure of the three selected models.

Figure 9 represents the results of the application of the NHPP model to the Windows Phone failure data. Once again, the RGA tool indicates the failure of the model.

One explanation of the failures of the models is the goodness-of-fit tests that failed for each model application. Two tests were used: (1) the Cramer-Von Mises (CVM) test and (2) the Chi-Square test. For both tests, the statistic (test value) has to be less than the critical value in order to have a successful fit, otherwise it is a failure, as was the case

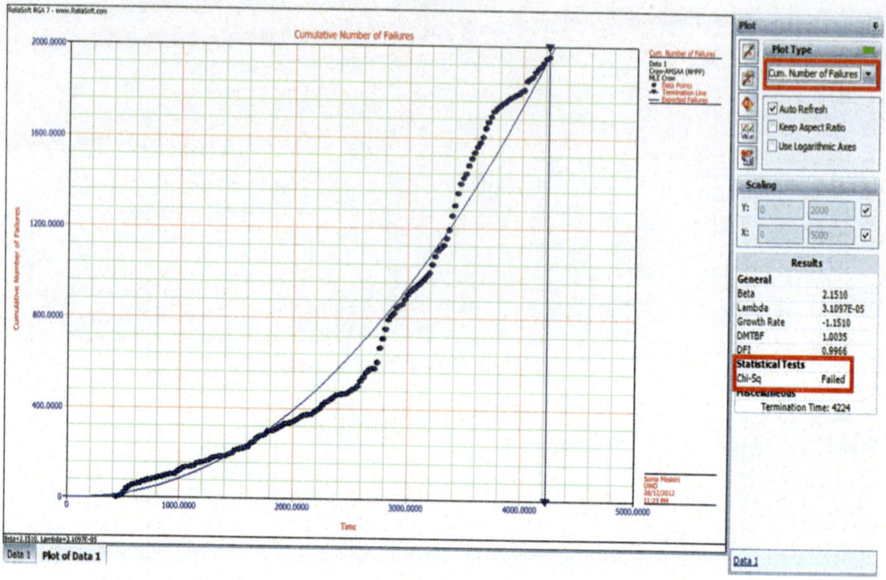

Fig. 9. Cumulative number of failures per time (Windows Phone data).

in our experiments (this was confirmed in Fig. 4 by performing the CVM test, where the model failed because the statistic (0.3588) is greater that the critical value (0.173)). We had the same result, shown in Fig. 9, when performing the Chi-Square test. The statistic (1282.0059) is higher than the critical value (180.0942), and the model did not accurately fir the data [15].

Thus, the most successful reliability models failed to fit all the data and to predict reliability in the mobile area for smartphone applications. This failure can be traced back to the main differences between the desktop area and smartphones. One of the mobile application failure characteristics is that they are application dependent, in the sense that they are dynamic and non-homogenously spread in time. Moreover, they are unpredictable; sometimes they decrease and sometimes they increase. One possible explanation is that reliability depends on how the application is used (Fig. 3 previous section), where it is used, and when it is used. The usage may differ from one person to another, from one country to another, from one condition and time to another, etc.; this explains the uncertainty of usage of the application in the execution and release time because all these factors play an important role in the reliability of the application.

Another reason is that the DLC of a mobile application is short (up to 90 days) and the programmer aims to develop the application as fast as possible to satisfy the time to market constraint, which leads to skip phases from the DLC. The phase most often skipped is the design phase, which is the most important phase in the DLC of the application [10]. Thus, it would be difficult to identify the causes of errors during the execution time and to find a convenient solution to fix them. Besides that, the failure or unreliability of the application may be caused by the technology used during the development process. The skills of the developer and the tester also play a huge role in the reliability of the application.

Moreover, the device itself and its hardware characteristics – such as the size of the screen, the performance, the keyboard, etc. – can have a direct effect on the reliability of the application [9]. For example to adjust the map size to a certain zoom level, a zoom in/out function is needed. However, to ensure a perfect use of this function, the performance of the device must be taken into consideration [9]. Based on different surveys and studies, reliability was identified as one of the most important quality attributes of the application software [6]. Thus, the reliability of smartphone applications needs to be ensured because individuals are using their own smartphones for daily life activities and tasks more than PCs. Our study confirms that a reliability growth model adapted to smartphone applications is needed, since the traditional reliability models turned out to be inefficient.

4 Failure Data Analysis Using Model Distributions

The preceding section was devoted to the application of the three most used SRGMs to two common smartphone applications, Skype and Vtok, and one private Windows phone application. The inputs to these models were the instantaneous failure data, i.e., the failure number and its exact time of occurrence. Those models failed to describe adequately the failure data. One possible reason for this is that on a real time scale, the failure data of smartphone applications fluctuate highly. Having tried several non-linear models to better fit the failure data, we found that Weibull and Gamma distributions can be used to model new collected failure data of the same application after sorting them by version number and grouping them in different time periods [16]. Therefore, we used the two mentioned distributions and their particular cases, the Rayleigh and the S-Shaped models, and compared their performances for each application. This study was carried out in two steps: (1) the failure data for each application were sorted by version number and (2) the data were grouped by larger time scales (days, weeks, and months). An estimation of the total number of defects in each smartphone application version was obtained.

As many users from different regions have responded to our call for failure data collection of smartphone applications, we have received new enriched data for Skype and Vtok applications. The data were collected synchronously and come from different versions of the applications. When we plotted the raw data on the real timescale, the obtained curves were highly fluctuating and no regularity could be detected. But after sorting the data by version number and grouping the data using a larger time scale (days, weeks, months), the relationship between failure counts and time (days, weeks, months) was represented by a non-linear graph. Each application version shows an early "burst of failures" followed by a decrease where the failures became less and less frequent. For the Windows phone application, we saw in the preceding section that the failure count curve was highly fluctuating (Fig. 3), but when plotted with larger time periods, it also presented the Weibull shape.

Based on this observation, we conducted a thorough study of the collected data [16]. Each version of each application was studied separately, when we had sufficient data for it. The versions with very few failure data were not considered, and they are evidently the most stable. For the Skype application we collected enough data for three versions whereas for Vtok we collected enough data for two versions. For the Windows phone application, we just grouped the failure data in larger time scales and modeled the failure count curves.

The Weibull and the Gamma distributions were characterized by two model parameters. Both contained the exponential distribution as a particular case. Excepting this case, their general shape went through a maximum and then continuously decreased. The Rayleigh and S-Shaped models, very often used in many reliability investigations, are particular cases of the Weibull and Gamma distributions, respectively. They were also used in the modeling of the observed failure data, grouped as previously indicated. A comparison between the models was carried out in each case based on the error criteria.

The Weibull distribution [24] is a two parameter function whose expression is given by:

$$f(t) = \mathrm{wblpdf(t\ ,\ a,\ b)} = \frac{\mathrm{b}}{\mathrm{a}} * \left(\frac{\mathrm{t}}{\mathrm{a}}\right)^{\mathrm{b}-1} \exp\left(-\left(\frac{t}{a}\right)^{b}\right). \tag{4}$$

The parameters a and b take positive values as well as the variable t. If we define $A = 1/a^b$ and $B = b$, the expression simplifies to:

$$\mathrm{f(t) = B\ A\ t^{\ B-1} exp\left(-A\ t^{\ B}\right)}. \tag{5}$$

A maximum for this function occurs at time $t = T_{max}$, such that

$$\mathrm{T_{max}^{b}} = \frac{\mathrm{B-1}}{\mathrm{A\ B}}. \tag{6}$$

The Gamma distribution is a two parameter function whose expression is given by:

$$\mathrm{f(t) = gampdf(t\ ,\ a,\ b)} = \frac{1}{\mathrm{b^a \Gamma(a)}} \mathrm{(t)^{a-1} exp}\left(-\frac{\mathrm{t}}{\mathrm{b}}\right). \tag{7}$$

for a, b and t taking positive values.

The maximum of this function occurs at $t = T_{max}$, such that:

$$\mathrm{T_{max} = b\ (a-1)}. \tag{8}$$

4.1 Results

This section presents a comparison and an evaluation of the use of the above mentioned distributions to model the failure fata of three versions of Skype (V1, V2, V3), two versions of Vtok (V1, V2), and the Windows phone application, based on the usual evaluation criteria: RMSE, Ad-R-Square, and MRE. V1, V2, and V3 are major versions, which mean that we collected the sub-versions 1.x in one major version 1. Due to space limitation, only Skype V1 and Vtok V2 will be presented from the versions we studied as well as the Windows phone application. The full and detailed results are found in [7].

For each application, the four distributions used were compared on the basis of their Root-Mean-Squared-Error (RMSE) and their Adjusted R-Square. The results of the estimated total number of defects were evaluated using the Magnitude of Relative Error (MRE).

- Root Mean Squared Error: the RMSE is the square root of the mean of the square of the differences between the actual and the predicted values and is expressed:

$$\mathrm{RMSE} = \sqrt{\sum_{i=1}^{N} \frac{\left(\mathrm{y_i} - \hat{\mathrm{y}}_{\mathrm{i}}\right)^2}{\mathrm{N}}}. \tag{9}$$

  where y_i and $\hat{y}_i$ are the actual and predicted number of failures respectively, N is the number of observations.
- The Magnitude of Relative Error (MRE) for each observation i can be obtained as follows:

$$\mathrm{MRE} = \frac{\left|\mathrm{C} - \hat{\mathrm{C}}\right|}{\mathrm{C}}. \tag{10}$$

  where C and $\widehat{C}$ are the actual and predicted cumulative number of failures respectively.
- The Adjusted R-Square: measures the proportion of the variation in the dependent variable accounted for by the explanatory variables. Unlike R-Square, Adjusted R-Square allowed for the degrees of freedom associated with the sums of the squares. Therefore, even though the residual sum of squares decreased or remained the same as new explanatory variables were added, the residual variance did not. For this reason, Adjusted R-Square is generally considered to be a more accurate goodness-of-fit measure than R-Square. The Adjusted R-Square can be negative.

Tables 1, 2, 3 and 4 give a compilation of all model parameters *(a, b)* along with the predicted or estimated T_{max} (time of maximum failure rate) and the expected proportion $(Y(t = T_{max})/C)$ of encountered failures by T_{max}. It also gives results of RMSE, As-R-Square, the estimated cumulative number of failures *C*, and the MRE shown by each model. Only the best and the second best model distributions are given for each application version. Likewise, Figs. 10, 11, 12 and 13 display a graphical comparison of all model distributions for the selected application version.

Table 1. Skype Version 1 – Error evaluation and model comparison.

Skype V1	Weibull	Gamma
Model parameters and deduced estimated values	a = 6.17 (5.26, 7.09) b = 2.82 (1.81, 3.84) Observed $T_{max} = 6$ Estimated $T_{max} = 5.98$ Estimated $(Y(t = T_{max})/C)$ = 47%	a = 6.14 (1.84, 10.44) b = 0.97 (0.21, 1.73) Observed $T_{max} = 6$ Estimated $T_{max} = 5.01$ Estimated $(Y(t = T_{max})/C)$ = 38.5%
RMSE	2.1966	2.2305
Ad-R-Square	0.6374	0.6262
C: Estimated cumulative number of failures or defects	50.54 (34.51, 66.58)	51.81 (34.32, 69.31)
MRE(%)	6.4	4

Note that the value of each parameter is given in a 95% confidence interval – particularly the estimated total number of failures *C*, which has a lower and an upper bound. Only the optimum value is used to calculate the MRE. For example, for Skype Version 1, 54 failures are collected. It can be conducted from Table 1 that Weibull (with parameters: a = 6.179 and b = 2.826) is the best distribution that models the failure data, compared to the other distributions. It has the lowest RMSE: 2.1966 associated with the highest Ad-R-Square: 0.6374. The lowest MRE belongs to the Gamma distribution (with parameters a = 6.141 and b = 0.975) with a value of 4% but based on the other evaluation criteria it is the second best distribution that fitted the data

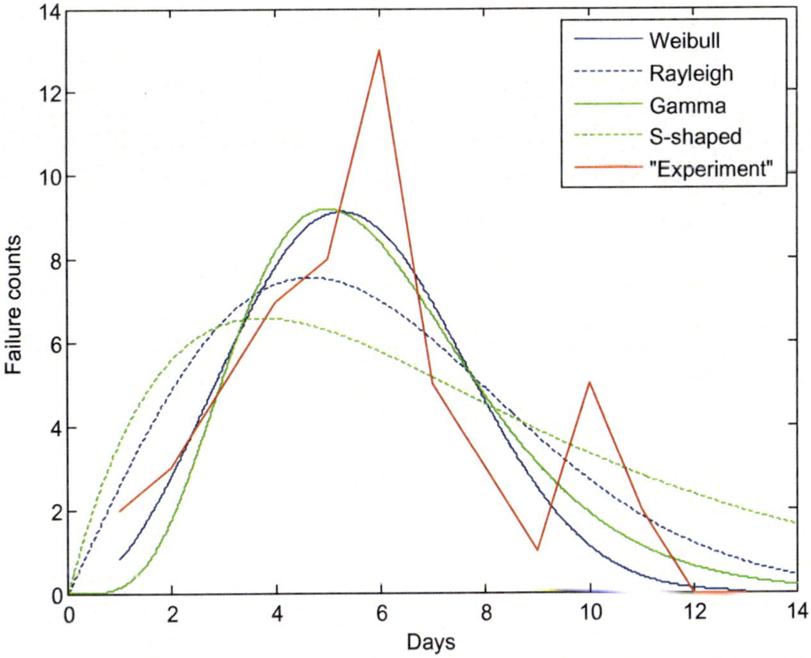

Fig. 10. Skype Version 1 – Model comparison.

Figure 10 portrays the results reported in Table 1. It can be noted from this figure that the Weibull distribution is the closest to the actual behavior curve of the application followed by the Gamma distribution. The difference between the actual curve and the modeling curves is explained by (1) the random nature of the failure event, which gives the spiky feature of the observed data, and (2) the size effect. Had we collected more data, the real failure curves would be smoother, but we still expect the general shapes to be explained by the chosen distributions because they capture the main behavior of the failure data of each smartphone application version. Besides that, the eventual modifications made to the application, its usage that differs from one environment to another and user to user [12] may play an important role in the behavior of the collected failure data.

For Vtok Version 2, 80 failures were collected in 13 weeks. Based on Table 2, it can be concluded that Gamma (with parameters a = 2.696 and b = 1.996) is the best distribution that models the failure data of this application version. It has the lowest RMSE: 2.2120 and the highest Ad-R-Square: 0.7665. The lowest MRE belongs to the Weibull distribution (with parameters: a = 5.75 and b = 1.793) with a value of 0.3%. However, based on the other evaluation criteria it is the second best distribution that models the data with an Ad-R-Square value of 0.7386 and an RMSE of 2.3401. According to the Gamma model distribution, the estimated total number of defects in the worst scenario is 100.

Table 2. Vtok Version 2 – Error evaluation and model comparison.

Vtok V2	Weibull	Gamma
Model parameters and deduced estimated values	a = 5.75 (4.63, 6.86) b = 1.79 (1.32, 2.26) Observed $T_{max} = 3$ Estimated $T_{max} = 3.64$ Estimated $(Y(t = T_{max})/C)$ = 39%	a = 2.69 (1.55, 3.83) b = 1.99 (0.89, 3.09) Observed $T_{max} = 3$ Estimated $T_{max} = 3.38$ Estimated $(Y(t = T_{max})/C)$ = 31%
RMSE	2.3401	2.2120
Ad-R-Square	0.7386	0.7665
C: Estimated cumulative number of failures or defects	80.28 (60.58, 99.99)	80.93 (61.63, 100.2)
MRE(%)	0.3	1.1

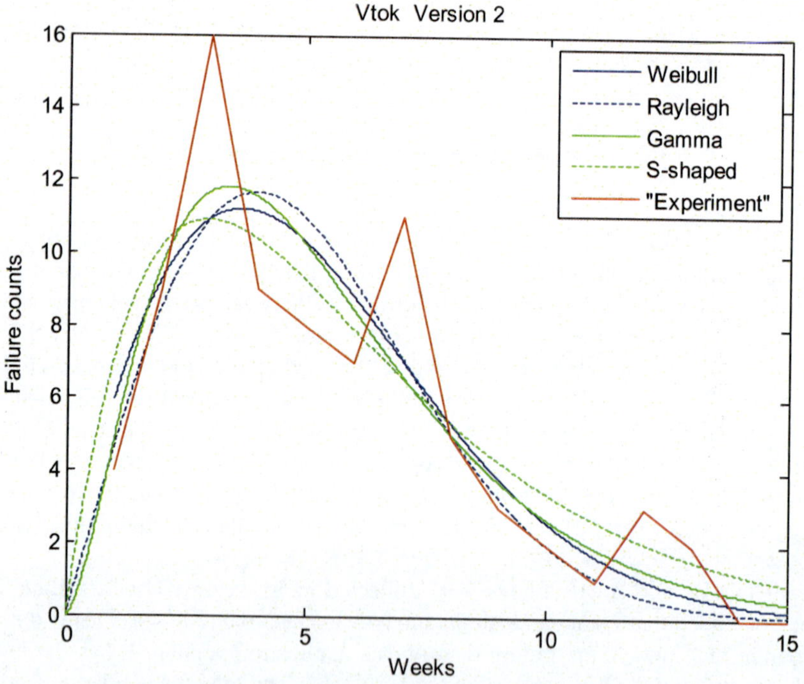

Fig. 11. Vtok Version 2 – Model comparison.

Figure 11 illustrates the results summarized in Table 2. It can be concluded that the Gamma distribution models adequately the failure data. It is to be noted that a small "burst" of failures occurred in the seventh week followed by a steady decrease.

As mentioned earlier, when plotted on the real time scale, the failure data of the Windows phone application are highly fluctuating and the time-to-event models could not adequately describe the data. Two time scales were adopted: the failure data are first grouped by week and in a second step are grouped by month. In each case, the same modeling carried previously was performed.

For the Windows phone application, it can be concluded that the Weibull distribution (with parameters: a = 22.11 and b = 6.248) performs better than Gamma (with parameters: a = 29.64 and b = 0724) but they do not differ markedly in modeling the failure data.

Table 3. Windows phone application (per weeks) – Error evaluation and model comparison.

Windows phone application	Weibull	Gamma
Model parameters and deduced estimated values	a = 22.11 (21.09, 23.14) b = 6.24 (4.47, 8.02) Observed $T_{max} = 22$ Estimated $T_{max} = 21.5$ Estimated $(Y(t = T_{max})/C)$ = 56.83%	a = 29.64 (10.49, 48.7) b = 0.72 (0.24, 1.20) Observed $T_{max} = 22$ Estimated $T_{max} = 20.75$ Estimated $(Y(t = T_{max})/C)$ = 45%
RMSE	44.2987	47.5420
Ad-R-Square	0.6112	0.5521
C: Estimated cumulative number of failures or defects	1722 (1315, 2130)	1721 (1245, 2198)
MRE(%)	12	12.05

Table 4. Windows phone application (per months) – Error evaluation and model comparison.

Windows phone application	Weibull	Gamma
Model parameters and deduced estimated values	a = 5.20 (4.94, 5.46) b = 5.42 (3.95, 6.88) Observed $T_{max} = 5$ Estimated $T_{max} = 5.01$ Estimated $(Y(t = T_{max})/C)$ = 55.76%	a = 25.48 (2.74, 48.21) b = 0.19 (0.01, 0.37) Observed $T_{max} = 5$ Estimated $T_{max} = 4.82$ Estimated $(Y(t = T_{max})/C)$ = 44.65%
RMSE	64.7462	106.89
Ad-R-Square	0.9363	0.8264
C: Estimated cumulative number of failures or defects	1831 (1420, 2243)	1767 (1081, 2454)
MRE(%)	6.43	9.7

Figure 12 depicts the results summarized in Table 3. It is to be noted that, even portioned in weeks, the Windows phone application failure data are still not smooth but the general trend is reproduced by the two models.

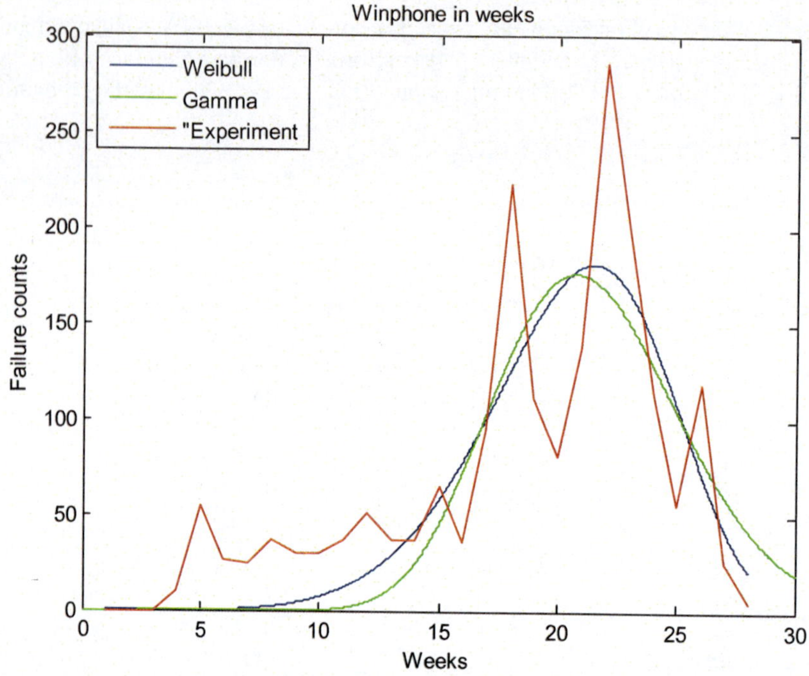

Fig. 12. Windows phone application failure data per week – Model comparison.

A further grouping of the Windows phone application failure data by months and the above analysis was carried out. Based on Table 4, it can be concluded that, again, the Weibull (with parameters: a = 5.20 and b = 5.42) model distribution performs better than Gamma (with parameters: a = 25.48 and b = 0.196).

Figure 13 illustrates the results summarized in Table 4. It can be concluded that the Weibull distribution adequately models the failure data.

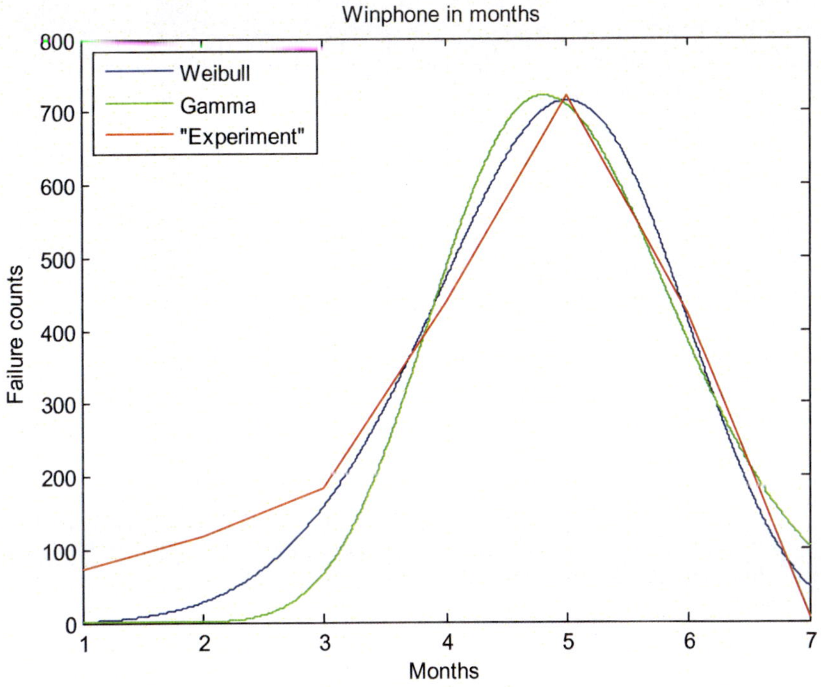

Fig. 13. Windows phone application failure data per month – Model comparison.

5 Discussion

According to the preceding section, and as an answer to the first research question raised in the abstract, it can be concluded that the most successful reliability models [17, 18] failed to account for the failure data and to predict the reliability of mobile applications. This failure can be traced back to the following main reasons:

5.1 Operational Environment and Usage Profiles of Smartphones Applications

One of the mobile application failure characteristics is that they are application dependent in the sense that they are dynamic and non-homogenously spread in time [19]. Moreover, they are unpredictable, happening in "bursts". One possible explanation is that reliability depends on the type of the application used (example the Windows phone application mentioned earlier), and on its operational environment and usage profile (where and when and how the application is used). The usage may differ from one user to another, from one country to another, from one condition to another, etc. [12], which explains the uncertainty of usage of the application in the execution and release time [10]; all of these factors play an important role in the reliability of the applications.

Another reason is that the DLC of a mobile application is short (up to 90 days) and the programmer aims to develop the application as fast as possible to satisfy the time to market constraint; this can lead to skipping phases from the DLC. The phase most often skipped is the design phase, which is the most important phase in the DLC of the application [3]. Thus, it is difficult to identify the causes of errors during the execution time and to find a convenient solution to fix them. Besides that, the failure or unreliability of the application may be caused by the technology used during the development process. Also, the skills of the developer and the tester play a huge role in the reliability of the application.

5.2 Hardware and Software Limitations

The device itself and its hardware characteristics – such as the size of the screen, the performance, the keyboard, etc. – can have a direct effect on the reliability of the application. For example, to adjust the map size to a certain zoom level, a zoom in/out function is needed. However, to ensure a perfect use of this function the performance of the device has to be taken into consideration [10].

Other reasons that may explain the dynamic aspect of smartphone applications and the various possible causes of their observed crashes are detailed in [20], which gives an idea about the different causes, external and internal, of the unreliability of the application.

5.3 Need to Re-examine the Standard Assumptions

Software managers usually base their assessment of the reliability of software and its future evolution (after release) on a simple extrapolation in time of a reliability model, based on the failure data collected in the testing phase [21]. They implicitly assume that the field operational use of the application will not differ greatly from that of the lab testing phase [22, 23]. This is the first standard assumption on which most of the reliability models are based.

This assumption is no longer valid in the mobile area as there are large variations and uncertainties in the operational profile of smartphone applications. There are as many possible operational profiles, as there are millions of users. If an application is used in different environments, its reliability may be different for each environment [12]. Therefore, the mobile feature has to be included in the initial assumptions of any reliability model suited to the mobile area in order for its assessments and predictions to be taken seriously.

5.4 The Answers to the Research Questions

Assuming all of these uncertainties, at a second stage, and in order to answer the second research question, we collected data from many users in different regions of the world, sorted them by application versions, and grouped them in different time periods (days, weeks, and months). Each application version failure data, when plotted in time periods, shows the same pattern: an early "burst of failures", due probably to the most evident defects, followed by a step decrease in failure rate. After trying several non-linear models to fit the failure data, we found that the observed behavior is better modeled by the Weibull or Gamma distributions. To answer the third research question the main features of this approach can be summarized as follows:

- For each application version, the model distributions are in fact distinguished by tiny differences in the calculated errors RMSE and Ad-R-Squared. Nevertheless, it can be concluded that no one single distribution can fit the data of all applications or even the different versions of the same application.
- As the parameters are given along with their 95% confidence intervals, it is to be noted that parameter b of the Weibull distribution, which is fixed to the value $b = 2$ for the particular case of Rayleigh distribution, has confidence intervals that include the value $b = 2$. The same can be noted for parameter a of the Gamma distribution, which is fixed to the value $a = 2$ in the particular case of the S-shaped model distribution. But most of the time, the general distribution models fit the failure data better than the particular cases.
- Similar to the famous 40% rule of the Rayleigh distribution, and independent of any application, the S-shaped distribution has a 26.4% rule. This means that by T_{max}, only 26.4% of the defects in a smartphone application will be uncovered. This can be tested on larger datasets and across many applications.
- For all the applications, the estimated and observed T_{max} are in good agreement. For the particular case of the Windows phone application only the Weibull and Gamma distributions can model the data because the value of parameter b for Weibull and that of parameter a for Gamma are greater than 2.
- It can be noted that the Gamma distribution, along with its particular S-shaped case, model more frequently the observed failure data.

6 Threats to Validity

As studying the reliability of smartphone applications is a new approach and no previous work or techniques were suggested, some limitations can become a threat to the validation of our approach, such as:

- The data collection was a challenge and insufficient data were collected due to the privacy policies of smartphone application developers and users who believed that someone could control their phone from the crash report of their applications. Therefore, the limited datasets of our investigations make a generalization of the results difficult.
- Experimentation with more elaborate data and other model distributions could be investigated. On the other hand, data from different platforms, like Android and its users, will be useful to get more accurate results.
- Only a few versions were used in this study. A study of more versions of the applications would help get a better idea whether the version failure data are independent of each other or correlated in a particular way.

- It is not guaranteed that the modifications made to the application from one version to another would not introduce more bugs or complexity. Moreover, it cannot be promised that the correction of old bugs would not result in other bugs.
- The data were collected during the operational phases of the applications. The early development stages of each application and its architecture, whether simple or complex, are unknown. Besides that, as mentioned earlier, the operational conditions and usage profiles are also unknown. All of these factors may create defects than expected.

7 Conclusions

Our work is a step toward the application and evaluation of traditional Software Reliability models in the mobile area. We selected three of the most used models that are known for their efficiency in the desktop area: the NHPP, Musa-Basic, and Musa-Okumoto models. We examined two iPhone applications, Skype and Vtok, which were used and tested differently to evaluate the models under different conditions, and one Windows phone application that whose name we didn't mention because of the company's confidentiality policies. It turned out that none of the selected SRGMs was able to account for the failure data satisfactorily.

Our study also highlighted the causes of the failure of the models and the need for a meticulous SRGM for Smartphone applications, because the existing software reliability approaches were developed for traditional desktop software applications that are static and stable during their execution. This is not the case for smartphone applications, which have an unknown operational profile, a highly dynamic configuration, and changing execution conditions. On a continuous background, the smartphone failures come in relatively short bursts from time to time, which explain the abrupt in the observed cumulative failure number curves. This particular feature cannot be accommodated by the SRGMs that were used. Thus, in order to evaluate the reliability of smartphone applications, new models, principles, and tools are needed to incorporate the underlying uncertainties of such applications [19].

Our investigation of smartphone application reliability through the use of well-known available growth models suited primarily to desktop applications is twofold: (1) highlight the versatile nature of mobile applications, their dynamic configuration, unknown operational profile, and varying execution conditions in contrast to the static and stable desktop ones, and (2) stress the need for the design of new reliability models suited for mobile applications that take into account the inherent versatility of such applications [19].

As is well known, reliability is one of the most important features of an application and great efforts have been devoted to tailor and predict it through the study of recorded failure data. A non-reliable application means dissatisfied customers, loss of market share, and significant costs to the supplier. For critical applications, such as banking or health monitoring, non-reliability can lead to great damage. Therefore, it is of great importance to ensure early detection and resolution of reliability issues in desktop applications as well as, increasingly, in mobile applications.

Our future work will focus on analyzing these selected SRGMs in more depth and trying to modify the closest one to the data and adapt in to smartphone applications. Moreover, we will check to find out if we need to have a specific model for each type of applications or if one model is applicable to all the categories, of Smartphone applications, taking into consideration the severity of the failure.

Another future project would be to evaluate the possibility of applying more than one model on the same failure data, such as the Windows phone crash count, by dividing the data into two or more categories and applying the convenient model to each category to predict the reliability of the application. Further investigations of Android failure data are also underway.

Funding. This study has been funded by the National Science and Engineering Research Council (NSERC) of Canada, grant number RGPIN-2020-04325.

Conflict of Interest. The authors declare that they have no conflict interest (financial or non-financial) related to this research.

References

1. Verkasalo, H., Lopez-Nicolas, C., Molina-Castillo, F.J., Bouwman, H.: Analysis of users and non-users of smartphone applications. Telematics Inform. **27**(3), 242–255 (2010)
2. Perera, U. D.: Reliability of mobile phones. In: Proceedings of Annual Reliability and Maintainability Symposium, Washington, DC, USA, pp. 33–38 (1995)
3. Pan, J.: Software reliability. Carnegie Mellon University (1999). http://www.ece.cmu.edu/~koopman/des_s99/sw_reliability/
4. Janicki, M., Katara, M., Paakkonen, T.: Obstacles and opportunities in deploying model-based GUI testing of mobile software: a survey. J. Softw. Test. Verification Reliab. **22**(5), 313–341 (2012)
5. Luy, M.R.: Handbook of Software Reliability Engineering. McGraw-Hill, New York, NY (1996)
6. Tamura, Y., Yamada, S.: Reliability assessment based on hazard rate model for an embedded OSS porting-phase. J. Softw. Test. Verification Reliab. **23**, 77–88 (2013)
7. Meskini, S.: Reliability Models Applied to Smartphone Applications, Master Thesis, Western University, London, Ontario, Canada (2013)
8. Meskini, S., Nassif, A.B., Capretz, L.F.: Can we rely on smartphone applications? In: Mazzara, M., Bruel, J.-M., Meyer, B., Petrenko, A. (eds.) TOOLS 2019. LNCS, vol. 11771, pp. 305–312. Springer, Cham (2019). https://doi.org/10.1007/978-3-030-29852-4_25
9. Jang, S., Lee, E.: Reliable mobile application modeling based on open API. In: Ślęzak, D., Kim, T.-H., Kiumi, A., Jiang, T., Verner, J., Abrahão, S. (eds.) Advances in Software Engineering, pp. 168–175. Springer Berlin Heidelberg, Berlin, Heidelberg (2009). https://doi.org/10.1007/978-3-642-10619-4_21
10. Wasserman, I.: Software engineering issues for mobile application development. In: Proc. of FSE/SDP Workshop on the Future of Software Engineering Research (FoSER'10), Santa Fe, NM, USA, pp. 397–400 (2010)
11. Luan, S.P., Hurang, C.Y.: An improved Pareto distribution for modeling the fault data of open source software. J. Softw. Test. Verification Reliab. **24**(6), 416–437 (2014)

12. Malaiya, Y.K., Denton, J.: What do the software reliability growth model parameters represent? In: Proceedings of 8th International Symposium on Software Reliability Engineering (ISSRE'97), Albuquerque, NM, USA, pp. 124–135 (1997)
13. Lai, R., Garg, M.: A detailed study of NHPP software reliability models. J. Softw. **7**(6), 1296–1306 (2012)
14. Capretz, L.F., Lee, P.A.: Reusability and life cycle issues within an object-oriented methodology. In: Proceedings of 8th International Conference on Technology of Object-Oriented Languages and Systems (TOOLS USA), Santa Barbara, USA, pp. 139–150 (1992)
15. ReliaSoft: Reliability Growth & Repairable System Data Analysis Reference, http://rga.reliasoft.com/ (2010)
16. Meskini, S., Nassif, A.B., Capretz, L.F.: Reliability prediction of smartphone applications through failure data analysis. In: Proceedings of 19th IEEE Pacific Rim International Symposium on Dependable Computing, Vancouver, BC, Canada, pp. 124–125 (2013)
17. Jha, K.: A Risk Catalog for Mobile Applications, Master Thesis, Florida Institute of Technology, Melbourne (2007)
18. Xu, J., Ho, D., Capretz, L.F.: An empirical validation of object-oriented design metrics for fault prediction. J. Comput. Sci. **4**(7), 571–577 (2008)
19. Malek, S., Roshandel, R., Kilgore, D., Elhag, I.: Improving the reliability of mobile software systems through continuous analysis and proactive reconfiguration. In: Proceedings of 31st IEEE International Conference in Software Engineering (ICSE-Companion 2009), Vancouver, BC, Canada, pp. 275–278 (2009)
20. Brinks, E.: Mobile application development lifecycle. http://www.socialcubix.com/blog/mobile-app-development-life-cycle (2012)
21. Geist, R., Offutt, A.J., Harris, F.C., Jr.: Estimation and enhancement of real-time software reliability through mutation analysis. IEEE Trans. Comput. **41**(5), 550–558 (1992)
22. El Yamany, H.F., Capretz, M.A.M., Capretz, L.F.: A multi-agent framework for testing distributed systems. In: Proceedings of 30th IEEE International Computer Software and Applications Conference (COMPSAC), Chicago, IL, USA, vol. 2, pp. 151–156 (2006)
23. Musa, J.D., Iannino, A., Okumoto, K.: Software Reliability – Measurement, Prediction, Applications. McGraw-Hill, New York, NY (1987)
24. Murthy, D.N.P., Xie, M., Jiang, R.: Weibull Models. Wiley, Hoboken, NJ (2004)

Non-square Multivariable System Control Case Study – Static Optimal Compensator Design and Application

Dominik Varga(✉), Daniel Honc, and František Dušek

Faculty of Electrical Engineering and Informatics,
University of Pardubice, Pardubice, Czech Republic
st49529@student.upce.cz

Abstract. By multivariable decentralized control, changing one set-point in result acts as a disturbance to other control loops. This can be solved by using multivariable controller or compensator. In this paper, a novelty approach to control non-square eighth-order system with four inputs and three outputs is demonstrated using a static compensator that guarantees autonomy in the steady state (changing one input, affects one output) and also optimal solution for non-square overdetermined systems (systems with more manipulated variables than controlled variables). To evaluate the control quality of this method, the system is also controlled without static compensator for comparison.

Keywords: Static compensator · Non-square system · MATLAB · Control

1 Introduction

Multivariable (Multi Input Multi Output) systems can be generally classified as square and non-square. The term square systems refers to systems with identical number of inputs and outputs. Non-square systems have different numbers of inputs and outputs. In this paper, a non-square system with less outputs than inputs are treated – so called over actuated system [1, 2].

The general procedure for controlling MIMO systems is to find suitable input-output variable pairs such that the manipulated variable has the maximum effect on the given output variable. However, finding optimal input-output pairs is not always trivial issue, since a single input can affect multiple outputs simultaneously [2–5].

One method of controlling non-square systems is to prepend a static compensator to the controlled system, which converts the non-square system into a pseudo-square system with the same number of inputs and outputs. There is no necessary to miss or combine some of the inputs and therefore it is possible to use the same number of controllers as the number of outputs of the controlled system. The compensator then automatically recalculates the controller outputs to the inputs of the system, based on the system gain information, and this operation minimizes the deviations of the applied inputs from the required inputs. By using a compensator, where the compensated closed-loop control system will be diagonally dominant, the output variable will be affected by

R. Silhavy et al. (Eds.): CoMeSySo 2021, LNNS 232, pp. 389–396, 2021.
https://doi.org/10.1007/978-3-030-90318-3_32

just one input variable. In this way, the multivariable system is converted into a series of one-dimensional systems that can be controlled much more easily, for example, by using single loop PID controllers [2, 3, 6].

2 Controlled System and Compensator Design

A multivariable eighth-order controlled system with four inputs and three outputs has the following block diagram. The blocks F_x represent the transfers functions of each input to the output of the system, *u* are the inputs and *y* the outputs of the system (Table 1 and Fig. 1).

Table 1. Transfer functions

$F_{1a} = \frac{0.6}{3s+1}$	$F_{1b} = \frac{0.625}{6s+1}$	$F_{2b} = \frac{0.7}{4.5s+1}$
$F_{3b} = \frac{0.5}{5.5s+1}$	$F_{4a} = \frac{0.8}{5s+1}$	$F_{4b} = \frac{0.556}{4s+1}$
$F_c = \frac{2.6315\cdot(15s+1)}{(3.5s+1)^2}$		

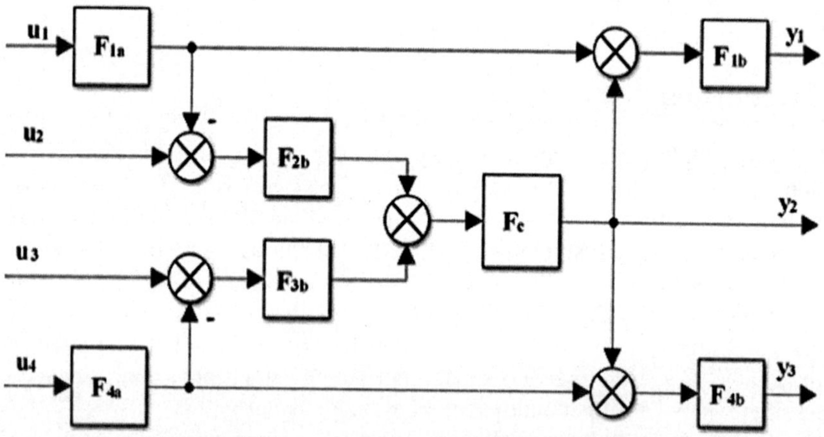

Fig. 1. Block diagram of the controlled system

For the steady-state holds:

$$y_\infty = Z \cdot u + z_0 \tag{1}$$

where Z is the gain matrix of the system and z_0 is the offset vector – we consider common practical situation that the steady-state characteristics does not pass through the origin.

There are an infinite number of different combinations of inputs that lead to the desired outputs. To solve this, we introduce a new vector u_w representing the preferred desired combinations of inputs to the system. The solution can be obtained by an optimization problem where it is desirable that the real inputs differ as little as possible from the desired inputs u_w, which means that this difference is minimal in the presence of given constraints. The optimization problem is described by equation as published in [3, 7]

$$minJ(u), J(u) = (u - u_w)^T \cdot M \cdot (u - u_w);\ constraints\colon y_\infty = Z \cdot u + z_0 \tag{2}$$

where u is the vector of inputs, u_w is the vector of desired inputs and M is the weighting matrix.

From mathematical point of view this is a minimization with bounded extreme, so we introduce the Lagrange multipliers λ and obtain modified equation in form.

$$J(u, \lambda) = (u - u_w)^T \cdot M \cdot (u - u_w) + (Z \cdot u + z_0 - y_\infty)^T \cdot \lambda \tag{3}$$

We solve this equation in the usual way, calculating the partial derivatives and setting these derivatives equal to zero. We obtain a system of linear equations which can be written in matrix form.

$$\begin{bmatrix} u \\ \lambda \end{bmatrix} = \begin{bmatrix} M^T & \frac{1}{2}Z^T \\ Z & 0 \end{bmatrix}^{-1} \cdot \begin{bmatrix} M^T \cdot u_w \\ y_\infty - z_0 \end{bmatrix} \tag{4}$$

We mark the inverse matrix as S and with two submatrices R and K.

$$\begin{bmatrix} u \\ \lambda \end{bmatrix} = S \cdot \begin{bmatrix} M^T \cdot u_w \\ y_\infty - z_0 \end{bmatrix} \quad S = \begin{bmatrix} R & K \\ X_1 & X_2 \end{bmatrix} \tag{5}$$

In solution we are only interested in the inputs of the system u, so we skip the row λ and multiply the resulting matrices. We obtain the matrix equation for the static compensator.

$$u = R \cdot M^T \cdot u_w + K \cdot (y_\infty - z_0) \tag{6}$$

The static compensator therefore converts the variables coming from the PID controllers (vector y_∞) into inputs to the controlled system (vector u) according to Eq. (6). The procedure takes into account offset parameters z_0 and the vector of desired inputs u_w [3] (Fig. 2).

3 Control Without Compensator

In the case of control without a static compensator, it is necessary to find such combinations of manipulated variables and controlled variables that lead to a sufficient process of control. Since the system has only three controlled variables and therefore only three controller outputs, only three inputs to the system can be used as manipulated variables, leaving the remaining input unused see Fig. 3.

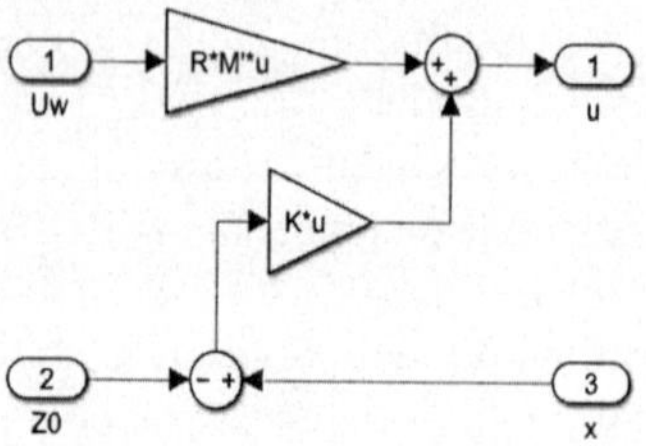

Fig. 2. Blok diagram of the compensator

The advantage of this solution is possibility to use PI controllers and classical methods of their setting. However, control without the use of a static compensator is considerably complicated, because it is necessary to find combinations of control loops and parameters of the individual PI controllers that additionally interact with each other in order to make the control system work at least acceptably. Trial-error iterative method was used to find PI controllers parameters.

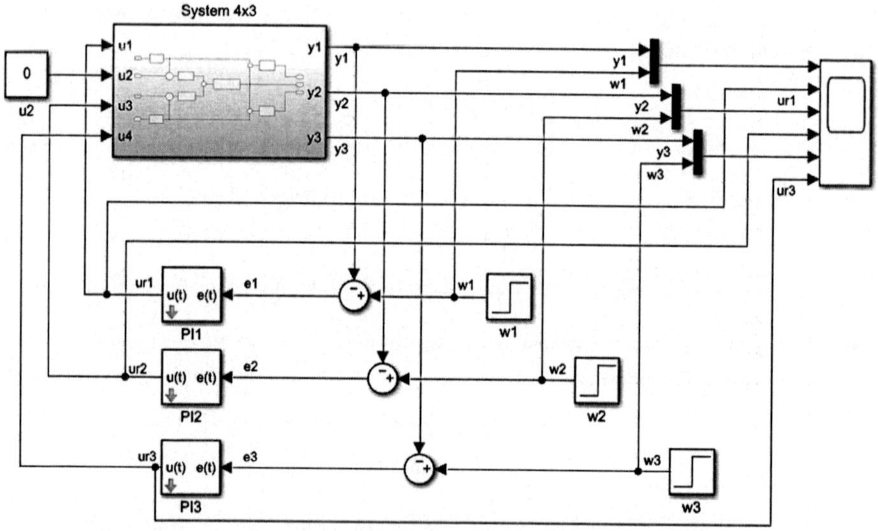

Fig. 3. Block diagram of control system without compensator

The control of the system, see Fig. 3, was simulated in MATLAB SIMULINK with the settings of individual PI controllers - parameters are shown in Table 2.

The individual control responses for controlling of the system without compensator can be seen in Fig. 4.

Integral absolute control error criterion ITAE is used to evaluate the quality of the control – see the Table 3.

Table 2. Parameters of PI controllers without compensator

	r_0	T_i
PI_1	3	9
PI_2	3	5
PI_3	3	10

Table 3. I_{TAE} integral criterion values without compensator

	The value of ITAE
PI1	2729
PI2	3570
PI3	3699
Total	9998

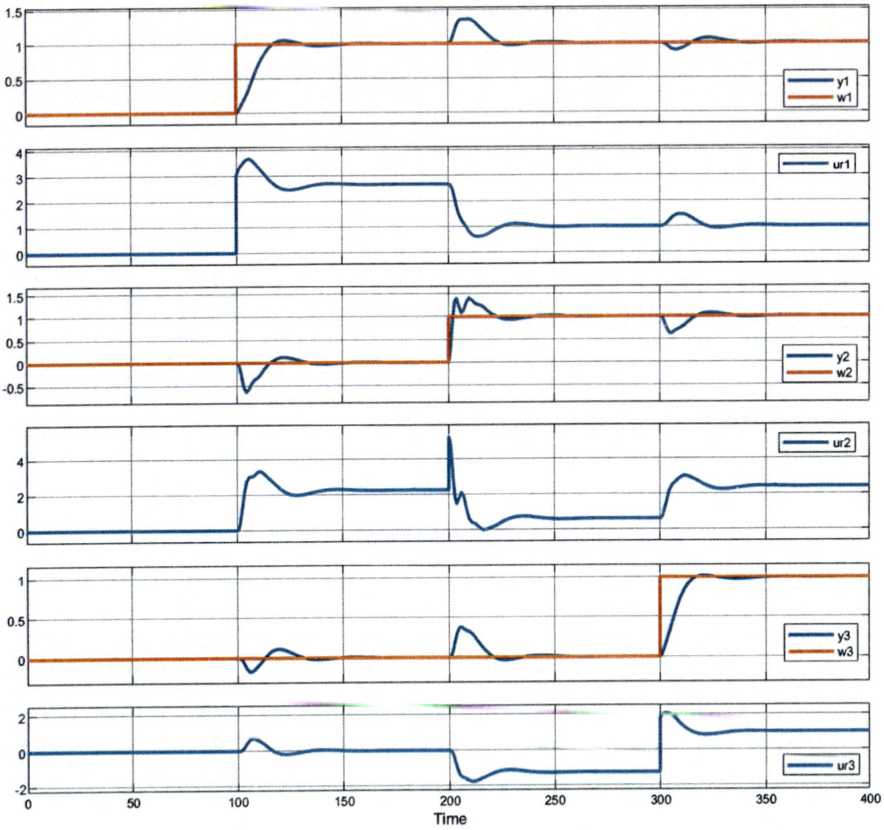

Fig. 4. Control responses for individual PI controllers without compensator

4 Control with Static Compensator

As already mentioned, static compensator greatly simplifies the control design of a multivariable system because it converts a non-square system into a pseudo-square system, therefore is not necessary to omit or combine any of the system inputs.

The compenzator generates propper number of manipulated variables obtained from the variables coming from the PI controllers. That simplifies the variables pairing for the control loops. As with the previous method without the compensator, decentralised control of the system and trial-and-error tuning of the PI controllers was used to achieve suitable control responses.

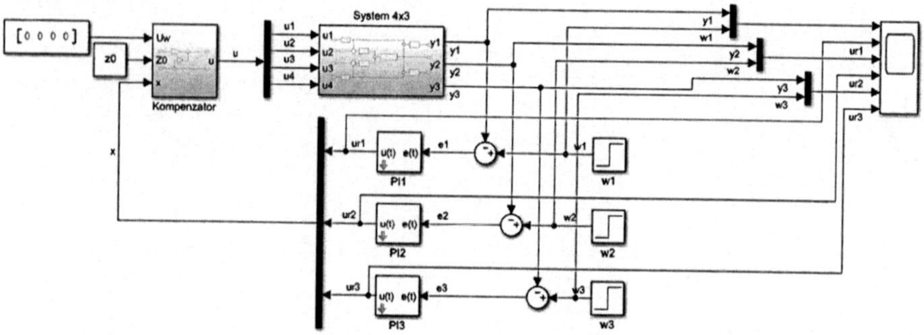

Fig. 5. Block diagram of control system with compensator

The control of the system, see Fig. 5, was simulated in MATLAB SIMULINK with the settings of individual PI controllers - parameters are shown in Table 4.

Table 4. Parameters of PI controllers with compensator

	r_0	T_i
PI_1	0,7	6
PI_2	0,07	1,1
PI_3	0,3	5

I_{TAE} criterion for individual variables is summarized in Table 5 (Fig. 6).

Table 5. ITAE integral criterion values with compensator

	The value of I_{TAE}
PI_1	3561
PI_2	9798
PI_3	6748
Total	20107

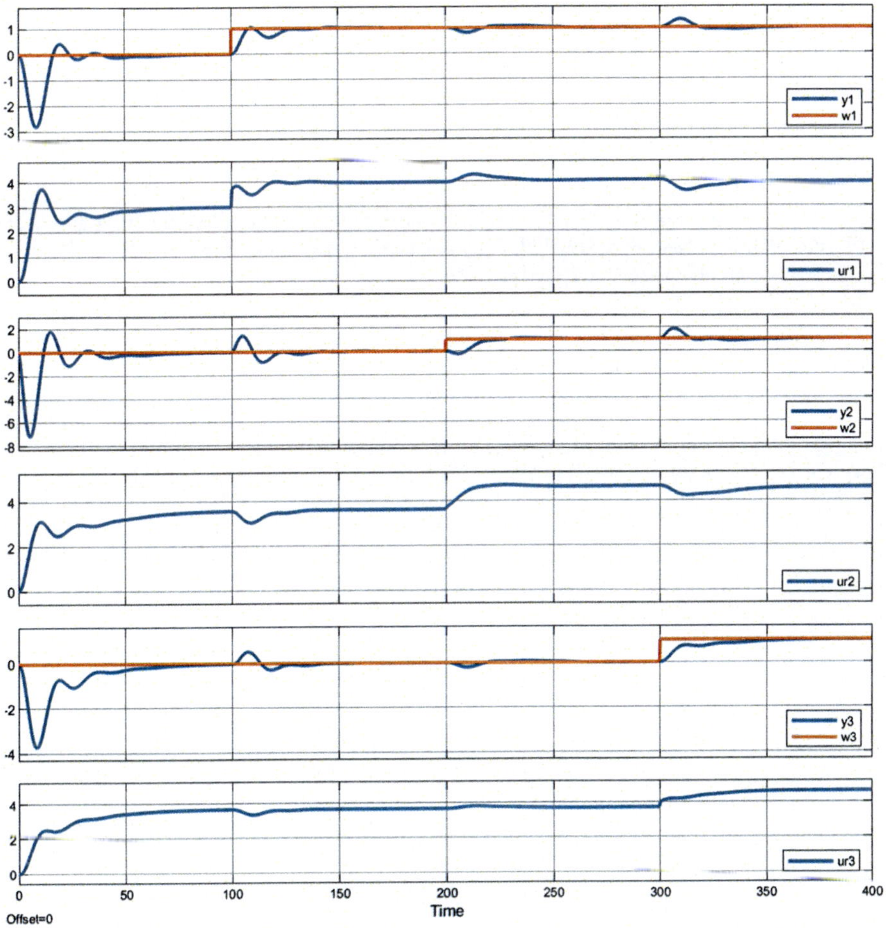

Fig. 6. Control responses for individual PI controllers with compensator

5 Conclusion

In this paper, novelty approach for non-square over actuated system was demonstrated on system with three controlled variables and four manipulated variables and static compensator. Steady-state offsets and also desired inputs are considered as a part of the optimal formulation of the problem solution control responses without and with compensator are simulated to be able to evaluate compensator benefits.

It was shown that the control of a given multivariable non-square system using a static compensator can simplify the problem of finding control loops. The controller also eliminates cross couplings in steady state, normalize the gains of the controlled system and allows to specify desired inputs in steady-state. The control process without the compensator exhibit less aperiodic response than with the static compensator. This should solve dynamical compensator or predictive controller with terminal state penalization which is authors' plan for their future research.

References

1. Kolavennu, S., Cocburn, J.C., Palanki, S.: Nonlinear control of nonsquare multivariable systems Chem. Eng. Sci. **56**(6), 2103–2110 (2001)
2. Goodwin, G.C., Graebe, S.F., Salgado, M.E.: Control System Design (Chapter 20. Analysis of MIMO Control Loops). Prentice-Hall, Upper Saddle River (2001)
3. Dušek, F., Honc, D.: Řízení soustavy se dvěmi vstupy a jedním výstupem. Perner's Contacts **3**(5), 107–112 (2008)
4. Nordfeldt, P., Hägglund, T.: Decoupler and PID controller design of TITO systems. J. Process Control **16**(9), 923–936 (2006)
5. Balátě, J.: Automatické řízení. BEN – technická literatura, Praha (2003)
6. Vázquez, F., Morilla, F.: Tuning decentralized PID controllers for MIMO systems with decouplers. IFAC **35**(1), 349–354 (2002)
7. Bequette, B.W.: Process Control: Modeling, Design, and Simulation. Prentice Hall PTP, Upper Saddler River (2003)

An Essential Intelligent Framework for Regulatory Compliance Management in the Public Sector: The Case of Healthcare Insurance in Egypt

Alsayed Abdelwahed Mohamed[1](✉), Nashwa El-Bendary[2], and A. Abdo[1]

[1] Faculty of Computers and Artificial Intelligence, Helwan University, Helwan, Egypt
[2] College of Computing and Information Technology, Arab Academy for Science, Technology, and Maritime Transport, Aswan, Egypt
nashwa.elbendary@aast.edu

Abstract. In process-centric public enterprises, the business process (BP) is at the center of value-delivery of public services to the citizens. Governing public enterprise BPs requires the ability to control and guide BP activities according to applied laws. Ensuring compliance of processes with legal regulations becomes a critical government requirement. This paper proposes an ontology-based framework for regulatory-compliance management that provides an integrated view of the law architecture model, public enterprise model, and BP lifecycle to enable the design of compliance solutions that cover the full business process compliance lifecycle, i.e., design time, runtime, and offline monitoring and analysis. The proposed framework is based on an integrated semantic model representing the business process model and administrative laws concepts which provide a shared conceptualization of the BP activities and the related legal rules. This work seeks to lay down the requirements of an integrated architecture for semantic regulatory-compliance management in the context of the BP management lifecycle. The focus of this paper is the design-time compliance approach. Our proposed design-time compliance solution was validated using a running example from the Universal Health Insurance program of Egypt.

Keywords: Business process modeling · Regulatory compliance · Law ontology · Enterprise model · Business process ontology

1 Introduction

In recent years governments and public administrations aim to digitally transform their internal and external processes and services. During digital transformation efforts of governmental services, the support of government operations and public services with Information systems is regarded as highly important. To fully develop the digital transformation of Government operations insight into the process is needed. What is the usual structure of BP? What are their input and output? What are the interdependencies between the different elements and applied laws? What should one consider when

R. Silhavy et al. (Eds.): CoMeSySo 2021, LNNS 232, pp. 397–409, 2021.
https://doi.org/10.1007/978-3-030-90318-3_33

changing these processes or input data, e.g., when introducing new IT systems? These aspects are dealt with Business Process Management (BPM) [1], enterprise architecture, and more specifically in business process architectures [2, 3] which include descriptions of the business processes and their interdependencies, or also meta-models containing an abstract view of all process elements of an enterprise. Public organizations are subject to regulations that govern their entire BPs that should be also identified and modeled. The compliance of BPs with regulations is crucial for public enterprises to achieve transparency and increase the added value that business processes present for the citizens.

To address emergent regulatory compliance needs, many public organizations develop tailored compliance solutions based on a case-by-case manner resulting in non-reusable and inconsistent solutions. Moreover, regulations require legal domain experts to clarify the ambiguities that may be impeded in the legal text which are not understandable by technical people. As a result of these problems the automation of compliance solutions at any of the phases of the BPM lifecycle; design-time, runtime, and offline monitoring is an onerous task.

BP modeling is the representation of the sequence and interactions of process activities. The core business of the public enterprise is decomposed of a set of interacting BPs that can be modeled using BP modeling approaches. On the other hand, administrative laws that represent BP compliance sources are decomposed into a set of legal policies and rules that can be modeled using ontology approaches. BP compliance can be achieved by relating BP modes with its related law ontology.

The management approach focusing on end-to-end BPs by aligning the organizational objectives and strategy with the objectives and values of its customers is known as BPM [4].

BPM addresses business strategies, objectives, vision, organizational structures, stakeholders, policies, and IT resources to understand, model, execute, control, govern, and enhance end-to-end processes. BPM is based on a defined lifecycle that consists of four main phases. First, a business process definition and design, process implementation, process execution, and finally, process evaluation. Great work has been done to provide solutions to the problem of business process compliance however, these solutions lack approaches that cover the whole BP management life cycle and organizational aspects in terms of enterprise models. Thus, in this paper, we propose an ontology-based regulatory compliance management framework that addresses the complete business process life cycle. The idea of linking law architecture, enterprise model, and BP life cycle will contribute to:

- Consistency between law models, enterprise models, and process models, as a result, law ontology, and business process ontology can be validated against each other.
- Analysis of the completeness of BP models.
- Monitoring of changes introduced to enterprise model triggered by new laws amendments and trace their effects on BP Models
- Establish semantic consistency and interoperability between law models, enterprise models, and process models domain ontology.
- Gaining a better understanding of laws, processes, and reasoning capabilities.

In the next section, we will present a state-of-the-art review of the related works. A detailed description of the proposed ontology-based BP compliance framework components including our proposed design-time compliance approach will be presented in Sect. 3. Section 4 presents the applicability of the proposed design-time approach to real-life cases; a running example from the Universal Health Insurance program (UHI) of Egypt is demonstrated. Finally, in Sect. 5, a brief analysis of our conclusions and future work are presented.

2 State of the Art

Our review of related work proposed for achieving business process compliance classifies them into three approaches namely, Design-Time compliance checking, Run-Time compliance checking, and offline compliance& analytics checking. Our classification approach is based on either the compliance checking approach using a proactive or reactive manner. Compliance Checking using Design-Time techniques has preventive nature aiming for ensuring that BP will comply with the related regulations. In this regard, during the modeling phase, some approaches assist in the elimination of non-compliant features of the process model, while others employ model testing techniques to validate the quality of the previously developed models. In [5] a language independent methodology was proposed for using labeling assistance functionalities to identify and annotate relevant ontology concepts to process model elements using a model consists of four steps. While this methodology provides automatic annotation of business processes with concepts from an ontology, it is limited to the presence of a terminologically standardized ontology. Moreover, ontologies need to be understandable to foster a common understanding of domain concepts. [6] proposed a proactive way to address the necessity for design time preventive compliance verification rather than an aftereffect runtime detective approach. The authors utilized a real-world case study to demonstrate how compliance must be examined and the benefits of including compliance checks throughout the process design stage. However, the scope of the demonstrated use case was regarding the data privacy and data access tailored for compliance with the GDPR only. [7] developed a meta-model that shows the interconnections of alternative conformity patterns with compliance processes that fulfill the same criteria. In addition, the authors proposed a method to automatically provide ideas for business process adaption to further assure compliance through the integration of alternative compliance methods. The proposed approach lacks the criteria for the selection of an alternative compliance process, in case of the existence of more than one appropriate compliance process. Moreover, the method lacks the representation of the enterprise model perspective. In [8] an approach that uses the viewpoint of data while examining structural features of a business process model was offered as a lightweight verification approach. The suggested technique creates models capturing the associated implementing states to the compliance conditions. The proposed method offered the ability for process compliance and enabled a formulation of the conditions that should and should not exist in the process. The method was implemented in Java and validated using synthetic datasets and a real-life case from the Australian telecom sector. The implementation of the proposed method lacks a visual user interface with graphical feedback. [9] addressed case management scenarios based on the fragment-based case management approach, the proposed approach divides a process into a set of fragments. These fragments can be synchronized through shared data and in addition to dynamically executed. The case models were formalized using Petri nets and demonstrated for design-time and run-time compliance checking implementation. The empirical evaluation of the method showed that rules can be checked in less than a second over a set of case models, however, the proposed method lacks the support of users eliciting, modeling, and managing compliance rules.

On the other side, Run-Time Compliance Checking approaches focus on executable BP instances by annotating BP models with compliance assertions that are aimed to be used either with executable BPs or by compliance checking engines for verification. In this regard, [10] presented a business rules compliance checking model and architecture for small and medium scale enterprises to monitor the BP process models at design time and at run-time. The method involved a logical classification of compliance requirements and analysis depending on a goal-based requirement engineering method before the verification in design time. The method also introduced the utilization of network analysis for runtime monitoring of BPs. The approach was tested on a financial institution in Nigeria. The approach needed to be demonstrated using other case studies rather than a financial business process to fully assess its usefulness in other business domains. [11] presented present time-aware business process model and discussed how to add temp features to the elements of the process, by evaluating the impact of events on temp constraints management. The proposed method classified decisions according to when they were taken and used in the process. The authors used two types of decisions based on how their results are treated to ensure proper business process execution. However, at the application level of the proposed method, it is critical to consider BPs only that can be executed in response to the previously executed activities, events, and decisions. [12] proposed a subjective logic-based predictive process monitoring approach. The proposed method facilitates the customize of both the reliability and sensitivity of the predictive system. The method was evaluated on synthetic data, and the results were compared with previous work. However, more exhaustive experiments using real-world datasets are needed to be performed as well as testing other discretization techniques are required to assess the validity of the proposed method. [13] presented an overview of the problems and the requirements for handling time in business processes and the solutions proposed based on temporal representation and reasoning. The authors discussed the challenges in temporal representation and reasoning for addressing advanced requirements of the management of business processes. [14] proposed a way of working to develop an application with known organizational flexibility for a small business domain, building on the expounded method. The proposed approach was executed in an agile way, allowing for iteration, continuous improvement, and consists of four elements: (1) creation and validation of the models of the business domain, using existing materials such as product catalogs, business process models. (2) identify the needed organizational variability for the business domain in terms of Organizational Implementation Variability. (3) mapping the essential models to the low code platform. (4) establishing the achieved organizational flexibility. However, the applicability of the proposed method did not demonstrate in the public sector domain.

Finally, Offline Compliance & Analytics approaches can be used to detect that non-compliant behavior of the business process happened by analyzing the historical execution instances of the BP and thus is a non-preventive approach. In this regard, [15] discussed several business process data-intensive operations along the process lifecycle. Moreover, the authors discussed Big Data systems-based solutions, and further improvements for such approaches were suggested. [16] highlighted the challenges of the GDPR implementation and compliance issues, especially the challenges associated with privileging the right to be forgotten to the data subjects. The process mining perspective to the problem was illustrated. The scenario was illustrated using an automotive lead generation use case. [17] presented a prescriptive business process monitoring method that translates the next activities in a key performance indicator into the next best actions. To ensure compliance with the recommended actions, the proposed method used business process simulation. The approach has been assessed with two real-life event logs and showed that the next best actions of the proposed method can overcome the next predictions of activity regarding optimization and distance from the current processes. However, the proposed technique cannot recommend an activity as the best action if the machine learning models were not observed during the training phase. [18] proposed a conceptual framework and architecture design for a data-driven Computer-Aided Process Planning solution, the study presented an individual production domain uses case and discusses the Manufacturing Execution Systems based on data-driven analytics integration scenarios details. The authors recommended the need to include in the proposed analytical process the building design figures that have been created in CAD systems. They have also disused the scientific and practical implications of process analysis in various production analysis scenarios. However, field studies are recommended to evaluate the proposed method. None of the approaches presented above address compliance checking in the context of the entire BPM lifecycle. Regulatory compliance for BP is a crucial organizational concern that requires considering the entire BPM layers including the design, execution, and offline analysis in addition to also considering laws architecture, enterprise model, design-time, run-time, and reactive compliance approaches. This point is highlighted when natural-language regulations are viewed as the starting point for any compliance checking effort. Moreover, this view will lead to a fully integrated compliance approach where the complexity of required algorithms and solutions can be reduced.

3 Proposed Framework

3.1 Framework Architecture

The proposed framework consists of four layers namely, law architecture model, enterprise model, business with each other as shown in Fig. 1. The operations of government agencies are defined and governed and by a set of administrative laws. Administrative laws define the requirements that specify which activities are allowed and which must be restricted. The structure of the administrative law is decomposed of a set of legal articles that implies a set of legal policies and a set of related rules. The law architecture framework [19] can be used for analyzing the administrative laws and extracting the set of legal policies and rules including their relationships with the business architecture,

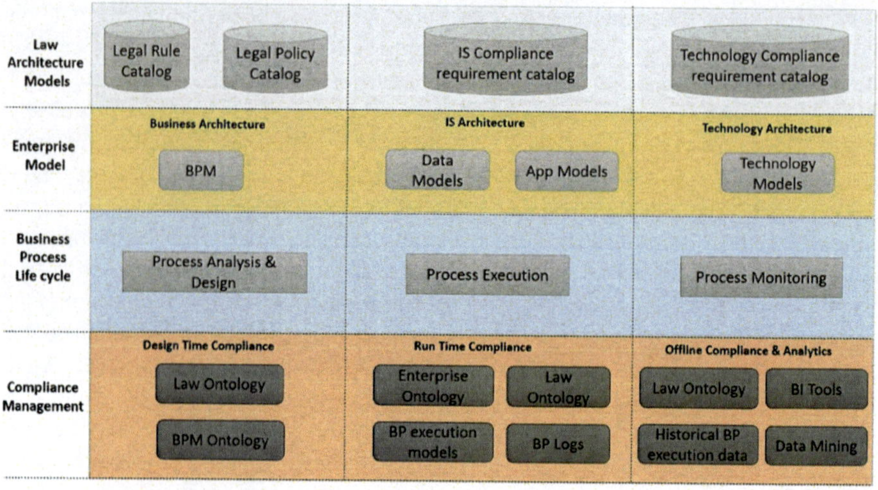

Fig. 1. An essential intelligent framework for regulatory compliance management in the public sector

information systems, and technology. The core concepts and artifacts of the law architecture are the legal policy catalog and legal rule catalog. A legal article consists of a set of statements that define a set of governmental activities and the preconditions and postconditions required to achieve these activities. These legal statements can be mapped to a legal policy. The intend of a legal policy is to accomplish a business objective. A legal policy's activities are performed by a stakeholder(s) that we call the subject of the policy, the legal policy activities are intended to deliver a certain value to a certain stakeholder which is called the target. Business capabilities or resources are required to perform the activities or rules defined in a legal policy. The set of activities and their related conditions defined in a legal policy can be decomposed into a set of legal rules. A legal rule is a set of conditional activities that belongs to one or more legal policies. We call these conditional activities actions. Actions can be classified into right, obligation, and anti-right based on the responsibility of performing or receiving the result of the action. If a stakeholder is responsible for performing an action, then this action can be classified as an obligation. If the actions implied in a legal rule tend to deliver a benefit to a certain stakeholder, then this action is classified as right, and if the action restricts a stakeholder to perform an activity, then the action is classified as ant-right. These concepts and their relationships are defined in the law architecture meta-model in a formal structure to ensure consistency within the law architecture framework and to facilitate the development of the law architecture within a public enterprise. Also, these concepts and their relationships were captured in our proposed law ontology which will be illustrated in the next section. The law architecture framework also has a process for identifying the compliance requirements of the business, IS, and technology architectures based on extracted legal policies and rules. The artifacts that can be extracted from the laws using the law architecture framework can be represented in terms of legal policy catalog, legal rule catalog, catalog of business compliance requirements, catalog of IS compliance requirements, and catalog of technology compliance requirements.

Enterprise architecture frameworks such as TOGAF [20] are widely used to define and guide the development of public enterprise models in terms of three architecture domains namely business architecture, IS architecture, and technology architecture. An enterprise model is a process-centered approach as in ARIS [4], and as such, we seek to link the BP model that is a core artifact of the enterprise model which is governed by law models using an integrating semantic model. The proposed framework is composed of four layers namely, law architecture model, enterprise model, BP lifecycle, and compliance management, that interacting with each other as follow:

- The analysis & design is the first phase of the BP lifecycle before starting this phase it is recommended to start with the law architecture framework to identify the legal policies and rules governing these processes in terms of law models. Then the analysis of the AS-IS processes and the design of the To-Be processes considering performing a gap analysis and suggesting a roadmap for transforming from the current state to the target state. The later activities are performed during the enterprise model definition activities and the complete business process model is defined as a core part of the enterprise model (Business Architecture). At this stage, compliance management practices emphasize linking the law models with business process models to achieve the compatibility between the two models based on a semantically enabled approach, as a design-time compliance approach.
- Once these business process models are verified and approved, they are deployed and executed in Information Systems (IS). IS has technical models and specifications that are defined in the enterprise models and has compliance requirements that are defined in the law architecture model and impact the way BP is executed. At this stage compliance management practices incudes, the analysis of BP execution ontology, BP logs, enterprise ontology, and law ontology models to ensure that BP is executed in a compliant mode as a run-time compliance approach
- The Business Process execution is latterly monitored by collecting variant statistics that help in analyzing the state of the executed instance of the business process. After a long time of executing BP, historical data of BP execution instances are created and stored on technology architecture, which is a part of the enterprise model and has compliance requirements. The regulatory compliance requirements of the technology architecture also are defined as an artifact of the law architecture model. The offline compliance approaches in this regard seek to utilize data analysis techniques and data mining tools to analyze historical BP execution instances combined with law ontology to discover the root cause of uncompliant BP instances.

We argue that integrating law architecture models with enterprise models and BPM Lifecycle in such a manner will achieve an integrated intelligent model that guarantees data quality of the models and compliance management across all public enterprise model domains.

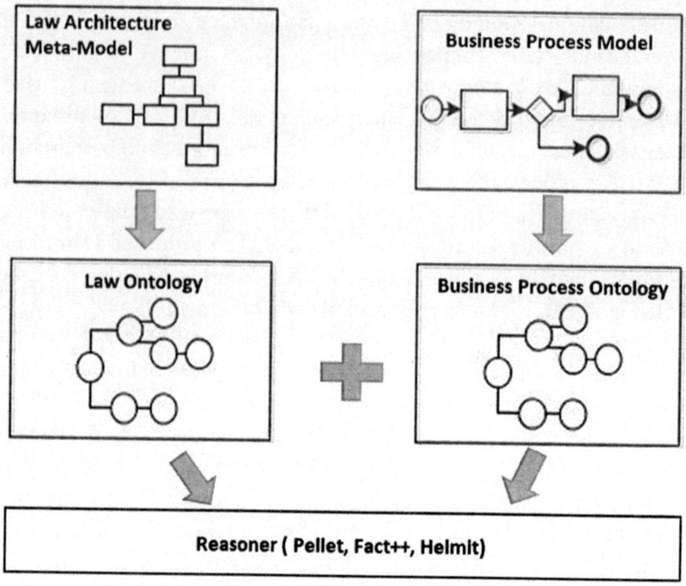

Fig. 2. Otology-based approach for design-time compliance

3.2 Design-Time Regulatory Compliance Approach

Since compliance management is an enterprise-wide problem, it is hard to cover all its aspects in a single study. Thus, this paper's focus will be the design-time checking approach aspects. Administrative laws that govern the business processes executed in the government domain are written in natural language and ambiguous. As a result, they require legal domain experts to analyze and extract the compliance requirements in terms of organization-specific requirements. A few approaches are available to guide organizations to undertake this refinement practice. We introduced in the previous section core concepts related to the legal policies and legal rules organized in the administrative laws. The below Fig. 2 depicts our proposed approach for design-time compliance that depends on transforming the core law concepts into an ontology model then extending the model to include the BP ontology model. The law ontology provides a generic and consistent structure for compliance concepts including the relationships between each other in addition to their relation to the business processes. We argue that making use of semantic technologies for representing compliance legal rules is necessary to deal with semantically process models. It also helps to make compliance checking more precise by allowing to model compliance rules at higher abstraction levels combined with related BP semantic models to cope with the ambiguity inherent to regulations. Below Fig. 3 depicts the main concepts of the proposed law ontology and their relationships.

We used a basic model for the business process domain rather than using a particular business process modeling notation. In this regard, the business process consists of a set of basic elements such as activity or task, role, organization unit, and software systems that integrating during process execution to accomplish process objectives. These BP elements are transferred into the Business Process Ontology (BPO) that represents the

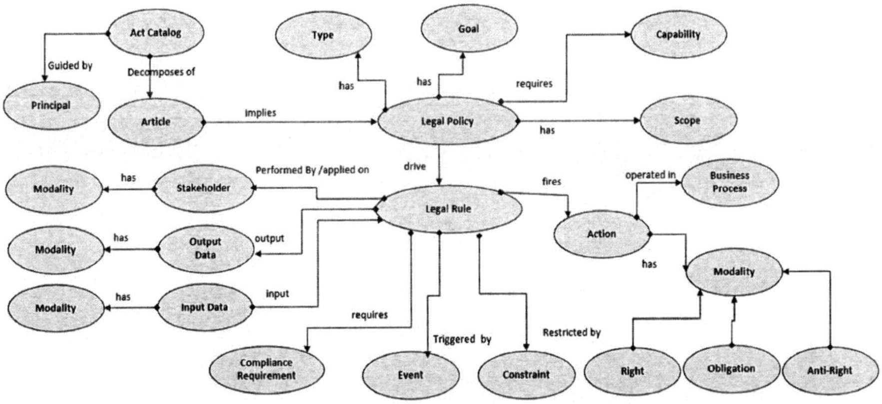

Fig. 3. Law ontology meta-model

main concepts in the BP model and their relationships. We merged law ontology and BPO into one ontology in order to check the business process compliance with legal rules provided that *pre-condition* in BPO and *action* in law ontology are defined as equivalent concepts and linked together. The consistency between the two merged ontologies is then checked by a reasoned (Fact++, Helmit, Pellet).

4 A Running Example

4.1 Example Overview

To demonstrate our proposed design-time approach for achieving BP compliance, a real process example from the Universal Health Insurance (UHI) program of Egypt was selected. The UHI project was selected as the corresponding author of this manuscript participated in the implementation of phase1 of the project, and the project also has a published law that regulates and controls each business process and activity of the project. Every single task within a business process could have different legal rules and responsibilities that should be undertaken by different governmental entities within the Egyptian government. Thus, analyzing the UHI and representing its concepts and their relationships into an ontological model then linking each mandated legal policy and rule with the related BP becomes a crucial objective of the project. Figure 4 illustrates our selected BP model for "Prepare Beneficiary Premium Invoice". The trigger of that business process is fired when a periodic timetable (monthly) is due for scheduled invoicing. The UHI officials calculate the premium amounts on factors such as family structure and income as mandated by law. The business process includes the following steps:

a) Retrieving beneficiary information to identify the family structure, household employment status, and basic salary.
b) Calculate the premium based on family structure and basic salary rules.
c) Finally, produce the invoice.

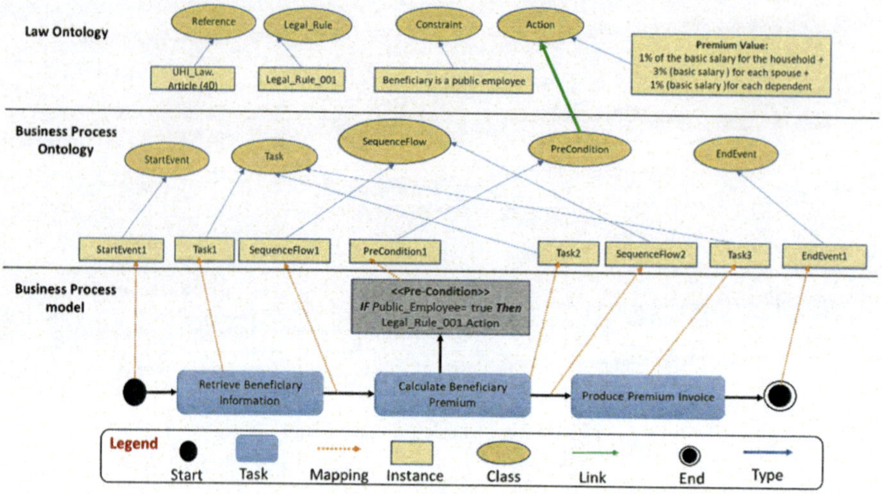

Fig. 4. Ontolog based compliance model for a real industry sample business process

In the bottom layer of Fig. 4, the process model based on UML is illustrated. On top of the process model, the figure shows the instance of the business process ontology, where each task in the BPM was mapped to the equivalent ontology class, and the relationships between these classes also are defined in the ontology model. At the top of the figure, the business process ontology is extended to include the UHI law model based on the law ontology meta-model illustrated in Fig. 3.

On the other side, the UHI project has a published law that was formally issued in 2018. The mandate of the law is to regulate the healthcare services in Egypt and organize the UHI services. According to article number (40), Table (1) of the UHI law, based on the employment status of the household the amount of the premium has a defined legal rule. For example, if the household is a public sector employee, then the amount of his premium that should be cut from his salary structured as follow:

- 1% of (the basic salary of the household), as a contribution of himself
- Plus 3% of (the basic salary of the household), for each spouse in his family
- Plus 1% of (the basic salary of the household), for each dependent of his family

For each employment status such as (private sector employee, self-employed, pensioner), the law stated different legal rule that defines a different premium amount structure. These legal rules are modeled in our proposed law ontology, each rule has a constraint and action that hold the premium structure defined in the law. To link the BP model with the PBO and law ontology, we added a pe-condition element to the BP model that can be assigned to any task of the process based on the need. The premium calculation task in the business process model has a pre-condition to check the employment status of the beneficiary and based on the status the link with the relevant legal rule from the law ontology is established based on the *has_legalAction* relation assigned to the pre-condition. Each pre-condition instance in the BPO is mapped to an action instance in the law ontology where the constraint of the legal rule of the law ontology satisfies the pre-condition in the BPO, so the consistency is obtained for the merged ontology of BRO and law ontology.

4.2 Results and Analysis

We used Protégé [21] to build our ontology models (Law and BPO) and obtain a machine-readable model that can be easily automated and analyzed as follow:

Reasoning
Description Logics (DL), one of the major Web Ontology Language (OWL) representation facilities. DL allows the use of proven inference and query techniques to analyze the consistency of the models. DL-reasoner performs various inferring services. The reasoning is used to verify inconsistencies in the models and to deduce various dependence between different business process model elements and legal models. To validate our integrated ontology, we utilized the following competency question:

> What Business *task* belongs to a business process, and has a *pre-condition* that is related to a legal rule's *action*?

It has been checked by a reasoned (Fact++, Helmit, Pellet).

Change Impact Analysis
Change impact analysis is about computing the effects of change in any part of a model on the rest of the related objects. For example, changes in laws might have numerous impacts on the related business process. Change impact analysis is mainly used to figure out what would happen if a change occurred in a proactive manner. This is particularly relevant to the link between BPO and law ontology in our running example. As indicators to assess the change impact analysis, for example, we formulated the below competency question as follow:

> Which business process and process *task* will be affected by updating *Article (40)* of the UHI law?

To answer this question, we can implement a SPARQL query that will be executed over our developed ontology model to retrieve the related business processes and tasks to the article (40) of the UHI law.

5 Conclusions and Future Work

This paper conducted a state-of-the-art review of current approaches to compliance checking within the BPM research area. According to this review, surveyed studies have made good efforts in this regard, but no approach is yet capable of covering all BPM life cycle phases. Thus, an ontology-based regulatory compliance management approach was proposed. The approach integrating law architecture models with enterprise models and BPM lifecycle while establishing their relation and traceability. In each dimension of the integrated models, several compliance solutions can be implemented i.e., design time, run time, and offline analytics. This integrated view of the public enterprise model contributes to the consistency between law models, enterprise models, and BP lifecycle, and achieves the data quality of the models. Moreover, an Otology-Based Approach for Design-Time Compliance was presented in this paper. The approach utilized the reasoning capabilities of ontology to represent complex legal policies and rules and their related actions along with its declarative semantics to validate BP compliance with the related laws, based on establishing the relationship and traceability between the two separate domains of business process and law. Traceability management primarily entails tracking compliance requirements back to their origins or forward to the BPs that implement them. Bidirectional traceability is crucial because it aids in detecting the impacts of changing regulations on BP. It also enables the analysis of why a specific decision in a business process was taken and what would happen to the process if the regulations changed. The proposed design-time compliance checking approach was illustrated using a running example from the universal health insurance program of Egypt. Future work will include proposing runtime and offline compliance solutions.

References

1. Weske, M.: Business Process Management - Concepts, Languages, Architectures, 2nd edn. Springer, Heidelberg (2012). https://doi.org/10.1007/978-3-662-59432-2_5
2. Dijkman, R., Vanderfeesten, I.I., Reijers, H.: The Road to a Business Process Architecture: An Overview of Approaches and their Use (2011)
3. Eid-Sabbagh, R.-H., Dijkman, R., Weske, M.: Business process architecture: use and correctness. In: Barros, A., Gal, A., Kindler, E. (eds.) BPM 2012. LNCS, vol. 7481, pp. 65–81. Springer, Heidelberg (2012). https://doi.org/10.1007/978-3-642-32885-5_5
4. Scheer, A.-W.: ARIS—Business Process Frameworks. Springer, Heidelberg (1999). https://doi.org/10.1007/978-3-642-58529-6
5. Riehle, D.M., Jannaber, S., Delfmann, P., Thomas, O., Becker, J.: Automatically annotating business process models with ontology concepts at design-time. In: de Cesare, S., Frank, U. (eds.) ER 2017. LNCS, vol. 10651, pp. 177–186. Springer, Cham (2017). https://doi.org/10.1007/978-3-319-70625-2_17
6. Kasse, J., Xu, L., Devrieze, P., Bai, Y.: The need for compliance verification in collaborative business processes. In: 19th Working Conference on Virtual Enterprises (PRO-VE), Cardiff, UK, pp.217–229 (2018). https://doi.org/10.1007/978-3-319-99127-6_19
7. Seyffarth, T., Kuehnel, S., Sackmann, S.: Business process compliance despite change: towards proposals for a business process adaptation. In: Cappiello, C., Ruiz, M. (eds.) Information Systems Engineering in Responsible Information Systems: CAiSE Forum 2019, pp. 227–239. Springer, Cham (2019). https://doi.org/10.1007/978-3-030-21297-1_20
8. Groefsema, H., van Beest, N.R.T.P., Armas-Cervantes, A.: Efficient conditional compliance checking of business process models. Comput. Ind. **115** (2020). https://doi.org/10.1016/j.compind.2019.103181
9. Haarmann, S., Holfter, A., Pufahl, L., Weske, M.: Formal framework for checking compliance of data-driven case management. J. Data Semant. **10**(1–2), 143–163 (2021). https://doi.org/10.1007/s13740-021-00120-3
10. Mustapha, A.M., Arogundade, O.T., Vincent, O.R., Adeniran, O.J.: Towards a compliance requirement management for SMSEs: a model and architecture. IseB **16**(1), 155–185 (2017). https://doi.org/10.1007/s10257-017-0354-y

11. Posenato, R., Zerbato, F., Combi, C.: Managing decision tasks and events in time-aware business process models. In: Weske, M., Montali, M., Weber, I., vom Brocke, J. (eds.) BPM 2018. LNCS, vol. 11080, pp. 102–118. Springer, Cham (2018). https://doi.org/10.1007/978-3-319-98648-7_7
12. Genga, L., Di Francescomarino, C., Ghidini, C., Zannone, N.: Predicting critical behaviors in business process executions: when evidence counts. In: Hildebrandt, T., van Dongen, B.F., Röglinger, M., Mendling, J. (eds.) BPM 2019. LNBIP, vol. 360, pp. 72–90. Springer, Cham (2019). https://doi.org/10.1007/978-3-030-26643-1_5
13. Eder, J., Franceschetti, M.: Time and business process management: problems, achievements, challenges. In: Leibniz International Proceedings in Informatics, LIPIcs. Schloss Dagstuhl-Leibniz-Zentrum fur Informatik GmbH. Dagstuhl Publishing (2020). https://doi.org/10.4230/LIPIcs.TIME.2020.3
14. Op't Land, M., Krouwel, M.R., Gort, S.: Testing the concept of the RUN-time adaptive enterprise. In: Aveiro, D., Guizzardi, G., Pergl, R., Proper, H.A. (eds.) EEWC 2020. LNBIP, vol. 411, pp. 228–242. Springer, Cham (2021). https://doi.org/10.1007/978-3-030-74196-9_13
15. Sakr, S., Maamar, Z., Awad, A., Benatallah, B., van der Aalst, W.M.P.: Business process analytics and big data systems: a roadmap to bridge the gap. IEEE Access **6**, 77308–77320 (2018). https://doi.org/10.1109/ACCESS.2018.2881759
16. Zaman, R., Hassani, M.: Process mining meets GDPR compliance: the right to be forgotten as a use case. In: van Dongen, B., Claes, J. (eds.) ICPM Doctoral Consortium 2019. Proceedings of the ICPM 2019 Doctoral Consortium co-located with 1st International Conference on Process Mining (ICPM 2019). CEUR-WS.org (2019)
17. Weinzierl, S., Dunzer, S., Zilker, S., Matzner, M.: Prescriptive business process monitoring for recommending next best actions. In: Fahland, D., Ghidini, C., Becker, J., Dumas, M. (eds.) BPM 2020. LNBIP, vol. 392, pp. 193–209. Springer, Cham (2020). https://doi.org/10.1007/978-3-030-58638-6_12
18. Mayer, L., Mehdiyev, N., Fettke, P.: Manufacturing execution systems driven process analytics: a case study from individual manufacturing. Procedia CIRP **97**, pp. 284–289 (2021)
19. Abdelwahed, A., El-Bendary, N., Abdo, A.: Law architecture for regulatory-compliant public enterprise model: a focus on healthcare reform in Egypt. J. Adv. Comput. Sci. Appl. (IJACSA) **12** (2021)
20. The Open Group: The Open Group Architecture Framework (TOGAF). Version 9.2 (2018)
21. Musen, M.A.: The Protégé Project: A Look Back and a Look Forward, pp. 4–12. Association for Computing Machinery (ACM) (2015)

Empowering Adaptive Learning Technology: Practical Insights from Social Network Site (SNS)

Miftachul Huda[1](✉), Arieff Salleh Rosman[2], Ahmad Kilani Mohamed[2], and Nurazmallail Marni[2]

[1] Universiti Pendidikan Sultan Idris, Tanjong Malim, Malaysia
miftachul@fsk.upsi.edu.my
[2] Universiti Teknologi Malaysia, Skudai, Malaysia

Abstract. The emerging trends of technology advancement have been clearly arisen into many sectors, one of which is in education setting. This context is substantially adopted with the wider progression through online based instruction in enabling the process in a distance basis. The need to have a sufficient knowledge in expanding the potential value of such initiative should be taken into consideration in a particular way in ensuring the process within a well organised platform and system. This article is attempted to examine the strategic approach of adaptive learning technology through getting the facilitation of Social Network Site (SNS). Several literatures were critically reviewed referring to the digital technology, adaptive learning and SNS platform. The finding reveals that there are three main points namely comprehensive way for sharing information across learning enhancement, social network site-based learning strategy enhancement, connecting significant interplay of professional and moral responsibility. This study is supposed to contribute into supporting the knowledge enhancement on digital-based learning innovation, and also may become one of the useful sources in enabling the stakeholder to adopt as the element considered for their decision making in the upgrade process of learning.

Keywords: Digital learning · Adaptive learning technology · Social Network Site (SNS) · Learning strategy

1 Introduction

In the last decades, utilising technology advancement could be clearly viewed into the number of sectors in societies at any platform initiative for instance smartphone usage in classroom [1, 2]. The particular enhancement of adopting the digital device through such platform enabled the process with an appropriate approach to transfer the knowledge in a smooth basis. With this regard, it is clear that developing the instruction tool through the technology enhancement could be deployed in a wider context in the sense that is properly transmitted along with an appropriate way [3, 4]. The further development on digital device means played a key role in distributing the wider opportunity in advancing

R. Silhavy et al. (Eds.): CoMeSySo 2021, LNNS 232, pp. 410–420, 2021.
https://doi.org/10.1007/978-3-030-90318-3_34

the extent of effectiveness together with disseminating the practices efficiency in the instruction process.

In particular to the context of teaching and learning in higher education (HE), the human interaction across digital technology advancement is widely bridged with supplying the integrated initiative of big data approach to carry out promoting the proper direction within the guideline [5, 6]. Bringing along with the digital learning enhancement through getting well prepared to facilitate the knowledge and information, attempts to incorporate the strategic practices within an appropriate approach refer to point out the worldwide to improve the quality of learning process. In supporting the integrated practices of learning enhancement in digital age through adopting the number of electronic platform, the adaptive strategy in underlying the appropriate way of instruction process might bring into following the procedure of the online learning [7, 8]. In considering this initiative to have a sufficient understanding in adopting the strategy of learning instruction, getting ready to absorb the multiple sources approached in facilitating the knowledge resources refers to incorporate the particular way in the process of learning and teaching in an online platform [9, 10]. With this regard, fulfilling the demand and supply balanced to the extent of inquiry on achieving the more properly engaged strategy refers to basically enlarged with sources in both online and print version.

In line with enhancing the facilitation of acquainting the sources in electronic sources specifically to gather the information distribution, it is necessary to point disseminating the learning and teaching (T&L) material resources through cooperating with the advantage of digital devices. The particular expansion in having the accessibility of online sources should be associated with instruction in the Higher Education (HE) context. Many studies were deployed in this area such as smartphone usage in classroom, big data in teaching, big data for innovative learning where all these refers to enhance the initiative point in facilitating the platform of online learning basis [11, 12]. While, not many tried to explore in a particular way in looking at the strategic approach on adaptive learning technology through Social Network Site (SNS) enhancement. As such, this study is attempted to examine the strategic approach of adaptive learning technology through getting the facilitation of Social Network Site (SNS). Several literatures were critically reviewed referring to the digital technology, adaptive learning and SNS platform.

2 Literature Review

2.1 Between Digital Learning and Adaptive Learning Technology

With the authentic style of implementing the learning instruction, the basis of genuine point in the activities of learning refers to make them to enable in practicing the competence skills in the environmental sphere. In similar, the competence skills in delivering the learning commitment need to transmit in expanding the accurate actions into the actual responsibilities in ensuring the individual ability in expecting what to do in line with the expertise and experience to go further in the related job [13, 14]. Moreover, the social concern in determining the way on taking a note in the communication basis refers to enhance the accessibility of learning enhancement with SNS approach. With this regard, it is necessary to conduct the subsequent phase in deliberating the social media as an attempt to encourage the commitment in enabling the accuracy for instructional tool in the learning process.

As the significant factor for medium in communication, SNS with its featured platform is potentially determined to enable the authentic learning through peers, supervision, coaching and monitoring appointment. As such, the need to incorporate the attitude in ensuring the learning program by providing the continuous access to the relevant information about the research finding and also applied models of teaching and learning enhancement [15]. In terms of the forms of simulated data within the real-world assigned into the appropriate style of format, the various categories of advice in transmitting the improved involvement in encompassing the formal instructional learning basis to informal instructional learning across the classroom.

In addition, providing the flexibility in the accommodation across instructional tool in the learning enhancement should bring along with the active involvement in participation. With the greater flexibility in participating to join collaboration in the learning process, digital device might become an instrumental point in accommodating the raise of interaction within the team management performance [16]. In facilitating the learning enhancement, the attempts of collaboration linked amongst other peers could be deployed within the digital approach in enhancing the communication sharing and engaged interaction between learners and instructors [17]. As the digital platform within the digital tool, providing the notices together with updates on the way to expand the strength across interactive relations could be transmitted to connect within online basis [18]. Attempts to connect the multiusers in online events might play a key role in driving the true practice in promoting the interactive atmosphere engaged into enhancing commitment and motivation.

Through engaged awareness on expertizing the social and individual concern, promoting the active involvement on SNS platform is entirely pointed out expanding the learning style in supporting the community basis. In using digital online basis, the number of plans assigned into the team based learning activities could give the mutual support amongst the groups in making available in the basis of creating the learning enhancement in complementing the accessibility to achieve the planned goal [19]. Moreover, encouraging the feedback through credibility of information and openness basis needs to take into account about the real-time access such as video, links, audio.

In line with providing the continued support of SNS in determining the setting of certain approach through social media platform as the instructional tool, the need to yield appointment of such a way in teaching basis should be taken into consideration with the digital integration skills. Providing the continued support together with encouraging the commitment in sharing the work in social basis refers to perform the digital expertise including multimedia presentation, blogging adoption and digital-based slide presentation [20]. In this view, there are some other platforms which can be performed in attaining the digital online-based instruction where the users can use it properly for their learning and teaching aid, apart from some challenging issues [21]. As an attempt to enhance the initial value of working together in determining the variety of learning functionalities, it might be incorporated in promoting the interactive, creative and informative basis on digital-based instruction.

With this regard, the feeling of having appropriate essence into adopting some virtual instruction including the smartphone, web-based instruction platform, and also online mode should be properly engaged into the social and personal interest on certain subject of material [22]. As such, the preparation among the teachers in their pedagogy approach needs to have the similar accessibility together with the committed awareness on utilising the internet access.

2.2 Towards Social Network Site (SNS)-Based Learning Platform

The initial value of SNS platform as the learning instruction tool is potentially an online basis in enabling all the users to have their own public profile. In this view, SNS platform is usually created with allowing the user feedback from their interaction member so that taking the benefit and avoiding the harmful point could be achieved in particular way [23]. Through incorporating the SNS platform, it is necessary to bring along with getting involved into the users in sharing and connecting amongst the other online members. At this point, the SNS platform pertaining to the online based-media interaction needs to take into account in creating the public-based profile properly within the standard context of digital humanity. With this regard, SNS refers to enhance the online platform in allowing the other users' member to have a simultaneous connection in sharing knowledge and also distributing the information [24]. As an attempt to incorporate SNS with its online platform in distributing the users' connection amongst others, it is important to take a note in determining the proper way on creating the simultaneous link with others. With having such SNS platform, the linked enhancement of the users refers to point out disseminating the information along with confirming the feedback in looking into the particular connection of applying for the electronic tool.

In addition, the connection of gaining the SNS with its particular aim in transmitting the beneficial value amongst the users across the number of platforms such as education sector, business platform, and social interaction network refers to play a significant role in distributing the sites usually bringing together with a new user input through the list of people with whom they share a connection and then allow the people on the list to confirm or deny the connection [25]. Through developing the SNS platform, the adoption of mobile application could be deployed into looking at the whole context whether positive and negative impact is potentially engaged to the digital platform framework.

In this view, transmitting the web-based application could become the initial trial-based tool in referring to perform the mutual basis in determining the direct instruction through the electronic tool provided through framework basis [26]. On the basis of gaining the natural outline providing the direct usage to deploy the electronic device, this approach can give insight into communicating towards the accessibility of data with this platform. The next approach to give insights into accessing the browser content refers to enhance the mobile-based application combined with the website basis.

In terms of incorporating the basis of crossline-integrated software applied for linking to others in sharing information, the approach of SNS with its theoretical framework on the way to enlarge the mobile basis should bring along with expanding the integration context about the certain information including venue, commitment and instruction [27]. With this regard, the nature point in enabling the users to adopt the native software of a mobile social networking framework refers to give an insight into the mobile-based socially enhanced line.

In expanding the possible inclusion on enabling the overall incorporation of material linked between usages between instructions, enhancing the attainment of using data based on pointing out the number of platform like Twitter, Facebook, LinkedIn, etc. With aiming at sufficient contribution to facilitate in building the communication platform in the attempts to share information into others, the purpose of social and personal basis refers to incorporate the intelligent system through the development of mobile-based community line [28]. The way of incorporating the basis of SNS in using the communication purpose should be transmitted through the set of nodes to represent the users together with its particular phase of interaction platform.

Through enhancing the SNS platform in disseminating the information along with the framework of social network, the general representation is used commonly in embodying the display of transmitting the set of stylish package on representing the number of users in determining the particular category in interacting with stating the explicit points into the other users within the mutual link for instance in the joining group, friendship appeal, commenting the certain post in social media and soon [29]. In terms of connecting with others determined to enhance the particular connections on implicit platform, analyzing the basis on mutual relation to transform the approach of interaction amongst others needs to take into consideration with embodying the digital human expert.

3 Methodology

Examining the adaptive learning technology should be enhanced with the further progression of digital platform of social media. The initial attempt to incorporate the exploration of SNS refers to expand the further development mainly in the learning process in enabling the students and teachers are in a widely engaged sustainability. Through attempting to investigate the advanced progress of SNS into the number of platform, such medium will give easily to the instructor in transferring the knowledge in an online basis. As such, several literatures were conducted by selecting the number of resources on adaptive learning technology and SNS platform. Moreover, Deployed with reviewing the critical literatures to have a look towards the particular contribution for strategic learning innovation, expanding the particular value of gaining the significant contributions of adaptive learning refers to give insights into playing a widely significant platform in enhancing the upgraded progress from the social media usage amongst the learners and instructors. The further investigation was then forwarded to point out the technology advancement adopted using the digital device, together with the social media. The approach of this investigation was substantially based on descriptive analysis through a critical discourse towards the recent works related to the proposed research issue and main topic. Moreover, investigation method organized in line with deploying the highlight of the main topic has been assigned into the main topic to structure the finding specification.

4 Analysis and Discussion

4.1 Comprehensive Way for Sharing Information Across Learning Enhancement

The potential basis of SNS in giving access on users' tools in connecting the users across multiple backgrounds refers to build the awareness on the important responsiveness towards the challenging issues, since the way to do wisely is significant to have sufficient information related to the subject. In this regard, the enhancement process should be empowered in particular basis through transmitting SNS platform in disseminating the users' competence skills, sharing and connecting with others' information together with running business. The number of factors to lead to enhance the continuance of adopting SNS refers to expand the particular extent of communication skills, conversation experts, and sharing information through the number of digital devices [30]. It is necessary to note that the commitment to use SNS platform needs to bring along with continuing the accessibility in providing the several services such as games, videos, music, and pictures. The influential continuance on enjoying SNS platform refers to incorporate the accessibility of sharing information, entertainment, and social connection. With providing the virtual interaction continued into the use of SNS adoption, generating the information content and news dissemination needs to go through providing the active collaborations to exchange with the round acquaintance.

In addition, looking at the way on expanding the information contents needs to have a confirmation from the sources credibility assigned into updating the moral responsibility together with gaining the professional accountability. In making the form of information credibility, ensuring the source of what to read along with reflecting inside is attempted to have a sufficient adoption of SNS in a proper way on generating the extensive point of content, news, and information to give insight more significant into the users [31]. The role of SNS in giving the platform for conducting the online communication plays a significant role in enhancing the various users with the wide range of background, since then it could make possible to create the collaboration opportunities as an effort to lead to the understanding level towards the certain information shared amongst its users. Apart from that, the significant point in the learning context may be prepared with acknowledging the potential value like business sectors, distributors, potential customers, and also other services.

4.2 Social Network Site-Based Learning Strategy Enhancement

In the enhancement process of pointing out the social network-based community engagement, it is necessary to have a sufficient contribution to advance the users' skills to enable them in maximizing their digital device. Amongst such content through the number of messages, attempts to have a mutual contribution on how to operate wisely should bring along with providing the appropriate manner referring to the actual context. In this view, the information source could be achieved through getting an appropriate way in following the significant point of online communication as an attempt to incorporate the necessity in determining the collaborative assessment [32]. In particular, the number of users across diverse background may have a similar point in running the digital communication such as running the business, by searching for their customers with digital marketing. Moreover, promoting such attempts need to have a sufficient knowledge on how to deliver their product or services into their customers properly within a wise manner. Apart from this, the skills on learning what to do and how to do wisely within the appropriate manner should bring along with promoting the information relevant to giving such understanding in order to maintain the information [33]. With this regard, the necessity to promote in monitoring the information relevant to the communities in the basis of demand has to go through looking at the potential contributions in understanding the way to do in line with the condition and circumstance. In terms of the particular promotion enhancement in underlining the online transaction, both knowledge understanding and experiential basis should be balanced in the harmonious sphere by accommodating the featured condition.

In the attempts to promote the way of learning enhancement together with its significance for online learning sources, the need to upgrade what to do in terms of developing the context of knowledge understanding through generating the technology experts together with professional and moral responsibility is a priority amongst the users in transmitting the information [34]. As the potential value in promoting the also promote a means of learning for users underlining its significance to accommodate the features of social networks through developing online learning resources. In enabling the users mainly amongst the students to have a sufficient access in generating knowledge information, attempts to incorporate the technology enhancement in controlling the benefit of SNS should bring along with advancing the skills enables the students to generate contents of knowledge to extend advancing technology to maximise their benefits together with gaining the control of operating smart-based learning activity [35]. In particular, critical analysis of social network could point the sufficient consideration in driving the right path on giving the pattern in underlying the interaction across the environmental condition. In exchanging the knowledge understanding and experienced skills, the communication basis assigned into the social support has to go through bringing along with professional capabilities and also moral responsibilities.

4.3 Connecting Significant Interplay of Professional and Moral Responsibility

The mutual link of SNS platform in connecting with the significant interplay of professional practice is related to the trend of search engines used into the online experts such as google meet. With enhancing the sources primarily into incorporating academic information, the online task may need to expand the accessibility of promoting the authentic basis of learning enhancement [18, 36]. It is true that getting the authentic basis on the way to run the process of online practice should bring along with the professional and moral responsibility as an attempt to ensure the wise basis amongst the users, in this case amidst the learners. With this regard, the wide connection in communication basis should refer to the engagement of online platform in accessing the information relevant to the subject course, so that the target of learning in enabling the good flow with the process itself could be achieved in particular way [20]. In this view, the extent of online platform in an effort to perform learning enhancement needs to have sufficient information in the way to promote the SNS basis on supporting the academic practice for instance research [37]. In the valuable insight of attempts in expanding the SNS performance, the wide perspective of providing the access with the proper format could be viewed into research outcomes, applied models, and also additional procedures of simulated data within the proper arrangement.

In addition, the authentic performance of learning enhancement needs to have a mutual line in expanding the accurate access on getting involved into the outstanding tasks in directing the need to implement coming from the adoption, transformation, and interpretation towards the certain information [38]. The requirement on using the online platform in this era is significantly necessary in mastering what to do in terms of knowing the way of learning. In this view, the transmission of knowledge understanding about the online practice is widely performing the initial basis of applying for the information gained through sharing information together with interacting and connecting into the professional expertise of course fitted into the relevant fields [39]. Moreover, the learning activities with performing the SNS basis may have an initial value in empowering the knowledge information about the certain issues combined with strengthening the contextual background, and thus this might need to bring along with creating the adoption and sharing the perspective through variety of idea converted into presentation, tagging and also blogging basis [40]. In this view, the natural expansion on achieving the available access in ensuring the learning experiences to go well through SNS platform should bring along with enhancing the structured skills of learning experiences. It is believed that attempts to perform the learning enhancement together with expanding the SNS platform might need to incorporate the experiential basis through the service basis about the teaching mode, since the responsibility engagement is potentially raising into the learning experience.

In the attempts to advance the digital learning enhancement, disseminating the numbers of individuals in using SNS needs to upgrade in updating the information acknowledged from the public basis where the general users can have a similar access in dealing their daily routine activities [41]. In particular, getting update about the information in distributing the online transaction refers to maintain the regulations and also people keep using SNS due to get updated information from the public. SNS for community help them in dealing with uncertainty that takes place in society [42]. The uncertainty motivates people to seek that information, and currently people tend to find the updated information from SNS because SNS allows community to gain access to point on the ground information about risk related issues or uncertainty very quickly [42]. Consequently, the SNS has become a heavily used source of information in reduction of risk-related uncertainty, information source in disaster and risk times. SNS patterns can be used to provide an understanding what kinds of information, objects, communications are exchanged among network actors, and how the flow of these exchanges supports overall goals.

5 Conclusion

This study focuses on addressing the learning enhancement with the SNS adoption basis to have a principal framework to establish the nation worldwide of technology in expanding the learning environment. In this study, the necessary act in incorporating the reflection aligned into pointing out the variant of pathways of learning motivation through SNS adoption is discussed on the learning motivation. It has to do with expanding the knowledge understanding amongst the digital users and also instruction basis in ensuring the flow of knowledge transfer. Moreover, addressing the behavioral management should be committed with SNS adoption. In looking at the particular contribution for learning enhancement, critical review of several literatures are deployed in the attempts to expand the initial value of SNS to play a significant role in developing the recent trends of technology development and transmitted into the social media basis. Amongst the committed awareness as the basic point of learning enhancement within the digital learning environment, the performance quality assigned into the strategic effort in building the comprehensive empowerment about the significant factors refers to propose the mutual combination of digital adoption behavior together with utilization procedure. The particular point could be viewed into the following. Those are empowering strategic foundation for learning enhancement, enhancing learning commitment with SNS platform, making personal and social balance of SNS users, expanding comprehensive way for sharing information across learning enhancement, connecting significant interplay of professional and moral responsibility.

References

1. Al-Dheleai, Y.M., Tasir, Z., Jumaat, N.F.: Depicting students' social presence on social networking site in course-related interaction. SAGE Open **10**(1), 2158244019899094 (2020)
2. Anshari, M., Almunawar, M.N., Shahrill, M., Wicaksono, D.K., Huda, M.: Smartphones usage in the classrooms: learning aid or interference? Educ. Inf. Technol. **22**(6), 3063–3079 (2017). https://doi.org/10.1007/s10639-017-9572-7
3. Baron, C., Sklarwitz, S., Blanco, M.Y.: Assessment of teachers' gains across multiple historic site-based professional development programs. Teach. Teach. Educ. **93**, 103077 (2020)
4. Baskaran, C.: Scholarly information share through Social Networks (SNs) and medias among social science scholars in selected State Universities in Tamilnadu. Int. J. Libr. Inf. Stud. **9**(3), 83–92 (2019)
5. Behera, R.K., Gunasekaran, A., Gupta, S., Kamboj, S., Bala, P.K.: Personalized digital marketing recommender engine. J. Retail. Consum. Serv. **53**, 101799 (2020)
6. Boland, J.K., Anderson, J.L.: The role of personality psychopathology in social network site behaviors. Pers. Individ. Differ. **151**, 109517 (2019)
7. Broekhuizen, T.L.J., Emrich, O., Gijsenberg, M.J., Broekhuis, M., Donkers, B., Sloot, L.M.: Digital platform openness: drivers, dimensions and outcomes. J. Bus. Res. (2019)
8. Closson, L.M., Bond, T.A.: Social network site use and university adjustment. Educ. Psychol. **39**(8), 1027–1046 (2019)
9. Deng, R., Benckendorff, P., Gannaway, D.: Progress and new directions for teaching and learning in MOOCs. Comput. Educ. **129**, 48–60 (2019)
10. Fong, K.C.H., Au, C.H., Lam, E.T.H., Chiu, D.K.: Social network services for academic libraries: a study based on social capital and social proof. J. Acad. Librarianship **46**(1), 102091 (2020)
11. Frost, R.D., Matta, V., MacIvor, E.: Assessing the efficacy of incorporating game dynamics in a learning management system. J. Inf. Syst. Educ. **26**(1), 6 (2019)
12. Hsieh, Y.J., Wu, Y.J.: Entrepreneurship through the platform strategy in the digital era: Insights and research opportunities. Comput. Hum. Behav. **95**, 315–323 (2019)
13. Huang, C.: Social network site use and Big Five personality traits: a meta-analysis. Comput. Hum. Behav. **97**, 280–290 (2019)

14. Huda, M.: Empowering application strategy in the technology adoption: insights from professional and ethical engagement. J. Sci. Technol. Policy Manag. **10**(1), 172–192 (2019)
15. Huda, M., et al.: Empowering technology use to promote virtual violence prevention in higher education context. In: Intimacy and Developing Personal Relationships in the Virtual World, pp. 272–291 (2019)
16. Huda, M., Maseleno, A., Jasmi, K.A., Mustari, I., Basiron, B.: Strengthening interaction from direct to virtual basis: insights from ethical and professional empowerment. Int. J. Appl. Eng. Res. **12**(17), 6901–6909 (2017)
17. Huda, M., et al.: Big data emerging technology: insights into innovative environment for online learning resources. Int. J. Emerg. Technol. Learn. **13**(1), 23–36 (2018). https://doi.org/10.3991/ijet.v13i01.6990
18. Huda, M., et al.: Understanding Modern Learning Environment (MLE) in big data era. Int. J. Emerg. Technol. Learn. **13**(5), 71–85 (2018). https://doi.org/10.3991/ijet.v13i05.8042
19. Huda, M., Hashim, A.: Towards professional and ethical balance: insights into application strategy on media literacy education. Kybernetes (2021). https://doi.org/10.1108/K-07-2017-0252
20. Huda, M., et al.: Towards divine values' strategic application: critical insights into management of change. In: Handbook of Research on Disruptive Innovation and Digital Transformation in Asia, pp. 542–566. IGI Global (2021)
21. Huda, M., et al.: Promoting environmental sustainability governance: a critical exposure from cross-sector partnership. In: Handbook of Research on Environmental Education Strategies for Addressing Climate Change and Sustainability, pp. 1–21. IGI Global (2021)
22. Ioannou, M., Ioannou, A.: Technology-enhanced embodied learning. Educ. Technol. Soc. **23**(3), 81–94 (2020)
23. Juhaňák, L., Zounek, J., Rohlíková, L.: Using process mining to analyze students' quiz-taking behavior patterns in a learning management system. Comput. Hum. Behav. **92**, 496–506 (2019)
24. Kim, C.Y.: SNS utilization of youth and use as a learning tool. J. Korea Multimed. Soc. **23**(1), 93–101 (2020)
25. Kim, C., Shen, C.: Connecting activities on Social Network Sites and life satisfaction: a comparison of older and younger users. Comput. Hum. Behav. **105**, 106222 (2020)
26. Kümpel, A.S.: The Matthew Effect in social media news use: assessing inequalities in news exposure and news engagement on Social Network Sites (SNS). Journalism **21**(8), 1083–1098 (2020)
27. Law, K.M., Lee, V.C., Yu, Y.T.: Learning motivation in e-learning facilitated computer programming courses. Comput. Educ. **55**(1), 218–228 (2010)
28. Lytras, M.D., Visvizi, A., Daniela, L., Sarirete, A., Ordonez De Pablos, P.: Social networks research for sustainable smart education. Sustainability **10**(9), 2974 (2018)
29. Norhailawati, M., et al.: The power of social networking sites: student involvement toward education. Int. J. Eval. Res. Educ. **8**(3), 549–556 (2019)
30. Salleh, U.K.M., Zulnaidi, H., Rahim, S.S.A., Bin Zakaria, A.R., Hidayat, R.: Roles of self-directed learning and social networking sites in lifelong learning. Int. J. Instr. **12**(4), 167–182 (2019)
31. Sánchez, R.M., et al.: Towards social justice through arts and language-based learning. In: Integrating Social Justice Education in Teacher Preparation Programs, pp. 101–123. IGI Global (2020)
32. Scagnoli, N.I., Choo, J., Tian, J.: Students' insights on the use of video lectures in online classes. Br. J. Edu. Technol. **50**(1), 399–414 (2019)
33. Simonson, M., Zvacek, S.M., Smaldino, S.: Teaching and Learning at a Distance: Foundations of Distance Education 7th edn. IAP (2019)

34. Solmaz, O.: Pre-service language teachers' use of social networking sites for language learning: a quantitative investigation. Eurasian J. Appl. Linguist. **5**(3), 423–439 (2019)
35. Sousa, M.J., Rocha, Á.: Digital learning: developing skills for digital transformation of organizations. Futur. Gener. Comput. Syst. **91**, 327–334 (2019)
36. Stallkamp, M., Schotter, A.P.: Platforms without borders? The international strategies of digital platform firms. Glob. Strategy J. (2019)
37. Trabucchi, D., Buganza, T.: Fostering digital platform innovation: from two to multi-sided platforms. Creativity Innov. Manag. **29**(2), 345–358 (2020)
38. Tranos, E.: Social network sites and knowledge transfer: an urban perspective. J. Plan. Lit. **35**(4), 408–422 (2020)
39. Wang, X.: Mobile SNS addiction as a learned behavior: a perspective from learning theory. Media Psychol. **23**(4), 461–492 (2020)
40. Yablonsky, S.: A multidimensional platform ecosystem framework. Kybernetes (2020)
41. Zatsarinny, A.A., Gorshenin, A.K., Kondrashev, V.A., Volovich, K.I., Denisov, S.A.: Toward high performance solutions as services of research digital platform. Procedia Comput. Sci. **150**, 622–627 (2019)
42. Zulkanain, N.A., Miskon, S., Abdullah, N.S., Ali, N.M., Bahari, M., Ahmad, N.: Social Network Sites (SNS) issues in learning landscape – systematic literature review. In: Badioze Zaman, H., et al. (eds.) IVIC 2019. LNCS, vol. 11870, pp. 592–600. Springer, Cham (2019). https://doi.org/10.1007/978-3-030-34032-2_52

Digital Learning Enhancement Through Social Network Site (SNS)

Khalili[1], Miftachul Huda[2](✉), Arieff Salleh Rosman[3], Ahmad Kilani Mohamed[3], and Nurazmallail Marni[3]

[1] State Islamic University of Sunan Kalijaga Yogyakarta, Depok, Indonesia
[2] Universiti Pendidikan Sultan Idris, Tanjung Malim, Malaysia
miftachul@fsk.upsi.edu.my
[3] Universiti Teknologi Malaysia, Johor Bahru, Malaysia

Abstract. In the last decade, the wide range of advanced technology development could be viewed into the number of recent electronic devices, and also services. Those include education development, business sector, and also public service. Moreover, the number could be seen into social media platform in facilitating global cross interaction. In this paper, the aims to examine the further development of social network site (SNS) are deployed in looking at the contribution for learning enhancement. Reviewing several critical literatures was deployed in the attempts to expand the initial value of SNS to play a significant role in developing the recent trends of technology development and transmitted into the social media basis. The finding points out that SNS with its valuable insight of developing adaptive learning technology is potential to play a significant role in facilitating the learning enhancement. This study is expected to contribute to give insights into enhancing the insightful contribution for learning technology together with the adaptive technology enhancement.

1 Introduction

In the last decades, amongst such significant elements in distributing the learning, there has been the number of the way to work with fully enlarging the supply and demands in technology development. As a significant contribution of the recent technology development into the number of fields such as education sector, business engagement, and also social interaction platform, the particular approach of utilizing such attainment should bring along with gaining the potential value of understanding what to do wisely within the moral ethics [1, 2]. In the attempts to incorporate the important way to signify such sectors, it is necessary to enhance the knowledge understanding in recognizing the true benefit apart from its harmful impact. Among such comprehension in the way to enlarge the beneficial value of social network site (SNS) together with handling the challenges are digital abuse, social media addiction, and also human behavior misconduct [3, 4]. In this view, the need to have a sufficient knowledge in enabling the users to perform well in certain task through platform has to be taken into consideration in particular way.

In addition, incorporating such attainment towards the significant contributions of SNS in enabling the users with its particular point in enhancing the certain tasks including

R. Silhavy et al. (Eds.): CoMeSySo 2021, LNNS 232, pp. 421–431, 2021.
https://doi.org/10.1007/978-3-030-90318-3_35

learning motivation refers to point out disseminating the wide range of achievement on the way to do wisely. In the learning context for example, the necessary point of enhancing the digital human behavior refers to the actual contribution in which the sufficient approach could be deployed through expanding the professionality together with morality [5, 6].

As an attempt to incorporate the way of building learning enhancement, adopting SNS basis into the number of tasks lies on going through the well-organized practices with the procedure in enlarging the learning motivation amongst the users [7]. Through incorporating the necessary points of building digital human community in underlying the way to go forward, the extensive point of learning enhancement needs to take into account in transmitting the professional skills together with the moral ability standard [8]. At this point of view, the use of SNS in the learning context should bring along with building the multifarious aspects of the significant approach, mainly the way on how to build the human behavior [9]. The necessary act with an extensive point of getting the enthusiasm in the learning context needs to incorporate the expansion of professional skills to be better interacted in the classroom basis.

In further, the particular contribution of SNS with its particular point of elements played a role in distributing the attempts to encourage the learning environment together with motivating the classroom atmosphere. In this view, it can make the wide range of learning enhancement to be more interested in and also motivated together with gaining the effectiveness on driving the right path of learning process [10–12]. In the attempts to incorporate the research development on the digital learning enhancement, this paper aims to examine the particular way on looking into the actual points of SNS with its featured relevance to bring along with having a sufficient contribution in digital learning environment context. Thus, the literature on SNS context is widely deployed together with making a point on giving insights into learning enhancement.

2 Literature Review

2.1 Social Network Site (SNS) in Education Setting

The initial value of SNS platform as the learning instruction tool is potentially an online basis in enabling all the users to have their own public profile. In this view, SNS platform is usually created with allowing the user feedback from their interaction member so that taking the benefit and avoiding the harmful point could be achieved in particular way [13, 14]. Through incorporating the SNS platform, it is necessary to bring along with getting involved into the users in sharing and connecting amongst the other online members. At this point, the SNS platform pertaining to the online based-media interaction needs to take into account in creating the public-based profile properly within the standard context of digital humanity [15].

With this regard, SNS refers to enhance the online platform in allowing the other users' member to have a simultaneous connection in sharing knowledge and also distributing the information [16, 17]. As an attempt to incorporate SNS with its online platform in distributing the users' connection amongst others, it is important to take a note in determining the proper way on creating the simultaneous link with others. With having such SNS platform, the linked enhancement of the users refers to point out disseminating the information along with confirming the feedback in looking into the particular connection of applying for the electronic tool.

In addition, the connection of gaining the SNS with its particular aim in transmitting the beneficial value amongst the users across the number of platforms such as education sector, business platform, and social interaction network refers to play a significant role in distributing the sites usually bringing together with a new user input through the list of people with whom they share a connection and then allow the people on the list to confirm or deny the connection. Through developing the SNS platform, the adoption of mobile application could be deployed into looking at the whole context whether positive and negative impact is potentially engaged to the digital platform framework [18].

In this view, transmitting the web-based application could become the initial trial-based tool in referring to perform the mutual basis in determining the direct instruction through the electronic tool provided through framework basis [19]. On the basis of gaining the natural outline providing the direct usage to deploy the electronic device, this approach can give insight into communicating towards the accessibility of data with this platform. The next approach to give insights into accessing the browser content refers to enhance the mobile-based application combined with the website basis.

2.2 SNS for Digital Learning Enhancement

In terms of incorporating the basis of crossline-integrated software applied for linking to others in sharing information, the approach of SNS with its theoretical framework on the way to enlarge the mobile basis should bring along with expanding the integration context about the certain information including venue, commitment and instruction [20]. With this regard, the nature point in enabling the users to adopt the native software of a mobile social networking framework refers to give an insight into the mobile-based socially enhanced line.

In expanding the possible inclusion on enabling the overall incorporation of material linked between usages between instructions, enhancing the attainment of using data based on pointing out the number of platform like Twitter, Facebook, LinkedIn, etc. With aiming at sufficient contribution to facilitate in building the communication platform in the attempts to share information into others, the purpose of social and personal basis refers to incorporate the intelligent system through the development of mobile-based community line [21, 22]. The way of incorporating the basis of SNS in using the communication purpose should be transmitted through the set of nodes to represent the users together with its particular phase of interaction platform.

Through enhancing the SNS platform in disseminating the information along with the framework of social network, the general representation is used commonly in embodying the display of transmitting the set of stylish package on representing the number of users in determining the particular category in interacting with stating the explicit points into the other users within the mutual link for instance in the joining group, friendship appeal, commenting the certain post in social media and soon [23]. In terms of connecting with others determined to enhance the particular connections on implicit platform, analyzing the basis on mutual relation to transform the approach of interaction amongst others needs to take into consideration with embodying the digital human expert.

3 Methodology

This study is an attempt to examine the further development of social network site (SNS) deployed in looking at the contribution for learning enhancement. Reviewing several critical literatures are deployed in the attempts to expand the initial value of SNS to play a significant role in developing the recent trends of technology development and transmitted into the social media basis. Through placing the method of investigation approach towards the recent articles, the systematic approach is deployed in highlighting the research procedure with the structural model of the proposed research basis. In developing the research context, the descriptive analysis was conducted through generalizing the finding specification.

4 Analysis and Discussion

4.1 Empowering Digital Learning Enhancement Through SNS Platform

As the online basis in allowing the users to have their own profile in the social media platform, creating website-based application referring to the possible inclusion in interacting into others refers to point out getting the input with the list to be confirmed or denied. It is true that the users may need to adopt the website basis in determining the SNS in linking and sharing with others in the attempts to give the enthusiastic atmosphere with being more interested [24]. As such, this platform could be also deployed into the context of learning enhancement in referring to incorporate the interesting basis with transmitting the motivation consciousness to be deployed amongst instructors, instructional designers, and learners. In the effective enhancement to transfer the knowledge understanding with learning enhancement [25], the variety of significant contributions of highlighting the online learning framework refers to enhance the motivation basis in the teaching and learning process [26].

In the attempts to highlight the learning motivation through incorporating the digital framework, the need to have a sufficient knowledge understanding in battling the emerging challenges such as less motivation, dropping out from the schools, digital abuse and soon should be taken into consideration in maximizing the environmental atmosphere [27, 28]. Moreover, the social concern is entirely needed to increase the engaged motivation in the basis of continuous support on online learning enhancement.

In line with incorporating the learning enhancement to handle the challenging issues, the well-organized instruction should bring along with expanding the accessibility in transmitting the knowledge and also competence skills in determining the digital communication tools in online learning basis. In this view, it is necessary to have a look at the whole context in transforming the traditional model of face-to-face setting to the digital modern basis [29]. In pointing out the major concern in handling the possibility of unexpected happen for instance, the balance between professional and ethical engagement should bring along with the competence skills in responding the environmental atmosphere and social sphere. As such, the necessity to give insight into enhancing the motivation basis, the particular approach in the social isolation may give the potentials on raising the issue of emerging challenges within the digital learning framework, since it can make less interaction in physical basis between learners and instructors or teachers [30, 31]. In offering the digital interaction, the basic implementation guideline on online educational settings needs to concern widely in the learning management systems (LMS) setting. The significant role of this attainment should bring along with enhancing the primary typicality of giving feedback to be achieved through online basis such as website.

In addition, the extensive point on expanding the technology integration in the learning setting for instance refers to advance the initial task in adopting LMS framework including Absorb, Moodle, Schoology, D2lbrightspaceand, Grovo, etc. [32]. The various usage in this context in the learning purpose should be determined in making LMS framework to have a significant contribution in handling the negative impact, such as less motivation, lack handwriting skill, and also teaching professionality commitment. The committed awareness in LMS framework has in particular played a role in transmitting the new model as the platform guide in the online course. Moreover, the motivation basis in LMS framework could adopt the massive open online course (MOOC) as the initiative to transfer knowledge in the basis of online learning course [33]. The number of this educational setting could be adopted in both institutions and universities apart from being suffered due to the variety of problems which might happen. At the same time, it is necessary to collaborate with the digital instruction platform to have a widely linked engagement along with the users.

4.2 Expanding Online-Based Strategic Approach for Learning Enhancement

Attempts to perform the strategic foundation to enhance learning basis are in line with focusing on the commitment of SNS to be utilized in further referring to the particular condition. Through empowering the learning context, strategic foundation in enhancing the role to play along with the behavioral management should bring along with the experienced calculation assessed in giving insight to drive in a right path. The users of digital learning environment need to take into consideration in particular in disseminating the content together with the context. In learning enhancement, it is necessary to take a look at the whole condition whether the users are in the ready condition with facilitation [34]. It means that the mutual commitment to ensure the learning facility in the digital environment has to bring along with the internet network in enhancing the stability between professional skills and moral responsibility.

As an important factor in giving direction of SNS in the learning environment [35, 36], the quality improvement needs to enhance in particular basis to have a sufficient commitment towards comprehending the benefit on technology use within wise and proper basis [37]. In addition, the direction on SNS as a learning environment tool refers to especially encourage the sufficient adoption of digital platform where this requirement should fulfill the practical basis in optimizing the technology-based current practice.

In the attempts to expand the new conceptual model basis on learning motivation with digital adoption, the technology use wisely and properly in the learning process could be deployed in following the procedure of operation within the virtual basis [38, 39]. Amongst such tools, there could be viewed into the model of digital learning environment where the instruction tool could be prepared in the basis of acceptance amongst the users. In pointing out this specification in committing with the learning motivation,

SNS adoption is particularly referred to raise the technology acceptance model (TAM) approach. In order to achieve the particular Outcome-based instruction, the capabilities of sustaining the model approach on digital environment should bring along with performing an effort to enhance the social concern [40]. Since the determinants of SNS adoption are part of adopting the consistency in the learning process, the relevance of having commitment to work together in expanding the assessment process whether to prove its significance or to ensure the procedure within the requirement context should bring along with responsiveness, consequence and approval in digital learning environment. It is true that both motivation and confidence played a key role in combining the digital-based learning enhancement together with maximizing the assessment process in ensuring the significant contribution in the learning sustainability.

In line with the attempts to empower the performance skills of learning motivation, enhancing the factors on the performance basis of strategic cooperation in the learning process refers to have a sufficient expansion of attitude commitment together with teaching instruction. Moreover, performance point was significantly determinant with the feedback in adopting the contribution of knowledge understanding about the SNS as instructional tool in the learning [41, 42]. In particular, developing the model of digital learning environment should bring along with the commitment of starting the procedural context on what to do in the learning process together with the teaching pedagogy skills [43, 44]. In adopting the strategic foundation in the learning enhancement, it is necessary to have a sufficient commitment in the process itself in ensuring the positive feedback to be achieved in contributing into the digital learning environment mainly in the university context.

In the attempts to incorporate the learning enhancement with digital environment context, the strategic foundation itself should be empowered with expanding the commitment awareness to collaborate with teaching instruction assigned into the tool to give insights into the contribution of knowledge transfer wisely within the purpose and goal [45, 46]. With this regard, it is necessary to point out committing with the instructional adoption through digital basis in order to achieve the outcome-based learning enhancement. Moreover, the strong commitment in the digital adoption in the learning process should bring along with advancing the assessment on driving the right path of learning purpose with directly related activities in the daily life basis.

4.3 Incorporating Professional and Ethical Balance in Learning Process

In terms of making balance between personal and social concern on digital-based instruction, it is necessary to point out disseminating what to do properly as a response for their learning process together with digital pedagogy, assigned into the Islamic based approach [47–49]. In maintaining the proper way to establish the theoretical basis on SNS platform, the balance between professional and ethical engagement would lead to build the access on what to do and how to achieve within the wise basis [50, 51]. In this view, the engagement process should bring along with providing the mutual access on certain technological device to the required technological devices to access the internet and therefore social networking sites including phones, tablets, and laptops [52]. In the attempts to describe the variety of digital platforms of social media such as Twitter,

Facebook, WhatsApp, and soon, SNS particularly adopted in making initiative to maintain the establishment of performing the mutual connection across academic purpose together with professional development is widely incorporated into the online activities and transaction.

In the effort to have a sufficient adoption between ethical and professional engagement amongst SNS users, it is necessary to point out disseminating the learning instruction together with connecting the initial networking basis in the online activities. In order to apply for the initiative performance on securing the interaction along with the digital platform basis, the maintenance procedure in the certain point of what to do in the context of interconnection aims to determine the online experts in searching for the resources related to the course of subject in the learning instruction [32, 53]. As an attempt to enhance the online source from the web-based library provided by the instructor, the multiple site of engaging the certain information referring to the authentic appointment of SNS might be empowered with facilitating the innovative instruction in supporting the academic activities together with encouraging the communication performance.

In terms of making available on enhancing the network site, the wider context of delivering the information along with the multiple user background needs to perform in an appropriate basis linked into obtaining the number of academic information. In order to achieve the appointment of authentic learning with enhancing SNS basis, it is clear to take a note that empowering the number of users across the variety of background is potentially enlarged into providing the accessibility of achieving the relevant information [30, 33]. In facilitating the connection together with interaction in the basis of professional and ethical balance, getting clear interpretation towards the certain issues regarding on workplace, or even in the family basis is potentially considered to transform the application procedure within the guideline prepared in the related study material [35, 54]. As such, using digital tool in SNS basis would enable in transmitting the interconnection and interaction in both learning and teaching basis.

In accessing the SNS platform to collaborate with the users' digital tool, it is necessary to point out disseminating the personal and social basis, since the consideration to take point of networking basis refers to enhance the digital devices with the wise application procedures [55, 56]. With this regard, the regular access of online basis might give an appointment in maximising the availability and accessibility towards the cyber space interaction [57, 58]. As such, the users in delivering information with accessing the internet basis have an appropriate basis in enlarging the digital devices including laptop, tablet and also smartphone platform [59]. In transmitting the online platform in the online learning basis, promoting the balance between personal and social concern has to be involved within SNS platform in enabling the users to get mutual access amongst the users.

The extent of accessibility on the way to transmit the information along establishing the connection across the multiple users should be enhanced in particular way so that the learning program is widely engaged into social networking basis. With this regard, the digital device assigned into the internet connection might be involved to uphold the accessibility of SNS-based instruction. As a result, the availability of personal and social concern is necessary to conduct the digital device including mobile phones, computer lab, laptop computers, and tablets together with online performance skills in enabling the users in responding the information demand [60, 61]. At this point of view, the accessibility of SNS performance using the digital device refers to catch up in disseminating the information linked into connecting interpersonal behaviour.

5 Conclusion

This study focuses on addressing the digital learning enhancement through the SNS adoption basis to have a principal framework to establish the nation worldwide of technology in expanding the learning environment. In this study, the necessary act in incorporating the reflection aligned into pointing out the variant of pathways of learning motivation through SNS adoption is discussed on the learning motivation. It has to do with expanding the knowledge understanding amongst the digital users and also instruction basis in ensuring the flow of knowledge transfer. Moreover, addressing the behavioral management should be committed with SNS adoption. In looking at the particular contribution for learning enhancement, critical review of several literatures are deployed in the attempts to expand the initial value of SNS to play a significant role in developing the recent trends of technology development and transmitted into the social media basis. Amongst the committed awareness as the basic point of learning enhancement within the digital learning environment, the performance quality assigned into the strategic effort in building the comprehensive empowerment about the significant factors refers to propose the mutual combination of digital adoption behavior together with utilization procedure. The particular point could be viewed into the following. Those are empowering strategic foundation for learning enhancement, enhancing learning commitment with SNS platform, making personal and social balance of SNS users, expanding comprehensive way for sharing information across learning enhancement, connecting significant interplay of professional and moral responsibility.

References

1. Huda, M.: Empowering application strategy in the technology adoption: insights from professional and ethical engagement. J. Sci. Technol. Policy Manag. **10**(1), 172–192 (2019)
2. Anshari, M., Almunawar, M.N., Shahrill, M., Wicaksono, D.K., Huda, M.: Smartphones usage in the classrooms: learning aid or interference? Educ. Inf. Technol. **22**(6), 3063–3079 (2017). https://doi.org/10.1007/s10639-017-9572-7
3. Behera, R.K., Gunasekaran, A., Gupta, S., Kamboj, S., Bala, P.K.: Personalized digital marketing recommender engine. J. Retail. Consum. Serv. **53**, 101799 (2020)
4. Boland, J.K., Anderson, J.L.: The role of personality psychopathology in social network site behaviors. Pers. Individ. Differ. **151**, 109517 (2019)
5. Broekhuizen, T.L.J., Emrich, O., Gijsenberg, M.J., Broekhuis, M., Donkers, B., Sloot, L.M.: Digital platform openness: drivers, dimensions and outcomes. J. Bus. Res. **122**, 902–914 (2019)
6. Closson, L.M., Bond, T.A.: Social network site use and university adjustment. Educ. Psychol. **39**(8), 1027–1046 (2019)
7. Deng, R., Benckendorff, P., Gannaway, D.: Progress and new directions for teaching and learning in MOOCs. Comput. Educ. **129**, 48–60 (2019)

8. Frost, R.D., Matta, V., MacIvor, E.: Assessing the efficacy of incorporating game dynamics in a learning management system. J. Inf. Syst. Educ. **26**(1), 6 (2019)
9. Hsieh, Y.J., Wu, Y.J.: Entrepreneurship through the platform strategy in the digital era: Insights and research opportunities. Comput. Hum. Behav. **95**, 315–323 (2019)
10. Huang, C.: Social network site use and Big Five personality traits: A meta-analysis. Comput. Hum. Behav. **97**, 280–290 (2019)
11. Huda, M., Kartanegara, M.: Islamic Spiritual Character Values of al-Zarnūjī's Ta 'līm al-Muta 'allim. Mediterr. J. Soc. Sci. **6**(4S2), 229–235 (2015)
12. Huda, M., Teh, K.S.M.: Empowering professional and ethical competence on reflective teaching practice in digital era. In: Dikilitas, K., Mede, E., Atay, D. (eds.) Mentorship Strategies in Teacher Education, pp. 136–152. IGI Global, Hershey (2018). https://doi.org/10.4018/978-1-5225-4050-2.ch007
13. Huda, M., et al.: Strengthening quality initiative for organization stability: insights from trust in cyberspace-based information quality. In: Cases on Quality Initiatives for Organizational Longevity, pp. 140–169. IGI Global, Hershey (2018). https://doi.org/10.4018/978-1-5225-5288-8.ch006
14. Huda, M., et al.: Innovative teaching in higher education: the big data approach. Turk. Online J. Educ. Technol. **15**(Special issue), 1210–1216 (2016)
15. Huda, M., Haron, Z., Ripin, M.N., Hehsan, A., Yaacob, A.B.C.: Exploring innovative learning environment (ILE): big data era. Int. J. Appl. Eng. Res. **12**(17), 6678–6685 (2017)
16. Huda, M., et al.: Empowering technology use to promote virtual violence prevention in higher education context. In: Intimacy and Developing Personal Relationships in the Virtual World, pp. 272–291. IGI Global, Hershey (2019). https://doi.org/10.4018/978-1-5225-4047-2.ch015
17. Huda, M., Jasmi, K.A., Alas, Y., Qodriah, S.L., Dacholfany, M.I., Jamsari, E.A.: Empowering civic responsibility: insights from service learning. In: Burton, S. (ed.) Engaged Scholarship and Civic Responsibility in Higher Education, pp. 144–165. IGI Global, Hershey (2017). https://doi.org/10.4018/978-1-5225-3649-9.ch007
18. Huda, M., Jasmi, K.A., Basiran, B., Mustari, M.I.B., Sabani, A.N.: Traditional wisdom on sustainable learning: an insightful view from Al-Zarnuji's Ta'lim al-Muta'allim. SAGE Open **7**(1), 1–8 (2017)
19. Huda, M., Maseleno, A., Jasmi, K.A., Mustari, I., Basiron, B.: Strengthening interaction from direct to virtual basis: insights from ethical and professional empowerment. Int. J. Appl. Eng. Res. **12**(17), 6901–6909 (2017)
20. Huda, M., Mulyadi, D., Hananto, A.L., Nor Muhamad, N.H., Mat Teh, K.S., Don, A.G.: Empowering corporate social responsibility (CSR): insights from service learning. Soc. Responsib. J. **14**(4), 875–894 (2018)
21. Huda, M., Sabani, N., Shahrill, M., Jasmi, K.A., Basiron, B., Mustari, M.I.: Empowering learning culture as student identity construction in higher education. In: Shahriar, A., Syed, G. (eds.) Student Culture and Identity in Higher Education, pp. 160–179. IGI Global, Hershey (2017). https://doi.org/10.4018/978-1-5225-2551-6.ch010
22. Huda, M., Shahrill, M., Maseleno, A., Jasmi, K.A., Mustari, I., Basiron, B.: Exploring adaptive teaching competencies in big data era. Int. J. Emerg. Technol. Learn. **12**(3), 68–83 (2017)
23. Huda, M., Teh, K.S.M., Nor, N.H.M., Nor, M.B.M.: Transmitting leadership based civic responsibility: insights from service learning. Int. J. Ethics Syst. **34**(1), 20–31 (2018). https://doi.org/10.1108/IJOES-05-2017-0079
24. Huda, M., et al.: Adaptive online learning technology: trends in big data era. In: Diverse Learning Opportunities Through Technology-Based Curriculum Design. IGI Global, Hershey (2019). https://doi.org/10.4018/978-1-5225-5519-3.ch008
25. Huda, M., Yusuf, J.B., Jasmi, K.A., Nasir, G.A.: Understanding comprehensive learning requirements in the light of al-Zarnūjī's Ta'līm al-Muta'allim. SAGE Open **6**(4), 1–14 (2016)

26. Juhaňák, L., Zounek, J., Rohlíková, L.: Using process mining to analyze students' quiz-taking behavior patterns in a learning management system. Comput. Hum. Behav. **92**, 496–506 (2019)
27. Kim, C.Y.: SNS utilization of youth and use as a learning tool. J. Korea Multimed. Soc. **23**(1), 93–101 (2020)
28. Kim, C., Shen, C.: Connecting activities on social network sites and life satisfaction: a comparison of older and younger users. Compute. Human Behav. **105**, 106222 (2020)
29. Law, K.M., Lee, V.C., Yu, Y.T.: Learning motivation in e-learning facilitated computer programming courses. Comput. Educ. **55**(1), 218–228 (2010)
30. Maseleno, A., et al.: Hau-Kashyap approach for student's level of expertise. Egypt. Inform. J. **20**, 27–32 (2018). https://doi.org/10.1016/j.eij.2018.04.001
31. Maseleno, A., et al.: Combining the previous measure of evidence to educational entrance examination. J. Artif. Intell. **10**(3), 85–90 (2017)
32. Maseleno, A., et al.: Mathematical theory of evidence to subject expertise diagnostic. ICIC Exp. Lett. **12**(4), 369 (2018). https://doi.org/10.24507/icicel.12.04.369
33. Maseleno, A., Sabani, N., Huda, M., Ahmad, R., Jasmi, K.A., Basiron, B.: Demystifying learning analytics in personalised learning. Int. J. Eng. Technol. **7**(3), 1124–1129 (2018)
34. Miller, B.J., Mundey, P.: Emerging SNS use: the importance of social network sites for older American emerging adults. J. Youth Stud. **23**(5), 613–630 (2020)
35. Moksin, A.I., Shahrill, M., Anshari, M., Huda, M., Tengah, K.A.: The learning of integration in calculus using the autograph technology. Adv. Sci. Lett. **24**(1), 550–552 (2018)
36. Othman, R., Shahrill, M., Mundia, L., Tan, A., Huda, M.: Investigating the relationship between the student's ability and learning preferences: evidence from year 7 mathematics students. New Educ. Rev. **44**(2), 125–138 (2016)
37. Petre, A.L.: Stimulating learning motivation of early education and primary school students through alternative methods of assessment. J. Plus Educ. 303–312 (2014)
38. Salleh, U.K.M., Zulnaidi, H., Rahim, S.S.A., Bin Zakaria, A.R., Hidayat, R.: Roles of self-directed learning and social networking sites in lifelong learning. Int. J. Instr. **12**(4), 167–182 (2019)
39. Scagnoli, N.I., Choo, J., Tian, J.: Students' insights on the use of video lectures in online classes. Br. J. Edu. Technol. **50**(1), 399–414 (2019)
40. Simonson, M., Zvacek, S.M., Smaldino, S.: Teaching and Learning at a Distance: Foundations of Distance Education 7th Edition. IAP (2019)
41. Sousa, M.J., Rocha, Á.: Digital learning: developing skills for digital transformation of organizations. Futur. Gener. Comput. Syst. **91**, 327–334 (2019)
42. Stallkamp, M., Schotter, A.P.: Platforms without borders? The international strategies of digital platform firms. Glob. Strategy J. (2019)
43. Trabucchi, D., Buganza, T.: Fostering digital platform innovation: from two to multi-sided platforms. Creat. Innov. Manag. **29**(2), 345–358 (2020)
44. Wang, X.: Mobile SNS addiction as a learned behavior: a perspective from learning theory. Media Psychol. **23**(4), 461–492 (2020)
45. Yablonsky, S.: A multidimensional platform ecosystem framework. Kybernetes (2020)
46. Zatsarinny, A.A., Gorshenin, A.K., Kondrashev, V.A., Volovich, K.I., Denisov, S.A.: Toward high performance solutions as services of research digital platform. Procedia Comput. Sci. **150**, 622–627 (2019)
47. Zulkanain, N.A., Miskon, S., Abdullah, N.S., Ali, N.M., Bahari, M., Ahmad, N.: Social network sites (SNS) issues in learning landscape – systematic literature review. In: Zaman, H.B., et al. (eds.) Advances in Visual Informatics: 6th International Visual Informatics Conference, IVIC 2019, Bangi, Malaysia, November 19–21, 2019, Proceedings, pp. 592–600. Springer, Cham (2019). https://doi.org/10.1007/978-3-030-34032-2_52

48. Zakirillah, Noorminshah, A.I., Huda, M., Fathoni, Heroza, R.I.: Design of a mobile based academic cyber counselling application in higher education. J. Inf. Syst. Res. Innov. **10**(3), 1–9 (2016)
49. Adela, A., Jasmi, K.A., Basiron, B., Huda, M., Maseleno, A.: Selection of dancer member using simple additive weighting. Int. J. Eng. Technol. **7**(3), 1096–1107 (2018)
50. Aminin, S., Huda, M., Ninsiana, W., Dacholfany, M.I.: Sustaining civic-based moral values: insights from language learning and literature. Int. J. Civ. Eng. Technol. **9**(4), 157–174 (2018)
51. Amin, M.M., Nugratama, M.A.A., Maseleno, A., Huda, M., Jasmi, K.A.: Design of cigarette disposal blower and automatic freshner using MQ-5 sensor based on atmega 8535 microcontroller. Int. J. Eng. Technol. **7**(3), 1108–1113 (2018)
52. Assahubulkahfi, M., Sam, Y.M., Maseleno, A., Huda, M.: LQR tuning by particle swarm optimization of full car suspension system. Int. J. Eng. Technol. **7**(2.13), 328–331 (2018)
53. Atmotiyoso, P., Huda, M.: investigating factors influencing work performance on mathematics teaching: a case study. Int. J. Instr. **11**(3), 391–402 (2018)
54. Putra, D.A.D., et al.: Tactical steps for e-government development. Int. J. Pure Appl. Math. **119**(15), 2251–2258 (2018)
55. Rosli, M.R.B., Salamon, H.B., Huda, M.: Distribution management of Zakat Fund: recommended proposal for Asnaf Riqab in Malaysia. Int. J. Civ. Eng. Technol. **9**(3), 56–64 (2018)
56. Sugiyarti, E., Jasmi, K.A., Basiron, B., Huda, M., Shankar, K., Maseleno, A.: Decision support system of scholarship grantee selection using data mining. Int. J. Pure Appl. Math. **119**(15), 2239–2249 (2018)
57. Susilowati, T., et al.: Determination of scholarship recipients using simple additive weighting method. Int. J. Pure Appl. Math. **119**(15), 2231–2238 (2018)
58. Susilowati, T., et al.: Getting parents involved in child's school: using attendance application system based on SMS gateway. Int. J. Eng. Technol. **7**(2.27), 167–174 (2018)
59. Susilowati, T., et al.: Learning application of Lampung language based on multimedia software. Int. J. Eng. Technol. **7**(2.27), 175–181 (2018)
60. Abadi, S., et al.: Application model of k-means clustering: insights into promotion strategy of vocational high school. Int. J. Eng. Technol. **7**(2.27), 182–187 (2018)
61. Aminudin, N., et al.: The family hope program using AHP method. Int. J. Eng. Technol. **7**(2.27), 188–1 (2018)

Measuring Success of Information System (IS) in Small and Medium Enterprises (SMEs) in Gauteng

Lungile Willie[1] and T. Zuva[2](✉)

[1] Vaal University of Technology, 1911 Andries Potgieter Blvd, Vanderbijlpark, South Africa
[2] Department of ICT, Vaal University of Technology, 1911 Andries Potgieter Blvd, Vanderbijlpark, South Africa
tranosz@vut.ac.za

Abstract. The impacts of information systems (IS) are often indirect and influenced by human, organizational, and environmental factors, therefore measurement of information systems (IS) success is both complex and elusive. Researchers have created models for success emphasizing the need for better and more consistent success metrics. This study endeavoured to measure IS success in SMEs in Gauteng, South Africa, in order to find out its impact on these businesses. This research paper proposed a comprehensive multidimensional model of measuring success of information system in SMEs which suggests that system quality, information quality, service quality, intention to use, user satisfaction, use, self-efficacy, individual benefit/impact, use and organizational benefits/impact are success variables to be used. Online survey questionnaire was distributed and hundred and twenty-six (126) participants retuned usable questionnaires. The data collection instrument fulfilled the reliability and validity tests. The hypotheses were tested using regression analysis. The results enabled the derivation of a generic formal measure of IS success in SMEs with organizational benefits/impact variable as the subject of the formula. The formula can be used by SMEs to measure the success of IS in their respective organizations.

Keywords: Small and medium enterprises · Information system · Information technology · DeLone and MacLean Model · Information quality · System quality · Service quality · User satisfaction · Organization benefit/impact · Use · Intention to use · Self-efficacy · Individual benefit/impact

1 Introduction

Almost each and every organisation we know of nowadays started out as an SME. SMEs internationally have a critical commitment to the arrangement of merchandise and enterprises to the general public. Without SMEs, huge organizations will most likely be unable to satisfy the need for products and ventures in an extending client base [1]. At present there is no acknowledged overall meaning of SMEs [1]. An assessment of the meaning of a SMEs demonstrates that it depends upon on who is characterizing it and

R. Silhavy et al. (Eds.): CoMeSySo 2021, LNNS 232, pp. 432–446, 2021.
https://doi.org/10.1007/978-3-030-90318-3_36

from where they are characterizing it. SMEs are depicted through wide assortment of workers utilized, capital utilized and deals turnover [2]. SMEs are thus classified via the scope of workers and by utilizing the expense of their resources. The grouping of SMEs by utilizing estimation is pertinent to area [2]. An organization of a given measurement ought to be little corresponding to one quarter the spot the market is enormous and there are numerous contenders, while a relationship of practically identical extents may also be seen monster in some other territory with less players or regularly more modest firms inside it [1]. It can likewise be fantastic to characterize size by means of the scope of workers in certain areas anyway extra astonishing to utilize turnover in others. Across governments, it is generally customary to quantify estimation in understanding to quantities of full-time workers or their equivalent [1]. Various nations diagram SMEs in an unexpected way, for instance, in Canada the time-frame SME alludes to offices with less than 500 representatives [1]. An organization that has more noteworthy workers than these shorts yet less than 500 representatives is named a medium-sized business [1]. By and large, in Canada a SMEs is any business foundation with 1 to 499 workers and considerably less than $50 million in net incomes. In Germany a SMEs has a limitation of 250 representatives while in Belgium it has a restriction of 100 workers [1]. In New Zealand a little venture has 19 representatives or less. In the US (U.S.A) a little business venture alludes to those with less than 100 representatives, while medium-sized business endeavour alludes to these with less than 500 work force [1].

Small and medium enterprises (SMEs) cover a wide spectrum of industries and play an important role in both developed and developing economies [1]. South Africa is no exception and SMEs occupy a prominent position in the planned development of the South African economy [1]. Traditionally the success of Information Systems (IS) has been studied in the context of large organizations with a focus on systems that have been developed in-house [2]. The South African SMEs have been consistently outperforming large industries on crucial parameters such as global competition, this has placed a considerable pressure on SMEs to improve on cost and efficiency to provide value-added services to meet market demand leading to a need for them to reexamine their competitive strategies [3].

In 2008, organizations continued to increase spending on information technology (IT) and their budgets continued to rise, even in the face of potential economic downturns [4]. However, fears about economic conditions and increasing competition create pressure to cut costs which require organizations to measure and examine the benefits and costs of technology [4]. Naturally, organizations are interested in knowing the return on these investments. The impacts of information technology (IT) are often indirect and influenced by human, organizational, and environmental factors, therefore measurement of information systems (IS) success is both complex and elusive. Researchers have created models for success emphasizing the need for better and more consistent success metrics [4]. This study endeavours to measure IS success in SMEs in Gauteng, South Africa in order to find out its impact on these businesses. This paper is arranged as follows:

2 Literature Review

Because of the popularity of the Delone & Maclean model (D&M) in the academic literature, it seemed appropriate to organize the studies of IS success measure in SMEs using the taxonomy of [5]. The literature will help the peruser to get comfortable with the significant estimates used to quantify the achievement of IS in SMEs. The part gives a short outline of the exploration foundations. Small and medium enterprises (SMEs) are vital engines of employment, innovation and growth. There is consensus on differentiating characteristics for SMEs from the larger counterparts which have important implications for the success of IS implementation initiatives. These include organizational structure, management and decision-making processes [6]. The presence of SMEs in all sectors of the economy would signify their critical role in steering development. However, according to [7], there is very little information from literature on specific roles and contributions of SMEs toward economic growth. This could be associated to the fact that SMEs are visible in all sectors and it is difficult to single them out against a few large businesses [7].

Information Systems (IS) has become a very important software applications with a significant influence on the business world and thus have attracted high levels of investment [8]. Despite the anticipated benefits few researches examined the success of such systems to ensure that benefits do materialize. However measuring such success is challenging due to the lack of consensus on contributing factors [9]. Firstly, the impact of IS, is indirect and is influenced by many factors such as human, organizational and environmental ones. The mixture of the technical and social aspects of an IS make such measurements complex and confusing [10]. Secondly, IS and work practices are very entangled thus making it difficult to identify their discrete influence on success [8]. The third point is related to the methodological perspectives that are used to measure IS success in which identifying dependent variables is difficult [11].

It is generally agreed that SMEs contribute significantly to economic development. They are associated with discovering new markets and exploiting them to their advantage. Similarly, they are the heart of founding new ventures and a source of income and employment for millions of Africans [12]. This means that SMEs are central to wealth creation by stimulating demand for goods, investment, and trade [12]. Without SMEs, many African governments will experience financial and developmental constraints, all of which would only worsen the living standards of low income persons often served by the sector [13]. Another important role played by SMEs is that of inventing, innovation of new ideas and technology. The businesses provide room for pre-incubating, incubating, introducing and commercializing new products [12].

3 Methodology

The study used a model derived from [5] and [14] in Fig. 1. The model consists of the following variables: system quality, information quality, service quality, intention to use, user satisfaction, use, self-efficacy, individual benefit/impact, use and perceived organizational benefits/impact, are success variables in SMEs. These variables are assumed to contribute passively to the organizational benefits as in Fig. 1.

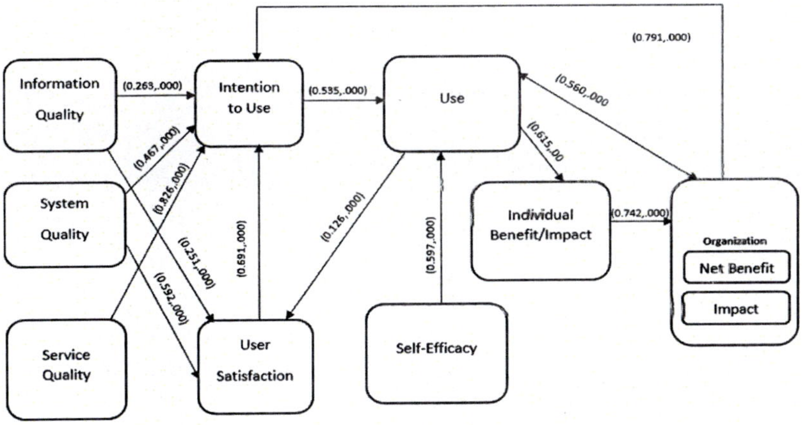

Fig. 1. Proposed model of IS success derived from ([5] and [14])

3.1 Data Collection

Data collection is a set of questions deliberately designed to elicit responses from respondents for the purpose of collecting data or information [15]. Data was collected using a set of an online survey questionnaire appropriate to each sample group. The research used a Likert-scale question perceived importance ranked from 1 to 5, where 1 = strongly disagree, 2 = disagree, 3 = neutral, 4 = agree, 5 = strongly agree. The questionnaires were grouped into two-part's, demographic and survey questionnaires. The survey questionnaire is viewed as the best instrument for gathering a large number of responses. According to [16] this tool gathers information about people's attitudes, facts, behaviour, activities and responses to events, and usually consists of a list of written questions. Introductory letters were sent to sampled emails of the survey followed by the link to the online survey questionnaire. Table 1 below represent number of questionnaires that were used to collect data. Population is a number of people or units from which research information will be obtained [17]. The target population of this study was SMEs owners/managers/employees operating in Gauteng province, South Africa. Sampling size is defined as a technique of electing the number of observations to include in a sample [18]. Additionally, the sample size is an important feature of any study or investigation in which the aim is to make inferences about the population from a sample [18].

Table 1. Questionnaires

Item description	Item code	Variable	Author(s)
Section 1			
IQ1. Information from the system is relevant to my work	IQ	Information quality	[14]
IQ2. Information I get from the system is accurate			
IQ3. Information is easy to understand information from the system			
IQ4. The information is presented in a useful format			
Section 2			
SysQ1. The system is easy to use	SysQ	System quality	[14]
SysQ2. The system is useful			
SysQ3. The system is easy to learn			
SysQ4. I can retrieve information I need easily			
Section 3			
SerQ1. The support services for the system are dependable	SerQ	Service quality	[14]
SerQ2. The support services give me individual attention			
SerQ3. Overall, the support services meet my needs			
Section 4			
US1. Overall, I am satisfied with the system	US	User satisfaction	[19]
US2. This information system is more useful than I had expected			[14]
US3. This information system assists me in performing my tasks better			
US4. This information system is extremely useful			
US5. Using this information system enables me to accomplish tasks more quickly			
US6. This information system makes it easier to do my tasks			
US7. This information system improves the quality of my decision making			

(*continued*)

Table 1. (*continued*)

Item description	Item code	Variable	Author(s)
US8. Use of the information system enables me to make better decisions			
US9. This information system assists me in making decisions more effec tively			
US10. Use of the information system enables me to set my priorities in decision making			
Section 5			
IU1. You intend to use any IS application	IU	Intention to use	[19]
IU2. You will reuse the IS applications in the future			[20]
IU3. You will use the IS applications frequently in the future			
IU4. Learning to use this information system was easy for me			
IU5. I found it easy to get this information system to do what I want it to do			[21]
IU6. My interaction with this information system was clear and understandable			[22]
IU7. It would be easy for me to become skilful at using this information system			
Section 6			
U1. How many minutes per shift do you spend on the system?	U	Use	[14]
U2. How many times a shift do you log on to the system?			
U3. How many functions in the system have you used?			
Section 7			
II1. Management relies a great deal on me to ensure proper operation or processing when I use the system	II	Individual impact	[23]
II2. Much is left to my discretion to ensure proper operation or processing when I use the system			
II3. I have considerable autonomy in deciding how to carry out my work			

(*continued*)

Table 1. (*continued*)

Item description	Item code	Variable	Author(s)
II4. The procedures to carry out a task are spelled out very clearly			
II5. If there is an error, it is very easy for my supervisor to trace when, where, and by whom it was committed through the IS system			
Section 8			
OI1. I know where to turn to when I need any assistance with our IS success	OI	Organizational impact	[23]
OI2. In my company we get good technical support for our IS			
OI3. We have extensive support to help with problems related to our IS			
OI4. Management is aware of the benefits that can be achieved with the use of IS			
OI5. Management provides most of the necessary help and resources to enable people to use IS			
OI6. Management is really keen to see that people are happy with using IS			
Section 9			
NB1. The product service of the IS application system is a good value for the money	NB	Net benefit	
NB2. The price of the product or service of the IS application system is acceptable			[20]
NB3. The time spent in the IS application system is appropriate			
Section 10			
SE1. When I enter data into the computer, I feel confident about what I am doing	SE	Self-efficacy	[14]
SE2. I feel comfortable to use the system			

4 Results

The response rate is the number of participants who actually completed survey out of all the invited survey takers [24]. In this study, a response rate was calculated on the basis of the primary data collection instrument i.e., the number of questionnaires distributed. Out of one hundred and sixty-seven questionnaires (167) distributed, one hundred and thirty-six (136) were returned, the number of usable returned questionnaires was one hundred and twenty-six (126), and the number of unusable questionnaires was ten (10). The returned questionnaires represent a response rate of 75%. The response rates were considered admissible given the recommendations by [25] that a response rate of 50% is adequate for analysis and reporting, a rate of 60% is generally good while a response rate of above 70% is excellent. This is also the same position taken by [26] who adds that a response rate of above 70% is deemed to be very good. Additionally, one hundred and twenty-six (126) out of one hundred and sixty-seven (167) respondents targeted for key informant online surveys fully participated, this represented a response rate of 75%. If the response rate is below 30 percent of the expected response then the validity methods used, and results will be questionable [27]. In this research study the percentage of 75% was above the proposed threshold. The sample size of one hundred and sixty-seven participants made it manageable with regards to time and resource constraints and it also provided critical analysis of the contents under study. Based on these assertions, this implies that the response rate for this study was adequate and increases confidence for generalization.

The assemble validity of the estimation comprises each convergent validity and discriminant validity [29]. Factor analysis is performed to discover whether or not the things are certainly estimating the builds. As referenced beforehand, the factor stacking suggests the degree of meeting with the assemble validity of the estimation comprises each convergent validity and discriminant validity [29]. Factor analysis is performed to discover whether or not the things are certainly estimating the builds. As referenced beforehand, the factor stacking suggests the degree of meeting with a priceless degree equivalent to or above 0.5. whilst AVE is a synopsis pointer for the combination, which want least estimation of 0.5 to set up focalized legitimacy [29]. After the adjustment, all element loading, and AVE values obtained are above the minimal level of 0.5. Therefore, the convergent validity was once established. A Cronbach's Alpha worth which is extra outstanding or equal to 0.6 is for the most phase precious for exhibiting inward unwavering high-quality [30]. The greater the value, the greater accurate the variable estimations are. The Cronbach's Alpha fee for all constructs exceeded the minimum value of 0.6. The assemble Intention to Use (IU) (0.978) featured the biggest fee for Cronbach Alpha, then Use (U) (0.957), Organizational benefit/impact (OBI) (0.934), User Satisfaction (US) (0.930), System Quality (SysQ) (0.913), Information Quality (IQ) (0.906) Service fine (SerQ) (0.830), Individual benefit/Impact (IBI) (0.864), and finally Self-Efficacy (SE) (0.909). Therefore, it can be concluded that the inner reliability is supported.

4.1 Regression Analysis

The hypothesises in this study had been tested using regression analysis. The regression analysis was once carried to take a look at the degree of impact of the unbiased variables on established variables. Table 2 shows the normal model summary, along with the relevant R values, R rectangular (R^2), adjusted R square, Std. Error of the estimate, and F-test results. (R^2) is a multiple coefficient of willpower which indicates the degree of goodness-of-fit of the regression model and adjusted (R^2) value determines also the model fitness with adjustment relies upon on the wide variety of variables in the mannequin [29]. In this study, the result from the model summary shows that adjusted (R^2) = 0.611, meaning 61.1% of the pattern variant in measuring the success of IS in SMEs is defined by means of the model. This suggests the model has a quite excellent sufficient goodness-of-fit and predictive strength [29]. 'F' statistic test is used to evaluate the utility of the regression model [29]. In the F-test result, it indicates $p = 0$ which is decrease than the large degree at $p<0.05$. This implies that the standard regression model is statistically beneficial for predicting the measuring success of IS in SMEs, i.e. at least one of the variables in the mannequin are useful [29].

Table 2. Regression model summary

Model	R	R Square	Adjusted R Square	Std. error of the estimate	F	Sig.
Regression	0.771	0.617	0.611	1.15	65.76	0.000

$^*p < 0.05$; $^{**}p < 0.01$; $p < 0.001$

Then in the Table 3, below, T statistic are used to assess the ten constructs (information quality, device quality, provider quality, user satisfaction, company advantage/impact, use, intention to use, self-efficacy, character benefit/impact) one at a time to determine whether or not they are contributing to the predictive relationship and substitution of values to formulate a new equation (Sub two into 1, Sub 1 into 3, Sub 3 into 4, Sub 4 into 5) was once calculated. In different words, it helps to take a look at the ten hypotheses in our study. In Table 3 below, it represents the speculation check consequences of assemble based on t-test, such as statistics of Beta (B) coefficients, values for well-known deviation of the sampling distribution and t-test results. According to [29], Beta coefficients suggests the relationship between the impartial variable and the based variable, i.e., to what extend the dependent variable is defined via the unbiased variable. A greater value of beta implies a stronger relationship. When a beta cost is > 0, it shows a nice relationship, vice versa (i.e., <0, bad relationship).

The hypothesis test results in this study that are supported or accepted at level significance of ≤ 0.05 are listed below:

Supported Study Hypothesis

H1. Information quality influence positively intention to use.
H2. Information quality influence positively user satisfaction.
H3. System quality influence positively intention to use.
H4. System quality influence positively user satisfaction.
H5. Service quality influence positively intention to use.
H6. User satisfaction influence positively intention to use.

Table 3. Regression measure

Coefficients	Unstandardized coefficients		Standardized coefficients	t	Sig.
Model	B	Std. Error	Beta		
Information Quality	0.25	.168	.263	8.464	.000, supported
System Quality	0.34	.196	.467	2.410	.000, supported
Service Quality	0.31	.138	.826	5.589	.000, supported
User Satisfaction	0.28	.170	.691	4.100	.000, supported
Organization Benefit/ Impact	0.31	168	.791	9.312	.000, supported
a. Dependent Variable: Intention to Use IU = 0.263(IQ) + 0.467(SysQ) + 0.826(SerQ) + 0.691(US) + 0.791(OBI) (1)					
Information Quality	0.23	.126	.251	1.879	.000, supported
System Quality	0.32	.132	.592	4.400	.000, supported
Use	0.21	.188	.126	1.847	.000, supported
b. Dependent Variable: User Satisfaction US = .251(IQ) + .592(SerQ) + .126(U) (2)					
Intention to Use	0.24	.125	.535	4.822	.000, supported
Self-Efficacy	0.26	.102	.597	6.315	.000, supported
Organization Benefit/Impact	0.26	.124	.560	5.363	.000, supported
c. Dependent Variable: Use U = 0.535(IU) + 0.593(SE) + 0.560(OBI) (3)					
Use	0.22	.088	.615	6.440	.000, supported
d. Dependent Variable: Individual Benefit/Impact IBI = 0.615(U) (4)					
Individual Benefit/Impact	0.28	.080	742	8.708	.000, supported
Use	0.26	.124	.560	5.363	.000, supported

(continued)

Table 3. (*continued*)

Coefficients	Unstandardized coefficients		Standardized coefficients	t	Sig.
Model	B	Std. Error	Beta		

e. Dependent Variable: Organization Benefit/Impact
OBI = 0.742(IBI) + 0.560 (U) (5)

H7. Intention to use influence positively the use.
H8. Use influence positively user satisfaction.
H9. Self-efficacy influences positively the Use.
H10. Use influence positively Individual benefits/impact.
H11. Individual benefits/impact influence positively organization benefit/impact.
H12. Use influence positively organization benefit/impact and Organization net benefit/impact influence positively the Use.
H13. Organization net benefit/impact influence positively intention to use (Fig. 2).

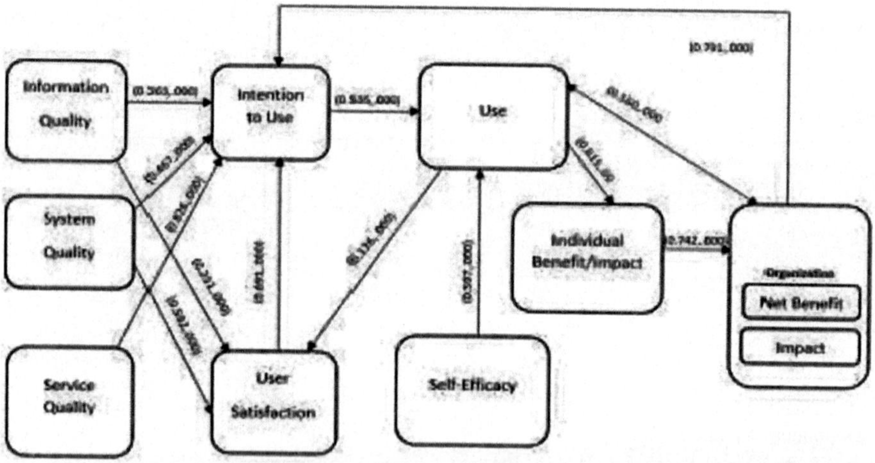

Fig. 2. Confirmed model of regression derived from ([5] and [14])

5 Formal Representation of the Confirmed Regression Model

The objective of this study was to measure the success of IS in SMEs. The main variable used to measure the success of IS in SMEs is organization benefit/impact. Measurement of the value of organizational benefit /impact from the model variables is illustrated below, the research used a system of equations derived from regression analysis that is 1 to 5:

$$IU = 0.263(IQ) + 0.467(SysQ) + 0.826(SerQ) + 0.691(US) + 0.791(OBI \quad (1)$$

$$US = .251(IQ) + .592(SerQ) + .126(U) \quad (2)$$

$$U = 0.535(IU) + 0.593(SE) + 0.560(OBI) \quad (3)$$

$$IBI = 0.615(U) \quad (4)$$

$$OBI = 0.742(IBI) + 0.560(U) \quad (5)$$

The measurement of success of IS in SMEs we calculate the value of the variable organizational benefit/impact. Its value is attributed to the success of IS in SMEs. The substitution is done in the following manner:

Sub 2 into 1,
Sub 1 into 3,
Sub 3 into 4,
Sub 4 into 5 as shown below.

Substitute 2 into 1
Intention to Use (IU) = 0.263(IQ) + 0.467(SysQ) + 0.826(SerQ) + 0.691(US) + 0.791(OBI)

$$IU = 0.263(IQ) + 0.467(SysQ) + 0.826(SerQ) + 0.691(US) + 0.791(OBI)$$

$$IU = 0.263(IQ) + 0.467(SysQ) + 0.826(SerQ) + 0.691*(0.251(IQ) + 0.592(SysQ) + 0.126(U)) + 0.791(OBI)$$

$$IU = 0.263(IQ) + 0.467(SysQ) + 0.826(SerQ) + 0.173(IQ) + 0.409(SysQ) + 0.087(U) + 0.791(OBI)$$

$$IU = 0.463(IQ) + 0.876(SysQ) + 0.826(SerQ) + 0.087(U) + 0.791(OBI) \quad (6)$$

Substitute 1 into 3

$$(U) = 0.535(IU) + 0.593(SE) + 0.560(OBI)$$

$$U = 0.535*(0.463(IQ) + 0.876(SysQ) + 0.826(SerQ) + 0.087(U) + 0.791(OBI)) + 0.593(SE) + 0.560(OBI)$$

$$U = 0.247(IQ) + 0.468(SysQ) + 0.441(SerQ) + 0.046(U) + 0.423(OBI) + 0.593(SE) + 0.560(OBI)$$

$$U = 0.247(IQ) + 0.468(SysQ) + 0.441(SerQ) + 0.046(U) + 0.983(OBI) + 0.593(SE) \quad (7)$$

Substitute 3 into 4

$$\text{Individual Benefit/Impact(IBI)} = 0.615(U)$$

$$(IBI) = 0.615*(0.247(IQ) + 0.468(SysQ) + 0.441(SerQ) + 0.046(U) + 0.983(OBI) + 0.593(SE))$$

$$(IBI) = 0.151(IQ) + 0.287(SysQ) + 0.271(SerQ) + 0.028(U) + 0.604(OBI) + 0.364(SE) \quad (8)$$

Substitute 4 into 5

$$\text{Organization Benefit/Impact(OBI)} = 0.742(IBI) + 0.560(U)$$

$$(OBI) = 0.742{*}(0.151(IQ) + 0.287(SysQ) + 0.271(SerQ) + 0.028(U) + 0.604(OBI) + 0.364(SE)) + 0.560(U)$$

$$(OBI) = 0.112(IQ) + 0.212(SysQ) + 0.201(SerQ) + 0.020(U) + 0.448(OBI) + 0.270(SE) + 0.415(U)$$

$$(OBI) = 0.112(IQ) + 0.212(SysQ) + 0.201(SerQ) + 0.435(U) + 0.448(OBI) + 0.270(SE)$$

$$(OBI) - 0.448(OBI) = 0.112(IQ) + 0.212(SysQ) + 0.201(SerQ) + 0.435(U) + 0.448(OBI) + 0.270(SE)$$

$$1 - 0.448(OBI) = 0.112(IQ) + 0.212(SysQ) + 0.201(SerQ) + 0.435(U) + 0.448(OBI) + 0.270(SE)$$

$$0.552(OBI) = 0.112(IQ) + 0.212(SysQ) + 0.201(SerQ) + 0.435(U) + 0.270(SE)/0552(OBI)$$
$$= 0.112(IQ)/0.552 + 0.212(SysQ)/0.552 + 0.201(SerQ)/0.552 + 0.435(U)/0.552 + 0.270(SE)/0.552$$

$$(OBI) = 0.202(IQ) + 0.384(SysQ) + 0.364(SerQ) + 0.788(U) + 0.489(SE) \quad (9)$$

The final generic equation is given in (10)

$$Y = 0.202(x) + 0.384(z) + 0.364(t) + 0.788(a) + 0.489(b) \quad (10)$$

where $y = OBI$, $x = IQ$, $z = SysQ$, $t = SerQ$, $a = U$, $b = SE$.

The value of organization benefit/impact is found using the Eq. (7) which indicates the value of success of IS in SMEs.

6 Conclusion

The study has an important effort towards a deeper understanding of measuring the success of IS in SMEs. This study succeeded in validating the proposed model and the supporting relationships among the key variables within the local SMEs context. The purpose of the research was to measure the success of IS in SMEs in Gauteng South Africa. Literature review was done, we were able to measure the success of IS in SMEs using a proposed model derived from [5, 14] Model. The hypothesis test results in this study were supported or accepted at a level significance of $= < 0.05$. Regression analysis was done we were able to come up with a confirmed model to find the value of organizational benefit/impact using a systematic of equations which indicate the value of success of IS in SMEs.

References

1. Levy, M., Powell, P.: Information systems strategy for small and medium sized enterprises: an organisational perspective. J. Strat. Inf. Syst. **9**(1), 63–84 (2000)
2. Sharma, M.K., Bhagwat, R.: Performance measurements in the implementation of information systems in small and medium-sized enterprises: a framework and empirical analysis. Meas. Bus. Excell. **10**(4), 8–21 (2006)

3. Sharma, M.K., Bhagwat, R., Dangayach, G.S.: Performance measurement of information systems in small and medium sized enterprises: a strategic perspective. Prod. Plan. Control **19**(1), 1–9 (2008)
4. Peter, S., McLean, E., DeLone, W.: Measuring information systems success: models, dimensions, measures and interrelationships. Eur. J. Inf. Syst. **17**(3), 236–263 (2008). https://doi.org/10.1057/ejis.2008.15
5. McLean, E.R., Delone, W.H.: The DeLone and McLean model of information systems success: a ten-year update. J. Manag. Inf. Syst. **19**(4), 9–30 (2003)
6. Alshardan, A., Goodwin, R., Rampersad, G.: Measuring the benefits of IS in small organizations in developing countries. Int. J. Concept. Comput. Inf. Technol. **1**(2), 12–17 (2013)
7. Muriithi, M.: African small and medium enterprises (SMEs) contributions, challenges and solutions. Eur. J. Res. Reflect. Manag. Sci. **5**(1), 1–3 (2017)
8. Alshardan, A., Goodwin, R., Rampersad, G.: A benefits assessment model of information systems for small organizations in developing countries. Comput. Inf. Sci. **9**(1), 1–8 (2016)
9. Irani, Z.: Editorial Information systems evaluation: what does it mean? J. Enterp. Inf. Manag. **17**(4), 276–282 (2008)
10. Petter, S., DeLone, W., McLean, E.R.: Measuring information systems success: models, dimensions, measures, and interrelationships. Eur. J. Inf. Syst. **17**, 236–263 (2008). https://doi.org/10.1057/ejis.2008.15
11. Seddon, P.B., Staples, S., Patnayakuni, R.: Dimensions of information systems success. Commun. Assoc. Inf. Syst. **2**(20), 1–42 (1999)
12. Broembsen, M.V., Wood, E., Herrington, M.: Global entrepreneurship monitor South African report. In: Global Entrepreneurship Monitor South African Report'. UCT Centre for Innovation and Entrepreneurship, 2005 edn (2005)
13. Santarelli, E., Vivarelli, M.: Entrepreneurship and the process of firms' entry, survival and growth. Ind. Corp. Chang. **16**(3), 455–488 (2007)
14. Yu, P., Qian, S.: Developing a theoretical model and questionnaire survey instrument to measure the success of electronic health records in residential aged care. PLoS One Measur. Success EHR Residential Aged Care **13**(1), 1–18 (2018)
15. Marajos, A., Michael, E.I., Rosemary, N.O.: Teaching agricultural science with instructional materials in secondary schools; what the standard deviation explains about the mean achievement scores. Creat. Educ. **7**(14), 1929–1934 (2016)
16. Ponto, J.: Understanding and evaluating survey research. J. Adv. **2015**(6), 168–171 (2015)
17. Chto, P., Tengeh, R.K., Iwi, C.G.: 'Daring to survive or to grow? The growth aspirations and challenges of survivalist entreprenuers in South Africa. Environ. Econ. **2014**(5), 93–101 (2016)
18. Singh, A.S., Masuku, M.B.: 'Sampling techniques and determination of sample size in applied statistics research: an overview. Int. J. Econ. Commer. Manag. **2**(2), 1–22 (2014)
19. Kim, Y.J., Garrity, E.J., Sanders, G.L.: Success measures of information systems. Inf. Resour. Manag. J. **2016**(4), 1–16 (2016)
20. Bahaddad, A.A.: Evaluating M-commerce systems success: measurement and validation of the DeLone and McLean Model of IS success in Arabic society (GCC case study). J. Bus. Theory Pract. **5**(3), 2329–2644 (2017)
21. Yakubu, M.N., Dasuki, S.I.: Assessing elearning systems success in Nigeria: an application of the Delone and Mclean information systems success model. J. Inf. Technol. Educ.: Res. **17**, 182–202 (2018)
22. Meriouh, Y.A.l., Roky, H.: 'Evaluation by users of an industrial information system (XPPS) based on the DeLone and McLean model for IS success. In: 4th World Conference on Business, Economics and Management, WCBEM, vol. 26, pp. 903–913 (2015)

23. Rajan, C.A., Baral, R.: Adoption of ERP system: an empirical study of factors influencing the usage of ERP and its impact on end user. IIMB Manag. Rev. **27**(2), 105–117 (2015)
24. Privitera, G.J.: Research Methods for the Behavioural Sciences. Sage Publications (2013)
25. Mugenda, O.M., Mugenda, A.G.: Research methods: quantitative and qualitative approaches. In: Proceedings of Research Methods. Acts Press. Nairobi (2012)
26. Kothari, C.: Research methodology; methods and techniques. In: Proceedings of Research Methodology; Methods and Techniques (2011)
27. Manfreda, K.L., Petric, G., Petrovcic, A.: The effect of email invitation elements on response rate in a web survey within an online community. Comput. Human Behav. **56**(56), 320–329 (2016)

Evolving Technological Advancements and Its Effectivity Study in Intelligent Traffic System

A. Mari Kirthima(✉) and S. K. Pushpa

Department of Information Science and Engineering, BMS Institute of Technology and Management, Bengaluru, India
{krithi.a,pushpask}@bmsit.in

Abstract. With the modernization of various technologies, the road and traffic system has also undergone significant growth in adopting smart and intelligent technologies. In this regard, Intelligent Traffic System is also known as ITS, where the term traffic is also interchangeable called transportation. At present, various literature has presented various contributions towards modernizing the concept of ITS by addressing various challenges associated with it. However, it is still open end issue to understand the degree of effectiveness of these studies. Hence, this manuscript contributes towards discussing the briefs of the most frequently exercised approaches in ITS. The core focus of the study is to understand the applicability of these over-tracking emergency vehicles and understand the strength and weakness of existing literature in ITS with the study of research trends.

Keywords: Intelligent transportation system · Vehicular network · Traffic management · Intelligent traffic

1 Introduction

Intelligent Traffic System (ITS) includes technological advancement towards managing various traffic operations [1]. An ITS offers the installation of various devices that can capture real-time traffic feeds, and these feeds of information are stored and analyzed for carrying out different operations [2–5]. The idea of this data collection and analysis in ITS is mainly to evolve up with better solutions towards various problems associated with traffic management [6, 7]. ITS is targeted to be used a wide range of applications viz. parking management, collision avoidance, sequence of the traffic light in dynamic patter, variable limits of speed, freight transportation system with radio frequencies, speed alerts, enforcement of speed over the road, notification system of the emergency vehicle, toll collection, etc. [8]. It should be noted that ITS mainly targets safety and mobility towards transportation systems, and this is accomplished by considering the participation of various modernized technologies that deal with the infrastructure of transportation. There are various studies carried out towards addressing various problems in ITS [9, 10]. Different techniques and methodologies have evolved up in ITS. However, none of the review papers has a simpler discussion of evolving techniques in ITS. Hence, all the review papers offer a similar discussion of theoretical approaches where it is challenging

R. Silhavy et al. (Eds.): CoMeSySo 2021, LNNS 232, pp. 447–457, 2021.
https://doi.org/10.1007/978-3-030-90318-3_37

to conclude any information of trending technologies. It is also challenging to precisely understand the contribution carried out towards investigating existing approaches of ITS. Therefore, the proposed system contributes to evolving a discussion paper that briefs frequently evolved technologies towards modernizing the ITS system. The organization of this manuscript is as follows: Sect. 2 discusses the essentials of ITS followed by a briefing of existing approaches of ITS concerning different categories in Sect. 3. Section 4 discusses the research trend, followed by highlights of the research gap in the existing approach in Sect. 5, and the conclusion is presented in Sect. 6.

2 Essential of ITS

A conventional ITS framework considers the presence of many essential operations; however, essential processes are only of 4 types, i.e., traffic, travellers, emergency, and transit [11]. The traffic information is requested and forwarded to the traveller as well, as it is linked with emergency vehicles and transit points [12]. The essential process will involve collecting data from the camera, sensor, global positioning system while the transmission is carried out in real-time via the traffic management center [13]. This is further followed by series of analytical operations, e.g., data synthesis, rectifying the error, and adaptive analysis [14]. Finally, intelligent information is obtained in different contexts, e.g., speed of the vehicle, accident, route change, condition of the work zone, etc. This information is forwarded to the traveller via highway advisory radio, variable message sign, etc. [15] (Fig. 1).

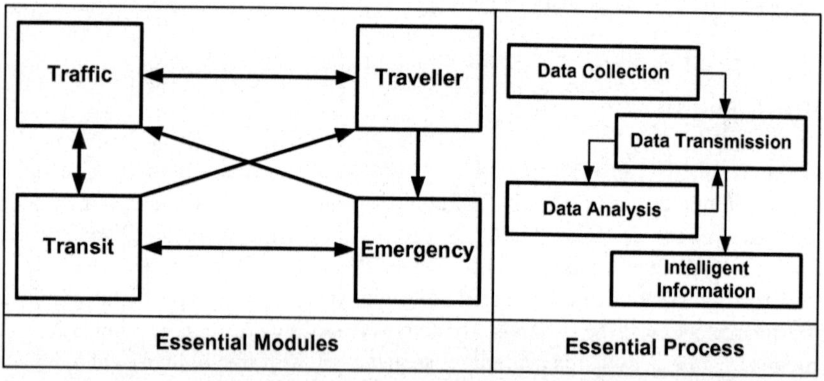

Fig. 1. Essential modules and process in ITS

Some of the significant and standard developments being witnessed in ITS are in Traveler Information System (ITS), Transit Signal Priority (TSP), Traffic Signal Coordination (TSC), Red Light Camera (RLC), Electronic Toll Collection (ETC) [16]. The next section discusses various categories of existing approaches towards modernizing ITS.

3 Existing Approaches

At present, there are various categories of approaches that are used for designing an ITS mechanism. The study papers are collected within publication years between 2016–2021, considering only journals from reputed publishers. Apart from this, the main inclusion criteria for choosing the papers are their connection with the emergency vehicle tracking system. This section discusses 4 significant taxonomies of some of the frequent exercises and recent approaches as follows:

3.1 Routing-Based Approach

This type of approach deals with a data forwarding scheme in ITS where various methodologies are associated with the route, distance, time, and artifacts associated with it. At present, various approaches deal with the routing-based approach as it enables direct communication-based issues in ITS. Din et al. [17] have developed a geographical routing to control overhead and manage various constraints associated with an environment. The study has used an approach of routing that doesn't use any form of the beacon; however, such an idea is in the very nascent stage of development. The work carried out by Luo et al. [18] has presented a solution for issues associated with controlling vehicles in a vehicular network. The work carried out by Paricio et al. [19] have developed a technique that uses traffic weights for constructing multi-maps for making a reliable decision for drivers as well as yield a better form of plan of contingency articulation. The work carried out by Hara et al. [20] has developed a scheme of ITS which extracts traffic information for overcoming the traffic congestion state in selfish routing. The model was assessed using both practical and hybrid road networks to find that it offers reduced travel time as its outcome. Table 1 highlights the summarized outcome of the routing-based approach in ITS.

Table 1. Summary of routing-based approach

Authors	Presented solution	Result	Limitation
Din et al. [17]	Geographic routing scheme	- Applicable for the sparse environment - Reduces delay, a drop of the packet, etc	Absence of benchmarking
Luo et al. [18]	Dynamic routing, multiple agents	Reduced computation time	The scheme is not interoperable
Paricio et al. [19]	Multi-maps using traffic weights	Improved global travel time	It doesn't consider the vehicle and environmental constraint
Hara et al. [20]	Optimal routing (selfish)	Reduce average travel time	Absence of benchmarking
Rezaei et al. [21]	Route guidance system	Reduced overhead & latency	Absence of benchmarking

3.2 Congestion Avoidance Approach

This type of approach uses unique computing resources and techniques to identify the traffic density followed by determining the state of congestion. The secondary aim of this approach is also to find out an alternative route to avoid the current state of congestion. Li et al. [22] have developed a model for predicting the state of congestion in traffic using fuzzy logic and machine learning. The work carried out by Qi et al. [23] has presented a controlling mechanism of the traffic light to prevent traffic congestion. The technique makes integrated use of warning and traffic lights for determining its accuracy with the aid of a two-leveled strategy by Petri net.

Further, predictive modelling has been carried out by Shin et al. [24], where a Short-Term Long Memory (LSTM) is used. The model can also be fine tuning any form of deviation created by spatial and temporal parameters missing. The adoption of machine learning has been further seen in the study of Sun et al. [25], where a supervised learning scheme is used along with computer vision to identify the state of traffic congestion. Wang et al. [26] have used traffic images and subject them to a deep neural network to detect traffic congestion. Table 2 presents a summarized snapshot of congestion avoidance approaches.

Table 2. Summary of congestion avoidance approach

Authors	Presented solution	Result	Limitation
Li et al. [22]	Fuzzy logic, machine learning	Satisfactory predictive outcome	higher iteration leads to complexity
Qi et al. [23]	Controlling traffic light	Timed Petri net	Make use of extensive resources
Shin et al. [24]	LSTM	Potentially reduced error score	Dependent on increased iteration
Sun et al. [25]	Convolution Neural Network, feature extraction from video frames	Higher accuracy of congestion detection	Poor stability and consistency of detection for all frames
Wang et al. [26]	Traffic image, deep neural network	Higher detection accuracy	Higher involvement of processing resources

3.3 Priority-Based Approach

This kind of methodology is applied for the ITS framework, where the priority is set for certain inbound and outbound vehicles from a defined threshold transit point. Koehler et al. [27] have presented a model where a strategy for priority control is being developed to control the vehicle's transit from one point. With the help of correcting the headway and the priority of a vehicle via intersection point, the idea is to reduce cumulative delay. Similar work is also presented by Long et al. [28], where priority is found from multiple routes of vehicle networks. The model presented by Seredynski et al. [29] where it was shown that requests for signal control could be reduced by a driver advisory system within a vehicle. A prioritized model is designed by Truong et al. [30] using stochastic modelling where downstream intersection signal is considered for bus progression, and upstream intersection signal is considered for the study. Zeng et al. [31] have developed a model where transit signal priorities have been developed with adaptive features with significant improvement in delay factors. Table 3 summarizes the priority-based approaches in ITS.

Table 3. Summary of priority-based approach

Authors	Presented solution	Result	Limitation
Koehler et al. [27]	Use of holding and priority for controlling vehicle transit point	Reduced waiting time	Absence of benchmarking
Long et al. [28]	Multi-route transit scheme of signal priority	Delay time reduced	The priority value is fixed
Seredynski et al. [29]	Driver advisory system	Minimize modification frequencies of the signal	Uses signals with fixed time
Truong et al. [30]	Stochastic modelling of arrival time	Reduced delay	Study subjective to delay minimization
Zeng et al. [31]	Optimization of signal timing	Reduced delay	Study subjective to delay minimization

3.4 Interference-Mitigation Approach

The existing system also identifies that communication among the vehicular nodes is significantly affected by interference, and hence it should be mitigated. Hence, this technique is used to ensure error-free communication in ITS. The work carried out by Angskog et al. [32] has developed a secure mechanism to resist antagonist attacks in the vehicular network considering interference caused intentionally by the electromagnetic. Hu et al. [33] have discussed outage probability associated with the vehicular networks and interference in them. This model adopts arbitrary multiple access protocols to control vehicular Adhoc Network interference (VANET). Koufos and Dettmann [34] have implemented a unique distance-based approach between two communicating vehicles to control interference. Yang et al. [35] have developed a model where multi-hop communication is carried out by optimizing spectrum and managing interference in VANET. Zelikman and Segal [36] have presented a model using Medium Access Control (MAC) along with clustering protocols to resist interference. A model is designed considering varied types of dissemination of message and its characteristics where the dual-slot is used along with clustering for mitigating interference. Table 4 summarizes the interference mitigation approaches in ITS.

Table 4. Summary of interference mitigation approach

Authors	Presented solution	Result	Limitation
Angskog et al. [32]	Security via intentional electromagnetic interference	Simplified architecture	No extensive analysis is carried out
Hu et al. [33]	Random multiple access protocol in VANET	Reduction in outage probability	Absence of benchmarking
Koufos and Dettmann [34]	Distance-based interference avoidance	Better interference reduction	Model doesn't have the inclusion of resource factors
Yang et al. [35]	Optimizing spectral efficiency	Maximization of throughput	Not scalable for heterogeneous nodes
Zelikman and Segal [36]	MAC, clustering	Supportive of broadcasting and collision avoidance	Absence of extensive test cases

4 Research Trend

All the associated core problems associated with ITS have been considered published between 2016–2021, as shown in Fig. 2. It show cases those studies towards ITS have been mainly focused on security and congestion avoidance. There has been significant availability of security approaches in ITS, and hence, it is no more the most significant concern in ITS. The degree of security threats in ITS is equivalent to that in VANET and other wireless networks. Hence, there is no significant trend towards this.

The outcome in Fig. 2 also show cases those fewer studies were carried out using a priority-based and interference-based approach. In reality, the existing approaches for addressing this problem in sensor networks, ad-hoc networks, and other wireless networks are less applicable in ITS. Hence, this can be looked upon as a significant point that calls for more research in this area. Routing-based approaches are increasing in ITS. It is consistently under attention with the evolution of different techniques. More or fewer policies for data forwarding are also incorporated in other approaches (priority-based, interference-mitigation, security, congestion avoidance). Hence, it can be said that the routing-based approach is one the most frequently used scheme in ITS.

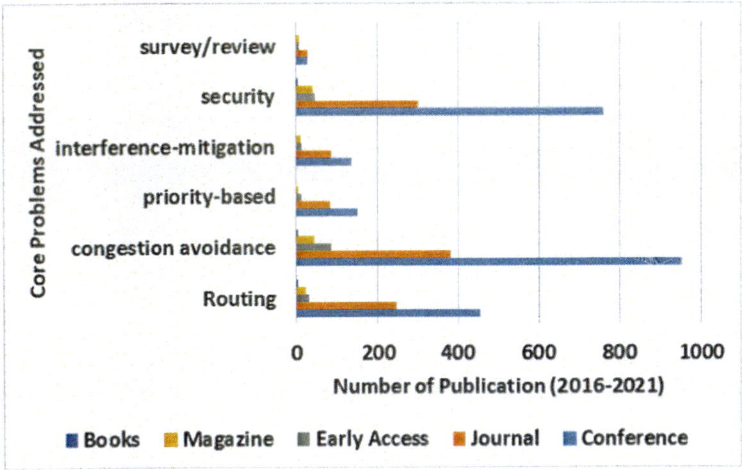

Fig. 2. Graphical Outcome of Research Trend

5 Research Gap

After reviewing the existing approaches associated with the ITS system, it is noticed that there are the following research gaps:

- *Biased Constraint Formulation*: At present, fewer studies are related to constraints of the vehicular nodes, and more constraint formulation is towards the environment. Apart from this, connectivity among constraints is not studied upon, leading to the non-evolution of potential computational modelling of ITS.
- *Non-Inclusion of Smart Feature*: Existing models extract traffic-related features from various forms and types of signals, but problems associated with data quality and reliability are not considered. The information collected does not offer more contextual information, which reduces the scope of successful implementation of even better predictive approaches.
- *Unrealistic Modelling Design*: Majority of existing study has been carried out in highly controlled research environment where the various practical aspect of vehicular network and traffic scene is not considered. Various conditions associated with emergency conditions are required to be dynamically designed.
- *Lesser Studies towards Emergency Vehicle Tracking*: All present studies consider emergency a fixed parameter for a vehicle, whereas, in reality, the state of emergency is stochastic and dynamic. None of the existing models is capable of tracking vehicles with dynamic emergencies with satisfactory performance parameters.
- *Lesser Benchmarking*: Despitethe large number of studies being carried out, the test environment is not found to be compared with other standard techniques, or a benchmarked test environment is ever considered with extensive test cases to prove the technical adoption of the presented ITS model.

Hence, all the research mentioned above is required to be addressed to evolve up with a potential computational model of an ITS. The next section briefs about the study's contribution.

6 Conclusion

This paper has offered a very compact insight into the evolution of research approaches in ITS, which is the futuristic trend of the traffic management system. The contribution of this paper is in the form of its findings as follows:

- The present studies towards modernizing ITS are mainly towards routing, security, and congestion avoidance; however, this emphasis is not towards emergency vehicle classification.
- The emergency factor is fixed in the existing model, which cannot deal with the dynamic situation where the emergency attribute must be changed. Hence, there is a severe need to redefine and implement the dynamic prioritization factor in ITS.
- The majority of the existing approaches in ITS has never been benchmarked or assessed for computational complexity. The lack of this evidence doesn't offer evidence of technology adoption.
- ITS's existing approaches are much focused on developing application-based models, which area highly scattered domain of research. There is no computational framework that can address the primary underlying common issues of ITS focus on scalability and cost-effectiveness.

Apart from this, it has been noticed that Software Defined Network (SDN), when integrated with ITS, offers some interesting outcomes [37–46]. The studies carried out in existing times show that there is a good scope of involving SDN. Hence, future work will be emphasized over developing a computational framework that can use SDN to offer more insight over the traffic-related information in its network plane that can be streamlined with ITS system for emergency vehicle tracking.

References

1. Zhu, L., Yu, F.R., Wang, Y., Ning, B., Tang, T.: Big data analytics in intelligent transportation systems: a survey. IEEE Trans. Intell. Transp. Syst. **20**(1), 383–398 (2018). https://doi.org/10.1109/TITS.2018.2815678
2. Wang, Y., Zhang, Y., Piao, X., Liu, H., Zhang, K.: Traffic data reconstruction via adaptive spatial-temporal correlations. IEEE Trans. Intell. Transp. Syst. **20**(4), 1531–1543 (2019). https://doi.org/10.1109/TITS.2018.2854968
3. Alemazkoor, N., Meidani, H.: Efficient collection of connected vehicles data with precision guarantees. IEEE Trans. Intell. Transp. Syst. **21**(11), 4637–4645 (2020). https://doi.org/10.1109/TITS.2019.2942568
4. Jin, Y., Tan, E., Li, L., Wang, G., Wang, J., Wang, P.: Hybrid traffic forecasting model with fusion of multiple spatial toll collection data and remote microwave sensor data. IEEE Access **6**, 79211–79221 (2018). https://doi.org/10.1109/ACCESS.2018.2884933

5. Han, L., Zheng, K., Zhao, L., Wang, X., Wen, H.: Content-aware traffic data completion in ITS based on generative adversarial nets. IEEE Trans. Veh. Technol. **69**(10), 11950–11962 (2020). https://doi.org/10.1109/TVT.2020.3007025
6. Nie, W., Lee, V.C.S., Niyato, D., Duan, Y., Liu, K., Nutanong, S.: A quality-oriented data collection scheme in vehicular sensor networks. IEEE Trans. Veh. Technol. **67**(7), 5570–5584 (2018). https://doi.org/10.1109/TVT.2018.2818190
7. Li, T., Tian, S., Liu, A., Liu, H., Pei, T.: DDSV: optimizing delay and delivery ratio for multimedia big data collection in mobile sensing vehicles. IEEE Internet Things J. **5**(5), 3474–3486 (2018). https://doi.org/10.1109/JIOT.2018.2847243
8. Rose, G., Gunasekaran, R., Aishwarya, G.: Mobility management for critical time and delay tolerant applications in vehicular networks. In: Tenth International Conference on Advanced Computing (ICoAC), pp. 344–348 (2018). https://doi.org/10.1109/ICoAC44903.2018.8939075
9. Tian, B., et al.: Hierarchical and networked vehicle surveillance in ITS: a survey. IEEE Trans. Intell. Transp. Syst. **18**(1), 25–48 (2017). https://doi.org/10.1109/TITS.2016.2552778
10. Chen, B., Cheng, H.H.: A review of the applications of agent technology in traffic and transportation systems. IEEE Trans. Intell. Transp. Syst. **11**(2), 485–497 (2010). https://doi.org/10.1109/TITS.2010.2048313
11. Alsrehin, N.O., Klaib, A.F., Magableh, A.: Intelligent transportation and control systems using data mining and machine learning techniques: a comprehensive study. IEEE Access **7**, 49830–49857 (2019). https://doi.org/10.1109/ACCESS.2019.2909114
12. Mershad, K., Artail, H.: A framework for secure and efficient data acquisition in vehicular Ad hoc networks. IEEE Trans. Veh. Technol. **62**(2), 536–551 (2013). https://doi.org/10.1109/TVT.2012.2226613
13. Suciu, G., Vochin, M., Vulpe, A., Fratu, O.: Vehicular mobile data collection platform to support the development of Intelligent Transportation Systems. In: 24th Telecommunications Forum (TELFOR), pp. 1–4 (2016). https://doi.org/10.1109/TELFOR.2016.7818753
14. Liu, Y., Chen, X., Chen, C., Guan, X.: Traffic big data analysis supporting vehicular network access recommendation. In: IEEE International Conference on Communications (ICC), pp. 1–6 (2016). https://doi.org/10.1109/ICC.2016.7510775
15. Al-Najada, H., Mahgoub, I.: Real-time incident clearance time prediction using traffic data from the internet of mobility sensors. In: IEEE 15th International Conference on Dependable, Autonomic and Secure Computing, 15th International Conference on Pervasive Intelligence and Computing, 3rd International Conference on Big Data Intelligence and Computing and Cyber Science and Technology Congress(DASC/PiCom/DataCom/CyberSciTech), pp. 728–735 (2017). https://doi.org/10.1109/DASC-PICom-DataCom-CyberSciTec.2017.127
16. ITS Research Fact Sheets - Benefits of Intelligent Transportation Systems. https://www.its.dot.gov/factsheets/benefits_factsheet.htm. Accessed 06 July 2021
17. Din, S., Qureshi, K.N., Afsar, M.S., Rodrigues, J.J.P.C., Ahmad, A., Choi, G.S.: Beaconless traffic-aware geographical routing protocol for intelligent transportation system. IEEE Access **8**, 187671–187686 (2020). https://doi.org/10.1109/ACCESS.2020.3030982
18. Luo, R., van den Boom, T.J.J., De Schutter, B.: Multi-agent dynamic routing of a fleet of cybercars. IEEE Trans. Intell. Transp. Syst. **19**(5), 1340–1352 (2018). https://doi.org/10.1109/TITS.2017.2710480
19. Paricio, A., Lopez-Carmona, M.A.: Urban traffic routing using weighted multi-map strategies. IEEE Access **7**, 153086–153101 (2019). https://doi.org/10.1109/ACCESS.2019.2947699
20. Hara, T., Sasabe, M., Kasahara, S.: Selfish yet optimal routing by adjusting perceived traffic information of road networks. IEEE Open J. Intell. Transp. Syst. **1**, 120–133 (2020). https://doi.org/10.1109/OJITS.2020.3019935

21. Rezaei, M., Noori, H., Mohammadkhani Razlighi, M., Nickray, M.: ReFOCUS+: multi-layers real-time intelligent route guidance system with congestion detection and avoidance. IEEE Trans. Intell. Transp. Syst. **22**(1), 50–63 (2021). https://doi.org/10.1109/TITS.2019.2952524
22. Li, L., Lin, H., Wan, J., Ma, Z., Wang, H.: MF-TCPV: a machine learning and fuzzy comprehensive evaluation-based framework for traffic congestion prediction and visualization. IEEE Access **8**, 227113–227125 (2020). https://doi.org/10.1109/ACCESS.2020.3043582
23. Qi, L., Zhou, M., Luan, W.: A two-level traffic light control strategy for preventing incident-based urban traffic congestion. IEEE Trans. Intell. Transp. Syst. **19**(1), 13–24 (2018). https://doi.org/10.1109/TITS.2016.2625324
24. Shin, D.-H., Chung, K., Park, R.C.: Prediction of traffic congestion based on LSTM through correction of missing temporal and spatial data. IEEE Access **8**, 150784–150796 (2020). https://doi.org/10.1109/ACCESS.2020.3016469
25. Sun, Z., Wang, P., Wang, J., Peng, X., Jin, Y.: exploiting deeply supervised inception networks for automatically detecting traffic congestion on freeway in china using ultra-low frame rate videos. IEEE Access **8**, 21226–21235 (2020). https://doi.org/10.1109/ACCESS.2020.2968597
26. Wang, P., et al.: Regional detection of traffic congestion using in a large-scale surveillance system via deep residual TrafficNet. IEEE Access **6**, 68910–68919 (2018). https://doi.org/10.1109/ACCESS.2018.2879809
27. Koehler, L.A., Seman, L.O., Kraus, W., Camponogara, E.: Real-time integrated holding and priority control strategy for transit systems. IEEE Trans. Intell. Transp. Syst. **20**(9), 3459–3469 (2019). https://doi.org/10.1109/TITS.2018.2876868
28. Long, K., Wei, J., Gu, J., Yang, X.: Headway-based multi-route transit signal priority at isolated intersection. IEEE Access **8**, 187824–187831 (2020). https://doi.org/10.1109/ACCESS.2020.3030686
29. Seredynski, M., Laskaris, G., Viti, F.: Analysis of cooperative bus priority at traffic signals. IEEE Trans. Intell. Transp. Syst. **21**(5), 1929–1940 (2020). https://doi.org/10.1109/TITS.2019.2908521
30. Truong, L.T., Currie, G., Wallace, M., De Gruyter, C., An, K.: Coordinated transit signal priority model considering stochastic bus arrival time. IEEE Trans. Intell. Transp. Syst. **20**(4), 1269–1277 (2019). https://doi.org/10.1109/TITS.2018.2844199
31. Zeng, X., Zhang, Y., Jiao, J., Yin, K.: Route-based transit signal priority using connected vehicle technology to promote bus schedule adherence. IEEE Trans. Intell. Transp. Syst. **22**(2), 1174–1184 (2021). https://doi.org/10.1109/TITS.2020.2963839
32. Ängskog, P., Näsman, P., Mattsson, L.G.: Resilience to intentional electromagnetic interference is required for connected autonomous vehicles. IEEE Trans. Electromagn. Compat. **61**(5), 1552–1559 (2019). https://doi.org/10.1109/TEMC.2018.2866178
33. Hu, D., Wu, J., Fan, P.: On the outage probability of interference-limited multi-hop linear vehicular ad-hoc network. IEEE Access **6**, 75395–75406 (2018). https://doi.org/10.1109/ACCESS.2018.2883648
34. Koufos, K., Dettmann, C.P.: Moments of interference in vehicular networks with hardcore headway distance. IEEE Trans. Wirel. Commun. **17**(12), 8330–8341 (2018). https://doi.org/10.1109/TWC.2018.2876241
35. Yang, F., Han, J., Ding, X., Wei, Z., Bi, X.: Spectral efficiency optimization and interference management for multi-hop D2D communications in VANETs. IEEE Trans. Veh. Technol. **69**(6), 6422–6436 (2020). https://doi.org/10.1109/TVT.2020.2987526
36. Zelikman, D., Segal, M.: Reducing interferences in VANETs. IEEE Trans. Intell. Transp. Syst. **16**(3), 1582–1587 (2015). https://doi.org/10.1109/TITS.2014.2348498
37. Abugabah, A., Alzubi, A.A., Alfarraj, O., Al-Maitah, M., Alnumay, W.S.: Intelligent traffic engineering in software-defined vehicular networking based on multi-path routing. IEEE Access **8**, 62334–62342 (2020). https://doi.org/10.1109/ACCESS.2020.2983204

38. Bao, K., Matyjas, J.D., Hu, F., Kumar, S.: Intelligent software-defined mesh networks with link-failure adaptive traffic balancing. IEEE Trans. Cogn. Commun. Netw. **4**(2), 266–276 (2018). https://doi.org/10.1109/TCCN.2018.2790974
39. Bhatia, J., Kakadia, P., Bhavsar, M., Tanwar, S.: SDN-enabled network coding-based secure data dissemination in VANET environment. IEEE Internet Things J. **7**(7), 6078–6087 (2020). https://doi.org/10.1109/JIOT.2019.2956964
40. Bi, Y., Han, G., Lin, C., Peng, Y., Pu, H., Jia, Y.: Intelligent quality of service aware traffic forwarding for software-defined networking/open shortest path first hybrid industrial internet. IEEE Trans. Ind. Inform. **16**(2), 1395–1405 (2020). https://doi.org/10.1109/TII.2019.2946045
41. Lin, C., Han, G., Du, J., Xu, T., Shu, L., Lv, Z.: Spatiotemporal congestion-aware path planning toward intelligent transportation systems in software-defined smart city IoT. IEEE Internet Things J. **7**(9), 8012–8024 (2020). https://doi.org/10.1109/JIOT.2020.2994963
42. Lin, C., Han, G., Du, J., Xu, T., Peng, Y.: Adaptive traffic engineering based on active network measurement towards software-defined internet of vehicles. IEEE Trans. Intell. Transp. Syst. **22**(6), 3697–3706 (2021). https://doi.org/10.1109/TITS.2020.3028990
43. Luo, G., Li, J., Zhang, L., Yuan, Q., Liu, Z., Yang, F.: sdnMAC: a software-defined network inspired MAC protocol for cooperative safety in VANETs. IEEE Trans. Intell. Transp. Syst. **19**(6), 2011–2024 (2018). https://doi.org/10.1109/TITS.2017.2736887
44. Meng, Q., Wei, J., Wang, X., Guo, H.: Intelligent routing orchestration for ultra-low latency transport networks. IEEE Access **8**, 128324–128336 (2020). https://doi.org/10.1109/ACCESS.2020.3008721
45. Wang, J., He, B., Wang, J., Li, T.: Intelligent VNFs selection based on traffic identification in vehicular cloud networks. IEEE Trans. Veh. Technol. **68**(5), 4140–4147 (2019). https://doi.org/10.1109/TVT.2018.2880754
46. Xiaolong, X., Yun, C., Liuyun, H., Anup, K.: MTSS: multi-path traffic scheduling mechanism based on SDN. J. Syst. Eng. Electron. **30**(5), 974–984 (2019). https://doi.org/10.21629/JSEE.2019.05.14

User-Configurable Structure of Decision Support System for Clinical Risk Management

Elena Kleymenova[1,3,4], Mikhail Dronov[2,5], Liubov Yashina[1,3(✉)], Svetlana Payushchik[1], Maria Nigmatkulova[1,4], and Vitaly Otdelenov[1,4]

[1] General Medical Center of the Bank of Russia, 66, Sevastopolsky pr., Moscow 117593, Russia
[2] Faculty of Physics, M.V. Lomonosov Moscow State University, Leninskie Gory, GSP-1, Moscow 119991, Russia
[3] Federal Research Center "Computer Science and Control" of the Russian Academy of Sciences, 44, Vavilova Street, Moscow 119333, Russia
[4] Russian Medical Academy of Continuous Professional Education, 2/1, Barrikadnaya Street, Moscow 125993, Russia
[5] Information Technology Department, Moscow City University, 4, Vtoroy Selskohoziajstvenny proezd, Moscow 129226, Russia

Abstract. The paper describes an application of relational structure-independent database for the design of user-configurable information system aimed at the intellectual support of clinical risk management in a hospital. The system represents a clinical decision support system (CDSS) and includes three modules: prevention of hospital-acquired adverse events (HA-AE), HA-AE prevention processes and audit of HA-AE cases. Thanks to flexible system configuration and graphic user interface, the clinical experts are able to compose complex knowledge artifacts directly working with the system, without engagement of knowledge engineers to support knowledge formalization or highly skilled professionals to ensure the physical layer performance. The successful practical approval of CDSS in a medical facility opens up prospects for a new level of cognitive support for diagnostic and treatment processes, patient safety improvement and healthcare costs reduction. #COMESYSO1120.

Keywords: Decision support system · Clinical risk management · Structure-independent database

1 Introduction

Patient safety is fundamental to the provision of health care in all settings. However, avoidable adverse events, errors and risks associated with health care remain major challenges for patient safety globally. Available evidence suggests that hospitalizations in low- and middle-income countries lead to 134 million adverse events annually, contributing to 2.6 million deaths [1]. Estimates indicate that in high-income countries, about one in 10 patients is harmed while receiving hospital care [2].

R. Silhavy et al. (Eds.): CoMeSySo 2021, LNNS 232, pp. 458–468, 2021.
https://doi.org/10.1007/978-3-030-90318-3_38

The clinical risk management framework in a general hospital is usually based on risk management standards (such as ISO 31000: 2018) and includes the following elements: 1) setting goals and objectives - reducing the risk of preventable hospital-acquired adverse events (HA-AE); 2) identifying risks in various information sources; 3) risk analysis and evaluation: assessing the rate, severity and preventability of HA-AE via standard approaches; 4) risk treatment: reducing the rate of HA-AE via process standardization and medical error reduction; 5) risk monitoring with measurable indicators of HA-AE prevention.

In the era of digital transformation in healthcare, clinical decision support systems (CDSS) and software tools based on artificial intelligence technologies (AIT) are becoming more and more compelling. The effectiveness and viability of such systems directly depends on their adaptability and flexibility, as well as on the applicability of their configuration for solving highly specialized and changeable tasks. The need to reduce the rate of HA-AE risk realizations and associated costs makes the use of AIT-applications for clinical risk management an urgent task. However, modern electronic medical records (EMR) usually lack specialized functions for identifying, analyzing and monitoring clinical risks.

The aim of the study was to develop and test a user-configurable information system intended for the intellectual support of clinical risk management in a general hospital.

2 Methods

2.1 The Structure of the Decision Support System

Based on the clinical risk management framework, a multifunctional decision support system for clinical risk management (SACRAL-CRM) was developed including three modules:

1. IHA-AE prevention: decision support for risk assessment, choice of preventive measures and their timeline for various types of HA-AE;
2. HA-AE prevention audit: current control of HA-AE prevention processes;
3. Analysis of realized risks: audit of HA-AE cases, assessment of HA-AE severity and preventability.

The first module is a knowledge-based CDSS for clinicians. It consists of computerized clinical guidelines, related to HA-AE prevention. Special scenario for preventing each HA-AE type includes decision-making algorithms based on recommendations from clinical guidelines and patient's individual risk factors. These data are translated into questions and answers. A branched tree of questions and answers comprises an epicrisis. Medical reasoning is based on if… then… rules combined into decision tables. A protocol with recommendations for specific preventive measures to be prescribed for a patient is generated after manual data entering. The module inherited the underlying principles of its creation from the previous system – a multifunctional CDSS (System for Automated Clinical Recommendations and Audit of treatment – SACRAL) [3].

The second module is designed for clinical experts (auditors), who control HA-AE prevention processes. The main functionality of the module is to support the process of checking the accuracy of preventive epicrisis completion by the doctor and preventive measures prescription in EMR according to the protocol.

The third module supports the process of standardized evaluation of adverse events. Its scenarios contain criteria for the in-hospital nature and severity of the events, possible factors contributing to the event. Comparison of the recommended patient management strategy and practical implementation of the CDSS recommendations allows auditor to draw conclusions about the preventability of complications, effectiveness of preventive strategy, and to determine measures for effective treatment of clinical risks.

Audit results are accumulated in a database to calculate the frequency of various HA-AE and the cumulative harm from each HA-AE type, to analyze common factors contributing to the HA-AE development, and to evaluate the effectiveness of corrective measures (see Fig. 1).

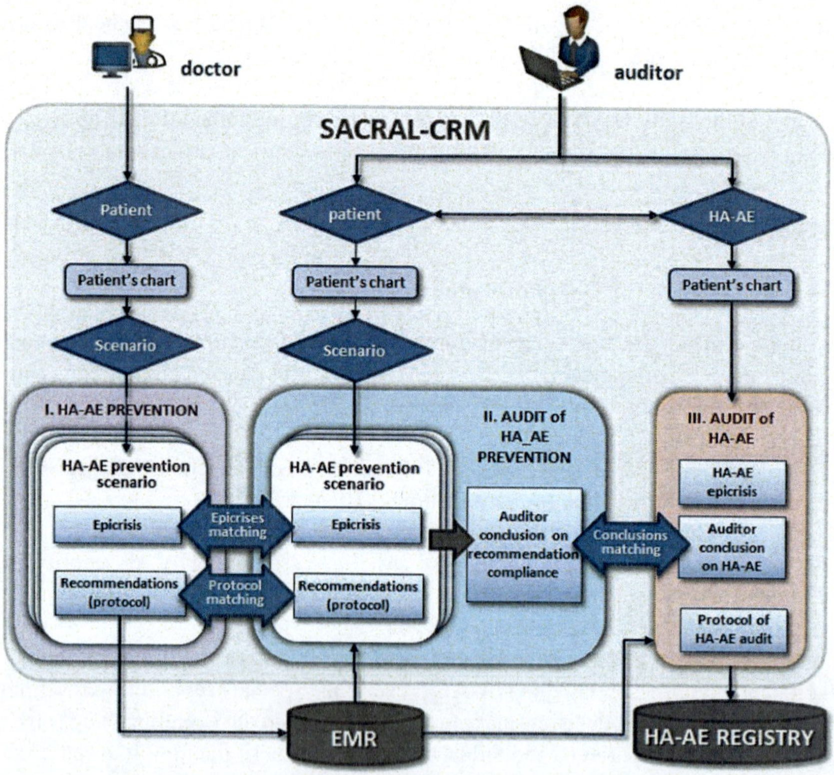

Fig. 1. The structure of user modules in SACRAL-CRM (abbreviations in the text).

One of the main principles of SACRAL-CRM design was to provide clinical experts the possibility of creating scenarios without programmers or even knowledge engineers. This possibility was accomplished as a special technology for generation of templates, calculators, questions and answers, to group questions, to link questions with predetermined sets of answers, as well as to provide references to full-text clinical guidelines.

Therefore, SACRAL-CRM has two user subsystems:

1) expert-analytical subsystem, intended for clinical experts who generate scenarios for the HA-AE prevention and audit,
2) medical subsystem, designed for clinicians who enter patient data and download decision protocols with recommended preventive measures.

Thus, Sacral-CRM provides the following main functions:

- transformation of textual clinical guidelines into the format of automated algorithms;
- configuring algorithms in the form of question-answers sets ("Epicrises"), auxiliary algorithms ("Calculators"), and output forms ("Protocols");
- manual input of patient data required for decision making according to the algorithms;
- the possibility of updating the knowledge base (configuration) of the system;
- automatic calculation of scenarios, calculators and protocols;
- storage of the patients data, scenarios, protocols and other objects of the system;
- recording and storage of the history of all system events in the event log;
- access control based on the predetermined role model.

2.2 User-Configurable Principles in the SACRAL-CRM

Current practical implementation of guideline-based CDSS is hampered by rapid updating of the subject area. It is known that 50% of knowledge in clinical guidelines becomes outdated in 5.8 years, and 90% of guidelines remain relevant for an average 3.6 years [4]. In this case, implemented guideline-based CDSS may soon become partially or completely inapplicable.

To meet this challenge SACRAL-CRM was implemented as a user-configurable information system (UCIS). This special class of information systems appeared as a flexible tool, able to change along with the expanding knowledge base of the subject domain and end-user demands [5]. In contrast to classical information systems, UCIS permits to configure not only interfaces and security policy rules, but also database structure (subject area) and rules of information resources processing (business logic) [6].

A static structure of UCIS relational database is isolated from the logical model of the stored data. Thus, UCIS can operate on data with a variable structure. This solution involves the creation of universal database to store the user data as well as the information about the subject area in the form of its logical data model. Such solutions are called structurally independent databases (SIDB). Their characteristic feature is that changes at the conceptual or logical level of the domain data model do not affect the physical structure of the database [7].

The principle of SACRAL-CRM operation can be described as follows:

- the logical data model of the subject area, defined by one of the classical types, is converted by deterministic transformation procedures into the form of its storage in the SIDB;

- the logical level of the SIDB is filled with the appropriate references, hierarchies, key values, etc.;
- the logical model of the SIDB is converted into a relational form by means of deterministic procedures. Metadata and data are distributed in appropriate tables.
- the process of reverse transformation is also based on deterministic procedures and has no fundamental differences.

The SACRAL-CRM conceptual model contains the basic entities required for the database design (see Fig. 2). A real model can contain a larger number of tables, according to the system tasks and requirements for the relational normalization.

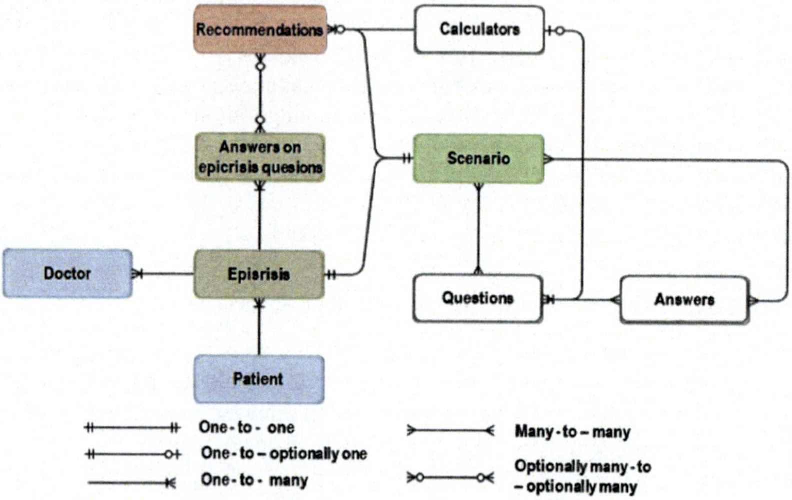

Fig. 2. Entities and their relationships in SACRAL-CRM.

The following stored basic components of SIDB are defined for SACRAL-CRM:

- objects of the subject area;
- characteristics of objects of the subject area;
- connections of characteristics with objects;
- connections of objects with objects;
- instances of objects;
- links of object instances;
- data stored as values of object's characteristics.

For relationships between schemes in the conceptual model, the following methods are proposed:

- registration of metadata for recording;
- adding metadata for recording;
- extracting an array with metadata keys from a record.

The use of the relationships and combined indexes in the model minimizes the number of database queries and execution time for typical calls to the CDSS. Examples of such calls are:

- editing the script by a clinical expert;
- data input by a doctor (clinical expert);
- receiving system recommendations based on data in the epicrisis.

3 Results

An open source platform MariaDB was used as a database management system (DBMS) for SACRAL-CRM. The program code of the DBMS was implemented in the declarative query language SQL.

During creating and editing scenarios, clinical experts can generate an unlimited number of questions and answers. There is a possibility to select type for each question: single choice, multiple choice, single or multi-line text, number, calculator, date, time. Questions in the scenario can be create de novo or picked up from the storage of questions created previously for other scenarios. Questions can be merged into groups, linked to the answer to another question or a set of answers. This mechanism implemented by means of graphic user interface (GUI) substitutes the use of standard logical operators, as far as clinical experts lack the knowledge of expression languages, such as GELLO or Clinical Quality Language (CQL) (see Fig. 3).

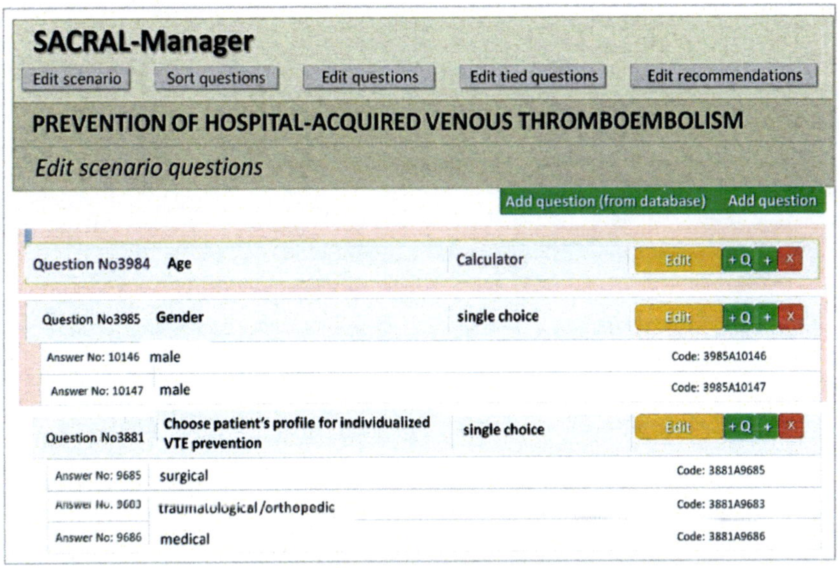

Fig. 3. Graphic user interface for generating a set of questions and answers in a scenario.

The ergonomics of the interface for end-users (doctors) can be improved by linking questions with the answer(s) to another question (tied questions). For example, personalized prevention of hospital-acquired venous thromboembolism (HA-VTE) is based on the calculation of risk scores that differ for patients of surgical, orthopedic, medical or oncological profiles. After selecting a patient profile, the system offers only relevant risk scales (see Fig. 4).

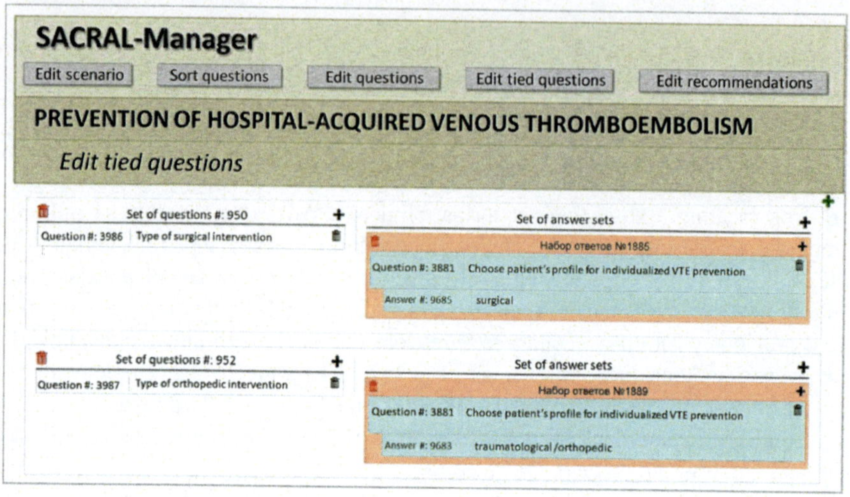

Fig. 4. Graphic user interface for binding questions (the left column) to the answers on another question (the right column).

Internal logic for recommendations can be represented not only in GUI, but also in human readable expressions. Example 1 below presents a constructed expression of internal system logic for a decision support recommendation:

Example 1:

$$
\begin{aligned}
& IF(4193A10508\ AND\ NOT\ 4183A10480\ AND\ 4192A10526) \\
& OR(4193A10508\ AND\ 4183A10480\ AND\ 4192A10530) \\
& OR(4193A10508\ AND\ NOT\ 4183A10480\ AND\ 4192A10527) \\
& THEN\ R3809
\end{aligned}
$$

This expression can be extended to more human readable form as:

$$
\begin{aligned}
& IF\,(Anticoagulant\ taken\ is\ warfarin\ \ldots) \\
& THEN\,(Perioperative\ interruption\ of\ warfarin\ is\ required)
\end{aligned}
$$

One of the challenging tasks in the SACRAL-CRM development was to embed the creation of calculators in the GUI. There are three types of calculators in the SACRAL-CRM:

- *formula calculators* process a formula using the numerical values of the answers (or calculators), support math functions, comparison operations and logical functions;
- *informational calculators* are used for algorithmization of risk scales; they can assign points to specific answers and to sum them up (they can additionally process formulas when the answer is the result of formula calculation);
- *temporal calculator* calculates the specified time intervals.

User-friendly interface for calculator construction allows clinical experts to select the questions involved in the calculation and assign codes to them, if required. Questions can be selected from the patient chart or scenarios. Figure 5 presents the example of the formula calculator for the glomerular filtration rate:

$$\begin{aligned}&Pow(0.993, B) * if\,(Gender = 1.141\\&* Pow(A/88.4/0.9, if\,(A <= 80, -0.411, -1.209)), 144\\&* Pow(A/88.4/0.7, if\,(A <= 62, -0.329, -1.209)))\end{aligned}$$

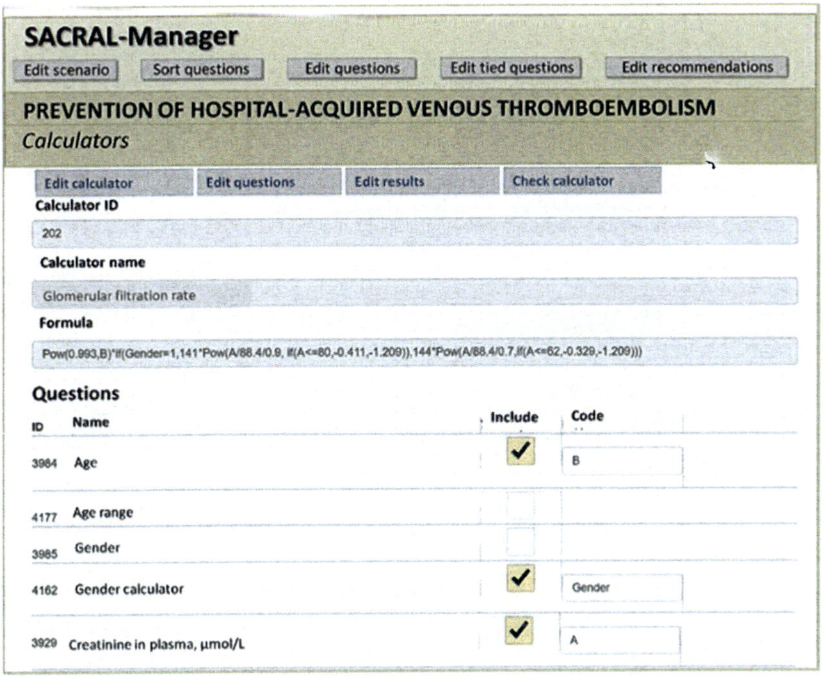

Fig. 5. Graphic user interface for constructing a formula calculator (glomerular filtration rate).

A convenient interface with a drop-down hierarchical tree has been created to generate recommendations. Recommendations can be organized into groups ("recommendation sets"). Ranks assigned to groups and individual recommendations determine the order of their presentation in the resulting protocol. Recommendations are textual fragments of clinical guidelines; appropriate field contains links to them together with the grade of evidence and strength of the recommendations (see Fig. 6).

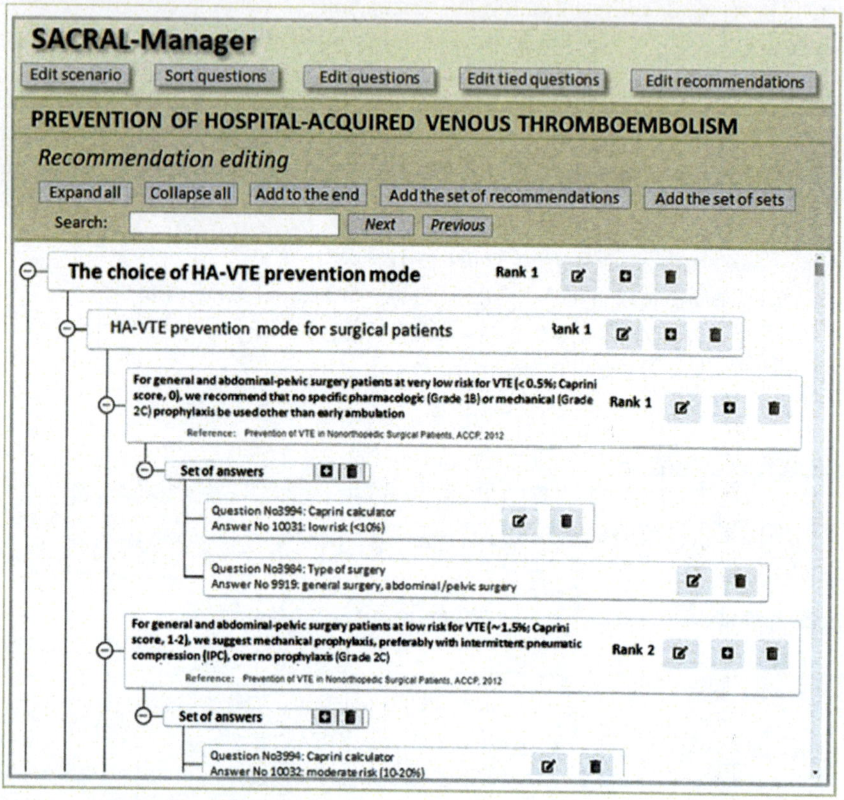

Fig. 6. Fragment of the decision tree for recommendations on prevention of hospital-acquired venous thromboembolism (HA-VTE).

4 Discussions

Guideline-based CDSS are often mentioned as an invaluable tool for improvement of healthcare quality and safety [8]. Nevertheless, their implementation in clinical practice is far from widespread, mainly due to the problems with creation and updating of the CDSS knowledge base [9]. Difficulties in keeping the knowledge base up to date are rooted in the variety of scientific information sources, in the determining an optimal level of knowledge structuring and formalization, in the need to combine and regularly update them, as well as in responsibility for the possible disastrous consequences of CDSS errors.

The main feature of our approach to the knowledge presentation in the CDSS is the absence of translating structured knowledge into a fully formalized language. We believe that the possibility to composed complex knowledge artifacts via GUI is one of strongest advantages of the SACRAL-CRM. While the majority of classical CDSS require the direct use of GELLO or CQL, SACRAL-CRM has no such requirements. Owing to this, clinical experts are able to input the scenario content straightforwardly into the system without any programming skills or additional conversions of input data. Besides, it is quite simple to check if a decision algorithm works properly. This technology was successfully implemented in the medical facility for creating more than a dozen scenarios for HA-AE prevention and audit.

On the other hand, this approach hampers the integration of such CDSS with other information systems.

SIDB of the SACRAL-CRM in couple with graphic user interface also simplifies the process of creating and modifying user database structures. Replacement of the development process with configuration process provides the dynamism both for the database and CDSS as a whole. Therefore, there is no need to engage highly skilled professionals to support the physical layer performance in case of a flexible CDSS with user-friendly GUI, thus both time and money can be saved.

5 Conclusion

CDSS creation as a user-configurable information system based on structurally independent database showed its applicability in healthcare domain, which is characterized by the high rate of knowledge update. Despite the lack of programming skills, clinical experts managed to create a range of decision support modules in the field of clinical risk management, including risk identification, analysis, treatment and continuous monitoring.

The described approach can provide a new level of intellectual information support for diagnostic and treatment processes. Its implementation in healthcare organizations can contribute to the improvement of patient safety and reduction of healthcare costs.

Acknowledgements. This work was supported by the Russian Foundation for Basic Research, grant №. 19-29-01132.

References

1. National Academies of Sciences, Engineering, and Medicine: Crossing the global quality chasm: improving health care worldwide. The National Academies Press, Washington (2018). https://www.nap.edu/catalog/25152/crossing-the-global-quality-chasm-improving-health-care-worldwide. Accessed 18 Oct 2020
2. Slawomirski, L., Auraaen, A., Klazinga N.: The economics of patient safety: strengthening a value-based approach to reducing patient harm at national level. OECD Health Working Papers, 96. OECD Publ., Paris (2017). http://dx.doi.org/10.1787/5a9858cd-en. Accessed 20 May 2018
3. Nazarenko, G., Kleymenova, E., Payushchik, S., Yashina, L.: The role of guideline-based decision support systems in execution of complex standard operating procedures. In: Williams, E. (ed.) Patient Safety and Management: Perspectives, Principles and Emerging Issues, pp. 37–66. Nova Science Publ., Hauppauge (2017)

4. Shekelle, P., Ortiz, E., Rhodes, S., et al.: Validity of the agency for healthcare research and quality clinical practice guidelines: how quickly do guidelines become outdated? JAMA **286**, 1461–1467 (2001)
5. Kucherov, S., Samoylov A., Grishchenko A.: Flexible database for configurable information systems: the relational approach. In: 8th IEEE International Conference on Application of Information and Communication Technologies, AICT 2014 - Conference Proceedings, p. 7035982 (2014)
6. Rogozov, Y., Sviridov, A.: The approach to the configurable system construction based on a methodological approach. In: 15th International Multidisciplinary Scientific Geoconference, SGEM 2015, pp. 353–360 (2015)
7. Rogozov, Y., Sviridov, A., Belikov, A.: Method of a structure-independent databases design in configurable information systems. In: Proceedings of 3rd International Conference on Data Management Technologies and Applications, pp. 139–145 (2014)
8. Varghese, J., Kleine, M., Gessner, S.I., et al.: Effects of computerized decision support system implementations on patient outcomes in inpatient care: a systematic review. JAMIA **25**(5), 593–602 (2018)
9. Koutkias, V., Bouaud, J.: Computerized clinical decision support: contributions from 2015. Yearb. Med. Inform. **1**, 170–177 (2016)

Heterogeneous Dynamic Hybrid Algorithm in Cloud Computing to Load Balance to Improve Cloud Server Speed Efficiency

Abdualmajed A. G. Al-Khulaidi[1](✉), Mansoor N. Ali[1], Muneer A. S. Hazaa[2], and A. A. Mohammed[3]

[1] Sana'a University, Sana'a, Yemen
[2] Thamar University, Dhamar, Yemen
[3] Researcher in Sana'a University, Sana'a, Yemen

Abstract. Load balancing is important in distributed operations in distributed environments and systems, and since cloud computing has grown and expanded rapidly in the recent period, and with the increase in its expansion, customers' requirements for more services and better results have increased, balancing the burden in cloud computing has become an important and expanding research topic for researchers and scholars in the field of cloud computing. In this paper, a group of algorithms used in balancing the burdens in cloud computing are discussed. Where a hybrid algorithm is designed to distribute burdens in cloud computing. The complexity of the proposed algorithm is O(n). Experiments proved that the proposed hybrid burden distribution algorithm in cloud computing is more efficient in performance compared to the PSO algorithm. The performance was better when the number of user requests was 200, which is 1.02 times. The hybrid algorithm works effectively with more orders and has a profit rate of 2.03%. The results of the experiments showed that the average execution time of the proposed algorithm for burden distribution is reduced by 4.73% compared to the PSO algorithm.

Keywords: Load balancing · Cloud computing · Cloud service

1 Introduction

Load unbalancing problem is a multi-variant, multi-constraint problem that degrades performance and efficiency of computing resources. Load balancing techniques cater the solution for load unbalancing situation for two undesirable facets- overloading and under-loading. In contempt of the importance of load balancing techniques to the best of our knowledge, there is no comprehensive, extensive, systematic and hierarchical classification about the existing load balancing techniques. Further, the factors that cause load unbalancing problem are neither studied nor considered in the literature. Cloud Computing is an internet based network technology that shared a rapid growth in the advances of communication technology by providing service to customers of various requirements with the aid of online computing resources. It has provisions of both hardware and software applications along with software development platforms and

R. Silhavy et al. (Eds.): CoMeSySo 2021, LNNS 232, pp. 469–488, 2021.
https://doi.org/10.1007/978-3-030-90318-3_39

testing tools as resources. Such a resource delivery is accomplished with the help of services. While as the former comes under category of Infrastructure as a service (IaaS) cloud, the latter two comes under headings of Software as a service (SaaS) cloud and platform as a service (PaaS) cloud respectively [3]. The cloud computing is an on-demand network enabled computing model that share resources as services billed on pay-as-you-go (PAYG) plan [4]. Some of the giant players in given technology are Amazon, Microsoft, Google, SAP, Oracle, VMware, Sales force, IBM and others [1, 2]. Majority of these cloud providers are high- tech IT organizations. The cloud computing model is viewed under two different headings. The first one is the service delivery model, which defines the type of the service offered by a typical cloud provider. Based on this aspect, there are popularly following three important service models SaaS, PaaS and IaaS [5, 6]. The other aspect of cloud computing model is viewed on its scale of use, affiliation, ownership, size and access. The official 'National Institute of Standards and Technology' (NIST) definition for cloud computing outlines four cloud deployment models namely private, public, community and hybrid clouds [7]. A cloud computing model is efficient if its resources are utilized in best possible way and such an efficient utilization can be achieved by employing and maintaining proper management of cloud resources. Resource management is achieved by adopting robust resource scheduling, allocation and powerful resource scalability techniques. These resources are provided to customers in the form of Virtual Machines (VM) through a process known as virtualization that makes use of an entity (software, hardware or both) known as hypervisor [8]. The greatest advantage of cloud computing is that a single user physical machine is transformed into a multiuser virtual machines [9, 10]. The Cloud Service Provider (CSP) plays a crucial role in service delivery to users and is a complex task with given available virtual resources. While serving user requests, some VMs will get a heavy traffic of user tasks and some will get a lesser traffic. As a result, the Cloud Service Provider (CSP) is left with unbalanced machines which have a huge gradient of user tasks and resource utilization [11]. The problem of load unbalancing is an undesirable event in the CSP side that degrades the performance and efficacy of the computing resources along with guaranteed Quality of Service (QoS) on agreed Service Level Agreement (SLA) between consumer and provider. Under these circumstances there arises need for load balancing (LB) and is a peculiar topic of research interest among researchers. The load balancing in cloud computing can be done at physical machine level or VM level [2]. This paper presents a hybrid algorithm for burden balancing in cloud computing.

2 Problem Statement

Time delay in responding to user requests in the server during processing in cloud computing through overload, which in turn reduces system performance efficiency in cloud computing. This problem is a load balancing problem.

3 Existing Load Balancing Algorithms and Method

Cloud computing technology is one of the modern technologies used and helps to accomplish human tasks, there are many users who deal in large organizations and companies with this technology [5, 7]. Several algorithms have been used to improve the performance of cloud computing systems such as Ant Colony Optimization, First Come First Serve, Round Robin, and Min-Min Algorithms to improve and speed up performance to complete tasks in less time [12].

3.1 Existing Load Balancing Algorithms and Method

The load balancing is based on the system's information and its previous behaviors and does not depend on the current system information, it is not possible to control the transfer of information from one node to another node [3]. Type of Load Balancing algorithms is:

- Static algorithms: Static algorithms are appropriate for systems with low variations in load [11, 12]. In static algorithm the traffic is divided evenly among the servers. This algorithm requires a prior knowledge of system resources the performance of the processors is determined at the beginning of the execution, therefore the decision of shifting of the load does not depend on the current state of system. However, static load balancing algorithms have a drawback in that the tasks are assigned to the processor or machines only after it is created and that tasks cannot be shifted during its execution to any other machine for load balancing.
- Dynamic Algorithms: In dynamic algorithm the lightest server in the whole network or system is searched and preferred for balancing a load. For this real time communication with network is needed which can increase the traffic in the system. Here, current state of the system is used to make decisions to manage the load. Dynamic algorithms respond to the actual current system state in making load transfer decisions. Since current state of the system is used to make dynamic load balancing decisions, processes are allowed to move from an over utilized machine to an under utilized machine in real time dynamically [12].

3.1.1 Static Algorithms

Following are the static load balancing algorithms that are currently common in simple cloud computing environments.

1 Round-Robin Load Balancing Algorithm:
The round-robin load balancing algorithm uses the round-robin scheme for allocating jobs. It selects the first node randomly and then allocates jobs to all other nodes in a round robin fashion. Processors are assigned to each process in a circular order without any sort of priority and hence there is no starvation. This serves the advantage of fast response in the case of equal workload distribution amongst processes. However, different processes have different processing times, therefore at any point of time some nodes may be heavily loaded while others remain idle and under-utilized [5, 12].
2 Weighted Round-Robin Load Balancing Algorithm:
Weighted round-robin was developed to improve the critical issues with round robin algorithm [5]. In weighted round robin algorithm, each server is assigned a weight

and according to the values of the weights, jobs are distributed. Processors with greater capacites are assigned a larger value. Hence the highest weighted servers will receive more tasks. In a situation where all weights become equal, servers will receive balanced traffic.

3 Opportunistic Load Balancing Algorithm:
Opportunistic load balancing algorithm attempts to keep each node busy [5, 12]. Therefore it does not consider the present workload of each computer. OLB dispatches unexecuted tasks to currently available nodes in a random order regardless of the node's current workload. Since OLB does not calculate the execution time of the node, the task to be processed will be processed in a slower manner resulting in bottlenecks despite some of the nodes being free.

4 Min-Min Load Balancing Algorithm:
Min = min load balancing algorithm begins by finding the minimum completion time for all tasks. Then among these minimum times, the minimum value is selected which is the minimum time amongst all tasks on any resource [14]. According to that minimum time, the task is then scheduled on the corresponding machine. The execution time for all other tasks is updated on that machine and that task is removed from the list. This procedure is followed until all the tasks are assigned the resource. In scenarios where the number of small tasks is more than the number of large tasks, this algorithm achieves better performance [5]. However, this approach can lead to starvation [12].

5 Max-Min Load Balancing Algorithm:
Max-min load balancing algorithm is similar to the min-min algorithm except the following: after finding out the minimum execution times, the maximum value is selected which is the maximum time amongst all tasks on the resources [12]. Then according to the maximum time, the task is scheduled on the corresponding machine. The execution time for all other tasks is updated on that machine and the assigned task is removed from the list of tasks that are to be assigned to the machines. Since the requirements are known beforehand, this algorithm is expected to perform well.

3.1.2 Dynamic Algorithms

Decisions on load balancing are based on current state of the system. No prior knowledge is required. Thus it is better than static approach [1, 6, 12].

Components of Dynamic Load Balancing Algorithms:

A load balancing algorithm has five major components [9, 12]:

1 Transfer Policy: In this policy it is responsible to determine when a task should be transferred from one node to the other node.
2 Selection Policy: In this policy it focuses on choosing the processor for load transfer so that the overall response time and throughput may be improved.
3 Location Policy: In this policy it determines the availability of essential resources for providing services and makes a selection based on location of resources.
4 Information Policy: In this policy it acquires workload related information about the system such as nature of workload and the average load on each node.

5 Load Estimation Policy: In this policy it determines the total workload of a node in a System.In cloud computing environment, there are various Load Balancing Algorithms which are closely analyzed and compared on the bases of some predefined metrics, including throughput, response time, Overhead, performance, fault tolerance, migration time, resource utilization, and scalability. Some of the commonly known Load Balancing Algorithms are:

   1 Biased Random Sampling:
   Biased Random Sampling algorithm [10] is a distributed and scalable load balancing approach that uses random sampling of the system domain to achieve self-organization thus balancing the load across all nodes of the system. In this algorithm the load on a server is represented as a virtual graph having connectivity with each node. Each server is symbolized as a node in the graph, with each in degree directed to the free resources of the server. Whenever a node executes a job, it deletes an incoming edge, which indicates a reduction in the availability of free resource. After completion of a job, the node adds on an incoming edge, indicating an increase in the availability of free resource. Random sampling is used for the increment and decrement processes [17]. The last node in the walk is selected for allocation of load; instead any other node based on certain criteria could also be preferred. A node on receiving a job, will execute it only if its current walk length is equal to or greater than the threshold value. Else, the walk length of the job under consideration is incremented and another neighbour node is selected randomly. Again a new directed graph is formed and load balancing is achieved in a fully decentralized manner, thus making it suitable for large network systems like cloud [3].
   2 Active Clustering: Active Clustering algorithm [11, 12] works on the principle of grouping the similar nodes and work together on the available groups. A set of processes is iteratively executed by each node on the network. Initially any node can become an initiator and selects another node from its neighbours to be the matchmaker node satisfying the criteria of being a different type than the former one [3]. The matchmaker node then forms a connection between neighbours of it which are similar to the initiator. The matchmaker node then removes the connection between itself and the initiator [3, 12].
   3 Honey Bee Foraging: Honey Bee Foraging algorithm [18] is derived from the behaviour of honey bees for finding and reaping food. In order to check for fluctuation in demand of services, servers are grouped under virtual servers, having its own virtual service queues. Each server processing a request from its queue calculates a profit or reward on basis of CPU utilization, which is corresponds to the quality that the bees show in their waggle dance and advertise on the advert board. Each of the servers takes the role of either a forager or a scout. A server serving a request, calculates its profit and compare it with the colony profit, if profit was high, then the server stays at the current virtual server and if it was low, then the server returns to the forager or scout behaviour, thus balancing the load with the server [12].
   4 Join Idle Queue: Join Idle Queue load balancing algorithm [14] is applied for dynamically scalable web services. This technique involves a dispatcher to whom

processors informs at the time of their idleness, without interfering with job arrivals. Thus removing the load balancing work from the critical path of request processing, system load is reduced; no communication overhead at job arrivals and no increment in actual response time [33] .

5 Load Balance Min-Min (LBMM): LBMM scheduling algorithm [15] and new optimized Load Balancing Max-MinMax (LB3M) [16] had main objective to minimize execution time of each task, also avoid unnecessary replication of task on the node thereby minimizing overall completion time. Opportunistic Load balancing algorithm when combined with LBMM (OLB + LBMM) keeps every node in working state to achieve load balance. Similar to LBMM, LB3M [16] also calculate average completion time for each task for all nodes. Then mark the task with maximum average completion time. After that it dispatches the task of marked node to the unassigned node with minimum completion task, thus balancing the workload evenly among all nodes [1, 12].

6 Ant Colony Optimization (ACO): ACO algorithm [17] is mainly proposed for load balancing of nodes and aims efficient distribution of workload among the nodes [3]. The ant will start to move towards the source of the food from the head node when the request is initialized. Ant records their data for future decision making and it keeps records for every node and it makes a visit to the record. Every ant is build with their own individual result set and further built for giving the complete solution. It makes to update continuously with a single result set rather than own result set is updating. This ant works in searching of new sources food with the use of existing food sources to shift the food back to the nest. This mainly aims that efficient distribution of the load among the nodes. It does not encounter the dead end of the movement to the node for building an optimum solution set. In ACO [18] two types of pheromones are used Foraging Pheromone (FP) used to explore overloaded node by forward movement of ants while Trailing Pheromone (TP) used to discover its path back to the under loaded node. In order to limit the number of ants in the network, they would commit suicide once it finds the target node [4, 12].

7 ACCLB (Load Balancing mechanism based on Ant Colony and Complex network theory): In [19] author proposed a load balancing mechanism based on ant colony and complex network theory in an open cloud computing federation. It uses small world and scale-free characteristics of a complex network to achieve better load balancing. This technique overcomes heterogeneity, is adaptive to dynamic environments, is excellent in fault tolerance and has good scalability hence helps in improving the performance of the system.

8 Exponential Smooth Forecast based on Weighted Least Connection (ESWLC): ESWLC algorithm [20] is improved form of Weighted Least-Connection (WLC) along with its features; it also takes into account time series and trials. WLC counts the connections of each server and reports the appropriate server based on the multiplication of a server weight and its count of connections. ESWLC algorithm concludes assigning a certain task to a node only after getting to know about the node capabilities. ESWLC builds the decision based on the experience of the node's CPU power, memory, number of connections and the amount of disk

space currently being used. ESWLC then predicts which node is to be selected based on exponential smoothing [33].

9 Honey Bee Behaviour inspired Load Balancing (HBB-LB): According to [2, 21], HBB-LB is a technique, which helps to achieve even load balancing across virtual machine to maximize throughput. It considers the priority task waiting in queue for execution in virtual machines. After that, the work load on VM calculated decides whether the system is overloaded, under load or balanced and based on these VMs are grouped. According to the load on VM the task is scheduled on VMs, which is removed earlier. To find the correct low loaded VM for the current task, tasks which are removed earlier from over loaded VM are helpful. Forager bee is used as a Scout bee in the next steps [3, 12].

10 Power Aware Load Balancing for Cloud Computing (PALB): In PALB algorithm [12, 13], the utilization percentages of each compute node is estimated. This helps in deciding the number of computing nodes which must keep operating while other nodes completely shut down. The algorithm has three sections: Balancing section, Upscale section and Downscale section. Balancing section is responsible for determining where virtual machines will be instantiated based on utilization percentages. The upscale sections power-on the additional compute nodes and the downscale section shutdowns the idle compute nodes. This algorithm is guaranteed to decrease the overall power consumption while maintaining the availability of resources as compared to other load balancing algorithms.

11 Equally Spread Current Execution (ESCE): According to [20], ESCE is a dynamic load balancing algorithm, which handles the process with priority. It determines the priority by checking the size of the process. This algorithm distributes the load randomly by first checking the size of the process and then transferring the load to a virtual machine, which is lightly loaded. The load balancer spreads the load on different nodes, and hence, it is known as spread spectrum technique.

12 Throttled Load Balancer (TLB): Throttled load balancer is a dynamic load balancing algorithm [12] in which the client first requests the load balancer to find a suitable virtual International Journal of Computer Applications (0975 – 8887) Volume 129 – No. 6, November 2015 17 machine to perform the required operation. In Cloud computing, there may be multiple instances of virtual machine. These virtual machines can be grouped based on the type of requests they can handle. Whenever a client sends a request, the load balancer will first look for that group, which can handle this request and allocate the process to the lightly loaded instance of that group.

13 Genetic Algorithm (GA): According to, Genetic Algorithm has been used as a soft computing approach, which uses the mechanism of natural selection strategy. A simple Genetic Algorithm is composed of three operations: genetic operation, selection, and replacement operation. The advantage of this technique is that it can handle a vast search space applicable to complex objective function and can avoid being trapped in locally optimal solution. A generation is a collection of artificial creatures (strings). In every new generation, a set of strings is created using information from the previous ones. Occasionally, a new part is effort for

good measure. According to [2, 12] Genetic Algorithms are randomized, but they are not simple random walks. They adept exploit historical information to speculate on new search points with expected improvement. The effectiveness of the GA depends in appropriate mix of exploration and exploitation.

14 Particle Swarm Optimization (PSO) Algorithm: Particle Swarm Optimization (PSO) as a meta-heuristics method is a self-adaptive global search based optimization technique introduced by Kennedy and Eberhart [7, 12]. The PSO algorithm is alike to other population-based algorithms like Genetic algorithms (GA) but, there is no direct recombination of individuals of the population. The PSO algorithm focuses on minimizing the total cost of computation of an application workflow. The objective is to minimize the total cost of execution of application workflows on Cloud computing environments. Results show that PSO based task-resource mapping can achieve at least three times cost savings as compared to Best Resource Selection (BRS) based mapping for application workflow. PSO balances the load on compute resources by distributing tasks to available resources.

4 Proposed Hybrid Load Balancing Algorithm in Cloud Computing

The demand rates that the data center receives are not constant all the time. Sometimes a large number of orders are accumulated in a short period of time which leads to an explosion. This affects the performance of the load balancing algorithm as it increases the processing time and response time in the data center. The performance of many load balancing algorithms varies according to the frequency of users' demand. For example, some algorithms work efficiently under a low workload while their performance deteriorates under a high workload and vice versa. To overcome the problem of rush and take advantage of the advantages of different load balancing algorithms, he proposed a new hybrid algorithm to improve load balancing.

In order to start with the hybrid algorithm, we must know that the characteristics of the nodes in the cloud consist of the following:

- Data processing power.
- storage capacity.
- Faster performance and high efficiency.

4.1 System Modeling

Consider M as non-VMs with RRVM resource requirements (CPU, Memory, Network, Bandwidth, Storage), which will be placed on N number of PMs with RCPM (CPU, Memory, Network, bandwidth, and storage) as resource capacity and scattered in the total number of C clusters. Consider PM1, PM2, PM3,.......... PMn as members of the PM in the cloud group, where n is the total number of PMs with the range 1 <= i <= n.

Likewise, the set of VM_s on PMi is VM_i1, VM_i2, VM_{im} wherein is the number of VMs on PM_i. If we place any VM to any PM, a load of the CPU, RAM, NW Bandwidth, and Storage has to be calculated individually. The CPU, RAM, Network Bandwidth and Storage utilization of the PM_i during time 'td' is denoted as below:

$$PM_i(CPU,\ td)_{=}\sum j = 1\ \text{to}\ \text{m}\ VM_{ij}(CPU, td) \tag{1}$$

$$PM_i(RAM,\ td)_{=}\sum j = 1\ \text{to}\ \text{m}\ VM_{ij}(RAM, td) \tag{2}$$

$$PM_i(NWB,\ td)_{=}\sum j = 1\ \text{to}\ \text{m}\ VM_{ij}(NWB, td) \tag{2}$$

$$PM_i(STO,\ td)_{=}\sum j = 1\ \text{to}\ \text{m}\ VM_{ij}(STO, vtd) \tag{4}$$

Where:
PMi - here refers to land.
C - The physical device of the group.
VMij - indicates CPU VM PMim indicates CPU usage, RAM indicates RAM usage.
NWB - indicates network bandwidth usage.
STO - Defines storage usage by all PMi VMs.

The average weighted workload of cluster C during time 'td' can be calculated as below:

$$PMC_c(AWL,\ TD) = \sum j = 1\ \text{to}\ \text{n}\ PMi(AWL, td) \tag{5}$$

Where PMC_c denotes the Cth cluster of the cloud, *AWL* denotes the average weighted load of Cluster during time 'td' and PMi denotes the ith PM of cluster C. Also, the load of all VM_S of any PM should be less than the capacity of PM at any point of time.

4.2 Objective Function

An objective Function is a key component of any optimization algorithm. Hence, the value of an objective function is used to determine its performance. When the value is high for maximization optimization problem, the performance can be considered as better. The termination or existence of an optimization algorithm is dependent on the objective function's fitness value.When there are $_n$*physical machines in cluster of cloud with* $_m$ number of VM_s in each machine, we can calculate REM remaining resource capacity as below:

$$PM_i(Rem\ CPU,\ td) = PM_i(TotCPU,\ td) - \sum j = 1\ to\ m$$
$$VM_{ij}(ReqCPU,\ td) \tag{6}$$

$$PM_i(Rem\ RAM,\ td) = PM_i(TotRAM,\ td) - \sum j = 1\ to\ m$$
$$VM_{ij}(ReqRAM,\ td) \tag{7}$$

$$PM_i(Rem\ NWB,\ td) = PM_i(TotNWB,\ td) - \sum j = 1\ to\ m$$

$$\mathrm{VM_{ij}(ReqNWB,\ td)} \tag{8}$$

$$\mathrm{PM_i(Rem\ STO,\ td) = PM_i(TotSTO,\ td)} - \sum j = 1\ to\ m$$
$$\mathrm{VM_{ij}(ReqSTO,\ td)} \tag{9}$$

Where $\mathrm{PM_i}$(RemCPU, td), $\mathrm{PM_i}$(Rem RAM, td) $\mathrm{PM_i}$ (TotNWB, td) and $\mathrm{PM_i}$ (Rem-STO, td) represents the remaining CPU, RAM Network Bandwidth and Storage of the $_{ith}$ physical Machine(PM)during time 'td' respectively. TotCPU, TotRAM, TotNWB and TotSTO represent the total CPU, RAM, Network Bandwidth and storage capacity of $_{ith}$ PM respectively. *VM $_{ij}$(ReqCPU, td), VM $_{ij}$(ReqRAM, td), VM $_{ij}$(ReqNWB, td) and VM $_{ij}$(ReqSTO, td), represents the requested CPU, RAM, Network Bandwidth and Storage of $_{jth}$ Virtual Machine (VM) of $_{ith}$ PM during time 'td'.*

$$PMCC\ \mu Rem\ CPU = \sum x = 1\ to\ n(PMi\ Rem\ CPU)/n \tag{10}$$

$$PMCC\ \mu Rem\ RAM = \sum x = 1\ to\ n(PMi\ Rem\ RAM)/n \tag{11}$$

$$PMCC\ \mu Rem\ NWB = \sum x = 1\ to\ n(PMi\ Rem\ NWB)/n \tag{12}$$

$$PMCC\ \mu Rem\ STO = \sum x = 1\ to\ n(PMi\ Rem\ STO)/n \tag{13}$$

Where $\mathrm{PMC_C}$ $\mu Rem\ CPU$, $\mathrm{PMC_C}$ $\mu Rem\ RAM$, $\mathrm{PMC_C}$ $\mu Rem\ NWB$ and $\mathrm{PMC_C}$ $\mu Rem\ STO$ denotes the mean value of the remaining CPU, RAM, Network Bandwidth and Storage respectively of physical cluster.Following fitness function is used to calculate the degree of load imbalance (DOI) in a cloud datacenter. By having Doi, the load difference of $\mathrm{PM_S}$ with $_{ith}$ mean is identifed, which makes it easy to identify the type and size of $\mathrm{VM_S}$ to be moved to achieve load balancing and maintaining Service Level Agreement (SLA) standards. $DOI_{minimize} = min$

$$\begin{bmatrix} \sqrt{\sum_{i=1}^{n} \frac{1}{N}(PMiRemCPU - PMCc\mu RemCPU)^2} & \sqrt{\sum_{i=1}^{n} \frac{1}{N}(PMiRemRAM - PMCc\mu RemRAM)^2} \\ \sqrt{\sum_{i=1}^{n} \frac{1}{N}(PMiRemNWB - PMCc\mu RemNWB)^2} & \sqrt{\sum_{i=1}^{n} \frac{1}{N}(PMiRemSTO - PMCc\mu RemSTO)^2} \end{bmatrix} \tag{14}$$

Here, the entire physical cluster is considered to calculate the workload instead $\mathrm{VM_s}$ in PM. To find the load imbalance metric, fitness function (Eq.) is applied, which attempts to minimize the imbalance load between each PM in the Cluster.

4.3 Proposed Hybrid Algorithm KNN-PSO

Though PSO is a good for global optimization, it suffeps from the trapping into his local minima. K-nn is powerful in terms of exploration, exploitation, local optimal avoidance, and convergence. However, its ability is still somewhat dependent or limited on some of the mechanisms in the balance between exploration and exploration and exploitation.

Hence, to achieve optimal global minima without trapping into the local minima, here combined both the approaches. In the proposed hybrid K-nn-PSO algorithm, first K-nn will be executed and will give the best position as alpha X_a. Then PSO will be executed with alpha position X_α instead of personal best position (pdi) for the calculation of the velocity of a particle. Because of frequent changes in VM load at runtime, initial VMs placement and static load balancing are insufficient in the cloud. Hence, dynamic load balancing is required to revise VMs placement and satisfy QoS constraints. VMs should be reallocated in such a way that the imbalance level between PMs of the cloud is minimized. Hence, the objective function (lL) (Eq. 14) is used to minimize an imbalance level in terms of remaining available resource capacity of the PMs in the cloud using hybrid algorithm K-nn-PSO. In this research, proposed KNN-PSO a hybrid algorithm that takes advantages of the both K-Nearest Neighbors and Particle swarm optimization algorithms. The K-Nearest Neighbor algorithm which randomly select to process the received tasks, does not need complex computation to make a decision. On the other hand, Particle swarm optimization algorithm selects the best node to handle the received task. The algorithm adopts the characteristics of K-Nearest Neighbors and Particle swarm optimization to make an efficient load balancing and covers their disadvantages. The algorithm considers the current resource information and the CPU capacity factor to achieve the objectives. In the proposed algorithm, the hybrid algorithm is used in order to decide the balancing degree of node, depending on the processor speed and the assigned load on the node. The hybrid algorithm, used to improve load balance in cloud calculators, focuses on the least loaded nodes to distribute users' requests from the most loaded nodes but not the least loaded, so the main goal is to improve the load balance in the use of cloud calculators. The hybrid algorithm strategy improves user request processing time.KNN-PSO algorithm is a load balancing algorithm used by the data center to distribute the received tasks efficiently over the node over workload by swapping between two policies depending on the request rates. When the data center receives a request, the burst detector determines the workload state (Normal or over). Depending on the burst detector decision, the data center will select the appropriate load balancing policy for that state. After that the selected load balancing algorithm will assign the received task to a suitable node, depending on the information supplied by the nearest neighbor. The service provided detector is responsible for detecting the variation in the workload, and determines whether the state of the workload is burst or not using a specific threshold. Depending on the detector decision, the hybrid algorithm will select the proper load balancing policy. The proposed hybrid algorithm uses two load balancing algorithms, one can search for set of the node and the other one can choose the best node. This KNN-PSO hybrid algorithm work efficiently in normal cases and the other one can work efficiently in burst cases. In this research, propose a KNN-PSO algorithm to improve load-balancing efficiency in cloud computing that deals with the advantages of the algorithms chosen to make the hybrid algorithm. Which aims to improve the efficiency of response speed to user requests cloud computing by balancing the load in cloud computing as some of the existing nodes are overloaded and some are under loaded.In the case of a loaded node, it has a pressure in the processors and storage capacities, and consequently it leads to full power consumption. In this case, the temperature in this node increases and results in environmental and server damage. The work of the KNN-PSO algorithm balances the loads in the loaded nodes and distributes them to the other less loaded nodes, as the idea of the search is based on choosing the less load some nodes to help the other loaded nodes.

4.3.1 Idea of a KNN-PSO Algorithm

The K-PSO algorithm is composed of two algorithms, K-NN and PSO. The K-PSO hybrid algorithm helps the decision maker in making the decision, noting that the Service Manager does not have the decision-making ability to move between nodes located in the same fie.The choice between a group of nodes is difficult, so the hybrid algorithm facilitates decision-making in choosing the nodes based on the least-loaded, most efficient and suitable nodes most of the time, the transition from the used node to the selected node is confirmed based on a fitness comparison using the fitness function in the hybrid algorithm for each node in the same field, and this grouping of existing nodes using the K-nn algorithm is choosing a set of random nodes, and based on the comparison between nodes, one of the nodes is chosen, thus increasing processing speed user requests and reducing the time to respond to user request (Figs. 1 and 2).

Step of Proposed Hybrid Algorithm to Improving Load Balancing in Cloud Computing

Step (1) search nodes randomly.
Step (2) Each node has an appropriate value (fitness value), as this value indicates the suitability of the solution of this node compared to the rest of the nodes. The process is carried out by a suitable function (fitness value) and the nodes have speeds based on (memory - processor - storage capacity) that lead them in their search for data, and the information is processed by using the following variables:
- P_{best}: The best fit value I have ever recorded.
g_{BEST}: The best value recorded within the contract.
L_{best}: Better node positioning compared to neighboring nodes.

Step (3) Calculate the values of each node in the current location and make a comparison through the following transactions:

1. $E(T) < E(P - i_{best})$
2. $E(P - i_{best}) = E(T)$. $E - i_{best} = si(t)$

Step (4) Memorizing the values of each node.
Step (5) Calculate the geographical distance between the nodes.
Step (6) Making a comparison between nodes in the cloud based on the lowest value in terms of the year and the minimum distance between the contract.

- IF case the contract was not obtained, go back to step 1.
- If the event that a positive value is obtained during the comparison between the nodes, it proceeds to step 7.

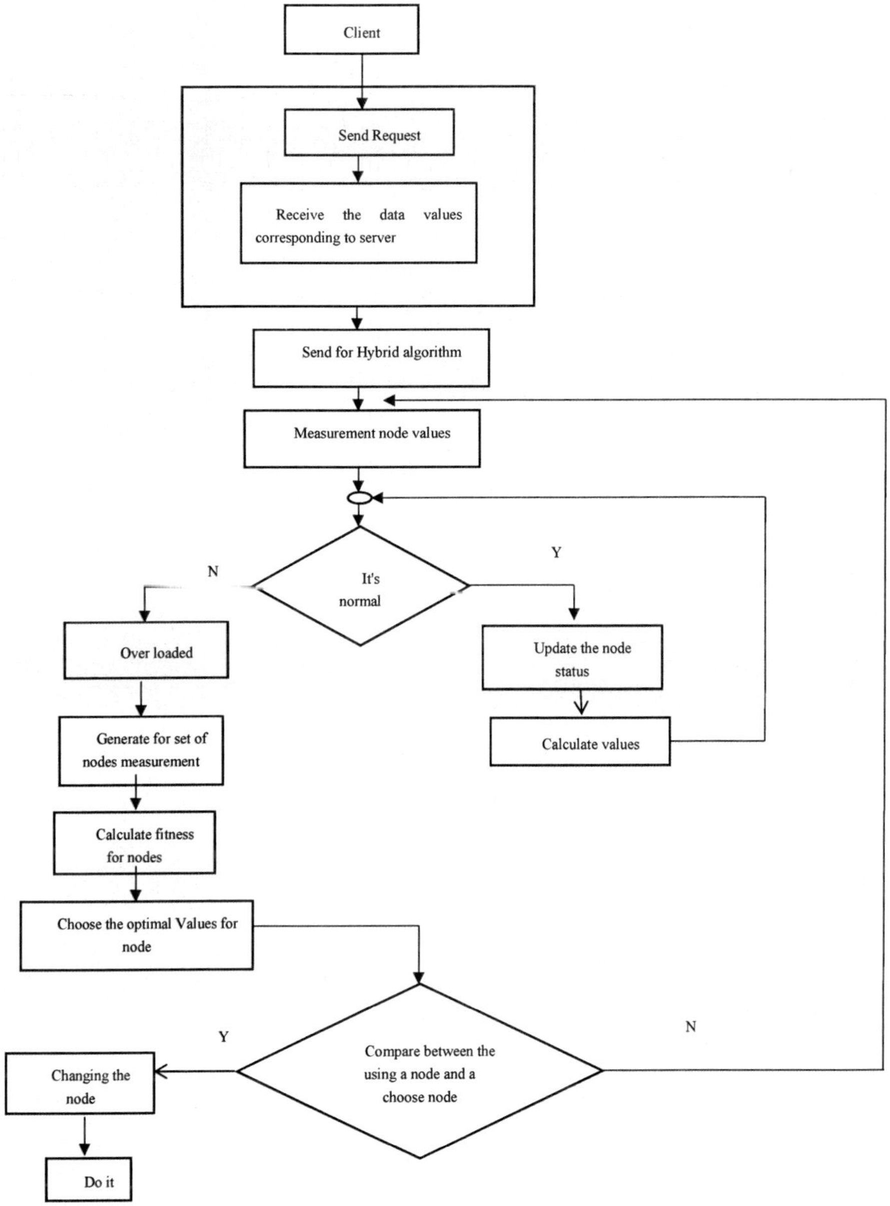

Fig. 1. Idea of the Hybrid algorithm

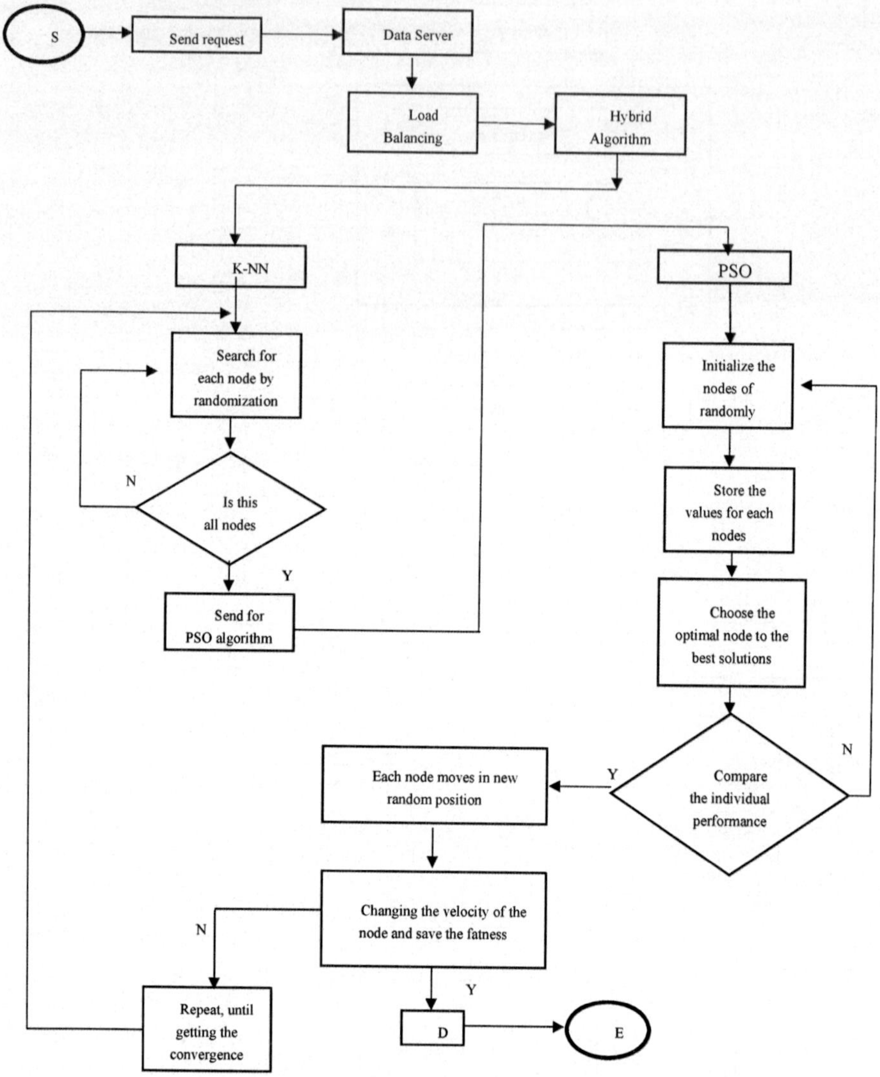

Fig. 2. How the Hybrid algorithm work

Step (7) Choosing the best node in the cloud based on the lowest load value and minimum distance.
Step (8) Sending a signal to the service controller in the presence of a qualified node.
Step (9) The service controller loads the selected node with a load from the neighboring node and the least load.
Step (10) We get a new contract value.
Step (11) We repeat step 6.
Step (12) Steps 6 to 9 are repeated until the loads are distributed on the contract in a balanced manner.
Step (13) Termination Benefits of the proposed hybrid algorithm for load distribution in cloud computing.

The hybrid K-PSO algorithm is one of the smart algorithms and one of the machine learning algorithms, which is more flexible and efficient in decision making because of its flexible combination between K-NN and PSO algorithm, which searches for nodes and selects the most efficient, closest and least overhead node.Finding the nodes using the characteristics of the K-NN algorithm is random by relying on the nodes closest to the base node. As well as through the qualities that the PSO algorithm possesses through the nodes of the least load. Therefore, it was found that the complexity of the hybrid Knn-PSO algorithm used to contribute to improving the load balance in cloud computing is O(n). Where: n- The number of customer requests. This insults that the complexity of the hybrid Knn-PSO algorithm used for load balancing in the cloud is better compared to the complexity of the algorithms used for load balancing in the cloud. An advantage of using the proposed hybrid algorithm is that it monitors the data server load balancing functions and is based on iterative random search and selection of the least loaded and most efficient nodes in the cloud (Fig. 3).

5 The Experiments Using the Proposed Hybrid Algorithm for Improving Load Balancing in Cloud Computing

One of the main problems with cloud computing is load balancing because overloading the system can lead to poor performance, which in turn leads to technology failure. Therefore, we propose a dynamic heterogeneous hybrid algorithm for the cloud computing environment to balance the burden to improve the efficiency of the cloud server speed to respond to user requests based on the nearest and least overburdened nodes. Therefore, we decided to conduct experiments and tests to find out the efficiency and performance of the hybrid algorithm for load balancing in cloud computing. Where two different algorithms were combined to form the hybrid Knn-PSO algorithm, one based on the traveling salesman problem and the other based on the question of how data is represented and applied in life. To find out the efficiency of the proposed algorithm, we compare the proposed hybrid algorithm with the PSO and K-NN algorithm through processing time and response time, as well as the size of the tasks to be balanced in the cloud computing environment. The objective of conducting experiments in this paper is to improve performance in cloud computing through the proposed hybrid algorithm used in the load balancing process. Experiments were conducted by virtualization with

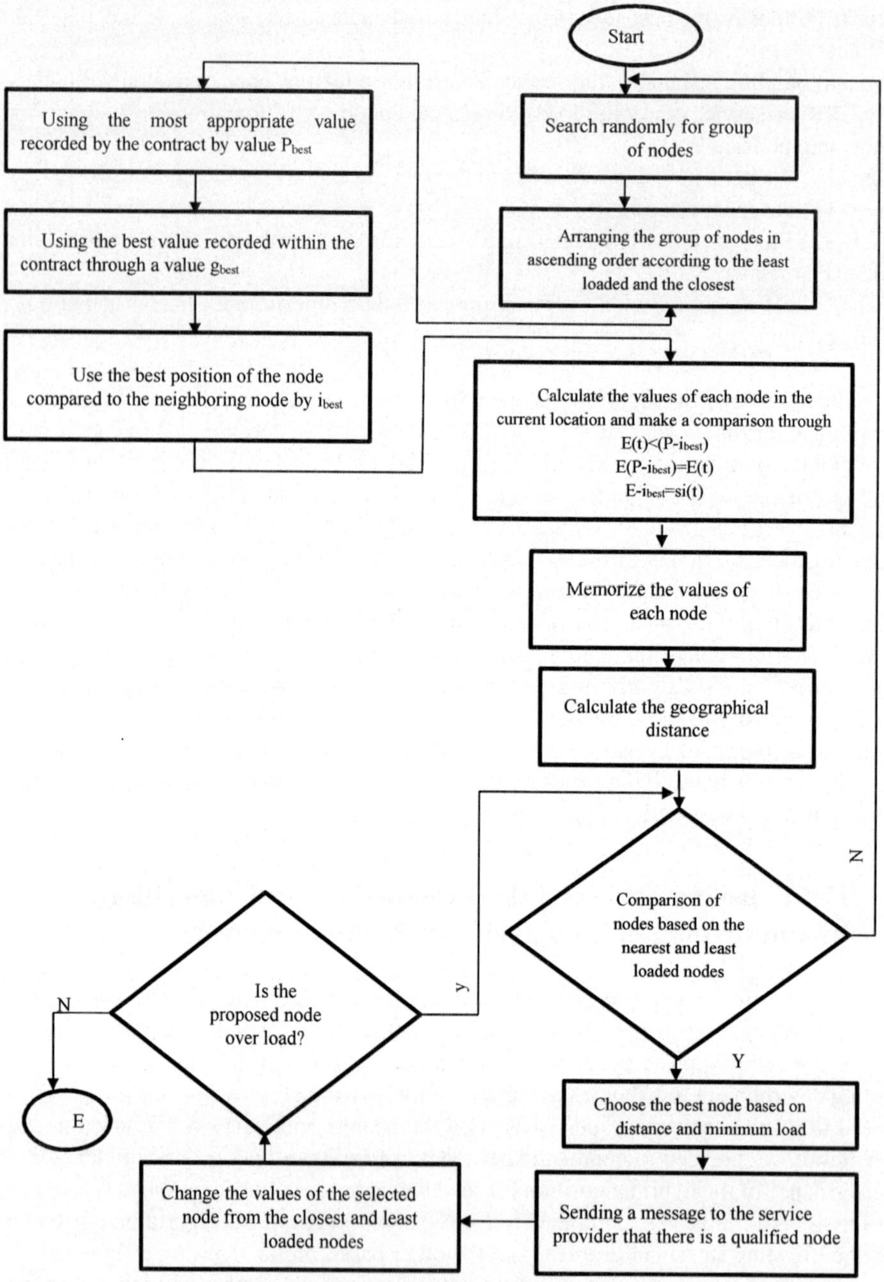

Fig. 3. Proposed K-PSO algorithms

a virtual computer connected to the cloud (internet) through a hypervisor that manages a group of virtual nodes (machines) connected to each other. These measurements have been made on a set of different default nodes to distribute loads to improve the efficiency of the cloud server to respond to users 'requests depending on the number of users' requests and the closest contract and less load.

5.1 The Results of the Knn-PSO Algorithm Performance Check Experiments for the Number of User Requests

In order to check the performance efficiency, the proposed Knn-PSO algorithm was compared with PSO algorithms in the process of load balancing in cloud computing based on the number of user requests (Table 1 and Fig. 4).

Table 1. Comparison of waiting time for number of requests for Knn-PSO Hybrid algorithm and PSO algorithm.

Number of requests	Waiting Time_PSO algorithm	Waiting Time_Knn-PSO_Hybrid Algorithm
100	2.37	2.13
150	3.38	3.25
200	4.42	4.33

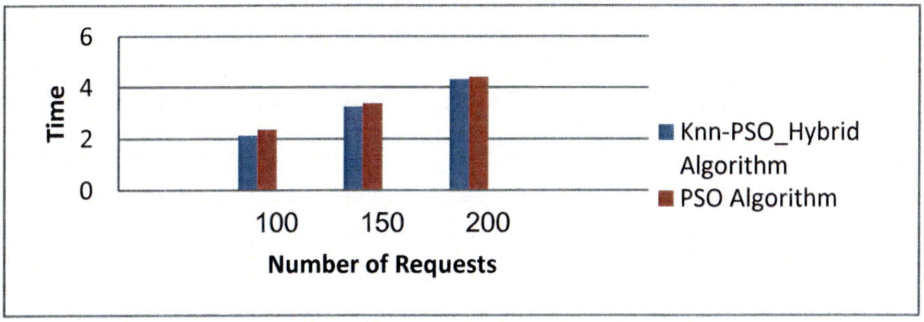

Fig. 4. Scheme of the dependence of the algorithm execution time on the number of requests

We conclude from the previous table and in terms of performance when the number of 200 applications have become marked for hybrid algorithm. The acceleration can be calculated by the following equation:

$$Y = \frac{T_{PSO}}{T_{\text{Knn-PSO\_Hybrid Algorithm}}} \tag{15}$$

Where: T_{PSO} - Time to implement the PSO algorithm.

$T_{\text{Knn-PSO\_Hybrid Algorithm}}$ - The implementation of the proposed Hybrid algorithm.

Y = 1.02 times in the case of the number of requests 200.

The algorithm's gain rate is calculated as follows:

$$G = \frac{T_{PSO} - T_{\text{Knn-PSO\_Hybrid Algorithm}}}{T_{PSO}} * 100\% \tag{16}$$

The algorithm works effectively as the number of requests increases, and a gain ratio of 2.03% is obtained in the case of 200 requests. The calculation of the error is defined as follows:

$$\Delta = \frac{T_{\text{Experiment}} - T_{\text{Theory}}}{T_{\text{Experiment}}} * 100\% \tag{17}$$

Where: T_T - The theoretical time of the proposed algorithm, T_E - Experimental time of the proposed algorithm.

The average implementation time of the proposed algorithm is calculated as follows:

$$\text{Average\_time} = \left| \frac{\sum T_{\text{Knn-PSO\_Hybrid Algorithm}} - \sum T_{\text{PSO}}}{\sum T_{\text{Knn-PSO\_Hybrid Algorithm}}} \right| * 100\% \tag{18}$$

$$\text{Average\_time} = \left| \frac{9,71 - 10,17}{9,71} \right| * 100\% = 4,73\%$$

The results of the experiments showed that the average execution time of the proposed algorithm for burden distribution is 4.73% less compared to the PSO algorithm. Through the experiments through which the hybrid algorithm was tested, we found a set of advantages and it controls the load balancing mechanism between geographically separated nodes. Study fit each node and see if the node is overloaded or underloaded on the set of nodes. It is used for nodes scattered over large distances between them in the cloud. A group of nodes is searched, so that it is easy to find the nearest nodes and tasks are distributed to the nearest nodes, and the hybrid K-PSO algorithm improves in terms of saving time to respond to user requests. The proposed algorithm prevents full redundancy which in turn leads to a high cost due to the large size of the data, but it leads to partial redundancy to save the data procedure in each node according to the capacity of each node in terms of capacity and processing power, which in turn leads to a better representation of the storage. The proposed hybrid K-PSO algorithm is less complex to perform operations in cloud computing. Whereas, the hybrid K-PSO algorithm is less complex than PSO and K-NN to use specific tasks when combining the algorithms. The proposed hybrid K-PSO algorithm takes into account system failures and avoids controller failures for nodes that in turn maintain the system, in terms of backing up the task transferred to the node in order to protect against failure. The proposed hybrid K-PSO algorithm sends requests to the nodes at the lowest cost. The proposed algorithm divides a single task into several tasks so that each part of the task is processed in a single node. The hybrid algorithm divides the file to be uploaded into parts according to the number of nodes available in cloud computing, and each node downloads the transferred part. The hybrid algorithm backs up the central console which in turn solves the problem when the system crashes. Ability to deal with dynamic budget problems.

6 Conclusion

Experiments proved that the proposed hybrid burden distribution algorithm in cloud computing is more efficient in performance compared to the PSO algorithm. The performance was better when the number of requests from users was 200, which is 1.02 times. The hybrid algorithm works effectively with more orders and has a profit rate of 2.03%. The results of the experiments showed that the average execution time of the proposed algorithm for burden distribution is reduced by 4.73% compared to the PSO algorithm.

References

1. Junaid, M., et al.: Modeling an optimized approach for load balancing in cloud. IEEE Access **8**, 173208–173226 (2020)
2. Devaraj, A.F.S., Elhoseny, M., Dhanasekaran, S., Lydia, E.L., Shankar, K.: Hybridization of firefly and improved multi-objective particle swarm optimization algorithm for energy efficient load balancing in cloud computing environments. J. Parallel Distrib. Comput. **142**, 36–45 (2020)
3. Junaid, M., Sohail, A., Ahmed, A., Baz, A., Khan, I.A., Alhakami, H.: A hybrid model for load balancing in cloud using file type formatting. IEEE Access **8**, 118135–118155 (2020)
4. Weng, W., Zhou, X., Srikant, R.: Optimal load balancing with locality constraints. Proc. ACM Meas. Anal. Comput. Syst. **4**(3), 1–37 (2020)
5. Afzal, S., Kavitha, G.: Load balancing in cloud computing–a hierarchical taxonomical classification. J. Cloud Comput. **8**(1), 22 (2019)
6. Adhikari, M., Amgoth, T.: Heuristic-based load-balancing algorithm for IaaS cloud. Future Gener. Comput. Syst. **81**, 156–165 (2018)
7. Moghaddam, J.S.M., O'Sullivan, M., Walker, C., Piraghaj, S.F., Unsworth, C.P.: Embedding individualized machine learning prediction models for energy efficient VM consolidation within cloud data centers. Future Gener. Comput. Syst. **106**, 221–233 (2020)
8. Mohanty, S., Patra, P.K., Ray, M., Mohapatra, S.: A novel meta-heuristic approach for load balancing in cloud computing. Int. J. Knowl.-Based Organ. (IJKBO) **8**(1), 29–49 (2018)
9. Puthal, D., Ranjan, R., Nanda, A., Nanda, P., Jayaraman, P.P., Zomaya, A.Y.: Secure authentication and load balancing of distributed edge datacenters. J. Parallel Distrib. Comput. **124**, 60–69 (2019)
10. Xiao, Z., Tong, Z., Li, K., Li, K.: Learning non-cooperative game for load balancing under self-interested distributed environment. Appl. Soft Comput. **52**, 376–386 (2017)
11. Li, Z., He, Z.: Load balance of cloud computing center based on energy awareness. In: Huang, C., Chan, Y.W., Yen, N. (eds.) Data Processing Techniques and Applications for Cyber-Physical Systems (DPTA 2019). Advances in Intelligent Systems and Computing, vol. 1088, pp. 667–675. Springer, Singapore (2020). https://doi.org/10.1007/978-981-15-1468-5_79
12. Randles, M., Lamb, D., Taleb-Bendiab, A.: A comparative study into distributed load balancing algorithms for cloud computing. In: Proceedings of IEEE 24th International Conference on Advanced Information Networking and Applications Workshops (WAINA), Perth, Australia, April 2010
13. Tong, L., Li, Y., Gao, W.: A hierarchical edge cloud architecture for mobile computing. In: IEEE INFOCOM 2016-The 35th Annual IEEE International Conference on Computer Communications, pp. 1–9. IEEE (2016)

14. Chaturvedi, Y., Kumar, S., Bansal, P., Yadav, S.: Comparison among APSO, PSO & GA for performance investigation of SEIG with balanced loading. In: 2019 9th International Conference on Cloud Computing, Data Science & Engineering (Confluence), pp. 459–463. IEEE (2019)
15. Jena, U., Das, P., Kabat, M.: Hybridization of meta-heuristic algorithm for load balancing in cloud computing environment. J. King Saud Univ.-Comput. Inf. Sci. (2020)
16. Gond, S.: Load balancing in cloud computing: a survey on comparison of two algorithms PSO and SJF-MMBF. IEEE (2018)
17. Singh, A.N., Prakash, S.: WAMLB: weighted active monitoring load balancing in cloud computing. In: Aggarwal, V., Bhatnagar, V., Mishra, D. (eds.) Big Data Analytics, vol. 654, pp. 677–685. Springer, Singapore (2018). https://doi.org/10.1007/978-981-10-6620-7_65
18. Sajjan, R., Yashwantrao, B.R.: Load balancing and its algorithms in cloud computing: a survey. Int. J. Comput. Sci. Eng. **5**(1), 95–100 (2017)
19. Issawi, S.F.: Efficient adaptive load balancing algorithm for cloud computing under bursty workloads by Sally Fouad Issawi supervised by: Dr. Alaa (2015)
20. Agarwal, D.A.: Efficient optimal algorithm of task scheduling in cloud computing environment. Int. J. Comput. Trends Technol. (IJCTT) **9** (2014)
21. Upadhyay, S.K., Bhattacharya, A., Arya, S., Singh, T.: Load optimization in cloud computing using clustering: a survey. Int. Res. J. Eng. Technol **5**(4), 2455–2459 (2018)
22. Ebadifard, F., Babamir, S.M.: A PSO-based task scheduling algorithm improved using a load-balancing technique for the cloud computing environment. Concurr. Comput. Pract. Exp. **30**(12), e4368 (2018)
23. Uma, J., Ramasamy, V., Kaleeswaran, A.: Load balancing algorithms in cloud computing environment-a methodical comparison. Int. J. Eng. Res. Technol. **3**(2), 272–275 (2014)
24. Alakeel, A.M.: A guide to dynamic load balancing in distributed computer systems. Int. J. Comput. Sci. Inf. Secur. **10**(6), 153–160 (2010)
25. Deepa, T., Cheelu, D.: A comparative study of static and dynamic load balancing algorithms in cloud computing. In: 2017 International Conference on Energy, Communication, Data Analytics and Soft Computing (ICECDS), pp. 3375–3378. IEEE (2017)
26. Pathania, B., Sharma, A.: improved hybrid DLBS artificial bee colony optimization algorithm based on parallel computing environment. Int. J. Comput. Appl. **164**(3) (2017)
27. Dobale, R.G.: Review of load balancing for distributed systems in cloud. Int. J. Adv. Res. Comput. Sci. Softw. Eng. **5**(2) (2015)
28. Xu, M., Tian, W., Buyya, R.: A survey on load balancing algorithms for virtual machines placement in cloud computing Concurr. Comput. Pract. Exp. **29**(12), e4123 (2017)
29. Choudhury, R., George, T., Kedia, M., Sabharwal, Y., Saxena, V.: Method for improving the performance of high performance computing applications on cloud using integrated load balancing. Ed: Google Patents (2015)

A Machine Learning Whale Algorithm Applied to the Resource Allocation Problems

Lorena Jorquera[1], Paola Moraga[1], Francisco Altimiras[2](✉), Pamela Valenzuela[1], and José-Miguel Rubio[3]

[1] Escuela de ingeniería en construcción, Pontificia Universidad Católica de Valparaíso, Valparaíso, Chile
{lorena.jorquera,paola.moraga,pamela.valenzuela}@pucv.cl
[2] Facultad de Ingeniería y Negocios, Universidad de las Américas, Santiago 7500975, Chile
faltimiras@udla.cl
[3] Facultad de Ingeniería, Ciencia y Tecnología, Universidad Bernardo O'Higgins Santiago, Metropolitana 8370993, Chile
josemiguel.rubio@ubo.cl

Abstract. Combinatorial optimization problems appear frequently in the areas of engineering and science. A significant number of these problems are of the NP-hard type, therefore when the problem grows, it is difficult to be addressed by complete techniques. This motivates the design of binary or discrete algorithms using continuous metaheuristic techniques. Particularly in the area of swarm intelligence, there are a large number of algorithms that work in continuous spaces and which can be adapted to solve binary or discrete problems. In this article, we use a general binarization mechanism based on the k-means technique to solve the multidimensional knapsack problem (MKP). Design experiments to demonstrate the practicality of the k-means technique in binarization.

Keywords: Combinatorial optimization · Knapsack · Metaheuristics · K-means

1 Introduction

The combinatorial problems have great relevance at industrial level, we find them in various areas such as Civil Engineering [1–5], Operational Research [6–9], resource allocation [10–12], scheduling problems [13–16], routing problems [17–20] among others. On the other hand, in recent years, algorithms inspired by nature phenomena to solve optimization problems have been generated. As examples of these algorithms we have cuckoo search, black hole, bat algorithm, and whale algorithms among others. Many of these algorithms work naturally in continuous spaces and therefore must be adapted to solve combinatorial problems. In the process of adaptation, the mechanisms of exploration and exploitation of the algorithm can be altered, having consequences in the efficiency of the algorithm.

R. Silhavy et al. (Eds.): CoMeSySo 2021, LNNS 232, pp. 489–498, 2021.
https://doi.org/10.1007/978-3-030-90318-3_40

Several binarization techniques have been developed to address this situation. In a literature search, the main binarization methods used correspond to transfer functions, angle modulation and quantum approach, for more detail refer to [21]. In this article we present a new binarization method that uses the clustering k-means technique to group the solutions and then perform the binarization process. To verify the efficiency of our method, we used the whale algorithm.

To check our binary k-means whale algorithm (BKWH), we use the well-known knapsack problem. Experiments were developed using a random operator to validate the contribution of the k-means technique in the binarization process of the whale algorithm. In addition, local search operator is used to strengthen the results.

2 KnapSack Problem

The multidimensional knapsack problem (MKP), models resource allocation situations. The goal is to find the subset of objects that produce the greatest benefit by satisfying a set of constraints. MKP is one of the most studied NP-hard class combinatorial problems, however it remains as a challenge due to the difficulties encountered in solving medium and large size instances. A search for the last few years in the literature, we find that MKP has been solved for example in [22] using a quantum binarization technique, in [23] applied the k-means clustering technique to perform binarization, in [24] used a differential algorithm with transfer functions, and in [25] a modification of the PSO equations was used. MKP can be set as:

$$\text{maximize} \sum_{j=1}^{n} p_j x_j \tag{1}$$

$$\text{subjected to} \sum_{j=1}^{n} c_{ij} x_j \leq b_i \ , \ i \in \{1, ..., m\}. \tag{2}$$

With $x_j \in \{0, 1\}$, $j \in \{1, ..., n\}$. p_j corresponds to the profit of element j. c_{ij} represents a cost associated with dimension i and element j. The constraints in each dimension i are represented by b_i. The solution can be modelled using a binary representation, in this representation a 0 means the element is not included in the knapsack.

3 Binary Whale Algorithm

The application of statistics and data mining, can be found in many areas such as transports, smart cities, agriculture and computational intelligence [23,26–31]. In this section, we explore the application of the k-means technique to the binarization of continuous metaheuristics.

The binary whale algorithm(BKWH) uses 4 operators to achieve its goal. An initialization operator, A local search operator, the binary operator using the k-means technique, and a repair operator for the case of solutions that do not meet any of constraints. The general scheme of the algorithm is shown in Fig. 1. In the first step, BKWH initializes the solutions, this is detailed in Sect. 3.1. Once we have all the solutions initialized, we check the compliance of the stopping criterion of the algorithm associated to the maximum number of iterations. The whale algorithm is then run in conjunction with the k-means operator, this is described in Sect. 3.2. Subsequently once the binarization was performed, the results are compared with the best solution obtained so far. In the case of finding a superior solution, it replaces the previous solution and is submitted to a local search operator. In the following subsection we will explain in detail the initialization method, the binary operator and the repair operator.

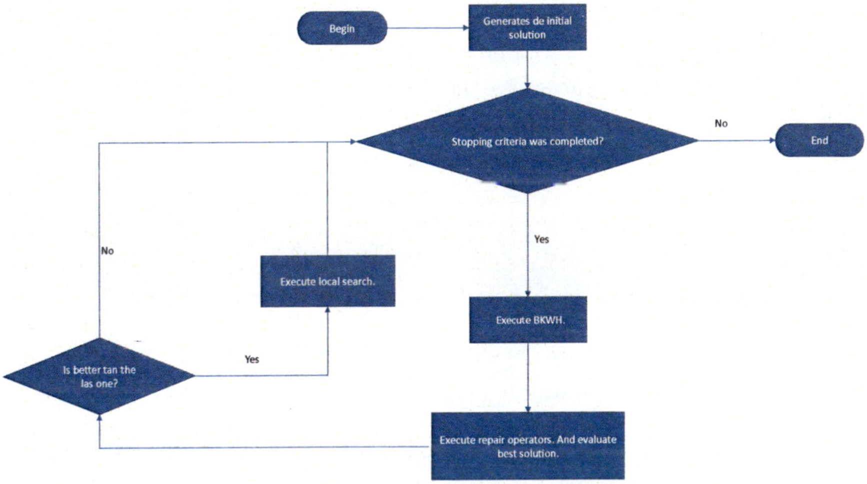

Fig. 1. Flowchart of the binary whale algorithm.

3.1 Initialization and Element Weighting

As BKWH is a swarm algorithm, to begin the exploration and exploitation of the search space, the list of solutions must be initialized. For the generation of each solution, an element is randomly chosen first. The next step is to check if other elements can be incorporated, for this we must evaluate the constraints of our problem. To select the new element, we generate a list of possible elements that comply with the constraints. Each element is calculated a weight and the best weight element is selected. This procedure is repeated until no additional element can be incorporated.

To calculate the weight of each element, several methods have been developed. In [32] a pseudo-utility in the surrogate duality approach was proposed.

The way to calculate it is shown in Eq. 3. In this equation the variable w_j corresponds to the surrogate multiplier whose value is between 0 and 1. This multiplier can be interpreted as a shadow prices of the j-th constraint.

$$\delta_i = \frac{p_i}{\sum_{j=1}^{m} w_j c_{ij}} \tag{3}$$

A more intuitive measure focused on the average resource occupancy was proposed by [33]. It is shown in Eq. 4.

$$\delta_i = \frac{\sum_{j=1}^{m} \frac{c_{ij}}{m b_j}}{p_i} \tag{4}$$

In this article we used a variation of this last measure focused on the average occupation and proposed in [23]. this variation considers the elements that exist in knapsacks to calculate the average occupancy. In each iteration depending on the selected items in the solution the measure is calculated again. The expression of this new measure is shown in Eq. 5.

$$\delta_i = \frac{\sum_{j=1}^{m} \frac{c_{ij}}{m(b_j - \sum_{i \in S} c_{ij})}}{p_i} \tag{5}$$

3.2 K-Means Transition Ranking Operator

Consider our metaheuristic is a continuous and population based. Due to its iterative nature, it needs to update the position of particles at each iteration. When the metaheuristic is continuous, this update is performed in $\mathbb{R}^n$ space. In Eq. 6, the update position is presented in a general manner. The x_{t+1} variable represents the x position of the particle at time t+1. This position is obtained from the position x at time t plus a Δ function calculated at time $t+1$. The function Δ is proper to each metaheuristic and produces values in $\mathbb{R}^n$. For example in Cuckoo Search $\Delta(x) = \alpha \oplus Levy(\lambda)(x)$, in Black Hole $\Delta(x) = \text{rand} \ \times (x_{bh}(t) - x(t))$ and in the whale, Bat and PSO algorithms Δ can be written in simplified form as $\Delta(x) = v(x)$.

$$x_{t+1} = x_t + \Delta_{t+1}(x(t)) \tag{6}$$

In the k-means transition ranking operator, a model for transitions in a discrete space is proposed. This model is based on considering the movements generated by the metaheuristic in each dimension for all particles. $\Delta^i(x)$ corresponds to the magnitude of the displacement $\Delta(x)$ in the i-th position. Subsequently these displacement are grouped using the magnitude of the displacement $\Delta^i(x)$. This grouping is done using the k-means technique where k represents the number of clusters used. Finally, a generic P_{tr} function given by the Eq. 7 is proposed to assign a transition probability.

$$P_{tr} : \mathbb{Z}/k\mathbb{Z} \rightarrow [0, 1] \tag{7}$$

A transition probability through the function P_{tr} is assigned to each group. Naturally, this P_{tr} function is modelled as a cumulative probability function. For the case of this study, we particularly use the linear function given in Eq. 8. In this equation, $\mathrm{N}(x^i)$ indicates the location of the group to which $\Delta^i(x)$ belongs. The α coefficient, corresponds to the transition probability and β to the transition separation coefficient. Both parameters are estimated in each metaheuristic. For our particular case, $\mathrm{N}(x^i) = 0$ corresponds to elements belonging to the group that has the lowest Δ^i values. $\mathrm{N}(x^i) = 7$ corresponds to the group of elements that have the greatest Δ^i values.

$$P_{tr}(x^i) = P_{tr}(N(x^i)) = \alpha + \beta N(x^i)\alpha \tag{8}$$

The k-means transition ranking operator starts calculating Δ^i for each of the particles. This step is specific in each metaheuristic. Subsequently the particles are grouped using k-means clustering technique and the magnitude of Δ^i as distance. With the group assigned to each particle we obtain the probability of transition using Eq. 8. Afterwards the transition of each particle is performed. In the case of firefly algorithm the rule 9 is used to perform the transition, where $\hat{x}^i$ is the complement of x^i. Finally, each solution is repaired using the repair operator.

$$x^i(t+1) := \begin{cases} \hat{x}^i(t), \text{ if } rand < P_{tg}(x^i) \\ x^i(t), \quad \text{otherwise} \end{cases} \tag{9}$$

3.3 Repair Operator

Local search and binary k-means technique can generate infeasible solutions. There are different mechanisms to address these infeasible solutions. In this article, the repair of the solutions was considered. To perform the repair, the measure described in the Eq. 5 was used. If the solution requires repair, the element with the maximum measure is chosen and it is removed from the solution, this process is iterated until a valid solution is obtained. The solution is then improved by exploring the possibility of incorporating new elements. This stage of improvement is iterated until there are no elements that can be incorporated without violating the constraints.

4 Results

4.1 Insight of BKWH Algorithm

This section aims to find out the contribution of the binary k-means operator to the performance of the algorithm. To carry out the comparison problems cb.5.250 from the OR-library were selected. Violin charts and the Wilcoxon non-parametric signed-rank test were used to perform the statistical analyzes. In the charts, the X axis identifies the studied instances and the Y axis the % -Gap described in the Eq. 10. The Wilcoxon test is run to determine if the results obtained by BKWH have significant difference with respect to other algorithms. The parameter settings and browser ranges are shown in Table 1.

$$\% - Gap = 100\frac{BestKnown - SolutionValue}{BestKnown} \tag{10}$$

Table 1. Setting of parameters for Binary whale algorithm.

Parameters	Description	Value	Range
N	Number of solutions	30	[20, 25, 30]
G	Number of clusters	5	[4, 5, 6]
Iteration number	Maximum iterations	1000	[1000]

Evaluation of k-Means Binary Operator. A random operator was designed to evaluate the contribution of the k-means binary operator. This random operator has the peculiarity of performing the transitions with a fixed probability of 0.5 without considering in which cluster the variable is located. Two configurations were studied: The first one where the local search operator is included and the second where the local operator is not considered. BKWH corresponds to our standard algorithm. *random.ls* is the random variant that includes the local search operator. *BKWH.wls* corresponds to the version with k-means binary operator without local search operator. Finally *random.wls* describes the random algorithm without local search operator.

When we compared the Best Values between BKWH and *random.ls* which are shown in Table 2. BKWH outperforms to *random.ls*. However the Best Values between both algorithms are very close. In the Average comparison, BKWH outperforms *random.ls* almost in all problems. The comparison of distributions is shown in Fig. 2. We see the dispersion of the *random.ls* distributions are bigger than the dispersion of BKWH. Therefore, the k-means binary operator together with local search operators, contribute to the precision of the results. Finally, the BKWH distributions are closer to zero than *random.ls* distributions, indicating that BKWH has consistently better results than *random.ls*. When we evaluate the behaviour of the algorithms through the Wilcoxon test, this indicates that there is a significant difference between the two algorithms.

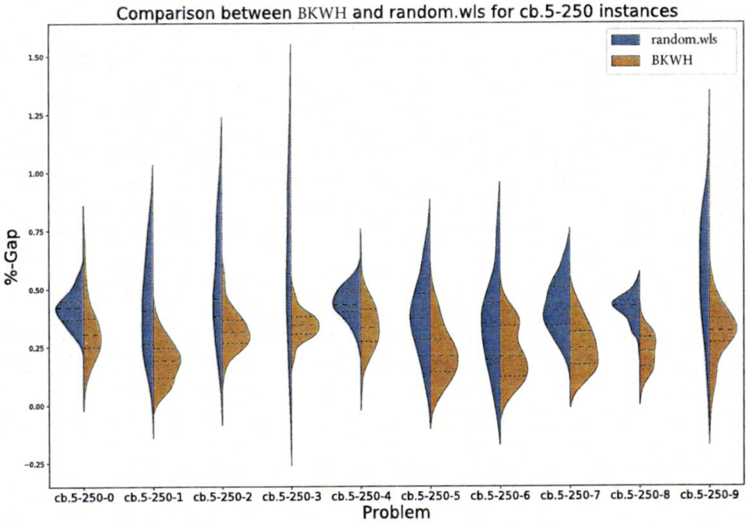

Fig. 2. Evaluation of k-means binary operator with Local Search operator

Table 2. Evaluation of k-means binary operator

Set	Best Known	Best *random.ls*	Best BKWH	Best *random.wls*	Best *wls*	Avg *random.ls*	Avg BKWH	Avg *random.wls*	Avg *wls*
cb.5.250-0	59312	59211	59225	59158	59175	59132.1	59143.2	59071.8	59132.1
cb.5.250-1	61472	61435	61435	61409	61409	61324.6	61384.1	61288.3	61370.7
cb.5.250-2	62130	62036	62036	61969	61969	61894.4	61975.5	61801.6	61924.6
cb.5.250-3	59463	59367	59446	59365	59349	59257.8	59321.2	59136.1	59269.3
cb.5.250-4	58951	58914	58951	58883	58930	58725.6	58821.6	58693.6	58758.5
cb.5.250-5	60077	60015	60056	59990	60015	59904.6	59952.3	59837.8	59945.9
cb.5.250-6	60414	60355	60355	60348	60349	60208.2	60337.2	60230.6	60311.1
cb.5.250-7	61472	61436	61436	61407	61407	61290.8	61331.9	61233.9	61342.3
cb.5.250-8	61885	61829	61885	61790	61782	61737.1	61781.2	61644.9	61738.1
cb.5.250-9	58959	58832	58832	58822	58787	58769.1	58782.3	58653.7	58770.4
Average	60413.5	60343	60365.6	60314.1	60317.2	60224.4	60282.7	60150.8	60256.3
p-value								1.1 e-04	3.2 e-04

5 Conclusions

In this article, the k-means technique is used to perform the binarization of the BKWH algorithm. It should be noted that the k-means technique can be applied in the binarization of any continuous swarm-intelligence algorithm. We used the knapsack problem to evaluate our binary algorithm. The contribution of the k-means binary operator was studied, observing that this operator contributes to the precision and quality of the solutions obtained.

As a future line of research, it is interesting to compare the performance of the k-means operator with the transfer functions, all applied in the binarization of the whale algorithm. Also, it is interesting to use the k-means operator to binarize other algorithms in addition to solving other NP-hard problems. From the theoretical point of view, it would also be interesting to understand how the exploration and exploitation of space are altered when we introduce the k-means operator. Another interesting research line is to explore adaptive techniques to automate the selection of parameters.

References

1. Yepes, V., Martí, J.V., García, J.: Black hole algorithm for sustainable design of counterfort retaining walls. Sustainability **12**(7), 2767 (2020)
2. García, J., Martí, J.V., Yepes, V.: The buttressed walls problem: an application of a hybrid clustering particle swarm optimization algorithm. Mathematics **8**(6), 862 (2020)
3. Crawford, B., et al.: Investigating the efficiency of swarm algorithms for bridge strengthening by conversion to tied-arch: a numerical case study on san Luis bridge. Iran. J. Sci. Technol. Trans. Civ. Eng. 1–13 (2020). https://doi.org/10.1007/s40996-020-00531-5
4. García, J., Yepes, V., Martí, J.V.: A hybrid k-means cuckoo search algorithm applied to the counterfort retaining walls problem. Mathematics **8**(4), 555 (2020)
5. Martínez-Muñoz, D., Martí, J.V., García, J., Yepes, V.: Embodied energy optimization of buttressed earth-retaining walls with hybrid simulated annealing. Appl. Sci. **11**(4), 1800 (2021)
6. Crawford, B., Soto, R., Monfroy, E., Astorga, G., García, J., Cortes, E.: A meta-optimization approach for covering problems in facility location. In: Figueroa-García, J.C., López-Santana, E.R., Villa-Ramírez, J.L., Ferro-Escobar, R. (eds.) WEA 2017. CCIS, vol. 742, pp. 565–578. Springer, Cham (2017). https://doi.org/10.1007/978-3-319-66963-2_50
7. García, J., Crawford, B., Soto, R., Astorga, G.: A percentile transition ranking algorithm applied to binarization of continuous swarm intelligence metaheuristics. In: Ghazali, R., Deris, M.M., Nawi, N.M., Abawajy, J.H. (eds.) SCDM 2018. AISC, vol. 700, pp. 3–13. Springer, Cham (2018). https://doi.org/10.1007/978-3-319-72550-5_1
8. García, J., Maureira, C.: A KNN quantum cuckoo search algorithm applied to the multidimensional knapsack problem. Appl. Soft Comput. **102**, 107077 (2021)
9. Crawford, B., et al.: Q-learnheuristics: towards data-driven balanced metaheuristics. Mathematics **9**(16), 1839 (2021)
10. García, J., Crawford, B., Soto, R., Astorga, G.: A percentile transition ranking algorithm applied to knapsack problem. In: Silhavy, R., Silhavy, P., Prokopova, Z. (eds.) CoMeSySo 2017. AISC, vol. 662, pp. 126–138. Springer, Cham (2018). https://doi.org/10.1007/978-3-319-67621-0_11
11. García, J., Altimiras, F., Peña, A., Astorga, G., Peredo, O.: A binary cuckoo search big data algorithm applied to large-scale crew scheduling problems. Complexity **2018**, 8395193 (2018). https://doi.org/10.1155/2018/8395193
12. García, J., Lalla-Ruiz, E., Voß, S., Droguett, E.L.: Enhancing a machine learning binarization framework by perturbation operators: analysis on the multidimensional knapsack problem. Int. J. Mach. Learn. Cybern. **11**(9), 1951–1970 (2020)
13. García, J., Crawford, B., Soto, R., García, P.: A multi dynamic binary black hole algorithm applied to set covering problem. In: Del Ser, J. (ed.) ICHSA 2017. AISC, vol. 514, pp. 42–51. Springer, Singapore (2017). https://doi.org/10.1007/978-981-10-3728-3_6

14. García, J., Crawford, B., Soto, R., Astorga, G.: A clustering algorithm applied to the binarization of swarm intelligence continuous metaheuristics. Swarm Evol. Comput. **44**, 646–664 (2019)
15. Crawford, B., Soto, R., Astorga, G., García, J.: Constructive metaheuristics for the set covering problem. In: Korošec, P., Melab, N., Talbi, E.-G. (eds.) BIOMA 2018. LNCS, vol. 10835, pp. 88–99. Springer, Cham (2018). https://doi.org/10.1007/978-3-319-91641-5_8
16. García, J., Astorga, G., Yepes, V.: An analysis of a KNN perturbation operator: an application to the binarization of continuous metaheuristics. Mathematics **9**(3), 225 (2021)
17. García, J., Peña, A.: Robust optimization: concepts and applications. In: Nature-inspired Methods for Stochastic, Robust and Dynamic Optimization, p. 7 (2018)
18. Crawford, B., Soto, R., Monfroy, E., Astorga, G., García, J., Cortes, E.: A meta-optimization approach to solve the set covering problem. Ingeniería **23**(3), 274–288 (2018)
19. García, J., et al.: A Db-scan binarization algorithm applied to matrix covering problems. Comput. Intell. Neurosci. **2019**, 3238574 (2019). https://doi.org/10.1155/2019/3238574
20. Tapia, D., et al.: A q-learning hyperheuristic binarization framework to balance exploration and exploitation. In: Florez, H., Misra, S. (eds.) ICAI 2020. CCIS, vol. 1277, pp. 14–28. Springer, Cham (2020). https://doi.org/10.1007/978-3-030-61702-8_2
21. Crawford, B., Soto, R., Astorga, G., García, J., Castro, C., Paredes, F.: Putting continuous metaheuristics to work in binary search spaces. Complexity **2017**, 8404231 (2017). https://doi.org/10.1155/2017/8404231
22. Haddar, B., Khemakhem, M., Hanafi, S., Wilbaut, C.: A hybrid quantum particle swarm optimization for the multidimensional knapsack problem. Eng. Appl. Artif. Intell. **55**, 1–13 (2016)
23. García, J., Crawford, B., Soto, R., Castro, C., Paredes, F.: A k-means binarization framework applied to multidimensional knapsack problem. Appl. Intell. **48**(2), 357–380 (2018)
24. Liu, J., Wu, C., Cao, J., Wang, X., Teo, K.L.: A binary differential search algorithm for the 0–1 multidimensional knapsack problem. Appl. Math. Model. **40**(23–24), 9788–9805 (2016)
25. Bansal, J.C., Deep, K.: A modified binary particle swarm optimization for knapsack problems. Appl. Math. Comput. **218**(22), 11042–11061 (2012)
26. García, J., Pope, C., Altimiras, F.: A distributed K-means segmentation algorithm applied to Lobesia botrana recognition. Complexity **2017**, 5137317 (2017). https://doi.org/10.1155/2017/5137317
27. Graells-Garrido, E., García, J.: Visual exploration of urban dynamics using mobile data. In: García-Chamizo, J.M., Fortino, G., Ochoa, S.F. (eds.) UCAmI 2015. LNCS, vol. 9454, pp. 480–491. Springer, Cham (2015). https://doi.org/10.1007/978-3-319-26401-1_45
28. Peredo, O.F., García, J.A., Stuven, R., Ortiz, J.M.: Urban dynamic estimation using mobile phone logs and locally varying anisotropy. In: Gómez-Hernández, J.J., Rodrigo-Ilarri, J., Rodrigo-Clavero, M.E., Cassiraga, E., Vargas-Guzmán, J.A. (eds.) Geostatistics Valencia 2016. QGG, vol. 19, pp. 949–964. Springer, Cham (2017). https://doi.org/10.1007/978-3-319-46819-8_66
29. Graells-Garrido, E., Peredo, O., García, J.: Sensing urban patterns with antenna mappings: the case of Santiago, Chile. Sensors **16**(7), 1098 (2016)

30. García, J., Moraga, P., Valenzuela, M., Pinto, H.: A db-scan hybrid algorithm: an application to the multidimensional knapsack problem. Mathematics **8**(4), 507 (2020)
31. Maureira, C., Pinto, H., Yepes, V., García, J.: Towards an AEC-AI industry optimization algorithmic knowledge mapping: An adaptive methodology for macroscopic conceptual analysis. IEEE Access **9**, 110842–110879 (2021)
32. Pirkul, H.: A heuristic solution procedure for the multiconstraint zero? one knapsack problem. Naval Res. Logistics **34**(2), 161–172 (1987)
33. Kong, X., Gao, L., Ouyang, H., Li, S.: Solving large-scale multidimensional knapsack problems with a new binary harmony search algorithm. Comput. Oper. Res. **63**, 7–22 (2015)

Inference Methods for Evaluable Knowledge Bases

Alexander Sakharov(✉)

Synstretch, Framingham, MA, USA
mail@sakharov.net

Abstract. This paper investigates inference from evaluable knowledge bases without reasoning by contradiction. These knowledge bases are comprised of non-Horn rules with partial predicates and functions, some of them are defined as recursive functions. This inference corresponds to model elimination without the reduction rule, which is equivalent to forward and backward chaining extended with rule contrapositives, to input resolution, and to constrained resolution without factoring including its ordered variant.

1 Introduction

The facts and rules of knowledge bases (KB) are usually expressible in the language of first-order logic (FOL) [19]. Typically, KB facts are atoms or literals, Atoms are expressions $P(t_1, ..., t_k)$ where P is a predicate and $t_1, ..., t_k$ are terms. Literals are atoms or their negations. Quantifier-free expressions $A \Leftarrow A_1 \wedge ... \wedge A_k$, where $A, A_1, ..., A_k$ are literals, are a common form of rules in KBs. The sets of such facts and rules are also called logic programs. Rule sets are treated as conjunctions. In normal logic programs, A is an atom. In Prolog, A and all A_i are atoms. These rules are equivalent to disjunctions of literals in classical FOL. These disjunctions are known as non-Horn clauses [25], and the disjunctions corresponding to Prolog rules are called Horn clauses. Their source rules and facts will also be called non-Horn and Horn, respectively.

Inference for KBs and logic programs differs significantly from inference in FOL. The outcome of this inference is literals as opposed to arbitrary FOL formulas. Efficiency is more important for logic programming and for KB inference than for theorem provers for FOL. The number of rules and especially the number of facts involved in KB inference may be huge. The interpretation of negation in logic programs may be different from FOL [7]. KBs and logic programs may include computable (evaluable) functions and predicates [16].

Forward and backward chaining are widely used as inference methods for Horn KBs with goals (queries) that are atoms. Backward chaining is a goal-directed method whereas forward chaining is data-driven [19]. These two chaining methods can be applied to non-Horn KBs or non-Horn logic programs and to goals that are literals [20]. In this case, backward chaining can be treated as a resolution strategy [5] in which one of any two resolved literals is a rule head or a fact. In this interpretation of backward chaining, every resolvent corresponds

R. Silhavy et al. (Eds.): CoMeSySo 2021, LNNS 232, pp. 499–510, 2021.
https://doi.org/10.1007/978-3-030-90318-3_41

to a goal list comprised of complementary literals. This strategy is similar to SLD resolution in application to Horn clauses. Prolog inference is implemented as backward chaining. Forward chaining can be treated as hyperresolution [5].

Backward chaining is a complete inference strategy for Horn rules but it is incomplete for non-Horn rules as regards to FOL. Resolution-based procedures are dominant inference methods for KBs with non-Horn rules. A variety of resolution methods are complete for FOL [5]. Complete inference procedures for FOL are much slower than backward chaining. Incomplete inference procedures are common for KBs because they are supposed to scale to large KBs. The use of incomplete procedures is also justified by the fact that KBs evolve, and thus they are incomplete in nature.

Many tasks are most conveniently specified by a combination of facts, rules, and evaluable functions. Evaluable functions are built-in or user-defined recursive functions. The lack of such functions is an impediment to the success of the KB systems that are solely based on FOL. The computational nature of KBs and logic programs is lost in the FOL treatment of rules and facts. In classical FOL, it is assumed that all functions and predicates are total. In a more computational environment, functions may be partial and their values may be undefined. Combining recursive functions and FOL has been extensively researched [4,10,18,24]. We assume that evaluable functions may occur in terms that are predicate arguments. Predicates may be defined as boolean recursive functions as well. KBs with non-Horn rules and facts and containing evaluable functions and predicates will be called evaluable KBs.

It is important for KB systems to support the derivation of negative literals. Negation is explicit in non-Horn KBs. It is classical negation with an intuitive interpretation. Formal semantics of normal logic programs are usually based on the concept of negation as failure to prove. Negation as failure makes inference nonmonotonic. In author's opinion, nonmonotonic reasoning is risky in application to evolving KBs because any modification may cancel previous results. Various attempts to formalize negation as failure have led to two conflicting views of logic programs [7]. In the autoepistemic view, negation is not treated as the classical one. In the definitional view, logic programs are treated as simultaneous inductive definitions of all predicates. In the presence of evaluable predicates, the entire sets of KB rules cannot be viewed as simultaneous definitions of all predicates.

Forward and backward chaining are based on Modus Ponens (MP) as the sole inference rule. They can be extended by additionally employing Modus Tollens (MT). We collectively refer to all these methods as chaining inference. It will be discussed later that adding one more rule - Reductio Ad Absurdum (RAA) - gives complete FOL inference in application to KBs, i.e. for the derivation of literals from facts and rules. RAA states that if A is deduced from hypothesis $\neg A$, then A is derivable. All functions and predicates in evaluable KBs are considered partial by default. In author's opinion, the legitimacy of RAA is questionable in the presence of partial predicates or functions. It seems faulty to apply RAA when the truth value of the hypothesis is undefined.

The aim of this paper is to investigate inference from evaluable KBs without using RAA, i.e. without reasoning by contradiction. Unlike MP, MT makes use of negation properties, which is significant for non-Horn rules. Chaining inference employing MP and MT corresponds to resolution without factoring and with a constraint on using goal descendants, to input resolution, and to model elimination without the reduction rule. Chaining as well as resolution including model elimination have been extensively tested and optimized in a multitude of software systems.

2 Evaluable Knowledge Bases

We assume that evaluable functions are defined by means that are external to KB rules. These functions could be defined as recursive functions in a functional programming language or as algorithms in a procedural programming language. The same assumption applies to evaluable predicates. The latter are boolean functions. Integration of KBs and recursive functions is also possible in the opposite direction, that is, the definitions of recursive functions may contain predicates specified by KB rules. This full integration of inference and computation may be beneficial for the specification of complex tasks.

A literal is called ground if it does not contain variables. A substitution is a finite set of mappings of variables to terms. The result of applying a substitution to a formula or set of formulas is called its instance. Recursive functions implementing evaluable functions and predicates may yield a value or may not terminate for some arguments.

We assume that occurrences of evaluable functions/predicates with constant arguments are evaluated as soon as they appear in derivations. If all goals in a backward chaining derivation are literals of evaluable predicates and these literals contain variables, then the literals are evaluated by iterating over the domains of the variables until they yield true. Any complete search strategy for backward chaining inference from evaluable KBs should continue and-or search [19] simultaneously with evaluations. If A is an evaluable predicate and the evaluation of ground literal $A(...)$ yields true, then $A(...)$ is considered a fact. If the evaluation of literal $A(...)$ yields false, then $\neg A(...)$ is considered a fact. These facts are not physically stored in KBs. There could be an infinite number of them.

Non-evaluable functions and predicates should also be considered partial by default. Rule sets are often incomplete and some atoms with constant arguments are not derivable in FOL or another formal system from KB rules and facts, and their negations are not derivable either. A KB is called consistent if no atom is a fact or is derivable from this KB along with its negation being derivable or a fact.

Evaluable KBs can be made more expressive if the closed world assumption [19] is adopted at the level of individual non-evaluable predicates. If all facts of predicate P are positive and have constant arguments, then P can be identified as closed by facts. Intuitively, P facts constitute a complete specification of the predicate. For instance, if $E(x, y)$ specifies edges of a graph, it is assumed that all edges of this graph are represented by KB facts. Predicate E can be identified as closed by facts by a KB developer. Negative literals of closed predicates with constant arguments are considered facts if their complementary atoms are absent in the KB. Closed predicates can represent infinite collections of facts.

A predicate can be identified as closed by rules (or simply closed) if all its facts and the rule heads in which it occurs are positive, these heads contain all variables from the respective rules, and predicate arguments in these heads are either variables or ground terms. The same intuition applies here: these rules and facts are assumed to fully specify the predicate. If predicate R is identified by a KB developer as closed by rules, then it can be automatically completed. Predicate completion begins with converting R facts and rules with predicate R in the head to rules of the following form [20]:

$$R(x_1...x_k) \Leftarrow Q_{11} \wedge ... \wedge Q_{1u} \qquad ... \qquad R(x_1...x_k) \Leftarrow Q_{q1} \wedge ... \wedge Q_{qv}$$

where $x_1...x_k$ are variables. It is done by renaming variables, generating rules for R facts, and transforming rules with predicate R in the head. Every fact of R is turned into a rule by adding atom $x_k = t_k$ to the rule body for every ground term t_k from the fact. For every rule with predicate R in the head, atom $x_k = t_k$ is added to the rule body for every ground term t_k from the head. The conjunction of all these new or transformed rules is equivalent to

$$R(x_1...x_k) \Leftarrow (Q_{11} \wedge ... \wedge Q_{1u}) \vee ... \vee (Q_{q1} \wedge ... \wedge Q_{qv})$$

Let $-A$ denote the complement of A, i.e. it is the negation of atom A, and the atom of negative literal A. This notation will also be used for sets of literals. Since the above rules do not contain other variables than $x_1...x_k$, automatic completion adds the following rules:

$$\neg R(x_1...x_k) \Leftarrow -Q_{1s} \wedge ... \wedge -Q_{qt} \qquad ... \qquad \neg R(x_1...x_k) \Leftarrow -Q_{1y} \wedge ... \wedge -Q_{qz}$$

where $s, y \in [1...u]$, $t, z \in [1...v]$ and $(s, ..., t)$, ..., $(y, ..., z)$ are distinct sequences of indexes [20]. The conjunction of these additional rules is equivalent to

$$R(x_1...x_k) \Rightarrow (Q_{11} \wedge ... \wedge Q_{1u}) \vee ... \vee (Q_{q1} \wedge ... \wedge Q_{qv})$$

Finally, rules containing complementary literals in their bodies are removed, and duplicate literals are removed from the bodies.

For example, suppose predicate R does not occur in facts, it occurs in the heads of the following rules, and it is marked as closed by rules:

$$R(x) \Leftarrow P(x) \wedge Q(x) \qquad\qquad R(x) \Leftarrow \neg P(x) \wedge \neg Q(x)$$

The completion procedure generates these two rules:

$$\neg R(x) \Leftarrow P(x) \wedge \neg Q(x) \qquad\qquad \neg R(x) \Leftarrow \neg P(x) \wedge Q(x)$$

Of course, incautiously marking predicates as closed may lead to KB inconsistencies, but the same is true about any rule sets.

3 Chaining Inference and Resolution

Chaining methods are formalized as inference based on Generalized MP as the sole inference rule [19]. If substitution θ is a unifier of literals A'_i and A_i, i.e. $A'_i\theta = A_i\theta$, for $i = 1...k$,

$$\frac{A'_1 \quad ... \quad A'_k \quad A_1 \wedge ... \wedge A_k \Rightarrow A}{A\theta} \ GMP$$

For facts, $k = 0$. Backward and forward chaining apply GMP in opposite directions. A forward chaining step derives $A\theta$ given that $A'_1, ..., A'_k$ are derived literals. Given goal list $L = \{...G...\}$ and such substitution θ that $G\theta = A\theta$, every step of backward chaining replaces goal G with $A_1\theta, ..., A_k\theta$ in L and also applies θ to the other goals in L. We will refer to these two method as MP chaining.

Negation is redundant in MP chaining. New predicate P^- can be introduced for every predicate P, and negative literals of P are replaced by atoms of P^-. Clearly, this transformation does not affect the inference power of MP chaining. MT makes it possible to utilize negation properties in KB inference. MT is generalized in the same manner as MP. If $A'_i\theta = A_i\theta$ for $i = 1...j-1$, $i = j+1...k$, and $A'\theta = -A\theta$,

$$\frac{A' \quad A'_1 \quad ... \quad A'_{j-1} \quad A'_{j+1} \quad ... \quad A'_k \quad A_1 \wedge ... \wedge A_k \Rightarrow A}{-A_j\theta} \ GMT$$

GMT can be used in chaining derivations similar to GMP. Adding GMT to GMP may significantly increase the range of derivable literals. For instance, if literals A and $\neg D$ are facts, then literal $\neg B$ is derivable by forward and backward chaining with GMT from rules $C \Leftarrow A \wedge B$, $D \Leftarrow C$. This derivation is not possible without GMT. The utility of GMT is demonstrated by rules defining total orders:

$$a = b \Leftarrow a \preceq b \wedge b \preceq a \qquad a \preceq c \Leftarrow a \preceq b \wedge b \preceq c \qquad a \preceq b \Leftarrow \neg(b \preceq a)$$

Other properties of total orders are derivable with using GMT but not with using GMP alone. For example, GMT makes it possible to derive literal $\neg(m \preceq n)$ from facts $n \neq m$, $n \preceq m$ and the aforementioned rules. The same literal is also derivable using GMT from these two facts: $\neg(m \neq k)$, $n \preceq k$. Not all FOL derivations are enabled by GMP and GMT. P is derivable from rules $P \Leftarrow Q$ and $P \Leftarrow \neg Q$ in FOL but it is not derivable from these rules by chaining with GMP and GMT. We will refer to forward and backward chaining with both GMP and GMT as MPT chaining or simply chaining. In general, MPT chaining is less efficient than MP chaining and more efficient than FOL inference.

For any rule $A \Leftarrow A_1 \wedge ... \wedge A_k$, the following implications $-A_i \Leftarrow -A \wedge A_1 \wedge ...A_{i-1} \wedge A_{i+1} ... \wedge A_n$ are called contrapositives to this rule [25]. Rules entail their contrapositives in classical FOL. Clearly, adding GMT to GMP is equivalent to adding contrapositives for all rules to the KB. This simulation of GMT by GMP with contrapositives shows a similarity of MPT and MP chaining, but it is more efficient to use GMT as opposed to multiplying the size of KBs by storing contrapositives.

Proposition 1. *A literal is derivable by MPT forward chaining if and only if it is derivable by MPT backward chaining.*

It is well-known that forward and backward MP chaining have the same inference power in application to Horn rules and facts. It is proved by a straightforward induction on the depth of derivations. Clearly, the same proof can be used to show that these two methods have the same inference power in application to non-Horn KBs. Since forward and backward MPT chaining can be simulated by using GMP with contrapositives, the two methods have the same inference power.

The resolution calculus [5] has two rules: resolution (left) and factoring (right).

$$\frac{A_1 \vee ... \vee A \vee ... \vee A_k \quad B_1 \vee ... \vee B \vee ... \vee B_m}{A_1\theta \vee ... \vee A_k\theta \vee B_1\theta \vee ... \vee B_m\theta} \qquad \frac{A_1 \vee ... \vee A \vee ... \vee A' \vee ... \vee A_k}{A_1\theta \vee ... \vee A\theta \vee ... \vee A_k\theta}$$

In the resolution rule, substitution θ is the most general unifier (MGU) of A and $-B$, that is, any other unifier of the two literals is a composition of θ and another substitution [19]. In the factoring rule, substitution θ θ is the MGU of A and A'. In input resolution, one of two premises of the resolution rule is a clause from the input set [5]. The resolution calculus is a complete inference method for FOL whereas input resolution is incomplete for FOL.

Suppose $\succ$ is a simplification order on formulas [15], i.e. it is:

- well-founded: there is no infinite sequence of formulas $t_0 \succ t_1 \succ ...$
- monotone: if s, l, r are formulas and $l \succ r$, then $s[l] \succ s[r]$;
if r is a subformula of l and $l \neq r$, then $l \succ r$
- stable: if $l \succ r$, then $l\theta \succ r\theta$ for any substitution θ

Literal A is maximal with respect to clause Σ if $B \succ A$ does not hold for any other literal $B \in \Gamma$. In ordered resolution [1], only maximal literals of the respective clauses are resolved or factored. This restriction dramatically reduces inference search space. The majority of modern theorem provers utilize ordered resolution due to its superior efficiency [15]. Ordered resolution is complete for FOL in the case of simplification orders.

The PTTP model elimination procedure [25] has two rules.

$$\frac{A \Leftarrow A_1, ..., A_k \quad B, B_1, ..., B_m}{A_1\theta, ..., A_k\theta, [B\theta], B_1\theta, ..., B_m\theta} \qquad \frac{B, B_1, ..., [B'], ..., B_m}{B_1\theta, ..., [B'\theta], ..., B_m\theta}$$

The left rule is a semi-ordered variant of the resolution rule. This rule does not eliminate the resolved literal. It is retained as a framed literal in square brackets. $A \Leftarrow A_1, ..., A_k$ is a rule or contrapositive, $B, B_1, ..., B_m$ is a goal list, substitution θ is the MGU of A and B. Leftmost framed literals are removed immediately. The right rule is called the reduction rule. $B_1\theta, ..., [B'\theta], ..., B_m\theta$ and $B, B_1, ..., [B'], ..., B_m$ are goal lists, substitution θ is the MGU of $-B$ and B'. The PTTP procedure is a complete inference method for FOL. The PTTP resolution rule applied to a KB rule corresponds to GMP. This resolution rule corresponds to GMT when it is applied to a contrapositive.

The reduction rule corresponds to RAA. This rule eliminates a hypothesis in the case that the hypothesis implies its complement. The derivation of P from rules $P \Leftarrow Q$ and $P \Leftarrow \neg Q$ by the PTTP procedure is an example of reasoning by contradiction [25]. This derivation involves the reduction rule. Chaining, input resolution, and model elimination can be implemented using stack-based memory.

4 Inference Without Reductio Ad Absurdum

The PTTP model elimination procedure [25] gives a characterization of proofs by contradiction. They are the PTTP derivations involving the reduction rule since this rule is essentially RAA.

Theorem 1. *Literal L is derivable by chaining if and only if –L is refutable by input resolution in which KB rules/facts are input clauses and –L is used in the first resolution step only.*

Proof. Factoring is admissible in resolution refutations satisfying the condition of this theorem. If the last application of the factoring rule eliminates A' from $A_1 \vee ... \vee A \vee ... \vee A' \vee ... \vee A_k$ using substitution θ, then each of $A_1\theta, ..., A\theta, ..., A_k\theta$ is refutable, and their refutations satisfy the condition of this theorem and do not involve the factoring rule. The lifting lemma [5] guarantees that if clause B is an instance of B' and C is the resolvent of B and rule/fact R, then there is such clause C' that C is its instance and C' is the resolvent of B' and R. By induction on the depth of resolution refutations, there exist refutations of $A_1, ..., A, ..., A', ..., A_k$ satisfying the condition of this theorem and not involving the factoring rule. By induction on the number of applications of the factoring rule, all them can be eliminated.

Consider a resolution refutation satisfying the condition of this theorem and not using the factoring rule. Every application of the resolution rule to a KB rule (treated as disjunction) or to a KB fact maps to a MPT backward chaining step. By induction on the depth of resolution refutations, the entire refutation corresponds to a chaining derivation.

Consider a backward chaining derivation. Let us ground it. The outcome of grounding is a legitimate backward chaining derivation. Suppose a step of this derivation produces goal list H from goal list G and rule/fact R. This step corresponds to an application of the resolution rule to $–G$ and R. Due to the lifting lemma, if G is an instance of another goal list G', then there is such goal list H' that H is its instance and $–H'$ is the result of applying the resolution rule to $–G'$ and R. By induction on the depth of backward chaining derivations, for any such derivation of goal list G which is an instance of G', there exists such resolution refutation that a KB rule/fact is used in every step and $–G'$ is used in its first step. Consequently, there exist a resolution refutation satisfying the condition of this theorem. □

Theorem 2. *Literal L is derivable by the PTTP model elimination procedure without the reduction rule if and only if –L is refutable by input resolution in which L is used in the first step only.*

Proof. Clearly, any PTTP derivation not using the reduction rule is a resolution refutation satisfying the condition of this theorem. As proved earlier, we can consider an input resolution refutation without factoring. The only difference between such input resolution refutations and PTTP derivations is that the order in which literals are resolved is fixed in PTTP derivations. This order is determined by the order of literals in the bodies of rules and contrapositives.

Let us transform the resolution refutation under consideration into another input resolution refutation in which only the first literals of resolvents are resolved. Suppose the last step not satisfying this requirement resolves A_i from $A_1 \vee ... \vee A_k$ where $i > 1$. All the following steps until the resolution of A_1 resolve descendants of A_i. Due to the lifting lemma [5], these steps can be moved right before the step that resolves A_{i+1}. This transformation reduces the number of steps resolving non-first literals by one. By repeatedly applying this transformation, we can get a resolution refutation in which first literals are resolved before others. This refutation directly maps to a PTTP derivation. □

Theorem 3. *Literal L is refutable by resolution in which the factoring rule is not used and at least one premise of every step is not L or its descendant if and only if it is refutable by input resolution in which L is used in the first step only.*

Proof. The 'if' part follows from the fact that factoring is admissible in the input resolution of L satisfying the condition of this theorem. Consider a resolution refutation of L in which at least one premise of every step is not L or its descendant and the factoring rule is not used. As usual, let us ground this refutation. Let $n(\Delta)$ denote the number of resolution steps in the resolution with end clause Δ and $h(\Delta)$ denote the sum of $n(\Psi)$ for left premises Ψ of all steps in this resolution. Input resolution is characterized by $h = 0$.

Without loss of generality, we assume that L and its descendants are right premises of resolution steps. Left premises are never descendants of L then. Consider the last resolution step whose left premise $\Gamma \vee B \vee \Pi$ is not an input clause. Depending on whether B occurs in the left or right premise of the resolution step with conclusion $\Gamma \vee B \vee \Pi$, we can apply one of the two following transformations to these two steps:

$$\dfrac{\dfrac{\Gamma \vee A \quad -A \vee B \vee \Pi}{\Gamma \vee B \vee \Pi} \quad -B \vee \Phi}{\Gamma \vee \Pi \vee \Phi} \quad \rightarrow \quad \dfrac{\Gamma \vee A \quad \dfrac{-A \vee B \vee \Pi \quad -B \vee \Phi}{-A \vee \Pi \vee \Phi}}{\Gamma \vee \Pi \vee \Phi}$$

$$\dfrac{\dfrac{\Gamma \vee A \vee B \quad -A \vee \Pi}{\Gamma \vee B \vee \Pi} \quad -B \vee \Phi}{\Gamma \vee \Pi \vee \Phi} \quad \rightarrow \quad \dfrac{-A \vee \Pi \quad \dfrac{\Gamma \vee A \vee B \quad -B \vee \Phi}{\Gamma \vee A \vee \Phi}}{\Gamma \vee \Pi \vee \Phi}$$

In these transformations, Greek letters denote clauses which are treated as multisets, i.e. they are not ordered. For the original (left) resolution in the first transformation,

$$h(\Gamma \vee \Pi \vee \Phi) = h(\Gamma \vee A) + h(-A \vee B \vee \Pi) + h(-B \vee \Phi) + 2n(\Gamma \vee A) + n(-A \vee B \vee \Pi) + 1$$

For the transformed (right) resolution in the same transformation,

$$h(\Gamma \vee \Pi \vee \Phi) = h(\Gamma \vee A) + h(-A \vee B \vee \Pi) + h(-B \vee \Phi) + n(\Gamma \vee A) + n(-A \vee B \vee \Pi)$$

For the original (left) resolution in the second transformation,

$$h(\Gamma \vee \Pi \vee \Phi) = h(\Gamma \vee A \vee B) + h(-A \vee \Pi) + h(-B \vee \Phi) + 2n(\Gamma \vee A \vee B) + n(-A \vee \Pi) + 1$$

For the respective transformed (right) resolution,

$$h(\Gamma \vee \Pi \vee \Phi) = h(\Gamma \vee A \vee B) + h(-A \vee \Pi) + n(\Gamma \vee A \vee B) + n(-A \vee \Pi)$$

Both transformations reduce the h value at least by one. By induction on h, every ground resolution without factoring and with the constraint on using L descendants can be transformed into another ground resolution with the same end clause and in which the left premise of every step is an input clause. By induction on the depth of resolution trees and applying the lifting lemma [5] in each step, these transformed input resolution refutations can also be carried out with variables. □

Theorem 4. *Literal L is refutable by resolution without the factoring rule if and only if it is refutable by ordered resolution without factoring and using a simplification order provided that at least one premise of every resolution step is not L or its descendant.*

Proof. Obviously, the 'if' part holds. Consider a resolution refutation of L without factoring and with the constraint on using L descendants. Let us ground this refutation. Let $\mathcal{S}$ be the set of literals in the ground refutation, and $\mathcal{R} = \emptyset$. We can pick up minimal literal A from $\mathcal{S}$ with respect to simplification order $\succ$, i.e. $P \succ C$ for no other literal $C \in \mathcal{S}$.

Consider the last step resolving A. If any of its descendant steps resolves another literal from $\mathcal{S}$, then one of the transformations from the proof of Theorem 3 is applied to this step. If both premises of the step resolving A are not descendants of L, then at least one premise of every resolution step is not a descendant of L after these transformations. The same is true in the case that one of the premises of the step resolving A is a descendant of L. We continue applying one of these transformations to the following resolution steps until no descendant step resolves another literal from $\mathcal{S}$.

Let us repeat the same transformation series for other occurrences of A. By induction on the number of these occurrences, all of them can be moved to the bottom, i.e. they are not followed by steps resolving other literals from $\mathcal{S}$. After that, A is moved from $\mathcal{S}$ to $\mathcal{R}$.

Now we can pick up a minimal literal from $\mathcal{S}$ again and repeat the same procedure for it. $B \succ A$ never holds for $B \in \mathcal{R}$, $A \in \mathcal{S}$. By induction on the number of distinct literals in the refutation, $\mathcal{S} = \emptyset$ at the end. For any step with premises $\Gamma \vee A$ and $-A \vee \Pi$, all literals from $\Gamma \vee \Pi$ are resolved in the following steps, and hence, $B \succ A$ does not hold for any $B \in \Gamma \vee \Pi$, i.e. the transformed resolution refutation is ordered.. By inductively applying the lifting lemma, this transformed ordered ground refutation can also be done with variables. Every resolvent in the ground refutation is an instance of the respective resolvent in the refutation with variables. Due to the stability property of simplification orders, order relations among literals are preserved in the refutation with variables. □

5 Related Work

Logics with more limited inference capabilities than FOL are sensible for KBs. Description logics [19] are a well-known example of that. Logics with limited inference capabilities also play an important role in argumentation. Simple logic has one inference rule - MP - and no logical axioms [2]. Simple logic corresponds to MP chaining. Direct derivations are defined in [14] as natural deduction done without using RAA. Direct derivations correspond to MPT chaining. Extending backward chaining by employing rule contrapositives is suggested in [21].

Integration of logic and functional programming has a long tradition. Recursive functions can be translated into Prolog rules [18]. Some languages like Curry [10] add logic programming capabilities to a functional language by means of non-deterministic functions. The implementation of inference capabilities of Curry is based on narrowing. Nondeterministic functional programs can be transformed to Prolog programs [10], and thus, MP chaining supports non-deterministic computations. Other languages such as Mercury [24] extend a logic programming language with features of functional languages. Ciao [11] combines more than two programming paradigms including logic and functional programming.

Designing a language supporting both evaluable KBs and recursive functions is not a goal of this paper. We simply recognize the necessity of allowing recursively-defined functions and predicates in KB rules and facts. Any existing functional or procedural programming language can be integrated with a KB language for expressing non-Horn rules and facts. Such programming language should support external calls for the incorporation of predicates defined by rules into the language.

In addition to the FOL interpretation of rules, they can also be treated as inductive definitions. Such rules in FO(FD) [12] rely on the notation of FOL. The paper [12] considers inference in FO(FD) for finite domains only. There exist other formalizations of inductive definitions as extensions of FOL [3,6]. The practicalities of inference in the respective proof systems are not addressed in these papers. Clearly, inference in these systems is more complicated than inference in FOL. They incorporate cyclic proofs. No concrete inference procedures are presented in these papers.

Stable and well-founded semantics [22] are arguably the two most-researched semantics of normal logic programs. They are based on the negation-as-failure paradigm [7]. The semantics of normal logic programs are closely related to the semantics of inductive definitions [8]. Inference procedures for the stable and well-founded semantics are more complex than those for FOL [23,26]. Their efficiency has been neither investigated nor tested in practice as much as that of inference procedures for FOL and its fragments.

Studies of some subsystems of FOL employ non-standard logical connectives [9,13]. This paper is concerned about standard logical connectives as implication, negation, and conjunction occur in KB rules. Standard connectives are intuitive and well understood by software developers. In authors opinion, having non-standard connectives in KBs that are developed and debugged by non-logicians is a recipe for problems, and those connectives should be avoided in logic programs and in KB rules and facts.

There exist a variety of MP chaining optimization techniques [19]. They are applicable to MPT chaining. Many resolution optimization strategies [5, 15] are applicable to resolution methods implementing reasoning without RAA. Investigation of inference without RAA in FOL with equality [17] is a topic for future research.

6 Conclusion

Evaluable KBs are quite versatile. Availability of rules with both positive and negative heads in these KBs eliminates the need in negation as failure and thereby helps avoid controversies stemming from it [7]. Still, negation properties are exploited in MPT chaining. These KBs naturally integrate computable functions into KBs. Recursively enumerable collections of facts can be represented in evaluable KBs. In author's opinion, it is crucial to have both reasoning and computation instruments handy in software systems.

This paper establishes a correspondence among MPT chaining, model elimination without the reduction rule, input resolution, and constrained resolution without factoring including ordered resolution. Since the reduction rule characterizes reasoning by contradiction, all of these methods are complete for FOL derivations without using RAA. The completeness of constrained ordered resolution without factoring is especially important because ordered resolution is recognized as one of the most efficient inference methods.

References

1. Bachmair, L., Ganzinger, H.: Resolution theorem proving. In: Handbook of automated reasoning, vol. 1, pp. 19–99. Elsevier, Amsterdam (2001)
2. Besnard, P., Hunter, A.: A review of argumentation based on deductive arguments. Handbook of Formal Argumentation, pp. 437–484 (2018)
3. Brotherston, J., Simpson, A.: Sequent calculi for induction and infinite descent. J. Logic Comput. **21**(6), 1177–1216 (2010)
4. Casas, A., Cabeza, D., Hermenegildo, M.V.: A syntactic approach to combining functional notation, lazy evaluation, and higher-order in LP systems. In: Hagiya, M., Wadler, P. (eds.) FLOPS 2006. LNCS, vol. 3945, pp. 146–162. Springer, Heidelberg (2006). https://doi.org/10.1007/11737414_11
5. Chang, C.L., Lee, R.C.T.: Symbolic Logic and Mechanical Theorem Proving. Academic Press, Cambridge (1973)
6. Cohen, L., Rowe, R.N.S.: Uniform inductive reasoning in transitive closure logic via infinite descent. In: 27th EACSL Annual Conference on Computer Science Logic, vol. 119, pp. 17:1–17:16. LIPICS (2018)
7. Denecker, M., Truszczynski, M., Vennekens, J.: About negation-as-failure and the informal semantics of logic programming. Association for Logic Programming (2017)

8. Denecker, M., Vennekens, J.: The well-founded semantics is the principle of inductive definition, revisited. In: 14th International Conference on the Principles of Knowledge Representation and Reasoning, AAAI (2014)
9. Girard, J.Y.: Linear logic. Theor. Comput. Sci. **50**(1), 1–101 (1987)
10. Hanus, M.: Functional logic programming: from theory to curry. In: Voronkov, A., Weidenbach, C. (eds.) Programming Logics. LNCS, vol. 7797, pp. 123–168. Springer, Heidelberg (2013). https://doi.org/10.1007/978-3-642-37651-1_6
11. Hermenegildo, M.V., et al.: An overview of Ciao and its design philosophy. Theor. Pract. Logic Program. **12**(1–2), 219–252 (2012)
12. Hou, P., De Cat, B., Denecker, M.: FO (FD): extending classical logic with rule-based fixpoint definitions. Theor. Pract. Logic Program. **10**(4–6), 581–596 (2010)
13. Jeřábek, E.: Admissible rules of Łukasiewicz logic. J. Logic Comput. **20**(2), 425–447 (2010)
14. Kakas, A.C., Mancarella, P., Toni, F.: On argumentation logic and propositional logic. Studia Logica **106**(2), 237–279 (2018)
15. Kovács, L., Voronkov, A.: First-order theorem proving and VAMPIRE. In: Sharygina, N., Veith, H. (eds.) CAV 2013. LNCS, vol. 8044, pp. 1–35. Springer, Heidelberg (2013). https://doi.org/10.1007/978-3-642-39799-8_1
16. McCune, W.: Otter 3.3 reference manual and guide. Technical report, Argonne National Lab. (2003)
17. Nieuwenhuis, R., Rubio, A.: Paramodulation-based theorem proving. Handbook Autom. Reasoning, vol. 1, pp. 371–443. Elsevier (2001)
18. Rodríguez-Hortalá, J., Sánchez-Hernández, J.: Functions and lazy evaluation in Prolog. Electron. Notes Theor. Comput. Sci. **206**, 153–174 (2008)
19. Russell, S., Norvig, P.: Artificial Intelligence: A Modern Approach, 3rd edn. Prentice Hall Press, Hoboken (2009)
20. Sakharov, A.: Hierarchical resolution for structured predicate definitions. In: 11th Hellenic Conference on Artificial Intelligence, pp. 202–210 (2020)
21. Sakharov, A.: Heuristic backward chaining based on predicate tensorization. In: Silhavy, R. (ed.) CSOC 2021. LNNS, vol. 229, pp. 573–586. Springer, Cham (2021). https://doi.org/10.1007/978-3-030-77445-5_52
22. Schlipf, J.S.: The expressive powers of the logic programming semantics. J. Comput. Syst. Sci. **51**(1), 64–86 (1995)
23. Shen, Y.D., Yuan, L.Y., You, J.H.: SLT-resolution for the well-founded semantics. J. Autom. Reasoning **28**(1), 53–97 (2002)
24. Somogyi, Z., Henderson, F., Conway, T.: The execution algorithm of mercury, an efficient purely declarative logic programming language. J. Logic Program. **29**(1–3), 17–64 (1996)
25. Stickel, M.E.: A Prolog technology theorem prover: a new exposition and implementation in Prolog. Theor. Comput. Sci. **104**(1), 109–128 (1992)
26. Yamasaki, S., Kurose, Y.: A sound and complete procedure for a general logic program in non-floundering derivations with respect to the 3-valued stable model semantics. Theor. Comput. Sci. **266**(1–2), 489–512 (2001)

A Review of e-Learning System Success Models

M. E. Rankapola(✉) and T. Zuva

Vaal University of Technology, Vanderbijlpark, Republic of South Africa
rankapolame@tut.ac.za

Abstract. This paper presents a review of e-learning system success models published between 1992 & 2019. Prominent models that focused on e-learning systems success, information systems success, e-learning system quality and user satisfaction from mainly open online journal articles, conferences and book chapters were identified and critiqued. The multidimensionality and interdependence of Information system success factors teethed arguments amongst various role players in the field. We observed that early research in the field of information system success attempted to combine a process model and a variance model into one framework thus leading to confusion in interpretations and implementation. On the other hand, most researchers focused on a single "chosen" variable than to comprehensively test associations of all the model constructs. Therefore, we believe that there are vital gaps that need to be filled in the e-learning system success field; notably, the previous success measurements did not deal with aspects such as voluntary system use versus mandatory system use. Other factors such as culture and cognition may also have a bearing on the e-learning system success. The paper concludes by highlighting relevant success measures from the existing e-learning system success models and suggests new perspectives that we believe are crucial system success measures. Future research work will concentrate on collating relevant system success measures to construct a more comprehensive e-learning system success model.

Keywords: E-learning · E-learning system success · Measuring e-learning system success · Information system success

1 Introduction

The vital terms and concepts that describe the paradigm of the contemporary society are information, knowledge and communication. The information and knowledge society in the digital age provides new prospects to the didactic discourse in higher education environments. The need to re-strategize and re-design new methods of teaching, learning and assessing students have become an imperative in order to keep up to speed with the technological advancements within the higher educational environments. Information and Communication Technology (ICT) based instruction or e-learning has been widely and intensively advertised to be more advantageous over the traditional teaching and learning modality. The e-learning systems literature argues that the ubiquitous presence of computing devices and increasing availability of the internet broadband provide a platform for the Higher Education Institutions (HEIs) and instructors to extend

R. Silhavy et al. (Eds.): CoMeSySo 2021, LNNS 232, pp. 511–526, 2021.
https://doi.org/10.1007/978-3-030-90318-3_42

their teaching and learning activities beyond the physical space. Consequentially, it has become extremely important to measure the success of the e-learning system to determine its impact and benefit to the academic discourse. It is also vital to justify investments made into the e-learning systems or to determine the return on investment (ROI) thereof. In an attempt to measure the e-learning systems success, various models & frameworks have been advanced to guide the adoption & use of e-learning systems in the educational environments.

2 Models Measuring the Success of the e-Learning Systems

In the quest to analyse and measure the success of the e-learning system, the study embarked on the models that concentrated on information systems success, e-learning systems success, user satisfaction and e-learning quality. The following models were identified in the literature completed between 1992 and 2019:

Model/Framework	Source
1992 IS success model	[1]
A re-specification of the IS success model	[2]
Revised IS success model	[1],
Lin's virtual communities model	[3]
Anita Lee-Post Model	[4]
Wang & Chiu model	[5]
The MELSS model	[6]
The Dorobat ELSS model	[7]
The Chuoa et. al model	[8]
The Mohammadi e-learning success model	[9]
The Findink-Conskuncay et. al e-learning success model	[10]
Kurt's e-learning system success model	[11]

2.1 The DeLone and McLean IS Success Model

The most prolific model that have been used for measuring the success of an information system (IS) is the D&M model, which was first presented in 1992 by [1]. The D&M model has served as a blueprint for measuring IS success hence it was reportedly used by over 300 scientific research papers by the year 2003 [7]. The original 1992 D&M IS success model includes six components i.e. system quality, information quality, use of the system, user satisfaction, individual impacts and organisational impacts. The model and the relationship between its components are depicted in Fig. 1. In the D&M IS Success Model, "systems quality" measures technical success; "information quality" measures semantic success; and "use, user satisfaction, individual impacts", and "organizational impacts" measure effectiveness success [1].

Since its inception in 1992, the D&M model has been widely supported and validated by some researchers but also criticised by some scientific research papers. Researchers,

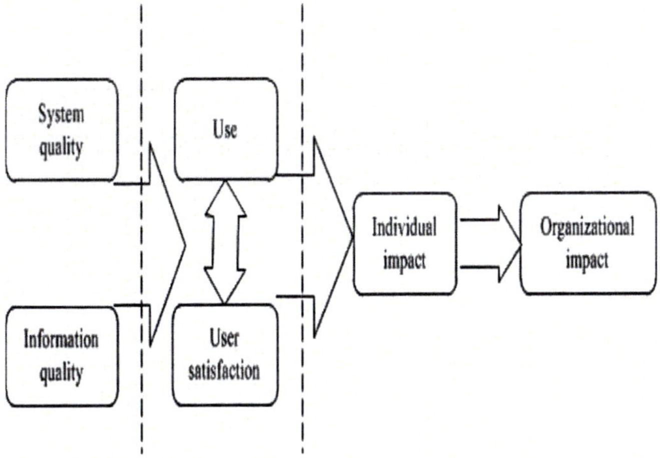

Fig. 1. The 1992 D&M IS success model

[6] argued that the D&M model cannot completely measure the IS success or the e-learning system success since it does not take into consideration aspects of culture, the loyalty of the system user etc. [2] argued that the D&M model's inclusion of both variance and process interpretations in one model leads to potentially confusing meanings that the value of the D&M model is diminished.

2.2 A Re-specified D&M Model

According to [1] most of the scientific articles that were presented between 1992 and 2003 tended to justify their own empirical measurement of IS success by citing the D&M IS Success Model, but failed to heed the appropriate cautions of the D&M model of 1992. The D&M further reported that some researchers have used the model to support their "chosen" success variable rather than to inform the development of a more comprehensive success construct thereby overlooking the main conclusion of the 1992 article which stipulated that IS success is a multidimensional and interdependent construct and that it is therefore necessary to study the interrelationships amongst, or to control for, those dimensions. According to [1], researchers should systematically combine individual measures from the IS success categories to create a comprehensive measurement instrument (Fig. 2).

In 2003, the D&M model was revised taking into consideration valuable inputs and critique of the original model from [2] who amongst other issues pointed out that intention to use and use of the system represent attitude and behaviour and therefore should be measured individually. Seddon [2] has also pointed out crucial issues regarding voluntary use of the system versus mandatory use of the system that may have a bearing on the system success measurements. The 2003 D&M model includes six components; i.e. system quality, information quality, service quality, intention to use/use of the system, user satisfaction and net benefits. The model and relationship between its components is illustrated in Fig. 3 below.

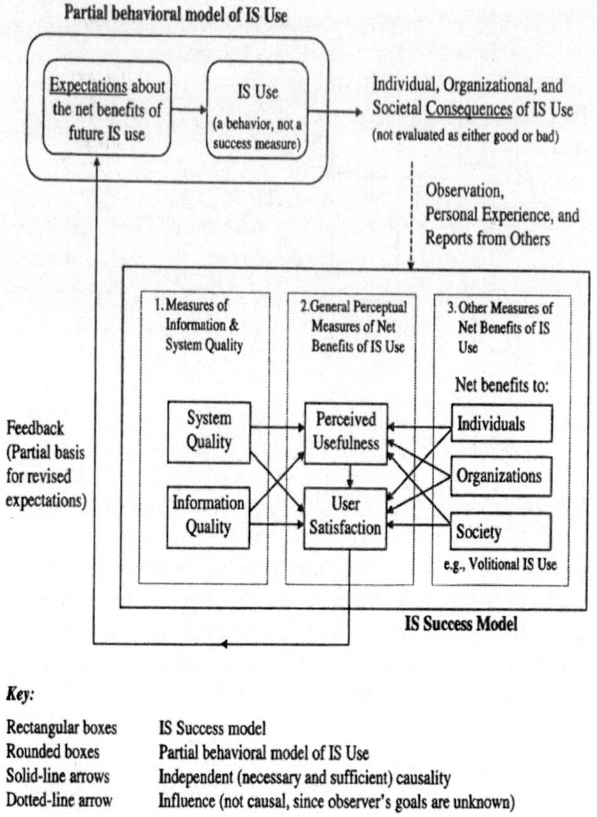

Fig. 2. A Re-specification of the D&M Model 1997

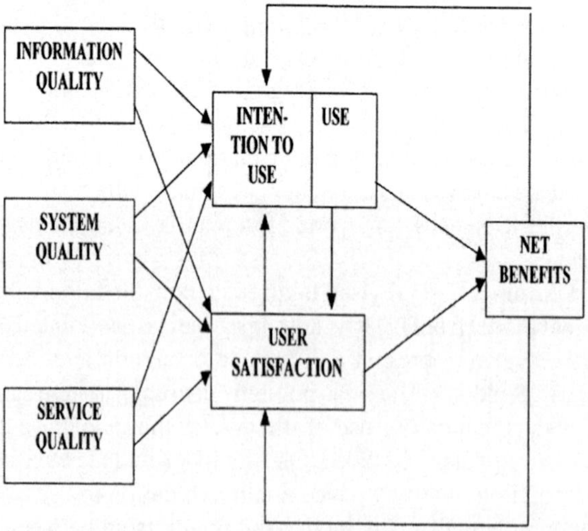

Fig. 3. IS success model: DeLone & McLean (2003:33)

In the 2003 model, service quality construct was added to the model and the individual impacts as well as the organisational impacts were collapsed into one construct called net benefits. The updated D&M IS Success Model includes arrows to demonstrate proposed associations among success dimensions in a process sense, but does not show positive or negative signs for those associations in a causal sense. According to [1], the nature of these causal associations should be hypothesized within the context of a particular study.

2.3 The Lin's Virtual Communities Model

In [3] the D&M model was magnified by examining the impact of system characteristics (namely, information and system quality) and social factors (namely, trust and social usefulness) on success in virtual communities. The Lin 2008 study contended that creating and sustaining the virtual community would promote knowledge sharing, with significant and positive impact on organisations. As depicted in Fig. 4 below, Lin's 2008 model made the following adjustments to the D&M model; the construct "use" was replaced with "sense of belonging" and the construct "net benefits" was replaced with "member loyalty". The study discovered that member satisfaction and a sense of belonging were determinants of member loyalty in the virtual community. Additionally, system characteristics (namely, information and system quality) were found to affect member satisfaction, whereas trust influenced the members' sense of belonging to the community.

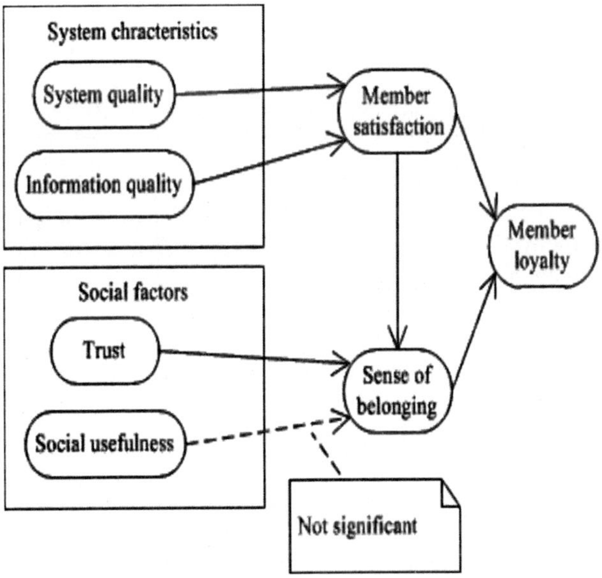

Fig. 4. Lin's virtual communities model

Lin's virtual communities model highlighted the significance of social factors such as "trust" on "member satisfaction" or "user satisfaction" which would impact on developing successful online/virtual communities. This suggests that institutions must take social factors into consideration when initiating or adopting the e-learning systems.

2.4 The Lee-Post Model

The Lee-Post model is an adaptation of the D&M model to measure the success of the e-learning system. This model encompasses a series of matrixes that that are depicted in Fig. 5.

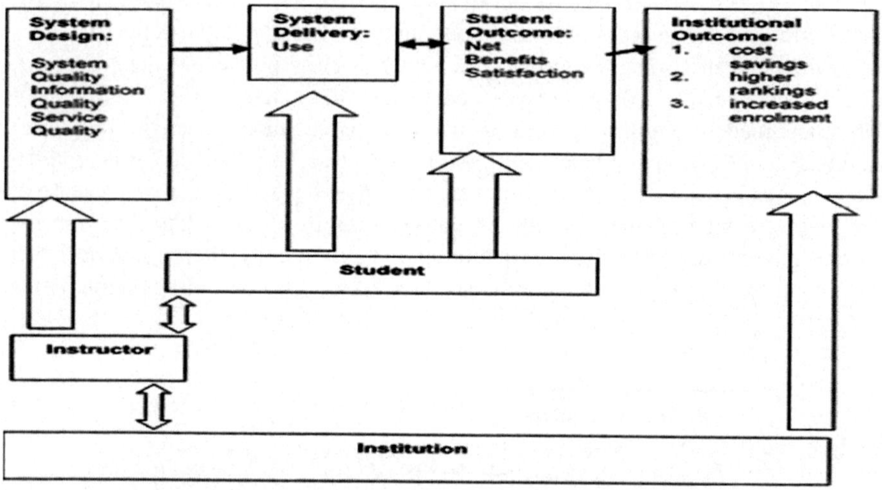

Fig. 5. E-learning success model: Anita Lee-Post, (2009:68).

The Lee-Post model purports to address the question of how to guide the design, development, and delivery of successful e-learning initiatives guided by theories of a user-centred information systems development paradigm. In [4], it was postulated that the process approach to e-learning success is fundamentally a continuous improvement process seeking to raise ratings of six success dimensions that includes system quality, information quality, service quality, use, user satisfaction, and net benefits in three stages: system design, delivery and outcome. It can therefore be deduced from this premise that the interdependences within the three stages as illustrated in Fig. 4, denote that attempts to improve a success rating should begin with raising the three quality ratings in the design stage and then proceed to advancing the use rating in the delivery stage and lastly advance the user satisfaction and net benefits ratings in the outcome phase. Although the Lee-Post model made an important contribution to the field of e-learning system success measurement by assisting us to guide the design, development and delivery of the system in order to attain desired outcomes, the model has its shortcomings.

Firstly, the model assumes that the instructors are knowledgeable and skilled system developers and e-learning experts. Different contexts would have brought about different datasets in e.g. instructors may not be skilled in system development. Secondly, the model is student-centred and depends heavily on students' e-learning experience as feedback for e-learning success. Thirdly, the Lee-Post model showed us how to demonstrated how action research can be applied in increasing e-learning quality. However, the model needs further testing and validation. Therefore, it cannot as yet be used as a comprehensive framework to measure the e-learning system success.

2.5 The Wang and Chiu Model

In a traditional e-learning system, instructors provide knowledge for learners but are unable to use a student's learning experiences to benefit the group of students as a whole [5]. Therefore, traditional e-learning systems support a one way kind of communication. The Wang & Chiu study explored e-learning success factors through the design and evaluation of an e-learning 2.0 system. A theoretical model that was advanced, evaluated user satisfaction and loyalty intentions to an e-learning system through communication quality, information quality, system quality & service quality as illustrated in Fig. 6 below:

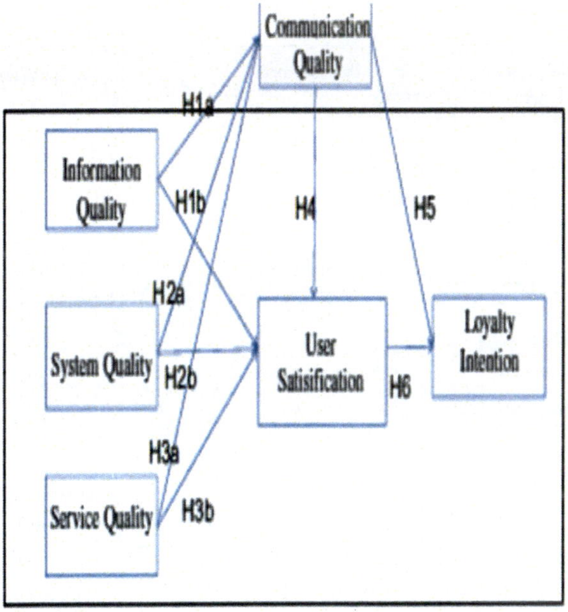

Fig. 6. The Wang & Chiu e-learning success model

The empirical results of the study revealed that communication quality, information quality, and service quality significantly positively affect user satisfaction and loyalty intentions to use an e-learning system. It is therefore apparent that loyalty intentions to use the system is underpinned by online communities' willingness to share experience, communicate with one another and obtaining feedback from each other. The Wang &

Chiu study applied the D&M IS success model to explore how to retain users and how to motivate them to continue using an e-learning 2.0 system. Although the study made positive contributions, it should be noted that the study focused only on one dimension of research participants who were mainly users of the system (students). The study also only used a web 2.0 system. It is therefore justifiably arguable that a variety of dimensions in this study would have broadened the dataset and affected the results of the study to a certain extent.

2.6 The MELSS Model

The MELSS model presented in Fig. 7 below was introduced in 2012 by [6]. This model also utilised the D&M model as its foundation to measure the e-learning system success. The MELSS model consists of 10 components that are used to measure the e-learning system success which includes: technical system quality, educational system quality, content and information quality, service quality, user satisfaction, benefits, intention to use, system use, loyalty to system and goals achievement. The MELLS model attempted to eliminate the shortcomings of the [1] model and the [4] model by incorporating three more components into the MELLS model.

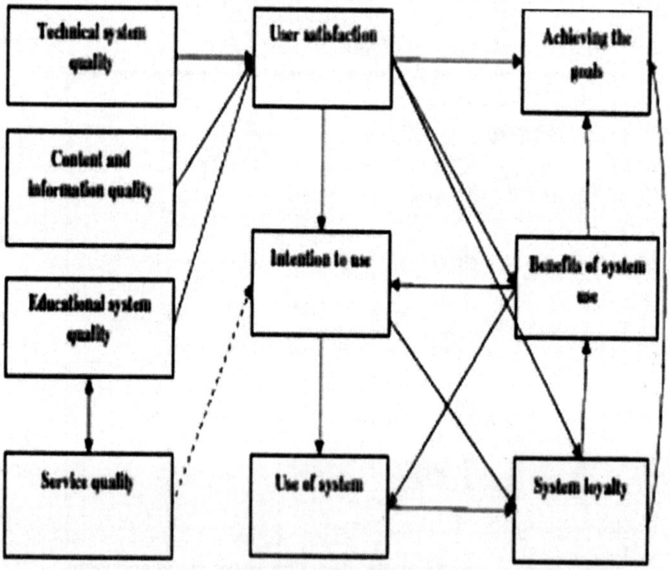

Fig. 7. The MELSS model

[6] attempted to make the MELSS model more comprehensive by surveying two categories of the research population i.e. e-learning experts (those who have knowledge and experience in the subject) and e-learning system users (including, students, alumni, and instructors).

In analysing the model [6] reported that (i) technical system quality has a direct effect on user satisfaction. This implies that the more technical quality, the higher the user satisfaction will be. Higher user satisfaction will lead to increased chances of the e-learning system success. (ii) Educational system quality has an influence on user satisfaction. However, the influence of educational quality on user satisfaction is less than that of the technical system quality. (iii) Educational quality and service quality are interrelated to one another implying that service quality through educational system quality can influence user satisfaction and intention to use. (iv) Content and information quality has the most direct effect on user satisfaction. This implies that if the quality of content and information is more, the users will be more satisfied with the use of the e-learning system. (v) User satisfaction leads to achieve personal and educational goals. (vi) User satisfaction has a direct impact on benefits which eventually encourages the users to re-use the system. (vii) When the user of the e-learning system is more satisfied, the loyalty to system will increase. (viii) Finally, intention to use the e-learning system has a direct effect on system usage. According to the MELSS model, even when the user's intention to use the system has not actualised the user will be loyal to the system and still suggest or recommend it to others. Although the MELSS model attempted to comprehensively measure the e-learning system success by incorporating additional components to the D&M model, just like any other scientific research the MELSS model has its strengths and weakness. An important contribution made by the MELSS model is that a number of stakeholders' perspectives were taken into consideration in constructing the model and not only the students. Users (students, instructors and alumni) and experts (experienced and knowledgeable e-learning practitioners) participated in the study. Finally, the model confirmed that components such as technical system quality, educational system quality, content and information quality, service quality, user satisfaction, intention to use, user loyalty to the system, benefits of using the system and goals achievement are suitable for measuring the e-learning system success. Nevertheless, the model needs further empirical testing and validation in different contexts. It may also be important to test the model from the opinions of different groupings used in this study both individually and collectively in order to increase the strength of the model.

2.7 The Dorobat's ELSS Model

The ELSS Model as presented in Fig. 8 below was proposed in 2014 by [7]. The ELSS model is founded on four perspectives namely, overall system quality, user perceived control, usefulness and user satisfaction, user attitude, social factors and benefits of using the e-learning system. This model combined constructs from various models notably, Technology acceptance model (TAM) that was first proposed by [12] and extended by various authors over the years including [13].

This hybrid model, attempted to explain the relationships between the e-learning system and the students' social contexts. However, the ELSS model has not been empirically tested and validated. The causal relationships and the interdependences between the constructs of the model need to be explained and validated.

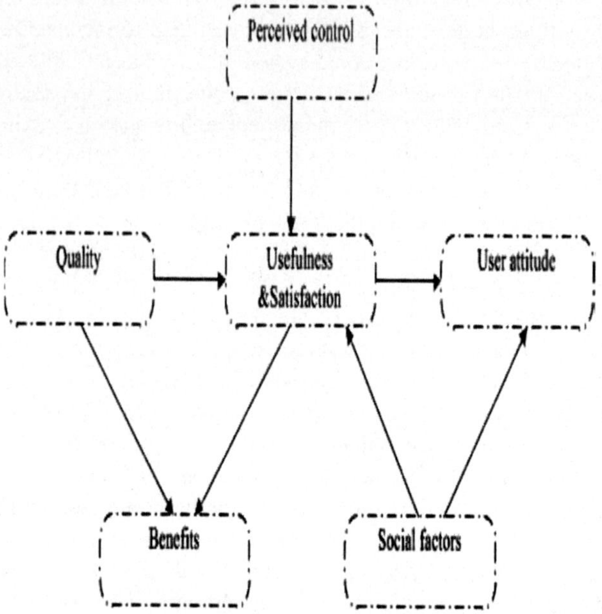

Fig. 8. ELLS model: Dorobat, (2014:86)

2.8 The e-Learning Success Model

The [8] study applied the IS success model as a foundation to investigate the factors influencing the effectiveness of e-learning in medical personnel. They explored how the constructs; system quality, information quality & service quality affect e-learning satisfaction. The study found that information quality, system quality & service quality exhibited the highest effect on e-learning satisfaction and that e-learning satisfaction exerted a significant influence on the effectiveness of e-learning as indicated on Fig. 9 below.

The study further demonstrated that learning climate and convenience have a considerable effect on student's e-learning satisfaction. This study affirms and endorses the notion that e-learning systems can remove the limitations of time and space and provide convenience to students because they can access learning material anytime, anywhere (on-demand learning).

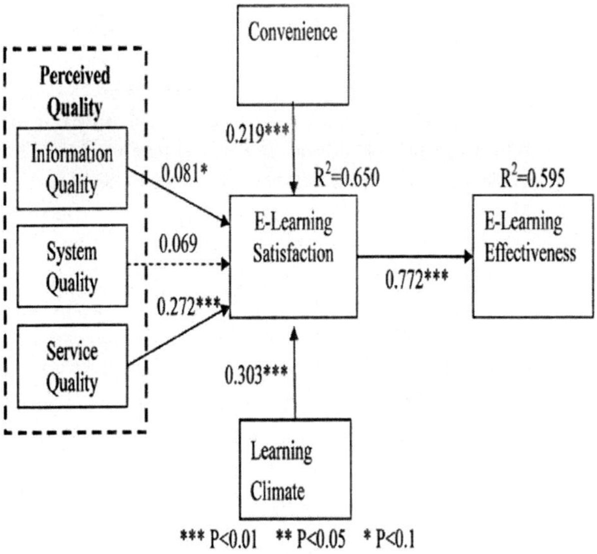

Fig. 9. Chuo, Liu & Tsai, (2015:S159)

2.9 The Integrated e-Learning Success Model

Similar to [7], the [9] used the D&M and the TAM lenses to explore the effects of quality features, perceived ease of use, perceived usefulness on users' intention and satisfaction alongside the mediating effect of usability towards use of e-learning systemes.

The Mohammadi model as depicted in Fig. 10 below included 9 constructs namely; educational quality, service quality, technical system quality, content and information

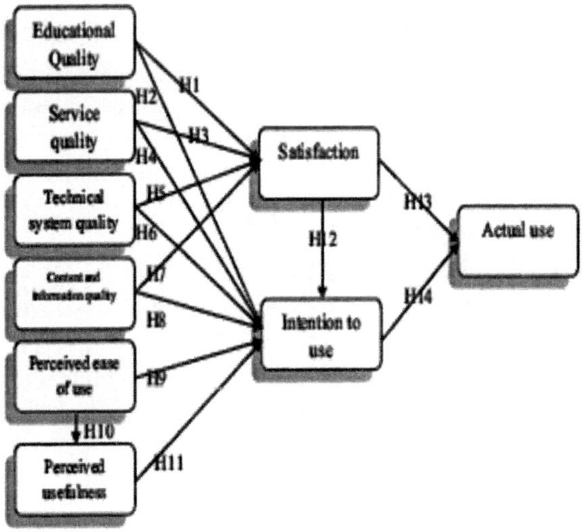

Fig. 10. The Mohammadi integrated e-learning success model

quality, perceived ease of use, perceived usefulness, satisfaction, intention to use and finally actual use.

The study found that intention to use and user satisfaction had positive effect on the actual use of the e-learning system. System quality and information quality were reported to be the main determinants influencing users' intentions to use the system and users' satisfaction towards the e-learning system. Perceived usefulness was reported to have had a positive effect on ease of use and users' intentions to use the system.

Despite positive contributions made by this study notably, (1) attempts to implement an integrated model of IS and TAM in the context of e-learning (2) comparison of past studies and support of literature review [6, 14] and (3) the study did not only examine the variables but also ivestigated the effects of variables on actual use and intention to use. It should be noted however that the study did not clearly and conclusively explain how the rest of the quality dimensions e.g. educational quality, content and information quality as well as service quality influence user satisfaction and intention to use and or actual use of the e-learning system. This argument is sustained by the conclusion of the study where it is claimed that all quality dimesions with significant effects may indirectly affect users' actual use. The study did not clearly and emprically indicated the interrelatedness of all quality dimensions in the model with "user satisfaction", "intention to use" and "actual use".

2.10 The Structural Model for Students' Adoption of LMSs and Their Behavioural Intention to Use the Systems

This study intended to identify factors affecting higher education students' behavioural intention to use Learning Management Systems (LMS). The study used the theoretical lens of the Technology Acceptance Model (TAM) with the following constructs; Perceived Usefullness (PU), Perceived Ease of Use (PEOU) and external factors which included self-efficacy, enjoyment, subjective norm, satisfaction, interactivity and control as seen in the Fig. 11 below.

The study found that the main predictors of student's behavioural intention to use the LMS were Perceived Usefulness, Perceived Ease of Use, enjoyment, subjective norm, satisfaction, interactivity and control. Thus the study proposed the structural model "LMS-TAM" as shown above which was assessed and validated in the this study. This study results support the results of [8, 9, 14]. User satisfaction has been found to be the most significant and positive predictor of students' behavioural intention to use an e-learning system which is mediated by PU & PEOU.

However, this model is also not sufficiently comprehensive. The self-efficacy construct which formed part of the proposed model was not analysed. Its proposed relationships with other constructs was not explored. There is therefore, a need to modify and extent this model to increase its generisability and comprehensiveness. It is also worth-noting that this study, only used the theoretical lense of TAM which can also limit the model's power to predict the student's behavioural intention to use the e-learning systems and the attainment of user satisfaction.

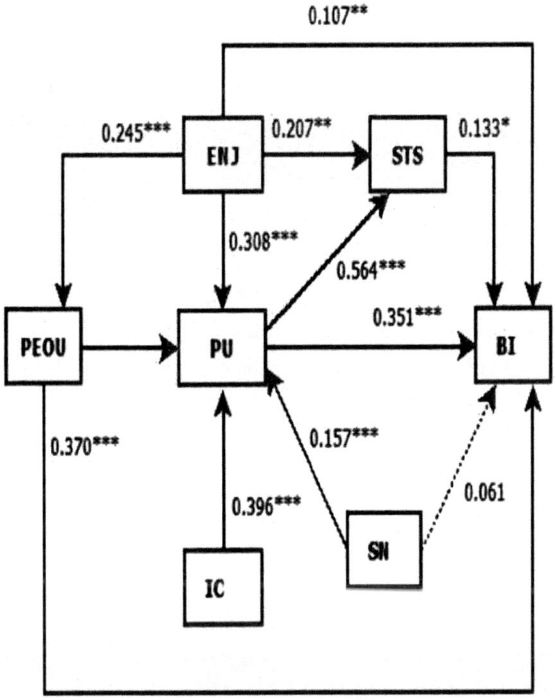

Fig. 11. The structural model predicting students' behavioural intention to adoption of LMS-TAM

2.11 Kurt's e-Learning System Through the Lens of IS Success Model

Kurt [11] study intended to examine an e-learning system based on students' perceptions through the lens of the the information systems success model. The study found that system quality had positive impact on system usage and user satisfaction whilst information quality only had significant impact on user satisfaction. User satisfaction and system usage both have positive impact on the system success. It can be said in summary that the study suggests that system quality and information quality afftect system use and user satisfaction which in turn leads to system success as depicted below (Fig. 12);

Although the model reported positive and significant findings, it is worth-noting that certain important aspects were excluded from the study. Notably, investigating readiness of students to e-learning, study discipline and other cognitive factors is important to understand the reasons for preferring e-learning systems by students. Again, instructor and content developers could have also been included in research sample to deeply understand the factors affecting system success from different perspectives. Other quality dimensions such as service quality and net benefits have been excluded in the model. It is also important to investigate the role of cultural context in e-learning experiences.

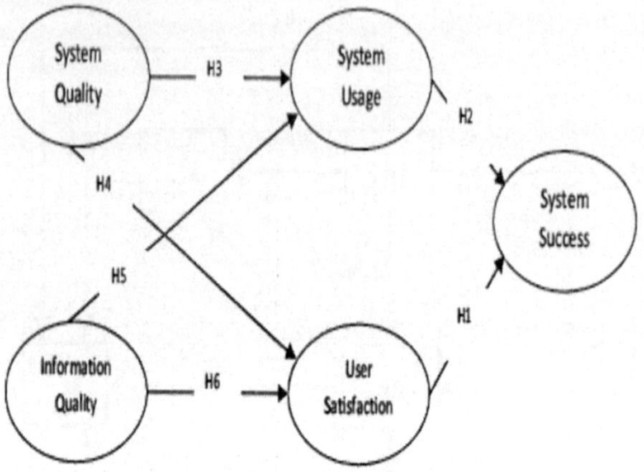

Fig. 12. Kurt's e-learning system success model

3 Discussion on Issues Emerging from the Literature Review

It is apparent that the DeLone & McLean model of Information Systems Success also known as the D&M IS success model served as a blueprint for the evaluation of information systems success including the e-learning system success. Since its inception in 1992, although criticised and refined over the years, the D&M IS success model has been applied by many researchers as a theoretical lens to measure the IS success. Since e-learning is a specific type of IS [4–6] were amongst the first researchers who used the D&M IS success model to measure the success of e-learning systems.

Some researchers [10, 11] have investigated the effectiveness and success of the e-learning systems through the theoretical lens of the Technology Acceptance Model (TAM). The argument advanced in this research studies is that when system users accept the e-learning system based on its perceived usefulness (PU) and perceived ease of use (PEOU), user satisfaction will be increased and the intention to use/use of the system will also be increased thus leading to the success of the e-learning system (increased use of the e-learning system).

Since e-learning system is a multi-faceted and a complex phenomenon, a number of researchers [7, 9] have attempted to examine e-learning system success through the dual theoretical lenses of the D&M IS success model and TAM. These studies have made a significant contribution in the field of e-learning systems because users of the systems are social beings whose social & cultural circumstances have a direct effect on the adoption/acceptance and use of the e-learning systems. However, these studies still need empirical testing and validation. On this note, it is clear that the e-learning system success measures are subject to the ever changing nature of the information system over time. Therefore, there is still a room for improvements and enhancements of the current e-learing systems success models. A more comprehensive e-learning success model that takes into account issues of culture, cognition and other factors such as whether or not the e-learning system success measurement considered e.g. mandatory use of the system or voluntary use of the system to determine its effectiveness and success. These aspects are crucial in determining the e-learning system success.

4 Contribution to the Body of Knowledge

First and foremost, this paper provides a scholarly review of prominent e-learning success models published between nearly three decades. This record will be of great assistance to scholars who are interested in the development and modifications of e-learning system success models. Secondly, it is of utmost importance to observe the e-learning system success patterns over time. Thirdly, this paper critiqued previously presented models in terms of their strengths and weaknesses. Isolating the weaknesses of different models and consolidating their strengths is likely to assist scholars in this field to develop, modify or augment various success models to achieve desired results.

5 Conclusions and Future Research Work

In this paper, we presented scholarly reviews of e-learning systems success. Remarkably, all e-learning system success models used a theoretical lens of the D&M model and TAM. Only a few studies examined the e-learning system success from the integrated perspectives of the D&M model and TAM. The literature study proved that even if attempts to improve these e-learning system success models have been made over time, there is still room for improvements and extensions.

For future research purposes, we intend to collate all relevant quality dimensions along with relevant dependent variables to construct a more comprehensive e-learning system success model. We believe that factors such as culture and cognition may have a bearing on the e-learning system success measurement. Instead of focusing on one success measure as other researchers did, we intend to holistically explain the causal relationships between all the model constructs. We also aim to formulate hypotheses and validate these hypotheses at a university of technology in the republic of South Africa. On completion of hypothesis validation, We intend to compare our research results with the results of other e-learning success models presented in this study.

References

1. DeLone, W., McLean, E.: The DeLone and McLean model of information systems success: a ten-year update. J. Manag. Inf. Syst. **19**(4), 9–30 (2003)
2. Seddon, P.B.: A respecification and extension of the DeLone and McLean model of IS success. Inf. Syst. Res. **8**(3), 240–253 (1997)
3. Lin, H.F.: Determinants of successful virtual communities: contributions from system characteristics and social factors. Inform. Manage. **45**(8), 522–527 (2008)
4. Lee-post, A.: e-Learning success model: an information systems perspective. Electron. J. e-Learn. **7**(1), 61–70 (2009)
5. Wang, H.C., Chiu, Y.F.: E-learning system success model. Comput. Educ. **57**(2011), 1790–1800 (2011)
6. Hassanzadeh, A., Kanaani, F., Elahi, S.: A model for measuring e-learning systems success in universities. Expert Syst. Appl. **39**, 10959–10966 (2012)
7. Dorobat, I.: Models for measuring E-learning success in universities: a literature review. Inform. Econ. **18**(3), 77–90 (2014)

8. Chuoa, Y., Liub, C., Tsai, C.: Effectiveness of e-learning in hospitals. Technol. Health Care **23**, S157–S160 (2015)
9. Mohammadi, H.: Investigating users' perspectives on e-learning: an integration of TAM and IS success model. Comput. Hum. Behav. **45**(2015), 359–374 (2015)
10. Findink-Conskuncay, D., Alkis, N., Ozkan-Yildirim, S.: A structural model for students' adoption of learning management systems: an empirical investigation in the higher education context. Educ. Technol. Soc. **21**(2), 13–27 (2018)
11. Kurt, O.E.: Examining an e-learning system through the lens of information systems success model: empirical evidence from Italy. Educ. Inf. Technol. **24**(2019), 1173–1184 (2019)
12. Davis, F.D.: Perceived usefulness, perceived ease of use, and user acceptance of information technology. MIS Q. **13**(3), 319–340 (1989)
13. Venkatesh, V., Bala, H.: Technology acceptance model 3 and a research agenda on interventions. Decis. Sci. **39**(2), 273–315 (2008)
14. Motaghian, H., Hassanzadeh, A., Karimzadegan, M.D.: Factors affecting university's instructors adoption of web-based learning system: Case study of Iran. Comput. Educ. **61**(2013), 158–167 (2013)

Distributed Flooding Algorithm for Sensor Fusion in Synchronous/Asynchronous Wireless Sensor Networks

Martin Kenyeres[1](✉) and Jozef Kenyeres[2]

[1] Institute of Informatics, Slovak Academy of Sciences, Dubravska cesta 9, 845 07 Bratislava, Slovakia
martin.kenyeres@savba.sk
[2] EBCONT proconsult GmbH, Millennium Tower, Handelskai 94-96, 1200 Vienna, Austria
jozef.kenyeres@ebcont.com
http://www.ui.sav.sk/w/en/dep/mcdp/

Abstract. Over recent decades, mechanisms for sensor fusion have been becoming a crucial complement in wireless sensor networks since they can optimize the executed applications in these networks in numerous aspects (e.g., data volume, energy consumption, quality of service, etc.). In this paper, we address the distributed flooding algorithm, a distributed straightforward method applicable to determining the exact value of various aggregate functions. We examine this approach in the synchronous and the asynchronous mode over random graphs with a varied graph order and connectivity by applying two metrics. Our goal is to identify in which mode the algorithm performs better in terms of the number of the clock ticks and the number of the sent messages as well as how the network size and connectivity can affect the performance of this algorithm. Finally, the distribution of the sent messages number is analyzed in both the examined modes.

Keywords: Distributed algorithms · Flooding algorithms · Data aggregation · Data fusion · Information fusion · Sensor fusion · Wireless sensor networks · Synchronous communication · Asynchronous communication

1 Introduction

Wireless sensor networks (referred to as WSNs), an infrastructure-less and self-organizing technology with several decades of history, are formed by tiny affordable autonomous entities (referred to as sensor nodes) for monitoring or supervising a wide range of physical and environmental circumstances [1–3]. Over recent years, this technology has been gaining popularity, thereby finding the application in many various areas (e.g., traffic monitoring, security, disaster monitoring, healthcare, intelligence surveillance) [4–6]. Typically, the sensor nodes operate

R. Silhavy et al. (Eds.): CoMeSySo 2021, LNNS 232, pp. 527–539, 2021.
https://doi.org/10.1007/978-3-030-90318-3_43

in an ad-hoc mode and consist of four main components, namely a sensing unit, a processing unit, a communication unit, and a power supply unit [7,8]. Besides, some variants of the sensor nodes can also be equipped with additional units providing additive functionalities. The purpose of the sensing unit is to sense physical quantities from the adjacent area and to digitize the sensed data. In general, we distinguish between three categories of the sensors - physical, chemical, and biological [9]. The processing unit control other units of a sensor node as well as processing the sensor data [7,8]. In order to ensure high financial affordability of the sensor nodes, the computational and memory capabilities of this unit are usually significantly constrained [7]. The goal of the communication unit is to send/receive messages to/from other sensor nodes in the network [7,8]. The last component, the power supply unit, acts as an energy source for all the other components [7,8]. Additional units can be formed by hardware components to ensure the mobility of the sensor nodes (a location finder, a mobilizer) or a power generator for recharging the power supply [7,8]. The general architecture of the sensor nodes is shown in Fig. 1 [7].

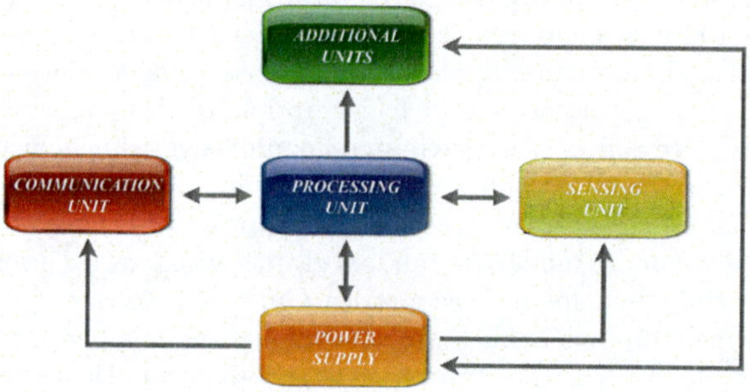

Fig. 1. General architecture of sensor node in wireless sensor networks

In WSN-based applications, the sensor nodes are required to record and transmit a specific data type to either a fusion node or their peers in the network via a wireless communication medium [10–12]. Their geographical position does not have to be predetermined; therefore, the sensor nodes can be randomly deployed (often densely) over the monitored area [13]. Frequently, the sensor nodes are situated in hard-accessible, inhospitable terrains and operate in real-time with no human interaction [14]. Thus, WSN-based applications are required to be robust to negative environmental effects and not to be prone to potential hardware/software lapses [12,14]. Compared to traditional IT technologies, the application of WSNs results in numerous benefits, e.g., flexibility, easy deployment, financial affordability, etc. [15]. However, on the other hand, the sensor nodes are vulnerable to many negative factors such as pressure variations, temperature, noises, radiation, data correlation, etc. [16]. As a consequence, the precision of sensor measurements can be significantly decreased whereby the quality of service (QoS) is worsened, and the overall energy consumption increases [16]. Therefore, in many modern WSN-based applications, sensor fusion mechanisms are applied to suppress (or even overcome) the impact of the mentioned factors, thereby increasing the QoS of the executed applications and extending the network lifetime [16].

Sensor fusion (also known as multi-sensor fusion) can be understood as a process of integrating a large amount of data from multiple sensor nodes into a more applicable summary form [17,18]. The goal of this process is to generate more reliable information about the observed phenomena and to decrease uncertainty whereby the constraints of the sensor nodes can be suppressed or even overcome [19]. The general architecture of sensor fusion in WSNs is shown in Fig. 2.

Fig. 2. General architecture of sensor fusion in wireless sensor networks

Based on where the sensor fusion is carried out, the authors of [20] define two categories of the sensor fusion methods - the centralized and the distributed one. In the first category, the centralized sensor fusion, the raw data measured by all the sensor nodes are transported to a fusion center, which performs data fusion [20]. This category is characterized by a large amount of bandwidth (due to massive data transfers), a significant time delay, the necessity to implement routing mechanisms, etc. [21]. Thus, it is not too optimal for real-world systems and so has been losing its popularity over recent years. In the other category, the distributed sensor fusion, each sensor node is fully responsible for data fusion in the network [20]. The sensor nodes exchange data with their neighbors in order to be capable of estimating/determining the value of the desired aggregate function. As the literature review shows, this approach has found a wide range of applications recently [19,22,23].

In this paper, we address the distributed flooding algorithm for determining various aggregate functions. The algorithm is analyzed over synchronous and asynchronous WNSs in order to identify in which communication mode it is better performing. Its performance is evaluated over random graphs (RGs) with a varied graph order (n) and connectivity by applying two metrics in order to examine the performance of the algorithm over various networks. In the last part, the distribution of the sent messages number is analyzed for both the modes.

The structure of this paper is organized as follows: in Sect. 2 (Distributed Flooding Algorithm for Sensor Fusion), we introduce the examined algorithm, its principle, and the analyzed synchronous/asynchronous mode. Section 3 (Applied Research Methodology) is concerned with modeling WSNs, the applied research methodology, and the metrics used to evaluate the performance of the distributed flooding algorithm. Section 4 (Experiments and Discussion) consists of the experimental results and a discussion about them. Section 5 (Conclusion) briefly summarizes the outputs of the research presented in this paper.

2 Distributed Flooding Algorithm for Sensor Fusion

As stated earlier in this paper, we examine the distributed flooding algorithm for sensor fusion in WSNs. The mentioned algorithm is a straightforward method for determining the exact value of various aggregate functions [20]. In contrast to many other distributed algorithms for sensor fusion, the distributed flooding algorithm can be applied to determine an extensive range of aggregate functions (e.g., the arithmetic mean, the sum, the extreme, the median, the mode, etc.) with no need to reconfigure the algorithm. In this paper, we consider sensor networks where each node performs local measurement, and its goal is to determine the value of the desired aggregate function given by the measurements of all the sensor nodes in the network. A sensor node stores, updates, and broadcasts a table (with n-slots) containing only the information measured by this node before the algorithm begins. All the other slots (n-1 slots) are empty, and each of them is reserved for the local measurement of all the other sensor nodes in the network. Once a sensor node receives the measured information of another node, it stores this information in its table unless it is already there (see Fig. 3 for an example of how a table is updated - in this case, the sensor node no. 1 receives two tables, specifically from the sensor node no. 2 and no. 4).

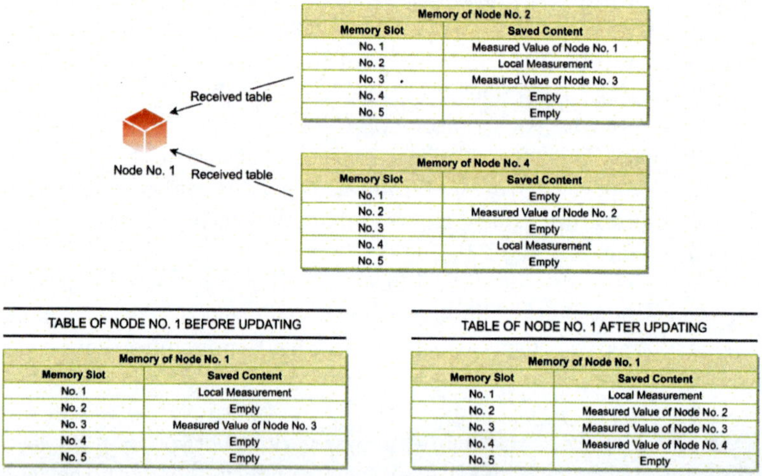

Fig. 3. Distributed flooding algorithm - example of table update

After the table of all the sensor nodes is filled in, the distributed flooding algorithm is said to have been completed (see Fig. 4 for an example of a table before and after the algorithm is applied). Thus, the sensor nodes can determine the exact value of the desired aggregate function (for example, the measured values of all the sensor nodes are added up whereby the sum of all the measurements is determined). Therefore, each sensor node finally acts as a fusion node, meaning that it has the measured information of all the sensor nodes and can easily determine the exact value of various aggregate functions.

As already mentioned, we analyze and compare two communication modes, namely [24,25]:

- *synchronous* – in this mode, all the sensor nodes follow the same global clock to synchronize communication in the network. Either the global clock is managed by a synchronization sensor node, or the sensor nodes perform synchronization by themselves. The principle of this mode can be described as follows: at each clock tick (the clock ticks periodically), each sensor node sends its table (respectively all the available data) to all of its neighbors (i.e., the sensor nodes further one-hop away) as well as receiving the table from all these sensor nodes. According to the received information, a sensor node updates its table for the next clock tick in such a way that the missing measured values are added to the table. In general, this communication mode is considered to be very prone to asynchronicity.
- *asynchronous* – in this case, each sensor node is equipped with its own local clock independent from the clocks of the other sensor nodes in the network. Once its local clock ticks (the clocks tick randomly and independently from each other), the sensor node broadcasts its current table (respectively all the available data) to all its neighbors, which update their current table according to the received information. This mode is a more robust solution compared to the previous one, however, more difficult to realize.

BEFORE

Memory of Node No. 1	
Memory Slot	**Saved Content**
No. 1	Local Measurement
No. 2	Empty
No. 3	Empty
⋮	⋮
No. n	Empty

AFTER

Memory of Node No. 1	
Memory Slot	**Saved Content**
No. 1	Local Measurement
No. 2	Measured Value of Node No. 2
No. 3	Measured Value of Node No. 3
⋮	⋮
No. n	Measured Value of Node No. n

Fig. 4. Comparison of table before and after distributed flooding algorithm is applied

In the following part, we demonstrate the difference between these two modes on a simple topology shown in Fig. 5, where we focus on the green node.

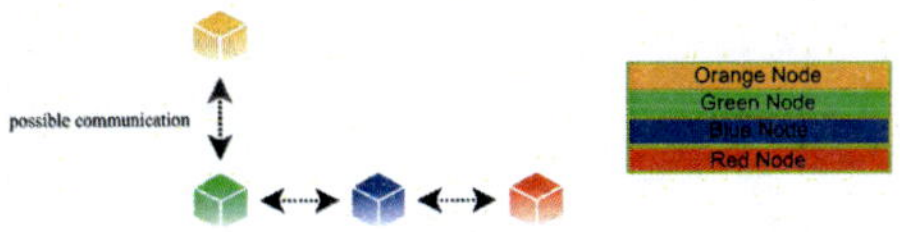

Fig. 5. Example network of line topology formed by four vertices

Synchronous mode

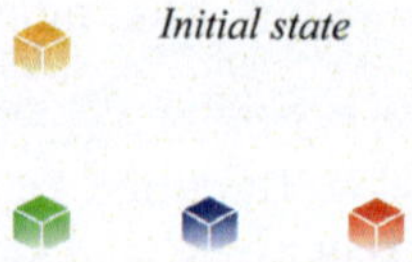

Memory of Green Node	
Memory Slot	Saved Content
No. 1	Local Measurement
No. 2	Empty
No. 3	Empty
No. 4	Empty

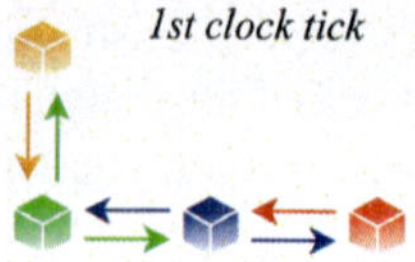

Memory of Green Node	
Memory Slot	Saved Content
No. 1	Local Measurement
No. 2	Measurement of Blue Node
No. 3	Measurement of Orange Node
No. 4	Empty

Memory of Green Node	
Memory Slot	Saved Content
No. 1	Local Measurement
No. 2	Measurement of Blue Node
No. 3	Measurement of Orange Node
No. 4	Measurement of Red Node

Fig. 6. Example of two clock ticks - synchronous mode

Asynchronous mode

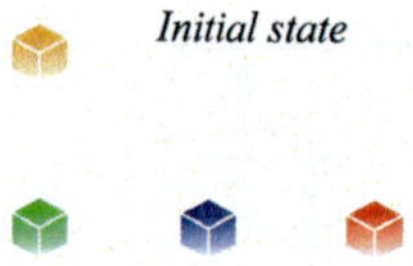

Memory of Green Node	
Memory Slot	Saved Content
No. 1	Local Measurement
No. 2	Empty
No. 3	Empty
No. 4	Empty

Memory of Green Node	
Memory Slot	Saved Content
No. 1	Local Measurement
No. 2	Measurement of Blue Node
No. 3	Empty
No. 4	Empty

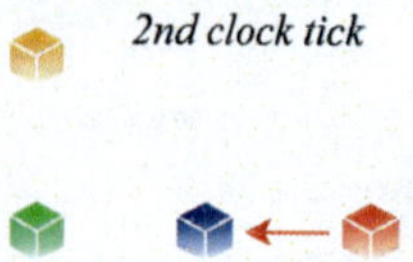

Memory of Green Node	
Memory Slot	Saved Content
No. 1	Local Measurement
No. 2	Measurement of Blue Node
No. 3	Empty
No. 4	Empty

Fig. 7. Example of two clock ticks - asynchronous mode

In Fig. 6 and Fig. 7, we demonstrate how these two modes differ from each other at the first two clock ticks. From Fig. 6 (the synchronous mode is analyzed), it can be seen that at each clock tick, all the sensor nodes broadcast their current table as well as receiving the table from all their neighbors. However, in the asynchronous mode (see Fig. 7), it can be observed that per a clock tick, only one sensor node (chosen randomly) broadcasts its current table, and only its neighbors can, therefore, update their table.

3 Applied Research Methodology

This section is concerned with modeling WSNs, the applied research methodology, and the metrics chosen for evaluating the performance of the examined algorithm. In our experiments, we model WSNs as simple finite undirected unweighted and connected graphs G determined by two sets: the vertex set $\mathbf{V}$ and the edge set $\mathbf{E}$ [26,27]. The vertex set $\mathbf{V}$ is formed by all the graph vertices, which are representatives of all the sensor nodes in WSN. The size of WSN is represented by the graph order n, which is equal to the cardinality of the vertex set $\mathbf{V}$. The other set, the edge set $\mathbf{E}$, contains all the edges in a graph. An edge linking two vertices represents a direct connection between these two vertices (these two nodes can directly communicate with one another - their distance is one-hop). In this paper, we apply RGs to model WSNs, meaning that connectivity between the vertices from $\mathbf{V}$ is conditioned by the probability p [28]. Thus, an edge between each pair of vertices exits with the probability p, and its existence is not dependent on the presence of the other edges in the graph. Higher connectivity of a graph is ensured by increasing the value of p. In our experiment, it takes the following values:

- $p = \{5\%, 10\%, 20\%, 30\%\}$

Also, we vary the graph order n of the generated graphs as follows:

- $n = \{30, 100, 200, 500\}$

For each pair p and n, we generate 100 graphs with unique topology in order to ensure the high credibility of the presented results. To evaluate the performance of the algorithm in both the modes, we apply these two metrics:

- *the number of clock ticks*: poses the minimal number of the local/global clock ticks required for completing the algorithm.
- *the number of the sent messages*: represents the minimal number of the sent messages required for each sensor node to obtain the measured information of all the sensor nodes in the network

In Fig. 8, 9, 10 and Fig. 11, we just show the average of the corresponding metric over 100 graphs for each pair p and n. Besides, in the asynchronous mode, the algorithm is repeated 100 times for each graph (as both the applied metrics differ over runs - one repetition of the algorithm is referred to as a run), and the average value is chosen as a representative of the corresponding metric in a graph.

In the last experiment, we analyze the distribution of the sent messages number for both the examined modes in four graphs with $n = 30$ and with $p = 5\%/10\%/20\%/30\%$. The algorithm is repeated 10 000 times in both the modes in this experiment.

4 Experiments and Discussion

In this section, we turn our attention to the numerical experiments executed in MatlabR2018a. At first, we deal with the number of the clock ticks required for the algorithm to be completed, i.e., all the sensor nodes are capable of determining the exact value of the desired aggregate function. In Fig. 8, we show the results for the case that the algorithm is executed in the synchronous mode. From the results, it can be seen that an increase in p (i.e., the graph connectivity is increased as a consequence) results in a decrease in the number of the required clock ticks regardless of the network size. However, for greater values of n, an

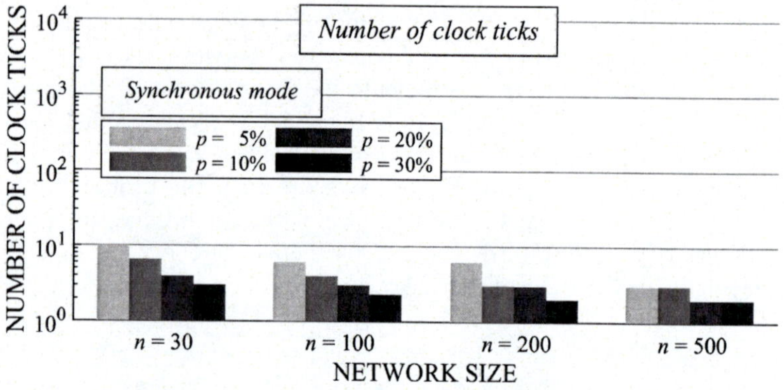

Fig. 8. Averaged number of clock ticks - synchronous mode

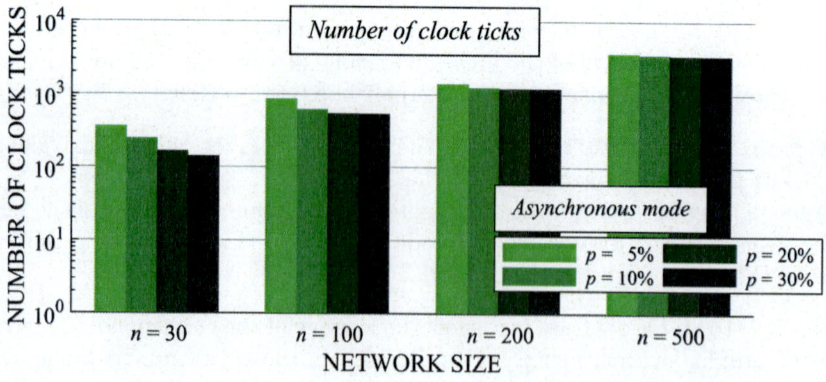

Fig. 9. Averaged number of clock ticks - asynchronous mode

increase in p causes either a marginal or no decrease in the number of the clock ticks. As further seen from the results, the algorithm needs only a small number of the clock ticks to be completed for each pair p and n. Also, it is observed that the number of the clock ticks is smaller in larger networks than in smaller ones. Thus, it is identified that the lowest number of the clock ticks in our experiments is required in the networks formed by 500 sensor nodes. The following part of this section is concerned with a performance analysis of the distributed flooding algorithm in the asynchronous mode. As seen from Fig. 9, an increase in p ensures that the number of the clock ticks required for completing the algorithm is smaller just like in the synchronous mode. Besides, an increase in n causes that the number of the clock ticks to increase, unlike the synchronous mode. As further seen from the depicted results, the asynchronous mode is outperformed by the synchronous one in terms of the number of the clock ticks for each value of n and p. Moreover, it can be observed that the difference between these two modes significantly increases with the growth of n.

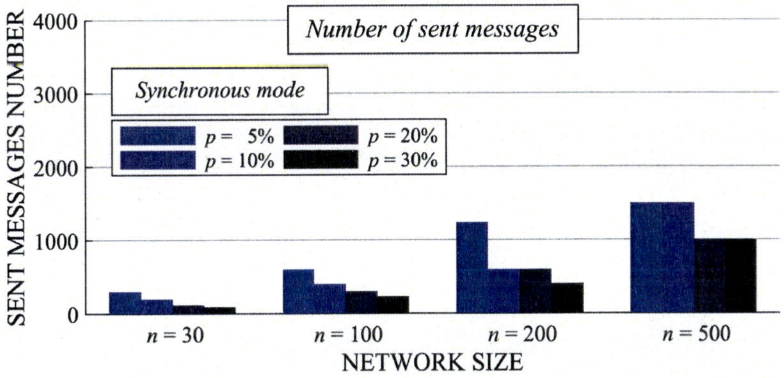

Fig. 10. Averaged number of sent messages - synchronous mode

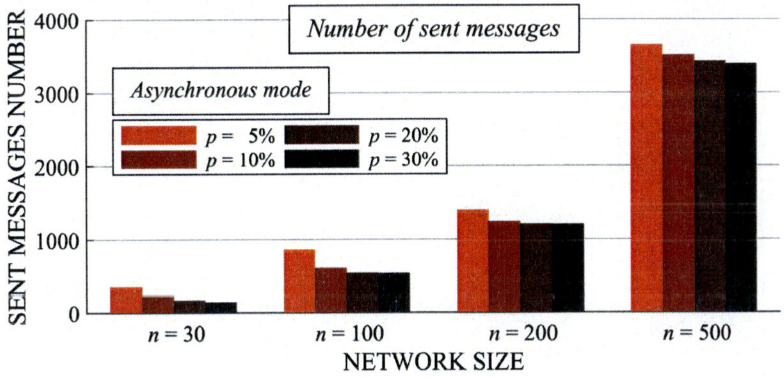

Fig. 11. Averaged number of sent messages - asynchronous mode

In this part, we focus on an analysis of the number of the sent messages required for completing the algorithm in both the examined modes. In Fig. 10, we show the results for the synchronous mode - it can be seen that an increase in p results in a decrease in the number of the sent messages for each value of n (except for several cases in larger networks when the number of the sent messages does not change even though p is increased). Moreover, it is identified that the algorithm achieves greater performance in smaller networks than in greater ones in contrast to the previous analysis, where the number of the clock ticks for completing the algorithm in the synchronous mode is smaller in larger networks. In the next figure (Fig. 11), we show the number of the sent messages in the case of the asynchronous mode. From the results, we can see that an increase in p and a decrease in n ensure that the algorithm performs better like in the synchronous mode. However, the algorithm achieves worse performance than in the synchronous mode again. Also, it is observed that this difference is greater for higher values of n like in the first analysis.

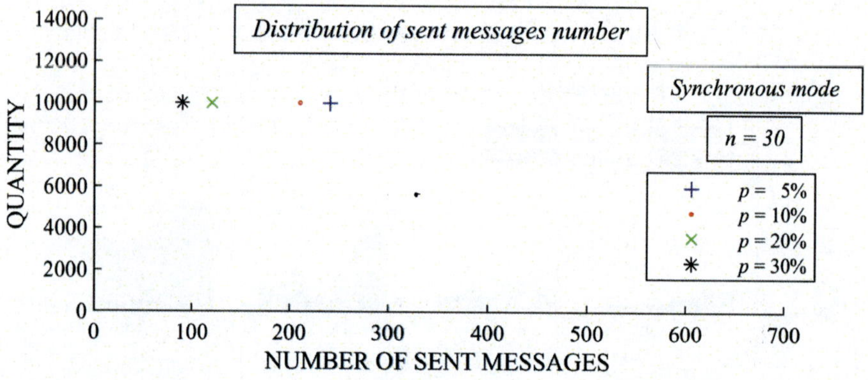

Fig. 12. Distribution of number of sent messages - synchronous mode

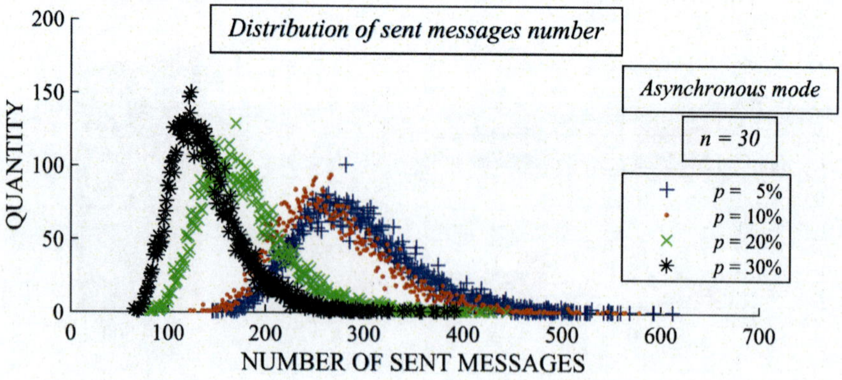

Fig. 13. Distribution of number of sent messages - asynchronous mode

In the last experiment, we pay our attention to the distribution of the sent messages number of the distributed flooding algorithm in both the synchronous and the asynchronous mode (see Fig. 12 and Fig. 13 for the results). As mentioned earlier in this paper, the algorithm is repeated 10 000 times in four RGs with different connectivity and with the graph order $n = 30$. From the results, we can see that the number of the sent messages does not change over runs, provided that the algorithm is executed in the synchronous mode (Fig. 12). This phenomenon is generally observable when deterministic distributed consensus algorithms are analyzed [19]. However, in the case of the asynchronous mode (Fig. 13), the number of the sent messages required for completing the algorithm has a skewed Gaussian distribution for each graph connectivity. The same distribution type can be seen in the case of distributed stochastic consensus algorithms as shown in [19]. Besides, it is identified that an increase in p ensures that the data are less scattered.

5 Conclusion

In this paper, we address the distributed flooding algorithm for sensor fusion over WSNs represented as RGs with various connectivity and graph order. The algorithm is executed in either the synchronous or the asynchronous mode in order to identify in which mode the performance of the examined algorithm quantified by two metrics (namely, the number of the clock ticks and the number of the sent messages) is better. Also, how the network parameters affect the performance of the algorithm is examined in our research. From the experimentally obtained results presented in this paper, it is seen that an increase in p generally results in a decrease in both the number of the clock ticks and the number of the sent messages in both the examined modes. It is also observed that an increase in n ensures that the number of the clock ticks is decreased in the synchronous mode. However, when the asynchronous mode is analyzed, greater values of n cause that the number of the clock ticks is increased. Moreover, it can be seen that an increase in n results in an increase in the number of the sent messages in both the investigated modes. As further identified by our research presented in this paper, the distributed flooding algorithm performs better in the synchronous mode than in the asynchronous one in terms of both the applied metrics. In the last experiment, we analyze the distribution of the sent message number for both the modes. From the results, it is seen that the number of the sent messages required for completing the algorithm does not differ over runs in the case of the synchronous mode. However, the number of the sent messages has a skewed Gaussian distribution, provided that the algorithm is executed in the asynchronous mode. Moreover, it is identified that the data are less scattered for greater values of p in this mode.

Acknowledgment. This work was supported by the VEGA agency under the contract No. 2/0155/19 and by the Slovak Research and Development Agency under the contract No. APVV-17-0116. Since 2019, Martin Kenyeres has been a holder of the Stefan Schwarz Supporting Fund.

References

1. Khalaf, O.I., Sabbar, B.M.: An overview on wireless sensor networks and finding optimal location of nodes. Period. Eng. Nat. Sci. **7**, 1096–1101 (2019). https://doi.org/10.21533/pen.v7i3.645
2. Stamatescu, G., Stamatescu, I., Popescu, D.: Consensus-based data aggregation for wireless sensor networks. Control. Eng. Appl. Inform. **19**, 43–50 (2017)
3. Pranathi, T.Y.S.S., Dhuli, S., Aditya, V.M.V.S., Charisma, B., Jayakrishna, K.: A hybrid routing protocol for robust wireless sensor networks. In: 12th International Conference on Computational Intelligence and Communication Networks (CICN 2020), pp. 102–106. IEEE Press, New York (2020). https://doi.org/10.1109/CICN49253.2020.9242627
4. Balasubramanian, D.L., Govindasamy, V.: Study on evolutionary approaches for improving the energy efficiency of wireless sensor networks applications. EAI Endorsed Trans. Internet Things **5**, 1–13 (2020). https://doi.org/10.4108/eai.13-7-2018.164856
5. Seda, P., Seda, M., Hosek, J.: On mathematical modelling of automated coverage optimization in wireless 5G and beyond deployments. Appl. Sci. **10**, 8853 (2020). https://doi.org/10.3390/app10248853
6. Sharma, H., Haque, A., Blaabjerg, F.: Machine learning in wireless sensor networks for smart cities: a survey. Electronics **10**, 1012 (2021). https://doi.org/10.3390/electronics10091012
7. Kenyeres, M., Kenyeres, J.: Average consensus over mobile wireless sensor networks: weight matrix guaranteeing convergence without reconfiguration of edge weights. Sensors **20**, 3677 (2020). https://doi.org/10.3390/s20133677
8. Anastasi, G., Conti, M., Di Francesco, M., Passarella, A.: Energy conservation in wireless sensor networks: a survey. Ad Hoc Netw. **7**, 537–568 (2009). https://doi.org/10.1016/j.adhoc.2008.06.003
9. Jino Ramson, S.R., Jackuline Moni, D.: Applications of wireless sensor networks - a survey. In: 2017 IEEE International Conference on Innovations in Electrical, Electronics, Instrumentation and Media Technology (ICIEEIMT 2017), pp. 325–329. IEEE Press, New York (2017). https://doi.org/10.1109/.2005.1467103
10. Marjanovic, M., Prijic, A., Randjelovic, B., Prijic, Z.: A transient modeling of the thermoelectric generators for application in wireless sensor network nodes. Electronics **9**, 1015 (2020). https://doi.org/10.3390/electronics9061015
11. Kenyeres, J., Kenyeres, M., Rupp, M., Farkas, P.: Connectivity-based self-localization in WSNs. Radioengineering **22**, 818–827 (2013)
12. Elsayed, W., Elhoseny, M., Sabbeh, S., Riad, A.: Self-maintenance model for wireless sensor networks. Electr. Comput. Eng. **70**, 799–812 (2018). https://doi.org/10.1016/j.compeleceng.2017.12.022
13. Lopez, J., Zhou, J.: Wireless Sensor Network Security. IOS Press, Amsterdam (2008)
14. Al-Karaki, A., Kamal, E.: Routing techniques in wireless sensor networks: a survey. IEEE Wirel. Commun. **11**, 6–28 (2004). https://doi.org/10.1109/MWC.2004.1368893
15. Gungor, V.C., Lu, B., Hancke, G.P.: Opportunities and challenges of wireless sensor networks in smart grid. IEEE Trans. Ind. Electron. **57**, 3557–3564 (2020). https://doi.org/10.1109/TIE.2009.2039455
16. Izadi, D., Abawajy, J.H., Ghanavati, S., Herawan, T.: A data fusion method in wireless sensor networks. Sensors **15**, 2964–2979 (2015). https://doi.org/10.3390/s150202964

17. Sasiadek, J.Z.: Sensor fusion. Annu. Rev. Control **26**, 203–228 (2002). https://doi.org/10.1016/S1367-5788(02)00045-7
18. Valcher, M.E., Parlangeli, G.: On the effects of communication failures in a multi-agent consensus network. In: 23rd International Conference on System Theory, Control and Computing (ICSTCC 2019), pp. 709–720. IEEE Press, New York (2019). https://doi.org/10.1109/ICSTCC.2019.8885721
19. Kenyeres, M., Kenyeres, J.: Comparative study of distributed consensus gossip algorithms for network size estimation in multi-agent systems. Future Internet **13**, 134 (2021). https://doi.org/10.3390/fi13050134
20. Xiao, L., Boyd, S., Lall, S.: A scheme for robust distributed sensor fusion based on average consensus. In: 4th International Symposium on Information Processing in Sensor Networks (IPSN 2005), pp. 63–70. IEEE Press, New York (2005). https://doi.org/10.1109/IPSN.2005.1440896
21. Castanedo, F.: A review of data fusion techniques. Sci. World J. **2013**, 1–20 (2013). https://doi.org/10.1155/2013/704504
22. Lacatusu, M., Ionita, A.D.: Metamodeling environment in cloud. U. Politeh. Buch. Ser. C **2020**, 27–34 (2020)
23. Mustafa, A., Islam, M.N.U., Ahmed, S.: Dynamic spectrum sensing under crash and byzantine failure environments for distributed convergence in cognitive radio networks. IEEE Access **9**, 23153–23167 (2021). https://doi.org/10.1109/ACCESS.2021.3053254
24. Wang, T., Ding, H., Xiong, H., Zheng, L.: A compensated multi-anchors TOF-Based localization algorithm for asynchronous wireless sensor networks. IEEE Access **7**, 64162–64176 (2019). https://doi.org/10.1109/ACCESS.2019.2917505
25. Sahoo, P.K., Pattanaik, S.R., Wu, S.-L.: A novel synchronous MAC protocol for wireless sensor networks with performance analysis. Sensors **19**, 5394 (2019). https://doi.org/10.3390/s19245394
26. Chen, L., Huang, X., Zhang, Z.: A simpler PTAS for connected k-path vertex cover in homogeneous wireless sensor network. J. Comb. Optim. **36**, 35–43 (2018). https://doi.org/10.1007/s10878-018-0283-9
27. Suriya Praba, T., Saravanan, S., Sethukarasi, T.: An efficient energy aware semigraph-based total edge domination routing algorithm in wireless sensor networks. J. Comb. Optim. **117**, 2423–2439 (2021). https://doi.org/10.1007/s11277-020-07982-z
28. Frieze, A., Karonski, B.: Introduction to Random Graphs. Cambridge University Press, Cambridge (2016). https://doi.org/10.1017/CBO9781316339831

An Adaptive Network Model for Sleep Paralysis: The Risk Factors and Working Mechanisms

Willem Huijzer and Jan Treur(✉)

Social AI Group, Department of Computer Science, Vrije Universiteit, Amsterdam, The Netherlands
w.d.huijzer@student.vu.nl, j.treur@vu.nl

Abstract. In this paper, an adaptive network model is presented for working mechanisms and risk factors of sleep paralysis with and without hallucinations. Sleep paralysis is a mysterious phenomenon where a person is awake but not able to move. Correspondingly, people can experience hallucinations and if so, often in a horrifying manner. The introduced network model covers the potential role of serotonin, temporoparietal junction, amygdala, 5-HT2AR receptor and covers the influence of risk factors as sleep deprivation, alcohol use, and the HT2AR gene.

Keywords: Adaptive network model · Sleep paralysis · Hallucinations · Alcohol · Sleep deprivation · Serotonin · 5-HT2AR receptor · Pimavanserin

1 Introduction

Many functions and working mechanisms of sleep have been addressed by researchers, although some phenomena of sleep are still a bit mysterious (Frank 2006). Sleep paralysis is one of those mysterious phenomena. During this state a person is awake, but can't move his or her muscles (Sateia 2014). Some people correspondingly experience hallucinations and if so, often in a horrifying manner. Not being able to move and having terrifying hallucinations can cause a lot of fear and distress for an individual. It has been hypothesized by researchers that sleep paralysis could be an explanation for people experiencing extraordinary events throughout history. In culturally specific contexts an occurrence of sleep paralysis could lead to the interpretation of a spiritual or supernatural event (Hufford 2005). Sleep paralysis normally occurs during the transition between sleep and wakefulness. In a systematic review paper of Sharpless and Barber (2011), the life time prevalence of experiencing at least one episode of sleep paralysis has been estimated to be 7.6% for the general population. Students and psychiatric patients are estimated to be more vulnerable, having a prevalence of respectively 28.3% and 34.6%. Researching the causes and working mechanisms of sleep paralysis could contribute to the understanding of sleep and might provide ways of coping for people that are troubled by sleep paralysis.

In the study reported here, the Network-Oriented Modeling approach for adaptive networks (Treur 2020) was applied to model possible working mechanisms and risk

R. Silhavy et al. (Eds.): CoMeSySo 2021, LNNS 232, pp. 540–556, 2021.
https://doi.org/10.1007/978-3-030-90318-3_44

factors of sleep paralysis. The proposed adaptive network model based on prior biological and neurological literature covers the potential role of serotonin level, risk factors, genes, medicine and more. Four different scenarios have been simulated to evaluate the model for: a normal sleep-wake cycle, the occurrence of sleep paralysis without hallucinations, sleep paralysis with hallucinations due to gene vulnerability and again without hallucinations due to medicine intake against the gene vulnerability.

2 Neurological Background

In this section the underlying biological and neurological background of regular sleep, sleep paralysis and possible hallucinations during a sleep paralysis will be discussed. The model proposed in the current study will be based upon the following neurological background.

2.1 Regular Sleep

Different internal and external factors have influence on regulating the sleep-wake cycle. Firstly, the human body itself has multiple complex systems that contribute to regulating the sleep-wake cycle. These systems can be narrowed down to sleep promoting circuits and wakefulness promoting circuits; between these circuits is a mutual inhibition to ensure stability. The switching between the circuits function analogously to an electronic 'flip-flop' switch (Fuller et al. 2006). Additionally, various external factors influence these the sleep-wakefulness regulation e.g., blue light (Lawrenson et al. 2017) and caffeine (Snel and Lorist 2011). The wakefulness promoting systems use a number of neurotransmitters to activate the brain; one of these neurotransmitters is serotonin. Serotonin promotes wakefulness (Brown et al. 2012), but also inhibits the REM sleep (Ursin 2002).

The Global Workspace Theory from Baars (1997, 2002) describes that an individual experiences a single flow of consciousness, but in underlying processes there are multiple potential sources of consciousness competing to be experienced. The comparison can be made with a theater stage on which the potential consciousnesses are competing and where the 'spotlight' of attention can only shine on one flow of consciousness. The idea is that a winner-takes-it-all competition makes up which consciousness will be experienced by the individual. It is put forward by Hesslow (2012), Damasio (1999) and Goldman (2006) that flows of consciousness can be related to the concept of internal (mental) simulation. In this internal simulation, sensory representation states (for example mental images) are activated and in response preparation states are activated for actions or bodily changes. The preparations state itself can trigger (as a way of generating possible predicted effects) other sensory representation states and the bodily changes can eventually trigger the amygdala through the as-if body loop (Damasio 1999) and make the individual experience (feel) certain emotions. In the internal simulation, the prefrontal cortex has a regulating role towards the amygdala and the sensory representation states (Morgan et al. 1993).

Mental images during wakefulness are built up through input from the visual system. Levin and Nielsen (2007) discuss how mental images during dreams get created by combining elements of different memories and how memory elements could be selected driven by emotions. Sleep paralysis has overlap of subjective descriptions and underlying brain mechanisms with dreams (Waters et al. 2016), therefore the theory on the emergence of dreams by Levin and Nielsen (2007) could be considered similarly for hallucinations.

2.2 Sleep Paralysis

Sleep paralysis is a state where a person is awake but experiences characteristics from REM sleep, namely not being able to move (Sharpless and Barber 2011). Jalal (2018) proposes a crucial role for an increased level of the neurotransmitter serotonin when experiencing a sleep paralysis. Prior research has suggested several predisposing factors for sleep paralysis, among which sleep deprivation (Stefani and Högl 2020) and alcohol use (Munezawa et al. 2011). Both sleep deprivation (Hery et al. 1970; Peñalva et al. 2003) and alcohol use (Chastain 2006) can increase the serotonin levels in the brain. Serotonin has numerous functions in the brain and body; one of the functions is regulating sleep. Serotonin promotes wakefulness and inhibits REM sleep (Ursin 2002). Taking the above description into account makes the role of serotonin plausible.

For sleep, four different stages are distinguished. The last stage, rapid eye movement (REM) sleep, is characterized by random rapid eye movements, motor atonia and the propensity of the sleeper to dream vividly. To prevent dangerous movements from being made when dreaming, REM sleep shuts down the motor neurons. The individual cannot use his or her muscles to talk or move; this is called motor atonia (Sharpless and Barber 2011). Furthermore, during REM sleep the prefrontal cortex gets stimulated (Vandekerckhove and Cluydts 2010). There are various theories about the mechanism and the involved neurotransmitters that cause the motor atonia during REM sleep. Brooks and Peever (2012) suggest a model where reduced motor neuron excitation strengthens the paralysis, but the inhibition of the motor neurons is the driving force behind the paralysis. They suggest that among others GABA neurons trigger the paralysis by directly inhibiting motor neurons during REM sleep. Alcohol intake (Kumar et al. 2009) and sleep deprivation (Modirrousta et al. 2007) could also stimulate GABA neurons.

2.3 The Hallucinations

In a survey under students in Poland, Wróbel-Knybel et al. (2020) found that 66% of their sample experienced hallucinations during a sleep paralysis: 37% visual, 31% auditory and 25% tactile hallucinations. Jalal (2018) proposes two processes that cause the hallucinations during a sleep paralysis. The first process starts with the motor-execution being blocked while at same time the sensory input from the body still is intact. This causes a disturbance at the temporoparietal junction (TPJ). The TPJ gets input from many brain areas among which are visual, auditory and somatosensory systems. The area is a crucial structure for the conscious experience of the normal self (Blanke et al. 2005). Research from Arzy et al. (2006) shows that stimulation of the TPJ can trigger the uncanny feeling that someone is nearby, out of body experiences and visual hallucinations of human-like figures.

The second process proposed by Jalal (2018) has a central role for the serotonin 2A receptor (5-HT2AR). Serotonin can bind to many different receptors and 5-HT2AR is one of them (Nichols and Nichols 2008). This receptor is present in different regions of the brain and there are several reasons to assume that this receptor is involved with the hallucinations. Firstly, Parkinson (Huot et al. 2010) and schizophrenic (González-Maeso et al. 2008) patients with visual hallucinations show an increased number of 5-HT2AR receptors in the visual cortex compared to patients without visual hallucinations. Secondly, when a highly selective 5-HT2A inverse agonist medicine Pimavanserin is given to a Parkinson patient the severity of hallucinations decreases (Meltzer et al. 2010). Furthermore, the activation of the serotonin 2A receptors also mediates visual hallucination induced by Psilocybin (Kometer et al. 2013). The serotonin 2A receptor is also present in frontoparietal cortical regions, among which is the TPJ (Tagliazucchi et al. 2016). Taking the above into account, it is plausible that the 5-HT2A receptor influences visual hallucinations.

Moreover, 5-HT2A receptors are densely concentrated in the amygdala; therefore, it could activate and increase activation the intensity of emotions felt (Frokjaer et al. 2008). The amygdala modulates among other things fear responses in humans (Ressler 2010). Increased intense negative emotions could contribute to retrieval of negative memory elements and while an individual already is in fear due not being able to move, this could explain why the hallucinations during a sleep paralysis are mostly terrifying.

Genes and medicine can also have an influence on the hallucinations experienced during sleep paralysis. The gene coding for HTR2A is found to be located on the long arm of chromosome 13 (Badner and Gershon 2002). Jalal (2018) proposes the medicine Pimavanserin to reduces visual hallucinations and fear during a sleep paralysis. This medicine is a highly selective 5-HT2A inverse agonist. As noted above, the medicine is used successfully to reduce hallucinations among Parkinson patients (Meltzer et al. 2010).

3 The Modeling Approach Used

In this section a brief summary will be given of the adaptive network modelling approach developed by Treur (2020). This approach was used to design the model that is introduced in this paper. Network-Oriented Modelling uses nodes and vertices; vertices are the connections between the nodes. Nodes are states with values (usually between 0 and 1) that vary over time, while the connections indicate the impact that states have on each other. The latter can be interpreted as the causal relationships between the nodes. In an adaptive network model, besides the states also the causal relationships can change over time. Table 1 summarizes the main concepts of network-oriented modelling. Every connection has a connection weight, which is a numerical value indicating the connection strength. The connections and their weights define the network's *connectivity characteristics*.

In addition to the latter, every state has a combination function that describes the manner in which the incoming impacts per connection are combined to form an aggregated impact. This defines the network's *aggregation characteristics*. The combination function can be chosen from the available Combination Function Library. The choice on which combination function is used, depends on the application and can also be node-specific. To define the network's *timing characteristics*, every state has a speed factor that determines how fast a state changes because of its received causal impact.

Table 1. An overview of the concepts in the conceptual component of temporal-causal networks.

Concepts	Notation	Explanation
States and connections	X, Y $X \rightarrow Y$	Denotes the nodes and edges in the conceptual representation of a network model
Connection weights	$\boldsymbol{\omega}_{X,Y}$	A connection between states X and Y has a corresponding *connection weight.* In most cases $\boldsymbol{\omega}_{X,Y} \in [-1,1]$
Aggregating multiple impacts on a state	$\mathbf{c}_Y(..)$	Each state has a *combination function* and is responsible for combining causal impacts of all states connected to Y on that same state
Timing of the effect of causal impact	$\boldsymbol{\eta}_Y$	The *speed factor* determines how fast a state is changed by any causal impact. In most cases: $\boldsymbol{\eta}_Y \in [0,2]$

The numerical representation derived from the network characteristics is summarized in Table 2. The last row of this Table 2, shows the difference equation. Adaptive networks are networks for which some of the characteristics $\boldsymbol{\omega}_{X,Y}$, $\mathbf{c}_Y(..)$, $\boldsymbol{\eta}_Y$ change over time. To model this, extra states are added that represent the adaptive characteristics. For example, for an adaptive connection weight $\boldsymbol{\omega}_{X,Y}$ a new state $\mathbf{W}_{X,Y}$ is added representing the dynamic value of $\boldsymbol{\omega}_{X,Y}$; this new state is called a self-model or adaptation state for $\boldsymbol{\omega}_{X,Y}$.

Table 3 shows an overview of the combination functions used in the designed network model. The first is the identity function, **id(.)**, which is commonly used when a state only has one incoming connection. The advanced logistic sum function, $\mathbf{alogistic}_{\sigma,\tau}(..)$, is used to aggregate impact for each state that has multiple incoming connections; it has as parameters steepness $\boldsymbol{\sigma}$ and threshold $\boldsymbol{\tau}$.

Table 2. Numerical representations of temporal-causal networks.

Concepts	Notation	Explanation
State value at time t	$Y(t)$	For every time t a state Y has a value in [0,1]
Single causal impact	$\mathbf{impact}_{X,Y}(t) = \omega_{X,Y}X(t)$	At any time t a state X, when connected to Y, impacts Y via connection weight $\omega_{X,Y}$
Aggregating multiple impacts on a state	$\mathbf{aggimpact}_Y(t) = \mathbf{c}_Y(\mathbf{impact}_{X_1,Y}(t), \ldots, \mathbf{impact}_{X_K,Y}(t)) = \mathbf{c}_Y(\omega_{X_1,Y}X_1(t), \ldots, \omega_{X_K,Y}X_k(t))$	The combination function $\mathbf{c}_Y$ determines the aggregated causal impact of states X_i on Y
Timing of the effect of causal impact	$Y(t+\Delta t) = Y(t) + \eta_Y [\mathbf{aggimpact}_Y(\mathrm{t}) - Y(t)]\,\Delta \mathrm{t} = Y(t) + \eta_Y\,[\mathbf{c}_Y(\omega_{X_1,Y}X_1(t), \ldots, \omega_{X_K,Y}X_k(t)) - Y(t)]\,\Delta t$	The speed factor Y determines how fast a state Y is changed by the aggregated causal impact of states X_i

Table 3. Overview of the combination functions used.

Combination function	Description	Formula $\mathbf{c}_Y(V_1,\ldots,V_k) =$
id(.)	Identity	V
$\mathbf{alogistic}_{\sigma,\tau}$(..)	Advanced logistic sum	$\left[\frac{1}{1+\mathrm{e}^{-\sigma(V_1+\cdots+V_k-\tau)}} - \frac{1}{1+\mathrm{e}^{\sigma\tau})}\right](1+\mathrm{e}^{-\sigma\tau})$

4 The Designed Adaptive Network Model

In the current section, an adaptive network model covering risk factors and working mechanisms of sleep paralysis will be introduced. In Table 4 all states are presented with a corresponding explanation. The fixed states remain at the initial value and do not change during a simulation; they make use of the **id(.)** function, while the other states make use of the $\mathbf{alogistic}_{\sigma,\tau}$**(..)** function. The last four states in the table are self-model states (adaptation states) and the other states are base level states. In the Appendix at URL https://www.researchgate.net/publication/353206807 the initial values (**iv**) of the states and the full specification by role matrices (Treur 2020) are presented: base connections (**mb**), connection weights (**mcw**), speed factors (**ms**), combination function weights (**mcfw**) and combination function parameters (**mcfp**).

In Fig. 1, a graphic conceptual representation is given of the connectivity of the introduced adaptive network model. It is shown that the wakefulness promoting circuits and sleep promoting circuits work analogously to a flip-flop switch. They inhibit each other and alternate by stimulating each other's activators; in this manner both promoting circuits will take turns in being active. One of the tools of the wakefulness promoting systems to activate the brain is serotonin. The serotonin level gets stimulated by the wakefulness promoting circuits and helps to stimulate wakefulness. Both external factors sleep deprivation and alcohol use can also stimulate the serotonin level. REM sleep gets stimulated by sleep promoting circuits and is inhibited by serotonin. During REM sleep the motor atonia increases; this connection is adaptive. The state context REM-Atonia stimulates the adaptive weight state W-REM-Atonia for a base level; alcohol use and sleep deprivation could strengthen the connection. When an individual is under motor atonia, the disturbance of TPJ and the activity of the amygdala could increase. Increased amygdala activity could lead to the feeling of fear, which in its turn could lead to retrieval of negative memory elements. Motor atonia can only have this effect when a person is awake; the outgoing weights states of the motor atonia to the amygdala and the disturbance of TJP are fully dependent of input from the wakefulness state.

In the model, the competition for the 'spotlight' of consciousnesses is represented by the states of dream, hallucinations and reality inhibiting each other. Each of those states gets stimulated by a corresponding sensory representation state and each sensory representation state has a feedback loop to the preparation state.

Furthermore, each sensory representation state gets negative feedback via a corresponding control state. The sensory representation state of dream further gets input from REM sleep, negative memory ER, and 5-HT2AR receptor. The sensory representation state of hallucinations further gets input from wakefulness, 2-HT2AR receptor, negative memory ER and disturbance TPJ. The sensory representation state of reality further gets input from wakefulness. The amygdala has a (negative) feedback loop with each of the control states and the preparation state. The 5-HT2AR receptor gets stimulated by the serotonin level and the threshold is represented by the T - 5-HT2AR receptor state and gets a base line input from the context receptor and further negative input from the HT2AR gene and positive input from the medicine Pimavanserin.

Table 4. All states of the introduced adaptive network model with explanations.

	State	Explanation
X_1	Sleep deprivation	**Fixed** state, indication of whether an individual is sleep deprived (1) or not (0)
X_2	Alcohol use	**Fixed** state, indication of whether an individual has used alcohol (1) or did not (0)
X_3	Serotonin level	The level of the neurotransmitter serotonin in the brain
X_4	Wakefulness	The likelihood of an individual being awake; a brain state where an individual is conscious and engages in mental processes with the external world
X_5	REM sleep	The likelihood that an individual is in a REM sleep phase
X_6	5-HT2AR receptor	The intensity of serotonin binding to the 5-HT2AR receptor
X_7	Motor atonia	Motor atonia is a condition where an individual can't move his or her muscles due to inhibition by the brain
X_8	Disturbance TPJ	This state represents potential disturbance in the temporoparietal junction
X_9	Amygdala	The activity of the amygdala, brain area that is responsible for emotional responses
X_{10}	Control dream	Activity of the part of prefrontal cortex inhibiting the amygdala and the sensory representation state for dreaming
X_{11}	Control hallucinations	Activity of the part of prefrontal cortex inhibiting the amygdala and the sensory representation state for hallucinating
X_{12}	Control reality	Activity of the part of prefrontal cortex inhibiting the amygdala and the sensory representation state for experiencing reality
X_{13}	Fear	Likelihood of an individual experiencing fear, an unpleasant emotion often caused by potential danger or harm
X_{14}	Negative memory ER	The probability that memory elements with a negative association will be retrieved
X_{15}	Wakefulness PCA	The activity of wakefulness promoting circuits activators
X_{16}	Sleep PCA	The activity of sleep promoting circuits activators
X_{17}	Wakefulness PC	The activity of wakefulness promoting circuits
X_{18}	Sleep PC	The activity of sleep promoting circuits
X_{19}	Preparation state	The activity of the preparation state
X_{20}	SRS dream	The activity of the sensory representation state of a dream
X_{21}	SRS hallucinations	The activity of the sensory representation state of hallucinations
X_{22}	SRS reality	The activity of the sensory representation state of a reality
X_{23}	Dream	The likelihood of an individual experiencing a dream

(*continued*)

Table 4. (*continued*)

	State	Explanation
X_{24}	Hallucinations	The likelihood of an individual experiencing hallucinations
X_{25}	Reality	The likelihood of an individual experiencing the reality
X_{26}	HT2AR gene	**Fixed state**, indication of whether an individual has a genetic predisposition for the creating 5-HT2AR receptors (1) or not (0)
X_{27}	Pimavanserin	**Fixed state**, indication of whether an individual has taken the medicine Pimavanserin (1) or not (0)
X_{28}	Context receptor	**Fixed state**, this state represents the influence of context factors on the threshold of the 5-HT2AR receptor
X_{29}	Context REM-atonia	**Fixed state**, this state represents the influence of context factors on the connection weight between REM sleep and motor atonia
X_{30}	$\mathbf{W}_{\text{Atonia-TJP}}$	This self-model state represents the value of the weight between the states motor atonia and temporoparietal junction
X_{31}	$\mathbf{W}_{\text{Atonia-Amygdala}}$	This self-model state represents the value of the weight between the states motor atonia and amygdala
X_{32}	$\mathbf{W}_{\text{REM-Atonia}}$	This self-model state represents the value of the weight between the states REM sleep and motor atonia
X_{33}	$\mathbf{T}_{\text{Receptor 5-HT2AR}}$	This self-model state represents the value of the threshold of the 5-HT2AR receptor state

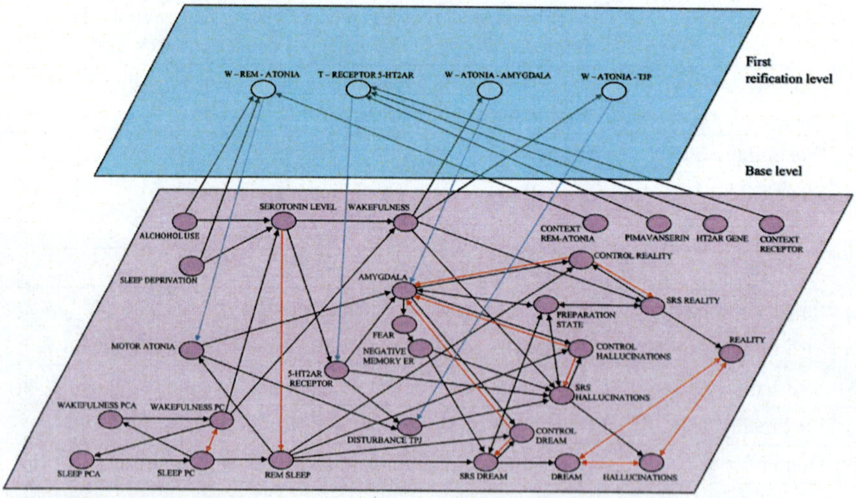

Fig. 1. Graphical representation of the connectivity of the introduced model.

5 Simulation Experiments

In this section, the simulation experiments performed are presented. The different scenarios were created merely by changing the initial values of sleep deprivation, alcohol use, HT2AR gene and Pimavanserin. The sleep and wakefulness promoting circuits are not influenced by other states and therefore remain identical for all the scenarios. Figure 2 shows how the sleep and wakefulness promoting circuits alternate with each other. When one system is active, the activator of the other system is stimulated and so they rotate in a loop. The x-as is an indication of time in hours.

5.1 Simulation of a Regular Sleep Cycle

In the first scenario a regular sleep cycle is simulated. For this scenario, the initial values of the following states are all set to 0: sleep deprivation, alcohol use, HT2AR gene and Pimavanserin.

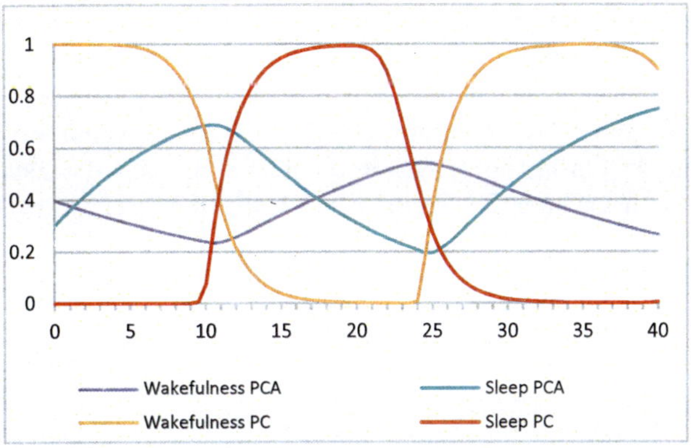

Fig. 2. Simulation of sleep-wake cycle.

Figure 3 left shows how the sleep-wake cycle is not interrupted and how wakefulness doesn't overlap with REM sleep nor motor atonia; there is no sleep paralysis. Figure 3 right shows how the activity of the dream, hallucinations and reality states with corresponding control and sensory representation states fluctuate over time.

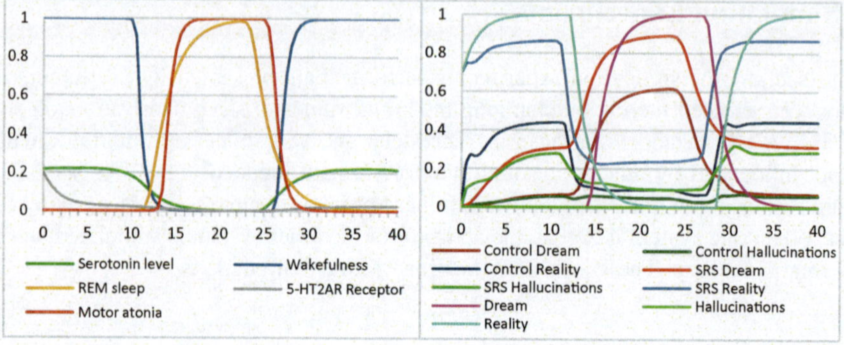

Fig. 3. Simulation of serotonin, wakefulness and REM sleep related states (left) and of dream, hallucinations and reality states with corresponding sensory representation and control states (right) during the first scenario.

It can be seen that the possibility of experiencing reality alternates with the possibility of experiencing a dream; the possibility of hallucinations does not surface in this scenario. As wakefulness does not overlap with motor atonia, the connection between REM sleep and motor atonia isn't strong when the motor atonia is active. As a result, the disturbance of TPJ and the amygdala only increases slightly during the transition between sleep and wakefulness (Fig. 4). Furthermore, the serotonin level remains low and the threshold of the 5-HT2AR remains high. The consequence is that the 5-HT2AR receptor has low activity.

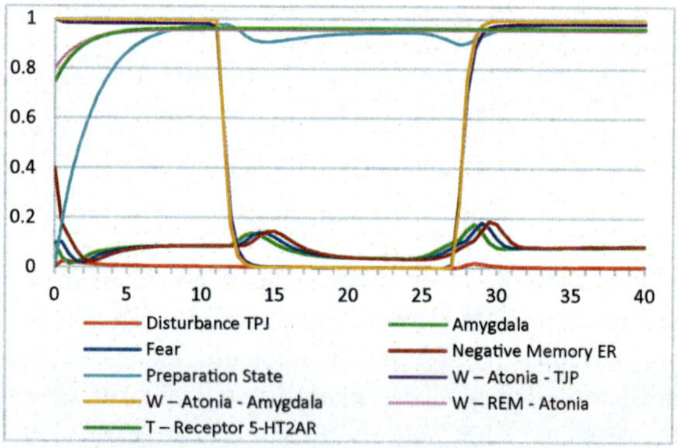

Fig. 4. Simulation of the disturbance of TPJ, emotion related states and adaptive states during the first scenario.

5.2 Simulation of Sleep Paralysis

In this scenario, the states alcohol use and sleep deprivation were set to 1. The other fixed states HT2AR gene and Pimavanserin remain 0. The effect of alcohol use and sleep deprivation is that the serotonin level rises (Fig. 5 left) and that connection between REM sleep and motor atonia get strengthened (Fig. 6). This causes the sleep-wake cycle to be disturbed; wakefulness, REM sleep and motor atonia now overlap during the transitions between sleep and wakefulness. This means that both at the beginning and at the end of the sleep cycle the probability of sleep paralysis is high. Similar to the previous scenario the likelihood of experiencing a dream and reality alternate each other. Only in this scenario experiencing a dream takes up less time, which is substituted by experiencing the reality for a longer period of time. The likelihood of hallucinations increases, but doesn't overrule the likelihood of experiencing a dream nor experiencing reality (Fig. 6 right). The amygdala, fear, negative memory retrieval and disturbance of TPJ all increase considerably around the sleep-wake transition areas. But this apparently doesn't stimulate the sensory representation state of hallucinations enough to rise above the other sensory representation states. An explanation for this could be found in a rather low activity of the 5-HT2AR receptor.

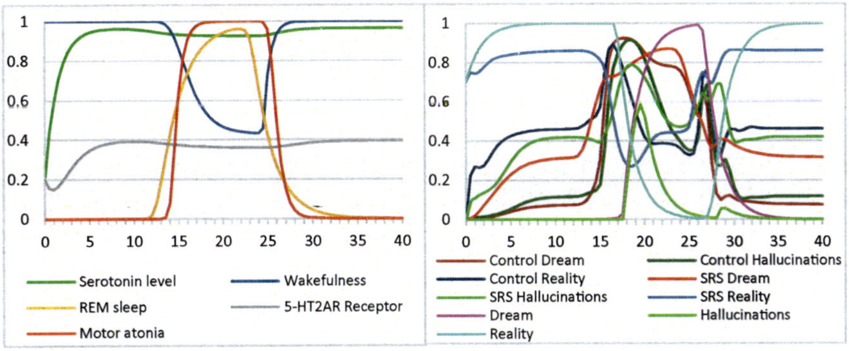

Fig. 5. Simulation of serotonin, wakefulness and REM sleep related states (left) and of dream, hallucinations and reality states with corresponding sensory representation and control states (right) during the second scenario.

5.3 Simulation of Sleep Paralysis with Genetic Vulnerability for Hallucinations

For this scenario, the states alcohol use, sleep deprivation and HT2AR gene were all set to 1. The fixed state Pimavanserin remains 0. Identical to the previous scenario, the probability of sleep paralysis is high during the transitions between sleep and wakefulness. The difference with the previous scenario is that during the current scenario hallucinations have a high likelihood of being experienced with potential sleep paralyses during the transitions between sleep and wakefulness (Fig. 7 left).

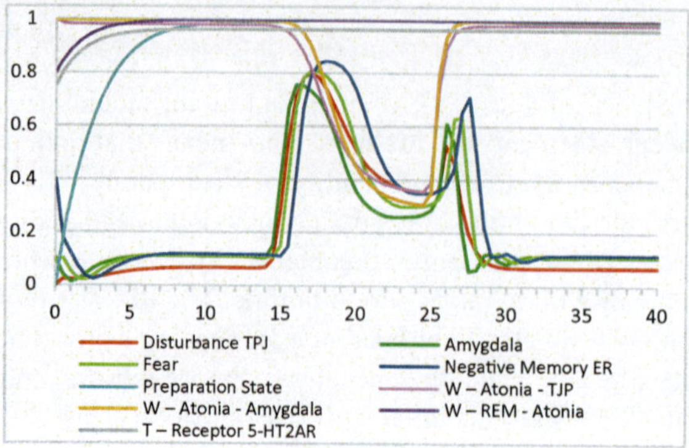

Fig. 6. Simulation of the disturbance of TPJ, emotion related states and adaptive states during the second scenario.

This is caused by the onset of the HT2AR gene; the gene is responsible for a greater number of 5-HT2AR receptors in the brain. In Fig. 7 right is shown how the threshold of the 5-HT2AR receptor decreases. As a result, the activity of 5-HT2AR receptor itself increases. The increased 5-HT2AR receptor activity has a direct influence on the sensory representation state of hallucinations. Additionally, it has an indirect influence on hallucinations through stimulating the disturbance of the TPJ and amygdala.

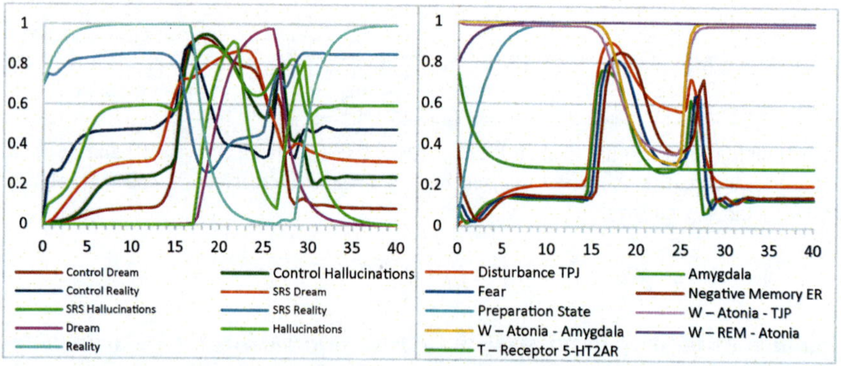

Fig. 7. Simulation of and of dream, hallucinations and reality states with corresponding sensory representation and control states (left) and of the disturbance of TPJ, emotion related and adaptive states (right) during the second scenario.

5.4 Simulation Sleep Paralysis Without Hallucination Due to Pimavanserin

For this scenario the initial values of the following states were all set to 1: sleep deprivation, alcohol use, HT2AR gene and Pimavanserin. Again, the probability of sleep paralysis is high during the transitions between sleep and wakefulness. During the current scenario the medicine Pimavanserin binds to the 5-HT2AR receptor, counteracting the regular effect. The threshold of the 5-HT2AR receptor eventually increases slightly, instead of decreasing as in the previous scenario. The outcomes of the current scenario are almost identical to the sleep paralysis scenario with no genetic vulnerability.

6 Discussion

In the study reported here, an adaptive network model was introduced that addresses the risk factors and working mechanisms of sleep paralysis. The presented network model is based on prior literature and validated by the simulations. The model has shown how risk factors such as alcohol use and sleep deprivation may increase the serotonin level and thereby disrupting the sleep-wake cycle. This may cause a sleep paralysis as wakefulness, REM sleep and most importantly motor atonia could overlap during the transition between wakefulness and sleep. This overlap causes fear and disturbance of the TPJ but doesn't necessarily cause hallucinations. The likelihood of experiencing hallucinations strongly increases if an individual has a genetic predisposition for creating a greater number of 5-HT2AR receptors. The receptor can directly stimulate hallucinations, but also indirectly by stimulating disturbance at the TPJ. Levin and Nielsen (2007) discuss how the selection of memory elements for creating mental images can be driven by emotions. And while the individual is likely to experience a lot of fear during a sleep paralysis, this could be an explanation for the fact that hallucinations often occur in a terrifying manner. Additionally, stimulation of the 5-HT2AR receptor can cause increased amygdala activity. The model further shows how the medicine Pimavanserin proposed by Jalal (2018) can counteract the genetic vulnerability and strongly decrease the likelihood of experiencing hallucinations.

The model is based on the latest literature available anno 2021; many components and working mechanisms of sleep are currently still under discussion and are subject of further research. For the model introduced, it has been attempted to find as much empirical literature as possible. However, for many components of the model there was no numerical empirical data available and qualitative empirical information had to be used. Based on the latter the current model has to be interpreted with a certain degree of prudence.

The model introduced leaves much room for further exploration in future studies. Both the risk factors and the working mechanisms could be refined or extended. Several predisposing factors for sleep paralysis have not yet been included in the current model, for example, jetlag (Stefani and Högl 2020) and the use of antidepressants (Sandman et al. 2015). Furthermore, Jalal and Ramachandran (2017) propose a role for mirror neurons as working mechanism for hallucinating during sleep paralysis.

References

Arzy, S., Seeck, M., Ortigue, S., Spinelli, L., Blanke, O.: Induction of an illusory shadow person. Nature **443**(7109), 287 (2006). https://doi.org/10.1038/443287a

Baars, B.J.: In the Theater of Consciousness: The Workspace of the Mind. Oxford University Press, Oxford (1997)

Baars, B.J.: The conscious access hypothesis: origins and recent evidence. Trends Cogn. Sci. **6**(1), 47–52 (2002). https://doi.org/10.1016/s1364-6613(00)01819-2

Badner, J.A., Gershon, E.S.: Meta-analysis of whole-genome linkage scans of bipolar disorder and schizophrenia. Mol. Psychiatry **7**(4), 405–411 (2002). https://doi.org/10.1038/sj.mp.4001012

Blanke, O., et al.: Linking out-of-body experience and self processing to mental own-body imagery at the temporoparietal junction. J. Neurosci. **25**(3), 550–557 (2005). https://doi.org/10.1523/JNEUROSCI.2612-04.2005

Brooks, P.L., Peever, J.H.: Identification of the transmitter and receptor mechanisms responsible for REM sleep paralysis. J. Neurosci. **32**(29), 9785–9795 (2012). https://doi.org/10.1523/JNEUROSCI.0482-12.2012

Brown, R.E., Basheer, R., McKenna, J.T., Strecker, R.E., McCarley, R.W.: Control of sleep and wakefulness. Physiol. Rev. (2012). https://doi.org/10.1152/physrev.00032.2011

Chastain, G.: Alcohol, neurotransmitter systems, and behavior. J. Gen. Psychol. **133**(4), 329–335 (2006). https://doi.org/10.3200/GENP.133.4.329-335

Damasio, A.R.: The Feeling of What Happens: Body and Emotion in the Making of Consciousness. Houghton Mifflin Harcourt, Boston (1999)

Frank, M.G.: The mystery of sleep function: current perspectives and future directions. Rev. Neurosci. **17**(4), 375–392 (2006). https://doi.org/10.1515/revneuro.2006.17.4.375

Fuller, P.M., Gooley, J.J., Saper, C.B.: Neurobiology of the sleep-wake cycle: sleep architecture, circadian regulation, and regulatory feedback. J. Biol. Rhythms **21**(6), 482–493 (2006). https://doi.org/10.1177/0748730406294627

Frokjaer, V.G., et al.: Frontolimbic serotonin 2A receptor binding in healthy subjects is associated with personality risk factors for affective disorder. Biol. Psychiat. **63**(6), 569–576 (2008). https://doi.org/10.1016/j.biopsych.2007.07.009

González-Maeso, J., et al.: Identification of a serotonin/glutamate receptor complex implicated in psychosis. Nature **452**(7183), 93–97 (2008). https://doi.org/10.1038/nature06612

Goldman, A.I.: Simulating Minds: the Philosophy, Psychology, and Neuroscience of Mindreading. Oxford University Press, Oxford (2006). https://doi.org/10.1093/0195138929.001.0001

Hery, F., Pujol, J.F., Lopez, M., Macon, J., Glowinski, J.: Increased synthesis and utilization of serotonin in the central nervous system of the rat during paradoxical sleep deprivation. Brain Res. **21**(3), 391–403 (1970). https://doi.org/10.1016/0006-8993(70)90419-1

Hesslow, G.: The current status of the simulation theory of cognition. Brain Res. **1428**, 71–79 (2012). https://doi.org/10.1016/j.brainres.2011.06.026

Hufford, D.J.: Sleep paralysis as spiritual experience. Transcult. Psychiatry **42**(1), 11–45 (2005). https://doi.org/10.1177/1363461505050709

Huot, P., et al.: Increased 5-HT2A receptors in the temporal cortex of Parkinsonian patients with visual hallucinations. Mov. Disord. **25**(10), 1399–1408 (2010). https://doi.org/10.1002/mds.23083

Jalal, B., Ramachandran, V.S.: Sleep paralysis, "the ghostly bedroom intruder" and out-of-body experiences: the role of mirror neurons. Front. Hum. Neurosci. **11**, 92 (2017). https://doi.org/10.3389/fnhum.2017.00092

Jalal, B.: The neuropharmacology of sleep paralysis hallucinations: serotonin 2A activation and a novel therapeutic drug. Psychopharmacology **235**(11), 3083–3091 (2018). https://doi.org/10.1007/s00213-018-5042-1

Kometer, M., Schmidt, A., Jäncke, L., Vollenweider, F.X.: Activation of serotonin 2A receptors underlies the psilocybin-induced effects on α oscillations, N170 visual-evoked potentials, and visual hallucinations. J. Neurosci. **33**(25), 10544–10551 (2013). https://doi.org/10.1523/JNEUROSCI.3007-12.2013

Kumar, S., et al.: The role of $GABA_A$ receptors in the acute and chronic effects of ethanol: a decade of progress. Psychopharmacology **205**(4), 529–564 (2009). https://doi.org/10.1007/s00213-009-1562-z

Lawrenson, J.G., Hull, C.C., Downie, L.E.: The effect of blue-light blocking spectacle lenses on visual performance, macular health and the sleep-wake cycle: a systematic review of the literature. Ophthalmic Physiol. Opt. **37**(6), 644–654 (2017). https://doi.org/10.1111/opo.12406

Levin, R., Nielsen, T.A.: Disturbed dreaming, posttraumatic stress disorder, and affect distress: a review and neurocognitive model. Psychol. Bull. **133**(3), 482 (2007). https://doi.org/10.1037/0033-2909.133.3.482

Meltzer, H.Y., et al.: Pimavanserin, a serotonin 2A receptor inverse agonist, for the treatment of Parkinson's disease psychosis. Neuropsychopharmacology **35**(4), 881–892 (2010). https://doi.org/10.1038/npp.2009.176

Modirrousta, M., Mainville, L., Jones, B.E.: Dynamic changes in $GABA_A$ receptors on basal forebrain cholinergic neurons following sleep deprivation and recovery. BMC Neurosci. **8**(1), 1–9 (2007). https://doi.org/10.1186/1471-2202-8-15

Morgan, M.A., Romanski, L.M., LeDoux, J.E.: Extinction of emotional learning: contribution of medial prefrontal cortex. Neurosci. Lett. **163**(1), 109–113 (1993). https://doi.org/10.1016/0304-3940(93)90241-C

Munezawa, T., et al.: Nightmare and sleep paralysis among Japanese adolescents: a nationwide representative survey. Sleep Med. **12**(1), 56–64 (2011). https://doi.org/10.1016/j.sleep.2010.04.015

Nichols, D.E., Nichols, C.D.: Serotonin receptors. Chem. Rev. **108**(5), 1614–1641 (2008). https://doi.org/10.1038/nature06612

Ressler, K.J.: Amygdala activity, fear, and anxiety: modulation by stress. Biol. Psychiat. **67**(12), 1117–1119 (2010). https://doi.org/10.1016/j.biopsych.2010.04.027

Peñalva, R.G., Lancel, M., Flachskamm, C., Reul, J.M., Holsboer, F., Linthorst, A.C.: Effect of sleep and sleep deprivation on serotonergic neurotransmission in the hippocampus: a combined in vivo microdialysis/EEG study in rats. Eur. J. Neurosci. **17**(9), 1896–1906 (2003). https://doi.org/10.1046/j.1460-9568.2003.02612.x

Sandman, N., Valli, K., Kronholm, E., Revonsuo, A., Laatikainen, T., Paunio, T.: Nightmares: risk factors among the Finnish general adult population. Sleep **38**(4), 507–514 (2015). https://doi.org/10.5665/sleep.4560

Sateia, M.J.: International classification of sleep disorders. Chest **146**(5), 1387–1394 (2014). https://doi.org/10.1378/chest.14-0970

Snel, J., Lorist, M.M.: Effects of caffeine on sleep and cognition. Prog. Brain Res. **190**, 105–117 (2011). https://doi.org/10.1016/B978-0-444-53817-8.00006-2

Sharpless, B.A., Barber, J.P.: Lifetime prevalence rates of sleep paralysis: a systematic review. Sleep Med. Rev. **15**(5), 311–315 (2011). https://doi.org/10.1016/j.smrv.2011.01.007

Stefani, A., Högl, B.: Nightmare disorder and isolated sleep paralysis. Neurotherapeutics **18**(1), 100–106 (2020). https://doi.org/10.1007/s13311-020-00966-8

Tagliazucchi, E., et al.: Increased global functional connectivity correlates with LSD-induced ego dissolution. Curr. Biol. **26**(8), 1043–1050 (2016). https://doi.org/10.1016/j.cub.2016.02.010

Treur, J. (ed.): Network-Oriented Modeling for Adaptive Networks: Designing Higher-Order Adaptive Biological, Mental and Social Network Models. Studies in Systems, Decision and Control, vol. 251. Springer, Cham (2020). https://doi.org/10.1007/978-3-030-31445-3

Ursin, R.: Serotonin and sleep. Sleep Med. Rev. **6**(1), 55–67 (2002). https://doi.org/10.1053/smrv.2001.0174

Vandekerckhove, M., Cluydts, R.: The emotional brain and sleep: an intimate relationship. Sleep Med. Rev. **14**(4), 219–226 (2010). https://doi.org/10.1016/j.smrv.2010.01.002

Waters, F., et al.: What is the link between hallucinations, dreams, and hypnagogic–hypnopompic experiences? Schizophr. Bull. **42**(5), 1098–1109 (2016). https://doi.org/10.1093/schbul/sbw076

Wróbel-Knybel, P., et al.: Prevalence and clinical picture of sleep paralysis in a Polish student sample. Int. J. Environ. Res. Public Health **17**(10), 3529 (2020). https://doi.org/10.3390/ijerph17103529

Localization of the Radio Signal with the Unknown Non-power Parameter

Yury Korchagin[1], Oleg Chernoyarov[2,3,4(✉)], Mikhail Trifonov[1], and Alexander Makarov[2]

[1] Voronezh State University, Universitetskaya sq. 1, 394018 Voronezh, Russia
[2] National Research University "MPEI", Krasnokazarmennaya Street 14, 111250 Moscow, Russia
chernoyarovov@mpei.ru
[3] National Research Tomsk State University, Lenin Avenue 36, 634050 Tomsk, Russia
[4] Maikop State Technological University, Pervomaiskaya Street 191, 385000 Maikop, Russia

Abstract. Application of the maximum likelihood method for estimating the moments of appearance and disappearance of a radio signal with a free-form envelope and the unknown initial phase is considered. Three approaches are proposed to overcome the prior uncertainty about the initial phase. In the first approach, instead of the unknown phase value, its expected value is used. The estimates of the times of appearance and disappearance of the signal obtained in this case by maximizing the decision determining statistics are the quasi-likelihood ones. In the second approach, the decision determining statistics are maximized by all the unknown parameters. However, the resulting maximum likelihood measurer has a complex multichannel structure. In the third approach, the three-dimensional decision determining statistics are represented as a sum of the two-dimensional random fields, each of them depending upon the unknown initial phase and the moment of appearance or disappearance of the signal. By the further separate maximization of the specified two-dimensional fields by the initial phase, the two-channel quasi-optimal measurer of the signal time parameters is obtained. One of the principal tasks is the performance testing of the introduced measurers. For this purpose, analytical formulas are found for the biases and the variances of the estimates of the moments of appearance and disappearance. It is shown that as the SNR increases, the characteristics of the quasi-optimal estimates and the maximum likelihood estimates demonstrate convergence. Thus, in conditions of high a posteriori accuracy, a two-channel quasi-optimal measurer can be used instead of a multichannel maximum likelihood measurer without any performance losses.

Keywords: Free-form signal · Moments of appearance and disappearance · Unknown phase · Maximum likelihood estimate · Quasi-optimal reception · Local Markov approximation method · Bias and variance of estimate

1 Introduction

The problem of estimating the moments of appearance and disappearance of the signal against noise is of considerable interest for many practical applications of statistical radio

R. Silhavy et al. (Eds.): CoMeSySo 2021, LNNS 232, pp. 557–571, 2021.
https://doi.org/10.1007/978-3-030-90318-3_45

physics, radio- and hydrolocation, navigation, seismology, etc. It has been repeatedly considered in the specialized literature (see, for example [1–4]). In [1], the problem of the discrete-time estimate of signal abrupt change points is studied. In [2–4] the algorithms for the continuous-time estimation of the unknown moments of appearance and disappearance of the quasi-deterministic low-frequency free-form signal against additive Gaussian white noise are examined. However, in many cases the high-frequency signals (radio signals) are used which often have an unknown initial phase due to the propagation features. Therefore, there is a good reason to consider the algorithms of localization of a radio signal in the conditions of parametric a priori uncertainty. In the present paper, the algorithms for estimating the moments of appearance and disappearance of a radio signal with a free-form envelope and the unknown initial phase are developed based on the maximum likelihood (ML) method. For the introduced algorithms, the asymptotic formulas for the characteristics of their performance are determined. The accuracy of these formulas increases with a signal-to-noise ratio (SNR).

2 The Problem Statement

The model of a radio signal with unknown initial phase and moments of appearance and disappearancc can be represented as follows [5, 6]

$$s(t, \varphi, \theta_1, \theta_2) = \begin{cases} f(t)\cos(\omega t - \varphi), & \theta_1 \le t \le \theta_2, \\ 0, & t < \theta_1, t > \theta_2. \end{cases} \tag{1}$$

Here the known continuous function $f(t)$ describes the form of the radio signal envelope; $\varphi \in [0, 2\pi]$ is the initial phase; ω is the frequency; θ_1 and θ_2 are the moments of appearance and disappearance, respectively, taking the values from the prior intervals $\theta_i \in [\theta_{i\,\min}, \theta_{i\,\max}]$, $i = 1, 2$.

In order that the signal does not disappear earlier than it appears, the following condition should be satisfied: $\theta_{1\,\max} < \theta_{2\,\min}$. It is also presupposed that the function describing the form of the radio signal envelope does not become zero at rising and falling edges so that $f(\theta_i) \neq 0$ and the radio signal (1) is discontinuous [5].

One considers that the random process

$$\xi(t) = s(t, \varphi_0, \theta_{01}, \theta_{02}) + n(t) \tag{2}$$

observed within the time interval $[0, T]$ is the additive mix of the useful signal $s(t, \varphi_0, \theta_{01}, \theta_{02})$ and a Gaussian white noise $n(t)$ with the one-sided spectral density N_0. In (1), the real values of the initial phase and the moments of appearance and disappearance moments are indicated through φ_0, θ_{01}, θ_{02} and they are unknown on the receiving side. Based on the processing of the observable realization (2), the receiver should produce the estimates of the moments of appearance and disappearance of the useful signal (1).

3 The Method of Study

If the initial phase of the useful signal (1) is a priori known, then the ML estimation algorithm can be applied. According to this algorithm, the estimates of the moments of appearance and disappearance coincide with the coordinates of the position of the largest maximum of the decision determining statistics that is the logarithm of the functional of likelihood ratio (FLR) and can be represented as the function of the current values θ_1, θ_2 of the unknown parameters θ_{01}, θ_{02} in the form [5, 6]

$$L(\theta_1, \theta_2) = \frac{2}{N_0} \int_{\theta_1}^{\theta_2} \xi(t) f(t) \cos(\omega t - \varphi_0)\, dt - \frac{1}{N_0} \int_{\theta_1}^{\theta_2} f^2(t) \cos^2(\omega t - \varphi_0)\, dt.$$

Hereinafter the first summand is the Ito integral.

However, when the moments of appearance and disappearance, together with the initial phase, are unknown, the FLR logarithm depends on the three variables [5, 6]

$$L(\varphi, \theta_1, \theta_2) = \frac{2}{N_0} \int_{\theta_1}^{\theta_2} \xi(t) f(t) \cos(\omega t - \varphi)\, dt - \frac{1}{N_0} \int_{\theta_1}^{\theta_2} f^2(t) \cos^2(\omega t - \varphi)\, dt. \quad (3)$$

By substituting some values of the initial phase into (3), instead of its unknown value, a number of estimation algorithms of the moments of appearance and disappearance are obtained both optimal and non-optimal ones. These values for the initial phase can be fixed or determined by the observable realization (2). The resulted estimation algorithms considered below differ by their efficiency and simplicity in hardware or software implementation.

4 The Quasi-likelihood Estimation Algorithm

One of the ways to overcome the prior parametrical uncertainty concerning the initial phase is to apply the quasi-likelihood (QL) estimation algorithm as it is described in [7]. The QL receiver should generate the FLR logarithm (3) for some expected value of the initial phase φ^* and for all the possible values of the moments of appearance and disappearance:

$$L^*(\theta_1, \theta_2) = \frac{2}{N_0} \int_{\theta_1}^{\theta_2} \xi(t) f(t) \cos\left(\omega t - \varphi^*\right) dt - \frac{1}{N_0} \int_{\theta_1}^{\theta_2} f^2(t) \cos^2\left(\omega t - \varphi^*\right) dt. \quad (4)$$

Then, the QL estimates of the moments of appearance and disappearance are determined as the position of the largest maximum of the decision determining statistics (4):

$$(\theta_1, \theta_2) = \arg\sup\ L^*(\theta_1, \theta_2). \quad (5)$$

According to (5), the receiver should produce the two-dimensional random field (4) for all the possible values of the unknown moments of appearance and disappearance. Therefore, in general case, its hardware implementation is complex enough. Indeed, determining the values (5) presupposes the construction of the system that is multi-channel for both unknown estimated parameters. However, difficulties with hardware implementation of the QL estimation algorithm (5) can be partially overcome by the expansion of the random field (4) into the sum $L^*(\theta_1, \theta_2) = L_1^*(\theta_1) + L_2^*(\theta_2)$ of two random processes, as it is carried out in [2–4]. The first of these random processes depends upon the moment of appearance θ_1, while the second – upon the moment of disappearance θ_2:

$$L_1^*(\theta_1) = \frac{2}{N_0}\int_{\theta_1}^{\theta} \xi(t)f(t)\cos(\omega t - \varphi^*)\,dt - \frac{1}{N_0}\int_{\theta_1}^{\theta} f^2(t)\cos^2(\omega t - \varphi^*)\,dt, \quad (6)$$

$$L_2^*(\theta_2) = \frac{2}{N_0}\int_{\theta}^{\theta_2} \xi(t)f(t)\cos(\omega t - \varphi^*)\,dt - \frac{1}{N_0}\int_{\theta}^{\theta_2} f^2(t)\cos^2(\omega t - \varphi^*)\,dt, \quad (7)$$

where θ is an arbitrary point within the interval $(\theta_{1\,\max}, \theta_{2\,\min})$.

It follows from (6), (7) that the random processes $L_1^*(\theta_1)$ and $L_2^*(\theta_2)$ are statistically independent, as they are the integrals from the white noise within the nonoverlapping intervals. Therefore, the maximum position of the random field $L^*(\theta_1, \theta_2)$ by the variable θ_1 coincides with the maximum position of the random process $L_1^*(\theta_1)$, while by the variable θ_2 – with the maximum position of $L_2^*(\theta_2)$. Consequently, the QL estimates of the moments of appearance and disappearance can be determined as

$$\theta_j^* = \arg\sup L_j^*(\theta_j), \quad \theta_j \in [\theta_{j\,\min}, \theta_{j\,\max}], \quad j = 1, 2 \quad (8)$$

The block diagram of the QL measurer of the moments of appearance and disappearance is shown in Fig. 1. Here the notations are the following: G is the generator that reproduce the signal $f(t)\cos(\omega t - \varphi^*)/2$; I1 is the integrator over the interval $[\theta_{1\,\min}, t]$, $t \in [\theta_{1\,\min}, \theta]$; I2 is the integrator over the interval $[\theta, t]$, $t \in [\theta, \theta_{2\,\max}]$; DL is the delay line for the time $t = \theta - \theta_{\min}$; E1 and E2 are the extremators which carry out searching the positions of the signal maxima within the time intervals $[\theta, \theta + \theta_{1\,\max} - \theta_{1\,\min}]$ and $[\theta_{2\,\min}, \theta_{2\,\max}]$, respectively.

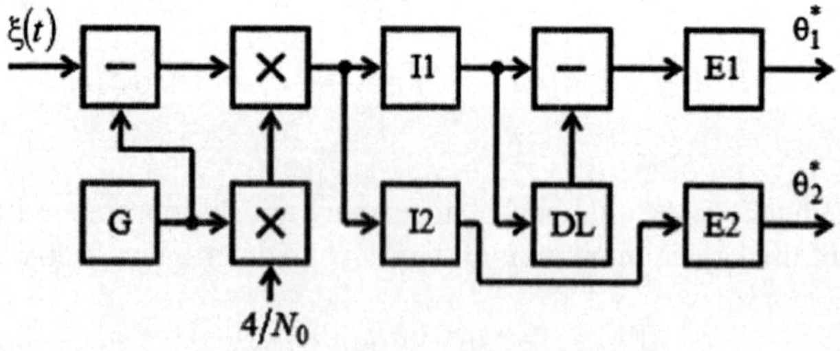

Fig. 1. The block diagram of the quasi-likelihood measurer of the moments of appearance and disappearance.

Representation of the decision determining statistics as the sum of two statistically independent random processes allows us both to introduce the simplified hardware QL measurer implementation and to analyze the QL estimation algorithm, according to the procedure presented in [3]. For the full statistical description of the decision determining statistics, it is enough to calculate the first two moments of the Gaussian independent random processes (6) and (7). By carrying out averaging of the expressions (6), (7), the following expressions can be obtained for the mathematical expectations

$$S_1^*(\theta_1) = \langle L_1^*(\theta_1)\rangle = Q[\max(\theta_{01}, \theta_1), \theta]\cos\Delta\varphi - Q(\theta_1, \theta)/2,$$

$$S_2^*(\theta_2) = \langle L_2^*(\theta_2)\rangle = Q[\theta, \max(\theta_{02}, \theta_2)]\cos\Delta\varphi - Q(\theta, \theta_2)/2,$$

and the correlation functions

$$B_1^*(\theta_{11}, \theta_{21}) = \langle\left[L_1^*(\theta_{11}) - \langle L_1^*(\theta_{11})\rangle\right]\left[L_1^*(\theta_{21}) - \langle L_1^*(\theta_{21})\rangle\right]\rangle = Q[\min(\theta_{11}, \theta_{21}), \theta],$$

$$B_2^*(\theta_{12}, \theta_{22}) = \langle\left[L_2^*(\theta_{12}) - \langle L_2^*(\theta_{12})\rangle\right]\left[L_2^*(\theta_{22}) - \langle L_2^*(\theta_{22})\rangle\right]\rangle = Q[\theta, \min(\theta_{12}, \theta_{22})].$$

Here

$$Q(\theta_1, \theta_2) = \frac{1}{N_0}\int_{\theta_1}^{\theta_2} f^2(t)\,\mathrm{d}t \tag{9}$$

is the signal-to-noise ratio (SNR) at the ML receiver output for the incoming signal with the moments of appearance θ_1 and disappearance θ_2, $\Delta\varphi = \varphi^* - \varphi_0$ is the intial phase mismatch of the QL measurer. Hereinafter it is presupposed that the signal (1) is narrow-band, so the integrals from the functions oscillating with the doubled frequency 2ω can be neglected.

Let the function $f(t)$ can vanish in the parts of interval $[\theta_{1\min}, \theta_{2\max}]$ having the null measure only. Then $Q(\theta_1, \theta)$ is the strictly decreasing function of the argument θ_1 while $Q(\theta, \theta_2)$ is the strictly increasing function of the argument θ_2, and, in addition, the following equalities hold:

$$Q(x, \theta) = -Q(\theta, x), \quad Q[\max(x, y), \theta] = \min[Q(x, \theta), Q(y, \theta)],$$
$$Q[\theta, \min(x, y)] = \min[Q(\theta, x), Q(\theta, y)].$$

Application of the function (9) properties specified above allows rewriting the mathematical expectations and the correlation functions of the processes (6), (7) in the form

$$S_j^*(\theta_j) = \min\left[(-1)^j Q(\theta, \theta_{0j}), (-1)^j Q(\theta, \theta_j)\right]\cos\Delta\varphi - (-1)^j Q(\theta, \theta_j)/2,$$
$$B_j^*(\theta_{1j}, \theta_{2j}) = \min\left[(-1)^j Q(\theta, \theta_{1j}), (-1)^j Q(\theta, \theta_{2j})\right]. \tag{10}$$

Hereinafter it is set that $j = 1, 2$. It is easy to verify that for $|\Delta\varphi| < \pi/2$ the mathematical expectations (10) top in the points coinciding with the real values θ_{0j} of the unknown moments of appearance and disappearance.

Further, in (6), (7), the variables θ_1, θ_2 are changed by the new variables $\lambda_j = (-1)^j Q(\theta, \theta_j)$, $\lambda_j \in [\lambda_{j\min}, \lambda_{j\max}]$, $\lambda_{1\min} = Q(\theta_{1\max}, \theta)$, $\lambda_{1\max} = Q(\theta_{1\min}, \theta)$, $\lambda_{2\min} = Q(\theta, \theta_{2\min})$, $\lambda_{2\max} = Q(\theta, \theta_{2\max})$. Then, for the random processes (6), (7), one gets

$$L_j^*(\theta_j) = L_j^*[g_j(\lambda_j)] = \mu_j^*(\lambda_j) = \min(\lambda_j, \lambda_{0j})\cos\Delta\varphi - \lambda_j/2 + v_j(\lambda_j), \qquad (11)$$

where $\lambda_{01} = Q(\theta_{01}, \theta)$, $\lambda_{02} = Q(\theta, \theta_{02})$, $g_j(\lambda_j)$ are the solutions of equations $(-1)^j Q(\theta, \theta_j) = \lambda_j$ with respect to θ_j, and $v_j(\lambda_j)$ are the statistically independent Gaussian random processes with zero mathematical expectations and correlation functions

$$B_j^*(\lambda_{1j}, \lambda_{2j}) = \min(\lambda_{1j}, \lambda_{2j}). \qquad (12)$$

According to the properties of the function (9), the maxima positions of the random processes (11)

$$\lambda_j^* = \arg\sup \mu_j^*(\lambda_j) \qquad (13)$$

are related to the estimates of the moments of appearance and disappearance by one-to-one transformations. Therefore, the conditional probability densities $W_{\theta j}^*(\theta_j|\theta_{0j})$ of the QL estimates (8) of the moments of appearance and disappearance can be expressed through the probability densities $W_{\lambda j}^*(\lambda_j|\lambda_{0j})$ of the random variables (13) as follows

$$W_{\theta j}^*(\theta_j|\theta_{0j}) = W_{\lambda j}^*\left[(-1)^j Q(\theta, \theta_j)\middle|(-1)^j Q(\theta, \theta_{0j})\right] |\partial Q(\theta, \theta_j)/\partial\theta_j|, \qquad (14)$$

Similarly to studies [5, 8, 9], the probability densities $W_{\lambda j}^*(\lambda_j|\lambda_{0j})$ can be represented in the form

$$W_{\lambda j}^*(\lambda_j|\lambda_{0j}) = \int_{-\infty}^{\infty} \frac{\partial}{\partial\lambda_j}\left[\left.\frac{\partial F_{2j}(u, v, \lambda_j)}{\partial u}\right|_{u=v}\right] du, \qquad (15)$$

where

$$F_{2j} = P\left[\sup_{\lambda_{j\min} \le \lambda < x} \mu_j^*(\lambda_j) < u, \sup_{x \le \lambda \le \lambda_{j\max}} \mu_j^*(\lambda_j) < v\right], \qquad (16)$$

are the two-dimensional distribution functions of absolute maxima of the random processes $\mu_j^*(\lambda_j)$. According to (11), (12), $\mu_j^*(\lambda_j)$ are the statistically independent Gaussian Markov (Wiener) random processes [10] with the drift k_{1j} and diffusion k_{2j} coefficients of the form

$$k_{1j} = \frac{1}{2}\begin{cases} 2\cos\Delta\varphi - 1, & \lambda_{j\min} \le \lambda \le \lambda_{0j}, \\ -1, & \lambda_{0j} < \lambda \le \lambda_{j\max}, \end{cases} \quad k_{2j} = 1. \qquad (17)$$

Therefore, using the results of studied [5, 8, 9], the function (16) can be expressed as follows:

$$F_{2j}(u, v, x) = \int_{-\infty}^{v} W_j(y, \lambda_{j\max})dy \qquad (18)$$

where $W_j(y, \lambda_j)$ are the solutions of the Fokker-Planck-Kolmogorov equation [10]

$$\frac{\partial W_j(y, \lambda_j)}{\partial \lambda_j} + \frac{\partial}{\partial y}\left[k_{1j} W_j(y, \lambda_j)\right] - \frac{1}{2}\frac{\partial^2}{\partial y^2}\left[k_{2j} W_j(y, \lambda_j)\right] = 0 \tag{19}$$

with the coefficients k_{1j}, k_{2j} (17), while the starting condition is

$$W_j(y, \lambda_{j\min}) = \exp\left[-\left(y + \lambda_{j\min}(\cos\Delta\varphi - 1/2)\right)^2 / 2\lambda_{j\min}\right] / \sqrt{2\pi\lambda_{j\min}}$$

and the boundary conditions are $W_j(y = u, \lambda_j) = W_j(y = -\infty, \lambda_j) = 0$, if $\lambda_j \in [\lambda_{j\min}, x]$, and $W_j(y = v, \lambda_j) = W_j(y = -\infty, \lambda_j) = 0$, if $\lambda_j \in [x, \lambda_{j\max}]$.

Using a reflection method with sign inversion [8], the separate solutions of Eq. (19) with the coefficients (17) can be obtained for the cases when $\lambda_j \in [\lambda_{j\min}, \lambda_{0j}]$ and $\lambda_j \in [\lambda_{0j}, \lambda_{j\max}]$, as it is described, for example, in [5, 8, 9]. By substituting the resultant solutions into (18), and then (18) into (15), similarly to [3], after appropriate transformations for the probability density of the random variable (13) one gets

$$W^*_{\lambda j}(\lambda_j|\lambda_{0j}) = \begin{cases} c_1\Psi\left[c_1(\lambda_{0j} - \lambda_j), c_1(\lambda_{0j} - \lambda_{j\min}), c_2(\lambda_{j\max} - \lambda_{0j}), 1/R\right], & \lambda_j \le \lambda_{0j}, \\ c_2\Psi\left[c_2(\lambda_j - \lambda_{0j}), c_2(\lambda_{j\max} - \lambda_{0j}), c_1(\lambda_{0j} - \lambda_{j\min}), R\right], & \lambda_j > \lambda_{0j}, \end{cases} \tag{20}$$

where

$$\Psi(y, y_1, y_2, y_3) = \left[\Phi\left(\sqrt{\frac{y_1 - y}{2}}\right) + \frac{1}{\sqrt{\pi(y_1 - y)}}\exp\left(-\frac{y_1 - y}{4}\right)\right]\frac{1}{2|y|^{3/2}\sqrt{\pi}}$$

$$\times \int_0^\infty x\exp\left[-\frac{(x + y)^2}{4y}\right]\left[\Phi\left(\frac{y_2 + y_3 x}{\sqrt{2y_2}}\right) - \exp(-y_3 x)\Phi\left(\frac{y_2 - y_3 x}{\sqrt{2y_2}}\right)\right]\mathrm{d}x, \tag{21}$$

$c_1 = R^2/2$, $c_2 = 1/2$, $R = 2\cos\Delta\varphi - 1$. It is now obvious that the probability density (20) does not depend on the value θ selection in (6), (7). With reference to (14) and (20), the expressions for the conditional biases and variances of the estimates of the moments of appearance and disappearance (5) are determined as follows:

$$B\left(\theta_j^*\middle|\theta_{0j}\right) = \int_{\theta j\min}^{\theta_{j\max}} (\theta_j - \theta_{0j}) W^*_{\lambda j}\left[(-1)^j Q(\theta, \theta_j)\middle|(-1)^j Q(\theta, \theta_{0j})\right]\left|\frac{\partial Q(\theta, \theta_j)}{\partial \theta_j}\right|\mathrm{d}\theta_j, \tag{22}$$

$$V\left(\theta_j^*\middle|\theta_{0i}\right) = \int_{\theta j\min}^{\theta_{j\max}} (\theta_i - \theta_{0j})^2 W^*_{\lambda j}\left[(-1)^j Q(\theta, \theta_j)\middle|(-1)^j Q(\theta, \theta_{0j})\right]\left|\frac{\partial Q(\theta, \theta_j)}{\partial \theta_j}\right|\mathrm{d}\theta_j, \tag{23}$$

Presupposing in (17) $\Delta\varphi = 0$ and therefore $y_3 = 1$ in (21), it can be seen that the expressions (20), (22), (23) move to the exact formulas obtained in [3] for the characteristics of the ML estimates of the moments of appearance and disappearance of the signal with a priori known initial phase.

The asymptotic behavior of the probability densities (20), the biases (22) and the variances (23) with SNR increasing and for $\Delta\varphi = 0$ is examined in [3]. Carrying out similar transformations for the normalized variables (or the generalized QL estimates of the moments of appearance and disappearance)

$$\kappa_j = \begin{cases} R^2(\lambda_j - \lambda_{0j})/2, & \lambda_j \le \lambda_{0j}, \\ (\lambda_j - \lambda_{0j})/2, & \lambda_j > \lambda_{0j}, \end{cases} \tag{24}$$

the limiting (with SNR increasing) probability density can be obtained in the form

$$W(\kappa_j) = \begin{cases} W_0(-\kappa_j, 1/R), & \kappa_j \le 0, \\ W_0(\kappa_j, R), & \kappa_j > 0, \end{cases} \tag{25}$$

$$W_0(x, y) = \Psi(x, \infty, \infty, y) = (2y + 1)\exp(y(y + 1)|x|)[1 - \Phi((2y + 1)\sqrt{|x|/2})] + \Phi(\sqrt{|x|/2}) - 1.$$

It is known [5] that with SNR increasing the QL estimates converge in mean square to the maxima positions of the mathematical expectations of the decision determining statistics (6), (7), coinciding with the real values θ_{0j} of the moments of appearance and disappearance, if $\Delta\varphi < \pi/2$. One develops the function $(-1)^j Q(\theta, \theta_j)$ as Taylor series in variable θ_j in the neighborhood of point θ_{0j} and reduces the serial expansion by terms of the first infinitesimal order:

$$(-1)^j Q(\theta, \theta_j) = (-1)^j Q(\theta, \theta_{0j}) + (-1)^j \rho_j^2 (\theta_j - \theta_{0j})/T_{\max},$$

where $\rho_j^2 = 2f^2(\theta_{0j})\, T_{max}/N_0$, and $T_{\max} = \theta_{2\max} - \theta_{1\min}$ is the maximum possible signal duration. From here, one gets

$$\lambda_j - \lambda_{0j} = (-1)^j \rho_j^2 (\theta_j - \theta_{0j})/T_{\max}. \tag{26}$$

Substituting (26) into (24) the following is obtained:

$$\kappa_j = (-1)^j \begin{cases} R^2 \rho_j^2 (\theta_j - \theta_{0j})/2T_{\max}, & \theta_j \le \theta_{0j}, \\ \rho_j^2 (\theta_j - \theta_{0j})/2T_{\max}, & \theta_j > \theta_{0j}. \end{cases} \tag{27}$$

Taking into account (25) and (27), the asymptotic values of the bias (22) and the variance (23) of the QL estimates of the moments of appearance and disappearance take the form

$$B_a\left(\theta_j^* | \theta_{0j}\right) = (-1)^j 2T_{\max}\left(R^2 - 1\right)/\rho_j R^2, \tag{28}$$

$$V_a\left(\theta_j^* | \theta_{0j}\right) = 8T_{\max}^2\left[R^5\left(2R^2 + 6R + 5\right) + 5R^2 + 6R + 2\right]/\rho_j^4 R^4 (R + 1)^3. \tag{29}$$

If $\Delta\varphi = 0$ and $R = 1$, then the expressions (28) and (29) coincide with the bias and the variance of the ML estimates of the moments of appearance and disappearance of the signal with a priori known initial phase, which appeared in [2], namely

$$B_{0j} = 0, \quad V_{0j} = 26T_{\max}^2/\rho_j^4. \tag{30}$$

Influence of prior ignorance of the initial phase on accuracy of the QL estimates of the moments of appearance and disappearance can be characterized by the normalized biases $b(\Delta\varphi) = B_a\left(\theta_j^*\middle|\theta_{0j}\right)/\sqrt{V_a\left(\theta_j^*\middle|\theta_{0j}\right)}$ and variances $v(\Delta\varphi) = V_a\left(\theta_j^*\middle|\theta_{0j}\right)/V_{0j}$ of the QL estimates. These values (accurate to a sign) are identical for the QL estimates of the moments of appearance and disappearance, and therefore they can be used as the characteristic of the loss in accuracy of the QL estimates, in comparison with the ML estimates of the moments of appearance and disappearance of the signal with a priori known initial phase.

In Fig. 2 and Fig. 3, there are presented the dependences of the normalized bias $b(\Delta\varphi)$ and variance $v(\Delta\varphi)$ upon the mismatch $\Delta\varphi$. It is obvious that the non-zero intial phase mismatch leads to appearance of the bias of the estimates of the moments of appearance and disappearance and, as a consequence, – to possible ten times increase of the estimates variance.

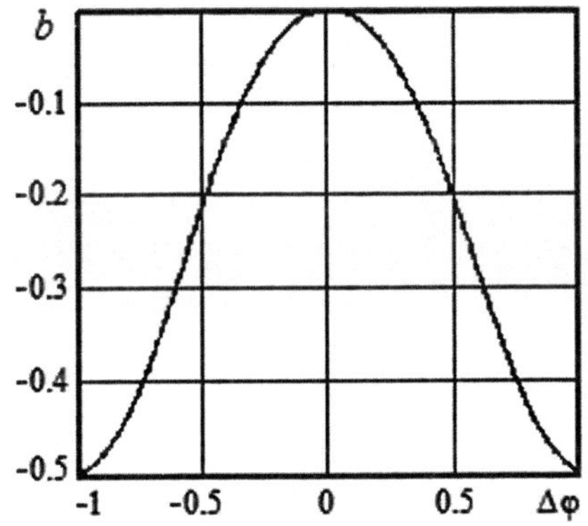

Fig. 2. The normalized bias of the estimates of the moments of appearance and disappearance.

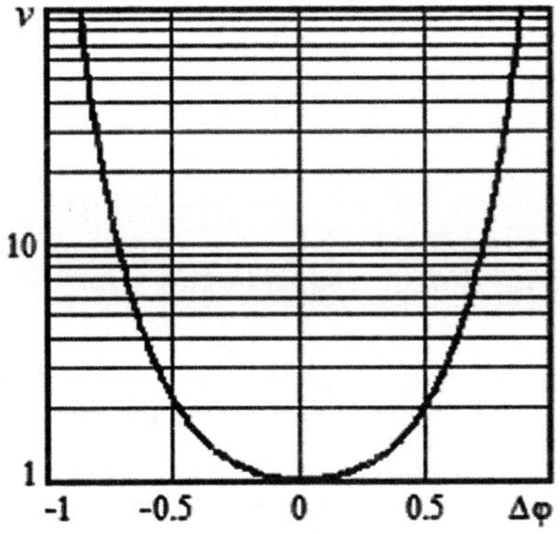

Fig. 3. The normalized variance of the estimates of the moments of appearance and disappearance.

5 The Maximum Likelihood Estimation Algorithm

In order to improve the accuracy of estimating the moments of appearance and disappearance, the ML algorithm can be applied. According to this algorithm, the unknown initial phase in (3) should be changed by its ML estimate φ_m that is equivalent to maximization of the FLR logarithm (3) by the current value of the initial phase φ:

$$L_\varphi(\theta_1, \theta_2) = L(\varphi_m, \theta_1, \theta_2) = \sup_\varphi L(\varphi, \theta_1, \theta_2). \quad (31)$$

The ML estimates of the moments of appearance and disappearance are determined as the position of the largest maximum of decision determining statistics (31):

$$(\theta_{m1}, \theta_{m2}) = \arg\sup L_\varphi(\theta_1, \theta_2). \quad (32)$$

Maximization of the FLR logarithm (3) by variable φ can be carried out analytically. For this purpose, by substituting the signal (1) into (3) the FLR logarithm is represented in the form

$$L(\varphi, \theta_1, \theta_2) = X(\theta_1, \theta_2)\cos\varphi + Y(\theta_1, \theta_2)\sin\varphi - \frac{1}{2N_0}\int\limits_{\theta_1}^{\theta_2} f^2(t)\,dt. \quad (33)$$

Here it is stated that

$$X(\theta_1, \theta_2) = \frac{2}{N_0}\int\limits_{\theta_1}^{\theta_2} \xi(t)f(t)\cos(\omega t)dt, \quad Y(\theta_1, \theta_2) = \frac{2}{N_0}\int\limits_{\theta_1}^{\theta_2} \xi(t)f(t)\sin(\omega t)dt$$

and the integrals from the terms oscillating with the doubled frequency 2ω are neglected. Further, the derivative of the function (33) by φ should be obtained and equated to zero:

$$\partial L(\varphi, \theta_1, \theta_2)/\partial\varphi = -X(\theta_1, \theta_2)\sin\varphi + Y(\theta_1, \theta_2)\cos\varphi = 0.$$

Solving the last equation in relation to φ:

$$\varphi = \mathrm{arctg}[Y(\theta_1, \theta_2)/X(\theta_1, \theta_2)] \quad (34)$$

and substituting (34) into (33), one gets

$$L_\varphi(\theta_1, \theta_2) = \sqrt{X^2(\theta_1, \theta_2) + Y^2(\theta_1, \theta_2)} - \frac{1}{2N_0}\int\limits_{\theta_1}^{\theta_2} f^2(t)\,dt. \quad (35)$$

In terms of (35), the ML measurer structure can be constructed. But, it should be noted that in practice it is not possible to produce the decision determining statistics (35) as continuous function of the moments of appearance and disappearance. In this case the measurer should generate the samples $L_{mg} = L_\varphi(\theta_{1\,\min} + m\Delta\theta_1, \theta_{2\,\min} + g\Delta\theta_2)$, $m = \overline{1, n_1}$, $g = \overline{1, n_2}$ of the random field (35) for a discrete value set of the moments of appearance and disappearance with the steps $\Delta\theta_1 = \{(\theta_{1\,\max} - \theta_{1\,\min})/n_1\}$, $\Delta\theta_2 = \{(\theta_{2\,\max} - \theta_{2\,\min})/n_2\}$, where $\{\cdot\}$ is an integer part. Therefore, the measurer should include $n_1 n_2$ channels.

In Fig. 4, the block diagram of one channel of the ML measurer that generates the FLR logarithm (35) for the fixed values of the moments of appearance $\theta_{1\,\min} + m\Delta\theta_1$ and disappearance $\theta_{2\,\min} + g\Delta\theta_2$ is represented. Here it is denoted: G1 is the generator that reproduce the signal $f(t)$; G2 is the cosine generator; I is the integrator over the time interval $[\theta_{1\,\min} + m\Delta\theta_1, \theta_{2\,\min} + g\Delta\theta_2]$. The ML estimates of the moments of appearance and disappearance are determined by numbers of the channel with the maximum output signal.

It must be emphasized that, besides complexities of practical implementation of the ML algorithm (32) caused its multichannel structure, there are also great difficulties in definition of the estimates characteristics by the position of the absolute maximum of the FLR logarithm (35).

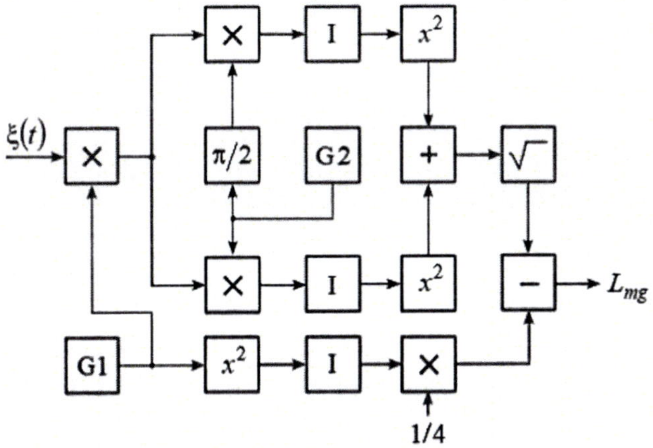

Fig. 4. The block diagram of one channel of the maximum likelihood measurer of the moments of appearance and disappearance.

6 The Quasi-optimal Estimation Algorithm

In order to simplify the hardware and software implementation of the ML measurer and a determination of its characteristics, it is possible to make use of the quasi-optimal (QO) estimates. For this purpose, one represents the FLR logarithm (3) as the sum $L(\varphi, \theta_1, \theta_2) = L_1(\varphi, \theta_1) + L_2(\varphi, \theta_2)$ of the two summands

$$L_1(\varphi, \theta_1) = \frac{2}{N_0} \int_{\theta_1}^{\theta} \xi(t) f(t) \cos(\omega t - \varphi)\, dt - \frac{1}{N_0} \int_{\theta_1}^{\theta} f^2(t) \cos^2(\omega t - \varphi)\, dt, \tag{36}$$

$$L_2(\varphi, \theta_2) = \frac{2}{N_0} \int_{\theta}^{\theta_2} \xi(t) f(t) \cos(\omega t - \varphi)\, dt - \frac{1}{N_0} \int_{\theta}^{\theta_2} f^2(t) \cos^2(\omega t - \varphi)\, dt, \tag{37}$$

where, as above, θ is an arbitrary point within the interval $(\theta_{1\max}, \theta_{2\min})$. Then the following estimates can be introduced:

$$\theta_{mj}^* = \arg\sup L_{j\varphi}(\theta_j), \quad L_{j\varphi}(\theta_j) = \max_{\varphi} L_j(\varphi, \theta_j). \tag{38}$$

The QO estimates (38), generally speaking, are not the ML ones, but, as it is shown below, with SNR increasing their efficiency asymptotically coincides with the efficiency of the ML estimates of the moments of appearance and disappearance of the signal with a priori known initial phase. The maximization of the functions (36), (37) by the variable φ is then carried out, similar to the one presented in (33)–(35), and thus one gets

$$L_{1\varphi}(\theta_1) = \sup_{\varphi} L_1(\varphi, \theta_1) = \sqrt{X_1^2(\theta_1) + Y_1^2(\theta_1)} - \frac{1}{2N_0} \int_{\theta_1}^{\theta} f^2(t) dt, \tag{39}$$

$$L_{2\varphi}(\theta_2) = \sup_{\varphi} L_2(\varphi, \theta_2) = \sqrt{X_2^2(\theta_2) + Y_2^2(\theta_2)} - \frac{1}{2N_0} \int_{\theta}^{\theta_2} f^2(t)\mathrm{d}t, \tag{40}$$

where

$$X_1(\theta_1) = \frac{2}{N_0} \int_{\theta_1}^{\theta} \xi(t) f(t) \cos(\omega t)\mathrm{d}t, \quad Y_1(\theta_1) = \frac{2}{N_0} \int_{\theta_1}^{\theta} \xi(t) f(t) \sin(\omega t)\mathrm{d}t,$$

$$X_2(\theta_{21}) = \frac{2}{N_0} \int_{\theta}^{\theta_2} \xi(t) f(t) \cos(\omega t)\mathrm{d}t, \quad Y_2(\theta_2) = \frac{2}{N_0} \int_{\theta}^{\theta_2} \xi(t) f(t) \sin(\omega t)\mathrm{d}t.$$

In Fig. 5, the block diagram of the device that generates the QO estimates (38) of the moments of appearance and disappearance is presented in terms of the expressions (39), (40). Here the notations are the following: G1 is the generator that reproduce the signal $f(t)$; G2 is the cosine generator; I1 and I2 are the integrators over the time intervals $[\theta_{1\,\min}, \theta]$ and $[\theta, \theta_{2\,\max}]$, DL is the delay line for the time $\theta - \theta_{1\,\min}$, E1 and E2 are extremators fixing the positions of absolute signal maxima on time intervals $[\theta, \theta + \theta_{1\,\max} - \theta_{1\,\min}]$ and $[\theta_{2\,\min}, \theta_{2\,\max}]$, respectively. Thus, the application of the estimates (38) allows considerable simplification of the technical measurer implementation. Indeed, while for the realization of the ML estimation algorithm (31) the multichannel receiver constructed on the appearance and disappearance moments is required, only the two-channel scheme is needed in order to determine the estimates (38).

Now it is time to analyze the estimation algorithm (38). It should be noted that the components (36), (37) of the FLR logarithm represent the realizations of the two-dimensional random fields. These fields are differentiable on the variable φ and nondifferentiable by the variables $\theta_j, j = 1, 2$. Hence, the initial phase is the regular parameter of the signal (1) while the moments of appearance θ_1 and disappearance θ_2 are discontinuous parameters [5, 11]. Thus, regularity conditions are partially violated. One denotes the SNR for the signal with the moments of appearance θ_{01} and disappearance θ as

$$z_{01}^2 = \frac{1}{N_0} \int_{\theta_{01}}^{\theta} f^2(t)\,\mathrm{d}t \tag{41}$$

and the SNR for the signal with the moments of appearance θ and disappearance θ_{02} as

$$z_{02}^2 = \frac{1}{N_0} \int_{\theta}^{\theta_{02}} f^2(t)\,\mathrm{d}t. \tag{42}$$

It is presupposed that SNRs (41) and (42) are rather great, so

$$z_{01}^2 \to \infty, \quad z_{02}^2 \to \infty.$$

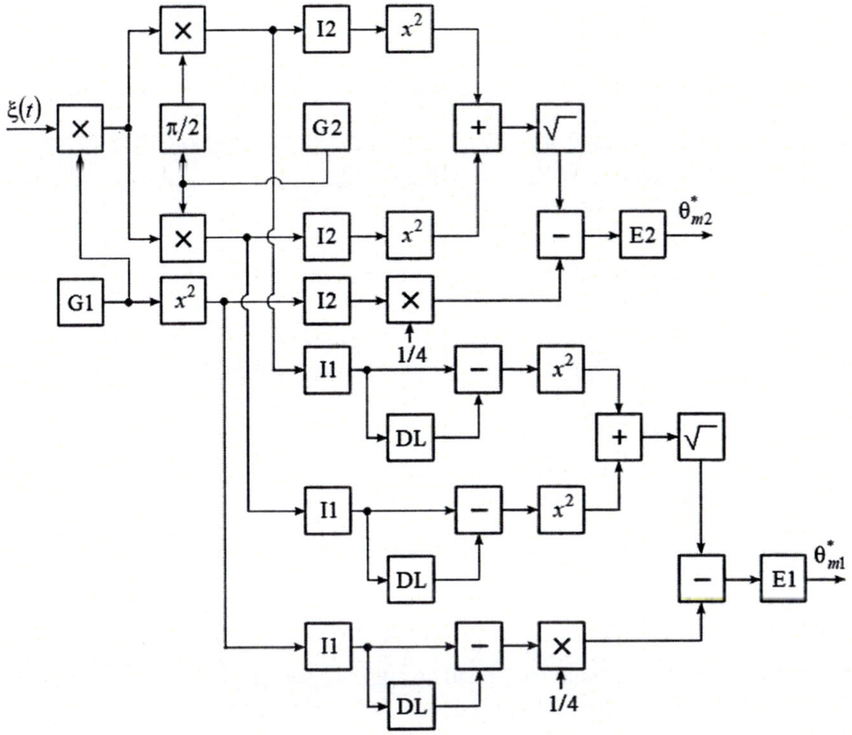

Fig. 5. The block diagram of the quasi-optimal measurer of the moments of appearance and disappearance.

The QO estimates (38) can be considered as a result of joint estimation of the initial phase and either the moment of appearance of the signal with SNR (41), or the moment of disappearance of the signal with SNR (42), respectively:

$$\left(\theta_{mj}^{*}, \varphi_{mj}\right) = \arg\sup L_j\left(\varphi, \theta_j\right)$$

In [11] it is shown that the accuracy of the estimates of discontinuous parameters (the moments of appearance and disappearance) does not asymptotically depend on the presence of a priori unknown regular parameter (the initial phase), with SNR increasing. It means that, according to [2], the biases and the variances of the QO estimates of the moments of appearance and disappearance (38) are of the form

$$B_{\varphi}\left(\theta_{mj}^{*}|\theta_{0j}\right) = 0, \quad V_{\varphi}\left(\theta_{mj}^{*}|\theta_{0j}\right) = 26\mathrm{T}_{\max}^{2}/\rho_j^4. \tag{43}$$

that coincides with (30). As it is obvious from (43), while the SNR increases, the accuracy of the QO estimates of the moments of appearance and disappearance (38) asymptotically coincides with the accuracy of the ML estimates of the moments of appearance and disappearance of the radio signal with a priori known initial phase. Hence, the estimates (38) are asymptotically the ML ones, and their application allows considerable simplification of the technical implementation of the ML measurer of the moments of appearance and disappearance (32).

Thus, the asymptotic values of the variances of the ML estimates of the moments of appearance and disappearance of the signal with a priori known initial phase (30) coincide with the asymptotic values of the variances (43) of the QO estimates (38). This implies that the asymptotic values of the variances of the ML estimates (32) of the moments of appearance and disappearance of the signal with a priori known initial phase are also defined by the formulas (30), (43). Indeed, the variances of the estimates (32) of the moments of appearance and disappearance of the signal with a priori unknown initial phase cannot be less than the variance (30) of the ML estimates of the moments of appearance and disappearance of the signal with a priori known initial phase. At the same time, the variance of the ML estimates (32) cannot be more than the variance of the QO estimates (43).

Therefore, the dependences shown in Fig. 2 and Fig. 3 characterize the accuracy gain of the ML estimates (32) or the QO estimates (38) in comparison with the QL ones (5). So, under $|\Delta\varphi| > 1/2$, the variance of the ML estimates (32) and the QO estimates (38) is more than two times less than the variance of the QL estimates (5).

7 Conclusion

The quasi-likelihood, maximum likelihood and quasi-optimal algorithms for estimating the moments of appearance and disappearance of the radio signal with free-form envelope and unknown initial phase are synthesized. The asymptotic characteristics for the performance of the introduced algorithms are obtained. It is noted that the maximum likelihood algorithm provides the best estimating accuracy among all other considered algorithms. However, its hardware or software implementation is the most complex. The accuracy of the simpler quasi-optimal estimation algorithm asymptotically coincides with the accuracy of the maximum likelihood estimation algorithm. It is shown that for the great signal-to-noise ratios the prior ignorance of the signal initial phase does not asymptotically influence the accuracy of the maximum likelihood and quasi-optimal estimates of the signal moments of appearance and disappearance. The obtained results allow making an informed choice of the estimation algorithm, depending upon the available prior information concerning the signal phase, as well as the estimation accuracy and the measure of simplicity of algorithm implementation required.

Acknowledgements. This work was financially supported by the Ministry of Education and Science of the Russian Federation (research project No. FSWF-2020-0022).

References

1. Mal'tsev, A.A., Silaev, A.M.: Optimum estimates of the times of random irregular variations in signal parameters. Sov. J. Commun. Technol. Electron. **34**(15), 116–124 (1989)
2. Trifonov, A.P., Korchagin, Y.E.: Optimal signals reception with unknown times of their appearance and disappearance. Problemy Peredachi Informatsii **37**(1), 52–71 (2001)

3. Trifonov, A.P., Korchagin, Y.E.: Exact formulas for computation of reception characteristics of a signal with unknown appearance and disappearance times. Probl. Inf. Transm. **45**(3), 158–167 (2009)
4. Chernoyarov, O.V., Shahmoradian, M.M., Marcokova, M., Korchagin, Y.E.: The secrecy of communications in the data transmission by the impulses with the unknown moments of appearance and disappearance. Int. J. Eng. Trans. A **32**(5), 548–554 (2019)
5. Trifonov, A.P., Shinakov, Y.S.: Joint Discrimination of Signals and Estimation of Their Parameters Against Hindrances. Radio i Svyaz', Moscow (1986).(in Russian)
6. Van Trees, H.L., Bell, K.L., Tian, Z.: Detection, Estimation, and Modulation Theory, Part I. Detection, Estimation, and Filtering Theory, 2nd edn. Wiley, New York (2013)
7. Mudrov, V.I., Kushko, V.L.: Methods of Measurement Processing. Radio i Svyaz', Moscow (1983)
8. Chernoyarov, O.V., Min, S.S.T., Salnikova, A.V., Shakhtarin, B.I., Artemenko, A.A.: Application of the local Markov approximation method for the analysis of information processes processing algorithms with unknown discontinuous parameters. Appl. Math. Sci. **8**(90), 4469–4496 (2014)
9. Chernoyarov, O.V., Zakharov, A.V., Trifonov, A.P., Salnikova, A.V., Makarov, A.A.: The common approach to calculating the characteristics of the signal parameters joint estimates under the violation of the decision statistics regularity conditions. Eng. Lett. **28**(3), 492–503 (2020)
10. Dynkin, E.B.: Theory of Markov Processes. Dover Publications, Mineola (2006)
11. Trifonov, A.P., Korchagin, Y.E.: Joint estimation of parameters of a signal with unknown moments of appearance and disappearance. Izvestiya Vysshikh Uchebnykh Zavedenij. Radioelektronika **43**(6), 34–43 (2000)

Fast Digital Devices for Accumulating and Measuring Parameters of Deterministic and Random Signals

Oleg Chernoyarov[1,2,3](✉), Serguei Dachian[4], Alexey Glushkov[5], Vladimir Litvinenko[6], and Yuliya Litvinenko[6]

[1] National Research University "MPEI", Krasnokazarmennaya str. 14, 111250 Moscow, Russia
chernoyarovov@mpei.ru

[2] National Research Tomsk State University, Lenin Avenue 36, 634050 Tomsk, Russia

[3] Maikop State Technological University, Pervomaiskaya str. 191, 385000 Maikop, Russia

[4] UMR 8524 - Laboratoire Paul Painlevé, University of Lille, CNRS, 59000 Lille, France

[5] Zhukovsky-Gagarin Air Force Academy, Mar. Nedelin street 133, 394052 Voronezh, Russia

[6] Voronezh State Technical University, Moscow Avenue 14, 394026 Voronezh, Russia

Abstract. Fast digital signal integrators operating over a specified time interval are introduced and the options of their application in measuring equipment are considered. They produce a sliding integration of the sampling pattern of a specified size by means of either the rectangle method or Simpson's rule while only a minimum number of simple arithmetic operations for the processing of the each subsequent sample of the received signal is required. The block diagrams and characteristics of the integrators are presented, the requirements to their parameters are studied and specified, and the possible errors resulting during their operation are estimated. The influence of the effects of the signal sample quantization on the integration accuracy is studied. The operability and efficiency of the proposed devices is tested using the simulations. It is shown that the integrators can be implemented in the software mode based on a microprocessor system, as well as in hardware set constituted of the field-programmable gate arrays. As an example, digital meters of the effective value of alternating signals and the statistical characteristics of the random processes are demonstrated. #COMESYSO1120.

Keywords: Digital integration · Rectangle rule · Simpson's rule · Fast processing · Root-mean-square signal measurement · Measurement of parameters of random process

1 Introduction

The integration of a signal in the time domain is widely used for its correlation processing in radio engineering devices [1–4], in electronic control systems [5, 6], in measuring equipment [7] and during physical experiments. For example, it is important while designing the devices for measuring magnetic fields when the acceptable margin of error shall not be greater than 10^{-5} [8]. Analog integrators are widely used in electronic

R. Silhavy et al. (Eds.): CoMeSySo 2021, LNNS 232, pp. 572–583, 2021.
https://doi.org/10.1007/978-3-030-90318-3_46

engineering [9], mainly in control devices, but they are of little use while measuring the integral within a specified time interval. Digital recursive integrating circuits [4, 5] implemented in measuring the integral within a specified time interval require external synchronization (reset) and do not provide continuous measurement. The common methods based on the conversion of the signal voltage into the frequency [10] are suitable for processing the relatively low-frequency signals and they do not completely ensure the accuracy of the result produced.

Digital integrators can be designed using digital averaging filters that can be the Dirichlet window based [11] or finite impulse response (FIR) filters [12]. Their disadvantage is the complexity of implementation in the case when a large signal sampling is to be processed.

The method of direct digital integration [8] focuses on the accumulation of the signal samples and allows implementing the computing units providing the continuous integration over a specified time interval. However, in this case, to ensure high accuracy, high computational and time costs are inevitable. Below, the fast digital algorithm is introduced for real-time processing of a large signal sampling that includes a minimum number of the arithmetic operations to generate the desired integral value.

2 Signal Integration

The input signal $s(t)$ is quantized by an analog-to-digital converter (ADC) with the frequency $f_s = 1/\tau$ generating a sampling of size N. Here τ is the time interval between the adjacent samples s_i while i is a sampling sequence number.

At the current moment t_i $(i > N)$, according to the quadrangle formula (the rectangle method) [13, 14], the value of the integral over the time interval from $t_i - N\tau$ to t_i can be found as

$$g1_i = \int_{t_i - N\tau}^{t_i} s(t)\ \mathrm{d}t \approx \tau \sum_{j=0}^{N-1} s_{i-j}. \tag{1}$$

The absolute error R of this approximation satisfies the inequality

$$R \le (N\tau)^2 F/2N, \tag{2}$$

where

$$F = \max_{t\in[t_i - N\tau, t_i]} |ds(t)/dt|, \tag{3}$$

A fast computational procedure (1) can be implemented by choosing

$$N = 2^n, \tag{4}$$

where n is an integer.

The same integral can be calculated using the parabola formula (Simpson's rule) [13, 14] as follows:

$$g2_i = \int_{t_i-(N-1)\tau}^{t_i} s(t)\ \mathrm{d}t \approx \frac{\tau}{3}\left(s_i + 4\sum_{j=1}^{2^{m+1}} s_{i-2j+1} + 2\sum_{j=1}^{2^{m+1}-1} s_{i-2j} + s_{i-N+1}\right)$$

$$= \frac{\tau}{3}\left(s_i + 4\sum_{j=1}^{2^{m+1}} s_{i-2j+1} + 2\sum_{j=1}^{2^{m+1}} s_{i-2j} - s_{i-N+1}\right), \quad (5)$$

while the calculation error does not exceed

$$R \leq (N\tau)^5 A/180[2(N-1)]^4, \quad F = \max_{t\in[t_i - N\tau, t_i]} \left|d^4 s(t)/dt^4\right| \quad (6)$$

and the sample size is such that $N = 2^{m+2} + 1 > 9$. In (5), m is an integer and is calculated as $m = \log_2(N-1) - 2$.

For the specified integration interval $N\tau$, under increasing N, the error R, in accordance with Simpson's rule (5), decreases significantly faster than in the case when the rectangle method (1) is applied. In this way, increasing integration accuracy is provided, but, at the same time, the number of computational operations involved increases also.

In the field of digital signal processing, the sums in (1) and (4) are the output signals of the averaging FIR filters [4]. To ensure a small integration error R, it is necessary to use the large sample sizes: $N >> 100$. However, direct calculation of the sums in (1) or in (4) require significant costs. Thus, it is urgent to produce such fast computational averaging algorithms that would require the minimum number of addition operations. It would allow simplifying the software or hardware implementation of the integrator.

3 Fast Integrator Designed Based on the Rectangle Method

To determine the integral (1) by the current sampling of N discrete signal samples s_i, the value z_i is calculated as follows

$$z_i = \sum_{j=0}^{N-1} s_{i-j}.$$

Then, referring to (1), one gets

$$g1_i = \tau z_i. \quad (7)$$

The block diagram of the integrator designed in view of the rectangle method is shown in Fig. 1. The samples of the input signal s_i are generated by the ADC under the control of a clock generator (CG), and then they are fed to the first unit for the accumulation of samples UAS_1. This unit includes multibit one cell shifter MR_1 in which the previous sample s_{i-1} has been saved before and the summator SUM_1. To the inputs of the summator SUM_1, the samples s_i and s_{i-1} are passed, while at its output the sum $s_i + s_{i-1}$ is produced. After that, the next value s_i is poked into MR_1. The sums of the pairs of samples are moved to UAS_2, in whose multibit two cell shifter the sums $s_{i-1} + s_{i-2}$ and $s_{i-2} + s_{i-3}$ have been previously sequentially saved. Then the last value is received at the input of the summator SUM_2, where it is added to the value $s_i + s_{i-1}$ producing the sum of the four samples $s_i + s_{i-1} + s_{i-2} + s_{i-3}$ at the UAS_2 output. Further, the data stored in MR_2 are shifted while the value $s_i + s_{i-1}$ is poked into its vacated first cell. Similarly, the sum of the eight samples is produced in UAS_3 that includes four cell shifter MR_3.

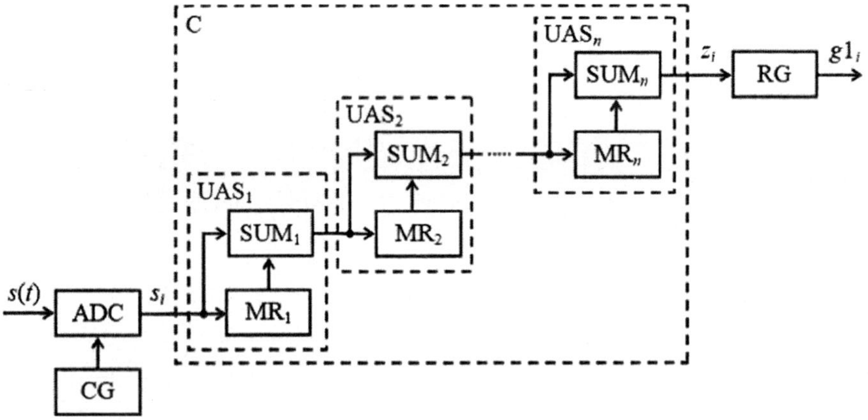

Fig. 1. The block diagram of the integrator designed based on the rectangle method.

In the last UAS$_n$, the sum $\sum_{j=0}^{N/2-1} s_{i-j}$ and the value $\sum_{j=N/2}^{N-1} s_{i-j}$ from $N/2$ cell shifter MR$_n$ are fed to the first and second inputs of the summator SUM$_n$ and thus the value (7) is generated at its output. The chain of the cascade-connected units UAS$_1$, UAS$_2$, ..., UAS$_n$ is denoted here as the calculator C. The obtained code passes to the result generator RG and then arrives to the integrator output in a\the specified form.

Thus, to calculate the sum of N signal samples, a minimum number of $n = \log_2 N$ summators and shifters of multidigit codes are required, and that can be conveniently implemented based on the main memory unit of N cells.

4 Fast Integrator Designed Based on Simpson's Rule

The block diagram of the integrator designed based on Simpson's rule is shown in Fig. 2. Here the signal samples s_i from the ADC output are poked into the multibit shifter MR5, in which the last 5 received samples $s_1 = s_i, s_2 = s_{i-1}, s_3 = s_{i-2}, s_4 = s_{i-3}, s_5 = s_{i-4}$ are stored. The pair of samples s_2 and s_4 is added in the summator SUM1 and then this sum is accumulated in m cascade-connected units UAS$_{11}$, UAS$_{12}$,..., UAS$_{1m}$ (calculator C$_1$) in the same way as it is performed by the calculator C in the circuit shown in Fig. 1. As a result, the value

$$z1_i = \sum_{j=1}^{2^{m+1}} s_{i-2j+1}$$

is produced at the output of the calculator C$_1$. Similarly, the pair of samples s_1 and s_3 is added in the summator SUM2 and then this sum is accumulated in m cascade-connected units UAS$_{21}$, UAS$_{22}$,..., UAS$_{2m}$ (calculator C$_2$) generating the value

$$z2_i = \sum_{j=1}^{2^{m+1}} s_{i-2j}$$

at the output of the output of the calculator C$_2$.

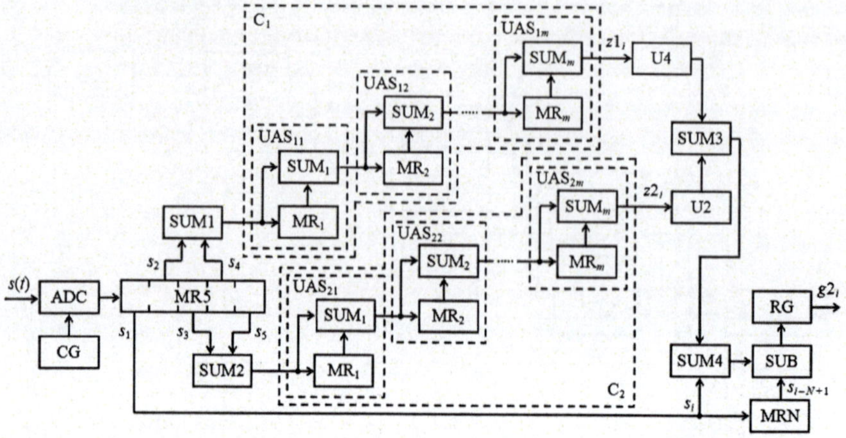

Fig. 2. The block diagram of the integrator designed based on Simpson's rule.

The values $z1_i$ are multiplied by 4 in the unit U4 (by shifting their binary code by 2 bits), while the values $z2_i$ are multiplied by 2 in the unit U2 (by shifting their binary code by 1 bit), respectively. Then they are added in the summator SUM3 and thereafter the current sample $s_1 = s_i$ is added to this last obtained sum in the summator SUM4 while the last sample s_{i-N+1} stored in multibit N cell shifter MPN is removed in the subtractor SUB. Thus, the SUB output signal is

$$z_i = s_i + 4\sum\nolimits_{j=1}^{2m+1} s_{i-2j+1} + 2\sum\nolimits_{j=1}^{2m+1} s_{i-2j} - s_{i-N+1}$$

It is passed to the result generator RG which produces the value $g2_i = \tau z_i/3$ at the output of the integrator.

5 Integrator Simulation

The study and testing of the integrator operation can be performed by the method of statistical simulation using the test signal, for which the exact value of the integral within the time interval $T_1 = N\tau$ from $t - T_1$ to t is known. As a test signal, the harmonic signal of the form $s(t) = \cos(2\pi ft)$ is chosen with the frequency f (period T) and the unit amplitude. For such the signal, one gets

$$U(t) = \int\limits_{t-T_1}^{t} s(x)dx = \int\limits_{t-T_1}^{t} \cos(2\pi fx)dx = \frac{1}{\pi f}\sin(\pi fT_1)\cos\left[2\pi f\left(t - \frac{T_1}{2}\right)\right]. \quad (8)$$

In this case, the values F (3) and A (7) are not greater than 1 and that allows using the error estimates R from (2) or (6). It should be noted that the oscillating behavior of the harmonic function is "unfavorable" for the integrator from the point of view of the accuracy of the result.

Now it is time to consider the integrator designed based on the rectangle method (Fig. 1) without the signal quantization by the level. In Figs. 3, by solid lines, there are shown the dependences of the response $g1_i$ (7) of the integrator (Fig. 3a) and its error margin (Fig. 3b)

$$\Delta_i = g1_i - U(i\tau + \tau/2) \tag{9}$$

upon the normalized sample number i/N, while $N = 256$ samples, $f = 10$ kHz, the length of the integration interval $T_1 = 1.5T$, the sampling interval $\tau = T_1/N$, and $N_1 = 1/f\tau = 171$ samples are generated within the signal period. The dotted line in Fig. 3a describes the change in the exact value of the integral $U(i\tau + \tau/2)$ calculated according to (8). As it can be seen, at the start, during the N samples, the multibit shifters are filled (Fig. 3a) and then the integrator proceeds with the operating mode in which the values $g1_i$ practically coincide with the value $U(i\tau + \tau/2)$. It follows from the graph presented in Fig. 3b that the maximum value of the error (9) is equal to $\Delta_i = 1.8 \cdot 10^{-9}$ while the maximum relative error is

$$\delta = \max_i \Delta_i / \max_i g_i = 5.7 \cdot 10^{-5}. \tag{10}$$

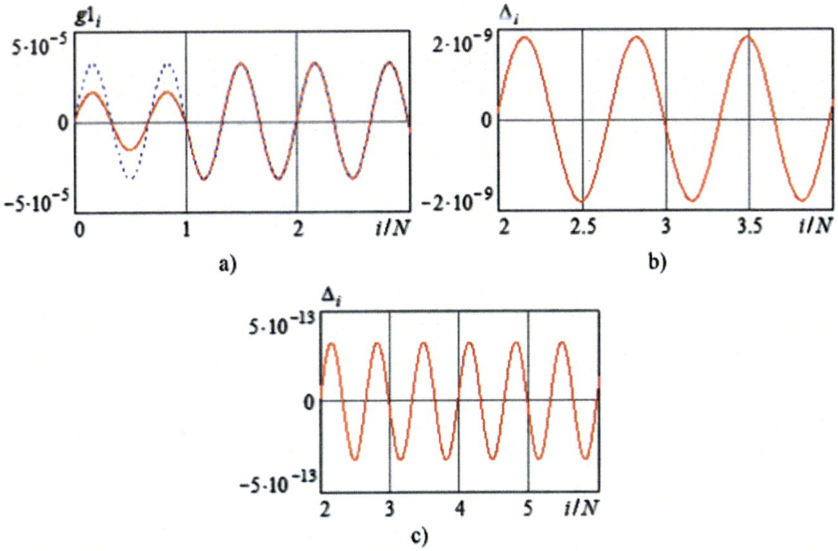

Fig. 3. The harmonic signal integration results: a) exact and calculated (based on the rectangle rule) integral value; b) error in calculating the integral based on the rectangle method; c) error in calculating the integral based on Simpson's rule.

The integrator designed based on Simpson's rule (Fig. 2) provides the same results, but with a smaller error [15] shown in Fig. 3c, while the value R (6) is practically zero.

It is obvious that the numerical integration error depends on the signal waveform, the values N and τ, as well as on the location of the integration interval at the time line. Under certain conditions, it can be zero as it can be seen from Fig. 3b or Fig. 3c. Under a small error (9), the deviation of the calculated value of the integral from the exact one is substantially influenced by the choice of the time points $i\tau + \tau/2$ for which the exact value of the integral is calculated using the analytical expression.

6 Relative Integration Error

Now one focuses on the relative error δ (10) of a digital integration of a harmonic signal. In Fig. 4a, one presents the dependences of the relative error upon the ratio of the signal period T and the length of the integration interval T_1, in the case when the sample size is $N = 33$ or $N = 256$. The dashed lines indicated by the number 1 correspond to the application of the rectangle method for the numerical integration, while the solid lines marked by the number 2 – to the application of Simpson's rule. Here the influence of the ADC-induced quantization effect is not taken into account. As it can be seen, the numerical integration error increases with the integration interval. Simpson's rule provides a significantly lower integration error, but its use leads to a more complex integrator structure.

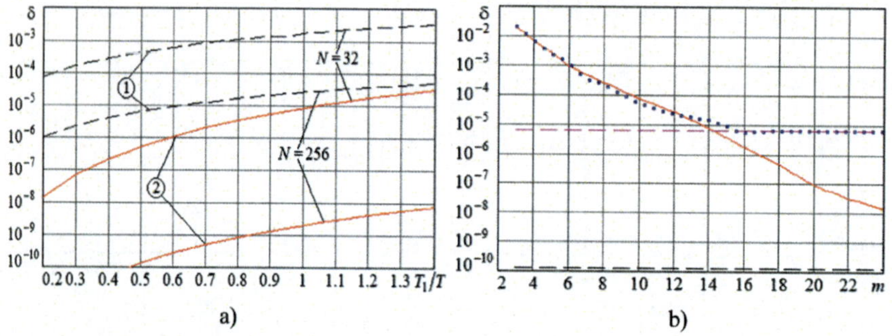

Fig. 4. The numerical integration error: a) excluding quantization effects; b) taking into account the quantization of signal samples.

In Fig. 4b, the results of simulating the integrator operation are demonstrated, taking into account the signal quantization by m-bit ADC, if $N = 256$ and $T_1 = 0.5T$. The dotted line corresponds to the application of the rectangle method, while the solid line – to the application of Simpson's rule. And here the dashed lines 1 and 2 describe the error δ values for the rectangle method and Simpson's rule, respectively, excluding the influence of the ADC (that is, these values are obtained under the high-precision calculations of the samples coded in floating-point extended format). It is obvious that they correspond to the values of δ presented in Fig. 4a and calculated for $N = 256$ and $T_1/T = 0.5$.

It follows from Fig. 4b that, under $m \leq 14$, the numerical integration error is determined by the ADC width and it does not depend upon the integration rule. To provide high integration accuracy, high-bit ADCs should be used. The shape of curves presented in Fig. 4b and recommendations for choosing the ADC width depend upon the shape of the processed signal.

When processing a signal over several its periods, the numerical integration error increases. In Fig. 5, the dependences of the relative error δ (10) upon the sample size

N are shown, while the integrators designed based on the rectangle method (curves 1) and Simpson's rule (curves 2) are used and $T_1 = 10.5T$. The solid lines correspond to the processing of the samples under the absence of the influence of the ADC-induced quantization effects (in the format of floating point numbers) and the dashed lines – to the processing of the quantized samples while the ADC width is equal to m.

As it can be seen from a comparison of the curves in Fig. 4a and Fig. 5, under $N = 256$, the numerical integration error increases with the number of the oscillating function integration periods. Under precise calculations, increasing the number of samples N while the value of T_1 is fixed provides an exponentially decreasing error (the solid lines in Fig. 5). In addition, the use of Simpson's rule reduces the error by several orders of magnitude in comparison with the results of a rectangle method application.

The use of m-bit ADC leads to the increasing integration error (that follows from the comparison of the solid curves and the dashed curves plotted under $m = 14$ and $m = 24$ in Fig. 5), while increasing the sample size in the range of $N > 2000 \div 4000$ samples does not lead to its decrease, and both integrators provide the same results approximately. The efficient way to decrease this error is to increase the ADC width up to $m = 24$ (it is possible, for example, with an analog-to-digital convertor AD7134). In this case, a relative error of the order of 10^{-8} (in fact, $10^{-6} \div 10^{-7}$) can be provided, since in (10) the maximum, but not the roof-mean-square deviation from the exact value of the integral is used. Such an error is feasible, for example, while measuring magnetic fields [8]. For the high-precision measurements, it is advisable to use an integrator designed based on Simpson's rule.

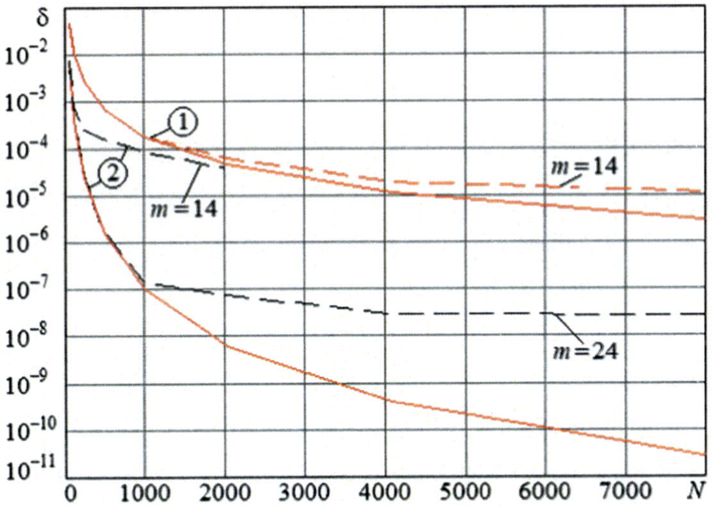

Fig. 5. The numerical integration error when processing several periods of a harmonic signal.

In the tasks of digital signal processing [1–6], very high integration accuracy is not required, and when $m = 8 \div 12$ (that is the width of ADC), it is acceptable. A digital integrator designed based on the rectangle method can be effectively used in the implementation of various measuring devices. Below, it is demonstrated by the examples of implementing the digital meters of both the root-mean-square signal with an arbitrary shape and the unknown period and the statistical characteristics of the random signals.

7 Digital Root-Mean-Square Signal Meter

The root-mean-square value of an alternating waveform $s(t)$ is determined as

$$S_{RMS} = \sqrt{\frac{1}{T_0}\int_{t_0}^{t_0+T_0} s^2(t)\,dt}, \quad (11)$$

where T_0 is the waveform period, t_0 is an arbitrary time point when the integration starts, and on the latter the value of the integral (11) does not depend. Under an unknown waveform period, the estimate of S_{RMS} by an arbitrary interval of integration T_{INT} can be represented in the form

$$\tilde{S}_{RMS} = \sqrt{\frac{1}{T_{INT}}\int_{t_0}^{t_0+T_{INT}} s^2(t)\,dt} \approx \sqrt{\frac{\tau}{T_{INT}}\sum_{k=0}^{N-1} s_{i-k}^2} = \sqrt{\frac{1}{N}\sum_{k=0}^{N-1} s_{i-k}^2}, \quad (12)$$

where N is the amount of samples in the signal sampling pattern and i is the number of the last received sample.

Technically, the values (11) and (12) coincide, if the integration interval T_{INT} is a multiple of the waveform period T_0. However, for many signals that are frequently encountered in practical applications, the estimate (12) converges to the real value (11) under N increasing. As it is specified in [16], the inequality

$$N >> 1000 \quad (13)$$

must be satisfied to provide an acceptable error in estimating the root-mean-square waveform according to (12).

The block diagram of the meter is shown in Fig. 6. Here the samples of the input signal s_i are squared by the multiplier M and fed to the calculator C (Fig. 1) that produces the value

$$z_i = \sum_{j=0}^{N-1} s_{i-j}^2.$$

The sum z_i arrives at the non-linear converter (NC) that generates the value $\tilde{S}_{RMS} = \sqrt{z_i/N}$ at its output. The easiest way for calculating $\tilde{S}_{RMS}$ by hardware is to use the storage device: at the address input of the storage device the z_i binary code is fed while the $\tilde{S}_{RMS}$ binary code has been pre-recorded in its corresponding memory cell.

The meter presented in Fig. 6 allows measuring the root-mean-square value of both an alternating waveform and a random process with an error less than 1% in a wide frequency range and in fulfilling (13). It should be noted that, in this case, the use of an integrator designed based on Simpson's rule is impractical.

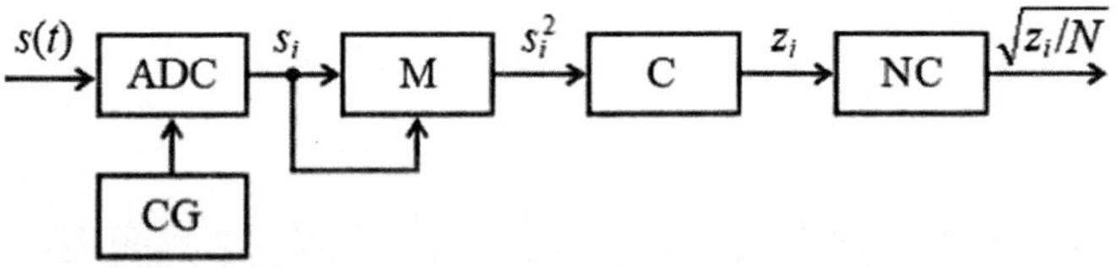

Fig. 6. The block diagram of the root-mean-square signal meter.

8 Digital Meter of Statistical Characteristics of Random Signals

A digital integrator designed based on the rectangle method can be used to implement a digital meter of the statistical characteristics of the random signal $x(t)$ including the mean value (the first initial moment) X_{mean}, the mean square (the second initial moment) M_2, the dispersion (the second central moment) D, etc. According to [17], the estimates $\tilde{X}_{mean}$, $\tilde{M}_2$, $\tilde{D}$ of the values X_{mean}, M_2, D by the sampling $\{x_i\}$ of the size of N are determined as follows

$$\tilde{X}\frac{1}{N}\sum\nolimits_{i=1}^{N} x_{i_{mean}}, \qquad \tilde{M}_2 = \frac{1}{N}\sum\nolimits_{i=1}^{N} x_i^2, \qquad \tilde{D} = \tilde{M}_2 - \tilde{X}_{mean}^{2\,\widetilde{\frac{1}{N}\sum_{i=1}^{N} x_{i\,mean}^{2}}^{2}}. \tag{14}$$

The block diagram of the meter of the parameters (14) is presented in Fig. 7. Here the following signals are generated: at the output of the multiplier M, one gets the square of the input sample; at the output of the chain including the calculator C_1 and the divider DIV by N, the value of $\tilde{X}_{mean}$ is produced; at the output of the chain including the calculator C_2 and the divider DIV by N, the value of $\tilde{M}_2$ is introduced; and at the output of the subtractor SUB, one gets the value of $\tilde{D}$. It can be observed that when it is binary coded, the DIV unit is actually a n-bit right shift register, where $n = \log_2 N$.

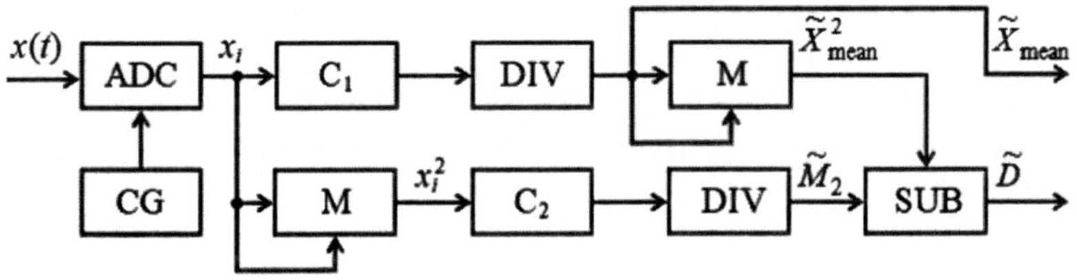

Fig. 7. The block diagram of the meter of the first two moments of the random process.

As it is shown in [16], in the presented meter the resulting measurement error of less than 1% can be provided, if the sample size N is chosen according to (13). In a similar way, the meters of the initial and central moments of high orders can be implemented.

9 Conclusion

Digital signal integrators designed based on the rectangle method and Simpson's rule are considered. The characteristics of the introduced integrators are determined. The numerical integration error is estimated and the influence on it of the signal and integrator parameters is studied. It is shown that the integrator using Simpson's rule provides the error that is several orders of magnitude lower than the one based on the rectangle method, but its structure is more complex. The best numerical integration results are obtained when integrating "smooth" signals, but when the oscillating signals are processed, then the numerical integration error increases significantly. In this case, to reduce it to the values of $10^{-5} \div 10^{-8}$, it is necessary to increase the signal sample size within a specified time interval. The ADC width has significant influence on the integration accuracy. Application of the eight-bit ADC makes it possible to provide the relative integration error $10^{-3} \div 10^{-4}$. This accuracy is feasible for the commonly used measuring instruments and thus, in this case, it is advisable to use the rectangle method. With 24-bit ADC, the error is reduced by 3–4 orders of magnitude, which is necessary for the implementation of the specialized measuring equipment, but an integrator designed based on Simpson's rule should then be used.

Based on the considered integrators, various devices for measuring signal parameters and statistical characteristics of random processes can be implemented. And finally, it is demonstrated that in the case of really large sample size (that is much more than a thousand) high measurement accuracy can be ensured.

Acknowledgements. This work was financially supported by the Ministry of Education and Science of the Russian Federation (research project No. FSWF-2020-0022) as well as RFBR and CNRS (research project number 20-51-15001).

References

1. Sklar, B.: Digital Communications: Fundamentals and Applications. Prentice Hall, Hoboken (2001)
2. Proakis, J., Salehi, M.: Digital Communications. McGraw-Hill, New York (2007)
3. Haykin, S., Moher, M.: Introduction to Analog and Digital Communications, 2nd edn. Wiley, New York (2007)
4. Lyons, R.G.: Understanding Digital Signal Processing. 3rd edn. Pearson India, Chennai (2011)
5. Miroshnik, I.V.: Automatic Control Theory. Linear systems (in Russian). Piter, St. Petersburg (2005)
6. Bolton, W.: Mechatronics: Electronic Control Systems in Mechanical and Electrical Engineering, 6th edn. Pearson, New York (2015)
7. Mazda, F.F.: Electronic Instruments and Measurement Techniques. Cambridge University Press, Cambridge (1987)
8. Batrakov, A.M., Il'yin, I.V., Pavlenko, A.V.: Precision digital signal integrators with accurate synchronization. Optoelectron. Instrum. Data Process. **51**(1), 51–57 (2015)
9. Peyton, A., Walsh, V.: Analog Electronics with Op-amps: A Source Book of Practical Circuits, 1st edn. Cambridge University Press, Cambridge (1993)
10. Precision Digital Integrator PDI5025. https://www.metrolab.com/wp-content/uploads/2015/07/PDI5025_manual_v22_rev20.pdf. Accessed 10 July 2021
11. Aggoun, L.: Measure Theory and Filtering: Introduction with Applications, 1st edn. Cambridge University Press, Cambridge (2004)
12. Van de Vegte, J.: Fundamentals of Digital Signal Processing. Pearson, New York (2001)

13. Korn, G.A., Korn, T.M.: Mathematical Handbook for Scientists and Engineers: Definitions, Theorems, and Formulas for Reference and Review. Dover Publications, New York (2000)
14. Burden, R.L., Faires, J.D.: Numerical Analysis, 9th edn. Cengage Learning, Boston (2010)
15. McCracken, D.D., Dorn, W.S.: Numerical Methods and Fortran Programming with Applications in Engineering and Science, 2nd edn. Wiley, New York (1965)
16. Chernoyarov, O.V., Glushkov, A.N., Lintvinenko, V.P., Matveev, B.V., Faulgaber, A.N.: Digital root-mean-square signal meter. Int. J. Eng. Trans. B **33**(11), 2201–2208 (2020)
17. DeCoursey, W.J.: Statistics and Probability for Engineering Applications (with Microsoft Excel). Newnes, Boston (2003)

Check for updates

A Method of Improving the Reliability of the Nodes Containing Ledger Replicas

Anna B. Klimenko[1(✉)] and Eduard V. Melnik[2]

[1] Scientific Research Institute of Multiprocessor Computer Systems of Southern Federal University, 2, Chekhov Street, 347922 Taganrog, Russian Federation

[2] Federal Research Center, The Southern Scientific Center of the Russian Academy of Sciences, 41, Chekhov Street, 344006 Rostov-on-Don, Russian Federation

Abstract. The fog- and edge-robotics emerged on the intersection of the fog-computing, MEC and robotics and enhance the cloud robotics to the edge of the network, improving the latency, QoS and network load. The distributed ledgers as data storages are used to provide almost the same: reducing the distance between the end-point devices and the data, the system reaction time is improved and the central storage "bottleneck" is eliminated. Yet, to the best of our knowledge, very few studies are devoted to the question of the reliability of those nodes, which contain the ledger replicas. Considering the fact that the data stores are redundant, the reliability degrading affects the latency and QoS, as well as other applications of the nodes are failed.

In this paper an issue of the reliability of the devices with the distributed ledger replicas is in question. We propose a new method of the reliability improving. The key idea is that the decrease of the distance between the data sources and the data processing nodes (i.e., ledger replicas) improves the latency system features. The usage of this emerging extra-time allows to decrease the computational workload of the nodes and so to improve their reliability function.

Considering the problem in this angle, some models have been developed and some selected simulation results are presented. The results are quite promising: the reliability improvement is achievable by means of the usage of additional time, which emerges because of the data transfer time decrease. Also, due to the extra-time usage the low performance nodes demonstrate the reliability function comparable with the one of the higher performance devices.

Keywords: Fog robotics · Edge robotics · Distributed ledger · Reliability · Latency

1 Introduction

Nowadays a new research field has emerged on the basis of the robotics, fog- and edge-computing. Considering that fact that this research area is situated in the intersection of the mentioned fields, fog- and edge-robotics attract the attention of multiple researches

This research is supported by the RFBR (projects19-07-00907 and 18-29-22086).

R. Silhavy et al. (Eds.): CoMeSySo 2021, LNNS 232, pp. 584–592, 2021.
https://doi.org/10.1007/978-3-030-90318-3_47

all over the world. To the best of our knowledge, the following studies consider the fog- and edge-robotics in the various angles [1–6]. Indeed, the concept of the Internet of Robotic Things requires those benefits, which the fog- and edge- computing can deliver.

The application of the distributed ledger technologies to the fog- and edge- robotics are considered also, in the most cases, from the following angles:

- As a mechanism of data sharing between the groups of robots [3, 7];
- As a delivery of data storage near the edge with more predictable time of data exchanges [8];
- As a mechanism to deliver the service foot-print in case of mobile robots and their migration relatively to the fog-nodes [8].

Yet, very few publications are devoted to the question of the reliability of the nodes/devices with the ledger replicas, though it is quite relative due to the following reasons:

- high failure rates of the nodes with data storages obviously affects the predictability of the response time of the hole system;
- the degrading of the node reliability affects its performance if the node process some applications excluding the ledger replica.

The problem to consider is how the DLT usage affects the nodes reliability. It is topical for the fog- and edge-computing-based systems, including fog- and edge-robotics.

We presuppose that the overall reliability of the system depends on the every node reliability. So, the degrading of one node's reliability is not desirable. Also we presuppose that the workload shift to the edge is quite useful due to the decrease of transit workload, which is considered in works [9, 10], and of system latency. Yet, the workload shift to the edge-devices with relatively low performance leads to their overloading and, as a result, to the reliability function deterioration.

In this paper a novel method of the nodes, containing the ledger copy, reliability is presented. The key idea of the method is as follows: the node reliability can degrade, when we try to place the ledger replica to the node of a low performance (in the fog or on the edge of the network), yet, utilizing the extra-time, which emerges due to the data transfer time reducing, the reliability of the node can be improved.

The following sections of this paper contain:

- a brief analytical overview of the reliability issue of the DLT and the fog- and edge-based systems;
- a method to improve the reliability of the nodes containing the ledger replicas by means of the extra-time usage;
- some selected simulation results;
- conclusions.

2 Reliability Issues of the Edge- and Fog- Processing

As is known, fog-computing concept was proposed in 2012 by Cisco to solve some latency and network loading issues under conditions of enormous growth of data exchange between sensors and data processing applications in the field of the Internet of Things. It has been proposed to set some data pre-processing procedures near sensors. Within the scope of the fog-computing data preprocessing functions are supposed to be placed to the industrial controllers, switches, routers, embedded servers, etc. The basic mechanism presupposes that the fog node makes a decision where to process the data – either on the fog-node or in the cloud. Later the idea was transformed to the edge-computing, where the data processing is performed on the edge devices, without the decision about the place of data processing.

Fog- and edge- robotics emerge later, in particular, as a development of the "Smart space" concept, in which edge-computing and robotics complement each other by computational and manipulation capabilities [11].

A huge amount of research was conducted in the field of fog-computing, including the following areas [12]:

- end-user application provisioning;
- resource management;
- communication issues;
- cloud and fog federation;
- algorithms for computing;
- algorithms for content storage and distribution;
- energy consumption.

Another field of published works relates to the fog- and edge- robotics, including the distributed ledger applications. These works are mostly devoted to the:

- distributed ledger federation for edge robotics [8];
- data sharing and integration in the fog-robotics systems [13–15];

Yet, to the best of our knowledge very little attention has been paid to the nodes reliability issue, which is interconnected with the nodes workload. The problem is that when the workload is shifted to the edge or to the fog, latency of the system decreases. Yet, placing the computations on the devices with relatively low computational performance increases nodes relative workload. So the reliability function degrades, and the question rises, where to set the data preprocessing to minimize the reliability degrading. The works which, to the best of our knowledge, pays attention to the nodes reliability problem, are the studies [16, 17], yet, the question considered is devoted to the interplay between the latency and the reliability of the data processing, to the questions of data transfers, etc., and does not consider the problem of nodes reliability generated by the workload shift.

The same can be said about the application of the distributed ledger, which receives, sends and synchronizes data from the end users. As is known, there is a wide range of ledger types, which use different data storage methods and different consensus ones. For example, there is a lot of consensus methods for blockchain-based ledgers: Proof of Work, Proof of Stake, Proof of Authority (the competitive consensus), PBFT (cooperative). Some consensus methods have been developed for ledgers with other storage types, e.g., Nano (Blocklattice), SwirldsHashgraph consensus algorithm (HederaHashgraph) and others.

Comprehensive classifications of the ledger types are presented in the following studies [18, 19].

Yet, the general feature of the ledger functioning can be determined as follows: ledger usage presupposes that users (or devices) can read from ledger, write to ledger, and the ledger is synchronized by the data transmission through the network. The generic functional structure of the ledger is presented in Fig. 1.

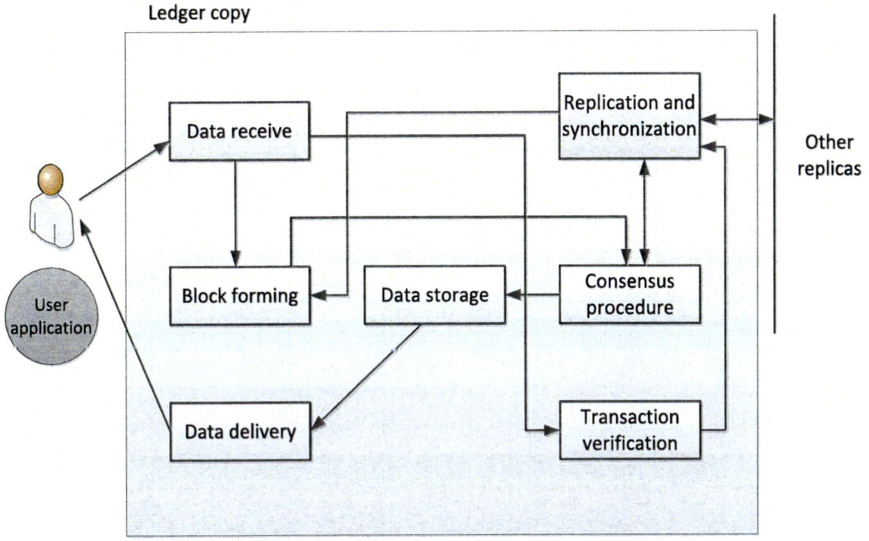

Fig. 1. Functional ledger structure.

So, considering the ledger as a computational task with particular data transfers, and taking into account the effect of the load shift, the question emerges: where to place the ledger replica and how to improve the node reliability, when the ledger replica is on the device with a low performance.

3 A Method of Improving the Reliability of the Nodes with Ledger Replicas

Consider the reliability function of the device such as the following equation:

$$P(t) = e^{-\lambda_0 \cdot 2^{dL/10} t}, \tag{1}$$

where λ_0 – a failure rate of the unloaded node; L – a node workload; d – a ratio between the workload and the temperature increase of the node.

Some previous studies of ours [9, 10] contain the quantitative analysis of the transit workload effect on the optimal application placement in the fog, including the distributed ledger placement.

Briefly, the results are as follows: there are particular cases, when the transit load is so considerable, that it is expedient to place the data processing on the edge. If the transit load is not of a large volume, there is an expedience to place it to the nodes with a high performance.

Yet, our previous works do not present the models which take into account the profit which can be got in case of latency decrease under conditions of computational workload increase. The process is shown in the causal diagram (Fig. 2), where blue arrows mean the negative relation, while the red arrows mean the positive one.

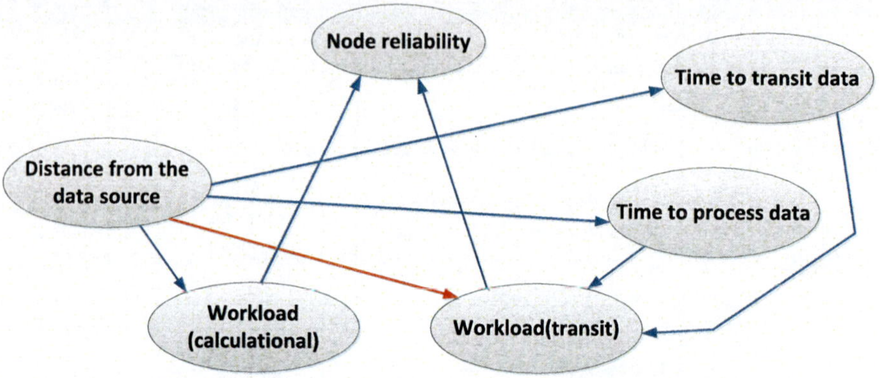

Fig. 2. The causal scheme of the parameters interrelations.

Firstly, with the reduce of the distance between the edge of the network (sensors, robots) and the data processing application, workload grows on those nodes, which process the data (transit workload reduces). In particular cases the increase of the computational workload degrades the reliability function, but it must be mentioned that in a wide range of cases there is a time constraint for the tasks to be solved to deliver the appropriate level of quality. So, we have the declared time period to perform some computations, and it is not necessary to shorten the time of the tasks solution. The dependency between the data processing placement and the time is shown in the diagram with the blue arrows (negative relations), which means that with the distance increase less and less time is available to perform some computations. This dependency plays an important role in the node reliability improvement: though we increase the computational workload, the additional time got by the decrease of distance allows to set that node workload such as to meet the time constraint and to minimize the computational workload. A possible effect of the workload distribution through the additional time is presented in Fig. 3.

Consider x as a share of operation time T_{op}, which is taken to solve the computational task. Then, 1-x is a share of T_{op}, when the node performs the data transfer.

The reliability function of the node is described by the following equation:

$$P(t) = e^{\int_{o}^{t} -\lambda(t)dt};$$

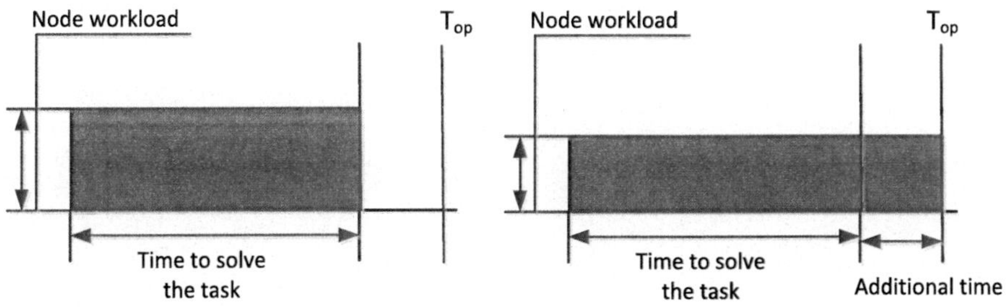

Fig. 3. The extra-time emerging due to the decrease of distance between the end devices and the ledger node.

While $\lambda(t)$ is a piecewise function here, the integral can be calculated as the sum of squares under the $\lambda(t)$ function:

$$P(t) = e^{\int_0^t -\lambda(t)dt} = e^{\int_0^{xTop} -\lambda_1 dt + \int_{xTop}^{Top} -\lambda_2 dt} = e^{-\lambda_1 xTop} \cdot e^{-\lambda_2 (1-x)Top}.$$

Hence,

$$P\left(T_{op}\right) = e^{-\lambda_1 xTop} \cdot e^{-\lambda_2 (1-x)Top} = e^{-\lambda_0 \cdot 2^{\frac{dW}{xTop*p*10}} * xTop} * e^{-\lambda_0 (1-x)Top}.$$

So, we presuppose that the node reliability can be improved by the replica placement near the end-point users, what consequently adds the additional time to solve some computational tasks of the ledger replica. In case of the low performance of the node, its reliability can be improved as well due to the workload distribution through the extra-time.

4 Simulation Results

Firstly, we checked one of the main ideas of this paper, which is illustrated in Fig. 4 and supposes that it is better for the device to solve the task through the longer time period with the smaller workload than to solve the same task for the shorter time period but with the higher workload. Under the conditions of the same computational complexity and node performance we changed the parameter of the share x, leaving larger and larger time for computational task. The following graph has been conducted, as is shown in Fig. 4.

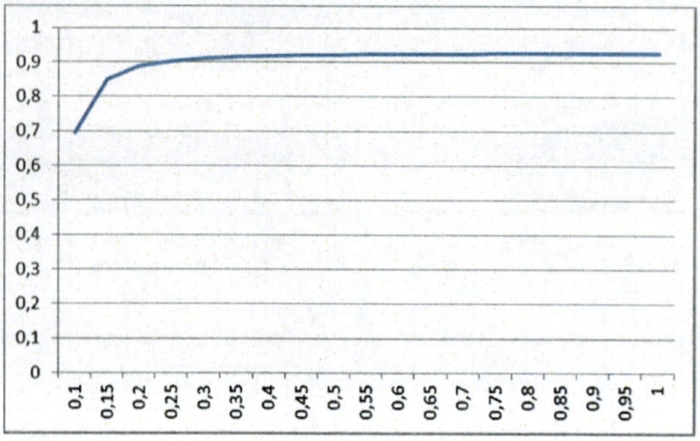

Fig. 4. The reliability function values increase with the extra-time share growth.

Figure 4 illustrates that the usage of additional time for computational task solving with the spreading the computational workload improves the reliability function of the node.

In the Fig. 5 the effect of the additional time usage is demonstrated for the nodes with different performance. It is seen that the workload spreading is highly relevant for the nodes with low performance.

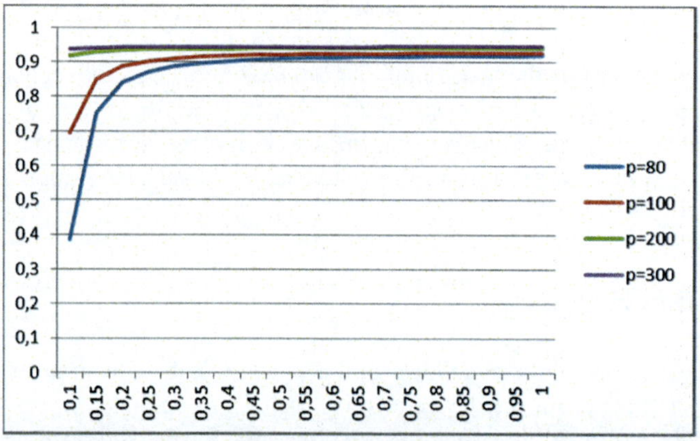

Fig. 5. The reliability functions for various performances.

Placing the ledger on the weak node, but at a smaller distance, one can achieve a reliability function comparable to the one obtained by means of a devices featuring higher performance, but located at a longer distance from the edge (Fig. 6).

The figure demonstrates an example of reliability functions, when at $x = 0{,}6\ T_{op}$ the device featuring $p = 100$ achieves the same reliability as that of a device featuring $p = 200$, located further from the network edge at $x = 0{,}2\ T_{op}$.

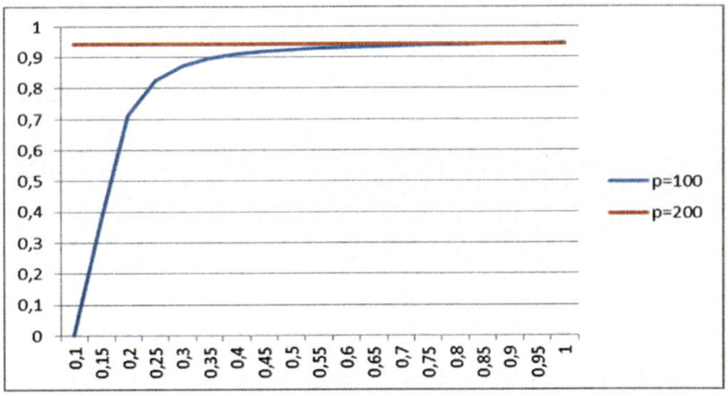

Fig. 6. The achievement of the comparable reliability function values for the nodes with performances $p = 100$ m.u. and $p = 200$ m.u. by means of extra-time usage

5 Conclusion

In this paper a question of ledger copy placement and its effect on the nodes reliability is considered. When an application (a ledger copy) is placed on the edge of the network, the transit workload decreases, but the growth of the computational workload on the node with low performance degrades its reliability dramatically. Such situation has been considered in our previous works, yet, in this paper the effect of workload shift to the edge is considered in the angle of the data transmission time decrease and the additional time emerging. This additional extra-time can be used to decrease the node workload and so to improve the reliability.

The main contribution of this paper is a Method of improving the reliability of the nodes containing the ledger replicas which bases on the possibility to improve the reliability by the usage of additional time for computational tasks solving. Also, we developed the causal diagram, which shows the complex dependencies between the model parameters. Simulation results allow to make the following conclusions:

1. the reliability improvement is achievable by means of the usage of additional time, which emerges because of the data transfer time decrease;
2. due to the extra-time usage the low performance nodes demonstrate the reliability function comparable with the one of the higher performance devices.

References

1. Gudi, S., Ojha, S., Clark, J., Johnston, B., Williams, M.-A.: Fog robotics: an introduction. In: IEEE/RSJ International Conference on Intelligent Robots and Systems, pp 1–2 (2017). (Electronic Materials: http://hdl.handle.net/10453/120748)
2. Fadelli, I.: Fog Robotics: A New Approach to Achieve Efficient and Fluent Human-Robot Interaction EE World online, pp. 1–3 (2018). (Electronic Materials: https://techxplore.com/news/2018-11-fog-robotics-approach-efficient-fluent.html)

3. Tanwani, A.K., Mor, N., Kubiatowicz, J., Gonzalez, J.E. and Goldberg, K.: A fog robotics approach to deep robot learning: application to object recognition and grasp planning in surface decluttering. In: Proceedings of the 2019 International Conference on Robotics and Automation (ICRA), pp. 4559–4566 (2019)
4. Tian, N., et al.: A fog robotic system for dynamic visual servoing. In: Proceedings of the 2019 International Conference on Robotics and Automation (ICRA), pp. 1982–1988 (2019)
5. Cao, H., Chen, R., Gu, Y., Xu, H.: Cloud-assisted tracking medical mobile robot for indoor elderly. In: Proceedings of the2017 IEEE 3rd Information Technology and Mechatronics Engineering Conference (ITOEC), pp. 927–930 (2017)
6. Antevski, K., Groshev, M., Cominardi, L., Bernardos, C.J., Mourad, A.Y., Gazda, R.: Enhancing edge robotics through the use of context information (2018). In: Workshop on Experimentation and Measurements in 5G (EM-5G 2018), pp. 7–12. ACM (2018)
7. Klimenko, A.B., Ivanov, D.Y.: J. Phys.: Conf. Ser. **1661**, 012049 (2020)
8. Antevski, K., Groshev, M., Baldoni, G., Bernardos, C.J.: DLT federation for Edge robotics (2020)
9. Klimenko, A.B., Melnik, E.V.: An experimental study of the fog-computing-based systems reliability. In: Silhavy, R. (ed.) CSOC 2020. AISC, vol. 1225, pp. 438–449. Springer, Cham (2020). https://doi.org/10.1007/978-3-030-51971-1_36
10. Melnik, E., Klimenko, A.: A condition of reliability improvement of the system based on the fog-computing concept. J. Phys. Conf. Ser. **1661**, 012007 (2020)
11. Afanasyev, I., et al.: Towards the Internet of Robotic Things: Analysis, Architecture, Components and Challenges (2019)
12. Hu, P., Dhelim, S., Ning, H., Qiu, T.: Survey on fog computing: architecture, key technologies, applications and open issues. J. Netw. Comput. Appl. **98**, 27–42 (2017)
13. Klimenko, A., Ivanov, D.: Data sharing by means of multiple fog robot servers. J. Phys: Conf. Ser. **1661**, 012049 (2020). https://doi.org/10.1088/1742-6596/1661/1/012049
14. Queralta, J.P., Qingqing, L., Zou, Z., Westerlund, T.: Enhancing Autonomy with Blockchain and Multi-Access Edge Computing in Distributed Robotic Systems (2020). https://doi.org/10.1109/FMEC49853.2020.9144809
15. Uddin, M.A., Stranieri, A., Gondal, I., Balasubramanian, V.: A survey on the adoption of blockchain in IoT: challenges and solutions. Blockchain. Res. Appl. 100006 (2021). https://doi.org/10.1016/j.bcra.2021.100006. ISSN 2096-7209
16. Inaltekin, H., Gorlatova, M., Mung, C.: Virtualized control over fog: interplay between reliability and latency. IEEE Internet Things J. **5**(6), 5030–5045 (2017)
17. Yao, J., Ansari, N.: Fog resource provisioning in reliability-aware IoT networks. IEEE Internet Things J. **6**(5), 8262–8269 (2019)
18. Park, S., Kim, H.: DAG-based distributed ledger for low-latency smart grid network. Energies **12**(18), 3570 (2019)
19. Teo, E.: Consensus for Blockchain and Distributed Ledger Technologies. Blockchain and Smart Contracts: Design Thinking and Programming for FinTech, pp. 99–119 (2021)

A Hybrid Approach for Protecting Clinical and Genomic Data Using Lossless Stego-DNA Compression

Mohammed A. Farahat[1](✉), A. Abdo[1,2], and Samar K. Kassim[3]

[1] Faculty of Computers and Artificial Intelligence, Department of Information Systems, Helwan University, Helwan, Egypt
amani.abdo@aou.edu.eg
[2] Faculty of Computing, Arab Open University in Egypt, Cairo, Egypt
[3] Faculty of Medicine, Department of Biochemistry and Molecular Biology, Ain Shams University, Cairo, Egypt
Samar_kassim@med.asu.edu.eg

Abstract. Clinical data is the most valuable and sensitive asset about humans. While there are a variety of effective policies, processes, and regulations in place to secure patient information, privacy and security vulnerabilities in electronic health record (EHR) systems remain major concerns. In the domain of communication and digital security, cryptography and steganography techniques are the most commonly used to secure confidential data from unauthorized access. New domains for DNA-based cryptography and steganography are developed to enable data protection using DNA as a cover medium by utilizing its bio-molecular computational features and capacity. In this paper, many DNA-based steganography related studies have been discussed with the basic DNA-based steganography methodologies according to crucial security factors. Furthermore; an approach (Lossless Stego-DNA Compression for Securing Clinical Data), LSDC has been proposed that can improve most of the previous security vulnerabilities and achieving a greater level of security. The approach consists of three stages: DNA steganography, DNA compression, and image or video hiding. It could hide the sensitive clinical data in the DNA sequence without any modification on it. It could achieve a higher level of accuracy and performance on both directions (cipher and decipher). The experimental results showed that LSDC could successfully achieve an acceptable security and compression ratio (from 53.3 to 62%).

Keywords: Clinical data privacy · DNA compression · Invisible communication · DNA steganography · Cryptography · Data hiding · Information security

1 Introduction

Safety and privacy of individual data are crucial in the healthcare field, protecting the confidentiality, integrity, and availability of clinical records is a serious concern. Clinical

R. Silhavy et al. (Eds.): CoMeSySo 2021, LNNS 232, pp. 593–608, 2021.
https://doi.org/10.1007/978-3-030-90318-3_48

data is typically regarded as the most delicate and confidential of all personal data [1]. Policies concerning protection and privacy must be adopted and made public. Despite several privacy and security issues, the healthcare sector achieved an extreme progress in the recent years. There are already a number of privacy and security techniques in place, and the advantages of these procedures should take place when applying healthcare information systems [2].

However, the expanding usage of Electronic Health Records (EHRs) and clinical data warehouses, as well as recent privacy violations, have boosted the demand for new data security solutions and ideas. The growing amount of clinical data from numerous sources is fragmentally processed in numerous locations and systems. Because security concerns in all of these systems may result in the release of data to unauthorized persons or entities, health information must be protected from unauthorized access and tampering [3].

Cryptography and steganography are both used to secure confidential data and information. The main variance between them in which with cryptography, a third party can see and realize that the two parties are communicating and interacting in secret, and the ciphered data can be easily noted; whereas with steganography, a third party cannot see and realize that the two parties are communicating and interacting in secret [4].

Steganography is derived from the Greek words “stegos” and “grafia” which mean “cover” and “writing”. So, it’s defined as “covered writing.”

Because of the rapid growth of the internet use, it has been utilized for sharing and transferring huge amounts of data [15]. The impact of lowering the risk of data being discovered or recognized during the spread is a critical concern nowadays. Steganography technique is considered as one of the most secure techniques for sharing secret data and information between parties since it hides the information in another medium, making it impossible for an attacker to observe or detect the message delivered [16, 17].

After transforming the data to disguise its existence, the data concealing methodology should keep the original cover media’s features as much as possible [18]. The key goal of steganography is to embed secret data with the least variation ratio between the original cover-medium and the altered one that cannot be noticed with the naked eye; as a result, the appeal reduces while the security rises. As a result, the tracker is unable to reveal the confidential information.

Recently, the most effective way for securing data has been to erase all traces of its existence. The best way to do this is to use steganographic techniques to hide data within the cover media. In addition to a new cover medium, robust algorithms are required to improve the security of steganography systems. Due to their high information density and capacity, deoxyribonucleic acid (DNA) sequences are now commonly used as cover mediums [5].

DNA Steganography is protected and often preferable more than cryptography for two major reasons: cryptography is insufficiently secure when sending data over an insecure, public communication channel; and cryptography is insufficiently secure when transferring data over an unsecured, public communication channel. It is simply considered as the art of concealed writing, whereas steganography is hidden writing that concentrates on concealing the message’s existence. Aside from that, DNA is being offered for utilizing in a variety of computational applications [10, 11].

Personal genomes are now available in unprecedented quantities, providing a rich resource for data mining and integration in the pursuit of a better knowledge of human biology and the translation of research findings into healthcare approach. By revealing personal identities or private features of individuals, attacks could theoretically be used to strategically breach genetic privacy [6].

With the emergence of biological characteristics of DNA sequences, researchers have proposed new data embedding methods based on DNA for achieving extreme privacy and good protection with large capacity and low rate of change. However, all of these studies focused on ways to reduce the pace of alteration. Aside from that, the DNA sequence should be kept private [12].

The rest of this paper is ordered as follows. First, a discussion of the related researches, and the authors' discussion are presented. In Sect. 3, the proposed approach and its phases are explained. Then, the material and methodology are presented in Sect. 4. In Sect. 5, the experimental results of the proposed approach. Finally, the conclusion of this paper and an outlook to the future work are presented in Sect. 6.

2 Related Works

A lot of studies and researches have gone into exploiting DNA sequences as a cover-medium to hide data.

The authors applied crucial security criteria in [5] to compare different DNA-based steganography. They went over the fundamentals of DNA-based steganography. Most additional steganographic structures, such as insertion, replacement, and complementary pair-based algorithms, use each approach as a building block. They proposed ideas to help future researchers build and enhance DNA storage systems to improve data security by using more effective, efficient, high-capacity, and biologically preserved DNA steganography approaches.

Zicheng Wang et al. investigated how to overcome DNA steganography's protection vulnerability and produced a DNA steganography-based message concealing technique in [7]. They created many ways to encrypt a message and then breakdown the cipher text into two portions. The sender utilizes DNA steganography to convey one portion. Whether the microdot is believed or contaminated, the message will be encrypted and deconstructed repeatedly until the microdot is not assumed or corrupted. The corresponding portion will be made public if it is not believed or corrupted with the microdot. The proposed solution is unique in that it includes a decomposed message to address the issue of DNA steganography protection. Based on the proposed system, the security issue of DNA steganography may be overcome, and a message can be transmitted while remaining hidden from an opponent.

The authors of [8] concentrated on the idea of deploying a speculative single-stranded DNA (ssDNA) retrieved from an unremarkable cover image as a primary cover. On the same basis, they investigated the current algorithm's security and efficiency vulnerabilities and proposed an improved approach. Both algorithms' efficiency is tested against a variety of visual and statistical threats, and both security and capability are parameterized. The comparison shows that the anticipated enhancements will provide better overall defense. The proposed study seeks to uncover concealed information in several layers of media coverage. The DNA is owing to such the pixel characteristics of the image. As a result, it was more reliable than approaches that relied on public database reference DNAs. The algorithm was made even more impenetrable by the various 2D logistic map settings. Multiple keys were required throughout the process, from DNA construction through data embedding, and their transfer between sender and receiver required a secure key exchange system.

The authors of [9] employed LSB, RSA, DNA cryptography, and DNA steganography techniques to address existing steganographic concealment methods, which are intended to protect movies, audios, and images from being copied, falsified, or unauthorized access. Existing approaches have disadvantages in terms of key size, computational cost, speed, and input size. As a result, Hyperelliptic Curve Cryptography (HECC) based on DNA steganography is recommended, which provides a greater level of image file security while simultaneously ensuring digital media security. In comparison to traditional cryptographic algorithms, the proposed HECC-based DNA steganography takes 30% and 42% longer to encrypt and decrypt data, respectively. Finally, the suggested system is compared to existing standard steganographic techniques in terms of Mean Square Error, Peak Signal to Noise Ratio (PSNR), and embedding capacity. The simulation findings show that the suggested steganographic method has a reduced embedding power, a high MSR, and a low PSNR value when compared to other steganographic approaches.

The authors of [13] proposed using DNA steganography for highly secure data transfer. This was indicated by displaying the results of an image steganography simulation utilizing a DNA sequence. The vast storage capacity of the DNA molecule qualifies it for usage in wireless security applications. The embedding procedure is divided into three stages. To begin, the cover and concealed message pixel values are converted to their respective DNA triplet values using characters from the DNA triplet conversion table. Second, utilizing DNA triplets, the binary value conversion table is used to convert the triplet values to binary values. In the third and final point, the two binary-valued matrices corresponding to cover and concealed messages are Ex-OR. At the conclusion of the reception, the extraction process is used.

The authors of [14] proposed a cryptographic DNA approach that used dynamic principles to promote data secrecy. Although encryption is used, first-level secrecy is guaranteed by utilizing the dynamic sequence table containing the substitution cypher features. They employed an asymmetric cryptosystem that provided second-level confidentiality. Finally, they improved its security by using dynamic DNA encoding, which retains the features of a product cypher while offering third-level secrecy. Furthermore, the role of mathematical series, iterations, and the DNA base sequences of the NCBI bank database, among other things, are linked to its concealment in the development of dynamic mechanisms. The NIST statistical test suite and the attack analysis ensured that the cipher text generated was pseudorandom and significantly secure.

Animesh Hazra et al. [19] gave a brief outline of DNA cryptography techniques and real-time implementation of those approaches. First is an arbitrary one-time pad based on DNA. This encryption approach employs a set of non-repeating, randomly ordered characters. These act as the input's cipher text. It's called a one-time pad since it's only utilized once when an input cipher text is used. The second method is based on DNA chips. The introduction of DNA chip technology has been a significant advancement in the field of DNA cryptography. It's handy for managing a huge number of genomic sequences as well as storing and handling biological data. The third method is DNA steganography. Because the procedure involves biological strands, the data may become tainted as a result of the abrupt change in the environment. Fragmentation of DNA is the fourth step. This method is used to build libraries from DNA sequences. It is used to break down the DNA sequence into little chunks. Many encryption techniques employ this as the second layer of protection. The key's encryption is also enforced. The encryption methods utilized in the four cryptography approaches, namely conventional cryptography, DNA cryptography, elliptic curve cryptography, and quantum cryptography, are being interchanged, resulting in the creation of new hybrid techniques.

To increase algorithm performance, the authors in [20] presented DNA steganography as one layer and symmetric second layer DES and AES cryptography with parallel programming. DNA-based steganography and cryptography use less time to encrypt and safeguard data from an interceptor than suggested parallel AES and DNA steganography (a 20–30% reduction in time). The proposed algorithms secure knowledge in double-fold techniques. In comparison to the Binary XOR operation, the DNA XOR operation has more stable resistance due to the 8 potential combinations for DNA Steganography operation. They used DNA steganography to turn a message into binary numbers, then transformed binary numbers into DNA sequences in the second phase; this was used as an input for step 2 of DNA steganography using human DNA reference to generate cipher-based indexing as an output. As one layer for translating a message into an index-based DNA sequence, these two stages are referred to as a middle cipher text. Both the AES and 3DES algorithms use the output of algorithm 1 as input. To produce the cipher text in AES, they used the preceding DNA steganography cypher and the AES keys. They created an array of cipher text using the same input and keys as 3DES after performing an XOR operation between the original DNA sequence and a human DNA reference (DNACIPER).

Amal Khalifa developed a steganographic methodology in which DNA is employed as a hidden information carrier that may be used for secret communication in [21]. The suggested system's concealment procedure featured two major phases. The first phase used an 8×8 play-fair ciphering technique to encrypt the concealed message into a DNA sequence. On the other hand, the extraction method involved breaking the stego-sequence into bits in order to distinct the message from cover sequences. The original message was then revealed after a deciphering process. Experimental results showed that the new method performed exceptionally well in terms of both blindness besides ability. The offered strategy attained a maximum concealing capacity of 2 bits/nucleotide, which was the uppermost of all other approaches.

Because of the presence of biological characteristics and features of DNA and RNA sequences, researchers have developed novel data embedding techniques based on DNA to provide maximal safety and powerful security with large capacity and low variation rate. However, all of these studies focused on ways to reduce the rate of alteration in terms of hiding capabilities and security concerns.

3 Proposed Approach

The aim of this research is to develop an approach for encoding confidential clinical data into genomic data objects like DNA or RNA. As illustrated in the figure below, the proposed approach (Lossless Stego-DNA Compression for Securing Clinical Data), LSDC consists of three stages: DNA steganography, compression, and image or video hiding (Fig. 1). The proposed approach can be thought of as a multi-layered security solution for hiding clinical data in a genomic data object, compressing the Stego-Genomic object, and finally hiding it in a cover-medium.

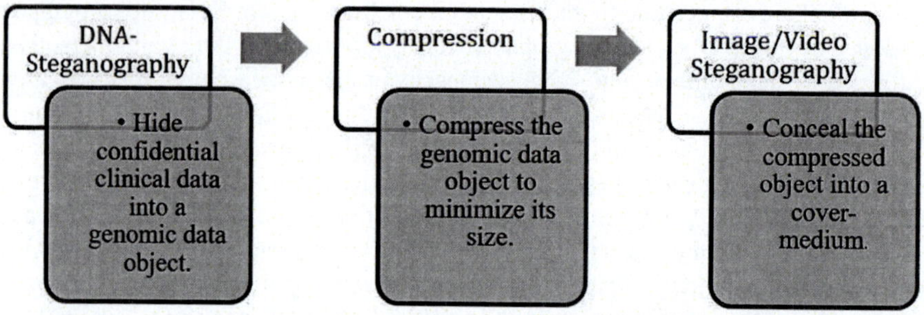

Fig. 1. LSDC approach to conceal and compress the clinical and genomic data.

4 Material and Methodology

DNA is a double-helix molecule made up of two chains that wrap around each other. The two DNA strands are made up of nucleotides, which are smaller monomeric units. Each nucleotide is made up of one of four nucleobases that contain nitrogen. (cytosine [C], guanine [G], adenine [A] or thymine [T]).

Sample of research: DNA dataset from real human tissues which are provided by H3ABioNet (Pan African Bioinformatics Network for the Human Heredity and Health in Africa (H3Africa) consortium) [24] and The National Center of Biotechnology Information (NCBI) databases [25].

4.1 Sender Side

In Step1, creating two functions to convert the genomic data into 2 binary bits to create the four possible binary pairs [(0,0), (0,1), (1,1), (1,0)] and vice versa as shown in (Fig. 2, Fig. 3, Fig. 4 and Fig. 5).

Converting DNA Nucleobase Letter into 2 Bits of Binaries

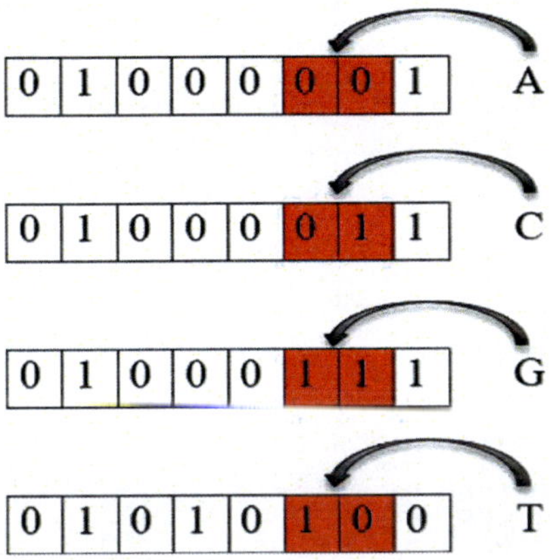

Fig. 2. Converting DNA Sequence into 2 bits of binaries

```
algorithm ConvertDNATo2Bits(string)
    Pre: DNA Sequence Letter (C, G, A or T)
    Post: Binary File (2bits of Zeros and Ones)
OSeq → input: DNA Sequence file
BinSeq → output: DNA Sequence in 2bits of binaries
 while OSeq.Length > 0
        for 1 ≥ i ≥ OSeq.Length
        OSeq= OSeq[i].ToBinary()
            string BinSeq = OSeq[i].Substring(5,2)
      end for
 end while
return BinSeq
end ConvertDNATo2Bits
```

Fig. 3. Pseudocode for Converting DNA Sequence into 2 bits of binaries

Converting 2 Bits of Binaries into a DNA Nucleobase Letter

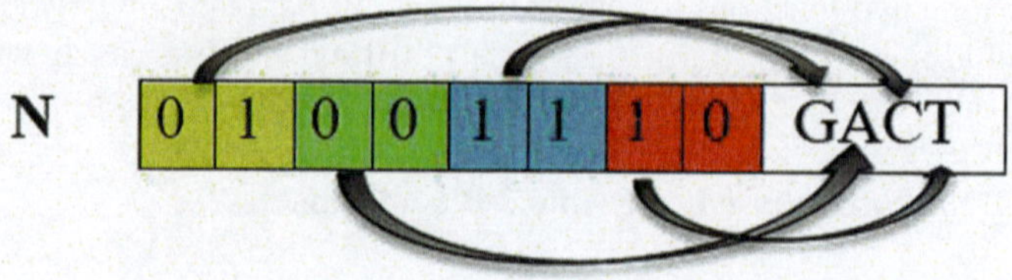

Fig. 4. Converting one character into four DNA bases

```
algorithm Convert2BitsTo1Letter(string of binaries)
    Pre: DNA Sequence in 2 bits of binaries (Streams of Zeros and Ones)
    Post: DNA Sequence letter (C, G, A or T)
BinSeq → input: DNA Sequence in 2bits of binaries
OSeq → output: DNA Sequence file
  while BinSeq.Length > 0
        for 1 ≥ i ≥ BinSeq.Length
        string c = BinSeq[i].Substring(i,2)
             switch (c)
             case "00":
             c = "A";
             case "01":
             c = "C";
             case "10":
             c = "T";
             case "11":
             c = "G";
      end switch
      OSeq,Append(c)
      end for
  end while
return OSeq
end Convert2BitsTo1Letter
```

Fig. 5. Pseudocode for converting one character into four DNA bases

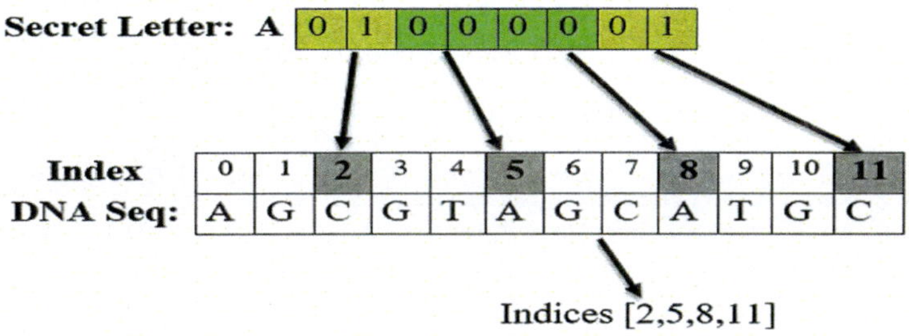

Fig. 6. Locating a string's indices on the DNA Sequence

In Step2, embedding the clinical report into the DNA sequence and locate its indices in the DNA sequence file and creating a file with these indices separated by comma "," as shown in (Fig. 6 and Fig. 7).

```
1) algorithm EmbedClincalReport(string,string)
2)     Pre: Clinical Report is shown and readable
3)     Post: Clinical Report has been embedded into DNA Sequence and Indices
       of the corresponding letters will be located.
4) OSeq → input: DNA Sequence file
5) CReport → input: Clinical report in string format file
6) Indices → output: an array of indices
7) Data → StringToBinary(CReport)
8) counter = 0
9)       while Data.Length >0
10)          for 1 ≥ i ≥ Data.Length
11)              string c = data.Substring(i, 2)
12)        c = Convert2BitsTo1Letter(c)
13)              for counter > j > OSeq.Length
14)                  string cr = OSeq. Substring(j, 1);
15)                      if (cr = c)
16)                    Indices.Append(j+",");
17)                    counter += 1;
18)                    goto Line 21;
19)                     end if
20)                  counter = counter +1
21)              end for
22)               end for
23) return Indices
24) end EmbedClincalReport
```

Fig. 7. Pseudocode for embedding clinical report

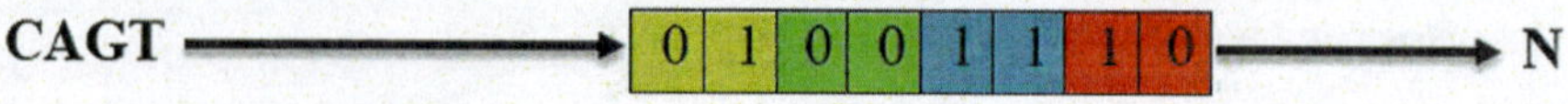

Fig. 8. Compress each four DNA letters into 1 letter

```
algorithm CompressDNASeq(string)
    Pre: DNA Seq is in its normal size and readable
    Post: DNA has been compressed and unreadable.
OSeq → input: DNA Sequence file
CompSeq → output: string file of unreadable letters
    while OSeq.Length >0
        for 1 ≥ i ≥ OSeq.Length
          BinSeq.Append = ConvertDNATo2Bits(OSeq)
        end for
    end while
CompSeq = BinaryToString(BinSeq)
return CompSeq
end CompressDNASeq
```

Fig. 9. Pseudocode for compressing DNA sequence

In Step3, Compressing the DNA Sequence into 25% of its size as shown in (Fig. 5). The aim of this step is to minimize the required storage, in addition to converting the DNA sequence into unreadable format as shown in (Fig. 8 and Fig. 9).

After applying the compression algorithm exactly while converting each 8 bits of binaries into one char, it has been noticed that there are several ASCII codes which represent nulls or white spaces. So, determining which byte was represented by this null or white space value is important. According to ASCII table and the extended table [26], the first 32 characters in the ASCII-table are unprintable control codes and are used to control peripherals such as printers, ASCII (127) represents delete button, and ASCII (129,141,143,144) represent white spaces. These Values must be determined correctly to be able to come backward again to the original text. Therefore, the BinaryToString algorithm is developed, which can be used as a function in the previous algorithm with respect of these exceptions as shown in (Fig. 10).

By now, the clinical report has been concealed into the DNA sequence file, then the sequence file has been compressed and minimized its size to 25% of its original size and converted it to unreadable format.

```
algorithm BinaryToString(string)
    Pre: Stream of zeros and ones in string format
    Post: Stream of strings.
inData → input: Stream of zeros and ones in string format
outData → output: Stream of strings
List<Byte> byteList
//Convert each 8 binary numbers to 1 bye in the byteList
    while inData.Length >0
        for 1 ≥ i ≥ inData.Length
    byteList.Add(Convert.ToByte(data.Substring(i, 8));
      end for
    end while
// Loop the byteList for the Null Chars or not
  while byteList.Count >0
    for 1 ≥ i ≥ byteList.Count
    int chr = (byteList[i].ToString()).ToInt32()
      if (0 ≤ chr < 32 or chr in[127,129,141,143,144])
  outData.Append("chr("+ byteList[i].ToString()+")");
      else
    outData.Append(byteList.GetString())
      end if
    end for
  end while
return outData
end BinaryToString
```

Fig. 10. Pseudocode for converting binaries to string

Then, a concatenation of the strings of the compressed DNA sequence with the indices array which were generated by EmbedClincalReport algorithm using any specific symbols or word will be applied.

In Step4, image or text steganography 3-3-2 Least Significant Bit technique (LSB) will be applied on the compressed file which contains the compressed DNA and the clinical report indices to conceal it into an image or video file resulting a stego-image/video file which contains the compressed DNA sequence and the clinical report in unreadable format. 3-3-2 LSB [22, 23] is a popular technique for embedding information in a cover file. The hidden message must be embedded inside the cover image/text, which necessitates changing LSB values (which is the carrier). So that the attackers cannot discover it, the secret message is deconstructed and hidden in the last bit of a cover image/text. The first three bits of the message are encoded in the last three bits of the Red component; the second three bits are encoded in the last three bits of the Green component; and the last two bits are embedded in the last two bits of the Blue component. Because blue fluctuation is more visible to the human eye than red and green variation, the researchers decided to use only two bits in the Blue component.

4.2 Receiver Side

The receiver will receive a normal image or video file which cannot be noticed as a secret communication between two parties. To get the original DNA sequence and the clinical report, the receiver should go backward by the following steps:

Step1, the compressed DNA included the clinical report will be extracted from the Stego-image/text using LSB extract technique [23].
Step2, the receiver should separate the compressed DNA and the indices array from the specific symbols or word, then decompress the compressed DNA sequence file to get the original DNA sequence back as shown in (Fig. 11).

```
algorithm DeCompressDNASeq(string)
    Pre: DNA has been compressed and unreadable
    Post: DNA Seq is in its normal size and readable
CompSeq → input: string file of unreadable letters
OSeq → output: DNA Sequence file
string BinSeq = StringToBinary(CompSeq)
      while BinSeq.Length >0
         for 1 ≥ i ≥ BinSeq.Length
    OSeq.Append = Convert2BitsTo1Letter(BinSeq)
                   end for
            end while
return OSeq
end DeCompressDNASeq
```

Fig. 11. Pseudocode for Decompressing DNA sequence

Step3, the receiver will extract the clinical report from the original DNA sequence using the indices array with indicate to the location the four 2bits which each character in the report as shown in (Fig. 12).

```
1)  algorithm ExtractCReport(string,string[])
2)      Pre: Original DNA Sequence file and Indices array
3)      Post: Clinical file has been extracted.
4)  OSeq → input: DNA Sequence file
5)  Indices[] → input: Indices array
6)  CReport → output: Clinical Report file
7)     while Indices.Length >0
8)          for 1 ≥ i ≥ Indices[].count
9)     string Index = Indices[i].ToString()
10)       for 1 ≥ j ≥ OSeq[j]
11)       string c = j.ToString()
12)          if (c = Index)
13)           ConvertDNATo2Bits(c)
14)          CReport.Append(c)
15)          goto line 18
16)       end if
17)     end for
18)    end for
19)     end while
20) CReport = BinaryToString(CReport)
21) return CReport
22) end ExtractCReport
```

Fig. 12. Pseudocode for extracting the clinical report

5 Experimental Results

The real genome sequences for four known viruses which are Influenza A Virus, Corona Virus, Hepatitis C Virus and FlaviVirus have been used in this research to test the proposed approach efficiency. The NCBI reference number of each virus as its clinical report is used to make the reports' size equal through the four test cases. Table 1 shows that after combining the clinical report indices and the original DNA sequence and compressing it, LSDC achieved a compression percentage from 53.3% to 62% of original sequence size. The algorithm compression ratio up to 25% depending on two main factors:

1) The length of the string on the clinical report.
2) The number of the ASCII codes which represent nulls or white spaces on the sequence, which will not be compressed to can convert it backward to its original value as mentioned before in BinaryToString algorithm.

Because of the presence of biological characteristics and features of DNA and RNA sequences, researchers have developed novel data embedding techniques based on DNA to provide maximal safety and powerful security with large capacity and low variation rate. However, all of these studies focused on ways to reduce the rate of alteration in terms of hiding capabilities and security concerns. The message was encrypted using the Playfair cypher algorithm in [9, 21]. The most prominent benefit mentioned in [9] is that

Table 1. Applying LSDC on four different viruses.

NCBI ID	Virus name	Original size	Clinical report size	Compressed DNA Size + Indices	Compression percentage
NC_002023.1	Influenza A virus	2.35 KB	11 bytes	1.45 KB	61.7%
NC_004102.1	Hepatitis C virus	9.68 KB	11 bytes	5.16 KB	53.3%
NC_021069.1	FlaviVirus	10.9 KB	11 bytes	6.25 KB	57.3%
NC_045512.2	CoronaVirus	30.0 KB	11 bytes	18.6 KB	62%

the suggested algorithm allowed the message and key to be of any kind and size. While [21] considers their approach to be a blind approach because the sender and receiver do not need to transmit the original carrier sequence because the extraction procedure may be completed entirely without mentioning to it. Despite their capacity to achieve a sufficient level of security. The secret key must be communicated, and both the sender and the receiver must agree on it. The authors of [8, 14] utilized a DNA picture as a cover-medium and employed the LSB approach to embed it. Multiple keys are required in [8] for the whole process, from DNA synthesis to data embedding, and their transfer between the sender and receiver needs a secure key exchange system, despite the possibility to attain acceptable levels of distortion metrics. Rafiul Biswas et al., improved data security by providing a second level of secrecy in [14]. To boot-up the security, they utilized dynamic DNA encoding, which provides third-level secrecy. Although some researchers have attempted to merge regular encryption techniques with steganography, security concerns such as exchanging cypher keys may still arise. Furthermore, employing DNA sequences as a cover-medium necessitates a significant amount of storage memory, and it is considered sensitive data in and of itself. As a result, it should be protected as well. Despite the fact that the three suggested cryptosystems passed various attack evaluations in [13], The ciphertext is many times longer than the plaintext, which is easy to guess considering that the major goal of steganography-based approaches is to reduce the attractiveness of the plaintext. The suggested methodology is strong enough in [20], since it uses DNA steganography as one layer, symmetric second layer DES and AES cryptography with parallel programming, and gene sequences as key values for DNA XOR operation, which is more stable resistance than Binary XOR operation. The major limitation of this approach is that it takes longer to encrypt small files.

The LSCD approach doesn't need a shared key between parties. There are no sensitive files or encrypted files to be shared. The LSCD approach could successfully locate the clinical report indices on the original DNA sequence. The challenge of hiding the clinical report in the DNA sequence without any modification has been achieved successfully. It could compress both of indices file and the original sequence into a compressed file, concealing this file into a text/image or video. On the other side, the compressed file could be extracted and decompressed to get the original sequence file correctly without any errors or modifications, then the clinical report can be extracted very easily and successfully. After comparing the original sequence and the clinical report with the extracted sequence and report, both of them are 100% typically the same. So, LSDC accuracy ratio is 100%, with respect to all issues and odd scenarios.

6 Conclusion and Future Work

Steganography-related research concentrated on ways and methods to reduce the rate of alteration in terms of hiding capabilities and security concerns. Using genomic data as a cover-medium necessitates a large amount of storage memory, and genomic data is a secret data itself.

In this research; a multi-layered security approach (LSDC) had been implemented to hide secret clinical data in a DNA sequence without any modification on the original DNA sequence, then hiding the DNA sequence in another cover-medium after compressing it to make it smaller and unreadable. Because of the genomic data privacy, the compressed DNA object will be hidden in another cover-medium, such as an image or video, using an optimization technique like Particle Swarm, Coco Search, Firefly, or Flower Pollination Algorithms.

References

1. Fernández-Alemán, J.L., Señor, I.C., Lozoya, P.Á.O., Toval, A.: Security and privacy in electronic health records: a systematic literature review. J. Biomed. Inform. **46**(3), 541–562 (2013)
2. Gunter, C.A.: Building a smarter health and wellness future: privacy and security challenges. ICTs and the Health Sector: Towards Smarter Health and Wellness Models, pp. 141–157 (2013)
3. Mellado, D., Fernández-Medina, E., Piattini, M.: Security requirements engineering framework for software product lines. Inform. Softw. Technol. **52**(10), 1094–1117 (2010)
4. Darshni, P., Ghanekar, U.: A hybrid data hiding scheme to enhance the capacity of one-third probability embedding method. In: Computational Intelligence & Communication Technology (CICT), 2015 IEEE International Conference on, pp. 269–272, 13–14 February 2015
5. Hamed, G., Marey, M., El-Sayed, S.A., Tolba, M.F.: Comparative study for various DNA based steganography techniques with the essential conclusions about the future research. In: 2016 11th International Conference on Computer Engineering & Systems (ICCES), Cairo, 2016, pp. 220–225 (2016)
6. Shi, X., Xintao, W.: An overview of human genetic privacy. Ann. N. Y. Acad. Sci. **1387**(1), 61–72 (2017)
7. Wang, Z., Zhao, X., Wang, H., Cui, G.: Information hiding based on DNA steganography. In: 2013 IEEE 4th International Conference on Software Engineering and Service Science, Beijing, 2013, pp. 946–949 (2013)
8. Das, P., Deb, S., Kar, N., Bhattacharya, B.: An improved DNA based dual cover steganography. Procedia Comput. Sci. **46**, 604–611 (2015)
9. Marwan, S., Shawish, A., Nagaty, K.: DNA-based cryptographic methods for data hiding in DNA media. Biosystems **150**, 110–118 (2016)
10. Sajisha, K.S., Mathew, S.: An encryption based on DNA cryptography and steganography. In: 2017 International conference of Electronics, Communication and Aerospace Technology (ICECA), Coimbatore, 2017, pp. 162–167 (2017)

11. Vinodhini, R.E., Malathi, P., Kumar, T.G.: A survey on DNA and image steganography. In: 2017 4th International Conference on Advanced Computing and Communication Systems (ICACCS), Coimbatore, 2017, pp. 1–7 (2017)
12. Mohammed, M.H., Taloba, A.I., Ali, B.H.: DNA-based steganography using neural networks. In: 2018 International Japan-Africa Conference on Electronics, Communications and Computations (JAC-ECC), Alexandria, Egypt, 2018, pp. 79–82 (2018)
13. Vijayakumar, P., Vijayalakshmi, V., Rajashree, R.: Increased level of security using DNA steganography. Int. J. Adv. Intell. Paradigms **10**, 74–82 (2018)
14. Biswas, M.R., Alam, K.M.R., Tamura, S., Morimoto, Y.: A technique for DNA cryptography based on dynamic mechanisms. J. ISA **48** (2019)
15. Yadav, P., Mishra, N., Sharma, S.: A secure video steganography with encryption based on LSB technique. In: Computational Intelligence and Computing Research (ICCIC), 2013 IEEE International Conference on, pp.1–5, 26–28 December 2013
16. Paul, R., Acharya, A.K., Yadav, V.K., Batham, S.: Hiding large amount of data using a new approach of video steganography. In: Confluence 2013: The Next Generation Information Technology Summit (4th International Conference), pp.337–343, 26–27 September 2013
17. Abbas, S.A., et al.: Optimized video steganography using Cuckoo Search algorithm. In: Intelligent Computing and Information Systems (ICICIS), 2015 IEEE Seventh International Conference on IEEE (2015)
18. Hanna, W.: Improving image steganography using a proposed mutated levy-flight firefly algorithm. Int. J. Comput. Appl. **182**, 23–29 (2018)
19. Hazra, A., Ghosh, S., Jash, S.: A review on DNA based cryptographic techniques. Int. J. Netw. Secur. **20**, 1093–1104 (2018)
20. Vantigaru, S., Basavaraj, R.: Two-layer encryption schemes for symmetric algorithms using DNA sequences. Int. J. Math. Comput. Sci. **14**(4), 953–964 (2019)
21. Khalifa, A.: A blind DNA-steganography approach using ciphering and random sequence splicing. In: 2020 10th International Conference on Information Science and Technology (ICIST), Bath, London, and Plymouth, United Kingdom, 2020, pp. 86–90 (2020)
22. Dasgupta, K., Mondalb, J.K., Duttac, P.: Optimized video steganography using genetic algorithm (GA). In: International Conference on Computational Intelligence: Modeling, Techniques and Applications (CIMTA), 2013 by Elsevier Ltd., Procedia Technology, pp.131–137 (2013)
23. Eltahir, M.E., Kiah, L.M., Zaidan, B.B.: High rate video streaming steganography. In: Information Management and Engineering, 2009. ICIME 2009. International Conference on 2009, pp. 550–553 (2009)
24. Pan African Bioinformatics Network for the Human Heredity and Health in Africa (H3Africa) consortium. https://www.h3abionet.org/
25. ASCII Code - The extended ASCII table. https://www.ascii-code.com/
26. National Center of Biotechnology Information. https://www.ncbi.nlm.nih.gov/

Analysis of Counterfeits Using Color Models for the Presentation of Painting in Virtual Reality

Irena Drofova(✉) and Milan Adamek

Faculty of Applied Informatics, Department of Security Engineering, Tomas Bata University in Zlin, Zlín, Czech Republic
drofova@utb.cz

Abstract. The black market for counterfeit works of art is the third most lucrative in the world. The digitalization of art and introduction into the virtual environment allows for a high rate of these counterfeit works. The issue of verifying original works in the online environment currently brings new challenges in the field of forensic science, in particular fields of color vision and working with light, colors, and color models, thus opening new approaches of art verification and presentation. This article deals with image processing and digitalization in color fields and their definition in color spaces. In order to determine a color model of a digitally processed image and for a specific color used, authentic artwork is analyzed by graphic software. This text aims to find and define the same color tone with the actual color tone of the original artwork in the digital environment and color spaces. The definition of color can be an attribute for verifying the originality of a work of art in a digital environment. At present, a precise definition of the color of a painting can help to achieve more accurate reproduction of artwork in a virtual reality environment and its visual perception by the user.

Keywords: Forensic art · Counterfeits · Color model · Virtual reality

1 Introduction

Forensic science is a multidisciplinary field that is an integral part of criminology, and it plays an important role in investigating crime and detecting potential perpetrators, primarily because the information and evidence that individual fields of forensic science can provide are based on detailed research. Crimes are reconstructed; crimes committed are detected; the personalities of suspects are analyzed; the evidence is presented in court proceedings [1]. One of the disciplines of forensic science is forensic art, which significantly contributes, among other things, to the possible disclosure of a perpetrator's identity. The appearance of a crime can be detected through the reconstruction of evidence fragments, evidence demonstration, victim identification (in case of an accident), or body decay. At last but not least forensic art can be used in the detection of counterfeit products, counterfeit artworks, and other valuables. Especially in the art field, the gradual

R. Silhavy et al. (Eds.): CoMeSySo 2021, LNNS 232, pp. 609–620, 2021.
https://doi.org/10.1007/978-3-030-90318-3_49

advent of digital technologies has created new challenges in this area. New methodologies and procedures for the detection of counterfeit works have been developed. Digital image diagnostics are used to evaluate the attributes of artworks. Computer imaging technologies can obtain evidence of possible manipulation or various information about the origin of the artwork. The possibility of applying modern forensic algorithms to the original painter's work for a diagnostic purpose can help detect and document the potential forgery of an artwork [2]. The use of the theories of light, colors, and color models can complement the detection and documentation of some attributes of artwork. Achieving the most exact representation of colors and structures of artwork depends on the fidelity of the digital reproduction of the image.

This article deals with the possibilities of classification and documentation of the original artwork in terms of colors used in combination with color models and areas for the most accurate reproduction of the image in a digital environment [3]. Experimental scanning and digitalization of an object are performed in order to identify the physical color in a digital environment and define the values of the color. A clearly defined color tone value is necessary for other color applications, especially in other color spaces or in subsequent artistic reproduction printing techniques, for digital archiving of artworks and reuse of color values in further digital image processing.

In the case of precise capture of artworks for further processing, evaluation of a definite attribute, and digital archives, it is desirable to have a quality device with an appropriate optical system for capturing the whole and partial specific parts. Primarily, if the goal is to present a work of art in virtual reality, the user may have an inaccurate visual perception of the object due to the light and textures in the virtual environment scene. By optimizing the elements that make up the virtual scene, the colors, and the display device, faithful reproduction of a work of art in virtual reality can be achieved.

2 Methods

Counterfeiting artworks are simple, and often even the author of an artwork cannot distinguish between counterfeit and original. Counterfeits often appear in the depositories of museums and galleries. Currently, there are modern and complementary methods for distinguishing counterfeits from the original. Options and methods for analyzing the original image may be combined. The choice of the methods also depends on the type of the painting and the purpose of reproduction.

2.1 Visual Cues Method

The initial *macroscopic examination* of the original work evaluates the features that are visible to the human eye. Differences in color tone and painting style may be evident in an artwork. We can also see, for example, the effort to obtain a credible patina that damaged the layer of painting, as well as the canvas to obtain the necessary old look of the artwork. *This stratigraphic method* can be used to evaluate used color pigment as well as the age of drawing. Some pigment can be identified by *polarizing microscopy* and scanning *electron microscopy* (SEM) or *electron microscopy* in conjunction with an *electron dispersion spectrometer* (EDS). The pigment can also be identified by a *microchemical test*.

The style of visual art is usually determined after the creation of the artwork. The style should be significantly different from other styles. Styles can overlap or complement each other. An artist can use more styles in his artwork. There is no specific definition of a visual style. However, we can apply some visual cues from paintings, such as color palette, scene, composition, lighting, contour, and brush strokes [4].

2.2 RGB to HSL Color Model Conversion

The RGB and CMYK color models are two models used in computer graphics and are directly related. Their abbreviation represents the primary colors; by mixing, we achieve other colors of the color spectrum. These are the example of the primary and secondary colors and their complementary. The RGB and CMYK color models are shown in Fig. 1.

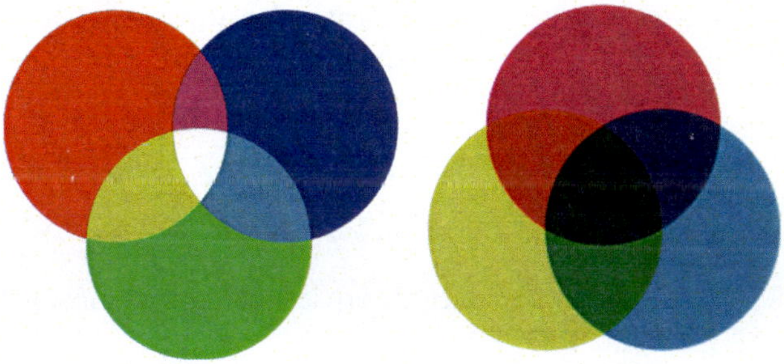

Fig. 1. RGB color model/CMYK color model.

All-digital display devices (such as scanners and monitors) using the RGB (red/green/blue) model, while the CMYK (cyan/magenta/yellow/contrast) color space works with real colors and outputs those colors to different materials. For this reason, it is necessary to recalculate the colors of the individual rooms according to the nature of the input and output information to minimize the loss of color information or to make the resulting display as faithful as possible to the original input information [4]. The HSL color model is very similar to the RGB model but works with different values. This model describes color relationships more accurately and is easier to work with. HSL color model works separately with tones and corresponds better to the color perception of the human eye. The HSL color model can provide other valuable information such as color grade, color saturation, and color brightness [5].

The advantage is a smooth change of tone. This color model is a favorite in digital graphic art. The color perceived by the human eye depends on the external environment and reaction to changes in it. Mainly lighting conditions significantly affect the resulting color perception.

2.3 Color Measuring Equipment

There are more or less accurate methods of measuring color. The following instruments are used for the measurement of color.

- *The photometer* is mainly used for display devices such as monitors.
- *The fluorometer* (fluorescence spectrometer) is used to measure photoluminescence.
- *The colorimeter* is used for direct measurement of color coordinates or *xy* from *trichromatic values*. This instrument is more accurate than the densitometer below and checks precisely specified and specific colors.
- *The spectrophotometer* is practically the most accurate of all the above instruments. It measures the spectral characteristics of direct or reflected light. With this device, it is also possible to find out the change in hue color depending on the type of lighting used.
- *The densitometer* measures the optical density (density) both in reflection and in the passage of light. These devices are especially suitable for determining the color thickness of layers. The densitometer determines:

  - Transparency (T), i.e., material permeability
  - Opacity (O), i.e., is impermeability rate. It is inverted transparency [6]

Density calculation for individual CMYK process colors using image analysis:

$$D_{cyan} = log_{10}\frac{I_W}{I_R} \tag{1}$$

$$D_{magenta} = log_{10}\frac{I_W}{I_G} \tag{2}$$

$$D_{yellow} = log_{10}\frac{I_W}{I_B} \tag{3}$$

$$D_{black} = log_{10}\frac{I_W}{(I_R + I_G + I_B)/3} \tag{4}$$

where $D_{cyan}, D_{magenta}, D_{yellow}, D_{black}$ are the optical densities of the individual process colors, I_W is the intensity of the comparative white, I_R is the intensity of the pixel red, I_G the intensity of the pixel green, I_B the intensity of the pixel blue.

The experimental study focused on specific artwork and worked with colors and color models to obtain plausible color reproduction in a digital environment. The artistic image belongs to a serial of paintings called Contrasts. The author of the picture, Michal Pasma, likes to use many different styles in his artworks.

The artwork was a painting with the acrylic technique on canvas, where one color is known. The material structure, stroke direction, and length are captured, as shown by the enlargement of the image in Fig. 2. The color measuring devices are not used in the experimental study.

Fig. 2. Structure of painting acrylic on canvas/structure detail enlarged +100%/structure in detail enlarged +200%.

3 Image Capture an Original Artwork

An artistic painting chosen for the experimental work was an artwork painted with acrylic paints on canvas. Three direct acrylic paints from Koh-i-Noor have been used: black 0700, white 0100, ocher 0600. The focus was only on the black paint, shown in Fig. 3 [7].

Fig. 3. Artwork on canvas painted with Koh-i-Noor acrylic paint black 0700 [7].

At the time, it was not possible to precisely define the values and characteristics of the light in the room. This artistic acrylic paint for painting is dilutable with water and fixed to the substrate. It is a high-quality dispersion paint with a high number of pigments and colorfastness. The paints can be used on any non-greasy substrate and, after drying, create an age-resistant color film. From the three acrylic colors mentioned, only black was used for the experiments because the other two colors were mixed from other color shades in the artwork. The black color can be theoretically related to the CMYK color model as a direct color.

The art object was photographed in not very good conditions, at dusk, with the spotlight at the space, without the proper lighting of the object. There was no color reference scale for the use of a DNG profile or unification. As expected, the scanned image had to be calibrated manually in a graphic program. The subject was captured with a Pentax K-50 mirror camera with the basic lens without the use of automatic or external flash. The properties of the reference image and the scanning device are shown in Table 1.

Table 1. Properties of the reference image.

Image size	19,3 MB
Image dimensions	4928 × 3264 px
Resolution	96 dpi
Bit depth	24 bit
Color range	sRGB
Aperture shuter	f/5
Exposure distance	1/200 s
Focal distance	35 mm
ISO	177

An important aspect when working with color and image digitization is the color calibration of input and output devices. This was used to define them and create their color scales to further refine colors and tones similar to original images. For these purposes, it is easiest to use a standardized color combination, and each sub-photo with the DNG (digital negative) profile needs to be created. It is reusable between sets of photos and between cameras. RAW files used to archive images may generate digital cameras differently, and specifications are not always available. DNG format is suitable for editing, simple archiving, and providing easy access to archived files.

The DNG format is supported by software vendors and used by manufacturers of cameras and 3D imaging devices. The classic standardized reference scale with 24 color fields for creating a DNG profile is used nowadays. The standardized reference color scale is a part of the chromatic triangle (chromaticity diagram) *xy* CIE 1931, as shown in Fig. 4. However, this color scale was not available when capturing the image. It was incorporated by subsequent image processing and onerous editing. Image processing, calibration, and conversion of colors and profiles were thus made only in graphics programs.

This article is focused on specific works of art and artworks with colors, color models, and spaces to obtain reliable reproduction and definition of colors in a digital environment. When an image of an object is captured under less favorable conditions, without proper illumination, a calibrated scanning device, or a basic standardized color gamut, it provokes several other questions about the image in terms of color processing. Above all, how to define black color paint in a digital environment. The following chapter describes color calibration, working with color models and spaces to approximate the tone of color spots and define the color and its properties in digital form.

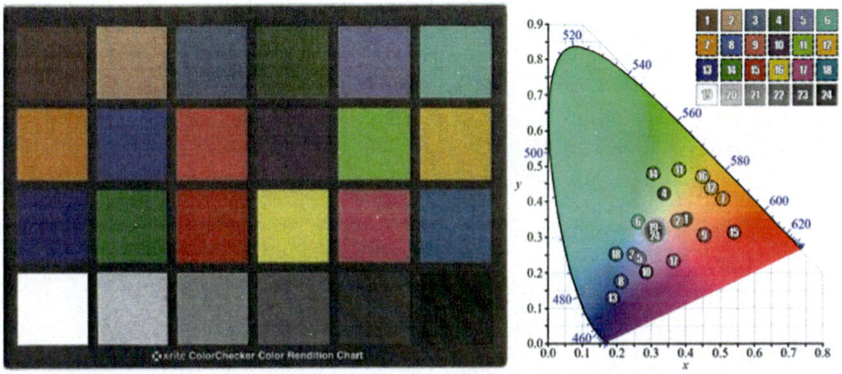

Fig. 4. Scale of 24 color fields and their position in the chromatic diagram.

4 Digital Processing of the Scanned Image

The selected artwork contains contrasting transitions and geometric elements. It was possible to work better with the image and the primary color. The light color on the paintings was mixed with ocher and white for the final shade. The painter used three colors. The color black was chosen for the experiment. This color could be theoretically related to the CMYK model as a primary color.

The original RAW format was loaded into the Camera Raw 13.1 [8] software workspace and then made basic adjustments to the digital image. Basic adjustment and color calibrations were adjusted in the RGB model, as well as the hue and saturation. Subsequently, the curve of the light and shadow values was corrected. This process is shown in Fig. 5.

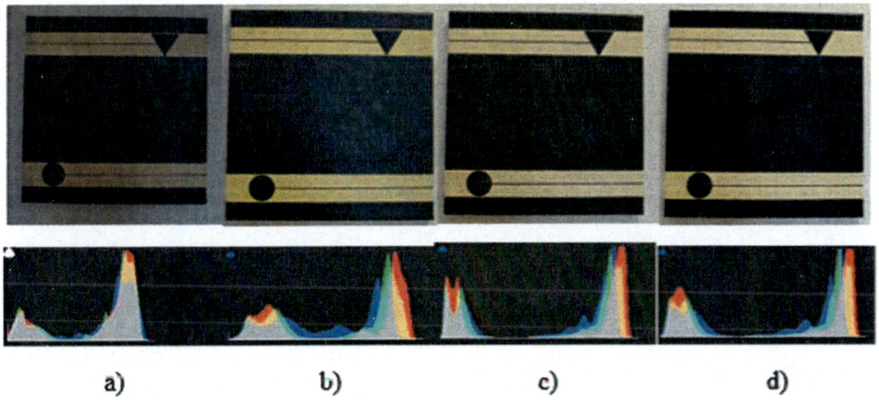

Fig. 5. Images and histograms after individual adjustments of the digitized image: a) of the original image in the RAW format before the adjustments; b) basic adjustments of the digital image. The basic parameters values adjusted from the initial values: hue; exposure; contrast, lights; shadows; and saturation; c) color calibration R, G, B; d) adjustment of the image curve values: light; shadows; white and dark values.

After basic adjustments, there were further image adjustments with emphasis on color transitions and color mixers in connection with the color model HSL. This color model is the most widely used in the digital graphics environment. Its advantage is the smooth transition of colors in tones compared to a similar model HSV. Further image adjustments are shown in Fig. 6.

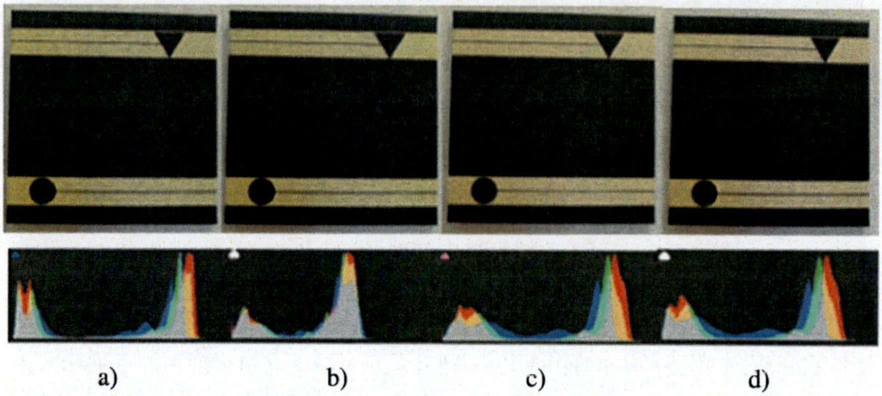

Fig. 6. Images and histograms after further adjustments of the digital image: e) adjustment of the amount and hue of the purple and green components in digital image reproduction; f) adjustment of color transitions in mid-tones, shadows, and light; g) color mixing and adjustment of individual colors RGB, CMYK, purple and orange to the HSL model; h) adjustment of color transitions in the HSL model.

After these adjustments, the digital image was evaluated and was recalculated as the object into the Adobe Photoshop [8] work environment. This image was center by rotating the canvas 2° to the left. Then the image was cropped to clean the background. In this case, it was walls. Further calibration of the RGB color values was applied, and the area with the highest possible black value was selected. The goal was to find the same shade of black that matches the color from this place. There were defined the color value of the painting's black color in the CMYK color model. The hue of the corresponding color was select from many color scales and spaces that the graphic program offers.

Figure 7 shows a comparison of two colors from the Pantone scales [9]. The first is the color Pantone Process Black C. This color is a process color from the colors of the CMYK Coated swatch. The second color is Pantone Black 3 CP from the Pantone Color Bridge Coated color range. This color was the same as the black tone on the image area. In the CMYK color model, it corresponds to the color values C = 67; M = 44; Y = 67; K = 95. Figure 8 shows both shades of black from the Pantone scales. The difference between them is obvious to the human eye.

Table 2 gives the exact numerical definition of the black tone on artwork in color models.

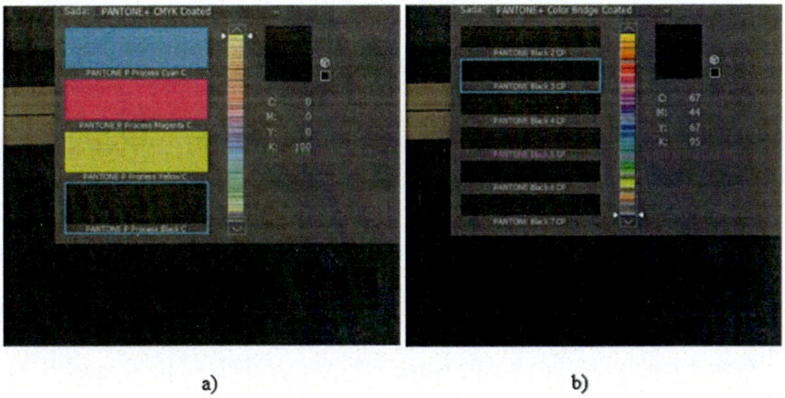

Fig. 7. Comparison of the color Pantone scale with the color in the image area: a) the color Pantone Process Black C; b) the color Pantone Black 3 CP.

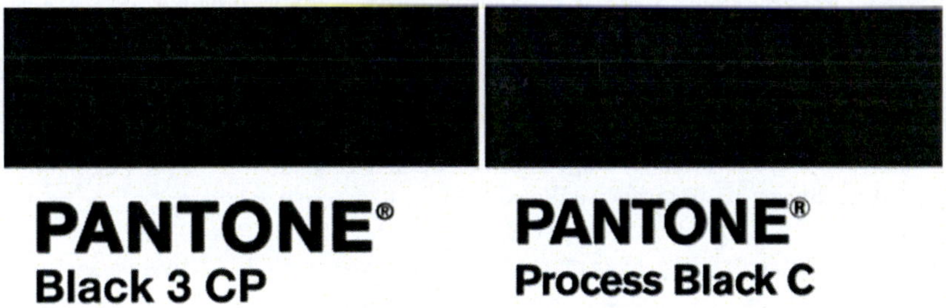

Fig. 8. Comparison of the color Pantone Process Black C and the color Pantone Black 3 CP [9].

Table 2. Numerical definition of the black tone.

Color model	Black color numerical definition
Hex	#262d26
RGB	rgb(38, 45, 38)
CMYK	cmyk(67, 44, 67, 95)
HSL	hsl(120,8.43%, 16, 27%)
RGBa	rgba(38, 45, 38, 1.00)
Lab Xyz	xyz(2.0876, 2.4288, 2.1927)
HSV	hsv(120,15.56%, 17, 65%)
HSVa	hsva(120,15.56%, 17.65%, 1)

5 Visual Comparison of Artwork in Virtual Reality

Currently, the virtual reality environment offers many possibilities. Virtual reality is a new opportunity for the presentation of artists and their artistic work. Galleries and museums are already using this technology. The exhibits can thus be presented to the public indirectly virtually in an online environment. Accurate reproduction of the object in a virtual environment is necessary. The virtual environment that is created affects the object and its presentation, primarily working with light in software to create virtual environments.

The artwork was embedded in a 2D virtual reality software environment. Primary virtual 3D space has also been created. In this study, Unity software [10] was used to create virtual reality. Figure 9 shows the effect of software on the quality of the reproduced image. The original image and the color-optimized image are placed side by side in the software's 2D and 3D graphics environment.

a) b) c) d) e) f)

Fig. 9. Comparison of the original image and the optimized image of the works of art: a) the initially captured image; b) an image calibrated in color in a graphics program; c) the captured initially image displayed in a 2D environment; d) a color-calibrated image displayed in a 2D software environment; e) the initially captured image displayed in the 3D software environment; f) a color-calibrated image displayed in a 3D software environment.

Subsequently, the work of art was inserted into a virtual environment with the wall's texture behind the painting and the wooden floor. The textures were created from the free Unsplash database [11]. Light first affects the perception of image and space in virtual reality. The white spotlight is used to illuminate both images at the center point between them. This light first affects the visual perception of the colors of both images. Figures 10 and 11 show the effect of light on the scene and the color perception of the two types of images.

Both the original and the optimized artwork look in color, as do all the types of images in Fig. 9. The images differ in the scene created to present the artwork in virtual reality. At first glance, the images and their colors are different in Figs. 10 and 11 compared to the images in Fig. 9. The images are also different when observing scenes with and without lighting. The background and its structure influence the perception of works of art in the scene. The light and lighting of scenes must be optimized for the reproduction of works of art in virtual reality. Color analysis in color spaces to clearly define colors in virtual reality will include the following studies as well.

Fig. 10. The scene without the spotlight.

Fig. 11. The scene with the spotlight.

6 Conclusion

Colors and color models are very narrow and have specific issues. In the case of artworks and counterfeits, colors play an important role. Their chemical and optical properties, as well as other properties, significantly varies not only in terms of time but also environmental influence on color sustainability. In the case of original artwork, its quality in digital reproduction can not only be preserved but its artistic value can also be transferred to other presentation channels and thus reach people all over the world. In terms of color reproduction, this is the most faithful approximation to a reused dye in a physical environment. There are many ways to achieve this, not only in terms of new uses but also for defining individual colors and outputs in different channels.

In this article, an experiment was demonstrated concerning a process of color definition for its further use in a digital graphic environment. The subject of the research was the black acrylic paint Koh-I-Noor 007. The original painting was painted using the acrylic technique on canvas. This paint is characterized by a high number of pigments, and its coating is fixed. The experiment included transferring artwork to an online environment, adjusting and calibrating colors, and further eliminating other aspects that have arisen in image capture, where there are no other high-level professional scanning techniques, no other accessories for working with colors and their properties. As can be observed, working with color and color spaces, and mixing channels thus requires considerable patience and time. The above-mentioned color was defined specifically in the form of a shade in the Pantone color swatch as Pantone Black 3 CP, and then it was specified for use in other color spaces.

The color defined in this way can serve as a basis for a faithful printed reproduction or further reproduction of the original in an online environment or in virtual reality with a clear specificity. In the case of two more colors, ocher and white, from the same manufacturer, the most appropriate methodology would probably be to perform direct measurements by instruments designed for this purpose and presented in this article. The main focus of this research was on using spectrophotometer or densitometer as the two most important methods concerning debated issue.

Acknowledgment. This research was based on the support of the author of the painting, Mr. Michal Pasma, and of the International Grand Agency of Thomas Bata University in Zlin,

IGA/CebiaTech/2021/004, and the Department of Security Engineering, Faculty of Applied Informatics.

References

1. Siegel, J.A., Saukko, P.J.: Encyclopedia of Forensic Sciences. Academic Press, Cambridge (2013). ISBN 978-0-12-382165-2
2. Pelagotti, A., et al.: Forensic imaging for art diagnostics. What evidence should we trust?, IOP Conf. Ser.: Mater. Sci. Eng. **949**, 012076 (2020). https://doi.org/10.1088/1757-899X/949/1/012076, ISSN 17578981
3. Gultebpe, E., Thomas, E., Conturo, M.M.: Predicting and grouping digitized paintings by style using unsupervised feature learning. J. Cult. Herit. **31**, 13–23 (2018). https://doi.org/10.1016/j.culher.2017.11.008, ISSN 1296-2074
4. Dohnal, M.: Barevne videni. Kolorimetrie, Univerzita Pardubice (2019). ISBN 978-80-7560-246-6
5. Ehkan, P., Siew, S.V., Zakaria, F.F., Warip, M.N.M., Ilyas, M.Z.: Comparative study of parallelism and pipelining of RGB to HSL colour space conversion architecture on FPGA. IOP Conf. Ser.: Mater. Sci. Eng. **767**, 012054 (2020)
6. Dohnal, M.: Fyzikalni zaklady reprodukce obrazu, Univerzita Pardubice (2007). ISBN 978-80-7194-945-9
7. Koh-i-Noor, the official website 2021. https://www.koh-i-noor.cz/
8. Adobe, the official website (2021). https://www.adobe.com/
9. Pantone, the official website (2021). https://www.pantone.com/eu/en/color-finder
10. Unity, the official website (2021). https://docs.unity3d.com/Manual/index.html
11. Unsplash (2021). https://unsplash.com/backgrounds/art/texture

Application of a Process Automation Approach in Investigative Activity Flows of Investigation Units: An Experience

Gleidson Sobreira Leite(✉) and Adriano Bessa Albuquerque

Universidade de Fortaleza, Fortaleza, CE, Brazil
adrianoba@unifor.br

Abstract. Preventing and combating crime is a global concern where, in this scenario, the strengthening of the performance of specialized units that act in investigations and combating crime is essential due to the economic and social impacts caused by the increase of crime rate. With the application of information technology and process automation, an opportunity arises to contribute to the treatment of the high volume of requests directed at these units, which tend to increase more and more as a result of the evolution of crime rates. This paper presents the execution of an experience in a real case scenario of investigation units, aiming at the application of a process automation approach and construction of a tool that aims to automate various investigative activity flows carried out by investigators. After conducting interviews with investigation units, a list of investigation activity flows was obtained and categorized. After applying the approach to define and prioritize the identified processes, the tool was developed and a comparative impact analysis was carried out. Results point out the benefits achieved, and discussions on challenges and limitations are presented.

Keywords: Information technology · Investigation units · Process automation

1 Introduction

Due to the economic and social impacts, as well as other aspects, the evolution of crime rates and the actions to combat it are a concern shared by several governments.

The evolution and use of technology not only provided benefits and advantages to society, but also created opportunities for greater performance and organization of criminal groups.

Several studies also point to concerns about the evolution and expansion of crime such as, for example, a statistic survey presented by [1] in October 2019, that showed the cost to business as a result of violence and crime (international comparison), presenting the top countries with the highest costs.

Figure 1 illustrates the cost index of the Top 10 countries where values close to 1 represent significant costs and close to 7, no costs. As shown in Fig. 1, Brazil, the country in which the experience of this study was carried out, is also on the list.

R. Silhavy et al. (Eds.): CoMeSySo 2021, LNNS 232, pp. 621–633, 2021.
https://doi.org/10.1007/978-3-030-90318-3_50

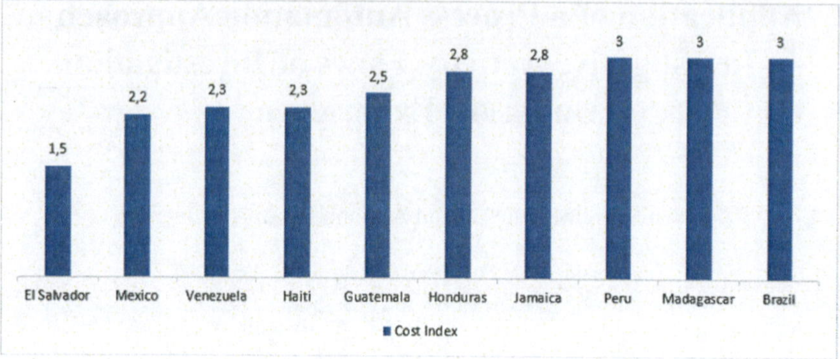

Fig. 1. Business costs of organized crime – Top 10 countries with the highest expenditure in 2019.

To this end, and due to the diversity and volume of existing crime practices, the importance of the role of institutions specialized in combating and investigating crimes is essencial, such as the FBI or EUROPOL, which have units specialized in combating diverse types of crimes, for example, criminal organizations, financial crimes, among others [2, 3].

However, due to the evolution of crime rate and the volume of requests directed at these specialized units, an opportunity arises to adopt solutions that contribute with greater agility or effectiveness of the work of these units, among other benefits such as simplification or optimization of processes, as well as the reduction of resources necessary to carry out operational activities.

Being a practice adopted in the market for many years, concepts of process automation can also be used to contribute to these guidelines when using integrated systems and data to improve the progress of workflows and, when possible, replacing manual activities with automated ones [4].

Motivated by this scenario and seeking to contribute to the body of knowledge related to studies regarding the use of information technology in the fight against crime, this study presents the execution of a real case scenario experience aiming the application of a process automation approach in investigation activity flows of investigation units.

The experiment was carried out in five specialized units where the objective was to identify a list of investigative activity flows carried out by investigators, and to implement a tool to automate these flows.

After the execution of the experience, a comparative impact analysis was carried out and discussions on challenges and limitations are presented.

It is also expected that this study can be adopted to help both practitioners and researchers in order to identify possible applications trends and improvements.

This paper is organized as follow: the methodology and procedures performed in the work (Sect. 2); background and related work (Sect. 3); the experience and discussions (Sect. 4), followed by the final considerations (Sect. 5).

2 Research Methodology

To accomplish the objective of this work, four specific actions were carried out. They were:

1) Bibliographic research in academic studies regarding the application of process automation in different domains.
2) Selection of the approach related to the application of process automation in the scenario of investigation units that will be adopted in the experience.
3) Definition of the scope and delimitation of the experience, as well as its planning
4) Conduction of the experience in the context of units that perform or support investigation activities aimed at fighting crime, and present discussions about results, challenges and limitations.

3 Background and Related Work

According to [5], investigation activities cover the collection of information and evidence to identify, apprehend and convict suspected criminals where operational activities are performed such as, collecting, grouping, information analysis and result presentation in order to achieve the expected objectives.

To achieve each objective in the investigation scenario, activity flows or processes are performed. According to [6–8], processes are composed of a set of activities ordered in a sequential way and organized with the participation of different individuals, resources and tools.

Regarding automation, [9–11] point out that it focuses on the application of techniques, software and/or equipment in an industrial process in order to increase its efficiency, reduce human effort or maximizing production with the lowest consumption of resources.

In this context, the concept of process automation can be understood as a factor that allows the integration of technology into the routine of organizations, strengthening and/or creating elementary values such as, efficiency, greater productivity, reduction of costs and time, among others [6–12].

Regarding the application of process automation, several studies and practices have been applied in organizations, government and the academy over the time.

[13] for example, performed an analysis of the historical perspectives on management automation and identified the impact in sectors like manufacturing, shipping, home care, food preparation, transportation and others, and a survey performed by [14], focused on the automation and control applications in developing regions considering an industry perspective of emerging technologies and challenges on projects regarding e-citizen services and smart city.

Regarding the application of process automation in the government, [15] designed a synthesized rural E-government system for use at the municipal to village levels where daily information management functions are included in the system contributing to office automation and provide an easy-to-use management platform for data management.

Regarding public sector, [16] pointed out that Robotic Process Automation (RPA), which uses software-based robots to perform structured tasks that require manual labor

[17], can contribute to the reduction of time allocation by the workers in repetitive and routine activities, allowing more time to be allocated in public attendance or more complexes jobs. In this context, [18] adopted RPA in a trade tax assessment scenario using deep convolution neural networks (CNN) for automated data extraction and processing. Findings pointed out that RPA has potential for the improvement of the efficiency of administrative work processes and for administrative modernization.

Despite the existence of several applications of process automation, it is essencial to adopt a strategy or an approach for its implementation to contribute to a higher success rate, as pointed out by [19], witch presented a principle named as the USA (Understand, Simplify and Automate) to serve as the basic strategy to implement process automation in a organization.

According to USA principal, before applying automation, it is first necessary to understand the process well, simplify it or optimize it (excluding unnecessary steps or adopted resources), and finally start the automation [19].

Considering the USA principle and taking into account the application of processes in the scenario of intelligence and investigation units, [20] presented an approach for the application of automation taking into account the specificities of these units, in order to propose criteria for the application of automation including the steps to be followed and calculations to be made regarding the inclusion or not of the processes for automation (and its prioritization) in the scenario of these units.

Taking into account the complexity of the investigative activity flows of these units, the proposed approach presents which procedures can be adopted for the application of automation in this scenario, scoring mandatory and optional steps to be followed in accordance with each scenario and limitations existing in the places where users plan to execute the approach [20].

In this study, the approach proposed by [20] was adopted in an experience of use in a real scenario of investigation units in a Brazilian public institution.

4 The Experience

4.1 Scenario

Considering a case of action in the fight against crimes, to carry out the experience proposed in this study, a public institution in Brazil (Public Prosecutor's Office of Ceará) that, among its attributions, also acts in the fight against crime, was selected.

In the institution, five units that perform or support intelligence or investigation activities that carry out or support complex criminal investigations participated in the experience.

Figure 2 shows the institution's organizational chart with respect to the five participating units. All units operate independently in the institution. Within the Intelligence and Technical Support Center a sub-sector composed of specialists who work in investigative activities on large volumes of data, called the Technology Laboratory Against Money Laundering, operates.

Among the motivations that led to the choice of this institution, is that it has an intensive performance in the fight against crime, and has in its structure units that carry out

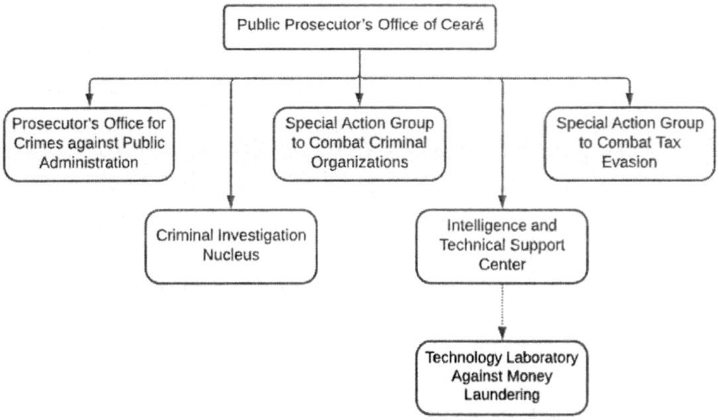

Fig. 2. Public Prosecutor's Office of Ceará org chart (only selected intelligence and investigation units)

or support intelligence and investigation activities and a technology laboratory against money laundering.

One of the reasons that motivated the institution's adherence to this study was due to the fact that the institution invested in and researched technological solutions that contribute in the fight against crime. However, the selected institution still faces difficulties in meeting the current high volume of specialized requests, and has limitations on financial and personnel resources. Another important factor would be the importance that the institution gives regarding the agility to combating criminal organizations.

4.2 Experience Planning

To carry out the experiment, the following actions were performed:

- Selection of the institution, specialized units and identification of participants investigators (prosecutors, analysts, among others)
- Interviews and information collection inherent to the process activities (investigative activity flows), aiming at its modeling and simplification before the start of automation (execution of the steps of the selected approach with the necessary adaptations)
- Implementation of an information system developed with the objective of automating the selected investigative activity flows (processes)
- Analyze the benefits/results reached after the execution of the experiment and the development of the system

4.3 Experience Execution

As already mentioned, for this experience investigators from five specialized units and a Technology Laboratory Against Money Laundering were selected.

A total of 12 specialists (two from each unit) participated in interviews carried out in order to identify the flows of investigative activities performed by the units, taking into account only flows in which the investigations involved the adoption of technological solutions and the use of different databases (to enable automation). The specialists also participated in the approach steps performed for the automation of the identified investigative activities flows.

The experts participated in an alternative way during the experience to not impact the execution of the daily work of the selected units, which also motivated the limited number of participants.

Table 1 shows the profile of the selected specialists regarding the academic degree, experience (in years) in intelligence or investigation activities, and time (in years) in the organization.

Table 1. Specialist's profile who participated in the experience

Participant	Academic degree	Domain experience (Years)	Time in the organization (Years)
P1	Specialization	1	5
P2	Specialization	1	6
P3	Specialization	3	4
P4	Specialization	4	4
P5	Specialization	4	9
P6	Specialization	7	8
P7	Specialization	9	11
P8	Specialization	9	12
P9	Master	4	9
P10	Master	9	13
P11	Graduate	2	4
P12	Graduate	1	3

The execution of the experiment was carried out following the flow of activities of the approach proposed by [20] (with the necessary adaptations), whose steps are illustrated in Fig. 3 and detailed below.

Initially, interviews and information collection were carried out with the selected investigators from each specialized unit, in order to obtain a list of investigative activities flows (processes) carried out, and the expected results.

From the obtained list, verification was performed to identify which processes are suitable for automation or not.

Following the premises pointed out by [20], for the analysis of whether a process is suitable for automation or not, two criteria were analyzed: maturity, which considers the process stability and the predictability of results, and standardization.

To analize the maturity, the participants where asked to answer to questions: If the same results are expected if a process is run multiple times, and if the process environment

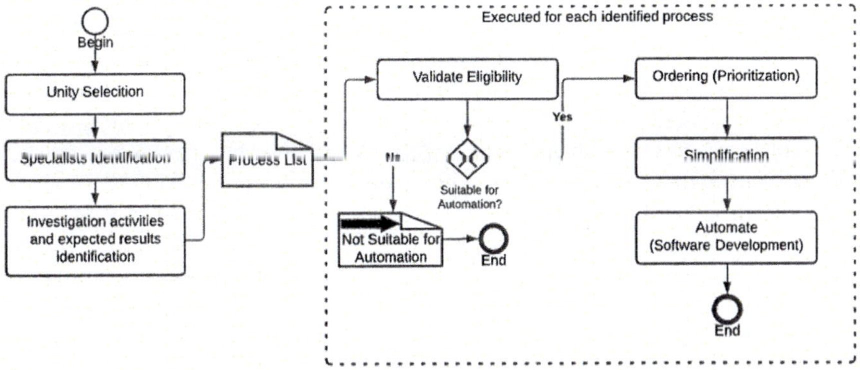

Fig. 3. Carrying out the steps of the adapted approach

is stable. If in any of the questions the answer is negative, the process is not suitable for automation.

Regarding the process standardization, specialists are asked to perform a validation if the process perspectives pointed out by [8] can be extracted. According [17], if its possible to extract the perspectives clearly from the process definition, it is also possible to adopt modeling tools in the process specification. The perspectives indicated by [8] are categorized into Functional, Behavioral, Organizational and Informational.

In the Functional perspective, which considered the activities performed and their interdependencies, the participants extracted basic information from the processes, answering questions such as, which activities were performed and expected input and output parameters, or which dependencies existed between them [8, 20].

The Behavioral perspective represents when the process elements are performed, where the order of execution of the activities and how the execution take place is considered, as well as how the information transition between activities occur [8, 20].

In the Organizational perspective, those responsible for executing the activities of the processes are identified, as well as the iterations between them. And in the Informational perspective, entities that are manipulated by the process are identified (including how iterations between databases and systems happen), as well as what standards/methodologies/techniques are used and how the information is used [8, 20].

Thus, if the process is mature and it is possible to extract the information provided in the perspectives proposed by [8], the process is suitable for automation, and the ordering (prioritization) step starts, as illustrated in Fig. 3.

Differently from the approach proposed by [20], in the experience it was decided not to perform the steps that involve the calculation of prioritization and classification indexes. The reason would be that the experiment aims to automate (in a single tool) all the flows of investigative activities identified and that have been suitable for automation, adopting only three criteria to define which processes to automate first.

The criteria adopted and chosen by the specialists of the selected units were: i) process with the highest volume of requests; ii) identical or similar processes that are being carried out by the largest number of specialized units; iii) processes with a longer execution time.

To facilitate the construction of the ordered list of processes for automation, for each criterion the experts determined a score from 0 to 1. After the sum of the criteria scores, the ordered list of investigative activity flows was obtained (tiebreakers were subjectively resolved by the experts).

Then, in the process simplification/optimization step, some activities were reordered, and unnecessary steps or procedures were eliminated.

With regard to the flows of investigative activities identified, there was a certain pattern in their execution. The vast majority of the performed activities involved the collection, crossing and analysis of data and information from different sources, followed by the application of investigative techniques and methodologies to then present the results in customized reports. Figure 4 presents and illustration of a generall workflow of the selected processes activities.

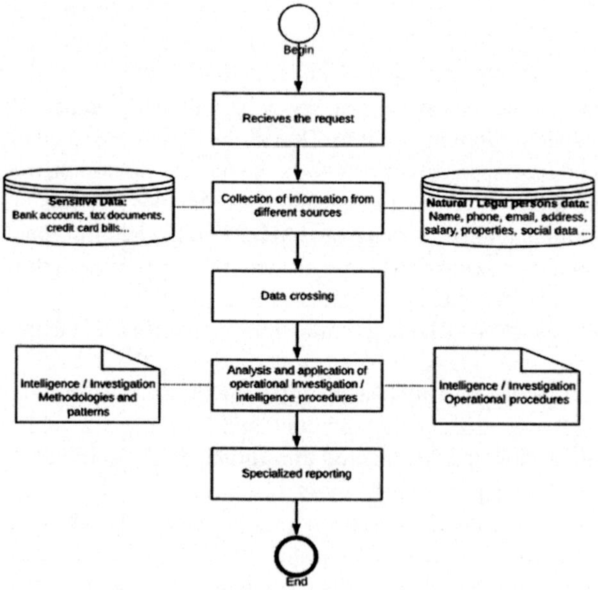

Fig. 4. General workflow illustration of the selected processes activities for automation

Finally, after obtaining the ordered list of processes, the implementation of the tool began (automation of the identified process). For the implementation of the solution, no investments were needed by the institution, since free tools or tools already adopted by the institution were used (e.g. programming languages: PHP, Javascript, HTML and CSS; application server: Apache; database management system: SQL Server).

4.4 Results and Discussions

After completing the development of the tool, 57 flows of investigative activities (process) were automated, which were categorized into 4 categories: Indicators (9 processes), investigative analysis (18 processes), linkage analysis (8 processes) and banking analysis (22 processes).

Although the flows of investigative activities have a certain pattern in their execution, as illustrated in Fig. 4, they differed from each other by the expected results, accessed data bases, objectives to be achieved and applied investigation methodologies/techniques (in all processes is possible to generate customized reports).

The Indicators category includes processes focused in proactive investigation actions by the specialized units (for example, without the institution having been called to act), which seek to obtain a list of individuals (investigated persons) depending on objectives of the investigations, and whose information is made available in graphs and reports.

Bank analysis aims to apply a set of methodologies and techniques for investigation and analysis focusing mainly on bank transaction statements of investigated people, also crossing the information with other databases.

Link analysis focuses on the application of investigative techniques in order to identify links between investigated individuals and/or companies through the use and crossing of different databases in order to achieve different objectives depending on the investigations.

And with regard to the category of investigative analysis, it includes flows of general investigative activities that did not fit into the other categories, but which are commonly performed in investigations carried out by specialized units.

In order to analyze the benefits reached after the execution of the experiment and the application and use of the developed tool, three comparative analyzes were carried out.

In the first comparative analysis, 3 investigative linkage analysis that contained a set of people to be investigated were selected.

Initially, the execution of the investigative work (activities flows) using systems, databases, techniques and methodologies was carried out by the specialized analysts (one analyst for each process).

The execution time of the works until the elaboration of the customized reports lasted about two weeks, in which few links were found and reported (in one of the works, no links were found in the manual analysis).

After obtaining the results of the work performed manually by the specialized analysts, the same work was performed by the developed tool in order to later enable the comparison of results.

The result provided by the tool (custom reports) was already made available immediately after processing the requests, in which it was found that the tool found a considerably higher amount of links (it even found links in one of the works in which the analyst had not identified any link). Figure 5 illustrates the execution of the first comparative analysis.

One of the reasons for the divergence of results obtained in relation to the quantity of links found (apart from the time the work was carried out), would be the fact that manual analyzes are limited and, with the use of systems that automate this analysis, they can be more in-depth or precises. Thus, not only is possible to achieve faster results, but also to perform deeper analyses/investigations achieving results that were previously unfeasible in the time available to carry out the investigations.

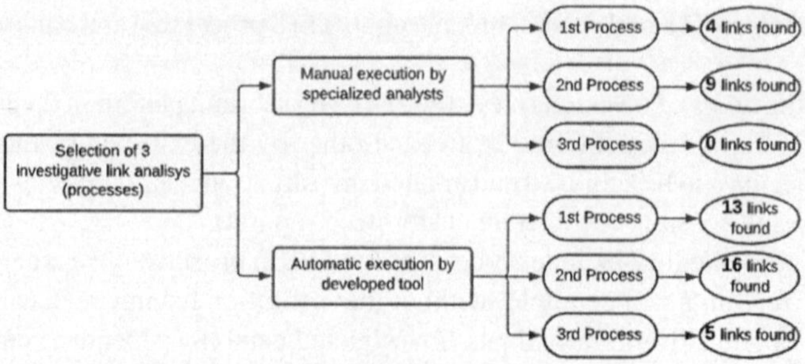

Fig. 5. Illustration of the first comparative analysis

Regarding the second comparative analysis, two bank transactions analysis were selected for the study. Following the same procedures as in the first comparative analysis, specialized analysts performed the works and then the same works were performed by the tools. The achievement of results by the analysts was obtained in an average of 15–20 days. The tool, on the other hand, made the results available automatically. Using the tool, it was also possible to provide more detailed results with more graphics and tables.

Finally, an overall comparative analysis was carried out with respect to overall results after application and use of the tool. Three criteria were used in the comparative analysis, namely: i) average time for the execution of the processes until the results are made available (customized reports); ii) number of processes executed per specialist (monthly); and iii) number of specialists involved in the execution of each process (average).

Table 2 presents the comparative analysis before and after the implementation of the developed tool, where the information contained in the "Before" column was provided by the specialized analists of the units, and the information in the "After" column were collected from the LOG history records obtained from use of the tool.

Table 2. Comparison analysis (before and after the adoption of the tool)

	Before	After
Quantity of processes executed per investigator (average 6 months)	12 to 18 per month	60 to 80 per month
Average time to finish process execution	6 to 18 h	2 to 8 min
Number of specialists involved (average)	1 to 3	*

It should be added that the average time to finish the execution of the processes by the tool varies as it depends on the internet connection speed (because it is a WEB system accessed by a browser), on the resources of the user's machine, among other factors.

Regarding the number of specialists involved in the execution of the processes, with the automation provided by the tool, which generated the expected results, the allocation of specialists in other specialized investigation activities was possible.

Finally, despite the identification of positive benefits obtained after the experience and the adoption of the developed tool, some considerations and limitations must be pointed out.

Because the experience was performed in a limited way (only one institution) in a restricted group of specialists of specialized units, it is not allowed to generalize the results. Due to the limited number of participants, there is a limitation in the conclusion of the results, which are considered evidence and not conclusive.

One possible threat identified in the realization of the studies is the participation of an author of this paper in the execution of the processes and implementation of the tool. Prior knowledge of the approach and its objectives may have positively influenced the results. Although the execution and application of the approach were carried out with the participation of the specialists (measure adopted in a way to reduce bias), it would have been for an unbiased third-party to conduct the approach to apply automation. However, the author from this work executed and managed the project at the request of the organization's senior management.

Regarding existing challenges in the application of automation in these scenarios, we can highlight the issue that the information treated and modeled (including investigative practices, techniques and methodologies) are considerably confidential, which makes access and assistance by third parties difficult, limiting the adoption of automation practices to domain experts (not all team members have extensive knowledge of information technology or software development).

Another challenge to be highlighted is the information security issue regarding the digital artifacts generated during the execution of a process automation approach, an aspect that was not considered in the experience performed in this study.

5 Conclusion

Considering the concern with fighting crime and the importance of strengthening the performance of institutions/units that act to fight crime, such as investigation units, studies that contribute to strengthening the performance of these units are essential.

Motivated by this scenario and seeking to contribute to the body of knowledge of studies regarding the use of technology in the fight against crime, this study presented the execution of a real case scenario experience where the application of a process automation approach was performed in investigation activity flows of investigation units.

After performing the experiment, developing and adopting the tool, benefits of applying process automation in the studied scenario were analyzed, and a comparative analysis was presented.

In addition to the benefits already mentioned, it should also be added that, after the experiment was carried out, several digital artifacts (such as documents, models, among others) were built, which enabled a greater standardization of the activities carried out, as well as their documentation and optimization/simplification.

Also as a benefit obtained, specialists can also reallocate their efforts in other activities (since it was possible to automate a good amount of investigative activity flows) indirectly contributing to the increase in the results of these units.

In addition to the limitations pointed out, as future works, it is recommended that the scope of this study be expanded to a greater number and variety of specialized units, as well as taking into account aspects related to information security due to the high degree of confidentiality of the information treated in these units.

References

1. STATISTA. Business costs of organized crime - Top 20 countries with the highest expenditure in 2019. In Statista - The Statistics Portal, October 2019. https://www.statista.com/statistics/268296/cost-index-for-business-cost-as-a-result-of-crime-and-violence/. Accessed 1 July 2021
2. FBI. What We Investigate. Federal Bureau of Investigation Department of Justice, United States. https://www.fbi.gov/investigate. 16 June 2021
3. EUROPOL. Crime Areas – Fighting Crime on a Number of Fronts. European Union Agency for Law Enforcement Cooperation. https://www.europol.europa.eu/crime-areas-and-trends/crime-areas. 16 June 2021
4. Sharma, K.L.S.: Overview of Industrial Process Automation, 2nd edn., pp 1–14. Elsevier (2017). https://doi.org/10.1016/B978-0-12-805354-6.00001-3
5. Osterburg, J.W., Ward, R.H.: Criminal Investigation: A Method for Reconstructing the Past. Lexis Nexis, New Providence (2010). ISBN 9781422463284
6. Weske, M.: Business Process Management: Concepts, Languages, Architectures, 2nd edn., pp. 1–24. Springer, Heidelberg (2012). https://doi.org/10.1007/978-3-642-28616-2
7. Scheel, V.H., Rosing, V.M., Scheer, A.W.: The Complete Business Process Handbook, pp. 1–9 (2015). ISBN 978-0-12-799959-3
8. Curtis, B., Kellner, M.I., Over, J.: Process modeling. Commun. ACM **35**, 75–90 (1992). https://doi.org/10.1145/130994.130998. Special issue on analysis and modeling in software development
9. Love, J.: Process Automation Handbook: A Guide to Theory and Practice (2007). https://doi.org/10.1007/978-1-84628-282-9
10. Goldberg, K.: What is automation? IEEE Trans. Autom. Sci. Eng. **9**(1), 1–2 (2012). https://doi.org/10.1109/tase.2011.2178910
11. Nof, S.Y.: Springer Handbook of Automation. Springer, Heidelberg (2009). https://doi.org/10.1007/978-3-540-78831-7
12. Jämsä-Jounela, S.-L.: Future trends in process automation. Annu. Rev. Control. **31**(2), 211–220 (2007). https://doi.org/10.1016/j.arcontrol.2007.08.003
13. Brown, A., Safford, H., Sperling, D.: Historical perspectives on managing automation and other disruptions in transportation. In: Empowering the New Mobility Workforce, pp. 3–30 (2019). https://doi.org/10.1016/b978-0-12-816088-6.00001-8
14. Aliu, A.: Automation and control applications in developing regions: an industry perspective of emerging technologies and challenges. IFAC-PapersOnLine **52**(25), 568–572 (2019). https://doi.org/10.1016/j.ifacol.2019.12.607
15. Wang, J., Dong, J., Li, L., Wang, Y.: Design and implementation of an integrated office automation/geographic information system rural E-government system. In: 2010 World Automation Congress, Kobe, Japan, 19–23 September 2010, pp. 377–384 (2010)
16. Willmer, A., Duhan, J., Gibson, L.: The new machinery of government - Robotic Process Automation in the Public Sector. Deloitte Development LLC (2017). https://www2.deloitte.com/content/dam/Deloitte/uk/Documents/Innovation/deloitte-uk-innovation-the-new-machinery-of-govt.pdf. 2 June 2021

17. Leshob, A., Bourgouin, A., Renard, L.: Towards a process analysis approach to adopt robotic process automation. In: 2018 IEEE 15th International Conference on e-Business Engineering (ICEBE), Xi'an, China, 12–14 October 2018, pp. 46–53 (2018). https://doi.org/10.1109/ICEBE.2018.00018
18. Houy, C., Hamberg, M., Fettke, P.: Robotic Process Automation in Public Administrations. Digitalisierung von Staat und Verwaltung, Lecture Notes in Informatics (LNI), Gesellschaft für Informatik, pp. 67–74 (2019)
19. Kapp, K.M.: A framework for successful E-technology implementation: understand, simplify, automate. J. Organ. Excell. **21**(1), 57–64 (2001). https://doi.org/10.1002/npr.1119
20. Leite, G.S., Albuquerque, A.B., Pinheiro, P.R.: Process automation and blockchain in intelligence and investigation units: an approach. Appl. Sci. **10**(11), 3677 (2020). https://doi.org/10.3390/app10113677

Sizing of Hybrid Renewable Energy System Based on Nature-Inspired Search Method and LPSP Model

Israel Edem Agbehadji[1](✉), Samuel Ofori Frimpong[2], and Richard C. Millham[2]

[1] Faculty of Accounting and Informatics, Durban University of Technology, P O Box 1334, Durban 4000, South Africa
[2] ICT and Society Research Group, Department of Information Technology, Durban University of Technology, Durban, South Africa
richardm1@dut.ac.za

Abstract. The energy sector is imperative for the development of every country. Companies in the energy sector often are challenged in terms of how to provide affordable energy within a climatically conducive manner, and with the ability to provide energy from the primary sources that meet the demand on the energy market. Developing countries often find it difficult to resolve these issues either because of a lack of technical expertise or financial resources. Although renewable energy sources can help address the challenges such as energy security, energy affordability and energy sustainability, it has not been fully exploited. In this paper, an approach for optimum sizing and minimizing Loss of Power Supply Probability (LPSP) of hybrid renewable energy systems (HRES) under the minimum cost of energy. The method underlying the HRES is based on a novel nature-inspired behaviour of the kestrel bird. The proposed algorithm was evaluated experimentally to compare its performance against a benchmarked algorithm that is WSA-MP. The result shows that by combining different sources of energy such as wind and solar, the proposed algorithm (KSA) guaranteed optimal cost over WSA-MP.

Keywords: Kestrel-based search algorithm (KSA) · Hybrid renewable energy sources (HRES) · Nature-inspired search method · Loss of Power Supply Probability (LPSP) · Wolf search algorithm with minus previous (WSA-MP)

1 Introduction

Currently, with a growing population and increasing demand, energy consumption worldwide is expected to rise. Providing access to cheap energy is a major challenge confronting many countries. The Internet of Things and edge computing could provide solutions to the global energy challenge by enabling clean energy technologies, creating better energy market dynamics, and optimizing the efficiency of existing energy production systems [1].

Companies in the energy sector from developing countries are often challenged in terms of energy supply to meet the demand for its' final consumers [2]. Mostly, consumers

R. Silhavy et al. (Eds.): CoMeSySo 2021, LNNS 232, pp. 634–649, 2021.
https://doi.org/10.1007/978-3-030-90318-3_51

who live in isolated locations have the least chances of connecting to the national energy grid are minimal due to the economic and technical constraints, the renewable energy system is the most preferred source of energy to meet the energy demand of consumers. In general, the energy market is faced with what is termed energy trilemma which refers to energy security, energy sustainability, and energy affordability. Energy security relates to how to manage the primary energy supply from domestic and external sources and the ability to provide energy to meet demand. Energy sustainability is the tendency to damage the environment by emitting CO_2 and greenhouse gases. Energy affordability is the cost of energy supply to the growing population. The security, sustainability and affordability are factors which cannot be isolated in energy management because it forms the basis to test new energy projects and policies and make the necessary adjustments.

Generally, there is a rise in energy demand in most developing economies which is a result of population growth, urbanization and industrialization. As the demand for energy increases, there is a need to look for alternative sources of energy that can minimize the cost of energy (that is, affordability in terms of price of electricity to residents), improve on the energy (that is security in terms of energy storage, system stability and recovery capacity) generation capacity, and provide environmentally friendly solutions (that is, sustainability) without CO_2 emission [3]. Research shows that individual renewable energy sources such as the sun, biomass and wind, which are found in the natural environment have the potential to produce affordable energy. However, the weather conditions are unpredictable which makes it difficult for reliable energy generation at any given time. Individual renewable energy systems can be combined to support each other, and when such a system is equipped with storage or backup devices such as battery bank and diesel generator, it can ensure reliable energy generation to meet peak-hour demand [4]. The batteries are used in the HRES as a storage device. The proposed nature-inspired algorithm will optimize the size of HRES and provide reliable power generation. The method of sizing considers the number of batteries, the PV modules, and wind generation capacity based on the reliability of the supply of power [4]. The probability of loss of power is applied to determine the reliability of the power supply system [4]. In this paper, the proposed algorithm combines the kestrel-based behaviour and characteristics, and the loss of power supply probability (LPSP) technique. The advantage of bio-inspired behaviour is the ability to self-adapt under different uncertainty from individual renewable energy sources.

The remaining sections of this paper are organized as follows: Sect. 2 presents the related works; Sect. 3 is the model proposed. Section 4 shows the algorithm to implement the proposed model; Sect. 5 presents the underlying parameter setting for the model. The experiment results are presented in Sect. 6 and the conclusion is in Sect. 7.

2 Related Work

Generally, the challenge faced in the energy market is termed as energy trilemma, which refers to energy security, energy equity and environmental sustainability. Energy security refers to how to ensure reliable production, energy equity is how affordable in terms of the price of electricity to residents. Environmental sustainability is how to provide energy such that it does not damage the environment by emitting CO_2. In this regard, the energy trilemma index provides insights into the energy performance trends of countries and helps to identify the most effective policy intervention for energy production [5].

Hybrid renewable energy systems (HRES) are becoming popular in most developing countries because the cost of photovoltaic (PV) and wind generation is lower compared to other forms of energy generation. The sizing of HRES system is based on finding the least cost when all generation factors are combined in terms of cost, radiation on the tilted surface and the ambient [4]. Currently, wind energy is an emerging source of energy because it is environmentally friendly. Wind is produced by irregular solar heating of the earth's land and sea level [6]. In wind power generation, the wind turbine has sensors attached and the sensors are controlled by algorithms that also analysis data for energy production optimization [1]. Mostly, because of uncertainties in energy production, it is ideal to combine the different renewable energy sources to form an HRES which might ensure reliable and consistent power generation, possibly at a cheaper cost. Thus, parameter tuning is imperative to ensure adaptability during uncertainties in energy production. Some of these uncertainties include economic, technical and design objectives [7].

A nature-inspired algorithm can play a significant role in HRES optimization due to its self-regulating ability that ensures some level of randomization in finding an optimal solution in terms of sizing of systems, without getting stuck in exploiting and exploring a search space.

The authors of [8] opine that the sporadic output and cost from renewable energy sources require an efficient way to self-control and self-manage the cost of energy production and other related cost components. For example, fuzzy logic and bio-inspired methods (e.g., grey wolf) have been applied to find efficient use of batteries on renewable energy systems [8]. Particle Swarm Optimization algorithm (PSO) was applied to optimize power [9]. Other bio-inspired methods include the Genetic Algorithm (GA) (Darwin, 1868) as cited by [10], Ant Colony Optimization (ACO), Wolf Search Algorithm (WSA) [11] and Social Spider Optimization algorithm (SSO).

Genetic algorithm, as an evolutionary method, is based on natural selection (Darwin, 1868 as cited by[10] where species considered to be weak and cannot adapt to the conditions of the habitat are eliminated while species considered as strong and can adapt to the habitat survive. Thus, natural selection is based on the notion that strong species have a greater chance to pass their genes to future generations, while weaker species are eliminated by natural selection. Sometimes, there are random changes that occur in genes due to changes within the external environments of species, which will cause new future species that are produced to inherit different genetic characteristics. At the stage of producing new species, individuals are randomly selected from the recent population within the habitat to be parents and use them to produce the off-springs for the subsequent iterations, thus successive iterations adapt to the habitat in respect of time.

Particle swarm is a bio-inspired method based on swarm behaviour such as fish and bird schools in nature [12]. The authors of [9] proposed an optimization of renewable hybrid energy systems for cost reduction using Particle Swarm Optimization (PSO).

Firefly produces short and rhythmic flashes [13] to attract other fireflies within its habitat. The flashing light attracts mating partners, warn other fireflies and attract potential prey. Other nature-inspired algorithms include Ant Colony Optimization (ACO) which is based on the foraging behaviour of ants in their natural environment. These ants used pheromone deposits to demarcate their path to the source of food; Wolf Search Algorithm (WSA) applies the preying behaviour of wolf and the ability to merge with wolves in a better position [14, 15]; Bat algorithm applies the behaviour of bats such as echolocation and pulse rate to control their search for prey [16]; Social Spider Optimization (SSO) algorithm is based on the social behaviour of the spider. Social spiders are known to cooperate with their neighbours using their web building and preying foraging ability. By constructing a social web, spiders provide a habitat that serves as a trap to hunt down prey. Any prey that is found/trapped on the web causes some vibration which can be detected by the neighbouring spider. As an entangled prey tries to detach itself from the web it amplifies the intensity and frequency of the vibration which intends to trigger spiders' awareness of a prey's presence on the social web. Thus, spiders become more aware of the presence of prey and inform neighbouring spiders to cooperate in hunting [17]. The social spider algorithm has been applied to the optimal sizing of power systems [18], the optimal operation of energy storage units [19], etc. A fitness function was used to predict the optimal size of the design of wind-PV under minimized cost. In applying the SSO algorithm, the author of [18] considered the number of PV, wind and battery with their economic conditions for an optimal size selection.

In this paper, a nature-inspired algorithm that finds the optimal size and reliability of a hybrid renewable energy system is proposed. The hybrid renewable system configuration is then optimized by employing the KSA, which searches for the optimal configurations of the HRES that minimize the cost function. In our approach, the LPSP model, economic conditions, the number of PV, number of wind turbine, number of houses and battery charge/discharge were considered.

3 Proposed Model

The basic structure underlying the proposed HRES with edge computing capability is shown in Fig. 1. The structure consists of the CIR component (that is, Check, Reduce and Improve), the Edge layer and the Fog-Cloud layer.

The CIR component provides a way to optimize the renewable energy resources by checking the climatic conditions at any given time and then receives the signal on the amount of energy that can meet the load demand at an optimal cost. Different kinds of data namely technical data (e.g., manufacturer's data for the PV modules, etc.), cost data (maintenance and operating cost, etc.) and load demand data are fed into the CIR component. Also, the CIR component provides an intelligent way to balance the load demand at any given time.

The Edge layer is the layer where devices are connected to sense real-time data on climatic conditions (e.g., ambient temperature, wind speed), and then sends a signal to the CIR component. The use of innovative technologies such as edge computing devices, sensor-enabled devices or IoT enabled devices, helps to monitor and collect timely data on the environmental conditions. Data includes weather data on solar radiation and air temperature are acquired from the edge devices.

The Fog-Cloud layer stores data for further processing on the Cloud computing environment.

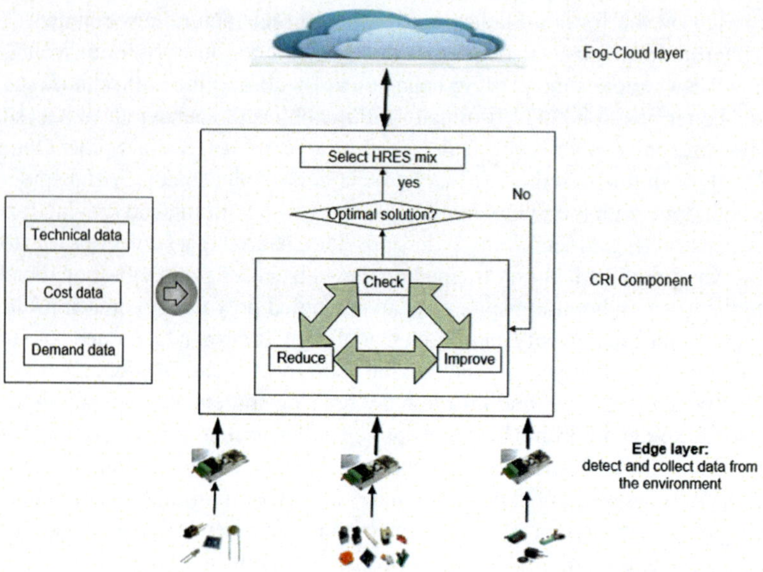

Fig. 1. Model proposed for a hybrid renewable energy system with edge computing capability

3.1 Proposed Mathematical Model of the Kestrel-Based Search Algorithm

In this paper, an algorithm is proposed for cost optimization of energy generation in the CIR component. The proposed algorithm uses the characteristics of Kestrel bird. The KSA is characterized by random encircling, use of the eye to detect the position of preys, position and velocity of kestrels to capture its prey, and possibility that the trail of prey will evaporate which is addressed based on the concept of the half-life of substances [20]. KSA achieves optimization through random encircling, randomly changed position, move with velocity and utilizing trail evaporation characteristics [21, 22]. Initially, the KSA starts by generating a set of random kestrels and then finds an optimal parameter in a search space [22]. The position of KSA is expressed in Eq. (1):

$$x_{i+1}^k = x_i^k + \beta_o e^{-\gamma r^2}\left(x_j - x_i\right) + f_i^k \tag{1}$$

Where x_{i+1}^k is the current better position of kestrel; x_i^k is the previous position based on random encircling formulation [21, 23]; $\beta_o e^{-\gamma r^2}$ is how attractive a light reflection is from a trail that varies between 0 and 1, the variable β_o represents initial attractiveness, r represents Minkowski distance [23], γ is light intensity variation between [0, 1]; x_j is kestrel in better position; f_i^k is the frequency of bobbing to detect its prey within sight measurement [23]. When Kestrels get to a better position, the frequency of bobbing is applied to exploit the frequently changing conditions at the time. The random encircling is mathematically expressed as follows in Eq. (2):

$$\vec{x}(t+1) = \vec{x_p}(t) - \vec{A} * \vec{D} \tag{2}$$

Thus:

$$\vec{A} = 2 * \vec{z} * \overrightarrow{r2} - \vec{z} \tag{3}$$

$$\vec{D} = \left| \vec{C} * \overrightarrow{x_p}(t) - \vec{x}(t) \right| \tag{4}$$

$$\vec{C} = 2 * \overrightarrow{r1} \tag{5}$$

Where $\vec{A}$ is coefficient vector, $\vec{D}$ is the encircling value obtained, $\overrightarrow{x_p}(t)$ is the position vector of the prey, $\vec{x}(t+1)$ is the previous position of Kestrels. Where $\vec{C}$ is the coefficient vector, $\vec{x}(t)$ is the position vector of a Kestrel, *r1* and *r2* are random numbers generated between 0 and 1, and as Kestrels move the centre to encircle a prey, it maximizes the chances of locating its prey hence, thus, the constant value of 2 was assigned. $\vec{z}$ represent a parameter to control the active mode with $\vec{z}_{hi}$ as the parameter for flight mode and $\vec{z}_{low}$ as the parameter for perched mode, which linearly decreases from 2 (high active mode value) to 0 (low active mode value) respectively during the iteration process. This is mathematically expressed as follows in Eq. (6):

$$\vec{z} = \vec{z}_{hi} - (\vec{z}_{hi} - \vec{z}_{low}) \frac{itr}{Max\_itr} \tag{6}$$

Where *itr* is current iteration whereas *Max_itr* is the total number of iterations, performed during the search, and other kestrels will update their position based on the best position of the leading Kestrel.

Finally, the velocity of Kestrel is updated expressed in Eq. (7):

$$v_{t+1}^k = v_t^k + x_t^k \tag{7}$$

Where v_{t+1}^k is the current best velocity that kestrels have to move, v_t^k represents the initial velocity, whilst x_t^k is the current best position that is occupied by a kestrel.

i. Trail evaporation

Trails are used to search for sources of food [24]. In our approach, trails are rather deposited by preys instead of kestrel [21]. Kestrel will not follow a depleted source of food and they prefer to explore new areas of food sources that are fresh [23, 25]. [24] suggested that the diminishment of the trail indicates the unstable nature of trail substances. Conceptually, if there are N unstable nodes, then the rate at which a substance decay with time t is expressed in Eq. (8):

$$\frac{dN}{dt} = -\gamma N \tag{8}$$

Thus, the decay rate (γ) with time (t) is simplified in Eq. (9):

$$\gamma_t = \gamma_o e^{-\varphi t} \tag{9}$$

Where γ_o represents initial value, t is the time of decay. The decay constant φ which indicates the duration of decay can be written in Eq. (10):

$$\varphi = \frac{ln\, 0.5}{-t_{\frac{1}{2}}} \tag{10}$$

Where φ is the decay constant and $t_{1/2}$ is the half-life period. If the value of decay constant is greater than 1, then the trail is new otherwise it is old, which is expressed in Eq. (11):

$$if\ \varphi \rightarrow \begin{cases} \varphi > 1, trail\ is\ new \\ 0, otherwise \end{cases} \tag{11}$$

Where φ is the decay constant [22].

In this paper, the model for the solar power plant and the wind power plant is considered independent and the cost of energy will be estimated separately.

3.2 Mathematical Model to Compute Power Generated from the Solar Plant

The mathematical model to compute the power output of PV modules is expressed in Eq. (12). The input variables are the solar radiation, the temperature per degree Celsius and the manufacturer's information. The time of the day influences the location of the sun from a reference point (that is the sky), which can be described by an hour angle in degree Celsius [26]. The power of the PV generator, P_{pv}, is computed using Eq. (12):

$$P_{pv}(t) = \eta_g A_s G_t \tag{12}$$

Where, η_g is the PV generation efficiency, $A_s(m^2)$ is the area of the panel used in a solar system, G_t is the solar irradiance that hits the surface of the tilted plane (W/m^2). The PV generation efficiency η_g is computed using Eq. (13):

$$\eta_g = \eta r * \eta pt[1 - \beta_t(Tc - Tr)] \tag{13}$$

Where ηr is the reference module efficiency and ηpt is the efficiency of power tracking equipment is equivalent to 1 for maximum power. β_t represents the temperature coefficient of efficiency that is between 0.004 to 0.006 per degree Celsius for silicon cells. Tr is reference cell temperature in degree Celsius. Tc is the temperature of PV cell in degree Celsius using Eq. (14):

$$Tc = Ta + \frac{NOCT - 20}{800} * Gt \tag{14}$$

Where Ta represents the ambient temperature in °C, $NOCT$ is nominal operating cell temperature in °C, and G_t is solar irradiation on tilted module plane (W/m^2). The value of A_s, ηpt, β_t, $NOCT$ of a cell is available on the manufacturer datasheet.

3.3 Mathematical Model to Compute Power Generated from Wind Plant

The computing power is given by [4] in Eq. (15):

$$P_w(t) = \frac{1}{2} C_p(\lambda, \beta) \rho_{pair} A_w V_w^3(t) \tag{15}$$

Where C_p is the power coefficient (where C_p is 0.593 refers to as Betz's coefficient) [4], ρ_{pair} density of the air (kg/m^3), A_w is rotor swept area (m^3) given by $A = \pi r^2$ where radius, r is equal to the blade length, V_w^3 is the average speed of the wind (m/s). λ is the tip speed ratio expressed in Eq. (16):

$$\lambda = \frac{\omega_r R}{V_w} \tag{16}$$

Where ω_r is the angular velocity of the turbine rotor in rad/s and V_w is the speed of the wind in m/s. The rotational speed n (r/min) and angular velocity ω_r is expressed in Eq. (17):

$$\omega_r - \frac{2\pi n}{60} \tag{17}$$

The adjustments of the height of the wind turbine is expressed in Eq. (18):

$$V = V_i \left(\frac{H}{H_i} \right)^{\alpha} \tag{18}$$

Where V is the speed of wind at hub height H, V_i is the speed of the wind that is measured from the reference height H_i and α refers to the power law exponent.

3.4 Model of Battery Storage

The battery bank can be both charged or discharged state. When the battery storage is in the discharge state, the load demand is higher than the energy generated. Thus, the battery bank capacity $C_{battery}$ at hour t is given as:

$$C_{battery}(t) = C_{battery}(t-1) * (1-\sigma) - \left(\frac{E_L(t)}{\eta_{inv}} - (E_s(t) + E_w(t)) \right) \tag{19}$$

Where $C_{battery}(t)$ and $C_{battery}(t-1)$ are the capacity of the battery bank (Wh) at time t and t − 1 respectively. η_{inv} is inverter efficiency. $E_L(t)$ is the load demand at hour t. $E_s(t)$ and $E_w(t)$ are the energy of the generating units from the PV module and wind generator respectively. Sigma (σ) is the self-discharge rate at 25% over six months at a temperature of 20 °C which indicates 0.14% per day [27]. Furthermore, when the battery storage is in the charged state, the total output of the generation units is greater than the load demand. Thus, the capacity of the battery bank at time t in an hour is expressed in Eq. (20)

$$C_{battery}(t) = C_{battery}(t-1) * (1-\sigma) - \left(E_s(t) + E_w(t) - \frac{E_L(t)}{\eta_{inv}} \right) * \eta_{battery} \tag{20}$$

Where $\eta_{battery}$ represents the battery efficiency. At any time t, the battery storage capacity should be within the minimum and maximum battery capacity such that:

$$C_{min\,battery} \le C_{battery} \le C_{max\,battery} \tag{21}$$

Where $C_{min\,battery}$, $C_{max\,battery}$ represents the permissible minimum and maximum storage capacity respectively. $C_{max\,battery}$ is the nominal storage capacity $C_{nominal\,battery}$, and $C_{minbattery}$ can be given by

$$C_{min\,battery} = DOD * C_{nominal\,battery} \tag{22}$$

Where DOD (%) is the maximum depth of battery discharge.

3.5 Model of Reliability

The model of reliability is based on the LPSP concept. The LPSP is the probability of loss of power supply when the HRES (a combination of PV module, wind turbine and battery storage) is unable to meet the load demand. LPSP value is either 0 or 1 where the value of 1 shows the load is never satisfied and the value of 0 signifies the load is always satisfied. For a specified period *T*, *LPSP* is expressed in Eq. (23)

$$LPSP = \frac{\sum_{t=1}^{T}(E_L(t) - \{E_s(t) + E_w(t) + C_{battery}(t-1) - C_{min\,battery}\} * \eta_{inv})}{\sum_{t=1}^{T} E_L(t)} \tag{23}$$

where T refers to the operation time within one year, when energy is generated then:

$$LPSP \le LPSP_{max} \tag{24}$$

where $LPSP_{max}$ is the maximum LPSP set by the user.

3.6 Mathematical Model on the Cost of Power from the Hybrid RES

The Total Annualized Life Cycle cost consists of capital cost and operating cost which is expressed as:

$$C_{an} = C_{anS} + C_{anw} \tag{25}$$

Where:

$$C_{ans} = C_{cs} * CRF_s + C_{os} \tag{26}$$

$$C_{anw} = C_{cw} * CRF_w + C_{ow} \tag{27}$$

Where:

- C_{an} C_{anS} C_{anw} represents the annual life cycle cost of energy for solar and wind generation.
- C_{cs}, C_{cw} represent the capital cost of investment for solar and wind.

- C_{os}, C_{ow} represent operating and maintenance costs for solar and wind
- CRF_s, CRF_w is the capital recovery factor for solar and wind

The unit costs of electricity C_{us}, C_{uw} of the solar and wind plants is expressed as:

$$C_{us} = \frac{C_{ans}}{E_s} \tag{28}$$

$$C_{uw} = \frac{C_{anw}}{E_w} \tag{29}$$

The cost of generated electricity by HRES in time T is given in Eq. (30)

$$CE(T) = a_s C_{us} \eta_g A_s GT + a_w C_{uw} \frac{1}{2} C_P \rho_{pair} A_w V_w^3 T + C_{battery}(T) \tag{30}$$

Where a_s, a_w are random parameters for a solar plant and wind power plant respectively. C_{us}, C_{uw} represent the unit costs of electricity generated by solar plant and wind power plants respectively. η_g is the PV generation efficiency, $A_s(m^2)$ is the area of a single solar panel used in a solar system and G_t is the solar irradiance when the plane is in a tilted position (W/m^2). C_p is the power coefficient [4], ρ_{pair} air density (kg/m^3), A_w is rotor swept area (m^3), V_w^3 is the wind speed (m/s), and T is the given time. Power generated should meet the following constraints:

1. The power generated $P_g(t)$ should satisfy the load demand constraint at time T which is expressed in Eq. (31):

$$P_g(t) = \alpha_s * P_s(t) + \alpha_w * P_w(t) \geq P_d(t) \tag{31}$$

2. The total power generated should be within the minimum P_{min} and maximum P_{max} power, this can be expressed in Eq. (32):

$$P_{min} \leq \alpha_s * P_s(t) + \alpha_w * P_w(t) \leq P_{max} \tag{32}$$

3. Variables should be within bounds as follow:

$$0 \leq n\alpha_s \leq N_s \tag{33}$$

$$0 \leq n\alpha_w \leq N_w \tag{34}$$

$$G_{min} \leq g\alpha_s \leq G_{max} \tag{35}$$

$$V_{wmin} \leq v\alpha_w \leq V_{wmax} \tag{36}$$

Where N_S represents the total number of solar plants and N_W represents the total number of wind power plants of equal capacity respectively in existence. G_{min}, and G_{max} represent the minimum and maximum irradiation, V_{wmin}, and V_{wmax} represent the minimum and maximum wind velocity respectively, $n\alpha_s$, $n\alpha_w$, $g\alpha_s$ and $v\alpha_w$ are random values within the respective bounds of the variables.

3.7 Economic Model

The Annualized Cost of System (ACS) considers both the Annualized Capital Cost (C_{acap}) and Annualized Maintenance Cost (C_{amain}) is derived by:

$$ACS = C_{acap} + C_{amain} \tag{37}$$

Thus

$$C_{acap} = C_{cap} * \frac{i * (1 + i)^{Yproj}}{(1 + i)^{Yproj} - 1} \tag{38}$$

Where C_{cap} represents the initial capital cost of components, i is the annual interest rate, Y_{proj} represents the lifetime of components. Mostly, the cost of maintenance is estimated between 20% to 25% of the total cost for the wind turbine, and that of PV module is 5%, and the battery is 10% [28].

3.8 Objective Function

The fitness objective function is expressed in term cost CE, penalty factor M and the LPSP which are expressed as:

$$fitness = CE(T) + M * [\max(0, LPSP - 0.5\%)]^2 \tag{39}$$

The proposed KSA searches for the optimal configuration that satisfies the fitness function.

4 Algorithmic Step to Implement the Proposed Model

The algorithmic step is expressed as follows:

Step 1: Set the model parameters.
Step 2: Get input data on load demand $P_d(t)$, wind velocity A(t),
solar irradiation G(t) and the economic cost variables per individual sources.
Step 3: Initial guess of the upper and lower bound of number of wind turbines,
solar panels, batteries and houses.
Step 4: Start iteration (loop until final criteria is met)
Compute position for each Kestrel using equation (1)
Evaluate the fitness function:
1) LPSP <= LPSP target,
2) ***min*** CE(T) using equation (30)
If $CE_i < CE_j$ then
Move Kestrel *i* towards *j*
End if
Update position of kestrel
Optimal configuration of the wind turbine, solar plant, batteries, houses
Step 5: End loop
Step 6: Display results of optimal size and cost

5 Parameter Setting for the Model

In a comparison of the unique characteristics of the KSA and WSA-MP with minus previous step, the following can be stated: In WSA-MP with minus the previous step, the behaviour is based on attractiveness, collective behaviour and escape from its enemy. In KSA, the search parameters for perched mode is 0 and flight mode 2, this ensures adequate search space [23]. In this paper, the KSA is evaluated against WSA-MP with the minus previous step. In WSA-MP with the minus previous step, the parameter for randomness (σ) is 0.2 and escape parameter is 0.25.

These nature-inspired algorithms are applied firstly to check the ambient temperature, wind data and load demand. Secondly, we attempt to reduce/minimize the economic cost of combining the different power sources. In the process, we perform 100 iterations to improve on the cost of combining different renewable energy sources and to produce an optimal size on the number of solar plants, wind turbines, batteries and houses.

The underlying assumption for the HRES includes but is not limited to the lifetime of the HRES [29]. An optimal combination is selected based on the satisfied LPSP and the minimum cost of the HRES [27].

For solar, the level of efficiency n is 0.986. The input data $G(t)$ is the solar irradiation which is measured in kWh/m^2, and $A(t)$ is wind velocity (m/s). $Cp = 0.593$ [4],$\beta_t = 0.004$ [6], the unit cost of power/electricity from PV and wind are 0.95 US dollars and the power is measured in \$/kilowatt-hour (kWh)). The area A_s of the solar panel (m^2) and panel efficiency are based on the design variable for the type of panel. The standard atmospheric condition that is air density (ρ) is approximately 1.225 kg/m^3 at sea level.

The cost variables considered in this simulation includes the initial operating and maintenance cost, interest rate, lifetime cost of devices and equipment, battery, diesel, cost of PV, wind turbine, the unit cost of power from solar, wind, etc. The annualized maintenance cost can either be estimated or constant fixed value. The considered penalty values is 0.9. The initial cost of wind turbine is 2000 US dollar, PV 3400 US dollars, and the battery is 280 US dollars, diesel is 1000 US dollars. The interest rate is 12%.

To test our proposed model on actual data, we use solar radiation data, ambient temperature, data on battery and load demand from rural household in Mersing, Malaysia. The power generated by PV modules is computed using the monthly input data on the solar radiation on the tilted surface and the ambient temperature per degree celsius in Table 1. The temperature at the reference condition was 25 °C.

Specification of wind turbine includes cut-in-wind speed (0.5 m/s), cut-out-wind speed (40 m/s), rated wind speed (9.5 m/s), rated power (5 kW), wind speed (1.2 m/s), H_{low} (43.6 m), H_{high} (70 m), blade diameter is 4 m and lifetime (24 months).

Depth of battery discharge is 0.8 and the state of battery charge is 0.2. Battery capacity is 40 kWh, nominal power is 4 kW, and based on manufactures data the lifetime of the battery is 12 months. Table 2 shows the hourly load demand per day for 24 h.

The candidate solution of each algorithm consists of six variables in the form of Npv, Nw, Nbat, Nhouse, LPSP and Cost.

Table 1. Monthly solar radiation and ambient temperature

Solar radiation (kWh/m^2)	Ambient temperature °C
4.12	25.5
4.97	25.7
5.02	26.1
5.08	26.5
4.84	26.6
4.66	26.3
4.61	26
4.7	26
4.83	26.1
4.7	26.3
4.05	26.1
3.62	25.7

Table 2. Hourly load demand per day

Hour	Load demand	Hour	Load demand
1	1.5	13	4.4
2	1	14	4
3	0.5	15	3.43
4	0.5	16	3
5	1	17	1.91
6	2	18	2.48
7	2	19	3
8	2.5	20	3
9	2.5	21	3.42
10	3	22	3.44
11	3	23	2.51
12	5	24	2

6 Experiment Results

The KSA and WSA-MP algorithms are applied to find the optimal configuration for the HRES model. The HRES system was sized for loss of power supply probability (LPSP) of zero, which shows the load will always satisfy the demand. The cost and

LPSP are computed for each combination of a renewable energy source. We present the experiment results in Tables showing the Npv, Nw, Nbat, Nhouse, LPSP and the cost.

Table 3, in respect of solution#1, it shows that the number of Npv, Nw, Nbat, Nhouse which are 19, 7, 9 and 60 satisfy LPSP value of zero with the minimum cost of 0.045 US dollars. In respect of solution#3, it shows that the number of Npv, Nw, Nbat, Nhouse which are 10, 3,1 and 84 do not satisfy LPSP value of zero with the minimum cost of 0.0276 US dollars.

Table 4 shows the optimal sizing of HRES using WSA-MP. It is observed that, in respect of solution#1, the number of Npv, Nw, Nbat, Nhouse which are 37, 10, 3 and 4 satisfy LPSP value of zero with the of cost 0.781 US dollars. However, in solution#7, it is observed that the number of Npv, Nw, Nbat, Nhouse which are 6, 19, 3 and 61 do not satisfy LPSP value of zero with the cost of 0.3832 US dollars.

Table 3. Optimal sizing of the HRES using KSA

Solution#	N_{pv}	N_w	N_{bat}	N_{house}	LPSP	Cost ($)
1	19	7	9	60	0	0.045
2	8	22	9	50	0	0.075
3	10	3	1	84	7.37e–05	0.0276
4	41	23	2	27	0	0.18797
5	14	8	10	56	1.92e–06	0.0434
6	15	4	5	77	0	0.06521
7	31	7	9	21	0	0.1189

Tables 3 and 4 shows that the change in generation units have an impact on the power generated in terms of the load demand and the cost. In comparison, KSA produced optimum size at minimal costs than WSA-MP.

Table 4. Optimal sizing of the HRES using WSA-MP

Solution#	N_{pv}	N_w	N_{bat}	N_{house}	LPSP	Cost ($)
1	37	10	3	4	0	0.78176
2	39	17	1	6	0	0.62814
3	31	24	16	88	0	0.20161
4	11	9	13	44	0	0.0488
5	19	12	31	69	0	0.3024
6	31	5	10	5	0	0.5988
7	6	19	3	61	0.18894	0.3832

7 Conclusion

In this paper, a nature-inspired algorithm is proposed to optimize the size of wind-solar HRES with battery storage. The economic factors and the LPSP model are used to determine the optimal configuration that ensures the reliability of the power supply and at the optimal cost of energy. We tested and evaluated the performance of KSA by comparing with WSA-MP. The simulation result shows that there is a minimum cost in combining a different source of power generation to meet load demand using KSA. Whereas, in WSA-MP, there is a high cost in combining different sources of energy. Future work will focus on collecting real-time data from IoT based devices connected to the proposed HRES model with edge computing capability.

Acknowledgement. The authors are thankful for the research supported grant by both the National Research Foundation of South Africa with grant number 117799 and the National Research Foundation of Korea (NRF) grant funded by the Korean government (MSIP) (NRF-2018K1A3A1A09078981).

References

1. Castro, D., Misra, J.: The Internet of Things (2013)
2. Nayanatara, C., Divya, S., Mahalakshmi, E.K.: Micro-Grid Management Strategy with the Integration of Renewable Energy Using IoT, pp. 160–165
3. El Kafazi, I., Bannari, R., Adiba, E.B.E.I., Nabil, H., Dragicevic, T.: Renewable energies: modeling and optimization of production cost. Energy Procedia **136**, 380–387 (2017)
4. Bhandari, B., Lee, K.-T., Lee, G.-Y., Cho, Y.-M., Ahn, S.-H.: Optimization of hybrid renewable energy power systems: a review. Int. J. Precis. Eng. Manuf. Green Technol. **2**(1), 99–112 (2015)
5. World Energy Council: World Energy Trilemma Index (2019)
6. Kajela, D., Manshahia, M.S.: Optimization of Renewable Energy Systems: A Review. vol. 3, no. 8 (2017)
7. Ghofrani, M., Hosseini, N.N.: Optimizing hybrid renewable energy systems: a review. In: Zobaa, A.F., Afifi, S.N., Pisica, I. (eds.) Sustainable Energy – Technological Issues, Applications and Case Studies. InTech (2016)
8. El-Bidairi, K.S., Nguyen, H.D., Jayasinghe, S.D.G., Mahmoud, T.S.: Multiobjective intelligent energy management optimization for grid-connected microgrids. IEEE, p. 6 (2018)
9. Amer, M., Namaane, A., M'Sirdi, N.K.: Optimization of hybrid renewable energy systems (HRES) using pso for cost reduction. Energy Procedia Mediterr. Green Energy Forum **42**(2013), 318–327 (2013)
10. Agbehadji, I.E., Millham, R., Fong, S., Hong, H.-J.: Kestrel-based search algorithm (KSA) for parameter tuning unto long short term memory (LSTM) network for feature selection in classification of high-dimensional bioinformatics datasets, In: Proceedings of the Federated Conference on Computer Science and Information Systems, pp. 15–20 (2018)
11. Tang, R., Fong, S., Yang, X.-S., Deb, S.: Wolf search algorithm with ephemeral memory. IEEE (2012)
12. Kennedy, J., Eberhart, R.C.: Particle swarm optimization. In: Proceedings of IEEE International Conference on Neural Networks, pp. 1942–1948. Piscataway, NJ (1995)

13. Yang, X.-S.: Firefly algorithms for multimodal optimization. In: Watanabe, O., Zeugmann, T. (eds.) Stochastic Algorithms: Foundations and Applications, pp. 169–178. Springer Berlin Heidelberg, Berlin, Heidelberg (2009)
14. Tang, R., Fong, S., Yang, X.-S., Deb, S.: Wolf search algorithm with ephemeral memory. pp. 165–17.
15. Agbehadji, I.E., Millham, R., Fong, S., Hong, H.-J.: Wolf search algorithm for numeric association rule mining, pp. 146–151
16. Yang, X.: A new metaheuristic bat-inspired algorithm. In: Nature Inspired Cooperative Strategies for Optimization (NICSO 2010), pp. 65–74 (2010)
17. Darchen, R., Delage-darchen, B.: American arachnological society societies of spiders compared to the societies of insects. J. Arachnol. **14**(2), 227–238 (2016)
18. Sandeep, S.R., Nandihalli, R.: Optimal sizing in hybrid renewable energy system with the aid of opposition based social spider optimization. J. Electr. Eng. Technol. 10 (2019)
19. Tabasi, M., Asgharian, P.: Optimal operation of energy storage units in distributed system using social spider optimization algorithm. AIMS Electron. Electr. Eng. **3**(4), 309–327 (2019)
20. Agbehadji, I.E., Millham, R.C., Fong, S.J., Yang, H.: Bioinspired computational approach to missing value estimation. Math. Probl. Eng. **2018**, 1–16 (2018)
21. Agbehadji, I.E., Frimpong, S.O., Millham, R.C., Fong, S.J., Jung, J.J.: Intelligent energy optimization for advanced IoT analytics edge computing on wireless sensor networks. In: International Journal of Distributed Sensor Networks, Nature Inspired Optimization for Sustainable Wireless Sensor Networks, SAGE Publishing, vol. 16, no. 2, pp. 1–18 (2020)
22. Agbehadji, I.E., et al.: Bio-inspired energy efficient clustering approach for wireless sensor networks. In: 7th International Conference on Wireless Networks and Mobile Communications (WINCOM'19), p. 8. Fez, Morocco (2019)
23. Agbehadji, I.E., Millham, R.C., Fong, S.J., Yang, H.: Bioinspired computational approach to missing value estimation. Math. Probl. Eng. **2018**, 16 (2018)
24. Agbehadji, I.E., Millham, R.C., Fong, S.J., Jung, J.J., Bui, K.H.N., Abayomi, A.: Multi-stage clustering algorithm for energy optimization in wireless sensor networks. In: Berry, M.W., Yap, B.W., Mohamed, A., Köppen, M. (eds.) Soft Computing in Data Science: 5th International Conference, SCDS 2019, Iizuka, Japan, August 28–29, 2019, Proceedings, pp. 223–238. Springer Singapore, Singapore (2019). https://doi.org/10.1007/978-981-15-0399-3_18
25. Agbehadji, I.E., Millham, R.C., Fong, S.J., Yang, H.: Integration of Kestrel-based search algorithm with artificial neural network for feature subset selection. Int J Bio-Inspired Comput. **13**(4), 222 (2019)
26. Hassan, Q., Jaszczur, M., Przenzak, E.: Mathematical model for the power generation from arbitrarily oriented photovoltaic panel. E3S Web Conferences **14**, 01028 (2017)
27. Tiwari, A.N., Dubey, N.: A methodology of optimal sizing for wind solar hybrid system. Asian Rev. Mech. Eng. **4**(1), 11–16 (2015)
28. Li, D., Liu, Y., Chen, Y. (eds.): CCTA 2010. IAICT, vol. 344. Springer, Heidelberg (2011). https://doi.org/10.1007/978-3-642-18333-1
29. Ming, M., Wang, R., Zha, Y., Zhang, T.: Multi-objective optimization of hybrid renewable energy system using an enhanced multi-objective evolutionary algorithm. Energies **10**(674), 1–15 (2017)

Development of a Simulation Model of Personnel Logistics of a Mining and Chemical Cluster

D. N. Khaliullina[1], S. N. Malygina[1,2], E. O. Neupokoeva[1], and V. V. Bystrov[1(✉)]

[1] Institute for Informatics and Mathematical Modeling Kola Science Centre of the Russian Academy of Sciences, Apatity, Murmansk region, Russia
{khaliullina,malygina,neupokoeva,bystrov}@iimm.ru

[2] Apatity Branch of Murmansk Arctic State University, Apatity, Murmansk region, Russia

Abstract. The work is devoted to the development of applied decision support systems based on the application of simulation modeling. The paper presents the intermediate results of the research focused on the development of methodological and software tools to support the management of personnel security of the regional mining and chemical cluster. Some aspects of creating a component of the applied decision support system, which is responsible for predicting the state of personnel security of the production cluster, are considered. The developed component is a poly-model complex, built on the basis of system dynamics, discrete-event and agent-based modeling. The article focuses on the issues of simulating the processes of labor movement within the cluster enterprises. The authors consider in more detail the software and algorithmic implementation of internal and external personnel logistics processes for one single enterprise. Within the framework of the work, an analysis of the difficulties encountered in the implementation of the component of forecasting the state of personnel security of a production cluster is made.

Keywords: Decision support system · Simulation modeling · Agents · Discrete-event modeling · Personnel security · Production cluster

1 Introduction

In recent years, the Institute for Informatics and Mathematical Modeling of the KSC RAS has been working on creating a decision support system in the field of personnel security management of the regional mining and chemical cluster. The paper [1] considers the architecture of this software system and provides a general algorithm for its functioning. In this article, we would like to consider in more detail the issues related to the development of the forecasting component of this decision support system. In particular, we focus on the development of the poly-model complex of personnel logistics of the regional production cluster as a fundamental element of the forecasting component.

It should be noted that modeling socio-economic processes is a rather difficult task. If we approach the issue of personnel logistics of the country or region, it is necessary to

R. Silhavy et al. (Eds.): CoMeSySo 2021, LNNS 232, pp. 650–667, 2021.
https://doi.org/10.1007/978-3-030-90318-3_52

take into account that the modeling of such processes will inevitably have a generalizing nature. Narrowing the scope of consideration to a separate enterprise or cluster will allow for modeling to take into account in more detail the features of the movement of personnel.

At the ongoing stage of the study, the following basic logistical processes in the cluster staffing are implemented:

1 Hiring unemployed persons.
2 Releasing staff unit.
3 Internal migration of personnel – transfer from one position to another within one enterprise.
4 External migration of personnel – transfer of an employee to another organization.
5 Changing staff list of the enterprise as a result of its functioning.
6 Demographic processes in the region.

Within the framework of modeling, special attention is paid to the reproduction of real processes: the search for vacancies for unemployed persons, the search for applicants for an open vacancy by an enterprise. It is proposed to carry out these procedures on the basis of competitive selection with the implementation of algorithms of multi-criteria choice.

2 Background

The area of current research is the study of modern trends in the field of human resource management of large socio-economic entities, for example, a regional production cluster. In particular, the authors focused on the creation of special software tools for information and analytical support for personnel security management of such systems.

In general, studies of the problems of labor resources management in socio-economic systems are diverse. Various researchers around the world are studying certain aspects of providing personnel to enterprises both individually and in different combinations of them. Having carried out the analysis of open sources of information on the given scientific subjects it is possible to tell that the majority of works correlate itself with such scientific direction as "management of human resources". This direction is multifaceted and includes the study from different positions of any issues related to the provision of human resources of enterprises. In recent years, a sub-direction, which is commonly called information systems for human resource management, has been actively developing. From the point of view of this article, it is most correlated with the ongoing research. Within the framework of this scientific and practical sub-direction the questions how modern information technologies and systems can help in human resource management are solved. Conventionally, software solutions in this area can be divided into the following categories:

- universal tools to support human resource management processes within one company. Such solutions usually allow solving a wide range of practical tasks on interaction with enterprise human resources, from search of candidates and hiring to strategic

development of enterprise human resources potential, including planning and implementation of individual career trajectories of employees. One of the reviews of this category of software systems can be found in work [2].
- specialized means that focus on individual processes of staffing of a single enterprise. For example, in the processes of recruiting or certifying employees in order to confirm their professional qualifications. As an overview of the most relevant online solutions can be found in the online publication [3].
- decision support systems for socio-economic macro systems. Within the framework of this category, software solutions are created that operate with the concepts of staffing at a higher level – territorial entity (city) or on the scale of a separate region, country or set of countries, as well as individual sectors of the economy. The traditional tasks of such software systems have become forecasting of supply and demand in the labor market, identifying trends in labor migration and personnel policy of territorial entities, etc. As an example of such a category of software solutions, we can consider the work [4], in which the authors propose an algorithm for optimal human resource planning based on the concept of a smart city. Another example is the work [5], which focuses on computer modeling of staffing processes for the healthcare system.

In Russian science, most of the research in the field of labor resource management can be attributed to the direction called "labor economics". This scientific and practical direction is also wide and includes the study of different aspects of staffing in various socio-economic systems. As part of the current study, the authors focused their attention on such a sub-direction as personnel logistics.

Personnel logistics of an organization is a section of logistics, which studies the optimization of the flow of human resources of enterprises and industry as a whole [6]. Personnel are characterized by high mobility, so the concept of the flow of labor resources fully reveals the need for constant monitoring and management of changes in human resources. Like any other type of resources, personnel must enter logistics systems (be hired), develop and be used in them (fulfill their job duties, train, move to other positions) and go beyond (to be dismissed). Thus, the personnel logistics of the organization has four main directions [7]:

1. optimization of input flows in accordance with the needs of the company;
2. using personnel;
3. development of personnel;
4. releasing personnel.

We adhered to this point of view when developing a poly-model complex of personnel logistics for a regional production cluster.

Within the framework of this section of the article, only a grain of scientific knowledge of the world community devoted to the study of various problems of human resource management is considered. We tried to show only the immediate vicinity of the scientific direction that we are engaged in the framework of the ongoing study.

3 Methods and Tools

Within the framework of this article, aspects of the practical implementation of a poly-model complex of personnel logistics of a regional production cluster are considered. The poly-model complex is a set of interconnected computer models, each of which implements certain socio-economic processes. A distinctive feature of this kind of complex is that any model included in its composition is developed using different approaches and modeling methods.

The following simulation methods were used in the practical implementation of the poly-model complex:

- system dynamics of J. Forrester's;
- agent-based modeling;
- discrete-event modeling.

System dynamics is a rather popular method of studying complex dynamic systems, proposed by J. Forrester in the middle of the last century [8, 9]. A feature of this method is the representation of the object of research in the form of a system of levels and flows that change the values of these levels, as well as their interaction by feedbacks. System dynamics has become quite widespread in the study of socio-economic processes due to its relative simplicity and ease of practical application.

Agent-based modeling is a relatively young direction in simulation modeling. It is associated with the transfer of ideas about representing the work of complex systems in the form of a network of interacting independent components [10, 11]. Thus, from the standpoint of this approach, the object of study is presented in the form of a set of interacting proactive elements, each of which has its own behavior algorithm.

Discrete-event modeling, as its name suggests, is associated with two basic concepts: discrete time and event [12, 13]. From the point of view of this approach, the object of research is considered as a system of any nature, the behavior of which can be described as a sequence of certain events occurring at discrete time intervals. Thus, discrete-event modeling emphasizes algorithmization rather than developing the structure of the system itself.

In this work, the poly-model complex is implemented in the software tool system Anylogic University 8.7, designed for open research [14]. Anylogic is a flexible and powerful tool for creating various computer models. It allows developing models on the basis of integration of all simulation modeling methods listed above.

4 Results

4.1 Architecture of the Poly-model Complex

The architecture describes the composition and interrelation of computer models, each of which reflects one of the components of the cluster's personnel security or auxiliary socio-economic systems (Fig. 1). The architecture is focused on the joint application of the method of system dynamics, discrete-event and agent-based modeling and consists of five structural blocks:

- "Regional economy" model,
- "Production cluster" model,
- "Regional labor market" model,
- "Regional training system" model,
- "Regional demography" model.

There are several types of relationships between structural blocks, which are determined depending on the type of objects passed through them. In Fig. 1, they are represented by two types of connections: the solid line is the flows of labor resources (personnel), the dotted line is other types of flows (information, financial, material).

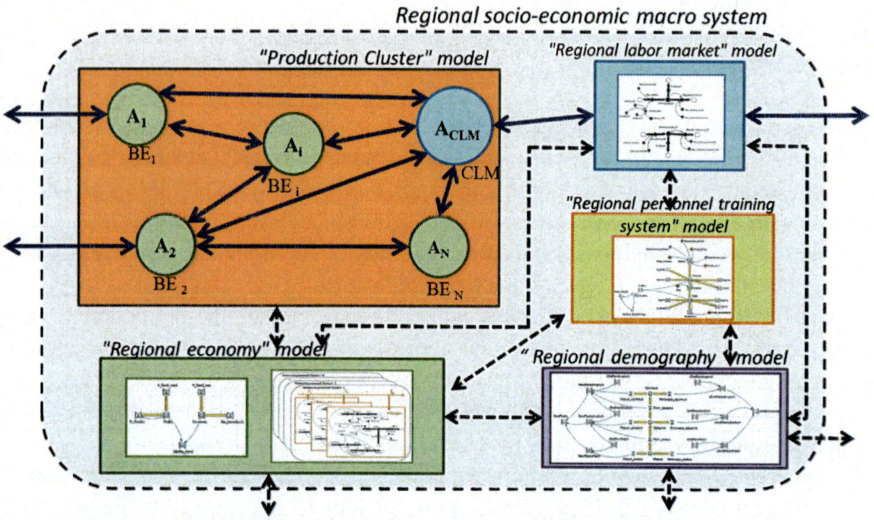

Designations: ⟷ - labor resource flows; ⇠⇢ - other types of interaction

BE_i - business entity; A_i – model agent of BE_i

CLM - cluster labor market; A_{CLM} - cluster labor market

Fig. 1. Architecture of the poly-model complex of personnel security of a regional production cluster.

The architecture of the poly-model complex presented in Fig. 1 reproduces a variety of processes occurring in the regional socio-economic macrosystem in the field of staffing. This article focuses on modeling the processes of labor resources movement in the production cluster.

4.2 Formal Description of Agent-based Models of the Movement of Labor Resources of the Cluster

Within the framework of work, the concept of "movement of personnel" means a change in the place and scope of application of labor, type of activity and production functions of the employee. The process of personnel movement is an orderly and legally regulated movement of personnel from one structural unit to another, from one position to another, both within the organization and outside it [15].

Traditionally, the movement of personnel includes:

1. Employment (filling of vacant positions). It can be carried out in two different ways:

    a. The company is looking for applicants for a vacant position.
    b. A person independently searches for a vacancy.

2. Dismissal (vacating position). It can happen for several reasons:

    a. Dismissal due to retirement (there is an opportunity for employment for another position, if the person is suitable for the parameters).
    b. Dismissal on their own request with the transition to unemployed person.
    c. Dismissal due to transfer to another enterprise.
    d. Dismissal due to reorganization or redundancy.
    e. Death of a person.

Modeling Objects

For the convenience of understanding the structure of the developed complex, we give a formal description of some of its components responsible for the movement of labor resources in the cluster. The following types of agents are considered as the main objects of modeling:

1. Able-bodied population.
2. Enterprises.
3. Positions.

The able-bodied population (W) in the model are people aged 18 years and older. This type of agents is described by the following parameters:

$$W = \langle name, sex, age, wage, exp, edu, spec, org, maxage \rangle,$$

where

- *name* – agent's name (double);
- *sex* – agent's gender (double);
- *age* – agent's age (double);
- *wage* – agent's salary (double);
- *exp* – experience (double);
- *edu* – education (double);
- *spec* – specialization (string);
- *org* – name of organization (string).
- *maxage* – retirement age (double).

In the implemented model, all agents of the type "Able-bodied population" (W) are divided into two populations ("Workers" (ws) and "Unemployed" (ums)). depending on whether a person is employed or not.

The type of agents "Position" (St) is described as follows:

$$St = \langle name, level, salary, sex, flag, minage, maxage, exp, edu, spec, org \rangle,$$

where

- *name* – title of the position (string);
- *level* – position rank (double);
- *salary* – salary (double);
- *sex* – required gender (double);
- *flag* – position employment parameter (double);
- *minage* – minimum age (double);
- *maxage* – maximum age (double);
- *exp* – required level of experience (double);
- *edu* – required level of education (double);
- *spec* – required level of specialization (string);
- *org* – name of organization (string).

Agents of this type, depending on the value of the *flag* parameter, can refer either to "Vacancies" ($flag = 0$) or to "Occupied Positions" ($flag = 1$).

In addition to the populations of agents of workers, unemployed and vacancies, the model implements the agent "Enterprise" (Work), the description of which has the following form:

$$Work = \left\langle nameWo, free, busy, total, ves_{age}, ves_{edu}, ves_{exp}, \{\text{ws}\}, \{\text{stes}\} \right\rangle,$$

where

- *nameWo* – company name (string);
- *free* – number of vacancies (double);
- *busy* – number of positions occupied (double);
- *total* – total positions (double);
- *ves_age* – the criterion that determines the value of the age of the applicant for the vacancy (double);
- *ves_edu* – the criterion that determines the value of an applicant's education for a vacancy (double);
- *ves_exp* – the criterion that determines the value of the applicant's work experience for a vacancy (double);
- {ws}– the "Workers" population, consisting of agents of the "Able-bodied population" type;
- {stes} – the "Staffing" population, consisting of agents of the "Position" type.

From the point of view of technical implementation, the employee's communication with the enterprise is carried out through the agent "Position", to which he is tied when hiring for work. The behavior of agents is implemented using the appropriate statechart diagrams (Figs. 2, 3) through the exchange of messages between agents such as "Able-bodied population" and "Position".

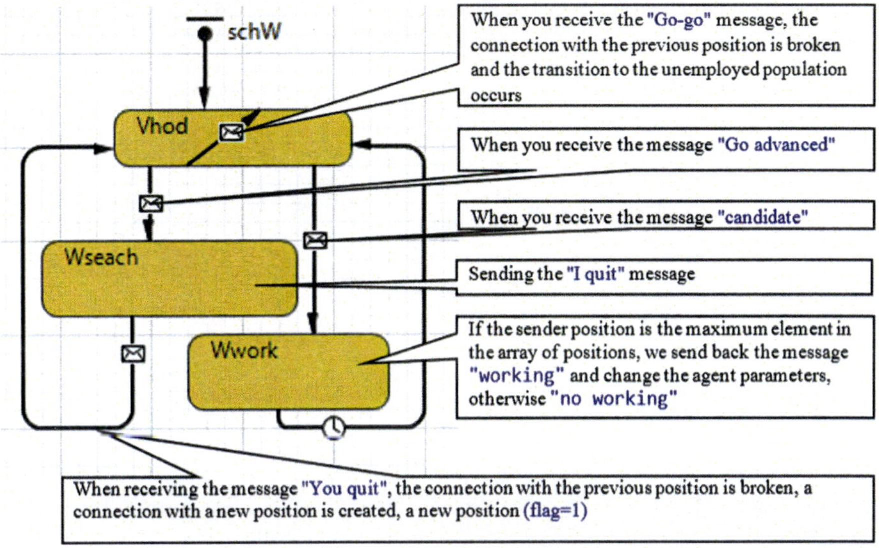

Fig. 2. Statechart diagram of the "Able-bodied population" agent

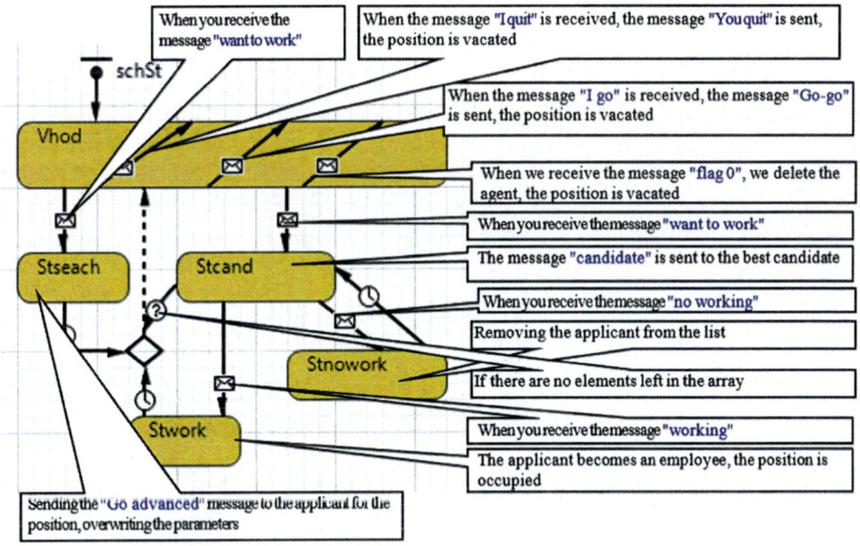

Fig. 3. Statechart diagram of the «Position» agent

A graphical representation of the model implementation of the proposed agents is presented in the Fig. 4. The upper level of the model (Main) contains two populations:

"Unemployed" (ums), consisting of agents of the "Able-bodied population" type, and "Enterprises" (works). Each agent "Enterprise" contains two populations "Workers" (ws) and "Staffing" (stes).

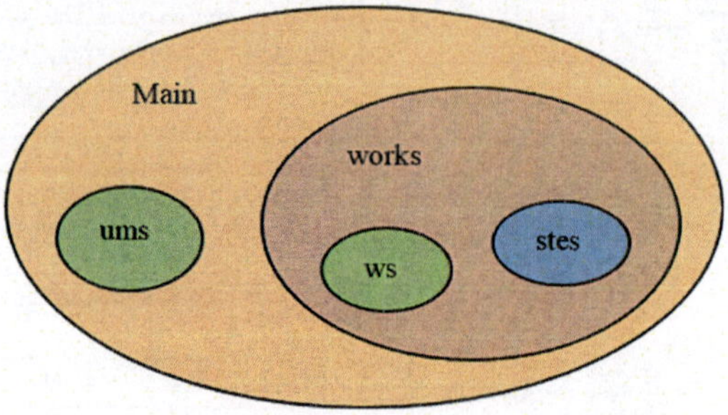

Fig. 4. Nesting of agent populations in the model.

All the initial information about enterprises, vacancies and specializations is stored in Excel file.

Implementation of a Complex of Simulation Models of Personnel Logistics of the Cluster

When the model is started, initialization occurs: as a result of calling the CreateWorld() function, the source data is read from an Excel file and the populations are sequentially created and populated by agents (Fig. 5).

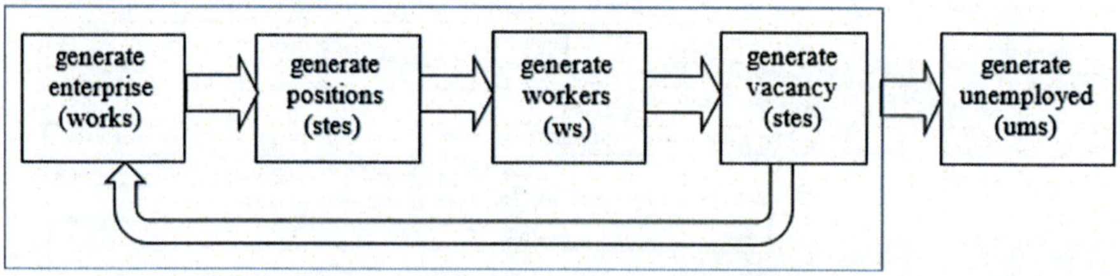

Fig. 5. Initialization of a set of models

The main processes are implemented in the model by the following functions:

death() – the aging of an agent of the "Able-bodied population" type. If the agent belongs to the "Unemployed" population, then when the retirement age is reached, it is removed from the model. If an agent belongs to the "Workers" population, then two processes are implemented for him: "Dismissal due to retirement" and "Dismissal at his own request with the transition to the unemployed population".

BornUms() – the emergence of new unemployed as a result of growing up and migration of the population.

WorkIn() – promotion to positions within the company.

GoToWork() – search for vacancies by unemployed persons.

The above functions are performed as a result of triggering the "Event" event at each step of the simulation (1 year):

```
for (int j=0; j<ums.size(); j++)
    {
      ums(j).cleanMassW(); // calling the array cleanup
function
      ums(j).death();  // calling the aging function of
unemployed persons
    }
colUms=begColUms;
BornUms(); // calling the function of appearance of new
unemployed persons
for (int j=0; j<works.size(); j++)
  {
    for (int i=0; i<works(j).ws.size(); i++)
      {
        works(j).ws(i).death(); // calling the aging work-
ers function
      }
    works(j).WorkIn(); // calling the promotion function
  }
GoToWork(); // calling the function of hiring unemployed
persons
```

Dismissal Due to Retirement

At the initial stage, to verify the functioning of the model, changing parameters of agents over the course of one year is laid. The age characteristics of agents are a key parameter in the implementation of this process. The algorithm of dismissal due to retirement is presented in Fig. 6.

It should be taken into account that if the age of a person retiring according to the age restrictions of the position allows him to get a job for another position, then in the model such an agent passes into the "Unemployed" population upon dismissal. Otherwise, he is simply removed from the model.

For the first case, when there is an opportunity to find a job, the program code has the following form:

```
if ((age>maxage) &&
(((sex==1)&&(age<65))||((sex==2)&&(age<60))))
  {
    vak.send ("I go");
  }
disconnectFromAll(); // breaking the connection with the
Position
goToPopulation(main.ums); // transfer to the "Unemployed"
population
```

In the second case, the agent in the model is destroyed:

```
if ((age>maxage) &&
(((sex==1)&&(age>65))||((sex==2)&&(age>60))))
  {
    vak.send ("flag 0");
  }
```

Dismissal at his Own Request with the Transition to the Unemployed Population

It is implemented similarly to the "Dismissal due to retirement" algorithm, except that sending a message from the agent of the "Worker" population is performed not according to a condition, but according to an event with some probability.

Search for a Vacancy (External Search)

The search for a vacancy is carried out once a year by the agent "Unemployed" according to the algorithm presented in Fig. 7.

When performing the function of hiring, agents are transferred from one population to another (from the "Unemployed" population to "Workers"). This process is carried out on request, while establishing a connection between each occupied position and the worker.

The Company is Looking for Applicants for a Vacant Position

When implementing this function, we suggest that all employees of the enterprise are ready for career growth, and accept the offer of promotion unconditionally. In the developed model, vacancies are offered first to employees of the current enterprise, and in the event that no one can apply for this position according to the parameters, the vacancy is transferred for consideration to all agents of the "Able-bodied population" type. The algorithm for searching for applicants consists of two parts: the first part is a search for its employees; the second is similar to the search for a vacancy (external search). Figure 8 shows the first part of the search (internal search for employees).

This section presents the implementation of only some of the declared processes of labor movement. The rest of the processes are implemented similarly and slightly differ from those considered. The processes are implemented as they take place in reality in accordance with the established rules of the Russian legislation in the field of employment.

4.3 Preliminary Results of Computational Experiments

For a preliminary assessment of the operability and adequacy of the complex of developed models, 100 computational experiments were conducted. The results of the computational experiment were exported to an excel file and further processed using the built-in tools of this tabular editor. The graphs show the average results of the experiments on the following indicators: the number of hired persons (Fig. 9), dismissed persons (Fig. 10) and the average annual number of employees (Fig. 11) in the cluster.

Verification of the adequacy of the results was carried out by comparing statistical data [16] with the average results of computational experiments in the period from 2014 to 2019. On average, the deviation according to these indicators was 12%, 8% and 6%, respectively. These quantitative estimates are crude for the following reasons:

1. Incompleteness of the poly-model complex itself.

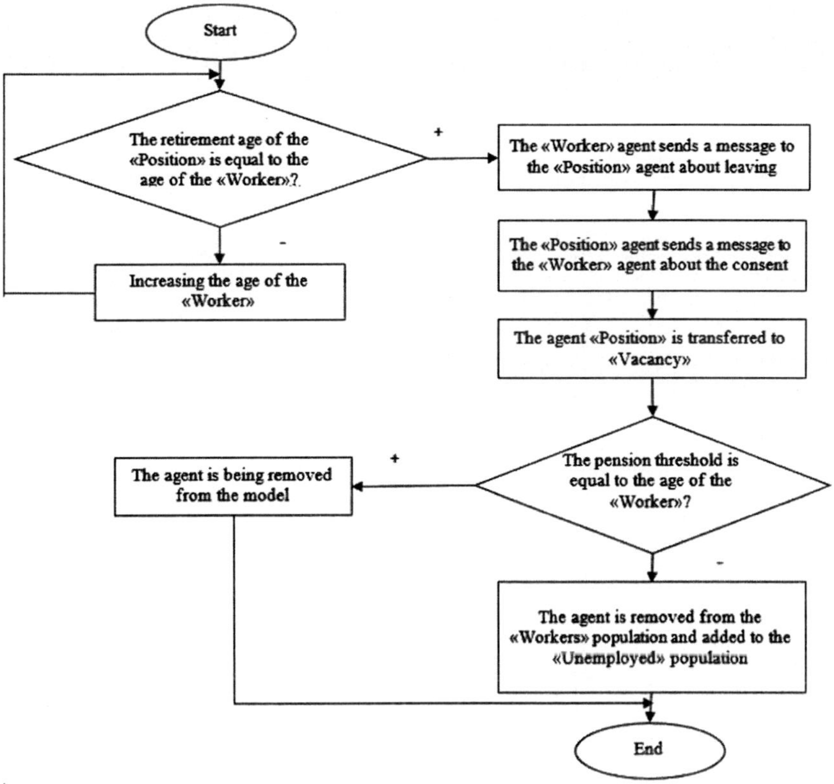

Fig. 6. Implementation of the "Dismissal due to retirement" process

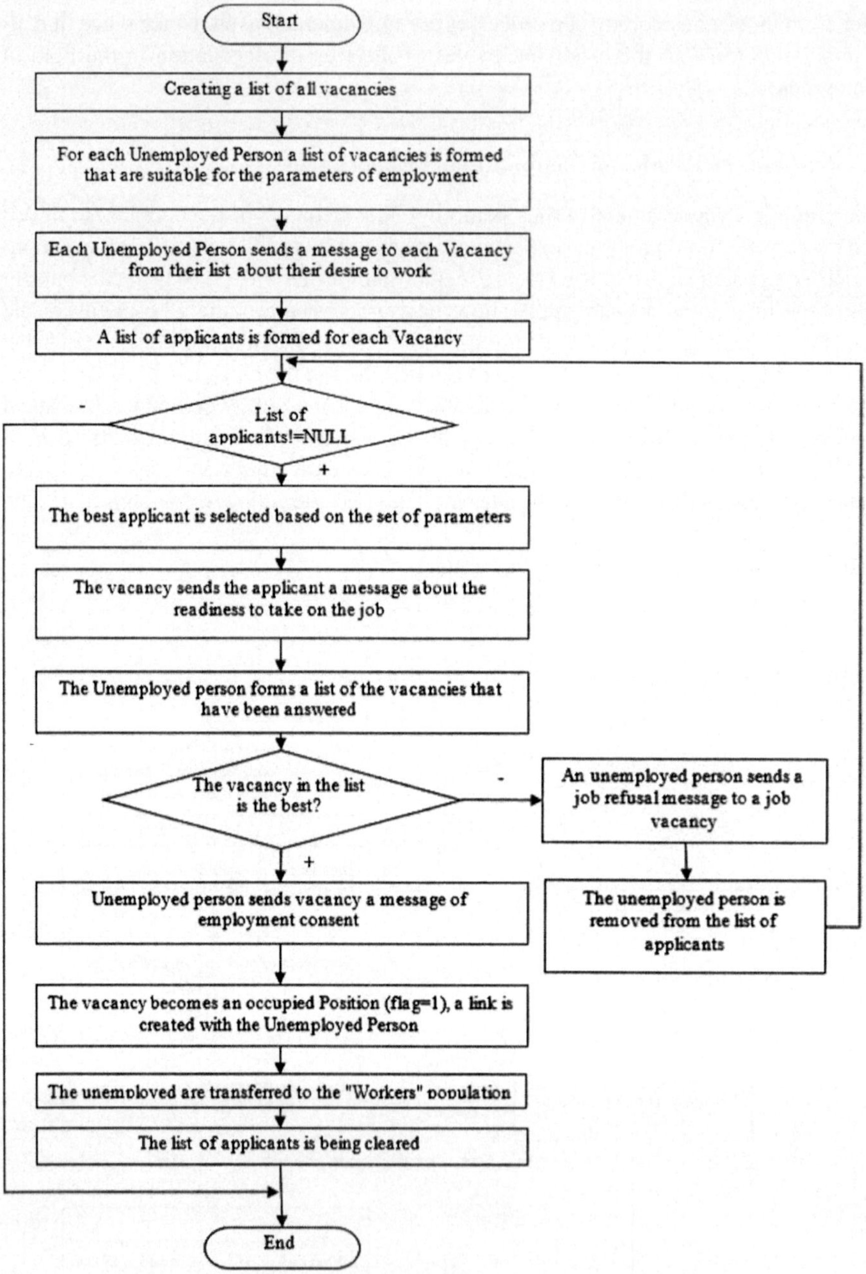

Fig. 7. Algorithm "Search for a vacancy" (external search)

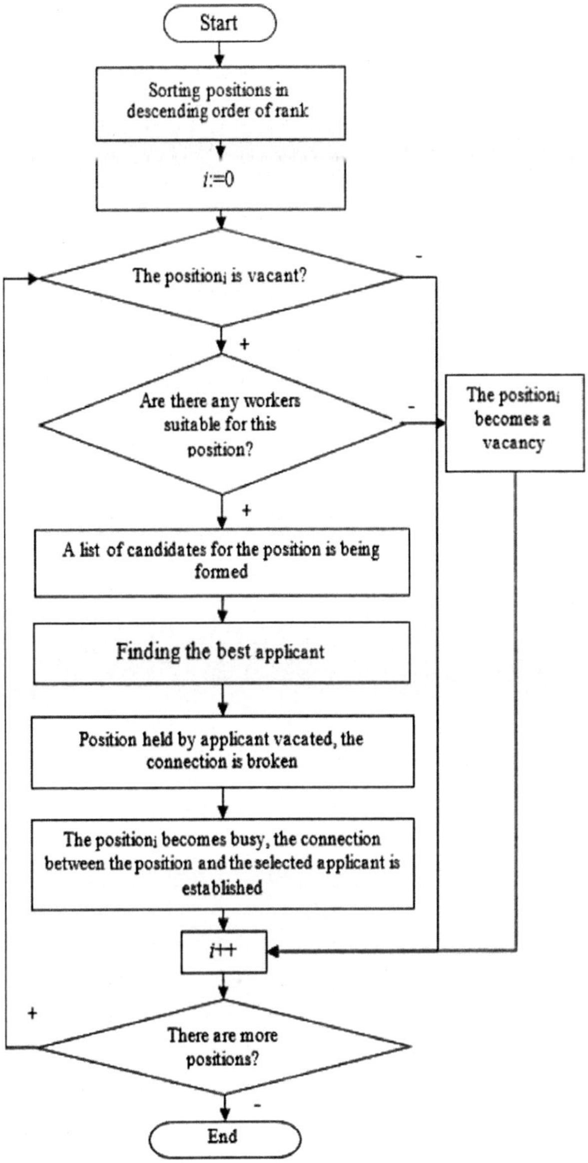

Fig. 8. Search for applicants within the enterprise (internal search).

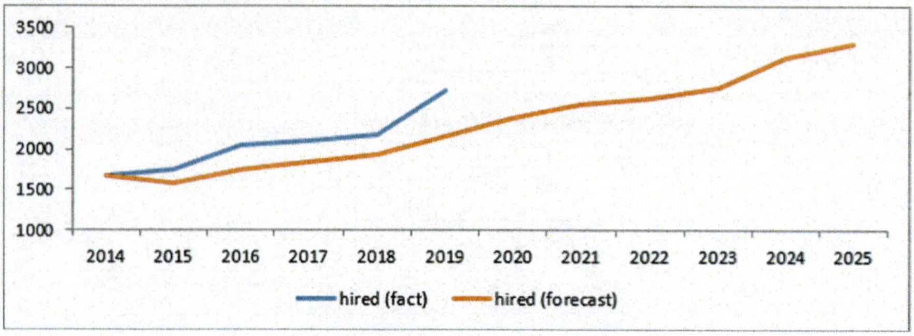

Fig. 9. Number of employees hired

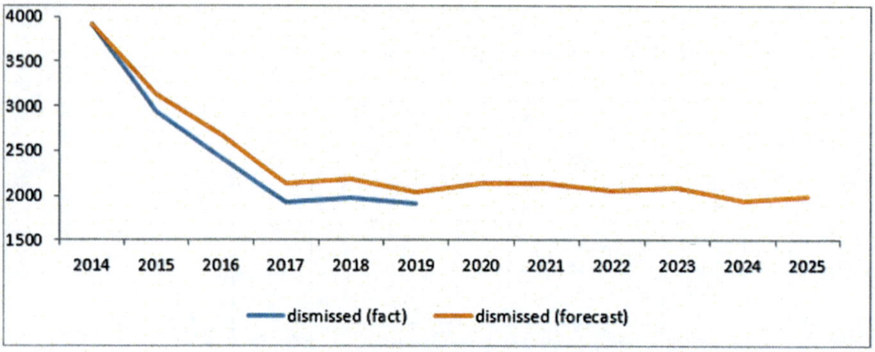

Fig. 10. Number of dismissed persons

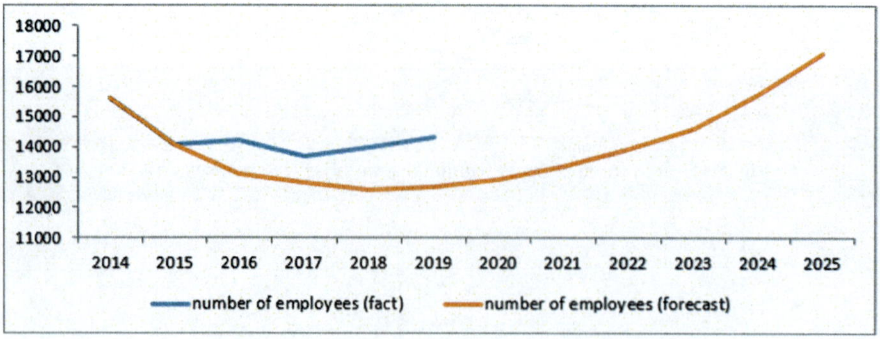

Fig. 11. Average annual number of employees

2. As initial statistical data, information from state statistics on the above indicators was taken. It should be borne in mind that there is a problem in the procedures for collecting and recording data from enterprises.
3. In connection with the peculiarities of the Russian state accounting of statistical data of the activities of enterprises, additional intermediate manipulations had to be carried out, which could also affect the purity of the initial data.

5 Discussion

The paper presents intermediate results on the implementation of the poly-model complex of personnel logistics of the regional production cluster. The main directions of continuation of work are:

- Adding a full-fledged model to the complex that reproduces the main processes of the enterprise functioning and operates with such concepts as production volumes, profit, capacity, efficiency. For this purpose, it is planned to use econometric regularities, for example, through the application of the Cobb-Douglas production function. It is planned to use the works on demo-economics by Y.S. Popkov [17] as a basis for application of such mathematical apparatus in the form of realization of computing procedures.
- A more detailed representation of demographic processes in the complex under development. In particular, external labor migration procedures for both the cluster and the region as a whole can be added.

It is worth noting separately the problem that we faced during the implementation of this research project. This problem is related to the openness and availability of the initial data for computational experiments:

- open reporting materials present only aggregated data, which require additional science-intensive operations to be used as input information for forecasting;
- aggregation of data is related to the peculiarities of the organization of the collection of official national statistics, as well as the organizational structure of the economic entities included in the production cluster. So, at the current moment, almost all enterprises of the mining and chemical cluster of the Murmansk region are part of larger international and national corporations. This leaves its mark on the formation of reports on the activities of such companies;
- reluctance of commercial companies to disclose detailed information about their actual economic activity indicators, which is also caused by competition and the desire to maintain commercial secrecy.

6 Conclusion

The paper presents only a fragment of a poly-model complex, which is directly related to the simulation of the processes of the movement of labor resources both within and between the enterprises of the production cluster. The authors have proposed a formal description of the components of the poly-model complex of the movement of labor resources in the production cluster. Aspects of the program implementation of individual processes related to personnel logistics, both internal and external to an individual enterprise, are given. Preliminary estimates of the adequacy of the developed complex of simulation models are made on the basis of comparison with the data of official state statistics for individual macro indicators.

The work will continue in several directions:

- simulation of production processes and their impact on the flows in the personnel logistics of the cluster;
- formation of values of certain parameters of models based on the analysis of additional sources of information, in particular, using automated data analysis procedures for general and specialized social networks;
- adding simulation of personnel training processes, in particular professional development.

The developed component for forecasting personnel logistics of a regional production cluster will make it possible to form forecasts for the provision of personnel for cluster enterprises in the study of different scenarios of the socio-economic development of the region.

Acknowledgements. The research was supported by the Ministry of Science and Higher Education of the Russian Federation (project No. 0226-2019-0035 «Configuration models and methods of the adaptive multi-level network-centric systems for regional security control in Arctic region of Russian Federation») and granted by the Russian Foundation for Basic Research (RFBR grant No. 19-07-01193-a «Methods and means of information support of personnel security management of the regional mining and chemical cluster»).

References

1. Bystrov, V.V., Khaliullina, D.N., Malygina, S.N.: Architecture of the decision support system for personnel security of the regional mining and chemical cluster. In: Silhavy, R., Silhavy, P., Prokopova, Z. (eds.) CoMeSySo 2020. AISC, vol. 1294, pp. 442–463. Springer, Cham (2020). https://doi.org/10.1007/978-3-030-63322-6_36
2. Subhi, R.M.Z., Hanan, M.S., Bzar Kh, H.: Human resource management systems for enterprise organizations: a review. Period. Eng. Nat. Sci. **7**(2), 660–669 (2019). https://doi.org/10.21533/pen.v7i2.428
3. Alyson, T.: The 10 best HRMS for human resource management in 2021. People managing people. https://peoplemanagingpeople.com/tools/hrms-human-resources-management-system/ (2021)
4. Zhang, L., Yang, W.: Simulation of enterprise human resource scheduling algorithm optimization in the context of smart city. Complexity **2020**, 1–10 (2020). https://doi.org/10.1155/2020/8830335
5. MacKenzie, et al.: Hum. Resour. Health (2019). https://doi.org/10.1186/s12960-019-0376-2
6. Serbin, V.D.: Fundamentals of Logistics: Manual. Taganrog, Russia (2004).(In Russian)
7. Kormin, N.G.: The application of personnel logistics in the personnel management of the organization. Bull. ChelSU **7**, 138–142 (2008). (In Russian)

8. Forrester, J.W.: World Dynamics, 2nd edn. Wright-Allen Press, p. 146 (1973). https://monoskop.org/images/d/dc/Forrester_Jay_W_World_Dynamics_2nd_ed_1973.pdf
9. Sterman, J.D.: Business Dynamics: Systems Thinking and Modeling for a Complex World, p. 1008. McGraw-Hill, Boston (2000)
10. Bonabeau, E.: Agent-based modeling: methods and techniques for simulating human systems. Proc. Natl. Acad. Sci. U.S.A. **99**, 7280–7287 (2002)
11. Railsback, S.F., Grimm, V.: Agent-Based and Individual-Based Modeling: A Practical Introduction, 2nd edn., p. 352. Princeton University Press (2019)
12. Bernd, P., Nicolas, K., Kruse, A.: A Discrete Event Simulation Framework for Agent-Based Modelling of Logistic Systems. INFORMATIK 2007: Informatik trifft Logistik. Band 1. Beiträge der 37. Jahrestagung der Gesellschaft für Informatik e.V. (GI), pp.24–27 (2007). https://www.researchgate.net/publication/221387171_A_Discrete_Event_Simulation_Framework_for_Agent-Based_Modelling_of_Logistic_Systems
13. Banks, J., Carson, J., Nelson, B., Nicol, D.: Discrete-event system simulation, – 4th edn., p. 271. Pearson (2005)
14. Anylogic: Homepage, https://www.anylogic.com/
15. Vertakov, Y.V.: Economics and organization of production: textbook, p. 381. INFRA-M, Moscow (2014) (in Russian)
16. Territorial agency of the Federal State Statistics Service for the Murmansk Region: https://murmanskstat.gks.ru/folder/36664. Accessed 12 June 2021
17. Popkov, Y.S.: Mathematical Demoeconomics: Integrative Demographic and Economic Approach. De Gruyter, Berlin, 500 p. (2014). DOI: https://doi.org/10.1515/9783110339161

MobileNetV2 in the Classification of Avian Influenza and CRD in Chickens

My Vo Dang Uyen, Nha Truong Thanh, Anh Nguyen My, Hang Lam Thi Khanh, Lan Le Thi Thu, and Luyl-Da Quach(✉)

Information Technology Department, FPT University, Can Tho, Vietnam
{MyVDUCE130115,NhaTTCE130121,AnhNMCE140573,HangLTKCE130393, lanltt11}@fpt.edu.vn

Abstract. The neural network architecture used on mobile devices is increasingly popular, especially the improvement of the MobileNet model. In this study, the MobileNetV2 model with many improvements compared to the previous version was used to evaluate the classification of chicken disease images expressed on a specific part of the body (chicken head) to evaluate the using the model on the phone, integrated into smart IoT systems for medium and large chicken farming. Research results are used to evaluate the best accuracy and accuracy of 85.25%, respectively. This shows the effectiveness of applying the MobileNetV2 model, however, there are still many errors in the model, which is important in perfecting the identification system applied in the disease identification problem chickens in industrial captivity.

Keywords: MobileNetV2 · Identify the disease in chickens · Image classification · Detect CRD · Detect avian influenza

1 Introduction

Mobile devices evolve rapidly has led to the development of neural network architecture models running on it, typically MobileNet. In this study, we use MobileNetV2 with improvements over MobileNetV1 version. MobileNetV2 inherits the advantages of a small, low-power model as it is parameterized to meet resource constraints on phones, using depth-separable aggregates as effective building blocks [1]. This will be convenient to use on phones, on IoT devices, etc. to identify diseases in chickens, especially avian influenza and chronic respiratory disease (CRD).

Avian influenza and chronic respiratory disease (CRD) are two common diseases in chickens. In particular, avian influenza (FluChicken) is caused by influenza A virus (IAV) of the Orthomyxoviridae family, which has appeared since 1997 [2]. Meanwhile, CRD, published since the 1980s, is caused by the Mycoplasma gallisepticum bacterium [3]. Both diseases lead to serious economic losses in the livestock industry such as reduced egg production, reduced meat quality, reduced fertility, and possibly death of chickens.

CRD in chickens, manifested when chickens are 4–8 weeks old, is usually associated with E.Coli disease with reduced appetite, runny nose, wheezing, conjunctivitis eyes,

R. Silhavy et al. (Eds.): CoMeSySo 2021, LNNS 232, pp. 668–678, 2021.
https://doi.org/10.1007/978-3-030-90318-3_53

swollen face, moody, picky eaters, etc. The main route of infection for CRD is through the air and through eggs [4]. In addition to CRD, avian influenza caused by AIV, AIV is classified as a low pathogenic but highly pathogenic virus that can cause large outbreaks and mortality in chickens, affecting quality of eggs and production [5]. Compared with CRD, FluChicken is dangerous to humans as it can cause infectious diseases from animals and humans, in which there are harmful substances to humans are H5N1 and H7N9 [6].

Realizing this, there have been many studies in the field of artificial intelligence in detecting diseases in poultry through phenomena and manifestations in chickens. Typically, [7] has implemented avian influenza identification through avian method according to the spectrum difference of sound with CSCNN method and focuses on avian influenza virus H9N2, the disadvantage of this method is the influence of acoustic elements in the environment of captive chickens. In contrast to this study, the study presented in [8, 9] has brought about image identification of sick chickens based on the use of VGGNet and ResNet-50 models, in which with the use of the Speeded up robust features (SURF) algorithm and K-means increase the accuracy of the model up to 93.56%. However, the disadvantage of this method is that the identification of diseases in chickens has not been focused on solving different parts such as head, body, legs, etc. because different diseases will be expressed differently on the chicken body.

From the above, the research takes an approach to solve the problems of identifying diseases in chickens on a specific part of the chicken head, especially CRD and avian influenza diseases (Fig. 1). Research suggested using the model MobileNetV2 [1] this makes a premise in embedding IoT devices to support the closed industrial chicken farming model of medium and large scale. Compared to MobileNetV1, MobileNetV2 is tested faster with accuracy across the entire latency spectrum, requiring fewer parameters and faster speeds. MobileNetV2 is used in many research, typically including [10, 11] with an accuracy of over 80%, this shows that the use of this model is appropriate in the study.

The research process was carried out according to the following steps:

- Step 1: Collect and process image data related to poultry diseases, focus on images of disease manifestations on chickens to distinguish them from normal chickens.
- Step 2: Use the MobileNetV2 model to evaluate the use of identity.
- Step 3: Evaluate the MobileNetV2 model on the dataset through accuracy, confusion matrix.

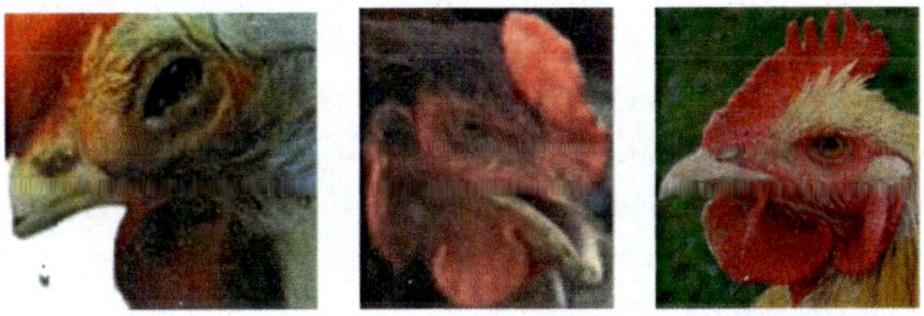

Fig. 1. Image of sick chicken flu and normal chickens (from left to right are CRD, FluChicken).

2 The Poposed Methodology

MobileNetV2 [9] introduced at CPVR (The IEEE Conference on Computer Vision and Pattern Recognition) 2018 conference aims to improve the performance of mobile models through multi-tasking and benchmarking as well as on various model sizes. The MobileNetV2 architecture has improved over the previous version thanks to the use of linear bottlenecks, depthwise separable convolution and inverted residual block.

2.1 Linear Bottlenecks

The add shortcut is adjusted so that the input and output depth of each residual block is narrowed. That's why it's called bottleneck layers (Fig. 2).

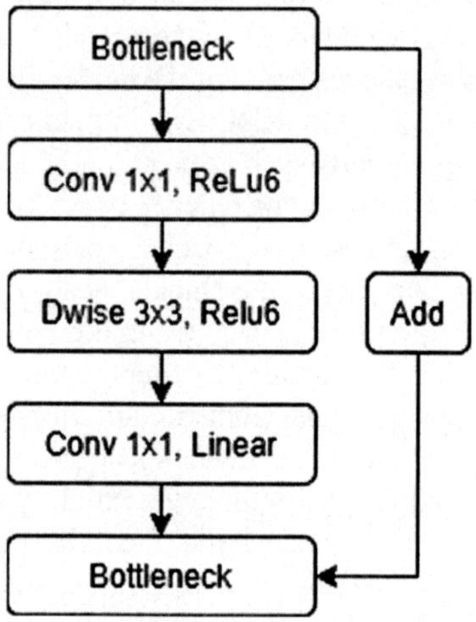

Fig. 2. Building block.

Each block has three different layers:

- 1×1 Convolution with ReLu6: The first convolution layer is a 1×1 combinator whose purpose is to expand the number of channels in the data before entering deep convolution. The expanded data are given by the expansion factor. The default expansion factor is 6.
- Depthwise Convolution: The depthwise convolution layer used to filter the final input is a convolution layer 1×1 making the number of channels smaller, also known as the projection layer or bottleneck.
- 1×1 Convolution without any linearity: It takes data with a large number of dimensions into a scale with a much lower number of dimensions. This layer also reduces the data flowing through the network.

2.2 Depthwise Separable Convolution

MobileNet has computational complexity and reduced the amount parameter is less than about nine times by using Separable Depthwise Convolution. Specifically, the process will be divided into two sequential steps: Depthwise Convolution and Pointwise (Fig. 3).

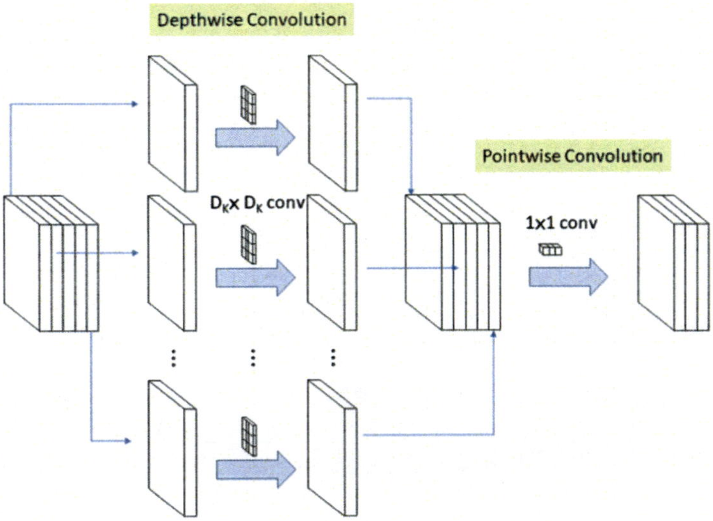

Fig. 3. The structure of Depthwise Separable Convolution.

Depthwise convolution is a type of convolution where we apply a unique convolution filter to each input channel. In conventional 2D convolution performed over multiple input channels, the filter is as deep as the input and allows us to freely mix the channels to produce each element in the output. In contrast, the variation in depth keeps each channel separate. Implementation steps:

1. Split input and filter into channels.
2. We transform each input with the corresponding filter.
3. We stack the transformed outputs together.

The complexity of Depthwise Convolution is:

$$D_k \cdot D_k \cdot M \cdot D_f \cdot D_f \tag{1}$$

Pointwise Convolution is a type of convolution that uses a 1 × 1 kernel, a kernel that iterates over every single point. This kernel has a depth equal to as many channels as the input image.

The complexity of Pointwise convolution is:

$$M \cdot N \cdot D_f \cdot D_f \tag{2}$$

The complexity of Depthwise Separable Convolution is:

$$D_k \cdot D_k \cdot M \cdot D_f \cdot D_f + M \cdot N \cdot D_f \cdot D_f \tag{3}$$

2.3 Inverted Residual Block

The intermediate layers in a block will do the nonlinear transformation, so they need to be thicker to create more transformations. Shortcut connections between blocks are performed on input and output bottlenecks, not on intermediate layers. Therefore, the input and output bottleneck layers only need to record the result and do not need to perform the nonlinear transformation (Fig. 4).

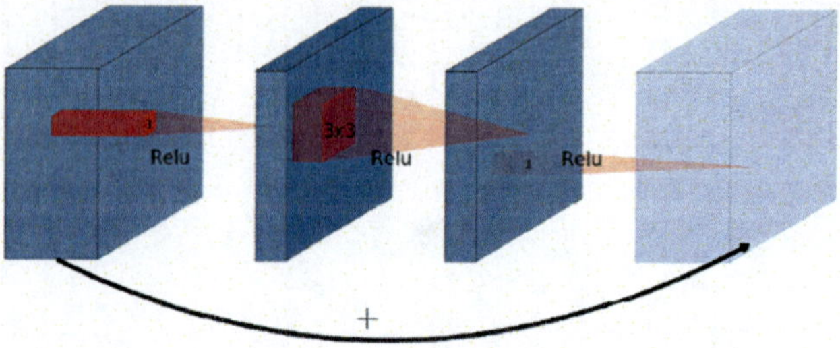

Fig. 4. A residual block.

Blocks in the previous layer are added directly to the next layer. If we consider the previous layer as x, after going through the two-way convolutional processing we get the result F(x), the final output is a residual block with the value x + F(x) (Fig. 5a).

In between layers in an inverse residual, we also use depth-separated convolutional transforms to minimize the number of model parameters. This is also the solution to help MobileNetV2 models have a reduced size, lighter computational weight than traditional models.

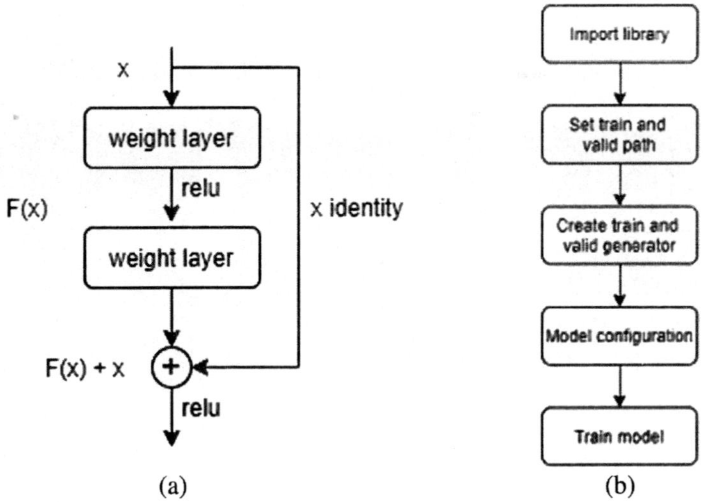

Fig. 5. a) Residual learning: a building block; and b) model training process.

3 Experiments and Results

3.1 Data Preparation

The image data set in the study was collected on the internet, the feature of the data set is the images of the disease in poultry with full body parts from head to toe (Fig. 6) with 3 types: normal chickens, CRD chickens and flu chickens. Then, we use the tool to perform chicken head image separation as input for the identification process using MobileNetV2 (Fig. 1) with the input size of the network is 244 × 244 × 3, the number of images The images are statistically specified as Table 1.

Table 1. Statistical data set of chicken diseases.

No	Label	Total species
1	CRD	114
2	Avian influenza	108
3	Normal	248
Total		470

3.2 Training

The data set was divided into training set and validation set in the ratio 67 and 33 (Fig. 5b) and pre-trained with the ImageNet dataset. We train MobileNetV2 on Google Colab with GlobalAveragePooling2D in global average aggregation for spatial data and

Fig. 6. Illustrations are originally collected.

dense layer is the trigger. The network job is trained with a batch size of 32,1,000 epochs and the learning rate is set to default. The model is compiled with the Adam optimization algorithm, the loss function is categorical_crossentropy and the metric is accuracy.

3.3 Evaluation Metrics

Research on identification of CRD and influenza in chickens using the assessment method based on accuracy, loss function, accuracy curves, confusion matrix [12]. Details are as follows:

- Accuracy refers to the degree of difference between the measured values and the actual or reference values. In which, n is noted to be the number of samples correctly predicted with m being the total number of samples included in the prediction.

$$Accuracy = \frac{n}{m} \tag{4}$$

- The Confusion matrix is used to compare which classes are classified the most, and which class data is often misclassified into another. Using the confusion matrix as a research will evaluate the inaccurate prediction value, which helps to better improve the MobileNetV2 model in classifying diseased chicken images (Fig. 7).

Predicted values	Actual Values	
	Positive (1)	Negative (0)
Positive (1)	True Positive	False Positive (Type 1 Error)
Negative (0)	False Negative (Type 2 Error)	True Negative

Fig. 7. Confusion matrix.

- o True Positive: Number of correct predictions.
- o True Negative: The number of indirectly correct predictions.
- o False Positive (Type 1 Error): Number of false predictions.
- o False Negative (Type 2 Error): The number of indirectly false predictions.
- o Precision: The metric is correct in all the given Positive predictions. This metric is calculated by the formula:

$$precision = \frac{TruePositive}{TruePositive + FalsePositive} \tag{5}$$

- o Recall: The case was predicted correctly in all the Positive cases. This index is calculated according to the formula:

$$recall = \frac{TruePositive}{TruePositive + FalsePositive} \tag{6}$$

- Accuracy: From all the classes (positive and negative), how many of them we have predicted correctly. This index is calculated according to the formula:

$$Accuracy = \frac{TruePositive + TrueNegative}{Total} \tag{7}$$

3.4 Result

In this research, the team repeatedly tested the appropriate parameters for the training process is the number of epochs and batch. The data set was scaled 67 to 33 to get the best results. We found that the results analyzed on curves appeared overfit, when the model is trained with 1000 epochs it is clearly visible. The best result we got was 85.25% with the parameters stated earlier (Fig. 8). Looking at the histogram (Fig. 9), the data are properly classified for other diseases as high as CRD (32/38), Fluchicken (24/30), and normal chicken (77/88).

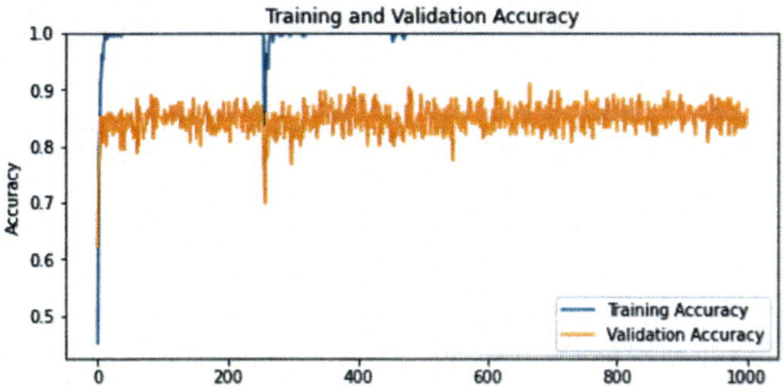

Fig. 8. Accuracy curves.

As stated in the previous section, we divide the data set into two parts: the training set and the validation set. In which, the training set is used to train the model and the validation set is used to evaluate the performance of the model. The contents on the training set show how the model is developing in terms of training, the contents on the validation set are used to evaluate the model's performance. In which, training accuracy is a measure of accuracy on training set and validation accuracy is a measure of accuracy on validation set.

Basically, the Confusion Matrix represents how many data points actually belong to a class, and are expected to fall into a class. We can immediately infer that the sum of the elements in this entire matrix is the number of points in the test set. The elements on the diagonal of the matrix are the number of correctly classified points of each data class. A good model will give a Confusion Matrix in which has the elements on the main diagonal have large values, the remaining elements have small values. In other words, when represented by colors, lines that are darker in color compared to the rest will net superior results.

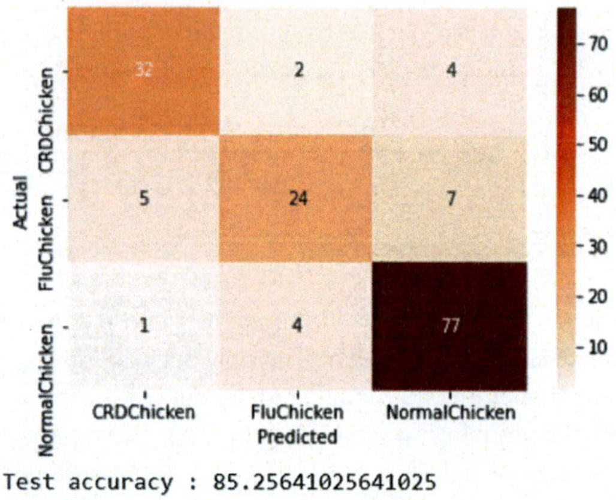

Fig. 9. Confusion matrix (without normalization)

4 Discussion

Based on the results shown in Fig. 8, it illustrates that there are errors in the identification process leading to overfitting. Nevertheless, the identification of the relative accuracy for each class depicts that the processing ability of the model has achieved high accuracy (Fig. 9). This represents the scalability of the MobileNet model in detecting diseases in chickens by overcoming the errors in the model.

5 Conclusions

Looking at the results of identifying two diseases CRD and FluChicken, which are common diseases in chickens, the research reveals the potential to use the typical classification model used on portable devices. high (Mo-bileNetv2). This opens up research directions in the use of IoT devices in quality monitoring of industrial chicken coops in medium and large scale. To further clarify how to do this, research should address the following issues:

- Using deep learning models to identify chickens in medium and large-scale barns with a stocking density of over 1,000 chickens/barn.
- Based on the obtained results, additional data on abnormal chicken images can be collected and this leads to being able to monitor individual chickens in large-scale barns.
- The comparison of models can track and classify diseased chickens based on high accuracy and consistent images on IoT monitoring devices such as Raspberry, Asus Tinker Board, etc.

Acknowledgment. This work is (partly) supported by the research project No.1 – DHFPT/2021/01 granted by FPT University.

References

1. Sandler, M., Howard, A., Zhu, M., Zhmoginov, A., Chen, L.-C.: MobileNetV2: inverted residuals and linear bottlenecks. IEEE Conf. Comput. Vis. Pattern Recognit. 4510–4520 (2018)
2. Haji, R.M., Babakir-Mina, M., Gubari, M.I.M.: Epidemiological parameter changes of global Avian Influenza H5N1 over two decades, from 1997 to 2017. Kurdistan J. Appl. Res. **2**(2), 128–136 (2017). https://doi.org/10.24017/science.2017.2.18
3. Nouh, M.S.A., Abd Alnaser, A.H.T., Zyan, K.A.M., Hashem, Y.H.M.: Molecular and histopathological investigation of mycoplsma gallisepticum isolates from chickens. Anim. Health Res. J. **7**(3), 31–41 (2019)
4. Ishfaq, M., et al.: The effect of Mycoplasma gallisepticum infection on energy metabolism in chicken lungs: through oxidative stress and inflammation. Microb. Pathog. **138**, 103848 (2020). https://doi.org/10.1016/j.micpath.2019.103848
5. Alkie, T.N., Yitbarek, A., Taha-Abdelaziz, K., Astill, J., Sharif, S.: Characterization of immunogenicity of avian influenza antigens encapsulated in PLGA nanoparticles following mucosal and subcutaneous delivery in chickens. PLoS ONE **13**(11), e0206324 (2018). https://doi.org/10.1371/journal.pone.0206324
6. Qu, Z., et al.: Identification of a key amino acid in hemagglutinin that increases human-type receptor binding and transmission of an H6N2 avian influenza virus. Microbes Infect. **19**(12), 655–660 (2017). https://doi.org/10.1016/j.micinf.2017.09.008
7. Cuan, K., Zhang, T., Huang, J., Fang, C., Guan, Y.: Detection of avian influenza-infected chickens based on a chicken sound convolutional neural network. Comput. Electron. Agric. **178**, 105688 (2020). https://doi.org/10.1016/j.compag.2020.105688
8. Quach, L.-D., Pham-Quoc, N., Tran, D.C., Fadzil Hassan, M.: Identification of chicken diseases using VGGNet and ResNet models. In: Vo, N.-S., Hoang, V.-P. (eds.) INISCOM 2020. LNICSSITE, vol. 334, pp. 259–269. Springer, Cham (2020). https://doi.org/10.1007/978-3-030-63083-6_20

9. Quach, L.-D., Quoc, N.P., Thi, N.H., Tran, D.C., Hassan, M.F.: Using SURF to improve resnet-50 model for poultry disease recognition algorithm. In: 2020 International Conference on Computational Intelligence (ICCI), pp. 317–321. Bandar Seri Iskandar, Malaysia (2020). https://doi.org/10.1109/ICCI51257.2020.9247698
10. Xiang, Q., Wang, X., Li, R., Zhang, G., Lai, J., Hu, Q.: Fruit image classification based on MobileNetV2 with transfer learning technique. In: Proceedings of the 3rd International Conference on Computer Science and Application Engineering – CSAE 2019, pp. 1–7. Sanya, China (2019). https://doi.org/10.1145/3331453.3361658
11. Buiu, C., Dănăilă, V.-R., Răduţă, C.N.: MobileNetV2 ensemble for cervical precancerous lesions classification. Processes **8**(5), 595 (2020). https://doi.org/10.3390/pr8050595
12. Buhmann, M.D., et al.: ROC analysis. In: Sammut, C., Webb, G.I. (eds.) Encyclopedia of Machine Learning, pp. 869–875. Springer US, Boston, MA (2010). https://doi.org/10.1007/978-0-387-30164-8_733

Diverging Mobile Technology's Cognitive Techniques into Tackling Malaria in Sub-Saharan Africa: A Review

Elliot Mbunge[1(✉)], Richard C. Millham[2], Maureen N. Sibiya[3], and Sam Takavarasha Jr.[4]

[1] Department of Computer Science, Faculty of Science and Engineering, University of Eswatini, Kwaluseni, Manzini, Eswatini

[2] Department of Information Technology, Faculty of Accounting and Informatics, Durban University of Technology, PO Box 1334, Durban 4000, South Africa
richardml@dut.ac.za

[3] Division of the Deputy Vice-Chancellor, Teaching and Learning, Durban University of Technology, PO Box 1334, Durban 4000, South Africa
nokuthulas@dut.ac.za

[4] Faculty of Management and Entrepreneurial Sciences, Women's University of Africa, 549 Arcturus Road, Harare, Zimbabwe

Abstract. Malaria is one of the most widespread infectious diseases caused by plasmodium parasites that continues to decimate populations globally. Several initiatives have been set as commitments towards malaria elimination and eradication. The initiatives have been rolled out to implement and monitor malaria control measures and interventions such as mosquito control, indoor residual spraying, intermittent preventive prophylaxis, insecticide-treated nets, prompt and effective treatment, and behavioural change education. To alleviate the catastrophic effects of malaria, some countries developed and deployed mobile health applications but failed to yield desired results especially in malaria-endemic countries in sub-Saharan Africa due to several emerging impediments. Therefore, we applied a rapid review to comprehensively review digital technologies and mobile applications deployed in various application domains to fight malaria while introspecting their success and challenges. The study revealed that past malaria mobile applications failed because of the digital divide, limited infrastructure, lack of internet access, among others. To strengthen malaria interventions, we propose the integration of cognitive computing techniques when developing cognitive malaria mobile health interventions. Such cognitive mobile applications could facilitate remote real-time collection of malaria data from isolated populations, disseminate malaria information, monitoring and predicting malaria prevalence and vector behaviours. This presents avenues for reconsidering malaria elimination strategies that subsequently reduce health service costs, increase accessibility, and facilitate vector control measures, contingent upon the success of countries having reached the malaria elimination phase.

Keywords: Mobile technology · Malaria elimination · Africa · Strategies

R. Silhavy et al. (Eds.): CoMeSySo 2021, LNNS 232, pp. 679–699, 2021.
https://doi.org/10.1007/978-3-030-90318-3_54

1 Introduction

Malaria remains a public health problem despite the concerted efforts made towards malaria elimination by the World Health Organization (WHO). Through the member states, the WHO set targets to eliminate malaria by 2035 in malaria-endemic countries. To achieve this, several initiatives including sustainable development goals (SDGs), WHO's Global Technical Strategy for Malaria (GTS), Roll Back Malaria (RBM) Partnership's Action and the Investment to defeat Malaria (AIM) have been set as commitments towards malaria elimination and eradication [1]. These initiatives have been rolled out to implement and monitor malaria control measures and interventions such as mosquito control, indoor residual spraying (IRS), intermittent preventive prophylaxis, insecticide-treated nets, prompt and effective treatment, and behavioural change education across the malaria-endemic regions. Such as measures partially achieved set milestones in the region through the WHO African Region between the year 2000 and 2019 [2]. For instance, there were an estimated 229 million malaria cases in 2019, and 409 000 deaths reported globally, and the WHO African Region accounted for 94% of both the 229 million malaria cases and 409,000 malaria deaths reported globally [3]. A report from the WHO (2020) indicated that malaria cases declined between 2000 and 2015, global malaria case incidence declined by 27%, and between 2015 and 2019 it declined by less than 2%, indicating a slowing of the rate of decline since 2015 [3]. African countries like Algeria, Cabo Verde are certified malaria-free since 2018 and 2019, respectively. While South Africa, Gambia, Ethiopia, Morocco, Ghana and Namibia achieved the 2020 milestones of reducing malaria incidence and deaths by 40% compared to 2015 [3]. This raises interesting concerns on how to manage malaria and the sustainability of the achieved millstones so that other countries in the region could learn. How did these countries achieve malaria-free status and 2020 milestones, and how is it sustained? What can be done differently to attain the 2030 targets despite the catastrophic impact of COVID-19 on malaria incidences? Though this is beyond the scope of this investigation, it may be worth further investigations.

Notably, the outbreak of coronavirus 2019 (COVID-19) tremendously derailed some countries in the region in their capacity to realize their milestones due to factors such as insufficient and reduced malaria budget, COVID-19 movement restrictions, substantial disruptions on the delivery of routine malaria preventive interventions and IRS activities [4], the limited number of malaria field workers, shortages of malaria commodities and poor access to health care especially for populace living in resource-constrained communities. COVID-19 restrictions subsequently affect the delivery and utilization of essential health services, including malaria services [5]. Gavi et.al [2] alluded that due to weaker health systems, the disproportionate burden of infectious diseases, including COVID-19 and Ebola; limited financial and healthcare professionals, more devastating effects of malaria are inevitable. For instance, in West Africa, during the Ebola outbreak, malaria deaths outnumbered Ebola-related deaths [6], and tuberculosis (TB) as well as human immunodeficiency virus (HIV) infections also surged [7]. Interestingly, the WHO predicted that the malaria mortality rate is likely to double if malaria interventions and operations are suspended and access to malaria services is restricted [8].

However, there is a growing demand to strengthen malaria control measures and interventions by diverging emerging digital technologies into tackling malaria especially in malaria-endemic countries in sub-Saharan Africa [9]. Developing digital malaria interventions using innovative digital technologies such as mobile technology cognitive computing, mobile health, artificial intelligence, Internet of Things (IoT), Internet of Medical Things, deep learning and machine learning. Other forms of technologies deployed to support malaria healthcare delivery are telemedicine, telehealth, digital health, telecare, virtual health, and others. These technologies are integrated and communicate together through a global positioning system (GPS), wireless fidelity and Bluetooth technology can significantly strengthen malaria control measures and ultimately improve the delivery of malaria healthcare services. Considering the tremendous amount of malaria data generated from various national health information systems, district health information systems (DHIS) and ubiquitous mobile phones, especially frugal smartphones, the application of mobile technology cognitive techniques are increasingly becoming imminent especially in malaria-endemic countries. Mobile technology cognitive computing involves the application of cognitive computing techniques in the development of cognitive mobile applications and devices. Such applications are developed by integrating different cognitive computing technologies, which include machine learning, deep learning, soft computing, the Internet of Things, big data and other platforms that can process huge malaria datasets to obtain patterns and insights. Cognitive mobile applications and devices can adapt to the environment, classify malaria vector species, identify and treatment of malaria patients, real-time mapping malaria hotspots. Mobile health technologies have been recognized as a low-cost approach to provide healthcare services and in many African countries with its fragile health systems, high prevalence of tropical diseases, high rate of infectious diseases, insufficient funding, high mortality rate, and many others [10]. There is a rise in the integration of mobile health applications into the existing electronic health services in many developing countries to fight diseases such as malaria, TB, Ebola, HIV/AIDS and COVID-19. This is necessitated by the high penetration of mobile phones and network coverage which is drastically shifting the general use of mobile phones into supporting healthcare services delivery through short message service (SMS) and voice calls. Adopting mobile phones for healthcare services could reach more people in resource-limited settings than the traditional forms of disease control [10].

However, the next-generation mobile technologies need to address malaria-specific challenges, especially in vulnerable communities. This can be achieved through the adoption of mobile technology cognitive computing which presents unprecedented opportunities to tackle malaria. For instance, cognitive mobile applications could facilitate remote collection of data from isolated populations, disseminate malaria information, monitor and predict malaria prevalence and vector behaviour. To accomplish this, this engine could utilize machine learning to detect patterns of malaria from the given outbreaks. To accomplish this, it requires training data to help develop a model and testing data to test the derived model against real-world data. However, the architectures of machine learning such as artificial neural networks are single and often consist of a single hidden layer, between the input and output layers, which transforms the raw input signals into a problem-specific feature space. The limitation of these architectures is that they are designed to solve only simple problems Another group of machine learning algorithms, deep learning, utilizes multiple hidden layers of transformation to uncover complex relationships among data. This presents avenues for reconsidering malaria elimination strategies that subsequently reduce health service costs, increase accessibility, and facilitate vector control measures, contingent upon the success of countries reached the malaria elimination phase. Notably, several malaria mobile applications have been developed and deployed but failed to yield desired results especially in malaria-endemic countries in sub-Saharan Africa. Thus, in this paper, we provide a comprehensive review of digital technologies and mobile applications that have been deployed in various application domains to fight malaria while introspecting their success and challenges. Since technology is ever-changing, we, therefore, propose a framework for possible integration of cognitive computing techniques in the development of malaria cognitive applications to address malaria-specific challenges in defined communities in Africa.

2 Methodology

We applied rapid review, as a methodology to search, analyse and synthesize published literature such as malaria policies, guidelines, frameworks, and articles. According to Tricco et al., a rapid review is a form of knowledge synthesis in which components of the systematic review process are simplified to produce information promptly [11]. The systematic literature review follows Preferred Reporting Items for Systematic Reviews and Meta-Analyses (PRISMA) steps namely, identification, screening, eligibility and synthesizing of literature included in the study. The first phase was to define the research questions, followed by defining search strategy, study selection, and lastly, study eligibility criteria and quality assessment. Therefore, this review aimed to address the following research questions:

- Which digital technologies were adopted to develop malaria mobile health interventions and applications?
- Why malaria mobile applications (interventions) have been developed and deployed but failed to yield desired results?
- How cognitive computing techniques be integrated into the development of malaria cognitive mobile applications, which supports malaria data collection, storage, and analysis?

2.1 Search Strategy

In conducting a review, a well-planned search strategy is paramount to guide the literature search and to ensure that all relevant digital technologies, mobile applications, and cognitive computing techniques are considered until July 2021(the time when the search was done). Before conducting the literature search, research questions were clearly defined. After that, online electronic databases such as Scopus, Science Direct, Google Scholar, IEEE Xplore Digital Library, Wiley Library, ACM Digital Library, Web of Science, and Springer Link were explored. The literature search was guided by the following Boolean search string: ((“malaria” OR “health interventions”) AND (“cognitive computing” OR “mobile phone applications” OR “digital technology” OR “SMS” OR “health information systems” OR “USSD”)). We further performed a citation chain for additional studies for each retrieved article to make sure that all relevant articles are considered for screening.

2.2 Inclusion and Exclusion Criteria

The search identified a total of 1,278 papers from electronic databases, as shown in Fig. 1. We excluded opinion pieces, non-peer-reviewed articles, incomplete articles, and studies in other languages with no English translation. We also discarded those studies which were not well-presented, contains unclear methodology, dataset, or contribution.

2.3 Eligibility Criteria

All authors double-screened all articles for quality assessment and eligibility. At the initial phase, we selected 1,278 articles by reading the titles and abstracts relevant to the topic. After perusing these articles, irrelevant studies were removed from the list of selected studies. We then created a list of eligible articles that initially passed the initial phase. In the second eligibility phase, we removed all duplicates and we had 515 articles. To further check the eligibility of literature, we read the full-text and to determine whether the contribution of the study is relevant to the objectives of this study. This assist to further eliminate studies that did not pilot or implement any of the malaria digital health interventions. Only 21 articles were considered in this study. The study designs included were mainly digital technologies applied to develop malaria mobile health interventions and applications in Africa. Steps followed to report the number of published articles are presented in a PRISMA flow diagram in Fig. 1.

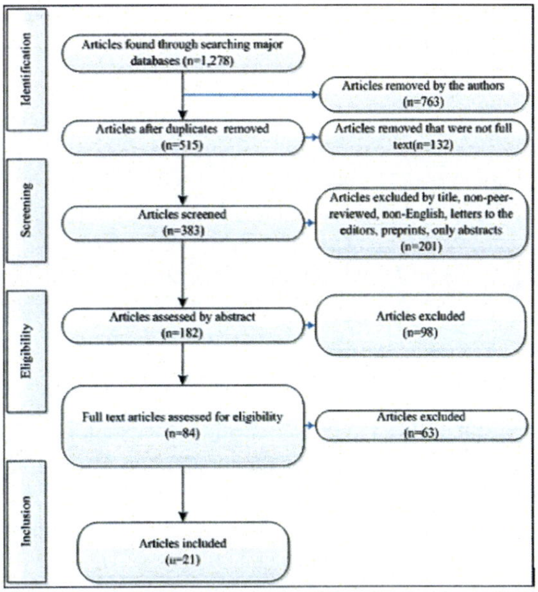

Fig. 1. PRISMA flow diagram

3 Results Analysis

All the relevant data from the selected articles were extracted and synthesized in response to the research questions. The results of mobile technologies applied to tackle malaria are shown in Table 1. Also, from Table 1, we counted the number of articles based on the year of publication as shown in Fig. 2.

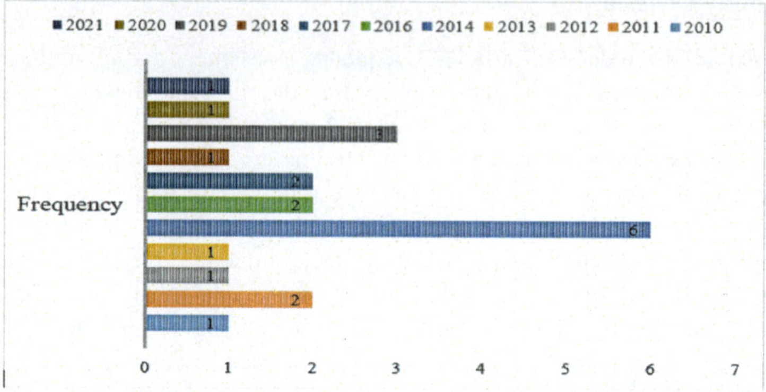

Fig. 2. Distribution of articles based on the year of publication.

Figure 2 shows that as of July 2021, most mobile technologies applied to tackle malaria were published in 2014 (n = 6).

3.1 Digital Technologies Utilized to Develop Malaria Applications

Different digital technologies have been utilized to develop malaria health interventions and applications in Africa. The study included malaria interventions rolled out at the national, district and ward level. Out of twenty-one studies, SMS has been prominently used (see Fig. 3) by different countries to deliver malaria services including malaria surveillance, reporting, monitoring, treatment adherence, health promotion, case management, among others.

SMS Technology

Studies in Table 1 shows that SMS technology is the most prominent technology used to tackle malaria in Africa. SMS technology is the most low-cost feasible communication technology that could be useful to enhance communication in rural areas and potentially leading to improved delivery of malaria services because of the ubiquitous of feature phones and existing infrastructure supporting text messaging and voice calling [18]. SMS uses communication protocols standardized in the Global System for Mobile communications, which support the exchange of messages between computing devices such as mobile phones, tablets, laptops and computers. Text messaging is the most affordable and preferred technology in most parts of the African continent due to the lower bandwidth required to transmit the message [24]. SMS technology requires minimal digital

Table 1. Mobile technologies applied to tackle malaria.

Ref.	Malaria mobile health interventions	Technology used	Purpose	Country	Implementation challenges and limitations
[12]	mSpray mobile-based application	GPS, SMS	Document IRS spray operations	South Africa	Logistical and technical challenges—missed mSpray entries, too few workers were equipped with phones, poor cellular coverage or upload speeds, uncharged cell phones (short battery life), cell phone failure, exhausted data bundles, cell phone malfunctions (included freezing and network problems), poor visibility of cell phone screens. Human error-inadequate GPS accuracy,
[13]	mHAT app-mobile phone-based web application	Image processing	Malaria rapid test result analysis and data aggregation	Zambia	Poor network coverage, poor coordination of data aggregation activities in satellites clinics
[14]	m-Health	Geographic information systems	Malaria diagnosis and mapping geo-localized data transmission	Uganda	Low-end phones with limited capabilities, Data transmission can be difficult and slow without a 3G connection. Very remote locations lacking healthcare
[15]	Timelier notification	SMS	Reporting and notification of malaria cases to improve diagnosis	South Africa	Poor connectivity (impossible to send data from one of the three clinics), technical failure of the system with receiving the forms at the central server for six weeks, users were unused to touch screen devices, a small number of malaria cases covered
[16]	SMS text message	GPS, SMS	Malaria case detection	Zambia	Lack of sensitive detection methods and past control programmes that were unable to detect the occurrence of reservoirs of infection were prone to a rapid return to high endemicity when the intervention foundered

(*continued*)

Table 1. (*continued*)

Ref.	Malaria mobile health interventions	Technology used	Purpose	Country	Implementation challenges and limitations
[17]	MalariaConnect-24-h mobile reporting tool	USSD technology	Real-time tracking of case trends and reporting	South Africa	Some health workers fail to report malaria cases through MalariaConnect, delays in both data collection and recording
[18]	Mobile Phone Text Messaging	SMS	Malaria health communication, monitoring and education	Africa (country not specified)	Implementation costs, technological illiteracy, and the digital divide
[19]	Mobile phone text-message reminders	SMS	Reminder and adherence to malaria treatment	Kenya	Undelivered messages in resource-limited settings without network coverage
[20]	Mobile phone text-messaging	SMS	Improve malaria treatment adherence and post-treatment reviews	Kenya	Poor mobile phone network, access, ownership and usage among health workers and patients. Charging of the phones outside of the households is a major prohibiting barrier for timely access to the text message
[21, 22]	SMS for Life-SMS and web-based reporting tool	SMS	Reducing Stock-Outs of Life Saving Malaria Commodities	Kenya, Tanzania, Uganda	Absence of a randomized controlled design to account for temporal trends in non-intervention areas
[23]	RapidSMS	SMS	Monitoring malaria in Remote Areas	Uganda, Nigeria, Rwanda	Network coverage delayed message delivery in remote areas. Intermittent internet access and power cuts reported as major setbacks, especially in rural areas
[24]	Mobile phone text messaging	SMS	Malaria control tool	Ghana	Lack of random assignment of participants to either an intervention or a control group affected the internal validity of the study design. Poor data recording and limited resources were reported by the study. The study also noted cluster effect and the confounding effects of other competing factors like social and behavioural change programs that were not addressed adequately

(*continued*)

Table 1. (*continued*)

Ref.	Malaria mobile health interventions	Technology used	Purpose	Country	Implementation challenges and limitations
[25]	Mobile phone text message reminders	SMS	Improve malaria case management and treatment guidelines	Malawi	Lack of support interventions alongside SMS reminders interventions affected the performance of the Mobile phone text message reminders. Structural barriers and lack of supervisory support may have affected the prevention of full implementation of SMS reminders
[26]	SMS reminder system	SMS	Strengthening malaria diagnosis and appropriate treatment	Namibia	The system was affected by the limited resources, limited time allocated to register data collection process and health workers' perceptions towards the system and malaria
[27]	WeChat	Social media	Improve malaria health literacy	Niger	The study reported that distance to access health services, as well as security and privacy of patients and health workers, are major drawbacks to adopt WeChat in Niger's health system
[28]	Mobile phone-based disease surveillance and treatment for malaria (ComDSTM) platform	GPS, Web technology	Real-time reporting of malaria and fever cases	Tanzania	Limited internet connectivity and high server downtime The study also noted that the deployed prototype does not capture data in real-time hence the need for a parallel paper-based system
[29]	District Health Information Software (DHIS)	Health Information Systems	Malaria surveillance, reporting, monitoring	Botswana, Zimbabwe, South Africa, Kenya, Tanzania, Nigeria, Rwanda, Ghana, Malawi, Uganda, etc	The challenges of utilising DHIS differ from country to country. The study classified challenges into different groups like operational challenges, leadership and management challenges, political, cultural, social and structural challenges, data security and privacy challenges

(*continued*)

Table 1. (*continued*)

Ref.	Malaria mobile health interventions	Technology used	Purpose	Country	Implementation challenges and limitations
[30]	Community-based surveillance systems	SMS	Surveillance and reporting malaria based on several indicators	Zambia	Some cases got lost in the process and were not entered into the database especially when messages failed to be delivered. The study reported human errors in the form of missing or duplicated data and transcription errors
[31]	Malaria sentinel surveillance system	SMS, GPS	Malaria surveillance and reporting	Togo	The study revealed that the timeliness of data and the generalization of the system deters the progress
[32]	Sentinel malaria surveillance system	SMS, GPS	Reporting and surveillance of malaria	Ethiopia	The study states that poor data recording practices and lack of supervision, incomplete data at health facilities have an unknown effect on routine system data quality
[33]	Malaria web-based application	Web technology	Forecasting malaria epidemics	Madagascar	The system is still in the pilot phase and further analysis is required

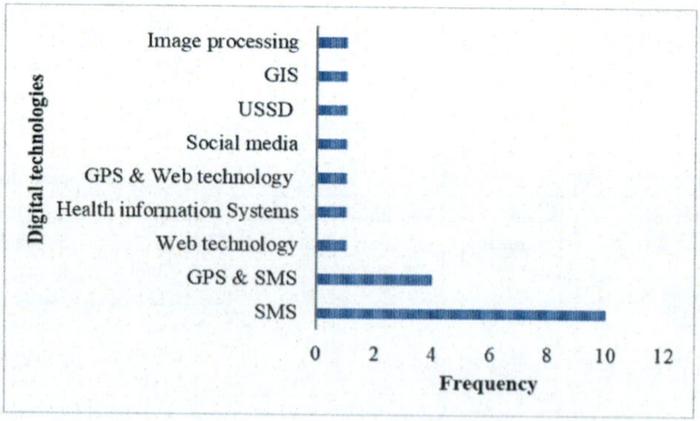

Fig. 3. Digital technologies utilized to develop malaria applications.

skills and increasingly became attractive because of its pervasiveness and ubiquitous. In Africa, most malaria health interventions have been evolving and delivering health services through SMS technology. For instance, a cross-sectional study conducted by [25] in Malawi utilized SMS technology to send one-way communication reminders to improve malaria case management and adherence guidelines to healthcare professionals such as medical assistants, clinical officers, nurses, among others. SMS reminders received high acceptability among healthcare professionals in Kenya [19] and Malawi [25] to improve patients and health workers' adherence to malaria guidelines. However, SMS technology is not immune to challenges and limitations. Even though SMSes have been greatly utilized to tackle malaria, bilateral communication is limited. Also, a retrospective study by [24] asserted that lack of support interventions and poor network coverage delayed SMSs and, in some instances, undelivered messages in some rural clinics impact the effectiveness of SMS in some parts of Africa.

Unstructured Supplementary Service Data (USSD) Technology

USSD technology is a global system for mobile (GSM) communication technology that allows users to send plain text between mobile phones through an application program hosted by the mobile phone service provider [34]. USSD technology is handset Independent and faster than SMS technology. Just like SMS technology, USSD technology provides a cost-effective and flexible mechanism for offering various interactive and non-interactive mobile services to a wide subscriber base. This technology supports the development of menu-driven applications that facilitate more user interactions and experience. Unlike SMS technology, USSD technology provides greater privacy that makes it more suitable for developing malaria interventions. For instance, MalariaConnect, a mobile surveillance tool for reporting malaria outbreaks in South Africa, was developed using USSD technology, SMS and supports web-based platforms [17]. However, USSD uses session-based services, which means if a session expires that user should restart. This restarting consumes time as evidenced with MalariaConnect, where some health workers noted duplicate reporting through the MalariaConnect system. Some health workers fail to report malaria cases using MalariaConnect because of poor network coverage [17].

Geographic Information Systems

A study conducted by Larocca et al. in Uganda shows that geographic information systems have been utilized to developed malaria apps that assist in diagnosis malaria and mapping geo-localized data transmission [14]. However, the study revealed that GIS applications require fast internet and smartphones which subsequently affects low-end phones with limited capabilities. The study also noted that data transmission can be difficult and slow without a 3G connection especially in remote locations where network signal is poor.

Geographical Positioning System (GPS)

This study also noted that several studies in Table 1 used GPS technology to develop malaria interventions. In the context of malaria, GPS technology provides the capability of capture the location of susceptible households or health facilities for mapping purposes. For instance, in South Africa, GPS technology was used to locate and document

IRS spray operations, through mSpraymobile-based applications. In Togo and Ethiopia, GPS technology was used to develop a sentinel malaria surveillance system for malaria surveillance and reporting. However, GPS technology consumes more phone battery power. This might affect the effectiveness of the interventions. A study by Eskenazi et al. alluded to challenges associated with GPS technology in malaria applications, which include poor cellular coverage and week data upload speeds, uncharged cell phones with active GPS results in short battery life, cell phone failure, exhausted data bundles, cell phone malfunctions caused by freezing and network problems, poor visibility of cell phone screens which led to human errors and wrong GPS accuracy [12]. Human errors and poor data recording practices have an unknown effect on routine system data quality.

Web Technology

This involves the use of web authoring tools and programming languages to develop web-based malaria applications. For instance, a malaria predicting and forecasting web-based application deployed in Madagascar was developed using web technologies [33]. Also, a pilot study conducted by [28] deployed mobile phone-based disease surveillance and treatment for malaria (ComDSTM) system developed using web technologies such as PHP, HTML and JavaScript. A web application was implemented using PHP and the mobile applications were implemented using the Android operating system. Apache HTTP web server and MySQL were implemented to manage the database. However, the study also noted that the deployed prototype does not capture data in real-time due to limited internet connectivity and high server downtime hence the need for a parallel paper-based system. In Niger, a social media platform, WeChat used to improve malaria health literacy among Chinese expatriates working in the country [27]. However, the study reported that distance to access health services, network coverage as well as security and privacy of patients are major drawbacks to adopt WeChat in Niger's health system.

4 Discussion of the Results

After synthesizing the literature in Table 1, we categorized malaria interventions based on the purpose and reporting flow within various health systems from the periphery of the health systems to the national information systems. The study revealed that malaria health interventions and applications are utilized for various purposes such as malaria surveillance, management, reporting, monitoring and patient compliance/adherence.

Mobile Health Applications for Malaria Surveillance and Reporting

Several studies in Table 1 shows that malaria mobile health interventions including SMS and applications promise to yield positive results in reducing case reporting, diagnosis and transmission delays, time-saving as well as error rates encountered when using conventional paper-based systems. However, in the study included, there is neither empirical evidence nor statistical comparison made in the same research area or groups to substantiate the claim. However, some studies including [16] and [15] state that reporting through SMS is easier, faster, affordable, and more reliable as compared to a paper-based system. A study by Déglise et al. noted that malaria surveillance and reporting through SMS technology is vulnerable to duplication and over-reporting of cases which ultimately affect the quality and validity of malaria data [35]. However, malaria surveillance and reporting are affected by the cost of SMS, data and mobile phones, mobile network coverage, access to electricity and high personnel turnover. For instance, a cross-sectional study conducted in South Africa by Quan et al. alludes that it was impossible to send data from one of the three rural clinics because of poor network connectivity [15]. In Zimbabwe, malaria surveillance and reporting are affected by the turnover of highly skilled health experts in malaria epidemiology, especially in malarious areas [2].

Mobile Health Applications for Malaria Mapping, Data Management and Aggregation

Malaria health interventions such as district health information systems (DHIS) and paper-based tally sheets report on a weekly, monthly and quarterly basis. These systems have been successfully implemented to manage malaria data in many African malaria-endemic countries like Zimbabwe, Nigeria South Africa, Kenya, Tanzania, Nigeria, Rwanda, Ghana, Malawi, Uganda, among others. DHIS tremendous assist in data capturing, aggregation, consolidation as well as establishing communication between remote health workers, and malaria experts as well as regulators. For instance, DHIS systems are most used to register malaria patients, malaria suspected cases (tested by rapid diagnostic test or blood slide) and confirmed cases (tested positive) in many African countries like Zimbabwe, Zambia, Uganda, and Congo. However, in some instances, data inaccuracy, incompleteness [17] and inconsistency were major setbacks for effective malaria data management and aggregation [13].

Mobile Health Applications for Creating Malaria Awareness and Prevention

This study revealed that SMS technology has been greatly used to create awareness by sending reminders to both health workers and patients. For instance, text messages have been used to strengthening malaria diagnosis and adherence to treatment in Namibia [26]. While, in Malawi and Kenya text messages were used to improve malaria case management and treatment guidelines [20, 25]. A study conducted by Quan et al. in South Africa reported that text messages were utilized for reporting and notification of malaria cases to improve diagnoses [15]. However, this study revealed that many Africa countries intensify malaria awareness through mass media digital platforms such as radio and television programmes. For instance, TalkBack, a Botswana weekly television program for HIV prevention, malaria, is broadcasted live, to stimulate interactivity with teachers and viewers through mobile phone lines, SMS, emails, and letters [36]. In the Central African Republic, UNICEF has been using SMS, radio, and television to encourage vaccination, use of long-lasting insecticidal nets, and handwashing. In Uganda, the SMS and television program, Text to Change, creates awareness through SMS quizzes designed to help resolve key issues around HIV transmission and prevention [37]. Nigeria adopted RapidSMS to create awareness and to help to manage the deployment of bed nets by tracking commodities from state stores to distribution points by monitoring coupon distribution and sending SMS reminders about distribution times and locations for beneficiaries. However, despite the progress made using text messages, radio and television programmes. A study by Dèglise et.al [37] noted that language barriers, difficulty in monitoring adherence and receiving real-time feedback for patients and health workers, as well as confidentiality and data protection concerns are major setbacks for effective reporting, monitoring adherence, and creating awareness [18]. There is a need for mass media digital awareness campaigns, behavioural change communication tailored for and to address malaria-specific challenges in defined communities in Africa.

4.1 Implementation Challenges Faced by Malaria Mobile Interventions and Applications in Africa

Despite the impressive progress towards the adoption of digital technologies to tackle malaria in Africa in the past years, emerging challenges continue to rise. For instance, Leon et al. posit that setting up m-health interventions in South Africa face some challenges such as weak health systems, organizational weaknesses and a gap between policy and implementation (which includes the introduction and use of information technology) and poor capacity of provinces and districts to use information systems [38]. In addition, a cross-sectional quantitative study conducted by Davies et al. also states that lack of infrastructure, shortage of healthcare workers, and lack of training of healthcare workers at healthcare facilities deterred malaria reporting rates and utilization of malaria interventions in KwaZulu Natal, Limpopo and Mpumulanga Province in South Africa [17]. Also, Eskenazi et al. piloted mSpray malaria mobile-based application to document IRS spray operations in South Africa [12]. However, their study reported that logistical and technical challenges affect the implementation of the intervention which led to missed mSpray entries, poor cellular coverage or upload speeds, uncharged cell phones (short battery life), cell phone failure, exhausted data bundles, cell phone malfunctions (included freezing and network problems), poor visibility of cell phone screens. Low-end phones with limited capabilities were also reported by Davies et al. when implementing geographic information systems to enhance malaria diagnosis and mapping geo-localized data transmission in Uganda [17].

Moore et al. [13] implemented mHAT mobile phone-based application to perform malaria rapid test result analysis and data aggregation in Zambia. However, the study reported that poor network coverage, poor coordination of data aggregation activities in satellites clinics tremendously impact the performance of the intervention. In Kenya, a study conducted by Barrington et al. noted that the SMS for Life system provides real-time malaria data reporting through SMS but its implementation challenges including poor usage and acceptance by healthcare professionals which subsequently lead to low accuracy of SMS reported surveillance data [22]. This is exacerbated by the lack of healthcare professionals which are generally overloaded with many tasks and malaria reporting is not considered a high priority [21]. However, in Malawi, Kaunda-Khangamwa et al. reported that low acceptance and usage of mobile technologies, especially SMS reminders, are associated with the lack of financial incentives and support interventions, structural barriers, and lack of supervisory support [25].

Many African countries including Botswana, Zimbabwe, South Africa, Kenya, Tanzania, Nigeria, Rwanda, Ghana, Nigeria, Malawi, and Uganda among others adopted district health information software to enhance malaria surveillance, reporting and monitoring. DHIS presents unprecedented opportunities for effective malaria monitoring However, several countries reported that several impediments regard the utilization of DHIS. These challenges may differ from country to country. A study by Dehnavieh et al. classified DHIS challenges into operational challenges, leadership and management challenges, political, cultural, social, and structural challenges, data security and privacy challenges [29]. Operational challenges include insufficient funding [26], inappropriate communication infrastructure, lack of training, difficulty integrating with existing malaria prototypes and systems [10].

5 All-Inclusive Generic Framework for Developing Data-Driven Malaria Digital Interventions in Africa

Despite the milestones achieved through text messaging, the next generations of mobile cognitive health interventions should support the development of adaptive data-driven solutions to predict malaria outbreaks. This is because fighting elusive malaria requires persistence, the development of resilient strategies and, most importantly, knowledge to understand vector behaviour, and other factors influencing malaria outbreaks. Some countries with once uniformly high levels of malaria transmission may now have a varied malaria landscape: low transmission in some areas, malaria "hotspots" in others [39]. Other countries that used to have low transmission can now begin moving toward the elimination of malaria. In some countries with changed malaria landscapes, national malaria control programmes and partners are now beginning to adjust strategies into predicting malaria outbreaks [40]. Modelling and predicting malaria incidences requires the adoption of adaptive technologies like cognitive computing technologies. Therefore, as a contribution to malaria control measures and elimination strategies, the following subsection presents a generic framework for developing smart and data-driven malaria digital interventions. The framework consists of four modules, including data acquisition, data management, cognitive computing engine, and lastly, a knowledge presentation module, as shown in Fig. 4.

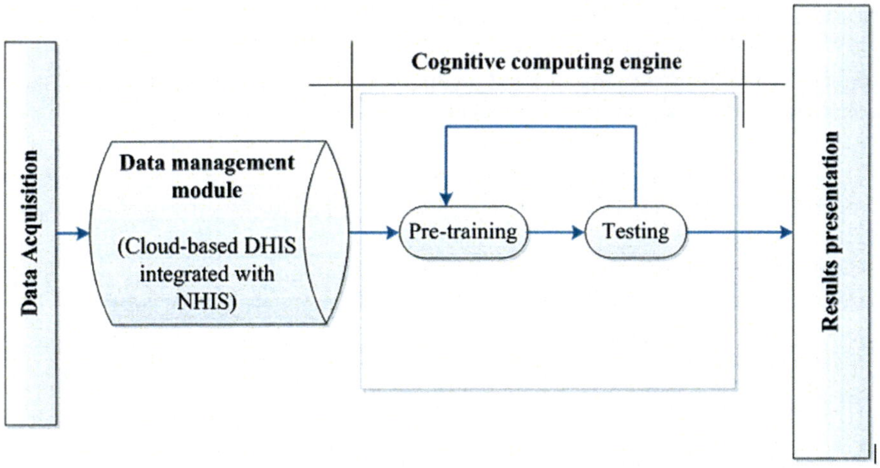

Fig. 4. Generic framework for developing data-driven malaria digital interventions.

a) **Data Acquisition**

This module consists of input data (variables) that are relevant to and associated with malaria outbreaks. Input variables could be malaria climatic and environmental risk factors and malaria incidences from health facilities (weekly and monthly). Automating data acquisition of environmental data such as temperature, rainfall, precipitation, normalized difference vegetation index, enhanced vegetation index among others will assist to reduce human errors.

b) **Data management module**

The data management module stores malaria input data from the data acquisition module. The data management module should be able to upload data and transfers the local data to an online repository for further analysis and record-keeping. Such data could be integrated and stored in the country-specific DHIS as it tends to increase its size and capacity. However, the latest DHIS v2.36 and Android v2.4 include many new features such as storing malaria data on the cloud, enhanced scalability to handle the increased load, supports concurrent access and significant performance improvements, especially for large-scale tracker implementations.

c) **Mobile technology cognitive computing engine**

The cognitive computing engine processes data from the data management layer to extract insights. The environmental risk factors such as temperature, rainfall, precipitation, normalized difference vegetation index, enhanced vegetation index and previous malaria cases from the DHIS are considered as inputs to predict malaria. Environmental data can be aggregated as a monthly frequency since malaria cases in some countries are recorded on either a weekly or monthly basis [13] to reduce the dimensionality of data and to speed up the cognitive processing time.

To accomplish this cognitive model, machine learning can be effectively utilised. Machine learning is used to manage and predict diseases often incorporate models that are specifically designed for efficient use of limited resources available on embedded devices such as smartphones [41–43]. The proposed model utilizes deep learning to find patterns among complex relationships of data and complex patterns as our input data are much varied with complex relationships that need to be discovered to achieve proper malaria prevention, monitoring, and surveillance. This will involve much training time and data in the cloud and can yield accurate results. Consequently, the cognitive model focuses on using environmental risk factors [44] and data sources as input and producing outputs to assist malaria prevention, prediction, surveillance in resource-limited areas.

d) **Results presentation**

The results presentation displays processed data from the cognitive engine. The output information might be in form of text or graphical representations based on the problem. Such information could be used for real-time reporting of malaria outbreaks to either district office or provincial health office, aid in decision making, resource allocation, predict malaria outbreaks, case management, tracing vector behaviour, automate alerts and warnings, mapping malaria hotspots at micro-level cascading down to endemic wards. Thus, incorporating mobile technology's cognitive techniques into malaria elimination strategies may advocate for better policymaking that addresses malaria-specific challenges in defined malaria-endemic communities in Africa.

Identified Gaps for Future Work to Tackle Malaria Through Mobile Health Interventions and Applications in Africa
This study noted that there are many parallel and recurring prototypes (malaria mobile applications) for tackling malaria. For instance, several malaria mobile applications such as the mSpraymobile-based application [12], Timelier notification [15] and MalariaConnect-24-h mobile reporting tool [17], DHIS [29] have been implemented to improve malaria services in South Africa. In Kenya, SMS for Life [21], mobile phone text messaging [20], Mobile phone text-message reminders [19] were rolled to remind healthcare professionals and patients about malaria adherence guidelines at different intervals. The integration of these interventions would to the development of resilient malaria interventions. Therefore, the future should focus on strategies ways of minimizing duplication or concurrent deployment of malaria interventions that intent to achieve relatively similar objectives.

This study revealed that there is a lack of adaptive technologies in malaria mobile interventions in Africa. Adaptive malaria interventions support the development of automated malaria early warnings and automated malaria screening using aggregated data from various malaria data sources. Such data-driven interventions prompt regulators and malaria experts to conduct corrective actions involving real-time mapping of malaria, malaria outbreak investigations, automated restocking alerts of malaria commodities. The future work may focus on developing malaria data-driven interventions designed to address malaria-specific challenges in defined vulnerable communities in Africa.

Lack of standardized mobile health policy [24, 45], implementation standards as well as guidelines for the development of malaria health interventions [13] have been noted as a major concern that consequently leads to the development of relatively similar malaria prototypes and pilot projects concurrently. There is a need to come up with prototype development standards guiding malaria data management, security and privacy of patients' data.

6 Conclusion

Malaria mobile interventions present opportunities in creating malaria awareness campaigns to enhance behavioural change communication and adherence to IRS and treated nets. The study revealed that most countries use DHIS to enhance malaria surveillance, monitoring [29] and reporting support for remote health workers by facilitating patient registration and clinical decision making as well as for patient compliance. Many countries in the African region widely use SMS technology to send reminders and adherence to malaria treatment, document IRS spray operations [22], reporting and notification of malaria cases to improve diagnosis, health communication, monitoring and education, improve malaria case management and treatment guidelines. Owing to emerging challenges and outbreaks of infectious diseases, it is time to reconsidering strategies to fight elusive malaria.

Technology is evolving; therefore, the next-generation mobile health interventions should support the development of adaptive data-driven malaria solutions tailored to addresses malaria-specific challenges in defined malaria-endemic communities in Africa. To achieve that we propose a framework to incorporate mobile technology cognitive technologies in developing data-driven malaria solutions. However, the proposed framework might face some hindrance especially when most of the population use feature phones instead of smartphones. In such circumstances, frugal computing would be the ideal technology to adopt. The future work of this study will focus on the development and implementation of malaria mobile applications guided by the proposed framework.

References

1. Zawawi, A., Alghanmi, M., Alsaady, I., Gattan, H., Zakai, H., Couper, K.: The impact of COVID-19 pandemic on malaria elimination. Parasite Epidemiol. Control **11**, e00187 (2020). https://doi.org/10.1016/j.parepi.2020.e00187
2. Gavi, S., Tapera, O., Mberikunashe, J., Kanyangarara, M.: Malaria incidence and mortality in Zimbabwe during the COVID-19 pandemic: analysis of routine surveillance data. Malar. J. **20**, 233 (2021). https://doi.org/10.1186/s12936-021-03770-7
3. Health Organization, W.: World malaria report 2020: 20 years of global progress and challenges (2020)
4. Penjor, K., Tobgyal, T.Z., Clements, A.C.A., Gray, D.J., Wangdi, K.: Has COVID19 derailed Bhutan's national malaria elimination goal? A commentary. Malar. J. (2021). https://doi.org/10.1186/s12936-020-03562-5
5. Nokuthula Sibiya, M., Ramlucken, L.: Willingness and feasibility of utilising short messaging services (SMS) as reminders of follow-up care to mental health care users in the uMgungundlovu District. KwaZulu-Natal. Int. J. Afr. Nurs. Sci. **14**, 100294 (2021). https://doi.org/10.1016/J.IJANS.2021.100294
6. Rogerson, S.J., et al.: Identifying and combating the impacts of COVID-19 on malaria. BMC Med. (2020). https://doi.org/10.1186/s12916-020-01710-x
7. Walker, P.G.T., White, M.T., Griffin, J.T., Reynolds, A., Ferguson, N.M., Ghani, A.C.: Malaria morbidity and mortality in Ebola-affected countries caused by decreased health-care capacity, and the potential effect of mitigation strategies: a modelling analysis. Lancet Infect. Dis. **15**, 825–832 (2015). https://doi.org/10.1016/S1473-3099(15)70124-6
8. Diptyanusa, A., Zablon, K.N.: Addressing budget reduction and reallocation on health-related resources during COVID-19 pandemic in malaria-endemic countries. Malar. J. **19**, 1–4 (2020). https://doi.org/10.1186/s12936-020-03488-y
9. Mbunge, E., Sibiya, M.N., Millham, R.C., Takavarasha, S.: M-health framework for improving malaria information dissemination in Buhera rural district amid COVID-19 and beyond. In: 2021 Conference on Information Communications Technology and Society, ICTAS 2021 – Proceedings, pp. 59–66. Institute of Electrical and Electronics Engineers Inc. (2021). https://doi.org/10.1109/ICTAS50802.2021.9395020
10. Osei, E., Mashamba-Thompson, T.P.: Mobile health applications for disease screening and treatment support in low-and middle-income countries: a narrative review. Heliyon **7**(3), e06639 (2021). https://doi.org/10.1016/j.heliyon.2021.e06639

11. Tricco, A.C., et al.: A scoping review of rapid review methods. BMC Med. **13**, 1–15 (2015). https://doi.org/10.1186/s12916-015-0465-6
12. Eskenazi, B., et al.: MSpray: a mobile phone technology to improve malaria control efforts and monitor human exposure to malaria control pesticides in Limpopo. South Afr. Environ. Int. **68**, 219–226 (2014). https://doi.org/10.1016/j.envint.2014.03.003
13. Moore, C., et al.: mHAT app for automated malaria rapid test result analysis and aggregation: a pilot study. Malar. J. **20**, 237 (2021). https://doi.org/10.1186/s12936-021-03772-5
14. Larocca, A., Moro Visconti, R., Marconi, M.: Malaria diagnosis and mapping with m-Health and geographic information systems (GIS): evidence from Uganda. Malar. J. **15**, 1–12 (2016). https://doi.org/10.1186/s12936-016-1546-5
15. Quan, V., Hulth, A., Kok, G., Blumberg, L.: Timelier notification and action with mobile phones-towards malaria elimination in South Africa. Malar. J. **13**, 1–8 (2014). https://doi.org/10.1186/1475-2875-13-151
16. Kamanga, A., Moono, P., Stresman, G., Mharakurwa, S., Shiff, C.: Rural health centres, communities and malaria case detection in Zambia using mobile telephones: a means to detect potential reservoirs of infection in unstable transmission conditions. Malar. J. **9**, 96 (2010). https://doi.org/10.1186/1475-2875-9-96
17. Davies, C., et al.: Effectiveness of 24-h mobile reporting tool during a malaria outbreak in Mpumalanga Province. South Afr. Malar. J. **18**, 45 (2019). https://doi.org/10.1186/s12936-019-2683-4
18. Zurovac, D., Talisuna, A.O., Snow, R.W.: Mobile phone text messaging: tool for malaria control in Africa. PLoS Med. **9**, e1001176 (2012). https://doi.org/10.1371/journal.pmed.1001176
19. Zurovac, D., et al.: The effect of mobile phone text-message reminders on Kenyan health workers' adherence to malaria treatment guidelines: a cluster randomised trial. Lancet **378**, 795–803 (2011). https://doi.org/10.1016/S0140-6736(11)60783-6
20. Otieno, G., Githinji, S., Jones, C., Snow, R.W., Talisuna, A., Zurovac, D.: The feasibility, patterns of use and acceptability of using mobile phone text-messaging to improve treatment adherence and post-treatment review of children with uncomplicated malaria in western Kenya. Malar. J. **13**, 1–7 (2014). https://doi.org/10.1186/1475-2875-13-44
21. Githinji, S., et al.: reducing stock-outs of life saving malaria commodities using mobile phone text-messaging: SMS for life study in Kenya. PLoS ONE **8**, 5 (2013). https://doi.org/10.1371/journal.pone.0054066
22. Barrington, J., Wereko-Brobby, O., Ward, P., Mwafongo, W., Kungulwe, S.: SMS for Life: a pilot project to improve anti-malarial drug supply management in rural Tanzania using standard technology. Malar. J. **9**, 1–9 (2010). https://doi.org/10.1186/1475-2875-9-298
23. Asiimwe, C., et al.: Use of an innovative, affordable, and open-source short message service-based tool to monitor malaria in remote areas of Uganda. Am. J. Trop. Med. Hyg. **85**, 26–33 (2011). https://doi.org/10.4269/ajtmh.2011.10-0528
24. Mohammed, A., Acheampong, P.R., Otupiri, E., Osei, F.A., Larson-Reindorf, R., Owusu-Dabo, E.: Mobile phone short message service (SMS) as a malaria control tool: a quasi-experimental study. BMC Public Health **19**, 1–11 (2019). https://doi.org/10.1186/s12889-019-7336-6
25. Kaunda-Khangamwa, B.N., et al.: The effect of mobile phone text message reminders on health workers' adherence to case management guidelines for malaria and other diseases in Malawi: lessons from qualitative data from a cluster-randomized trial. Malar. J. **17**, 1–13 (2018). https://doi.org/10.1186/s12936-018-2629-2
26. Lourenço, C., Kandula, D., Haidula, L., Ward, A., Cohen, J.M.: Strengthening malaria diagnosis and appropriate treatment in Namibia: a test of case management training interventions in Kavango Region. Malar. J. **13**, 1–8 (2014). https://doi.org/10.1186/1475-2875-13-508

27. Li, W., Han, L.Q., Guo, Y.J., Sun, J.: Using WeChat official accounts to improve malaria health literacy among Chinese expatriates in Niger: an intervention study. Malar. J. **15**, 1–13 (2016). https://doi.org/10.1186/s12936-016-1621-y
28. Francis, F., et al.: Deployment and use of mobile phone technology for real-time reporting of fever cases and malaria treatment failure in areas of declining malaria transmission in Muheza district north-eastern Tanzania. Malar. J. **16**, 308 (2017). https://doi.org/10.1186/s12936-017-1956-z
29. Dehnavieh, R., et al.: The district health information system (DHIS2): a literature review and meta-synthesis of its strengths and operational challenges based on the experiences of 11 countries. Health Inf. Manage. J. **48**(2), 62–75 (2018). https://doi.org/10.1177/1833358318777713
30. Hamainza, B., Killeen, G.F., Kamuliwo, M., Bennett, A., Yukich, J.O.: Comparison of a mobile phone-based malaria reporting system with source participant register data for capturing spatial and temporal trends in epidemiological indicators of malaria transmission collected by community health workers in rural Zambia. Malar. J. **13**, 1–13 (2014). https://doi.org/10.1186/1475-2875-13-489
31. Thomas, A., Bakai, T.A., Atcha-Oubou, T., Tchadjobo, T., Voirin, N.: Implementation of a malaria sentinel surveillance system in Togo: a pilot study. Malar. J. **19**, 330 (2020). https://doi.org/10.1186/s12936-020-03399-y
32. Yukich, J.O., et al.: A description of malaria sentinel surveillance: a case study in Oromia regional state. Ethiopia. Malar. J. **13**, 1–13 (2014). https://doi.org/10.1186/1475-2875-13-88
33. Girond, F., et al.: Analysing trends and forecasting malaria epidemics in Madagascar using a sentinel surveillance network: a web-based application. Malar. J. **16**, 1–11 (2017). https://doi.org/10.1186/s12936-017-1728-9
34. Perrier, T., Derenzi, B., Anderson, R.: USSD: The third universal app. In: ACM DEV-6 2015 – Proceedings of the 2015 Annual Symposium on Computing for Development, pp. 13–21. Association for Computing Machinery, Inc. (2015). https://doi.org/10.1145/2830629.2830645
35. Déglise, C., Suggs, L.S., Odermatt, P.: SMS for disease control in developing countries: a systematic review of mobile health applications. J. Telemed. Telecare **18**(5), 273–281 (2012). https://doi.org/10.1258/jtt.2012.110810
36. Tsheko, G.N.: HIV/AIDS and education in Botswana. Int. Perspect. Educ. Soc. **18**, 141–159 (2012). https://doi.org/10.1108/S1479-3679(2012)0000018009
37. Déglise, C., Suggs, L.S., Odermatt, P.: Short message service (SMS) applications for disease prevention in developing countries. J. Med. Internet Res. **14**(1), e3 (2012). https://doi.org/10.2196/jmir.1823
38. Leon, N., Schneider, H., Daviaud, E.: Applying a framework for assessing the health system challenges to scaling up mHealth in South Africa. BMC Med. Inform. Decis. Mak. **12**, 1–12 (2012). https://doi.org/10.1186/1472-6947-12-123
39. Slutsker, L., Kachur, S.P.: It is time to rethink tactics in the fight against malaria. Malar. J. (2013). https://doi.org/10.1186/1475-2875-12-140
40. Tizifa, T.A., Nkhono, W., Mtengula, S., van Vugt, M., Munn, Z., Kabaghe, A.N.: Leveraging phone-based mobile technology to improve data quality at health facilities in rural Malawi: a best practice project. Malar. J. **20**, 203 (2021). https://doi.org/10.1186/s12936-021-03742-x
41. Hartanto, C.A., Wibowo, A.: Development of mobile skin cancer detection using faster R-CNN and MobileNet v2 model. In: 7th International Conference on Information Technology Computer and Electrical Engineering ICITACEE 2020 – Proceedings, pp. 58–63 (2020). https://doi.org/10.1109/ICITACEE50144.2020.9239197
42. Abiodun, O.I., Jantan, A., Omolara, A.E., Dada, K.V., Mohamed, N.A., Arshad, H.: State-of-the-art in artificial neural network applications: a survey. Heliyon **4**(11), e00938 (2018)

43. Li, D., Hinton, G., Kingsbury, B.: New types of deep neural network learning for speech recognition and related applications: an overview. In: 2013 IEEE International Conference on Acoustics, Speech and Signal Processing, pp. 8599–8603. IEEE (2013)
44. Mbunge, E., Sibiya, M.N., Millham, R.C., Takavarasha, S.: Micro-spatial modelling of malaria cases and environmental risk factors in Buhera rural district, Zimbabwe. In: 2021 Conference on Information Communications Technology and Society ICTAS 2021 – Proceedings, pp. 52–58 (2021). https://doi.org/10.1109/ICTAS50802.2021.9394987
45. Mbunge, E., Millham, R., Sibiya, M.N., Takavarasha, S.: Impact of COVID-19 on Malaria Elimination: Juxtaposing Indoor Residual Spraying and Mobile Phones in Buhera Rural District, Zimbabwe. https://doi.org/10.21203/rs.3.rs-173130/v2

Corporative Financing of Innovative Activity in the Russian Regions

Iuliia S. Pinkovetskaia[1(✉)], Anton V. Lebedev[2], Natalia E. Bobrova[2], Diego Felipe Arbelaez Campillo[3], and Magda Julissa Rojas Bahamon[3]

[1] Ulyanovsk State University, Ulyanovsk 432000, Russia
[2] National Research Ogarev Mordovia State University, Saransk 430005, Russia
[3] University Amazonia, Florencia, Colombia

Abstract. Aim of the article is in the assess the share of the cost of financing innovation in the income received from the sale of manufactured goods, works and services by innovative organizations located in each of the Russian Federation constituent entities. Initial information used in our research is the official data of the Federal Statistical Service of Russia on corporate financing of innovation activities for all 82 continent entities of the country for 2017–2019. In the course of study we consider the indicators characterizing the share of the cost of financing innovation in the amount of income received from the sale of manufactured products by organizations located in the constituent entities. The assessment of the levels of three indicators used economic-mathematical modeling based on density functions of normal distribution. The use of these functions allows to define medium values and ranges changes in considered indexes typical for most constituent entities. In addition, the research identifies regions, territories and republics which are characterized by the maximum and minimum values of three indicators. The scientific novelty and originality of this research is as follows: we developed models describing the distribution of indicators according to data for 2017, 2018, 2019; It is proved that there were no significant changes in the value of the cost shares of financing innovation in the amount of income received from the sale of manufactured products by organizations located in the regions of the country in 2017–2019; it is shown that the average cost of financing innovation by organizations amounted to a small share (less than 2%) in the amount of income received from the sale of manufactured products.

Keywords: Innovative activity of organizations · Assessment of innovation costs · Financing of innovations · Russian constituent entities · Functions of normal distribution

1 Introduction

According to [1, 2], innovations include the new goods and services creation or modernization of the existing ones, as well as the production processes transformation. Two types of innovation are distinguished: product and technological innovation. Innovation of the twenty-first century is a critical element in the national economies' development.

R. Silhavy et al. (Eds.): CoMeSySo 2021, LNNS 232, pp. 700–711, 2021.
https://doi.org/10.1007/978-3-030-90318-3_55

They ensure the progress in engineering and technology [3], the development of enterprises activity [4], implementing new ideas [5], and applied creativity [6]. The article [7] draws attention to the improvement of companies' efficiency and their survival due to the use of innovation. The development and implementation of innovative technologies are often driven by increased competition [8–10]. In addition, innovation contributes to improving the quality of people life and their comfort [11]. Thus, innovation makes a significant contribution to the socio-economic development of countries.

As shown by the analysis of the academic publications, the effectiveness of innovation has grown dramatically in recent decades as a result of increased consumer demand for various goods and services [12–14]. A number of studies have noted the innovation importance for efficiency and business development [15, 16]. One must admit that innovation has evolved in different sized organizations, both small and large enterprises [17–19]. Most innovations require significant financial costs [20].

Given the above, for the majority of developed and developing economies the urgent problem is to assess the financial costs, providing the organizations innovation activity. As organizations here and below we mean various types of companies, firms and enterprises. This issue is in the focus of our research. The study of the current level of innovation development financing is significant both for individual enterprises and organizations, as well as for government agencies that regulate innovation.

Despite the extensive studies in the field of innovation, little attention has been paid to the innovation activities financing. Therefore, the given article responds to calls for evaluation of needed costs level associated with the innovation development. These ideas have been formulated, in particular, in the works [21–23].

In Russia, there are high expectations for the growth of innovation in the economy of each of the country's constituent entities. The innovations' introduction ensure appearing of new goods and services, and also the transition to new more technological and effective forms and methods of activity of enterprises and organizations. To increase the level of innovation activity, it is vital to provide its financing. Therefore, the urgent problem is to consider the financing for the innovation development in modern conditions. Consequently, major attention should be paid to the study of regional aspects of the mentioned scientific problem, i.e., the peculiarities typical for different constituent entities of the Russian Federation (republics, territories and regions).

The study aims at assessing the share of the cost of financing innovation in the income received from the sale of manufactured goods, works and services by innovative organizations located in each of the Russian Federation constituent entities. Our article is aimed to obtain a certain empirical and methodological contribution to the knowledge of innovation financing. Specifically, the authors propose the methodology for modeling the level of innovation financing in modern countries using the normal density functions. The empirical contribution is related to determining the average values and standard deviations for the Russian Federation constituent entities of such indicator as the share of the cost of financing innovation in the amount of income received from the sale of manufactured products by organizations. In addition, the study determines the Russian Federation constituent entities with the maximum and minimum values of this indicator.

The structure of this work is as follows: Sect. 2 provides a review of the literature on the problem of financing innovation; Sect. 3 describes the methodology, data and design used to conduct the research. The results of the model building are presented in Sect. 4. The conclusions are featured in Sect. 5. The final section contains bibliographic references.

2 Review of Literature

According to the study [24], innovation accounts for about 50% of economic growth in the developed countries. The importance of innovation in both developed and developing countries is highlighted in the articles [25, 26]. Innovations have improved production efficiency not only in high-tech industries such as electronics, engineering, information and biological technology, and chemistry [27], but also in health care, most consumer goods manufacturing, retail and education [28]. Much attention in scientific publications is paid to the peculiarities of innovation activity in different regions. For example, the work [29] indicates that innovation activity depends on such factors as the experience in forming an environment of innovation, the involvement of most organizations and enterprises in innovation, as well as the relevant activities of scientific structures located in the regions. Therefore, the differentiation of innovation performance by regions is significant [30].

The study [31] concentrates on two main sources of financing for innovation. External sources are related to obtaining loans from banks or other legal entities or individuals. Internal sources of financing are based on the use of retained earnings and the business owners' own capital. The articles [32, 33] conclude that organizations and enterprises tend to use financial resources obtained when manufacturing their products or providing services to innovate. Doubtless, these resources are limited, but, according to the authors of the mentioned articles, outside fund raising may not be cost-effective. A similar conclusion is drawn from [34].

When considering the problem of innovation, it is necessary to take into account that the outcomes on innovation within a market economy can be skew-distributed [35]. This imposes certain constraints on the costs associated with financing innovation, as described in [36–40]. An empirical analysis by the authors [41] shows that in the U.S. from 1994 to 2004, the share of spending on financing innovation ranged from 1.4% in 1994 to 1.9% in 2004 of total output in seven high-tech industries.

A number of scientific works published in 2019–2020 are devoted to the problem of financing innovative activity in Russia. A brief description of these publications is given in Table 1.

Based on the data presented in Table 1, it can be stated that the problem of researching the financing of innovation activities is relevant in Russia. Most of the reviewed studies analyzed the financing of innovation for the country as a whole, or for individual types of economic activity. It can be noted that Russian studies pay little attention to the comparative analysis of the funds spent by the organizations carrying out technological innovations in the Russian Federation constituent entities. Various publications, as a rule, assess the absolute values of the costs of innovative activities, which seems illogical for a comparative analysis by the constituent entities of the country, since they differ significantly in the number of enterprises and organizations, population, size and location. Taking this into account, it seems advisable to analyze the costs of organizations for

Table 1. Scientific publications on financing innovative activities in Russia.

Authors	Subject matter	Period, years	Objects of innovation	Indicator type
1	2	3	4	5
Marenkov [42]	Assessment of the state of innovation and investment in Russian industry	2005–2016	Industrial enterprises of Russia	Absolute and specific
Semenova [43]	Main sources of funding for innovation	2000–2017	Organizations in Russia	Absolute and specific
Krasova [44]	Structure of sources of funding for innovation	2010–2017	Budget expenditures in Russia	Absolute and specific
Makarov et al. [45]	Financing innovative projects	2017–2018	PJSC Rostelecom	Absolute
Sokolova, Kabannik [46]	Financing innovative companies	2013–2018	Organizations in Russia	Qualitative
Barkova [47]	Sources of funding for innovation at different stages	–	Innovative enterprises	Indices
Sheina [48]	Venture funding for innovation	2013–2018	Organizations in Russia	Absolute
Raiskaya [49]	Innovation through leasing	2015–2019	Leasing market	Absolute
Bykovskaya et al. [50]	Financing innovation in the digital economy	2018	Organizations in Russia	–
Leshchinskaya et al. [51]	Funding for technological innovation	2014–2019	Oil industry	Indices
Denisova [52]	Venture financing for innovative small businesses	–	Small businesses	–

Source: The table based on the information provided in the Russian Science Citation Index was compiled by the authors.

innovation activity in terms of specific indicators in each of the constituent entities of the country.

3 Materials and Methods

It should be noted that the financing of innovation includes the costs associated with the following activities:

- introduction of innovative equipment, tools and components;
- change in production technologies and the creation of new products (goods and services);
- change in logistics, transportation, distribution operations related to the supply of organizations and sales of finished products;
- improvement of technology and organization of information processes;
- application of more efficient methods of production activities, their management, including accounting and control issues;
- development of interaction between organizations and the external environment;
- improving the efficiency of personnel policy;
- improvement of methods and forms of marketing and pricing.

From the analysis of previous studies, it can be concluded that to assess the costs of financing innovations, it is advisable to use indicators that characterize the share of these costs in the income received by innovative organizations from the sale of manufactured goods, works and services.

During the research, three stages were implemented. The first stage concerned the formation of initial empirical data describing the costs of innovation in the income received by organizations from this activity for each of the country's constituent entities. The second stage involved assessing the distribution of the values of the share of costs for financing innovation in the volume of products sold. The third stage was related to the comparative analysis, which allowed to identify the constituent entities of the country in which the indicators were minimum and maximum.

The study is based on the use of official statistical data for 2017–2019, characterizing the costs of financing innovations by organizations located in 82 constituent entities of the Russian Federation [53].

The economic and mathematical modelling, which was used for distribution assessment of the expenses for innovation financing in the share of income from the products by the country's areas, applied the functions of normal distribution. The articles [54, 55] present the methodological approach to the development and use of such functions to determine the average values of indicators for the constituent entities of the country, as well as their variation range.

The study involved testing the following three hypotheses:

- hypothesis 1 – the costs of financing innovations by organizations make up a small share in the amount of income received from the sale of manufactured their products;
- hypothesis 2 – there are significant differences in the values of each of the indicators sets of organizations located in constituent entities of the country;

– hypothesis 3 – the geographical position of the country's constituent entities does not significantly affect whether the values of the indicators for them belong to the maximum and minimum.

4 Results of the Calculation Experiment

The calculation experiment consisted in the development of economic and mathematical models describing empirical information. Models characterizing the frequency distribution (y) of three indicators (x, %) for different years for 82 constituent entities of the country are shown below:

– the share of the cost of financing innovation in the amount of income received from the sale of manufactured products by organizations located in the constituent entities of the country in 2017

$$y_1(x_1) = \frac{75.17}{1.21 \times \sqrt{2\pi}} \cdot e^{-\frac{(x_1 - 1.53)^2}{2 \times 1.21 \times 1.21}} \quad (1)$$

– the share of the cost of financing innovation in the amount of income received from the sale of manufactured products by organizations located in the constituent entities of the country in 2018

$$y_2(x_2) = \frac{88.83}{1.18 \times \sqrt{2\pi}} \cdot e^{\frac{-(x_2 - 1.41)^2}{2 \times 1.18 \times 1.18}} \quad (2)$$

– the share of the cost of financing innovation in the amount of income received from the sale of manufactured products by organizations located in the constituent entities of the country in 2019

$$y_3(x_3) = \frac{68.33}{1.09 \times \sqrt{2\pi}} \cdot e^{-\frac{(x_3 - 1.38)^2}{2 \times 1.09 \times 1.09}} \quad (3)$$

The second stage of the work allowed to identify patterns arising from the analysis of the developed models from the first to the third. The results of this analysis are summarized in Table 2. The table presents the data for each of the years under consideration, characterizing both the average values of the cost shares of financing innovations in the amount of income received from the sale of manufactured products by organizations located in the constituent entities of the country (second column), and the ranges changes in the corresponding values for most subjects of the country (third column).

The average values of the cost share of financing innovation in the amount of income received from the sale of manufactured products by organizations located in the constituent entities of the country (Table 2), were in the range from 1.53% to 1.38% in 2017–2019. At the same time, over the years under consideration, there was a slight (by 0.15%) decrease in the indicators under consideration. In 2019, the values of the indicators under consideration at a level below the average (1.38%) were observed in 47 constituent entities of the country. The intervals of change in indicators for most constituent

Table 2. Percentage of expenses on innovation activity, %

Indicators	Average values	Values for most constituent entities of the country
1	2	3
The share of the cost of financing innovation in the amount of income received from the sale of manufactured products by organizations located in the constituent entities of the country in 2017	1.53	0.32–2.74
The share of the cost of financing innovation in the amount of income received from the sale of manufactured products by organizations located in the constituent entities of the country in 2018	1.41	0.23–2.59
The share of the cost of financing innovation in the amount of income received from the sale of manufactured products by organizations located in the constituent entities of the country in 2019	1.38	0.29–2.47

Source: Indicators are determined by the authors based on models (1)–(3).

entities over these years were as follows: the least value was 0.23%, and the greatest was 2.74%. Namely, the costs for most constituent entities did not exceed one thirty-sixth of the total cost of products and services rendered by innovative organizations. Thus, the above shows that the first hypothesis has been confirmed.

To test hypothesis 2 on the differentiation of indicators by the constituent entities of the country, we analyzed the range of variation of each of the indicators presented in Table 2. The indices of variation were: for the first indicator – 79%, for the second indicator – 84%, for the third indicator – 79%. This confirms the second hypothesis about significant differences in all indicators across the constituent entities of the country.

The third stage made it possible to determine the lists of the country's constituent entities with the maximum and minimum values of the shares of costs for financing innovation in the amount of income received from the sale of manufactured products by organizations located in the respective constituent entities. In this respect, the maximum and minimum values are ones that respectively exceed the upper limit of the range in 2019 and are less compared to the lower limit of the range (the third column). These indicators are presented in the second and third columns of Table 3. In addition to the lists of the country's constituent entities, these columns also contain the titles of federal districts and the shares of innovative organizations in the country's constituent entities, which are given in brackets.

Table 3. Russian constituent entities with the maximum and minimum values of indicators in 2019

Indicators	Maximum values	Minimum values
1	2	3
The share of the cost of financing innovation in the amount of income received from the sale of manufactured products by organizations located in the constituent entities of the country in 2019	Nizhny Novgorod Region (8.0%), Tula Region (6.9%), Sakhalin Region (5.5%), Republic of Tyva (5.3%), Khabarovsk Territory (5.0%), Lipetsk Region (3.8%), Moscow (3.6%), Sevastopol (3.5%), the Republic of Buryatia (3.5%), Chuvash Republic (3.5%), Tambov Region (3.5%), Republic of Tatarstan (3.3%), Republic of Mordovia (3.1%), Samara Region (2.9%). They are located in the Central (four constituent entities), Privolzhsky (five constituent entities), Southern (one constituent entity), the Far Eastern and Siberian federal districts, two constituent entities each	Republic of Dagestan (0.25%), Vologda region (0.20%), Republic of Kalmykia (0.19%), Republic of Khakassia (0.10%), Pskov region (0.10%), Oryol region (0.10%), Ivanovo region (0.10%), The Karachay-Cherkess Republic (0.08%), the Republic of North Ossetia (0.05%), the Chechen Republic (0.04%), the Republic of Ingushetia (0.02%). Located in the Central (two constituent entities), North-West (two constituent entities), South (one constituent entity), North Caucasian (five constituent entities), Siberian (one constituent entity) federal districts

Source: The analysis was carried out by the authors based on the data in Table 2 and statistical information.

Thus, Table 3 indicates the amount of financing for innovation activities in organizations located in the constituent entities of the country with the maximum (column 2) and minimum (column 3) values of the share of costs. Analysis of information on the territorial location of the country's constituent entities showed that there was no connection with their territorial location in federal districts. Thus, we can state the confirmation of the third hypothesis.

5 Conclusions

The goal set in the study, i.e., to assess the share of the cost of financing innovation in the amount of income received from the sale of manufactured products by organizations located in the constituent entities of the country for 2017–2019, was achieved. Results of research which have scientific novelty and authenticity are gathered below. The study presented the methodological approach to assessing the share of costs for financing innovation in the amount of income received from the sale of manufactured products by organizations located in the constituent entities of the Russian Federation. Economic and mathematical models describing the distribution of indicators according to data for 2017, 2018, 2019 have been developed. It is proved that there were no significant changes in the value of the cost shares of financing innovation in the amount of income received from the sale of manufactured products by organizations located in the regions of the country in 2017–2019. It is shown that the average cost of financing innovation by organizations in 2017–2019 amounted to a small share (less than 2%) in the amount of income received from the sale of manufactured products. It is shown that the percentage of the costs of financing innovation was significantly differentiated by the constituent entities. The research identified the constituent entities of the country with the maximum and minimum values in 2019 of the cost shares for financing innovations.

In Russian constituent entities with a high level of financing of innovations, the corresponding costs range from 2.9% to 8% of the amount of income received from the sale of manufactured products by organizations located in these entities. As for the constituent entities with low values of indicators, they are in the range from 0.02% to 0.25%. The findings of the study indicate the unevenness of the innovative development of organizations in Russian regions. This situation contradicts the tasks of the significant increase in the innovative activity of enterprises and organizations in all regions without exception, determined by decisions of the Government of the Russian Federation. Therefore, in regions with a low level of funding for innovations, it is necessary to implement measures to stimulate investments in the production of innovative products.

The results of this study are of theoretical and practical importance. The tools presented in the article for assessing the share of costs of financing innovation in the amount of income received from the sale of manufactured products by organizations located in the constituent entities of the country can be used in further research. The proposed tools can be applied to monitor the costs of financing innovative activities in organizations. The results of the work can be implemented in the current activities of government institutions and public organizations, when justifying measures for the development of innovation activities. In addition, the information obtained can be used to solve the problems of increasing the share of innovative organizations in the constituent entities of the country where such organizations are not well-developed. The results of the work are critical for leasing companies that ensure the introduction of new equipment and advanced technologies.

Further research can be carried out to assess industry specificities for financing innovation. In the course of the study, there were no restrictions on empirical data, since the research considered the data on all 82 constituent entities of the Russian Federation.

References

1. OECD. Eurostat. Oslo Manual: Guidelines for Collecting, Reporting and Using Data on Innovation. The Measurement of Scientific, Technological and Innovation Activities, OECD Publishing, Paris (2018)
2. OECD: Innovation Strategy, OECD Publishing, Paris (2020)

3. Cramond, B., Fairweather, E.: Future problem solving as education for innovation. In: Shavinina, L. (ed.) Routledge International Handbook of Innovation Education, pp. 212–222. Routledge, London, United Kingdom (2013)
4. De Clercq, D., Pereira, R.: Resilient employees are creative employees, when the workplace forces them to be. Creativity and Innovation Management, pp. 329–342 (2019)
5. Soleas, E.K.: Leader strategies for motivating innovation in individuals: a systematic review. J. Innov. Entrepreneurship **9**(1), 1–28 (2020)
6. Horkoff, J., Maiden, N., Asboth, D.: Creative goal modeling for innovative requirements. Inf. Softw. Technol. **106**, 85–100 (2019)
7. Leiponen, A., Helfat, C.: Innovation objectives, knowledge sources, and the benefits of breadth. Strateg. Manage. J. **31**(2), 224–236 (2010)
8. Hu, Y., Aziz, E., Chassapis, C.: Creativity-based design innovation environment in support of robust product development. Int. J. Interact. Des. Manuf. **10**(4), 335–353 (2014)
9. Dereli, D.: Innovation management in global competition and competitive advantage. Procedia Social Behav. Sci. **195**, 1365–1370 (2015)
10. Damanpour, F., Aravind, D.: Managerial innovation: concepts, processes, and antecedents. Manage. Organ. Rev. **8**(2), 423–454 (2012)
11. Kotsemir, M.N., Meissner, D.: Conceptualizing the innovation process – trends and outlook. SSRN Electron. J. (2013). https://doi.org/10.2139/ssrn.2249782
12. Schaarschmidt, M., Kilian, T.: Impediments to customer integration into the innovation process: a case study in the telecommunications industry. Eur. Manage. J. **32**(2), 350–361 (2014)
13. Tamayo, M., Huergo, E.: Determinants of internal and external R&D offshoring: evidence from spanish firms. Ind. Innovation **24**(2), 143–164 (2017)
14. Klewitz, J., Hansen, E.: Sustainability-oriented innovation of SMEs: a systematic review. J. Cleaner Prod. **65**, 57–75 (2014)
15. De Jong, M., Marston, N., Roth, E.: The eight essentials of innovation. McKinsey Q. **2**, 1–12 (2014)
16. Teirlink, P., Spithoven, A.: Formal R&D management and strategic decision making in small firms in business services. R&D Manage. **43**, 37–51 (2013)
17. Yadollahi, F., Toghraee, M.: Identification the main challenges of small and medium sized enterprises in exploiting of innovative opportunities. J. Global Entrepreneurship Res. **4**(4), 1–15 (2014)
18. Diallo, O.: Small and medium enterprises (SMEs) as drivers of productive capacity and job creation in Africa. Background Paper for Regional Preparatory Meeting for Africa. ECOSOC, New York (2012)
19. Lee, S., Park, G., Yoon, B., Park, J.: Open innovation in SMEs: an intermediated network model. Res. Policy **39**, 290–300 (2010)
20. Xie, X., Zeng, S., Peng, Y., Tam, C.: What affects the innovation performance of small and medium-sized enterprises in China? Innovation: Management. Policy Pract. **15**(3), 271–286 (2013)
21. Carlino, G., Kerr, W.R.: Agglomeration and Innovation. Harvard Business School Working Paper 15-007 (2014)
22. Lerner, J.: The Architecture of Innovation: The Economics of Creative Organizations. Harvard Business School Press, Boston, MA (2012)
23. Kerr, W.R., Ramana, N.: Financing Innovation Working Paper 15-034. Harvard Business School Press, Boston, MA (2014)
24. Kayal, A.: National innovation systems: a proposed framework for developing countries. Int. J. Entrepreneurship Innovation Manage. **8**(1), 74–86 (2008)
25. Audretsch, D., Frisch, M.: Linking entrepreneurship to growth: the case of West of Germany. Ind. Innovation **10**(1), 65–73 (2003)

26. Galindo, M., Mendez, M.: Emprendedores y objetivos de polltica economica. Inf. Comercial Esp. **841**, 29–40 (2008)
27. Blake, M., Hanson, S.: Rethinking innovation: context and gender. Environ. Plann. **37**(4), 681–701 (2005)
28. Robb, A., Coleman, S.: Gender differences in innovation among US entrepreneurs. Chapter 6. In: Lewis, K., Henry, C., Gatewood, E.J., Watson, J. (eds.) In Women's Entrepreneurship in the 21st Century, pp. 117–137. An International Multi-Level Research Analysis. Edward Elgar Publishing, United Kingdom (2014)
29. de Oliveira Sousa, S.R., da Silva, W.V., da Veiga, C.P., Zanini, R.R.: Theoretical background of innovation in services in small and medium-sized enterprises: literature mapping. J. Innovation Entrepreneurship **9**(1), 1–26 (2020)
30. Savrula, M., Incekara, A.: The effect of r&d intensity on innovation performance: a country level evaluation. Procedia Social Behav. Sci. **210**, 388–396 (2015)
31. Hottenrott, H., Peters, B.: Innovative capability and financing constraints for innovation more money, more innovation? Discussion paper 09-081, ZEW, Zentrum fur Europaische Wirtschaftsforschung GmbH. Center for European Economic Research (2012)
32. Bougheas, S., Gorg, H., Strobl, E.: Is R&D financially constrained? Theory and evidence from Irish manufacturing. Rev. Ind. Organ. **22**(2), 159–174 (2003)
33. Czarnitzki, D., Hottenrott, H.: R&D investment and financing constraints of small and medium-sized firms. Small Bus. Econ. **36**(1), 65–83 (2011)
34. Peeters, C., van Pottelsberghe, B.: Measuring innovation competencies and performances: a survey of large firms in Belgium, pp. 3–16. Institute of Innovation Research Working Paper, Hitotsubashi University (2003)
35. Scherer, F.M., Harhoff, D.: Technology policy for a world of skew-distributed outcomes. Res. Policy **29**(4–5), 559–566 (2000)
36. Kaplan, S., Zingales, L.: Investment-cash flow sensitivities are not valid measures of financing constraints. Q. J. Econ. **115**(2), 707–712 (2000)
37. Avdogan, A.: How sensitive is investment to cash flow when financing is frictionless? J. Finance **58**(2), 707–722 (2003)
38. Moven, N.: Investment-cash flow sensitivities: constrained versus unconstrained firms. J. Finance **59**(4), 2061–2092 (2004)
39. Dhanani, A.: Corporate dividend policy: the views of British financial managers. J. Bus. Finance Acc. **32**(7 & 8), 1625–1672 (2005)
40. Brown, J.R., Martinsson, G., Petersen, B.C.: Do financing constraints matter for R&D? Eur. Econ. Rev. **56**(8), 1512–1529 (2012)
41. Brown, J.R., Fazzari, S.M., Petersen, B.C.: Financing innovation and growth: cash flow, external equity, and the 1990s R&D boom. J. Finance **64**(1), 151–185 (2009)
42. Marenkov, I.M.: Innovations in the Russian industry: state and prospects of development. Intell. Innovation Investment **1**, 35–42 (2019)
43. Semenova, N.N.: Financing of innovations as a condition for the formation of a new model of economic development of Russia. Financial Life **1**, 77–82 (2019)
44. Krasova, E.V.: State financing of innovations in Russia: dynamics and specifics. The Territory of New Opportunities, pp. 47–58. Vestnik VSUES 1 (2019)
45. Makarov, V.V., Slutsky, M.G., Alexandrova, N.A.: Results of implementation of innovative projects in PJSC "Rostelecom." Eurasian Sci. Association **5–4**(51), 286–289 (2019)
46. Sokolova, A.P., Kabannik, E.A.: Economic conditions and priority directions of innovative development of the Russian economy. Bull. Altai Acad. Econ. Law **1**, 86–94 (2020)
47. Barkova, E.E.: Development of an algorithm for selecting a source of financing for innovative projects at the early stages of the life cycle. Bull. Russ. Univ. Cooperation **1**(39), 11–18 (2020)
48. Sheina, E.G.: Venture financing of innovative activity of business entities in the conditions of formation of the economy of advanced development. Russ. Econ. Bull. **3**(2), 206–214 (2020)

49. Raiskaya, M.V.: Scope of application, state and prospects of innovative leasing development. Econ. Vector **1**(20), 72–79 (2020)
50. Bykovskaya, E.N., Gafiyatullina, Y., Kharchilava, G.P.: Analysis of sources and mechanisms of financing innovation activities in the digital economy. Stand. Living Popul. Reg. Russ. **16**(2), 36–41 (2020)
51. Leshchinskaya, A.F., Ivolgina, N.V., Stepanova, D.I., Akimova, N.A.: Problems of financing technological innovations in the oil production market. Econ. Ind. **13**(2), 233–243 (2020)
52. Denisova, O.N.: Venture financing of innovative small business in Russia. Innovations Investments **6**, 3–7 (2020)
53. Federal State Statistics Service: Science and Innovation. Available at https://rosstat.gov.ru/folder/14477?print=1. Accessed 19 June 2021
54. Pinkovetskaia, I., Slepova, V.: Estimation of fixed capital investment in SMEs: the existing differentiation in the Russian federation. Bus. Syst. Res. **9**(1), 65–78 (2018)
55. Pinkovetskaia, I., Nuretdinova, Y., Nuretdinov, I., Lipatova, N.: Mathematical modeling on the base of functions density of normal distribution. Rev. Universidad del Zulia **12**(33), 34–49 (2021)

Mathematical Modeling of Various Physical Processes in Technical Systems Described by Polynomial Matrix Equations

Alexander G. Madera(✉)

Scientific Research Institute of System Analysis of Russia Academy of Sciences, 36, k.1, Nachimovskiy prosp, 117218 Moscow, Russia

Abstract. This article presents an analytical method for the solving matrix polynomial equations which in many cases are the mathematical models for a lot of various nonlinear physical processes (thermal, electrical, mechanical) in the different technical systems. The developed method makes it possible to reduce a complex system of nonlinear interconnected equations to a system of independent unconnected algebraic equations, each of which can be easily solved. The method is based on similarity transformations and spectral decomposition of the system matrix. The application of the developed method is considered by the example. Comparison of the results obtained by the developed and numerical methods showed their complete agreement.

Keywords: Mathematical modeling · Nonlinear matrix equation · Polynomial functions · Eigendecomposition of the matrix · Spectrum of similarity transformation

1 Introduction

Mathematical modeling of various processes of different physical nature [1–7] (mechanical, electrical, thermal, etc.) that occur in various technical systems in the course of their operation often leads to systems of algebraic equations with nonlinear polynomial right-hand side that depends on the sought vector x, namely:

$$Hx = F(x), \tag{1}$$

where H is a $n \times n$ square matrix of a system with constant coefficients; $x = (x_1, x_2, \ldots, x_n)^T$ is the sought n-vector; $F(x) = (f_1(x_1), f_2(x_2), \ldots, f_n(x_n))^T$ is the vector-function where functions $f_i(x_i)$ depend nonlinearly on elements of the sought vector x.

Mathematical model of this type (1) is, for instance, the result of mathematical modeling of thermal processes in technical systems where the consumed power depends nonlinearly on the temperature [7].

Since any nonlinear function $f_i(x_i)$ can be approximated with some polynomial of the k-th degree, or can be expanded into a Taylor series, it is reasonable to assume that

R. Silhavy et al. (Eds.): CoMeSySo 2021, LNNS 232, pp. 712–719, 2021.
https://doi.org/10.1007/978-3-030-90318-3_56

nonlinear functions on the right-hand side of the equation are polynomials of the k-th degree, namely:

$$F(x) = \begin{pmatrix} f_1(x_1) \\ f_2(x_2) \\ \vdots \\ f_n(x_n) \end{pmatrix} = \begin{pmatrix} f_{01} + f_{11}x_1 + f_{21}x_1^2 + \cdots + f_{k1}x_1^k \\ f_{02} + f_{12}x_2 + f_{22}x_2^2 + \cdots + f_{k2}x_2^k \\ \vdots \\ f_{0n} + f_{1n}x_n + f_{2n}x_n^2 + \cdots + f_{kn}x_n^k \end{pmatrix}. \tag{2}$$

Matrix Eq. (1) with polynomial right-hand side, due to lack of mathematical methods of solving, is normally solved by assuming that nonlinear functions $f_i(x_i), i = 1, 2, \ldots, n$ can be approximated by linear ones, or by leaving only first-order terms in the Taylor series expansion of nonlinear functions. All these assumptions produce inadequate modelling results and errors in design of technical systems.

The existing literature [8–14] considers a nonlinear polynomial matrix equation

$$A_0 + A_1X + A_2X^2 + \cdots + A_kX^k = 0,$$

in which an unknown $n \times n$ square matrix X has to be defined, where A_l, $l = 0, 1, 2, \ldots, k$, are the given $n \times n$ square matrices. Such mathematical problems as solving the equation $X^m = A$, arising from extracting the m-th root from nonsingular matrix X, or determination of nilpotent ($X^m = 0$) or idempotent ($X^2 = X$) matrix X, result in polynomial matrix equations.

Solution of a nonlinear polynomial matrix equation is based upon transformation of similarity $X = TDT^{-1}$, where matrix D is considered known, and T is a nonsingular matrix [8], or is based on numerical methods [15]. Since $X^m = TD^mT^{-1}$, $m = 1, 2, \ldots, k$, then by putting matrices X^m, $m = 1, 2, \ldots, k$, in the original polynomial matrix equation and by multiplying it by matrix T on the right-hand side, an equation in the unknown matrix T is obtained, namely

$$A_0T + A_1TD + A_2TD^2 + \cdots + A_kTD^k = 0.$$

As it is mentioned in [8], the only way to find a matrix solution T of this equation is to substitute the obtained matrix equation for a system of homogenous linear scalar equations in the elements of the sought matrix T. Each nonsingular solution T plugged in the transformation $X = TDT^{-1}$ will yield a solution of the original nonlinear polynomial matrix equation.

At the same time, the above polynomial matrix equation with the unknown matrix X is not extensively applied in mathematical modelling, since it does not describe any real-life physical process, at least the existing literature contains no such mathematical models. At the same time, vector-matrix nonlinear Eqs. (1), (2) with polynomial right-hand side, where an unknown vector x (but not matrix X) has to be found, quite often occur during mathematical modelling of real-life physical processes in technical systems. However, methods of solving nonlinear polynomial matrix-vector Eqs. (1), (2), as well as the constructive theory of solving such equations, are currently missing.

This article develops an general method of solving nonlinear matrix-vector equations of type (1) with polynomial right-hand side (2). The method has been tested for a polynomial matrix equation with the right-hand side in the form of a fifth-degree polynomial. The obtained results were verified by numerical methods and have been shown high accuracy. The application of developed in this article general method in relation to mathematical modeling of thermal processes in electronic systems with thermal feedback has been discussed in detail in the article by A. Madera and E. Grebennikova published in this journal.

2 Method of Solving Polynomial Matrix-Vector Equation

This section describes a method of solving nonlinear matrix equations with a polynomial right-hand side (1), (2). In the suggested method, the original matrix-vector Eq. (1) for the unknown vector x is first reduced to a matrix equation for the unknown matrix X, that relate to each other as $x = Xj$. It allows to apply the similarity transformation of matrix $X = TDT^{-1}$ where matrix D is the unknown. Then matrices of the system H of Eq. (1) is subjected to eigendecomposition and is reduced to a diagonal identity matrix. As a result, a matrix equation that defines the diagonal eigenvalue matrix of matrix D, which expands to independent algebraic equations of the k-th degree. Determination of roots of obtained algebraic polynomial equations allows finding eigenvalues of matrix D, which in turn help finding the unknown vector of the nonlinear matrix-vector Eq. (1), (2).

2.1 Reduction of Original Matrix-Vector Equation to Matrix Equation and Diagonalization of Matrix of System of Eqs. (1)

Note that in order to apply similarity transformation to the original equation it is first of all necessary to switch from the matrix-vector Eq. (1), where vector x is unknown, to a matrix equation, where the matrix is unknown.

To do so, let us write vector x as a product $x = Xj$ of $n \times n$ diagonal matrix $X = diag\{x_1, x_2, \ldots, x_n\}$ and unit vector $j = (11\ldots 1)^T$, while X^m $(m = 1, 2, \ldots k)$ is also a diagonal matrix equal to $X^m = diag\{x_1^m, x_2^m, \ldots, x_n^m\}$, namely [16]:

$$x = \begin{pmatrix} x_1 \\ x_2 \\ \vdots \\ x_n \end{pmatrix} = \begin{pmatrix} x_1 & & & 0 \\ & x_2 & & \\ & & \ddots & \\ 0 & & & x_n \end{pmatrix} \begin{pmatrix} 1 \\ 1 \\ \vdots \\ 1 \end{pmatrix} = Xj,$$

$$\begin{pmatrix} x_1^m \\ x_2^m \\ \vdots \\ x_n^m \end{pmatrix} = \begin{pmatrix} x_1^m & & & 0 \\ & x_2^m & & \\ & & \ddots & \\ 0 & & & x_n^m \end{pmatrix} \begin{pmatrix} 1 \\ 1 \\ \vdots \\ 1 \end{pmatrix} = X^m j.$$

After introducing vectors

$$F_0 = \begin{pmatrix} f_{01} \\ f_{02} \\ \vdots \\ f_{0n} \end{pmatrix}, F_1 = \begin{pmatrix} f_{11} \\ f_{12} \\ \vdots \\ f_{1n} \end{pmatrix}, F_2 = \begin{pmatrix} f_{21} \\ f_{22} \\ \vdots \\ f_{2n} \end{pmatrix}, \cdots, F_k = \begin{pmatrix} f_{k1} \\ f_{k2} \\ \vdots \\ f_{kn} \end{pmatrix},$$

instead of matrix-vector Eq. (1), (2) in the unknown vector x, we obtain a matrix equation in the unknown matrix X, namely,

$$HXj = F_0 + XF_1 + X^2F_2 + \cdots + X^kF_k. \tag{3}$$

Let us find all solutions X of the matrix Eq. (3) among nonsingular $n \times n$ matrices in the similarity transformation $X = TDT^{-1}$. Since $X = TDT^{-1}$, $X^2 = TD^2T^{-1}$, ..., $X^k = TD^kT^{-1}$, then Eq. (3) will be written as follows:

$$HTDT^{-1}j = F_0 + TDT^{-1}F_1 + TD^2T^{-1}F_2 + \cdots + TD^kT^{-1}F_k. \tag{4}$$

Let us reduce matrix H of the equation system (4) to a diagonal form.

Let us assume that matrix H in Eq. (1) is nonsingular and positively defined. In this case, there is an eigendecomposition $U^{-1}HU = \Lambda_H$ with $n \times n$ similarity transformation matrix U and $n \times n$ diagonal eigenvalue matrix of matrix H [10, 11].

Let us write the transformation matrix T in Eq. (4) as $T = U\Lambda_H^{-\frac{1}{2}}V^{-1}$, where V is the yet-unknown $n \times n$ similarity transformation matrix, whose significance will be explained below, and let us multiply Eq. (4) on the left-hand side by matrix $V\Lambda_H^{-\frac{1}{2}}U^{-1}$. The resulting equation will be as follows:

$$V\Lambda_H^{-\frac{1}{2}}U^{-1} \cdot H \cdot U\Lambda_H^{-\frac{1}{2}}V^{-1}DT^{-1}j =$$

$$= V\Lambda_H^{\frac{1}{2}}U^{-1}F_0 + DT^{-1}F_1 + D^2T^{-1}F_2 + \cdots + D^kT^{-1}F_k. \tag{5}$$

Since

$$V\Lambda_H^{-\frac{1}{2}}\underbrace{U^{-1} \cdot H \cdot U}_{\Lambda_H}\Lambda_H^{-\frac{1}{2}}V^{-1} = VV^{-1} = I,$$

the matrix H of the system of equations is thus reduced to a diagonal unit form $I = diag\{1, 1, \ldots, 1\}$, and Eq. (5) will be as follows:

$$V\Lambda_H V^{-1}DT^{-1}j = V\Lambda_H^{\frac{1}{2}}U^{-1}F_0 + DT^{-1}F_1 + D^2T^{-1}F_2 + \cdots + D^kT^{-1}F_k. \tag{6}$$

2.2 Determination of Equations to Which Elements of Matrix *D* in Similarity Transformation TDT^{-1} Must Adhere

Considering that $T^{-1} = V\Lambda_H^{\frac{1}{2}}U^{-1}$ and introducing to Eq. (6) representations of vectors.

$$U^{-1}j = \alpha = (\alpha_1, \alpha_2, \ldots, \alpha_n)^T, U^{-1}F_l = \beta_l = (\beta_{l1}, \beta_{l2}, \ldots, \beta_{ln})^T, \quad l = 0, 1, 2, \ldots, k,$$

we obtain the equation

$$V\Lambda_H V^{-1}DV\Lambda_H^{\frac{1}{2}}\alpha = V\Lambda_H^{\frac{1}{2}}\beta_0 + DV\Lambda_H^{\frac{1}{2}}\beta_1 + D^2V\Lambda_H^{\frac{1}{2}}\beta_2 + \cdots + D^kV\Lambda_H^{\frac{1}{2}}\beta_k. \quad (7)$$

We will require that matrix V is a similarity transformation matrix for matrix D, so that its eigendecomposition is written as

$$V^{-1}DV = \Lambda_D, \quad (8)$$

where $\Lambda_D = diag\{\lambda_{D1}, \lambda_{D2}, \ldots, \lambda_{Dn}\}$ is the diagonal eigenvalue matrix of matrix D.

Elements of matrix Λ_D are eigenvalues of matrix D in similarity transformation and at the same time must satisfy the specific polynomial Eq. (1), (2).

Let us find an equation to determine elements of the diagonal matrix Λ_D. By multiplying Eq. (7) on the left-hand side by matrix V^{-1}, considering eigendecomposition (8) of matrix D, as well as diagonal structure of matrices Λ_D and Λ_H, we obtain the final matrix polynomial equation, which must be satisfied by the eigenvalue matrix Λ_D

$$\beta_0 + \Lambda_D(\beta_1 - \Lambda_H\alpha) + \Lambda_D^2\beta_2 + \cdots + \Lambda_D^k\beta_k = 0. \quad (9)$$

Matrix Eq. (9), due to diagonality of matrices Λ_H and Λ_D, decomposes into n independent polynomial equations of k-th degree in the unknown eigenvalues λ_{Di}, $i = 1, 2, \ldots, n$, namely

$$\beta_{0i} + (\beta_{1i} - \lambda_{Hi}\alpha_i)\lambda_{Di} + \beta_{2i}\lambda_{Di}^2 + \ldots + \beta_{ki}\lambda_{Di}^k = 0, \quad (10)$$

$$i = 1, 2, \ldots, n.$$

The resulting Eqs. (1) are, in essence, conditions to which eigenvalues of matrix D in similarity transformation TDT^{-1} must obey.

2.3 Determination of Unknown Solution Vector *x* of Nonlinear Eq. (1) with Polynomial Right-Hand Side (2)

Unknown vector x, which is the solution of matrix Eq. (1) with polynomial right-hand side (2), is determined by the following relationships (see items 2.1, 2.2):

$$x = Xj = TDT^{-1}j = U\Lambda_H^{-\frac{1}{2}}\underbrace{V^{-1}DV}_{\Lambda_D}\Lambda_H^{\frac{1}{2}}U^{-1} = U\Lambda_D U^{-1}j. \quad (11)$$

In Eq. (11), similarity transformation matrix U is known from the eigendecomposition of matrix H of the original system of Eqs. (1), and elements λ_{Di}, $i = 1, 2, \ldots, n$, of diagonal matrix Λ_D are determined by solutions of algebraic polynomial equations of the k-th degree (10).

The developed method which allows to obtain vector solution x of the matrix nonlinear Eq. (1) with polynomial right-hand side (2) is applied by performing the following stages:

- eigendecomposition $U^{-1}HU = \Lambda_H$ of matrix H of the original system of Eqs. (1), $n \times n$ similarity transformation matrix U and $n \times n$ diagonal eigenvalue matrix $\Lambda_H = diag\{\lambda_{H1}, \lambda_{H2}, \ldots, \lambda_{Hn}\}$ of matrix H are determined [10, 11]. Since matrix H is nonsingular and positively defined, its eigendecompostion exists;
- n independent algebraic polynomial Eqs. (10) of k-th degree are formed, and their solutions are used to determine eigenvalues λ_{Di}, $i = 1, 2, \ldots, n$, of matrix D;
- k roots found by solving algebraic equations of the k-th degree (10) generally contain both complex and real roots, and while complex roots are set aside and are not considered, real roots of the equation must be analyzed and screened based on conceptual physical significance of the problem being solved;
- unknown vector of the solution x of the original nonlinear Eq. (1) with polynomial right-hand side (2) is determined in accordance with Eq. (11).

3 Example of Method Application

Let us consider matrix polynomial equation $Hx = F(x)$ consisting of two equations with the right-hand side in the form of a fifth degree polynomial, namely

$$Hx = \begin{pmatrix} h_{11} & h_{12} \\ h_{21} & h_{22} \end{pmatrix} \begin{pmatrix} x_1 \\ x_2 \end{pmatrix} = \begin{pmatrix} f_{01} + f_{11}x_1 + f_{21}x_1^2 + f_{31}x_1^3 + f_{41}x_1^4 + f_{51}x_1^5 \\ f_{02} + f_{12}x_2 + f_{22}x_2^2 + f_{32}x_2^3 + f_{42}x_2^4 + f_{52}x_2^5 \end{pmatrix}$$

The last equation, using transformation $x = Xj$ (see item 2.1) with $n \times n$ diagonal matrix $X = diag\{x_1, x_2\}$ and vectors $F_l = (f_{l1}, f_{l2})^T$, $l = 0, 1, 2, 3, 4, 5$, and $j = (11)^T$ can be written as follows (see (3)):

$$HXj = F_0 + XF_1 + X^2F_2 + X^3F_3 + X^4F_4 + X^5F_5.$$

In terms of its physical meaning, the considered system of polynomial equations models, for example, a steady-state thermal process in an electronic systems. In more detail the case of thermal process in real electronic systems with thermal feedback is considered in this journal in the article "Thermal processes with thermal feedback in electronic systems and their modeling by nonlinear matrix equations

Input data of the considered system are as follows:

- $h_{11} = 0{,}31422$, $h_{12} = h_{21} = -0{,}0065$, $h_{22} = 0{,}66675$,
- $f_{01} = 11{,}05$, $f_{11} = 9{,}86 \cdot 10^{-2}$, $f_{21} = 2{,}9 \cdot 10^{-3}$, $f_{31} = -1{,}06 \cdot 10^{-4}$, $f_{41} = 8{,}07 \cdot 10^{-7}$, $f_{51} = -1{,}34 \cdot 10^{-9}$

- $f_{02} = 6{,}65, f_{12} = -0, 12, f_{22} = 6, 24 \cdot 10^{-3}, f_{32} = -3, 5 \cdot 10^{-5}, f_{42} = -4, 12 \cdot 10^{-7}, f_{52} = 2, 9 \cdot 10^{-9}$.

According to the method application procedure (see item 2.3), it is necessary to first determine eigendecomposition of matrix $H = U\Lambda_H U^{-1}$, which equals to

$$\begin{pmatrix} 0, 31422 & -0, 00605 \\ -0, 00605 & 0, 66675 \end{pmatrix} =$$

$$= \begin{pmatrix} 58, 3 & -0, 0172 \\ 1 & 1 \end{pmatrix} \begin{pmatrix} 0, 314 & 0 \\ 0 & 0, 667 \end{pmatrix} \begin{pmatrix} 0, 0171 & 0, 000294 \\ -0, 0171 & 1, 0 \end{pmatrix},$$

and of matrix U и Λ_H.

$$U = \begin{pmatrix} 58, 3 & -0, 0172 \\ 1 & 1 \end{pmatrix}, \Lambda_H = \begin{pmatrix} 0, 314 & 0 \\ 0 & 0, 667 \end{pmatrix},$$
$$U^{-1} = \begin{pmatrix} 0, 0171 & 0, 000294 \\ -0, 0171 & 1, 0 \end{pmatrix}.$$

Then it is necessary to form algebraic polynomial equations (in the considered example – equations of the fifth degree) that determine elements λ_{D1}, λ_{D2} of diagonal matrix Λ_D, namely:

$$-2, 21 \cdot 10^{-11}\lambda_{D1}^5 + 1, 37 \cdot 10^{-8}\lambda_{D1}^4 - 1, 82 \cdot 10^{-6}\lambda_{D1}^3$$
$$+5, 14 \cdot 10^{-5}\lambda_{D1}^2 - 0, 004\lambda_{D1} + 0, 19 = 0,$$

$$2, 93 \cdot 10^{-9}\lambda_{D2}^5 - 4, 26 \cdot 10^{-7}\lambda_{D2}^4 - 3, 32 \cdot 10^{-5}\lambda_{D2}^3$$
$$+0, 006\lambda_{D2}^2 - 0, 78\lambda_{D2} + 6, 46 = 0.$$

These equations have five pairs of roots, three of which are complex roots (not considered) and two pairs are real roots, namely: 1st pair of roots $(\lambda_{D1}, \lambda_{D2})_1 = (46, 63; 8, 85)$ and 2nd pair of roots $(\lambda_{D1}, \lambda_{D2})_2 = (156, 26; 185, 32)$. Estimate calculations of thermal processes show that the actual physical significance of the considered mathematical model is satisfied by the 2nd pair of roots.

The unknown solution vector x of the considered nonlinear mathematical model equals to (see (11))

$$x = U\Lambda_D U^{-1} j =$$
$$= \begin{pmatrix} 58, 3 & -0, 0172 \\ 1 & 1 \end{pmatrix} \begin{pmatrix} 156, 26 & 0 \\ 0 & 185, 32 \end{pmatrix} \begin{pmatrix} 0, 0171 & 0, 000294 \\ -0, 0171 & 1, 0 \end{pmatrix} \begin{pmatrix} 1 \\ 1 \end{pmatrix} = \begin{pmatrix} 155, 33 \\ 184, 87 \end{pmatrix}.$$

Results obtained using the developed method have been verified by solving the original equations with the Newton's method. Comparison of results yielded by the developed method and the numerical method proved them to be consistent (relative error not exceeding 0.6%).

4 Conclusion

We have considered a method of solving a matrix nonlinear Eq. (1) with a nonlinear function on the right-hand side of Eq. (2) defined as a polynomial of the k-th degree in the unknown vector. Mathematical models of various physical processes occurring in technical systems of various applications are reduced to the considered nonlinear equation. The right-hand side of the equation represented as a polynomial of the k-th degree is the most common, since any nonlinear function that has a physical significance can be expanded as a Taylor series, or can be approximated by a polynomial.

The method developed in the article is generic and does not have any limitations in terms of polynomial type and degree. The method allows finding matrix solutions of matrix nonlinear equations without using numerical methods.

Acknowledgment. Publication is made as part of national assignment for SRISA RAS (fundamental scientific research 47 GP) on the topic No.0580-2021-0001 (121031300047-6).

References

1. Chua, L.O., Lin, P.-M.: Computer-aided Analysis of Electronic Circuits. Prentice-Hall, Englewood Cliffs, New Jersey (1975)
2. Wellstead, P.E.: Introduction to Physical System Modelling. Electronic Publisher, Control Systems Principles, www.control-systems-principles.co.uk (2005)
3. Ellison, G.N.: Thermal Computations for Electronics. Conductive, Radiative, and Convective Air Cooling. CRC Press, New York (2011)
4. Frazer, R.A., Duncan, W.J., Collar, A.R.: Elementary Matrices and Some Applications to Dynamics and Differential Equations. Cambridge University Press, New York (1960)
5. Madera, A.G.: Interval-stochastic thermal processes in electronic systems: analysis and modeling. J. Eng. Thermophys. **26**(1), 17–28 (2017)
6. Madera, A.G., Kandalov, P.I.: Thermal processes in electronic equipment at uncertainty. J. Eng. Thermophys. **29**(1), 170–180 (2020)
7. Madera, A.G.: Modeling thermal feedback effect on thermal processes in electronic systems. Comput. Res. Model. **10**(4), 483–494 (2018)
8. Gantmacher, F.R.: The Theory of Matrix. Chelsea Publishing Company, New York (1960)
9. Bellman, R.: Introduction to Matrix Analysis. McGraw-Hill, New York (1960)
10. Lankaster, P.: Theory of Matrix. Academic Press, New York – London (1969)
11. Horn, R.A., Johnson, C.R.: Matrix Analysis. Cambridge University Press, Cambridge (1986)
12. Bini, D.A., Latouche, G., Meini, B.: Solving matrix polynomial equations arising in queueing problems. Linear Algebra Appl. **340**(1–3), 225–244 (2002)
13. Williams, M.P.: Solving polynomial equations using linear algebra. J. Hopkins APL Tech. Dig. **28**(4), 354–363 (2010)
14. Jonsson, G.F., Vavasis, S.A.: Accurate solution of polynomial equations using Macaulay resultant matrices. Math. Comput. **74**(249), 221–262 (2004)
15. Han, Y.-H., Kim, H.-M.: Solving a matrix polynomial by Newton's method. J. Korean Soc. Ind. Appl. Math. **14**(2), 113–124 (2010)
16. Madera, A.G.: A method for explicitly solving matrix differential equations in second-order ordinary derivatives based on diagonalizing the matrices of the equation using spectral decomposition and Kronecker matrix algebra. Differencialnie Uravnenia i Protsesy Upravlenia (Differential Equations and Control Processes) **3**, 10–24 (2021)

Light Rail Transit (LRT) Passenger Management System Using Image Processing Algorithm with Raspberry Pi and Multiple Cameras

Decerry Millenna Paulme(✉), Marielle Perez, and Julius Sese

Mapua University, Manila, Philippines
{dmspaulme,mcperez}@mymail.mapua.edu.ph, jtsese@mapua.edu

Abstract. The Covid 19 pandemic has brought challenges to transportation as social distancing measures were hard to implement on overcrowded public vehicles. This paper presents an LRT passenger management system using image processing algorithms with a Raspberry Pi and multiple cameras. This works by counting and limiting the passengers inside the train. The model also includes an early warning system and an alarm to alert passengers when the vehicle has passed a particular threshold limit. After testing on different scenarios, the system obtained 96.5% accuracy on single entry and exit of passengers and 1 to 2 s to count. Meanwhile, the results have shown 88.22% for simultaneous entry and 93% for simultaneous exits. For simultaneous entry and exits, the system has obtained 94.44%. Overall, this study was able to develop a low-cost passenger management system with high accuracy and fast time to count. Moreover, we were able to determine the parameters to vary to improve the overall accuracy of the system.

Keywords: Passenger management · Light rail transit · Image processing

1 Introduction

According to the World Health Organization (WHO), "coronavirus disease, or COVID-19, is an infectious disease that has infected millions of people around the globe." As a preventive measure for this virus, physical distancing is highly recommended. With that, the Covid-19 has brought new challenges in public transport as social distancing measures were difficult to implement in overcrowded public vehicles. Hence, the government has proposed the 'new normal' bill wherein new norms of physical distancing are imposed in public places to avoid new surges in cases. In consideration of 'new normal', major train line stations, specifically the Light Rail Transit-1, or LRT-1, have implemented passenger limits to encourage social distancing. In the simulation done by the transport lines, the passenger limiting was managed by the security guards and personnel which is impractical as it is time-consuming and prone to errors.

Overcrowding in the public mass rapid transports is already a common problem in our society even before the pandemic has started. Since there were a lot of commuters

R. Silhavy et al. (Eds.): CoMeSySo 2021, LNNS 232, pp. 720–748, 2021.
https://doi.org/10.1007/978-3-030-90318-3_57

in the Philippines, LRT-1 is one of the most used public transports. In previous studies, several crowd control systems were made possible with the use of advanced technology. In the Philippines, one study developed simulation software to monitor the crowds in train stations. In the Netherlands, an RFID-based ticketing system was used to count the number of people and use it to catch fare-dodgers. Another study has evaluated the crowd dynamics of commuters through static floor fields to plan effective crowd control schemes. Different techniques for counting people were also developed including video counting, infrared sensors, treadle mats, and several image processing techniques.

However, the past papers were made for different purposes such as crowd monitoring, infrastructure preparedness, and crowd dynamic models. Hence, they do not control and limit the passengers per train as they focused on monitoring the crowd for later interventions which are not suitable for the "new normal". Aside from that, existing people counters, especially infrared sensors, and video cameras are mostly made for buses and several establishments such as auditoriums and malls. These people counters are not suitable for train cars as they are not capable of counting multiple passengers simultaneously. In the Philippines, manual counting and limiting of passengers in railway stations are solely done by selected staff and personnel.

In this paper, the researchers aim to design an LRT passenger management system by implementing passenger limits on trains and by giving an immediate response for the new normal adjustment. Specifically, the objectives of this paper are (1) to develop a passenger counter circuit using multiple cameras programmed with OpenCV Python on Raspberry Pi (2) to implement image processing algorithms for passenger detection and movement analysis (3) to create an alarm system maintaining the suggested capacity of passengers for the new normal adjustment, and (4) to determine the speed and accuracy of the counter in different scenarios.

This study will help the Philippines regarding its current problem of social distancing on public transportation. This will help the rail operators, specifically of LRT, to measure and limit the crowd density to aid in implementing social distancing. This study will benefit society as it helps in ensuring the health and safety of all the personnel, operators, and passengers during this pandemic. This could also be used for future studies about managing the existing problem of overcrowding in public transport.

The scope of this study is to manage the crowd to follow social distancing for 'new normal' adjustment. This will focus on developing a passenger counter and alarm to limit passenger numbers inside the train only. Though it might help at some point, this will not cover other common LRT station problems such as ticket queue and platform overcrowding. This device will not directly implement social distancing but will only aid in its protocol implementation by limiting the crowds inside the train. As of now, LRT train cars have plastic seat barriers and seat marks to avoid direct contact among the passengers. Along with the existing LRT train car policies and our passenger management system, social distancing protocols will be better implemented inside the train overall. Another limitation of this paper is the sample size for testing. Considering the pandemic situation, the number of people for testing is at max five persons per door.

2 Review of Related Literature

2.1 Covid-19 and Its Mode of Transmission

According to the World Health Organization (WHO), "coronavirus disease, or COVID-19, is an infectious disease caused by a newly discovered coronavirus" [1]. In December 2019, the outbreak that occurred in the province of Hubei, Wuhan spread rapidly into entire China. The virus was initially known as the 2019 novel coronavirus (2019-nCoV), but it was renamed severe acute respiratory syndrome coronavirus 2 (SARS-CoV-2) by the WHO [2]. Since the virus mainly targets the respiratory system, those people with weak immune systems which most of the time are the elderly are the most prone to the virus [3]. Since it can be transmitted via respiratory droplets and touch paths, the number of confirmed cases has risen rapidly.

2.2 Social Distancing as a Preventive Measure to Covid-19

Due to the mode of transmission of the Covid-19 virus, social distancing measures have been implemented most especially in crowded places and high-risk areas like malls and public transportations [4]. Most of the places with confirmed cases decided to implement the new normal to be able to follow social distancing measures and to ensure the safety of everyone. With these sudden changes because of the pandemic, it is expected that traveling will be lessened especially for those who have no private vehicle and only use public transportation [5]. People with low income have no personal vehicles rely solely on public transport vehicles. In these trying times, the authorities must focus on creating a safe place for those people who do not have their vehicles or those who are physically disabled to move around to purchase necessities such as food and medication. Everyone, particularly during this crisis, needs to have access to necessities [6].

2.3 Crowd Monitoring Systems in Public Transportation

Overcrowding has always been a problem, especially in public transportations even in other countries. There are some cases in the other country that their trains are in an open space or gate-free so there is a higher possibility that it will have more people including the people the fare-dodgers. The Netherlands used RFID-based OV-Chip cards that can be used in different transport systems such as trains, trams, metros, and buses to monitor the number of passengers and to prevent fare-dodgers [7]. Another case is in India, overcrowding in the compartments of their trains is their problem. For them to solve the problem, they come up with a study on how to monitor the number of people in each compartment of the trains. It uses a video camera, and its pixel data will help the personnel to monitor whether one compartment is occupied or not and will let the passengers know where the compartment is not occupied or not overcrowded [8].

2.4 Current LRT Crowd Control System

According to the LRT authority, there are 13–14 million passengers every month with an average of almost 400,000 passengers daily in the Philippines [9]. With a huge volume of passengers, overcrowding has been a long-term problem on LRT stations. As of now, a Passenger Limit Per Platform (PLPP) was implemented on stations to prevent overloading on the train station. This scheme is carried out manually, which is inconvenient for employees and vulnerable to errors [10]. Despite the issue, only a few studies about overcrowding were done in the country. In the study of [11], a simulation tool was implemented to understand the behavior of the crowd in railway stations. This was designed for strategic planning and assessing infrastructure preparedness to mitigate overcrowding. Another study was done wherein a smart control management system for LRT 1 was developed. The system works by using IR sensors that will count the passenger in the platform. The passenger information is intended for measuring crowds.

Currently, the Department of Transportation (DOTR) has reduced the capacity of the train to ten percent of its original capacity [12, 13]. However, these measures were solely done by the security guards and personnel assigned. Manual counting is an impractical way to count passengers as it consumes a lot of time and is prone to errors. Moreover, human labor has a short attention span in analyzing a large amount of data, especially in crowded situations [14].

2.5 Image Processing for Passenger Counting

Manual counting of passengers includes estimating ridership from passenger revenue, distributing surveys, and using traffic checkers. One advantage of manual counting is that it does not require special technical knowledge or huge capital. However, it is very labor-intensive and time-consuming. The accuracy is also a concern as random human errors might occur in the data collection [15]. Image processing is a method to perform some operation on an image to get useful information and needed data such as the number of people and their direction. The accuracy of this method is just like the accuracy of video counting, it may be not so accurate when the entrance and area are wide, also when the place is crowded, and some people are blocking others from the camera [16].

2.6 Gaussian Blur

Gaussian blur occurs when an image is captured through a turbulent medium and the acquisition or exposure duration is significantly longer than the period of Brownian motion of the particles in the medium. With the finite size of the sampling pulse, Gaussian blur is also connected to sensor blur. However, this impact is usually of minor consequence. Furthermore, the Gaussian kernel is frequently employed as a rough approximation of other blurs that are too hard to work with precisely. Gaussian blur is intentionally applied to soften the image, decrease its additive noise, and apply a local averaging before performing the downscaling [17]. The Gaussian Blur is created by convolution the pixels of an image with a kernel specified by a Gaussian function. The formula for the Gaussian Kernel is as follows:

$$(G(xy) = \frac{1}{2\pi\sigma^2}e - \frac{x^2 + y^2}{2\sigma^2}) \tag{1}$$

The filter's implementation is defined by several symbols in the formula. The following is a list of the symbols that make up the Gaussian Kernel formula:

- G(x y) – The Gaussian Kernel formula is used to obtain this number. This value represents a single element that is part of a Kernel.
- σ – This symbol merely reflects a user-defined threshold or factor value.
- e – Euler's number **is** represented by the symbol. The value of Euler's number has been calculated to be 2.718 as a mathematical constant.
- x, y – The variables x and y refer to the coordinates of pixels in an image. The row is represented by y and the column is represented by x (Fig. 1).

Fig. 1. Effect of Gaussian blur

2.7 Thresholding Binary

One of the simplest, but most powerful tools for segmenting images is thresholding. It is useful for telling the difference between the foreground and the backdrop. The threshold operation is utilized when converting a grayscale image to a binary image [18]. Creating a binary image is advantageous as it simplifies the process of recognition and classification by reducing the complexity of the data. Selecting a single threshold value, T, is the most common approach to convert a gray-level image to a binary image. Then, any gray level values below T were identified as background black (0), while that above T will be categorized as objects white (1). Thresholding is a grey value remapping process that is defined as:

$$g(x, y) = 0 f f(x, y) < T \tag{2}$$

$$g(x, y) = 1 \ if \ f(x, y) \geq T \tag{3}$$

where (x,y) denotes a gray value and (x,y) denotes the threshold value's coordinates. T denotes the threshold value: g(x,y) is the threshold image and f(x,y) is the gray level image pixels or input image (Fig. 2).

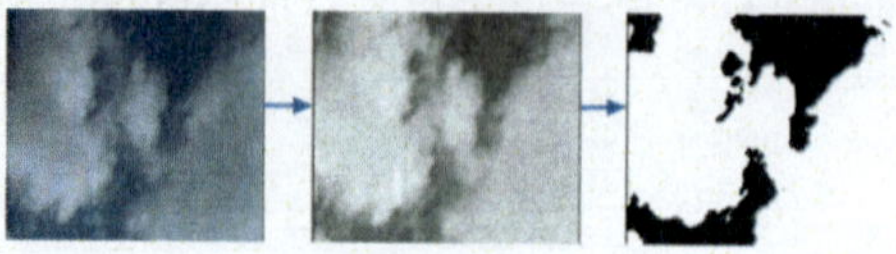

Fig. 2. Conversion from grayscale to binary

2.8 Image Dilation

Dilation is used for an element to be enlarged by creating additional structural elements to the pixels of the image [18]. These elements are also pixels added to the object edges. The maximum value of all the pixels in the vicinity determines the output pixel's value. If any of the nearby pixels have the value 1, the pixel is set to (Fig. 3).

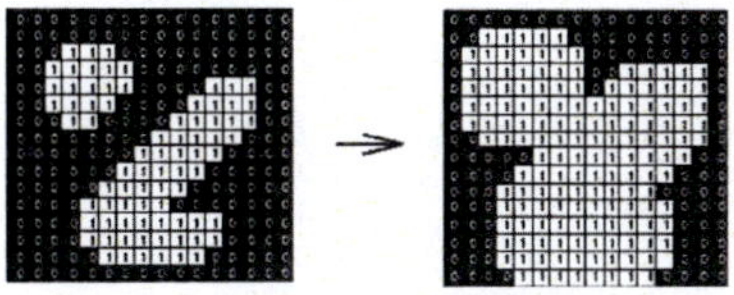

Fig. 3. Effect of image dilation

2.9 Image Contour

Image contour is an image processing process that identifies the structure of the objects in an image by outlining their shape. It is a curve joining continuous points in the image that has the same color [19]. In determining the shape of objects in an image, the contour properties are significant. These include the aspect ratio, extent, and solidity of the object. The aspect ratio refers to the width and height of the object whereas the extent is the area of the bounding box. The bounding box is an area that encloses the object. For better accuracy in image detection, binary images are used. This could be achieved by threshold application (Fig. 4).

Fig. 4. Effect of contoured image

2.10 Region of Interest

Region of interest, or ROI, is an effective method for visualizing segmentation results. ROI is the proposed region from the original image in which will be further processed [19]. This is done by creating a binary mask. A binary mask is created by a binary image with the same size as the selected image to process. The pixels defining the ROI are set to 1 and all the other pixels are set to 0. The objects under the ROI boundary will be processed using the contour outline.

2.11 Centroid Computation

The centroid of an image refers to the point where the values of the pixels surrounding are balanced. To find the centroid, the image moment must be determined [20]. Image moment is the weighted average of image pixel intensities. This is found through the radius, area, and such properties of an image. The centroid is given by the equation:

$$C_x = \frac{M_{10}}{M_{00}};\ C_y = \frac{M_{01}}{M_{00}} \tag{4}$$

where Cx and Cy are the x and y coordinates, respectively. Meanwhile, M denotes the image moment (Fig. 5).

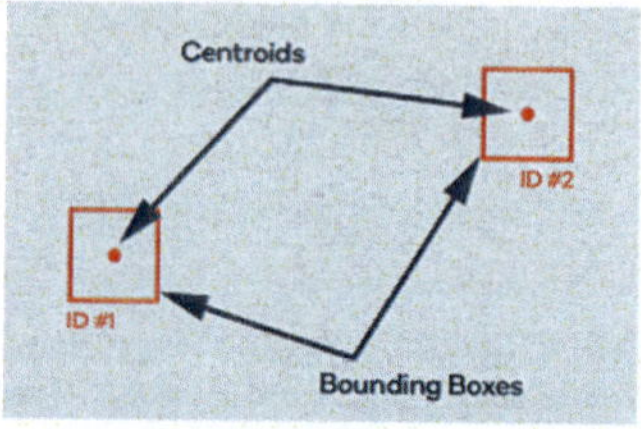

Fig. 5. Centroid of an image for tracking

2.12 Object Tracking

The tracking of the image is obtained from the initial object detection applied. To track an object, a unique ID was created from the object detection such as color thresholding, contouring, linear SVM, SSDs, and such. These IDs will maintain the tracking of the object throughout the video [21].

2.13 Real-Time Passenger Counting Using Cameras

In this study, they have used surveillance cameras and combinations of image processing algorithms, specifically the Canny Edge algorithm, to create a passenger counter for a bus. Different crowd conditions were simulated to test their passenger counter. Their results have shown an optimum Gaussian value of 21 for the Canny edge detection algorithm with a 5.18% of error in counting [22].

2.14 Multiple Webcam in a Raspberry Pi

A study conducted multiple live webcam streams and time-lapse videos on a single Raspberry Pi. The time-lapse videos were done using the Pi Camera module coded into the LXTerminal. This will take the images at a certain period. A similar method was done for the live streams of the webcam. They have connected three webcams on the Raspberry Pi. The optimum set-up of connection for the three webcams and Raspberry Pi is connecting one directly to the Pi and connecting the remaining through USBs [23].

3 Methodology

3.1 Project Design

The Passenger management system is designed to count and limit the passengers in every train car to encourage social distancing. Aside from counting and displaying the view count in each train, the early warning system and alarm will be activated when the threshold limits of passengers inside the train car are reached. These provide immediate control measures for the passengers to avoid overcrowding inside the train car and strictly follow the suggested train car capacity in line with the physical distancing protocols. Figure 6 shows the block diagram of the design of the project.

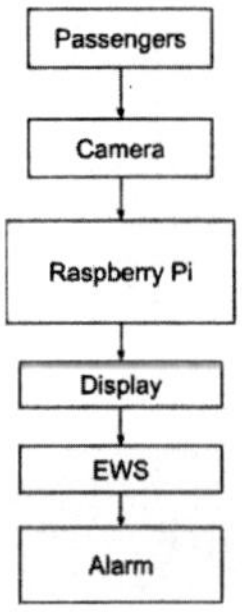

Fig. 6. Passenger management system design

3.2 Conceptual Framework

As seen in Fig. 7, the passengers from the frames will be the input for the system to start; then, the information will undergo the image processing and counting algorithm to determine the passenger counts. After that, the output will be seen at a screen display which is the number of people inside the train. Another output is the early warning system to avoid reaching the total capacity of the train. The final output is the alarm which will then be activated when the passenger limit is reached.

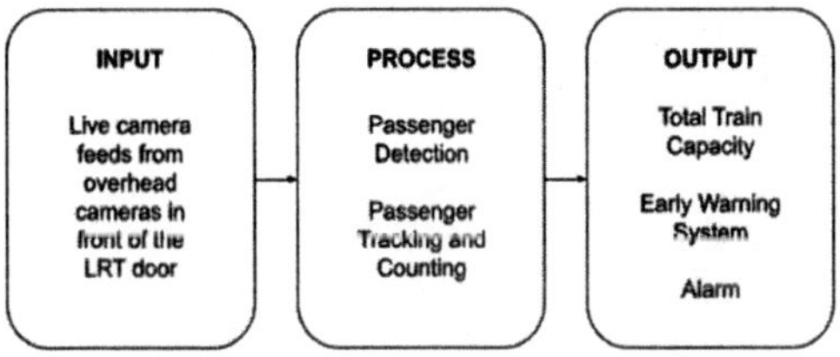

Fig. 7. Conceptual framework

3.3 Hardware Design

For the hardware components, the passenger management system consists of a passenger counter, display, and alarm. The passenger counter is designed using multiple cameras and Raspberry Pi. The Raspberry Pi will be used in the prototype as a microcontroller. The image processing and counting algorithms will be stored in this microcontroller. This was attached to the back of the screen display used in the system by the pins and an HDMI adaptor. This screen will show the GUI which displays the number of passengers inside the train. Figure 8 shows the connection of the hardware components.

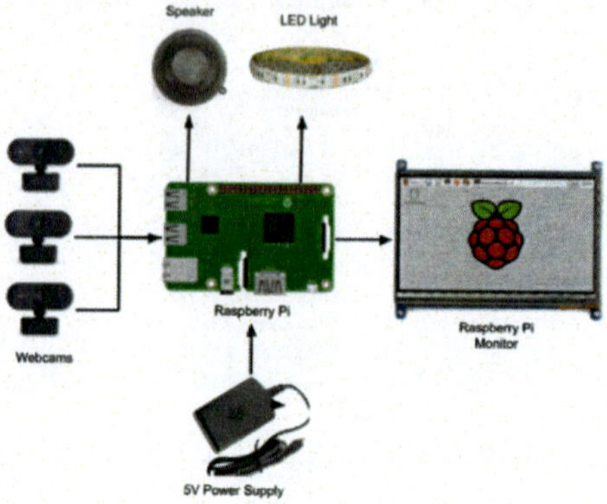

Fig. 8. Hardware components

3.4 Software Development

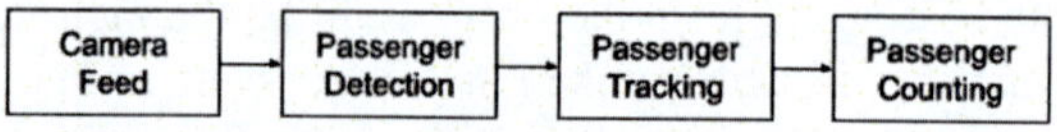

Fig. 9. Software development

Figure 9 shows the whole working of the software of the passenger counter. It starts with the passenger detection and will then proceed to the tracking and counting of the passengers. The image processing algorithms used in each stage are all encoded in the Raspberry Pi microcontroller through OpenCV python.

Passenger Detection. Once the passengers enter the frames of the camera installed, they will undergo multiple image processing techniques to determine the passenger counts. The initial stage of the counter is object detection. Figure 10 below shows the input frame from one of the cameras.

Fig. 10. Input image from the camera

With the same aspect ratio, the image will be specifically resized from its original resolution of 1088 × 1932 to 281 × 500 pixels. After that, the resized image will be converted to grayscale. This lessens the data which helps in making its processing faster. Figure 11 below shows the input frame resized and converted to grayscale.

Fig. 11. Grayscale image

After that, the gaussian blur will be used to soften the input image and reduce its additive noise. This blurs the whole image reducing its detail. The gaussian blur works as a lowpass filter that filters out the high-frequency information of the image to avoid aliasing. Figure 12 shows the grayscale image after the gaussian blur application.

Fig. 12. Grayscale image with gaussian blur

The next process was background subtraction which will separate the foreground objects from the background image in a sequence. The foreground is identified by subtracting the current frame from the background which is the static part of the frame.

In this case, the moving passengers will be separated from their background. The base reference used for this was the open-door frame. Figure 13 shows the grayscale image with gaussian blur after the background subtraction.

Fig. 13. Image after background subtraction

The thresholding will be applied next, which converts the grayscale image to binary. The advantage of starting with a binary image is that it makes the detection and classification process easier by minimizing the data's complexity. Figure 14 below shows the image after the thresholding process.

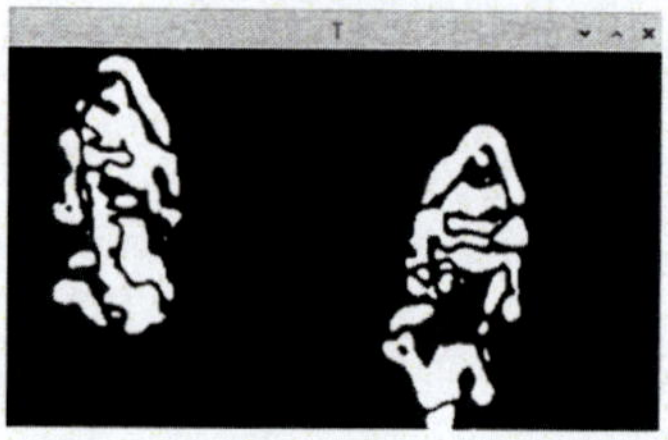

Fig. 14. Image after thresholding

Following that, dilation was used to connect the subtracted binary image by adding additional pixels surrounding it. This makes the image obtained from the thresholding more accurate in size. Figure 15 shows the image after dilation.

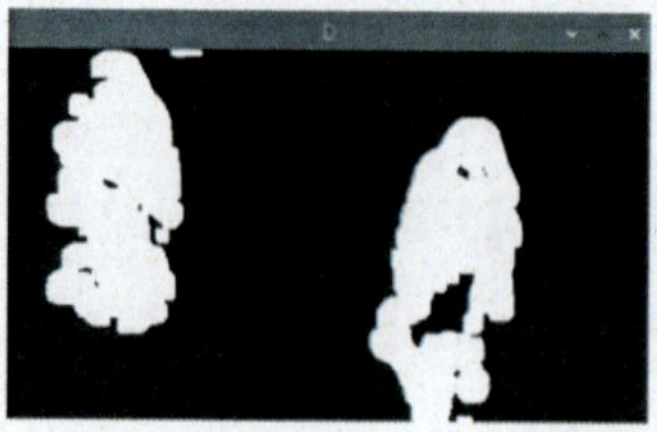

Fig. 15. Image after dilation

After the dilation, the objects on the image will be contoured. Through image contouring, the structure of the object, in this case, is the passenger, will be outlined. The image contour works more accurately on the binary image; hence, we have applied grayscale and thresholding earlier. Once the passenger shape is outlined by the contour, a bounding box will enclose the passenger which indicates that it is detected and within the bounds of ROI. This ROI is applied for the tracking of the passengers. ROI works by using correlation filters to identify the object (Fig. 16).

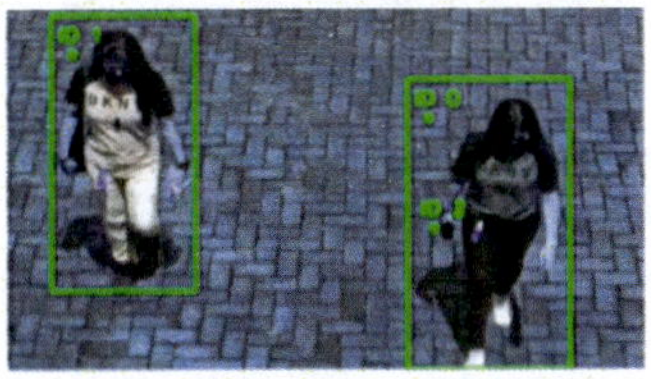

Fig. 16. Image after contour with applied ROI

Passenger Tracking. After the object detection, the system will now track and count the passengers. Once the bounding box or the ROI appears on the detected passenger, its centroid must be determined to track it. The centroid can be identified by finding the moment of the object. The moment of the image is the weighted average of pixels surrounding it. When the centroid is detected, centroid tracking will be applied to track the passengers. This will accept the x and y coordinates of the passengers detected and will later assign a unique ID to each. As the objects move in subsequent frames, the centroid tracking predicts that the pairs of centroids computed with the minimum distance in between them must have the same object ID. Therefore, the object must be within the bounds of the ROI applied. This enables the system to track the movement of the object. Once a new passenger enters, a new ID will then be assigned.

Passenger Counting. To count the passenger entrance and exits towards the LRT train door, we will consider the direction of movement of the passengers. They-coordinate of the passenger moving will be considered to determine its direction. Specifically, the direction will be computed by taking the difference between the current centroid location and the mean of all the previous centroid locations.

$$Y = PC - CC \tag{5}$$

$$\text{if Y is negative, the direction is upwards,}$$
$$\text{is positive, the direction is downwards,}$$

where

Y – Y coordinate
PC – mean of previous centroid location
CC – current centroid location

If the direction is positive, the direction is moving down which means that the passenger enters the train. If the direction is negative, the direction is moving up which means that the passenger exits the train car. The counts will be updated once the passenger moves across the crossing lines set on the frames. Since a train has multiple doors, the overall passenger counts for each train will be coming from the multiple cameras. The counting will be the addition of all the detected entries from all the doors minus the detected exits. This is equal to the current capacity of the train.

3.5 Process Flowchart

Figure 17 shows the flowchart of the overall process of the system.

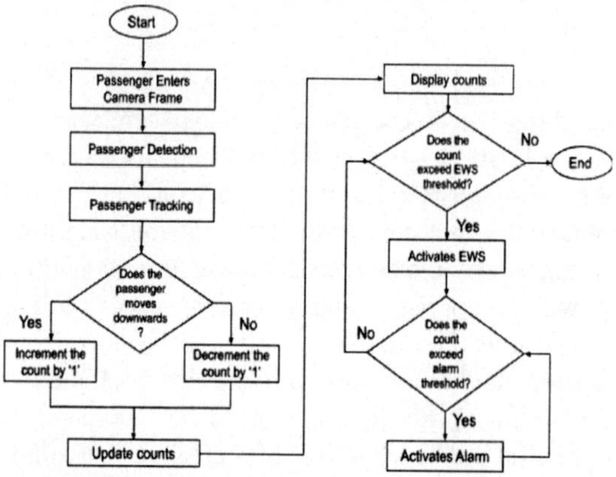

Fig. 17. Flowchart of the Passenger Counter

3.6 Early Warning System and Alarm

The Early Warning System (EWS) and an alarm will be activated, alerting the passengers through the lights and speaker when the train car's threshold limits are reached. The Early Warning System (EWS) will be activated when the train is close to its maximum capacity. A red light will illuminate as a warning that the train compartment is approaching its limit. The use of red light was chosen because it catches people's attention and is typically used to convey a warning or reminder. On the other hand, the alarm will be activated once the threshold limit has been reached, which means that a specific message will be heard once the train capacity has exceeded the limit.

Sample Message (for alarm): *"This train car has reached the maximum capacity. Only __ number of passengers are allowed. Kindly step out and please wait for the next train car instead. Thank you!"*.

3.7 Idle Time

In this system, idle time was implemented on the cameras and is triggered once the unit detects that the door is closed. This works in a way that when the door is closed, the contour that detects the object will overlap with the door. Hence, the estimated size set in the system will not be met which disregards the passenger detection. The passenger will only be detected once the door is open. On idle time, the device would not count the number of people identified by the camera to prevent a miscount of people.

3.8 Device Set-Up

The device's setup is critical to the overall operation of the passenger management system. The system of devices was placed in their optimal and most practical setting to ensure their effectiveness. Figure 18 shows the set-up of the devices inside the train car.

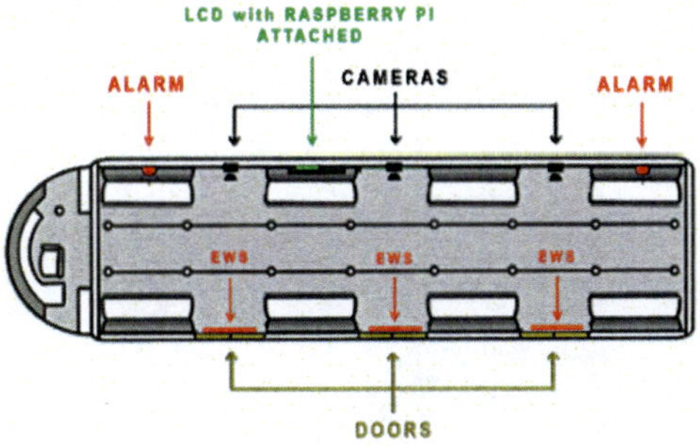

Fig. 18. Set-up of devices inside the train car

As seen from the figure, the devices shown inside the train car are the multiple cameras, microcontroller, display, and alarm. As seen, the cameras are strategically installed overhead, in front of the door entrance. This was done to gather clear images of the passengers as required on the object detection and image processing method. The Raspberry Pi is placed on the central part of the train where all the cameras in each train door are connected. On the other hand, two speakers to ensure that its sound will reach and inform all the passengers once it buzzes.

3.9 Graphical User Interface

Figure 19 shows the graphical user interface of the system that will be displayed on the monitor. As seen, this shows the live stream of the cameras. The display shows the total number of entries and exits, the total number of passengers on the train, and the threshold limits. The GUI also contains the start, stop, and reset buttons.

3.10 Actual Prototype

Figure 20 below shows the actual prototype for our passenger counter. As seen in the figure, all the components are mainly connected to the Raspberry Pi. The Raspberry Pi

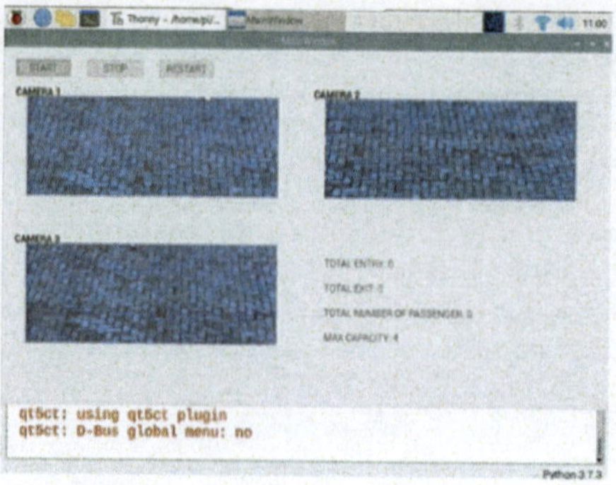

Fig. 19. Actual GUI of the prototype

was attached to the back of the screen display by connecting the pins on the back and by using the HDMI adapter. The cameras are directly connected to the Raspberry Pi. The alarm was also connected on the Raspberry Pi on the audio port located on the top. All the components have their respective ports making them easier to connect.

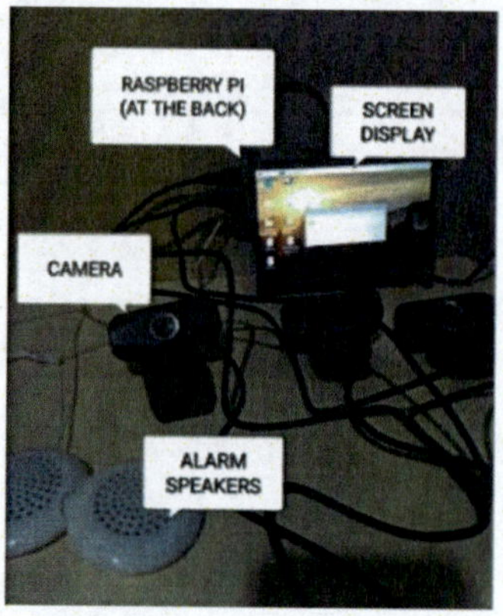

Fig. 20. Actual prototype

3.11 Calibration

The calibration of the set-up was done by testing the system itself. The area of counting was dependent on the camera range which in this case was 2 m. This was suitable for train car doors as they have a width of 1.5 m. The camera was mounted on the wall with a height of 2.5 m concerning the height of train car doors of 1.9 m (Table 1).

Specifically, the parameters we have calibrated are the angle of the camera, threshold values, dilation iterations, position of the crossing line, and the estimated size of the person or passengers. We have tested it with persons going in and out of the frame. We have observed which value of the certain parameters was the counting most accurate (Table 2).

Table 1. Positioning of cameras

	Camera	LRT train door
Width (camera range)	2 m	1.5 m
Height	2.5 m	1.9 m

Table 2. Calibration of the camera angles

Camera angles	Actual count	System count	Accuracy (%)
200°	1	0	0
220°	1	1	100
240°	1	0	0

The angle of the camera's tilt was calibrated as this affects the passenger detection performance. The line of sight of the camera plays a significant role in passenger detection since the detection will be based on the person size. The angle should show the whole body of the person entering for it to be accurate.

The cameras were calibrated into three different angles: 200°, 220°, and 240°. This was tested in detecting one passenger only since we only aim to determine at which angle was the person's whole body was seen clearly by the camera. Based on the results, the person was only detected at the 220° angle. Since this is a mid-range angle, this shows the whole body of the person completely making it accurate. For the 200°, the person was not detected since the view was already cutting the bottom part of the body. Meanwhile, the 240° angle was too low which cuts the upper part of the person's body in the frame (Table 3).

Table 3. Calibration of threshold values

Threshold values	Actual count	System count	Accuracy (%)
Entry			
1	2	4	50
5	2	2	100
100	2	2	0
Exit			
1	2	1	50
5	2	2	100
100	2	0	0

We have tested three threshold values, 1, 5, and 100, to see which will give the most accurate result to our image detection. For the threshold value 1, only 50% accuracy was obtained from both entry and exit detection. This happens since the image processing was too dark with the said value. As for the threshold of 100, no persons were detected on both entry and exit. This was due to the image being too light. The most accurate threshold value for the image was 5 as it detected all the entries and exits (Table 4).

Table 4. Calibration of the person size

Person size	Actual count	System count	Accuracy (%)
Entry			
50 × 90–100 × 400	2	2	100
80 × 120–100 × 400	2	2	100
80 × 300–120 × 160	2	0	0
Exit			
50 × 90–100 × 400	2	0	0
80 × 120–00 × 400	2	2	100
80 × 300–120 × 160	2	0	0

The person size was calibrated as it will be the basis for the image contour for passenger detection. The values for the estimated person size are in terms of pixel size. The pixel size chosen to be tested was obtained concerning the resolution of the frame size itself. As can be seen in the table, the system count on 50 × 90–100 × 400 pixel size has 100% accuracy on entries, but 0% accuracy on exits. Meanwhile, the system count for the 80 × 300–120 × 160 pixel size has an accuracy of 0% both in entry and exit because it was unable to detect any person passing through the door. The most accurate pixel size for persons was 80 × 120–100 × 400 pixels, which can count all of the people from entry to exit with 100% accuracy (Table 5).

The number of dilation iterations was also calibrated as this affects the image detection. As seen on the table, with 8 iterations on dilation, no entry and exit were detected by the system. This was because a lot of iterations make the size of the person bigger exceeding the person's size limit for detection. As for the 1 iteration, it was able to detect the exit but not the entry of passengers. With minimal iteration, the image for contouring might be too small to be detected by the size limit set by the user. The most accurate value of iteration for dilation was 2 (Table 6).

Table 5. Calibration of dilation iterations

Dilation iterations	Actual count	System count	Accuracy (%)
Entry			
1	2	1	50
2	2	2	100
8	2	0	0
Exit			
1	2	2	100
2	2	2	100
8	2	0	0

Table 6. Calibration of the crossing line positions

Crossing lines position	Actual count	System count	Accuracy (%)
Entry			
48% and 52%	2	3	66.67
10% and 90%	2	0	0
40% and 60%	2	2	100
Exit			
48% and 52%	2	1	50
10% and 90%	2	0	0
40% and 60%	2	2	100

The values of the crossing line positions are obtained concerning the height of the video frame. This crossing line will be the indicator for the system to count the passengers. As seen from the table, 48% and 52% crossing line positions had 66.67% and 50% accuracy for entry and exit, respectively. Meanwhile, 10% and 90% crossing line positions obtained 0% accuracy on both entry and exit. This happened since these crossing lines positions are near the door which disables the camera to capture the whole image of the person. The most accurate crossing line position was 40% and 60% which was placed near the center of the frame. This counted all the entries and exits (Table 7).

3.12 Testing of the Prototype

The accuracy of a passenger counter was determined by counting the number of entries and exits of passengers in the vehicle separately and simultaneously. The accuracy of the counter will be determined by the ratio between the number of people identified and the total number. Equation 7 below was used to get the overall accuracy of the counter

Table 7. Summary of the used calibrated condition

Parameters	
Angle	220°
Threshold value	5
Person size	80 × 120–100 × 400 pixel
Dilation iterations	2
Crossing line positions	40% and 60% of the frame size

for both the boarding and alighting of passengers.

$$n_n = \frac{P}{P_T} \times 100 \tag{6}$$

$$n_T = \frac{n_1 + n_2 + n_3}{3} \tag{7}$$

where

$$n_n - accuracy\ per\ door$$

$$P - number\ of\ people\ detected$$

$$P_T - actual\ number\ of\ people$$

$$n_T - total\ accuracy\ in\ one\ train\ car$$

This was tested under five different scenarios to fully evaluate the performance. In the first two cases, a simple accuracy test was done wherein several passengers will enter or the train one at a time. The next two cases were focused on the simultaneous counting capability of the counter wherein two or more people enter or exit. For the last case, the system will be tested if it can handle simultaneous entry and exits of people. The actual testing of our prototype was done using a controlled environment since the testing on an actual train car was prohibited due to the pandemic. For our testing set-up, instead of actual train, a designated area was just set for the people to go inside and outside the train car.

4 Results and Discussion

4.1 Passenger Counter

Case 1: Passengers Enters Train Car One at a Time

Figure 21 below shows the testing scenario for Case 1. As seen in the figure, one person is entering each frame.

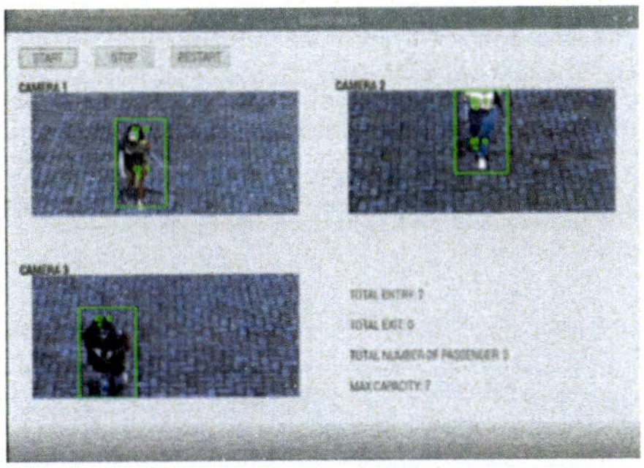

Fig. 21. Persons enters one at a time

Table 8 shows the first case of our simulated environment for testing where the people are entering one at a time. As seen from the table, all the passengers were detected from door 1 and door 3. In Door 2, there was one passenger that was not detected. The average accuracy of the system counts was 100%, 80%, and 100% for door 1, door 2, and door 3, respectively. This gives an average of 93% overall. As for the time to count, it was only around 1 to 2 secs which shows that there are minimal delays in counting.

Table 8. Passengers enters train car one at a time

	Door 1				Door 2				Door 3			
Trials	Actual Counts (P_T)	System Counts (P)	Time to count (sec)	Accuracy n_1 (%)	Actual Counts (P_T)	System Counts (P)	Time to count (sec)	Accuracy n_2 (%)	Actual Counts (P_T)	System Counts (P)	Time to count (sec)	Accuracy n_3 (%)
1	1	1	2	100	1	1	1	100	1	1	2	100
2	1	1	1	100	1	1	2	100	1	1	2	100
3	1	1	1	100	1	1	2	100	1	1	1	100
4	1	1	2	100	1	0	-	0	1	1	2	100
5	1	1	1	100	1	1	2	100	1	1	1	100
Ave				100				80				100
Total Accuracy, $n_{T\,(\%)}$: 93.00												

Case 2: Passengers Exits Train Car One at a Time

Figure 22 below shows the testing scenario for Case 2. As seen in the figure, there was one person that exits in each camera frame.

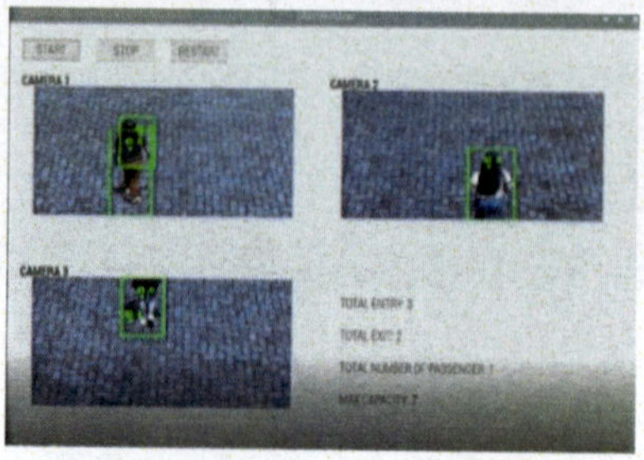

Fig. 22. Persons exits one at a time

Table 9 shows the results of testing the prototype as passengers are exiting the door one by one. We can see in the table that door 1, door 2, and door 3 have 100% accuracy since they count correctly all of the passengers who passed through that door. Combining the results for single entry and exit, the system has obtained 96.5% accuracy. The high accuracy shows that the system has no difficulty in detecting one person that enters or exits.

Table 9. Passengers exits train car one at a time

	Door 1				Door 2				Door 3			
Trials	Actual Counts (P_T)	System Counts (P)	Time to count (sec)	Accuracy η_1 (%)	Actual Counts (P_T)	System Counts (P)	Time to count (sec)	Accuracy η_2 (%)	Actual Counts (P_T)	System Counts (P)	Time to count (sec)	Accuracy η_3 (%)
1	1	1	1	100	1	1	2	100	1	1	2	100
2	1	1	1	100	1	1	2	100	1	1	2	100
3	1	1	2	100	1	1	2	100	1	1	1	100
4	1	1	1	100	1	1	1	100	1	1	1	100
5	1	1	1	100	1	1	1	100	1	1	2	100
Ave				100				100				100
Total Accuracy, η_T (%): 100.00												

Case 3: Passengers Enters Train Car Simultaneously

Figure 23 below shows the testing scenario for case 3. As seen in the figure, several people were entering each frame.

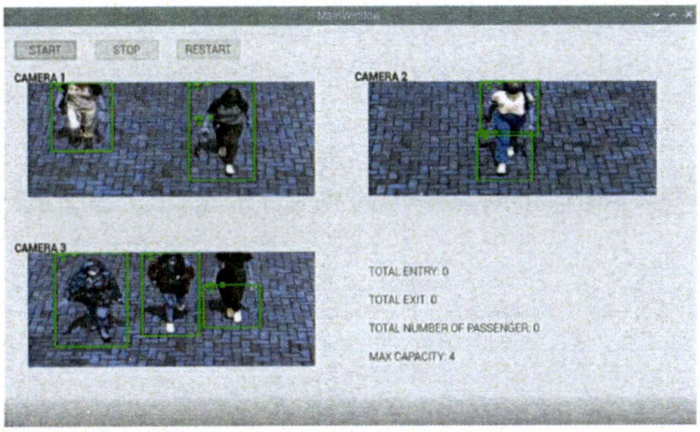

Fig. 23. Persons enters simultaneously (trial 1)

Table 10 shows the testing for simultaneous passenger entries. As seen from the table, the overall accuracy for this case was 88.22%. Door 1 has the highest accuracy because there were fewer people who entered this door compared to the other doors, reducing the number of errors. However, we can see in trial 5 of door 1 that it has low accuracy because many people passed through that door and the camera had a hard time counting them all, especially the people who were close to each other. Doors 2 and 3 have 84.33% and 88.33% accuracy, respectively. This was still considered high as there were only one to two people not counted per door. As for the time to count, it was observed that mostly took 2 s to count the passengers.

Table 10. Passengers enters train car simultaneously

	Door 1				Door 2				Door 3			
Trials	Actual Counts (P_T)	System Counts (P)	Time to count (sec)	Accuracy n_1 (%)	Actual Counts (P_T)	System Counts (P)	Time to count (sec)	Accuracy n_2 (%)	Actual Counts (P_T)	System Counts (P)	Time to count (sec)	Accuracy n_3 (%)
1	2	2	1	100	1	1	2	100	3	2	2	66.67
2	2	2	2	100	1	1	2	100	3	3	1	100
3	3	3	2	100	4	3	2	75	4	4	3	100
4	3	3	3	100	4	4	3	100	4	3	2	75
5	5	3	4	60	3	2	2	66.67	3	3	2	100
Ave				92				88.33				88.33
Total Accuracy, $n_{T\,(\%)}$: 89.55												

Case 4: Passengers Exits Train Car Simultaneously

Figure 24 below shows the testing scenario for Case 4. As seen in the figure, several people exit in each frame.

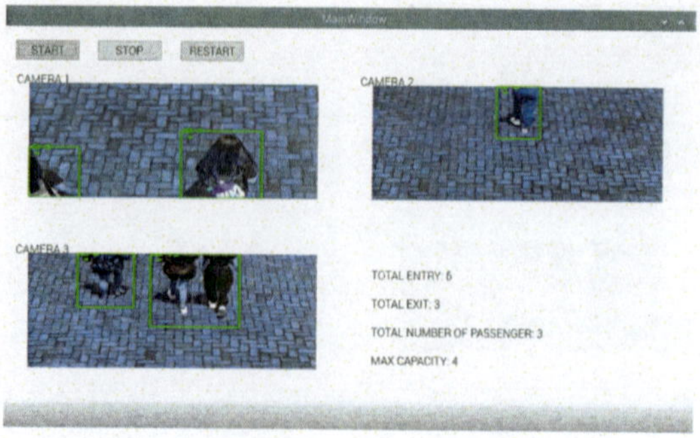

Fig. 24. Persons exits simultaneously (trial 1)

Table 11 depicts the results of evaluating the prototype when people exit the door simultaneously. The accuracy of Doors 1 to 3 was measured by having a maximum of 5 people exit each door at the same time. In Door 1, 100% accuracy was achieved by the system. Meanwhile, the third trial in Door 2 was only 75% accurate since only three out of four people are counted. This gives an average accuracy of 95% for door 2. Lastly, for door 3, trials 1 and 2 were both tested for three people exiting the door simultaneously. However, the system only identifies two that exit the door for the said trials resulting in only 66.67% accuracy. Nonetheless, the rest of the trial has maximum accuracy for door 3 obtaining an average accuracy of 86.67. The overall accuracy for this case was 93.89%

Table 11. Passengers exits train car simultaneously

	Door 1				Door 2				Door 3			
Trials	Actual Counts (P_T)	System Counts (P)	Time to count (sec)	Accuracy η_1 (%)	Actual Counts (P_T)	System Counts (P)	Time to count (sec)	Accuracy η_2 (%)	Actual Counts (P_T)	System Counts (P)	Time to count (sec)	Accuracy η_3 (%)
1	2	2	4	100	1	1	2	100	3	2	2	66.67
2	2	2	1	100	1	1	2	100	3	2	3	66.67
3	3	3	4	100	4	3	2	75	4	4	3	100
4	3	3	2	100	4	4	3	100	4	4	1	100
5	5	5	3	100	3	3	2	100	3	3	3	100
Ave				100				95				86.67
Total Accuracy, $\eta_{T\ (\%)}$: 93.89												

Case 5: Passengers Enters and Exits the Train Car Simultaneously

Figure 25 shows the testing scenario for Case 5. As seen in the figure, the people enter and exit the frame at the same time.

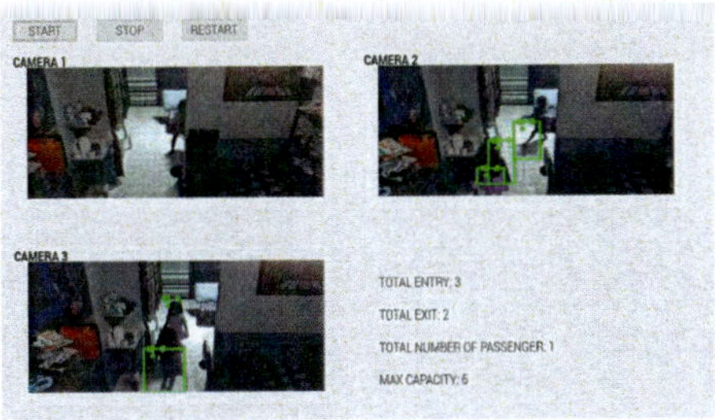

Fig. 25. Persons enters and exits simultaneously

Table 12 shows the results of the testing when several persons enter and exit the frame at the same time. As seen from the table, the system has an average accuracy of 90% for entry and 100% for exit on door 1 obtaining 95% accuracy for this door. For door 2, all the entries were detected obtaining 100%. For the exit, the results have shown 96.65% accuracy. The accuracy for door 2 is 98.32%. In Door 3, 100% accuracy was obtained from entries. Meanwhile, only 90% were detected for the exit resulting in 90% accuracy for door 3. The overall accuracy for this case is 94.44%.

4.2 Alarm System

The working of the alarm system will be done by intentionally reaching the passenger limits of the train car. The alarm should be activated once the limits are reached.

The alarm system's accuracy is shown in Table 13. The researchers only set the threshold limit to four people to measure the alarm's accuracy as this was just a controlled environment. As shown on the table, the device has 100% accuracy since the alarm activated when it reached the defined threshold limit. Trial 1 has a total of 6 people that can enter the door, but the table shows that only 5 people are counted. This was because the alarm is already activated when the device counts 4 people. The alarm was not triggered in trial 3 as seen in the table because the actual and system counts are 3 only, which does not meet the system's threshold limit. Figure 26 below shows the first trial of the alarm system. As seen, the GUI turned red which indicates that the alarm was activated.

4.3 People Sitting/Standing Near the Door

As the range of the camera was not exact concerning the width of the door, there might be people sitting or standing caught within the camera frames. This testing is included to test the accuracy of detection if people are sitting or standing near the door. Since this was a controlled environment, we instructed a person to stand near the door while others will go inside and outside the frame. The system should exclude the people sitting or

Table 12. Persons enter and exit simultaneously

	Door 1 (Enter)				Door 1 (Exit)			
Trials	Actual Counts (P_T)	System Counts (P)	Time to count (sec)	Accuracy (%)	Actual Counts (P_T)	System Counts (P)	Time to count (sec)	Accuracy (%)
1	1	1	1	100	1	1	1	100
2	1	1	2	100	1	1	1	100
3	2	1	2	50	1	1	1	100
4	2	2	2	100	1	1	1	100
5	1	1	1	100	1	1	2	100
Ave				90				100
Accuracy, a_1 (%)					95			
	Door 2 (Enter)				Door 2 (Exit)			
Trials	Actual Counts (P_T)	System Counts (P)	Time to count (sec)	Accuracy (%)	Actual Counts (P_T)	System Counts (P)	Time to count (sec)	Accuracy (%)
1	1	1	1	100	1	1	1	100
2	1	1	1	100	1	1	1	100
3	2	2	1	100	1	1	1	100
4	2	2	2	100	1	1	1	100
5	1	1	1	100	3	2	2	66.67
Ave				100				96.65
Accuracy, a_2 (%)					98.32			
	Door 3 (Enter)				Door 3 (Exit)			
Trials	Actual Counts (P_T)	System Counts (P)	Time to count (sec)	Accuracy (%)	Actual Counts (P_T)	System Counts (P)	Time to count (sec)	Accuracy (%)
1	1	1	1	100	1	1	1	100
2	1	1	2	100	1	1	1	100
3	1	1	1	100	1	1	1	100
4	1	1	1	100	1	1	1	100
5	1	1	1	100	1	0	-	0
Ave				100				90
Accuracy, a_3 (%)					90			
Total Accuracy, a_T (%)					94.44			

Table 13. Alarm system

Trials	Overall actual counts (P_T)	Overall system counts (P)	Threshold limit	Is the alarm activated?
1	6	5	4	Yes
2	5	5	4	Yes
3	3	3	4	No
Accuracy				100

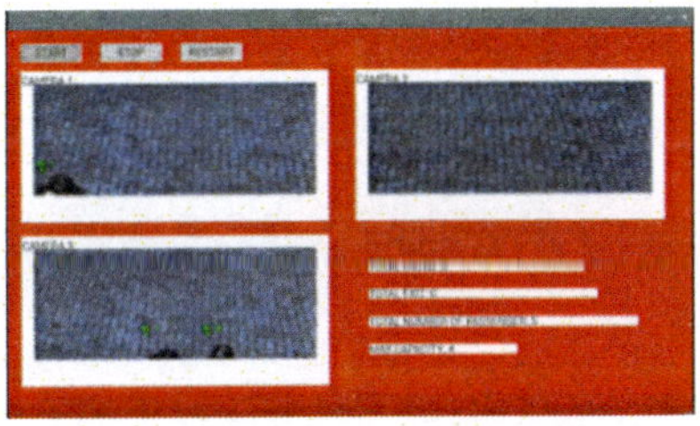

Fig. 26. Alarm activated

standing near the door from detection, and only detect the people entering or leaving the frames door. As seen on Fig. 27, there was no bounding box on the person standing near the door compared to the people entering the door. This shows that the system counts accurately even when there are people standing near the door.

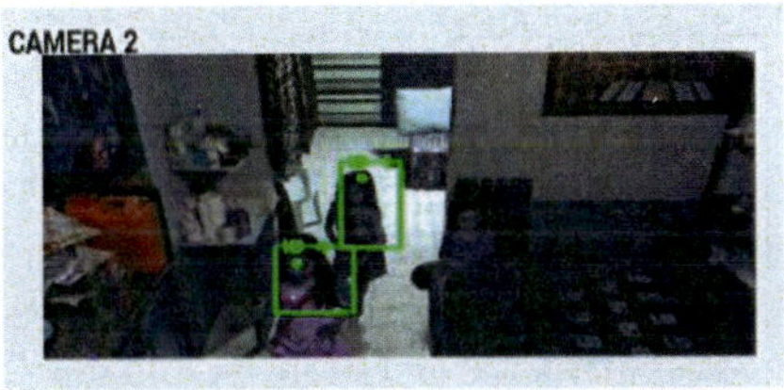

Fig. 27. Testing detection when a person is standing near the door

4.4 Idle Time Testing

Similar to the previous testing setup, the idle time was also tested in a controlled environment. Instead of LRT doors, we simulated the closed-door view using a curtain. We tested whether the person in front of the curtain will be counted by the system or not. The set-up for testing the idle time was shown in Figs. 28 and 29. It can be noticed that a bounding box for the person appears in Fig. 28 when the door is still open, and it disappears when the door is already closed, as seen in Fig. 29.

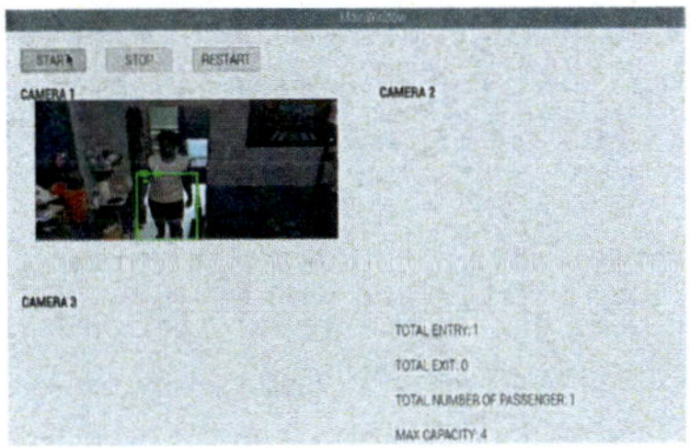

Fig. 28. Idle time testing (open door)

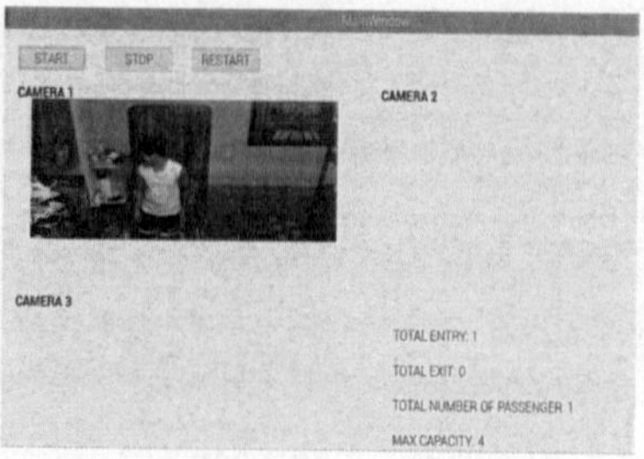

Fig. 29. Idle time testing (closed door)

5 Conclusion

The LRT passenger management system using Image Processing Algorithm and Multiple Cameras was successfully done using Raspberry Pi as the main device where all other components are integrated. With this system, the passengers on a train car will be counted and limited to the number in line with the social distancing protocols.

To achieve the highest system accuracy, we calibrated the system with different parameters. For the camera angle, the results have shown that the 220° angle is the most accurate. For the threshold, the value of 5 produces the most balanced lightness and darkness of the image. For the estimated size, the most accurate detection is at a range of 80 × 120–100 × 400. For the number of iterations in dilation, 2 iterations obtained the best result. For the position of crossing lines, 40% and 60% of the frame size produced the highest accuracy detecting all the entries and exits on the calibration.

For testing the overall accuracy of our system, different cases were simulated. Based on the results, the counting of passengers has the highest percentage of accuracy of 100% when passengers exit one at a time. It also shows high accuracy for single entries averaging 93% accuracy. For both the single entry and exit, the time to count recorded was around 1 to 2 s. On simultaneous entry, the average accuracy is 88.22%. Meanwhile, for the simultaneous exit, the average accuracy is 93.89%. The time to counts for both cases is around 2 to 4 s. As for the alarm, it was activated on the intended trials showing its accuracy.

Our research has shown the accuracy of using multiple image processing techniques, for passenger detection and counting as we have used three cameras with Raspberry Pi, we need to consider that the algorithms to be used will not cause a huge delay on the system. Compared to other algorithms, the multiple-image processing techniques used in the system reduces the complexity of data to fit the system's capability. Overall, this study was able to develop a low-cost passenger management system with high accuracy and fast time to count.

6 Recommendations

As this system uses image processing algorithms for passenger detection and tracking, it is recommended that more research into image processing techniques should be done to improve the system. As observed from the testing, the system has delays in detecting and counting the people. Having multiple cameras, the Raspberry Pi was not able to process all the data efficiently due to its hardware limitations. Therefore, the researchers recommend using a more capable microprocessor. Moreover, if a more advanced processor is to be used, researchers could further use deep learning. IoT could also be integrated for the passenger data to be sent to the rail operators. This could help them create better plans for handling the passengers, or for crowd analysis. Also, future research should test the prototype in an actual LRT station since the researchers only provided a controlled environment for the testing due to the pandemic. For the calibration, not all values for the parameters are explored. Thus, more values for each parameter could be calibrated in future studies.

References

1. WHO: Coronavirus. https://www.who.int/health-topics/coronavirus (2020)
2. Zheng, Y.-Y., Ma, Y.-T., Zhang, J.-Y., Xie, X.: COVID-19 and the cardiovascular system. Nat. Rev. Cardiol. **17**(5), 259–260 (2020)
3. Aslam, F.: COVID-19 and Importance of Social Distancing (Apr. 2020). https://doi.org/10.20944/preprints202004.0078.v1
4. Andersen, M.: Early Evidence on Social Distancing in Response to COVID-19 in the United States (6 Apr. 2020)
5. De Vos, J.: The effect of COVID-19 and subsequent social distancing on travel behavior. Transport. Res. Interdiscipl. Perspect. **5**, 100121 (May 2020). Accessed: 19 Jun. 2020 [Online]
6. Goldbaum, C., Cook, L.R.: They Can't Afford to Quarantine. So They Brave the Subway. The New York Times (30 Mar. 2020)
7. Khoeblal, R., Laohapensaeng, T., Chaisricharoen, R.: Passenger monitoring model for easily accessible public city trams/trains. In: 2015 12th International Conference on Electrical Engineering/Electronics, Computer, Telecommunications and Information Technology (ECTI-CON), pp. 1–6 (Jun. 2015)
8. Trivedi, U., Hari, S.N.: Intelligent crowd management system in Trains. Int. J. Comput. Sci. Electron. Eng. **4**(2) (2016)
9. Ganiron Jr., T.U.: Technological Evolution of Manila Light Rail Transit System (2016). https://doi.org/10.14257/ijast.201..0.
10. Goh, M.L.I., Goh, J.E.E.: Smart Crowd Control Management System For Light Rail Transit (LRT) 1. In: 2019 International Conference on Computational Intelligence and Knowledge Economy (ICCIKE), pp. 608–613 (Dec. 2019)
11. Samson, B.P.V., Aldanese IV, C.R., Chan, D.M.C., San Pascual, J.J.S., Sido, M.V.A.P.: Crowd dynamics and control in high-volume metro rail stations. Procedia Comput. Sci., **108**, 95–204 (2017)
12. Comms, D.M.: DOTr – Summary of the DOTr's Guidelines for railway-based public transport operations in areas under the General Community Quarantine (GCQ). http://www.dotr.gov.ph/55-dotrnews/1324-summary-of-the-dotr-s-guidelines-for-railway-based-public-transport-operations-in-areas-under-the-general-community-quarantine-gcq.html. Accessed 20 Jun.2020
13. CNN Philippines: COVID-19: LRT, MRT implement passenger limit on platforms as 'social distancing' encouraged. CNN Philippines. https://cnnphilippines.com/transportation/2020/3/13/MRT-LRT-social-distancing-coronavirus-COVID.html
14. Ashkanani, A.M., Roza, A.S.M., Naghavipour, H.: A design approach of automatic visitor counting system using a video camera. IOSR J. Electr. Electron. Eng., 2278–1676 (2015)

15. Oberli, C., Torres-Torriti, M., Landau, D.: Performance evaluation of UHF RFID technologies for real-time passenger recognition in intelligent public transportation systems. IEEE Trans. Intell. Transp. Syst. **11**(3), 748–753 (2010). https://doi.org/10.1109/TITS.2010.2048429
16. Chen, C., Chang, Y., Chen, T., Wang, D.: People counting system for getting in/out of a bus based on video processing. In: 2008 Eighth International Conference on Intelligent Systems Design and Applications, Kaohsiung, pp. 565–569 (2008)
17. Li, F., Yang, F., Liang, H., Yang, W.: Automatic passenger counting system for bus based on RGB-D video. In: 2016 International Conference on Electronics, Electrical Engineering and Information Science, vol. 117, pp. 209–2019 (2017)
18. Chato, P., Chipantasi, D.J.M., Velasco, N., Rea, S., Hallo, V., Constante, P.: Image processing and artificial neural network for counting people inside public transport. In: 2018 IEEE Third Ecuador Technical Chapters Meeting (ETCM), pp. 1–5. Cuenca (2018). https://doi.org/10.1109/ETCM.2018.8580287
19. Lodwick, G., et al.: A survey of preprocessing and feature extraction techniques for radiographic images. IEEE Trans. Comput. **20**(09), 1032–1044 (1971). https://doi.org/10.1109/T-C.1971.223399
20. Ma, Z., Chan, A.B.: Counting people crossing a line using integer programming and local features. IEEE Trans. Circuits Syst. Video Technol. **26**(10), 1955–1969 (2016). https://doi.org/10.1109/TCSVT.2015.2489418
21. Rong, W., Li, Z., Zhang, W., Sun, L.: An improved Canny edge detection algorithm. In: 2014 IEEE International Conference on Mechatronics and Automation, Tianjin, pp. 577–582 (2014). https://doi.org/10.1109/ICMA.2014.6885761
22. Calindas, R., Dimayayac, V., Reyes, F.: Real-time Passenger Counting System using Surveillance Camera with Applications of Image Processing Algorithms. Mapua University (Sept. 2020)
23. Berboucha, M., Najmudin, Z.: Multiple live webcam streams using a Raspberry Pi. SLAC National Accelerator Laboratory (September 2014). https://doi.org/10.13140/RG.2.1.3843.2809

Analysis of Intersection of Working Areas Within the Human-Robot Interaction in a Shared Workspace

Rinat R. Galin(✉), Roman V. Meshcheryakov, and Mark V. Mamchenko

V.A. Trapeznikov Institute of Control Sciences of Russian Academy of Sciences, 117997 Moscow, Russian Federation
grr@ipu.ru, mrv@ieee.org

Abstract. This paper addresses the issues of modeling of individual workspaces of a human and a collaborative robot, as well as their intersections, taking into account the requirements of the safety of their interaction without losses in the efficiency of collaboration. Methods of safe human-robot collaboration have been considered. A review of a human functioning in a workspace within both the easy-to-access, and the optimum area depending on his motor functions has been undertaken. The approaches to solving the problem of human movement using a discrete cellular automata method, and the problem of intersection of the working bodies of a robot and a human have been considered. A model in the MATLAB simulation environment has been formed, which allows to create clouds of points of workspaces for both a robot and a human, taking into account the anthropometric parameters of a human, the dimensional characteristics of a collaborative robot, and their relative locations. In addition, the initial stage of numerical modelling has been carried out in order to form a separate cloud of points of intersection of the working areas of a collaborative robot and a human worker, as well as to calculate its approximate volume.

Keywords: Collaborative robot · Human-robot interaction · Shared workspace · Workspace intersection · Collaboration efficiency

1 Introduction

The use of robots in various spheres of human activity is a current trend in the modern world. In a higher risk environment and in context of hazardous work the integration of robots is an effective solution to prevent injuries of human workers. However, full automation of a workplace using a robot is currently not possible. This is due to the difficulties in ensuring both the continuous and efficient functioning of a robotic system, and the absolute safety of human workers carrying out tasks with the robots in a single working space.

In modern robotic solutions, the concept of human-robot interaction is a common term, which characterizes the process of joint, purposeful and safe human actions in collaboration with a robot [1, 2]. The concept of "collaborative interaction" refers to

R. Silhavy et al. (Eds.): CoMeSySo 2021, LNNS 232, pp. 749–759, 2021.
https://doi.org/10.1007/978-3-030-90318-3_58

the process of safe cooperation between a human and a robot, and such robots are called “collaborative robots” (or “cobots”) [3, 4]. The use of cobots does not require additional safety measures (e.g., robot barrier walls); their operation in a single working space is defined by technical specifications and international standards [5]. Thus, partial automation through the use of cobots in technological processes and production facilities makes it possible to increase labor productivity.

This paper summarizes the current state of collaborative robotics and proposes an approach to improving the efficiency of collaborative robotic systems. This approach involves the identification, calculation and analysis of areas of intersection of a human and a cobot, taking into account the movements of their working bodies while performing collaborative tasks.

2 Related Work

An important component in the organization of an efficient working process in a collaborative robotic system is human worker’s safety (including while moving and changing posture) in terms of the direct interaction with a robot. Human’s possible deviations from the correct working posture may lead to the reduction of the intersection area (volume) of human’s working bodies and robot’s working zone in a single workspace. In turn, the robot must take into account possible deviations in human worker’s position in order to ensure that he will not be harmed during the collaboration.

There are many studies concerning human behavior in crowds, the search for an exit in limited and open spaces (purposeful movements), as well as human motion in terms of biomechanics. Such approaches lead to “linked” or interdependent movements, similar to Newtonian mechanics. However, there are important differences, for example, in general Newton’s third law is not enforced [6–10]. Paper [11] examines a special case of possible movement of a human on the plane in three-dimensional space. In developing a probabilistic model of human motion both basic structure (two-dimensional grid on the plane) and special conditions (the initial position of a person that occupies one cell in a grid corresponding to an area of 160 square centimeters) have been considered.

Studies [12–14] consider a discrete approach to modelling human motions in different spatial scenarios. The analyzed model is based on the theory of cellular automata, with the probabilities of transitions being calculated so that the motions are reproduced from random to strictly directed ones. A distinguishing characteristic of such a model is the consideration of the surrounding environment (in relation to the current cell). The field of a cellular automaton is defined by two matrices: the first one contains values corresponding to the presence or absence of a particle in the cell, while the second matrix has the values indicating the presence or absence of the obstacles in the cell. For the von Neumann neighborhood (number of sides is $N = 4$) the state of the cell is influenced by the adjacent sides and cells. Depending on the specific task, the Moore neighborhood may be considered when the number of the sides increases up to $N = 8$ [15]. The direction of the movement is determined on the basis of the transition probabilities in each direction for each particle in terms of discrete time and the rules of transition.

On the other hand, much of the published papers on collaborative robotics focuses on safe human-robotic interaction, with much less attention being paid to improving the efficiency of such interaction. However, a number of studies address, for example, the issues of establishing robots' working processes using group management and swarm methods and algorithms [16–19].

However, finding solutions to improve the efficiency of human-robot collaboration in ergatic robotic systems (collaborative robotic systems) is still a pressing challenge [20–22]. In addition to kinematic analysis, the influence of the distance between a human and a cobot is also considered. Like a robot, a human has its own working area, which can be expressed by the distance from his location to the end point of his arm. In this way human's working space is formed; and this area (or volume) can be defined as a circle (or a sphere) with a given radius. Our hypothesis is that maximizing the efficiency of human-robot interaction lies in the increase value of the area (the volume) of the intersection of human's and robot's workspaces – a shared workspace (given the safety of such collaboration).

In a shared workspace mutual beneficial collaboration between the robot and the human occur. For example, given the various human factors that may affect the human's ability to work, the robot is able to compensate the loss of the overall working capacity of considered tandem. In turn, the human worker is able to provide expert analysis of the processes of human-robot collaboration in order to ensure the overall supervision of joint tasks and improve the performance of the cobot through its fine tuning.

3 Simulation of Human and Robot Workspaces

In order to achieve the required level of safety of human-robot interaction without decreasing the efficiency of the overall work process, an analysis of the methods presented in [23] has been carried out. In particular, the main safety and management techniques in robotic systems are:

- *Movement control.* Controlling the motion of the robot before and after the contact with the human. Pre-collision methods comprise limiting the values of the basic parameters of the robot's operation (speed, energy capacity, etc.), as well as collision avoidance using methods that include the definition of safety zones, tracking the distance and directing the movement vector of the robot away from people. Post-collision methods include different ways of minimizing injuries by switching between different control techniques in the event of a collision. In addition, intentional and unintentional contact are distinguished for the purpose of efficient cooperation, and safe physical contact is allowed.
- *Movement planning.* The goal is to plan safe robot trajectories to avoid potential collisions. For this purpose, different parameters and values (including distance between the robot and the human, direction of view when planning the movement, etc.) should be taken into account.
- *Human motion prediction.* Such a prediction is based on possible human actions and movements, as well as various methods, including sequence matching, probabilistic plan recognition and motion analysis. The predictability of the robot should also be taken into account, since the role of the robot is as important as the role of the operator in multi-agent interaction.

The results of the analysis showed the relevance of considering the issue of modeling human motion and behavior in a shared workspace.

3.1 Human-Robot Motion Field During Collaboration

While modeling human worker's motion, difficulties arise at the stage of describing human's behavior. This is due to the fact that human actions are caused by a variety of rational and irrational factors. Basically, research may be carried out using two approaches: continuous and discrete. The continuous approach is based on solving a system of differential equations. The movement of a human is described using Newton's second law, taking into account his mass and motion velocity when interacting with obstacles. The discrete approach proposes the unification of physical parameters in the process of human motion. Spatial discretization allows to consider the movement of a human within a single time step.

The authors consider a discrete probabilistic model based on the theory of cellular automata within the framework of the study presented in the article. The workspace of a human is determined by the "reachable" area of the motion field in the horizontal plane, taking into account the length of his working bodies. Manual labor operations are carried out within the motion field area during human-robot collaboration. A workplace is designed to perform physical work of intermediate complexity, the size of the working area should be determined based on the general requirements of ergonomics [24].

Performing frequently repeated operations should be within the following easy-to-reach and optimal motion field zones (see Fig. 1):

- *Zone I.* Optimal motion field area for placing very frequently used and most important controls, tools and objects;
- *Zone II.* The easy-to-reach area of the motion field for placing frequently used controls, tools and objects;
- *Zone III.* Motion field reachable area for placing rarely used controls, tools and objects.

Human's movement can be either arbitrary or involuntary, and the effective workspace can be changed within the reach of the motion field. Thus, obtaining and processing predictive data about the motions of a human agent in a collaborative robotic system is an urgent problem, and forming a model of human movements within the workplace may be useful for a collaborative robotic system when making decisions. Such solutions, for example, may include: reducing the speed of motions, changing the trajectory of the robot's working bodies, and assessing the psychophysiological state of the human agent [25].

3.2 The Workspace of a Collaborative Robot

Collaborative robots have additional functions that regulate the safety of interaction. In particular, such functions include: speed control and limiting the motion power [26–28]. A human operator and a robot can simultaneously move in a single workspace, and the robot's movements are technically limited to ensure a safe value of the collision force.

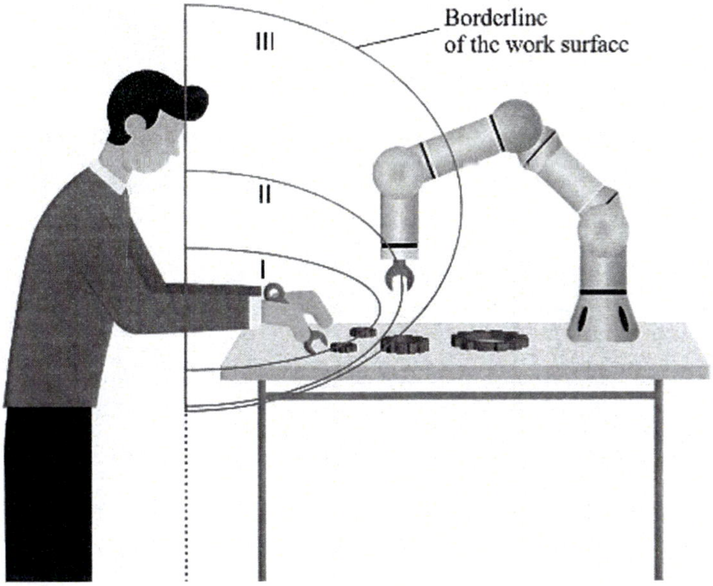

Fig. 1. Areas of performing manual operations in a vertical plane.

There are four embedded spaces in the area of interaction between a collaborative robot and a human [29]:

1. Maximum — the full amount of space, taking into account the volume of all possible parts that can be used by the robot;
2. Forbidden — a part of the maximum space that a human has limited access to when the robot is working;
3. Operational — a part of the forbidden space that the cobot uses to perform its actions;
4. Collaborative (shared) — a part of the operational space designed for safe joint human-robot activities.

The following principles (the so called "human-robot collaboration principles") are applied in the design of collaborative robots:

- the operations performed by the robot must be safe in relation to other agents when acting in a shared workspace, both to human operators, and the objects of external environment [30];
- the robot should adapt to the human, not vice versa (e.g., if the collaborative system does not function efficiently enough, it is necessary to tune the robot, not to replace the human operator);
- it is necessary to create such an environment in which any human operator in the working space can make the most of the cobot;
- it is necessary to increase labor productivity and reduce the number of errors at the stage of design of the collaborative system [31].

The key differences between collaborative robots and other types of robotic systems are mainly defined by the above-mentioned principles [32].

3.3 Finding the Intersection of Workspaces of Human and Robot Agents

We define the arms of a human operator as the main working bodies that are used by him to perform work together with a collaborative robot. In a simplified form, the motion of a human's arm can be described as two intersecting hemispheres with the same radius $R_{leftarm} = R_{rightarm} = R_{arm}$. They are located opposite to the volume of the collaborative robot's workspace at some distance L from the cobot's mounting position on the working surface, and these two hemispheres are located at some height h_{arm} from the working surface and at a distance of d_{arm} from each other (corresponds to the distance between the two shoulder joints). Thus, these two intersecting hemispheres represent the working space of a human. In turn, the workspace of a collaborative robot can also be described as a hemisphere with a radius R_{cobot} that is opposite to the workspace a human. This hemisphere of a robot corresponds to its working space with the maximum radius of motion of all its working bodies.

Using the well-known equation of a sphere centered at a point in three-dimensional space with certain coordinates $C(x_c; y_c; z_c)$ and radius R, the intersection of the volumes of human and robot workspaces V_{HRC} can be described via a set of solutions of a system of three following nonlinear inequalities:

$$V_{HRC} = \begin{cases} \left(x - x_{leftarm}\right)^2 + \left(y - y_{leftarm}\right)^2 + \left(z - z_{leftarm}\right)^2 \le R_{arm}^2 \\ \left(x - x_{rightarm}\right)^2 + \left(y - y_{rightarm}\right)^2 + \left(z - z_{rightarm}\right)^2 \le R_{arm}^2 \\ (x - x_{cobot})^2 + (y - y_{cobot})^2 + (z - z_{cobot})^2 \le R_{cobot}^2 \end{cases} \quad (1)$$

Taking the mounting point of the collaborative robot on the working surface as an origin point in the Cartesian coordinate system ($x_{cobot} = 0$; $y_{cobot} = 0$; $z_{cobot} = 0$), the points of the location of the shoulder joints of a human will have $(L;\ \frac{d}{2};\ h)$ and $(L;\ -\frac{d}{2};\ h)$ coordinates respectively (provided that the human is standing straight, and the Ox axis is directed towards the human).

MATLAB modeling framework was used to simulate the intersection of the working space of a human and a robot during their collaboration. The built-in model of the Universal Robots UR5 collaborative manipulator was used. The distance from the human to the mounting point of the cobot on the working surface was $L = 1$ m the height of the placement of the human shoulder joints above the working surface was $h_{arm} = 0.3$ m, the distance between the shoulder joints was $d_{arm} = 0.5$ m, the length of the human's outstretched arm $R_{arm} = 0.7$ m. Because of the calculations (homogeneous transformation using start and end configurations of the robot, as well as Cartesian representation of the translation vector from it) the radius of the sphere for the robot corresponding to its maximum possible length in the working space was $R_{cobot} = 0.8173$ m.

The construction of the point cloud of the maximum working space of the collaborative robot was carried out by consecutive turns of joint of its shoulder link (assuming the maximum length of all other links) by 180 degrees in the horizontal plane and turning the joint of its upper arm link to 1 degree in the vertical plane, while the initial position of the manipulator was chosen parallel to the working surface.

The simulation of the motion of human arms was carried out by creating two manipulators at a distance d_{arm} from each other and the height h_{arm} from the working surface and the distance L from the collaborative robot. The manipulator (simulating one hand) was constructed using three joints and three links respectively. The procedure of creating the cloud of points of the maximum workspace was similar to the cobot's one (the rotation of the main joint by 180 degrees in the horizontal plane, and the rotation of the neighboring joint by 1 degree in the vertical plane; the initial position is parallel to the plane of the working surface, see Fig. 2). The motion of the links was carried out using the built-in bidirectional Rapidly-exploring RandomTree (*manipulatorRRT*) functions and the optimal trajectory (plan) scheduler.

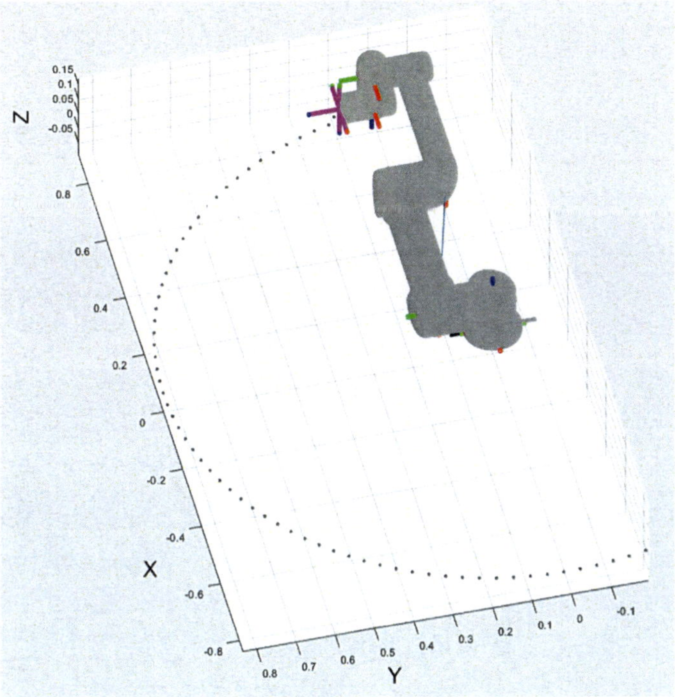

Fig. 2. The process of forming a cloud of points of intersection of working bodies of a robot and a human (MATLAB simulation).

Then the function was used to convert points from the source body frame to the target one with a specific configuration of the robot, as well as extract the Cartesian representation of the translation vector from a homogeneous transformation. As a result, an array of values of three-dimensional Cartesian coordinates of the "end-effector" location was obtained for all three "manipulators" (*body3* – for human's hands and *tool0* – for UR5 robot). The values were obtained, taking into account their minimal movement along a given trajectory, and rotation in the horizontal and vertical planes.

The calculation of the total volume for the above three hemispheres was carried out using the built-in *fmincon* function to solve a system of nonlinear equations with nonlinear constraints.

Thus, the calculation of the cloud of points belonging to the volume of the intersection of the three hemispheres allows creating a computer model for finding the areas of intersection of the working spaces of a human and a collaborative robot, their volume and shape. We assume that the efficiency of interaction between a human and a collaborative robot is directly proportional to the volume of the area of intersection of their working zones.

Figure 3 shows the cloud of points of the working spaces of a collaborative robot and a human operator, as well as the area of intersection of their working bodies.

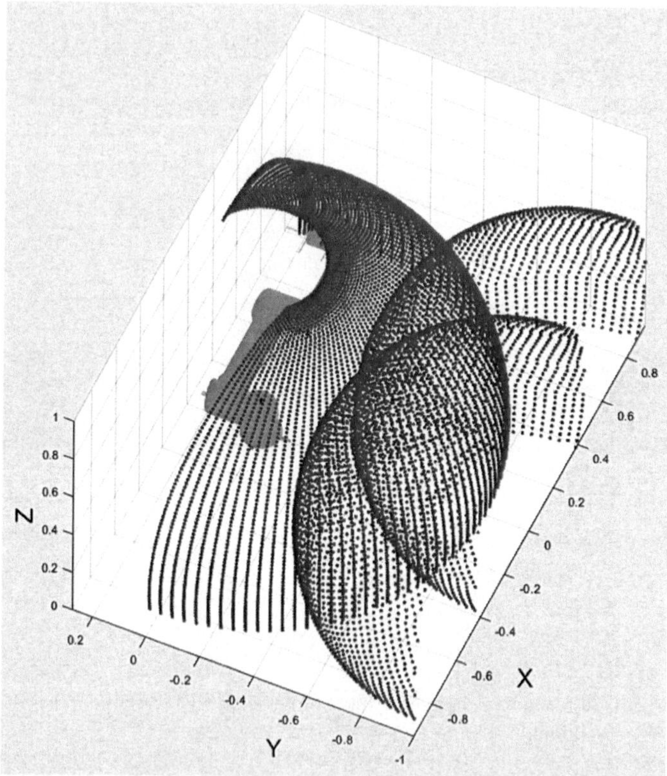

Fig. 3. Cloud of points of working spaces of a cobot and a human, and area of intersection of their working bodies (MATLAB simulation).

4 Conclusions and Future Work

Within the conducted research, the issue of determining the intersection of the areas of the working bodies of a human and a robot in a collaborative robotic system was considered. As a result, a computer model has been developed; it allows forming clouds of points of human-robot workspaces. The constructed model depends on the distance from the human to the mounting point of the collaborative robot, the anthropometric parameters of the human, as well as and the dimensional characteristics of the cobot. In addition, the initial stage of numerical modeling was performed to form a point cloud, and estimate the volume of intersection of the working areas of a collaborative robot and a human.

Further research is planned to use numerical methods to develop a universal model for finding the intersection volumes of the working zones of a cobot and a human, taking into account changes in their positions, working radius values, motion and changes in the human posture, etc. We also plan to develop mathematical software for the analytical solution of finding the intersection of these working spaces, as well as the corresponding volumes in multi-agent collaborative robotic systems.

Acknowledgements. The reported study was partially funded by RFBR according to the research project No. 19-08-00331.

References

1. Cherubini, A., Passama, R., Crosnier, A., Lasnier, A., Fraisse, P.: Collaborative manufacturing with physical human-robot interaction. Robot. Comput. Integr. Manufa. **40**, 1–13 (2016)
2. Galin, R., Meshcheryakov, R., Kamesheva, S., Samoshina, A.: Cobots and the benefits of their implementation in intelligent manufacturing. In: Proceedings of the 2nd International Conference "MIP: Engineering-2020: Modernization, Innovations, Progress", vol. 862 (032075), pp. 1–5. IOP Publishing Ltd, Bristol (2020)
3. Ermishin, K.V., Yuschenko, A.S.: Collaborative mobile robots – a new stage of development of service robotics. Robot. Tech. Cybern. **3**(12), 3–9 (2016)
4. Zanchettin, A.M., Ceriani, N.M., Rocco, P., Ding, H., Matthias, B.: Safety in human-robot collaborative manufacturing environments: metrics and control. IEEE Trans. Autom. Sci. Eng. **13**(2), 882–893 (2015)
5. Hoffman, G.: Evaluating fluency in human-robot collaboration. IEEE Trans. Hum.-Machine Syst. **49**(3), 1–10 (2019)
6. von Krüchten, C., Schadschneider, A.: Empirical study on social groups in pedestrian evacuation dynamics. Physica A **475**, 129–141 (2017)
7. Duives, D.C., Daamen, W., Hoogendoorn, S.P.: State-of-the-art crowd motion simulation models. Transport. Res. C: Emerg. Technol. **37**, 193–209 (2013)
8. Kirchner, A., Schadschneider, A.: Simulation of evacuation processes using a bionics-inspired cellular automaton model for pedestrian dynamics. Physica A **312**(1–2), 260–276 (2002)
9. Nazarov, V.T.: Analytical representation of the athlete's movements. Questions Theory Pract. Phys. Cult. Sports **14**, 121–122 (1984). (in Russian)
10. Nazarov, V.T.: Athlete's Movement, 1st ed. Polimya, Minsk (1984) (in Russian)
11. Weidmann, U.: Transporttechnik der Fussgänger. Transporttechnische Eigenschaften des Fussgängerverkehrs, Literaturauswertung. Schriftenreihe des IVT 90, pp. 1–84 (1992)
12. Breukelaar, R., Bäck, Th.: Using a genetic algorithm to evolve behavior in multi dimensional cellular automata: emergence of behavior. In: Proceedings of the 7th annual conference on Genetic and evolutionary computation (GECCO '05), pp. 107–114. Association for Computing Machinery, New York (2005).

13. Lobanov, A.I.: Models of cellular automata. Comput. Res. Model. **2**(3), 273–293 (2010) (in Russian)
14. Kirik, S.E., Krouglov, D.V., Yurgel'yan, T.B.: On discrete people movement model with environment analysis. J. Siberian Federal Univ. Math. Phys. **1**(3), 262–271 (2008) (in Russian)
15. Stepantsov, M.E.: Mathematical model of the directional movement of a group of people. Math. Model. **16**(3), 43–49 (2004) (in Russian)
16. Galin, R.R., Kamesheva, S.B.: Effective functioning of a collaborative robotic system in a shared workspace. The scientific peer-reviewed journal "Proceedings of the Kabardino-Balkar Scientific Center of the Russian Academy of Sciences" **1**(99), 5–14 (2021) (in Russian)
17. Vorotnikov, S., Ermishin, K., Nazarova, A., Yuschenko, A.: Multi-agent Robotic Systems in Collaborative Robotics. In: Ronzhin, A., Rigoll, G., Meshcheryakov, R. (eds.) ICR 2018. LNCS (LNAI), vol. 11097, pp. 270–279. Springer, Cham (2018). https://doi.org/10.1007/978-3-319-99582-3_28
18. Gaiduk, A.R., Kalyaev, I.A., Kapustyan, S.G.: Managing a team of intelligent objects based on pack principles. Bull. YNC RAS **1**(2), 20–27 (2005) (in Russian)
19. Melnik, E., Klimenko, A., Safronenkova, I.: An Estimation of Distributed Algorithms of the Fault-Tolerant Management in the Robot Groups. In: Ronzhin, A., Rigoll, G., Meshcheryakov, R. (eds.) ICR 2020. LNCS (LNAI), vol. 12336, pp. 212–221. Springer, Cham (2020). https://doi.org/10.1007/978-3-030-60337-3_21
20. Galin, R.R., Serebrenniy, V.V., Tevyashov, G.K., Shirokiy, A.A.: Human-robot Interaction in Collaborative Robotic Systems. In: Emelyanov, S.G., et al. (eds.) Proceedings of the Southwest State University, vol. 24(4), pp. 180–199. Southwest State University, Kursk (2020) (in Russian)
21. Kebria, P.M., Al-wais, S., Abdi, H., Nahavandi., S.: Kinematic and dynamic modelling of UR5 manipulator. In: 2016 IEEE International Conference on Systems, Man, and Cybernetics (SMC), pp. 004229–004234 (2016)
22. Galin, R., Meshcheryakov, R., Samoshina, A.: Mathematical modelling and simulation of human-robot collaboration. In: 2020 International Russian Automation Conference (RusAutoCon), pp. 1058–1062. IEEE, New York (2020)
23. Lasota, P.A., Shah J.A.: Analyzing the effects of human-aware motion planning on close-proximity human-robot collaboration. Human factors. J. Hum. Factors Ergon. Soc. **57**(1), 21–33 (2015)
24. GOST 12.2.033-78: Occupational safety standards system. Operator's location in a standing position. General ergonomic requirements. https://docs.cntd.ru/document/1200005187. Last accessed 1 July 2021
25. Korovin, I.S., Klimenko, A.B., Kalyaev, I.A., Safronenkova, I.B.: An experience of the cognitive map-based classifier usage in astronaut's emotional state monitoring. Acta Astronaut. **181**, 537–543 (2021)
26. Byner, C., Matthias, B., Ding, H.: Dynamic speed and separation monitoring for collaborative robot applications–concepts and performance. Robot. Comput. Integr. Manuf. **58**, 239–252 (2019)
27. ISO/TS 15066:2016: Robots and robotic devices – collaborative robots. https://www.iso.org/ru/standard/62996.html. Last accessed 1 July 2021
28. Galin, R.R., Meshcheryakov, R.V.: Human-Robot Interaction Efficiency and Human-Robot Collaboration. In: Kravets, A.G. (ed.) Robotics: Industry 4.0 Issues & New Intelligent Control Paradigms. SSDC, vol. 272, pp. 55–63. Springer, Cham (2020). https://doi.org/10.1007/978-3-030-37841-7_5
29. Mihelj, M., et al.: Collaborative robots. Robotics, 2nd ed. Springer International Publishing AG, Cham (2019)
30. Franklin, C.S., Dominguez, E.G., Fryman, J.D., Lewandowski, M.L.: Collaborative robotics: new era of human–robot cooperation in the workplace. J. Saf. Res. **73**, 153–160 (2020)

31. Konz, S., Johnson, S.: Work Design: Industrial Ergonomics, 5th ed. Holcomb Hathaway Pubs, Scottsdale (1999).
32. Paluch, S., Wirtz, J., Kunz, W.H.: Service robots and the future of services. In: Bruhn, M., Burmann, C., Kirchgeorg, M. (eds.) Marketing Weiterdenken, pp. 423–435. Springer Gabler, Wiesbaden (2020)

New Technologies in Solar Power Plant Study in the MATLAB Program Applied to the Conditions of Angarsk City

Yu V. Konovalov(✉) and A. N. Khaziev

Angarsk State Technical University, Angarsk, Russian Federation

Abstract. In order to study the solar power plant, data on temperature and insolation for Angarsk city were analyzed, a computer model was developed in the MATLAB software package, which gives opportunity to study the main characteristics of a solar power plant in given weather conditions. The model allows you to investigate the main output parameters, such as: total power generated by a set of panels, daily schedule of electricity generation, charts of battery charge, discharge and voltage for a given period of time. The proposed model, which implements the method of power supply of the receiver in a traditional power supply system together with solar photovoltaic power plants, allows minimizing capital costs and intake of usable areas, improving the quality of power supply and reducing the negative environmental impact of energy sources. #COMESYSO1120.

Keywords: Solar power plant · Modeling · Power supply

1 Introduction

The changes in the energy field are very significant and occur quickly. The priorities of energy sources and conditions for energy production are changing, as well as the ways and methods of transmission and distribution of electricity. Although traditional industries and services are becoming more energy efficient, the growth of the world's population and the emergence of new services leads to an increase in overall energy consumption. According to the forecast of world electricity production [1–3] in the period from 2016 to 2050, the annual growth rate of world electricity consumption will decrease from 2.2 to 1.1%, but at the same time, in 2015, world energy consumption amounted to 20300 trillion kWh, forecast for 2030–26400 trillion kWh, and by 2050 will increase to 34,500 trillion kWh.

Keeping pace with changing needs requires new technologies, cutting edge solutions, products and services that span the entire energy chain. In particular, tangible and digital technologies, products and services are required for the energy of the future.

Among great number of industries, electricity-producing industry has the most significant impact on human life. Today it is difficult to imagine meeting the majority of everyday human needs without the use of electricity. The work of machines in production, city lighting, the Internet is impossible without high-quality power supply.

R. Silhavy et al. (Eds.): CoMeSySo 2021, LNNS 232, pp. 760–770, 2021.
https://doi.org/10.1007/978-3-030-90318-3_59

As the consumption of electricity by mankind increases, its production increases, which, in the case of using classical energy, causes significant harm to the environment. In recent years, there has been a trend towards planning a gradual transition to alternative energy. A number of developed countries are developing normative and legislative acts regulating its commissioning and operation. Already now, a significant percentage of electricity generation in the countries of the European Union is occupied by power plants producing energy based on renewable energy sources (RES) [1–3].

A significant percentage of electricity produced by renewable energy sources falls on solar photovoltaic power plants (SPPP), the number of which is steadily growing every year. One of the reasons for this is the growth in the production of thin-film modules for such power plants, as well as an increase in the number of service companies involved in the installation and connection of SPPP [3, 4]. Solar PV power plants directly convert solar radiation into electrical energy. For the manufacture of SPPP, materials of different physical and chemical properties are used. In their composition, they can contain: silicon, germanium, gallium arsenide, selenium, complex polymers and organic compounds.

Analysis result of most silicon batteries common types in terms of their efficiency, advantages and disadvantages [5–8], is presented in Table 1.

Table 1. Comparative characteristics of silicon batteries types.

Batteries type	Single-crystal	Poly-crystal	Amorphous
Efficiency, %	22	17	8
Structure	Single-crystals of silicon	Poly-crystals of silicon	Amorphous silicon
Advantages	High efficiency	Use for production of waste and recycled materials	Waste-free production More efficient use of diffused sunlight Increase in the output of electricity when heating elements
Disadvantages	High price	Sensitivity to temperature changes	Low efficiency rate per unit area

2 Methods

In most cases, SPPP are used as part of hybrid power plants, in which the operation of SPPP is supposed to be combined with other renewable energy sources, such as wind turbines (WT), diesel power plants (DPP) and mini hydroelectric power plants. This approach assumes the presence of an automatic control system for a hybrid power plant. In this case, the following components are required:

- a system that determines the optimal use of a particular type of power plant: SPPP, WT, DPP, mini hydroelectric power plants, depending on many uncertain parameters (weather, temperature, etc.);
- automation relay protection systems that control the distribution and optimal control of the various sources included in the hybrid power plant;
- consideration of environmental component, which has recently been of great importance in the use of RES. For example, WTs occupy significant areas, produce low-frequency noise, change the landscape, etc. For diesel power plants, special premises are needed for the station itself, for storing fuel reserves, measures to solve the problems of fuel delivery, the possibility of pollution. For mini hydroelectric power plants – the availability of areas for the accumulation of water reserves, the solution of environmental issues.

A method is proposed for using new technologies in the form of SPPP together only with traditional sources of electricity as part of existing power supply systems (PSS).

In this case, only a system that controls the amount of electricity for the needs of the consumer, which is produced by a set of solar panels, is sufficient for functioning, or it is possible to use the stored electricity in storage batteries (SB) charged with a sufficient amount of solar insolation. If this is not enough, then the consumer switches to power from a traditional PSS. The algorithm for implementing this method is shown in Fig. 1: if the power generated by the SPPP P_S is greater than the electrical power of the consumer P_C, then the consumer is supplied from the SPPP. Otherwise, power is supplied from a traditional power supply system.

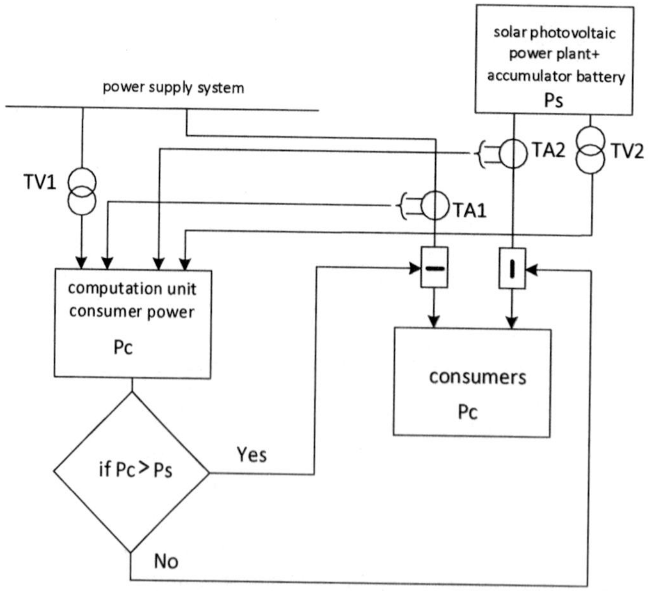

Fig. 1. Flow diagram of the algorithm implementation.

When implementing this method, there are significantly fewer capital costs for power supply to the consumer and negative, from the point of ecology view, factors of influence of hybrid power plants various components. At the same time, this approach has a number of significant advantages:

- losses during transmission of electricity according to the traditional power supply scheme of the receiver are reduced, because the electricity produced by the SPPP is generated in the immediate vicinity of its consumption. If long-term power outages are foreseen, then the issue of increasing the power of the SB and the power of solar panels is being resolved;
- roofs, walls and other areas not involved in economic activity can be used for solar panels;
- the possibility of passing the morning maximum load without exceeding it, because the morning maximum coincides mostly with active solar insolation period.

3 Results

The studies were carried out during modeling with solar panels, based on the LG405N2W-A5 module, the appearance of which is shown in Fig. 2, its current-voltage characteristics are shown in Fig. 3, and the main parameters are given in Table 2.

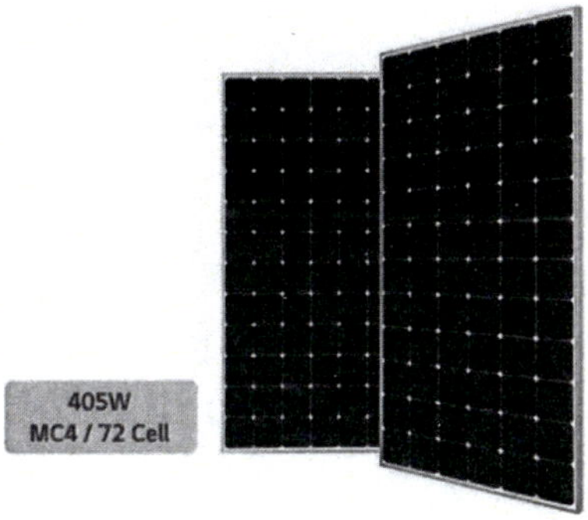

Fig. 2. Solar battery module brand LG405N2W-A5.

To conduct research on the use of solar photovoltaic power plants in relation to the conditions of Angarsk city, the MATLAB/Simulink software complex was used [9]. The developed computer model of SPPP allows to study the operation of photovoltaic cells with different input parameters, simulate the operation of a power plant with various kinds of load and build graphical characteristics describing the state of circuit objects.

The practical significance of the developed computer model is to consider the possibility of constructing a real SPPP based on the data obtained during the study of such a model in relation to specific weather and geographic conditions.

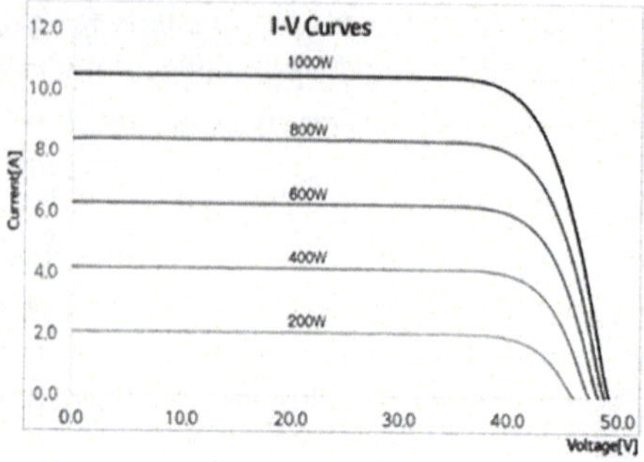

Fig. 3. Volt-ampere characteristics of the panel.

Table 2. Electrical parameters of the solar panel.

Maximum power	Maximum voltage	Maximum current flow	Closed-circuit voltage	Short-circuit current	Module efficiency
W	V	A	V	A	%
405	41.0	9.89	49.4	10.51	19.5

The used complex model of a solar power plant consists of two subsystems:

1) a photovoltaic module connected through an electric filter to a battery storing electricity, shown in Fig. 4;
2) a module consisting of a battery connected to a DC-to-AC converter and then to a transformer to match the voltage with specific consumers. The block diagram of this module is shown in Fig. 5.

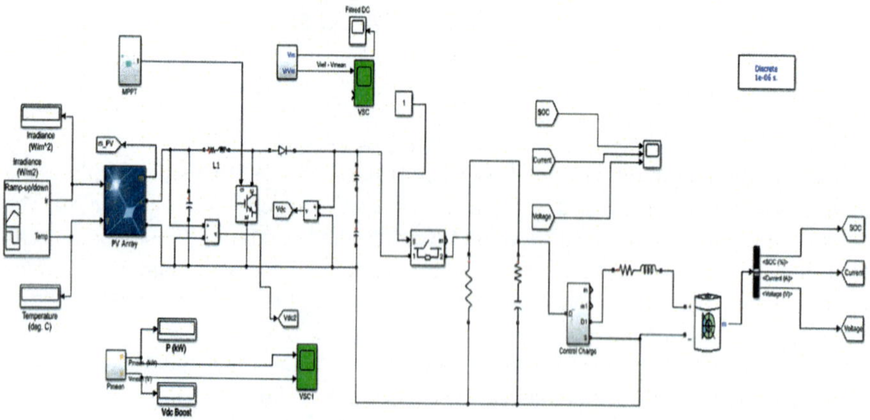

Fig. 4. Solar battery charging module.

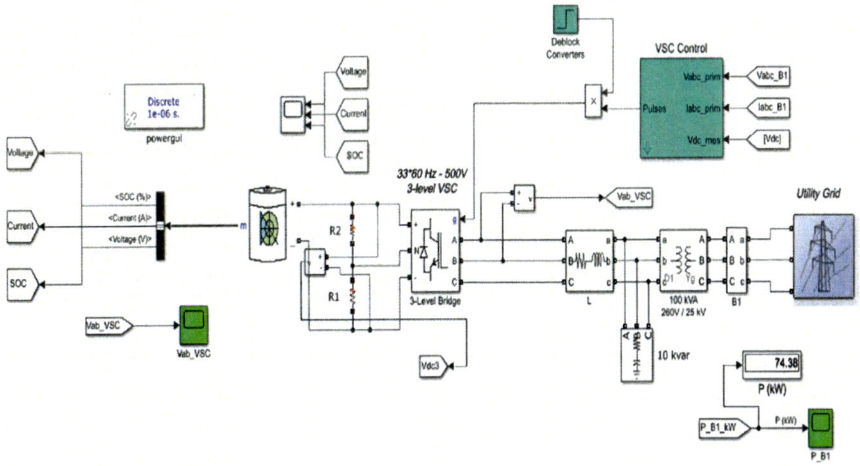

Fig. 5. Battery operation module for consumer network.

Both modules are based on 100 kW photovoltaic station connected to the consumer network [10]. The first one contains the following main elements: a set of solar panels implemented by the PV Array unit, an MPPT controller represented by an MPPT unit that controls IGBT, a DC filter network filter, a control charge subsystem and the battery itself.

The second module includes: a battery, a 3-Level Bridge DC-to-AC converter, a 3-phase step-up transformer and a load.

Studies of the solar power plant possibility were carried out in relation to Angarsk city, which is located in the southern part of Irkutsk region, its coordinates are 52° 34′ north latitude and 103° 55′ east longitude. The city is included in the zone with an increased number of sunshine hours, there are more than 2000 h of sunshine in the city per year [11, 12].

Information on solar insolation was taken from a resource specializing in providing climate data [13]. On the bases of obtained array, the following were investigated:

- the dependence of insolation for all year months with an hourly interval of insolation change, shown in Fig. 6. The graph of this dependence reflects how the illumination changes during the day for a certain month;
- histogram of annual temperature by months for four different hours of the day (see Fig. 7).

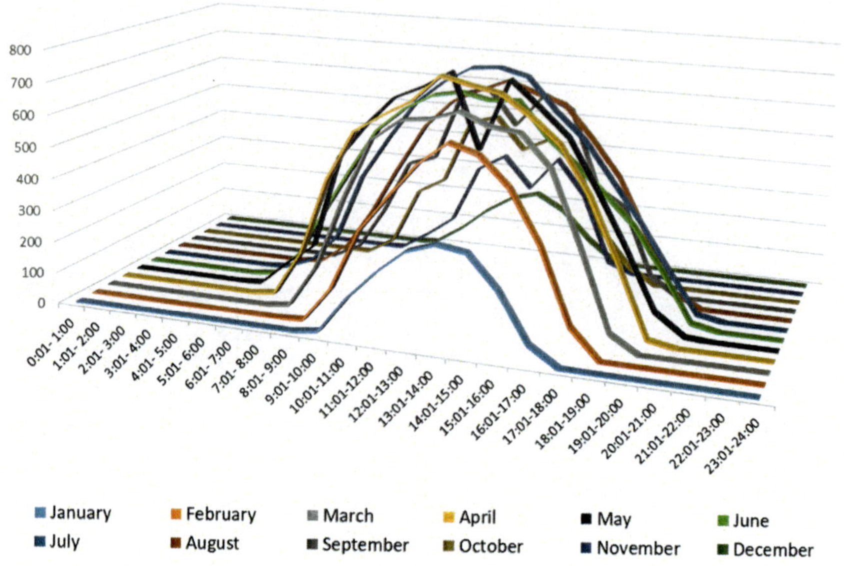

Fig. 6. Schedule of insolation for Angarsk city.

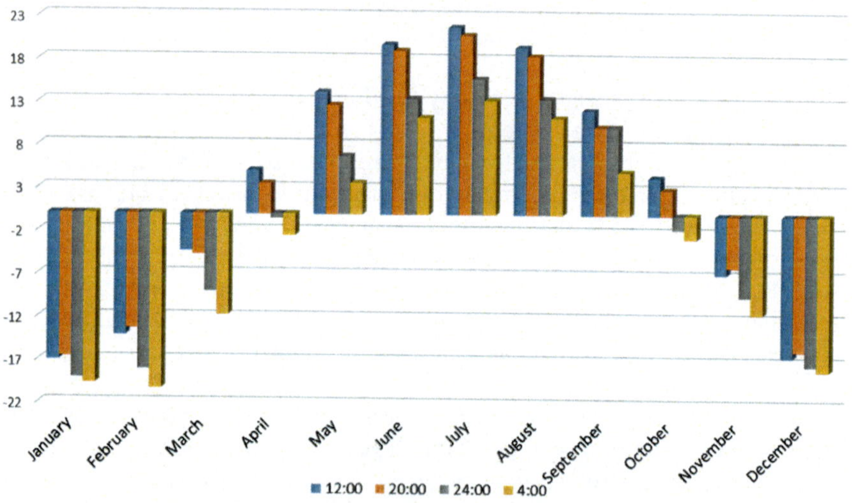

Fig. 7. Histogram of annual temperature for Angarsk city.

According to the results of a study of the available data for Angarsk city, it was found that the average daily solar insolation is 4341 Wh/m^2, or 131571 Wh/m^2 per month, with the established average daily duration of sunshine equal to 12.5 h.

The model considered in the study includes 250 solar panels of the LG405N2W-A5 brand. The number of panels affects the amount of energy produced. When operating the SPPP, it is necessary to take into account the angle of panel inclination in relation to the sun. In the simulated system, the resulting output corresponds to ideal conditions at an angle of 90°. This condition is implemented by the solar tracker control unit on the corresponding element base.

To implement the power supply of the consumer from batteries at night, a model has been created that includes 18 lithium-ion batteries connected in parallel-series, shown in Fig. 8.

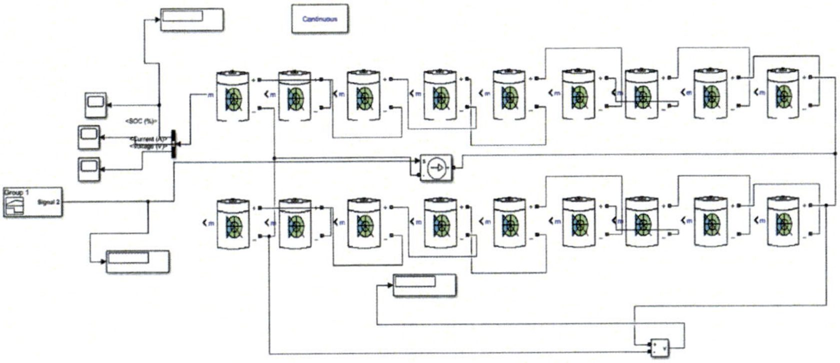

Fig. 8. Scheme of switching on solar panels of SPPP.

The electricity generation schedule for an average day in May, each division of which is equal to one hour, is shown in Fig. 9.

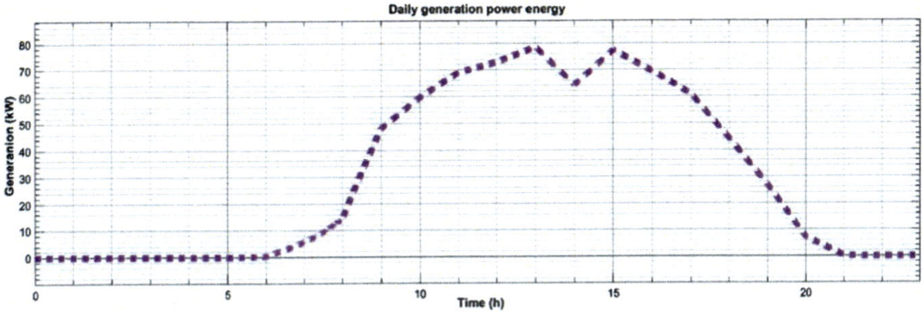

Fig. 9. Daily hourly electricity generation schedule.

As can be seen from the graph, in Angarsk conditions, the maximum total power of the panels reaches 80 kW, with 103 kW of potentially maximum output of the modules used.

The oscillogram of the inverted voltage is shown in Fig. 10.

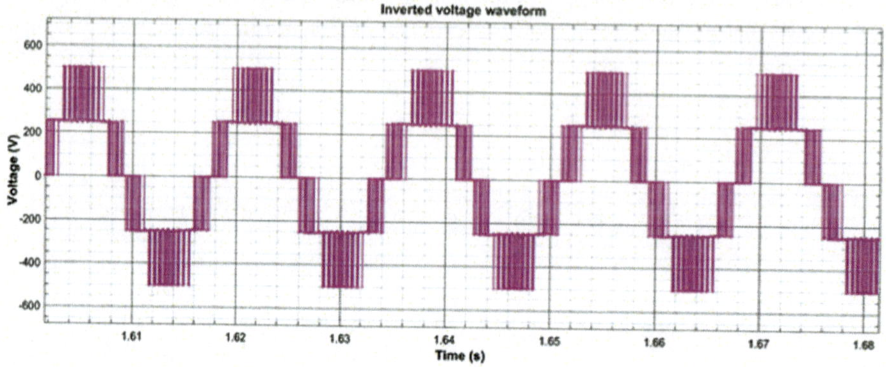

Fig. 10. Oscillogram of voltage in discrete simulation.

Figure 11 shows the oscillogram of the voltage at the solar battery output.

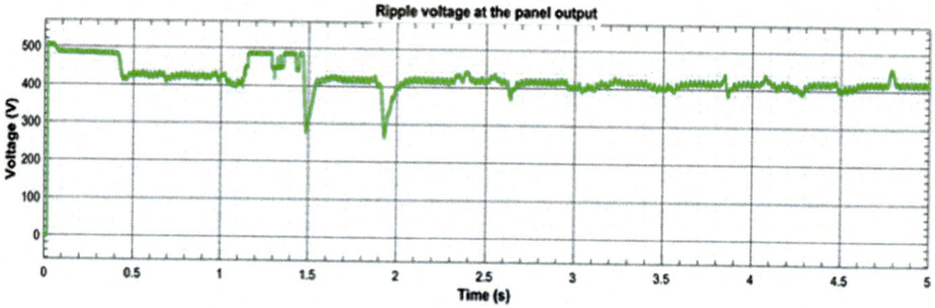

Fig. 11. Oscillogram of voltage at the battery output.

It should be noted that the model under consideration is a universal SPPP that can operate both directly into the network and in the mode of accumulating energy in batteries for subsequent return to the consumer. The work was simulated for May, when the number of sunshine hours prevails.

As known, the level of solar insolation is different during the day, therefore, the generation of electrical energy by panels at different hours can have a wide range of values. Added to this is the variability of the consumer load.

Graphs displaying the processes of charging and discharging the battery, which correspond to the universal type of SPPP, taking into account the above factors, are shown in Figs. 12 and 13.

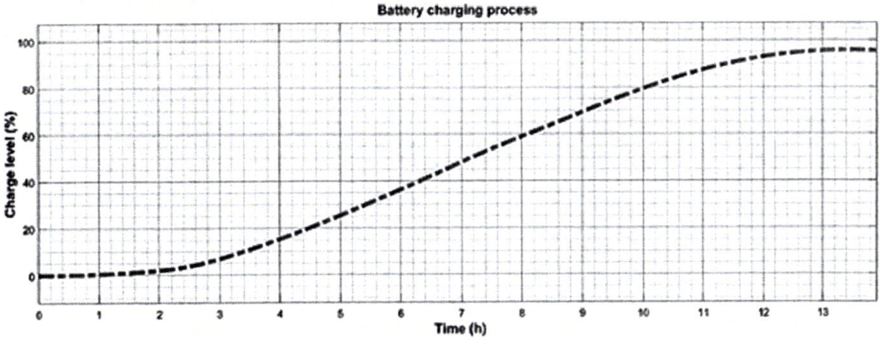

Fig. 12. Chart of battery charge for 14 h of operation (%).

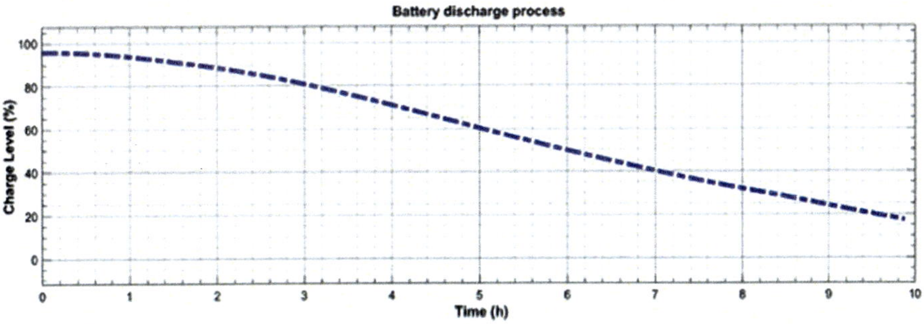

Fig. 13. Chart of battery discharge for 10 h of operation (%).

The graph in Fig. 14 shows voltage change at the battery output when operating at rated load.

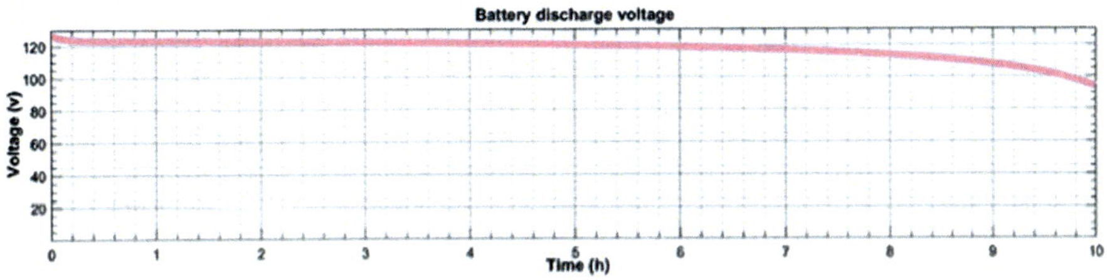

Fig. 14. Chart of battery voltage (V) for 10 h of operation without recharging from a solar panel.

4 Conclusion

The paper analyzes data on temperature and insolation for Angarsk, developed a computer model in the MATLAB software package using the Simulink extension, which allows to study the main characteristics of a solar power plant in given weather conditions. The model allows you to investigate the output parameters, such as: total power generated by a set of panels, daily schedule of electricity generation, charts of battery charge, discharge and voltage for a given period of time.

A model is proposed that implements the method of power supply to the consumer in the traditional power supply system together with SPPP, which minimizing capital costs and the use of useful areas, improving the quality of power supply and reducing the negative environmental impact of energy sources.

References

1. Nigmatulin, B.I.: Forecast of World Electrical Production at Nuclear Power Plants for the Period Up to 2050, http://www.proatom.ru/modules.php?name=News&file=article&sid=7390. Last accessed 28 July 2021
2. Marola, A.-M.: Share of Renewable Energy in the EU Up to 18.0%, pp. 1–3. https://ec.europa.eu/eurostat/documents/2995521/10335438/8-23012020-AP-EN.pdf/292cf2e5-8870-4525-7ad7-188864ba0c29. Last accessed 24 June 2021
3. Jones, D.: Global Electricity Review. https://ember-climate.org/wp-content/uploads/2021/03/Global-Electricity-Review-2021.pdf (2021). Last accessed 20 June 2021
4. Grigorash, O.V., Stepura, Y.P., Uskov, A.E., Kvitko, A.V.: Renewable energy sources: terms, definitions, advantages and disadvantages. In: Proceedings of KubGAU, Krasnodar 5, pp. 189–192 (2011)
5. Patrusheva, T.N.: Manufacturing technologies for components of oxide solar cells. Siberian Federal University, Krasnoyarsk. http://elib.sfu-kras.ru/handle/2311/139641 (2015). Last accessed 28 July 2021
6. Bessel, V.V., Kucherov, V.G., Mingaleeva, R.D.: Study of Solar Photovoltaic Cells, p. 90. Publishing Center of the Russian State University of Oil and Gas (NRU) named after I. M. Gubkin, Moscow (2016)
7. Afanasyev, V.P., Terukov, E.I., Sherchenkov, A.A.: Thin-Film Solar Cells Based on Silicon, p. 168. Publishing House of SPETU "LETI", St. Petersburg (2011)
8. Maronchuk, I.I., Sanikovich, D.D., Mironchuk, V.I.: Solar cells: current state and development prospects. Energy economics. Izv. Higher Study Instit. Energy Assoc. CIS **62**(2), 105–123 (2019)
9. Chernykh, I.V.: Simulation of Electrical Devices in MATLAB, p. 288 SimPowerSystems and Simulink. Peter, St. Petersburg (2008)
10. Detailed Model of a 100-kW Grid-Connected PV Array, https://www.mathworks.com/help/physmod/sps/ug/detailed-model-of-a-100-kw-grid-connected-pv-array.html. Last accessed 20 June 2021
11. Cities of Russia: An Encyclopedia, p. 559. Great Russian Encyclopedia, Moscow (1994)
12. Popel, O.S., et al.: Climatic Data for Renewable Energy in Russia (Climate Data Base): Study Guide, p. 56. Moscow (2010)
13. http://climate.onebuilding.org/WMO_Region_2_Asia/RUS_Russian_Federation/index.html#IDIRK_Irkutsk. Last accessed 22 June 2021

Methods for Functional Risks Assessment

V. B. Gisin, E. S. Volkova, I. Z. Yarygina(✉), and I. V. Lukashenko

Financial University under the Government of the Russian Federation, Leningradsky pr. 49, Moscow 125993, Russia

Abstract. The article reveals the features of the assessment and management of financial risks of enterprises. The purpose of the article is to propose new methods of quantitative analysis of functional risks that contribute to the effective management of financial resources of enterprises, including banks as enterprises of a special kind, the main purpose of which is to make a profit. In the framework of the work, the justification of functional risk as a source of losses is presented.

The research begins with a description of the trust distribution function on a binary frame consisting of two elements, meaningfully corresponding to the answers "yes" and "no". Trust distribution functions are associated with a static description of the information system. To describe the dynamics of the transfer of trust through the information system, the Dempster evidence combination rule is used.

The material of the study considers various examples of quantitative assessment of functional risks, mathematical expectation of losses and its reflection in information databases. The results contribute to the formation of the reasonable strategy for enterprises in the face of increasing threats of losses caused by functional risks.

Keywords: Functional risk · Soft computing technique · Measure of confidence · Measure of likelihood · Theory of evidence

1 Functional Risk as a Factor in the Occurrence of Losses

The Russian scientific school defines a bank as a special kind of enterprise that produces and sells banking products, provides services, organizes distribution channels, and manages relations with clients (consumers, buyers). In the current conditions, the key area of activity of a credit institution is risk management based on a quantitative and qualitative assessment of their occurrence threat. Russian enterprises, including banks, use a wide range of methods for assessing country, financial and functional risks based on the use of modern information technologies. In modern conditions of market relations, the correct quantitative and qualitative assessment of functional risks plays a special role. As they cover all areas of the production process and arise due to the inability to timely and fully control the financial and economic process, collect, and analyze the information required to ensure the reproductive process. It seems appropriate to adhere to the following classification of functional risks (Table 1).

R. Silhavy et al. (Eds.): CoMeSySo 2021, LNNS 232, pp. 771–783, 2021.
https://doi.org/10.1007/978-3-030-90318-3_60

Table 1. Functional risks classification.

Group	Risk class	Risk category
External risks	Risks of the operating environment	• Regulatory and legal risk • Competition risk • Economic risk • Country (sovereign) risk
Internal risks	Management risks	• Fraud • Ineffective administration
	Risks of goods/services delivery	• Technological • Operating • Implementation of new tools • Strategic

Practice has shown that enterprises widely use methods of qualitative assessment of functional risks, including the use of various control tools based on an independent assessment of risks, limiting possible losses, and expanding a conservative approach to forming an activity strategy, creating incentives for risk management, transferring risks to third parties, forming resources to cover potential losses. Since there is an extremely wide variety of possible manifestations of functional risks, one of the possibilities for managing them is dividing the business process into its constituent elements, forming a process flow chart, identifying, and evaluating possible manifestations of risk sources at each stage of the operation in order to prevent them.

The role of quantitative assessment of functional risks and the effectiveness of using the results to ensure the sustainable operation of enterprises of all forms of ownership are increasing in the context of the digitalization of the economy the expansion of cyber space and the information content of databases. Let us dwell on this issue in more detail.

2 Trust and Likelihood Measures on Finite Frames

Let us start with a description of the trust distribution function on a binary frame consisting of two elements, meaningfully corresponding to the answers "yes" and "no". The normalized trust function on such a frame can be specified by a pair of numbers such as $\mu, \nu \geq 0$ and $\mu + \nu \leq 1$. In relation to vulnerability, the first number can be interpreted as a subjective assessment of the arguments in favor of the vulnerability being exploited in an attack, the second as a subjective assessment of the arguments in favor of the vulnerability not being exploited. The "arguments" are assessed on a scale from 0 to 1. The value $\pi = 1 - \mu - \nu$ serves as an estimate of the uncertainty. These estimates can be made probabilistic: a value μ is a lower probability estimate that a vulnerability will be exploited, ν - a lower probability estimate that a vulnerability will not be exploited.

In general, the finite frame can be interpreted as a set of possible answers to some question (in the case of a binary frame, the answer must be "yes" or "no"). The evidence on the frame Θ is given by the baseline trust distribution function for which $\sum\limits_{X \subseteq \Theta} m(X) =$

1. The sets $X \subseteq \Theta$ for which $m(X) \neq 0$ are called foc. The expert, giving evidence, as it were, indicates the focal sets and sets the probability distribution between them. The probability of a focal set X can be interpreted as the probability that the correct answer is in the set X. If $m(\Theta) = 1$, the corresponding evidence does not carry any information. We will call such distribution and evidence empty. The number $m(\emptyset)$ is the probability that the frame does not contain the correct answer. If $m(\emptyset) = 1$, all non-empty subsets are assigned a zero probability. The corresponding evidence is called contradictory. The baseline probability distribution is considered m normalized if $m(\emptyset) = 0$. An arbitrary consistent baseline probability distribution is easy to normalize by going to a "conditional" distribution, for which.

$$m * (X) = \frac{m(X)}{1 - m(\emptyset)},$$

where $X \neq \emptyset$ and $m * (\emptyset) = 0$.

On a binary frame $\Theta = \{y, n\}$ the base distribution of trust (not necessarily normalized) can be given by four numbers.

$$\mu = m(y);\ \nu = m(n);\ \pi = m(y, n)$$

such that

$$\mu + \nu + \pi + \lambda = 1,\ \mu, \nu, \pi, \lambda \geq 0$$

(If this does not lead to confusion, we write $\mu = m(y)$ instead $\mu = m(\{y\})$ and similarly in other cases).

The basic trust distribution on an arbitrary finite frame Θ is associated with the set functions: the measure of confidence *bel* and the measure of likelihood *pl*, which in a sense replace the probability measure. For an arbitrary set $X \subseteq \Theta$, the values $bel(X)$ and $pl(X)$ are given by the following formulas:

$$bel(X) = \sum\nolimits_{Y \subseteq X, Y \neq \emptyset} m(Y),\ pl(X) = \sum\nolimits_{Y \cap X \neq \emptyset} m(Y).$$

It is easy to see that

$$bel(X) + pl(\overline{X}) = 1,$$

Where $\overline{X} = \Theta \backslash X$ — complement of the set X relative to the frame Θ.

Knowing the measure of trust, it is possible to restore the function of the basic distribution of trust:

$$m(X) = \sum\nolimits_{Y \subseteq X} (\ 1)^{|X|-|Y|}(1 + bel(Y) - bel(\Theta)),$$

Where $|S|$ denotes the number of elements in a finite set S. Similarly, the baseline trust distribution function is restored as the likelihood increases.

As an illustration, consider a dice roll situation. Suppose that the die is not completely symmetrical. Our confidence that i will fall out is slightly less than $\frac{1}{6}$, and is equal to

$\frac{1}{6} - \varepsilon$. We introduce the frame $\Theta = \{1, 2, 3, 4, 5, 6\}$ and define the basic distribution of trust on it, assuming

$$m(i) = \frac{1}{6} - \varepsilon \text{ for each } i = 1,2,\ldots,6 \text{ и } m(\Theta) = 6\varepsilon.$$

The only focal sets with such a distribution are one-point subsets and the frame itself. Then

$$bel(i) = \frac{1}{6} - \varepsilon \text{ и } pl(i) = \frac{1}{6} + 6\varepsilon$$

for all $i = 1, 2, ..., 6$

The mathematical expectation of confidence measures and similarity can be used as an estimate of the mathematical expectation of an unknown probability distribution.

Consider a general scheme leading to confidence and likelihood measures and an example of calculating the expected value.

Let $V = \{v_1, ..., v_n\}$ be a set of vulnerabilities with a given probability distribution. Next, let $A = \{a_1, ..., a_k\}$ be a set of attacks. We will assume that the relationship between attacks and vulnerabilities $R \subseteq V \times A$ is given. If v is a vulnerability, then $R(v) \subseteq A$ is a family of attacks related to the vulnerability v. On the set of attacks, we define the basic distribution of trust, assuming

$$m(X) = \sum_{v \in V} \{P(v) | R(v) = X\},$$

where X is an arbitrary set of attacks. Measures of confidence and likelihood are determined by the following formulas:

$$bel(X) = \sum\nolimits_{Y \subseteq X} m(Y);\ pl(X) = \sum\nolimits_{Y \cap X \neq \emptyset} m(Y)$$

Suppose now that each attack a has an associated loss size $L(a)$. Using the Lebesgue-Stieltjes integral, you can specify the upper and lower estimates of the expected losses:

$$E * (L) = \int_{-\infty}^{\infty} r \cdot dbel(\{a | L(a) \leq r\}) = \sum\nolimits_{X \subseteq A} m(X) \cdot \max_{v \in X} L(a);$$

$$E_*(L) = \int_{-\infty}^{\infty} r \cdot dpl(\{a | L(a) \leq r\}) = \sum_{X \subseteq A} m(X) \cdot \min_{v \in X} L(a)$$

Example 1. Let there be five vulnerabilities with probabilities 0.01; 0.02; 0.04; 0.02 respectively. Add to them a "fictitious" zero vulnerability, for which the probability is 0.9.

Next, let there be seven options for an attack, for which the losses are respectively 35000; 28000; 26000; 24000; 20, 000; 15000; 10000.

We add to them a zero attack, in which there are no losses.

Attacks are related to vulnerabilities as follows:

$$R(v_1) = \{a_2, a_3, a_4\};\ R(v_2) = \{a_3, a_5, a_6\};\ R(v_3) = \{a_1, a_3, a_4\};$$
$$R(v_4) = \{a_2, a_6, a_7\};\ R(v_5) = \{a_2, a_4\};\ R(v_0) = \{a_0\};$$

There are six focal sets in the attack space:

$$m(\{a_2, a_3, a_4\}) = 0{,}1;\ m(\{a_3, a_5, a_6\}) = 0{,}2;\ m(\{a_1, a_3, a_4\}) = 0{,}4;$$
$$m(\{a_2, a_6, a_7\}) = 0{,}2;\ m(\{a_2, a_4\}) = 0{,}1;\ m(\{a_0\}) = 0{,}9$$

Further

$$bel(\{a_0, a_4, a_5, a_6, a_7\}) = 0{,}9;\ bel(\{a_0, a_3, a_4, a_5, a_6, a_7\}) = 0{,}92;$$
$$bel(\{a_0, a_2, a_3, a_4, a_5, a_6, a_7\}) = 0{,}96;$$
$$bel(\{a_0, a_1, a_2, a_3, a_4, a_5, a_6, a_7\}) = 1;$$
$$pl(\{a_0\}) = 0{,}9;$$
$$pl(\{a_0, a_7\}) = 0{,}92;$$
$$pl(\{a_0, a_7\}) = 0{,}92;$$
$$pl(\{a_0, a_6, a_7\}) = 0, 94;$$
$$pl(\{a_0, a_4, a_5, a_6, a_7\}) = 1{,}00.$$

The loss function L is a step function with discontinuities at the points

35000; 28000; 26000; 24000; 20000; 15000; 10000.

With this in mind, we find:

$$E * (L) = 2572;\ E_*(L) = 2240$$

If we assume that all non-zero attacks are equally probable, the mathematical expectation of losses will be 2257. Thus, $E * (L)$ and $E_*(L)$ serve as the upper and lower bounds of the mathematical expectation of losses, respectively.

The latter reflects the following general fact. Let us denote the set of probabilistic measures enclosed in the interval from the confidence measure to the likelihood measure as Π, i.e., such that

$$bel(X) \leq P(X) \leq pl(X)$$

for any set of attacks $X \subseteq A$. Then.

$E * (L) = \sup\{E_P(L)|P \in \Pi\};\ E_*(L) = \inf\{E_P(L)|P \in \Pi\}$.

If there are alternative estimates of the probability distribution on a set of vulnerabilities and alternative combinations of vulnerabilities in an attack, a consistent result can be obtained by combining the corresponding Dempster trust distribution functions, as described in the next section.

3 Combining Confidence Building Measures

Trust distribution functions are associated with a static description of the information system. To describe the dynamics of the transfer of trust through the information system, the Dempster evidence combination rule is used.

Let there be some finite frame Θ. Information about the elements of the frame is represented by evidence, the formal expression of which is the basic distribution of trust

m_1. Suppose there is a second piece of evidence, and the baseline distribution of trust m_2 matches it. Dempster's Rule allows one to construct a combined normalized baseline distribution of confidence $m_1 \oplus m_2$ under the assumption that both m_1 and m_2 are normalized and compatible. The last condition means that

$$\sum_{X \cap Y = \emptyset} m_1(X)m_2(Y) \neq 1$$

If the subset $Z \subseteq \Theta$ is not empty, then

$$(m_1 \oplus m_2)(Z) = k \sum_{X \cap Y = Z} m_1(X)m_2(Y),$$

where

$$k = (1 - \sum_{X \cap Y = \emptyset} m_1(X)m_2(Y))^{-1}.$$

Consider, in particular, the combination of basic trust distributions on a binary frame $\Theta = \{y, n\}$. Assuming that m_i, $i = 1, 2$, is given by the values $\mu_i = m_i(y)$, $\nu_i = m_i(n)$, $\mu_i = m_1(y)$, for $\nu = (m_1 \oplus m_2)(n)$, $\mu = (m_1 \oplus m_2)(y)$ we have

$$\mu = 1 - \frac{(1 - \mu_1)(1 - \mu_2)}{1 - (\mu_1\nu_2 + \mu_2\nu_1)}; \quad \nu = 1 - \frac{(1 - \nu_1)(1 - \nu_2)}{1 - (\mu_1\nu_2 + \mu_2\nu_1)}$$

Note that the compatibility condition in the case of a binary frame means that $\mu_1\nu_2 + \mu_2\nu_1 \neq 1$.

The above rule has a fairly transparent interpretation. Suppose there are n experts who are presented with certain evidence related to the assessment of the truth for some statement S. Let us denote by M the set of experts whom evidence E convinced of the statement S truth, and by N - the set of those experts who were convinced by this evidence of the falsity of the statement in question. Let us denote by $\mu = \frac{|M|}{n}$ the fraction of experts who consider the statement S to be true, and by $\nu = \frac{|N|}{n}$ - the fraction of experts who consider it false (here and below, the sign of a module applied to a finite set denotes the number of its elements). It is assumed that the sets M and N do not intersect, so that $\mu + \nu \leq 1$. Thus, the basic trust distribution corresponding to the evidence E is set on the binary frame.

Now let us consider the situation when there are two evidences E_1 and E_2, associated with the assessment of the truth for the statement S. Designations $M_i, N_i, \mu_i, \nu_i, i = 1, 2$, will refer to the evidence with the corresponding index. The evidences are presented to the same set of experts. The multitudes $M_1 \cap N_2$ и $M_2 \cap N_1$ constitute a conflict zone of evidence: the first evidence convinced them of one thing, and the second of the exact opposite. The condition of the compatibility of the evidence means that not all experts have ended up in the conflict zone.

We will consider the evidences to be independent in the sense of the classical conditional probability, i.e., we will assume that the relation holds

$$\frac{|M_1 \cap M_2|}{|M_1|} = \frac{|M_2|}{n}$$

as other similar equalities too.

Let us find the share of experts who were convinced by at least one testimony among experts who did not get into the conflict zone. We get:

$$\mu = \frac{|M_1|+|M_2|-|M_1 \cap M_2|-|M_1 \cap N_2|-|N_1 \cap M_2|}{n - |M_1 \cap N_2|-|N_1 \cap M_2|}$$

whence

$$\mu = 1 - \frac{(1-\mu_1)(1-\mu_2)}{1-(\mu_1\nu_2+\mu_2\nu_1)}$$

follows.

The formula for ν is obtained in a similar way.

It is easy to verify that the operation of combining the trust distribution functions according to Dempster's rule is commutative and associative.

In the future, another variant of combining the trust distribution functions will be used:

$$(m_1 \odot m_2)(Z) = \sum_{X \cap Y = Z} m_1(X)m_2(Y), \quad (1)$$

which differs from the Dempster combination by the absence of a normalization factor. This operation will be called an unnormalized combination of evidences (or trust distribution functions).

4 Possibilistic Assessment of Attack Graph Nodes

In [11], it was proposed to build an attack graph from separate basic blocks to assess the vulnerability of an information network as a whole. Each of the blocks describes the logical connection of parent nodes with their direct descendants. The simplest communication option is a serial connection of nodes: vulnerability *A* creates a condition for exploiting vulnerability *B*, and sequential exploitation of vulnerabilities *A* and *B* allows an attacker to achieve the target state. More complex connections are described through conjunctions and disjunctions. For example, exploiting one of vulnerabilities *A* or *B* allows exploiting vulnerability *C*, etc. Moreover, there may be a dependency between parent nodes, leading to an adjustment in the evaluation of logical formulas.

When using the Bayesian approach, each of the nodes receives a probabilistic estimate in the form (μ, ν), where $\mu, \nu \geq 0$ and $\mu + \nu = 1$. The use of assessments with the use of confidence-building measures makes it possible to obtain a model describing the situation under conditions of greater uncertainty. The Bayesian computing engine provides the transfer of probability over the graph. By combining Dempster evidence, trust transfer can be achieved.

In this section, we describe the trust transfer mechanism using ideas from [9] and [13] as a starting point.

For convenience, we will further denote the elements of a binary frame by a^+ и a^-. Correspondingly, the normalized basic distribution of trust m on a binary frame $\Theta = \{a^+, a^-\}$ is given by the numbers $\mu = m(a^+)$, $\nu = m(a^-)$, and $\pi = m(\Theta) = 1 - \mu - \nu$.

Let A_i, $i = 1, 2, ..., n$, be the set of nodes in the attack graph that are parent to the node B. Let $U(X_1, ..., X_n)$ be a disjunctive normal form describing the logical connection of node A_i with a node B. The trust-likelihood score of a node A_i relies on evidence E_i with a basic trust distribution function m_i on a binary frame $\Theta_i = \{a_i^+, a_i^-\}$. A node A_i estimate is a pair of numbers $w(A_i) = (\mu_i, \nu_i)$, where $w(A_i) = (\mu_i, \nu_i)$. Accordingly, $m_i(\Theta_i) = \pi_i$ и $\mu_i + \nu_i + \pi_i = 1$.

Likewise, the node B score is given by a pair of numbers $w(B) = (\mu_B, \nu_B)$, based on evidence E_B with a trust distribution function m_B on a binary frame $\Theta_B = \{a_B^+, a_B^-\}$

Next, let $I \subseteq \{1, 2, ..., n\}$ be some set of indices. The dependence between evidence with numbers from a set of indices I will be represented by a number $p_I \in [0, 1]$ - the intensity of the evidence connection.

To describe the trust transfer mechanism, it is required to estimate the attack, the target of which is located directly behind the node B on the attack graph.

We start by describing the logical evaluation of a formula for $U(X_1, \ldots, X_n)$,using a triangle

$$T = \{(\mu, \nu)|\mu, \nu \geq 0, \mu + \nu \leq 1\}$$

as a logical scale (matrix, as they say in a similar case in mathematical logic).

Let us compose an auxiliary frame Θ'. Its elements are some words of the form

$$w = z^{\omega} a_1^{\omega_1} a_2^{\omega_2} \ldots a_n^{\omega_n} \tag{2}$$

where $\omega, \omega_1, \ldots, \omega_n \in \{+, -\}$. For a word w of the form (2), we define a binary vector $x = \phi(w) \in \{0, 1\}^n$ as follows:

if $\omega_i = +$, then $x_i = 1$; if $\omega_i = -$, then $x_i = 0$.

A word w of the form (2) is a frame element Θ' if one of the following conditions is satisfied for the vector $x = \phi(w)$:

1) $U(x_1, ..., x_n) = 1$ and $\omega = +$;
2) $U(x_1, ..., x_n) = 0$ and $\omega = -$.

For each $i = 1, 2, ..., n$, we denote by $m_i^{'}$ the cylindrical continuation of the basic distribution of confidence m_i from Θ_i to Θ':

$$m_i^{'}(\{w = z^{\omega} a_1^{\omega_1} a_2^{\omega_2} ... a_n^{\omega_n} \in \Theta' | \omega_i = +\}) = \mu_i;$$
$$m_i^{'}(\{w = z^{\omega} a_1^{\omega_1} a_2^{\omega_2} \ldots a_n^{\omega_n} \in \Theta' | \omega_i = -\}) = \nu_i;$$
$$m_i^{'}(\Theta') = \pi_i.$$

To take into account the relationship between the evidences E_i, $i \in I$, we select in the frame Θ' the subset W_I containing all words $w \in \Theta'$ with $\omega = +$, as well as those words $w \in \Theta'$ with $\omega = -$ that have the following property: if you change all the values x_i in the vector $x = \phi(w)$ to opposite values $1 - x_i$, the value assumed by the formula for U will change from 0 to 1. Let set the basic distribution of trust $m_I^{'}$ on the frame Θ', assuming.

$$m_I^{'}(W_I) = p_I; m_I^{'}(\Theta') = 1 - p_I.$$

We suppose.

$$m' = m'_l \odot m'_1 \odot \ldots \odot m'_n.$$

Let θ be the projection of the frame Θ' onto the binary frame $\Theta = \{z^+, z^-\}$, for which $\theta(w) = z^{\omega}$, if $w = z^{\omega} a_1^{\omega_1} a_2^{\omega_2} \ldots a_n^{\omega_n} \in \Theta'$. Let us denote by m'_{θ} the basic distribution of trust on the frame such that.

$m'_{\theta}(z^{\omega}) = \sum_{\theta(X)=z^{\omega}} m'(X)$; $m'_{\theta}(\Theta) = m'(\Theta')$.

The distribution m'_{θ} may turn out to be, generally speaking, unnormalized. It is its normalization that gives a logical assessment of the formula for U in terms of trust:

$$bel(U) = \frac{m'_{\theta}(z^+)}{m'_{\theta}(z^+) + m'_{\theta}(z^-) + m'_{\theta}(\Theta)};$$

$$pl(U) = 1 - \frac{m'_{\theta}(z^-)}{m'_{\theta}(z^+) + m'_{\theta}(z^-) + m'_{\theta}(\Theta)}$$

As an example, let us consider a situation when an attack through a node B is possible only when the parent nodes A_1 and A_2 are conjunctively connected. That is, the vulnerability corresponding to a node B can be used for an attack only if the attacker has succeeded in exploiting the vulnerabilities corresponding to nodes A_1 and A_2.

Thus, we are considering an example in which $U = X_1 \wedge X_2$. The frame Θ' in this case contains four words:

$$\Theta' = \{z^+ a_1^+ a_2^+, z^- a_1^+ a_2^-, z^- a_1^- a_2^+, z^- a_1^- a_2^-\}$$

Suppose that the link between the nodes A_1 and A_2 expresses a type of assertion: if one of the vulnerabilities was successfully exploited, then the other is likely to be exploited. Let the number $q \in [0, 1]$ indicate the strength of this connection. Formally, to represent this connection, we separate the subset into Θ'

$$\Theta_{12} = \{z^+ a_1^+ a_2^+, z^- a_1^- a_2^-\}$$

and define the distribution of trust on the frame Θ', assuming $m'_{12}(\{z^+ a_1^+ a_2^+, z^- a_1^- a_2^-\}) = q$; $m'_{12}(\Theta') = p$,

Where $p = 1 - q$.

The value $q = 0$ means the absence of connection (the independence of the testimonial confidence-likelihood estimates for the vulnerabilities under consideration). The value $q = 1$ means the presence of the maximum possible connection (if one vulnerability is used, the other will also be used).

Cylindrical extensions of the confidence distribution functions m_1 and m_2 are given by the following formulas:

$$m'_1(\{z^+ a_1^+ a_2^+, z^- a_1^+ a_2^-\}) = \mu_1;\ m'_1(\{z^- a_1^- a_2^+, z^- a_1^- a_2^-\}) = \nu_1;\ m'_1(\Theta') = \pi_1;$$

$$m'_2(\{z^+ a_1^+ a_2^+, z^- a_1^- a_2^+\}) = \mu_2;\ m'_2(\{z^- a_1^+ a_2^-, z^- a_1^- a_2^-\}) = \nu_2;\ m'_2(\Theta') = \pi_2$$

Now we calculate $m' = m'_{12} \odot m'_1 \odot m'_2$.
It is easy to see that

$$m'(Z) = \sum_{W \cap X \cap Y = Z} m_{12}(W) m_1(X) m_2(Y)$$

With this keeping in mind, we have:

$$m'(\Theta') = p\pi_1\pi_2;$$
$$m'(\{z^+a_1^+a_2^+, z^-a_1^+a_2^-\}) = p\mu_1\pi_2;\ m'(\{z^-a_1^-a_2^+, z^-a_1^-a_2^-\}) = p\nu_1\pi_2;$$
$$m'(\{z^+a_1^+a_2^+, z^-a_1^-a_2^+\}) = p\pi_1\mu_2;\ m'(\{z^-a_1^+a_2^-, z^-a_1^-a_2^-\}) = p\pi_1\nu_2;$$
$$m'(\{z^+a_1^+a_2^+, z^-a_1^-a_2^-\}) = q\pi_1\pi_2;$$
$$m'(\{z^+a_1^+a_2^+\}) = q\mu_1\mu_2 + p\mu_1\mu_2 + q\pi_1\mu_2 + q\mu_1\pi_2;$$
$$m'(\{z^-a_1^+a_2^-\}) = p\mu_1\nu_2;$$
$$m'(\{z^-a_1^-a_2^+\}) = p\nu_1\mu_2;$$
$$m'(\{z^-a_1^-a_2^-\}) = q\nu_1\pi_2 + q\pi_1\nu_2 + p\nu_1\nu_2 + q\nu_1\nu_2;$$
$$m'(\emptyset) = q\mu_1\nu_2 + q\nu_1\mu_2.$$

Now we find the marginal trust distribution m'_θ on the frame $\Theta = \{z^+, z^-\}$. Since $\theta(X) = z^+$ at $X = \{z^+a_1^+a_2^+\}$,
then

$$m'_\theta(z^+) = m'(\{z^+a_1^+a_2^+\}) = q\mu_1\mu_2 + p\mu_1\mu_2 + q\pi_1\mu_2 + q\mu_1\pi_2 = \mu_1\mu_2 + q(\pi_1\mu_2 + \mu_1\pi_2)$$

Further $\theta(X) = z^-$, at

$$X = \{z^-a_1^-a_2^+, z^-a_1^-a_2^-\};\ X = \{z^-a_1^+a_2^-, z^-a_1^-a_2^-\};$$
$$X = \{z^-a_1^+a_2^-\};\ X = \{z^-a_1^-a_2^+\};\ X = \{z^-a_1^-a_2^-\}.$$

It means, that

$$\begin{aligned} m'_\theta(z^-) &= p\nu_1\pi_2 + p\pi_1\nu_2 + p\mu_1\nu_2 + p\nu_1\mu_2 + q\nu_1\pi_2 + q\pi_1\nu_2 + p\nu_1\nu_2 + q\nu_1\nu_2 \\ &= \nu_1\pi_2 + \pi_1\nu_2 + p\mu_1\nu_2 + p\nu_1\mu_2 + \nu_1\nu_2 \end{aligned}$$

Finally, $\theta(X) = \Theta$, at

$$X = \{z^+a_1^+a_2^+, z^-a_1^+a_2^-\};\ X = \{z^+a_1^+a_2^+, z^-a_1^-a_2^+\};$$
$$X = \{z^+a_1^+a_2^+, z^-a_1^-a_2^-\};\ X = \Theta'$$

and $\theta(X) = \emptyset$ at $X = \emptyset$.
Hence,

$$m'_\theta(\emptyset) = q(\mu_1\nu_2 + \nu_1\mu_2);$$
$$m'_\theta(\Theta) = p\pi_1\pi_2 + q\pi_1\pi_2 + p\mu_1\pi_2 + p\pi_1\mu_2 = \pi_1\pi_2 + p(\mu_1\pi_2 + \pi_1\mu);$$

Suppose that

$$G = \mu_1\pi_2 + \pi_1\mu_2;\ H = \mu_1\nu_2 + \nu_1\mu_2$$

Let us denote by $m*$ the normalized trust distribution function associated with $m_{\theta}^{'}$. Then the normalizing factor can be represented as $(1 - qH)^{-1}$, and, therefore,

$$m * \left(z^{+}\right) = \frac{\mu_1\mu_2 + qG}{1 - qH}; \quad m * \left(z^{-}\right) = 1 - \frac{(1 - \nu_1)(1 - \nu_2)}{1 - qH} \tag{3}$$

They are formulas (3) that give a logical estimate of U in formulas:

$$bel(U) = \frac{\mu_1\mu_2 + qG}{1 - qH}; \; pl(U) = \frac{(1 - \nu_1)(1 - \nu_2)}{1 - qH}.$$

Assessment of an attack through a block of nodes A_1, A_2, B is obtained by combining the trust distribution functions $m*$ and m_B. Thus, the possibility of an attack $\mathrm{A} : (A_1 \wedge A_2) \rightarrow B$ is assessed as follows:

$$bel(\mathrm{A}) = \frac{\mu * \mu_B}{1 - (\mu * \nu_B + \nu * \mu_B)};$$

$$pl(\mathrm{A}) = \frac{(1 - \nu*)(1 - \nu_B)}{1 - (\mu * \nu_B + \nu * \mu_B)},$$

Where $\mu* = bel(U)$, $\nu* = 1 - pl(U)$, while μ_B и ν_B are the estimates for the node B.

Example 2. Suppose that interval estimates are obtained for the probabilities of exploiting vulnerabilities in nodes A_1, A_2, B:

$$0, 1 \leq \Pr(A_1) \leq 0, 15; \; 0, 2 \leq \Pr(A_2) \leq 0, 3; \; 0, 15 \leq \Pr(B) \leq 0, 25$$

Then, in terms of the trust functions, we can assume that

$$\mu_1 = 0, 1; \; \nu_1 = 0, 85; \; \mu_2 = 0, 2; \; \nu_2 = 0, 7; \; \mu_B = 0, 15; \; \nu_B = 0, 75.$$

Then.

$$\pi_1 = 0, 05; \; \pi_2 = 0, 15.$$

Next,

$$G = 0, 1 \cdot 0, 15 + 0, 05 \cdot 0, 2 = 0, 025;$$
$$H = 0, 1 \cdot 0, 7 + 0, 85 \cdot 0, 2 = 0, 24$$

Let $q = 0, 5$. Let us find an estimate of the conjunction $U = A_1 \wedge A_2$ in terms of trust:

$$bel(U) = \frac{0, 1 \cdot 0, 2 + 0, 5 \cdot 0, 025}{1 - 0, 5 \cdot 0, 24} = 0, 037;$$

$$pl(U) = \frac{0, 15 \cdot 0, 3}{1 - 0, 5 \cdot 0, 24} = 0, 051.$$

Thus, we can assume that, considering the dependence of the nodes A_1 and A_2, the probability of an attack through a conjunction $A_1 \wedge A_2$ lies in the range from 0.037 to 0.051. Now we can find the trust scores for the attack A:

$$bel(\mathrm{A}) = \frac{0,037 \cdot 0,15}{1 - (0,037 \cdot 0,75 + 0,949 \cdot 0,15)} = 0,0067;$$

$$pl(\mathrm{A}) = \frac{0,051 \cdot 0,25}{1 - (0,037 \cdot 0,75 + 0,949 \cdot 0,15)} = 0,0153.$$

The probability of an attack is 0.0063 if, instead of interval estimates for vulnerabilities A_1, A_2 and B, we use average values and do not consider the relationship between A_1 and A_2.

5 Conclusion

Disclosure of the features of quantitative assessment of functional risks showed that methods based on confidence and likelihood measures can be applied in banks and other enterprises affected by risks, including those associated with their functioning in a competitive environment. The proposed methods for quantitative risk assessment play the role of forming the boundaries of the loss probability. Analysis of the mathematical expectation of losses and its reflection in information databases contributes to the formation of the reasonable strategy for enterprises in the face of increasing threats of financial losses. It is important to note that confidence and likelihood measures are used to logically assess the likelihood of combined loss-causing attacks. The quantitative risk assessment, based on the proposed soft computing methodology contributes to effective risk management, probabilistic calculations including.

References

1. U.S. Supplement to the 2014 Global Economic Crime Survey, PricewaterhouseCoopers LLP. http://www.pwc.com/gx/en/economic-crime-survey/. Accessed 30 Apr 2021
2. Positive Technologies Homepage. https://www.ptsecurity.com/ru-ru/research/analytics/cybersecurity-threatscape-2020-q1/#id2. Accessed 30 Apr 2021
3. Positive Technologies news. https://www.ptsecurity.com/ru-ru/about/news/positive-technologies-v-kazhdoj-gosudarstvennoj-i-promyshlennoj-kompanii-obnaruzheny-vredonosnoe-po-i-narusheniya-reglamentov-kiberbezopasnosti/. Accessed 30 Apr 2021
4. Chockalingam, S., Hadžiosmanović, D., Pieters, W., Teixeira, A., van Gelder, P.: Integrated safety and security risk assessment methods: a survey of key characteristics and applications. In: Havarneanu, G., Setola, R., Nassopoulos, H., Wolthusen, S. (eds.) CRITIS 2016. LNCS, vol. 10242, pp. 50–62. Springer, Cham (2017). https://doi.org/10.1007/978-3-319-71368-7_5
5. Naik, N., Jenkins, P., Kerby, B., Sloane, J., Yang, L.: Fuzzy logic aided intelligent threat detection in cisco adaptive security appliance 5500 series firewalls. In: 2018 IEEE International Conference on Fuzzy Systems (FUZZ-IEEE), pp. 1–8. IEEE (2018)
6. Guzairov, M.B., Vulfin, A.M., Kartak, V.M., Kirillova, A.D., Mironov, K.V.: Comparative analysis of cognitive modeling algorithms in assessing information security risks analysis of the Russian Academy of Sciences. In: Proceedings of the Institute for Systems Analysis of the Russian Academy of Sciences, vol. 69. №. 4, pp. 62–69. (2019). https://doi.org/10.14357/20790279190408

7. Srivastava, R.P., Shafer, G.R.: Belief-function formulas for audit risk. In: Auditors Report, vol. 17, no. 1, pp. 8–12 (1993)
8. Fukukawa, H., Mock, T.J., Srivastava, R.P.: Assessing the risk of fraud at olympus and identifying an effective audit plan. In: The Japanese Accounting Review, vol. 4, pp. 1–25 (2014)
9. Srivastava, R.P., Mock, T.J., Turner, J.L.: Analytical formulas for risk assessment for a class of problems where risk depends on three interrelated variables. Int. J. Approx. Reason. **45**(1), 123–151 (2007)
10. Frigault, M., Wang, L., Singhal, A., Jajodia, S.: Measuring network security using dynamic Bayesian network. In: Proceedings of the 4th ACM workshop on Quality of protection, pp. 23–30 (2008)
11. Frigault, M., Wang, L.: Measuring network security using Bayesian network-based attack graphs. In: 2008 32nd Annual IEEE International Computer Software and Applications Conference, pp. 698–703. IEEE (2008)
12. Sabaliauskaite, G., Mathur, A.: Aligning cyber-physical system safety and security. In: Cardin, M.-A., Krob, D., Lui, P.C., Tan, Y.H., Wood, K. (eds.) Complex Systems Design & Management Asia, pp. 41–53. Springer, Cham (2015). https://doi.org/10.1007/978-3-319-12544-2_4
13. Bulyga, R.P., Gisin, V.B., Volkova, E.S.: Combining interrelated bodies of evidence in audit risk assessment. In: 2016 XIX IEEE International Conference on Soft Computing and Measurements (SCM), pp. 453–455. IEEE (2016)

Algorithms and Devices for Non-coherent Digital Processing of the Multi-level Amplitude Differential Phase-Shift Keyed Signals

Oleg Chernoyarov[1,2,3](✉), Alexey Glushkov[4], Kaung Myat San[1], Vladimir Litvinenko[5], and Yuliya Litvinenko[5]

[1] National Research University "MPEI", Krasnokazarmennaya str. 14, 111250 Moscow, Russia
chernoyarovov@mpei.ru
[2] National Research Tomsk State University, Lenin Avenue 36, 634050 Tomsk, Russia
[3] Maikop State Technological University, Pervomaiskaya str. 191, 385000 Maikop, Russia
[4] Zhukovsky-Gagarin Air Force Academy, Marshal Nedelin Street 133, 394052 Voronezh, Russia
[5] Voronezh State Technical University, Moscow Avenue 14, 394026 Voronezh, Russia

Abstract. The possibilities of implementing non-coherent demodulation of joint multi-level amplitude modulated and differential binary or quadrature phase-shift keyed signals are considered. As basic processing units, digital non-coherent demodulators of differential binary phase shift keyed and differential quadrature phase shift keyed signal are used. One of their major advantage is that it is not necessary to apply phase locking. The algorithms for processing signals and making decisions on the received code combinations are presented. The noise immunity of the introduced demodulators is evaluated both theoretically and experimentally. A comparative estimation of the energy efficiency of the considered signals is carried out. It is demonstrated that the proposed algorithms allow separate processing of the signal amplitude and phase on the receiving (in the modulator) and transmitting (in the demodulator) ends of the communication line. #COMESYSO1120.

Keywords: Phase shift keying · Signal demodulation · Non-coherent processing · Bit-error probability · Noise immunity · Statistical simulation

1 Introduction

In modern discrete data-transmission system, multi-level amplitude- and phase-shift keyed signals are widely used, providing high frequency and energy efficiency [1]. These include pulse amplitude modulation (PAM), phase-shift keyed (PSK), amplitude- and phase-shift keyed (APSK), quadrature amplitude modulation (QAM), and quadrature phase-shift keyed (QPSK) signals [2–5]. In most cases, the coherent demodulation of the received signals is used [3]. But, it requires phase synchronization of the reference and received signals, while the demodulator should provide high noise immunity and protection against "backward operation" effect.

R. Silhavy et al. (Eds.): CoMeSySo 2021, LNNS 232, pp. 784–795, 2021.
https://doi.org/10.1007/978-3-030-90318-3_61

Non-coherent processing of PSK signals is devoid of the specified disadvantages and can be implemented when using differential phase-shift keying (DPSK). Below, digital non-coherent demodulators of differential binary phase-shift keyed (DBPSK) [6, 7] and differential quadrature phase shift-keyed (DQPSK) [8] signals are introduced. The operation of these demodulators is based on the fast digital signal processing algorithms [9] that are supplemented by the procedures for estimating the amplitude of the received elements. This makes it possible to implement demodulators of amplitude differential binary phase-shift keyed (ADBPSK) and amplitude differential quadrature phase shift keyed (ADQPSK) signals. For example, an eight-level ADQPSK signal is used in directly modulated lasers (DMLs) to obtain a complex optical signal modulated by both amplitude and phase simultaneously [10, 11]. ADBPSK signals are being widely used in mobile radio communication [2].

2 The Signal Models

The received PSK signal can be represented as follows [3–5]

$$s(t) = S\cos(2\pi f_0 t + \psi(t) + \psi_0). \tag{1}$$

In (1), S is the amplitude, f_0 is the carrier frequency, ψ_0 is the initial phase, while $\psi(t) = a_k\pi$, if $a_k = 0$ or 1, for the DBPSK signal and $\psi(t) = a_k\pi/2$, if $a_k = 0, 1, 2$ or 3, for the DQPSK signal. Here a_k is the modulating signal of the k-th PSK signal element.

The signal (1) is sampled by an analog-to-digital converter (ADC) with the sampling rate $f_S = 4f_0$ as it is shown in Fig. 1, where the samples s_{1i}, s_{2i}, s_{3i} and s_{4i} obtained within the current i-th period $T_0 = 1/f_0$ are marked by dots. It is presupposed that the information symbol duration is equal to $T = NT_0$, where N is the number of periods.

The modulating signals a_k are defined by both the transmitted phase-shift coding information symbols b_k and the preceding values a_{k-1}. So, for the DBPSK signal with the number of phase shift levels $M_1 = 2$, one gets.

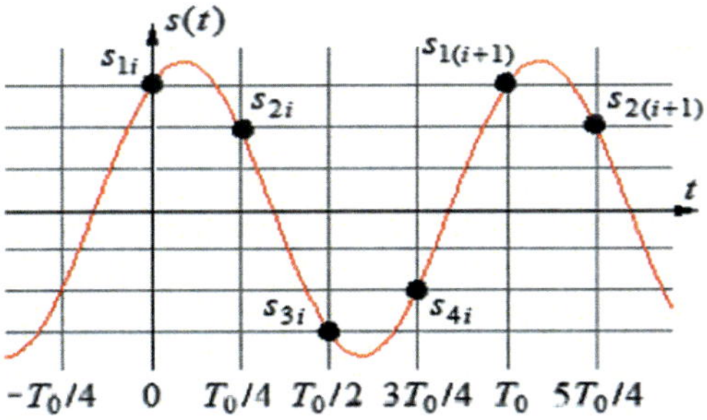

Fig. 1. The PSK signal quantization timing.

$$a_k = \mathrm{mod}(b_k + a_{k-1}, 2), \tag{2}$$

while for the DQPSK signal with the number of phase shift levels $M_1 = 4$, it can be written

$$a_k = \mathrm{mod}(b_k + a_{k-1}, 4). \tag{3}$$

In (2), (3), $\mathrm{mod}(x, y)$ is the integer value of x modulo y.

With amplitude shift keying of the DBPSK signal elements by the transmitted information symbols $c_k = \overline{1, M_2}$ (where M_2 is the number of amplitude levels), the S magnitude takes the discrete values.

$S = (2c_k - 1)U$.

Here U is the minimum value of the signal element amplitude.

In Fig. 2, there are drawn the spatial diagrams [5] or the ADBPSK signal "constellations", while in Fig. 3 – the similar ADQPSK signal "constellations". It is worth noting that they are the versions of the constellations with Gray labeling (Gray-APSK) [12]. Unlike the PSK signal diagrams [5], in these cases the angles relative to the X axis counterclockwise represent the phase shift of the current DPSK signal element relative to the previous one.

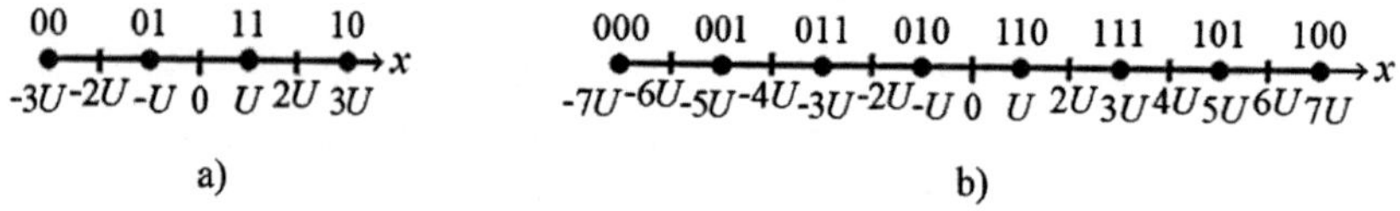

Fig. 2. The spatial ADBPSK signal diagrams: a) M1 = 2, M2 = 2; b) M1 = 2, M2 = 4.

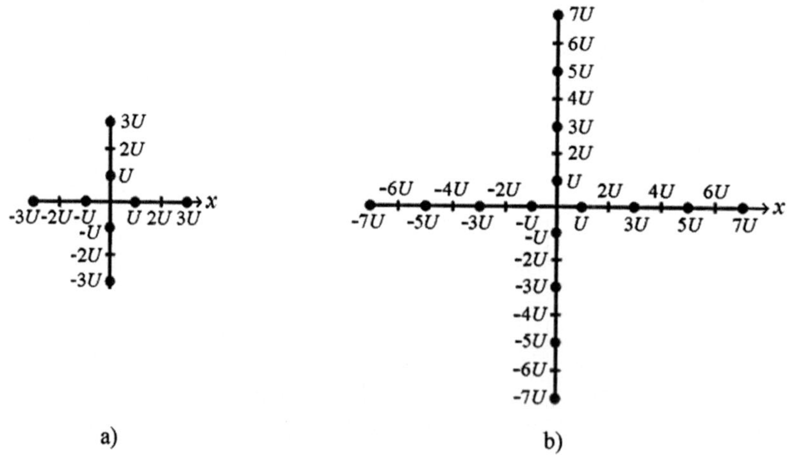

Fig. 3. The spatial ADQPSK signal diagrams: a) M1 = 4, M2 = 2; b) M1 = 4, M2 = 4.

The signal levels with the even values $2U$, $4U$, etc. specify the threshold values with which the amplitudes of the received symbols are compared to make a decision about the information symbol c_k. As a result, the number of levels for the ADBPSK signal is $M = 2M_2$, while for the ADQPSK signal it is $M = 4M_2$, respectively. Each signal value (the "constellation" point) is encoded by the transmitted binary code, for example, the Gray code, the combinations of which are shown in Fig. 2.

3 The Demodulation Algorithms

In the basic digital signal processing algorithm [9], the sums of differences of the odd and even samples for the current i-th period are calculated for the last N received periods:

$$y_{0i} = \sum_{j=0}^{N-1} \left(s_{1(i-j)} - s_{3(i-j)}\right),\; y_{1i} = \sum_{j=0}^{N-1} \left(s_{2(i-j)} - s_{4(i-j)}\right). \qquad (4)$$

In the DBPSK signal demodulator [6], the values

$$\begin{cases} z_{0i} = \sqrt{\left(y_{0i} + y_{0(i-N)}\right)^2 + \left(y_{1i} + y_{1(i-N)}\right)^2}, \\ z_{1i} = \sqrt{\left(y_{0i} - y_{0(i-N)}\right)^2 + \left(y_{1i} - y_{1(i-N)}\right)^2} \end{cases}$$

are determined by which a decision on the incoming information symbol b_k is made upon the completion of the reception of the k th signal element (when the clock synchronization pulse of the demodulator is generated) in accordance with the rule.

$$z_i = z_{0i} - z_{1i} <> 0.$$

Namely, if $z_i > 0$, then the phases of the current and previous DBPSK signal elements coincide (that is, the phase shift is 0) and the binary information symbol $b_k = 0$ is received, otherwise the specified phases are opposite (that is, the phase shift is π) and the symbol $b_k = 1$ is accepted.

In the four-level DQPSK signal demodulator [8], based on the responses y_{0i} and y_{1i} (4) of the basic algorithm, the following values are calculated:

$$\begin{cases} z_{0i} = y_{0i}y_{0(i-N)} + y_{1i}y_{1(i-N)} + y_{1i}y_{0(i-N)} - y_{0i}y_{1(i-N)}, \\ z_{1i} = y_{0i}y_{0(i-N)} + y_{1i}y_{1(i-N)} - y_{1i}y_{0(i-N)} + y_{0i}y_{1(i-N)}. \end{cases} \qquad (5)$$

Then, by the signs of z_{0i}, z_{1i} (5), upon the completion of the reception of the i-th DQPSK signal element, a decision on the incoming information symbol b_k (the phase shift between adjacent elements is equal to $b_k\pi/2$) is made according to rules presented in Table 1.

Using the responses y_{0i}, y_{1i} (4) of the basic algorithm, the signal amplitudes can be determined at the end of the reception of the k-th signal element as follows.

$$U_{ik} = \sqrt{y_{0i}^2 + y_{1i}^2}.$$

They are compared with the thresholds $U_{Tk} = 2c_k U$, and the estimate of the received value c_k is selected from the closest of them. In the result of the demodulation one gets the values b_k and c_k that determine the incoming information symbol.

Table 1. The rules of the four-level DQPSK signal demodulation according to the responses of the basic processing algorithm.

z_0	z_1	b_k	Binary code
$z_0 > 0$	$z_1 > 0$	0	00
$z_0 < 0$	$z_1 > 0$	1	01
$z_0 < 0$	$z_1 < 0$	2	11
$z_0 > 0$	$z_1 < 0$	3	10

4 The Probabilities of Erroneous Decisions

The noise immunity of the considered digital differential phase shift keyed signal demodulators is studied in [7], and there it is stated that when receiving the DBPSK signal the demodulation error probability is

$$P_{DPSK2} = \exp\left(-h^2\right)/2, \tag{6}$$

while when receiving the DQPSK signal it is

$$P_{DPSK4} = 1 - (1 - P_B)^2. \tag{7}$$

In (7), the bit-error probability P_B is determined by the expression.
$P_B = Q(\alpha, \beta) - \exp(-h^2) I_0\left(h^2/\sqrt{2}\right)/2$, $\alpha = \sqrt{2} h \sin(\pi/8)$, $\beta = \sqrt{2} h \cos(\pi/8)$, $Q(\alpha, \beta) = \int_\beta^\infty t \exp\left[-\left(\alpha^2 + t^2\right)/2\right] I_0(\alpha t)\, dt$ is the Marcum function [13], $I_0(x)$ is the zero-order modified Bessel function [14], while

$$h^2 = NS^2/\sigma_n^2 \tag{8}$$

is the signal-to-noise ratio (SNR) and σ_n^2 is the dispersion (mean power) of the Gaussian noise samples at the demodulator input.

It follows from (6) and (7) that the error probability when demodulating the DBPSK signal decreases rapidly when h^2 (or symbol amplitude S) increases. With the minimum amplitude $S = U$, the basic SNR is

$$h_0^2 = NU^2/\sigma_n^2 \tag{9}$$

and if h_0^2 corresponds to 8 dB (i.e., $20 \log_2 h_0 = 8$), then, according to (6), for the binary differential phase shift keying, one gets $P_{DPSK2} \approx 10^{-3}$. For the next signal level $S = 3U$, the value h^2 (8) increases by 9.5 dB, and the error probability drops to almost zero. The similar results are obtained for the DQPSK signal demodulation when applying (7). Thus, the noise immunity of the DBPSK signal demodulation is determined by the value h_0^2 taking into account that the minimum element amplitude is probable. Under the equiprobable information symbols, the average error probability is

$$P_{DPSK2av} = P_{DPSK2}/M_2 \tag{10}$$

or

$$P_{DPSK4av} = P_{DPSK4}/M_2 \tag{11}$$

when demodulating the DBPSK or DQPSK signals, respectively.

In the case when an additive Gaussian interference appears in a transmission channel, the probability density of the signal amplitude estimate takes the form [7].

$$w(V) = \exp\left[-(V - V_k)^2/2\sigma_V^2\right]/\sqrt{2\pi}\sigma_V,$$

where $\sigma_V^2 = 2N\sigma_n^2$ is the dispersion and $V_k = 2NS_k$ is the mathematical expectation of the estimated amplitude.

When estimating the amplitude V of the symbol in the k-th position, an error occurs, if the inequality $|V - V_k| > 2NU$ holds [7]. Then, for the error probability when demodulating the amplitude of the signal element, one obtains [5]

$$P_{AM} = 2(M_2 - 1)\Phi\left(\sqrt{2}h_0\right)/M_2, \tag{12}$$

where $\Phi(x) = \int\limits_x^\infty \exp\left(-z^2/2\right) dz/\sqrt{2\pi}$ while the SNR h_0^2 corresponding to the minimum amplitude U of the signal element at the input of the demodulator is defined from (9).

A demodulation error occurs when the position of the amplitude of the received signal or the signal phase shift are incorrectly determined. The specified error probabilities can be found according to (12) and (10) or (11). Therefore, the total error probability is determined as follows

$$P_{err} = 1 - (1 - P_{AM})(1 - P_{DPSKav}) \approx P_{AM} + P_{DPSKav}, \tag{13}$$

where P_{DPSKav} is one of the probabilities (10) or (11).

The multi-level ADBPSK signal elements have different amplitudes, and when they are equally probable to occur, under $M \geq 4$ the root-mean-square value of the amplitude is equal to.

$$S_{rms} = \sqrt{\frac{2}{M}\sum\nolimits_{k=0}^{M/2-1}(2k+1)^2U^2} = U\sqrt{\left(M^2 - 1\right)/3}.$$

Then, as it follows from (8) the root-mean-square SNR is

$$h_{rms} = h_0\sqrt{\left(M^2 - 1\right)/3}. \tag{14}$$

The multi-level signals have different information rates, therefore, for their correct comparison, the equivalent SNR h_2 for binary elements is used that is equal to [4, 5]

$$h_2 = h_{rms}/\sqrt{\log_2 M}. \tag{15}$$

Thus, taking into account (14), when using the ADBPSK signals one gets

$$h_2 = \mu h_0, \tag{16}$$

where $\mu = \sqrt{(M^2 - 1)/3\log_2 M}$.

Similarly, when using the ADQPSK signals the number of the amplitude levels is $M_2 = M/4$ and under $M \geq 8$ the root-mean-square value of the amplitude is determined as.

$S_{rms} = \sqrt{\frac{4}{M}\sum_{k=0}^{M/4-1}(2k+1)^2U^2} = U\sqrt{(M^2-4)/12}$,

while

$$h_{rms} = h_0\sqrt{(M^2-4)/12} \quad (17)$$

and it should be taken $\mu = \sqrt{(M^2 - 4)/12\log_2 M}$ in (16).

5 The Noise Immunity When Using the ADBPSK Signals

In Fig. 4a, for the ADBPSK signal ($M_1 = 2$), the solid line 1 shows the dependence of the error probability (12) under $M_2 = 2$ ($M = 4$), the dashed line demonstrates the dependence of the error probability (6), and the solid line 2 introduces the dependence of the total error probability (13) upon the basic SNR (9) expressed in decibels. The estimates of the error probabilities P_{AM}, P_{DPSK2}, P_{err} obtained by means of a statistical simulation are marked by circles, black triangles and crosses, respectively. In Fig. 4b, the similar results are presented for the case when $M_2 = 4$ ($M = 8$) and it can be seen that the experimental data are well described by the corresponding theoretical formulas in a wide range of SNR values.

In Fig. 5, for this signal, the dependences of the error probabilities P_{err} upon the SNR for binary element h_2 (15) are drawn by solid lines. The bottom curve is calculated using the formula (13) under $M_2 = 2$, while the upper curve – under $M_2 = 4$. By dash-dotted line, the dependence (7) of the demodulation error probability in the case of the non-coherent DQPSK signal receiving is plotted. As it can be seen, when $M_2 = 2$ the ADBPSK signal loses in terms of the reduced noise immunity to the four-level DPSK signal by 2 dB. This is due to the use of two values U and $3U$ of the symbols amplitude. In turn, while the eight-level ADBPSK signal with $M_2 = 4$ is used the equivalent noise immunity is additionally deteriorated by 2.5 dB. However, at the same time, the communications rate increases 1.5 times.

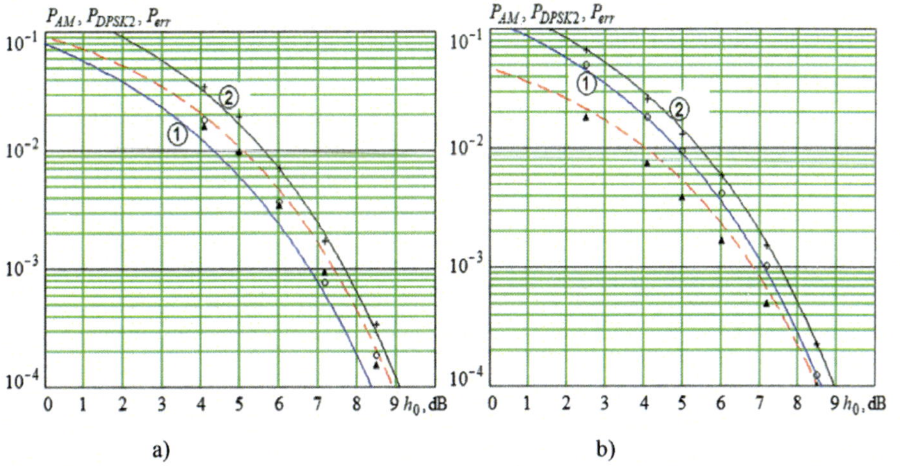

Fig. 4. The theoretical and experimental demodulation error probabilities when receiving the ADBPSK signal: a) M1 = 2, M2 = 2; b) M1 = 2, M2 = 4.

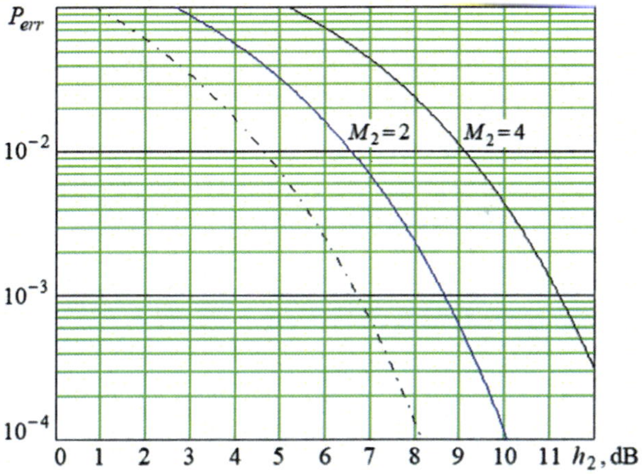

Fig. 5. The demodulation error probabilities when receiving the ADBPSK and DQPSK signals.

6 The Noise Immunity When Using the ADQPSK Signals

In Fig. 6, for the ADQPSK signal, the theoretical and experimental dependences of the error probabilities P_{DPSK4} and P_{AM} upon the SNR h_0 (9) expressed in decibels are shown. The upper solid line is calculated using the formula (7) under $M = 4$, while the blue circles mark the corresponding values obtained by means of a statistical simulation. These curve and circles describe the noise immunity of the DPSK4 (without AM) signal demodulation. Below, the dashed lines are calculated according to (7) and describe the dependences of the error probability when demodulating the DQPSK signal. By black triangles and squares, the corresponding simulation results are shown for the cases when $M = 8$ and $M = 16$. At the bottom of Fig. 6, the dependences of the error probability

(12) are drawn under $M = 8$ (dashed line) and $M = 16$ (solid line), while by pluses and crosses the experimental values of this probability are presented. It follows from Fig. 6 that a good coincidence of the theoretical and simulation results can be confirmed.

In Fig. 7, by solid lines, the theoretical dependences of the error probability of the ADQPSK signal demodulation P_{err} (13) upon the SNR h_2 (15) are shown for the cases when $M = 8$ and $M = 16$. By crosses (under $M = 8$) and squares (under $M = 16$) the values are marked of the demodulation error probability obtained by means of statistical simulations. The dashed and dotted lines calculated by the formulas [4].

$$P_{errQAM} \frac{1}{\log_2 \sqrt{M}}\left(1 - \frac{1}{\sqrt{M}}\right)\left(\sqrt{\frac{3\log_2\sqrt{M}}{M-1}}\sqrt{2}h_2\right),$$

$$P_{errMDPSK}\left(h_2\sqrt{2\log_2 M}\,\sin\left(\pi/M\sqrt{2}\right)\right)$$

describe the demodulation error probabilities when using the quadrature amplitude modulation (QAM) signal and the M-level differential phase-shift keyed (MDPSK) signal, respectively. Here the $\Phi(x)$ function is defined as in (12).

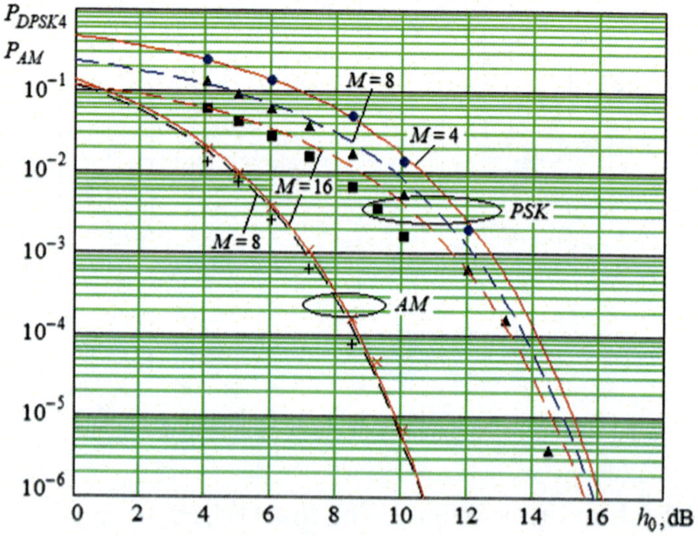

Fig. 6. The theoretical and experimental demodulation error probabilities when receiving the ADQPSK signal.

It follows from the presented results that an increase in the number of signal levels leads to a decrease in noise immunity, but this increases the information transfer rate. The ADQPSK signals are inferior in noise immunity to both the QAM signals by 8–9 dB and the MDPSK signals by 2–3 dB in the case when $M = 16$. However, they allow non-coherent demodulation.

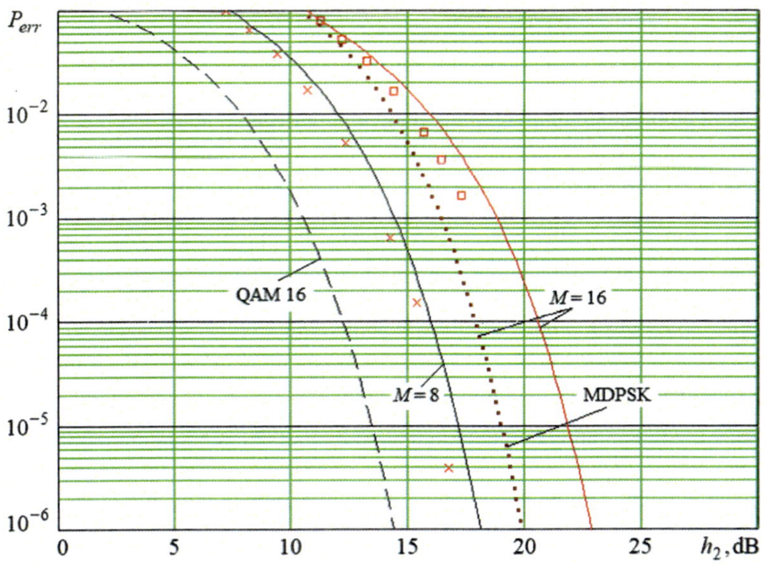

Fig. 7. The error probabilities when demodulating the ADQPSK, QAM and MDPSK signals.

7 The Energy Efficiency

The energy efficiency of multi-level signals is estimated by the value [1, 4]

$$\beta = 1/h_2^2, \tag{18}$$

where the root-mean-square SNR per binary symbol h_2^2 is determined by (15) or (16) and is calculated for a specified error probability P_0 as a solution to equation of the form [1].

$$P_{err}\left(h_2^2\right) = P_0.$$

In Fig. 8, by solid lines, there are shows the dependences (18) of the energy efficiency β expressed in decibels upon the error probability under $M = 4$ (Fig. 8a) and $M = 16$ (Fig. 8b) when demodulating the ADBPSK and ADQPSK signals. For comparison, by dashed and dotted lines, there are drawn the similar dependences when demodulating the DPSK and QAM signals.

In the case when $M = 4$, the ADQPSK ($M_2 = 1$), DPSK (or DQPSK) and QAM (one amplitude gradation) signals have the same constellations and their energy efficiency is almost the same, while the ADBPSK signal efficiency under $M_2 = 2$ (two amplitude gradations) is about 3 dB less. When using the sixteen-level signals (Fig. 8b), the energy efficiency decreases and the considered ADBPSK signals are inferior in energy efficiency to the QAM signals by 10 dB, while the ADQPSK signals – by 8.5 dB, respectively.

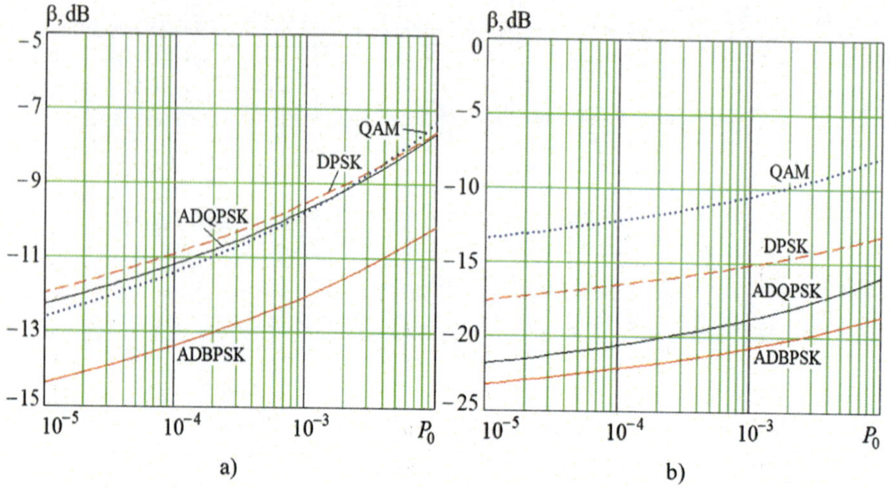

Fig. 8. The energy efficiency when using the ADBPSK, ADQPSK, DPSK and QAM signals: a) M = 4; b) M = 16.

8 Conclusion

The amplitude differential binary phase shift keyed signals are proposed, their processing can be effectively implemented based on the digital non-coherent demodulators of the binary (BPSK) and the four-level (DPSK) differential phase shift keyed signals with an additional modulation of the symbol amplitude. At the same information rate, the ADPSK signals are significantly inferior to QAM and DPSK signals in terms of a noise immunity and an energy efficiency. Their advantage is separate modulation of amplitude and phase, which is advisable, for example, in optical communication lines. In addition, under the non-coherent demodulation, phase synchronization is not required contrarily to the case of, for example, QAM signal demodulation. It greatly simplifies the requirements to the receiving equipment. As an element base for the implementation of the introduced digital algorithms, it is advisable to use field-programmable gate arrays.

Acknowledgements. This work was financially supported by the Ministry of Education and Science of the Russian Federation (research project No. FSWF-2020–0022).

References

1. Burachenko, D.A., Savischenko, N.V.: Bandwidth and limiting frequency-energy efficiency in systems with two-dimensional M-QAM, M-PSK and M-APSK signals (in Russian). Inf. Manag. Syst. **1**, 64–73 (2013)
2. Stuberg, G.L.: Principles of Mobile Communication. Springer, New York (2017)
3. Feher, K.: Wireless Digital Communications. Modulation and Spread Spectrum Applications. Prentice-Hall, Hoboken (1995)
4. Sklar, B.: Digital Communications: Fundamentals and Applications, 2nd edn. Prentice Hall, Hoboken (2017)

5. Proakis, J., Salehi, M.: Digital Communications. McGraw-Hill, New York (2007)
6. Chernoyarov, O.V., Glushkov, A.N., Litvinenko, V.P., Matveev, B.V., Makarov, A.A.: Algorithms and devices for noncoherent digital radio signal processing. Eng. Lett. **28**(4), 1238–1248 (2020)
7. Chernoyarov, O.V., Glushkov, A.N., Litvinenko, V.P., Litvinenko, Y.V., Matveev, B.V.: Fast digital algorithms for the noncoherent demodulation of the differential phase-shift keyed binary signals. Int. Rev. Electr. Eng. **13**(4), 334–341 (2018)
8. Chernoyarov, O.V., Glushkov, A.N., Litvinenko, V.P., Litvinenko, Y.V., Makarov, A.A., Matveev, B.V.: Digital noncoherent demodulator of four-position differential phase shift keyed signals. Eng. Lett. **28**(2), 306–311 (2020)
9. Glushkov, A.N., Litvinenko, V.P., Matveev, B.V., Chernoyarov, O.V.: Basic algorithm for the noncoherent digital processing of the narrowband radio signals. Appl. Math. Sci. **9**(95), 4727–4735 (2015)
10. Prat, J., Velasquez, J.C., Tabares, J.: Direct PSK-ASK modulation for coherent udWDM. In: 2019 21st International Conference on Transparent Optical Networks, ICTON, pp. 1–4. IEEE, France (2019)
11. Elfiqi, A.E., Morra, A.E., Khallaf, H.S., Shalaby, H.M.H., Hranilovic, S.: Two-level MPPM-MDPSK modulation for free-space optical channels. In: 2018 IEEE Global Communications Conference, GLOBECOM, pp. 1–6. IEEE, United Arab Emirates (2018)
12. Liu, Z., Xie, Q., Peng, K., Yang, Z.: APSK constellation with gray mapping. IEEE Commun. Lett. **15**(12), 1271–1273 (2011)
13. Meyer, D.P.: Radar Target Detection: Handbook of Theory and Practice. Academic Press, New York (1973)
14. Abramowitz, M., Stegun, I.A.: Handbook of mathematical functions with formulas, graphs and mathematical tables. National Bureau of Standards. Appl. Math. Ser. **55**, USA (1964)

Realization of an Interdependent Architecture for Sustainable Agro-Economic Industry 4.0 for Post-Covid-19 Pandemic Economic Era

Vijay Kumar Pandey[1(✉)] and Vikash Kumar Singh[2(✉)]

[1] Department of Computer Science, Indira Gandhi National Tribal University, Amarkantaka, Madhya Pradesh, India

[2] Faculty of Department of Computer Science, Indira Gandhi National Tribal University, Amarkantaka, Madhya Pradesh, India

Abstract. The impact of the COVID-19 pandemic accelerates the Industry 4.0 implementation in the various walk of business processes. The tightly and loosely coupled integration among machines and the processes through intelligent software systems and pervasive computing promises a higher level of automation based on real-time prediction-based control systems. Therefore, Industry 4.0 plays a vital role in maintaining a balance between social distancing compliance and production demand. The proposed system introduces a novel model where the agricultural process is considered automated in adherence to Industry 4.0 standards over rural areas complying with the situation of the COVID pandemic. The proposed model offers a sophisticated operation of sustainability and introduces a new analytical model for Industry 4.0. The simulated outcome shows that the proposed system offers a cost-effective solution towards achieving better sustainability in agricultural innovation.

Keywords: Sustainability · Industry 4.0 · COVID 19 · Agriculture

1 Introduction

The notion of automation starts with no human intervention to run a business process from a small scale to a large-scale manufacturing unit [1]. In the last decade, tThe technical fraternity has witnessed a tremendous innovation into various promising technologies that include i) Sensor Technology (ST) [2], ii) Embedded Systems [3], (ES), iii) Systems on Chip (SOC) [4], iv) Cloud Computing (CC) [5], v) Software Defined Network (SDN) [6], vi) Intelligent Computing Models (ICM) [7] – Artificial Intelligence [8], Machine Learning [9] & Blockchain [10] (AIMLB), vii) Robotics [11], viii) Internet of Things(IoT) [12], ix) Digital Twin [13] and x) Converged Internet with 5G Communication Standard(CI-5GCS) [14]. These technologies are enablers to realize the vision of Industry 4.0 or Industry X.0 in a true sense. The Goal of Industry 4.0 realization also complements the path of sustainable development (SD). The very basic objective of SD is to fulfill the adequate supply of the demand of the current generation without creating any challenges for future generations to come.

R. Silhavy et al. (Eds.): CoMeSySo 2021, LNNS 232, pp. 796–808, 2021.
https://doi.org/10.1007/978-3-030-90318-3_62

An explicit example is an invention of excessive production. The use of undegradable plastic fulfills the current generation's requirements but creates a substantial multi-dimensional hazard for the generations to come. Since the last century, the rapid development in the production processes of various products and population growth has exposed the environmental and human ecosystem to the Agro-food crisis, management of production waste, energy imbalances & water supply, among various other issues. Thus, it is an open research problem to synchronize sustainability and industry 4.0 vision together. One of the research studies of Gad-el-Hak et al. [15] provides a centrality flow control with different research dimensions, as in the Fig. 1.

The centrality flow logic in the figure-1 is interdependent on the various keys of a set, including the elements ∈ {**BD**: BigData, **ES**: Embedded System, **SM**: Sustainable Manufacturing, **IR**: Industrial Research, **SD**: Sustainable Development, **CPS**: Cyber-Physical System, **MI**: Manufacturing Industry, **LC**: Life-Cycle, **PL**: Product Life-Cycle, **SC**: Supply Chain, **DM**: Digital Manufacturing.}. This paper proposes a holistic architectural design of a model by collaborating and synchronizing the agricultural production process as a seed-to-plate with social distancing factors, sustainability, and industry 4.0 as "**SPSuSI4.0**".

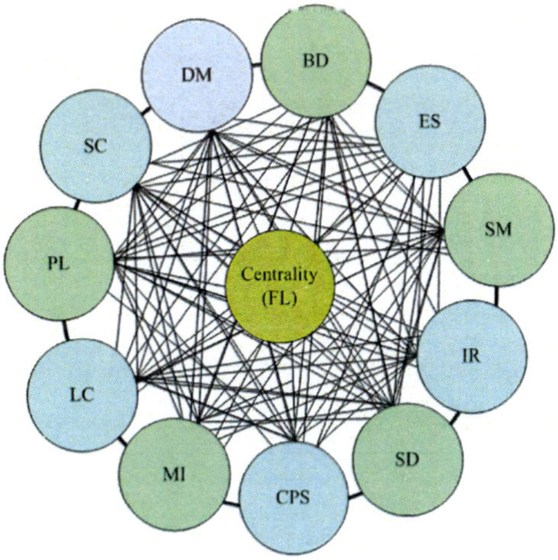

Fig. 1. Centrality flow control with connected elements for sustainable I.4.0. [15]

Unlike smart cities, progressive development is taking place for smart villages, and a sustainable Agri model as a seed to plate can be achieved with the adoption of Industry 4.0 technological advancements. There exists an ISO standard for smart cities with sustainable development as ISO 37122:2019. The collective efforts by the researchers and the real-time implementation of sustainable agriculture automation using Industry 4.0 will provide technical indicators simultaneously with the evolution. The majority of the research has focused on the urban regions, whereas a large population yet resides in the rural sections. Therefore, the automated Agri-system complements the smart village vision, brings a large economical boost, and improvises the well-being of the rural populations [16].

The rest of the paper is organized as Sect. 2, a literature review, Sect. 3, a proposed model and its architectural design, Sect. 4 results and analysis, and Sect. 5 conclusion and future research direction.

2 Review of Literature

Although many technical discussions and articles are in the way of the paving of Industry 4.0 in the research community, there is no one generally accepted standard definition of Industry 4.0 [17]. Rather it is a promised to automate the processes involved into various models of activities using advanced technologies. The penetration of the smart city is much higher than the smart villages and, in this direction, initiation by the European Union (EU) is witnessed recently in the year 2017 [18]. The post COVID19 economic era requires a new dimension of collaboration of the smart and intelligence network-enabled valuable systems to meet the suitable consumer quality of service and experience [19]. The balance of the urban and rural population growth over some time with the quality-of-life development is quite an imbalance, as shown in Fig. 2.

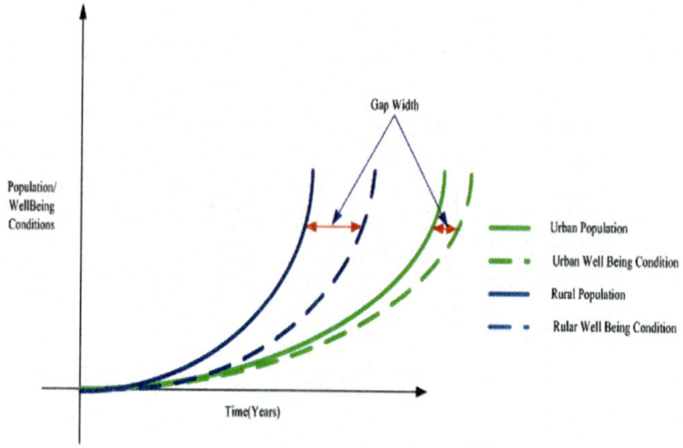

Fig. 2. Imbalance of urban-rural population their quality of life

It is indicated that the gap between the proportion of urban population growth and provisioning of adequate smart technologies and hence improvement into urban population well-being conditions is faster than the rural environment. Moreover, this gap motivates researchers, government agents, and industrialists to invest their respective efforts towards the research and development of incubation and implementation of the technology towards the Agro-economy. The following work has been carried out in this direction.

The issues of the correlation between the factors among 1) fossil fuel-to-environment 2) social to economic for achieving a pathway for sustainable-energy development as a model namely:" Smart-Rular" to promote renewable energy-based efficient smart grid system by (Poggi et al. 2017) to achieve a non-zero energy equilibrium at corporation level [20]. The logical discussion concerning the questions like whether the rural-automation with sustainability is all about is the requirement of the rural growth or collaborate the technology with the sustainability for rural growth, are well in the work of (Visvizi, Lytras, and Mudri (2019)) [21]. Apart from this, there are various studies carried out considering the rural area, e.g., Moschen et al. [22], Huovilla et al. [23], vanGenvelt et al. [24], and Sutriadi [25]. These studies have considered the adoption of various smart approaches with a focus on sustainable development towards agriculture. However, the existing approaches are just like a constructive guideline towards moving agriculture to a modernized version. However, they are potentially associated with various loopholes that should be addressed in order to make way for productive model formulation.

3 Key Research Gap Finding

The following are the key research gap found during the study of the literature:

- Many of the research studies are being carried out towards smart cities or urban areas, but a smaller number of studies are found for the rural developments suitable for the post-COVID-19 era economy for smart Agro-system to complement smart villages' economies.
- Much of the research towards the smart Agro-system and Smart villages is contextual to a region or country that is not applicable as it is for another context and regions; therefore, a standard universal architectural model is an open research problem.

4 Proposed Model

The statistics show that the world population is increasing by 3.5% year irrespective of dead rates, calamities, epidemics, and pandemics. The growing population is directly correlated to the land use for residential, industrial, sports, amusement, and government and private establishments for various services. As a result of additional land demand, the coverage of the agricultural land is declining/shrinking. Therefore, the widening gap between the yield and demand of food -pro leads to rising prices. The economic disparity and imbalance supply-demand system will seriously bring a home war-like situation. The recent episode of the Covid-19 pandemic has introduced a new social behavior called "social distancing." As a result, the entire processes of human society-related activities, including the way school &college, industrial units, offices, functions, etc., have to be operated within the social distancing norms, is a natural demand of research today. The correlation of social distancing, sustainability, and Industry 4.0 requires a new dimension of architecture. Therefore, the orientation of the proposed model is for the social distance - sustainable automation from seed to plate using I4.0 is quite significant for present and future endeavors, as shown in Fig. 3.

The typical modeling of seed to plate (Sp) is a case of supply-chain optimization, whereas sustainability (SuS) is a problem of resource optimization such that the factors that cause harm to the environment are minimized. The consumption of natural resources

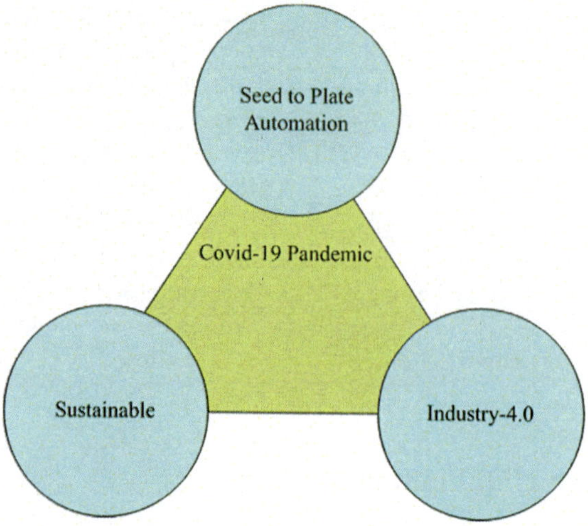

Fig. 3. Orientation diagram for the sustainable manufacturing using Industry 4.0 (SMI4.0)

is also minimized. However, the consideration of I4.0 is the problem of placement and collaboration of technologies. Therefore, the overall objective of the SPSuSI4.0 is to produce an optimal model that suites a practical case study of a system for Seed-to-Plate.

4.1 Modeling Social Distance Compliance Cultivation Process

The process of cultivation, transit and distribution of Agri related products involves many activities. Ac represents the activities related to cultivation, whereas At and Ad represent activities related to transit and distribution. The distributed and small piece of landforms a G(V, E) network as in Fig. 4.

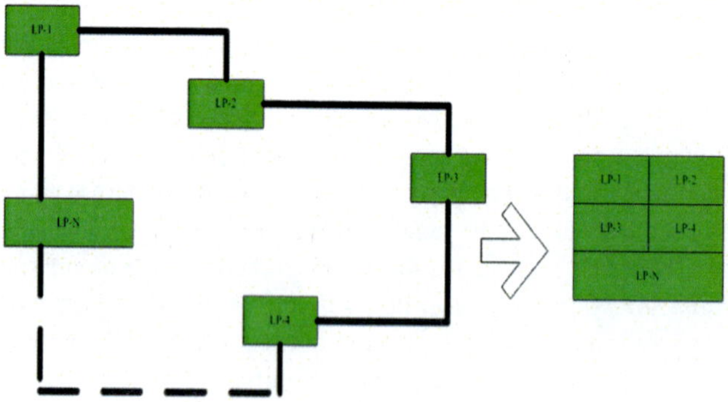

Fig. 4. Network of distributed small pieces of land.

Where V represents land pieces, and E represents the connecting roads or pathways. The sustainability factors like consumption of fuel(F), emission of CO_2 is directly proportional to the degree of distribution of $\forall$ LPi $\in$ {LP}. The convergence of distributed land system with cooperative and consolidated form minimizes the consumption of F and emission of CO_2, and not only that, the implementation of cyber-physical systems under I4.0 for all cultivation activities is realizable and feasible with most minor human interventions. Thus social distance compliance becomes possible. Therefore, the objective function for the social distance compliance in synchronization with SuS and I4.0 is as in Eq. 1.

$$\text{Sdcul} \leftarrow (\text{SuS},\ \text{I4.0},\ \text{G(LP)}) \tag{1}$$

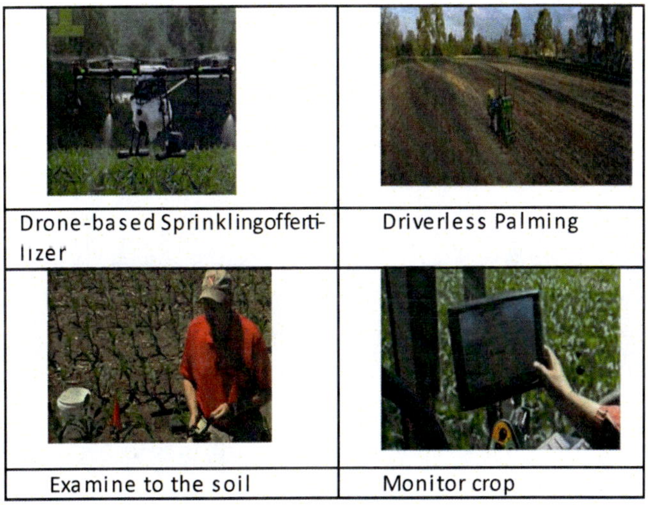

4.2 Modelling of Sustainability

In order to offer better sustainability in Agriculture 4.0, the proposed system uses the Original Equipment Manufacturer (OEM) associated with a seeds processing company, chemical fertilizer, financial institutions, and government institutions. It is also necessary to include dealers of the distributors. From a theoretical viewpoint, agriculture can only be rendered sustainable if it includes diverse technologies for eliminating various obstructions towards healthy crops and cultivation. It should also integrate crops and livestock using various sensing technologies in farming to cover up input supply and production. Further post-harvesting, processing, and distribution will include agents, transporters, and processers with final yield.

4.3 Modelling of Industry 4.0

Figure 5 highlights the proposed model of industry 4.0 used in improving the sustainability of agriculture. The model consists of mainly four different blocks of operation. The first block associates with the adoption of Industry 4.0, which uses IoT sensors, big data, analytics, cloud computing, and robotics. The second block consists of origin, production, soil quality, place of transformation, transport temperature, and packaging material. The third block is about production, transformation, packaging, delivery, and selling. The fourth block is about the agri-food business model. All these blocks are connected to an IT solution supported by the different application and website variants.

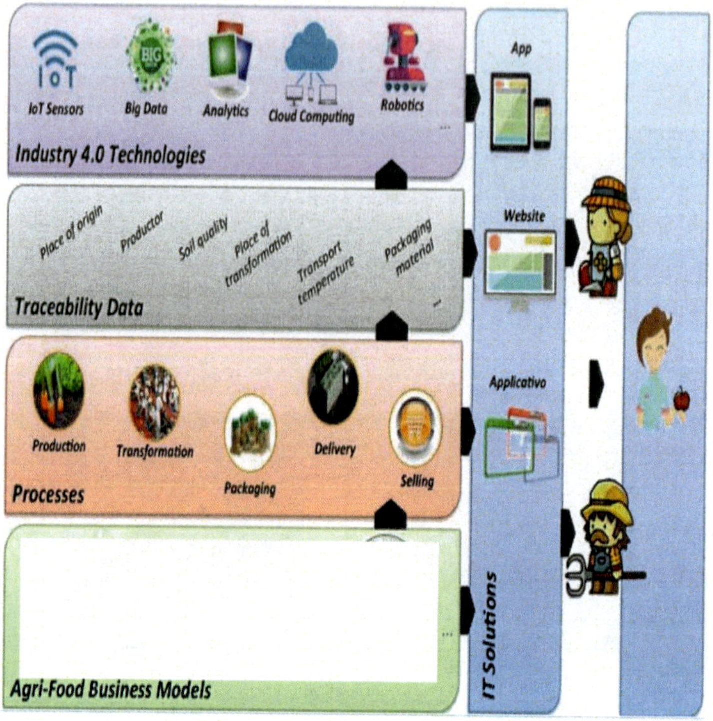

Fig. 5. Proposed model of industry 4.0 in agriculture

4.4 Collaborative Architecture of Social Distancing- Sustainability and Industry 4.0

Figure 6 showcases a mechanism of collaborative architecture of the proposed system, which offers social distancing while incorporating sustainability and applying industry 4.0 in agriculture over rural areas. The first block of *agriculture control* is meant for performing complete control over the production of the cultivation area. The complete operation of this block depends upon the input from real-time decisions obtained from sensing devices. The second important block of operation is that of renewable energy generation systems that extract various forms of renewable energy to facilitate crop cultivation. The third important block of operation is a sensor board crop management device that extracts vital statistics from crops using sensors. This block also integrates with artificial intelligence-based operations, and various other software-based platforms are also connected in this process.

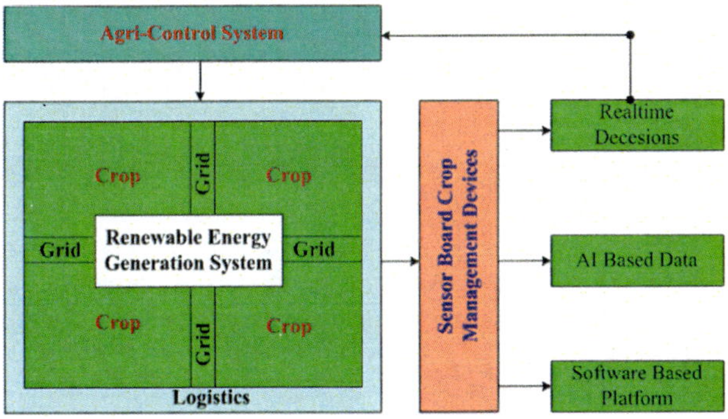

Fig. 6. Model for collaborative architecture

5 Results and Analysis

The result of the proposed system is mainly intended to assess the degree of sustainability factors obtained in the proposed system. The analysis of the proposed study is carried out using multiple performance parameters viz. energy, environment, economy, health hazard, housing, and Safety. The comparative analysis is carried out concerning the existing system where the agricultural aspect has been improvised using automation.

5.1 Analysis of Energy

Figure 7 highlights the comparative analysis to show that energy consumption for the proposed system is significantly less in contrast to the existing system where automation is not used in agriculture. Adopting Industry 4.0 ensures a correct determination of the demanded energy, and so it offers usage of specific energy enabling better energy conservation.

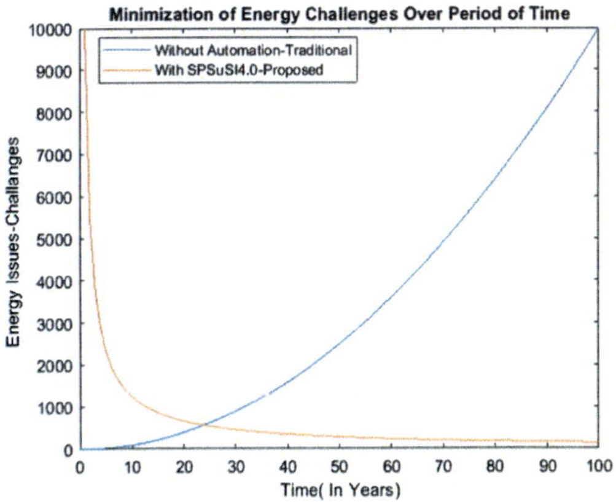

Fig. 7. Comparative analysis of energy

5.2 Analysis of Environment

Figure 8 highlights the comparative analysis of the environment, representing the adverse effect that the proposed system can have on the environment. A closer look into the outcome shows that without adoption of automation approach is incapable of identifying the loopholes that lead to a more significant number of challenges in the environment. However, adopting analytics in the proposed system can quickly determine the detrimental factor and perform effective decision-making using machine learning to find the best solution. This causes the proposed system to offer a highly reduced degree of environmental challenges.

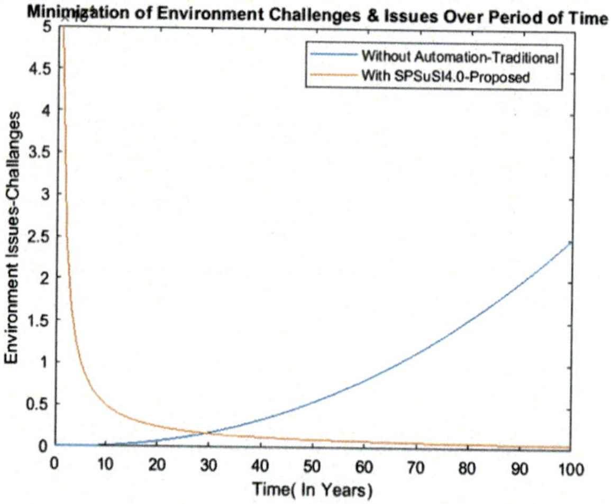

Fig. 8. Comparative analysis of environment

5.3 Analysis of Economy

Figure 9 highlights that the proposed system offers significantly reduced economic challenges in contrast to the existing traditional system. The adoption of cost-effective sensing tools and wireless components is powered by high-end cloud computing resources that do not lead to any form of economic challenges in the proposed system. Therefore, the proposed system offers cost-effective crop cultivation, which can be practically adopted in the existing system, which is in adherence to industry 4.0 norms.

5.4 Analysis of Health Hazard

Figure 10 showcases that the proposed system offers significantly less degree of health hazard. The prime reason for health hazards is to consider adopting any tools or machineries that could negatively affect the environment and health. There is no such adoption when agriculture is carried out considering simplified sensor and computational-based tools. More adoption towards tools can be adopted to reduce possibilities of any

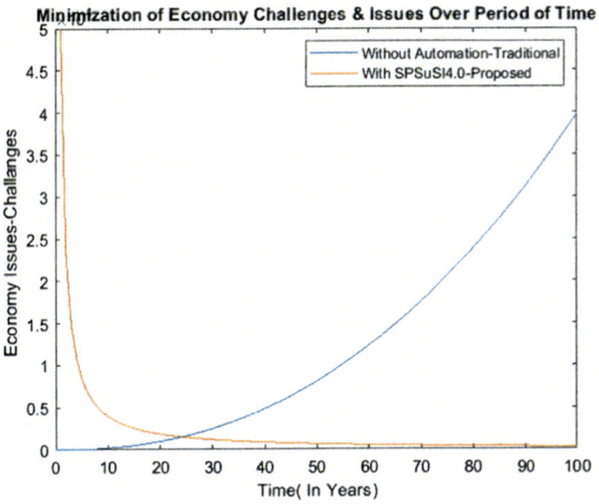

Fig. 9. Comparative analysis of economy

factors that could lead to an adverse condition in health. With artificial intelligence, adverse health conditions can be identified, and better solutions can be evolved using machine learning. Therefore, the adoption of an automation-based approach offers better agriculture outcomes.

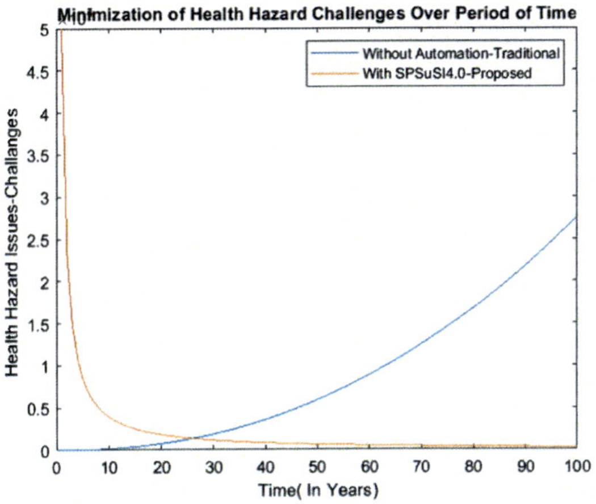

Fig. 10. Analysis of health hazard

5.5 Analysis of Housing

Figure 11 showcases that the proposed system offers no challenges over housing. Adopting wireless sensing leads to an effective analysis of the cultivation area, and it can consistently evolve to find the better location of cultivation, excluding the housing location. Hence, a better solution can always be extracted using data analytics, and hence, agriculture 4.0 has a less significant effect on housing.

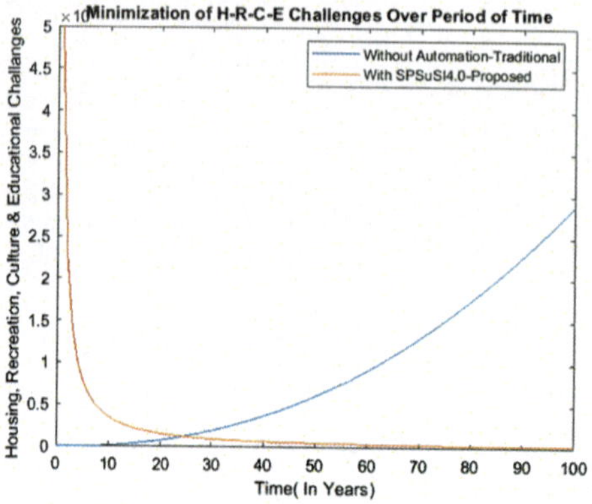

Fig. 11. Analysis of housing

5.6 Safety

A closer look into the proposed system showcase that the proposed system does not offer any form of safety-based challenges, as seen in Fig. 12. Environment and physical information collected by various sensing devices are seamlessly transmitted towards IT solutions where safety hazards are assessed using a predictive algorithm. Hence, there is a lesser probability of evolution of any safety hazard in the proposed system.

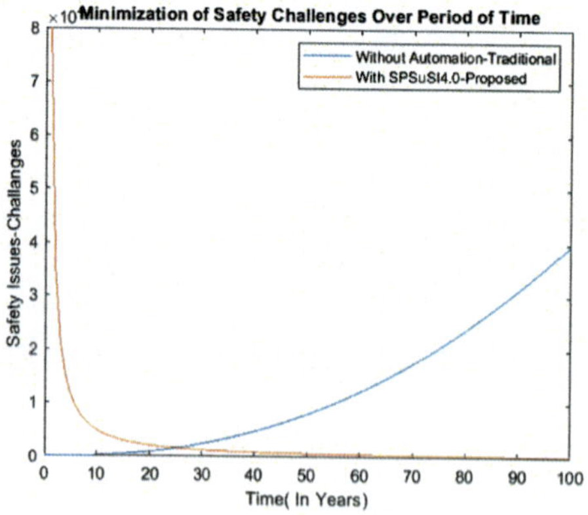

Fig. 12. Analysis of safety

6 Conclusion

This paper discusses a novel model where agriculture is considered a topic to be automated using Industry 4.0 standards. The study considers a case study of rural area cultivation where the pandemic situation has been considered. Hence, considering social distancing norms, the proposed system implements a model in adherence with Industry 4.0 to find out the proposed system offers a cost-effective solution in contrast to the existing system.

References

1. Parasuraman, R., Sheridan, T.B., Wickens, C.D.: A model for types and levels of human interaction with automation. IEEE Trans. Syst. Man Cybern. Part Syst. Hum. **30**(3), 286–297 (2000). https://doi.org/10.1109/3468.844354
2. Borghetti, M., Cantù, E., Sardini, E., Serpelloni, M.: Future sensors for smart objects by printing technologies in industry 4.0 scenario. Energies **13**(22), 5916 (2020)
3. Galkin, P., Umiarov, R., Grigorieva, O.: Design embedded system testbench based on FPGA and microcontrollers for TATU smart lab as education component of industry 4.0. In: 2019 IEEE 2nd Ukraine Conference on Electrical and Computer Engineering (UKRCON), Lviv, Ukraine, pp. 628–633 (2019). https://doi.org/10.1109/UKRCON.2019.8879996
4. Urbina, M., Astarloa, A., LáZaro, J., Bidarte, U., Villalta, I., Rodriguez, M.: Cyber-physical production system gateway based on a programmable SoC platform. IEEE Access **5**, 20408–20417 (2017). https://doi.org/10.1109/ACCESS.2017.2757048
5. Sokolov, B., Ivanov, D., Dolgui, A. (eds.): Scheduling in Industry 4.0 and Cloud Manufacturing. ISORMS, vol. 289. Springer, Cham (2020). https://doi.org/10.1007/978-3-030-43177-8
6. Tsuchiya, A., Fraile, F., Koshijima, I., Ortiz, A., Poler, R.: Soft-ware defined networking firewall for industry 4.0 manufacturing systems. J. Ind. Eng. Manag. (JIEM) **11**(2), 318–333 (2018)
7. Sanchez, M., Exposito, E., Aguilar, J.: Autonomic computing in manufacturing process coordination in industry 4.0 context. J. Ind. Inf. Integr. **19**, 100159 (2020)
8. Canito, A., Mota, D., Marreiros, G., Corchado, J.M., Martins, C.: Contextual adaptative interfaces for industry 4.0. In: Rodríguez González, S., et al. (eds.) Distributed Computing and Artificial Intelligence, Special Sessions, 17th International Conference. DCAI 2020. Advances in Intelligent Systems and Computing, vol. 1242, pp. 149–157. Springer, Cham (2021). https://doi.org/10.1007/978-3-030-53829-3_14
9. Weber, C., Hirmer, P., Reimann, P.: A model management platform for industry 4.0 – enabling management of machine learning models in manufacturing environments. In: Abramowicz, W., Klein, G. (eds.) Business Information Systems. BIS 2020. Lecture Notes in Business Information Processing, vol. 389, pp. 403–417. Springer, Cham (2020). https://doi.org/10.1007/978-3-030-53337-3_30
10. Leng, J., et al.: Blockchain-empowered sustainable manufacturing and product life-cycle management in industry 4.0: a survey. Renew. Sustain. Energ. Rev. **132**, 110112 (2020)
11. Kliestik, T., Nica, E., Musa, H., Poliak, M., Mihai, E.-A.: Net-worked, smart, and responsive devices in industry 4.0 manufacturing systems. Econ. Manag. Finan. Mark. **15**(3), 23–29 (2020)
12. Routray, S.K., Javali, A, Ghosh, A.D., Sarangi, S.: An outlook of narrowband IoT for industry 4.0. In: 2020 Second International Conference on Inventive Research in Computing Applications (ICIRCA), vol. 13, pp. 923–926. IEEE (2020)

13. Lu, Y., Liu, C., Kevin, I., Wang, K., Huang, H., Xu, X.: Digital twin-driven smart manufacturing: connotation, reference model, applications and research issues. Robot. Comput. Integr. Manufact. **61**, 101837 (2020)
14. Oyekanlu, E.A., et al.: A review of recent advances in automated guided vehicle tech-nologies: integration challenges and research areas for 5G-based smart manufacturing applications. IEEE Access **8**, 202312–202353 (2020). https://doi.org/10.1109/ACCESS.2020.3035729
15. G-Hak, M., Pollard, A., Bonnet, J-P.: Flow Control: Fundamentals and Practices. Springer Science & Business Media, vol. 53 (2003)
16. Nair, A.V., Nellippallil, A.B., Allen, J.K., Mistree, F.: Speculating a successful and relevant global development enterprise in the year 2035. Mississippi State Univ., Starkville, MN, USA, Tech. Rep. NSF/ASME Design Essay Competition (2019)
17. Lu, Y.: Industry 4.0: a survey on technologies, applications and open research issues. J. Ind. Inf. Integr. **6**, 1–10 (2017)
18. Klenova, T.V., Ivanov, A.S., Koneva, D.A.: Development of rural areas by means of smart village concept. In: Popkova, E.G., Sergi, B.S. (eds.) ISC 2020. LNNS, vol. 155, pp. 998–1006. Springer, Cham (2021). https://doi.org/10.1007/978-3-030-59126-7_110
19. Rowan, N.J., Galanakis, C.M.: Unlocking challenges and opportunities presented by COVID-19 pandemic for cross-cutting disruption in agri-food and green deal innovations: Quo Vadis?. Sci. Total Environ. 141362 (2020)
20. Poggi, F., Firmino, A., Amado, M.: SMART RURAL: a model for planning net-zero energy balance at municipal level. Energy Procedia **122**, 56–61 (2017)
21. Visvizi, A., Lytras, M.D., Mudri, G.: Smart Villages in the EU and Beyond. Emerald Publishing Limited, Bingley, U.K. (2019)
22. Moschen, S.A., Macke, J., Bebber, S., Da Silva, M.B.C.: Sustainable development of communities: ISO 37120 and UN goals. Int. J. Sustain. High. Educ. **20**(5), 887_900 (2019)
23. Huovila, A., Bosch, P., Airaksinen, M.: Comparative analysis of standardized indicators for smart sustainable cities: what indicatorsand standards to use and when?' Cities **89**, 141-153 (2019)
24. van Gevelt, T., et al.: Achieving universal energy access and rural development through smart villages. Energy Sustain. Develop. **43**, 139-142 (2018)
25. Sutriadi, R.: De_ning smart city, smart region, smart village, and technopolisas an innovative concept in indonesia's urban and regional development themes to reach sustainability. In: Proceedings IOP Conference Series Earth Environment Science, vol. 202, 012047 (2018)

Development of Text-to-Sign Language Translator for Bengali Sign Language

Arifa Ashrafi[1,2], Victor Sergeevich Mokhnachev[1,2](✉), Yuriy Nikolaevich Philippovich[1,2], and Tsilenko Lyubov Petrovna[1,2]

[1] Moscow Polytechnic University, Moscow 107023, Russian Federation
gagashaggy@inbox.ru
[2] Springer Heidelberg, Tiergartenstr. 17, 69121 Heidelberg, Germany

Abstract. The article focuses on the development of a text-to-sign language translator for Bengali sign language. The task includes problem statements, existing approaches, scopes for a better outcome in the future, proposed approaches, and recommendations for future work. The static and dynamic signs that have been used for advancing the translator are original and reviewed by the sign language experts for their accuracy. The existing open-source code aiming to create a text to sign language translator is being modified and improved to match the proposed methodology and further development. #COMESYSO1120.

Keywords: Bengali Sign Language · Text-to-Sign · Sign Language Dictionary · Static and dynamic signs · Computer vision · Machine translation · Russian Sign Language

1 Introduction

Bengali is currently the sixth most spoken language, and fifth most-spoken native language in the world according to the world factbook [1, 2]. It is the most widely spoken language of Bangladesh and the second most widely spoken language of India, after Hindi. With approximately 228 million native speakers and another 37 million as second language speakers [3, 4], Hearing Impairment is widely spread in both India, and Bangladesh. It occurs when there is a problem with or damage to one or more parts of the ear, and the degree of hearing impairment can vary widely from person to person [5]. An official study from 2004 showed that about 13 million people are suffering from variable degrees of hearing loss of which 3 million are suffering from severe to profound hearing loss leading to disability [6, 7]. Right now, the number of hearing impaired people has increased in Bangladesh because of constant sound pollution, aging, and various other reasons. To cope with this situation it is highly necessary to ensure modern technologies for hearing impaired people. Our research aims to create text-to-sign language translators for hearing impaired individuals. The project is divided into two parts - sign-to-text language translator, and text-to-sign language translator. This article outlines the development of a text-to-sign language translator only focusing on Bengali Sign Language.

R. Silhavy et al. (Eds.): CoMeSySo 2021, LNNS 232, pp. 809–819, 2021.
https://doi.org/10.1007/978-3-030-90318-3_63

2 Bengali Text-to-Sign Language Translator

Sign language, or hand-speak, has become a popular method of communicating for those who cannot verbally speak [8]. The usage of Sign Languages is officially supported by many countries, but in Bangladesh, sign language lacks official recognition. To provide a better lifestyle for hearing impaired people in Bangladesh, there is no alternative to proper learning and teaching of Bengali sign language.

Bengali sign languages are both one-handed and two-handed, shown in Fig. 1. According to the national center for special education, department of social service in Bangladesh, the method of teaching hearing-impaired children started almost six decades ago. Lip reading use of sign language and speech training has been followed in teaching them. Though the basis of sign language is the same, yet like every other country, Bangladesh has developed its techniques of sign language according to its language [9].

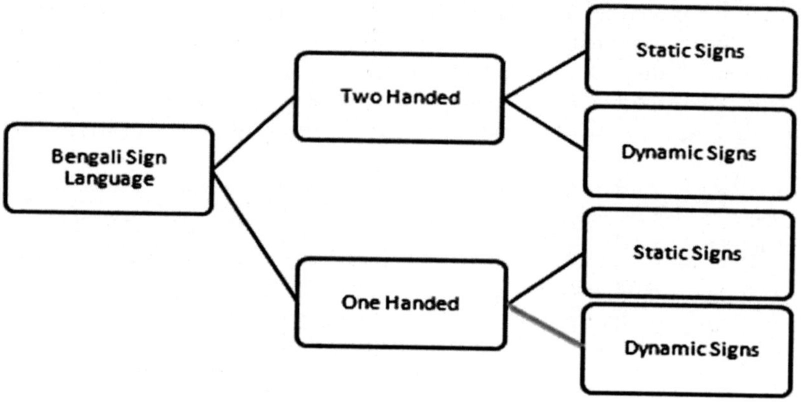

Fig. 1. Bengali sign language type

Nowadays, the widespread use of smart devices such as smartphones, tablets, and other personal communication devices has made it easy for people to access various learning applications including both the web and mobile applications. They can serve as a useful tool for people with hearing impairments.

2.1 Problem Statement

There has been much research available about text-to-sign language translators. Almost every Sign Language has its mobile or web application to reduce the communication gap with hearing-impaired people. An interactive sign language teaching program [10] was developed by Kadam et al. for American Sign Language (ASL) using gloves [11]. In order to detect the hand gesture, they used flex sensors in the fingers of the glove. Verma et al. invented a system [12] for detecting sign language gestures using Microsoft Kinect. A bidirectional translation system [13] for Japanese Sign Language (JSL) has been designed by Imagawa et al. They made a real-time system for tracking the hands of signers and finding the location of each hand. An application named'Gesture Audio Video Conferencing Application' proposed by Tarte et al. in [14].

But unfortunately, there are few studies related to Bengali sign language. For developing sign-to-text language translator and sign language recognition some new methods proposed recently. With the use of deep convolutional neural network and utilizing the learned features from a pre-trained network and fine-tuning the top layers of this network, authors have achieved a high overall recognition rate of 96.33% on the training dataset and 84.68% on the validation dataset [15]. Image-based Bengali Sign Language Alphabet Recognition for Deaf and Dumb Community also got an accuracy of 89.6% approximately [16].

At the moment, only a few textbook dictionaries of Bengali Sign Language are available. In addition, there is no convenient multilingual Bengali Sign Language related text-to-sign translation software or research that could meet the huge demand, so the development and improvement of text-to-sign Language translators for Bengali Sign Language are very essential. The project aims to develop a video-based text-to-sign language translator.

2.2 Steps for Developing Proposed Bengali Language Translator

There is no accurate image dataset or video dataset available for Bengali Sign Language. So the video datasets will be used here to create the web application, for both static and dynamic signs are unique and first of their kind. The video clips are evaluated by the Bangladesh government-approved Bengali Sign Language Dictionary edited by Bangladesh Sign Language Committee.

Building an accurate sign language translator depends on several things. Collecting accurate data sets is one of them. In Fig. 2, necessary steps are shown.

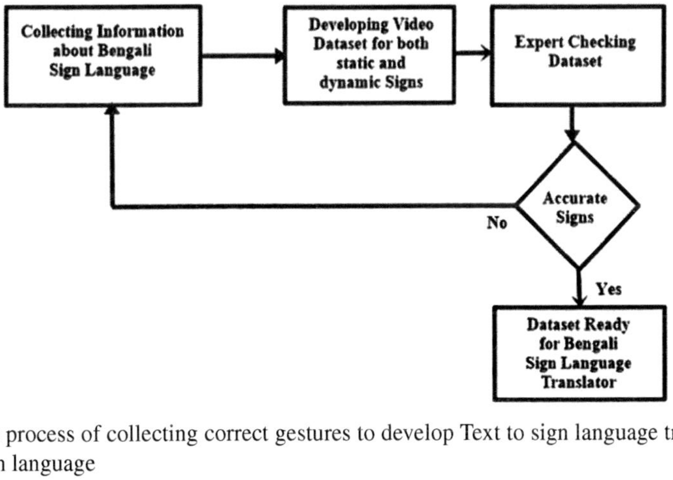

Fig. 2. The process of collecting correct gestures to develop Text to sign language translator for Bengali sign language

For our research, we choose to develop a web application. The proposed algorithm for developing the web application is defined in Fig. 3. At an early stage of the development of this translator, we will use 353 video clips for both static and dynamic signs. The datasets will be extended in order to improve the translator. The proposed algorithm can be updated in future work to cope up with user requirements and market analysis.

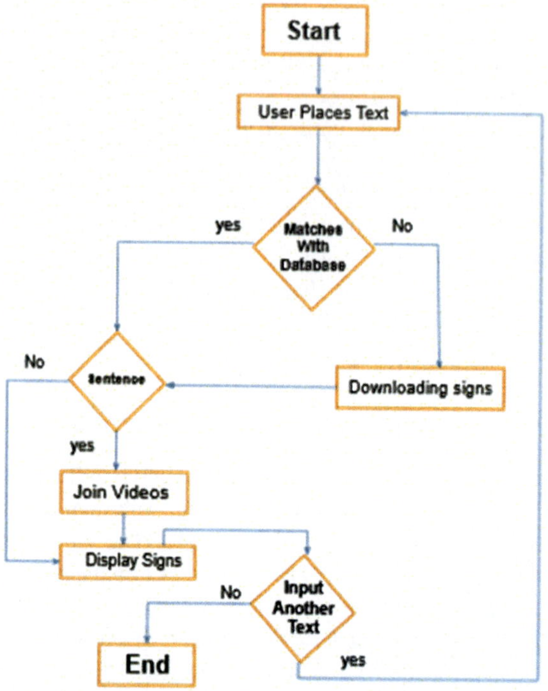

Fig. 3. Flowchart of proposed sign language translator

2.3 Understanding the Joining of Signs

Joining of different signs is necessary to complete a sentence, which the user wants to observe as an output. A research by Krape, Slavko & Solina, Franc. in 1999 discussed the problem of video clip joining. There it has been described that two digital video clips (each contained in its own file) should be done in such a way that the viewer perceives them as a single video sequence with smooth motion [17].

Two video clips joining the process are described with the aim of easily generalizing any number of video clips.

Suppose, the first video clip as a n_1 dimensional vector v_1 and second video clip as a n_2 dimensional vector v_2.

The First Video clip is represented as,

$$v_1 = (p_1, p_2, \ldots, p_{n1}).$$

Where, the elements of the vector represent the image that forms the video clip. The number of images in the video clip is n_1.

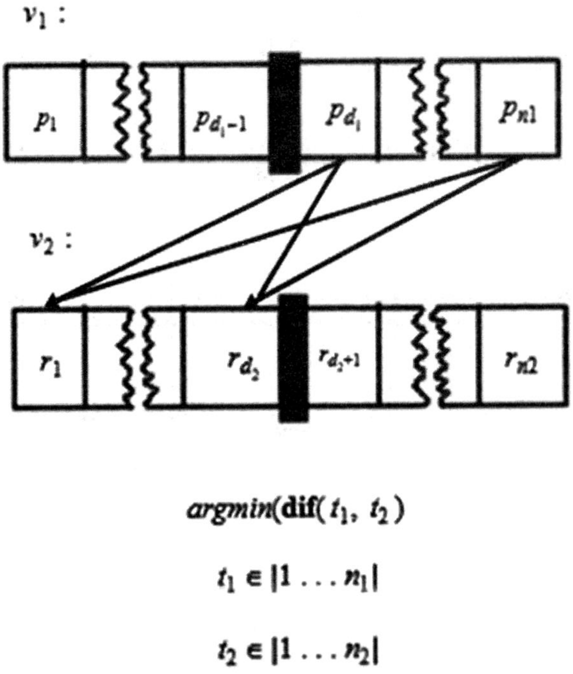

Fig. 4. Determination of the t_1 and t_2 values [17]

The second video clip is represented as,

$$v_1 = (r_1, r_2, \ldots, r_{n1}).$$

Where, the elements of the vector represent the image that forms the video clip. The number of images in the video clip is n_2.

The joining can occur on any image of the clips of v_1 and v_2. Suppose, the image in which joining should occur presents time t_1 and t_2, described in Fig. 4. They represent the indexes, which are, in the case of the first video clip, shown as the last and as the first in the case of the second video clip [17].

$$t_1 \in |1 \ldots n_1|$$

$$t_2 \in |1 \ldots n_2|$$

We need two more variables d_1 and d_2 to explain. d_1 is the index of the image in the first video clip, from which on, the joining is allowed. d2 is the index of the image in the second video clip till which the joining is allowed. With these constraints t_1 and t_2 can occupy the values given below:

$$t_1 \in |d_1 \ldots n_1|$$

$$t_2 \in |1 \ldots d_2|$$

The joining of video clips should be performed in such a way that the transition from one video clip to another would not be noticed. This function dif (i, j) calculates the difference between pictures p_i and r_j depending on a given criterion. The smaller the function value, the more both pictures resemble each other [17].

2.4 Dataset for Bengali Sign Language

For the Development of a text-to-sign language translator, largely focused on Bengali sign language it is necessary to create a dataset for both one-handed and two-handed signs. Also, we need to remember not all signs are static. Most signs are dynamic and this is why a video dataset is necessary. In Bengali scripts, there are a total of 50 letters, among them, the number of vowels is 11, and 39 consonants. Bengali script is written from left to right and has no distinct letter cases. It is recognizable, as are other Brahmic scripts, by a distinctive horizontal line known as, matra running along the tops of the letters that links them together. The Bengali script is however less blocky and presents a more sinuous shape [18]. For implementation into the translator, video clips have been selected from around more than a thousand videos. In Fig. 5, some random samples of datasets are visible.

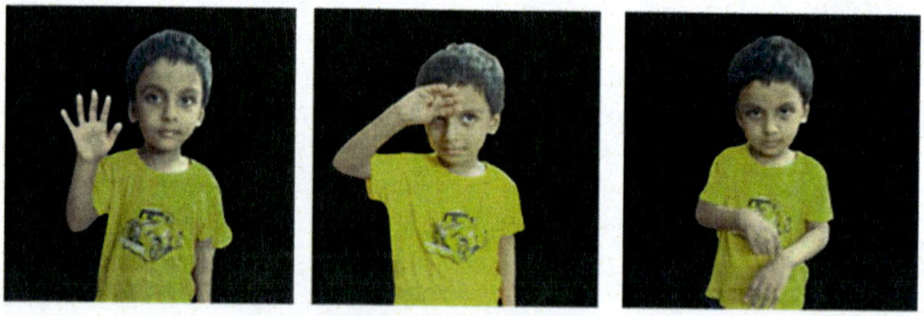

Id: 0 **Value : <দুপুর>** **Id: 1** **Value : <সালাম>** **Id: 2** **Value : <সময়>**

Fig. 5. Random samples of Bengali sign language dataset

2.5 Implementation of Proposed Translator

Our created dataset of Bengali sign language includes Bengali alphabets for both one handed and two handed signs, important words people use day to day life, and numbers for counting, etc. The video clips that we are using are in mp4 format. We used dark background, bright sunlight or flashlight, and dark to medium bright skin colors of the participants for most of the videos.

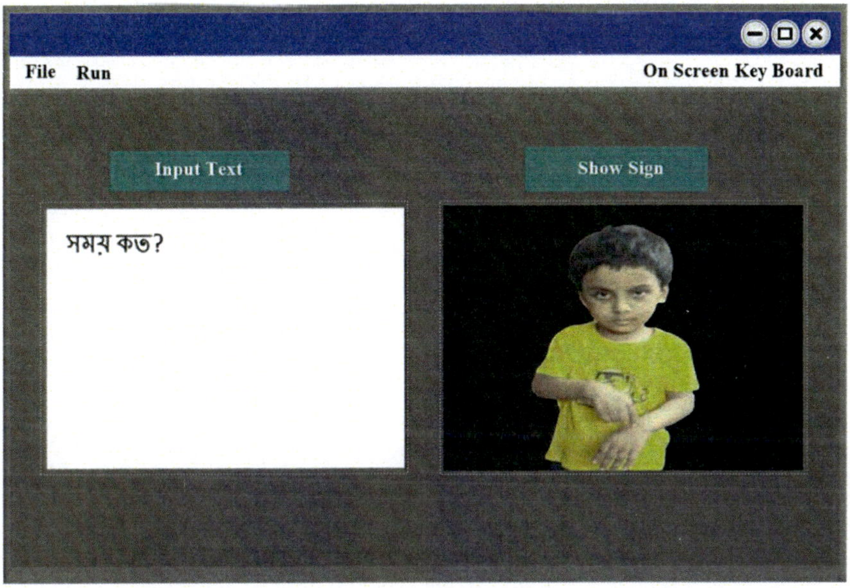

Fig. 6. The outlook of proposed Bengali sign language translator

For implementing the proposed algorithm for creating Bengali Sign Language translator, the Python programming language has been used with HTML considering its readability, maintainability, robust standard library, and many open source resources, tools, etc. An open source Text-to-Sign Language translator [19] has been modified and improved for developing the program. The user interface of Bengali sign language translator is represented in Fig. 6, which needs to be improved in order to make it more user friendly and more effective.

The user has to input a text, and every time the program encounters a word it does not currently have in the database, it will show an error, but if it matches with the database, it will check whether it is a sentence or a word.

If it is a sentence and all words are from the database, the videos will merge and display the signs as a video clip.

In case of a word, which matches with the database, the user will get an output.

If in an inputted sentence some words or a single word from the database but another words or a single word do not match with the database, then the program will show an error.

```
# Get text
text = get_input_text()
# Process text
words = process_text(text)
# If the user entered incorrect words that are not in the
database, show an error and break.
real_words = []
all_words_exist = True
for w in words:
  real_name = find_in_db(w)
  if real_name:
     real_words.append(real_name)
  else:
     all_words_exist = False
     break
if all_words_exist:
  words = real_words
  # Concatenate videos and save output video to folder
  merge_signs(words)
  # Play the video
  from os import startfile
  startfile(SIGN_PATH + "\\Output\\out.mp4")
else:
  show_error()
```

[Main function of the program].

For merging the video files we used the command-line tool FFmpeg with necessary parameters. Before running the FFmpeg, the program writes a txt file with paths to the essential videos, and sends it among parameters.

2.6 Results and Analysis

The developed translator shows good results of the sentences containing words matched with the database but there are some issues that need to be addressed in order to improve the web application.

The web application shows errors when spelling is wrong. For example, the word " সময়" in Bengali, which means "time", displays an output if the word is written correctly. But if the spelling is wrong with almost same pronunciation like " শময়", the program shows error. These spelling mistakes are common within Bengali users. This is why it needs to be solved for the future.

Secondly, if any word from the inputted sentence by users does not match the database, then the program shows an error. Recommendation for solving this problem is to improve the database by increasing more video datasets or at least displaying the words within a sentence, which match the database.

Most importantly, the program does not follow grammatical rules of Bengali Sign Language. It joins signs in the sequence of inputted data. Also the program shows error or it breaks when there is change in tense, cases, or something similar.

For analysing the merging time of video clips we checked from simple to complex sentences. In Fig. 7, the details are shown.

Using linear regression, we get the value of R square 0.2233. It ranges in value from 0 to 1. And it is usually interpreted as summarizing the percent of the variation in the

Number of Sentences Tested	Sentence's Word Count	Average Duration of video clips (seconds)	Average Time of Joining (seconds) for each sentence
5	2	2.0	0.0765
8	3	4.0	0.0845
3	4	7.0	0.0823
2	5	8.0	0.08275
4	6	11.0	0.08210

Fig. 7. Average duration vs average time of joining for each sentence

response that the regression model explains. So an R-squared of 0.2233 might mean that the model explains about 22.33% of the variation in our dependent variable [20].

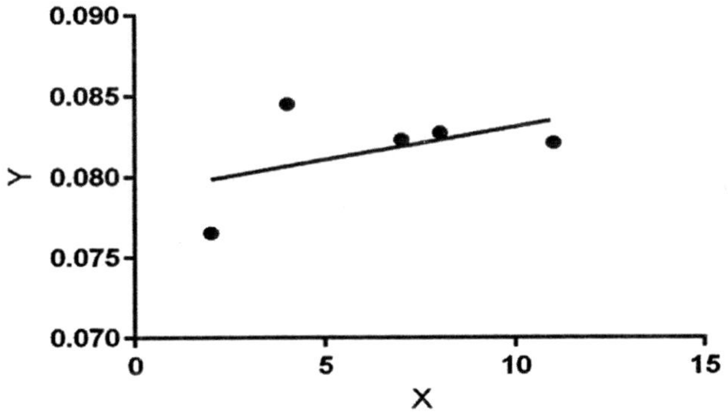

Fig. 8. Linear regression graph for average duration of video clips (x) vs merging time (y)

From the result of our tested data, it shows:

- No noticeable delay or error for the tested sentences which matched with database
- Length of sentence doesn't prolong the execution time
- Users had no problem to download signs which matches the database
- Only a few users were interested in looking for another text input after getting an error.

Most sign language translators can show video clips with normal speed or with lower or faster speed than normal [17]. But this program lacks this feature. With this feature, it can be adapted to the user's ability of sign language recognition [17].

3 Conclusion / Future Work

Sign language works as a bridge to reduce the communication gap with hearing-impaired people. Sign Language can be used also in many situations where there is no hearing impairment involved such as autism spectrum disorders, etc.

In our future work, we are aiming to develop a large dataset of Bengali Sign Language and evaluate its accuracy to build a more accurate and improved text-to-sign language translator.

In the future, we are looking forward to completing some tasks to make Bengali Sign Language easy to understand for all:

- Improvement of Proposed Bengali Text-to-Sign language translator;
- Development of a large video data set which will include almost all the words;
- Creation of a text-to-sign language translator that will display letter to letter, text to sign translation;
- Development of a sign-to-text Language Translator using deep learning.
- Collecting, understanding, and visualizing the reviews and improve the implementation for further development;
- Testing the developed Sign Language Translator with many users and receiving feedback from them.

A large number of people in the world suffer from various physical disabilities. A huge part of this population cannot hear or speak. Hearing impairment can not only be congenital, it can also occur in the elderly, patients with various neurological diseases, as well as a result of any accident, hearing loss or speech impairment can happen. And in such a situation, the help of sign languages is required. Most of the developed countries of the world have already come a long way in solving problems associated with hearing impairment using modern technology.

However, underdeveloped and developing countries lag far behind in this regard. Many countries do not even recognize sign language as an official language. Also, it is very difficult to find any information, even after a long search on the Internet. But the number of Internet users in some of these countries is significant. For example, both Bangladesh and India are currently in the top ten countries for the number of Internet users. But the use of technology for people with disabilities is very small there. Considering all these reasons for the development of Bengali sign Language Translator, both text-to-sign and sign-to-text are essential.

References

1. Central intelligence agency Bangladesh. In: The world factbook (2018). https://www.cia.gov
2. Summary by language size. Ethnologue: Language of the world (2018)
3. Top 10 most spoken languages (2021). https://www.ethnologue.com/guides/ethnologue200
4. Encyclopedia of World Cultures: South Asia, Macmillan Reference USA, p. 416 (1992)
5. Hearing Impairment. https://kidshealth.org/en/teens/hearing-impairment.html
6. Alauddin, M., Joarder, A.H.: Deafness in Bangladesh. In: Suzuki, J.I., Kobayashi, T., Koga, K. (eds.) Hearing Impairment. Springer, Tokyo (2004). https://doi.org/10.1007/978-4-431-68397-1_13

7. Amin MN: Prevention of Deafness and Primary Ear Care (Bengali) - Society for Assistance to Hearing Impaired Children (SAHIC), Mohakhali, Dhaka-1212, Bangladesh
8. Elmahgiubi, M., Ennajar, M., Drawil, N., Elbuni, M.S.: Sign language translator and gesture recognition. Glob. Summit Comput. Inf. Technol. (GSCIT) **2015**, 1–6 (2015). https://doi.org/10.1109/GSCIT.2015.7353332
9. Bengali Sign Language Dictionary Dhaka: National Centre for Special Education, Ministry of Social Welfare, Norwegian Association of the deaf, and Bangladesh National Federation of the Deaf, pp. 3–149 (1994)
10. Kadam, K., Ganu, R., Bhosekar, A., Joshi, S.: American sign language interpreter. In: 2012 IEEE Fourth International IEEE, Conference on Technology for Education (T4E), pp. 157–159 (2012)
11. Huda, M.N., Alam, I.: Automated Bangla Sign Language Conversion System: Present and Future (2020)
12. Verma, H.V., Aggarwal, E., Chandra, S.: Gesture recognition using kinect for sign language translation. In: 2013 IEEE Second International Conference on Image Information Processing (ICIIP), pp. 96–100. IEEE (2013). https://doi.org/10.1109/ICIIP.2013.6707563
13. Imagawa, K., Lu, S., Igi, S.: Color-based hands tracking system for sign language recognition. In: Proceedings of Third IEEE International Conference on Automatic Face and Gesture Recognition IEEE, pp. 462–467 (1998).https://doi.org/10.1109/AFGR.1998.670991
14. Tarte, N.T., Bhadane, S.N., Kulkarni, P.S.: A gesture audio-video conferencing application for the ease of communication between normal person and deaf and dumb person. Int. J. Sci. Eng. Technol. Res. **3**(5) (2014)
15. Hossen, M.A., Govindaiah, A., Sultana, S., Bhuiyan, A.: Bengali sign language recognition using deep convolutional neural network. In: 2018 Joint 7th International Conference on Informatics, Electronics & Vision (ICIEV) and 2018 2nd International Conference on Imaging, Vision & Pattern Recognition (icIVPR), pp. 369–373 (2018). https://doi.org/10.1109/ICIEV.2018.8640962.
16. Rafi, A.M., Nawal, N., Bayev, N.S.N., Nima, L., Shahnaz, C., Fattah, S.A.: Image-based bengali sign language alphabet recognition for deaf and dumb community. IEEE Glob. Hum. Technol. Conf. (GHTC) **2019**, 1–7 (2019). https://doi.org/10.1109/GHTC46095.2019.9033031
17. Krape, S., Franc, S.: Synthesis of the Sign Language of the Deaf from the Sign Video Clips. 66 (1999)
18. Cardona, G., Jain, D.: The Indo-Aryan Languages. Routledge (2003)
19. Text To Sign Language. github.com/MifletzetDigdoogim/TextToSignLanguage
20. Is R-squared Useless? https://data.library.virginia.edu/is-r-squared-useless/

A Conceptual Consumer Data Protection Maturity Model for Government Adoption: South African Context

Ileen E. Bredenkamp[1](✉), Elmarie Kritzinger[2], and Marlien Herselman[3]

[1] University of South Africa, Pretoria 0002, South Africa
[2] School of Computing, University of South Africa, Pretoria 0002, South Africa
kritze@unisa.ac.za
[3] Next Generation Enterprises and Institutions, CSIR, Pretoria 0002, South Africa
mherselman@csir.co.za

Abstract. In South Africa (SA), being a developing country, there is a significant gap between the level of trust consumers extend towards organizations they share personal information with whilst conducting everyday transactions and the degree to which that trust is justified by the Information Security (IS) compliance efforts of organizations. Consumers are lacking IS compliance awareness which influences their ability to make informed decisions on when to share or not to share their personal information with organizations. There is currently no government-led/sponsored IS compliance awareness training initiatives in SA and therefore no way to measure the progress/maturity of such framework implementation. This research problem called for the development of a Consumer Data Protection Framework (CDPF) in a prior study to assist the South African government with improving the IS compliance awareness of consumers through government-led awareness training initiatives [1]. This paper extends that work by proposing a Consumer Data Protection Maturity Model (CDPMM) to assist the South African government with evaluating the maturity of the aforementioned CDPF implementation. The evaluation results can be used to create an improvement plan which will guide the South African government to reach its target maturity level. The primary research objective of this paper is therefore to propose a CDPMM that can be used to measure the progress of the CDPF implementation in a South African government setting. The CDPMM is developed based on key components derived from literature and evaluated through expert reviews for improvement.

Keywords: Government · Consumer data protection · Information security · Compliance awareness · Maturity model · Design artefact

1 Introduction

Human factors continue to be regarded as a weak link in the protection of personal information [2–4]. Awareness, being a critical human factor, prevails as one of the most significant data protection measures. According to Hadlington [5], consumers are lacking awareness and knowledge and therefore are ignorant of the need to protect their personal

R. Silhavy et al. (Eds.): CoMeSySo 2021, LNNS 232, pp. 820–834, 2021.
https://doi.org/10.1007/978-3-030-90318-3_64

data. More importantly, consumers lack IS compliance awareness – awareness of the IS compliance status of organizations they transact/share personal information with. A lack of IS compliance awareness may negatively influence the IS behaviour of consumers as they may extend trust towards non-compliant/untrustworthy organizations unknowingly.

In a search to address consumer data protection in South Africa, being a developing country, Bredenkamp, Kritzinger and Herselman [1] proposed a Consumer Data Protection Framework (CDPF) to assist the South African government with improving IS compliance awareness amongst consumers through IS compliance awareness training initiatives. Another aspect to consider for addressing the problem is the ability to measure the progress/maturity of such awareness training initiatives once implemented. It came to the researchers' attention that such a CDPF implementation would require the development of a maturity model artefact to assist the government with measuring and tracking the progress/maturity of the CDPF implementation in the government setting for continuous improvement.

The goal of this paper is to propose a Consumer Data Protection Maturity Model (CDPMM) artefact, developed by utilizing the Design Science Research (DSR) methodology as it combines the practical dimension and the scientific dimension. This paper is structured in seven sections. First, the background to the study and are explained in Sect. 1. This is followed by Sect. 2, where the research methodology is outlined. Section 3 discuss the problem identification of the research and details the findings from a systematic literature review on existing information security/awareness-related models and a comparison between the existing maturity models from a human factors perspective. Then Sect. 4 discuss the maturity model design and the maturity model evaluation, the latter being an iterative approach through conducting expert reviews. Section 5 presents and discuss the proposed CDPMM. Section 6 provides a summary of the expert review results. Finally, Sect. 7 details the conclusions of the proposed CDPMM and suggestions on future work to be considered on this research topic.

2 Research Methodology

This research applied the DSR method as defined by Hevner et al. [6, 7] whilst adopting the maturity model development process of Becker et al. [8]. DSR is described as "a designer answering questions relevant to human problems via the creation of innovative artefacts, thereby contributing new knowledge to the body of scientific evidence. The designed artefacts are both useful and fundamental in understanding that problem" [9]. The major benefit of DSR is due to its ability to addresses real-world problems and simultaneously contributes to the body of knowledge [7]. The maturity model development process of Becker et al. [8] was applied as it is based on DSR and therefore provides a methodological artefact construction platform very suitable for application in this research approach. In addition, Becker et al. provide a strict and uniform development process, which is in line with the DSR guidance of Hevner et al. [6]. Becker et al. [8] define maturity models as artefacts that serve to solve the problem of appreciating capacity and obtain improvement measures. Hevner and Chatterjee [9] posit that design science allows the development of artefacts such as models, constructs, methods and instantiations which assist with improving problem-solving capabilities. Therefore, DSR is appropriate for the development of maturity models.

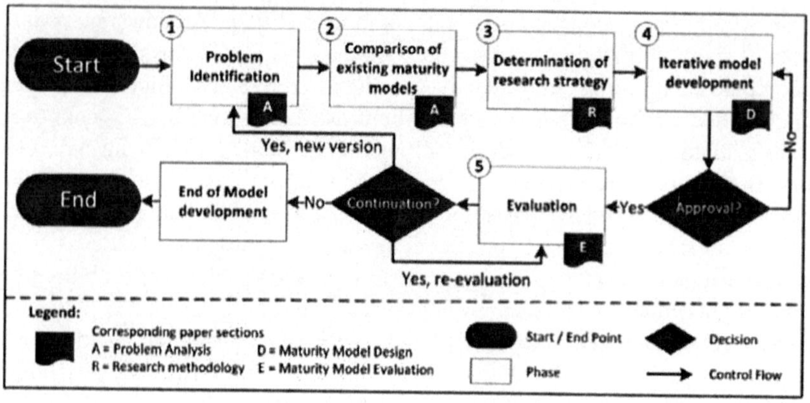

Fig. 1. Procedure model of the research approach adopted from Becker et al. [8]

The maturity model development procedure of Becker et al. [8] that is adopted for this study consists of five distinct and mandatory steps to be documented. As depicted in Fig. 1, the first step (Step 1) focus on problem identification. In Step 1 the research problem is identified and discussed, the relevance of the problem is articulated in practical terms and the value of the artefact is justified. Step 2 follows which is a comparison of existing maturity models. This second step is based on the problem identification of the first step and the analysis of existing information security/awareness-related maturity models, which leads to the identification of gaps in these models. A systematic literature review was conducted on the theme of information security maturity/awareness maturity that was derived from a prior systematic literature review. From this literature analysis, thirteen maturity models were uncovered (see Table 2). The next step tends to the determination of the research approach (Step 3) as discussed in this section of the paper. Step 3 is followed by maturity model development (Step 4). In Step 4, model adoption techniques were utilized, such as configuration, instantiation, aggregation, specialization and analogy [10] to incorporate relevant and useful aspects of the existing selected maturity models into the proposed maturity model for Consumer Data Protection. This allowed the creation of a rigorous maturity model regarding both structure and content which is grounded in literature. The last step of evaluation (Step 5) is to conduct expert reviews with subject-matter experts to evaluate (rate and rank) the CDPMM components for determining its relevance, value/importance and practicality (see Sect. 6).

3 Problem Identification and Analysis

As mentioned, in Step 1 the research problem is identified and detailed. The practical relevance of the problem is specified and the value of the artefact is justified. To provide a consistent and precise problem definition, the type of activities the proposed CDPMM must measure were identified. Table 1 presents the activities per phase of the PDCA cycle [11] and per layer of the CDPF [1].

Table 1. CDPMM activities for measurement

PDCA cycle	CDPMM activities		CDPF layers
	Maturity level 1: Initial		
	A1:	Obtain governmental management authorization to establish a project to implement and operate the CDPF	
Plan	**Maturity level 2: Planning**		
	A2:	Define the scope and boundaries of the CDPF within the government departmental setting	Strategic layer
	A3:	Define a CDP policy and update Government Gazette accordingly	
	A4:	Determine the government internal posture and governance model in terms of their readiness to perform IS compliance awareness training initiatives/programs (campaigns and incentivized training programs): • Develop an awareness training workforce department • Assign roles and responsibilities for IS compliance awareness and training initiatives • Control the workforce life cycle (resources and budget) • Internal training on IS compliance awareness aspects for Consumer Data Protection • Sharing of IS compliance status information and communication activities	

(continued)

Table 1. (*continued*)

	A5:	Acknowledge, make provision, and establish mediums through which IS compliance awareness and training initiatives will be realized: • Government incentivized awareness training and monitoring programs for participating organizations, including; • Government-led public IS compliance awareness campaigns	Tactical layer
	A6:	Acknowledge, make provision, and establish topics, content and tools necessary for the delivery of awareness and training programs, including public consumer IS compliance awareness campaigns	Preparation layer
Do	**Maturity level 3: Implementation**		
	A7:	Implement incentivized awareness training programs by liaising with participating organizations Implement IS compliance awareness campaigns Manage relationships with participating organizations	Delivery layer
Check	**Maturity level 4: Monitoring & Evaluation**		
	A8:	Execute monitoring and reviewing procedures and controls for training and awareness programs and IS compliance awareness campaigns	Monitoring & Evaluation layer
	A9:	Conduct internal audits (Government) to ensure CDPF processes and procedures are being followed	
	A10:	Undertake management review of the CDPF. Ensure the improvements achieve their objectives	
Act	**Maturity level 5: Improvement**		
	A11:	Implement the identified improvements in the CDPF based on the outcomes of the Monitoring & Evaluation Layer	Continuous improvement
	A12:	Take appropriate corrective actions	
	A13:	Repeat A8 and onwards to establish continuous improvement	

Table 2. CDP maturity model activities for measurement

Maturity models	Purposes/Strengths	Limitation to CDP from a human factors perspective	Maturity levels
CCSMM [12, 13]	Community effort and communication capability in communities	Focus more on cybersecurity and technical implementation thereof, rather than IS awareness and training evaluation	Initial, Advanced, Self-assessed, Integrated, Vanguard
COBIT [12–18]	IT governance maturity-provides enterprises with a repeatable, reliable, and robust methodology for assessing the capability of their IT processes	Focus more on IT governance maturity evaluation rather than IS awareness and training evaluation	Incomplete, Performed, Managed, Established, Predictable, Optimized
CSF CCM [12, 19, 20]	Improves federal critical infrastructure through a set of activities designed to develop individual profiles for operators	Non-technical security issues (i.e. IS awareness training evaluation)	Identify, Protect, Detect, Respond, Recover
C2M2 [12, 21–23]	Assessment of implementation and management in critical infrastructure	Focus more on cybersecurity and technical implementation thereof, rather than IS awareness training evaluation	Not Performed, Initiated, Performed, Managed
GSMM [24]	Information protection	Non-technical security issues (i.e. IS awareness and training evaluation)	Blind Trusting, Repeatable, Defined, Managed, Maintenance
GISMM [25]	Security awareness and risk management	Non-technical security issues	Blissful Ignorance, Awareness, Corrective, Operations Excellence
ISEM [12, 26]	Security awareness and evaluation	Non-technical security issues	Complacency, Acknowledgment, Integration, Common Practice, Continuous Improvement

(*continued*)

Table 2. (*continued*)

Maturity models	Purposes/Strengths	Limitation to CDP from a human factors perspective	Maturity levels
ISM2 [12, 27]	Provides a framework for review and measure the information security posture of an information security program	Non-technical security issues (i.e. IS awareness and training evaluation)	Policies, Procedures, Implemented, Tested, Integrated
ISM3 [12, 28, 29]	Prevent and mitigate incidents and optimize the use of information, money, people, time, and infrastructure. IS management, risk assessment and process integration	Organizational cultural issues	Undefined, Defined, Managed, Controlled, Optimized
NICE CCM [12, 13, 30]	Workforce planning for cybersecurity best practices	The requirement for understanding current staffing capabilities	Limited, Progressing, Optimized
SSE CMM [12, 31–33]	Evaluation of software security engineering processes	Focus more on software security and technical implementation thereof, rather than IS awareness training evaluation	Performed Informally, Plan and Track, Well Defined, Control, Continuous Improvements
SANS SAMM [34]	Identifies the current stage of a security awareness program	Focused on identifying the current stage of a security awareness program, but evaluation criteria are limited	Non-existent, Compliance Focused, Promoting Awareness & Behaviour Change, Long-term Sustainment & Culture Change, Robust Metrics Framework
HAIS-Q [35–38]	Measures an individual's knowledge, attitude, and self-reported behaviour relating to information security in the workplace	HAIS-Q model questions on the 7 specific focus areas are not completely aligned to measure consumer IS awareness	Each of these focus areas contain knowledge, attitude and behaviour statements which can be answered on a 5-point Likert scale, which ranges from strongly disagree to strongly agree

The activities are used as a reference baseline to assess the appropriateness of thirteen existing information security/awareness-related maturity models. Table 2 presents a summary of the thirteen maturity models that were identified and their shortcomings in terms of Consumer Data Protection from a human factors perspective in a government setting.

Six of these models were shortlisted based on relevant characteristics with the potential to support the activities. Useful aspects of each of the six shortlisted maturity models were identified and adapted for incorporation into the proposed CDPMM. These shortlisted maturity models are presented in Table 3, together with a description of the aspects considered for adaption/adoption.

Table 3. Shortlisted maturity models and associated aspects selected for adoption/adaption.

Maturity model	Adapted/Adopted aspects description
CCSMM	The concepts of the CCSMM geography dimension and scales (organization, community and nations) were adapted to be the government and the public consumer community for the CDPMM
COBIT	Adapted governance process aspects of COBIT for inclusion in the CDPMM as the CDPMM strategic and tactical layer activities needs to be supported by governance processes
C2M2	Adapted the concept of the C2M2 domains and associated activities to be performed to establish a mature capability with a focus on CDPMM domains. The CDPMM domains are Consumer IS awareness, IS Trust, IS behaviour. The CDPMM activities are listed in Table 1
ISM2	Considered the concept of the following ISM2 strategic aspects to address in the CDPMM: • Information security culture • Planning • Security awareness • Training and education • Budget and resources • Lifecycle management
NICE CCM	Adapted the concept of NICE CCM key activities for inclusion in the CDPMM, assisting with government workforce planning: • Process and analytics • Integrated governance • Trained professionals and enabling technology
SANS SAMM	Adapted awareness level identification aspects of the SANS SAMM to assist with activities in the CDPMM monitoring and evaluation layer
HAIS-Q	Adapted aspects of the HAIS-Q model for measuring consumer IS compliance awareness, extended to include measurement of consumer IS trust and IS behaviour as part of the monitoring and evaluation layer activities of the CDP-MM

From the above analysis, it can be concluded that a gap exists in terms of maturity models which cover the CDPMM activities with a focus on improving the CDPMM domains (IS awareness, IS trust, IS behaviour) of this study. The finding motivates the need to construct a new CDPMM to be able to measure the maturity of the CDPF implementation if the South African government decides to adopt the CDPF.

4 Maturity Model Design

According to the maturity model development approach of Becker et al. [8], a new maturity model has to be developed if no existing, or the progress of an existing one, is not able to address the identified problem. Therefore, based on the findings of the research analysis, there is no maturity model which satisfactory fulfil the entire activities baseline with a focus on the CDP domains. Therefore, a new maturity model is developed to address this gap. As mentioned, the proposed CDPMM adopts or adapts established structural elements, including relevant and useful aspects of seven maturity models, namely: CCSMM, COBIT, C2M2, ISM2, NICE CCM, SANS SAMM and HAIS-Q.

As noted in Table 3, five maturity levels are proposed: initial, planning, implementation, monitoring and improvement. These maturity levels are based on the PDCA cycle used within the ISO/IEC 27001 industry standard for information security [11]. Table 1 details the activities for each maturity level on which the proposed CDPMM is based. The CDPF consists of five layers namely, strategic layer, tactical layer, preparation layer, delivery layer and lastly, the monitoring and evaluation layer, followed up with a continuous improvement process based on the outcome/results obtained during the monitoring and evaluation layer. The CDPF layers and PDCA phases assisted with the logical grouping of CDPMM activities.

To improve from level X to level X + 1, the government must comply with all the criteria from level X, which makes this maturity model follow a "stages" approach. What the government can expect from progressing through the maturity levels is that their internal capabilities to establish and run with consumer data protection initiatives will become increasingly managed, defined and optimized.

The authors acknowledge that the evaluation step is an important element of Design Science Research as it is necessary to demonstrate the utility, quality and efficacy of a design artefact [7]. For this reason, iterative expert reviews were conducted with six subject-matter experts in the ICT, Finance/Banking, Law, Government and Education industries to inform the initial maturity model, refining its content and structure based on recommendations during each review cycle (see Sect. 6).

5 Proposed Consumer Data Protection Maturity Model (CDPMM)

Figure 2 presents the proposed CDPMM, a theoretical model which is grounded in literature. The CDPMM consists of five maturity levels: initial, planning, implementation, monitoring and improvement (L1 to L5). The CDPMM includes thirteen activities (A1 to A13 as listed in Table 1) and is depicted in Fig. 2. The CDPMM activities are logically grouped per PDCA life cycle phases which also coincides with the CDPMM maturity levels. To progress from one maturity level to the next, all activities per specific maturity level needs to be completed, starting with Level 1 (L1 - Initial) and systematically progressing through Level 2 to Level 5, chronologically completing the activities per maturity level. Therefore, the successfully completed activities (A1 to A13) as deliverables per each PDCA phase, constitute the evaluation criterion to progress from one maturity level to another. The CDPMM focuses on three domains for the improvement of Consumer Data Protection. The three CDP domains are IS Awareness Maturity (D1), IS Trust Maturity (D2) and IS Behaviour Maturity (D3). These CDP domains are incorporated into the CDPMM activities dealing with incentivized IS compliance awareness training program material development, including IS compliance awareness campaign material development. Lastly, the CDPMM consists of four dimensions namely, Government (Dim1), Public Consumers (Dim2), Regulatory Bodies (Dim3) and Public/Private Organizations (Dim4), which represents the stakeholder scope of the proposed CDPMM.

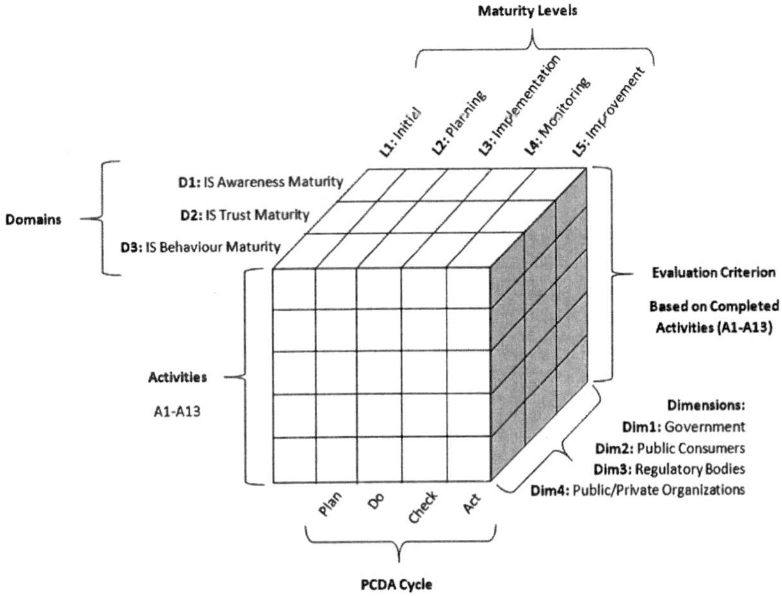

Fig. 2. Proposed consumer data protection maturity model (CDPMM)

6 Summary of Expert Review Findings

As mentioned, six expert reviewers, all of whom are knowledgeable in the field of personal data protection and compliance, and have significant practical experience in the ICT, Education, Finance/Banking, Legal and Government industries, were conveniently and purposefully selected to evaluate the components of the CDPMM via an online survey. The expert reviewers were tasked to firstly rate each CDPMM component in terms of its relevance, value and practicality, after which the researcher analyzed and compared the expert reviewer results to rank each CDPMM component from most to least relevant, valuable and practical for measuring the progress/maturity of the CDPF implementation. Table 4 summarizes the CDPMM component and sub-component average rating percentage scores of the expert evaluation, including the rankings values assigned for the CDPMM components and sub-components.

Table 4. Expert evaluation: summary of CDPMM component and sub-component rating and ranking scores

CDPMM Components	CDPMM Sub-components	Avg. Relevance % Rating	Relevance \ Ranking	Avg. Value % Rating	Value Ranking	Avg. Practicality % Rating	Practicality Ranking	Total Avg. Rating % per CDPMM Component	Value Ranking per CDPMM Component
Maturity Levels (L1-L5)	L1: Initial	83,33%	2	80,00%	1	66,67%	3		
	L2: Planning	86,67%	1	80,00%	1	76,67%	1		
	L3: Implementation	83,33%	2	80,00%	1	76,67%	1	78,67%	4
	L4: Monitoring	83,33%	2	80,00%	1	70,00%	2		
	L5: Improvement	83,33%	2	80,00%	1	70,00%	2		
	Total Avg. Rating Score for Relevance, Value & Practicality:	***84,00%***		***80,00%***		***72,00%***			
Evaluation Criterion/ Activities (A1-A13)	**A1:** Obtain governmental management authorization to establish a project to implement and operate the CDPF.	86,67%	1	83,33%	1	80,00%	1		
	A2: Define the scope and boundaries of the CDPF within the government departmental setting.	83,33%	2	80,00%	2	76,67%	2		
	A3: Define a CDP policy and update Government Gazette accordingly.	83,33%	2	80,00%	2	73,33%	3		
	A4: Determine government internal posture and governance model in terms of their readiness to perform IS compliance awareness and training initiatives/programs (campaigns and incentivized training programs)	83,33%	2	80,00%	2	70,00%	4		
	A5: Acknowledge, make provision and establish mediums through which IS compliance awareness and training initiatives will be realised.	83,33%	2	76,67%	3	76,67%	2		
	A6: Acknowledge, make provision and establish topics, content and tools necessary for the delivery of awareness and training programs, including public consumer IS compliance awareness campaigns.	86,67%	1	83,33%	1	80,00%	1	80,17%	3
	A7: Implement incentivised awareness training programs by liaising with participating organizations. Implement IS compliance awareness campaigns. Manage relationships with participating organizations.	83,33%	2	80,00%	2	76,67%	2		
	A8: Execute monitoring and reviewing procedures and controls for awareness training programs and IS compliance awareness campaigns.	86,67%	1	80,00%	2	73,33%	3		
	A9: Conduct internal Audits (government) to ensure CDPF processes and procedures are being followed.	83,33%	2	80,00%	2	76,67%	2		
	A10: Undertake management review of the CDPF.	86,67%	1	80,00%	2	80,00%	1		

(continued)

Table 4. (*continued*)

	Ensure the improvements achieve their objectives.								
	A11: Implement the identified improvements in the CDPF.	83,33%	2	80,00%	2	73,33%	3		
	A12: Take appropriate corrective actions.	83,33%	2	80,00%	2	73,33%	3		
	A13: Repeat A8 and onwards to establish continuous improvement.	83,33%	2	83,33%	1	73,33%	3		
	Total Avg. Rating Score for Relevance, Value & Practicality:	***84,36%***		***80,51%***		***75,64%***			
Domains (D1-D3)	D1: IS Awareness Maturity	83,33%	2	83,33%	2	73,33%	2		
	D2: IS Trust Maturity	86,67%	1	83,33%	2	73,33%	2	81,11%	2
	D3: IS Behaviour Maturity	83,33%	2	86,67%	1	76,67%	1		
	Total Avg. Rating Score for Relevance, Value & Practicality:	***84,44%***		***84,44%***		***74,44%***			
Dimensions (Dim1-Dim4)	Dim1: Government	83,33%	1	83,33%	2	76,67%	2		
	Dim2: Public Consumers	80,00%	2	86,67%	1	76,67%	2	81,67%	1
	Dim3: Regulatory Bodies	83,33%	1	86,67%	1	80,00%	1		
	Dim4: Public/Private Organizations	83,33%	1	83,33%	2	76,67%	2		
	Total Avg. Rating Score for Relevance, Value & Practicality:	***82,50%***		***85,00%***		***77,50%***			
PDCA Process Steps (Step1-Step4)	Step1: Plan	80,00%		80,00%		70,00%			
	Step 2: Do	80,00%		76,67%		86,67%		78,06%	5
	Step 3: Check	80,00%		80,00%		73,33%			
	Step 4: Act	76,67%		80,00%		73,33%			
	Total Avg. Rating Score for Relevance, Value & Practicality:	***79,17%***		***79,17%***		***73,03%***			

As indicated in Table 4, the CDPMM Dimensions component obtained the highest total average rating percentage of 81,67% (ranked 1st place) for its relevance, value and practicality, followed by the CDPMM Domains component which obtained a total average rating percentage of 81,11% (ranked 2nd place). The CDPMM Activities component obtained a total average rating percentage of 80,17% (ranked 3rd place), followed by the CDPMM Maturity Levels component which obtained a total average rating percentage of 78,67% (ranked 4th place). Lastly, the CDPMM PDCA Process Steps component obtained the lowest total average rating percentage of 78,06% (ranked 5th place). All the components of the CDPMM obtained a total average rating percentage over 75,00% and therefore it can be argued that all CDPMM components are considered relevant, valuable and practical for measuring the progress/maturity of a CDPF implementation.

7 Conclusions

The mass utilization of consumer personal information by non-compliant organizations poses a significant risk to consumer privacy. It would seem that consumers are largely unaware of these risks due to a lack of awareness regarding organizations' IS compliance statuses, which cause consumers to share their personal information with untrustworthy organizations, and as such putting themselves at risk. In recognition of this matter, the researchers wish to promote a culture of personal data protection amongst consumers within South Africa. Consumer IS compliance awareness training initiatives play a big role in cultivating such a culture. There is currently no government-led/sponsored IS compliance awareness training initiatives in SA and no means of measuring the progress/maturity of such implementation. To address the former gap, the researchers proposed a CDPF for IS compliance awareness in a previous study [39]. The primary research objective of this paper is to complement the CDPF by proposing a CDPMM to assist the South African government with measuring and tracking the progress/maturity of the CDPF implementation as a means of driving continuous improvement. It is anticipated that the inclusion of the CDPMM to accompany the CDPF implementation will assist the South African government to instil an improved Consumer Data Protection culture amongst South African citizens.

The CDPMM can be used to derive a roadmap towards an evolutionary improvement of the CDPF capabilities (activities) and its effectiveness and efficiency. The first part of the paper elaborates the CDPF activities requirements which were used as a reference baseline to investigate whether existing maturity models are capable of holistically assessing these activities. The findings revealed that existing maturity models do not cover the entire reference baseline sufficiently, since they only selectively address some of the activities, with limited focus on the CDP Domains. Hence, no existing maturity model can solve the identified problem. Finally, it was decided to design a new maturity model in line with the defined research strategy. In the second part of the paper, the development of the CDPMM for Consumer Data Protection was described, including the model itself as well as the evaluation techniques that may be employed. The developed model is based on existing maturity model structures and inherits concepts and aspects of these maturity models, whilst following the PDCA cycle of ISO/IEC 27001 [11].

The authors also evaluated the CDPMM components through an iterative expert review process to determine its relevance, value and practicality utilizing Likert scale measurements. The overall findings of the expert reviews indicated the components of the CDPMM to be relevant, valuable and practical for measuring the progress of the CDPF implementation. The authors suggest testing the implementation of the CDPF and CDPMM through case studies in the South African government, as well as in the government setting of other countries in future work as this could lead to a more generic CDPMM to enable cross-country benchmarking.

References

1. Bredenkamp, I.E., Kritzinger, E., Herselman, M.: A conceptual framework for consumer IS compliance awareness: South African government context. In: Silhavy, R. (eds.) Informatics and Cybernetics in Intelligent Systems. CSOC 2021. LNCS, vol. 228, pp. 682–701 (2021). Springer, Cham. https://doi.org/10.1007/978-3-030-77448-6_66
2. Connolly, L.Y., Lang, M., Gathegi, J., Tygar, D.J.: Organisational culture, procedural countermeasures, and employee security behaviour. Inf. Comput. Secur. **25**(2), 118–136 (2017)
3. Tsohou, A., Karyda, M., Kokolakis, S.: Analyzing the role of cognitive and cultural biases in the internalization of information security policies: recommendations for information security awareness programs. Comput. Secur. **52**, 128–141 (2015). https://doi.org/10.1016/j.cose.2015.04.006
4. Zhang, X., Ghorbani, A.A.: Human factors in cybersecurity: issues and challenges in big data. In: Security, Privacy, and Forensics Issues in Big Data, IGI Global, pp. 66–96 (2020)

5. Hadlington, L.: The human factor in cybersecurity: exploring the accidental insider. In: Research Anthology on Artificial Intelligence Applications in Security, IGI Global, pp. 1960–1977 (2021)
6. Hevner, A.R., March, S.T., Park, J., Ram, S.: Design science in information systems research. MIS Q. 75–105 (2004)
7. Hevner, A., vom Brocke, J., Maedche, A.: Roles of digital innovation in design science research. Bus. Inf. Syst. Eng. **61**(1), 3–8 (2018). https://doi.org/10.1007/s12599-018-0571-z
8. Becker, J., Knackstedt, R., Pöppelbuß, J.: Developing maturity models for IT management. Bus. Inf. Syst. Eng. **1**(3), 213–222 (2009). https://doi.org/10.1007/s12599-009-0044-5
9. Hevner, A., Chatterjee, S.: Design science research in information systems. In: Design Research in Information Systems. Integrated Series in Information Systems, vol. 22, pp. 9-22. Springer, Boston, MA (2010). https://doi.org/10.1007/978-1-4419-5653-8_2
10. vom Brocke, J.: Design principles for reference modeling: reusing information models by means of aggregation, specialisation, instantiation, and analogy. In: Reference modeling for business systems analysis, IGI Global, pp. 47–76 (2007)
11. Omoyiola, B.O.: The evolution of information security. IOSR J. Comput. Eng. **22**(3), 50–54 (2020). https://doi.org/10.1016/j.neunet.2012.06.003
12. Le, N.T., Hoang, D.B.: Capability maturity model and metrics framework for cyber cloud security. Scalable Comput. **18**(4), 277–290 (2017). https://doi.org/10.12694/scpe.v18i4.1329
13. Barclay, C.: Sustainable security advantage in a changing environment: the cybersecurity capability maturity model (CM2). In: Proceedings of 2014 ITU Kaleidosc. Academic Conference Living a Converg. World - Impossible Without Standards K 2014, pp. 275–282 (2014) https://doi.org/10.1109/Kaleidoscope.2014.6858466.
14. Pratiwi, A., Indah, D.R., Jauhari, J., Firdaus, M.A.: Security capability assessment on network monitoring information system using COBIT 5 for information security. In: Sriwijaya International Conference on Information Technology and Its Applications (SICONIAN 2019), pp. 167–171(2020)
15. Umar, R., Riadi, I., Handoyo, E.: Analysis security of SIA based DSS05 on COBIT 5 using capability maturity model integration (CMMI). J. Appl. Eng. Technol. Sci. **1**(2), 124–135 (2020)
16. ISACA, COBIT 5: A business framework for the governance and management of enterprise IT. Isaca (2012)
17. Mugobo, V.V., Baschiera, M.G.: Chapter3_The impact of personalized engagement with customers and efficient stock management software systems on customer service at a clothing retailer in cape town, South Africa. Mediterr. J. Soc. Sci. **6**(1), 40–49 (2015). https://doi.org/10.5901/mjss.2015.v6n1p40.15
18. El, S., Youssfi, K., Boutahar, J.: CAT5: a tool for measuring the maturity level of information technology governance using COBIT 5 framework. Int. J. Adv. Comput. Sci. Appl. **7**(2), 385–391 (2016). https://doi.org/10.14569/ijacsa.2016.070253
19. Almuhammadi, S., Alsaleh, M.: Information security maturity model for NIST cyber security framework. Comput. Sci. Inf. Technol. **7**(3), 51–62 (2017)
20. Barrett, M.: Framework for improving critical infrastructure cybersecurity. In: Proceedings Annual ISA Analysis Div. Symposium, vol. 535, pp. 9–25 (2018)
21. von Solms, S.H.B.: A maturity model for part of the African Union convention on cyber security. In: 2015 Science and Information Conference (SAI), pp. 1316 1320 (2015)
22. Adler, R.M.: A dynamic capability maturity model for improving cyber security. In: 2013 IEEE International Conference on Technologies for Homeland Security (HST), pp. 230–235 (2013)
23. Payette, J., Anegbe, E., Caceres, E., Muegge, S.: Secure by design: cybersecurity extensions to project management maturity models for critical infrastructure projects. Technol. Innov. Manag. Rev. **5**(6), 26–34 (2015). https://doi.org/10.22215/timreview904

24. Lessing, M.M.: Best practices show the way to information security maturity. In: 6th National Confernce Process Establish Assess Improvement Inforamtion Technology (ImproveIT 2008), pp. 1–9 (2008)
25. Dzazali, S., Sulaiman, A., Zolait, A.H.: Information security landscape and maturity level: case study of Malaysian public service (MPS) organizations. Gov. Inf. Q. **26**(4), 584–593 (2009). https://doi.org/10.1016/j.giq.2009.04.004
26. Group, C.: Secretary of defense Corporate fellows program (2000). http://www.ndu.edu/sdcfp/reports/Citigroup.doc. Accessed 26 Jun 2020
27. Karokola, G., Kowalski, S., Yngström, L.: Towards an information security maturity model for secure e-Government services: a stakeholders view. In: Proceedings 5th International Symposium Human Asp. Information Security Assurance HAISA 2011, pp. 58–73 (2011)
28. Hohan, A.I., Olaru, M., Pirnea, I.C.: Assessment and continuous improvement of information security based on TQM and business excellence principles. Procedia Econ. Financ. **32**(15), 352–359 (2015). https://doi.org/10.1016/s2212-5671(15)01404-5
29. Group, T.O.: The Open Group Releases Maturity Model for Information Security Management O-ISM3 Framework Ensures Security Management Processes Operate at a Level Consistent with Business Requirements (2011). https://www.opengroup.org/open-group-releases-maturity-model-information-security-management.
30. Newhouse, W., Keith, S., Scribner, B., Witte, G.: National Initiative for Cybersecurity Education Cybersecurity Workforce Framework NIST Spec. Publ. pp. 1–385 (2017)
31. Drivas, G., Chatzopoulou, A., Maglaras, L., Lambrinoudakis, C., Cook, A., Janicke, H.: A NIS directive compliant cybersecurity maturity assessment framework (2020). http://arxiv.org/abs/2004.10411
32. Rea-Guamán, A.M., Sánchez-García, I.D., San Feliu, T., Calvo-Manzano, J.A.: 16 Maturity models in cybersecurity: a systematic review. In: 2017 12th Iberian Conference on Information Systems and Technologies (CISTI), pp. 1–6 (2017)
33. Kurniawan, E., Riadi, I.: Security level analysis of academic information systems based on standard ISO 27002:2003 using SSE-CMM (2018). https://doi.org/10.13140/RG.2.2.20925.15840
34. SANS, The Rising Era of Awareness Training SANS Security Awareness Report (2019). https://www.knowbe4.com/hubfs/SANS-Security-Awareness-Report-2019.pdf
35. Parsons, K., McCormac, A., Butavicius, M., Pattinson, M., Jerram, C.: Determining employee awareness using the human aspects of information security questionnaire (HAIS-Q). Comput. Secur. **42**(May), 165–176 (2014). https://doi.org/10.1016/j.cose.2013.12.003
36. Parsons, K.M., Young, E., Butavicius, M.A., McCormac, A., Pattinson, M.R., Jerram, C.: The influence of organizational information security culture on information security decision making. J. Cogn. Eng. Decis. Mak. **9**(2), 117–129 (2015)
37. Parsons, K., Calic, D., Pattinson, M., Butavicius, M., McCormac, A., Zwaans, T.: The human aspects of information security questionnaire (HAIS-Q): two further validation studies. Comput. Secur. **66**, 40–51 (2017). https://doi.org/10.1016/j.cose.2017.01.004
38. Mccormac, A., Parsons, K., Zwaans, T., Butavicius, M., Pattinson, M.: Test-retest reliability and internal consistency of the human aspects of information security questionnaire (HAIS-Q), Australas. Conference Information System, pp. 1–10 (2016)
39. Bredenkamp, I.E., Kritzinger, E., Herselman, M.: A conceptual framework for consumer is compliance awareness: South African government context. In: Silhavy, R. (eds.) Informatics and Cybernetics in Intelligent Systems. CSOC 2021. Lecture Notes in Networks and Systems, vol. 228 (2021). Springer, Cham. https://doi.org/10.1007/978-3-030-77448-6_66

Analyzing Correlation of the Relationship Between Technical Complexity Factors and Environmental Complexity Factors for Software Development Effort Estimation

Ho Le Thi Kim Nhung(✉), Vo Van Hai, and Huynh Thai Hoc

Faculty of Applied Informatics, Tomas Bata University in Zlin, Nad Stranemi 4511, Zlin 76001, Czech Republic
{lho,vo_van,huynh_thai}@utb.cz

Abstract. In this paper, a new method called Correlation-based Feature Selection in Correction Factors is proposed. The method is based on the feature selection method used in software development effort estimation to reduce redundant correction factors. In this paper, the impact of correlation-based feature selection on the method's estimation accuracy is investigated. Multiple linear regression was used as the basic technique for the correction factors preprocessed by the feature selection method. The results were evaluated using six unbiased accuracy measures through the 5-fold cross-validation over the historical dataset. The proposed method leads to a significant improvement in estimation accuracy by simplifying the evaluation of correction factor values in the use case points method, thus increasing the usefulness of the proposed method in practice.

Keywords: Software development effort estimation · Use case points · Correlation-based feature selection · Multiple linear regression

1 Introduction

Software Development Effort Estimation (further only SDEE) is a critical factor in the early stages of the software development life cycle. SDEE can help project managers make early software development decisions, i.e., to prepare the necessary project plan and budget within the expected completion time [1, 2]. The application of appropriate SDEE methods determines the success or failure of a software project. In this context, various approaches have been proposed and evaluated to address this problem. SDEE approaches can be categorized as expert judgment, parametric techniques, and machine learning techniques [3].

The Use Case Points (further only UCP) method can be used as a functional size metric in the early stages of the software lifecycle [4]. The UCP method is based on a Use Case Model (further only UCM) structured scenario and actor analysis. The method estimates effort based on software size and fixed productivity factors (20 person-hours). However, the original UCP method was analyzed for its low precision [5–8].

R. Silhavy et al. (Eds.): CoMeSySo 2021, LNNS 232, pp. 835–848, 2021.
https://doi.org/10.1007/978-3-030-90318-3_65

Combining ML techniques to produce SDEE models based on the original UCP formula could be a method to improve accuracy. Some estimation approaches [9–18] have also explored variant models, especially the use of statistical regression or ML models to obtain better effort estimation based on historical data. The benefits of using UCP-based effort estimation methods are numerous. These methods have made many advances in reducing the influence of human error during UCM analysis and simplifying the original principles of UCP. We recently conducted a literature review on UCP-based SDEE methods in the context of system development [19]. This review highlighted that there are some challenges in estimating the components of UCP, especially the correction factors - which are thirteen technical complexity factors (TCF) and eight environmental complexity factors (ECF). According to Luis et al. [20], a small variation in the weighting value of the correction factors can dramatically affect the software size. Nassif et al. [21] also pointed out the need to refine correction factors that are directly related to estimates computed by the UCP method.

In the UCP method, not all correction factors are crucial for finding the hidden knowledge among the important data. In many cases, some of the factors under consideration are irrelevant and unnecessary. By selecting appropriate subsets of correction factors, we can reduce the complexity of the UCP method and UCP-based SDEE methods. The motivation for us to evaluate whether or not the influence of the feature selection method on correction factors improves the accuracy of the UCP method. To address these issues, this paper proposes the Correlation-based Feature Selection in Correction Factors (named CFSiCF) method for SDEE. We evaluate the construction of correction factors based on the machine learning technique preprocessed with a feature selection method. The method used is the Correlation-based Feature Selection (further only CFS) [22, 23]. The Multiple Linear Regression (further only MLR) technique is chosen as the base method for the selected correction factors to minimize human error's influence during the analysis of these factors. The research questions are answered as follows:

- RQ1: What is the correlation and benefit of the number of technical complexity and environmental complexity factors for size estimation?
- RQ2: Is the proposed method more estimation accurate than the UCP method and other tested methods?

The main contributions of this study are:

- Evaluating the construction of the best correction factors based on MLR models preprocessed with the CFS algorithm.
- The results achieved by the proposed method are compared with three tested estimation methods to verify their accuracy. The methods are run on the dataset of projects from three data donators. The project by each data donator differs in size (measured in UCP). Unbiased evaluation measures (8–13) were used to evaluate the accuracy of the methods.

The rest of the article is organized as follows: Sect. 2 introduces the background of the methods used. Section 3 describes the research methodology. Section 4 demonstrates the experimental evaluation. Finally, Sect. 5 concludes the paper and suggests future work.

2 Background

2.1 Use Case Points

The UCP method was introduced by Karner [4] to estimate the size of object-oriented software projects. The UCP method is calculated by computing four basic size metrics - Unadjusted Actor Weight (UAW), Unadjusted Use Cases Weight (UUCW), Technical Complexity Factors (TCF), and Environmental Complexity Factors (ECF), as shown in Eq. (1).

$$UCP = (UAW + UUCW) \times TCF \times ECF \tag{1}$$

The UAW is calculated by taking the weighted sum of the number of actors in each type, as shown in Eq. (2). The actors are classified based on their complexity (see Table 1).

$$UAW = \sum_{i=1}^{3} a_i \times w \tag{2}$$

where, a_i is the number of actors in the ith actor type and w_i is the associate complexity weight for each type.

Table 1. Actor classification and their weights.

Actor classification	Complexity weight
Simple	1
Average	2
Complex	3

The UUCW is calculated by taking the weighted sum of number of use cases as shown in Eq. (3). The use cases are classified based on the number of transactions in the use case (see Table 2).

$$UUCW = \sum_{j=1}^{3} uc_j \times w_j \tag{3}$$

where, uc_j is the number of use case in the jth use case type and w_j is the associate complexity weight for each type.

The TCF is calculated from thirteen factors representing the complexity of software projects, as shown in Eq. (4). Table 3 presents the technical factors as defined in the UCP.

$$TCF = 0.6 + \left(0.01 \times \sum_{i=1}^{13} t_i \times fw_i\right) \tag{4}$$

where, t_i is the value of complexity factor i, and fw_i is the weight of factor i.

The ECF is calculated from a set of eight factors that describe the non-functional requirements, as shown in Eq. (5). The environmental factors are listed in Table 4.

$$ECF = 1.4 - \left(0.03 \times \sum_{i=1}^{8} e_i \times ew_i\right) \tag{5}$$

where, e_i is the value of complexity factor i, and ew_i is the weight of factor i.

Table 2. Use case classification and their weights.

Use case classification	Number of transactions	Complexity weight
Simple	(0–4)	5
Average	<4–7>	10
Complex	(7 -∞)	15

Table 3. Technical complexity factors

Factor ID	Description	Weight
T1	Distributed system	2
T2	Response adjectives	2
T3	End-user efficiency	1
T4	Complex processing	1
T5	Reusable code	1
T6	Easy to install	0.5
T7	Ease of use	0.5
T8	Portable	2
T9	Easy to change	1
T10	Concurrent	1
T11	Security feature	1
T12	Access for third parties	1
T13	Special training required	1

2.2 Correlation-Based Feature Selection

Performing feature selection is considered a step of data preprocessing to determine the best subset of features to improve estimation accuracy [24, 25]. Feature selection techniques can be classified: Filter, Wrapper, and Hybrid algorithms. Filter methods select the most relevant features based on the properties of a dataset. In contrast, wrapper methods select the best feature subset based on assessing the effects of various subsets

Table 4. Environmental complexity factors

Factor ID	Description	Weight
E1	Familiar with RUP	1.5
E2	Application experience	0.5
E3	Object-oriented experience	1
E4	Lead analyst capability	0.5
E5	Motivation	1
E6	Stable requirements	2
E7	Part-time workers	– 1
E8	Difficult programming language	2

of features on the performance of an estimation system. Embedded or hybrid methods determine the best feature subset by performing the selection step and model building concurrently or combining filtering and wrapper techniques.

In this study, we use the CFS algorithm, which uses correlation to evaluate a feature subset derived from the Pearson correlation coefficient. The CFS evaluates different subsets of features according to the heuristic evaluation function and selects the best one.

$$Heuristic_{fs} = \frac{nr_{fcc}}{\sqrt{n + n(n-1)r_{ffi}}} \tag{6}$$

where, $Heuristic_{fs}$ is the heuristic value of a feature subset fs containing n features, r_{fcc} is the average feature-class correlation, and r_{ffi} is the average feature-feature intercorrelation [26]. To apply $Heuristic_{fs}$, a correlation matrix and heuristic search are computed to find a best subset of features. In this research, the CFS algorithm is used with Greedy stepwise forward search by the space of attribute subsets.

2.3 Multiple Linear Regression

MLR is one of the types of effort estimation methods used to determine how the dependent variable is related to a change in the independent variables and which independent variable is relevant to the dependent variable [27]. The regression models are built based on historical data. The regression models are then evaluated and compared with alternative models [9–18]. The MLR model is shown in Eq. (7):

$$Y = \alpha_0 + \alpha_1 X_1 + \alpha_2 X_2 + \ldots + \alpha_n X_n + \varepsilon \tag{7}$$

where Y is the dependent variable is related to the independent variables $X_1, \ldots, X_n$; α_0 is the intercept parameter; $\alpha_1, \ldots, \alpha_n$ are the regression coefficients; and ε_i are the error residuals.

3 Research Methodology

3.1 Data Description

The methods are evaluated on a dataset of 70 observations (each observation represents a description of a realized software system) from three data donators [27]. All data donators work in different government, health, and business sectors. The projects were installed in Java and C# programming languages. For comparison and estimation accuracy of the methods, a standard of 20 man-hours per UCP [4] is used, without considering the productivity factor (PF - man-hours per 1 UCP). The statistical properties of the dataset are described in Table 5. Figure 1 shows the boxplot of Real_P20 in the dataset. Real_P20 is the real effort in man-hours divided by productivity.

Table 5. Statistical characteristics of the dataset

	Median man-hours	Median Real_P20	Standard deviation	Minimum Real_P20	Maximum Real_P20	n
Dataset	6406	320.3	33.21	288.75	398.5	70

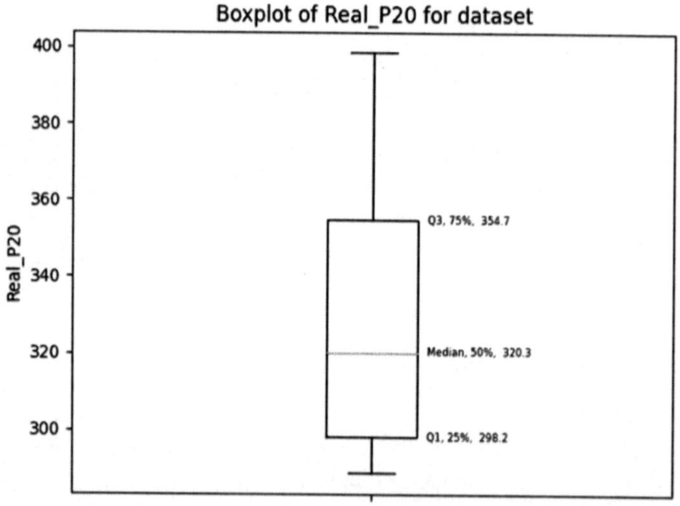

Fig. 1. Boxplot of real_P20 in the dataset

3.2 Evaluation Criteria

To evaluate the accuracy of the methods, we used evaluation criteria commonly used in the field of the software system size estimation [28]. The accuracy measure that used in this paper are Mean Absolute Residual (MAR), Mean of Magnitude of Relative Error

(MMRE), Percentage of prediction within x% (PRED(x)), Mean Absolute Percentage Error (MAPE), Sum of Squared Errors (SEE), and the Mean Squared Error (MSE). These measures were chosen because they behave very differently. The best method is the one where MAR, MMRE, MAPE, SSE and MSE are minimized and PRED (x) is maximized.

Mean absolute residual (MAR)

$$MAR = \frac{1}{n}\sum\nolimits_{i=1}^{n} \left|y_i - \hat{y}_i\right| \tag{8}$$

Mean magnitude of relative error (MMRE)

$$MMRE = \frac{1}{n}\sum\nolimits_{i=1}^{n} \frac{\left|y_i - \hat{y}_i\right|}{y_i} \tag{9}$$

Percentage of prediction within x% (PRED(x))

$$PRED(x) = \frac{1}{n}\sum\nolimits_{i=1}^{n} \{1 if \frac{|y_i - \hat{y}_i|}{y_i} \leq x; 0\, otherwise \tag{10}$$

Mean absolute percentage error (MAPE)

$$MAPE = \frac{1}{n}\sum\nolimits_{i=1}^{n} \frac{\left|y_i - \hat{y}_i\right|}{y_i} \times 100 \tag{11}$$

Sum of square error (SSE)

$$SSE = \sum\nolimits_{i=1}^{n} \left(y_i - \hat{y}_i\right)_i^2 \tag{12}$$

Mean squared error (MSE)

$$MSE = \frac{1}{n}\sum\nolimits_{i}^{n} \left(y_i - \hat{y}_i\right)_i^2 \tag{13}$$

where, n is the number of observations, y_i is the known real value, $\hat{y}_i$ is the estimated value.

3.3 Methodology Used

This section describes the procedure for investigating the accuracy of the proposed method. As described in Sect. 2, the CFS algorithm selects the most appropriate factors for the correction factors to be used as input to the MLR models. The methodology for the historical dataset is as follows:

Step 1: Using the CFS algorithm to determine the correlation and benefit of the number of technical complexity and environmental complexity factors for size estimation.
Step 2: Building the two MLR models based on the best correction factors preprocessed by the algorithm CFS. The regression model on CFS-based selected technical factors is called TCF_{CFS}. The regression model on CFS-based selected environmental factors is called ECF_{CFS}.

Step 3: The effort estimation result of the proposed CFSiCF method is calculated as the aggregate of four metrics - UAW (Eq. (2), UUCW (Eq. (3), TCF_{CFS}, and ECF_{CFS}

$$UCP_{CFSiCF} = (UAW + UUCW) \times TCF_{CFS} \times ECF_{CFS} \tag{14}$$

Step 4: Evaluate the methods using six performance measures (MAR, MMRE, PRED (0.25), MAPE, SEE, and MSE).

4 Results

This section evaluates and discusses the experimental results according to the methodology detailed in Sect. 3.

4.1 RQ1

The accuracies of experimental validation for the proposed CFSiCF method and other tested methods are listed at Fig. 2, 3, 4, 5, 6 and 7. As the results, we can confirm that the proposed CFSiCF method gives the best MAR, MMRE, MAPE, SSE, MSE, PRED (x) values in four selected different subsets of factors.

Specifically, the correlation and benefit of the number of technical complexity and environmental complexity factors in descending order of effectiveness (see Table 6): Case 1 - eight selected technical factors (T1, T2, T3, T4, T5, T7, T9, T10) and four selected environmental factors (ENV5, ENV6, ENV7, ENV8); Case 2 - eight selected technical factors (T1, T2, T3, T4, T5, T7, T9, T10) and five selected environmental factors (ENV4, ENV5, ENV6, ENV7, ENV8); Case 3 - eight selected technical factors (T1, T2, T3, T4, T5, T7, T9, T10) and seven selected environmental factors (ENV1, ENV3, ENV4, ENV5, ENV6, ENV7, ENV8); and Case 4 - eight selected technical factors (T1, T2, T3, T4, T5, T7, T9, T10) and eight selected environmental factors (ENV1, ENV2, ENV3, ENV4, ENV5, ENV6, ENV7, ENV8).

Based on the above results, RQ1 can be answered using the CFS method to lead different subsets of factors that build diverse CFSiCF methods.

Table 6. Selected factors on TCF and ECF

Case	Selected technical factors	Selected environmental factors
1	T1, T2, T3, T4, T5, T7, T9, T10	ENV5, ENV6, ENV7, ENV8
2	T1, T2, T3, T4, T5, T7, T9, T10	ENV4, ENV5, ENV6, ENV7, ENV8
3	T1, T2, T3, T4, T5, T7, T9, T10	ENV1, ENV3, ENV4, ENV5, ENV6, ENV7, ENV8
4	T1, T2, T3, T4, T5, T7, T9, T10	ENV1, ENV2, ENV3, ENV4, ENV5, ENV6, ENV7, ENV8

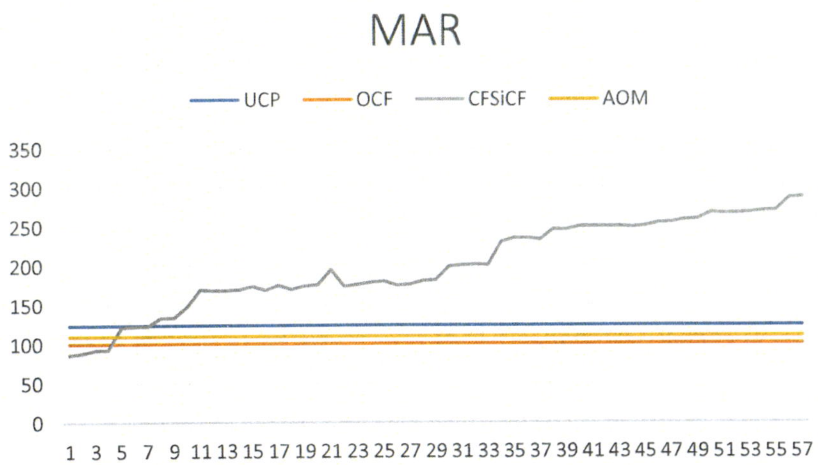

Fig. 2. The MAR results of the UCP, AOM and CFSiCF methods

4.2 RQ2

We measured the accuracy improvements achieved by the proposed CFSiCF method over the baseline method - UCP and other tested methods - AOM[14] and OCF[29]. As the results in Table 7, the proposed method outperforms all other methods with superior accuracy in MAR, MMRE, MAPE, SSE, MSE, PRED (x).

Specifically, the proposed CFSiCF method outperforms the UCP, OCF and AOM methods by 33.45–43.50%, 8.42– 16.58% and 18.48–27.41% respectively for MAR, by

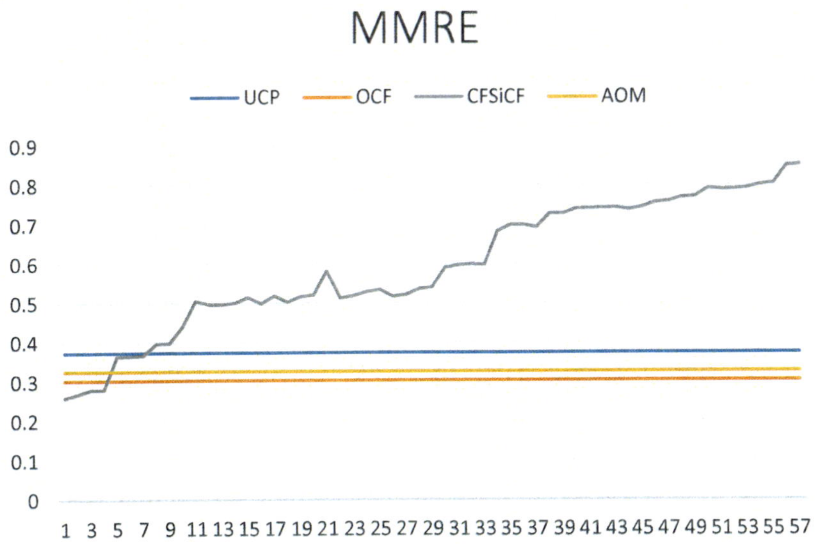

Fig. 3. The MMRE results of the UCP, AOM and CFSiCF methods

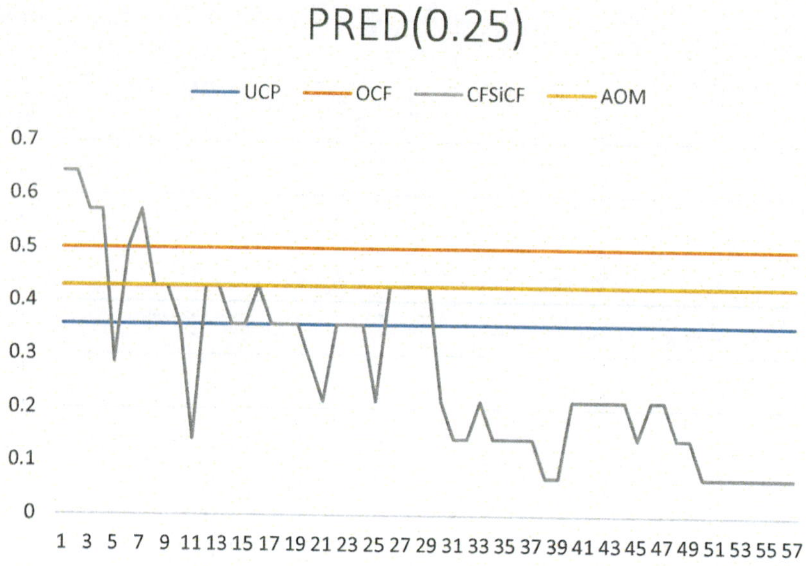

Fig. 4. The PRED(0.25) results of the UCP, AOM and CFSiCF methods

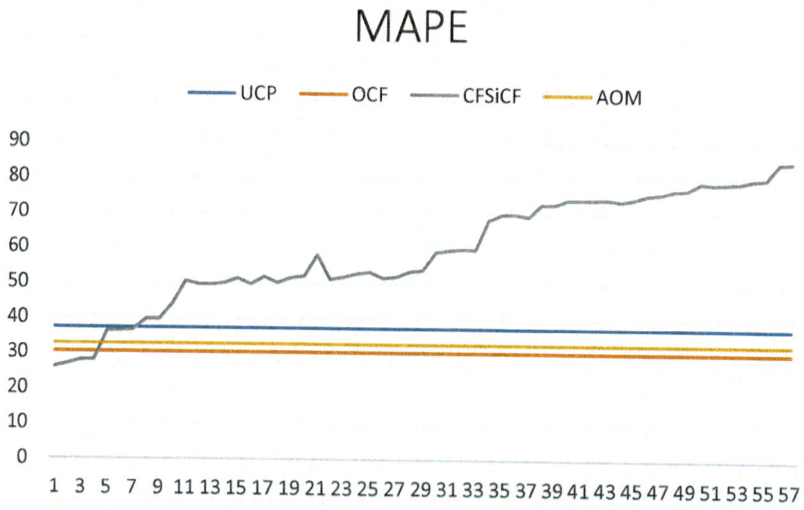

Fig. 5. The MAPE results of the UCP, AOM and CFSiCF methods

33.57 –43.85%, 8.57–16.92% and 16.78–25.77% respectively for MMRE, by 37.47–44.48%, 12.43–22.24% and 24.87– 33.28% respectively for PRED (0.25), by 33.43–43.72%, 8.43–16.79% and 16.66–25.66% respectively for MAPE, by 1.69–2.24 times, 1.19–1.57 times and 1.39–1.85 times respectively for SSE and MSE.

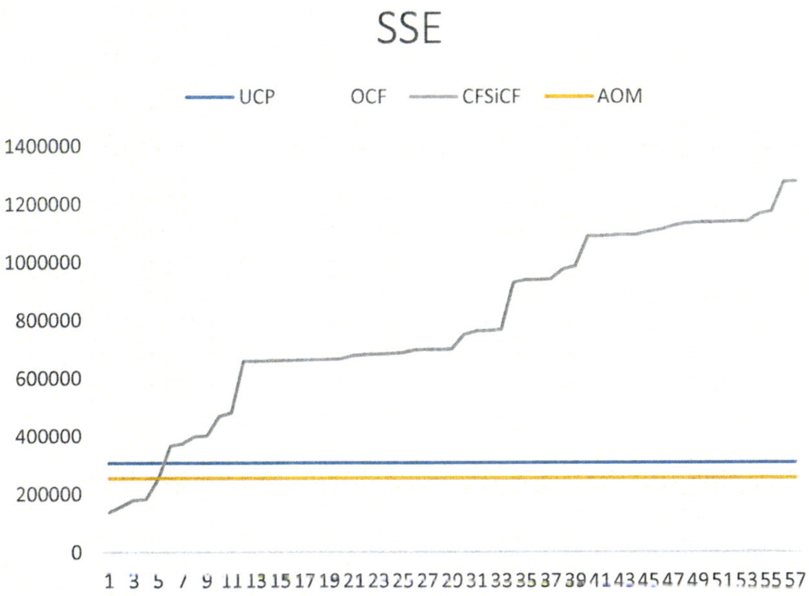

Fig. 6. The SSE results of the UCP, AOM and CFSiCF methods

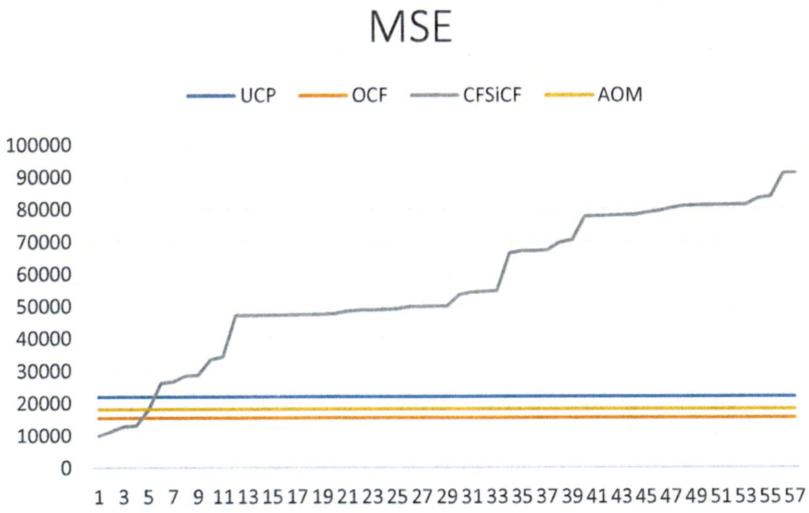

Fig. 7. The MSE results of the UCP, AOM and CFSiCF methods

Table 7. Comparison of the proposed method to other method variants

Method	MAR	MMRE	PRED(0.25)	MAPE	SSE	MSE
UCP	123.753	0.374	0.35	37.371	307472.9	21962.3
AOM	109.872	0.327	0.42	32.676	253954.5	18139.6
OCF	100.535	0.304	0.5	30.37	216512.7	15465.1
CFSiCF (Case 1)	86.234	0.26	0.64	26.002	130774.5	9791.04
CFSiCF (Case 2)	88.474	0.269	0.64	26.87	157356.1	11239.7
CFSiCF (Case 3)	92.454	0.28	0.57	27.954	178883.7	12777.4
CFSiCF (Case 4)	92.73	0.28	0.57	28.008	181691.4	12977.9

5 Conclusions

In this experimental study, we investigated the influence of the feature selection method on the correction factors to improve the estimation accuracy of the UCP method. For this purpose, the MLR method was chosen for correction factors preprocessed by the CFS algorithm. The proposed CFSiCF method was compared with the other tested methods on six accuracy measures using the 5-fold cross-validation method.

In the evaluation of RQ1, it can be concluded that the effect of the CFS algorithm on the correction factors was demonstrated. The most accurate estimation results were achieved with eight technical factors (T1, T2, T3, T4, T5, T7, T9, T10) and four environmental factors (ENV5, ENV6, ENV7, ENV8). It also follows from the analysis that in answer to RQ2, it can be said that the proposed CFSiCF method leads to a significant increase in estimation accuracy compared to the UCP, AOM, and OCF methods. It is also worth highlighting the simplification of the evaluation of the correction factor values, which increases the usefulness of the proposed CFSiCF method in practice.

Future research will further investigate other feature selection methods that can increase diversity and produce more estimation accuracy of the CFSiCF method. In addition, further research direction using clustering approaches will be considered a solution in our future work.

Acknowledgment. This study was supported by the Faculty of Applied Informatics, Tomas Bata University in Zlin, under Project IGA/CebiaTech/2021/001.

References

1. Kotonya, G., Sommerville, I.: Requirements Engineering: Processes and Techniques. Wiley, 1st edition (1998)
2. Trendowicz, A., Jeffery, R.: Software Project Effort Estimation. Springer, Foundations and Best Practice Guidelines of Success (2014)
3. Boehm, B., Abts, C., Chulani, S.: Software development cost estimation approaches - a survey. Ann. Softw. Eng. **10**, 177–205 (2000). https://doi.org/10.1023/A:1018991717352

4. Karner: Resource Estimation for Objector Projects. Objective Systems SFAB (1993)
5. Azzeh, M., Nassif, A.B., Attili, I.B.: Predicting software effort from use case points: a systematic review. Sci. Comput. Program. **204**, 102596 (2021)
6. Singh, Y., Bhatia, P.K., Kaur, A., Sangwan, O.P.: A review of studies on effort estimation techniques of software development. In: 2nd National Conference Mathematical Techniques: Emerging Paradigms for Electronics and IT Industries (2008)
7. Mahmood, Y., Kama, N., Azmi, A.: A systematic review of studies on use case points and experts-based estimation of software development effort. J. Softw. Evol. Process (2019)
8. Subriadi, A.P.S, Ningrum, P.A.: Critical review of the effort rate value in use case point method for estimating software development effort. J. Theor. Appl. Inf. Technol. **59** (2005)
9. Nassif, A.B., Capretz, L.F., Ho, D.: Enhancing use case points estimation method using soft computing techniques. J. Glob. Res. Comput. Sci. (2010)
10. Nassif, A.B., Capretz, L.F., Ho, D.: Estimating software effort based on use case point model using sugeno fuzzy inference system. In: 23rd IEEE International Conference on Tools with Artificial Intelligence, pp. 393–398 (2011)
11. Nassif, A.B., Ho, D., Capretz, L.F.: Regression model for software effort estimation based on the use case point method. In: International Conference on Computer and Software Modeling, IPCSIT, vol.14 (2011)
12. Jorgensen, M.: Regression models of software development effort estimation accuracy and bias. Empir. Softw. Eng. **9**, 297–314 (2004)
13. Ochodek, M., Nawrocki, J., Kwarciak, K.: Simplifying effort estimation based on use case points. Inf. Softw. Technol. **53**, 200–213 (2011)
14. Silhavy, R., Silhavy, P., Prokopova, Z.: Algorithmic optimisation method for improving use case points estimation. PLoS ONE, **10**(11), e0141887 (2015)
15. Nassif, A.B., Caprets, L.F.: Software effort estimation in the early stages of the software life cycle using a cascade correlation neural network model. In: ACIS International Conference on Software Engineering, Artificial Intelligence, Networking and Parallel/Distributed Computing (2012)
16. Iraji, M.S., Motameni, H.: Object oriented software effort estimate with adaptive neuro fuzzy use case size point (ANFUSP). Int. J. Intell. Syst. Appl. **4**, 14–24 (2012)
17. Khatibi Bardsiri, V., Jawawi, D.N.A., Hashim, S.Z.M., Khatibi, E.: A flexible method to estimate the software development effort based on the classification of projects and localization of comparisons. Empir. Softw. Eng. **19**(4), 857–884 (2013). https://doi.org/10.1007/s10664-013-9241-4
18. Chiu, N.H., Huang, S.J.: The adjusted analogy-based software effort estimation based on similarity distances. J. Syst. Softw. **80**, 628–640 (2007). https://doi.org/10.1016/j.jss.2006.06.006
19. Nhung, H.L.T.K., Hoc, H.T., Hai, V.V.: A review of use case-based development effort estimation methods in the system development context. In: Silhavy, R., Silhavy, P., Prokopova, Z. (eds) Intelligent Systems Applications in Software Engineering. CoMeSySo 2019. Advances in Intelligent Systems and Computing, vol. 1046. Springer, Cham (2019). https://doi.org/10.1007/978-3-030-30329-7_44
20. Luis, M.H., Sussy, B.O.: Factors affecting the accuracy of use case points. Trends Appl. Softw. Eng. Adv. Intell. Syst. Comput. **537** (2017)
21. Nassif, A.B., Ho, D., Caprets, L.F.: Towards an early software estimation using Log-linear regression and a multilayer perceptron model. J. Syst. Softw. **86**, 144–160 (2013)
22. Hall, M.A.: Correlation-based feature selection for machine learning. Citeseer **113**, 1–8 (1999)
23. Sánchez-Maroño, N., Alonso-Betanzos, A., Tombilla-Snaromán, M.: Filter methods for feature selection. a comparative study. In: Intelligent Data Engineering and Automated Learning (IDEAL), pp. 178–187 (2007).https://doi.org/10.1007/978-3-540-77226-2

24. Chandra Blessie, E., Karthikeyan, E.: Sigmis: a feature selection algorithm using correlation based method. J. Algor. Comput. Technol. **6**, 385–394 (2012)
25. Idri, A., Cherradi, S.: Improving effort estimation of fuzzy analogy using feature subset selection. In: IEEE Symposium Series on Computational Intelligence (SSCI), pp. 1–8 (2016). https://doi.org/10.1109/SSCI.2016.7849928
26. Hall, M.A., Holmes, G.: Benchmarking attribute selection techniques for discrete class data mining. IEEE Trans. Knowl. Data Eng. **15**, 1437–1447 (2003)
27. Silhavy, R., Silhavy, P., Prokopova, Z.: Analysis and selection of a regression model for the use case points method using a stepwise approach. J. Syst. Softw. **125**, 1–14 https://doi.org/10.1016/j.jss.2016.11.029(2017)
28. Silhavy, R., Silhavy, P., Prokopova, Z.: Using actors and use cases for software size estimation. Electronics **10**, 2021 (2021). https://doi.org/10.3390/electronics10050592
29. Le Thi Kim Nhung, H., Hoc, H.T., Van Hai, V.: An evaluation of technical and environmental complexity factors for improving use case points estimation. In: Silhavy, R., Silhavy, P., Prokopova, Z. (eds.) CoMeSySo 2020. AISC, vol. 1294, pp. 757–768. Springer, Cham (2020). https://doi.org/10.1007/978-3-030-63322-6_64

Entropy Modeling of Random Vectors in the Presence of Discrete Components

Alexander N. Tyrsin[1,2](✉) and Dariya V. Ostroushko[3]

[1] South-Ural State University (National Research University), Chelyabinsk, Russia
[2] Scientific-Engineering Center Reliability and Life of Large Systems and Machines, Ural Branch, Russian Academy of Science, Yekaterinburg, Russia
[3] Mother and Child National Research Center, Minsk, Republic of Belarus

Abstract. Entropy modeling is widely used in the study of open stochastic systems in various fields. However, when using differential entropy to model stochastic systems, all components of a random vector must be continuous random variables. In practice, the phenomena under study are usually continuous, and discreteness occurs when data is grouped or when moving to scoring indicators. The article describes the technique of entropy modeling of multidimensional stochastic systems on the example of groupings and score indicators. It is shown that differential entropy can be used in modeling discrete random variables for cases when they are obtained as a result of grouping data or when moving to scoring indicators. This is achieved by switching from discrete random variables to their approximations by continuous random variables having piecewise linear distribution functions. Examples are considered.

Keywords: Differential entropy · Model · System · Discrete random vector · Grouping · Score indicator

1 Introduction

Entropy is one of the fundamental properties of stochastic systems. Currently, it is quite common to use entropy to describe the behavior of open stochastic systems in various fields [1–6]. Common in these works is the use of the information entropy introduced by K. Shannon [7].

However, the use of information entropy as a model of multidimensional stochastic systems faces difficulties – it is necessary to estimate the probabilities of all possible states of the system (this requires large volumes of samples, in addition, some states may be unknown in advance), and it is also difficult to model the relationships between the elements of multidimensional systems.

Those disadvantages do not exist if differential entropy is used [8, 9]. It is based on the representation of the system as a random vector and the decomposition of its differential entropy into components – the entropy of chaotic and entropy of self-organization. However, all components of the vector must be continuous random variables. This significantly narrows the scope of entropy modeling, since in many applications, for example,

R. Silhavy et al. (Eds.): CoMeSySo 2021, LNNS 232, pp. 849–856, 2021.
https://doi.org/10.1007/978-3-030-90318-3_66

in medicine, economics, their grouped values are often used instead of the actual values of features or their point (rating) estimates are introduced [10–12].

The purpose of the article is to describe the methodology of entropy modeling of multidimensional stochastic systems, all or part of the components of which are scoring indicators or obtained using grouping, as well as its approbation on model and real data.

2 Methods

Let's imagine a multidimensional stochastic system as a random vector $\mathbf{Y} = (Y_1, Y_2, \ldots, Y_m)$. Its differential entropy is equal to

$$H(\mathrm{Y}) = \int_{-\infty}^{+\infty} \cdots \int_{-\infty}^{+\infty} p_{\mathrm{Y}}(x_1, x_2, \ldots, x_m) dx_1 dx_2 \ldots dx_m \tag{1}$$

where $p_{\mathbf{Y}}(x_1, x_2, \ldots, x_m)$ is the distribution density of a random vector $\mathbf{Y}$.

Formula (1) was proposed by C. Shannon in [7] as a formal analogue of the concept of information entropy for an m-dimensional continuous random vector $\mathbf{Y}$. Value (1) was later called differential entropy by A.N. Kolmogorov together with I.M. Gelfand and A.M. Yaglom [13].

$$H(\mathbf{Y}) = \sum\nolimits_{i=1}^{m} ln\sigma_{Y_i} + \sum\nolimits_{i=1}^{m} \kappa_i + \frac{1}{2}\sum\nolimits_{k=2}^{m} ln\left(1 - R^2_{Y_k/Y_1Y_2\ldots Y_{k-1}}\right) \tag{2}$$

where $\kappa_i = H\left(Y_i/\sigma_{Y_i}\right) = H\left(\widehat{Y}_i\right) = -\int_{-\infty}^{+\infty} p_{\widehat{Y}_i}(x)\ln p_{\widehat{Y}_i}(x)dx$ is an entropy indicator of the type of the distribution law of a random variable Y_i; $R^2_{Y_k/Y_1Y_2\ldots Y_{k-1}}$ are the indices of the determination of regression dependencies. The first two terms of $H(\mathbf{Y})_V = \sum_{i=1}^{m} ln\sigma_{Y_i} + \sum_{i=1}^{m} \kappa_i$ are called the entropy of randomness, and the third $H(\mathbf{Y})_R = \frac{1}{2}\sum_{k=2}^{m} ln\left(1 - R^2_{Y_k/Y_1Y_2\ldots Y_{k-1}}\right)$ is the entropy of self-organization.

The problem is that all components of Y_i in (1) must be continuous random variables, which will not allow us to determine entropy indicators κ_i for discrete random variables. This shall be briefly demonstrated. Consider some discrete random variable Z, which has a distribution series presented in Table 1.

Table 1. Random variable distribution series Z.

Z	z_1	z_2	z_3	...	z_{n-1}	z_n
$p_k = P(Z = z_k)$	p_1	p_2	p_3	...	p_{n-1}	p_n

Let's write the distribution function $F_Z(x)$ of a random variable Z

$$F_z(x) = \begin{cases} 0, & x \le z_1, \\ \sum_{i=1}^{k} p_i, & z_k < x \le z_{k+1}, k = 1, \ldots, n-1, \\ 1, & x > z_n. \end{cases}$$

It is obvious that the probability density $p_Z(x)$ of a random variable Z is zero everywhere except at points z_k, and at points z_k does not exist, i.e.

$$p_z(x) = \begin{cases} \lim\limits_{\Delta x \to 0} \frac{p_k}{\Delta x}, & x = z_k, k = 1, \ldots, n, \\ 0, & x \neq z_k. \end{cases}$$

Consider the differential entropy of a discrete random variable Z

$$H(Z) = -\int_{-\infty}^{+\infty} p_z(x) \ln p_z(x) \Delta x = -\sum\nolimits_{k=1}^{n} \lim_{\Delta x \to 0} (p_z(z_k) \ln p_z(z_k) \Delta x) =$$

$$= -\sum_{k=1}^{n} \lim_{\Delta x \to 0} \left(\frac{p_k}{\Delta x} \ln \frac{p_k}{\Delta x} \Delta x \right) = -\sum_{k=1}^{n} \lim_{\Delta x \to 0} \left(p_k \ln \frac{p_k}{\Delta x} \right).$$

Since $\forall k \; 0 < p_k < 1$ and $\lim\limits_{\Delta x \to 0} \ln \frac{1}{\Delta x} = +\infty$, the limit in each term diverges and tends to $+\infty$. Therefore, the differential entropy of a discrete random variable Z does not exist ($H(Z) \to -\infty$).

Thus, when using the entropy model (2), all components of the random vector $\mathbf{Y}$ must be continuous random variables. If some component Y_i is a discrete random variable, then it must be replaced by a continuous one. In general, this cannot be done, because depending on the type of continuous distribution function $F_W(x)$ approximating the function $F_Z(x)$, you can get almost any entropy value $H(W)$, from some constant to any arbitrarily large negative value (with increasing approximation accuracy). Thus, entropy (1) can be used for discrete random variables only if they are obtained from continuous ones by transformations (grouping, transition to point values, etc.). In this case, to determine it, it is necessary to restore the original distribution function of a continuous random variable X, which was replaced by a discrete random variable Z.

It is impossible to restore the true function $F_X(x)$. Therefore, we will limit ourselves to an approximate version in relation to common situations when people move from X to Z using grouping data and scoring indicators.

3 Results

In the results section, both cases shall be considered.

3.1 Data Groupings

Let the distribution series of a discrete random variable Z (Table 2) be obtained by grouping the values of some continuous random variable X.

We denote the midpoints of all the inner intervals of the groups as $z_{k,k+1} = (z_k + z_{k+1})/2$. Usually, when grouping data, the left border of $z_{0,1}$ of the first interval and the right border of the last interval are determined as follows [12]: $z_{0,1} = z_1 - (z_2 - z_1)/2$, $z_{6,7} = z_6 + (z_7 - z_6)/2$. As a result, we will move from the distribution series from Table 2 to the grouping (Table 3).

Table 2. Random variable distribution series Z.

z_k	4	8	10	14	19	22
$p_k = P(Z = z_k)$	0.1	0.15	0.2	0.25	0.2	0.1

Table 3. Grouping for a random variable Z.

Group	$(z_{0,1}, z_{1,2})$	$(z_{1,2}, z_{2,3})$	$(z_{2,3}, z_{3,4})$	$(z_{3,4}, z_{4,5})$	$(z_{4,5}, z_{5,6})$	$(z_{5,6}, z_{6,7})$
	(2; 6)	(6; 9)	(9; 12)	(12; 16,5)	(16,5; 20,5)	(20,5; 23,5)
p_k	0.1	0.15	0.2	0.25	0.2	0.1

Assuming that at each interval some continuous random variable Y is distributed uniformly, taking into account the given probabilities p_k, it is enough to simply restore the probability density $p_Y(x)$: at each k-th interval it will be constant and equal to $p_k/(z_{k,k+1} - z_{k-1,k})$. Figure 1 shows a graph of the "restored" probability density $p_Y(x)$.

Figure 2 shows the graphs of the distribution function $F_Z(x)$ of the original discrete random variable Z (dotted line) and the distribution function $F_Y(x)$ of the continuous random variable Y (continuous line).

In general, for a series of distributions from Table 1, the approximating probability density of a continuous random variable $p_W(x)$ will be equal to

$$p_W(x) = \begin{cases} 0, & x \in (-\infty;\ z_{0,1}], \\ p_k^*, & x \in (z_{k-1,k};\ z_{k,k+1}], \\ 0, & x \in (z_{n,n+1};\ +\infty), \end{cases} \tag{3}$$

where $p_k^* = p_k/(z_{k,k+1} - z_{k-1,k})$, $z_{0,1} = z_1 - (z_2 - z_1)/2$, $z_{k,k+1} = (z_k + z_{k+1})/2$, $z_{n,n+1} = z_n + (z_n - z_{n-1})/2$, $k = 1, 2, \ldots, n - 1$.

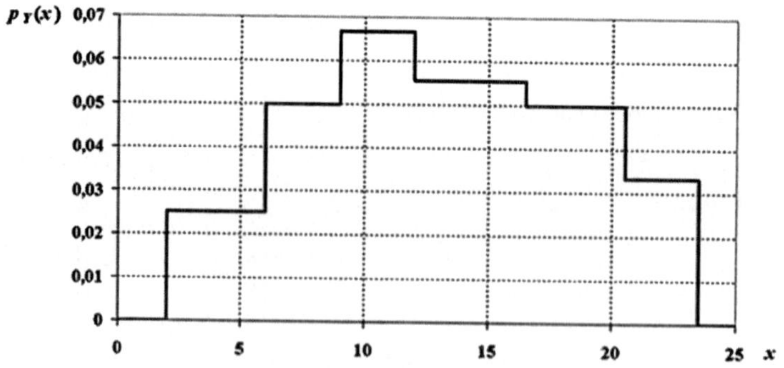

Fig. 1. Probability density graph $p_Y(x)$.

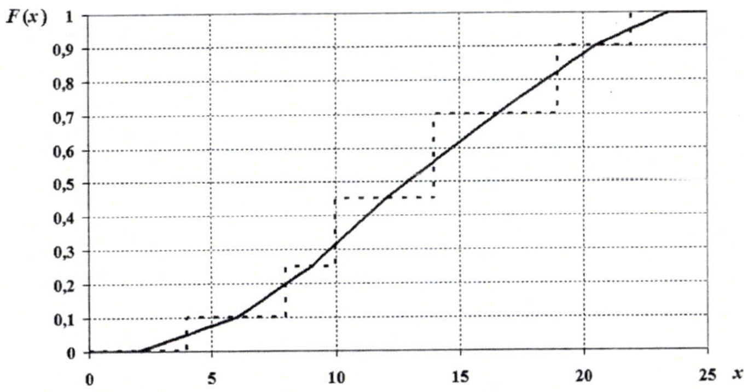

Fig. 2. Graphs of distribution functions: $F_Z(x)$ – dotted line, $F_Y(x)$ – continuous line.

Now we calculate the estimate of the differential entropy of the distribution (3) by the formula

$$H(W) = -\sum_{k=1}^{n} p_k^* \ln p_k^* \bullet \left(z_{k,k+1} - z_{k-1,k}\right).$$

3.2 Scoring Indicators

Let the investigated continuous random variable X be replaced by a number of point indicators as a result of some transformations (for certainty, we count points from 1 to M) (Table 4).

Table 4. Distribution series of a point random variable Z.

Z	1	2	...	$M-1$	M
$p_k = P(Z = k)$	p_1	p_2	...	p_{M-1}	p_M

Obviously, this is a special case of the grouping case discussed above, if we equate $z_k = k$, $k = 1, 2, \ldots, M$. Then, instead of (3), we get the formula for the approximating probability density of a continuous random variable $p_W(x)$

$$p_W(x) = \begin{cases} 0, & x \in \left(-\infty;\, z_{0,1}\right], \\ p_k, & x \in \left(z_{0,1};\, z_{M,M+1}\right], \\ 0, & x \in \left(z_{M,M+1};\, +\infty\right), \end{cases}$$

where $z_{k,k+1} = k + 0.5$, $k = 0, 1, \ldots, M$.

4 Discussions

At this point, examples using this technique and are based on real data will be discussed.

4.1 Example 1

We will generate a sample from the standard normal distribution X of the volume of 100 numbers. The sample mean square deviation was equal to $s = 0.928$. The differential entropy is equal to [14]: $H(X) = \ln(2\pi e)/2 = 1.344$. Now it will group the data into 7 intervals (Table 5). The width of the interval of each group was equal $\Delta = 0.59$.

Table 5. Grouping for a sample of 100 observations.

$(z_{0,1}, z_{1,2})$	$(z_{1,2}, z_{2,3})$	$(z_{2,3}, z_{3,4})$	$(z_{3,4}, z_{4,5})$
(–2.19; –1.6)	(–1.6; –1.01)	(–1.01; –0.42)	(–0.42; 0.17)
$p_1 = 0.06$	$p_2 = 0.13$	$p_3 = 0.18$	$p_4 = 0.16$
$(z_{4,5}, z_{5,6})$	$(z_{5,6}, z_{6,7})$	$(z_{5,6}, z_{6,7})$	–
(0.17; 0.76)	(0.76; 1.35)	(1.35; 1.94)	–
$p_5 = 0.29$	$p_6 = 0.12$	$p_7 = 0.06$	–

The differential entropy for the distribution given in Table 5 is equal to

$$H(Z) = -\int_{-\infty}^{+\infty} p_z(x)\ln p_z(x)dx = -\sum_{k=1}^{7} \frac{p_k}{\Delta}\ln\left(\frac{p_k}{\Delta}\right)\Delta = 1.291.$$

The difference between the values of $H(Z)$ and $H(X)$ was less than 4%, which indicates a fairly accurate estimate of the differential entropy.

4.2 Example 2

In [15], a method for assessing the severity of the condition of premature infants with a gestational age of less than 30 weeks is described. Two groups of preterm infants were considered – "severe" and "very severe"). Two of their nine informative signs turned out to be discrete variables (Y_1 – pregnancy: 0 – singleton, 1 – multiple; Y_2 – method of delivery: 0 – through the natural birth canal, 1 – planned cesarean section, 2 – emergency cesarean section). To conduct a comparative entropy analysis, we will use the described method of entropy modeling of discrete random vectors. The results of the entropy analysis for statistically significant features are shown in Table 6.

We see that in the "very severe" group, compared to the "severe" group, the entropy of randomness and self – organization is significantly greater (the entropy of randomness is 2.87 or 27.4% more, and the entropy of self-organization is 0.42 or 38.4% more). This means that in general, "very severe" newborns have a general variation of indicators, their variability is higher, and the closeness of the relationship between the indicators is lower. Accordingly, the total entropy of "very severe" is greater than that of "severe" by 3.29 or 35.4%. Thus, from the point of view of considering the organism as a complex system, the elements (informative diagnostic signs") in the group (system) "very severe" behave more chaotically and interact much less with each other compared to the elements of the group (system) "severe".

Table 6. Results of vector entropy analysis of the groups "severe" (IND = 0) and "very severe" (IND = 1).

IND	Entropy of randomness $H(\mathbf{Y})_V$	Entropy of self-organization $H(\mathbf{Y})_R$	Total entropy $H(\mathbf{Y})$
0	10.47	−1.11	9.37
1	13.34	−0.68	12.66

In conclusion, vector entropy analysis showed a significant difference in the behavior of groups of premature newborns "severe" and "very severe" as complex multidimensional systems. Informative diagnostic signs "in the "very severe" group behave more chaotically and interact much less with each other compared to the elements of the "severe" group.

5 Conclusion

It is shown that differential entropy cannot be used in modeling discrete random variables.

For cases when discrete random variables are obtained as a result of grouping data or switching to point indicators, it is possible to use differential entropy. This is achieved by switching from discrete random variables to their approximations by continuous random variables having piecewise linear distribution functions.

The technique of entropy modeling of multidimensional stochastic systems is described, all or part of the components of which are scoring indicators or obtained by grouping.

Funding. This work was supported by a joint Russian-Belarusian project of the Russian Foundation for Basic Research (grant no. 20-51-00001) and the BRFFR (grant no. M20R-008).

References

1. Bailey, K.D.: Social Entropy Theory. State University of New York Press (1990)
2. Deutscher, G.: The Entropy Crisis. World Scientific Publishing Co. Pte. Ltd. (2008)
3. Ellis, R.S.: Entropy, Large Deviations, and Statistical Mechanics. Springer (2006)
4. Haken, H.: Information and Self-Organization: A Macroscopic Approach to Complex Systems. Springer (2000)
5. Saridis, G.N.: Entropy in Control Engineering. World Scientific Publishing Co. Pte. Ltd. (2001)
6. Wilson, A.G.: Entropy in Urban and Regional Modelling. Pion, London (1970)
7. Shannon, C.E.: A mathematical theory of communication. Bell Syst. Tech. J. **27**(379–423), 623–656 (1948)
8. Tyrsin, A.N.: Entropy Modeling of Multidimensional Stochastic Systems. Science Book Publishing House, Voronezh (2016). (in Russian)
9. Tyrsin, A.N., Gevorgyan, G.G.: Population health assessment based on entropy modeling of multidimensional stochastic systems. In: Strijov, V.V., Ignatov, D.I., Vorontsov, K.V. (eds.) IDP 2016. CCIS, vol. 794, pp. 92–105. Springer, Cham (2019). https://doi.org/10.1007/978-3-030-35400-8_7

10. Siburina, T.A.: Basic assessment and practice of rating assessments in healthcare. Social Aspects of Public Health. **5**(27) (2012). (in Russian). http://vestnik.mednet.ru/content/view/427/30/
11. Orlov, A.I.: Organizational and Economic Modeling. Part 2: Expert Assessments. Publishing House of the Bauman Moscow State Technical University, Moscow (2011). (in Russian)
12. Efimova, M.R., Ganchenko, O.I., Petrova, E.V.: Workshop on the General Theory of Statistics, 3rd edn. Finance and Statistics, Moscow (2011). (in Russian)
13. Gelfand, I.M., Kolmogorov, A.N., Yaglom, A.M.: The amount of information and entropy for continuous distributions. In the book: Proceedings of the III All-Union Mathematical Congress. Academy of Sciences of the USSR, Moscow, vol. 3, pp. 300–320 (1958). (in Russian)
14. Cover, T.M., Thomas, J.A.: Elements of Information Theory. Wiley, New York (1991)
15. Tyrsin, A.N., Ulezko, E.A., Ostroushko, D.V., et al.: Predicting the severity of the condition of premature newborns with a gestation period of less than 30 weeks// Modern problems of science and education. № 6 (2020). https://doi.org/10.17513/spno.30288. http://www.science-education.ru/ru/article/view?id=30288

Calibrating Function Complexity in Enhancement Project for Improving Function Points Analysis Estimation

Vo Van Hai(✉), Ho Le Thi Kim Nhung, and Huynh Thai Hoc

Faculty of Applied Informatics, Tomas Bata University in Zlin,
Nad Stranemi 4511, 76001 Zlin, Czech Republic
{vo_van,lho,huynh_thai}@utb.cz

Abstract. Producing a good software product on time and within budget, the initial software estimation takes a significant role. Reality has shown that most software fails because the initial software estimation is not correct. Many researchers have proposed methods for software estimation. It has been developed since the 70s of the last century, but it is still of great interest until now. We also know that creating a new software product is difficult; it is even more difficult to innovate. In the framework of this paper, we propose an improved method based on the FPA method of IFPUG. We named this proposed model is Calibrating Function Complexity in Enhancement Project (CFCEP). This method is based on the Linear Regression technique to give coefficients of function complexity. The experimental results based on the ISBSG dataset show that the estimation based on this new coefficient gives much better results than using the coefficients of the standard FPA method.

Keywords: Software measurement · Function point analysis · Effort estimation · Multiple linear regression

1 Introduction

Software estimation is an essential issue in the process of developing a software system [7]. Many projects have not been completed on time or have ultimately failed, primarily because the effort was underestimated or overestimated [8]. For this reason, various software effort estimation models have been developed to manage the budget and schedule of software projects [5]. Accurate effort estimation is essential for project managers and customers [6]. It supports creating an application for bidding, contract negotiation, scheduling, monitoring, and controlling software development.

According to Moløkken [16], only about 30–40% of software projects are completed. The CHAOS team's report shows a partial or ultimately failure rate of software projects, which was 70% in 1994 [17], and this number increased to 83.9% in 2019 [10]. Several studies have been conducted to find out the reasons for the failure of software projects. Galorath et al. [18] indicate that the causes of this failure are inadequate engineering, poor project planning, sudden decisions at the early stages of the project, and incredibly

R. Silhavy et al. (Eds.): CoMeSySo 2021, LNNS 232, pp. 857–869, 2021.
https://doi.org/10.1007/978-3-030-90318-3_67

inaccurate estimates. Some researchers have also pointed out that the leading cause of project failure is incorrect cost estimates [7, 17, 19]. Therefore, software effort estimation methods need to be improved to evaluate different types of projects. One of the possible research directions could be to study methods based on Functional Point Analysis (further only FPA) [2]. An important aspect to be measured in software engineering projects is the functional size of the software [20]. FPA is a standardized method for determining the size of software based on its functional requirements. It is designed to be applicable regardless of programming language and implementation technology. Albrecht [2] recommends FPA to measure the size of the system that processes data from end-users.

The FPA method has been standardized by ISO/IEC 20926:2010 [21]. However, the FPA method is controversially discussed by different researchers regarding its advantages and limitations. Henderson et al. [22] studied the FPA method from a manager and developer perspective based on 13 attributes with three main findings: (1) Source Lines Of Code (further only SLOC) count is less complex than Functional Point (further only FP); (2) developers better understand FP profitability than managers, and (3) differences between managers and developers in values block the communication needed to make informed decisions. Kampstra et al. [23] and Kemerer [24] reported that the FPA method does not produce consistent results when applied by different metrics. Meli [25] showed a mismatch between the established complexity for Base Functional Components (further only BFC) and possible yield estimates.

Many studies have shown that the FPA for BFC scores is incorrect. For example, the same data function and/or the same transaction function can be classified with different DET and RET/FTR combinations with the same complexity. This results in the same function score for data functions and/or transactional functions. It was also found that in some situations, functions with very similar DET and RET/FTR can be classified with different complexity and therefore receive different weights according to the FPA.

Xia et al. [26] suspect that the unadjusted FP weighted value based on research on data processing systems at IBM (locally) cannot reflect software globally. In [27], they repeatedly point out that the classification is ambiguous, and that the original method may not adequately reflect the reality of software complexity in a particular software application. In [28], they demonstrated no clear boundary between the two classifications in FP counting. The authors propose combining three techniques (fuzzy logic, artificial neural network construction, and statistical regression) in a neural fuzzy function point calibration model to solve these problems.

According to Hajri et al. [29], the classification of function types into simple, average, and complex does not reflect the full extent of complexity required to develop a user system. The main innovative idea of this study is to introduce a new weighting system for measuring FP using the Artificial Neural Network (further only ANN) - using the backpropagation technique. In the first step, the original weighting system is used as a baseline to establish new weights. Next, they train one of the most popular techniques in Neural Networks to predict the value of new weights. Then, the new weights and the original weights are inserted into the FP model. Finally, the size of FP is calculated as a function of the original weights and the new weights.

Ya-Fang et al. [30] reported the IFPUG FPA barely distinguishing the complexity of adjacent components. If this situation occurs in a software program, the estimated results will not correspond to the actual situation. To address this shortcoming, they believe that fuzzy theory should be used to analyze component complexity. He also said that the weights of the BFCs, as set by the IFPUG, are supposed to reflect the functional size of the software, but software today is vastly different from before, so it no longer exists suitable anymore. Agree with Ya-Fang et al. [30] state that this inconsistency in a large number of BFCs, located over boundary regions of defined periods, is even worse. This is because an incorrect classification of the different functions of the system will skew its functional dimensions. Like Xia et al. [7, 19, 20], the authors also link fuzzy logic with artificial neural networks. The neural network can learn from the estimated data from the rules and linguistic terms defined by fuzzy logic.

In recent decades, scientists have done a lot of research on the implementation of localized learning using soft computational techniques and statistical techniques. Specifically, Šilhavý et al. [33] proposed a new categorical variable segmentation model based on dataset segmentation to estimate software development effort using a remarkable model trained on a particular data segment. In addition, Prokopová et al. [34] also determined the influence of the VAF factor on the accuracy of the software cost estimation process. Wena et al. [35] showed that the ANN technique is the most widely used and provides the most accurate results along with Support Vector Regression (further only SVR). Although soft computational techniques have shown positive results, there is a significant variation between results reported from different studies.

Since there are many of the above limitations, this study proposes a function complexity weight calibrated based on multiple linear regression for estimating the development effort of enhancement software projects when the IFPUG FPA method is used for baseline estimation. This study will answer the research question of whether the proposed method can significantly improve the prediction of the IFPUG FPA method. The methods are applied to the International Software Benchmarking Standards Group (ISBSG) dataset, which contains data on completed software projects [4]. Unbiased evaluation criteria (Eq. 9–11) were used to evaluate the estimation accuracy of the methods.

The remainder of the paper is organized as follows: Sect. 2 presents the background of the methods used. Section 3 describes the research methodology. Section 4 introduces the results and the discussion, and Sect. 5 the conclusions and suggestions for future work.

2 Background

2.1 Function Point Analysis

FP was first introduced by A. J. Albrecht [2] and widely accepted with many variants from both academics and practitioners [31]. The study in this area is also known as FPA or Function Size Measurement (FSM). The function points measurement could be classified into FP counting and FP estimation [32].

The FPA assumes that a software program consists of a set of functions. In turn, each function can be a data transaction or transactional function. The data transaction has consisted of Internal Logic Files (ILF) and External Interface Files (EIF), and the transaction function comprises External Input (EI), External Output (EO), or External

Inquiry (EQ). In turn, each of these is judged as simple, average, or complex and assigned a weight accordingly (see Table 1).

Table 1. Data and transactional function complexity

		Component				
		EI	EO	EQ	EIF	ILF
Functional Complexity	Low	3	4	3	5	7
	Average	4	5	4	7	10
	High	6	7	6	10	15

The following equation can obtain the total unadjusted function points (UFP)

$$UFP = \sum_{i=1}^{n} \sum_{j-1}^{m} (W_{ij} \times S_{ij}) \tag{1}$$

where W_{ij}'s are the complexity weights and S_{ij}'s are the counts for each function component, n is the number of types, and m is the number of complexity groups.

The Value Adjustment Factor (VAF) should be aimed at before calculating the function points. The VAF is based on the rate of 14 general system characteristics (GSCs - Data Communications, Distributed Data Processing, Performance, Heavily Used Configuration, Transaction Rate, Online Data Entry, End-User Efficiency, Online Update, Complex Processing, Reusability, Installation Ease, Operational Ease, Multiple Sites, Facilitate Change). The GSCs are rated based on their degree of influence or complexity on the 0 – 5 scale. The following table represents the significance of the influence factors rating.

Table 2. The GSCs Factor weights

System Influence	Rating
Not present or no influence	0
Incidental influence	1
Moderate influence	2
Average influence	3
Significant influence	4
Strong influence throughout	5

The VAF count is adjusted as

$$VAF = 0.65 + 0.01 \times \sum_{i=1}^{14} (F_i \times rating) \tag{2}$$

After all, the adjusted function points (AFP) can be reached by using the following Eq. (3)

$$AFP = UFP \times VAF \tag{3}$$

In the previous IFPUG, the function point was used to develop a new project but not enhance an existing project. However, in the latest version (4.3.1), this problem is changed. We can use FP for both Adjusted Development Project Function Point (aDFP) and Adjusted Enhancement Project Functional Size (aEFP).

Use the following formula to calculate aDFP:

$$aDFP = DFP \times VAF \tag{4}$$

where $DFP = AFP$. In aEFP, the following formula was used to calculate the function size:

$$aEFP = [(ADD + CHGA + CFP) * VAFA] + (DEL * VAFB) \tag{5}$$

where:

ADD - the size of the functions being added by the enhancement project.

CHGA - the size of the functions being changed by the enhancement project – as they are / will be after implementation.

CFP - the size of the conversion function.

VAFA - the value adjustment factor of the application after the enhancement project is complete.

DEL - the size of the functions being deleted by the enhancement project.

VAFB is the value adjustment factor of the application before the enhancement project begins.

2.2 The Multiple Linear Regression

In the data analysis, a common task is to study the dependence of a responding variable $Y \in R^1$ on many independent variables $X = [X_1, X_2, ..., X_n] \in R^n$. The model given in the form of a mathematical equation $Y = f(X)$ describing each relationship between Y and X is called the regression model. In the case of only one independent variable, it is called Simple Linear Regression. There are two or more independent variables called Multiple Linear Regression (MLR) [14]. The essential model has the following form [15]:

$$\mathrm{Y} = \beta_0 + \beta_1 \mathrm{X}_1 + .. + \beta_n \mathrm{X}_n + \varepsilon \tag{6}$$

where Y is the predicted dependent variable, n is the number of variables, β_0 is an intercept, X_i are the independent variables and $\beta_i i \in 1, n$ are called partial regression coefficients, ε is error residual. In this study, the β_0 and ε value was omitted to get the β_i which does not depend on the intercept and the error residual.

3 Research Methodology

3.1 Data Description

In this article, we use the ISBSG dataset version August 2020 R1 [4]. This dataset has been contributed by companies in the field of information technology for many years. In this experiment, we conduct data preprocessing according to the following criteria: First, we only select records with the counting approach of IFPUG (including IFPUG Old and IFPUG 4+). This study is for enhancement, so we choose the development type as an enhancement. Of course, the Data quality rating is selected as A and B. Since the records are related to functions, the data transaction and transactional function must not be empty.

The data to this point is basically complete; however, in some records with enough data, the value of VAF is missing; we proceed to recalculate this value based on the formula

$$VAF = \frac{AFP}{UFP} \tag{7}$$

Finally, due to some errors in data collection, we proceed to remove records whose sum of added, changed, and deleted functions is different from UFP. At this point, we have a dataset with 911 records ready for testing.

We notice that there is a possibility that the data may be noisy because some values of SWE are too far from the mean group, as shown in Fig. 1. In this study, we used the interquartile range (IQR) method to determine the outlier. According to this method, suppose that Q1 is the lower quartile, and Q3 is the upper quartile, we have Eq. (10).

$$\text{IQR} = \text{Q3} - \text{Q1} \tag{8}$$

Any value that is greater than Q3 + 1.5 × IRQ or less than Q1 − 1.5 × IRQ was considered an outlier. The simulation of this process was shown in Fig. 2. After removing outliers, our dataset can be seen in Table 3.

Table 3. Statistic characteristics of SWE

	count	mean	std	min	25%	50%	75%	max
Before remove outliers	911	3413.367	5055.922	16	812.0	1930.0	3867.5	61891.0
After remove outliers	637	3003.243	3327.755	26	929	1974	3705	25900

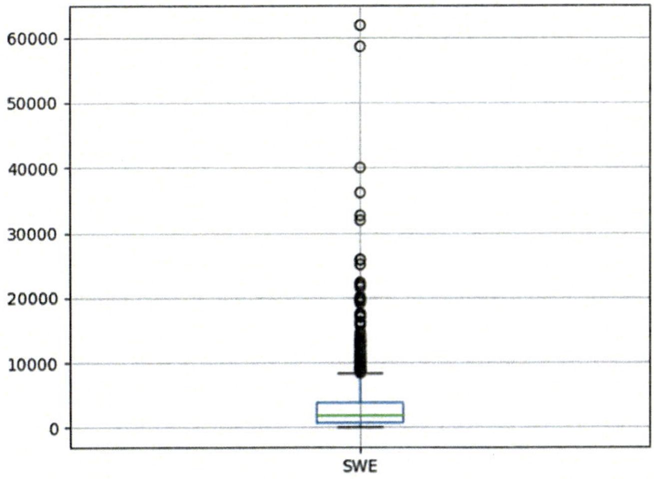

Fig. 1. Boxplot of SWE for dataset

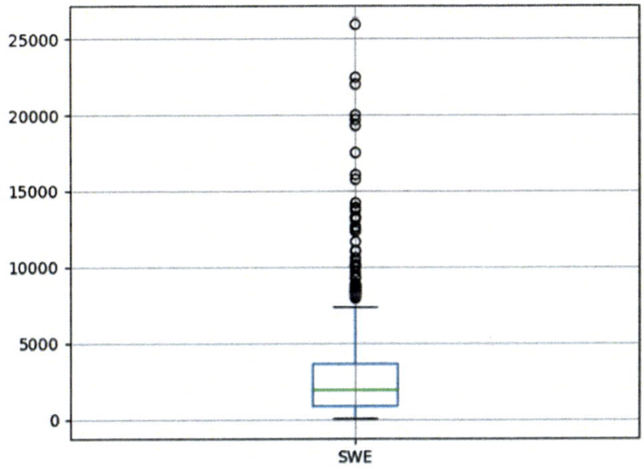

Fig. 2. Boxplot of SWE after removing outliers for dataset

3.2 Evaluation Criteria

Regarding the accuracy of the measurement, according to Foss et al. [11], there have been many studies and evaluations on the relevance of the error function that have been proposed and used so far. However, there is no silver bullet to choose the best predictive models among several alternatives. That means that every accuracy metric presents certain advantages over the rest. Kitchenham et al. [12] have shown that the most crucial factor in making meaningful comparisons between predictive models is determining each error function using actual measures. In this study, to evaluate the accuracy of the proposed model, we use the most popular and widely used measure to evaluate the predictive ability of the comparative models [13].

MAE (Mean Absolute Error). MAE is calculated as:

$$MAE = \frac{\sum_{i=1}^{n} \left| Effort_{actual} - Effort_{predict} \right|}{n} \tag{9}$$

where n is the number of test projects, the $Effort_{actual}$ is the actual effort and the $Effort_{predict}$ is the predicted effort. MAE is considered a good choice for measuring symmetric unbiased under or overestimates [9, 10]. A higher (better) predictive performance can be achieved with lower MAE.

MSE (Mean Squared Error) is the mean of the square of the differences between the actual and the predicted efforts.

$$MSE = \frac{\sum_{i=1}^{n} \left(Effort_{actual} - Effort_{predict} \right)^2}{n} \tag{10}$$

RMSE (Root Mean Square Error) is a frequently used measure that is just the square root of MSE. RMSE can be calculated by using the equation given below

$$RMSE = \sqrt{\frac{\sum_{i=1}^{n} \left(Effort_{actual} - Effort_{predict} \right)^2}{n}} \tag{11}$$

3.3 Methodology Used

The use of function complexity values shown in Table 1 has many limitations, as mentioned in part 1. In this study, we use MLR to propose changes on these function complexity values. Our proposed method can be visualized as follows: first, from the ISBSG dataset, we select the attributes and filter it according to the criteria described in the experiments. Each data column corresponding to EI, EO, EQ, ILF, and EIF will be transformed into three values : low, average, and High. Then the data will be removed outliers to reduce noise. The next step is to use the k-fold cross-validation technique to split the data into five folds and run each of these folds against the MLR. Finally, the regression results will be computed to obtain Low, Average, and High values for the function complexity. We named this proposed model is Calibrating Function Complexity in Enhancement Project (CFCEP). Figure 3 shows the steps described above in a more intuitive way.

For the purpose of this article to find function complexities, we divided the function points values of EI, EO, EQ, ILF, and EIF into 15 values corresponding to low, average, and high for each function point. These 15 values are obtained for each record by approximating the solutions to the 3-variable equation and then taking the mean value. Based on these 15 values, we will calculate the UFP_{CFCEP} value according to the MLR model. Then we calculate the AFP and the Effort values. The MLR model is described as follows:

$$\begin{aligned} UFP_{CFCEP} = {} & \beta_1 x_1 - \beta 2_2 x_2 - \beta_3 x_3 - \beta_4 x_4 + \beta_5 x_5 + \beta_6 x_6 + \beta_7 x_7 \\ & + \beta_8 x_8 - \beta_9 x_9 + \beta_{10} x_{10} + \beta_{11} x_{11} - \beta_{12} x_{12} + \beta_{13} x_{13} \\ & + \beta_{14} x_{14} - \beta_{15} x_{15} \end{aligned} \tag{12}$$

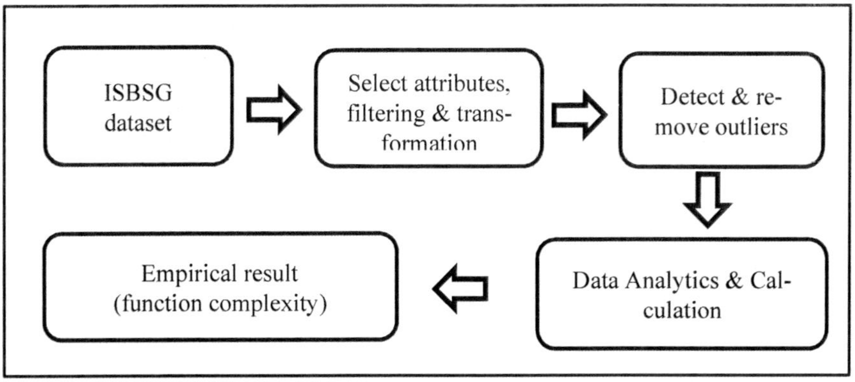

Fig. 3. CFCEP model

4 Result and Discussion

In this section, we will review the experimental results. This result estimated through MLR will be compared with using FPA for calculation. Actual results and estimated results will be evaluated based on error criteria MAE, MSE, RMSE.

Table 5 shows us the results of five experimental. As we can see, the result of effort estimation using MLR is much better than the FPA method. With each experiment as well as the average value, the estimated results based on our proposed method are always closer to the actual results than the FPA method. Specifically, the accuracy of this proposed method is 96.13%.

Table 4. Effort estimation results

	SWE	Effort using FPA	Effort using CFCEP
ex-1	3214.661	46599.3	2784.752
ex-2	2991.458	47375.3	2851.248
ex-3	2845.984	48915.68	2942.287
ex-4	3207.625	53258.45	2995.198
ex-5	3168.885	52608.05	3258.606
mean	**3085.723**	**49751.35**	**2966.418**

A smaller value represents a more accurate estimate for the evaluation criteria MAE, MSE, and RMSE [12]. With Table 5, it is clear that the proposed method has much higher accuracy than the FPA method.

Table 5. Evaluation criteria results

	MAE FPA	MAE MLR	MSE FPA	MSE MLR	RMSE FPA	RMSE MLR
exp_1	43442.04	2500.785	4.42E + 09	15246804	66496.63	3904.716
exp_2	44446.85	2017.467	5E + 09	10126571	70721.76	3182.227
exp_3	46083.04	2198.764	5.38E + 09	11095411	73340.03	3330.978
exp_4	50120.9	2459.952	5.25E + 09	15106096	72434.24	3886.656
exp_5	49502.85	2429.726	5.51E + 09	12530844	74244.84	3539.893
mean	**46719.14**	**2321.339**	**5.11E + 09**	**12821145**	**71447.5**	**3568.894**

The regression model obtained from our experiment is as the following

$$UFP_{CFCEP} = x_1 - 0.58x_2 - 1.41x_3 - 0.49x_4 + 0.26x_5 + 0.31x_6 + 1.02x_7 + 0.34x_8 - 1.20x_9 + 0.08x_{10} + 0.10x_{11} - 0.07x_{12} + 1.45x_{13} + 0.05x_{14} - 1.40x_{15} \quad (13)$$

And the function complexity we also obtained by calculating is shown in Table 6.

The research objective of this study is to propose a new function complexity value for the enhancement project. Table 6 is our answer to this problem. With this result, we can apply function complexity to enhancement projects with better accuracy.

5 Conclusion

This paper has achieved a new function complexity table for the enhancement project by applying the MLR model to the ISBSG dataset filtered by the suggested criteria. We can see from the experimental results that the average effort achieved with our proposed new method is much closer to the actual recording of SWE compared to the IFPUG FPA method. Specifically, the new method has an accuracy of 96.13%.

Future work will focus on a complete framework for both new development types and enhancement types. In addition, we will be using other methods like artificial neural networks and support vector machines in our model.

Table 6. Function complexity results

		std	exp_1	exp_2	exp_3	exp_4	exp_5	mean
EI	**L**	3	6.023	7.178	6.628	4.496	7.987	**6.463**
	A	4	-2.445	-1.654	-2.674	-2.859	-1.942	**-2.315**
	H	6	-7.338	-10.939	-8.239	-3.704	-12.093	**-8.463**
EO	**L**	4	-1.159	-1.588	-2.779	-2.856	-1.436	**-1.964**
	A	5	-0.775	2.467	2.507	1.615	0.694	**1.301**
	H	7	3.794	-0.305	1.874	3.298	2.062	**2.145**
EQ	**L**	3	2.468	3.059	2.298	3.927	3.49	**3.048**
	A	4	1.297	0.883	1.3	2.3	1.006	**1.357**
	H	6	-5.825	-6.703	-5.771	-10.336	-7.493	**-7.226**
ILF	**L**	5	0.067	0.433	0.473	0.586	0.474	**0.407**
	A	7	1.078	0.4	0.57	1.625	-0.186	**0.697**
	H	10	-1.049	0.192	-0.363	-2.601	0.208	**-0.722**
EIF	**L**	7	8.107	11.54	11.288	12.103	7.832	**10.171**
	A	10	0.732	0.48	0.374	-0.893	1.628	**0.464**
	H	15	-16.369	-24.64	-22.381	-24.527	-17.3	**-21.043**

Acknowledgment. This work was supported by the Faculty of Applied Informatics, Tomas Bata University in Zlín, under project IGA/CebiaTech/2021/001.

References

1. IFPUG: International Function Point Users Group. http://www.ifpug.org/. Accessed July 2021
2. Albrecht, A.J.: Measuring application development productivity. In: Proc. IBM Applications Develop. Symp., p. 83 (1979)
3. Stone, M.: Cross-validatory choice and assessment of statistical predictions. J. Royal Stat. Soc. Ser. B (Methodological) 111–147 (1974)
4. ISBSG, ISBSG Release 2020 R1
5. Albrecht, A.J., Gaffney, J.E.: Software function, source lines of code, and development effort prediction: a software science validation. IEEE Trans. Softw. Eng. 639–647 (1983)
6. Putnam, L.H.: A general empirical solution to the macro software sizing and estimation problem. IEEE Trans. Softw. Eng. 345–361 (1978)
7. Caper, J.: Estimating software cost. Mc-Graw-Hill Edition (2007)
8. Boehm, B.W.: Software Engineering Economics. Prentice-Hall, Englewood Cliffs, NJ, USA (1981)
9. Lokan, C., Mendes, E.: Investigating the use of the chronological split for software effort estimation. IET Softw. **3**(5), 422–434 (2009)
10. The Standish Group, https://www.standishgroup.com/. Accessed July 2021
11. Foss, T., Stensrud, E., Kitchenham, B., Myrtveit, I.: A simulation study of the model evaluation criterion MMRE. IEEE Trans. Softw. Eng. **29**(11), 985–995 (2003)

12. Kitchenham, B., MacDonell, S., Pickard, L., Shepperd, M.: What accuracy statistics really measure. IEE Proc. Softw. Eng. **148**(3), 81–85 (2001)
13. Briand, L.C., Emam, K.E., Surmann, D., Wieczorek, I., Maxwell, K.D.: An assessment and comparison of common software cost estimation modeling techniques. In: International Conference on Software Engineering, pp. 313–322 (1999)
14. Khatibi Bardsiri, V., Jawawi, D.N.A., Hashim, S.Z.M., Khatibi, E.: A flexible method to estimate the software development effort based on the classification of projects and localization of comparisons. Empir. Softw. Eng. **19**(4), 857–884 (2013). https://doi.org/10.1007/s10664-013-9241-4
15. Mendenhall, W.: A Second Course in Statistics: Regression Analysis. Pearson Education Inc, Boston, MA, USA (2012)
16. Moløkken, K., Jørgensen, M.: A review of surveys on software effort estimation. In: International Symposium on Empirical Software Engineering. Retrieved from ACM Digital Library database, pp. 223–231 (2003)
17. Moløkken – Østvold, K., et al.: Project Estimation in the Norwegian Software Industry. A Summary. Simula Research Laboratory (2004)
18. Galorath, D.D., Evans, M.W.: Software sizing, estimation, and risk management: When performance is measured performance improves. Auerbach, Boca Raton, FL (2006)
19. Kemerer, C.F.: An empirical validation of software cost estimation models. Commun. ACM, **30**(5), 416–429 (1987)
20. Ravichandran, T.: Organizational assimilation of complex technologies: an empirical study of component-based software development. IEEE Trans. Eng. Manag. **52**(2), 249–268 (2005)
21. ISO/IEC 20926:2009 (IFPUG), Software and systems engineering -- software measurement -- IFPUG functional size measurement method 2009
22. Marcos, F., Marcelo, F., Sun, V.: Sun, Improvements to the function point analysis method: a systematic literature review. IEEE Trans. Eng. Manage. **62**(4), 1–12 (2015)
23. Kampstra, P., Verhoef, C.: Reliability of function point counts — Department of Computer Science, VU University Amsterdam, Amsterdam, The Netherlands (2010)
24. Kemerer, C.: F: Reliability of function points measurement: a field experiment. Commun. ACM **36**(2), 85–97 (1993)
25. Meli, R.: Functional metrics: problems and possible solutions. Proc. 1st Eur. Software Meas. Conf., Antwerp, Belgium (1998)
26. Xia, W., Capretz, L.F., Ho, D.: Neuro-fuzzy approach to calibrate function points. Proc. 8th WSEAS Int. Conf. Fuzzy Syst., pp. 116–119 (2007)
27. Xia, W., Capretz, L.F., Ho, D., Ahmed, F.: A new calibration for function point complexity weights. Inf. Softw. Technol. **50**(7–8), 670–683 (2008)
28. Xia, W., Ho, D., Captrez, L.F.: A neuro-fuzzy model for function point calibration. WSEAS Trans. Inf. Sci. Appl. **5**(1), 22–30 (2008)
29. Hajri, M.A., Ghani, A.A.A., Sulaiman, M.N., Selamat, M.H.: Modification of standard function point complexity weights system. J. Syst. Softw. **74**(2), 195–206 (2005)
30. Ya-Fang, F., Xiao-Dong, L., Ren-Nong, Y., Yi-Lin, D., Yan-Jie, L.: A software size estimation method based on improved FPA. Proc. 2nd World Congr. Softw. Eng., pp. 228–233 (2010)
31. Gencel, C., Demirors, O.: Functional size measurement revisited. ACM Trans. Softw. Eng. Methodol. **17**(3), 15.1–15.36 (2008)
32. Meli, R., Santillo, L.: Function point estimation methods: a comparative overview. In: FESMA 1999 Conference Proceedings, Amsterdam, 4–8 October 1999
33. Silhavy, P., Silhavy, R., Prokopová, Z.: Categorical variable segmentation model for software development effort estimation. IEEE Access. 1 (2019). https://doi.org/10.1109/ACCESS.2019.2891878

34. Prokopova, Z., Silhavy, P., Silhavy, R.: VAF factor influence on the accuracy of the effort estimation provided by modified function points methods. In: Annals of DAAAM & Proceedings. vol. 29, pp. 0076–0084 (2018)
35. Wena, J., Lia, S., Linb, Z., Huc, Y., Huang, C.: Systematic literature review of machine learning-based software development effort estimation models. Inf. Softw. Technol. **54**(1), 41–59 (2012)

Analysis of «Leader – Followers» Algorithms in Problem of Trajectory Planning for a Group of Multi-rotor UAVs

Igor Lebedev and Valeriia Lebedeva(✉)

St. Petersburg Institute for Informatics and Automation of the Russian Academy of Sciences, St. Petersburg Federal Research Center of the Russian Academy of Sciences (SPC RAS), 39, 14th Line, St. Petersburg 199178, Russia

Abstract. Trajectory planning a group of robots is an urgent task of basic robotics. The paper evaluates the efficiency of using the path planning algorithms A*, RRT, RRT* and the visibility graph (VG) to solve the group flight problem using the «leader-follower» strategy. According to this strategy, it is important to provide a trajectory without collisions with obstacles for the robot leader, and the trajectories of the followers are determined relative to the position of the leader. The paper presents results of comparing of path planning algorithms, as well as solving the problem of moving a group of 7 UAVs in the environment with obstacles. The processing time for calculating the trajectory for the leader of the group does not exceed 1.3 s in the working area in the form of a square with a side of 100 m, the processing time for solving the flight task performed by the UAV group takes no more than 2 min, provided that the speed of the UAV is 11 m/s. The proposed strategy is flexible to modifications, therefore, further research of the leader-follower strategy will be aimed at modifying the A* algorithm when finding a path for a group leader in an environment with dynamic obstacles.

Keywords: Leader – Followers · Group control · Swarm robotics · UAV · Multi-rotor UAV · A* · RRT · RRT* · Visibility graph

1 Introduction

The flight of a group of unmanned aerial vehicles (UAVs) in a fixed formation allows solving many civil and military tasks [1]. When performing flight missions in real conditions, the UAV group must fly around obstacles [2, 3] in order to avoid a possible collision with them. In these conditions, methods to build coordinated trajectories for all agents in the group are required. Let us consider the known approaches to solving the problem of group movement of UAVs according to the "leader-follower" strategy in a fixed formation at a certain height. The movement of the formation is carried out in an environment with obstacles.

In [4], the concept of a virtual leader was introduced. A virtual leader is an artificial agent, represented as a point, which supports the formation, as if attracting individual agents to itself. This group management strategy allows you to influence the movement of

R. Silhavy et al. (Eds.): CoMeSySo 2021, LNNS 232, pp. 870–884, 2021.
https://doi.org/10.1007/978-3-030-90318-3_68

the entire group, controlling only one virtual leader. In paper [5], the author demonstrates the use of potential functions to overcome obstacles. A repulsive potential is generated around the obstacle, while the potentials around each agent are used to simulate a force similar to gravity to achieve the desired distance. Minimizing the potential function through gradient descent results in agents avoiding each other and obstacles, as well as reaching the desired formation.

In [6], a combination of the approaches described above is presented. Agents follow a virtual leader who follows a predetermined trajectory across the workspace. Moving a group of UAVs based on the virtual leader and the architecture of an artificial potential field may seem to be effective for solving the problem of overcoming obstacles. But this technique is poorly studied and until that moment was used to solve the problem of moving only small groups of UAVs.

Tuning the optimization function parameters for this type of system can be quite complex. From a numerical point of view, the authors of [6] found that the size of the simulation and optimization step plays a very important role in solving the group control problem. The simulation step size should be small to accurately represent the robot's behavior and to detect any potential collisions but cannot be too small because it can add to the computational burden.

The strategy for managing a group of UAVs through a virtual leader is similar to the strategy for managing a group with the help of a physical leader. Often in the works one can find methods of control of the formation on the principle of «leader – follower» in various variations [7], for example, using the exchange of information from onboard sensors to determine the location of the driven group [8]. Despite the status of a «leader» (virtual or physical), he needs to build a route of movement in space. In other words, solve the path planning problem.

The hybrid group control system proposed in [3] contains two parts of the subsystem: a module responsible for constructing a flight trajectory for the «leader», and modules for coordinated movement of the follower group, where all other agents repeat the trajectory of the «leader» corrected for their own trajectory, and the relative position is corrected by exchanging data with each other. Such a system is capable of solving many flight tasks assigned to the group. Thanks to such a system, a group of UAVs can carry out prospecting work over a wide area, map the area or process crops faster and more efficiently than a single UAV.

Traditionally, the problem of planning a trajectory for a robot is considered as a problem of finding a path on a graph in the space under study. The vertices of the graph correspond to different positions of the agent in space, the edges correspond to elementary trajectories, i.e. such trajectories, following which the agent is considered a trivial task (usually, these are line segments or curves of a certain radius). To model the environment, graphs of regular decomposition (grids) are usually used (the classic representative of the pathfinding algorithm on a graph model is the A* algorithm) [9], as well as visibility graphs [10], probabilistic terrain schemes (for example, RRT) [11], etc.

The purpose of this paper is to provide an overview of the literature on modern path planning algorithms and evaluate the performance of the four most commonly used path planning algorithms, namely A* [12], RRT [13], RRT * [14], and visibility graph pathfinding (VG) [15] in the context of two-dimensional planning of the trajectory of the UAV movement [16] for solving the problem of group movement «leader-follower». These algorithms were chosen for the study due to the fact that algorithms of the RRT are more efficient for simple maps with a few obstacles and a small number of path homotopies, while A* and VG -based algorithms are more efficient for more complex maps with a large number of obstacles and a large number of path homotopy [17]. The investigated parameters in the work are the length of the found path and the time for calculating the path. For UAVs, the path length determines the travel time, hence the time it takes for the UAV to reach the target point. Path planning time is of fundamental importance, especially in a dynamic environment in which some parameters may have changed towards the end of the path planning. In addition, this time will limit the response of the UAV control algorithm to changes both in the UAV model and in the environment. Performance analysis is important to determine the advantages and disadvantages, as well as the feasibility of using any of the algorithms for two-dimensional planning of the UAV trajectory in the group control problem.

The work is organized as follows. Section 2 introduces and discusses the theoretical aspects of the A*, RRT, RRT* and VG-algorithms, followed by a review of the literature on the above algorithms. Section 3 contains a description of the strategy for managing the "leader-follower" group. Section 4 presents a comparison of planning algorithms and the results of a group flight simulation in Gazebo.

2 Analysis of Path Planning Algorithms for Leader

This section discusses modern path planning algorithms and analyzes the performance of the four most commonly used path planning algorithms, namely A*, RRT, RRT* and VG for 2D UAV trajectory planning.

2.1 Principle of Operation of A* Algorithm

Algorithm A* is a heuristic algorithm for finding the shortest path on a graph [18]. Heuristic search is about finding the least cost route. In order to use the A* algorithm to build a route, it is necessary to present a map of the work area in the form of a graph. This is possible using the cellular decomposition of the workspace map [19]. In this case, the environment surrounding the robot is represented as a grid with the same mesh size. The center of each cell is the vertex of the graph, and the edges connecting the vertices are transitions to neighboring cells. Each cell has a cost to move through it and can also be «free» of obstacles or «occupied». The advantages of presenting a room map as a grid function map are the known distance from each cell to the nearest obstacle and taking into account the leader's dimensions.

The problem of finding a path using the A* algorithm is reduced to determining the sequence of cells through the centers of which the UAV will move in the shortest time. Each vertex is assigned a cost - a relative value of the complexity of the path. In the algorithm, the search starts from the target vertex and ends at the vertex where the leader is. The cost of the vertex is determined as the sum of the cost of moving from a given cell to the target point, following the already found path, and the approximate cost of moving from the starting point to the current one. As a result of performing several iterations of recursive vertex processing, there is a map, the cells of which contain the cost of the path to the target point, as well as directions that indicate the cells, the cost of which decreases as we move in these directions. Moving from the starting point along these directions, the leader reaches the target point - this will be the desired path.

2.2 Principle of Operation of RRT Algorithm

The RRT algorithm was developed by Stephen Lavelli [20]. RRT is a type of algorithm based on Probabilistic Roadmap Methods (PRM). When working with PRM algorithms, the input data are the start and target position and the obstacle map. The output data is the trajectory found for the UAV on this map. The algorithm has a strong probabilistic completeness property. This means that given enough time, it finds a solution, if one exists. However, there is no guarantee as to the exact time it will take for the algorithm to construct the path due to the random nature of the algorithm. For example, its performance can be very different in terms of the time to connect the start and target positions even in two tests with the same conditions (obstacle map), as in Fig. 1.

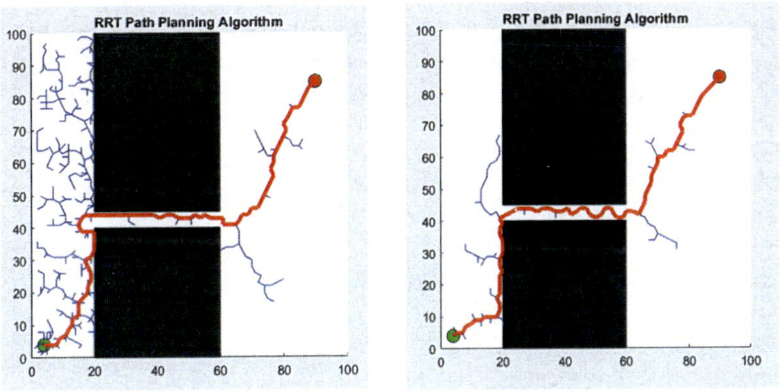

Calculation time 106.976978 s   Calculation time 7.407148 s

Fig. 1. An example of the performance of the RRT algorithm on the same card. The experiments were carried out on a PC with a 2.4 GHz Intel Core i7 processor and 16 GB of RAM.

2.3 Principle of Operation of RRT* Algorithm

RRT* is an optimized version of the RRT algorithm [21]. When the number of nodes approaches infinity, the RRT* algorithm still finds the shortest path to the goal. The basic principle of RRT* is the same as RRT, but two additions to the basic algorithm led to different results.

First, the RRT* records the distance traveled by the robot at each vertex relative to its parent vertex. This is called the *cost ()* of the vertex. After the nearest node is found in the graph, the neighborhood of the vertices of a fixed radius from the new node is examined. If a node with a lower *cost ()* than the previous node is found, then the node with the lowest cost replaces the previous node. The effect of this feature can be seen due to the fan-shaped structure of the constructed tree.

The second difference in RRT* is an additional tree check. After a vertex has been connected to a vertex of lesser value, these neighboring vertices are checked again. Neighboring vertices are checked if joining the newly added vertex will reduce their cost. If the value does decrease, the neighboring vertex is connected to the newly added vertex. This difference makes the planned path smoother than the standard RRT version. To find a path, especially in a dense field of obstacles, the trajectory constructed with the RRT* does not require additional smoothing. The constructed trajectory, as it were, smoothly bypasses obstacles, thereby reducing the length of the path. If the destination changes, the route found can still be used as it is the shortest route to most locations in the study area.

2.4 Principle of Operation of Visibility Graph Algorithm

A visibility graph is a graph of mutual visible locations [22]. Each node on the graph represents the location of a point, and each edge represents a visible connection between them. The vertices of the visibility graph correspond to points on the plane, which are the vertices of impassable polygons, as well as the start and target points. If the line segment connecting two vertices does not pass through the obstacle, then an edge is drawn between them. The collection of such edges and vertices is called the visibility graph (VG).

The construction of the VG from the known coordinates of the polygons bounding impassable regions on the plane is a classical problem in computational geometry. A description of the basic algorithm for its solution can be found, for example, in [23]. The complexity of the algorithm is estimated as $O(N^3)$, where N is the number of vertices of the polygons that bound the impassable regions. There are also more efficient algorithms for constructing the visibility graph. For example, in [24], an algorithm with a quadratic complexity of $O(N^2)$ is presented.

A detailed comparison of the presented algorithms by two criteria is presented below in Sect. 3.

3 Description of the «Leader-Follower» Strategy

To control a group of robots, a «leader-follower» strategy is considered in the simulation [25]. In this case, it is important to provide a path without collisions with obstacles for the robot leader, and the trajectories of the followers are determined relative to the position of the leader.

In solving the problem of controlling a UAV group, an option is proposed when the leader and the group directed by him, after takeoff and until the end of the mission, operate at an altitude that does not change over time. Therefore, we will accept the condition that UAVs move in a two-dimensional plane. The state of the *i*-th UAV (both the leader and the slave) can be described as $[x_i, y_i, \theta_i]^T$, where x and y denote the coordinates on the *Oxy* plane of the *i*-th UAV in the inertial coordinate system, and θ_i is the orientation of *i*-th UAV. Accordingly, such a UAV will move according to the following law of motion [26]:

$$\begin{cases} \dot{x}_i = v_i \cos\theta_i \\ \dot{y}_i = v_i \sin\theta_i \\ \dot{\theta}_i = \omega_i \end{cases}$$

where v_i and ω_i denote the translational and angular velocities of the *i*-th UAV, respectively. UAVs are subject to restrictions on input control signals: $0 < v_{min} \le v_i \le v_{max}$, where v_{max} and v_{min} are the maximum and minimum speed of the UAV.

A flight mission is described by a set of waypoints known as a list of waypoints, which collectively represent the path found by the planning algorithms from the starting position to the target. Starting from its current location, the lead UAV must fly sequentially to each waypoint in its waypoint list.

Description of the task of following the path (Fig. 2): point p_i is the projection of the UAV leader onto the directed path Γ_i.

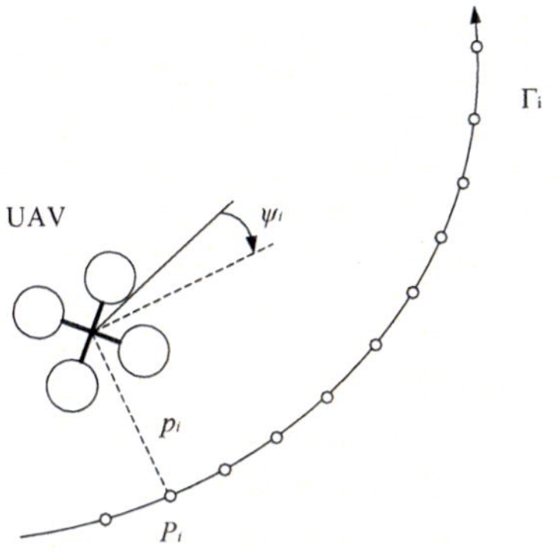

Fig. 2. Description of the task of following the path: point p_i is the projection of the UAV leader onto the directional path Γ_i.

According to the trajectory tracking control law proposed in [27], the trajectory-following error of the leading UAV tends to zero. A detailed solution to the problem of managing «leader-follower» coordination is also described in [27]. When the slave UAV follows the master UAV and changes its position in the formation, it forms a route based on its distance from the desired position. The direct induced route is a line segment from the current follower position to the desired position (Fig. 3).

The follower speed is adjusted depending on its position relative to the desired position [28]. Once the desired yaw angle and speed are obtained, they are fed to the input of the lower-level controller (flight controller) to correct the position and speed of the slave UAV.

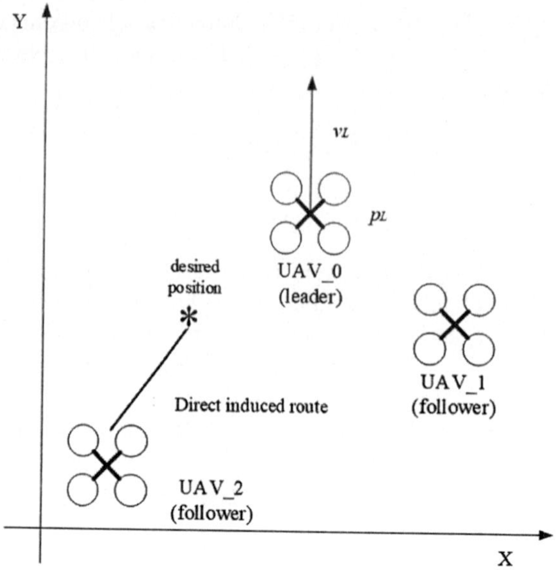

Fig. 3. Direct induced route. The leader's position and speed are denoted as p_L and v_L, respectively.

4 Results of Comparison of Planning Algorithms and Simulation of a Group Flight in Gazebo

Below the test environment and the results of experiments with the scheduling algorithms are presented. After evaluating the results obtained, it is possible to deduce the advantages and disadvantages of each algorithm for use on board the leader of the UAV group.

For the experiments, a working area of 100 by 100 m was chosen (Fig. 4). The planning problem was solved on three maps: map №1 without obstacles, map №2 with two obstacles, map №3 with five obstacles. Obstacles on the map are indicated by black areas. The start and target positions are fixed in all experiments. The start position with coordinates [9, 9] is highlighted with a red marker, the target position with coordinates [70, 90] is highlighted in green. A multi-rotor UAV in the experiments was represented as a circle with a radius of 1 m. The trajectories found by the described planning algorithms: A* (Fig. 4, a), RRT (Fig. 4, b), RRT * (Fig. 4, c), GV (Fig. 4, d) - are marked with a red line. Blue lines, if they are present in the figure, indicate auxiliary elements for constructing the trajectory.

In order to study the effectiveness of the described algorithms, 100 tests were carried out on each card. To study the efficient operation of the algorithms, two parameters were chosen: the average length of the found path and the average computation time of this path. The results of all experiments are presented in Table 1, where T_{cp} is the average computation time in seconds, L_{cp} is the average path length in meters.

According to the table, the average time for finding a path using VG does not exceed 1 s for the map with the largest number of obstacles. This is due to the fact that all the tops of the obstacles were written in the code of the algorithm that implements the path

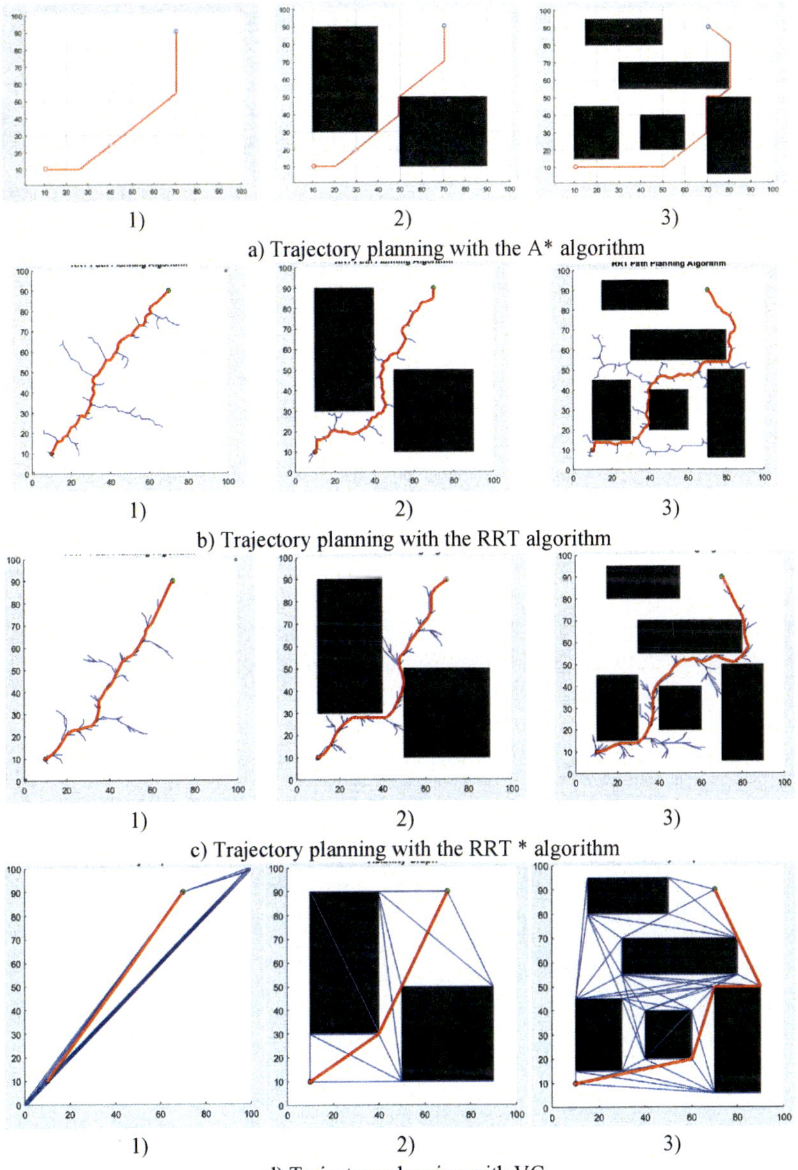

Fig. 4. The results of the experiments, presented in the form of the found path on the map of the workspace.

Table 1. Experimental results.

	A*		RRT		RRT*		VG	
	T_{cp}, s	L_{cp}, m	T_{cp}, s	L_{cp}, m	T_{cp},s	L_{cp}, m	T_{cp}, s	L_{cp}, m
Map №1	1,76914	116,28280	5,28412	112,42420	6,65924	105,45960	0,19187	100,00000
Map №2	1,33207	112,00000	3,79108	124,92140	7,66588	113,28050	0,22360	103,13760
Map №3	1,84623	141,13400	21,37331	151,4581	21,71970	138,60260	0,823636	147,33430

search by means of the VG, i.e. the algorithm fully knows the state of the work area. The time for finding a path using A* includes the decomposition of the work area and its analysis, i.e. the agent does not know anything about the location of obstacles on the maps. RRT and RRT* are also unknown about the work area, and the values of the time parameter are so high due to the probabilistic nature of the algorithms.

For clarity of presentation of the data obtained, diagrams were built on which it is convenient to compare the results of experiments (Fig. 5). Let us analyze the performance of A*, RRT, RRT* and VG in a two-dimensional environment from the point of view of solving the problem of planning the trajectory of the leader of the UAV group. According to the data obtained, using the VG algorithm, the time spent on generating paths turned out to be less than when using A*, RRT and RRT*.

However, the creation of VG -based trajectory planners and their subsequent use in UAV control systems is associated with a number of problems. One significant problem is associated with the fact that usually agents operating autonomously in the environment do not have information about the surrounding space in the form necessary for constructing a VG, i.e. there is no information about the coordinates of the polygons bounding impassable regions on the plane. This is due to the fact that the agent usually has a discrete digital map of the area, updated in real time using data from various sensors and sensors of the agent. Accordingly, to extract the necessary information (coordinates of the vertices of impassable polygons), it is necessary to develop a separate class of algorithms, which requires additional costs. Thus, it can be argued that the use of VG for solving the problem of planning the trajectory of the group leader is in practice impractical.

Estimating the obtained path length, we can say that RRT* is looking for the smoothest paths that are visually similar to real UAV trajectories. And their length was slightly less than the length of the path built by A*. In addition, A* only explored the areas needed to build the path, while the RRT* explored the environment evenly. But A* found paths in less time compared to RRT* and RRT.

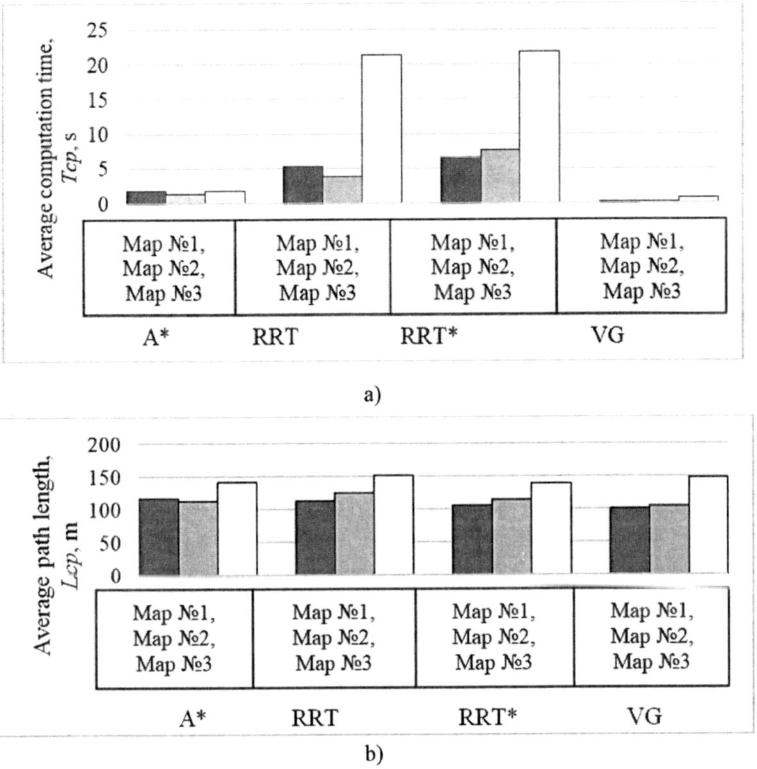

Fig. 5. Experimental results presented in the form of resulting diagrams.

Comparing RRT and RRT*, you can see that RRT finds the path faster than RRT*. However, the found RRT path cannot be used in practice and must be processed with additional anti-aliasing tools, for example, the Bézier curves method, which can affect the running time of the scheduler. The constructed trajectory RRT* smoothly bypasses obstacles, thereby reducing the length of the path. Also, the path found by RRT* does not need additional processing. Such a trajectory is more suitable for the movement of a real object than the other paths found.

In addition to avoiding obstacles, path creation time is of fundamental importance, especially in dynamic environments where the path needs to be constantly updated. While optimal path length is critical in long-distance travel situations, there are limitations on the UAVs onboard resources such as fuel and battery.

RRT and RRT* can perform better in terms of path construction time and length than A* if random nodes are chosen taking into account the shape, position and equation of motion of the obstacle, which is rather difficult due to the probabilistic nature of the algorithm. Therefore, A* stably finds a short path in a short time compared to RRT and RRT*.

As a result, the low computational requirement makes A* a key candidate for solving the problem of online planning of two-dimensional trajectories for the group leader in an environment with static and dynamic obstacles. Moreover, A* allows for different sampling of the environment in accordance with the requirements of different parts of the scenario, ensuring optimal use of the UAV resources.

Simulation of a group flight according to the proposed strategy and path planning algorithms was carried out in the Gazebo environment (Fig. 6). To simulate the task, the A* algorithm was chosen, which consistently produces an optimal solution with small spent time resources.

a) b)

Fig. 6. The result of modeling the task in Gazebo: a) a group of UAVs in a given formation («leader» in front of the group) before the start of the flight mission; b) a group of UAVs in a given formation in flight to the target point (led in the «line» formation, the leader is in front of the group).

Simulation in Gazebo was carried out with standard UAV quadrocopters. The used UAV «iris» has the following characteristics: dimensions 440 mm × 330 mm × 100 mm, weight 1.2 kg, maximum speed 18 m/s. For simulation, the UAV speed was set to 11 m/s.

The route planning module is written in Python and implements an A* based path finding algorithm. When searching for a path, the size of the group is taken into account: the surface (area) occupied by the group, which can be described by a circle. The module receives a map of the area (map №2 was chosen for modeling) and the radius of the circle where the UAVs are located (it is necessary so that the group can overcome the obstacle). The result of the work of the trajectory planning module for the «leader» of the group is a file with intermediate coordinates, which together constitute the path of the UAV from the start to the final coordinate. This coordinate file is sent to the leader's flight controller. The flight controller of the «leader» processes the resulting trajectory. Until takeoff, UAVs do not interact with each other. As soon as all UAVs take off, the «leader» interacts with the «followers». The «followers» UAVs, using the MAVLink data transfer protocol, receive the coordinates of the «leader». In accordance with the concept of «leader-followers» UAVs «followers» maintain their positions relative to the «leader» to perform a synchronous flight to the target point. The total time for completing the flight mission, including the takeoff of the entire group and the return of all agents to the starting point, took 2 min. The length of the traversed path is 118.7 m, which is slightly different from the length of the planned path. Figure 7 shows a graph of the trajectories of the «leader» UAV.

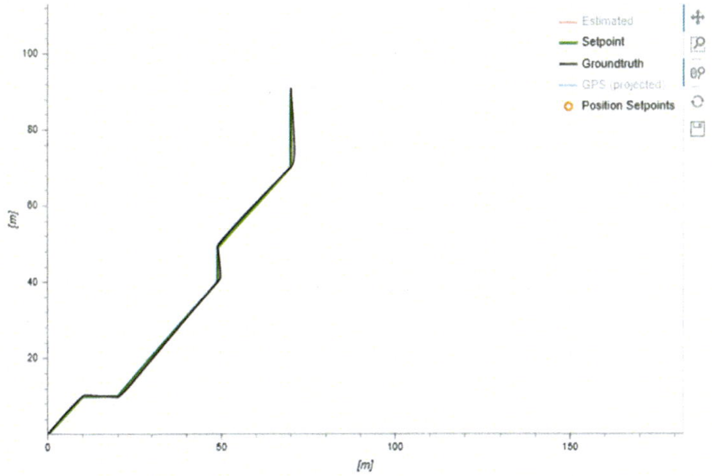

Fig. 7. The trajectory of movement of the «leader» of the group.

Figure 8 shows the parameters for roll, pitch and yaw angles.

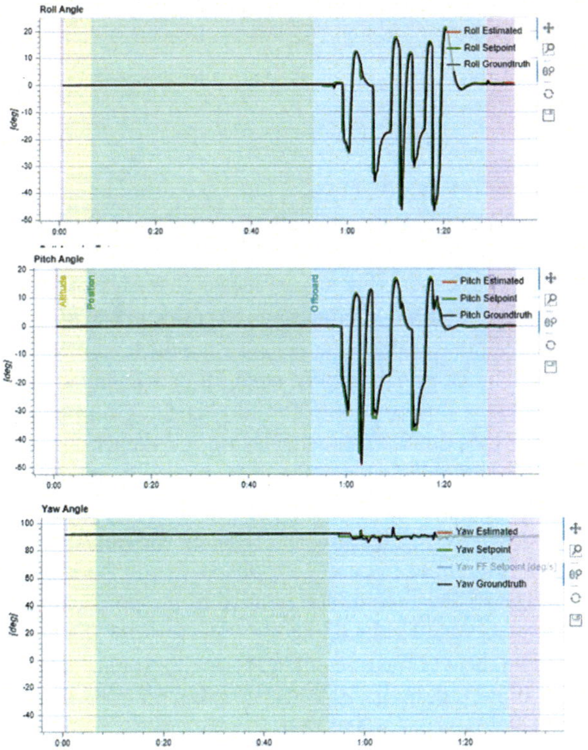

Fig. 8. Indicators of the angle of roll, pitch and yaw.

In Fig. 7, the found trajectory using the A* algorithm is indicated in green. The black line shows the real path along which the UAV was moving. The initial experiment was conducted without wind. The results of the flight logs show that the actual trajectories of the UAV almost coincide with the found trajectories, which were transmitted to the Pixhawk flight controller. Small deviations of the UAV from the trajectory were observed during the flight. In real conditions, in the presence of wind, the distance in the formation between UAVs is almost impossible to keep constant. For a safe flight, it is recommended to make the difference in altitude of adjacent UAVs. In addition to the wind, the simulation is influenced by the readings of on-board navigation equipment, for example, GPS. Therefore, UAVs located at intermediate points slightly deviated from the specified route. The accumulation of errors in GPS and large deviations from the trajectory can cause the model to fail. These problems can be solved by using RTK GPS (Real Time Kinematic GPS), which has high positioning accuracy [29].

The background color of the graphs (Fig. 8) is used to indicate the UAV flight mode. Each color corresponds to its own flight mode: yellow - altitude mode, green - position mode (these two modes are necessary when controlling from the remote control, therefore there are no changes on the graph and these modes can be considered as waiting modes for a flight task), blue - a mode for performing a flight task, sent from the ground device to the UAV via MAVLink, purple - landing mode (the command to land is also sent from the ground device).

As a result, the total simulation time for a group flight does not exceed 2 min. The length of the calculated path and the length of the traveled path are 112.00 m and 118.7 m, respectively.

5 Conclusion

Route planning when performing flight tasks most often occurs as follows: having an initial map and building a global route along it, the agent moves to the target and at the same time dynamically rebuilds its path when new obstacles are detected, using proximity sensors and other similar sensors. The study focuses on the use of global path planning algorithms A*, RRT, RRT* and VG. Comparing the performance of these algorithms, it was found that the paths found using A* and RRT* are most applicable in the problem of group leader movement. However, the low need for path calculations using A* is the main criterion for choosing this algorithm for solving the group movement problem. Also, the advantages include the fact that when using A*, the working environment can be divided into cells in accordance with different requirements, ensuring optimal use of UAV resources.

To study group flight, the «leader-follower» strategy was applied, where it is important to provide a path without collisions with obstacles for the robot-leader, and the trajectories of the followers are determined relative to the position of the main agent. The time spent on the process of calculating the trajectory for the leader of the group does not exceed 1.3 s on the working area in the form of a square with a side of 100 m, the time spent on the flight task by the UAV group takes no more than 2 min, provided that the speed of the UAV is 11 m/s (map № 2).

Experiments in the Gazebo simulation showed that when using the proposed strategy for controlling a group of UAVs, it is necessary to take into account the size of the group of UAVs, environmental parameters and a map of obstacles to the movement of the group in the working area. The required input data are the start and end coordinates. They are necessary to construct a trajectory based on any studied planning algorithm. The constructed path, represented as an array of coordinates, is sent to the leader's flight controller. And the further movement of the group occurs by keeping the given positions of the followers relative to the leader.

The proposed strategy is flexible to modifications, therefore, further research of the leader-follower strategy will be aimed at modifying the A* algorithm when finding a path for a group leader in an environment with dynamic obstacles. Also in the near future, it is planned to study the problem of combining global planning algorithms on a known map with local planning algorithms, for example, a potential fields algorithm or a joint planning algorithm based on methods of simultaneous localization and navigation (SLAM) or neural networks.

References

1. Stecz, W., Gromada, K.: UAV mission planning with SAR application. Sensors **20**(4), 1080 (2020)
2. Ruan, W.-Y., Duan, H.-B.: Multi-UAV obstacle avoidance control via multi-objective social learning pigeon-inspired optimization. Front. Inf. Technol. Electron. Eng. **21**(5), 740–748 (2020). https://doi.org/10.1631/FITEE.2000066
3. Wang, X., et al.: Coordinated flight control of miniature fixed-wing UAV swarms: methods and experiments. Sci. China Inf. Sci. **62**(11), 1–17 (2019)
4. Leonard, N.E., Fiorelli, E.: Virtual leaders, artificial potentials and coordinated control of groups. In: Proceedings of the 40th IEEE Conference on Decision and Control (Cat. No. 01CH37228), vol. 3, pp. 2968–2973 (2001)
5. Han, K., Lee, J., Kim, Y.: Unmanned aerial vehicle swarm control using potential functions and sliding mode control. Proc. Inst. Mech. Eng. Part G J. Aerospace Eng. **222**(6), 721–730 (2008)
6. Raj, S., Supreet, K., Chris, G.: Optimizing settling time for swarm UAV formation control in the presence of obstacles. In: Department of Mechanical and Aerospace Engineering University at Buffalo Buffalo, New York, 14261
7. Brunner, M., Brüggemann, B., Schulz, D.: Hierarchical rough terrain motion planning using an optimal sampling-based method. In: 2013 IEEE International Conference on Robotics and Automation, pp. 5539–5544 (2013)
8. Noreen, I., Khan, A., Habib, Z.: Optimal path planning using RRT* based approaches: a survey and future directions. Int. J. Adv. Comput. Sci. Appl **7**(11), 97–107 (2016)
9. Yap, P.: Grid-based path-finding. In: Conference of the Canadian Society for Computational Studies of Intelligence, pp. 44–55. Springer, Berlin, Heidelberg (2002)
10. Wooden, D.T.: Graph-based path planning for mobile robots (Doctoral dissertation, Georgia Institute of Technology) (2006)
11. Kavraki, L.E., Svestka, P., Latombe, J.C., Overmars, M.H.: Probabilistic roadmaps for path planning in high-dimensional configuration spaces. IEEE Trans. Robot. Autom. **12**(4), 566–580 (1996)
12. Jeauneau, V., Jouanneau, L., Kotenkoff, A.: Path planner methods for UAVs in real environment. IFAC-PapersOnLine, **51**(22), 292–297 (2018)
13. Ren, X., Tan, L., Jiaqi, S., Lian, X.: Multi-target UAV path planning based on improved RRT algorithm. J. Phys. Conf. Ser. **1786**(1), 012038 (2021)

14. Adiyatov, O., Varol, H.A.: A novel RRT*-based algorithm for motion planning in dynamic environments. In: 2017 IEEE International Conference on Mechatronics and Automation (ICMA), pp. 1416–1421 (2017)
15. Huang, S., Teo, R.S.H.: Computationally efficient visibility graph-based generation of 3D shortest collision-free path among polyhedral obstacles for unmanned aerial vehicles. In: 2019 International Conference on Unmanned Aircraft Systems (ICUAS), pp. 1218–1223 (2019)
16. Ghandi, S., Masehian, E.: Review and taxonomies of assembly and disassembly path planning problems and approaches. Comput.-Aided Des. **67**, 58–86 (2015)
17. Aggarwal, S., Kumar, N.: Path planning techniques for unmanned aerial vehicles: a review, solutions, and challenges. Comput. Commun. **149**, 270–299 (2020)
18. Nikolos, I.K., Valavanis, K.P., Tsourveloudis, N.C., Kostaras, A.N.: Evolutionary algorithm based offline/online path planner for UAV navigation. IEEE Trans. Syst. Man Cybern. Part B (Cybernetics), **33**(6), 898–912 (2003)
19. Le, A.V., Prabakaran, V., Sivanantham, V., Mohan, R.E.: Modified a-star algorithm for efficient coverage path planning in tetris inspired self-reconfigurable robot with integrated laser sensor. Sensors **18**(8), 2585 (2018)
20. LaValle, S.M.: Rapidly-exploring random trees: a new tool for path planning (1998)
21. Karaman, S., Walter, M.R., Perez, A., Frazzoli, E., Teller, S.: Anytime motion planning using the RRT. In: 2011 IEEE International Conference on Robotics and Automation, pp. 1478–1483 (2011)
22. Islam, F., Nasir, J., Malik, U., Ayaz, Y., Hasan, O.: RRT∗-smart: rapid convergence implementation of RRT∗ towards optimal solution. In: 2012 IEEE International Conference on Mechatronics and Automation, pp. 1651–1656 (2012)
23. Huang, H.P., Chung, S.Y.: Dynamic visibility graph for path planning. In: 2004 IEEE/RSJ International Conference on Intelligent Robots and Systems (IROS) (IEEE Cat. No. 04CH37566), vol. 3, pp. 2813–2818 (2004)
24. Drug, F., Sheimos, M.: Computational geometry: an introduction (1989) (In Russia)
25. Edelsbrunner, H.: Algorithms in combinatorial geometry, vol. 10. Springer Science & Business Media (2012)
26. Yun, B., Chen, B.M., Lum, K.Y., Lee, T.H.: A leader-follower formation flight control scheme for UAV helicopters. In: 2008 IEEE International Conference on Automation and Logistics, pp. 39–44 (2008)
27. Hao, C., Xiangke, W., Lincheng, S., Yirui, C.: Formation flight of fixed-wing UAV swarms: a group-based hierarchical approach. Chin. J. Aeronaut. **34**(2), 504–515 (2021)
28. Wu, J., Zhou, R., Dong, Z., Che, J.: Formation flight control method of multiple UAVs based onguidance route. J. Beijing Univ. Aeronaut. Astronaut. **42**(7), 1518 (2016)
29. Scientific Research Publishing, https://www.scirp.org/html/376.html. Accessed 20 July 2021

The Factors that Influence the Users' Adoption and Acceptance of Cloud Computing at a University of Technology in South Africa

Manche Aloycia Lieta[1](✉), Odilia Sehume[2], and Tranos Zuva[2]

[1] Vaal University of Technology, Andries Potgieter Blvd, Vanderbijlpark 1911, South Africa
[2] Faculty of Applied and Computer Science, Department of ICT, Vanderbijlpark 1900, South Africa

Abstract. Cloud computing can assist in overcoming the present boundaries in mobile learning (m-learning) with regard to the limited processing and storage capabilities of the mobile devices. This way, learning applications can run on students' mobile devices while the heaviest computing tasks take place in the cloud. For this to happen leaners have to accept the use of cloud computing in their studies. This study endeavours to identify factors that influence the user's adoption and acceptance of cloud computing at universities. A model derived from a unified theory of acceptance and use of technology (UTAUT) model was used to find factors that influence learners to adopt and accept cloud computing during their studies. The model consists of seven variables that is performance expectancy (PE), effort expectancy (EE), social influence (SI), facilitating conditions (FC), trust (TR), behaviour intention (BI) and user behaviour (UB). The result showed that all the variables influenced adoption and acceptance of cloud computing except EE. User behaviour was mostly influenced by trust, then facilitating conditions followed by performance expectancy and lastly social influence. In conclusion, trust seems to be the greatest influential factor when it comes to the adoption and usage of applications hosted on the internet.

Keywords: Cloud computing · Adoption and acceptance model · e-learning · m-learning

1 Introduction

Technology improvements have always had a significant impact on industry growth, affecting even the most traditional systems such as education [1]. There are new technological tools which can support or elevate the traditional learning systems. Cloud computing is an innovative technology that has evolved rapidly [2], and is among the innovations that guarantees to take the knowledge delivery procedures to a higher level [3]. Cloud computing is defined [4] as the provision of on-demand computing resources such as intelligence, networking, databases, analytics, software and storage on the internet, according to the method of pay-as-you-go. Users access computing resources in accordance with their needs, regardless of where the resources are hosted or how they

R. Silhavy et al. (Eds.): CoMeSySo 2021, LNNS 232, pp. 885–898, 2021.
https://doi.org/10.1007/978-3-030-90318-3_69

are provided. In other words, cloud computing is a highly scalable form of computing that uses virtualized resources that can be distributed among users. Users do not require a basic knowledge of services. Moreover, an Internet user may communicate with multiple servers at the same time and these servers exchange information with each other.

Higher education institutions play an important role in ensuring economic growth, especially for countries with growing economies such as South Africa. In the opinion of [5] emerging technology enables greater access and flexibility in the delivery of quality teaching and learning. Whilst e-learning is the process of receiving or giving systematic instruction, sharing of knowledge or skills and expertise over the internet [5], cloud computing on the other hand has greater scalability, flexibility and mobility in the utilisation of resources for teaching purposes [6]. Three cloud computing service models: - software as a service (SaaS): software is hosted in the cloud and can be used by any individual or organisation as per requirement e.g., Google, Twitter and Facebook [7]. Platform as a service (PaaS): suitable for users intending to deploy their own applications. Infrastructure as a service (IaaS): provides cloud-based infrastructure such as storage, processing and networking elements. Cloud computing tends to be beneficial over traditional computing due to the reduced infrastructure and IT costs, increased accessibility, enabled collaboration and flexibility and in a schooling environment it helps in reaching more and more diverse students [8].

Cloud computing offers students and staff numerous and different online applications which can be utilised as learning tools [9]. These applications are typically web-based, available anywhere, anytime on the internet, creating a good study environment for learners. They can access information in a very suitable way, even from home without purchasing product licenses, overpowering the physical limitations of the faculty and increasing their learning capabilities. It assists in overcoming the current boundaries in mobile learning (m-learning) regarding the limited processing and storage capabilities of the devices [10]. In this way, learning applications can run on students' mobile devices while the heaviest computing tasks take place in the cloud and cloud-based solutions can therefore be effective in supporting global collaboration and cooperative learning, by the use of virtual classrooms via online and video conferencing [11]. Cloud computing will enable learners to come out from the walled classrooms and explore the world [12].

Among its other benefits, cloud computing is cost effective because the hardware and software are shifted from local premises to cloud, thus reducing installation and maintenance costs [12], and institutions will consequently, only pay for what they use. The aim of this study was to propose a framework for understanding the factors that influence the user's adoption and acceptance of cloud computing in education at universities.

1.1 Problem Statement

Various studies investigated the adoption and benefits of cloud computing and found a low rate of cloud computing adoption exclusively in higher education [13]. There are limited numbers of studies which looked into the factors that influence cloud computing adoption in developing countries [14], of which South Africa is no exception. Furthermore, [15] added that most studies examined cloud computing adoption in various small, medium sized enterprises (SME) or government institution but very few studies were conducted under education area. In the study of [16], factors that influence cloud computing adoption by small, medium sized enterprises were determined, but not so much has been done to measure factors that influence individuals in education to adopt and accept cloud computing. There is a lack of results in factors that influence the adoption rate of cloud computing to be slower than expectations [17].

The users' trust towards cloud computing is one of the greatest socio-technical issues of concern today [18]. It can be said that one of the key factors that formulate the users' trust is the protection of their privacy, however, in the adoption and acceptance of cloud studies there has been a lack of attention given to issues such as trust and security.

This study attempts to bridge the knowledge gap through the identification of factors that influence the adoption, acceptance and the use of cloud computing and its benefits to the university in executing their day-to-day business of teaching and learning.

1.2 Research Question

What are the factors that influence the user to adopt and accept cloud computing at a University of Technology in South Africa?

Secondary questions

- What framework was proposed to measure a users' adoption and acceptance of cloud computing at a university of technology in South Africa?
- How was the influence of each variable of the model in the adoption and acceptance of cloud computing at university level measured?

1.3 Aim and Objectives

The goal of the research was to investigate the factors that would influence the adoption and acceptance of cloud computing at the University of Technology.

- To propose a model to use to measure the degree of influence of factors in user's adoption and acceptance of cloud computing.
- To measure the degree of influence of factors in the proposed model in the adoption and acceptance of cloud computing at University of Technology in South Africa.

2 Literature Review

The theory of reasoned action (ToRA or TRA) [19] targeted to explain how the action of human beings is formed by the connection of their attitudes and behaviours. Fundamentally, it is utilised to anticipate how people will act in the future based on how they reacted previously (pre-existing attitudes). A person's choice to take part in a specific conduct depends on the results the individual expects will come because of their participation. As per [20] the theory of reasoned action (TRA) postulates that an individual's behaviour is determined by their motive to perform the behaviour, people weigh up the inferences of their deeds before taking any decision of whether to partake or not to partake in a behaviour (their expectations). In the study of [21] TRA was joined with another model to investigate the behaviour of students towards the use of cloud computing innovations. The investigation was directed using an online survey, with the sample size of 478 individuals. The validity and reliability of the proposed model were checked. For assessment R2, Beta q2, and f2 were utilised. The outcomes affirmed that the multiple model approach had a sufficient informative capacity to clarify BI toward cloud computing classrooms.

The theory of technology acceptance model (TAM) was founded by Davis (1989), with the aim of explaining and predicting why users accept information systems and this model was derived from other adoption theory (theory of reasoned action). The model was later upgraded [22] to extend technology acceptance model TAM2. This model empowers the use of outside factors in order to explain how those factors affects users' attitude. In the investigation of [23], technology acceptance model 2 (TAM2) was used to investigate connections between factors related to Web 2.0 and the behaviour of the user. Information was gathered through questionnaires with sampling sizes of 400. The model tested for reliability and validity, the relationship between factors were investigated, and lastly the coefficient and t-test were dissected. The outcomes demonstrated that Web 2.0 sites were commonly acknowledged by clients.

Innovation is the way toward making another technology, gadget or system [24], whereas diffusion is the way toward spreading thoughts, ideas, abilities and learning through society. The innovation diffusion theory (IDT) portrays how developments or innovation become acknowledged and spread through the media [24]. The study of [25] investigated the adoption of cloud computing as a form of innovative IT by the South African organisations. The study tested institutional theory and diffusion of innovation theory (DOI) characteristics that influence the organisational adoption of cloud computing. The model was developed and tested using a survey methodology, with the sample size of 980 medium to large organisations. Data were tested for validity and reliability, correlation and regression were used for measuring. The results of the study indicated that the mimetic pressures construct drawn from institutional theory is more important than normative and coercive pressures in explaining adoption of cloud computing, and that the diffusion of innovation DOI factors of compatibility and relative advantage were also significant [25]. The study conducted by [26] reshaped theory of interpersonal behaviour (TIB) developed by trends in 1997. The aim of this adoption was to propose a model which predicts personal computer (PC) /IS usage behaviour instead of the intention. The Model of personal computer utilization (MPCU) examines the direct control of long-term consequences of use, social influences, complexity of job fit, affect towards use and facilitating conditions on behaviour. The model assumes that participants have required experience [27]. Social cognitive theory was founded in 1986 by Bandura, this is the theory of learning which consider human behaviour changes. The learning happens when individuals observe what others do and learn from them. In this theory the following (personal factors, environmental factors and behaviour) interacts and utilise reciprocal determinism to impact each other. Additionally, this theory is widely conducted in researches which investigates how mass media can trigger a change in someone's behaviour or attitude [28]. The study of [29] used social cognitive theory to examine the factors of the adoption of internet banking. The banks in Ghana were utilised for sampling. Structured equation modelling and confirmatory factor analysis was used for analysis. The findings of the study showed that the adoption of internet banking if significantly influenced by the following (websites' social feature, trust, compatibility with lifestyle and online customer services).

Already, different adoption models, and accepted theories were utilised to examine the acceptance of IT and IS technology. Lately unified theory of acceptance and use of technology (UTAUT) model was created from eight models (explained above) to be specific: theory of reasoned action (TRA), technology acceptance model (TAM), motivational model (MM), theory of planned behaviour (TPB), and CTAM-TPB, model of PC use (MPCU), innovation diffusion theory (IDT), and social cognitive theory (SCT). Furthermore, [30] established UTAUT by incorporating the past referenced eight models to remunerate their shortcomings and to frame a total and better model to foresee clients' aims on innovation reception. UTAUT comprise of six centre factors that are performance expectancy, effort expectancy, social influence, facilitating conditions, behaviour intention, and use behaviour. It additionally incorporates four mediators to be specific, sex, and age, involvement and wilfulness. The utilisation of this model in cloud computing adoption is still limited in number [31].

The examination led by [32] recognised elements that influence the users' adoption of cloud computing in education in Iraq, the study deployed unified theory of acceptance and use of technology (UTAUT) along with security and trust. The outcomes demonstrated that exhibition anticipation is the most significant component that influences the conduct goal to utilise cloud computing pursued by exertion hope, social impact, security, and trust. Conduct expectations and encouraging conditions had a solid impact on use conduct of cloud computing.

The research by [33] explored the key factors that impact client reception of cloud-based, community learning innovation in the instructional setting, the investigation was grounded in a unified theory of acceptance and use of technology (UTAUT), personalisation develop was connected to the conduct expectation, execution hope and exertion anticipation. Utilising survey questionnaire, data were collected from 150 students of four Malaysian public universities. The outcome of the SEM demonstrated that clients' conduct expectation was fundamentally impacted by execution anticipation, exertion hope, social impact and personalisation. In this manner, the UTAUT model is reasonable for understanding the use of cloud computing in education.

3 Methodology

After going through the literature review for previous studies. The researcher proposed revised unified theory of acceptance and use of technology (UTAUT) model which is derived from the following eight adoption models namely: Theory of Reasoned Action (TRA), technology acceptance (TAM), motivational model (MM), theory of planned behaviour (TPB), and CTAM-TPB, model of PC utilisation (MPCU), innovation diffusion theory (IDT), and social cognitive theory (SCT). The study followed [30] original UTAUT model as closely as possible with its measurements and analysis. Nevertheless, a revised version of UTAUT was utilised to fit the context of this research. The original model consists of 4 key independent constructs: performance expectations, effort expectancy, social influence, facilitating conditions, and two dependent constructs: behaviour intention and use behaviour. In addition, the original model consists of 4 control variables which includes: gender, age, experience, voluntariness of use.

For this study trust was incorporated to be one of the independent constructs, this was to make the model more relevant for cloud computing adoption and acceptance. This model was chosen because [32] used the same model in measuring the factors that influence the user to adopt cloud computing in Iraq and [34] utilised the same model to identify factors that explain customers' behavioural intention to adopt and use internet banking services in Nigeria, this model was found to be convenient for both studies. Hence this same model will now be used in South African university to measure the factors that influence the user to adopt and accept cloud computing (Fig. 1).

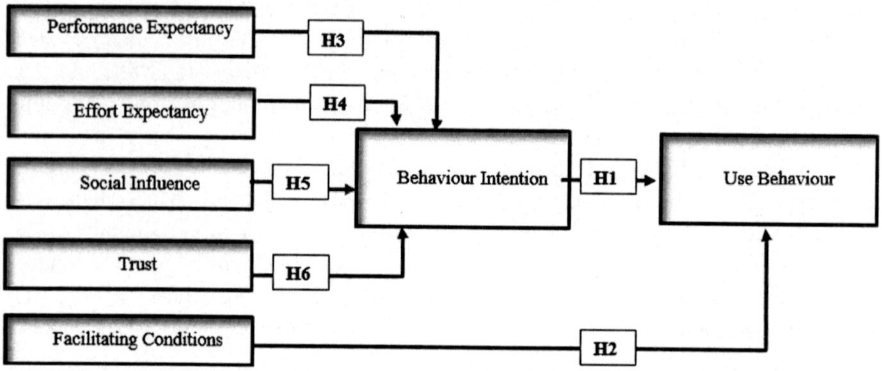

Fig. 1. The proposed model of the revised UTAUT Model derived from [30, 35].

Table 1. Measures and sources.

Construct	Number of items	Sources
Performance expectations (PE)	10	[30, 36]
Effort Expectation (EE)	3	[30, 36]
Social influence (SI)	6	[30, 36]
Facilitating conditions (FC)	4	[30, 36]
Behavioural intention (BI)	3	[30, 36]
User Behaviour (UB)	3	[30, 36]
Trust (TR)	7	[37]

3.1 Research Design

Quantitative approach was used together with survey questionnaires. This was chosen because a large number of related past studies used it and it was proven to be convenient (Table 1).

3.2 Population, Sample and Sample Technique

The study was limited to South African undergraduate and postgraduate students from a university of technology, who had qualities that were significant to this study. The study was conducted in English because it is a universal language in South Africa. The sample used was 349 participants. Both male and female from the age of 17 and above, qualified for participation. The convenience sampling was used because the study aimed to get participants from all over the university. This sampling strategy enables samples to be chosen based on availability and accessibility.

3.3 Ethical Consideration

According to [38] the rules of the International Development Research Centre stipulate that the research work, including human subject must be done with high morality. The consent form was sent together with the questionnaire to ensure that every member understood the motive behind the study, got clarification based on their voluntariness of participation and also to give them freedom to withdraw, lastly to ensure them of their data protection and confidentiality.

3.4 Data Collection and Instruments

The survey questionnaire was an appropriate tool to gather information about the factors that influence the adoption and acceptance of cloud computing in South Africa at a university of technology. The questionnaire was based on the proposed theoretical model. The study used the online form of survey distribution. The link to the survey was sent to participants via social media as it was convenient during the pandemic lockdown and cost-effective. Participants were asked for their wiliness to engage, once they agreed to participate, they were issued, a link which had a consent letter attached to the first page of the questionnaire. This was done to ensure that participants understood the guidelines of the questionnaire and for ethical purposes. About 377 questionnaires were distributed to students and 349 were completed. The questionnaire was divided into three parts:

Part 1: asked multiple-choice questions about participants' personal information such (age, gender and education level) and
Part 2: consisted of multiple-choice questions about participants' computer knowledge and cloud computing experience.
Part 3: consisted of technology adoption related questions, which in this study contained 36 questions from seven constructs of the UTAUT model. This part aimed at gathering information on the participants 'perception of factors that influence them to adopt and accept cloud computing. Respective items of the construct were measured in order to determine the degree at which the participant agreed with each item using a five-point Likert scale where 1 = strongly disagree and 5 = strongly agree.

3.5 Validity and Reliability

Reliability testing involved the testing of internal consistency of the study constructs. To assess the existence of some differences within the test items that explore the same construct. Validity testing ensured that the study instrument accurately measured what it intended to measure. The items on the questionnaire were taken from existing studies from the literature review. This was done to avoid problems of validity and reliability of the measures. Some questions were modified to fit in a cloud computing environment. Cronbach's alpha and exploratory factor analysis were utilised for validity and reliability.

4 Data Analysis

The study had a total of 349 respondents. Statistical Package for the Social Sciences software (SPSS) version 27 was found to be the appropriate data analysis tool.

4.1 Demographic Information

The usable data for study participants was 349, with a return rate of 93% of which 43.97% were male and 56.03% participants. The age group ranged from 21–30 old with 56.03%. The majority of the participants were undergraduates 60.34% and their knowledge of cloud computing was moderate 38.22%. Additionally, 32.47% have used cloud computing for the period of 1–3 years. See Table 2 below:

Table 2. Demographics.

Total Participants = 349	Labels	Percentages
Gender	Male	43.97
	Female	56.03
Age	20 or under	13.51
	21–30	56.03
	31–40	26.44
	41–50	2.87
	51+	1.15
Education Level	Postgraduates	39.66
	Undergraduates	60.34
Cloud Computing Knowledge	Very poor	7.18
	Poor	18.39
	Moderate	38.22
	Good	25.00

(continued)

Table 2. (*continued*)

Total Participants = 349	Labels	Percentages
	Very good	6.90
	Other	4.31
Cloud Computing Experience	Don't Use	20.98
	Less than 1 year	27.01
	1–3 years	32.47
	More than 3 Years	19.54

4.2 Validity and Reliability

Cronbach's alpha and exploratory factor analysis were utilised for validity and reliability. According to [39] the factor loadings should exceed 0.5, any loading less than this number should be eliminated because it contributes less towards other variables. In Table 4 the majority of the latent components had loadings greater than 0.5 and almost all other variables loaded strongly on these identified components. All the seven extracted components had Eigenvalues greater than one, hence they were considered for further analysis. The 36 items from the study questionnaire are clubbed together into 7 components (factors) which explained 56% of the total variance of the user adoption and acceptance of cloud computing. This confirms that the variables are valid and were measured (Table 3).

Table 3. Exploratory factor analysis.

Variables	Component	Factor loadings	Eigen value	Total variance explained
Performance Expectancy (PE)	PE8	.764	10,253	28,482
Effort Expectancy (EE)	EE3	.751	2,585	35,662
Social Influence (SI)	SI3	.699	1,834	40,755
Facilitating Conditions (FC)	FC4	.673	1,622	45,26
Behaviour Intention (BI)	BI2	.515	1,405	49,162
User Behaviour (UB)	UB3	.776	1,358	52,935
Trust (TR)	TR	.801	1,126	56,063

The summary of the results of Cronbach's Alpha values is shown in Table 4. Through Cronbach's Alpha analysis, the results demonstrated that 6 factors (PE, EE, SI, FC, UB and TR) are highly reliable with more than .7 reliability coefficient while 1 factor (BI) have an average alpha of .6, which can be interpreted as questionable, but, acceptable. The overall proposed model coefficient was found to be highly reliable at more than .9 which confirmed the reliability of the collected data is to fulfil the goal of this study (Table 4).

Table 4. Crobach's alpha.

No	Factors of the proposed research model	Cronbach's Alpha	Number of items
1	PE	.860	10
2	EE	.706	3
3	SI	.779	6
4	FC	.728	4
5	BI	.653	3
6	UB	.769	3
7	TR	.860	7
8	Cronbach's Alpha for overall model	.933	36

4.3 Hypothesis Testing

Prior to conducting multiple regression analysis the following assumptions had to be fulfilled: outliers were deleted, normality of data was checked, linear relationships between variables were confirmed (linearity) and lastly there were no multicollinearity issues.

The regression analysis results depicted in Table 5 demonstrate that performance expectancy is the highest predictor of behaviour intention towards cloud computing adoption with the highest B coefficient, followed by trust towards cloud computing adoption, then social influence. Thus, all these three variables have P-Value < 0.05 and are found to be significantly influential on behaviour intention. Additionally, facilitating condition and behaviour intention significantly affect the user behaviour with p-value less than .05, in this behaviour intention is the highest predictor of cloud computing acceptance, then facilitating conditions follows. Lastly effort expectancy has a p-value > 0.05, meaning it does not have a significant impact on behaviour intention (it cannot predict BI) in the adoption of cloud computing.

The following independent variables (PE, EE, SI, TR) are directly linked to behaviour intention and have an R squared off .354, meaning these independent variables explain 35% of the variance in BI towards the adoption of cloud computing. Furthermore BI and FC explain variation of 22% in user behaviour towards the acceptance of cloud computing (Tables 5 and 6).

Table 5. Multiple regression analysis.

Model		Unstandardized coefficients		Standard Beta Coefficient	T	Sig.
Dependent variables		B	Std. Error			
Behaviour Intention	(Constant)	.763	.254		3.001	.003
	PE	.357	.066	.305	5.374	.000
	EE	.065	.043	.075	1.489	.137
	SI	.138	.053	.140	2.584	.010
	TR	.281	.058	.241	4.891	.000
	(Constant)	1.206	.280		4.310	.000
User Behaviour	BI	.379	.069	.312	5.450	.000
	FC	.282	.074	.218	3.802	.000

Table 6. Summarised results of hypothesis testing.

	Constructs	P-value	Beta coefficient	Result
H1: Behavioural intention has a positive effect on use behaviour of cloud computing acceptance	BI	P-value < 0.05 .000	(.312)	Accepted
H2: Facilitating condition has a positive effect on use behaviour of cloud computing services	FC	P-value < 0.05 .000	(.218)	Accepted
H3: Performance expectancy has a positive effect on users' behavioural intention to use cloud computing services	PE	P-value < 0.05 .000	(.305)	Accepted
H4: Effort expectancy has a positive effect on users' behavioural intention to use cloud computing services	EE	P-value > 0.05 .137	(.075)	Rejected
H5: Social influence has a positive effect on users' behavioural intention to use cloud computing services	SI	P-value < 0.05 .010	(.140)	Accepted
H6: High trust has a positive effect on users' behavioural intention to use cloud computing services	TR	P-value < 0.05 .000	(.241)	Accepted

The final representation of regression analysis is calculated as follows:

$$BI = 0.357\,PE + 0.138SI + 0.281TR + 0.763 \tag{1}$$

$$UB = 0.379\,BI + 0.282FC + 1.206 \tag{2}$$

Substitute 1 in 2.

$$UB = 0.379(0.375PE + 0.138SI + 0.281TR + 0.763) + 0.282FC + 1.206$$

$$UB = 0.135\,PE + 0.052SI + 0.289TR + 0.289 + 0.282FC + 1.206$$

$$UB = 0.135\,PE + 0.052SI + 0.289TR + 0.282FC + 1.495 \tag{3}$$

Equation 3 is the final representation of the study. Use behaviour is mostly influenced by trust (.289) then facilitating conditions (.282) followed by performance expectancy (.135) and lastly social influence (.052). Our results are consistent with other researchers' findings that had the following variables supported performance expectancy, social influence and trust [40, 41]. Effort expectancy was not supported in our research study, which is also consistent with other researchers' findings [36, 42, 43].

5 Conclusion

The aim of this study was to investigate factors that influence users (learners) to adopt and accept cloud computing. To achieve this the study proposed revised UTAUT model. This model was used to examine the level of influence performance expectancy (PE), effort expectancy (EE), social influence (SI), facilitating condition (FC) and trust (TR) have on university students' adoption and usage of cloud computing. The testing of hypothesis was conducted through multiple regression analysis, the outcome proved that the following (PE, SI, FC and TR) have a positive effect on students' decision to use cloud computing. Furthermore the results demonstrated students' trust towards the use of cloud computing as the highest factor in their decision. This was followed by facilitating conditions, performance expectancy and then social influence. Nevertheless, effort expectancy did not have an impact on students' intention to use cloud computing, therefore the hypothesis was rejected. In conclusion, trust seems to be the greatest influential factor when it comes to the adoption and usage of applications hosted on the internet, this is consistent with [40, 41].

References

1. Wahl, H.: An essay about the impact of the digital revolution on higher education in art and design. In: International Conference on Human-Computer Interaction. Springer, pp. 317–333 (2018)
2. Buyya, R., Yeo, C.S., Venugopal, S., Broberg, J., Brandic, I.: Cloud computing and emerging IT platforms: vision, hype, and reality for delivering computing as the 5th utility. Future Generation Comput. Syst. **25**(6), 599–616 (2009)

3. Chibaro, N.: Adoption of cloud pedagogy by higher learning institutions in Southern Africa. Cape Peninsula University of Technology (2015)
4. Copeland, M., Soh, J., Puca, A., Manning, M., Gollob, D.: Microsoft Azure and cloud computing: microsoft Azure. Springer, pp. 3–26 (2015)
5. Kettle, M.: International student engagement in higher education (Multilingual Matters, 2017 (2017)
6. Alharthi, A., Yahya, F., Walters, R.J., Wills, G.: An overview of cloud services adoption challenges in higher education institutions (2015)
7. Mell, P., Grance, T.: Sp 800–145. the nist definition of cloud computing, Gaithersburg, MD (United States, Tech. Rep, 2011, edn.) (2011)
8. Ali, O., Soar, J., Yong, J.: An investigation of anticipated benefits of cloud computing adoption in Australian regional municipal governments. Serv. Trans. Cloud Comput. **3**(3), 13–24 (2015)
9. González-Martínez, J.A., Bote-Lorenzo, M.L., Gómez-Sánchez, E., Cano-Parra, R.: Cloud computing and education: a state-of-the-art survey. Comput. Educ. **80**, 132–151 (2015)
10. Chen, X., Liu, J., Han, J., Xu, H.: Primary exploration of mobile learning mode under a cloud computing environment. In: 2010 International Conference on E-Health Networking Digital Ecosystems and Technologies (EDT), pp. 484–487. IEEE (2010)
11. Pardeshi, V.H.: Cloud computing in higher education institutes: architecture, strategy and recommendations for effective adaptation. Procedia Econ. Finance. **11**, 589–599 (2014)
12. Munjal, M.N.: Cloud computing in higher education: opportunities, challenges and counter measures. Int. J. Adv. Res. Sci. Eng. **4**(1), 659–668 (2015)
13. Alghamdi, F., Sharma, D., Sathye, M.: Investigating the factors affecting the adoption of cloud computing in SMEs: a case study of Saudi Arabia. In: European, Mediterranean, and Middle Eastern Conference on Information Systems. Springer, pp. 158–176 (2018)
14. Hassan, H., Mohd Nasir, M.H., Khairudin, N.: Cloud computing adoption in organisations: review of empirical literature (2017)
15. Chen, S.-L., Chen, J.-H., Lee, Y.H.: A comparison of competing models for understanding industrial organization's acceptance of cloud services. Sustainability, **10**(3), 673 (2018)
16. Al-Dwairi, R.M., Al-Tweit, N., Zyout, K.: Factors influencing cloud-computing adoption in small and medium e-commerce enterprises in Jordan, pp. 73–78 (2018)
17. Raza, M.H., Adenola, A.F., Nafarieh, A., Robertson, W.: The slow adoption of cloud computing and IT workforce. Procedia Comput. Sci. **52**, 1114–1119 (2015)
18. Kalloniatis, C.: Increasing internet users trust in the cloud computing era: the role of privacy. J. Mass Commun. Journalism, 6(3) (2016)
19. Rossmann, C.: Theory of reasoned action-theory of planned behavior (Nomos Verlagsgesellschaft mbH & Co. KG, 2010. 2010)
20. Silverman, B.G., Hanrahan, N., Huang, L., Rabinowitz, E.F., Lim, S.: Artificial intelligence and human behavior modeling and simulation for mental health conditions. In: Artificial Intelligence in Behavioral and Mental Health Care, Elsevier, pp. 163–183 (2016)
21. Shiau, W.-L., Chau, P.Y.: Understanding behavioral intention to use a cloud computing classroom: a multiple model comparison approach. Inf. Manage **53**(3), 355–365 (2016)
22. Venkatesh, V., Davis, F.D.: A theoretical extension of the technology acceptance model: four longitudinal field studies. Manage. Sci. **46**(2), 186–204 (2000)
23. Wu, M.-Y., Chou, H.-P., Weng, Y.-C., Huang, Y.-H.: TAM-2 based study of website user behavior-using web 2.0 websites as an example. WSEAS Trans. Bus. Econ. **4**(8), 133–151 (2011)
24. Rogers, E.: Diffusion of innovations. Free Press, 551 (2003)
25. Trope, J.: Adoption of cloud computing by South African firms: an institutional theory and Diffusion of Innovation theory perspective (2014)
26. Thompson, R.L., Higgins, C.A., Howell, J.M.: Personal computing: toward a conceptual model of utilization. MIS Q. 125–143 (1991)

27. Momani, A.M., Jamous, M.: The evolution of technology acceptance theories. Int. J. Contemporary Comput. Res (IJCCR) **1**(1), 51–58 (2017)
28. Burney, V.H.: Applications of social cognitive theory to gifted education. Roeper Rev. **30**(2), 130–139 (2008)
29. Boateng, H., Adam, D.R., Okoe, A.F., Anning-Dorson, T.: Assessing the determinants of internet banking adoption intentions: a social cognitive theory perspective. Comput. Human Behav. **65**, 468–478 (2016)
30. Venkatesh, V., Morris, M.G., Davis, G.B., Davis, F.D.: User acceptance of information technology: toward a unified view. MIS Q. 425–478 (2003)
31. San Martín, H., Herrero, Á.: Influence of the user's psychological factors on the online purchase intention in rural tourism: integrating innovativeness to the UTAUT framework. Tourism Manage. **33**(2), 341–350 (2012)
32. Hashim, H., Hassan, Z.: Factors that influence the users' adoption of cloud computing services at iraqi universities: an empirical study. Australian J. Basic Appl. Sci. **9**(27), 379–390 (2015)
33. Yadegaridehkordi, E., Nasir, M.H.N.B.M., Noor, N.F.B.M., Shuib, L., Badie, N.J.A.S.C.: Predicting the adoption of cloud-based technology using fuzzy analytic hierarchy process and structural equation modelling approaches. Appl. Soft Comput. **66**, 77–89 (2018)
34. Okonkwo, I.: Behavoral intention to adopt internet banking (2012)
35. Alharbi, S.T.: Trust and acceptance of cloud computing: a revised UTAUT model. In: 2014 International Conference on Computational Science and Computational Intelligence, pp. 131–134. IEEE (2014)
36. Wu, Y.-L., Tao, Y.-H., Yang, P.-C.: The use of unified theory of acceptance and use of technology to confer the behavioral model of 3G mobile telecommunication users. J. Stat. manage. Syst. **11**(5), 919–949 (2008)
37. Gao, S., Krogstie, J., Siau, K.: Developing an instrument to measure the adoption of mobile services. Mobile Inf. Syst. **7**(1), 45–67 (2011)
38. Tansey, G.: The future control of food: a guide to international negotiations and rules on intellectual property, biodiversity and food security (Routledge, 2012. 2012)
39. Afthanorhan, W.: A comparison of partial least square structural equation modeling (PLS-SEM) and covariance based structural equation modeling (CB-SEM) for confirmatory factor analysis. Int. J. Eng. Sci. Innovative Technol. **2**(5), 198–205 (2013)
40. Akgül, Y.: Adoption of e-government services in Turkey. **9**(32), 7–22 (2018)
41. Lee, W.-I., Fu, H.-P., Mendoza, N., Liu, T.-Y.: Determinants impacting user behavior towards emergency use intentions of m-health services in Taiwan. In: Healthcare, p. 535 (2021)
42. Miladinovic, J., Hong, X.: A study on factors affecting the behavioral intention to use mobile shopping fashion apps in Sweden (2016)
43. Adel Ali, R., Rafie Mohd Arshad, M.: Empirical analysis on factors impacting on intention to use m-learning in basic education in Egypt. Int. Rev. Res. Open Distributed Learn. **19**(2) (2018)

Fuzzy Control Model to Determine the Score in Virtual Reality-Based Appendectomy Practices

Alieski Véliz Vega[1](✉), Omar Correa Madrigal[2], and Vlada Kugurakova[3]

[1] Universidad de Las Ciencias Informáticas, Carretera a San Antonio de los Baños Km 2 ½, Havana, Cuba
veliz@uci.cu

[2] Universidad de Las Ciencias Informáticas, Carretera a San Antonio de los Baños Km 2 ½, Havana, Cuba
ocorrea@uci.cu

[3] Kazan Federal University, Kremlyovskaya St, 18, 420008 Kazan, Tatarstan Republic, Russian Federation

Abstract. Virtual Reality simulators allow surgical training in a safe and controlled environment. Through simulation, procedures can be repeated until a certain level of skill is acquired. This paper presents a fuzzy logic model implemented separately from the Virtual Reality simulator so that it can be used from other simulators. Multiple linear regression is applied for training using a Python library. The model is part of a self-adaptive model for a Virtual Reality-based surgery simulator.

Keywords: Appendectomy · Fuzzy logic · Surgery · Virtual reality

1 Introduction

Virtual Reality (VR) simulators are an important resource for skills training, especially if the practice of the real activity is costly and risky. The use of VR simulators offers the advantage of repeating procedures in a safe and less costly environment. VR technologies share the common characteristics of "immersion, perception and interaction with the environment" [1, 2]. Immersive environments provide a multisensory, playful, multimedia and interactive stimulus, and allows to control the stimulation conditions and repeat actions as in the real world.

VR has proven its effectiveness, particularly with the development of immersive simulators for training in areas such as medicine, rehabilitation, biotechnology, military industry, among others [3–7]. Medical training and especially in surgery has been an area that has explored the advantages of simulation. The need for surgical simulation stems from the advantages of safety and repeatability of procedures and its advantages over the traditional training.

The traditional paradigm of surgical training is based on the performance of procedures supervised by experts and based on the principle of observing, assist and operate.

R. Silhavy et al. (Eds.): CoMeSySo 2021, LNNS 232, pp. 899–906, 2021.
https://doi.org/10.1007/978-3-030-90318-3_70

Despite the efficiency of this model, it has the disadvantage of being dependent on the technical level of the expert tutor and his or her ability to transfer skills to the trainee [8]. On the other hand, confronting students early on with highly complex procedures can lead to an unsatisfactory learning experience for both the student and the tutor, adding possible consequences for the patient [8].

For surgical training there are practices that include working on cadavers, whose advantage is the correct anatomy, however, they have a single use and no bleeding; plastic models, low cost and repeated use, but with lack of reality; computerized mannequins have multiple applications, repeated use, are very useful for learning routines, in individual or team training, with rehearsal of serious events but have a high cost with preset incisions; and VR simulators, have unlimited use, can handle multiple diverse and complex scenarios. The development of VR technologies has marked a growing interest in building skills in students and professionals through training in immersive simulators [9–11]. The use of this technology provides more confidence to perform surgical procedures from the realistic experience of immersion, adding that the training is performed in a safe, controlled, standardized environment and can be repeated until a sufficient level of skill is acquired.

With the aim of achieving higher degrees of motivation and better training outcomes, the educational context has seen the introduction of concepts such as adaptive learning commonly referred to as personalized learning. Adaptive learning has a number of important potential applications in smart devices, the rapid development of Artificial Intelligence (AI), VR, and other technologies [13]. This paper presents a model linked to Virtual Reality separating the AI implementation from the simulator implementation, different of related works to link fuzzy logic and Virtual Reality [13–15].

An adaptive learning system refers to its ability to produce changes in its environment for specific situations, when necessary. Adaptive learning allows to personalize learning for each student using the resources offered from a platform, covering the content in depth in a rational way and serving as a tool for both the student and the teacher [16]. These systems rely on data collection techniques to learn from each learner and provides an adaptation to the content they display.

A self-adaptive simulator must be able to capture the user's actions and react to them, being able to update and optimize the content. In this sense, the system must collect and classify all the information provided by the user, based on his emotions, behavior, skills and knowledge. Correct and incorrect actions must be taken into account for user classification according of its level of knowledge. In this way, when the user performs incorrect actions, the system provides instant feedback and offers suggestions to the user. Based on the previous considerations, the development of self-adaptive immersive simulators for training is a valuable application. This paper presents a fuzzy control model trained to score a VR-based surgery training for the open appendectomy procedure.

2 Methods

Appendectomy is a common surgical procedure. Some reference works point to the time of intervention that on average is performed in an interval of 30 min and up to 2 h with occurrence of complications [17–23]. Based on the average surgical intervention times for this procedure, and an approximate number of possible errors, a fuzzy control model was designed to determine the score to be awarded to a surgical student during practice in a simulator.

The general logic of the model follows the idea of awarding a better score for training with fewer errors and adequate time. On the other hand, it is considered that the appearance of patient complications may lengthen the surgical time, however, the occurrence of errors depends on the level of skill and knowledge achieved by the student to perform the procedure.

For the implementation of this model a Request-Response approach was followed, so that the model is queried through an API as it is not implemented in the simulator itself. In this way, the simulator only provides the input parameters and the model returns a response. Figure 1 shows the process performed for the development of the proposal divided into four stages.

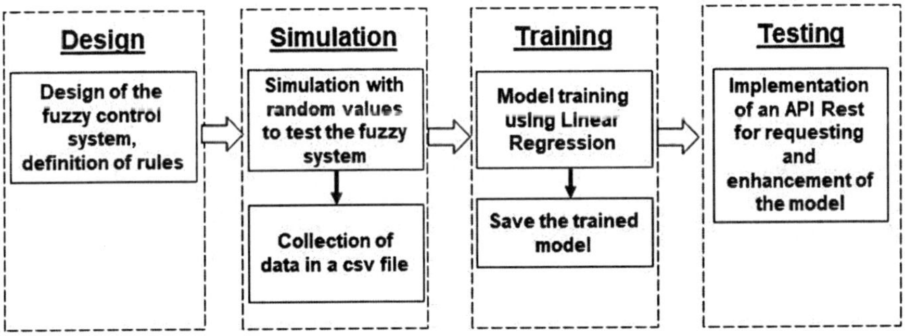

Fig. 1. Development stages of the model

First, a system of rules was created in the Python programming language. The specialized libraries skfuzzy and numpy were used for the first stage.

The error variable was modeled in an approximate interval of 0–10 (low, medium, high), and the time variable in an interval of 30–120 min following the average time references described above. Additionally, times below 30 min were considered as very low and very high above 90 min, being defined as "very low", "low", "average", "high" and "very high". On the other hand, the variable score was modeled on a scale of 0–5 points as shown in Fig. 2.

Was performed a simulation to fill a dataset.csv file. Random values for time, number of errors, user name, procedure, and date were entered. The values of time and number of errors as the main ones for training, the rest of the variables are part of a broader user model (MU) to propose feedback, adjustments, and to be able to classify users into groups according to the level of learning.

The extended MU referenced above is one of the three fundamental components designed for a VR-based self-adaptive simulator. This model groups user characteristics divided into three subcomponents: Skills and Knowledge, Emotions, and Behavior. From the interaction of the MU with the other components of a self-adaptive model for VR simulators, learning difficulties are identified during training. The variables associated

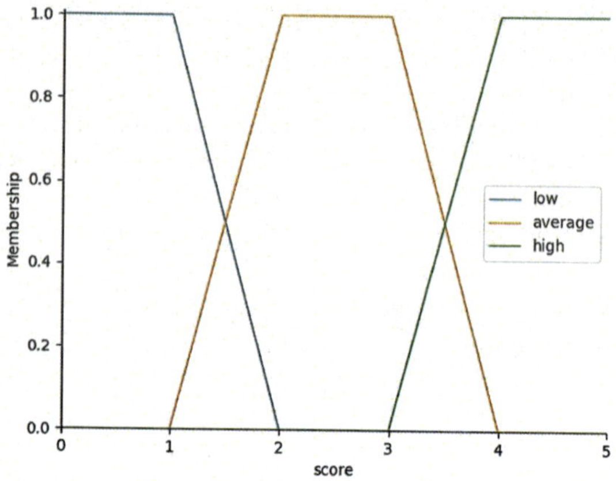

Fig. 2. Representation of the time variable

with Skills and Knowledge group data related to users' skills and knowledge from the performance of activities and levels of acquired knowledge. The variables associated with emotions provide a dataset related to the user's emotional states during training. Through interaction devices with emphasis on haptic devices, these variables are obtained. Can be identified some emotional states with the use of pressure sensors and heart rate monitors. With the support of these devices, it is possible to identify where the user is looking or how is moving to infer possible emotional states. These emotions can have an impact on the user's skills and/or knowledge, and can indicate whether he has acquired a level of knowledge about an activity or has achieved a skill. Behavioral variables provide a set of data related to the user's behavior and closely related to emotions.

Different times in the established interval as well as variation in the number of errors were considered so that the dataset would have a diverse data set and a sufficient number of entries for comparisons. Figure 3 shows a representation of data after simulation. The model was then trained using the *sklearn* Python library for Linear Regression, the *pandas* library and numpy for handling the *dataset.csv* data. Finally, the trained model was saved in a *trained_model.pickle* file so that it can be used again without the need for retraining.

Linear Regression [24–26] is a mathematical model used to approximate the relationship of dependence between a dependent variable Y, m independent variables X_i, con $m \in \mathbb{Z}^+$

$$Y = \beta_0 + \beta_1 \mathrm{X}_1 + \cdots \beta_\mathrm{m} \mathrm{X}_\mathrm{m} + \mathcal{E} \tag{1}$$

In this paper, a multiple linear regression model is applied, given that multiple predictors are used, having multiple X_i (*time, errors*) to predict the response, Y (*score*)

$$Y = \beta_0 + \beta_1 \mathrm{X}_1 + \beta_1 \mathrm{X}_1 + \mathcal{E} \tag{2}$$

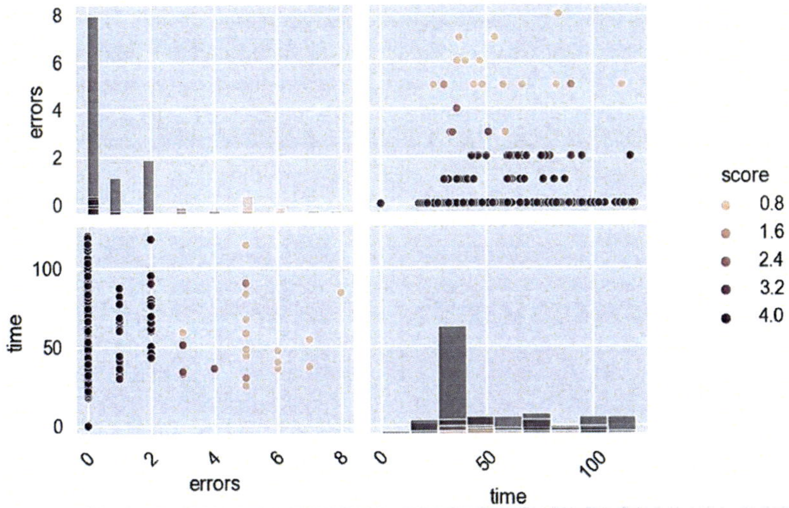

Fig. 3. Representation of data form dataset after the simulation

3 Results

Figure 4 shows the interaction between the technologies involved in the use of this model. Was implemented an Application Programming Interface (API) using the flask library to capture the data sent from the surgery simulator to the trained model. The simulator sent the information supported by hand controllers. These controllers obtain information of movements, gestures, pressure with which performed an incision, how are used the surgical instruments, the order of performed procedures, and all the interactions during

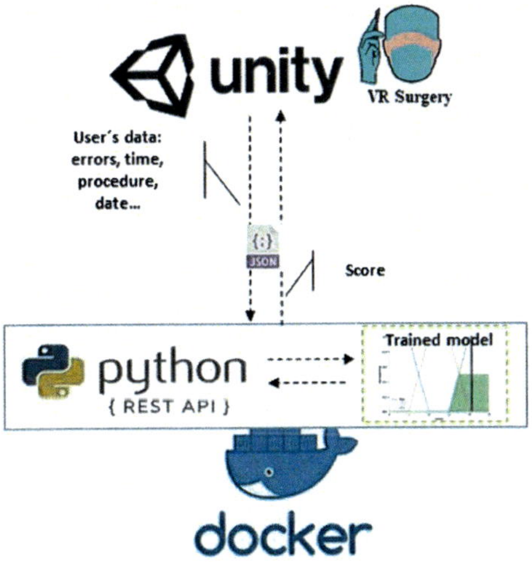

Fig. 4. General design of the proposal with interaction among technologies

the intervention so that the errors made, and the operation time. Finally, the application is deployed as a Docker container so that it can be queried through the API from a surgery simulator and specifically for the Appendectomy procedure.

The accuracy of the model was tested experimentally with the Rest Insomnia client. With this client a *Json* is sent with the indicated parameters and the score is obtained. Table 1 shows some results.

Table 1. Results in the test stage of the model

Time (X_1)	Errors (X_2)	Score (Y)
60	0	4.39
60	1	3.85
30	1	3.75
35	0	4.35
90	2	3.42
25	0	4.27
45	0	4.34
51	3	2.75
30	5	1.61

4 Discussion

The model was developed from a fuzzy control system following approximate values of a real situation. It was assumed that the error variable affects the score, taking into account that the higher the error, the lower the score.

This work introduces the idea of separating the AI implementation from the simulator, so that the simulator providing the necessary information for the trained model to provide a score. Considering these elements, a deeper research intends to go beyond scoring and supply possible difficulty settings to the simulator, classify students according to their learning level, and provide adequate feedback to improve student skills.

The trained model was based on a simulation with random values applying a system of fuzzy logic rules to fill the data file, however a more effective result of the model would be obtained by sending real data from the simulator during training. In this sense, it is considered as an advantage, that the implemented API adds new values to the data file to continue enriching the model and to be able to retrain it to improve its accuracy, also considering that as a student performs a training improves his skills, and as a result decreases the occurrence of errors and improves the time. The authors of this work, consider conducting a study with real students in a VR simulator to reaffirm this hypothesis and achieving self-adaptability in training.

References

1. Mendez, K.J.W., et al.: Virtual and augmented reality: Implications for the future of nursing education, Nurse Educ. Today. **93**, 104531 (2020). https://doi.org/10.1016/j.nedt.2020.104531
2. Hudson, S., Matson-Barkat, S., Pallamin, N., Jegou, G.: With or without you? Interaction and immersion in a virtual reality experience. J. Bus. Res. **100**, 459–468 (2019). https://doi.org/10.1016/j.jbusres.2018.10.062
3. Pallavicini, F., Argenton, L., Toniazzi, N., Aceti, L., Mantovani, F.: Virtual reality applications for stress management training in the military. Aerosp. Med. Hum. Perform. **87**, 1021–1030 (2016). https://doi.org/10.3357/AMHP.4596.2016
4. Sultanova, R., Sharaeva, R.: Virtual reality-based immersive simulation mechanics for invasive surgery training. In: Proc. - Int. Conf. Dev. ESystems Eng. DeSE (2019). https://doi.org/10.1109/DeSE.2019.00171
5. Siu, K.C., Best, B.J., Kim, J.W., Oleynikov, D., Ritter, F.E.: Adaptive virtual reality training to optimize military medical skills acquisition and retention. Mil. Med. **181**, 214–220 (2016). https://doi.org/10.7205/MILMED-D-15-00164
6. Labovitz, J., Hubbard, C.: The use of virtual reality in podiatric medical education. Clin. Podiatr. Med. Surg. **37**, 409–420 (2020). https://doi.org/10.1016/j.cpm.2019.12.008
7. Vaughan, N., Gabrys, B., Dubey, V.N.: An overview of self-adaptive technologies within virtual reality training. Comput. Sci. Rev. **22**, 65–87 (2016). https://doi.org/10.1016/j.cosrev.2016.09.001
8. León Ferrufino, F.: Simulación en cirugía laparoscópica, Cir. Esp. **93**, 4–11 (2015). https://doi.org/10.1016/j.ciresp.2014.02.011
9. Bernardo, A.: Virtual reality and simulation in neurosurgical training. World Neurosurg. **106**, 1015–1029 (2017). https://doi.org/10.1016/j.wneu.2017.06.140
10. Qian, K., Bai, J., Yang, X., Pan, J., Zhang, J.: Virtual reality based laparoscopic surgery simulation. In: Proceeding of ACM Symp. Virtual Real. Softw. Technol. VRST, Association for Computing Machinery, New York, NY, USA, pp. 69–78 (2015). https://doi.org/10.1145/2821592.2821599
11. Wang, X., Wang, X.: Virtual Reality training system for surgical anatomy. In: ACM Int. Conf. Proceeding Ser., Association for Computing Machinery, New York, New York, USA, pp. 30–34 (2018). https://doi.org/10.1145/3293663.329367.
12. Xie, H., Chu, H.C., Hwang, G.J., Wang, C.C.: Trends and development in technology-enhanced adaptive/personalized learning: a systematic review of journal publications from 2007 to 2017. Comput. Educ. **140**, 103599 (2019). https://doi.org/10.1016/j.compedu.2019.103599
13. Ota, D., Loftin, B., Saito, T., Lea, R., Keller, J.: Virtual reality in surgical education. Comput. Biol. Med. **25**, 127–137 (1995). https://doi.org/10.1016/0010-4825(94)00009-F
14. Toraño, J., Diego, I., Menéndez, M., Gent, M.: A finite element method (FEM) – Fuzzy logic (Soft Computing) – virtual reality model approach in a coalface longwall mining simulation. Autom. Constr. **17**, 413–424 (2008). https://doi.org/10.1016/J.AUTCON.2007.07.001
15. Pan, J.J., et al.: Virtual reality training and assessment in laparoscopic rectum surgery. Int. J. Med. Robot. Comput. Assist. Surg. **11**, 194–209 (2015). https://doi.org/10.1002/RCS.1582
16. Del, M., Morillo Lozano, C.: Aprendizaje adaptativo (2016). http://uvadoc.uva.es/handle/10324/21000. Accessed 7 April 2021
17. Fingerhut, A., Millat, B., Borrie, F.: Laparoscopic versus open appendectomy: time to decide. World J. Surg. **23**(8), 835–845 (1999). https://doi.org/10.1007/s002689900587
18. Hilaire, R., Fernández, Z.R., García, L.I.R., Sánchez, L.P.R.: Apendicectomía videolaparoscópica frente a apendicectomía convencional. Rev. Cuba. Cirugía. **53**, 30–40 (2014). http://scielo.sld.cu. Accessed 22 July 2021

19. Malhotra, L., Pontarelli, E.M., Grinberg, G.G., Isaacs, R.S., Morris, J.P., Yenumula, P.R.: Cost analysis of laparoscopic appendectomy in a large integrated healthcare system. Surg. Endosc. 2021, 1–8 (2021). https://doi.org/10.1007/s00464-020-08266-0
20. Ali, R., Anwar, M., Akhtar, J.: Laparoscopic versus open appendectomy in children: a randomized controlled trial from a developing country. J. Pediatr. Surg. **53**, 247–249 (2018). https://doi.org/10.1016/J.JPEDSURG.2017.11.022
21. Baumgarten, H.D., Brown, E.G., Russell, K.W., Adzick, N.S., Laje, P.: Laparoscopic appendectomy — outcomes of senior trainees operating without supervision versus experienced pediatric surgeons. J. Pediatr. Surg. **54**, 276–279 (2019). https://doi.org/10.1016/J.JPEDSURG.2018.10.092
22. Fortea-Sanchis, C., et al.: Apendicectomía laparoscópica frente al abordaje abierto para el tratamiento de la apendicitis aguda. Rev. Gastroenterol. México. **77**, 76–81 (2012). https://doi.org/10.1016/J.RGMX.2012.02.001
23. Apendicectomía por laparoscopia en la apendicitis aguda complicada, (n.d.). http://ve.scielo.org/scielo.php?pid=S0367-47622002000200007&script=sci_arttext&tlng=en. Accessed 29 July 2021
24. Linear Regression Analysis - George A. F. Seber, Alan J. Lee - Google Libros, (n.d.). https://books.google.com.cu/books?hl=es&lr=&id=X2Y6OkXl8ysC&oi=fnd&pg=PR5&dq=linear+regression+&ots=sejUF5nMpy&sig=81UH7LJS-e6O937G0ROukPeeDuQ&redir_esc=y#v=snippet&q=multiple&f=false. Accessed 1 August 2021
25. Uyanık, G.K., Güler, N.: A study on multiple linear regression analysis. Procedia - Soc. Behav. Sci. **106**, 234–240 (2013). https://doi.org/10.1016/J.SBSPRO.2013.12.027
26. Granados, R.M.: Modelos de regresión lineal múltiple, Granada. España Dep. Econ. Apl. Univ, Granada (2016)

Recommender System for Enterprise Resource Planning

Alexey Kurennykh[1(✉)] and Vladimir Sudakov[2]

[1] Moscow Aviation Institute (National Research University), Moscow, Russia
[2] Plekhanov Russian University of Economics, Moscow, Russia

Abstract. This paper deals with a major scientific and technical problem for many research and production enterprises that are faced with the need to carry out logistics activities. The authors describe combined algorithms for recommender systems, that take into account different ways to evaluate objects: content-based and collaborative filtering and knowledge-based approach as well. Such implementation allows to carry out efficient and exact ranking in order to make rational recommendations for users. In this research we propose four different kinds of criteria that could be obtained with the use of ERP system database and other data sources available via Internet connection. Developed algorithms are invariant to the subject area and criteria itself, thus designed software can be applied in different modern scientific problems.

Keywords: Recommender system · Enterprise resource planning · Combined approach

1 Introduction

Currently, we deal with the worldwide trend of digitalization of various spheres of human activity. Information systems of various classes are constantly developing, a wide range of production processes are being automated, the use of modern computing technologies has affected all types of business processes. The increase in the area of informatization is especially clearly visible at manufacturing enterprises, where electronic document management systems are actively distributed, and corporate information systems, product life cycle management systems, design documentation systems, automated workstations, etc. are being introduced. The use of corporate information systems has a number of advantages that go beyond simple process automation and facilitating the work of enterprise employees using these systems. In the majority of cases, the tasks of the corporate information system are reduced to the automation of accounting processes at the enterprise: warehouse accounting, accounting and others, the tasks of enterprise resource management are solved, for which there is a separate class of systems – ERP systems. However, the needs of some users of these systems go beyond accounting functions. Modern realities, high competition in the domestic and foreign market require enterprises to accurately plan their operations, procurements, costs and staffing requirements in the long-term. The current development of information systems makes it possible to build

R. Silhavy et al. (Eds.): CoMeSySo 2021, LNNS 232, pp. 907–912, 2021.
https://doi.org/10.1007/978-3-030-90318-3_71

production plans with high accuracy with respect to the range of manufactured products, the required material resources, professions, time periods and other analytics that are important for the enterprise. Another advantage of using corporate information is the ability to automatically collect statistics accumulated in the accounting system for more accurate planning. Resource planning tasks are really important and relevant for many research and production associations, and their solution requires the use of mathematical approaches in order to ensure high accuracy in predicting the need for certain types of resources. Thus, the authors of this research came up with the idea of implementing the module for developing recommendations as an addition to the main information system of the enterprise. The purpose of this module is not limited to resource planning tasks, but also finds application in various daily tasks, for example, in the task of selecting a supplier of material resources. Recommender systems, as well as corporate information systems, have their own classification based on the principle of their operation and the algorithmic approaches used to develop recommendations. The principle of operation of recommender systems is based on various approaches that have their own advantages and disadvantages:

- the first type of recommender systems is based on collaborative filtering, such systems have a sufficiently high accuracy in the presence of a sufficient amount of information, are well applicable in the search for films [1] and musical compositions [2], also there are attempts to use them in the healthcare [3];
- the second type of recommender systems is based on content - data collected about each object and other objects that the subject was already interested in. This approach has been successfully used in text recommendations [4] and news feeds [5]. These systems are quite effective, but highly dependent on the subject area in which they are applied;
- the third type of recommender systems is based on knowledge of the subject area, such recommendations are highly accurate, allow to take into account a large number of heterogeneous factors, relationships between objects. For example, such systems are used in the field of financial funds [6], corporations [7]. The core of such a system is the knowledge base about the subject area, which can be effectively represented in the form of a knowledge graph [8];
- the fourth type of recommender systems are hybrid. They are based on combinations of the above approaches, which allows to achieve high-quality results. The main disadvantage of such systems is the high complexity of development. Hybrid recommender systems may be an evolution of knowledge-based systems [9], but the most common and used example is a combination of collaborative and content-based systems [10]. In hybrid systems, various approaches are used, for example, fuzzy methods and artificial intelligence [11, 12], approaches of cluster data analysis as well [13].

2 Formulation of the Problem

The authors task in this study was to develop algorithmic and software for making recommendations using combined methods. The main idea of the developed algorithms is invariance to the subject area, which makes it possible to use the development not only in the field of logistics, which is considered in this article as an example, but also in many others.

In the process of executing the algorithm, two entities are processed: the compared objects (alternatives) themselves, as well as a set of their characteristics, which can be own attributes of alternatives, statistical data, external characteristics, and also express the subjective opinion of the user working with these objects (criteria). As alternatives in the process of solving one problem of developing recommendations, a finite set of objects of the same class can act, we denote them by a set

$$\boldsymbol{O} = \{o_i\}, i = \overline{1, k}$$

where o_i is the specific alternative considered at the moment, k is the number of alternatives of this class in the system at the time of solving the problem of making recommendations. In order for the developed algorithm to claim universality, as well as invariance with respect to the subject area, it is not stipulated in advance which class of objects forms a set $\boldsymbol{O}$. The filling of this set with specific alternatives should occur automatically at the time of solving a specific problem without the need to edit the source code of the software. The criteria characterizing the alternatives are also not subject to strict declaration in the program code, but are determined by the user of the application software when solving a specific problem. We denote the set of criteria for alternatives

$$\boldsymbol{C} = \left\{c_{ij}\right\}, i = \overline{1, k}, j = \overline{1, n}$$

where c_{ij} is a criterion with a number j for an alternative with a number i, n is the number of criteria for alternatives of a given class. The main task of the developed algorithm is the aggregation of vector criteria for individual alternatives with the subsequent ranking of objects in accordance with the optimization criterion in a specific problem of making recommendations, which can be formally written as follows

$$r(o_i) = f(\boldsymbol{c}_i) \quad (1)$$

In this expression, the following designations are introduced: $r(o_i) \in \mathbb{R}$ is the rank of the alternative with a number i, which is its final assessment in the problem of making recommendations; $f(\boldsymbol{c}_i)$ is an aggregating function, the parameter of which is the vector of criteria values for the given alternative. We write the optimization criterion in the following form

$$r(o_i) \xrightarrow[o_i]{} max \quad (2)$$

The minimization of this value is also not excluded, the optimization criterion is set by the user and is determined by the problem being solved.

A separate place in the formulation of the problem for algorithmic and software is the user interface, allowing one to work with alternatives and their criteria, the choice of aggregation methods, as well as input of initial data. This interface should provide the ability to select a class of objects from the system-wide directory of catalogs, the selection of criteria that are attributes of the object under consideration at the level of the system or its database, as well as the addition of external characteristics that do not have a direct connection with the considered objects within the enterprise information system. Also, the interface should provide the ability to customize the aggregation method for each task of developing recommendations and allow using the variability of the algorithm for each user, since each of them may have their own subjective views on the decision problem.

3 Development of a Combined Algorithm

Let us consider the implemented approaches to calculating and collecting criteria values for alternatives. All methods of calculating the criteria in these approaches to solving the problems of developing recommendations are divided into four types:

- stored directly in the database, i.e. they are criteria that do not require significant mathematical calculations to achieve their values. Their values can be taken directly from specific database tables. For example, such a criterion may be the entry of a supplier into the list of "black counterparties", or, for example, the price of any material or service in the current price list of the enterprise. Let us designate a set of these criteria as

$$\boldsymbol{SC} = \{sc_{ij}\}, \boldsymbol{SC} \subset \boldsymbol{C}$$

- calculated at the application level. For these criteria, values can be calculated by collecting statistical data from a database or by various mathematical calculations. For example, the average delay in the delivery of a material for a specific supplier or the frequency of detecting defective items. One can also consider as an example the estimated cost of the purchase, taking into account the volume of demand, as well as discounts provided by the supplier. Let us designate a set of these criteria as

$$\boldsymbol{CC} = \{cc_{ij}\}, \boldsymbol{CC} \subset \boldsymbol{C} \text{ and } \boldsymbol{SC} \cap \boldsymbol{CC} = \emptyset$$

- external criteria. The values of these criteria fall into the recommendation system exclusively from the outside world. These criteria are not significant for the main functionality of the enterprise information system, however, they can have a positive effect on the ranking of alternatives, increasing the accuracy of the results. Such criteria may include, for example, the number of employees of a supplier of products, or even such an indirect indicator as the value of his shares, if any are publicly available. Let us designate a set of these criteria as

$$\boldsymbol{OC} = \{oc_{ij}\}, \boldsymbol{OC} \subset \boldsymbol{C} \text{ and} \boldsymbol{SC} \cap \boldsymbol{CC} \cap \boldsymbol{OC} = \emptyset$$

- custom criteria. The values of these criteria can enter the system only through manual input and subsequent transformations. The criteria of this group can express the subjective views of the user in relation to the problem of choice being solved. For example, this criterion could be "supplier experience". Let us designate a set of these criteria as

$$\boldsymbol{UC} = \{uc_{ij}\}, \boldsymbol{UC} \subset \boldsymbol{C}, \boldsymbol{SC} \cap \boldsymbol{CC} \cap \boldsymbol{OC} \cap \boldsymbol{UC} = \emptyset \text{ and}$$
$$\boldsymbol{SC} \cup \boldsymbol{CC} \cup \boldsymbol{OC} \cup \boldsymbol{UC} =$$

Consideration of these groups of criteria in the problem of developing recommendations implements a hybrid approach in the recommender system. Indeed, the first, third and fourth groups of criteria implement a collaborative and content-based approach to develop recommendations, and the second group of criteria uses knowledge about the subject area. In accordance with (1), the values of the criteria for each of these groups are subject to aggregation; it is possible that the aggregation of each individual group is performed in a different way, for example, by the weighted sum method, Germeier convolution, multiplicative convolution, or any other method that can be configured in the system. Thus, the final rank of the alternative can be determined:

$$r(o_i) = f(f_1(\boldsymbol{sc}_i), f_2(\boldsymbol{cc}_i), f_3(\boldsymbol{oc}_i), f_4(\boldsymbol{uc}_i)) \quad (3)$$

f, f_1, f_2, f_3, f_4 are various aggregating functions that can be used in the multicriteria assessment of alternatives.

4 Conclusion

In this paper we considered the actual scientific and technical problem of developing combined algorithms for developing recommendations for scientific and industrial enterprises. Taking into account the application of various approaches to the assessment of alternatives for recommender purposes, as well as the possibility of changing the list of considered criteria and methods of their processing, depending on the specific task at the applied level, the developed algorithms can be considered universal and invariant to the subject area.

The prior direction for the continuation of this research is the allocation of the fifth group of criteria – the criteria for evaluating the simulation results, which will be the result of running multi-agent models that allows to get more information about the subject area. Such an extension of the approach to the development of recommendations will not require significant modernization of algorithms in terms of ranking alternatives, however, it is necessary to develop mechanisms for integration with systems that allow simulation experiments.

Potentially, the use of models and automated processing of their results significantly expands the boundaries of applicability of the described algorithms, for example, in the field of production planning, the military industry, as well as the field of healthcare.

Acknowledgments. This research is supported by RFBR, project 20-31-90043.

References

1. Yutian, H., Xiong, F., Dongyuan, L., Wang, X., Xiong, X., Chen, H.: Movie collaborative filtering with multiplex implicit feedbacks. Neurocomputing **398**, 485–494 (2020). https://doi.org/10.1016/j.neucom.2019.03.098
2. Sánchez-Moreno, D., Gil González, A.B., Dolores Muñoz Vicente, M., López Batista, V.F., Moreno García, M.N.: A collaborative filtering method for music recommendation using playing coefficients for artists and users. Expert Syst. Appl. **66**, 234–244 (2016). ISSN 0957-4174. https://doi.org/10.1016/j.eswa.2016.09.019

3. Mustaqeem, A., Anwar, S.M., Majid, M.: A modular cluster based collaborative recommender system for cardiac patients. Artif. Intell. Med. **102**, 101761 (2020). ISSN 0933-3657. https://doi.org/10.1016/j.artmed.2019.101761
4. Wang, D., Liang, Y., Xu, D., Feng, X., Guan, R.: A content-based recommender system for computer science publications. Knowl. Based Syst. **157**, 1–9 (2018). ISSN 0950-7051. https://doi.org/10.1016/j.knosys.2018.05.001
5. Li, J., Xu, H.: Suggest what to tag: recommending more precise hashtags based on users' dynamic interests and streaming tweet content. Knowl. Based Syst. **106**, 196–205 (2016). ISSN 0950-7051. https://doi.org/10.1016/j.knosys.2016.05.047
6. Tejeda-Lorente, Á., Bernabé-Moreno, J., Herce-Zelaya, J., Porcel, C., Herrera-Viedma, E.: A risk-aware fuzzy linguistic knowledge-based recommender system for hedge funds. Procedia Comput. Sci. **162**, 916–923 (2019). ISSN 1877-0509. https://doi.org/10.1016/j.procs.2019.12.068
7. Zhen, L., Huang, G.Q., Jiang, Z.: An inner-enterprise knowledge recommender system. Expert Syst. Appl. **37**(2), 1703–1712 (2010). ISSN 0957-4174. https://doi.org/10.1016/j.eswa.2009.06.057
8. Lully, V., Laublet, P., Stankovic, M., Radulovic, F.: Enhancing explanations in recommender systems with knowledge graphs. Procedia Comput. Sci. **137**, 211–222 (2018). ISSN 1877-0509. https://doi.org/10.1016/j.procs.2018.09.020
9. Tarus, J.K., Niu, Z., Yousif, A.: A hybrid knowledge-based recommender system for e-learning based on ontology and equential pattern mining. Future Gener. Comput. Syst. **72**, 37–48 (2017). ISSN 0167-739X. https://doi.org/10.1016/j.future.2017.02.049
10. Walek, B., Fojtik, V.: A hybrid recommender system for recommending relevant movies using an expert system. Expert Syst. Appl. **158**, 113452 (2020). ISSN 0957-4174. https://doi.org/10.1016/j.eswa.2020.113452
11. Kermany, N.R., Alizadeh, S.H.: A hybrid multi-criteria recommender system using ontology and neuro-fuzzy techniques. Electron. Commer. Res. Appl. **21**, 50–64 (2017). ISSN 1567-4223. https://doi.org/10.1016/j.elerap.2016.12.005
12. Paradarami, T.K., Bastian, N.D., Wightman, J.L.: A hybrid recommender system using artificial neural networks. Expert Syst. Appl. **83**, 300–313 (2017). ISSN 0957-4174. https://doi.org/10.1016/j.eswa.2017.04.046
13. Mohammadpour, T., Bidgoli, A.M., Enayatifar, R., Javadi, H.H.S.: Efficient clustering in collaborative filtering recommender system: hybrid method based on genetic algorithm and gravitational emulation local search algorithm. Genomics **111**(6), 1902–1912 (2019). ISSN 0888-7543. https://doi.org/10.1016/j.ygeno.2019.01.001

Applying an Agent-Based Approach to Modeling COVID-19

Vladimir A. Sudakov[1,2] and Tatiana V. Sivakova[2,3](✉)

[1] Moscow Aviation Institute (National Research University), 4 Volokolamskoe Shosse, 125993 Moscow, Russian Federation
[2] Keldysh Institute of Applied Mathematics, Miusskaya sq. 4, 125047 Moscow, Russia
[3] Plehanov Russian University of Economics, 36 Stremyanny lane, 117997 Moscow, Russia

Abstract. The paper discusses approaches for predicting the modeling of the epidemiological situation. To solve this problem, it is proposed to go to simulation modeling. The advantages of the agent-based approach are revealed, in which the main role is to take into account the interaction and mutual influence of individuals.

Keywords: Multi-agent modeling · Epidemic · Individuals · Expert judgments

1 Introduction

In recent years, humanity has faced another pandemic - COVID-19 [1, 2]. A difficult epidemiological situation gives rise to a crisis in which measures to ensure the preservation of human health and safety are reflected directly in the values of economic indicators. Therefore, the creation of methods and tools for predictive assessment of the consequences of decisions made in the context of the effectiveness of each of the specified criterion factors and the observance of their balance is an urgent task. A predictive assessment of the crisis can be obtained by modeling the spread of the epidemic process with the introduced measures and their impact on the functioning of economic entities.

To date, methods of modeling behavior in society are based on probabilistic approaches, however, it is difficult to obtain appropriate probabilistic estimates in practice. This was especially evident in the example of COVID-19. First of all, the lack of structured reliable statistical data does not allow modeling epidemics with the required accuracy. Therefore, it is proposed to use the mathematical apparatus of the membership functions of fuzzy sets, which will allow using the potential of expert judgments in the field of the spread of COVID-19.

Expert judgments, as a rule, have a fuzzy, vague nature and their modeling with fuzzy numbers ensures that this uncertainty is taken into account when making important decisions. These models can evaluate possible solutions in the context of epidemics. However, the vector criterion for the effectiveness of solutions has a high dimension, improving some criteria, we can worsen others. The criteria can be qualitative, quantitative and fuzzy, and it is necessary to investigate approaches to decision-making based on methods of aggregating high-dimensional criteria in relation to assessing the results of modeling socio-economic decisions in the context of the spread of coronavirus infection.

R. Silhavy et al. (Eds.): CoMeSySo 2021, LNNS 232, pp. 913–919, 2021.
https://doi.org/10.1007/978-3-030-90318-3_72

As one of the important directions of this work, one can single out the development of more advanced decision support tools for a quick assessment of the situation and forecasting its development, including the development of an appropriate strategy based on an assessment of the levels of effectiveness of alternative response options. The creation of such tools is associated with a paradigm shift in approaches to modeling epidemics in the context of the development of platform solutions to support decision-making by government bodies.

2 Literature Review

It is known that attempts to use the mathematical apparatus for assessing the consequences and forecasting of epidemics have almost three centuries of history, starting with the works of Daniel Bernoulli in the middle of the 18th century.

At the heart of most modern approaches, the "epidemiological part" of the model is based on the classical population SIR-Kermak-McMackendrick model.

$$\frac{dS}{dt} = -\propto IS, \frac{dI}{dt} = \propto IS - \beta I, \frac{dR}{dt} = \beta I \tag{1}$$

$$\frac{dS}{dt} + \frac{dI}{dt} + \frac{dR}{dt} = 0 \tag{2}$$

S is the number of healthy individuals, but susceptible to the disease.

I is infected individuals.

R is recovered.

$\propto$ and β are coefficients of intensity of contacts and recovery.

With the help of a system of integro-differential equations, the process of dynamics in a population of groups of individuals is described. This approach is accompanied by the setting of a certain set of initial conditions and a large number of parameters, which, as a rule, are determined very subjectively. This is what leads to a significant scatter of the results obtained by various researchers, which significantly reduces the predictive value of such models.

The above mathematical model of the spread of coronavirus has the following limitations and is not taken into account: population size; the age of the individuals; the possibility of vaccination; an indicator of the distance between individuals; time of contact with an infected person and many other indicators.

The current stage in the development of epidemic modeling is largely associated with the rapid progress in the field of computer technology. An overview of the current state of the most significant and common approaches to predicting infectious morbidity and the development of the epidemic process is given in the following works [3, 4]. It should be noted that the approaches are quite limited and there is a need to improve the efficiency of calculations and the further use of mixed forecasting approaches based on the joint use of various methods.

Despite the potential application of various techniques, the analysis of modern publications on the topic of epidemic modeling shows that the most frequently used models are based on the description of the processes considered in them by systems of nonlinear

ODEs [5, 6], with a set of initial conditions and a large number of parameters, which, as a rule, are determined very subjectively. It is the latter circumstance that leads to a significant scatter of the results obtained by various researchers, which significantly reduces the predictive value of such models. In addition, almost all the approaches considered are focused primarily on the description of the epidemic processes themselves, outside of their interdependence with the life of society.

The following conclusions can be drawn:

1. A qualitative forecast of the spread of the disease is achievable only on the basis of adequate mathematical models;
2. Effectively use several modeling techniques;
3. A significant scatter of the results obtained is associated with a large number of initial conditions, which are often determined subjectively;
4. Traditional models of the spread of diseases use assumptions (for example, the uniformity of individuals, their continuous uniform mixing in the simulated territory), which make the results of forecasting with their help insufficiently accurate;
5. There is no connection with the life of society;
6. It is necessary to comprehensively model and predict the economic and social consequences of epidemics and take protective measures at the level of management decisions.

Therefore, the authors of the study propose to use fuzzy methods of decision theory and multiagent modeling in predicting the spread of coronavirus infection.

3 Formalization of the Method

It is proposed to solve the problem on the basis of a multi-agent approach with fuzzy judgments of agents [7–9]. The multi-agent approach makes it possible to simulate the system as close as possible to reality. A great advantage of multi-agent modeling is the ability to conduct predictive studies to assess the impact of various options for changing the external environment (including infections) or the conditions for the functioning of agents (introducing some kind of restrictive measures). The use of the multi-agent approach in the development of simulation models has proven itself well in those problems in which the individual behavior of objects is essential in the system, and the dynamics of the entire system follows from these features.

To build this model, it is necessary to take into account as much of the communication between people as possible, while the very probability of contacts between individuals is predetermined by many parameters. People are divided by gender, age category, social status, income level, professional affiliation, place and living conditions, place of work, health status, etc. Each person can simultaneously perform in different societies in different roles (in a family, in a work collective, in public transport, at a retail outlet, etc.). Any of these factors can affect the likelihood that a person is at a particular time in a particular place. As a basic model, it makes sense to choose the normal mode of life of the society, based on the well-known fact of the "typical daily routine".

In ordinary everyday conditions, the entire system of relations is in a certain conditionally equilibrium state. The values of the characteristic quantities that determine these ratios can initially be set from some a priori considerations based on expert estimates obtained by generalizing the accumulated observations and experience. In the future, which is a distinctive advantage of the approach proposed here, in order to improve the accuracy characteristics of the model, its parameters must be corrected. Correction of the values of statistically significant parameters of the general model of the functioning of society is supposed to be carried out by establishing fuzzy consistency with the set of available factual data. Taking into account the trends in the development of a unified information environment, the model can consider the potential for obtaining direct or indirect data from many different sources, access to which is currently not available, but, if necessary, is potentially possible in the future.

To clarify the parameters of the movement activity of various groups of the population, data from video surveillance systems, registration of transactions at the cash desks of retail outlets, the dynamics of linking mobile phones to cell towers and other similar systems can be used.

Information about statistically significant deviations from typical trajectories can be obtained from processing passenger registration data (or tickets sold to them) at transport hubs (airport, railway or bus station), registration of visits to medical institutions, places of cultural events, etc.

In addition, dynamic data on resource consumption can be used as indirect sources of information - the volumes and dynamics of the supply of electricity, fuel, water, cargo, finance turnover and many others.

In the modern digital world, the possibilities for obtaining additional data to clarify the parameters of the created model are constantly expanding. And the wider the range of involved actual direct or indirect data from various registration systems used to calibrate the simulation model, the higher the level of its reliability and, accordingly, the predictive and research potential of such a model will be provided.

Within the framework of this approach, we will assume that each agent is an analogue of a person, is a separate entity and is endowed with a set of parameters describing its features, spatial location, social status and indicators related to the state of his health (Fig. 1).

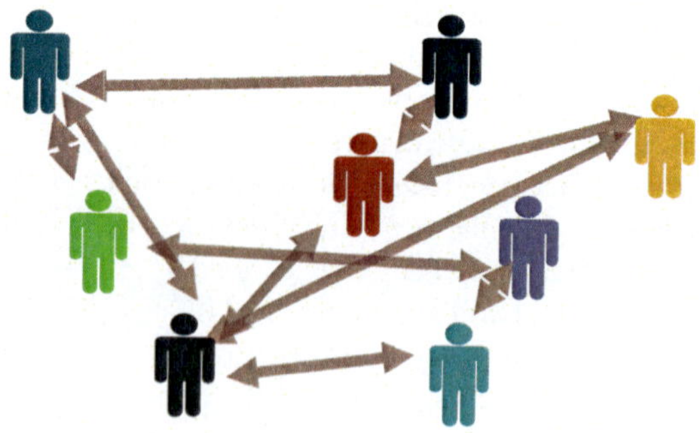

Fig. 1. Interaction between agents.

Let's endow our agent with the following properties: status, location (movement of the agent); age group.

The status of each person in the model can be divided into 4 types: vulnerable, infected, diseased, and acquired immunity. Moreover, at the stage of infection, it is necessary to take into account while the agent is not dangerous to others, and then when it can pose a danger to others (Fig. 2).

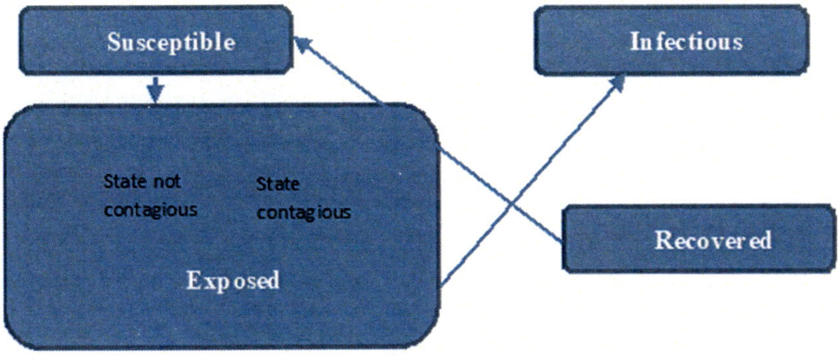

Fig. 2. Agent state stages.

Here, each agent of the system is characterized by a number of states [10]. An agent changes states through various transitions that can be triggered by certain events, for example, meeting with another agent.

Let's highlight the main events in the model that form the epidemic situation:

- "contact" between agents leading to infection;
- the course of the disease at the agent (at home or in the hospital);
- vaccination (usually in 2 stages);
- movement of agents;
- introduction of certain quarantine measures in specific locations (regulated by an expert);
- external influences (such as weather, increased movement of agents due to vacations, etc.).

4 Results

The use of an agent-based approach to simulation modeling provides an opportunity to study the course of epidemic and infectious processes at a detailed level, as well as to play all kinds of scenarios of epidemic outbreaks, test variable strategies to combat the epidemic and assess the impact of multicomponent intervention strategies on the dynamics of epidemics.

Age group helps predict diseases based on the location of the agent. For each age group, we will define the rule for staying in a certain place and the time of stay in it.

Such a model, with corrective refinement of its current parameters due to the permanently accumulated heterogeneous data, can serve the purposes of analyzing the situation in real time, be used to study possible changes in the situation during the transformation of external conditions (infections, weather anomalies, economic shocks), be used for predictive study of the consequences of implementation. various management decisions. The proposed method. allows you to most accurately investigate, predict and evaluate the effectiveness of control measures for the spread of coronavirus infection (as opposed to population-based approaches), but at the same time will require the involvement of significant computing resources.

All this helps to analyze the situation and take the necessary protective measures for managers, such as wearing a mask, social dropping, teleworking, restricting the operation of stores and a number of enterprises and many others.

5 Conclusion

The result of this approach is the system integration of multi-agent models and means of information and analytical support for research and expert activities to form scientific ideas about the development of effective management decisions in a pandemic.

One of the important aspects of modeling and forecasting is the preparation of recommendations for governing structures on certain measures to influence the spread of infection.

The applied significance of this approach makes it possible to solve many predictive problems, including, but not limited to:

- forecasting the dynamics of the spread of morbidity, including the study of the dependence on the characteristics of the causative agent of the infection (which can also play a role for the timely development of adequate measures in the event of other epidemics);
- an assessment of the additional load on the health infrastructure as the epidemic spreads;
- a comparative assessment of the expected effectiveness of various preventive and anti-epidemic measures;
- the possible impact of the epidemic and the measures taken on the change in the nature of life and behavior of people;
- a comprehensive assessment of the socio-economic consequences;
- in assessing the effectiveness of management decisions on the rational management of medical infrastructure, including the forecast of socio-economic consequences.

Acknowledgments. The reported study was funded by RFBR and CNPq, FASIE, DBT, DST, MOST, NSFC, SAMRC according to the research project №. 20-51-80002.

References

1. Anderson, R.M., May, R.M.: Infectious Diseases in Humans. Oxford University Press, Oxford (1992)

2. Park, J.Y., Han, M.S., Park, K.U., Kim, J.Y., Choi, E.H.: First pediatric case of coronavirus disease 2019 in Korea. J. Korean Med. Sci. **35**(11), 1–7 (2020)
3. Barlow, N.S., Weinstein, S.J.: Accurate closed-form solution of the sir epidemic model. Physica D Nonlinear Phenom. **408**, 132540 (2020)
4. West, R.W., Thompson, J.R.: Models for the simple epidemic. Math. Biosci. **141**, 29–39 (1997)
5. Nucci, M.C., Leach, P.G.L.: An integrable SIS model. J. Math. Anal. Appl. **290**, 506–518 (2004)
6. Ma, Z., Liu, J., Li, J.: Stability analysis for differential infectivity epidemic models. Nonlinear Anal. Real **4**, 841–856 (2003)
7. Jing, H., Zhu, D.: Global stability and periodicity on SIS epidemic models with backward bifurcation. Int. J. Comput. Math. **50**, 1271–1290 (2005)
8. Kampf, G., Todt, D., Pfaender, S., Steinmann, E.: Persistence of coronaviruses on inanimate surfaces and their inactivation with biocidal agents. J. Hosp. Infect. **104**(3), 246–251 (2020)
9. Osipov, V., Posadskii, A., Sivakova, T.: Methods of rational decision making in multi-agent systems for evaluating the effectiveness of innovations. In: Proceeding of the AIP Conference Proceedings, pp. 2181 (2019). https://doi.org/10.1063/1.5135691
10. Wolfram, C.: An agent-based model of COVID-19. Complex Syst. **29**(1), 87–105 (2020)

What Keeps You Going? Modelling Motivation and Lifestyle Change in Patients with Type 2 Diabetes

Karine van't Land[1], Sietze Kai Kuilman[2], Simon Laub[3], and Jan Treur[4](✉)

[1] Department of Public and Occupational Health, Amsterdam UMC, Amsterdam, The Netherlands
land.vant.k@menzis.nl
[2] Delft University of Technology, Delft, The Netherlands
S.K.Kuilman@tudeft.nl
[3] EAAA, Business Academy Aarhus (Dk), Viby, Denmark
sila@eaaa.dk
[4] Vrije Universiteit Amsterdam, Amsterdam, The Netherlands
j.treur@vu.nl

Abstract. Chronic lifestyle-related disease is one of the major health problems the world is facing in the twenty-first century. One of the ways to address the problem of chronic lifestyle-related disease is through programs which support lifestyle changes. Research shows the actual benefits of lifestyle changes for patients. However, in reality only a fraction of patients succeeds in changing their lifestyle. Their level and type of motivation appear to be important factors in shaping their chances for success. We devised a second order temporal-causal network for adoption of habit with regards to motivation, focusing on patients with type 2 diabetes who receive advice to exercise. Taking into account a variety of motivational forces, the model clearly shows the benefits of more and repeated interactions. They lead to a more long-lasting change, even though these interactions may also amplify negative emotions. These findings are in accordance with tentative findings in research.

Keywords: Diabetes · Temporal-causal networks · Lifestyle · Motivation · Exercise

1 Introduction

Chronic lifestyle-related disease is one of the major health problems the world is facing in the twenty-first century. Obesity, type 2 diabetes, hypertension and cardiovascular disease are not just problems of so-called Western countries. For instance, worldwide, 422 million people have type 2 diabetes, the majority living in low and middle-income countries[1]. The Covid-19 pandemic only stressed the severity of the problem, as it showed the vulnerability of people suffering from chronic disease. The world is in need

[1] Estimates taken from the World Health Organisation.

R. Silhavy et al. (Eds.): CoMeSySo 2021, LNNS 232, pp. 920–932, 2021.
https://doi.org/10.1007/978-3-030-90318-3_73

of effective, scalable and affordable solutions in order to prevent chronic lifestyle-related disease and support those suffering from these diseases.

One of the ways to address the problem of chronic lifestyle-related disease is formed by programs to support lifestyle change in patients. Research clearly points to the actual benefits lifestyle change has to offer to these patients. When patients with type 2 diabetes follow a suitable energy-restricted diet and exercise regularly, many of them can lower their medication or even completely stop using medication. However, in reality only a fraction of patients succeeds in successfully changing lifestyle [3].

Key factors for the adoption of new habits actually arise from self-esteem and intrinsic motivation. Especially the latter can be taken as indicative of long-term change. The question is of course: how can one boost such motivation? Next to these promoting factors, patients with type 2 diabetes encounter serious barriers to lifestyle change. It is important to take these barriers into account as well.

This paper presents a second order temporal-causal network for adoption of lifestyle habit, focusing on patients with type 2 diabetes who receive advice to exercise. Our aim is to provide caregivers of patients with type 2 diabetes with insights on the nature of promoting factors and barriers for the adoption of lifestyle change and the development of new habits. The temporal-causal network enables us to present findings from qualitative research in a new and visual manner. Furthermore, by attaching weights, speeds and adaptivity to the diverse connections, we can tentatively experiment with different types of care-giving and predict their effects. After devising our model, we compared Dutch standard care for type 2 diabetes patients with a typical blended care program. Dutch standard care is delivered by well-trained assistants of general practitioners, who see their patients in a face-to-face consultation once every three months. The blended care program has a strong digital component and consists of 100 days of daily interaction.

This paper consists of four sections: firstly, a discussion about the theoretical aspects of motivation and temporal-causal networks followed up with a general overview of our model, secondly, a few simulations that show the effectiveness of the model, thirdly a verification of the model based on findings in the literature, and lastly a discussion about how the model could be extended and a few of the possible objections.

2 Method

In this section we provide relevant theoretical notions on motivation. We present research findings on barriers and promoting factors to change in type 2 diabetes patients. Finally, we present theory on networks and explain the model we devised based on research findings.

Motivational Theory. Over the years many theories have been put forward to explain why we behave as we do, such as Maslow's theory of the hierarchy of needs [7], McClelland's Human Motivation Theory (needs for achievement, affiliation, or power) [9] and Skinner's operant conditioning theory (where response or behavior is strengthened by rewards) [15]. Sometimes we are moved by external forces such as rewards, grades or other peoples opinions, and sometimes we are moved by internal forces such as interests, curiosity and our own values. The interplay between these extrinsic forces and intrinsic motives and how they affect our ability to make choices and manage our lives is a central part of theories about self-determination [2]. Self-determination theory is the common construct nowadays to address and understand lifestyle change.

Ryan and Deci have suggested that types of motivation can be arranged according to the extent to which the motivation emanates from one's self [13]. Self-determination theory posits that more autonomous forms of motivation are associated with more sustainable behavior change [8]. Therefore, we arrive at a continuum of motivations from extrinsic to intrinsic. Concepts such as introjected motivation, identified motivation and integrated motivation are often used to explain different parts of the middle ground that sit between both extremes.

We have been guided by the need to include as many motivational concepts as necessary, in order to make the model relevant. At the same time, we felt the need to keep the model as simple as possible. Balancing these needs, we have included extrinsic, intrinsic motivation and identified motivation in a first version of the motivational part of our model.

According to research, the level and type of motivation of type 2 diabetes patients appear to be central factors in shaping their chances to successfully change their lifestyle. Extrinsic motivation leads them to work on lifestyle change in order to either avoid punishment or to receive praise, for instance in the interaction with caregivers who advised the lifestyle change. Qualitative research shows that patients for whom this type of motivation is predominant, only make slow progress in implementing lifestyle advice into their daily routine. Identified motivation makes patients feel responsible to follow the program, since it aligns with their own expectations to do so. After a period of changed behavior, some people develop intrinsic motivation to sustain their new lifestyle, actually enjoying their new routines and valuing them in themselves. Intrinsic motivation enables patients to more easily integrate new behavior with their daily routines and it sustains long-term lifestyle change and habit formation [6, 14, 16, 18].

Barriers and Resistance. Patients with type 2 diabetes suffer from real barriers in adopting and implementing lifestyle change. Many of them have a lower socialeconomic status, implying that they received lower education and/or have lowly paid jobs. This heightens their chance of being functionally illiterate and of possessing only low health literacy. Their social and physical surroundings may not support lifestyle change. They may experience financial difficulties. In our model, we have little assumptions about the social and physical environment or educational level of the patient, though considering the target group we can generally perceive them to be lowly educated, possess low health literacy and to be not particularly supported in changing their lifestyle by the people they live with[2].

Lifestyle related type 2 diabetes arises from diet and lack of exercise. Qualitative research shows that many patients positively value their diet and their lack of exercise, perceiving them as comfortable and a luxury. If they are advised to give up these comforts and to exchange them for a more moderate diet and regular exercise, they may experience

[2] Rodriguez-Saldana, Joel. The Diabetes Textbook Clinical Principles, Patient Management and Public Health Issues. 1st ed. 2019. Cham: Springer International Publishing, 2019. Web.

resistance. When actually going through the advised lifestyle changes, patients may feel grief and a experience a sense of loss [16]. Many patients appear to be unaware of the gravity of the health complications associated with type 2 diabetes. When they lack a full comprehension of their disease, this forms a strong barrier towards taking action for lifestyle change [4].

Barriers to exercise also arise from the disease itself, the associated physical states and comorbidities. Type 2 diabetes patients tend to be overweight or obese. These conditions in themselves can pose barriers to exercise. People with these conditions are prone to suffer from muscle of joint pain, and exercise might easily exacerbate these complaints. General tiredness is a common problem in patients with type 2 diabetes. Approximately twenty percent of type 2 diabetes patients also suffers from depression [10].

In our model, we presented the barriers as 'negative emotional states' and 'negative physical states'. Negative emotional states comprise emotions as guilt, grief, a sense of failure and feeling pressured. Extrinsic motivation tends to enhance these emotions, as does receiving a caregiver's advice to exercise. Negative physical states comprise tiredness, muscle and joint pain, obesity and depression. These states are taken to be enhanced by starting exercise, while they also function as a barrier to exercise. Yet, when patients persist in exercising, these negative physical states will go down [5, 16]. When patients do not perceive their disease as threatening their future health, this is taken as a strong barrier.

3 The Adaptive Network Model

In order to illustrate, how advice given to the patient first causes action, and eventually forms a habit (to exercise), a temporal-causal network model was constructed. Such a type of network provides a type of dynamic modelling that was apt to simulate the kind of repeated interactions a patient may have over time with their caregiver.

To better illustrate and analyze the network, we should view the model in the following way. A patient gets advice to exercise from their caregiver. To actually undertake such an action is dependent on a variety of factors as we have seen in the previous sections. Both stimulants and barriers to exercise exist in the case of patients with type 2 diabetes. Each connection between nodes can be seen as a directional influence. These connections can suppress or amplify the state, thus providing barriers or stimulants. Whereas the caregivers' advice to start exercising is given on one point in time, the patients' actual predictions of the effects of action, the preparations for action and adaptations or learning effects take place over a period of time, comprising complicated interactions with one another. A temporal-causal network provides a type of dynamic modelling that is apt to simulate the repeated interactions patients may have over time with their caregivers.

The Network-Oriented Modeling approach used here is based on ideas presented in Treur's book [17]. This gives us both a conceptual framework and a numerical representation. The former is the one we can see in Fig. 1. It involves the representation between certain factors found in the literature. The states have activation levels that may vary over time. Yet not all connections are created equally, thus also introducing a notion of strength. Furthermore whenever multiple relations affect a state, we require some way to aggregate these impacts. Lastly, a notion of speed is used to represent the malleability of a node. This gives us three important factors.

Connectivity: The strength of a connection between state X and Y is represented by some *connection weight* $\boldsymbol{\omega}_{X,Y}$.

Aggregation: Each state Y is also passed along some *combination function* $\mathbf{c}_Y$ which combine the impacts from other states via (multiple) incoming connections on Y.

Timing: Each state Y also takes along a *speed factor* $\boldsymbol{\eta}_Y$ which represents the speed by which state Y changes upon impact.

Using the notion of *self-modeling network* (also called *reified network*) introduced in (Treur 2020), every network characteristic can be made adaptive by adding a (self-model) state to the network that represents the value of this characteristic. This will be applied here to obtain a second-order adaptive network in which for some states X and Y: (1) first-order self-model states $\mathbf{W}_{X,Y}$ are included in the network that represent the value of connection weight $\boldsymbol{\omega}_{X,Y}$, and (2) second-order self-model states $\mathbf{H}_{\mathbf{W}_{X,Y}}$ are included in the network that represent the value of the speed factor of $\mathbf{W}_{X,Y}$ (learning rate).

In the model a number of links between states are assumed to be adaptive, meaning that they are learned over time. They are depicted as red arrows in Fig. 1. However to better represent the variety and adaptivity that takes place in habit formation we required a first- and second-order self-model (depicted in different planes) to describe the adaptive relations. In the full model, as shown by Fig. 2, some connection strengths are learned over time through Hebbian learning. The second-order self-model states represent the speed of said adaptation, following the second-order adaptation principle 'adaptation accelerates with increasing exposure' formulated in [11]. More details of the design of the model can be found in Linked Data at URL https://www.researchgate.net/publication/353654589.

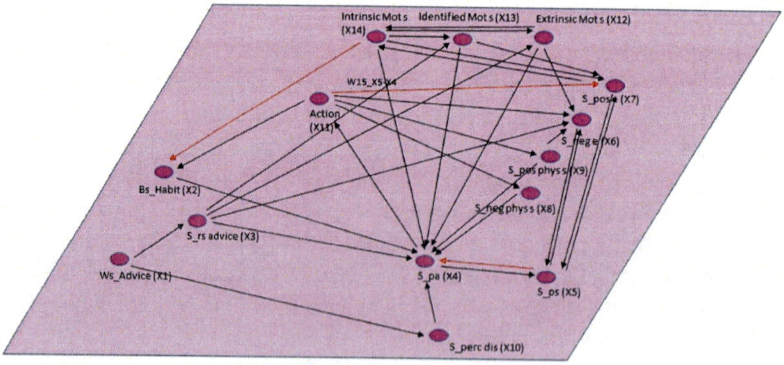

Fig. 1. Base level of model

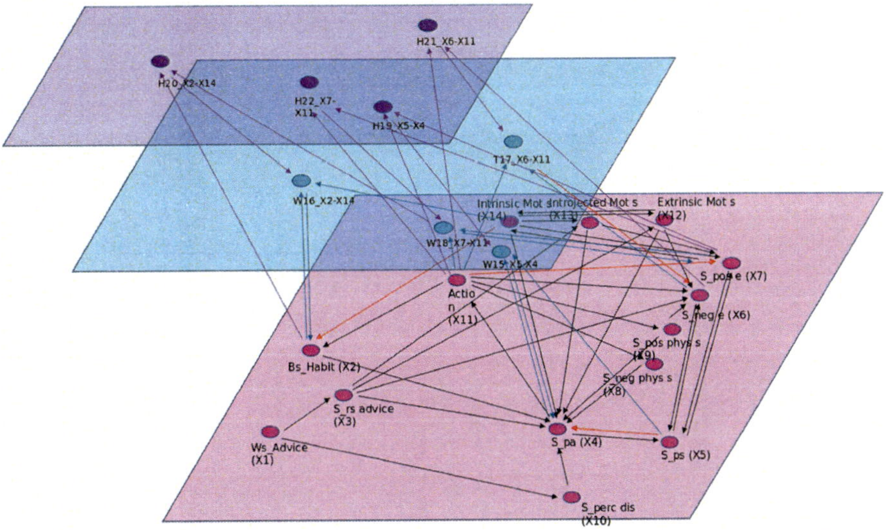

Fig. 2. Complete adaptive network model for formation of habits in patients with Diabetes

4 Simulation Results

While individuals may vary, we found that a significant impediment in habit formation arises when patients have infrequent interaction with their caregivers. By contrast, when running our model with the input frequency of advice every other day for a period of 100 days, we found that the patient does develop a habit to exercise (Fig. 6b). In order to comprehend this difference in outcomes, we need to focus on the complex interaction between advice given, physical body states, emotions and motivational drives.

Comparing the graphs arising from our model, we see that when patients receive advice every other three months, they do start to exercise right after the encounter with the caregiver. The action arises despite the barriers of negative emotions, negative physical states and sub-optimal motivational state we placed in the model. Yet, after exercise patients experience pain, tiredness and negative emotions, while they lack repeated stimulus for action. Their attempts to exercise therefore quickly fade away. Patients show no signs of habit formation[3].

We found that frequent advice prevents the attempts to exercise from fading away so quickly. Patients start exercising after receiving advice (Fig. 3b). This causes them to experience negative physical states and emotions, such as muscle pain and a sense of loss. Yet, because of the repetition of the advice, patients repeat their attempts to exercise. Pain and negative emotions gradually drop off with more exercise (Fig. 4a). After exercising for some time, positive effects start to emerge, such as more stamina, better physical shape, actual enjoyment of the exercise and a sense of well-being (Fig. 4b). These effects lead to fewer negative emotions (Fig. 5a). In time, the threshold for negative emotions goes up, making the new lower state for negative emotions more stable (Fig. 5b). As the

[3] This of course depends on the state of the individual, their basic intrinsic motivation, their perception of the disease and many of the other factors we modelled.

intrinsic motivation to exercise picks up, this increases the adaptive weight to the habit (Fig. 6a) that will later cause the habit to form (Fig. 6b).

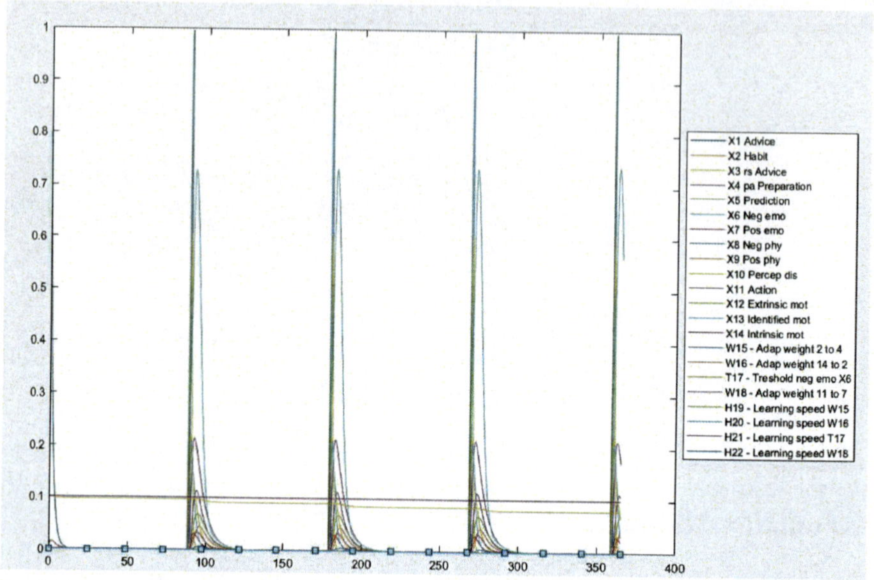

(a) When the advice (X1) is only given every 90 days, a habit (X2) for exercise is not formed.

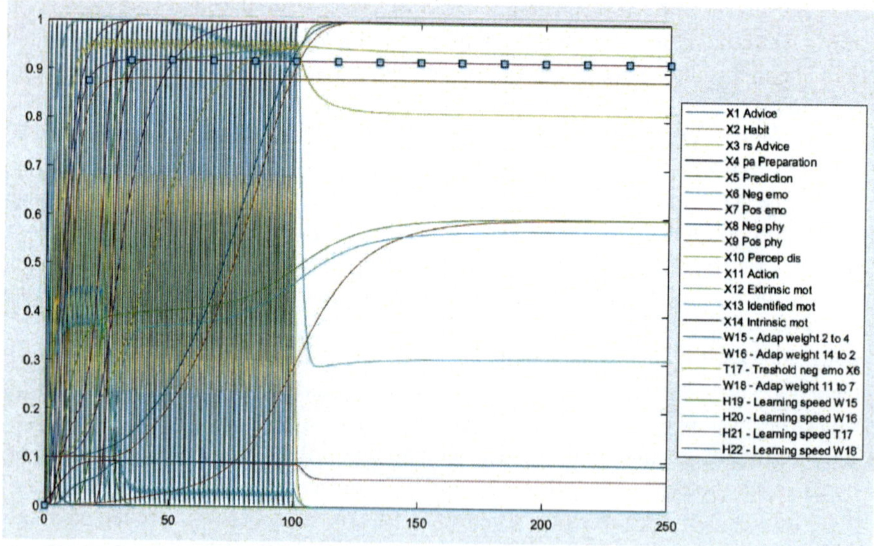

(b) Given frequent advice, until time step 100, the patient begins to exercise (X11 Action)

Fig. 3. Difference in advice

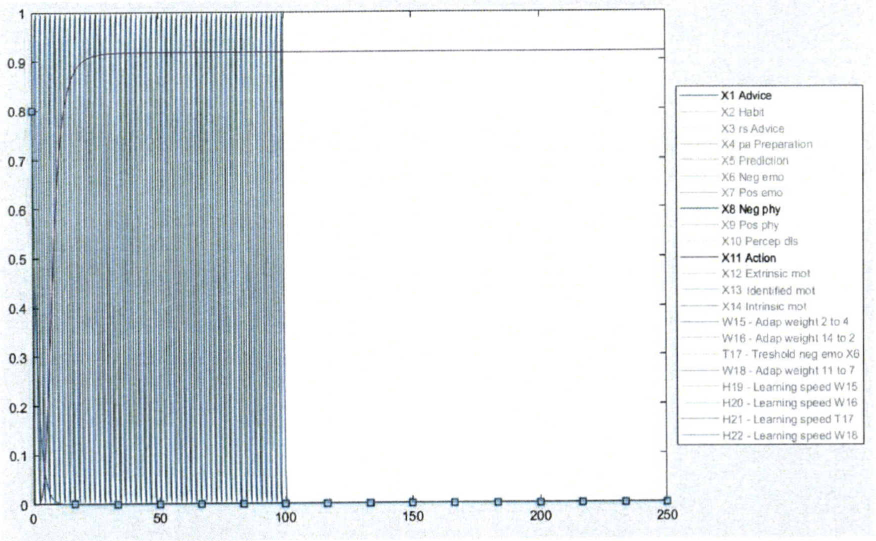

(a) As the patient begins to exercise (X11 Action), the negative physical body state (X8) begins to drop (from an initial 0.8 value)

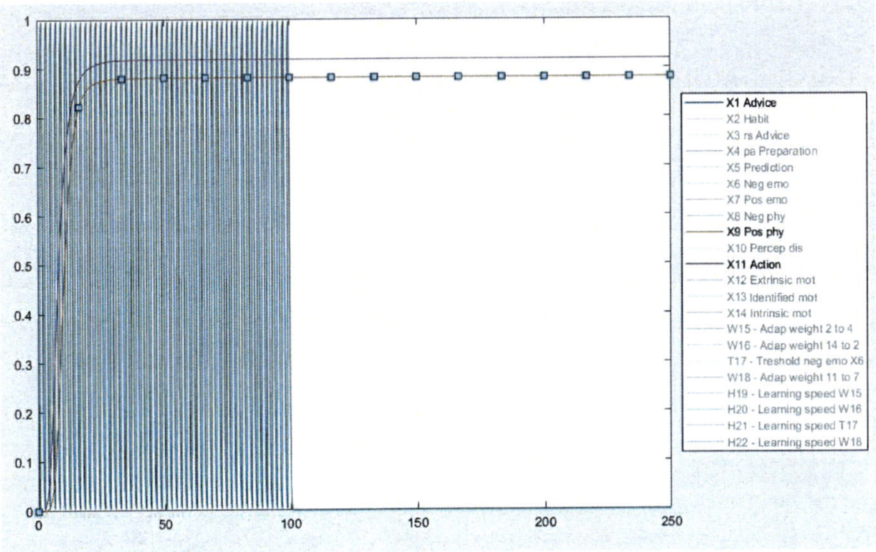

(b) The positive physical state (X9) begins to improve with more exercise

Fig. 4. Negative and positive physical states

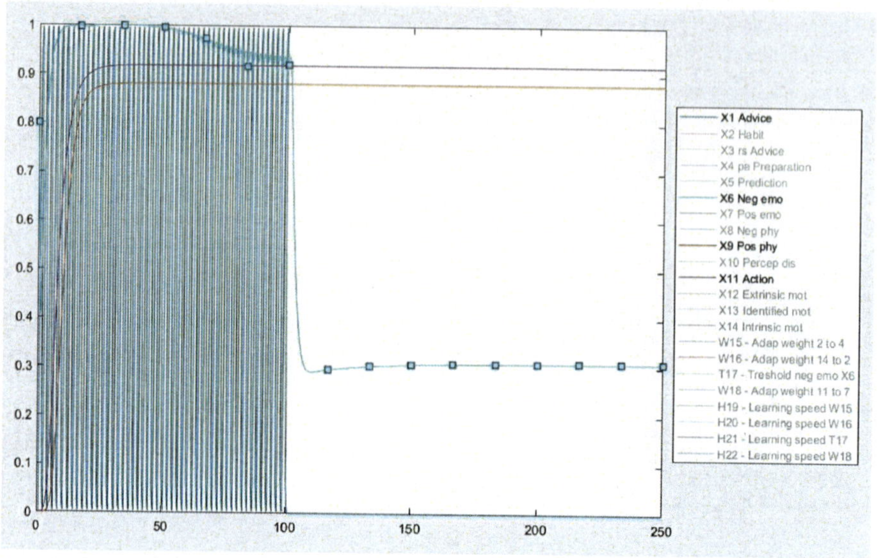

(a) As the positive physical body state (X9) improves and stays high, negative emotions (X6) goes down

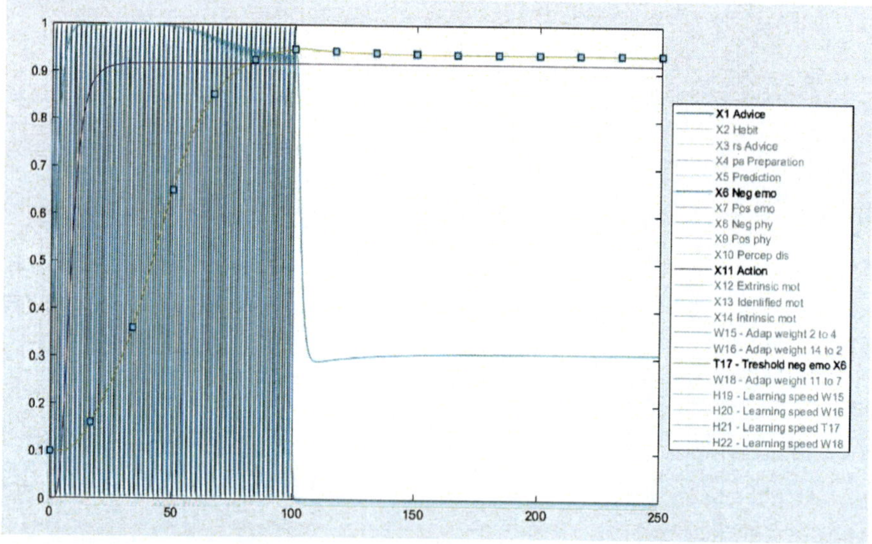

(b) As threshold (T17 neg emo) for changing the negative emotions goes up, the emotional state become more stable, and it becomes more difficult to push it (X6 negative emotions) upwards again

Fig. 5. Negative emotion and its threshold

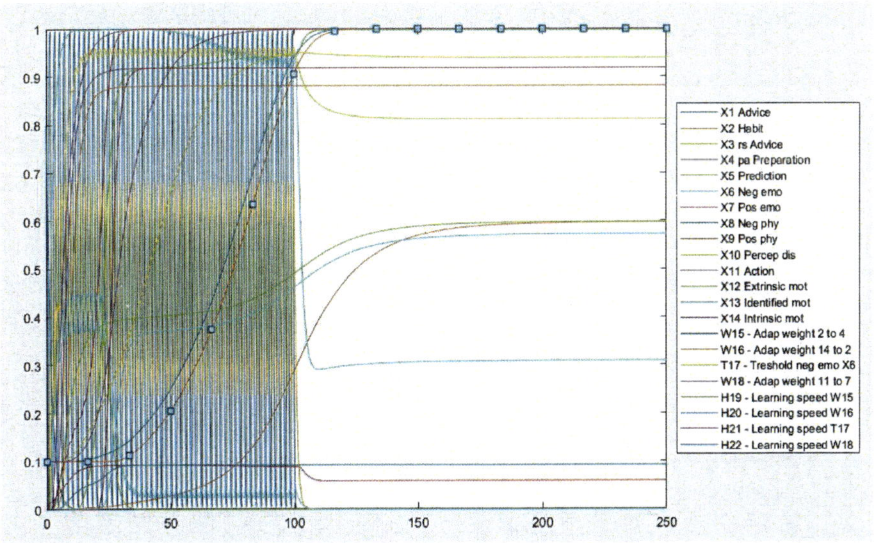

(a) As the intrinsic motivation (X14) picks up, this adjusts the adaptive weight (W16) to the habit, that will later cause the habit to form.

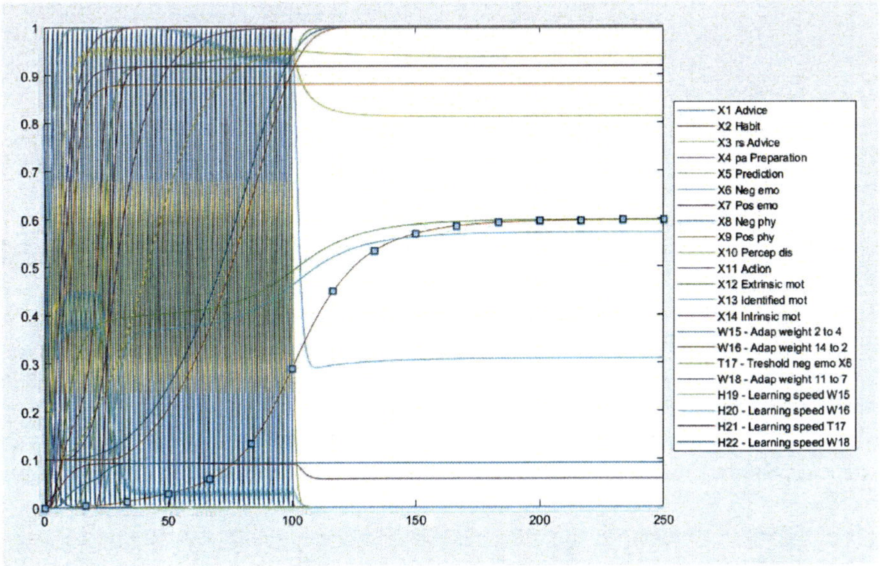

(b) As the strength of the W16 weight increases this then leads to the formation of the habit (X2)

Fig. 6. Habit formation

5 Discussion

In reality, more interaction will not easily and not always lead to habit formation. To put it bluntly: there are more factors at work here. We have not modeled all the personal and interpersonal factors which might be influential in shaping the caregiver-patient relation, for instance. The way one receives advice and from whom matters. The effect of advice might depend on personality of both patient and caregiver. The content and tone of the advice are important. Is the advice concrete or vague, tailored or general, fitting for the patient or wildly astray? Is the tone warm or cold, relaxed or hasty, understanding or accusatory, emphatic with the real problems or overly positive?

Furthermore, we only used three types of motivation, whereas other types may influence the effectiveness of advice and consequently habit formation as well. Literature on motivation theory points to a relation between habit formation and identified motivation [19]. However, in the case of type 2 diabetes the self-identification with say being a exercising person, does take ample amount of time. Moreover, literature [16] on type 2 diabetes specifically points to intrinsic motivation as a key factor to persevering in exercising. Identified motivation is presented as sustaining the difficult first phase of habit formation for exercise. We would advocate further research into the relation between identified motivation and habit formation for exercise.

Possibly even more importantly, the model has neglected the physical and social environment of patients. Modern societies created a highly obesogenic and therefore diabetogenic environment. Our built environment is often solely caroriented, our main hobbies comprise various forms of screen-watching, and many cities resemble food swamps, flooded with unhealthy options of highly processed and lowly nutritious food, which we have learnt to perceive as 'yummy' from our toddler days onward. It is tough to change your lifestyle when you live in an environment filled with cues to sit and to eat [9]. It is even tougher to change your lifestyle when you share this environment and the lifestyle it promotes with many others. Their aggregated social influence might be stronger than that of the most convincing caregiver, blended or otherwise. Furthermore, we should take into account that, as stated above, many patients with type 2 diabetes have a lower socioeconomic position, which makes them prone to struggle with other problems than their diabetes, such as poverty, unsafe neighborhoods, unemployment and chronic stress. Those patients who encounter such problems may well find it hard to direct their mental and financial resources towards lifestyle change [1].

Thus, chronic lifestyle-related disease such as type 2 diabetes presents itself as a complex and wicked problem in real life, not as a complicated yet neat model. Why did we endeavour to create such a model in the first place, and how do we think it can prove useful? Our aim in building and running the model was to provide caregivers with insights on the nature of promoting factors and barriers for lifestyle change for type 2 diabetes patients. Our model, we feel, might prove useful to enhance caregivers' understanding of their patients' difficult position and the barriers they encounter. Our results stress the limitations of standard care for inducing lifestyle change, and they could raise the awareness of blended care's potential.

We would advise to expand this first model. In next versions, we would advocate to consider the nature and content of the relation between caregiver and patient in more detail, possibly focusing on finding the right balance between face-to-face and digital contact in blended care. Other types of motivation could be taken into account. Barriers in the physical and social environment could be modeled, as well as pervasive problems such as poverty. We believe that step by step, next versions of the model will show opportunities to overcome barriers and to achieve better results for more people.

6 Conclusion

We constructed a second order temporal-causal network model on type 2 diabetes patients receiving caregivers' advice to exercise, taking into account motivational forces, emotions, physical states and habit formation. We based ourselves on qualitative research findings when we consciously constructed a model filled with many barriers. We found that the advice frequency of Dutch standard care of once every other three months did lead to attempts to exercise, which quickly faded away. According to our model, this advice frequency did not lead to habit formation or to lasting lifestyle change. When we increased the advice frequency to every other day, as can be achieved through blended care with a strong digital component, we found that the attempts to exercise were no longer extinguished by the barriers of physical pain en negative emotions. These barriers still rose after starting exercise. Yet, through the repetition of the advice and the ensuing repetition of attempts, negative effects of exercise went down and positive effects, such as stamina and well-being, started to emerge. Eventually, these positive changes sustained habit formation. We would advise to expand our model later on, adding relational factors, motivational states, barriers in the physical and social environment and socioeconomic problems.

As is well known, intrinsic motivation is crucial for lifestyle adaptation. In order to successfully treat chronic lifestyle-related disease like type 2 diabetes, caregivers require a good understanding of boosting this motivational state. The motivational model represented in this paper provides suggestions on how to improve adoption rate. Frequent and consistent interactions, while also garnering more negative emotions, showed to have a positive effect on the patient's lifestyle in the long run. With the established severity and pervasiveness of type 2 diabetes we should ensure that people are getting the best care that they can get and that most likely means better guidance for those in need.

References

1. Agardh, E., Allebeck, P., Hallqvist, J., Moradi, T., Sidorchuk, A.: Type 2 diabetes incidence and socio-economic position: a systematic review and meta-analysis. Int. J. Epidemiol. **40**(3), 804–818 (2011)
2. Deci, E.L., Ryan, R.M.: Intrinsic Motivation and Self-Determination in Human Behavior (Perspectives in Social Psychology). Plenum Press (1985)
3. Jacob, E., Avery, A.: Energy-restricted interventions are effective for the remission of newly diagnosed type 2 diabetes: a systematic review of the evidence base. Obes. Sci. Pract. (2021)
4. Jenkins, D.W., Jenks, A.: Exercise and diabetes: a narrative review. J. Foot Ankle Surg. **56**(5), 968–974 (2017)
5. Kang, H.J., Wang, J.C.K., Burns, S.F., Leow, M.K.S.: Is self-determined motivation a useful agent to overcome perceived exercise barriers in patients with type 2 diabetes mellitus? Front. Psychol. **12**, 2134 (2021)

6. Lakerveld, J., Palmeira, A., van Duinkerken, E., Whitelock, V., Peyrot, M., Nouwen, A.: Motivation: key to a healthy lifestyle in people with diabetes? Current and emerging knowledge and applications. Diabet. Med. **37**(3), 464–472 (2020)
7. Maslow, A.H.: A theory of human motivation. Psychol. Rev. **50**(4), 370–396 (1943)
8. Matsumoto, H., Takenaka, K.: Motivational profiles and stages of exercise behavior change. Int. J. Sport Health Sci. **2**, 89–96 (2004)
9. McClelland, D.C.: The Achieving Society. Simon and Schuster (1961)
10. Nicolaidis, S.: Environment and obesity. Metabolism **100**, 153942 (2019)
11. Owens-Gary, M.D., Zhang, X., Jawanda, S., Bullard, K.M., Allweiss, P., Smith, B.D.: The importance of addressing depression and diabetes distress in adults with type 2 diabetes. J. Gen. Int. Med. **34**(2), 320–324 (2019)
12. Robinson, B.L., Harper, N.S., McAlpine, D.: Meta-adaptation in the auditory midbrain under cortical influence. Nat. Commun. **7**(1), 1–8 (2016)
13. Ryan, R.M., Deci, E.L.: Intrinsic and extrinsic motivations: classic definitions and new directions. Contemp. Educ. Psychol. **25**(1), 54–67 (2000). https://www.sciencedirect.com/science/article/pii/S0361476X99910202
14. Sebire, S.J., et al.: "i've made this my lifestyle now": a prospective qualitative study of motivation for lifestyle change among people with newly diagnosed type two diabetes mellitus. BMC Public Health **18**(1), 1–10 (2018)
15. Skinner, B.: The Behavior of Organisms. Appleton-Century (1938)
16. Teixeira, P.J., Carraça, E.V., Markland, D., Silva, M.N., Ryan, R.M.: Exercise, physical activity, and self-determination theory: a systematic review. Int. J. Behav. Nutr. Phys. Act. **9**(1), 1–30 (2012)
17. Treur, J.: Network-oriented modeling for adaptive networks: designing higher-order adaptive biological, mental and social network models. Springer Nature (2020). https://link.springer.com/book/10.1007/978-3-030-31445-3
18. Vilafranca Cartagena, M., Tort-Nasarre, G., Rubinat Arnaldo, E.: Barriers and facilitators for physical activity in adults with type 2 diabetes mellitus: a scoping review. Int. J. Environ. Res. Public Health **18**(10), 5359 (2021)
19. Vlachopoulos, S.P., Karageorghis, C.I.: Interaction of external, introjected, and identified regulation with intrinsic motivation in exercise: relationships with exercise enjoyment. J. Appl. Biobehav. Res. **10**(2), 113–132 (2005)

Aggressive Action Recognition Using Pre-trained 3D CNNs with Attention

Mikhail Uzdiaev(✉)

St. Petersburg Federal Research Center of the Russian Academy of Sciences (SPC RAS), St. Petersburg Institute for Informatics and Automation of the Russian Academy of Sciences, Laboratory of Autonomous Robotic Systems, 39, 14th Line, 199178 St. Petersburg, Russia

Abstract. The article is devoted to the problem of improving the accuracy of aggressive action recognition on video. The application of the attention mechanism to improve the recognition accuracy in pre-trained 3D Convolutional Neural Networks (3D CNN) is considered. Inception 3D neural network (I3D), Residual 3D neural network (R3D), Residual (2+1)D neural network (R(2+1)D) are used as models to compare. The neural attention mechanism that distinguishes the most significant information on the RSL feature maps of video frames is described. The architectural features of the application of the attention mechanism in 3D CNNs are also given. The research was carried out on the RWF-2000 dataset. The considering neural networks were evaluated in terms of accuracy, precision, recall, f1-score metrics. Comparative analysis of the results of models with attention mechanism and models without attention showed an increase in recognition accuracy in neural networks with Attention Mechanism for all the considering 3D CNN architectures.

Keywords: I3D · R3D · R(2+1)D · Attention mechanism · Aggressive action recognition

1 Introduction

In recent years, video surveillance systems especially Close Circuit Television (CCTV) have become widespread. This poses the urgent task of analyzing the video received from these cameras. At the same time, it is necessary to highlight the analysis and identification of aggressive behavior on CCTV video due to CCTV often record human aggressive behavior. Fights, attacks, robberies, mass clashes, brawls, etc. can be distinguished as aggressive actions. It is necessary to respond promptly to aggressive actions in order to prevent human casualties. In addition, it is necessary to counteract the distribution of video content containing aggressive actions on the Internet [1]. A large amount of video material is difficult to analyze manually, therefore, it becomes relevant to develop automatic methods for recognizing aggressive human actions on video.

Aggressive action recognition on video is a particular instance of the general task of human action recognition. It is noted that 3D CNN architectures show high efficiency

R. Silhavy et al. (Eds.): CoMeSySo 2021, LNNS 232, pp. 933–944, 2021.
https://doi.org/10.1007/978-3-030-90318-3_74

in the task of classifying a large number of human actions [2]. Such neural networks perform spatio-temporal analysis of the feature dynamics on video stream frames and classify human actions. At the same time, training a neural network for a particular task such as aggressive action recognition from scratch may not be effective in terms of time, computational costs and the final recognition accuracy. To solve a particular problem of recognizing aggressive actions, the application of the transfer learning approach becomes promising. Transfer learning approach uses a neural network that was pre-trained on a large representative dataset to extract feature vectors with the most valuable information about human actions. These feature vectors are further used to fine-tune another neural network that performs the particular aggressive action recognition task. The use of transfer learning approaches has shown its effectiveness, in particular, in the problem of recognizing aggressive actions on video [3].

One of the most promising methods for increasing the efficiency of neural network recognition is the attention mechanism [4], which was originally developed for automatic text translation methods. Also, this method is used in other tasks, neural network processing of various information [5–7]. The essence of the attention mechanism is to extract the most significant intermediate features during the neural network data processing and selection the relevant information to solve an appropriate task. At the moment, the application of the attention mechanism in the tasks of recognizing aggressive actions on video has not been sufficiently disclosed.

This work reveals the results of a study of the application of the transfer learning approach and the attention mechanism in the problem of recognizing aggressive human actions on CCTV video. To obtain results reflecting the improvements that are made by the attention mechanism, a comparative study of the results of recognizing aggressive actions in I3D [8], R3D [9] и R(2+1)D [10] architectures with and without attention mechanisms was performed.

2 Prior Work

C. Ding et al. [11] for the first time proposed to use 3D CNNs for human aggressive actions recognition based on the analysis of grayscale frames of a video stream. This method has a recognition accuracy of 91% on the Hockey Fights [12] dataset.

The authors of the work [13] proposed a new method using convolutional-recurrent neural network architecture in the recognizing aggressive human actions. This architecture uses CNN to extract spatial features of visual objects on video frames. Temporal analysis of these features is performed by the Long-Short Term Memory (LSTM) neural network architecture [14] This method achieves 93.9% accuracy in human aggressive action recognition task on the Hockey Fights dataset.

S. Sudhakaran and O. Lanz [15] use the convolutional LSTM architecture [16] to analyze the sequence of spatial features of video frames. This method achieves 94.57% accuracy on the Violent Flow dataset [17], 97.1% accuracy on the Hockey Fights dataset and 100% on the Movies Fights dataset [12].

The authors of the approach [18] propose to use a bidirectional LSTM to recognize aggressive actions on a video stream from CCTV using only RGB video frames. The method achieves 83.19% aggressive action recognition accuracy on the Hockey Fights dataset, 88.7% recognition accuracy on the Movies Fights dataset, 75.73% recognition accuracy on the CCTV video dataset that was collected by the authors.

The FightNet architecture presented in [19], performs the classification of aggressive human actions using the C3D architecture to analyze in parallel RGB video frames, optical stream frames, as well as frames for accelerating objects in video. The accuracy of this method is 97% on the Hockey Fights dataset and 100% accuracy on the Movies Fights dataset.

W. Song et al. [20] presented a new algorithm for selecting key frames based on calculating the center of mass of moving objects on grayscale frames of a video. The selected keyframes are further processed using the C3D architecture. This approach achieves 99.6% recognition accuracy on the Hockey Fights dataset, 99.9% accuracy on the Movies Dataset, 93.4 on the Violent Flow dataset.

Work [3] considers transfer learning approach for aggression action recognition using 3D CNN that are pretrained on a representative Kinetics-400 [21] dataset. Transfer learning approach achieves 99.5% accuracy on an aggressive actions dataset that was combined from Hokey Fights, Movies Fights and Violent Flow datasets.

Liang Q. et al. [22] presented so-called two-cascade temporal shift module with attentions mechanism. This module performs channels shift in the temporal dimension. This module is used along with CNN to extract spatial features and LSTM+attention mechanism is used to process temporal information. Authors claim 98.9% accuracy on Hockey Fights dataset and 89.3% accuracy on RWF-2000 dataset [23].

3 Proposed Model

To solve the problem of recognizing aggressive actions, a neural network architecture was proposed, consisting of two blocks: a block for the extraction of spatio-temporal features and a fine-tuning block, which includes the attention mechanism. The proposed architecture is shown in Fig. 1.

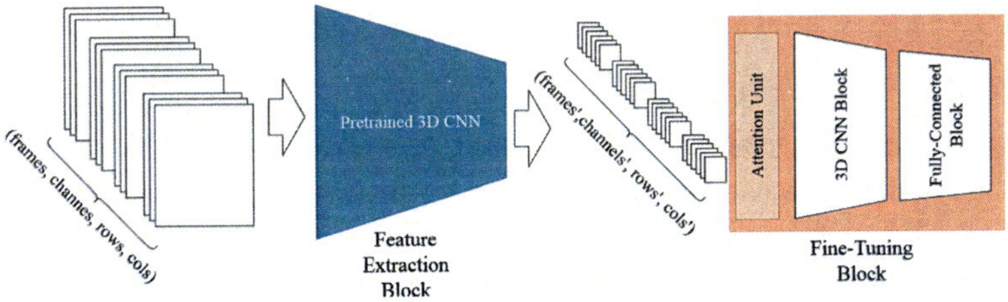

Fig. 1. 3DCNN+Attention architecture

The block for the extraction of spatio-temporal features of moving visual objects on frames of a video stream is implemented using a 3D CNN neural network, previously trained on a representative Kinetics [21] dataset to classify 400 classes of human actions. The weights of the initial layers of this neural network are frozen and do not change in the process of further fine-tuning by the specific task of recognizing aggressive actions. This feature extracting neural network processes RGB video frames and generates at the output a spatio-temporal feature map that reflects information about moving objects on the processed frames. The feature extraction block is highlighted in blue in Fig. 1. The extracted feature maps are fed to the fine-tuning block, which consists of a block that implements the attention mechanism [6], a 3D convolution block, and an output fully connected classifier. Fine-tuning block weights are initialized with random values and are changed during fine-tuning procedure for a specific task of aggressive action recognition. The fine-tuning block is highlighted in red in Fig. 1.

The attention mechanism is a method for highlighting the most significant information in the processed feature maps. This is done using one or more fully connected neural networks with a single output neuron. In this work, we use the attention mechanism, which highlights the most significant feature map within a single video frame by analyzing individual channels of each frame. The architecture of the used block that implements the attention mechanism is shown in Fig. 2.

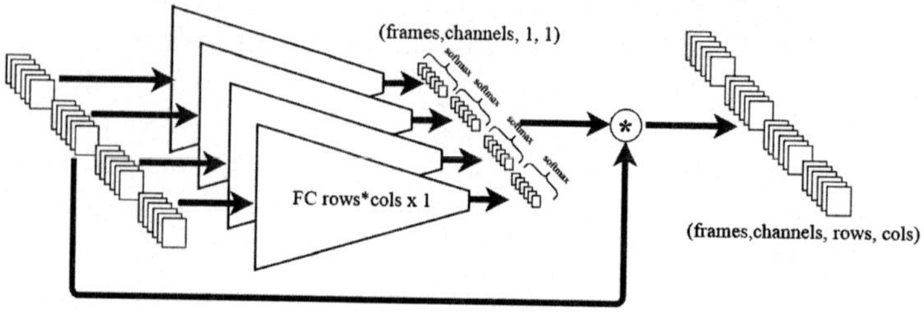

Fig. 2. Attention block

An attention unit consists of several fully connected neural networks with one output neuron. The number of neural networks is equal to the number of frames processed. Each of the fully connected neural networks processes flattened feature maps coming from the previous layer:

$$z_i^j = f^j(x_i^j) \tag{1}$$

where $\mathbf{x}_i^j$ is a flatetned feature map; f^j is fully-connected neural network; z_i^j is a score computed by f^j; i is an index of feature map; j is an index of a frame.

z_i^j scores are fed to the layer of the softmax activation function. The output values of this layer are the coefficients of the significance of a particular feature map:

$$a_i^j = softmax(z_i^j) = \frac{e^{z_i}}{\sum_{k=1}^{N} e^{z_k}} \tag{2}$$

where α_i^j is a coefficient of significance of the i-th feature map of the j-th frame; N is a number of feature maps in the frame. Then, all the elements of the processed feature maps are multiplied by the corresponding significance factor α_i^j of the feature map within one frame. The resulting tensor has the same dimension as the input tensor:

$$o_i^j = x_i^j a_i^j \tag{3}$$

where $\mathbf{o}_i^j$ is an output tensor.

In order to show the changes in recognition results of 3DCNN that use attention blocks, the 3DCNN+Attention results are compared with the results of the same neural networks without attention. Let us now consider each of the investigated 3DCNNs in more detail.

3.1 Inception 3D Neural Network

Let's take a closer look at the architectural features of the Inception 3D neural network (I3D) [8]. A distinctive feature of this architecture is the use of 3D Inception blocks [24] for the spatio-temporal analysis of signs of movement of visual objects in a video stream. The architecture of Inception block is shown in Fig. 3.

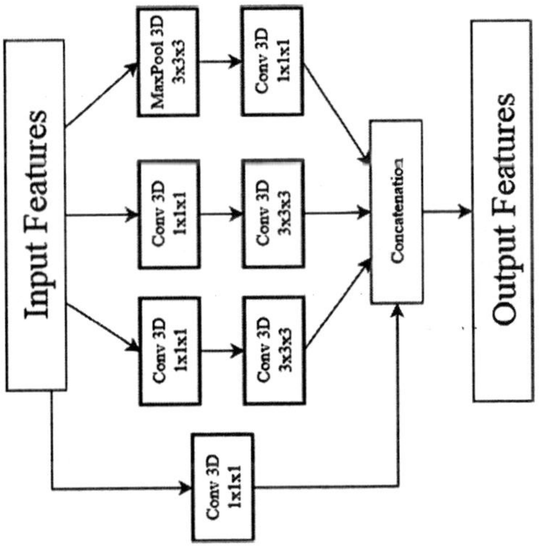

Fig. 3. Inception block

Inception blocks are analogs of blocks in the Inception-V1 neural network [24], where the feature map obtained on the previous layer is processed in parallel by four different branches. The first branch is a convolution with a kernel of size (1 × 1 × 1), the second and third branches each contain two consecutive convolutional layers with kernels of size (1 × 1 × 1) and (3 × 3 × 3), respectively, and the third branch contains two consecutive layers - Max-Pool with a field of size (3 × 3 × 3) and a convolutional layer of size (1 × 1 × 1). Next, the resulting feature map is formed using the concatenation of feature maps obtained using all four branches of the Inception module.

The I3D architecture is shown in Fig. 4.

The layers that are highlighted in blue in Fig. 4 compose feature extractor. The weights of the feature extractor are frozen and not updated during the training procedure. Fine-tune block is also presented on the Fig. 4. It is highlighted in red. The weights of the fine-tune block are randomly initialized and updated during the training procedure.

The I3D+Attention architecture is shown in Fig. 5.

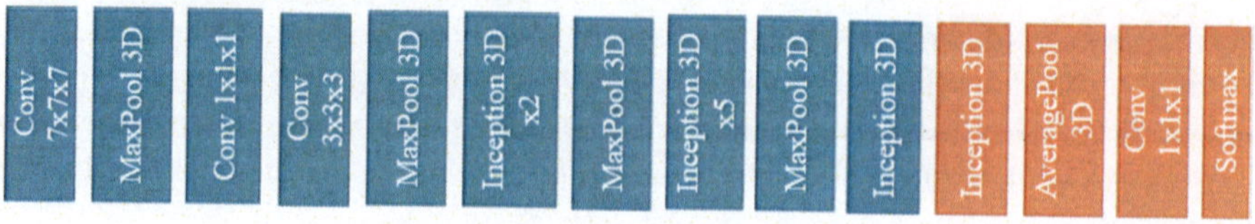

Fig. 4. I3D architecture

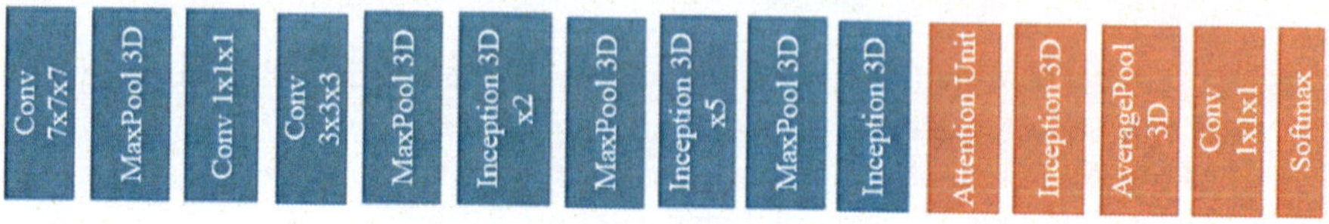

Fig. 5. I3D+Attention architecture

As well as plain I3D, the layers that are highlighted in blue in Fig. 5 compose feature extractor. Fine-tune block is also presented on the Fig. 5. It is highlighted in red. The attention unit is at the input of the fine-tune block.

3.2 Residual 3D Neural Network

Residual 3D Neural Network (R3D) [9,25] is analogous to 2D ResNet [26]. The main architectural component of this network is the so-called Residual Block, which is shown in the Fig. 6.

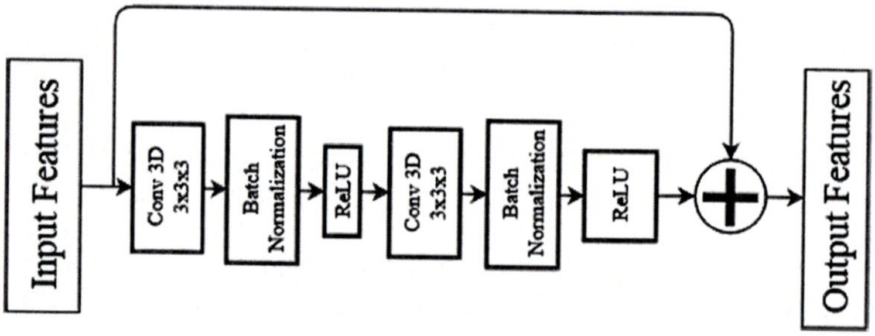

Fig. 6. Residual block

The main distinguishing feature of this block is the shortcut of the input features to the output of the Residual block. The convolutional part of the Residual block consists of three consecutive blocks, each of which consists of a convolutional layer, a batch normalization layer [27] and ReLU activation [28].

The 3D ResNet architecture is shown in Fig. 7.

R3D can process RGB video fragments of various lengths. The size of each frame is 112×112. The I3D architecture model was chosen as the feature extractor, which was pre-trained on the general task of classifying 400 human actions on the Kinetics dataset [21].

Fig. 7. R3D architecture

The layers that are highlighted in blue in Fig. 7 compose feature extractor. The weights of the feature extractor are freezed and not updated during the training procedure. Fine-tune block is also presented on the Fig. 7. It is highlighted in red. The weights of the fine-tune block are randomly initialized and updated during the training procedure.

The R3D+Attention architecture is shown in Fig. 8.

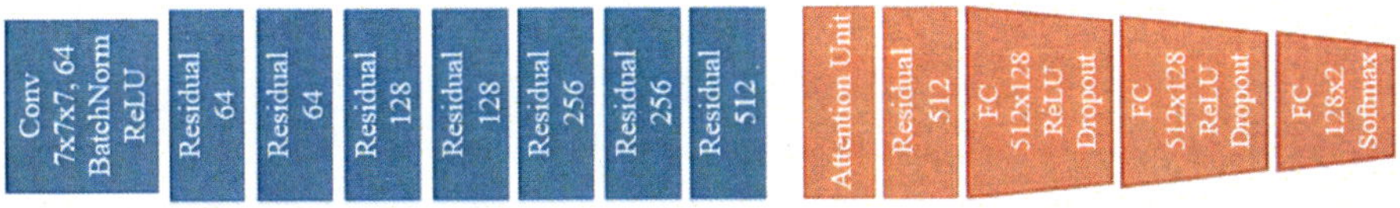

Fig. 8. I3D+Attention architecture

As well as plain R3D, the layers that are highlighted in blue in Fig. 8 compose feature extractor. Fine-tune block is also presented on the Fig. 8. It is highlighted in red. The attention unit is at the input of the fine-tune block.

3.3 Residual (2+1)D Neural Network

R(2+1)D neural network [10] has an architecture that almost completely reproduces the R3D architecture discussed above. A distinctive feature of this architecture is a different structure of convolutional layers that process the spatio-temporal features of visual objects presented in video frames. The structure of convolutional layers that process space-time information in R(2+1)D is shown in Fig. 9.

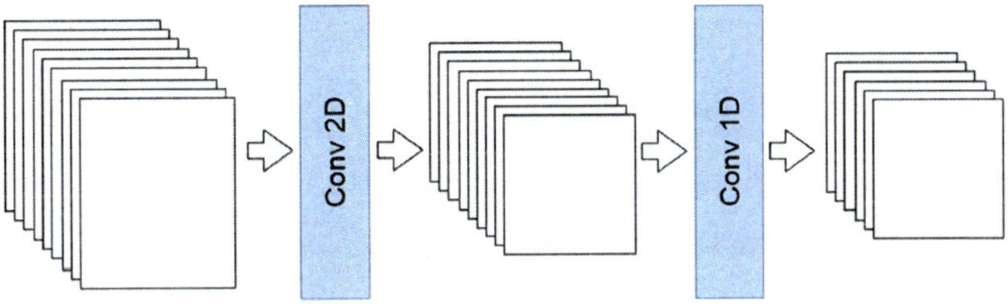

Fig. 9. Processing of spatio-temporal information using convolution (2+1)D

In plain 3D CNNs, the 3D convolution operation analyzes both the spatial and temporal components of the video stream. In (2+1)D convolution, spatio-temporal analysis is divided into a sequence of two operations: first, the spatial component of visual objects on video frames is analyzed using a 2D convolution, and then the temporal component is analyzed using 1D convolution. This separation is due to lower computational costs as well as lower memory costs. Feature extractor R(2+1)D fully reproduces the architecture of feature extractor R3D. Feature extractor R(2+1)D+Attention fully reproduces the architecture of feature extractor R3D+Attention.

4 Experimental Setup, Results and Discussion

To obtain results that reflect the improvements made by adding the attention mechanism, the results obtained by each of the models with the attention mechanism and without it were examined. To assess the effectiveness of the proposed approach, the RWF-2000 dataset [23] was used. Figure 10 shows the video frames of the RWF-2000 dataset.

(a) (b)

Fig. 10. Frames from the RWF-2000 dataset. Nonviolent (a) and violent (b) examples of video frames

This dataset consists of CCTV videos containing aggressive human actions. This dataset consists of 1000 videos containing aggressive behavior, as well as 1000 videos that do not contain manifestations of aggression. Each video is exactly 5 s long. Aggressive behavior is expressed in fights and brawls present in the video. One of the main factors in choosing this dataset is that this dataset contains videos only from one domain that is CCTV videos. This factor increases the representativeness of the data presented in the dataset. All videos in the dataset were additionally divided into fragments of 1 s or 30 frames. This was done to increase the number of instances in the dataset while keeping the proportions of the dataset labels unchanged. Thus, 5000 videos containing aggression and 5000 without aggression were made. These videos were used to teach the studied architectures.

Each model was trained for 200 epochs. Batch size is equal to 16. Adam optimizing algorithm was taken [32] to perform backpropagation learning procedure with learning rate 0.001. To prevent overfitting the following frame augmentation techniques was used: random horizontal and random vertical flipping. Binary cross-entropy was used as a loss function:

$$L(\widehat{y_i}, y_i) = -y_i \, log \, \widehat{y_i} - (1 - y_i) log(1 - \widehat{y_i}) \tag{4}$$

The best results were achieved at 43, 52, 46, 79, 92, 111 for I3D, R3D, R(2+1)D, I3D+Attention, R3D+Attention, R(2+1)D+Attention respectively. The training was fullfilled using the PyTorch [29] library.

In this work, the following metrics for assessing the recognition efficiency are used:

1. Accuracy

$$accuracy = \frac{TP + TN}{TP + TN + FP + FN} \tag{5}$$

2. Precision (Positive Predictive Value – PPV)

$$precision = PPV = \frac{TP}{TP + FP} \tag{6}$$

3. Recall (True Positive Rate – TPR)

$$recall = TPR = \frac{TP}{TP + FN} \tag{7}$$

4. F1-score:

$$F1 - score = 2\frac{precision \cdot recall}{precision + recall} = \frac{TP}{TP + \frac{TP+FN}{2}} \tag{8}$$

where TP is the number of true positive number of recognitions, TN – is the number of true negative number of recognitions, FP – is the number of false positive number of recognitions, FN – is the number of false negative number of recognitions. The results of the experiments are summarized in the Table 1.

During training and testing on the RWF-2000 dataset, the proposed new neural network architectures with the attention mechanism showed better results compared to the same neural networks without attention mechanism for all the considering metrics. The best accuracy result is 0.970 for I3D+Attention and R(2+1)D+Attention models. At the same time R3D+Attention has shown a comparable accuracy result that is equal to 0.969. I3D+Attention has shown the best results for the precsion metric for non-aggressive

Table 1. Results on the RWF-2000 dataset

Modal	Accuracy	Precision		Recall		F1-score	
		Aggr	Non aggr	Aggr	Non aggr	Aggr	Non aggr
I3D	0.956	0.945	0.966	0.965	0.946	0.955	0.956
R3D	0.958	0.959	0.956	0.954	0.961	0.956	0.959
R(2+1)D	0.954	0.961	0.946	0.943	0.964	0.952	0.955
I3D+Attention	**0.970**	0.962	**0.978**	**0.977**	0.963	**0.969**	**0.971**
R3D+Attention	0.969	0.961	0.976	0.975	0.962	0.968	0.969
R(2+1)D+Attention	**0.970**	**0.965**	0.973	0.972	**0.967**	**0.969**	0.970

videos (0.978) and R(2+1)D+Attention has shown the best precision results for aggressive videos (0.965). At the same time, I3D+Attention has shown the best results for the recall metric for aggressive videos (0.977) and R(2+1)D+Attention has shown the best recall results for non-aggressive videos (0.967). I3D+Attention and R(2+1)D+Attention have shown the best results for F1-score metrics (0.969) for aggressive videos. The best result of F1-score for non-aggressive videos achieved by I3D+Attention model. It is worth noting that the results of appropriate metrics that have achieved by different neural networks are very close to each other.

5 Conclusion

The implementation of the attention mechanism has shown an improvement in the results of recognizing aggressive actions for the considered 3D CNN architectures. Moreover, the improvement is observed for all the considering metrics. However, the results of models with attention mechanism are very close to each other and it is hard to distinguish one outstanding model. In further research, it is planned to expand the number of neural network architectures: 3D ResNext [25], T3D [30], CNN+LSTM, etc. In addition, it is planned to explore other versions for the mechanism of attention, for example, Multihead Attention [31]. It is also planned to apply the considered approach to the general human action recognition task using the datasets Kinetics [21], Sports-1M [32] for a wide range of different 3DCNN architectures.

References

1. Xu, Z., Liu, Y., Mei, L., Hu, C., Chen, L.: Semantic based representing and organizing surveillance big data using video structural description technology. J. Syst. Softw. **102**, 217–225 (2015)
2. Panzner, M., Cimiano, P.: Comparing hidden Markov models and long short term memory neural networks for learning action representations. In: Pardalos, P., Conca, P., Giuffrida, G., Nicosia, G. (eds.) MOD 2016. LNCS, vol. 10122, pp. 94–105. Springer, Cham (2016). https://doi.org/10.1007/978-3-319-51469-7_8

3. Saveliev, A., Uzdiaev, M., Dmitrii, M.: Aggressive action recognition using 3D CNN architectures. In: 2019 12th International Conference on Developments in eSystems Engineering (DeSE), pp. 890–895. IEEE (2019)
4. Bahdanau, D., Cho, K., Bengio, Y.: Neural machine translation by jointly learning to align and translate. arXiv preprint arXiv:1409.0473 (2014)
5. Xu, K., et al.: Show, attend and tell: neural image caption generation with visual attention. In: International Conference on Machine Learning, pp. 2048–2057. PMLR (2015)
6. Chorowski, J., et al.: Attention-based models for speech recognition. arXiv preprint arXiv:1506.07503 (2015)
7. Chu, Q., et al.: Online multi-object tracking using CNN-based single object tracker with spatial-temporal attention mechanism. In: Proceedings of the IEEE International Conference on Computer Vision, pp. 4836–4845 (2017)
8. Carreira, J., Zisserman, A.: Quo vadis, action recognition? A new model and the kinetics dataset. In: Proceedings of the IEEE Conference on Computer Vision and Pattern Recognition, pp. 6299–6308 (2017)
9. Hara, K., Kataoka, H., Satoh, Y.: Learning spatio-temporal features with 3D residual networks for action recognition. In: Proceedings of the IEEE International Conference on Computer Vision, 3154–3160 (2017)
10. Tran, D., et al.: A closer look at spatiotemporal convolutions for action recognition. In: Proceedings of the IEEE Conference on Computer Vision and Pattern Recognition, pp. 6450–6459 (2018)
11. Ding, C., Fan, S., Zhu, M., Feng, W., Jia, B.: Violence detection in video by using 3D convolutional neural networks. In: Bebis, G., et al. (eds.) ISVC 2014. LNCS, vol. 8888, pp. 551–558. Springer, Cham (2014). https://doi.org/10.1007/978-3-319-14364-4_53
12. Bermejo Nievas, E., Deniz Suarez, O., Bueno García, G., Sukthankar, R.: Violence detection in video using computer vision techniques. In: Real, P., Diaz-Pernil, D., Molina-Abril, H., Berciano, A., Kropatsch, W. (eds.) CAIP 2011. LNCS, vol. 6855, pp. 332–339. Springer, Heidelberg (2011). https://doi.org/10.1007/978-3-642-23678-5_39
13. Dong, Z., Qin, J., Wang, Y.: Multi-stream deep networks for person to person violence detection in videos. In: Tan, T., Li, X., Chen, X., Zhou, J., Yang, J., Cheng, H. (eds.) CCPR 2016. CCIS, vol. 662, pp. 517–531. Springer, Singapore (2016). https://doi.org/10.1007/978-981-10-3002-4_43
14. Hochreiter, S., Schmidhuber, J.: Long short-term memory. Neural Comput. **9**(8), 1735–1780 (1997)
15. Sudhakaran, S., Lanz, O.: Learning to detect aggressive videos using convolutional long short-term memory. In: 2017 14th IEEE International Conference on Advanced Video and Signal Based Surveillance (AVSS), pp. 1–6. IEEE (2017)
16. Xingjian, S., Chen, Z., Wang, H., Yeung, D., Wong, W., Woo, W.: Convolutional LSTM network: a machine learning approach for precipitation nowcasting. In: Advances in Neural Information Processing Systems, pp. 802–810 (2015)
17. Hassner, T., Itcher, Y., Kliper-Gross, O.: Violent flows: real-time detection of violent crowd behavior. In: 2012 IEEE Computer Society Conference on Computer Vision and Pattern Recognition Workshops, pp. 1–6. IEEE (2012)
18. Ditsanthia, E., Pipanmaekaporn, L., Kamonsantiroj, S.: Video representation learning for CCTV based violence detection. In: 2018 3rd Technology Innovation Management and Engineering Science International Conference (TIMES-iCON), pp. 1–5. IEEE (2018)
19. Zhou, P., Ding, Q., Luo, H., Hou, X.: Violent interaction detection in video based on deep learning. J. Phys. Conf. Ser. **844**(1), 012044 (2017)
20. Song, W., Zhang, D., Zhao, X., Yu, J., Zheng, R., Wang, A.: A novel violent video detection scheme based on modified 3D convolutional neural networks. IEEE Access **7**, 39172–39179 (2019)

21. Kay, W., et al.: The kinetics human action video dataset. arXiv preprint arXiv:1705.06950 (2017)
22. Liang, Q., et al.: Violence behavior recognition of two-cascade temporal shift module with attention mechanism. J. Electron. Imaging **30**(4), 043009 (2021)
23. Cheng, M., Cai, K., Li, M.: RWF-2000: an open large scale video database for violence detection. In: 2020 25th International Conference on Pattern Recognition (ICPR), pp. 4183–4190. IEEE (2021)
24. Szegedy, C., et al.: Going deeper with convolutions. In: Proceedings of the IEEE Conference on Computer Vision and Pattern Recognition, pp. 1–9 (2015)
25. Hara, K., Kataoka, H., Satoh, Y.: Can spatiotemporal 3D CNNS retrace the history of 2D CNNS and imagenet? In: Proceedings of the IEEE Conference on Computer Vision and Pattern Recognition, pp. 6546–6555 (2018)
26. He, K., et al.: Deep residual learning for image recognition. In: Proceedings of the IEEE Conference on Computer Vision and Pattern Recognition, pp. 770–778 (2016)
27. Ioffe, S., Szegedy, C.: Batch normalization: accelerating deep network training by reducing internal covariate shift. arXiv preprint arXiv:1502.03167 (2015)
28. Nair, V., Hinton, G.E.: Rectified linear units improve restricted Boltzmann machines. In: Proceedings of the 27th International Conference on Machine Learning (ICML-10), pp. 807–814 (2010)
29. https://pytorch.org/
30. Liu, K., et al.: T-C3D: temporal convolutional 3D network for real-time action recognition. In: Proceedings of the AAAI Conference on Artificial Intelligence, vol. 32, no. 1 (2018)
31. Vaswani, A., et al.: Attention is all you need. In: Advances in Neural Information Processing Systems, pp. 5998–6008 (2017)
32. Karpathy, A., et al.: Large-scale video classification with convolutional neural networks. In: Proceedings of the IEEE Conference on Computer Vision and Pattern Recognition, pp. 1725–1732 (2014)

Control of the Group of Mobile Robots Without Pre-planned Route for Monitoring and Diagnostic Purposes

S. G. Kapustyan, M. V. Orda-Zhigulina, and D. V. Orda-Zhigulina(✉)

Southern Scientific Center of the Russian Academy of Sciences, st. Chehova, 41, Rostov-on-Don 344006, Russia

Abstract. In this paper an approach is presented to solving the problem of control a heterogeneous group of mobile robots with a leader. In particular this approach has been applied for a group of unmanned aerial vehicles (UAVs) during monitoring (scanning) geographically distributed areas.

A description of the solution to the problem is given for the subtask for determining the number of UAV subgroups required for scanning areas; for the subtask for distribution of scanning sub areas between UAV subgroups; for the subtask for distribution of UAVs by subgroups; for the subtask for UAV distribution over scanning bands.

The main attention is paid to the issues of working out pre-planned trajectories of movement of individual UAVs in well-known works dedicated the problem of group control of UAVs in solving various target tasks. Unlike these works in this paper an approach proposes to autonomous planning of trajectories which allows minimizing the intersection of trajectories and thereby reduce the risk of collisions of UAVs with each other during operation.

Keywords: Data collection · Robotics · UAV · Processing and storage systems · Internet of Things · Decentralized decision support system · Distributed ledger · Mobile robotic complex · Hazardous phenomena

1 Introduction

The authors proposed the combined method for creating systems for predicting and monitoring hazardous processes and phenomena [1]. The group of robotic UAVs is used as terminal devices (sensors) for collecting information, as well as UAVs with separate sensors or analyzers on board, photo and video capabilities within the framework of this method.

Therefore, the task of control the UAV group becomes urgent. The following subtasks have been solved:

a) the subtask for determining the number of UAV subgroups required for scanning areas;
b) the subtask for distribution of scanning sub-areas between UAV subgroups;

R. Silhavy et al. (Eds.): CoMeSySo 2021, LNNS 232, pp. 945–953, 2021.
https://doi.org/10.1007/978-3-030-90318-3_75

c) the subtask for distribution of UAVs by subgroups;
d) the subtask for UAV distribution over scanning bands.

We need a formal statement to solve a problem of scanning multiple areas using a UAV. The formal statement is required and discussed in detail as it is described in the paper [2].

The following tasks were identified to solve the main problem:

1. The number of UAV subgroups is determined which is necessary to scan M areas in the first phase. Moreover, the several subgroups may be required to scan the same area. It depends from the size of this area.

   Each area is divided into parts, and each subgroup must scan its own part of the area (sub-area).
2. The distribution is carried out of scanning sub-areas between UAV subgroups in the second phase.
3. Scanning bands of width L are allocated in each sub-area in the third phase. Also, the number of UAV subgroups is determined. The UAV is distributed among subgroups R_j $(j \in [1, N])$.
4. Finally, it is necessary to determine the specific scanning lanes (trajectories) for each UAV of the subgroup in the fourth phase.

   Let us consider the solution of the listed subtasks in more detail.

2 The Subtask of Determining the Number of UAV Subgroups Required to Scan Areas

We need firstly to estimate the size of each area and determine the number of parts into which this area should be divided to solve this subtask. This can be done as follows.

If the width H_i of the area P_i is greater than 2D, this area should be divided into h_i parts. The number of pieces in width is defined as:

$$h_i = \left\lceil \frac{H_i}{2D} \right\rceil, \tag{1}$$

where H_i is rounding to the larger integer.

The result is several pieces of width:

$$\Delta H_i = \frac{H_i}{h_i}, \tag{2}$$

and long F_i.

Since all sub-areas can be scanned at the same time, it is possible to determine the scan time of their area:

$$t_i^{scan} = \frac{F_i}{V}. \tag{3}$$

If $t_i^{scan} > t_p$, then this area must be divided in length into f_i parts.

The number of parts (sub-areas) can be defined as

$$f_i = \left\lceil \frac{F_i}{V \cdot t_p} \right\rceil. \tag{4}$$

As a result, the area is divided into $h_i \cdot f_i$ parts with width ΔH_i and length ΔF_i, where ΔF_i is defined as

$$\Delta F_i = \frac{F_i}{f_i}. \tag{5}$$

Thus, the entire set of areas P_i is divided into S sub-areas, each area must be scanned by one subgroup of UAVs headed by the UAV-leader. The number of sub-areas is determined according to the required number of UAV subgroups in the following way

$$S = \sum\nolimits_{i=1}^{M} s_i, \tag{6}$$

where $s_i = h_i \cdot f_i$ – is the number of sub-areas in the P_i area.

If $S > K$, that is the number of UAV subgroups is less than required to scan the entire set $\{M\}$ of areas. It is mean the problem cannot be solved in a given time t_p. Therefore, it is necessary to reduce the number of sub-areas by increasing their width ΔH_i $(i = [1, M])$ to achieve the condition

$$S \leq K. \tag{7}$$

This can be done as follows.

We should select a subset of areas $\{\boldsymbol{M}^s\}$ $(|\boldsymbol{M}^s| = M^s)$ in the set of sub-areas $\{M\}$. The width of the subareas is less than 2D and. Also, we should select a subset of areas $\{\boldsymbol{M}^l\}$ $\left(\left|\boldsymbol{M}^l\right| = M^l\right)$. The width of the subareas is greater than 2D, that is $M = M^s + M^l$.

It is required $S^s = M^s \cdot f_i$ $(\forall P_i \in \{\boldsymbol{M}^s\})$ subgroups of the UAV to scan M^s areas, then M^l subgroups remain for scanning $K - S^s \cdot$ areas.

Further, it is determined the average width of the sub-area for the subset of areas $\{\boldsymbol{M}^l\}$

$$\Delta H^l = \frac{\sum_{\forall P_i \in \{\boldsymbol{M}^l\}} \Delta H_i \cdot h_i}{K - S^s}. \tag{8}$$

The width of the sub-area will exceed the range of the communication and data exchange subsystem of the UAVs performing scanning in this case. Consequently, UAVs communication and data exchange subsystem must provide relaying of data from remote UAVs - leaders A delay may occur in the transmission of information which may affect the process of coordinating the actions of the UAV of a subgroup by its leader in this case. This delay is incommensurably shorter than the scanning time of this sub-area and does not significantly affect this time. Analysis of this delay about controlling a subgroup of UAVs will be discussed in the next paper.

The number of sub-areas which are divided along the width of each area $P_i \in \{\boldsymbol{M}^l\}$ and is defined as

$$h_i^l = \left\lceil \frac{H_i}{\Delta H^l} \right\rceil \quad (\forall P_i \in \{\boldsymbol{M}^l\}). \tag{9}$$

The width of each sub-area of the domain $P_i \in \{\boldsymbol{M}^l\}$ is defined as

$$\Delta H_i = \frac{H_i}{h_i^l} \quad (\forall P_i \in \{\boldsymbol{M}^l\}), \tag{10}$$

and the number of sub-areas in the subset of areas $\{\boldsymbol{M}^l\}$ is defined as

$$S^l = \sum\nolimits_{\forall P_i \in \{\boldsymbol{M}^l\}} h_i^l \cdot f_i. \tag{11}$$

Then the total number of sub-areas for scanning is determined as

$$S = S^l + S^s = K. \tag{12}$$

Each sub-area is assigned a number $s \in [1, S]$ in a certain way.

It is possible to calculate the coordinates $<x_s^c, y_s^c>$ $(s \in [1, S])$ of the center of each sub-areas at the base of the calculated data on the sub-areas and the given orientation angles of the scanned areas. Since this subtask is trivial it is not considered in this paper.

The solution of this subtask is assigned to special services of the monitoring and diagnostic system. The calculation results are recorded in a distributed ledger, copies of which are available to each UAV-leader RL_k $(k = [1, K])$ and is used to distribute scanning sub-areas between UAV subgroups.

3 The Subtask of Distribution of Scanning Sub-Areas Between UAV Subgroups

The solution of this subtask is assigned to the UAV-leaders on-board control. The following algorithm can be proposed in the simplest case. The UAV-leader RL_1 chooses the closest sub-area S_s as the target that is the sub-area at the base of the data on the coordinates of the centers $<x_s^c, y_s^c>$ of all sub-areas $S_s\left(s = \overline{1, S}\right)$ and it is realized the follow condition:

$$\sqrt{(x_1 - x_s^c)^2 + (y_1 - y_s^c)^2} = min, \tag{13}$$

where x_1, y_1 are the current coordinates of the UAV RL_1. The UAV RL_1 enters the results of its choice in a distributed ledger which makes the information available to other UAVs-leaders who must take this choice into account.

Next UAV RL_2 makes the decision. It chooses as target sub-area S_q, for which the condition:

$$\sqrt{(x_2 - x_s^c)^2 + (y_2 - y_s^c)^2} = min. \tag{14}$$

UAVs-leaders carry out the sequential selection of target sub-areas in a similar way. In this case the next UAV makes a choice only among those sub-areas that have not yet been selected. This problem is discussed in detail here [3, 4].

4 The Subtask of UAV Distribution by Subgroups

First of all, the solution to this subtask consists in determining the number of UAVs R_j ($j \in [1, N]$), which can simultaneously scan the sub-area S_s. Obviously one UAV R_j ($j \in [1, N]$) can scan a certain band in a sub-area S_s of width L along its entire length as shown in Fig. 1. Since several UAVs scan this band simultaneously, they will interfere with each other.

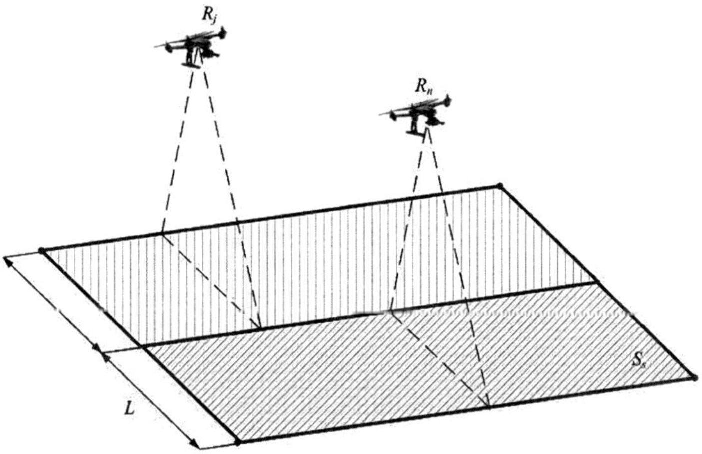

Fig. 1. The bands of the S_s sub-area scanned by UAVs R_j and R_n

It should be noted that the value of the scan bandwidth L is chosen less than the value that is provided by the onboard sensor systems, and the bands should overlap. Firstly, this is necessary there are no gaps between the bands due to the positioning error of the UAV. Secondly, they can slightly expand the band and solve the scanning task without losing information when exiting building at least one UAV if some UAVs performing scanning fail.

We can conclude based on these considerations that no more than N_s UAVs can simultaneously scan the S_s sub-area and

$$N_s = \left\lceil \frac{\Delta H_s}{L} \right\rceil, \quad (15)$$

where H_s is the width of the S_s sub-area;

L - UAV scanning bandwidth R_j ($j \in [1, N]$).

Consequently, the size of the subgroup intended for scanning the S_s sub-area is N_s UAVs (excepting the UAV-leader).

The next step is to determine which particular UAVs R_j ($j \in [1, N]$) will be included in the subgroups scanning the S_s sub-areas.

We will assume the current coordinates $<x_j, y_j>$ of the UAV R_j ($j \in [1, N]$) are known and information about them is entered in the distributed ledger and is available to the UAVs-leaders. We will also assume that the targets for the UAV R_j ($j \in [1, N]$) are the centers of the sub-areas S_s specified by the coordinates $<x_s^c, y_s^c>$. On-board control devices of the UAVs-leaders determine which UAVs R_j ($j \in [1, N]$) will be included in subgroups. These on-board control devices make decision based of the information contained in the distributed ledger about the UAV R_j ($j \in [1, N]$) and the results of the distribution of sub-areas between the UAVs-leaders at the previous phase.

Any of the mentioned above distribution algorithms can be used in this case. There is only one difference from the algorithm which is used at the previous phase for forming the matrix of assessing the effectiveness of the assignment. There are also taken into account the distances between the current UAV positions and the centers of the sub-areas in addition to the intersections of the UAV trajectories R_j ($j \in [1, N]$). The task is to minimize the intersections of trajectories and movements of the UAV.

5 The Subtask of UAV Distribution by Scan Bands

After the UAVs R_j ($j \in [1, N]$) groups are distributed over the scanning sub-areas it is necessary to determine a specific scan band for each UAV.

It lets N_s UAVs be allocated for scanning the S_s sub-area. That is the S_s sub-area is divided into N_s strips with a width of at most L.

You need firstly to define the so-called UAV entry-exit points for each band. Two such points can be defined for each band which is illustrated in Fig. 2.

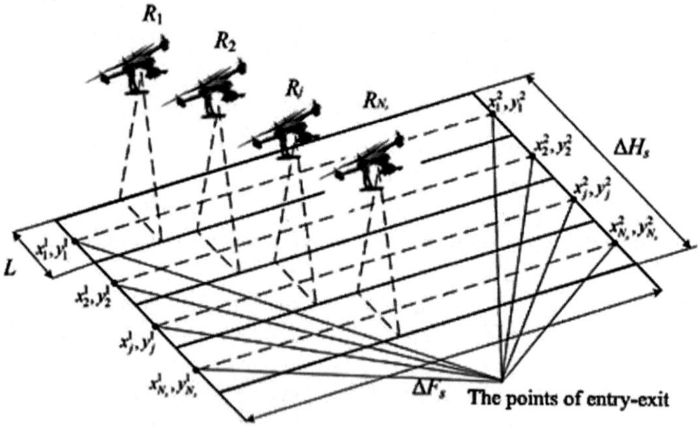

Fig. 2. The points of entry-exit of scanning bands

Coordinates of entry and exit points $<x_{j_s}^1, y_{j_s}^1>$ and $x_{j_s}^2, y_{j_s}^2>$ $(j = \overline{1, N_s})$ are defined by equations:

$$\begin{cases} \bar{x}_{j_s}^1 = x_s^c - \frac{\Delta F_s}{2}, \\ \bar{y}_{j_s}^1 = y_s^c - \frac{\Delta H_s}{2} + (j_s - 1) \cdot L + \frac{L}{2}, \\ x_{j_s}^1 = x_s^c + (\bar{x}_{j_s}^1 - x_s^c) \cdot \cos\varphi_s - (\bar{y}_{j_s}^1 - y_s^c) \cdot \sin\varphi_s, \\ y_{j_s}^1 = y_s^c + (\bar{x}_{j_s}^1 - x_s^c) \cdot \sin\varphi_s + (\bar{y}_{j_s}^1 - y_s^c) \cdot \cos\varphi_s. \end{cases} \quad (16)$$

And

$$\begin{cases} \bar{x}_{j_s}^2 = x_s^c + \frac{\Delta F_s}{2}, \\ \bar{y}_{j_s}^2 = y_s^c - \frac{\Delta H_s}{2} + (j_s - 1) \cdot L + \frac{L}{2}, \\ x_{j_s}^2 = x_s^c + (\bar{x}_{j_s}^2 - x_s^c) \cdot \cos \varphi_s - (\bar{y}_{j_s}^2 - y_s^c) \cdot \sin \varphi_s, \\ y_{j_s}^2 = y_s^c + (\bar{x}_{j_s}^2 - x_s^c) \cdot \sin \varphi_s + (\bar{y}_{j_s}^2 - y_s^c) \cdot \cos \varphi_s, \end{cases} \tag{17}$$

where ΔF_s is the length of the scanning area S_s;
ΔH_s is width of the scanning area S_s;
L is the UAV scanning bandwidth;
φ_s is angle of orientation of the scanning sub-area S_s;
$<x_s^c, y_s^c>$ are coordinates of the geometric centers of the scanning sub-areas;
$<\bar{x}_{js}^{1}, \bar{y}_{js}^{1}>$, $<\bar{x}_{js}^{2}, \bar{y}_{js}^{2}>$ are intermediate values of the coordinates of the centers of the bands before turning through the angle φ_s.

Each UAV-leader RL_s enters the obtained values of the coordinates of the entry-exit points of the lanes of its sub-area into the distributed ledger so that these values are available for the other UAVs-leaders.

It was determined the size of the subgroup corresponds to the number of scan bands into which the corresponding sub-area can be divided at the previous phase. It means each UAV R_{j_s} ($j_s \in [1, N_s]$) UAV-the leader RL_s assigned to this the sub-area defines the scan band using again one of the above assignment task solving algorithms.

However, it is necessary to define for each band of each sub-area the entry points $<x_{j_s}^1, y_{j_s}^1>$ or $<x_{j_s}^2, y_{j_s}^2>$. The movement of the UAV R_{j_s} ($j_s \in [1, N_s]$) was carried out in one direction in order to avoid emergency situations in the adjacent along the length sub-areas of the area P_i.

The corresponding UAVs-leaders make the decision on the choice of entry points for their subgroup, for example, according to the minimum of the total movements of the UAVs R_{j_s} ($j_s \in [1, N_s]$) assigned to these sub-areas.

The procedure can be based on the considerations presented in the work [4] for selecting entry points of scanning sub-areas for UAVs R_{j_s} ($j_s \in [1, N_s]$).

The corresponding UAVs can be sent to the group gathering area for subsequent tasks after completing the scanning procedure for each area.

The proposed approach is based on digital economy technologies, in particular, blockchain and distributed ledger technologies. This approach makes it possible to solve the problem of simultaneous scanning of several sufficiently large areas of space by a heterogeneous group of UAVs which is the basis of the multi-robotic complex of monitoring and diagnostics system in a time not exceeding a given one.

6 Conclusions

The methods and algorithms which are proposed in this work are simple and have low computational complexity for solving the problem of distributing scanning areas between a group of mobile robots (a group of UAVs).

Also, these methods and algorithms can be implemented by on-board control devices of individual UAVs of the group united by communication channels into a computer network.

In current paper we suggest the new approach to autonomous planning of trajectories which allows minimizing the intersection of trajectories and thereby reduce the risk of collisions of UAVs with each other during operation. It is new approach compared the suggested approaches from well-known works [5–11], where the main attention is paid to the issues of working out pre-planned trajectories of movement of individual UAVs in dedicated the problem of group control of UAVs in solving various target tasks.

The results are obtained in the current paper can be used for solving problems of creating monitoring and diagnostics systems using UAVs. Also, these results can be used for motion control systems for UAVs and UAV groups in different applications.

Acknowledgment. The publication was prepared as part of GZ SSC RAS N GR project AAAA-A19-119011190173-6 and RFBR projects 18-05-80092 and 18-08-01226.

References

1. Orda-Zhigulina, M.V., Melnik, E.V., Ivanov, D.Y., Rodina, A.A., Orda-Zhigulina, D.V.: Combined method of monitoring and predicting of hazardous phenomena. In: Silhavy, R. (eds.) CSOC 2019. AISC, pp. 55–61, vol. 984. Springer, Cham (2019). https://doi.org/10.1007/978-3-030-19807-7_6
2. Gaiduk, A.R., Kapustyan, S.G., Plaksienko, V.S., Kabalan, A.E.A.: Upravlenie gruppoi BLA pri neopredelennykh zapazdyvaniyakh v kanalakh svyazi. In: Nauchnyi vestnik Novosibirskogo gosudarstvennogo tekhnicheskogo universiteta, vol. 2–3, no. 79, pp. 37–56 (2020)
3. Kapustyan, S.G., Usachev, L.Zh., Dyachenko, A.A.: Sposob reshenija stroevoj zadachi dlja gruppy BLA samoletnogo tipa. Extreme Robot. **1**(1), 93–98 (2016)
4. Kaljaev, I.A., Kapustjan, S.G., Usachev, L.Zh.: Metod reshenija zadachi raspredelenija celej v gruppe BLA setecentricheskoj sistemoj upravlenija. Izvestija Juzhnogo federal'nogo universiteta. Tehnicheskie nauki **12**(185), 55–70 (2016)
5. Gajduk, A.R., Kapustjan, S.G., Merkulov, V.I., Besklubova, K.V.: Gruppovoe upravlenie BLA. Informacionno-izmeritel'nye i upravljajushhie sistemy **11**, 24–36 (2014)
6. Chandra, R.S., Breheny, S.H., D'Andrea, R.: Antenna array synthesis with clusters of unmanned aerial vehicles. Automatica **44**(8), 1976–1984 (2008)
7. Kingston, D., Beard, R.W., Holt, R.S.: Decentralized perimeter surveillance using a team of UAVs. IEEE Trans. Robt. **24**(6), 1394–1404 (2008)
8. Spry, S.C., Girard, A.R., Hedrick, J.K.: Convoy protection using multiple unmanned aerial vehicles: organization and coordination. In: Proceedings of the 2005, American Control Conference, pp. 3524–3529. IEEE (2005)
9. Chung, J.: Cooperative control of UAVs using a single master subsystem for multi-task multi-target operations. In: Kim, J.-H., Yang, W., Jo, J., Sincak, P., Myung, H. (eds.) Robot Intelligence Technology and Applications 3. AISC, vol. 345, pp. 193–212. Springer, Cham (2015). https://doi.org/10.1007/978-3-319-16841-8_19

10. Merkulov, V.I., Samodov, I.O.: Issledovanie jeffektivnosti kollektivnogo upravlenija bespilotnymi letatel'nymi apparatami v sostave lokal'noj seti. Informacionno-izmeritel'nye i upravljajushhie sistemy **12**(3), 39–42 (2014)
11. Verba, V. S., Merkulov, V. I., Salovskij, P. A.: Gruppovoe upravlenie bespilotnymi letatel'nymi apparatami. Vozmozhnosti i ogranichenija. Polet. Obshherossijskij nauchno-tehnicheskij zhurnal (8–9), 16–24 (2015)

CSMI-AW: Computational System for Medical Image Authentication Using Watermarking

Anusha Chacko[1,2](✉) and Shanty Chacko[3]

[1] Karunya Institute of Technology and Sciences, Coimbatore, Tamil Nadu, India
anushachacko@vjec.ac.in

[2] Department of Electronics and Communication Engineering, Vimal Jyothi Engineering College, Chemperi, Kerala, India

[3] Department of Electrical and Electronics Engineering, Karunya Institute of Technology and Sciences, Coimbatore, India

Abstract. The enormous growth of digital data in the era of information technology (IT) has provided broader scope towards improving the quality of life. However, the technological aspects in various data-driven solutions also resulted in data duplication and malevolent manipulation during the digital data distributions through networks, especially when it comes to images. Thereby the demand to protect the privacy of image data is highly envisioned and a crucial concern in the current scenario. Currently, medical facilities also require protecting medical image data ownership with copyright protection and content authentication to preserve the authenticity of the patient clinical details. The study in this regard introduces a computationally efficient technique, namely-computational system for medical image authentication using watermarking (CSMI-AW) to protect the ownership of the patient chest x-ray images. The system uses the potential factors associated with the spectral domain techniques, such as with Discrete Cosine Transformation (DCT) approach with non-blind detection. The optimized procedure of DCT-based significant technique for message image/watermarking and substitution methods for embedding process makes the CSMI-AW more robust to retain the medical image watermark and watermarked image perceptual quality. It also simplifies the algorithm steps in such a way where it optimizes the performance of CSMI-AW with a proper balance between robustness, computational cost, and embedding efficiency. However, the CSMI-AW system not only ensures a higher degree of security for digital x-ray images but also attains better computational performance with higher perceptual quality of the watermark. It attains approximately 40% better outcome of PSNR and 60% computational improvement. The comparative performance analysis for the traditional approaches shows its effectiveness in terms of computational efficiency and peak-signal-to-noise ratio (PSNR).

Keywords: Medical image watermarking · Spectral domain · Bit-plane extraction · Optimized security performance · Perceptual quality

R. Silhavy et al. (Eds.): CoMeSySo 2021, LNNS 232, pp. 954–965, 2021.
https://doi.org/10.1007/978-3-030-90318-3_76

1 Introduction

The modern era of high-speed computer networks has witnessed tremendous data explosion and growth irrespective of its type. The demand for sharing the data or content over various networking platforms has created different opportunities for businesses. It means an inclination of a broader spectrum in new business ideas correspond to scientific exploration, entertainment, and social opportunities [1, 2]. The prime reason behind the growth in various business opportunities also lies in demand for digitally formatted data, as digital media offers various advantages compared to analog media, especially when it comes to dynamic data of both textual and image formats [3]. However, usage of images is more than text in the collaborative networks, thereby protecting the transmitted digital image data plays a vital role from the security viewpoint as it can consist of essential information corresponding to one individual or one organization. Thereby to provide adequate copyright protection along with ownership identification in the context of data integrity image watermarking techniques are widely used over various internet and multimedia applications [4, 5]. Digital image watermarking refers to the embedding procedure of a stream of bits correspond to a signal into an image file. It has wide range of applications in broadcast monitoring, owner identification and proof of ownership etc. [6]. Here in the context of image watermarking the host file is referred as "asset" whereas the bit stream to be embedded is referred as "message".

However, its deployment also serves the purpose of content ownership when it comes to medical images such as chest x-ray, brain MRI etc. The prime underlying motive here is to protect the patient image data with proper content authorization and authentication so that it should not get mishandled with malicious intention [7]. Thereby in medical diagnostic systems also- scope of implementing digital image watermarking techniques restrict the data duplication and falsification by incorporating effective copyright protection aspect [8, 9]. The traditional approaches of digital image watermarking techniques in the existing system mostly adopts two popularly known methods such as either spatial domain watermarking or spectral domain watermarking, however both have got their respective strength factors and limitations. The prime advantage of Discrete Cosine Transform (DCT) in spectral domain based approach is- the watermarking approach with DCT computing factors employ easier steps of sequential work-flow modeling which enhances the computational efficiency of the system also balance the trade-off between robustness and perceptual quality of the watermark image. On the other hand frequency domain based spectral analysis oriented watermarking mostly lack efficiency when it comes to cost of computing also affects the invariance properties of the system. But DCT based watermarking approach outperforms the LSB based approach as LSB based approaches are not much robust against different form of attacks. Frequency domain watermarking involves computation of the DCT/DWT of the pixel matrix of both the image as well as the watermark to be embedded on it. Then both the DCT/DWT coefficients are scaled and added [10–12]. Finally the resulting DCT/DWT coefficients are subjected to IDCT/IDWT computation. It can be observed that in frequency domain watermarking, lot of complex logic is involved but in this proposed study the DCT based watermarking approach is simplified with ease of execution flow. Keeping the limiting factors of Discrete Coefficient Transformation (DCT) based approaches, the study designs and develops a system namely computational system for medical image authentication using watermarking (CSMI-AW) to perform efficient watermarking of medical images specially chest x-ray images in order to protect the integrity of the patient information correspond to the x-ray image data. The simplified execution work flow makes the system robust against various potential attacks such as noise, cropping and scaling and also lower the degradation of the image quality of both watermarked image and watermark image which is also referred as message image. The system design corresponds to CSMI-AW modeled with numerical modeling and computing and designed taking the spectral domain watermarking concept in reference, however the CSMI-AW system model from the design viewpoint differs from the traditional approach of spectral domain watermarking as it incorporates a modified version of DCT extraction and substitution phase with less execution steps. Its performance is assessed compared to the existing system of spectral-domain watermarking techniques. The outcome shows that the proposed solution of CSMI-AW accomplishes better image perceptual quality on the measure of PSNR with reduced computational effort and ensures robustness against different forms of security threats.

Section 2 discusses about the existing research work followed by problem identification in Sect. 3. Section 4 discusses about proposed methodology followed by elaborated discussion of algorithm implementation in Sect. 5. Comparative analysis of accomplished result is discussed under Sect. 6 followed by conclusion in Sect. 7.

2 Related Work

Information verification is one of the essential imperatives in present-day correspondence and networking systems. In the context of image processing, information verification is executed by utilizing watermarking procedures. Whatever the method of watermarking, one of the significant elements to be considered is robustness, which indicates the defensive factor against different forms of attacks. The study by (**Megalingam et al. 2010**) [13] examines an original vigorous technique for advanced watermarking in spectral domains. Here, the strategy manages an image in the spectral space, watermarked at various intensity-based subsections.

Furthermore, the study also scrutinized the performance of the proposed strategy by executing the technique utilizing Verilog HDL and Matlab. A correlation of the proposed novel spectral space strategy with the spectral method is likewise given here. They have similarly analyzed the PSNR that can be computed utilizing the proposed spectral space watermarking and the DCT/IDCT based watermarking. The study of (Goli and Naghsh 2017) [14] on the other hand, proposes an unique approach of watermarking which is robust against crop attack. The technique is designed and developed taking two-step approach of Sudoku into consideration. The experimental outcome shows that the defensive measure by the presented approach to mitigate the possibilities of crop attack increases by 98.8%. The study by (Zhen et al. 2012) [15] introduced a novel approach of DCT-based digital watermarking algorithm which ensures better defensive approach against various attacks and also poses better watermarking capability as compared to existing system. However, the study does not talk about its applicability on medical image watermarking aspects. There are many authors who have mostly emphasized on the spatial domain watermarking such as [16–21] but very few incorporated simplified techniques of embedding watermarking. However, out of these studies, the study by (Wong et al. 2013) [20] introduced a novel watermarking scheme for digital mammogram authentication.

3 Problem Description

Although the review and analysis of the conventional related literature show that a considerable effort has been laid towards improving the digital image watermarking in computational performance, security, and imperceptibility aspects, the contribution of the existing approaches limits the analysis to the usual images only. However, minimal literature talks about the medical image watermarking opportunities. Also, the traditional approaches based on spectral analysis involve complex procedural execution and do not incorporate optimized workflow modeling with simplified steps. On the other hand the traditional techniques of spatial watermarking lack basic robustness and vulnerable to various attacks such as noise cropping and scaling. It is also observed that most of the traditional approaches for digital image watermarking involve hardware dependencies which limit the scope of optimizing the computational performance of the spectral domain-based algorithms. Hence, the problem statement is "*Designing a spectral domain-based optimized medical image watermarking technique is challenging aspect for balancing the problems of computational performance efficiencies and perceptual quality over constraint resources and different forms of security attacks*".

4 Proposed Methodology

The underlying core design principle of the study exhibits an analytical methodology for privacy preservation by means of medical image watermarking. The study also adopts the core principle and strength factors of digital image watermarking where functionally a hidden stream of bits are embedded into another signal in the form of a file in a robust and invisible way.

4.1 CSMI-AW: Load $I_x \rightarrow I_{Asset}$: Chest x-Ray

In the proposed study context the analytical modeling and algorithm design for watermarking is implemented exclusively for medical chest x-ray images (I_x). It can be represented also as asset image for digital watermarking purpose and can be shown as below.

$$I_x \rightarrow I_{Asset} : \forall I_x \in I_m \quad (1)$$

Where the formulated approach targets to preserve the authenticity of I_x data while transmitted thorugh a medium/channel components. The proposed CSMI-AW is designed in a way where initially it locates the medical chest x-ray image I_x into a repository of patient database ($I_m \rightarrow pDB[]$) which consists of different set of I_m. After locating the corresponding I_x in a particular index of $pDB[]$), it returns two distinct attributes I_n, $I_p \in I_x$. Here I_n corresponds to the name of the particular x-ray image and I_p belongs to the path to locate the file within $pDB[]$. It also performs concatenation of $I_{loc} \leftarrow [I_n \| I_p]$ to generate the location of that particular asset image.

The system CSMI-AW here applies a function $f_1(x) \rightarrow I_{Asset}(I_{loc})$ which performs digitization of the asset image file I_x both spectrally and in amplitude. Further it computes its corresponding numerical matrix form for the ease of analysis. The I_{Asset} chest x-ray after the processing stage of digitization forms a matrix of pixel brightness values and can be represented as

$$I_{Asset} = \begin{bmatrix} I(1,1)\ldots I(1,c) \\ \vdots \quad \vdots \quad \vdots \\ I(r,1)\ldots I(r,c) \end{bmatrix} (r \times c \times d)$$

In numerical form of the asset image here $r \rightarrow$ the number of rows in the matrix and $c \rightarrow$ number of columns in the matrix and d represents the dimension of the image matrix. The system of CSMI-AW is mechanized in a way where it checks for the dimension of I_{Asset} chest x-ray image and if it is found to be more than 2-dimensional then to reduce the computational effort of the watermarking, the system applies another function $f_2(x) \rightarrow I_{Asset}$ to reduce the dimension of the image with the complexity upper bounded with $O(d^2)$ from$O(d^3)$. This approach basically speed up the process of bit-plane extraction in uint8 format and also watermarking which means the number of bits to preserve the numerical value of each pixel is : $2 \leq b_{depth} \leq 8$. The lowest pixel value of the matrix I_{Asset} is become 0 and highest pixel value becomes 255 for the ease of computation.

4.2 CSMI-AW: Load $W_m \rightarrow W_{message}$

The study considers another image in the form of message which is $W_{message}$ which could consist of unique data of coded or straight arrange of bits in the form of another image. Here both the asset and message images of chest x-ray are computed for size of $(r \times c)$ in the context of CSMI-AW. The system also computes the 8-bits form of both the I_{Asset} and $W_{message}$ for the approximation of bits of pixel attributes.

4.3 CSMI-AW: Embedding Procedure in Spectral Domain

The system initially computes the bit-planes correspond to both asset chest x-ray image and message image with bit depth of 8 bits. The study here design and develop another optimized functional module $f_{DCT2}(x) \rightarrow I_{Asset}$ which acts as a splitter object and compute planes for every corresponding bit of pixel entity. The computation and digitization of both I_{Asset} and $W_{message}$ takes place in such a way where for asset image 8-bit image the range of pixel ranges between 0 to 255. To handle the memory constraints here CSMI-AW initially converts the asset into grayscale image of size $(r \times c)$ and also computes the unsigned integer format sequence of the bit-sequence to handle the memory constraints associated with the asset and message image. In the spectral domain the formulated approach of CSMI-AW explores the frequency domain properties correspond to the asset image of chest x-ray where coefficients are slightly altered. This approach of CSMI-AW makes changes in the I_{Asset} and $W_{message}$ in a way where the modified changes cannot easily be visually perceived. This approach in the CSMI-AW also makes the system more defensive against different form of attacks as compared to the traditional spatial domain methods. The study applies $f_{DCT2}(x) \rightarrow I_{Asset}$ on the asset chest xray image and bring the image into frequency domain matrix form. The functional module basically implements the following Eq. (1) for the DCT computation of the asset image which is shown as follows.

$$y(k) = w(k) \sum_{i=1}^{N} x(i) \times \cos(\frac{\pi}{2N}(2i-1)(k-1)) \tag{2}$$

Here the N belongs to the length of the asset image and x and the y is of the same size. Here x is represented with digitized matrix of the asset image and DCT is applied here with column-wise orientation. Here k ranges between 1 to N. The system here also incorporates another functional term which is w (k) for the ease of computation of DCT in frequency domain. The function w (k) returns $1/\sqrt{N}$ when the k value is considered 1 else it returns $\sqrt{2/N}$ if the values range between 2 to N. The study computed DCT for total 5 number of test images as shown in the following Table 1.

Table 1. DCT Computation outcome of sample chest x-ray: asset Images

	Test Chest X-ray Images				
I_{Asset} : **Original form** (1024x1024)	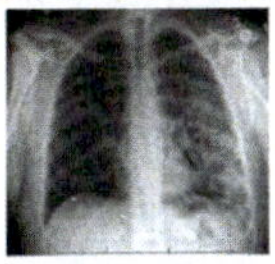	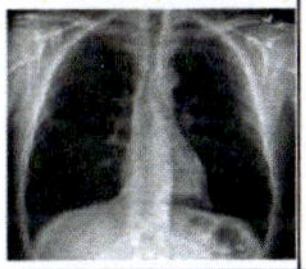	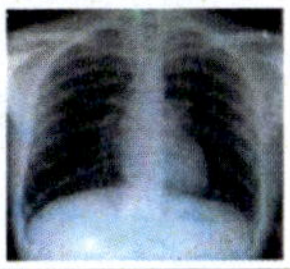	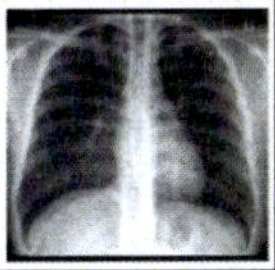	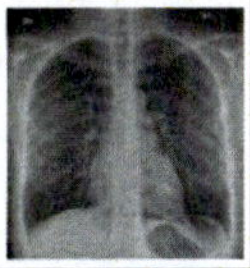
	Sample-1	**Sample-2**	**Sample-3**	**Sample-4**	**Sample-5**
DCT Outcome (1024x1024)					

After the DCT computation the output matrix of asset image result in a matrix of same size as the original form but the numerical values here correspond to the class of double. The DCT computation outcome interpretation shows that the absolute values related to the low frequency coefficients are higher and appear in the up-left corner of the $\boldsymbol{I_{Asset}}$, whereas high frequency coefficients appear in the down-right corner of the asset chest x-ray images. Here further the $\boldsymbol{W_{message}}$ is also coded into a spread spectrum sequence. The prime potential aspect of DCT computation is it makes the watermark message of spread spectrum sequence safe from many attacks where one of the primary attacks is JPEG compression. It also makes the message visually imperceptible. The study emphasizes on designing the watermarking procedure with simplified computational steps to execute the work-flow faster, so that it overcomes the limitations associated with the existing DCT based transformation procedures. The following Fig. 1 shows the embedding watermarking procedure involved in CSMI-AW where the asset image undergoes through DCT transformation as discussed above and produce DCT matrix of frequency coefficients which later embeds the $W_{message}$. Here the Inverse DCT basically produces the watermarked image with higher imperceptibility and with faster execution.

The proposed CSMI-AW block-based design of watermark embedding is shown as follows:

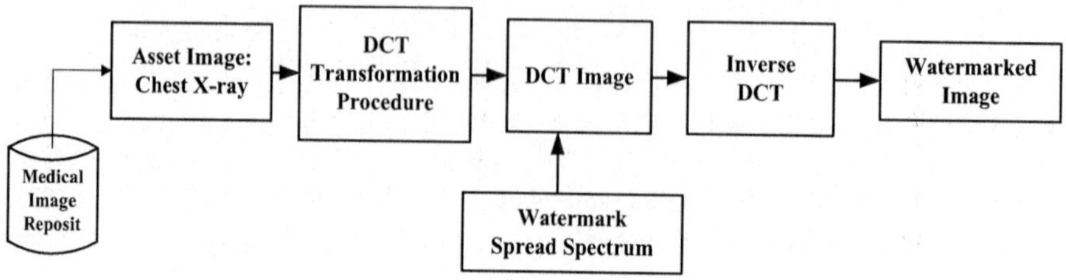

Fig. 1. CSMI-AW watermarked image generation procedure

Here the proposal for CSMI-AW is made in such a way where it also performs extraction of $W_{message}$ and performs a validation check **as** a part of authentication of the I_{Asset}. The extracted watermark is also further compared with the original watermark spread spectrum to analyze the performance pertaining to PSNR and the robustness against attacks. The process of $W_{message}$ extraction and validation step can be visualized with the following block based representation (Fig. 2).

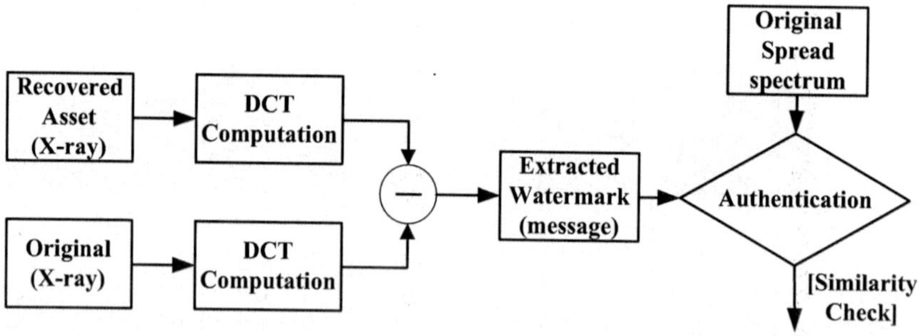

Fig. 2. CSMI-AW authentication procedure with security validation

The next part of the study shows the proposed algorithm workflow modeling with mathematical steps of execution flow. The study also performs experimental analysis process to observe the outcome of the CSMI-AW watermarking process in the consecutive phases.

5 Algorithm Implementation

The proposed algorithm is analytically modeled in a way where it considers a set of parameters for step-by-step execution flow in the design and modeling. The step by step algorithm modeling is presented as follows.

CSMI-AW: Algorithm for Watermarking Procedure

Input: $[I_{Asset}\,(r \times c)\,,\, W_{message}\,]$
Output: $W2_{message}$ []
Start

1. Initialize: $[I_{Asset}\,(r \times c)\,,\, W_{message}\,]$
2. Load → I_{Asset}
3. $f_1(x) \rightarrow I_{Asset}\;(I_{loc})$ where $I_{loc} \leftarrow [I_n \,||\, I_p]$
4. *Define:* w ←Size($W_{message}$)
5. Apply: $f_2(x) \rightarrow I_{Asset}$
6. Compute: D← $f_{DCT2}(x) \rightarrow I_{Asset}$
7. Reorder DCT values
8. $W_{message} \leftarrow f_3(1, w)$
9. Ix2 ← Ix(2,w+1)
10. For k ← 1 to w
    a. x ← Ix2(k)/r+1 // provides the associated Col numerical in image
    b. y ← mod(Ix2(k),r) // associated row based numerical in image
    c. Compute: row-column order
11. End
12. For k ← 1 to w
    a. $I_{Asset}\,(y, x) = I_{Asset}\,(y, x) + .1 \times I_{Asset}\,(y,x) \times w(k)$
13. End
14. $Inv(I_{Asset}) \leftarrow f_{IDCT2}(x) \rightarrow I_{Asset}$
15. *Perform* Optimized $W_{message}$ extraction and authentication process

End

The analytical modeling of the formulated CSMI-AW watermarking approach is exclusively carried out in a numerical computing environment which considers a set of medical chest x-ray images as asset image. The algorithm initially takes the asset and watermark spread spectrum as inputs and process the inputs for sequential operations. It basically initializes both the input parameters in (**Line-1**). After initializing the entities, the system further load the asset image by incorporating a functional module in (**Line-2 to Line-3**). Here the system basically performs the digitization of the image to normalize the pixel entities within o to 255 ranges for the ease of computation. The system further also computes the size of watermark spread spectrum and stores it into a variable w in (**Line-4**). And further it applies the dimensional reduction of the image to reduce the complexity of data with another function in (**Line-5**). The system further computes the DCT values correspond to the asset image and store it into a vector D[] in (**Line-6**) followed by reordering the DCT values and computation of the spread spectrum in (**Line-7 and Line-8**). Further (**Line-9–Line-11**) the system defines the workflow to compute the row-column order of vector coefficients for the image. Further the system model insert the values of the $W_{message}$ into I_{Asset} with the proposed embedding process and computation and generates the watermarked image of I_{Asset} with inverse IDCT2 (**Line-12 to Line-14**). Finally the CSMI-AW system here applies extraction process to extract the $W_{message}$ spread spectrum sequence from the watermarked image and compare that with the original watermark spectrum in the DCT domain to validate the authenticity of the spread spectrum embedded in the chest x-ray image with optimized computational steps. The similarity score is checked with the following Eq. (3)

$$S(W, W^*) = \frac{W^*.W}{\sqrt{W^*.W^*}} \tag{3}$$

The next segment of the study further discusses the experimental outcome obtained after simulating the CSMI-AW in a numerical computing environment.

6 Results Discussion

The simulation study has been performed considering five different chest x-ray images of different dimension and size which are further scaled to a specific dimension. The study perform comparison with the existing related system in terms of two distinct performance parameters which are PSNR in dB and Processing time in seconds. The study considers (Bamatraf et al. 2011) [22] as a baseline approach for comparison. In the first place the execution time of the watermarking procedure is compared for both the proposed and the existing approach which is further presented in the following Table 2 and the visualized with Fig. 3.

Table 2. Numerical outcome represented for computational time complexity

Images	Proposed	Existing system
Image-1	0.674773	4.1123
Image-2	0.295128	5.2342
Image-3	0.28592	1.23452
Image-4	0.297375	1.23111
Image-5	0.295656	1.14555

The numerical outcome correspond to the processing time shows that the proposed approach of CSMI-AW attains comparatively better outcome as compared to the existing approach during the embedding procedure of watermark spread spectrum. The outcome also visualized with the following Fig. 3.

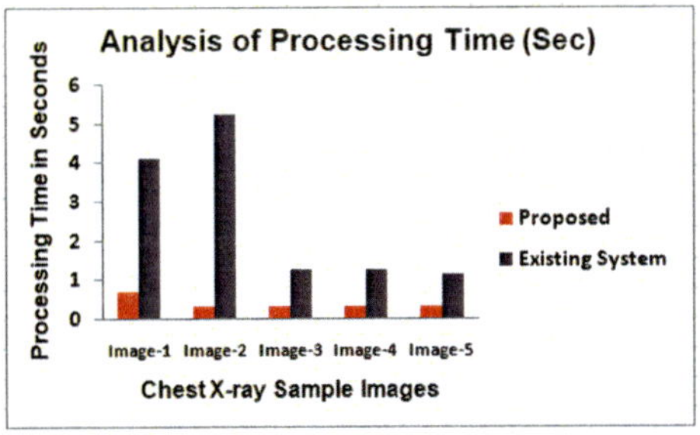

Fig. 3. Analysis of comparable outcome correspond to processing time

A closer interpretation of the outcome in Fig. 3 also shows that the system accomplishes very lesser processing time for five different medical chest x-ray images whereas in the case of existing system the computational time complexity is quite higher. The prime reason behind this is the proposed approach optimizes the computational steps in DCT domain and also reduces the size of the image in the domain of grayscale which result in better computational performance. The study also performs comparison on the basis of PSNR outcome for different images which are numerically represented as follows (Table 3):

Table 3. Numerical outcome represented for PSNR (dB)

Images	Proposed	Existing
Image-1	85.2186	54.5961
Image-2	83.5272	54.6673
Image-3	80.8397	54.6216
Image-4	81.1183	54.6899
Image-5	89.6421	54.695

The analysis of PSNR outcome corresponds to the watermarked image shows that it is quite higher in the case of the proposed system. The maximum value of PSNR obtained is 89.6421 in the proposed system and also the CSMI-AW outperforms the existing approach by approximately 40% performance improvement in the PSNR. The PSNR outcome is visualized with the following Fig. 4.

The interpretation of the outcome correspond to PSNR also shows that the proposed approach attains higher PSNR, the prime reason behind that is it applies the watermarking in DCT domain, on the other hand the existing system adopts LSB based approach in

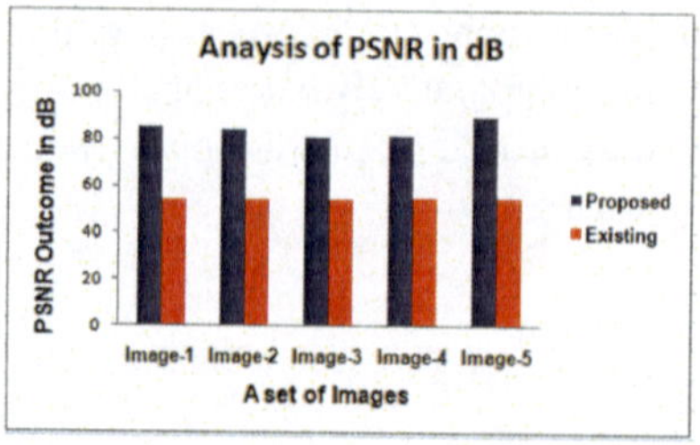

Fig. 4. Analysis of comparable outcome correspond to PSNR

the spatial domain and lacks efficiency not only in the context of PSNR but also do not ensure robustness against various kind of attacks. Thereby, based on the performance it can be said that the proposed watermarking solution of CSMI-AW not only attains better PSNR with higher imperceptibility but also ensure robustness against various forms of attacks with ease of execution.

7 Conclusion

Introduces a novel approach namely CSMI-AW to perform medical image watermarking in spectral domain with authentication procedure of the watermark spread spectrum sequence. The system is analytically modeled in a way which simplifies the execution workflow with optimized DCT computation and numerical analysis. The experimental outcome of the study shows that the proposed CSMI-AW is computationally efficient and its PSNR outcome is considerable to justify its applicability towards resisting various types of image attacks with high imperceptibility. The future work further aims to extend the authentication procedure with the extensive analysis of extracted watermark in the spectral domain.

References

1. Mousavi, M., Naghsh, A., Abu-Bakar, S.A.R.: Watermarking techniques used in medical image: a survey. J. Digit. Imaging **27**(6), 714–729 (2014)
2. Sunesh, H.K.: Watermark attacks and applications in aatermarking. In: National Workshop-Cum-Conference on Recent Trends in Mathematics and Computing (RTMC) (2011)
3. Shukla, D., Tiwari, N.: Survey on digital watermarking techniques. Int. J. Signal Process. Image Process. Pattern **8**(9), 6–11 (2016)
4. Nangedda, M., Sudharsan, R.A.: Medical image steganography with digital watermarking. Int. J. Adv. Res. Comput. Sci. Softw. Eng. **4**(7) (2014)
5. Gupta, V., Barve, A.: A review on image watermarking and its techniques. Int. J. Adv. Res. Comput. Sci. Softw. Eng. **4**(1) (2014)
6. Goli, M.S., Naghsh, A.: A comparative study of image-in-image steganography using three methods of least significant bit, discrete wavelet transform and singular value decomposition. Bull. de la Société Royale des Sci. de Liège **85**, 1465–1474 (2016)
7. Saneei, S., Naghsh, A.: Introducing a new method of robust digital image watermarking against cropping and salt & pepper noise using Sudoku. Majlesi J. Multimedia Process. **4**(4) (2015)

8. Khalid, S.K.A., Deris, M.M., Mohammad, K.M.: Anti cropping digital image watermarking using Sudoku. Int. J. Grid Utility Comput. **4**(2/3), 169–177 (2013)
9. Cox, I.J., Kilian, J., Leighton, T., Shamoon, T.: A secure, robust watermark for multimedia. In: Workshop on Information Hiding. University of Cambridge (1996)
10. Fang, Y.M., Huang, J.W., Wu, S.Q.: CDMA-based watermarking resisting to cropping. In: Proceedings of the ISCAS 2004, pp. 25–28 (2004)
11. Aggarwal, A., Singla, M.: Robust watermarking of color image under noise and cropping attack in spectral domain. Int. J. Comput. Sci. Inf. Technol. **2**(5), 2036–2041 (2011)
12. Rawat, S., Raman, B.: A new robust watermarking scheme for color images. In: Proceedings of the IEEE 2nd International Advance Computing Conference, pp. 206–209 (2010)
13. Megalingam, R.K., Nair, M.M., Srikumar, R., Balasubramanian, V.K., Sarma, V.S.V.: Performance comparison of novel, robust spectral domain digital image watermarking with the conventional frequency domain watermarking techniques. In: International Conference on Signal Acquisition and Processing, pp. 349–353 (2010)
14. Goli, M.S., Naghsh, A.: Introducing a new method robust against crop attack in digital image watermarking using two-step Sudoku. In: 3rd International Conference on Pattern Recognition and Image Analysis (IPRIA), pp. 237–242 (2017). https://doi.org/10.1109/PRIA.2017.7983054
15. Zheng, J.Y., Ling, D.H., Liang, J.Z., Jin, M.: A DCT-based digital watermarking algorithm for image. In: 2012 International Conference on Industrial Control and Electronics Engineering, pp. 1217–1220 (2012). https://doi.org/10.1109/ICICEE.2012.323
16. Basu, A., Roy, S.S., Chattopadhyay, A.: Implementation of a spatial domain salient region based digital image watermarking scheme. In: Second International Conference on Research in Computational Intelligence and Communication Networks (ICRCICN), pp. 269–270 (2016). https://doi.org/10.1109/ICRCICN.2016.7813669
17. Basu, A., Sinha Roy, S., Sarkar, S.: FPGA implementation of saliency based watermarking framework. In: 6th International Conference on CODEC. IEEE (2015). ISSN: 978-1-4673-9513-7/15
18. Kutter, M., Petitcolas, F.A.P.: A fair benchmark for image watermarking systems. J. Electron. Imaging **4**, 445–455 (2000)
19. Sinha Roy, S., Saha, S., Basu, A.: Generic testing architecture for digital watermarking. In: Proceedings of FRCCD-2015, pp. 50–58, December 2015. ISBN 978-93-8592-600-6
20. Wong, M.L.D., Lau, S.I.J., Chong, N.S., Sim, K.Y.: A salient region watermarking scheme for digital mammogram authentication. Int. J. Innov. Manag. Technol. **4**(2), 228–232 (2013)
21. Hong, W., Chen, T.S.: A novel data embedding method using adaptive pixel pair matching. IEEE Trans. Inf. Forensics Secur. **7**(1), 176–184 (2012)
22. Bamatraf, A., Ibrahim, R., Salleh, M., Mohd, N.: A new digital watermarking algorithm using combination of least significant bit (LSB) and inverse bit. arXiv preprint arXiv:1111.6727 (2011)

Development of Scientific-Methodological Approaches to Frame Analysis of the Effectiveness of Organizational Systems

S. V. Pronichkin[1](✉), Y. S. Chernyshova[2], A. V. Gayer[2], and I. B. Mamai[2]

[1] Federal Research Center "Computer Science and Control" of Russian Academy of Sciences, Moscow, Russia

[2] Federal Center of Theoretical and Applied Sociology of Russian Academy of Sciences, Moscow, Russia

Abstract. Evaluation of the effectiveness of organizational systems refers to semi-structured decision-making tasks. In such tasks, it is difficult to give numerical estimates of the criteria, as well as to assess the dependencies between them. As a methodological basis, it is proposed to use the verbal analysis of decisions based on frames. Algorithms for information-analytical processing of frames have been developed to assess the effectiveness of organizational systems. During the information and analytical processing of frames, the purpose of processing is set and, therefore, the composition and form of the situational framework, which reflects the vision of the effectiveness of the organizational system. The identification of frames is carried out by comparing the properties of the marked-up data on the activities of the organizational system with the properties of the elements of the approach-result frame. It is proposed to sequentially select from the general data array segments that include links and attributes. Then frames are selected that have associative and cause-and-effect relationships, then frames that include attributes without relationships. Removal of unconnected frames from the segment is carried out, and a semantic network is built on the basis of related frames. A knowledge base has been implemented for storing a set of frames and algorithms for their information and analytical processing to support decision making. #COMESYSO1120.

Keywords: Organizational system · Efficiency · Knowledge base · Frame analysis

1 Introduction

Efficiency issues are of decisive importance in the activities of various organizations [1, 2]. Averaging convolutions used in existing approaches [3, 4], a set of numerical coefficients, as well as estimated parameters, significantly complicate the analysis of the integral criterion of effectiveness and complicate the explanation of the calculated values for decision makers (DM). Therefore, the reliability of the results obtained for assessing the effectiveness of organizational systems cannot be considered sufficient

R. Silhavy et al. (Eds.): CoMeSySo 2021, LNNS 232, pp. 966–970, 2021.
https://doi.org/10.1007/978-3-030-90318-3_77

for program-targeted analysis [5, 6]. The program-target orientation of organizational systems, first of all, is manifested in the structure of the distribution of the scale of this activity by the objects of goal-setting [7, 8]. For scientific organizational systems, the distribution of the scale of research activities by goal-setting objects, the priority of costs by areas of activity, it is proposed to assess the appropriate structure of resources spent on research and development [9, 10].

An important characteristic of the process of production of scientific knowledge is the saturation of the acquired knowledge of potential and real consumers of this knowledge, that is, indicators of the intensity of knowledge production, indicating the achieved degree of provision of the external environment with scientific research. Methodological defects of existing quantitative approaches to multicriteria analysis of scientific organizational systems are one of the limitations on the path of effective predictive and analytical substantiation of management decisions, large-scale innovation projects and scientific and technical programs, as well as the formation of intellectual potential.

2 Methods

Multi-criteria analysis of the effectiveness of organizational systems consists of two stages. At the first stage, a decision-making model is built within the framework of identifying the preferences of the decision-maker. At the second stage, the structure of classes of organizational systems with boundary objects for each class of efficiency is determined. The task of assessing the effectiveness of organizational systems is a semi-structured decision-making task [11, 12]. In such problems, it is difficult to give numerical estimates of the criteria, as well as to assess the dependencies between them. Inaccuracies arise in the definition of classes, as well as in their number and boundaries. As a methodological basis, it is proposed to use the verbal analysis of decisions based on frames [13, 14]. It is obvious that a decision maker or an expert can make mistakes and contradict their preferences.

It is proposed to use methods for checking the consistency of verbal information received from a decision maker with previous answers at each iteration of the decision-making process [15]. Thus, multi-criteria analysis of the effectiveness of organizational systems becomes inter-active. The procedure for identifying the preferences of the decision maker is built into the methods and organized through a dialogue with the decision maker. At each iteration, the DM assigns one specially selected frame to a class. Based on the decision maker's answer, some organizational systems are classified automatically. The procedure is repeated until each frame is assigned to a certain efficiency class. The consistency of the information received from the decision maker with the previously received one is checked. In case of any contradictions, the decision maker revises his answers. At the end of the classification algorithm, the decision maker can obtain a meaningful information portrait of each organizational system in the efficiency class.

3 Results

Algorithms for information and analytical processing of frames have been developed to assess the effectiveness of organizational systems. In the process of information-analytical processing of frames, the purpose of processing is set and, consequently, the composition and form of the situational framework, which reflects the vision of the effectiveness of the activity of the organizational system. Identification of frames is carried out by comparing the properties of the marked data on the activities of the organizational system with the properties of the elements of the approach-result frame. It is proposed to sequentially select from the general data array segments that include links and attributes. Then frames are allocated that have associative and cause-and-effect relationships, then frames that include attributes without relationships. Removal of unconnected frames from the segment is carried out, and a semantic network is built on the basis of related frames.

The following designations are used: D - frame identifier; C - excellence of the approach; E - completeness of the approach; O - assessment and revision of the approach; L - associative and cause-and-effect relationships; P^O - achievements; P^L - completeness of coverage.

The algorithm consists in sequential selection of elements:

$S_1 \subseteq S \sim \langle L, \otimes \rangle$, where $\otimes$ - empty set;

$S_2 \subseteq (S \cap S_1) | S_2 \sim \langle \{\bar{L} \cup \bar{O}\}, R_{LS} \rangle;$

$S_3 \subseteq (S \cap S_2) | S_3 \sim \langle \{\bar{L} \cup \bar{O} \cup \bar{P}\}, \left\{R_{LS} \cup R_C \cup R_E \cup R_O^{O|R_{LS}(O,L)}\right\} \rangle;$

$S_4 \subseteq (S \cap S_3) | S_4 \sim \langle \{\bar{L} \cup \bar{O} \cup \bar{P}\}, \left\{R_{LS} \cup R_C \cup R_E \cup R_O^{O|R_{LS}(O,L)} \cup R_{PP}^{O|R_{LS}(O,L)} \cup R_O^L\right\} \rangle;$

$S_5 \subseteq (S \cap S_4) | S_5 \sim \langle \{\bar{L} \cup \bar{O} \cup \bar{P}\}, \left\{R_{LS} \cup R_C \cup R_E \cup R_O^{O|R_{LS}(O,L)} \cup R_{PP}^{O|R_{LS}(O,L)} \cup R_O^L \cup R_{LO}\right\} \rangle;$

$S_6 \subseteq (S \cap S_5) | S_6 \sim \langle \{\bar{L} \cup \bar{O} \cup \bar{P}\}, \left\{R_{LS} \cup R_C \cup R_E \cup R_O^{O|R_{LS}(O,L)} \cup R_{PP}^{O|R_{LS}(O,L)} \cup R_O^L \cup R_{LO} \cup R_O^{O|R_{LO}(O,L)} \cup R_{PP}^{O|R_{LO}(O,L)}\right\} \rangle,$

where $\bar{O}^{R_{LO}} \rightarrow P^{R_O^L};$

$S_6^b \subseteq (S \cap S_5) | S_6^b \approx \langle \{\bar{L} \cup \bar{O} \cup \bar{P}\}, \left\{R_{LS} \cup R_C \cup R_E \cup R_O^{O|R_{LS}(O,L)} \cup R_{PP}^{O|R_{LS}(O,L)} \cup R_O^L \cup R_{LO} \cup R_O^{O|R_{LO}(O,L)} \cup R_{PP}^{O|R_{LO}(O,L)}\right\} \rangle,$

where $\bar{P}^{R_O^L} \rightarrow O^{R_{LO}};$

$S_7 \subseteq S | S_7 \approx \langle \left\{\bar{O}^{O|\forall o \in O, o = \Theta} \cup \bar{P}\right\}, \left\{R_C \cup R_O^{O|\neg(R_{LO}(O,L) \vee R_{LO}(O,L))} \cup \neg R_{PP}^{O|\neg(R_{LO}(O,L) \vee R_{LO}(O,L))}\right\} \rangle,$

where Θ - empty object;

$S_8 \subseteq S | S_8 \approx \langle \{\bar{O} \cup \bar{P}\}, \left\{R_C \cup R_E \cup R_O^O \cup R_{PP}^{O|\neg(R_{LO}(O,L) \vee R_{LO}(O,L))}\right\} \rangle;$

$S_9 \subseteq S | S_9 \approx \langle \{\bar{O} \cup \bar{P}\}, \left\{R_C \cup R_E \cup R_O^O \cup R_{PP}^{O|\neg(R_{LO}(O,L) \vee R_{LO}(O,L))}\right\} \rangle,$

where the target element matches the property $\overline{O} \rightarrow P$.

The result of the algorithm is a frame $\overline{\overline{S}} = S_6 \cup S_6^a \cup S_6^b \cup S_7 \cup S_8 \cup S_9$.

In the absence of associative links, an algorithm has been developed that allows the absence of one of the properties. For a more detailed analysis of such components of the frame as the completeness of the approach, its assessment and revision, it is necessary to restructure the frames. The restructuring is carried out taking into account the links and the resulting attributes. The restructuring itself consists in the selection of new frames from the existing ones based on the common values of the attributes.

4 Discussions

In the problems of classification of organizational systems, decision makers are only interested in a certain set of effective systems. It is possible to reduce the number of calculations if the set of real frames is much smaller than the set of frames obtained as the Cartesian product of the criteria scales. However, frames obtained as a Cartesian product of the criteria scales are still used if they have a higher information content compared to the identified frames. Variation in the number of indirectly classified multi-attribute frames significantly reduces the number of computations. Only a given set of organizational systems is classified and the relative information content is taken into account, but not absolute. At the same time, an unresolved dilemma remains. The effectiveness of the method decreases significantly with an increase in the number of criteria or gradations, and with an increase in the number of classes, the effectiveness of the method increases.

It is proposed to use a multi-method technology for multicriteria assessment of organizational systems based on methods for reducing the dimension of big data. It was found that the main problem of the existing methods of dimensionality reduction is the sparseness of big data arrays. To reduce the size of big data arrays of semi-structured data, a more efficient approach based on typing is proposed, which takes into account the strengths and weaknesses of known methods of extraction and abstraction. Typification of mapping of visualized large amounts of data is carried out by introducing a reference multiset. The generalized characteristic of the proximity of each multiset to the reference multiset is calculated based on various optimality principles.

5 Conclusion

The developed algorithms for information and analytical processing of frames were used as a scientific and methodological basis for the implementation of expert mechanisms for assessing the effectiveness of the activities of scientific organizations. Software has been developed for assessing the effectiveness of the activities of scientific organizations. There are three categories of efficiency - process, internal and external. Process efficiency reflects approaches to ensuring the quality of activities, resource potential, its support and improvement. Internal and external effectiveness reflect how effectively the potential is used, respectively, by internal and external stakeholders.

A knowledge base has been implemented for storing a set of frames and algorithms for their information and analytical processing to support decision making. Using an ontology that describes a system of concepts and a declarative model that takes into account the relationships between concepts and constraints of integrity, a knowledge base model has been developed for assessing the activities of scientific organizations. The knowledge base consists of 40 tables, relationships between them and stored procedures that ensure the integrity of the knowledge base and algorithms for information-analytical processing of frames.

Acknowledgments. The article is prepared with the financial support of the Russian Science Foundation, project № 19-78-10055.

References

1. Konig, W.: Energy efficiency in industrial organizations - a cultural-institutional framework of decision making. Energy Res. Soc. Sci. **60**, 314–322 (2019)
2. Rayeni, M., Saljooghi, F.: Network data envelopment analysis model for estimating efficiency and productivity in universities. J. Comput. Sci. **6**, 1252–1257 (2010)
3. Inua, O., Maduabum, C.: Performance efficiency measurement in the Nigerian public sector: the federal universities dilemma. Mediterr. J. Soc. Sci. **5**, 838–847 (2014)
4. Gimenez, V., Martinez, J.: Cost efficiency in the university: a departmental evaluation model. Econ. Educ. Rev. **25**, 543–553 (2006)
5. Ramon, N., Ruiz, J., Sirvent, I.: Two-step benchmarking: setting more realistically achievable targets in DEA. Expert Syst. Appl. **92**, 124–131 (2018)
6. Ruiz, J., Sirvent, I.: Common benchmarking and ranking of units with DEA. Omega **65**, 1–9 (2016)
7. Aparicio, J., Pastor, J.: Closest targets and strong monotonicity on the strongly efficient frontier in DEA. Omega **44**, 51–57 (2014)
8. Dai, X., Kuosmanen, T.: Best-practice benchmarking using clustering methods: application to energy regulation. Omega **42**, 179–188 (2014)
9. Ruiz, J., Segura, J., Sirvent, I.: Benchmarking and target setting with expert preferences: an application to the evaluation of educational performance of Spanish universities. Eur. J. Oper. Res. **242**, 594–605 (2015)
10. Daraio, C., Bonaccorsi, A., Simar, L.: Rankings and university performance: a conditional multidimensional approach. Eur. J. Oper. Res. **244**, 918–930 (2015)
11. Lin, Y., Jun, Z., Zhanfang, F.: A method of extracting the semi-structured data implication rules. Procedia Comput. Sci. **131**, 706–716 (2018)
12. Aparicio, J., Cordero, J., Pastor, J.: The determination of the least distance to the strongly efficient frontier in data envelopment analysis oriented models: modelling and computational aspects. Omega **71**, 1–10 (2017)
13. Larichev, O., Asanov, A., Naryzhny, Y.: Effectiveness evaluation of expert classification methods. Eur. J. Oper. Res. **138**, 260–273 (2002)
14. Larichev, O., Kortnev, A., Kochin, D.: Decision support system for classification of a finite set of multicriteria alternatives. Decis. Support Syst. **33**, 13–21 (2002)
15. Larichev, O.: Ranking multicriteria alternatives: the method ZAPROS III. Eur. J. Oper. Res. **131**, 550–558 (2001)

The Technology of Spatial Relations Visualization Based on the Analysis of Natural Language Texts

A. V. Vicentiy[1,2](✉) and M. G. Shishaev[1,2]

[1] Institute for Informatics and Mathematical Modeling – Subdivision of the Federal Research Centre "Kola Science Centre of the Russian Academy of Science", 24A, Fersman Street, Apatity, Murmansk Region 184209, Russia

[2] Apatity Branch of Murmansk Arctic State University, Lesnaya Street 29, Apatity, Murmansk Region 184209, Russia

Abstract. This paper considers the problem of recognizing information about the spatial relations of geographical objects in natural language texts. The proposed technology of spatial relations visualization makes it possible to extract geographical information from unstructured texts and present it in a structured form suitable for use by modern geographical information systems. The extracted information can be used for automated filling and updating the database of geographic information system for generating cartograms that allow visual analysis of the spatial connectivity of arbitrary geographical objects. To get the best result, we combine neural network methods to recognize named entities (geographical objects), a rule-based approach to identify potential spatial relationships, and domain-specific lexical patterns. Since the information about recognized geographical entities and spatial relationships is stored in a standardized structured form, we can use standard geoservices for visualization without additional preparation of geodata in the last stage of the technology. The result of the visualization is a geographical image (cartogram) showing the spatial relations of geographical objects recognized as a result of the analysis of natural language texts.

Keywords: Natural language text analysis · Spatial relationship recognition · Named entity recognition · Geovisualization · Text-to-map conversion · Geographic information systems · Geo-attributed entity

1 Introduction

Natural language processing is currently a hot research area in both scientific and industrial fields of activity. The results of natural language text processing are used in such tasks as text generation, search systems, marketing and sociological research, knowledge extraction from texts, decision support, medical document analysis and others. [1–3] There are also works aimed at the analysis of natural language texts for the purpose of geographical object recognition. [4, 5] However, in our opinion, the topic of

R. Silhavy et al. (Eds.): CoMeSySo 2021, LNNS 232, pp. 971–980, 2021.
https://doi.org/10.1007/978-3-030-90318-3_78

geographical data extraction from texts is not paid enough attention, and the development of new ways to extract geodata from texts is a promising task that will significantly expand the capabilities of modern geographical information systems (GIS).

The tasks of geodata extraction from natural language texts belong to such classes of tasks as Information Retrieval tasks [6, 7], Named Entity Recognition (NER) tasks [8, 9] and Geographical Information Retrieval tasks [10]. As a result of solving such problems, structured information relevant to a certain problem or user query is extracted from a collection of unstructured texts.

The results of geographical context search and, above all, toponyms present in natural language texts can be useful for many different applications, such as geographical search engines, navigation systems, social network user behavior analysis systems and others. One well-known method of searching for toponyms is based on the presence of so-called influential terms in texts. [11] Influential terms are terms that can indicate the presence of geographical references in the text. This method can partially solve the problem of ambiguity in the recognition of place names in natural language texts through the use of a pre-generated list of influential terms in the process of information retrieval of geodata. However, creating heuristics based on the presence of influential terms for automatic toponym detection is challenging. Such heuristics are created manually based on human knowledge and experience. In addition, they are not universal and require modification when the text language changes or the subject area changes significantly.

In the field of geographical information retrieval different methods of geographical content recognition are used [12, 13]. In particular, to solve the problem of disambiguation of multiple locations with the same name, models of recognition of geographical objects based on sets of heuristics are used [10].

Virtually all modern software systems designed to recognize geographical objects in natural language texts are based on more general methods and approaches for the recognition of named entities developed earlier. [14] Currently, three main approaches can be distinguished in the field of named entity recognition. These are Rule-based approach (Rule-base NER), Machine Learning-base NER, Hybrid NER approach [15, 16].

In recent years, the predominant approach is based on the use of machine learning and neural networks. [17] However, rule-based approaches are also used in named entity recognition systems for languages with rich morphology [18, 19], for complex, subject-specific domains, [20–22] as well as in hybrid systems to improve the accuracy of named entity recognition [23].

It can be noted that almost all of the above-mentioned works related to the extraction of named entities from natural language texts deal with point or area objects. In this case, most of the problems are reduced to relatively simple tasks of extracting place names from unstructured text, identifying some location, and associating some text with a certain location by some identifiers, such as geotags. None of the reviewed works proposes methods or approaches for recognizing spatial relationships of named entities.

A comparison of the capabilities of modern frameworks and natural language processing libraries (SpaCy, Google Cloud NLP, Natasha, and others) has shown that they are also unable to recognize spatial relations in texts. Thus, it is currently impossible to perform automatic extraction of spatial relations from natural language texts and the construction of mappings showing the spatial connectivity of various geographical objects based on the use of standard named entity extraction tools.

To solve this problem, we propose an information technology for visualizing spatial relations based on the analysis of natural language texts. This technology allows to recognize both named geographical entities and spatial relations between them by combining several methods and approaches of natural language text processing.

2 Materials and Methods

The technology of spatial relations visualization based on the analysis of natural language texts described in this paper includes two main stages: 1) the geo-attributed entity extraction stage and 2) the geocoding and geo-visualization stage.

By geoassociated entity in this paper we mean both geographic named entities (toponyms) and spatial relations between them. For example, in the sentence "A new highway will be built between Murmansk and Belokamenka." we can distinguish three geoassociated entities: "Murmansk", "Belokamenka", "highway". Moreover, "Murmansk" and "Belokamenka" are named entities of location (toponym) type. Such entities are relatively easy to recognize with existing approaches and software tools. A geo-attributed entity " highway" is an entity of the spatial relationship type. Such entities cannot be recognized using existing approaches and software tools without additional effort. In this work, such additional effort is text processing using a combination of several natural language processing methods, including neural network approach, rule-based approach, lexico-syntactic approach based on original (developed by the authors) lexical patterns corresponding to the subject area.

As the initial data for the analysis we chose texts in Russian that contain references to real geographical objects and the connections between them. Most of these texts are texts devoted to the transport and logistics system of the Arctic Zone of the Russian Federation (AZRF). This choice of texts is related to the fact that the transport and logistics system of the AZRF has been actively developing in recent years, which is the reason for the emergence of new spatial relationships (including transitive ones) between geographical objects, which cannot always be promptly reflected on "official" maps.

In addition, it is necessary to note the seasonal variability of the transport and logistics system of the Arctic zone of the Russian Federation. For example, special winter roads (zimniki) are created in the northern regions of Russia in winter. Such roads can be used only in winter when the temperature is below zero. Therefore, winter roads cease to function normally in the spring. Specific dates of commissioning and termination of use of a particular winter road are usually published on the official websites of regional executive authorities and local governments in the form of text messages. Extracting such information from texts and visualizing it with the help of automated creation of operational connectivity maps of local geographical objects can be useful for notifying local residents about the changes.

Also of practical value in the complex natural and climatic conditions of the Arctic can be operational maps, taking into account the location and timing of ice crossings, seasonal air transportation (including regional helicopter and small aviation transportation), snowmobile and cross-country routes, roads "technical purpose", routes of river and coastal navigation and so on.

As sources of textual information we used texts from official sources (websites of executive authorities and local authorities), Internet news resources, social networks and other available sources. If necessary, technical preparation of the texts was carried out. In the process of technical preparation all texts were converted into.txt format without formatting. A single coding was set for all texts. The single format of the texts greatly simplified their automatic processing.

The choice of methods is due to several factors. One of them is the peculiarities of the Russian language. The Russian language belongs to the group of Slavic languages. Therefore, some methods of named entities recognition, which show good results for the English language, often do not work well for the texts in Russian. The quality of named entities recognition for the Russian language is affected by its features such as free word order, rich inflection, derivation, and others. Because of these features, the recognition of named entities and spatial relations of geoassociated entities is a difficult task for the Russian language [24, 25].

As the main tool for the recognition of named entities in Russian-language texts, implemented on the basis of the neural network approach, we have chosen the Natasha library. Natasha library allows solving all the basic tasks for the Russian language: tokenization, sentence segmentation, morphological tagging, lemmatization, NER tagging, fact extraction, and others. The Natasha library is implemented based on the distillation of Google's BERT language model [26] and shows quality no worse than current state-of-the-art solutions for Russian language news texts. [27] However, as with other neural network-based language models, the accuracy and completeness of named entity recognition using Natasha can decrease when the subject area or type of texts being analyzed changes [28].

In addition, the analysis of the capabilities of existing libraries and frameworks for the recognition of named entities showed that none of the libraries can cope with the task of recognizing the spatial relations of geographical objects. To solve this problem, we have added the lexico-syntactic analysis stage as the second stage of the described technology. This stage allows to improve the quality (accuracy and completeness) of the recognition of geoassociated entities, as well as to recognize indications in the text of the spatial relationships between different geographical objects. For lexico-syntactic analysis we use the Yargy-parser. Yargy-parser implements Earley's syntactic analysis algorithm [29] and is specifically designed to extract structured information from Russian texts based on the use of rules and dictionaries [30].

Despite the relatively high efficiency of the rule-based approach for a well-defined subject area, it is rather difficult to take into account in dictionaries and rules all possible variants of geoassociated entities relevant for visualization of spatial relations. That is why we also supplemented the technology with the stage of text analysis using lexical patterns.

We believe that by using lexical templates we can recognize with a high degree of probability in texts such lexical constructions that describe spatial relations between geographical objects. We distinguish two groups of lexical templates: 1) general lexical templates characteristic for the Russian language [31] and 2) subject-specific lexical templates characteristic for a specific subject area.

The primary test set of lexical templates was created heuristically, using subject matter experts as well as grammatical patterns specific to the Russian language [32] and taking into account the subcategorization frame for each selected lexeme of the lexical pattern [33].

3 Results and Discussion

The overall description of the technology of spatial relations visualization based on the analysis of natural language texts is shown in Fig. 1.

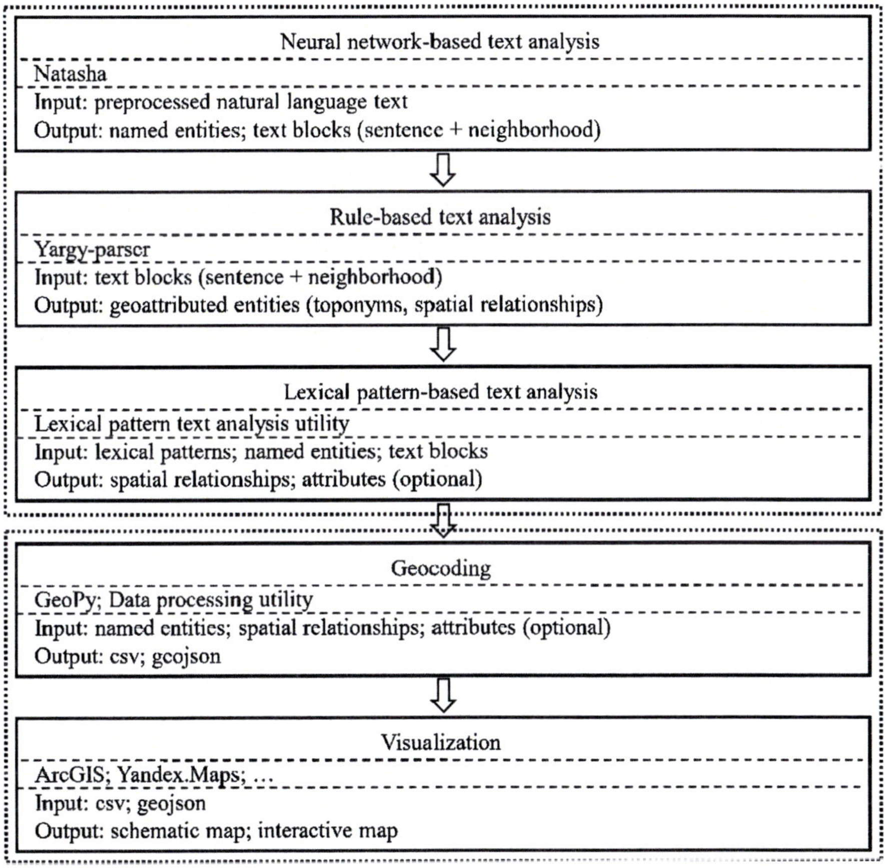

Fig. 1. The overall description of the technology of spatial relations visualization based on the analysis of natural language texts.

Preliminary experiments have shown that at the stage of neural network analysis of the text it is possible to recognize on average not less than 97% of the named entities (toponyms). This level of recognition corresponds to the declared capabilities of the Natasha library that we use.

As a result of the neural network stage of text analysis we not only obtain a list of toponyms, but also save for further processing the sentences in which the named entities were detected, and the neighborhood of these sentences. Such sentences and their neighborhoods are stored in separate text blocks.

By neighborhood, we mean several sentences to the right and/or left of a sentence containing a toponym. Consideration of neighborhood allows us to take context into account in order to recognize geoassociated entities in the text. The larger the neighborhood we choose, the more deeply we can take context into account. On the other hand, the larger the neighborhood we choose, the lower the recognition accuracy of geoassociated entities. We hypothesize that this is due to the fact that if the neighborhood is large enough, contexts that are not relevant to the sentence being analyzed are "captured".

Preliminary experiments have shown that the most effective neighborhood size for most cases is size $(-1; +1)$. This means that we keep one sentence to the left of the sentence with the toponym and one sentence to the right of it. For example, in the usual analysis of the test "*The project will build a new highway in 2021. It will connect Murmansk with the village of Belokamenka*" in the second sentence, the neural network recognizes two toponyms "*Murmansk*" and "*Belokamenka.*". But the second sentence does not contain information that a new road will be built between these settlements. Thus, without taking into account the context (in this case the sentence on the left) it is impossible to establish the type of spatial relationship between these named entities. However, if we set the size of the neighborhood (-1) for this sentence (that is, one sentence to the left), we can find that "*Murmansk*" and "*Belokamenka*" are connected by a spatial relationship expressed by the geoassociated entity "*road*". A similar example can be given for the neighborhood of size (+1): "*It will become easier to get from Murmansk to Teriberka. They will be connected by a new asphalt road.*" It should be noted that such descriptions, when named entities are mentioned in one sentence, and their relationship is mentioned in another sentence, are quite common for texts in Russian. Therefore, the use of neighborhoods for the recognition of geoassociated entities is very useful, provided that the choice of neighborhood size is adequate.

In the second stage we perform syntactic analysis of the text based on rules and dictionaries (gazetteers). Yargy parser, which we use to analyze the text, allows us to use custom grammars and gazettings. In this way we are able to increase the accuracy of the recognition of geoassociated entities. For example, we can recognize such complex toponyms as "Track Estate 9 km of the Luostari-Nikel railroad" and "Settlement 25 km of the Monchegorsk-Olenya railroad", which recognition with the neural network approach causes significant difficulties.

Each text block (sentence + neighborhood) obtained in the previous stage is analyzed separately, within a "local context". This improves the accuracy of the recognition of geoassociated entities of the spatial relation type, as well as identifying the type of spatial relation (road, flight, railroad, and so on.) and its attributes (optional).

In the third stage we perform text analysis based on lexical patterns. To do this we manually create a basic set of lexical patterns corresponding to the subject area in question. Creating patterns is a complex and time-consuming process, since the patterns are created heuristically involving expert knowledge in the subject domain. However, well-crafted lexical patterns allow us to recognize a large number of spatial relationships and their attributes in the text and provide high completeness and accuracy of recognition.

Given the fact that the second and third stages do not process the entire text, but only those blocks of text that were obtained in the first stage, the overall speed of text analysis does not depend much on its initial volume. The most time-consuming stage is the neural network analysis. Thus, despite the presence of three stages of text analysis, the overall analysis speed remains at an acceptable level and allows us to analyze large volumes of text.

To perform the geocoding stage, we use the GeoPy service and a specially developed utility for data transfer and processing. The result of this stage of the technology is a structured file in csv or geojson format containing geographic coordinates and a description of all identified geoassociated entities (toponyms and spatial relationships).

At the visualization stage we use the capabilities of modern geoservices that support csv or geojson formats of geodata transfer. We successfully used such geoservices as ArcGis, Yandex.Maps and Google Maps to perform test geovisualization.

To illustrate the possibilities of the technology of spatial relations visualization based on the analysis of natural language texts discussed in this paper, here is an example of visualization for a text fragment: "As part of the project, a new road will be built in 2021. It will connect Murmansk with Belokamenka village. In addition, the project provides for the development of railway infrastructure, including the construction of the railway line Vykhodnoy - Lavna. In the near future it will be easier to get from Murmansk to Teriberka. They will be connected by new asphalt road."

The analyzed text was divided into three blocks: 1) "As part of the project, a new roadwill be built in 2021. It will connect Murmansk with Belokamenka village." ; 2) "In addition, the project provides for the development of railway infrastructure, including the construction of the railway line Vykhodnoy - Lavna.";3) "In the near future it will be easier to get from Murmansk to Teriberka.. They will be connected by new asphalt road." Yellow color marks geoassociated entities of type location (toponyms), and green color marks geoassociated entities of type spatial relation. Based on the analysis we get the following data for visualization: 1) "Murmansk - Road – Belokamenka", 2) "Vykhodnoy - Railway – Lavna", 3) "Murmansk - Road – Teriberka". The result of geocoding and visualization in the form of a mapping is shown in Fig. 2.

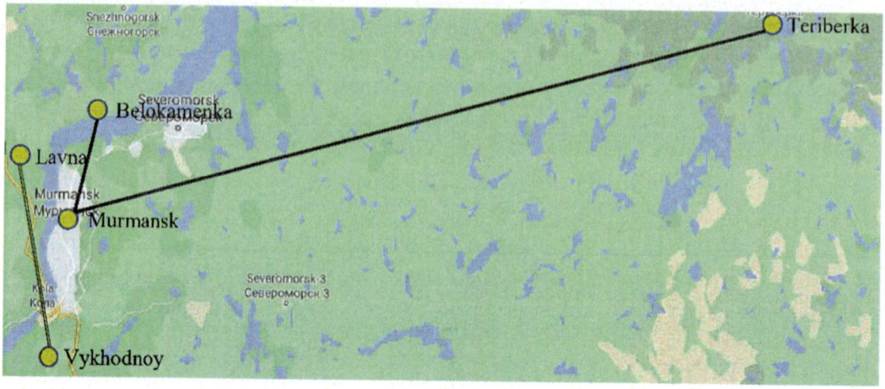

Fig. 2. The result of the technology of spatial relations visualization based on the analysis of natural language texts.

4 Discussion and Future Work

The technology presented in this paper allows the automatic generation of schematic maps based on the extraction of geographical information from unstructured natural language texts. The proposed technology is based on the combination of three methods of text processing (neural network approach, syntactic analysis and lexico-syntactic analysis based on lexical templates). Due to the three-stage text analysis the proposed technology allows to recognize the spatial relations of geoobjects, which is a difficult task for modern natural language text analysis systems.

The results of preliminary testing of the technology confirm the possibility of using the obtained schematic maps to quickly understand the geographical meaning of the text. However, the question of accuracy and the adequate level of detail of the visual representation of the data remains open. At the moment, the generated schematic maps are relatively simple. In particular, the current implementation of visualization does not take into account some important factors, such as relief features and others.

Also, this work has not considered the development of a reasonable hierarchy of spatial relationships and the formalization of attributes of different types of spatial relationships. These issues should be the subject of future research.

Also testing the technology has shown that the quality of syntactic analysis based on the use of rules and gazetteers strongly depends on the number of toponyms in the gazetteer. We are convinced that the inclusion of "local toponyms" (especially complex ones) in the gazettirs will significantly improve the quality of text analysis and increase the completeness of recognition of geoassociated entities.

Due to the fact that the described technology uses heuristic lexical patterns created empirically, in further research we plan to evaluate their applicability when used in other subject areas.

References

1. Agarwal, M.: An overview of natural language processing. Int. J. Res. Appl. Sci. Eng. Technol. (IJRASET) **7**, 2811–2813 (2019)

2. Pilecki, B.M., Vicentiy, A.V.: Development of a method for extracting spatial data from texts for visualization and information decision-making support for territorial management. In: IOP Conference Series: Earth and Environmental Science. Institute of Physics Publishing (2020). https://doi.org/10.1088/1755-1315/539/1/012087
3. Pedrosa, J., Oliveira, D.M., Meira, W., Ribeiro, A.L.: Automated classification of cardiology diagnoses based on textual medical reports. In: Proceedings of the 8th Symposium on Knowledge Discovery, Mining and Learning (KDMILE 2020), pp. 185–192 (2020)
4. Purves, R.S., Clough, P., Jones, C.B., Hall, M.H., Murdock, V.: Geographic information retrieval: progress and challenges in spatial search of text. In: Foundations and Trends in Information Retrieval, vol. 12, pp. 164–318. Now Publishers Inc. (2018)
5. Vicentiy, A.V., Dikovitsky, V.V., Shishaev, M.G.: Automated extraction and visualization of spatial data obtained by analyzing texts about projects of arctic transport logistics development. In: Silhavy, R., Silhavy, P., Prokopova, Z. (eds.) Intelligent Systems Applications in Software Engineering. AISC, vol. 1046, pp. 419–433. Springer, Cham (2019). https://doi.org/10.1007/978-3-030-30329-7_37
6. Grishman, R.: Information extraction. In: Clark, A., Fox, C., Lappin, S. (eds.): The Handbook of Computational Linguistics and Natural Language Processing, pp. 515–530. Wiley-Blackwell, Malden (2010)
7. Doddington, G.R., Mitchell, A., Przybocki, M., Ramshaw, L., Strassel, S., Weischedel, R.: The Automatic Content Extraction (ACE) Program – Tasks, Data, and Evaluation (2004). http://www.lrec-conf.org/proceedings/lrec2004/pdf/5.pdf
8. Nadeau, D., Sekine, S.: A survey of named entity recognition and classification. Int. J. Linguist. Lang. Resour. Investig. **30**, 3–26 (2007). https://doi.org/10.1075/li.30.1.03nad
9. Ratinov, L., Roth, D.: Design challenges and misconceptions in named entity recognition. In: Proceedings of the Thirteenth Conference on Computational Natural Language Learning, pp. 147–155. Association for Computational Linguistics (2009)
10. Campelo, C.E.C., de Souza Baptista, C.: A model for geographic knowledge extraction on web documents. In: Heuser, C.A., Pernul, G. (eds.) Advances in Conceptual Modeling - Challenging Perspectives. LNCS, vol. 5833, pp. 317–326. Springer, Heidelberg (2009). https://doi.org/10.1007/978-3-642-04947-7_38
11. Campelo, E.C., Baptista, C.S.: Mining influential terms for toponym recognition and resolution. Revista Brasileira de Cartografia. **68**, 1123–1132 (2016)
12. Zenasni, S., Kergosien, E., Roche, M., Teisseire, M.: Spatial information extraction from short messages. Expert Syst. Appl. **95**, 351–367 (2018)
13. Acheson, E., Volpi, M., Purves, R.S.: Machine learning for cross-gazetteer matching of natural features. Int. J. Geogr. Inf. **34**, 1–27 (2019)
14. Capineri, C., et al.: European Handbook of Crowdsourced Geographic Information, p. 474. Ubiquity Press, London (2016). https://doi.org/10.5334/bax
15. Yadav, V., Bethard, S.: A survey on recent advances in named entity recognition from deep learning models. In: Proceedings of the 27th International Conference on Computational Linguistics, pp. 2145–2158. Association for Computational Linguistics, Santa Fe (2018)
16. Stock, K.: Mining location from social media: a systematic review (2018). www.elsevier.com/locate/ceus, https://doi.org/10.1016/j.compenvurbsys.2018.05.007
17. Collobert, R., Weston, J., Bottou, L., Karlen, M., Kavukcuoglu, K., Kuksa, P.: Natural language processing (almost) from scratch. J. Mach. Learn. Res. **12**, 2493–2537 (2011)
18. Ivanitskiy, R., Shipilo, A., Kovriguina, L.: Russian named entities recognition and classification using distributed word and phrase representations. In: Proceedings of the 3rd Annual International Symposium on Information Management and Big Data - SIMBig 2016, Cusco, Peru, 1–3 September 2016, pp. 150–156. CEUR-WS.org (2016)

19. Alfred, R., et al.: A rule-based named-entity recognition for Malay articles. In: Motoda, H., Wu, Z., Cao, L., Zaiane, O., Yao, M., Wang, W. (eds.) Advanced Data Mining and Applications. LNCS (LNAI), vol. 8346, pp. 288–299. Springer, Heidelberg (2013). https://doi.org/10.1007/978-3-642-53914-5_25
20. Song, H.J., Jo, B.C., Park, C.Y., Kim, J.D., Kim, Y.S.: Comparison of named entity recognition methodologies in biomedical documents. Biomed. Eng. Online **17**, 158 (2018). https://doi.org/10.1186/s12938-018-0573-6
21. Eftimov, T., Seljak, B.K., Korošec, P.: A rule-based named-entity recognition method for knowledge extraction of evidence-based dietary recommendations. PLoS ONE **12** (2017). https://doi.org/10.1371/journal.pone.0179488
22. Işıklar, Y.E., Çiçekli, N.: A TV content augmentation system exploiting rule based named entity recognition method. In: Abdelrahman, O.H., Gelenbe, E., Gorbil, G., Lent, R. (eds.) Information Sciences and Systems 2015. LNEE, vol. 363, pp. 349–357. Springer, Cham (2016). https://doi.org/10.1007/978-3-319-22635-4_32
23. Sari, Y., Hassan, M.F., Zamin, N.: Rule-based pattern extractor and named entity recognition: a hybrid approach. In: Proceedings 2010 International Symposium on Information Technology - Engineering Technology, ITSim 2010, pp. 563–568 (2010). https://doi.org/10.1109/ITSIM.2010.5561392
24. Piskorski, J., Pivovarova, L., Šnajder, J., Steinberger, J., Yangarber, R.: The first cross-lingual challenge on recognition, normalization, and matching of named entities in Slavic languages. In: Proceedings of the 6th Workshop on Balto-Slavic Natural Language Processing, pp. 76–85. Association for Computational Linguistics, Stroudsburg (2017). https://doi.org/10.18653/v1/w17-1412
25. Piskorski, J., et al.: The second cross-lingual challenge on recognition, normalization, classification, and linking of named entities across Slavic languages. In: Proceedings of the 7th Workshop on Balto-Slavic Natural Language Processing, pp. 63–74. Association for Computational Linguistics, Stroudsburg (2019). https://doi.org/10.18653/v1/W19-3709
26. Official BERT github. https://github.com/google-research/bert. Accessed 17 Aug 2021
27. Official Natasha. https://natasha.github.io/. Accessed 17 Aug 2021
28. Brykina, M.M., Faynveyts, A.V., Toldova, S.Yu.: Dictionary-based ambiguity resolution in Russian named entities recognition: a case study, vol. 1, pp. 163–177 (2013)
29. Earley, J.: An efficient context-free parsing algorithm. In: Communications of the ACM, vol. 13, pp. 94–102 (1970)
30. Official Yargy parser. https://github.com/natasha/yargy. Accessed 17 Aug 2021
31. Naykhanova, L.V.: The technology of creating methods for automatically constructing ontologies using genetic and automatic programming, UlanUde, p. 244 (2008). (in Russian)
32. Nenausnikov, K.V., Kuleshov, S.V.: Algorithm of automatic selection of collocations from the text. J. Instrum. Eng. **62**(11), 976–981 (2019). (in Russian)
33. Apresyan, Y.D.: Three-level control theory: lexicographical aspect. In: Typology of Language and Theory of Grammar, pp. 17–21 (2007). (in Russian)

Fault Detection Model for Software Correctness and Reliability

Eze Nicholas[1(✉)], Ogochukwu Nwosu[1], Akuchinyere Chilaka[1], Peter Obichukwu[1], Stella Osondu[2], Chukwuemeka Adindu[1], Catherine Kanu[1], and Maureen Madu[1]

[1] University of Nigeria, Nsukka, Nigeria
{nicholas.eze,fidelia.nwosu,akuchinyere.chilaka, uzochukwu.obichukwu,emeka.adindu,catherine.kanu, maureen.madu}@unn.edu.ng

[2] Michael Okpara University of Agriculture, Umudike, Nigeria
stella.osondu@mouau.edu.ng

Abstract. We evaluate various technical developments in software testing and fault detection that are currently advanced as potential breakthroughs in determining software correctness and reliability. Testimatica Model was developed for testing the correctness of the entire structure of software for purpose of detecting software faults using phase-to-phase approach. Findings reveals that the test based on the design using the developed testing model when evaluated against other testing/ fault detecting approaches can be a more powerful tool for checking if the structures at the design level of many software systems are correct. The findings also reveal that the model improves customer satisfaction, speeds up the development process and improves productivity of software development teams. The model is very unique and novel as it deals with the entire structure of the design on a phase-wise basis; and not just the control structures of software as reported in literature thus radically implicating a novel pathway for software development firms to develop error free products to achieve software reliability. More quantitative facts on the impact of the developed model on software reliability can be substantially proposed and improved on.

Keyword: Correctness proofs · Fault detection model · Software reliability · Testing · Testmatica

1 Introduction

There is an old saying that software testing can show the presence of faults but definitely not the complete absence of it [1]. What this means is that, if a product undergoes some testing and the output did not meet the specification, then that product is wrong. On the other hand, if a product is tested and the output met the specification, the product may likely still have some faults in it. For this reason, there must be some behavioral attribute of the software which needs to be tested to ascertain whether the program still shows the presence of fault or not. Among these attributes are reliability and correctness [2, 3].

R. Silhavy et al. (Eds.): CoMeSySo 2021, LNNS 232, pp. 981–1000, 2021.
https://doi.org/10.1007/978-3-030-90318-3_79

Reliability of software products here refers to the measure of the frequency and criticality of software product failure which is an unacceptable effects or behavior under permissible operating condition that occurs as a consequence of a fault [4]. It is necessary to know how often software products can fail and how bad the effects of such failure can be. For example, when a software product fails, an important issue should not just be focused on how long it takes on average to repair it but often more important should be how long it takes to repair the results of the failure. This last point is frequently overlooked. However, to avoid software product failures requires that software faults should be detected as early as possible through the instrumentation of testing. Testing here refers to the means of showing that a software product runs correctly to the modern attitude that testing should be used to prevent requirements, analysis, design and implementation faults [5].

However, an alternative approach to testing to ensure a fault free software products that satisfies their specifications is to test the product step-wise and to check that each steps of the software development endeavor preserves correctness. Correctness here means that a software product satisfies its output specification independent of its use of computing resources, when operated under permitted condition [6]. What this further means is that if the inputs that satisfies the input specification of a product is provided and given all the resources it requires, then the product is correct if only the output satisfies the output specification. The two definitions of correctness and testing has some worrisome implications. For example, suppose a product has been tested successfully against a broad variety of test data. Does this mean that the product is acceptable? Unfortunately, it does not. However, if a product is correct, all that means is that it satisfies its specifications. But what if the specifications themselves are incorrect? It will lead to detection of incorrect products full of fatal errors when testing is conducted.

Research in software engineering has shown that the greatest threat to software development endeavor is the high amount of risks and errors detected when testing is conducted [7–9]. That is to say, the method by which software is developed often determines the time and how testing is carried out. For example, in the traditional waterfall model, testing is carried out only after the design of the entire software [10]. But, under an agile programming approach, requirements, programming, and testing are often carried out at the same time [8, 11, 12]. However, because testing is conducted only during the final stage of software development in a traditional waterfall process, problem is bound to occur especially when it takes a very long period of time before an error that occurred at the earlier phases of the process model is detected [13]. More so, reports have shown that project managers are always overanxious in placing targets on task duration times which do not guarantee implementation of projects without a problem when finally released [14]. This is because software test coverages are always low and only few testing are conducted with several faults detected [15].

However, even when there is enough time to test the software, not every fault is detected when the entire code base is examined [16]. This is because all the test codes are examined at once probably as a result of reluctance in making the necessary process changes that agile methodologies required or mere disagreements when prioritizing requirements and no one will know how faulty the implemented code is until the date and time it will be released [17, 18]. Then when the functionality of such software products are tested, a latent and reasonable number of faults may be detected leading software developers into heavy loss in terms of increased cost, lesser profit, delayed delivery of software product, client's dissatisfaction and project deviating from earlier budget [19].

However, the entire problem is not unsolvable [10, 20, 21]. For example, Charles W. et al. suggested a plan-based project approach and agility-based project approach as dominant paradigms for managing software development projects. The authors insist that software products are not to be tested only at the time of delivery to the client without carrying step-wise check of every step taken to develop it. This is because if the software developer makes a mistake while eliciting and documenting the requirements, then the remaining phases of software development (the specifications, the design, the code and so on), will all be affected [22]. The authors concluded that software faults should be detected as earlier as possible because the earlier a fault is detected; the cheaper it is to fix; though it depends on what has to be done to fix the fault. For example, let's say if a problem that is supposed to be discovered earlier at the requirement or the specification stage (as requirements, analysis, and design stage faults constitute over 60% of all faults) [23] is found only at the post-delivery stage, then it would consume huge amount of resources (money, time, etc.) to correct than if the fault had been detected at the earlier stages [19].

Testing for fault detection and proofing of correctness of software is therefore very important in software development not only because it help to detect faults earlier rather it also helps to reduce the cost and time of delivery of software products and increase software reliability. Software is developed by humans who because they are not perfect could make mistakes. However, without proper testing for fault detection and proofing of correctness of software, there will also be of less value in terms of reliability for any software development group to discover after several months of developing a software that a single mistake they made at the early phases of the development process will cause them to redesign the entire software. Because of this, phase-wise testing model is developed in this study as it requires over 70% of software development effort in other to achieve software that one can rely on [24]. To achieve this reliability will require that the requirement must be tested; the specification document must be tested; the design also must be tested; and so on; [23] and each of these phases must be certified error free by the software quality assurance team before moving to the next phase.

In sum, based on the literature so far reviewed, it is so glaring that because of when and how testing is conducted, there are always errors when the software is finally released [10]. Trying to correct such errors could lead to increased cost and budget and clients not being satisfied because there is always delay in delivering software products. All these problems have persisted and what researchers are only interested on is getting empirical data to know if software testing strategies are reliable or not [25]. In many ways these observations have inspired this study to first have a better understanding of the problem and develop a technical fault detection model for correctness of software reliability. It is therefore to the best of our knowledge that this study appears to be the very first to develop a model of this nature in the existing literature of software development and this is how our study contributes to the existing knowledge of software development, software testing and fault detection.

To start addressing the problem, this study first dealt with phase-wise testing of the design structures of software and steps on how the model developed can be used in every phase/iteration of software development process to achieve fault free software in proofing the correctness of the control structures of software. This is to support Brooks' in his article 'no silver bullet', that a major potential silver bullets in software development, is to prove its correctness [26].

2 Literature Survey

This section reviews previous research from different scholars on software testing in relation to fault detection and software reliability and consequences of lack of it. We examined the various testing methods and techniques employed at each phase of software development. The aim is to enable the readers gain more insight on software testing using the model so developed for proofing the correctness of software products to achieve reliability.

2.1 An Overview of Prior Research

So many success stories have been heard of software development. Of these are that software development has reduced radically the rate at which firms across the globe fail in business. Despite these, huge amounts of software products are still not delivered within the stipulated time frame and when it is delivered, a lot of errors are detected [19]. A very good example is a Standish Group study involving 8,380 development projects which was completed in 2006 with 365 respondents on the adoption of agile methodology in software development process as summarized in Fig. 1 [27].

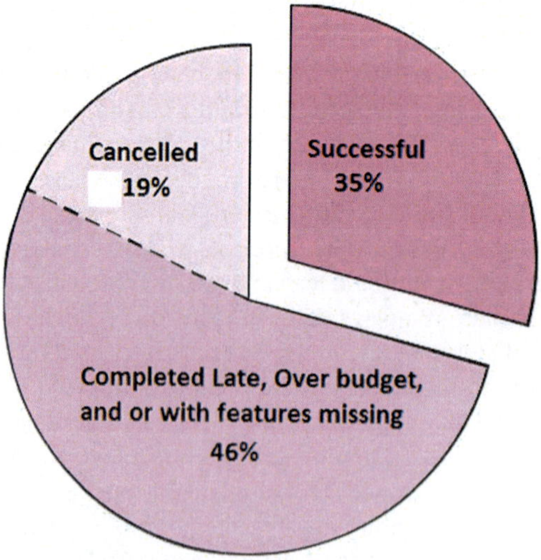

Fig. 1. The outcome of over 8,000 development project completed in 2006 (Rubenstein 2007)

The major findings of this study were that, of a hundred percent success rate with respect to timely completion and within budget successful delivery of projects, only a meager 35% of these projects were successfully completed and delivered within stipulated time and budget. The rest 65% were not even channeled on completing and delivering the entire projects; rather 19% of these projects were terminated at one point while the product is still being developed or were never implemented at all. Then only 46% of the software product were completed and handed over to the client. What this means was that the initially specified features and functionality of these projects were no longer maintained as projects were seen delivered over budget and late. Hence, in 2006, the success rate of software development endeavor was just one third with over 50% of the projects showing so many symptoms of the software crisis which was due to improper software testing. The financial implications of these failures and crises are horrendous on both the firms and the customer. For example, research has shown that late delivery of a software product could lead to serious legal action which is capable of sapping huge amount of resources from clients and firms [28]. However, Cutter Consortium in 2002 reported that:

1. A surprising 78% of software development firms have settled their disputes in court.
2. 67% of the issues lies in the fact that software developers failed to measure up with the functionality and performance of the software products as delivered.
3. In 56% of those cases, the actual dates for delivering of the product were not met.
4. In 45% of those cases, there were so many errors on the software that could not be managed hence not usable.

These reports are not healthy for successful software development considering its many success stories as they come with heavy cost implications. Corroborating, a study conducted by National Institute of Standards and Technology (NIST) in 2002, also shows that one of the greatest threats to the U.S. economy is software bugs [9]. According to the report, the United State loses a whopping $59.5 billion annually. The study however suggested that better software testing could save more than a third of this cost [29]. The above suggestion is in agreement with the result of the study as reported in [10, 30] that while developing an operating system, proper testing can bring down the amount of error by 85%. The Jet Propulsion Laboratory (JPL) also corroborated when they reported that over $25,000 is saved per inspection of software as every little inspection revealed at least 4 major faults and 14 minor faults [31]. Developers at NASA's Jet Propulsion Laboratory (JPL) know firsthand that a single defect, a poorly chosen design, or single untested component can result in missing project goals, mission failure, or worse, total loss of software reliability. Because the stakes are high, JPL has always followed the famous Russian advice, "*doveryai*, no *proveryai*", or "trust, but verify" when making critical software development decisions that depend on quality. However, an Australian group [32], conducted a survey involving 131 respondents of software development groups and firms that used testing methodologies in their software development projects and their report shows that, productivity was improved by 93%; development cost was reduced by 49%; 88% quality and customer satisfaction was achieved; and 83% business satisfaction was equally great. Further research also reveals that conducting a thorough testing on software can lead to an increase in the overall quality of the product and decrease the cost of the product [33].

However, a comparative study was conducted between 1997 and 2007 to determine the perceived influence of testing methods and techniques used by different software development firms on productivity gain [34]. Their findings as shown in Fig. 2; is the discovery that companies that use some degree of testing techniques have more satisfied employees and clients. The perception of University students is that use of testing methods avails them more experience and relevant training. On productivity gains, the study found that there is a greater increase in productivity gain if testing is thoroughly done while developing a software product.

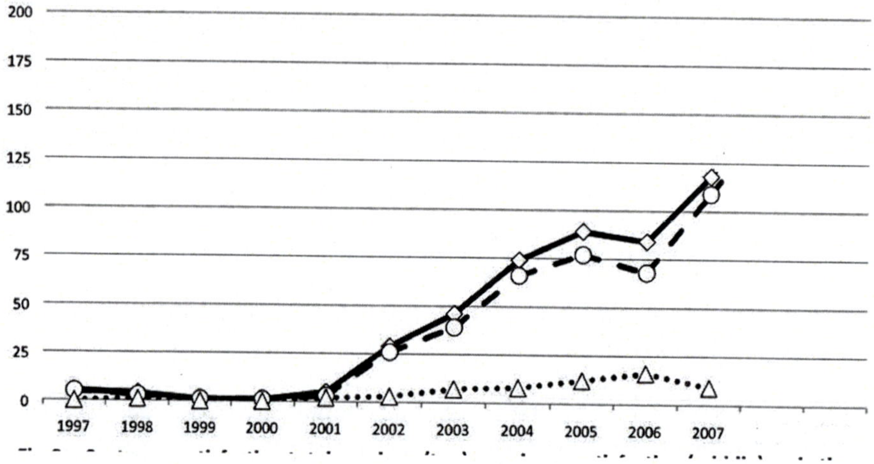

Fig. 2. Customer satisfaction total number (top), employee satisfaction (middle) and other user's satisfaction

Summing up the literature review, testing is a nice tool for software development. This is because it comes with a lot of benefits for the software development team, as well as business benefits to the client [32–34]. Secondly, testing should not be seen as a separate phase of software development. For example, consider Fig. 3.

In the preceding list of phases of software development in Fig. 3, notice there is no separate testing phase. This omission is deliberate for the purpose of this study. This study is advocating that testing rather should be an activity that should take place all the way through software production [5, 21] as in Fig. 4.

More emphases should be placed on testing more than other activities in software development. This normally should take place at the end of every phase or iteration (verification); and also before the product is finally handed down to the client (validation) [35]. This is to verify if system behaves according to specification, and to check the system correctness to ascertain if the clients' requirement as specified actually met the desired output.

Note that this verification could be achieved in two ways, 1) Test Mode, and 2) walk- through mode [36]. Under the test mode, we first generate the correct operational

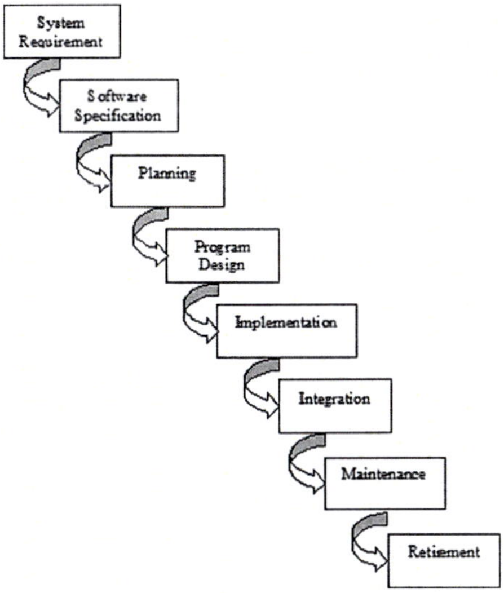

Fig. 3. Software development life cycle

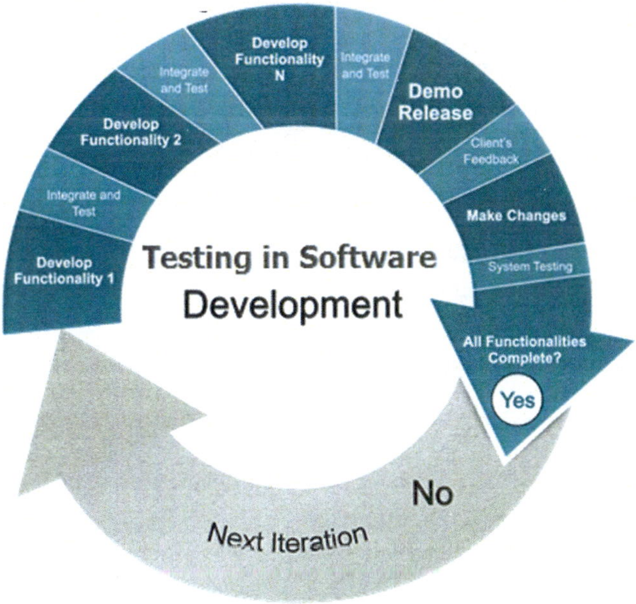

Fig. 4. Software testing diagram

sequences of the design according to the requirement specification and semantics of operations. Then the true input responses are generated and compared with the correct version as in Fig. 5.

```
1:    ignore;  1
*:    empty-ab; l: ignore
 *:   acc-ab;
 4): acc-ab;
 Case input of  * /empty-ab
(*:   acc-ab)  2 3
(1:   acc-ab)  *  acc-ab
```

Fig. 5. Input response in comparison with correct version

In walkthrough, input sequences from the design and their corresponding responses can be taken as "path programs" where its correctness could be achieved by a walk-through procedure based on the specification by applying the testmatica model as will been seen later in this study. During this step, one may discover that there may be ambiguity or incompleteness in the specification; and also, not all specification errors can be revealed. But the application of the testimatica model must establish that the input specification satisfies the output specification.

However, although there are times when testing predominate, there should never be times when no testing is carried out else it will have a high negative impact on productivity gain [11]. For these reasons, the testing of each phase of software development process must be certified error free by software quality assurance team of the firm [37]. before moving on to the next phase. This will make software development 'more agile' which is the best for addressing the problems of the traditional waterfall, incremental and spiral model of software development [11].

2.2 Strategies for Testing Software

To achieve good software testing, one needs to get acquainted with the best testing strategies capable of integrating various software test case design methods into a well-organized series of steps. Our earlier review of literature shows that software testing strategies are necessary for testing and these are developed by project managers, Software Quality Assurance Engineers and individuals who are specialist in testing. This study has identified and employed four software testing strategies in other to draw inference from reports across firms that use testing techniques as an agile method of software development. They include:

- **Unit testing**
  This type of testing is conducted only at the smallest level often referred to as "unit", "iteration", "module", or "component". Here, the entire program is broken down into separate, smaller sections called modules, units, subroutines, subprograms as they are interchangeably referred to. Each module has a specific job to do and is relatively easy to test each module to verify the functionality of a particular section of the code. This type of testing is generally grouped into a white box test class. However, in an object-oriented programming environment, this is usually at the class level in which the constructors and destructors constitute the unit test [38–40].
- **Integration Testing**
  In integration testing, program codes to be tested are organized within various control structures and then merged as one bigger entity which has direct interface with the control structures. Testing is then conducted on the interfaces in other to discover faults in the interfaces and interaction between the merged components (modules). The testing and integration of the larger groups of software components corresponding to elements of the architectural design continues gradually until the software works as a system [41] (See section C for more details).
- **System Testing**
  In system testing, both the functional and requirement specifications including the behavioral properties of the entire system are tested so as to know if the product functions correctly [42, 45]. The SQA group first look at these properties, in parts, run it with known input data, and examines the output [43]. The SQA simply does this by inputting intentionally erroneous data into the system to check the functionality of the product or if the mechanism for detecting faults is still working perfectly to detect faults incase bad data is inputted into the system. After this, the internal consistency of the product source code will be tested to know if the new inputted product will have any form of effect on the client's existing computer operations.
- **Acceptance Testing**
  Acceptance testing precedes Integration testing as it is the testing that stands to give complete assurance that the entire system is perfectly working without any residual fault [4]. This type of testing is done at the time when the complete design artifact is to be handed over to the client by the developer. In other words, without a software product passing through its acceptance testing, it can never be said to be correct or meet its specification or perceived to be useful [44].
  However, in the course of this testing the SQA must ensure that the portability of the system as well as working as expected should protect the systems operating environment from damage to avoid causing other processes within the environment to become inoperative [43].

2.3 Testing Process of Software Development

In this section the seven phases of the software life cycle, are carefully analyzed and the role played by software quality assurance team during testing of each phase to improve software quality by detecting software bugs were explored. The first phase is the requirement phase.

- **Requirement Phase Testing**

  Every software development organization has Software Quality Assurance (SQA) group [37]; who ensures that the quality of a software product that is delivered to the client met with the specification of the client. This is so because; the quality and the correctness of any software product depend on the extent to which it meets its specification [15]. In that case, the SQA is set up to enforce these standards and must play a role right from the beginning of the software development process. In particular, it is important that clients get satisfied when the product is finally handed over to them by first crosschecking with the client to know if the rapid prototype of the software reflects their current needs.

  Nevertheless, no matter how meticulously this is done, there is always the possibility that forces beyond the control of the development team will cause changes to happen during the design process. Further development will then have to be put on hold until the necessary modifications have been made to the partially completed products.

- **Specification Phase Testing**

  Before the specification phase can be deemed to be finished, the SQA group must carefully check the specifications, looking for contradictions, ambiguities, and any sign of incompleteness [37, 43]. In addition, the SQA group must also certify the capacity of the specified hardware and online disk storage in handling the new products.

  If a specification document is to be testable, then one of the properties it must have is traceability where every statement in the output specification document is traced back to every statement in the input specification document. If the requirement has been methodically presented, then the SQA group will have fewer jobs tracing through the specification documents. If rapid prototyping has been used in the requirement phase, then the relevant statements of the specification document should be traceable to the rapid prototype.

  However, a way out for checking the specification document is by means of a review where walkthrough or inspection could be used. During this, both the SQA team and the client meet to determine the correctness of a specification document. The specification documents are reviewed, ensuring that there are no misunderstandings about the documents.

- **Planning Phase Testing**

  In this phase, a Software Project Management Plan (SPMP) is drawn and carefully checked by the SQA team. The SPMP contains detailed plan on delivery date and an estimate of the cost for developing the software.

  To test the planning phase, software development firms obtain more than one independent quotes of both delivery date and cost at the commencement of the planning phase. If there is any difference in the two quotes, such differences are reconciled through the process of a review similar to the review of the specification document. This review will make it possible for software development cost and delivery date not to slip away.

- **Design Phase Testing**

  Testing the design phase of software requires traceability where every aspect of the design is traced to a statement in the specification document. This design review is similar to the specification reviews as it helps the SQA team to check and know if the actual design conforms to what is specified or that whatever is in the specification document reflects with what is in the design.

During the design review, clients need not be physically present because of the technical nature of the design. Here software development team and the SQA team work through the entire design and each separate module, looking for logic faults, interface faults, etc. all with the intent of achieving a correct and perfect design. In addition, it is important that the review team should know that some specification faults were not detected during the previous phase before the design and so pay more attention to discover such faults during the design phase.

- **Implementation Phase Testing**

  The modules should be desk checked by the programmer during implementation and also tested immediately they have been implemented and run against test cases. After this informal testing, the SQA team then performs methodical testing on the modules.

  Code review could be employed in detecting programming faults by the programmer and the SQA representatives. This procedure is similar to reviews of specifications and design described previously.

- **Integration Phase Testing**

  Integration phase testing ensures that various units or modules are combined correctly in other that the entire product meets with its specification. Here module interfaces are meticulously tested and checked by the compiler and the linker to ensure that number, order, and types of formal arguments matches with number, order, and types of actual arguments.

  The SQA team carries out product testing [43] at the completion of the integration testing during which the entire product is tested against its specification. In particular, the constraints listed in the specification documents must be tested to ascertain if the correct specifications have been implemented. Only after this are the test cases drawn up.

  In some cases, early versions (alpha version) of the complete product are sent to the client for testing on site. After correction, this corrected alpha version, now called the beta version is intended to be close to the final version. Alpha testing and beta testing are particularly important during software development [45, 46].

- **Maintenance Phase Testing**

  From the foregoing, once the software developer has satisfied that the desired changes have been implemented correctly, then the software product must undergo regression testing [47] to ascertain if the functionality of the rest of the product has been compromised or not. Here all the previous test cases are retained, together with the results of running those regression tests. If it is discovered that the product has been compromised, then fault could be easily traced and corrected.

3 Method

3.1 Proofing the Correctness and Reliability of Software System

A product is simply said to be correct if the input specification satisfies the output specifications [6]. But we should understand that there is no law that says a product must be accepted simply because the product has been successfully tested against a broad range of test data. If a product is correct and should be accepted, that simply means that the output specifications were met. But what if the specifications themselves are incorrect? Consider the following illustrations. Illustrating the above difficulty, one may think that the specification in Fig. 6 is correct simply because, may be when the product was tested against a given set of data, it runs correctly. But it is not.

Input specification:	p : array of n integers, n 0.
Output specification:	q :array of n integers such that q[0] q[1] ··· q[n 1]

Fig. 6. Incorrect specifications for a sort

The specification has only created a specification gap by showing that the input to the sort is an array p of n integers, and the output is another array q sorted in non-decreasing order; without recourse that the element of q and the output array are all a permutation of the element of the input array p.

The second method in Fig. 7, tricksort then capitalizes on this specification fault and tries to correct it by setting all n elements of array q to 0 as in Fig. 8.

```
void trickSort (int p[], int q[])
{
    int i;
    for (i    0; i    n; i    )
        q[i]  0;
}
```

Fig. 7. Method tricksort, which satisfies the specifications

Figure 8 is then the corrected specification for the sort. This example has set the record straight that it is no meaning claiming that a product is correct when its specification is incorrect.

In other words, it is better to look at what accounted for the correctness of a product rather than just showing that the product is correct. Therefore, it is necessary to prove that a product specification is correct. This proof which is a mathematical technique should be carried out in conjunction with step by step coding and the design [2], after which it is tested to know if the coding and design is equivalent to the specification and therefore is correct [4].

However, to prove the correctness of software [47] proposes an automated method that merely *reduces redundant test executions* for SPL *regression testing only for* the control structures of software. Their result only showed that the method is correct.

Input specification:	**p : array of n integers, n0**
Output specification:	**q : array of n integers such that q[0]q[1]-q[n1]** **The elements of array q are a permutation of the elements of array p, which are unchanged.**

Fig. 8. Corrected specification for the sort

However, in furtherance of their study, we propose here a similar method but based on TESTMATICA MODEL where the entire structure of the design is thoroughly checked; and tested for errors while satisfying some reasonable assumptions. The test based on the design is evaluated against the specification and comparison was made with other testing approaches to compare their error detecting capabilities.

The result shows that our proposed method "Testmatica model" testing strategy can be a powerful testing tool for checking if all the control structures at the design level of many software systems are correct. In achieving this, we identified the following steps; 1) the highest number of states in the design is estimated; for example, see an informal specification of a comment pointer in Fig. 9 below: 2) According to the coding and the design, the test sequence is generated; and finally, 3) the test sequence generated in step 2 is then checked. The flowchart equivalent of Fig. 9 is represented in Fig. 10.

Comment Pointer:

<u>**Specification**</u>: **input consists of characters n{1,2,3}**
print only comment. A comment is an
input sub-sequence enclosed by{*
on the left and *} on the right. (it may
contain other {*'s but not *{'s)
now looking at the specification;

```
int k, s;
int y[n];
k  0;
s  0;

while (k  n)
{
  s  s  y[k];
  k  k  1;
}
```

Fig. 9. Informal specification for a comment printing system

To prove that the code fragment and the corresponding flowchart in Figs. 9 and 10 respectively is correct, we set, the variable s to contain the sum of the n elements of the array y after the code has been executed.

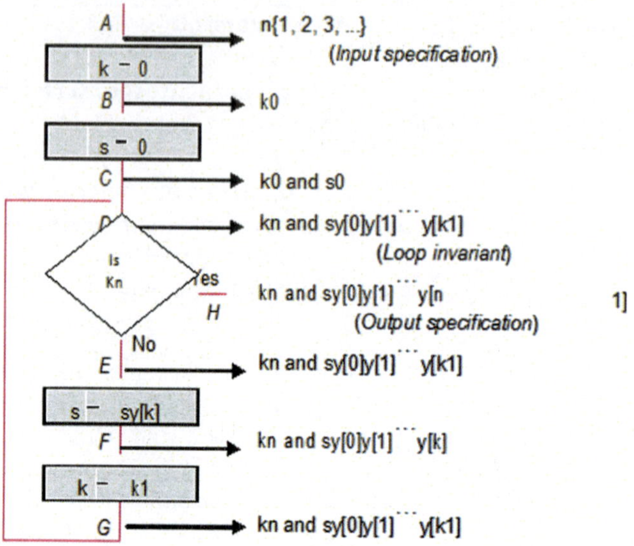

Fig. 10. A representation of figure with input specification output specification, loop invariant, and assertions added

However, observe that Fig. 10 contains a certain claim of a mathematical attribute from statement A to H; that is at the beginning and at the end of each statement. Having observed that, then each statement can be proven as thus.

The input specification, just at the beginning of the code execution holds at A and initializes n value as a positive integer; i.e.,

$$A: \; n \in \{1, 2, 3, \ldots\} \tag{1.1}$$

The output specification to this is that, when control gets to *H*, then the addition of the n value will be stored in the s value which will all be stored in array y, that is,

$$H: \; s \; y[0] + y[1] + \ldots + y[n-1] \tag{1.2}$$

Then a very strong output specification can only prove correct the code fragment as shown:

$$\mathrm{H}: \mathrm{k} \; \mathrm{n} \; \mathrm{and} \; \mathrm{s} \; y[0] + y[1] + \ldots + y[n-1] \tag{1.3}$$

After declaring the input and output specification, we move to prove the mathematical expression that holds at point D not minding if its loop invariant has been executed several times or not. That is;

$$D: \; k \, \# \, n \; \mathrm{and} \; s \; y[0] + y[1] + \ldots + y[k-1] \tag{1.4}$$

Next is to prove that the code fragment is correct by showing that the output specification in (1.3) holds at point H; and the input specification in (1.1) holds at point A.

Then the assignment statement k R 0 where the control is at B is first executed and this is what holds:

$$B: \text{k } 0 \tag{1.5}$$

That is to say that at point *B*, the statement holds as k 0 and n ∈ {1, 2, 3,...} thus making the input specification in (1.1) to hold at all points in the flowchart after which k 0 and n ∈ {1, 2, 3,...} is removed from the prove.

Then at point *C*, as a result of the statement s R 0, seen in the second assignment, the statement below holds true:

$$C: \text{k } 0 \text{ and s } 0 \tag{1.6}$$

Next is to prove by induction the correctness of the loop invariant in (1.4). While considering that statement (1.6) holds that C: k 0, and s 0; then k 0 by statement (1.6) and n $ 1 from input specification (1.1), will require k # n. Now since k 0, in statement (1.6) holds; then k − 1 − 1, will empty the sum in (1.4) and s 0 as required thereby proving that the Loop invariant (1.4) is true.

Next, the inductive hypothesis will be performed to ascertain if k is equivalent to some value k_0, 0 # k_0 # n, and execution is at point *D*. then the statement that holds is

$$D: k_0 \# n \text{ and s } y[0] + y[1] + \ldots + y[k_0 - 1] \tag{1.7}$$

Next if k_0 $ n, and k_0 # n by hypothesis, it means that k_0 n. Going by the inductive hypothesis in (1.7), the following statement implies

$$H: k_0 \text{ n and s } y[0] + y[1] + \ldots + y[n - 1] \tag{1.8}$$

The above assertion is the output specification as in (1.3). Now, supposing the test k_0 $ n? did not hold, then control will be passed from a point *D* to a point *E*. Again *since* k_0 *is neither* ≥ n, k_0 n, then the statement implies.

$$\begin{array}{l} E: \text{k n and s } y[0] + y[1] + \ldots + y[k - 1] \\ \quad\quad \text{O } 0 \end{array} \tag{1.9}$$

Now that the statement s R s + y[k_0] is executed, then from statement (1.9), at point *F*, the following statement must hold:

$$\begin{array}{l} k_0 \text{ n and s } y[0] + y[1] + \ldots + y[k_0 - 1] + y[k_0] \\ \quad\quad\quad\quad y[0] + y[1] + \ldots + y[k\,] \\ \quad\quad\quad\quad 0 \end{array} \tag{1.10}$$

After that, k_0 R k_0 + 1 will be executed. Now if we set the value of k_0 to 19 say before the statement is executed, then the last term in the sum in (1.10) will be y[19]. Notice that the value of k_0 is increased by 1 to 20 with the sum s unaltered, hence the last term in the sum still remains y[19], thereby bringing it to y[k_0 − 1]. Again, the incremental value of k_0 by 1 at point *F*, k_0 n presupposes k_0 # n should the inequality be hold at point G. *Now the outcome of* increasing k_0 by 1 is that at point G the following statement must hold:

$$G : k_0 \# n \text{ and } s\ y[0] + y[1] + \ldots + y[k_0 - 1] \tag{1.11}$$

Statement (1.11) that holds at point G above reveals its similarity with assertion (1.7) meaning that suppose (1.7) holds at D for k k_0, it will also hold at D with k $k_0 + 1$ proving the statement that the loop invariant in (1.4) holds for k 0 which also stand for all values of k, 0 # k # n.

Next is to prove the termination of the loop by observing that the loop adds 1 to the value of k at every iteration each time the statement k R k + 1 is executed. Remember also that statement (1.6) sets k to be equal to 0. This loop addition per iteration will therefore continue to increase by 1 until k reaches the value n thereby causing the exit of the loop and setting the value of s as in assertion (1.8), thereby satisfying output specification (1.3).

To conclude the review, observe that the input specification is given in (1.1), and that loop invariant in (1.4) holds irrespective of the number of times executed, which thereafter terminates after n iterations through an incremental value of 1. When this happens, the values of k and s will be seen to satisfy the output specification as in (1.3) hence proving correct the code fragment of Fig. 9 and the flowchart in Fig. 10.

The above model gives the programmer confidence that the software is correct and reliable since the correctness and reliability of any software simply means that input specification must satisfy the output specification. It also means that numbers of faults are reduced to the barest minimum and productivity gains are obvious.

4 Benefit of TESTMATICA Model

Thus far, having reviewed the literature and explored the testing techniques applied in software development, this section discusses the benefits of the developed model as tool for testing.

- **Improves Quality**
  Testimatica model improves quality. When testing and review is conducted regularly at every phase of software development, quality is improved since every fault detected will be exposed and fixed at once. This is done more easily and quickly when the entire project is broken down into units or modules. Then it will allow the project development team to focus on developing high quality products.
- **Allows for Change**
  During the process of software development, the developers tries to constantly shape and reshape the overall product backing so as to meet customers demand. However, this process can introduce new or changed backlog items in the process thereby providing the opportunity for making or creating change.
- **Predictable Cost and Schedule**
  Testing allows for cost of developing a product to be predicted. Owing to the fact that each scheduled time box is a fixed duration, the cost can be predictable. It will make clients to be well informed of the approximate cost which the product will take and also help to achieve good decisions on which feature should be considered first.

- **Early and Predictable Delivery**
  By using time box where jobs are scheduled to be delivered within a particular spate of time, testing makes it easier and possible for product delivery date and time to be highly predictable. Those faults that could have delayed the development process are detected earlier and fixed thereby making way for speedy development and delivery.
- **Transparency**
  Testing guarantees transparency because it includes the client in the development process so as to make input. This gives the client a good impression that they are seeing their work in progress. Clients follow the development process and tend to be satisfied with what they are observing.

5 Result

Predicting correctness and reliability of software system will require this model. This model will be widely used as a way to help engineers develop high-quality software. The model developed when applied as engine for software testing will indicate top benefits such as customer satisfaction, speeding up the development process and improved productivity of teams.

6 Conclusion and Recommendations

We have presented our model for the use of software developers. Our first recommendation for future studies suggests that the way software is produces should be changed in other to achieve comparable future breakthroughs. To improve cost and productivity gains also, software should be developed incrementally to enable it to be constructed phase-wise with testing carried out at each phase instead of trying to build the product as a whole. Some teams must begin phase-to-phase testing as necessary. Great and potential software designers should be encouraged to undergo more training on software development to be up-to-date with recent trends in software development. This will bring about greatest hope considering that Brooks's opined that great designers should be considered first if we wish to improve software production and cost. Considering the central role of the model in developers' daily activities, it can be expected that results considering developers' experiences while using the model play a prominent role in the development of software products that is correct and reliable. In relation to the concept of Testmatica, our paper contributes to increased understanding of software faults, and the key features of the model and its relation to software development.

References

1. Dijkstra, E.: The humble programmer. Commun. ACM **15**(10), 859–866 (1972)
2. Chandrasekar, A., Sudha, R., Rajesh, P.: A research study on software quality attributes. Int. J. Sci. Res. Publ. **4**(1), 14–19 (2014)
3. Gol Mohammadi, N., et al.: An analysis of software quality attributes and their contribution to trustworthiness. In: Proceedings of the 3rd International Conference on Cloud Computing and Services Science, pp. 542–552 (2013). https://doi.org/10.5220/0004502705420552

4. Schach, S.: Object Oriented and Classical Software Engineering, 15th edn. McGraw-Hill, New York (2015).ISBN: 970-0-07337618-9
5. Abid, J., Muhammad, A., Normi, S., Awang, A., Akhlaq, A.: Software testing techniques: a literature review. In: 6th International Conference on Information and Communication Technology for the Muslim World (ICT4M) (2016). https://doi.org/10.1109/ICT4M.2016.045
6. Felix, E., Lee, S.: Predicting the number of defects in a new software version. PLoS ONE **15**(3) (2020). https://doi.org/10.1371/journal.pone.0229131
7. Mamdouh, A., Sadiq, A.: Security risks in the software development lifecycle. Int. J. Recent Technol. Eng. **8**(3), 7048–7055 (2019). https://doi.org/10.35940/ijrte.C5374.098319
8. David, B., Pam, R.: Agile and secure software development: an unfinished story. Issues Inf. Syst. **20**(1), 144–156 (2019). https://doi.org/10.48009/1_iis_2019_144-156
9. NIST: National Vulnerability Database (2019). https://nvd.nist.gov/vuln/search/statistics?adv_search=false&form_type=basic&results_type=statistics&searchtype=all
10. Azeem, U., Abhineet, A.: Importance of software testing in the process of software development. Int. J. Sci. Res. Dev. **6**(12), 141–145 (2019)
11. Suryaatmaja, K., Wibisono, D., Ghazali, A., et al.: Uncovering the failure of agile framework implementation using SSM-based action research. Palgrave Commun. 6(8) (2020). https://doi.org/10.1057/s41599-019-0384-9
12. Malek, A., Mariam, B., Wael, E., Adnan, S.: Agile software development methodologies: survey of surveys. J. Comput. Commun. **5**(1), 74–97 (2017). https://doi.org/10.4236/jcc.2017.55007
13. Software Testing Lifecycle. https://www.softwaretestinghelp.com/what-is-software-testing-life-cycle-stlc/. Accessed 31 Aug 2018
14. Didar, Z., Francesca, D., Muneera, B.: Problems and challenges of user involvement in software development: an empirical study. In: International Conference on Evaluation and Assessment in Software Engineering (EASE), Nanjing, China (2015)
15. Eze, N.: Development process for controllable software. Int. J. Educ. Res. **5**(7), 37–52 (2017)
16. Evaluation of Software Testing tools (coursework). http://myassignmenthelp.info/assignments/master-thesis-title-evaluation-software-testing-tools-87710/. Accessed 15 July 2015
17. Gregory, P., Barroca, L., Sharp, H., Deshpande, A., Taylor, K.: The challenges that challenge: engaging with agile practitioners' concerns. Inf. Softw. Technol. **77**(3), 92–104 (2016). https://doi.org/10.1016/j.infsof.2016.04.006
18. Marques, R., Costa, G., Silva, M., Gonçalves, P.: A survey of failures in the software development process. In: Proceedings of the 25th European Conference on Information Systems (ECIS), Guimarães, Portugal, 5–10 June 2017, pp. 2445–2459 (2017). ISBN 978-989-20-7655-3, Research Papers
19. Certified Tester Foundation Level Syllabus. International Software Testing Qualifications Board. https://www.bcs.org/upload/pdf/ct-foundation-syllabus.pdf. Accessed 31 Mar 2017
20. Charles, W., Butler, L., Vijayasarathy, R., Nicholas, R.: Managing software development projects for success: aligning plan- and agility-based approaches to project complexity and project dynamism. Proj. Manag. J. **51**(3), 262–277 (2020). https://doi.org/10.1177/8756972819848251
21. Madeyski, L., Kawalerowicz, M.: Software engineering needs agile experimentation: a new practice and supporting tool. In: Madeyski, L., Śmiałek, M., Hnatkowska, B., Huzar, Z. (eds.) Software Engineering: Challenges and Solutions. AISC, vol. 504, pp. 149–162. Springer, Cham (2017). https://doi.org/10.1007/978-3-319-43606-7_11
22. Yusupbekov, N., Rustambekovicha, S., Gulyamovb, M., Usmanovac, N., Mirzaev, D.: Challenging the ways to determine the faults in software: technique based on associative interconnections. In: 9th International Conference on Theory and application of Soft Computing, Computing with Words and Perception, ICSCCW, Budapest, Hungary, 22–23 (2017)

23. Kamlesh, R., Arun, A., Khushboo, J., Singh, V.: A time-variant fault detection software reliability model SN Appl. Sci. 3(18) (2021). https://doi.org/10.1007/s42452-020-04015-z
24. Welman, C., Kruger, F., Mitchell, B.: Research Methodology, 3rd edn. Oxford University Press, London (2005)
25. Divyani, S.: Comprehensive study of software testing techniques and strategies: a review. Int. J. Eng. Tech. Res. 9(08) (2020). https://doi.org/10.17577/IJERTV9IS080373
26. Brooks, F.: No silver bullet. In: Kugler (ed.) Information Processing, pp. 10–19. Elsevier North-Holland, New York (1986)
27. Standish Group Report: There's Less Development Chaos Today. https://community.embarcadero.com/blogs/entry/chaos-report-2007--there-is-less-software-development-chaos-today-33392. Accessed 21 Mar 2007
28. Highsmith, J.: Agile Project Management: Principles and Tools, 2nd edn. Cutter Consortium, Arlington (2003)
29. The Economic Impacts of Inadequate Infrastructure for Software Testing. National Institute of Standards and Technology, pp. 1–309 (2002)
30. Fuqun, H., Bin, L.: Software defect prevention based on human error theories. Chin. J. Aeronaut. **30**(3), 1054–1070 (2017)
31. Dan, P., Joel, W.: The value proposition for assurance of JPL systems. Procedia Comput. Sci. **28**, 398–403 (2014). Conference on Systems Engineering Research (CSER 2014)
32. Shine Technologies. Agile Methodologies Survey Results. https://www.scirp.org/journal/PaperInformation.aspx?PaperID=75114. Accessed Mar 2017
33. Maria, P., Lassenius, C.: How does an agile coaching team work? A case study. In: Proceedings of the 2011 Sixth IEEE International Conference on Global Software Engineering, New York, pp. 29–38 (2011)
34. Dyba, T., Dingsoyr, T.: Empirical studies of agile software development: a systematic review. Inf. Softw. Technol. **50**(9), 833–859 (2008)
35. Verification/validation/certification. https://en.wikipedia.org/wiki/Verification_and_validation. Accessed: 07 Sept 2018
36. Rostislav, G., Igor, R., Alexander, R.: Methods of Software Verification. Science and Education of the Bauman MSTU (2015). https://doi.org/10.7463/1015.0823129
37. Gregory, J., Crispin, L.: More Agile Testing, 1st edn., pp. 23–39. Addison-Wesley Professional, Boston (2014)
38. Hadas, C., Dani, A., Dov Benyomin, S., Mark, G., Robin, P., Shlomo, M.: The unit test: facing CICD – are they elusive definitions? J. Inf. Technol. Manag. **19**(2), 40–54 (2018)
39. Passier, H., Bijlsma, L., Bockisch, C.: Maintaining unit tests during refactoring. In: Proceedings of the 13th International Conference on Principles and Practices of Programming on the Java Platform: Virtual Machines, Languages, and Tools. ACM (2016)
40. Daka, E., Fraser, G.: A survey on unit testing practices and problems. In: 2014 IEEE 25th International Symposium on Software Reliability Engineering. IEEE (2014)
41. Nahid, A., Susmita, K.: Review paper on various software testing techniques and strategies. Global J. Comput. Sci. Technol. C Softw. Data Eng. **19**(2), 43–49 (2019)
42. Pecorelli, F., Palomba, F., De Lucia, A.: The Relation of test-related factors to software quality: a case study on apache systems. Empirical Softw. Eng. **26**(2), 1–42 (2021). https://doi.org/10.1007/s10664-020-09891-y
43. Sowunmi, O.Y., Misra, S., Fernandez-Sanz, L., Crawford, B., Soto, R.: An empirical evaluation of software quality assurance practices and challenges in a developing country: a comparison of Nigeria and Turkey. Springerplus **5**(1), 1–13 (2016). https://doi.org/10.1186/s40064-016-3575-5
44. Eze, N., Obichukwu, P., Subodh, K.: Perceived usefulness, perceived ease of use in ICT support and use for teachers. IETE J. Educ. 62(1) (2021). https://doi.org/10.1080/09747338.2021.1908177

45. Emily, A., Dennis, W., Erin, A., Alan, M.: Alpha test results for a housing first eLearning strategy: the value of multiple qualitative methods for intervention design. Pilot Feasibility Stud. 3(46) (2017).https://doi.org/10.1186/s40814-017-0187-y
46. Nuraini, C.K., Mohda, C.K., Faaizah, S.: Personalized learning environment: alpha testing, beta testing & user acceptance test. Procedia Soc. Behav. Sci. **195**, 837–843 (2015). World Conference on Technology, Innovation and Entrepreneurship
47. Pilsu, J., Sungwon, K., Jihyun, L.: Efficient regression testing of software product lines by reducing redundant test executions. Appl. Sci. **10**, 8686 (2020). https://doi.org/10.3390/app10238686

Author Index

R. Silhavy et al. (Eds.): CoMeSySo 2021, LNNS 232, pp. 1001–1003, 2021.
https://doi.org/10.1007/978-3-030-90318-3